Introduction to Probability and Statistics for Data Science

Introduction to Probability and Statistics for Data Science provides a solid course in the fundamental concepts, methods, and theory of statistics for students in statistics, data science, biostatistics, engineering, and physical science programs. It teaches students to understand, use, and build on modern statistical techniques for complex problems. The authors develop the methods from both an intuitive and mathematical angle, illustrating with simple examples how and why the methods work. More complicated examples, many of which incorporate data and code in R, show how the method is used in practice. Through this guidance, students get the big picture about how statistics works and can be applied. This text covers more modern topics such as regression trees, large-scale hypothesis testing, bootstrapping, MCMC, time series, and fewer theoretical topics such as the Cramer–Rao lower bound and the Rao–Blackwell theorem. It features more than 250 high-quality figures, 180 of which involve actual data. Data and R code are available on the book's website so that students can reproduce the examples and complete hands-on exercises.

Steven E. Rigdon is Professor of Biostatistics at Saint Louis University. He is a fellow of the American Statistical Association and is the author of *Statistical Methods for the Reliability of Repairable Systems*, *Calculus*, 8th and 9th editions, *Monitoring the Health of Populations by Tracking Disease Outbreaks* (2020), and *Design of Experiments for Reliability Achievement* (2022). He has received the Waldo Vizeau Award for technical contributions to quality, the Soren Bisgaard Award, and the Paul Simon Award for linking teaching and research. He is also Distinguished Research Professor Emeritus at Southern Illinois University Edwardsville.

Ronald D. Fricker, Jr. is Vice Provost for Faculty Affairs at Virginia Tech, where he has served as head of the Department of Statistics, Senior Associate Dean in the College of Science, and, subsequently, interim dean of the college. He is the author of *Introduction to Statistical Methods for Biosurveillance* (2013) and, with Steve Rigdon, *Monitoring the Health of Populations by Tracking Disease Outbreaks* (2020). He is a fellow of the American Statistical Association, a fellow of the American Association for the Advancement of Science, and an elected member of the Virginia Academy of Science, Engineering, and Medicine.

Douglas C. Montgomery is Regents' Professor and ASU Foundation Professor of Engineering at Arizona State University. He is an Honorary Member of the American Society for Quality, a fellow of the American Statistical Association, a fellow of the Institute of Industrial and Systems Engineering, and a fellow of the Royal Statistical Society. He is the author of 15 other books, including *Design and Analysis of Experiments*, 10th edition (2020) and *Design of Experiments for Reliability Achievement* (2022). He has received the Shewhart Medal, the Distinguished Service Medal, and the Brumbaugh Award from the ASQ, the Deming Lecture Award from the ASA, the Greenfield Medal from the Royal Statistical Society, and the George Box Medal from the European Network for Business and Industrial Statistics.

Introduction to Probability and Statistics for Data Science

with R

Steven E. Rigdon
Saint Louis University

Ronald D. Fricker, Jr.
Virginia Polytechnic Institute and State University

Douglas C. Montgomery
Arizona State University

CAMBRIDGE
UNIVERSITY PRESS

CAMBRIDGE
UNIVERSITY PRESS

Shaftesbury Road, Cambridge CB2 8EA, United Kingdom

One Liberty Plaza, 20th Floor, New York, NY 10006, USA

477 Williamstown Road, Port Melbourne, VIC 3207, Australia

314-321, 3rd Floor, Plot 3, Splendor Forum, Jasola District Centre, New Delhi - 110025, India

103 Penang Road, #05–06/07, Visioncrest Commercial, Singapore 238467

Cambridge University Press is part of Cambridge University Press & Assessment,
a department of the University of Cambridge

We share the University's mission to contribute to society through the pursuit of
education, learning and research at the highest international levels of excellence.

www.cambridge.org
Information on this title: www.cambridge.org/highereducation/isbn/9781107113046

DOI: 10.1017/9781316286166

When citing this work, please include a reference to the DOI 10.1017/9781316286166

First published 2025

Printed in Mexico by Litográfica Ingramex, S.A. de C.V.

A catalogue record for this publication is available from the British Library.

A Cataloging-in-Publication data record for this book is available from the Library of Congress.

ISBN 978-1-107-11304-6 Hardback
ISBN 978-1-009-56835-7 Paperback

Additional resources for this publication at www.cambridge.org/ProbStatsforDS.

Steve Rigdon

To my wife Pat, who has always supported me and been at my side.

Ron Fricker

To my spouse, Christine: *Tu ventus sub alis meis es.*

And to my first statistics professor, Randy Spoeri: You introduced me to the subject and made it fun.

Doug Montgomery

To Cheryl, who has always supported and encouraged me. And to the memory of my first statistics professor, Ray Myers, mentor, colleague, collaborator, and friend.

Contents

Preface

This book is designed for students in statistics, data science, biostatistics, engineering, and mathematics programs who need a solid course in the fundamental concepts, methods, and theory of statistics. Our goal is to give students enough background in the methods and theory of statistics that they can understand modern techniques used in statistics and be able to apply them in the practice of data science.

We had to make some difficult choices regarding topic coverage. We do cover the important concepts of statistics, including maximum likelihood, the information matrix, power, etc., because these are needed for a student to be a successful statistician. When we cover maximum likelihood estimation, we specifically cover the method of approximating the maximum of the (log) likelihood function. Nowadays, data are so plentiful that we are often faced with testing multiple null hypotheses. Holm's method and the Benjamini–Hochberg method are derived and applied to real problems. There are a number of statistical methods that were developed in the late twentieth and early twenty-first centuries, including regression trees, large-scale hypothesis testing, methods of cross-validation, the bootstrap, Markov chain Monte Carlo, and others. We address the optimal selection of levels of a predictor variable to maximize the information we obtain; this leads to an introduction to the topic of optimal design. With some exceptions, these techniques have not found their way into introductory textbooks, especially those that emphasize theory. Throughout, we have tried to include topics that a statistician would use in the practice of statistics and to cover these thoroughly. We don't develop every aspect of statistical theory; for example, we cover very little of the limit theorems in statistics (convergence in probability, convergence in distribution, almost sure convergence, Slutsky's theorem, etc.). We don't cover the Cramer–Rao lower bound or the Rao–Blackwell theorem. We cover joint continuous distributions using multiple integration, but we do not go into great depth.

The emphasis is on modern methods of statistical inference. We develop enough theory so that students will understand these methods. If a statistician or data scientist is to work effectively with practitioners, it is up to the statistician to be the one to explain how methods work, what assumptions underlie the methods, what the limitations are, and how (or whether) the assumptions can be checked. Subject matter experts (i.e., the nonstatisticians) are not trained to do this. This is why it is important for students of statistics to understand the underlying theory behind the methods.

The flip side of our approach is that we do not develop theory for theory's sake. No theory is developed for the purpose that it might be usable in a future course. We have found that students who understand probability and the foundational concepts of statistical theory can understand and use advanced statistical methods. Without a solid grounding on the theory and concepts of statistics it is difficult to pick up new methods.

Calculus is used in a number of places in the book, so students will need at least one or two semesters of calculus. There are a few uses of multiple integrals when we discuss joint continuous distributions, and for these the third semester of calculus will be needed. An instructor can skip these topics or sidestep the use of multiple integrals. We use calculus when it is necessary, for example in getting expected values of continuous random variables. We use R throughout the book. Although we do cover an introduction to R, it would be helpful if students had some prior background in R.

We use data extensively throughout the book. Most of the data sets are real (although at times we give small data sets to introduce a method). Many of these data sets are large. In most cases, we have provided a csv (comma separated values) file for the data. We also provide the R code used in the book to analyze the data sets that we provide. This can be found at: www.cambridge.org/ProbStatsforDS

While the book's website contains information about getting R up and running, we offer the following advice about loading in data sets and packages. First, it is always good practice to set the working directory to the directory on your computer that contains your data files. You can do this with the setwd() command. For example,

```
setwd("C:/Users/Documents/Rfiles")
```

will force R to read (write) files from (to) this directory. Note two things: (1) the path must be enclosed in quotes, and (2) subdirectories are indicated by forward slashes, not backslashes. Second, many of the methods we apply in this book require special R packages to run. These packages are collections of functions, dataframes, etc. Before you can use a package you must (1) install it, and (2) load it in during each R session. To install a package, such as dplyr, type

```
install.packages("dplyr")
```

Then, every time you start a new R session, you will have to load this package using

```
library(dplyr)
```

You need only install a package once on your computer, but you must call library() each time you begin an R session.

If you type library for a package you haven't installed, you will get an error. For example, if you haven't installed the testassay package and if you type library(testassay), then you will get an error like this:

```
Error in library(testassay) : there is no package called 'testassay'
```

The remedy is to first install the package by typing install.packages("testassay") and then typing library(testassay). If you ever get an error like the following

```
Error in arrange(df, y) : could not find function "arrange"
```

there is a good chance you forgot to load the package that contains the function arrange(), which is in the dplyr package. The remedy is to first type library(dplyr).

Most two-semester courses will include a fairly standard first semester, which would likely cover the following chapters:

Semester 1

Chapter	Topic
1	Introduction
2	Data Visualization
3	Basic Probability
4	Random Variables
5	Discrete Distributions
6	Continuous Distributions
7	About Data and Data Collection
8	Sampling Distributions
9	Point Estimation
10	Confidence Intervals
11	Hypothesis Testing
12	Hypothesis Tests for Two or More Populations

The choice of topics for a second course would depend on the nature of the course. For example, our book could be used in a mathematical statistics course that emphasizes applications of statistics without sacrificing any of the underlying theory. Such a course could use the following material in the second term:

Semester 2

Chapter	Topic
13	Hypothesis Tests for Categorical Data
14	Regression
15	Bayesian Methods
17	The Jackknife and Bootstrap
18	Generalized Linear Models and Regression Trees
20	Large-Scale Hypothesis Testing

For a course that leans toward data science, the second semester coverage might include:

Semester 2

Chapter	Topic
13	Hypothesis Tests for Categorical Data
14	Regression
16	Time Series Methods
17	The Jackknife and Bootstrap
18	Generalized Linear Models and Regression Trees
19	Cross-Validation and Estimates of Prediction Error
20	Large-Scale Hypothesis Testing

A course for scientists or engineers could include selected topics in the above chapters, with additional methods from Chapter 15. For example, a course in biostatistics might emphasize the sections on logistic regression, discrimination, and classification since these are frequently used in medical and public health research. Such a course could minimize or skip material on regression trees. Instructors could also use this as a textbook for a one-semester course by selecting (and omitting) material in the early part of the book. For example, the following chapters could be covered in a one-semester course:

One-semester course emphasizing statistics

Chapter	Topic
1	Introduction
2	Data Visualization (omitting data visualization for survey data, geospatial data, and network data)
3	Basic Probability
4	Random Variables
5	Discrete Distributions (possibly omitting the hypergeometric and multinomial distributions)
6	Continuous Distributions (possibly skipping the Weibull, Beta distributions, and the sections on transformations, moment-generating functions, and QQ plots)
7	About Data and Data Collection (hitting just the main ideas)
8	Sampling Distributions (skipping the proof of the Central Limit Theorem)
9	Point Estimation
10	Confidence Intervals
11	Hypothesis Testing
12	Hypothesis Tests for Two or More Populations
13 or 14	Hypothesis Tests for Categorical Data/Regression

For situations where students have had a prior course on statistics (possibly one that did not use calculus), a course could be designed to emphasize data science:

One-semester course emphasizing data science

Chapter	Topics
4–6	Select topics in these chapters to bring students up to speed
7	About Data and Data Collection (hitting just the main ideas)
8	Sampling Distributions (skipping the proof of the Central Limit Theorem)
9	Point Estimation
10	Confidence Intervals
11	Hypothesis Testing
12	Hypothesis Tests for Two or More Populations
13	Hypothesis Tests for Categorical Data
14	Regression
17.	The Jackknife and Bootstrap
18.	Generalized Linear Models and Regression Trees
20.	Large-Scale Hypothesis Testing

This book was typeset in LaTeX using a modified version of The Legrand Orange Book template originally created by Mathias Legrand and modified by Vel and the authors.

We would like to thank Emily Rigdon for LaTeXing much of the material in the book and Gary Smith for his careful reading and editing of the manuscript. We would also like to thank the staff at Cambridge, especially Lauren Cowles, Maggie Jeffers, and Lucy Edwards for their help in molding this book into what it has become, and for their patience through the process.

Steven E. Rigdon

Ronald D. Fricker, Jr.

Douglas C. Montgomery

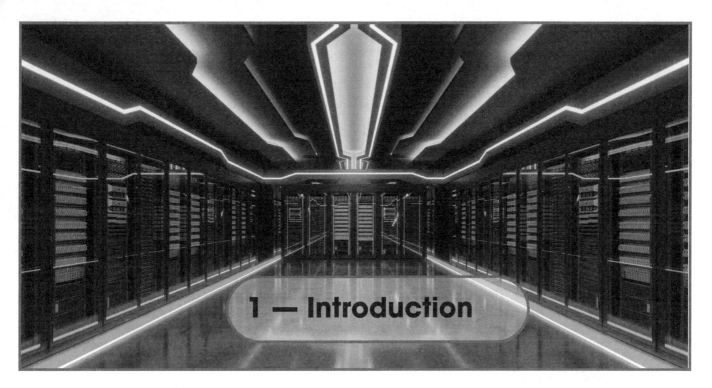

1 — Introduction

In the 1990s and before, most of the world's information was stored on paper and other analog media, such as film. However, with the proliferation of personal computers and the internet, by 2000 one-quarter of the world's information was stored digitally. Since that time, the amount of digital data has exploded, roughly doubling every couple of years, so that now *more than 98% of all stored information is digital*.

Much of this digital data is the result of the *datafication* of the world. Datafication is both the digitization of existing analog media and, more significantly, the collection of digital data on people, processes, and other things in ways that until recently were not possible. For example, the rise of social media has resulted in the generation of massive amounts of digital data by and about individuals throughout the world. More generally, the proliferation of smart sensors and ever-cheaper storage is driving the availability of data from all types of societal, commercial, and government processes and systems. The result is an exponentially increasing amount of data being collected and stored, much of which is in need of analysis so that useful information can be extracted from the data.

What does some of this data look like? In May 2018, Bernard Marr, writing for *Forbes* magazine, said

> The amount of data we produce every day is truly mind-boggling. There are 2.5 quintillion bytes of data created each day at our current pace, but that pace is only accelerating with the growth of the Internet of Things (IoT). Over the last two years alone 90 percent of the data in the world was generated. This is worth re-reading!

The information cited by Marr, which is now several years old, includes over 527,000 photos shared on Snapchat, 120 new professionals on LinkedIn, over 4 million YouTube videos, and over 450,000 tweets; all of these statistics are per minute! These numbers have increased since 2018 and are continuing to increase. We use Google to conduct 40,000 searches per second. One-fifth of the world's population, 1.5 billion people, use Facebook daily.

This flood of digital data contains valuable information that can be used to inform all types of decisions. Indeed, cutting-edge commercial organizations now focus their operations around the analysis and exploitation of knowledge gleaned from data. This is what data science is all about: turning data into useful information.

1.1   Data Science and Statistics

Both data science and statistics are concerned with the extraction of useful information from data. Let's start by defining the two fields.

> **Definition 1.1.1 — Statistics.** Statistics is the science of learning from data, including collecting, organizing, analyzing, interpreting, and presenting data, often with a particular focus on measuring, controlling, and communicating the uncertainty inherent in the data, associated analyses, and final results or conclusions.

Statistics traces its roots back to 2 CE, when the Han dynasty conducted a census of the Chinese population, where it counted 57.7 million people in 12.4 million households. As this early application illustrates, statistics is about both the collection of data and its analysis. Furthermore, as Definition 1.1.1 makes clear, statistics is also focused on determining the uncertainty in data and in the conclusions drawn from data. This is an important consideration when only a sample of data is observed because the results will be subject to sampling error, and classical statistics is generally predicated on the idea that it's not possible to observe all the data, either because it's too expensive or it's impossible to collect it all. We'll explore this concept more in Section 1.2, but briefly the idea is to use the data not only to summarize what was observed but also to quantify what can be said about *all* of the data, both observed *and* unobserved.

Statistics can also be split into two broad subfields: theoretical statistics and applied statistics. Theoretical statistics is concerned with the creation and development of methods and techniques to summarize and analyze data, including clearly defining how and when to use the methods and their associated pros and cons. Applied statistics, on the other hand, is the application of the methods to data and the process of conducting rigorous and principled analyses. In the statistical profession, the division between applied and theoretical statisticians is fuzzy, with most statisticians doing both, though perhaps with an emphasis on one or the other.

The theory and practice of statistics go hand-in-hand. A rigorous theoretical foundation is what keeps statistics from being a purely empirical exercise of sifting through data and a rich set of applied problems is what keeps statistics relevant to solving real-world problems. As David Bartholomew (1995), former president of the Royal Statistical Society, said, "There can be no statistics without data and no statistics with data alone."

When developing methods, statisticians seek to understand how the methods perform on particular types of data, including how efficiently they extract information from the data and how well they characterize the uncertainty inherent in a sample of data. That is, just as automobile designers seek to understand the performance characteristics of a new car, statisticians seek to understand how their methods perform. In particular, as Lindsay et al. (2004) say, "A distinguishing feature of the statistics profession, and the methodology it develops, is the focus on a set of cautious principles for drawing scientific conclusions from data."

> **Definition 1.1.2 — Data Science.** Data science is the study of how to extract useful information from data using quantitative methods and theories from many fields, including statistics, operations research, computer science, and various engineering disciplines. Data science often focuses on large data sets not originally designed or collected to address the question of interest.

In many ways data science is a modern extension of statistics and, to the extent they use statistical methods, data scientists can be characterized as applied statisticians. Indeed, while some trace the inception of data science to the 1960s, where it was then focused on data processing, modern data science originated with a lecture given by Professor C.F. Jeff Wu in 1998 entitled "Statistics = Data Science?" In that lecture, Professor Wu characterized statistical work as data modeling, analysis, and decision making, and he proposed that statistics be renamed data science and statisticians be called data scientists.

However, since the late 1990s, particularly with the explosion of massive, heterogeneous, and often unstructured data sets, the term data science has expanded to include the ability to collect, manage, and analyze such data. Today, data scientists are expected to be adept in both statistics and computer science, particularly as applied to extracting and manipulating large data sets, as well as to have a solid working knowledge of the field in which they are trying to answer questions. In particular, data scientists must be able to:

- find, manage, and interpret large and complex data sets;
- analyze the data, including building mathematical models; and
- present and communicate results.

An important distinction between data science and statistics is whether we wish to explain or predict.[1] The goal of a study may be to make predictions. For example, the goal might be to predict when the actual number of people with influenza is high, and the predictors might include internet searches. Here the goal is to make accurate *predictions*, not to explain what variables affect the response and the extent to which they have an effect. Other times, the goal might be to explain why something happens. For example, when a cluster of cancer cases is discovered, epidemiologists will use data to search for a cause. This is an example of using data to *explain*. Often data scientists make predictions and use methodology designed for this purpose; statisticians develop models that explain how the world works. This overgeneralizes, since both perform both tasks, but this is a useful way of looking at the differences between data science and statistics.

As a result, as Definition 1.1.2 makes clear, today's data scientists come from a variety of fields and academic backgrounds and they collect and analyze data using a variety of methods. Thus, data science now extends beyond the realm of traditional statistics that was generally focused on collecting and analyzing smaller and typically very structured number-based data sets. Yet, coming full circle, those who collected and collated the census data back in 2 CE for the Han dynasty – where collecting information on 57.7 million people was undoubtedly a huge undertaking that resulted in a massive amount of data for that era – could have been called data scientists!

1.2 More on Statistics

While most of the colloquial and popular media references to statistics concern the collection and summarization of a set of numbers (e.g., baseball statistics or stock market returns), real statistics is about much more than that. If the field of statistics were only concerned with describing data, that is, *descriptive statistics*, this book would conclude with Chapter 2.

Statistics is most fundamentally about methods for describing uncertainty. For example, uncertainty may arise if a data science question is about a particular *population* but data are only available on a subset of the population – a *sample*. Hence, there is uncertainty about how closely the results from the sample correspond to the results for the population. Similarly, uncertainty may arise if the data science question involves forecasting the future which, of course, can only be answered using data from the past and present.

For example, what if we wanted to know the average starting salary for a person obtaining a master's degree in data science in the United States? One way to find out would entail getting the salary information for every new data scientist in the United States and then calculating the average. The left side of Figure 1.1 illustrates this idea. However, obtaining the starting salary data for every single data scientist in the United States is probably impossible. Alternatively, we could collect the starting salary information from a *sample* of new data scientists with a master's degree and use it to *estimate* the average salary of the entire population. The right side of Figure 1.1 illustrates this idea, where the goal is to use

---

[1]This distinction is the topic of a paper by Shmueli (2010).

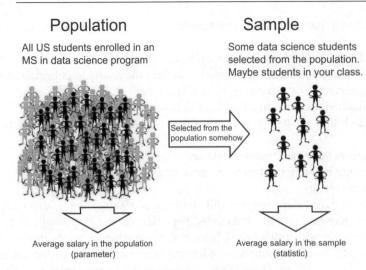

Figure 1.1 Calculating the average starting salary for the entire population of data scientists in the United States versus the average starting salary for a sample of data scientists. The average calculated from a sample is unlikely to be the same as the average calculated from the population.

the sample results to understand the population, and so it is clearly important to ensure the sample is representative of the population.

In either situation the natural question that arises is "How far off is the estimated or predicted average salary from the actual value?" After all, in Figure 1.1 the sample is *not* the population, so any analysis on sample data is likely to differ from the same analysis done on the complete population's data. Statistical methodology is designed to formally specify the precision and uncertainty inherent in any such estimate or prediction.

Of course, since it is often difficult or impossible to know the true result for the population, it can be easy for unprincipled analysts to fool those not skilled in statistics. *Good statistics is about defining mathematically rigorous ways to do estimation, hypothesis testing, and modeling combined with principled methods for quantifying how far off an estimate is likely to be from the true answer.* That may sound like a bit of magic, but you will learn how to do it in this book!

1.2.1 Populations and Samples

We used the terms *population* and *sample* to motivate what statistics is all about: quantifying uncertainty. A population is the set of all people or things that meet the criteria of a particular research study or data science question. A sample is a subset of the population upon which the study or analysis will actually be done. A *random sample* is a subset that is not drawn in any systematic way from the population. (We'll learn more about sampling in Chapter 7.)

For example, if we were interested in saying something about the average GRE scores for graduate students studying data science this year, then the population would be all students enrolled in a data science degree program this year. A sample of that population could be the students in your statistics class. That sample is not likely to be random, however, since it systematically excludes certain groups of students (such as students enrolled in data science programs at other schools).

If we are interested in the average height of students in your statistics class, the class is the population. A sample might be all the women in the class. Is that a random sample? If we used the average of the heights of the women in the class to estimate the average height of all students in the class, would we be making a good estimate?

Why sample? Often it is either impossible or financially prohibitive to observe an entire population. In fact, sometimes even with significant resources and extraordinary effort it is difficult to accurately measure an entire population. A good example is the US Census. Every decade the US government spends millions of dollars and puts forth significant effort trying to count every individual in the country. And, every decade, the Census is challenged for failing to accurately count certain segments of society.

Several governmental agencies collect large amounts of data in complex surveys. For example, the Centers for Disease Control and Prevention (CDC), part of the US Department of Health and Human Services, selects approximately 10,000 participants and provides thorough health data through a physical exam. The sampling design involves both stratification and cluster sampling, and analysis of the National Health and Nutrition Examination Survey (NHANES) data must account for this. The CDC also conducts the Behavioral Risk Factor Surveillance System, which interviews 400,000 adults regarding their health and risk behavior. The US Department of Justice administers the National Crime Victimization Survey twice per year. Each time about 50,000 households are selected for interview. All of these surveys involve a complicated sampling design which must be taken into account when doing an analysis. Interested readers are referred to Lohr (2010).

As it turns out, with good statistical practice we can often get as precise answers from a sample as we can from an attempt to collect data from the whole population. There are times when taking a sample can be *more* precise than trying to get the whole population. How can this be? Well, for the same amount of effort or cost one can either get precise data from the sample or imprecise data from the population. The idea is that under certain conditions it is preferable to allow for a moderate increase in *sampling error* in order to achieve a greater reduction in *measurement error.*

1.2.2 Descriptive versus Inferential Statistics

Descriptive statistics and data visualization are ways to numerically and graphically summarize data, whether the data are from a sample or a population. Why is this important? Think about the US Census, with its information on more than 300 million people. If we wanted to understand the economic status of people in the United States we would certainly not want to do so by looking at each and every Census record. Rather, we would use ways to describe the data in a more concise way, either through summary statistics or graphical plots of the data. That is, we would use descriptive statistics to summarize the data.

Most of the rest of this book is about *inferential statistics*, though we will have to spend quite a few chapters developing the probability tools we need to do statistical inference first. This is the machinery designed for using a *statistic* calculated from a sample of data to say something about the population. As illustrated in Figure 1.2, if it is impossible to obtain the starting salary for every data scientist in the United States, then we will have to use information from a sample to *infer* what it is for the population. However, inference is also more than using a sample average as an estimate for a population average; it

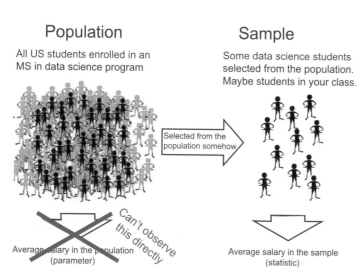

Population

All US students enrolled in an MS in data science program

Sample

Some data science students selected from the population. Maybe students in your class.

Selected from the population somehow

Can't observe this directly

Average salary in the population (parameter)

Average salary in the sample (statistic)

Figure 1.2 Statistical inferential methods are designed to estimate population *parameters* using *statistics* generated from a sample and to quantify the precision of the estimate.

is also about statistical methods to quantify how accurate the sample average is and thereby specify our uncertainty in our knowledge of the population.

1.3 An Introduction to R

R is cutting-edge, free, open-source statistical software. R runs on a wide variety of UNIX platforms, and on the Windows and MacOS operating systems. To download it, go to www.r-project.org and follow the downloading and installation instructions. There are many good tutorials and books on R coding. We assume that the reader knows the basics of R and how it runs. In this book we will explain how R is used to analyze data while assuming some familiarity with R.

A common file format for data is csv, which stands for comma separated variables. Given that your data are in csv format, they can be read using the read.csv() function, where the argument is the file name (including the appropriate path) enclosed in quotes. It is customary to set R's working directory through the setwd command; for example:

```
setwd("/Users/rdfricker/Desktop")
```

Once the directory is set, R looks for files in this directory. If you are a PC user, note the use of forward slashes rather than backslashes. Also note that the default option for the read.csv() function is that the first line of the file contains the variable names and each subsequent line contains the data, one line for each observation (row) in the data, where each item in the data is separated from the next by a comma. If your file begins with the data in row 1 – that is, there is no row for the variable names – you can set the header to be FALSE:

```
my.data <- read.csv("/Users/rdfricker/Desktop/data.csv",header=FALSE)
```

You can then assign names to the columns using the names() function:

```
names(my.data) = c("var1","var2","var3")
```

This will give names to the first three variables in your data frame.

Functions

One of the strengths of R is the ability to use and write functions. Functions are basically mini-programs, where you can create a function that even calls other functions. Just about everything you do in R involves applying a function, usually to data. And, as we'll discuss shortly, R users can write their own functions that can be published to the wider R community via packages that everyone can then download and use.

R has thousands of functions. We'll use many of them as we go through the text. For now, below is a list of those you need to know to get started.

- c() is the concatenate function, which is often used to create a vector, as in vector1 <- c(1,2,3) or to join two or more vectors, as in new.vector <- c(vector1,vector1).
- dim(*data*) returns the number of rows and columns respectively in *data* which can be either a data frame object or matrix object.
- help(*function.name*), ?*function.name*, and ??*text* are useful for getting help.
- is.na() is a function that returns a logical object indicating whether data are missing (i.e., NA) or not.
- length(*vector*) returns the number of elements in *vector*.
- library(*name*) loads the package called *name* so you can use/access its contents.
- ls() lists the objects in the workspace. No arguments are required to run the function, but you do have to type the parentheses because it is a function.

- `read.csv()` and `read.table()` are useful for reading data into R. For reading data in specialized formats, the `foreign` package is very useful.
- `rm(`*name*`)` deletes the object *name* from the workspace. It works on single objects or a series of objects separated by commas.
- `rm(list = ls(all = TRUE))` deletes all the files in your workspace. Be very cautious when using this – there is no undo option!
- `sin`, `cos`, `exp`, and `log` are, respectively, the sin, cos, exponential (i.e., e^x), and natural log functions. Most other mathematical functions you've encountered are built into R.

We can define and write our own functions in R using the `function` command. The following takes one argument x and returns $(x+1)^2$:

```
f = function(x)
{
        y = x + 1
        z = y^2
        return( z )
}
```

This code defines the function `f` and says that it takes one argument x. It then computes y = x+1 and finally squares it to obtain the variable z. The statement `return(z)` returns the computed value of z back to R. When we run the above code, nothing seems to happen, although after having run it we notice that the function is now in R's global environment; this can be seen in the upper right panel in R Studio. Once the function is executed, we can call it using the syntax `f(x)`, where x is some number or variable. For example, typing `f(9)` yields

```
f(9)
[1] 100
```

1.4   Descriptive Statistics

Descriptive statistics is all about summarizing data. In this chapter we will learn how to describe data numerically using statistics; in Chapter 2 we will learn how to graphically describe data. In fact, a *statistic* is simply a number calculated from data that summarizes something about the data. For example, the average is a statistic, and there are many other types that we will learn about in this chapter.

Descriptive statistics are becoming increasingly important as data scientists deal with ever-larger data sets and as data collection accelerates in our ever more computerized and interconnected world. They are important because the human mind is limited in its ability to assimilate individual facts; we simply aren't good at being able to synthesize lots of numbers. Indeed, human short-term memory capacity is only about seven digits – the length of a US telephone number *not* counting the area code. So, in our increasingly data-rich world, knowing how to appropriately summarize data is a critical skill.

Returning to our discussion of the US Census in Section 1.2.1, the only way to get some understanding of the US population using the Census is to apply descriptive statistics (as well as other methods we will talk about in later chapters) to summarize the data. Just looking at individual Census records would not provide us with much insight into the entire US population. And, while the 300 million plus Census records may sound like a lot, these days it is not a big data set: There are now more tweets sent *per day* than there are Census records.

Before we proceed further, let's formalize what we've just discussed with some definitions.

Definition 1.4.1 — Data. Information, often numerical but not necessarily so, collected from an experiment, a survey, administrative records, the internet, etc. The word "data" is plural. One piece of information is a "datum."

> **Definition 1.4.2 — Statistic.** A numerical fact, usually computed from a data set. Statistics can also be computed from subsets of the data and can even be just a single datum.

> **Definition 1.4.3 — Descriptive Statistic.** A statistic that usefully summarizes a data set, where the data can be either for an entire population or for a subset of the population.

Good descriptive statistics can help data scientists understand what the data are trying to say. They can highlight and bring out the underlying information in a data set, which might not (and probably will not) be evident by just inspecting the individual data elements.

1.4.1 Types of Data

We can divide data into two basic types, quantitative and qualitative. *Quantitative* data are data that can be measured or characterized with a numerical value; *qualitative* data cannot be so measured. For example, if we think about demographic data, height, weight, and age are all quantitative, while gender and eye color are qualitative.

We can also divide data into *cross-sectional* data and *longitudinal* (also known as *time series*) data. Cross-sectional data are data that occur either in one time period or are constant over time, while longitudinal data covers multiple time periods or varies over time. For example, referring back to the previous demographic data, gender and eye color are cross-sectional in the sense that they are unlikely to change over time. Height and weight data could be cross-sectional if they are recorded for one period of time and longitudinal if they are repeatedly measured over multiple time periods.

As shown in Figure 1.3, we can further describe quantitative data as either *continuous* or *discrete*. Data are discrete if there are gaps between the values the data can assume. For example, the number of people in a family can be 1, 2, 3, …. It cannot be 2.7. If there are no gaps between possible data values, then we say the data are continuous. Another way to think about continuous data is as if we had an infinitely accurate measuring device, then we could express the data to any number of decimal places and it would always make sense. For example, height is continuous: we can talk about someone being 6 feet tall, or 5.97 feet tall, or 5.9722683 feet tall.[2]

Unlike quantitative data, qualitative data cannot be measured or described numerically. As shown in Figure 1.3, qualitative data can be either *nominal* or *ordinal*. Ordinal qualitative data are data for which there is a natural ordering, but the data cannot be expressed on a numerical scale. For example, shirt size is ordinal: "large" shirts are bigger than "medium" shirts which are bigger than "small" shirts. In contrast, with nominal data there is no natural ordering to the data. For example, gender is a nominal type of data: each person can be classified as "male" or "female," but it does not make any sense to say that "male" is greater than "female."

Note that nominal data can be represented numerically for purposes of analysis, but care must be taken not to over-interpret the numerical labels. For example, we will sometimes use an *indicator*

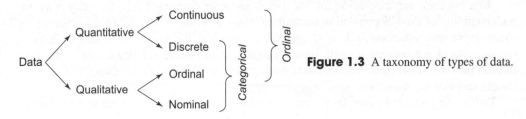

Figure 1.3 A taxonomy of types of data.

---

[2]In practice, height can be measured to a fixed degree of precision, probably about one-tenth of an inch. Thus, all data are discrete because we can measure to a fixed degree of precision. Despite this, it is often helpful to think of data such as height as continuous. All models, such as how we measure height, are *approximations* of reality.

variable to do analyses, say setting a variable called gender equal to 1 for men and 0 for women. But just because the numbers 0 and 1 are ordinal, this property does not carry over to the original qualitative variable, and so care must be taken not to over-interpret or misuse the indicator variable.

We use the term *categorical* to refer to data that are either discrete or qualitative because we can naturally categorize these types of data into groups. Continuous data are clearly ordinal since numeric data have an obvious ordering. Finally, note that we can turn continuous data into categorical data by defining ranges of values for each category. For example, for height, people may be categorized as "short" if they are less than 5 feet tall; "average" if they are between 5 and 6 feet; and "tall" if they are 6 feet or greater. The reason these distinctions are important is that the appropriate statistical analyses, and even the proper way to display data, will depend on the type of data.

1.4.2 Example Data: US Domestic Flights from 1987 to 2008

To help make the ideas and methods of this chapter concrete, we will illustrate them using a data set consisting of US domestic airline flight arrivals and departures and associated details for all commercial flights from October 1987 to April 2008. This is a large data set: there are nearly 120 million records for 3,376 airports. In this chapter we'll mainly focus on the data from years 1988, 1997, and 2007, where there are 5,202,096, 5,411,843, and 7,453,215 observations (i.e., flights), respectively, for those years.

See the book's website for instructions on how to download this and other data sets. As shown in Table 1.1, the data set contains information on the date, day, and time of each flight, the airline and particular airplane, flight origin and destination, and lots of information about the length of each flight and the types of delays, if any, experienced. Table 1.2 shows five randomly selected observations from the 2007 data.

Looking through Table 1.2 we see, for example, that the time an airplane is in the air (see the AirTime variable) and the distance flown (see the Distance variable) are continuous data, while the day of the month (see the DayofMonth variable) is an example of discrete data. The reason for flight cancellation (see the CancellationCode variable) and airport of origination (see the Origin variable) are examples of nominal data.

Some of the data that are numerically coded are actually qualitative. For example, as we see in Table 1.2, the day of the week (see the DayOfWeek variable) contains the integers 1 to 7, where the number 1 corresponds to Monday, 2 to Tuesday, etc. Though these data are numbers, they are not quantitative data. Instead, the numbers are only codes that represent ordinal qualitative data. Similarly, flight numbers (FlightNum) are also coded numerically, but they are really nominal.

1.5 Cross-Sectional Data

As we just discussed, cross-sectional data are collected during the same period of time. Statistics can then be used to summarize these data.

1.5.1 Measures of Location

Measures of location, also referred to as *measures of central tendency*, are typically used to quantify where the "center" or mass of the data is located. There are a number of common measures of central tendency, each of which quantifies the "center" in a different way. The most common measure is the *mean*, which is the average of a set of observations in either a sample or a population.

> **Definition 1.5.1 — Population Mean.** For data from a population, denoted $x_1, \ldots, x_N$, the population mean is calculated as
>
> $$\mu = \frac{1}{N} \sum_{i=1}^{N} x_i = \frac{x_1 + x_2 + \cdots + x_N}{N}.$$

Table 1.1 Aircraft data set variables and brief descriptions.

	Variable name	Description
1	Year	1987–2008
2	Month	1–12
3	DayofMonth	1–31
4	DayOfWeek	1 (Monday) – 7 (Sunday)
5	DepTime	Actual departure time (local, hhmm)
6	CRSDepTime	Scheduled departure time (local, hhmm)
7	ArrTime	Actual arrival time (local, hhmm)
8	CRSArrTime	Scheduled arrival time (local, hhmm)
9	UniqueCarrier	Unique carrier code
10	FlightNum	Flight number
11	TailNum	Plane tail number
12	ActualElapsedTime	In minutes
13	CRSElapsedTime	In minutes
14	AirTime	In minutes
15	ArrDelay	Arrival delay, in minutes
16	DepDelay	Departure delay, in minutes
17	Origin	Origin IATA airport code
18	Dest	Destination IATA airport code
19	Distance	In miles
20	TaxiIn	Taxi in time, in minutes
21	TaxiOut	Taxi out time, in minutes
22	Cancelled	1 = yes, 0 = no
23	CancellationCode	Reason for cancellation (A = carrier, B = weather, C = NAS, D = security)
24	Diverted	1 = yes, 0 = no
25	CarrierDelay	In minutes
26	WeatherDelay	In minutes
27	NASDelay	In minutes
28	SecurityDelay	In minutes
29	LateAircraftDelay	In minutes

Definition 1.5.2 — Sample Mean. For a sample of data, $x_1, \ldots, x_n$, the sample mean $\bar{x}$ is calculated as

$$\bar{x} = \frac{1}{n}\sum_{i=1}^{n} x_i.$$

The difference between the population mean μ and the sample mean $\bar{x}$ is subtle, but important. The population mean μ is the average across all units in the population. Usually μ is unknown, but it is important to have a symbol for it because estimating it is an important problem in statistics. The sample mean $\bar{x}$ is usually known, because it is the result of an observed sample. We often use $\bar{x}$ to estimate μ. More on this idea of estimation in Chapter 9.

■ **Example 1.1 — Calculating the sample mean.** Calculate the mean for the following sample of data: {2.3, 8.1, 5.5, 9.0, 7.8}.

Table 1.2 Five randomly selected observations from the 2007 aircraft data set.

	Year	Month	DayofMonth	DayOfWeek	DepTime
1	2007	3	30	5	1339
2	2007	4	2	1	NA
3	2007	4	30	1	1353
4	2007	12	27	4	2235
5	2007	1	20	6	805

	CRSDepTime	ArrTime	CRSArrTime	UniqueCarrier	FlightNum
1	1340	1548	1588	YV	7352
2	824	NA	1005	OO	6625
3	1348	1420	1410	AS	381
4	2135	143	33	B6	861
5	805	1045	1033	YV	2681

	TailNum	ActualElapsedTime	CRSElapsedTime	AirTime	ArrDelay
1	N516LR	69	78	49	-10
2	0	NA	101	NA	NA
3	N788AS	87	82	63	10
4	N588JB	188	178	146	70
5	N911FJ	160	148	139	12

	DepDelay	Origin	Dest	Distance	TaxiIn
1	-1	ORD	CAK	344	3
2	NA	DEN	BZN	525	0
3	5	BOI	SEA	399	8
4	60	JFK	TPA	1005	5
5	0	PWM	CLT	812	8

	TaxiOut	Cancelled	CancellationCode	Diverted	CarrierDelay
1	17	0		0	0
2	0	1	A	0	0
3	16	0		0	0
4	37	0		0	0
5	13	0		0	0

	WeatherDelay	NASDelay	SecurityDelay	LateAircraftDelay
1	0	0	0	0
2	0	0	0	0
3	0	0	0	0
4	0	10	0	60
5	0	0	0	0

Here we have $n = 5$, so

$$\bar{x} = \frac{1}{5} \sum_{i=1}^{5} x_i = \frac{2.3 + 8.1 + 5.5 + 9.0 + 7.8}{5} = \frac{32.7}{5} = 6.54.$$

■ ■ ■

A couple of additional points on notation:

- Using lower case Roman letters for data (i.e., x) indicates that this is a calculation for a particular sample of numbers. That is, the letter x represents a particular number. (In Chapter 3 we will talk about random variables which will be denoted by capital letters.)
- Similarly, it is convention to use the letter n to indicate the number of items in a sample and for N to denote the population size. We usually assume that the sample is drawn from the population *without replacement*, meaning each observation in the population can only be drawn and put into the sample once.
- In later chapters we will use the sample mean to estimate the population mean (for those times when it is not possible to observe the whole population). In such cases it is common to use the notation $\hat{\mu}$ instead of $\bar{x}$ to indicate that the sample mean is an estimate of the population mean μ. (This notation is used more generally, putting a "hat" over the symbol for a population quantity to indicate an *estimate* of that population quantity.)

In R, the `mean()` function can be used to calculate either a population or sample mean, depending simply on whether the data set is for a population or a sample. For example,

```
mean(c(2.3, 8.1, 5.5, 9.0, 7.8))
```

yields

```
[1] 6.54
```

The real power of R comes not from using it to calculate the mean of five numbers – we can do that easily on a calculator or in Excel – but in using it to do calculations for larger data sets such as the airline data. For example, let's look at airline departure delays (contained in the `DepDelay` variable), which is defined as the difference in minutes between the scheduled and actual departure times. Example 1.2 shows the average departure delay for all flights, which we can see has been increasing over the past 20 years.

■ **Example 1.2** Use R to calculate the means for the years 1988, 1997, and 2006.

Solution: The command `read.csv(file.choose())` will open up a dialog box and allow you to choose the file you wish to load. The `na.rm=TRUE` removes all missing values from the calculation.

```
data88 <- read.csv(1988.csv)
data97 <- read.csv(1997.csv)
data07 <- read.csv(2007.csv)
mean(data88$DepDelay, na.rm=TRUE)
mean(data97$DepDelay, na.rm=TRUE)
mean(data07$DepDelay, na.rm=TRUE)
```

R returns this output

```
[1] 6.706768
[1] 8.235566
[1] 11.39914
```

■ ■ ■

In Example 1.2, setting the `na.rm` option to `TRUE` tells the mean function to ignore all the missing observations. Thus, the example shows that the mean departure delay *for those flights that had data* was 6.7 minutes in 1988, 8.2 minutes in 1997, and 11.4 minutes in 2007. This raises the question of the fraction of flights for which we're missing data in each year.

■ **Example 1.3** Calculate the proportion of missing data in the `DepDelay` variable for each of the years 1988, 1997, and 2006.

Solution:

```
mean(as.numeric(is.na(data88$DepDelay)))
mean(as.numeric(is.na(data97$DepDelay)))
mean(as.numeric(is.na(data07$DepDelay)))
```

R returns

```
[1] 0.009642844
[1] 0.01806464
[1] 0.02156761
```

The function `is.na` returns a logical result, TRUE or FALSE. The `as.numeric` function transforms this into a numeric result with 1=TRUE and 0=FALSE, so when we take the mean of this, we are adding up the number of TRUEs and dividing by the number of observations; this is like taking the proportion of TRUEs. ■ ■ ■

Example 1.3 shows that in 1987 just under 1% of the flights did not have departure delay data. This increased to just under 2% in 1997 and just over 2% in 2007. Thus, overall, the amount of missing data is small, but is increasing across time.

Another measure of central tendency is the *median*. The median is simply the middle value of the *ordered* data. If there are an odd number of observations, there is a value in the middle of the ordered data. If there is an even number of observations, then the median is the average of the middle two points of the ordered data.

To express the median in a mathematical formula, we will use the notation $x_{(i)}$ which denotes the *i*th *ordered* observation. For example, returning to the data from Example 1.1, $x_{(1)} = 2.3$, $x_{(2)} = 5.5$, $x_{(3)} = 7.8$, $x_{(4)} = 8.1$, and $x_{(5)} = 9.0$. Then, using this notation, the median is defined below.

> **Definition 1.5.3 — Median.** For a set of data $x_1, x_2, \ldots, x_n$, defining the ordered values to be $x_{(1)}, x_{(2)}, \ldots, x_{(n)}$, the median $\tilde{x}$ is calculated as
>
> $$\tilde{x} = \begin{cases} x_{((n+1)/2)}, & \text{if } n \text{ is odd} \\ \dfrac{x_{(n/2)} + x_{(n/2+1)}}{2}, & \text{if } n \text{ is even.} \end{cases}$$

■ **Example 1.4 — Calculating the sample median.** Calculate the median for the data from Example 1.1: {2.3, 8.1, 5.5, 9.0, 7.8}.

Solution: Because $n = 5$ is odd, $\tilde{x} = x_{\left(\frac{n+1}{2}\right)} = x_{\left(\frac{5+1}{2}\right)} = x_{(3)} = 7.8$. ■ ■ ■

In R, the `median()` function can be used to calculate the median. For example,

```
median(c(2.3, 8.1, 5.5, 9.0, 7.8))
```

yields

```
[1] 7.8
```

While both the median and the mean are measures of central tendency, they are different. If, when we use the term *central tendency*, we mean a typical value in the middle of the data, then it often makes sense to use the median. If, on the other hand, we really want to know the average, then the mean is the appropriate measure. When the data contain one or more *outliers*, meaning data points that are unusually large or small when compared to the rest of the data, the median may be preferred. This is because, as the next example shows, the median is less affected by outliers.

■ **Example 1.5** Consider a sample of seven observations: $\{0,1,2,2,2,3,4\}$. Compute the mean and median.

Solution: For these data the mean equals the median:

$$\bar{x} = \frac{1}{7}\sum_{i=1}^{7} x_i = \frac{0+1+2+2+2+3+4}{7} = \frac{14}{7} = 2,$$

and

$$\tilde{x} = x_{\left(\frac{7+1}{2}\right)} = x_{(4)} = 2.$$

Yet, what if a mistake was made entering the data so that the first observation was accidentally recorded as 70 instead of 0? In that case, the median would be unchanged at 2, but the mean would be 12. This is clearly a big difference, and it's only because of a typographical error for one datum.　　　　　　　　　　　　　　　　　　　　　　　　　　　　　　　　　　■■■

> **Exercise 1.1 — Calculating means and medians.** Confirm that for the data with the typo, $\{70,1,2,2,2,3,4\}$, $\bar{x} = 12$ and $\tilde{x} = 2$.

If the data have one or more extreme outliers, then using the mean as a measure of central tendency could be questionable. For example, with the mistaken data set from Exercise 1.1, the resulting sample average is larger than all but one of the observations. However, the median is the same whether the outlier is included in the data or not. One way to describe this property is to say that the median is *robust* to outliers while the mean is *sensitive* to them. Whenever data are *skewed*, meaning one of the tails contains more observations than the other, then the median is likely to be a better measure of central tendency since it better describes a "typical" value in the data.

The *trimmed mean* is another measure of central tendency that can be a useful compromise between the mean and the median. For $0 \le p \le 0.5$, the $100p$ percent trimmed mean is calculated by first discarding the smallest $100p/2\%$ and the largest $100p/2\%$ of the data values, and then averaging the remaining observations. Mathematically, let $x_{(1)} \le x_{(2)} \le \cdots \le x_{(n)}$ be the order statistics of n observations. Then the trimmed mean $\bar{x}_{tr(p)}$ is defined below.

> **Definition 1.5.4 — Trimmed Mean.** For a sample of data, $x_1, \ldots, x_n$, and some specified fraction $0 \le p \le 0.5$, the $100p$ percent trimmed mean is calculated as
>
> $$\bar{x}_{tr(p)} = \frac{1}{n-2m}\sum_{i=m+1}^{n-m} x_{(i)},$$
>
> where $m = p \times n$. If m is not an integer, simply round it down to the nearest integer.

Note that the $100p$ percent trimmed mean actually deletes $2 \times 100p$ percent of the data. Also note that $\bar{x}_{tr(0)} \equiv \bar{x}$ and $\bar{x}_{tr(0.5)} \equiv \tilde{x}$, though for any other value of $0 < p < 0.5$ there is no guarantee that the trimmed mean will be between the mean and median.

■ **Example 1.6 — Calculating and comparing trimmed means.** Calculate the 20% trimmed mean for the data from Example 1.5, both with and without the typo.

Solution: First, because $p \times n = 0.2 \times 7 = 1.4$, $m = 1$. Then the original set of ordered data $\{0,1,2,2,2,3,4\}$ is trimmed to $\{1,2,2,2,3\}$. That is,

$$\bar{x}_{tr(0.2)} = \frac{1}{n-2m}\sum_{i=m+1}^{n-m} x_{(i)} = \frac{1}{5}\sum_{i=2}^{6} x_{(i)} = \frac{1+2+2+2+3}{5} = \frac{10}{5} = 2.0.$$

In a similar way, the data with the typo, $\{70,1,2,2,2,3,4\}$, is trimmed to $\{2,2,2,3,4\}$ so that

$$\bar{x}_{tr(0.2)} = \frac{2+2+2+3+4}{5} = \frac{13}{5} = 2.6,$$

which, while somewhat larger than $\bar{x} = 2$, is not nearly as large as $\bar{x} = 12$. In R, the trimmed mean is calculated using the mean() function with the trim option:

```
mean(c(0,1,2,2,2,3,4),trim=0.2)
mean(c(70,1,2,2,2,3,4),trim=0.2)
```

This yields

```
[1] 2.0
[1] 2.6
```

■ ■ ■

In the mid-1980s, the mean salary of geography majors at the University of North Carolina was over $150,000 (worth roughly $350,000 in 2020 dollars). Does that sound high? Does it make you rethink your decision to study statistics and explore other options? It turns out that basketball star Michael Jordan graduated in 1986 from the University of North Carolina with a degree in geography. His salary with the Chicago Bulls the following year was $845,000. This one outlier caused the mean salary to be high. The median salary was closer to $40,000. This is why you often hear of *median salary*, instead of the *mean salary*.

Note that the mean, median, and trimmed mean are all appropriate to use with either continuous or discrete data. Returning to the previous example of children in a family, while it does not make sense to say that any particular family has 2.5 children, because of the discrete nature of the data, it is perfectly fine to say that the average number of children per family is 2.5. However, when data are nominal none of these measures work. For nominal data, the *mode*, which is the category with the largest number of observations, is the appropriate measure.

Definition 1.5.5 — Mode. For a sample of data, $x_1,\ldots,x_n$, which could be quantitative or qualitative, the *mode* is the value that occurs most often.

For example, in the airline delay data, we previously noted that in spite of it being coded numerically, the DayOfWeek variable contains qualitative (ordinal) data. Because it's qualitative, it would not make sense to calculate the mean or median for DayOfWeek.

■ **Example 1.7** Use R to create a table of flights by day of the week and observe the mode.
Solution: The R code

```
table(data88$DayOfWeek)
table(data97$DayOfWeek)
table(data07$DayOfWeek)
```

yields

1	2	3	4	5	6	7
755898	757140	757963	753415	766364	697795	713521
1	2	3	4	5	6	7
790298	791617	802130	785731	786342	706198	749527
1	2	3	4	5	6	7
1112474	1078562	1088858	1097738	1101689	933338	1040556

The mode tells us which day of the week has the most flights: for 1988, it's Friday (DayOfWeek = 5); for 1997, it's Wednesday (DayOfWeek = 3), and for 2007 it's Monday (DayOfWeek = 1). ■ ■ ■

Exercise 1.2 Both manually and in R, calculate the mean, median, and 10% trimmed mean of the following sample of data: {98.6, 92.4, 97.4, 118.6, 84.1, 111.1, 84.1, 103.8, 90.5, 102.4, 101.0, 100.6, 109.6, 102.5, 106.0, 98.7, 79.5, 85.0, 114.1, 89.3}.
Answer: $\bar{x} = 98.465$, $\tilde{x} = 99.65$, and $\bar{x}_{\text{tr}(0.1)} = 98.3125$.

1.5.2 Measures of Variation

When describing or summarizing a set of data, providing measures of both location and variation is important. That is, while the mean, trimmed mean, and median provide information about the middle or typical observation in the data, they do not give any information about the variability of the data. But knowing whether the data tend to be close together or spread far apart can be useful and informative.

The most common measure is the standard deviation. It is based on the *sample variance*, s^2, which is defined below.

Definition 1.5.6 — Sample Variance and Standard Deviation. For a sample of data, $x_1, \ldots, x_n$, the sample *variance* s^2 is calculated as

$$s^2 = \frac{1}{n-1} \sum_{i=1}^{n} (x_i - \bar{x})^2.$$

The sample *standard deviation* is the square root of the variance:

$$s = \sqrt{s^2}.$$

The calculation of the sample variance, using Definition 1.5.6, requires two passes through the data. First, we have to calculate the sample mean ($\bar{x}$) and then, once we have that, we must go though the data a second time to calculate the squared differences from the mean for each of the data points. Theorem 1.5.1 provides an alternate "shortcut" formula, that is mathematically equivalent, for calculating the sample variance that only requires one pass through the data.

Theorem 1.5.1 — Sample Variance "Shortcut" Formula. For a sample of data, $x_1, x_2, \ldots, x_n$, a shortcut formula for sample variance is

$$s^2 = \frac{1}{n-1} \left[\sum_{i=1}^{n} x_i^2 - n\bar{x}^2 \right].$$

For a proof, see Problem 1.33.

The sample variance is nearly equal to the average squared deviation from the sample mean. (If the denominator were n instead of $n-1$, then the sample variance would be *exactly* the average squared deviation from the mean.) If most observations are close to the mean, then most of the deviations $x_i - \bar{x}$ will be near zero, so the squared deviations will be small, leading to a small variance. If there are values far from the mean, then $x_i - \bar{x}$ will be large for these values, and consequently the sum of squared deviations will be large. The sample variance measures how spread out the data values are: the larger the sample variance, the more the points are spread out around the mean.

■ **Example 1.8 — Calculating the sample variance.** Calculate the sample variance for the data from Example 1.1, {2.3, 8.1, 5.5, 9.0, 7.8}, both manually and in R.

Solution:

$$
\begin{aligned}
s^2 &= \frac{1}{n-1} \sum_{i=1}^{n} (x_i - \bar{x})^2 \\
&= \frac{1}{4} \sum_{i=1}^{5} (x_i - 6.54)^2 \\
&= \frac{1}{4} \left[(2.3 - 6.54)^2 + (8.1 - 6.54)^2 + (5.5 - 6.54)^2 + (9.0 - 6.54)^2 + (7.8 - 6.54)^2 \right] \\
&\approx 7.283.
\end{aligned}
$$

Now, in R using the `var()` function:

```
var(c(2.3,8.1,5.5,9.0,7.8))
```

yields

```
[1] 7.283
```

■■■

The calculation for the population variance, σ^2, is slightly different from the sample variance. As shown in Definition 1.5.7, the population mean replaces the sample mean in the formula and the sum of the squared differences is divided by N, not $n-1$.

Definition 1.5.7 — Population Variance. For population data, $x_1,\dots,x_N$, the population variance is calculated as

$$
\sigma^2 = \frac{1}{N} \sum_{i=1}^{N} (x_i - \mu)^2.
$$

It is important to note that in most realistic situations we get a look only at a sample of size n from the population of size N. Thus, we don't know, and we'll never know μ and σ^2. It is still worth conceptualizing and naming these variables and we will often want to estimate them from sample data. This is what Chapters 9 and 10 are about.

For both the sample and population calculations, the standard deviation is simply the square root of the variance. Typically the standard deviation is used to characterize variation because it is measured in the same units as the mean.

Definition 1.5.8 — Standard Deviation. The standard deviation is the square root of the variance. For a sample it is $s = \sqrt{s^2}$ and for a population it is $\sigma = \sqrt{\sigma^2}$.

As the next example shows, to calculate the sample standard deviation in R, we can use the `sd()` function or take the square root of the results of the variance. For example, returning to the airline delay data, the standard deviation of `DepDelay` is in minutes, just like the mean and the data, and thus it is interpretable. The variance, on the other hand, is in units of minutes-squared for these data.

■ **Example 1.9** Use R's `sd` function to compute the standard deviation of the airline delays in 1988, 1997, and 2008.

Solution: The R code

```
sd(data88$DepDelay, na.rm=TRUE)
sd(data97$DepDelay, na.rm=TRUE)
sd(data07$DepDelay, na.rm=TRUE)
```

produces

```
[1]  21.77714
[1]  28.47112
[1]  36.14189
```

∎∎∎

The *range, R*, is another measure of variation.[3] The range has the advantage that it is easy to calculate because it is just the difference between the largest and smallest observations. Using the order statistic notation, the range is defined below.

> **Definition 1.5.9 — Range.** For a sample of data, the range R is defined as $R = x_{(n)} - x_{(1)}$, where $x_{(1)}$ and $x_{(n)}$ are the smallest and largest values in the sample; for a population it is $R = x_{(N)} - x_{(1)}$, where $x_{(1)}$ and $x_{(N)}$ are the smallest and largest values in the population.

We can use the `min()` and `max()` functions to calculate the range in a sample. R also has a `range()` function, but it returns the minimum and maximum values for a set of data, not R. However, as Example 1.10 shows, we can use the `range()` function to calculate R in a couple of ways.

■ **Example 1.10** Use the `range()` function to calculate R for the data from Example 1.1: {2.3, 8.1, 5.5, 9.0, 7.8}.

Solution: The `range` function in R gives a vector of the smallest and largest values of the data set. This is a common understanding of the term "range," but it it not how statisticians interpret the range. In statistics, when we say range, we mean the difference between the largest and smallest values. Thus the range is a single number. Here is what the output from the range function looks like:

```
range(c(2.3,8.1,5.5,9.0,7.8))
```

yields

```
[1]  2.3 9.0
```

Using the bracket notation to extract the parts of the `range()` function output, we can calculate R via the expression:

```
range(c(2.3,8.1,5.5,9.0,7.8))[2] - range(c(2.3,8.1,5.5,9.0,7.8))[1]
```

This gives

```
[1]  6.7
```

∎∎∎

> **Exercise 1.3** Both manually and in R, calculate the variance, standard deviation, and range of the sample data from Exercise 1.2: {98.6, 92.4, 97.4, 118.6, 84.1, 111.1, 84.1, 103.8, 90.5, 102.4, 101.0, 100.6, 109.6, 102.5, 106.0, 98.7, 79.5, 85.0, 114.1, 89.3}.
>
> *Answer:* $s^2 = 115.821$, $s = 10.762$, and $R = 39.1$.

1.5.3 Measures of How Two Variables Co-vary

When looking at a data set with more than one variable, it is often natural to ask whether they seem to be related in some way. For two variables x and y, one such measure is the *sample covariance*.

---

[3]Notation confusion alert: The range is denoted by an italicized capital Roman letter *R*. Don't confuse this with the nonitalicized letter R, which refers to the R software program.

> **Definition 1.5.10 — Sample Covariance.** For a sample of data $\{(x_1, y_1), (x_2, y_2), \ldots, (x_n, y_n)\}$, the covariance is calculated as
>
> $$\text{cov} = \frac{1}{n-1} \sum_{i=1}^{n} (x_i - \bar{x})(y_i - \bar{y}).$$

As Definition 1.5.10 shows, covariance is a measure of how two variables co-vary about their means. Note that the data come in pairs: a pair consisting of one x observation and one y observation. Each pair is indexed from 1 to n. So, x_1 goes with y_1, x_2 goes with y_2, on up to x_n with y_n. The equation, then, essentially calculates the average of the product of the differences between each x in the pair from the mean of the xs and each y in the pair from the mean of the ys.

Covariance is a measure of both the strength and direction of the *linear* relationship between x and y. If the covariance is a number with a large magnitude (either positive or negative) then the strength of linear association is great. If the covariance is near zero, then the strength of linear association is weak or nonexistent. Similarly, if the sign of the covariance is positive, then the association is positive (meaning the x values tend to vary in the same direction as the y values); if the sign is negative then the association is negative (meaning the x values tend to vary in the opposite direction to the y values).

■ **Example 1.11 — Estimating the Covariance.** To illustrate the concept of covariance, let's look at data from the 2015 Napa Valley Marathon. Specifically, compute covariance between the variables age (in years) and finishing time (measured in minutes) for the men who participated in the race.

Solution: One might think that there is a relationship between these variables, with lower ages corresponding to shorter finishing times. A scatterplot of age versus finishing time, shown in Figure 1.4, indicates a slight positive relationship. You could predict that an older runner will have a longer finishing time, but there is a lot of variability in such a prediction. We can plot the graph and compute the covariance by saying

```
napa  <- read.csv("napa_marathon_fm2015")
ageM  <- napa$Age[napa$Gender=="M"]
timeM <- napa$Hours[napa$Gender=="M"]*60
plot( ageM , timeM , pch=19 , xlab="Age in Years" ,
      ylab="Finishing Time in Minutes" , xlim=c(10,80) , ylim=c(100,400) )
cov(  ageM , timeM )
```

The covariance is computed to be

```
[1] 83.89024
```

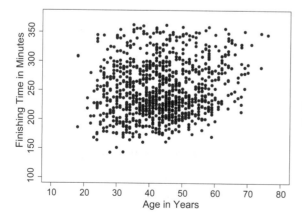

Figure 1.4 Scatterplot of age and finishing time for the 2015 Napa Valley Marathon.

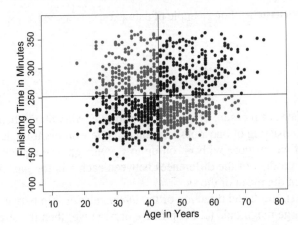

Figure 1.5 Scatterplot of age and finishing time for the 2015 Napa Valley Marathon with blue dots for those instances where both variables were above the mean or both were below the mean, and red where one variable was below the mean and the other was above.

What if the age or finishing time were measured in different units, for example age in months or finishing time in hours? Does this change the covariance? The code

```
cov( ageM*12 , timeM/60 )
```

yields

```
[1] 16.77805
```

so the covariance *does* depend on the scales being used to measure the variables.   ■ ■ ■

To get some insight into what the covariance measures, let's look at the terms in the summation $\sum_{i=1}^{n}(x_i - \bar{x})(y_i - \bar{y})$. Either, or both, of the differences $x_i - \bar{x}$ and $y_i - \bar{y}$ can be negative. If both x_i and y_i are above their respective means, then both will be positive so the product will be positive. If both x_i and y_i are below their respective means, then both will be negative and the product will be positive. If one is below its mean and the other above its mean, then the product will be negative. The numerator in the definition of the covariance is therefore a sum of numbers that can be positive or negative. If most of the points are either both above or both below their means, with few being discordant, the positives will dominate and the covariance will be positive. Conversely, if most points are such that one is above the mean and the other below, then the negatives will dominate and the covariance will be negative. In Figure 1.5 we show horizontal and vertical lines at the means of age and time, respectively. Points whose values are both above or both below the means are indicated in blue, whereas points for which one is above and the other below are in red. Notice that there are slightly more blue dots than red dots, so we would expect the covariance to be positive, and it is.

It is difficult to know when to call a covariance "large" because it depends on how the observation is measured. That is, changing the measurement units changes the value of the computed covariance. Example 1.11 illustrates this with the marathon data, where we see that the covariance between age and finishing time is different when we measure them in different units. This is troublesome, since changing the measurement units (inches instead of feet; grams instead of kilograms) does not change the association between x and y. And, if the association is the same, the numerical measure describing that association should be the same.

The *sample correlation* solves this scaling problem. Sample correlation is the sample covariance divided by the sample standard deviation of the x values and the sample standard deviation of the y values. Denoted by r, the definition of the sample correlation is shown below.

Definition 1.5.11 — Sample Correlation. For a sample of paired data $(x_1, y_1), (x_2, y_2), \ldots, (x_n, y_n)$, the correlation is calculated as

$$r = \frac{\text{cov}}{s_x \times s_y} = \frac{\sum_1^n (x_i - \bar{x})(y_i - \bar{y})}{\sqrt{\sum_1^n (x_i - \bar{x})^2}\sqrt{\sum_1^n (y_i - \bar{y})^2}}.$$

Dividing by the standard deviations of x and y makes the correlation independent of the measurement scale, so that the correlation is always between -1 and 1. That makes interpretation much easier. Example 1.12 revisits the marathon data from Example 1.11 and shows that the correlation is the same whether age is measured in years or months (or any other time unit for that matter).

■ **Example 1.12 — Correlation of age and finishing time in marathon data.** Compute the correlation coefficient between the age and finishing time using the cor() function.

Solution:

```
cor( ageM , timeM )
cor( ageM*12 , timeM/60 )
```

The output from this is

```
[1] 0.1672689
[1] 0.1672689
```

These are identical, indicating that the correlation between two variables does not depend on the scales used to measure them. ■ ■ ■

Figure 1.6 illustrates what various levels of correlation look like. A correlation near 1 is a strong positive linear association, and a correlation near -1 is a strong negative linear association, where a correlation of either $+1$ or -1 is a perfect linear relationship. For example, not surprisingly, the arrival delays are highly correlated with departure delays (see Problem 1.25), which says that if a plane departs late then it is very likely to arrive late as well. A correlation of zero means that there is no linear association between x and y.

The words "association" and "linear" are purposely used in the previous explanation. It is possible for two variables to be related in a nonlinear fashion yet have zero correlation. So, as the next example demonstrates, observing a correlation of zero does not mean there is no association between two variables, only that there is no *linear* relationship.

■ **Example 1.13 — Causation but zero correlation.** Here's a simple example in which there is clear causation between two variables, x and y, where y is a deterministic function of x (so if I tell you x then you also know y), and yet the correlation is (essentially) zero. For the data in the scatterplot of x and $y = x^2$ shown in Figure 1.7, compute the correlation.

Solution: The R code to compute the correlation is this:

```
x <- seq(-1,1,0.1)   # x is a sequence from -1 to 1 in steps of 0.1
y <- x^2             # y is a direct function of x
cor(x,y)
```

which yields

```
[1] 1.216307e-16 # essentially zero correlation
```
■ ■ ■

A nonzero correlation does not mean there is a causal relationship between the variables. Correlation between the two variables can occur for reasons not associated with direct causality, so the most that can be said is that there is an association between the variables. For example, there is a positive correlation between monthly ice cream sales and the number of shark attacks in a month. But eating ice cream does not cause an increase in shark attacks, nor do shark attacks cause an increase in ice cream sales. Rather, *both* tend to occur more often in the summer and less often in the winter.

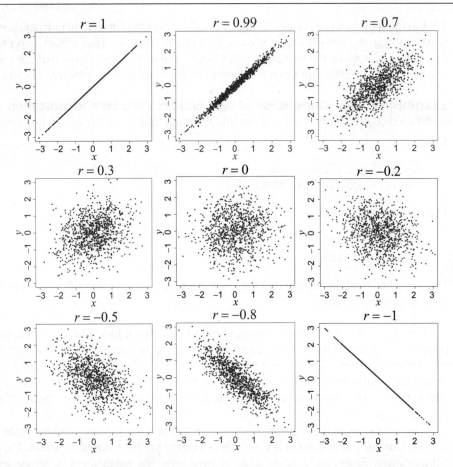

Figure 1.6: Scatterplots illustrating various levels of correlation, from $r = 1$ in the upper left, to $r = 0$ in the middle, to $r = -1$ in the lower right.

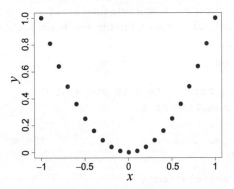

Figure 1.7 A deterministic (completely nonrandom) function $y = x^2$ for $x = -1, -0.9, \ldots, 1$ that yields a zero correlation.

Exercise 1.4 Both manually and in R, calculate the covariance and correlation of the following sample data:

$$x = \{10.9, 10.8, 11.2, 9.1, 11.5, 10.4, 10.5, 10.7, 9.9, 10.1\}$$
$$y = \{22.4, 23.6, 22.5, 20.6, 23.0, 21.6, 21.6, 21.9, 23.1, 22.2\}$$

Answer: cov $= 0.3561, r = 0.5912$.

1.5.4 Other Summary Statistics

There are other useful summary statistics in addition to measures of location and variation. One is the *percentile*. The pth percentile is the value of x such that p percent (where p is between 0 and 100) of the data are less than or equal to x. For a given x, the information communicated by its percentile is where x ranks with respect to the rest of the data. That is, it specifies how much of the rest of the data is less than or equal to x.

A standardized testing example will make the idea more concrete. If a person achieves a score of $x = 720$ on the mathematics section of the SAT, the number 720 is not particularly informative unless we have some idea of where it places the person among the rest of those who took the test. Now, if we're told that a score of 720 is the 96th percentile, then it's now clear what the score means: 96 percent of the other test takers scored 720 or less or, conversely, that only 4% of the test takers did better.

The percentile for a particular value of x in a data set can be calculated directly, and for other values that are not observed it can be interpolated. That is, for a particular order statistic $x_{(i)}$ in a sample of data with n observations, the percentile is calculated as described in Definition 1.5.12.

> **Definition 1.5.12 — Percentiles for Sample Data.** The ith order statistic $x_{(i)}$ is the p_ith percentile of a sample of data with n observations, where
>
> $$p_i = \left(\frac{i-1}{n-1} \right) \times 100.$$

There are some special percentiles. The 100th percentile is the largest (maximum) observation in the data, the median is the 50th percentile, and the 0th percentile is the smallest (minimum) observation. These follow directly from Definition 1.5.12. Also, from this, we can define the range R as the 100th percentile minus the 0th percentile.

■ Example 1.14 — Calculating the percentile values for sample data. Find the p_i values for the data from Example 1.1: $\{2.3, 8.1, 5.5, 9.0, 7.8\}$.

Solution:

i	p_i	Ordered Data
1	$(1-1)/(5-1) \times 100 = 0\%$	2.3
2	$(2-1)/(5-1) \times 100 = 25\%$	5.5
3	$(3-1)/(5-1) \times 100 = 50\%$	7.8
4	$(4-1)/(5-1) \times 100 = 75\%$	8.1
5	$(5-1)/(5-1) \times 100 = 100\%$	9.0

So, for example, 5.5 is the 25th percentile, while 8.1 is the 75th percentile.   ■ ■ ■

As we previously mentioned, linear interpolation can be used to estimate the percentiles of other values between $x_{(1)}$ and $x_{(n)}$. That is, to calculate the percentile for x, where $x_{(1)} < x < x_{(n)}$, find the two order statistics closest to x. Let's call them $x_{(i)}$ and $x_{(i+1)}$, where $x_{(i)} < x < x_{(i+1)}$. Then the percentile for x, let's call it p_x, is

$$p_x = p_i + (p_{i+1} - p_i) \left(\frac{x - x_{(i)}}{x_{(i+1)} - x_{(i)}} \right).$$

Also notice how, with a little algebra, we can rearrange the above expression so that for a given percentile p_x we can find the associated value of x. That is,

$$x = x_{(i)} + \left(x_{(i+1)} - x_{(i)} \right) \left(\frac{p_x - p_i}{p_{i+1} - p_i} \right).$$

■ **Example 1.15 — Calculating particular percentiles.** For the data from Example 1.1, {2.3, 8.1, 5.5, 9.0, 7.8}, find the 95th percentile.

Solution: Because $p_4 < 95\% < p_5$,

$$
\begin{aligned}
x &= x_{(i)} + \left(x_{(i+1)} - x_{(i)}\right)\left(\frac{p_x - p_i}{p_{i+1} - p_i}\right) \\
&= x_{(4)} + \left(x_{(5)} - x_{(4)}\right)\left(\frac{95 - p_4}{p_5 - p_4}\right) \\
&= 8.1 + (9.0 - 8.1)\left(\frac{95 - 75}{100 - 75}\right) \\
&= 8.1 + 0.9 \times 0.2/0.25 \\
&= 8.82.
\end{aligned}
$$

■ ■ ■

Quantiles are often used in the statistical literature instead of percentiles. Quantiles are simply percentiles divided by 100. That is, while percentiles are on a scale of 0 to 100 (%), quantiles are on a scale from 0 to 1. Thus, if we denote quantiles by q_i, then $q_i = p_i/100$.

Those quantiles that divide the data into fourths are called *quartiles*, where the first quartile is the same as the 25th percentile or the 0.25 quantile, the second quartile is the median, and the third quartile is the same as the 75th percentile or the 0.75 quantile. The *interquartile range* or IQR is defined as the 75th percentile (equivalently, the 0.75 quantile) minus the 25th percentile (equivalently, the 0.25 quantile). Appropriately enough, in R the quantile() function is used to calculate quantiles and IQR() calculates the IQR. The quantile function has two arguments: the observed data in the form of a vector and either a single q_i value or a vector of q_i values. The function then returns the quantiles corresponding to the q_i value or values.

Percentiles and quantiles can be useful as a way to understand how the data are distributed. While measures of location tell us what the typical or average value is, and measures of variation tell us something about how spread out the data are, quantiles and percentiles can provide detailed information about how the data are distributed. Example 1.16 illustrates this idea.

■ **Example 1.16 — Airline delay percentiles.** Returning to the airline delay data, find the 5th and the 95th percentiles of the 1988, 1997, and 2007 data.

Solution:

```
quantile(data88$DepDelay,c(0.05,0.95),na.rm=TRUE)
quantile(data97$DepDelay,c(0.05,0.95),na.rm=TRUE)
quantile(data07$DepDelay,c(0.05,0.95),na.rm=TRUE)
```

yields

```
5% 95%
-2  35

5% 95%
-4  48

5% 95%
-9  74
```

In 1988, 5% of the flights actually left two or more minutes early (where the negative number means that the actual departure time was before the scheduled departure time) and only 5% of the flights were delayed more than 35 minutes. By contrast, in 1997, the 95th percentile increased to 48 minutes and in 2007 it's up to 74 minutes. So, even though the 5th percentile indicates that some planes are leaving earlier, the delay time for planes at the 95th percentile more than doubled.   ▪ ▪ ▪

An alternative to the first and third quartiles are *hinges*.

> **Definition 1.5.13 — Hinges.** For a sample of size n, the lower and upper hinges are defined as the $x_{(j)}$ and $x_{(n-j+1)}$ order statistics for
>
> $$j = \frac{\left\lfloor \frac{n+1}{2} + 1 \right\rfloor}{2}.$$
>
> If j is not an integer then appropriately interpolate between the two closest order statistics. Here, $\lfloor x \rfloor$ is the floor function and it means that if x is not an integer then round down. If n is even, the hinges are the median values of the upper and lower halves of the sorted data. If n is odd, the hinges are the median values of the upper and lower halves of the sorted data, *where each half includes the median data point*.

The R function `fivenum()` returns the minimum, lower hinge, median, upper hinge, and maximum of the data. So we see that the hinges exactly match the first and third quartiles. This will always be true when n is odd. When n is even they may not match but will typically be quite close. R also has a `summary()` function that appropriately summarizes whatever object it operates on. The summary function run on a vector of numeric data returns the minimum, first quartile, median, mean, third quartile, and maximum of the data.

▪ **Example 1.17 — Airline delay summary() statistics.** In the airline delay data set, compute the results of the `summary()` function on the `DepDelay` variable.

Solution: The R code

```
summary(data07$DepDelay)
```

gives us

Min.	1st Qu.	Median	Mean	3rd Qu.	Max.	NA's
-305.0	-4.0	0.0	11.4	11.0	2601.0	160748

Note that the `summary` function does not need to be told whether or not to count NA values. It automatically excludes them and provides a count of the number of NA values in the vector.   ▪ ▪ ▪

1.6 Tabular Summaries of Data

For categorical data, sometimes the most useful way to summarize the data is in a table, either in terms of counts, or percentages, or both. Example 1.18 shows a "one-way" table, meaning a table that summarizes one categorical variable.

▪ **Example 1.18 — Tabulating airline cancellation codes for 2007.** The table below is a tabulation of the cancellation codes for the 2007 airline data. "A" means the airline canceled the flight, "B" means the flight was canceled due to weather, "C" means the flight was canceled by the National Air

System (NAS), and "D" means the flight was canceled for security reasons. Tabulate the cancellation codes for all canceled flights.

Solution: The table below shows the number of times flights were canceled as well as the percentage of canceled flights for each code.

	CancellationCode				
	A	B	C	D	Total
Count	66,779	61,936	31,995	39	160,749
%	41.542	38.530	19.904	0.024	100.0

The table shows that the greatest number of cancellations in 2007 were attributed to the airlines, followed closely by the weather. In comparison, cancellations due to the NAS were only about half as many as either the airlines or the weather, and the number canceled for security reasons was a very small fraction of all canceled flights.　　　　　　　　　　　　　　　　　　　　　　　　■ ■ ■

Tables of counts are easy to calculate in R using the `table()` function. For example, the counts in the above table were calculated using the code:

```
table(data07$CancellationCode[data07$CancellationCode!=""])
```

which yields

```
      A      B      C     D
0  66779  61936  31995    39
```

To interpret the R code, first note that `data07$CancellationCode!=""` produces a logical vector that contains `TRUE` and `FALSE` entries. It is true when the factor label of `data07$CancellationCode` is something other than blank and false otherwise.[4] Enclosing this in the square brackets after the vector `CancellationCode` means R will return all values of that vector with a `TRUE` value in the brackets, meaning in this case it will return all nonblank cancellation codes. Hence, in the above output, note there are 0 blank observations (the far left column in the output table).

Why bother with this complicated code? Well, if we just ran the code

```
table(data07$CancellationCode)
```

we would get a table, but the table would contain one more count corresponding to those flights that did not have a cancellation code (i.e., the number of flights that were not canceled). Try it.

This methodology is a useful and powerful approach for subsetting data in R. You can write virtually any logical statement in the square brackets and those observations for which the logical statement is true are then selected. For example,

```
table(data07$CancellationCode[data07$UniqueCarrier=="AA"])
```

would produce a table of cancellation codes for American Airlines flights (carrier code "AA").

---

[4]Note the subtle difference here between a blank factor label and missing data denoted by `NA`. Key is the concept of a factor, which is how the `CancellationCode` variable is coded, in which character vectors are more compactly stored by translating them into integers, one integer for each unique vector entry, along with "levels" that are the actual character values. For a variable containing long character strings this can be an efficient way to store the data since each character string is only stored once and integers, which are likely to be much shorter, are used as placeholders in the vector. This storage scheme is transparent if you print the vector to the screen because R automatically substitutes the factor labels for the integer placeholders before printing the results. So, when R read in the cancellation code data, it created `CancellationCode` as a factor variable. In so doing, all of the fields missing a cancellation code were assigned the integer 1 with corresponding label `""`. The result is that there are no missing values in the vector and, thus, to identify which flights were not canceled we must use the syntax `data07$CancellationCode!=""`.

Exercise 1.5 In R, tabulate the number of 2007 flights delayed by more than three hours by day of the week.

Answer:

	Monday	Tuesday	Wednesday	Thursday	Friday	Saturday	Sunday
Count	23,461	19,400	21,871	27,074	26,235	14,141	22,010

In addition to one-way tables, we can also create two-way or higher tables that give the counts or percentages (or both) for two or more variables. Example 1.19 is a two-way table of cancellation codes by day of the week.

■ **Example 1.19** Compute a two-way table of the number of cancellations by cancellation code and by day of the week for 2007.

Solution: Creating a two-way table in R just requires adding a second categorical variable into the `table()` function. For example,

```
cancellation.codes <- factor(data07$CancellationCode[
                  data07$CancellationCode != ""])
cancellation.dow   <- data07$DayOfWeek[data07$CancellationCode != ""]
table(cancellation.dow,cancellation.codes)
```

The table below shows the results:

	CancellationCode				
Day of the Week	A	B	C	D	Total
Monday	11,314	6,607	4,346	20	22,287
Tuesday	10,612	9,563	5,404	0	25,579
Wednesday	10,196	10,113	6,114	3	26,426
Thursday	9,607	8,891	6,634	14	25,146
Friday	8,923	10,013	5,281	0	24,217
Saturday	7,238	7,402	1,348	2	15,990
Sunday	8,889	9,347	2,868	0	21,104
Total	66,779	61,936	31,995	39	160,749

The table shows, for example, that overall the greatest number of cancellations occurs mid-week and the fewest on the weekend, but for carrier-related cancellations the most cancellations occurred on Monday. ■ ■ ■

1.7 Chapter Summary

This chapter described some important statistical concepts, particularly descriptive versus inferential statistics and populations versus samples. Chapter 2 will explore descriptive statistics in much more detail, and in Chapter 9 and onward we will learn about a host of inferential statistical methods. In Chapter 7 we will learn more about sampling, including different sampling methods, and gain some insight into the utility of sampling.

We covered an introduction to the R language, including functions, data frames, and objects. We also learned how to read data into R and how to get help. This material will be important for understanding the R applications presented throughout the rest of this text and, perhaps more importantly, R is becoming so ubiquitous that one cannot be an effective statistician without being functional in R.

As we discussed in the beginning of the chapter, the purpose of descriptive statistics is simply to summarize data, whether the data is from an entire population or just a subset of the population. And the whole point of these data summaries, as with the visualizations described in the next chapter, is to make it easier to understand the data and thus facilitate insight into the underlying phenomenon that gave rise to the data. Of course, the connection between the descriptive statistics derived from a particular set of data and the appropriate characterization of the underlying phenomenon is also dependent upon how the data were collected and whether it is appropriately representative – something we will learn more about in Chapter 7.

The question of which descriptive statistics to apply to a given data set depends on the type of data and the question or questions of interest. The most important principle, however, is to carefully use descriptive statistics to appropriately summarize the data. The goal is *not* to search for statistics in order to twist the data to fit a preconceived theory or story. Rather, the descriptive statistics should provide an honest and objective summary of the data. Good descriptive statistics, as well as good visualizations, should let the data speak for itself.

1.8 Problems

Problem 1.1 Each week, a company receives from its supplier a shipment of 200 units that are to be used in production. Each time a shipment is received, the receiving company selects 20 units and tests the strength of them. Testing requires destroying the unit. The company accepts the shipment if the strengths are sufficiently high among the 20 tested units.

(a) What is the population?
(b) What is the sample?
(c) Explain why testing *all* 200 units in the shipment is impractical.

Problem 1.2 A university runs a polling group to conduct surveys on various issues. One issue concerns a bond issue for upgrading the system of water lines in the city. A sample of 200 registered voters from the city, which contains 300,000 registered voters, is selected and asked whether they favor a bond issue for this upgrade. Responses are recorded as "Approve," "Don't Approve," or "Don't Know."

(a) What is the population?
(b) What is the sample?
(c) Is it possible to select all 300,000 registered voters and ask them? Is this feasible?

Problem 1.3 Using R, calculate the following:

(a) $1.2 + 3.4 + 5.6 + 7.8 + 9.0$
(b) 3.75×4.1
(c) $5.12 \div 8.45$
(d) $4.9 \times 0.3 + 7.1$

Problem 1.4 Using R, calculate the following:

(a) $(37.923 - 14.720) \times 2.5$
(b) $\left(\sqrt{833} + 47.5\right) / 275.8^2$
(c) $\log_{10}(50)$
(d) $\ln(50)$

Note that the `log()` function in R calculates the natural log and the `log10()` function in R calculates log base 10.

Problem 1.5 Using R, calculate the following:

(a) $(82.71 + 2.73)/(7.32 \times 9.58)$
(b) $\frac{72}{195} \times \frac{33}{94.7}$
(c) $\cos(20)$
(d) e^2

Note that the `cos()` function in R calculates the cosine and the `exp(x)` function in R calculates e^x.

Problem 1.6 Create a new data frame called `item` by inputting the following data directly into R using the concatenate function `c()`:

item number: 28300, 98723, 52038, 23908
weight (lbs): 1.4, 12.3, 3.9, 0.4
cost ($): 1.20, 4.99, 0.75, 0.07

Doing the calculation using the appropriate vectors from the data frame, calculate the cost per pound of each of the items.

Problem 1.7 Repeat the calculations in Problem 1.6 by first creating a file containing the data, say in Excel, and saving it as a csv file. Then read the csv file into R and calculate the cost per pound of each of the items.

Problem 1.8 Look up the R help page for the `read.table()` function. What is the difference between the `read.csv()` and `read.table()` functions?

Problem 1.9 Install the `wordcloud` packages on your computer.

Problem 1.10 Install the `bootstrap` package on your computer.

Problem 1.11 Using the data taxonomy of Figure 1.3, classify what type of data is contained in the following airline dataset variables: `AirTime`, `Origin`, `Cancelled`, `Month`, and `TailNum`.

Problem 1.12 Using the data taxonomy of Figure 1.3, classify what type of data is contained in the measurements on the units in the shipment described in Problem 1.1.

Problem 1.13 Using the data taxonomy of Figure 1.3, classify the type of data contained in the measurement on the survey respondents in Problem 1.2.

Problem 1.14 Using the data taxonomy of Figure 1.3, classify the type of data contained in the variables in Problem 1.6.

Problem 1.15 Manually calculate the mean, median, and 20% trimmed mean for the following sample data:

8, 10, 1, 7, 12, 25, 6, 12, 2,
47, 9, 12, 5, 11

Confirm your results using R.

Problem 1.16 Manually calculate the mean, median, and 10% trimmed mean for the following sample data:

−1.5, −3.3, −42.7, 2.9, 7.4,
0.9, −1.2, 0.3, 4.5

Problem 1.17 Manually calculate the mean, median, mode, variance, and standard deviation of the following sets of numbers:

(a) 0, 0, 0, 0, 1
(b) −0.6325, 0, 0, 0, 0.6325
(c) −0.4473, −0.4473, 0, 0.4473, 0.4473

Problem 1.18 Manually calculate the variance and standard deviation of the data from Problem 1.15. Confirm your results using R.

Problem 1.19 Manually calculate the variance and standard deviation of the data from Problem 1.16. Confirm your results using R.

Problem 1.20 Manually calculate the 25th and 75th percentiles and the hinges of the data from Problem 1.15. Confirm your results using R.

Problem 1.21 Manually calculate the 25th and 75th percentiles and the hinges of the data from Problem 1.16. Confirm your results using R.

Problem 1.22 Use R to calculate the covariance and correlation for the following data:
x: $-0.779, 0.253, 0.529, 1.632, 1.332, -2.167$
y: $-0.415, 1.482, 2.181, 4.240, 3.633, -3.332$.

Problem 1.23 Using R, calculate the 2007 aircraft flight arrival delay (the `ArrDelay` variable) mean, median, and 10% trimmed mean. Also calculate the variance, standard deviation, and range for the 2007 flight times.

Problem 1.24 Using R, calculate the 2007 aircraft flight time (the `AirTime` variable) mean, median, 10% trimmed mean, and standard deviation.

Problem 1.25 Using R, calculate the covariance and correlation between the 2007 aircraft flight arrival delay (the `ArrDelay` variable) and the aircraft flight time (the `AirTime` variable). Are the two variables highly correlated?

Problem 1.26 Using R, calculate the covariance and correlation between the 2007 aircraft flight time (the `AirTime` variable) and the distance flown (the `Distance` variable). Are the two variables highly correlated?

Problem 1.27 Using R, for the 2007 aircraft data construct a two-way table of counts of `Cancellation Code` versus `Diverted`. What does this tell us?

Problem 1.28 Using R, for the 2007 aircraft data construct a two-way table of `Diverted` versus `Month`. What does this tell us?

Problem 1.29 A number of data sets come with the base package in R. (Type `data()` to view them.) One is called `women` and it contains the heights and weights of 15 women. (Type `women` to view the data and `?women` to get some information about the data.) Describe and summarize the data using appropriate numerical descriptive statistics.

Problem 1.30 The `faithful` data set installed with the R base package has data on 272 eruptions of the Old Faithful geyser in Yellowstone National Park. The data set contains the length of each eruption and the time between eruptions, in minutes. Describe and summarize the data using appropriate numerical descriptive statistics.

Problem 1.31 The `cars` data set installed with the R base package has data on 50 observations of car speed (in miles per hour) and distance to stop (in feet). Describe and summarize the data using appropriate numerical descriptive statistics. What is the correlation between speed and distance?

Problem 1.32 The `mtcars` data set installed with the R base package has data on 32 cars from the early 1970s. Describe and summarize each of the variables in the data using appropriate numerical descriptive statistics. Which of the other 10 variables is most positively correlated with miles per gallon (`mpg`)? Most negatively correlated? (Hint: The `cor()` function run on the whole data frame will produce a "correlation matrix" that summarizes all pairwise correlations.) Are any of the cars unusual with respect to the other cars in the data set? How so?

Problem 1.33 Prove that the shortcut formula for the sample variance in Definition 1.5.1 is mathematically equivalent to the formula in Definition 1.5.6.

Problem 1.34 Clearly $s^2 \geq 0$. Prove that $s^2 = 0$ if and only if $x_i = \bar{x}$ for $i = 1, 2, \ldots, n$.

Problem 1.35 Some ask why in the sample variance formula the differences are squared. Prove that $\frac{1}{n-1} \sum_{i=1}^{n} (x_i - \bar{x}) = 0$ for any $x_1, x_2, \ldots, x_n$.

Problem 1.36 Clearly $\sigma^2 \geq 0$. Prove that $\sigma^2 = 0$ if and only if $x_i = \mu$ for $i = 1, 2, \ldots, N$.

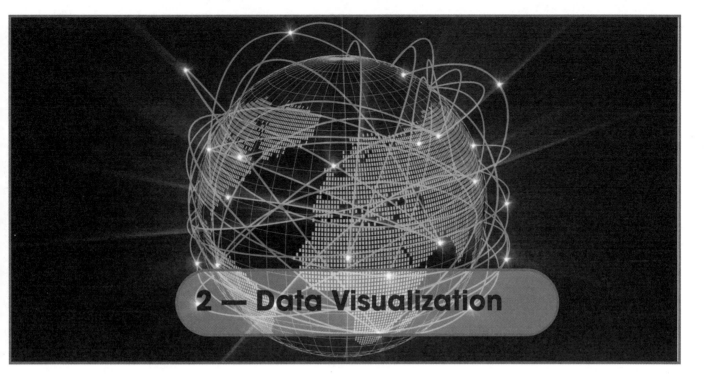

2 — Data Visualization

2.1 Introduction

Graphical plots are the means by which data are most easily visualized and understood. Indeed, there is no better tool for finding patterns in data than the human eye applied to appropriate displays of relevant data, particularly patterns that are ill-specified or unknown. Done correctly, good graphs can facilitate:

- *Insight*: Plotting data can enhance perception and comprehension of phenomena, particularly relationships among variables and trends over time, that may not be apparent otherwise. With large data sets or complicated data (or both), appropriate visualizations can make the data easier to understand.
- *Exploration*: With the appropriate analytical software (such as R) that facilitates interaction with data, the data can be examined and evaluated in a variety of ways. For subject matter experts and data scientists, interactive methods allow them to bring their expertise to bear in the visualization of the data.
- *Impact*: Data visualization is a powerful way to communicate information, relationships, and even the stories present in data. It's the old adage that a picture is worth a thousand words applied to data: A good graph is worth thousands of data points.

Good graphs of data can reveal relationships in the data that are simply not apparent via summary statistics. The classic example is Anscombe's data, plotted in Figure 2.1. Visually these four sets of data clearly show very different relationships between the *x* and *y* variables, yet:

- the means and standard deviations of each of the *x* variables are exactly the same;
- the means and standard deviations of the *y* variables are also the same;
- the correlations between *x* and *y* in each of the four cases are the same; and
- the best-fit lines are the same.[1]

In the absence of plotting the data, by just looking at some descriptive statistics one could be completely misled into thinking that there is little difference in the underlying phenomena.

In summary, there are clear benefits from graphing data and these benefits increase in direct relation to the size of the data, though, as we've just seen, visualization can provide unique benefits even with small data sets such as Anscombe's. But unlike Anscombe's data, which is small and simple to visualize via scatterplots, data scientists are frequently faced with large, complicated, and messy data sets. In this

---

[1] You will learn about regression and line fitting in Chapter 14.

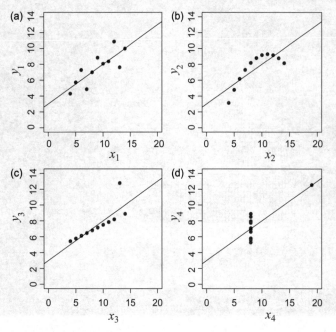

Figure 2.1 Plots of Anscombe's data, where all the descriptive statistics match, the regression fits (as shown by the lines) are the same, and yet the relationships between the x and y variables are clearly different in all four cases.

situation, good statistical graphs and other forms of data visualization are critical to understanding the data and underlying phenomena.

2.2 Traditional Statistical Graphics

Statistical graphics have traditionally been designed for displaying numerical data. Common to most of these methods is that the data are quantitative, typically with a small number of observations and variables (at least by today's standards), and the plots are mainly static. We'll branch out from traditional graphics in Section 2.3, but it's important to recognize that the graphics we discuss in this section are still in widespread use, are still very useful, and are quite powerful.

The graphical methods presented in this section are useful complements to the cross-sectional numerical descriptive statistics, particularly for further summarizing and displaying such cross-sectional data. As Anscombe's example illustrates in Figure 2.1, when appropriately designed, implemented, and presented, these graphical methods give the user intuitive insights into the data that may not be achievable otherwise.

2.2.1 Bar Charts

Bar charts are useful for summarizing categorical data, particularly for visually comparing the relative sizes of the various categories. For example, Figure 2.2 plots the percentage of 2007 canceled flights by cancellation code. This is a graphical representation of the percentages first tabulated in Example 1.18. As before, the "A" code means the airline canceled the flight, "B" means the flight was canceled due to weather, "C" means the flight was canceled by the National Air System (NAS), and "D" means the flight was canceled for security reasons. In the plot, we clearly see that flights were canceled by the carriers more often than any other reason, but it is followed closely by weather. We can also see that the number of NAS-related cancellations are roughly half as big as either carrier- or weather-related cancellations and that the number of security-related cancellations is very small.

Figure 2.3 shows plots equivalent to Figure 2.2 with counts instead of percentages. When the category names are long, it is often preferable to plot the bars horizontally so that the names can also be written out horizontally on the y-axis. Note that it is virtually impossible to discern the precise number

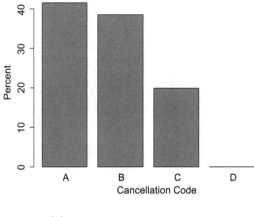

Figure 2.2 Bar chart for the percentage of flights canceled in 2007 by cancellation code. "A" means the airline canceled the flight, "B" means the flight was canceled due to weather, "C" means the flight was canceled by the NAS, and "D" means the flight was canceled for security reasons.

(a)

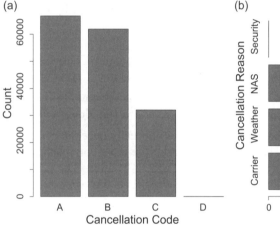

(b)

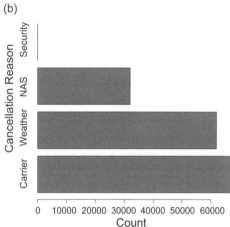

Figure 2.3 Examples of bar charts: number of flights that were canceled in 2007 by flight cancellation code. See Example 1.18 for the exact counts.

of flights canceled in each category in Figure 2.3, but that's not the point of the bar chart. Instead, the goal is to give some insight into the overall distribution of data by cancellation code, and for that the bar chart works quite well. If it's the precise counts we want, then a table such as Table 1.18 would be a better way to summarize the data.

As shown in Figure 2.4, bar charts can also display subgroupings, either by breaking the bars up to show the constituent subgroups in a stacked bar chart (a), or as a set of bars in a side-by-side bar chart (b). Stacked bar charts facilitate comparing the main groupings (cancellation code in the figures) while also allowing for comparison of the relative sizes of the subgroupings within each main group (day of the week in the figures). In contrast, side-by-side bar charts allow direct comparison of the sizes of the subgroupings within the main groupings, but this comes at the cost of not being able to directly compare the sizes of the main groups.

In R, bar charts are created using the `barplot()` function. Key to using the function is that it takes as inputs the aggregate counts or percentages for each category. Thus, you must calculate the counts or percentages ahead of time, typically using the `table()` function. Example 2.1 provides code similar to that used to generate Figures 2.2–2.4.

■ **Example 2.1 — Bar chart R code.** Use R to create plots similar to those in Figures 2.3 and 2.4.

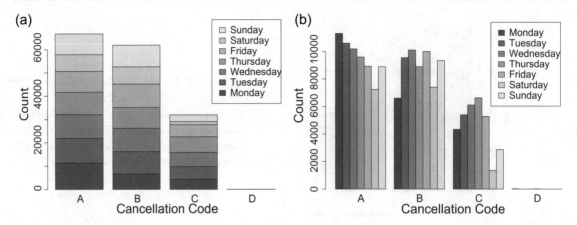

Figure 2.4 Examples of stacked and side-by-side bar charts: number of flights that were canceled in 2007 by flight cancellation code and day of the week. (a) A stacked bar chart; (b) a side-by-side bar chart. See Example 1.19 for the exact counts.

Solution: The easiest bar charts to plot are those based on counts. Continuing with the US flight data from Section 1.4.2, the code below generates the left plot in Figure 2.3(a).

```
cancellation.counts <- table(data07$CancellationCode)[2:5]
barplot(cancellation.counts,xlab="Cancellation Code",ylab="Count")
```

The creation of the `cancellation.counts` vector is not really necessary but is used just to make the example a little more readable. The use of `[2:5]` after the table command tells R to only assign the second through fifth entries of the table to `cancellation.counts`. The first entry in the table is the count of all flights without cancellation codes (i.e., the flights that were not canceled) which are not relevant to this particular plot.

The code below generates plot in Figure 2.3(b). Here we use the `horiz` option to plot horizontal bars and the `names.arg` option to replace the cancellation codes with the cancellation categories:

```
barplot(cancellation.counts,xlab="Cancellation Code",ylab="Count",
    horiz=TRUE, names.arg=c("Carrier",  "Weather", "NAS", "Security"))
```

Generating the bar charts in Figure 2.4 is a bit more complicated. First, let's extract the data that we will plot:

```
cancellation.codes = factor(
                data07$CancellationCode[data07$CancellationCode!=""] )
cancellation.dow   = data07$DayOfWeek[data07$CancellationCode!=""]
```

Again, the creation of the `cancellation.codes` and `cancellation.dow` vectors is not really necessary but they make the example a little more readable. In the first line above, the code extracts all the values in the `CancellationCode` vector that are not blank (i.e., for which there is a code entry). In a similar way, the second line extracts all the values in the `DayOfWeek` for which the associated `CancellationCode` entry is not blank. The result is two vectors of the same length, where each pair of entries in the two vectors corresponds to the same flight.

The code below then generates each of the plots in Figure 2.4:

```
barplot(table(cancellation.dow,cancellation.codes),legend.text=c("Monday",
        "Tuesday","Wednesday","Thursday","Friday","Saturday","Sunday"))
barplot(table(cancellation.dow,cancellation.codes),legend.text=c("Monday",
        "Tuesday","Wednesday","Thursday","Friday","Saturday","Sunday"),
        beside=TRUE)
```

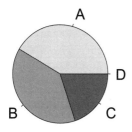

Figure 2.5 Sample pie chart for the flight cancellation codes of flights canceled in 2007. Compare this to the bar chart in Figure 2.2. The R code to generate this figure is shown in Example 2.2.

The key to generating the stacked and side-by-side bar charts is that the first argument in the `barplot()` function is a two-way table. The default is then a stacked bar chart, but a side-by-side plot can be obtained by setting the `beside` option to TRUE. Also, the legends in the plot are created using the `legend.text` option.   ■ ■ ■

2.2.2 Pie Charts

Pie charts are a commonly used graphic for displaying the relative sizes of categories within a set of data. Figure 2.5 is an example, where this particular plot is an alternative to Figure 2.2. Statisticians prefer bar charts to pie charts because research has shown that people are able to more accurately extract information from bar charts than pie charts. In particular, people are better able to compare the relative sizes of the categories with bar charts. For example, in Figure 2.2 it is clear that a larger percentage of cancellations are attributable to carriers than to weather, but the difference is not discernible in the pie chart of Figure 2.5.

■ Example 2.2 — Pie chart R code. Using the `cancellation.counts` vector created in Example 2.1, the code below generates the pie chart in Figure 2.5:

```
pie(cancellation.counts,col=c("light gray","gray","dark gray","black"))
```

The `col` option is used to recolor the pie slices. The default colors are shades of pastel.   ■ ■ ■

> **Exercise 2.1** Using the 2007 airline data, create a bar chart and pie chart showing the number of flights by month. Which plot makes it easier to identify the month with the greatest number of flights? The month with the fewest flights?

2.2.3 Histograms

A histogram is akin to a bar chart, but for continuous data. As just discussed, bar charts are for discrete data, and hence each bar in the chart corresponds to a distinct category in the data. In contrast, the histogram is applied to continuous data by dividing the real line into contiguous ranges (typically called "bins") for which the number of observations that fall into each bin are summed up. Then bars are plotted, where the height of each bar corresponds to the number (or fraction) of observed values that fall in the range of the bar.

Histograms are often used to gain insight into the distribution of the data. What is its shape? Where are most of the observations located? How spread out are the data? Are there any unusual outliers or concentrations of the data?

Figure 2.6 is a histogram of the distance of 2007 flights showing the number of flights within various mileage bands. The numerical summary statistics for these data are: $\bar{x} = 719.8$ and $\tilde{x} = 569.0$ miles with $s = 562.3$ and $R = 4,951$ miles. These statistics are visually evident in the histogram, which shows a bump around the median, with quite a bit of variation around it, and extreme values that extend from zero to almost 5,000 miles.

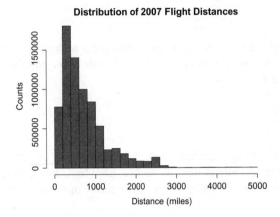

Distribution of 2007 Flight Distances

Figure 2.6 Example histogram: distribution of flight distances for all 2007 flights.

In Figure 2.6, the heights of the bars denote the number of flights in each mileage bin. Visually prominent in the plot is a spike at the 200–400 mileage range, as well as that all of the higher bars are in the 0–1,000 mile range, indicating that most flights in 2007 were 1,000 miles or less. The *skew* of the distribution is to the right, meaning that the right "tail" of the distribution is much longer than the left "tail," which is bounded by zero – distance must be non-negative. This skew is both visually evident in the plot and in the summary statistics, with the mean greater than the median ($\bar{x} > \tilde{x}$). Had the distribution been skewed to the left then the mean would have been less than the median.

■ **Example 2.3 — Histogram R code.** Construct a histogram using the `Distance` variable in the 2007 flight data.

Solution: The code below generates the histogram in Figure 2.6.

```
hist( data07$Distance , xlab="Distance (miles)" , ylab="Counts" ,
      main="Distribution of 2007 Flight Distances" , col="royalblue")
```

Note that the `xlab` and `ylab` options are used to label the axes, the `main` option adds the title at the top, and the `col` option is used to color the histogram bars. The default color is white and the bar's boundaries are black lines.   ■ ■ ■

Given the ubiquity of statistical software, we will ignore the details of how to manually construct a histogram. However, be aware that every software package makes certain choices when drawing a histogram, based on default values in the software. Two important choices that need to be made when drawing a histogram are: (1) how many "bins" should be used, and (2) where those bins should be located. A common choice is to begin the first bin at the data's minimum value and to end the last bin at the maximum value. In terms of the right number of bins, when n is relatively small a good choice is often $\sqrt{n}$, and when n is large a good choice is often $10\log_{10} n$.

■ **Example 2.4 — R code for the histograms in Figure 2.7.** Using the `AirTime` variable in the 2007 data, generate histograms with varying bin sizes.

Solution: The histograms in Figure 2.7 were created with the following code:

```
hist(data07$AirTime[data07$Origin=="MRY"],xlab="Flight Time (minutes)",
     ylab="Counts",main="",col="royalblue",breaks=8)
hist(data07$AirTime[data07$Origin=="MRY"],xlab="Flight Time (minutes)",
     ylab="Counts",main="",col="royalblue",breaks=15)
hist(data07$AirTime[data07$Origin=="MRY"],xlab="Flight Time (minutes)",
     ylab="Counts",main="",col="royalblue",breaks=30)
```

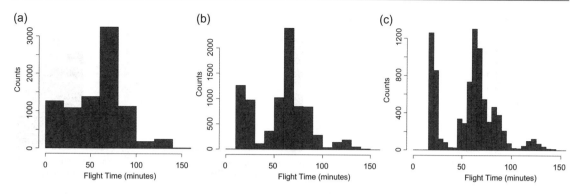

Figure 2.7 Histograms with varying numbers of "bins" showing the distribution of flight times for 2007 flights originating from the Monterey, California airport: (a) has 8 bins, (b) has 15 bins, and (c) has 30 bins.

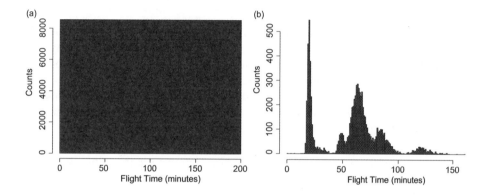

Figure 2.8 Extreme histograms for the same flight time data from Figure 2.7. (a) A trivial "histogram" with just one bar that shows the total number of flights for all flight times. (b) Basically just a bar chart since there is a distinct bar for each and every minute.

The breaks option allows user input to the number of histogram bins, but the input is only a suggestion and R may override it. ■ ■ ■

Regardless of the defaults, it is important to try out alternatives, particularly in terms of the number of bins, since the information communicated with the plot can vary with these choices. For example, Figure 2.7 shows three different histograms for the distribution of 2007 flights originating from the Monterey, California airport (airport code "MRY"). The difference is in the number of bins, where the histogram in (a) has 8 bins, (b) has 15 bins, and (c) has 30 bins (set using the breaks option of the hist() function). The figures show that the reader's perception of the shape of the distribution varies with the number of bins.

Now, while it's important to vary the number of bins in order to find the histogram that most appropriately communicates what is important in a given set of data, it's also important to be careful of extremes. Figure 2.8 shows the most extreme histograms for the same flight time data from Figure 2.7. Figure 2.8(a) is a trivial "histogram" with just one bar that shows the total number of flights for all flight times. This "histogram" says nothing about the underlying distribution of flight times and can be more efficiently summed up numerically: there were 8,554 flights in 2007 that originated in Monterey with flight times from 4 to 161 minutes. The right "histogram" is really just a bar chart with a distinct bar for every possible flight time.

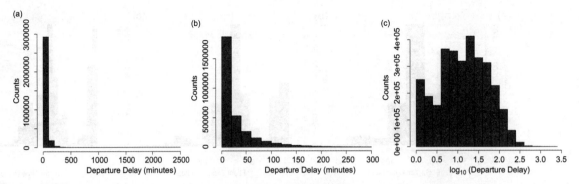

Figure 2.9 Histograms of 2007 departure delay times: (a) is for all of the data with delay times greater than zero; (b) is for departure delays between 1 and 300 minutes; and (c) is for the log-transformed departure delay times (for delay times greater than zero).

Sometimes, because of outliers, it can be useful to plot the data rescaled. For example, the left histogram in Figure 2.9 plots the 2007 delay times for all flights with delay times greater than zero. Because of a small number of very large delay times, most data values are compressed into the one bar on the far left. This doesn't provide much insight into the distribution of most of the data. Now, we can rescale the plot's *x*-axis to "zoom in" on the smaller times as shown in the histogram in Figure 2.9(b). This is better, but we've left some of the data out of the display. Another alternative is to transform the data and plot the transformed values. For example, the histogram in Figure 2.9(c) shows the log (base 10) transformation of delay times. The log transformation has the effect of compressing the larger delay times and spreading out the smaller delay times. Here we see that plotting data in units other than the natural units can be useful in picturing all the data. In such cases, transformations can reveal interesting aspects of the data.

Finally, note that the R `hist()` function defaults to *frequency histograms* that show counts (i.e., frequencies) on the *y*-axis. That's what we have been looking at in Figures 2.6–2.9. Setting the `freq` option to FALSE tells R to create a *density histogram*, where the areas of all the bars add up to 1. This type of histogram will become important when we start learning about probability distributions in later chapters.

Exercise 2.2 Using the 2007 airline data, create a histogram showing the elapsed time of flight (which is contained in the `ActualElapsedTime` variable).

2.2.4 Lattice (or Trellis) Plots

Lattice plots, also known as trellis plots, are an array of some type of statistical graph of one variable subset according to the values of one or more categorical variables. The categorical variables are also referred to as conditioning variables. The idea is to create a series of plots of one variable, say bar charts of the fraction of delayed flights in each cancellation category by separate levels of some categorical variable such as `Airline`. Figure 2.10 shows this lattice plot for the 2007 airline data. Here the conditioning variable is `Airline`, of which there are 20, which results in a lattice of 20 bar charts.

What Figure 2.10 shows is that the distributions of cancellation reasons differs by airline. For example, while cancellation code A (carrier-related cancellation) is often the most frequent reason for flight cancellation, for some airlines (American, American Eagle, Continental, Comair, and Jet Blue) cancellation code B (weather-related cancellation) is the most frequent reason, and for only two airlines (Expressjet and Skywest) is cancellation code C (NAS-related cancellation) most frequent. Interestingly, for Aloha and Hawaiian airlines, nearly all cancellations are carrier-related, presumably because of their routes' weather and NAS are less of an issue than for other airlines.

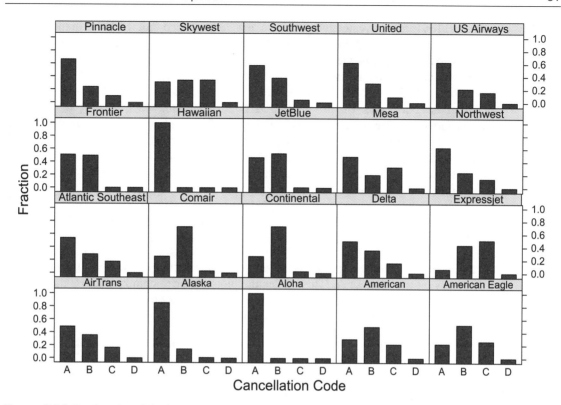

Figure 2.10 Lattice plot of the frequency of canceled 2007 flights by cancellation code conditioned on airline.

■ **Example 2.5 — R code for the lattice plot in Figure 2.10.** The code below generates the histograms in Figure 2.10. It is substantially more complicated than most of the R code we have used thus far, so below the code is a line-by-line explanation. Be sure to install the `lattice` package first.

```
library(lattice)
airline.names.alpha_order <- c("AirTrans","Alaska","Aloha","American",
        "American Eagle","Atlantic Southeast","Comair","Continental","Delta",
        "Expressjet","Frontier","Hawaiian","JetBlue","Mesa","Northwest",
        "Pinnacle","Skywest","Southwest","United","US Airways")
airline.codes.matching.alpha_order <- c("FL","AS","AQ","AA","MQ","EV",
        "OH","CO","DL","XE","F9","HA","B6","YV","NW","9E","OO","WN","UA","US")
UC <- factor(data07$UniqueCarrier,levels=airline.codes.matching.alpha_order,
            labels=airline.names.alpha_order)
CancelCode <- data07$CancellationCode
cnts <- table(CancelCode,UC)[2:5,]
pp <- as.data.frame.table(prop.table(cnts,margin=2))
barchart(pp$Freq ~ pp$CancelCode | pp$UC,xlab="Cancellation Code",
        ylab="Fraction")
```

The first line loads the `lattice` library so that we can use its `barchart()` function. In the next two lines we manually create a vector of airline names in alphabetical order and a vector of the airline codes corresponding to the alphabetical names. We do this so that the final plot will use the airline names rather than the codes, and so that the names will be in alphabetical order.

In the fourth command, using the two vectors we just created, we create a new vector that directly corresponds to the `UniqueCarrier` data vector, but our new UC vector now has airline names rather than codes and it knows to print them in alphabetical order. In the fifth command we extract the data from the `CancellationCode` vector and put it in a new vector called `CancelCode`. In the sixth command we create a table of counts (called `cnts`) from the two vectors, but we exclude the first row of counts that are for those flights not canceled. In the seventh command we create a new table called pp that converts the counts to fractions by airline.

Finally, in the last line we create the lattice plot. Besides the use of the `barchart()` function, the key part of the code is pp$Freq ~ pp$CancelCode | pp$UC, which says to plot the `Freq` variable for each category of the `CancelCode` variable and do this for each level of the `UC` variable. This corresponds to plotting the fraction of canceled flights for each of the cancellation categories by each of the airlines. To really understand the code, sequentially run each command above and take a look at the resulting object. That will clearly show you what each step is doing and you'll understand both the output of each step and thus what the inputs are for the next step.   ■■■

A lattice plot is nothing more than a set of repeated graphs, one for each category of the conditioning variable. Three things make lattice plots more useful than simply manually repeating some plot by variable levels. The first is that good software facilitates exploring the data by making it easy to generate lattices. Second, and more importantly, the graphs in the lattice are all plotted with the horizontal and vertical axes on the same scale. This makes the plots easier to compare across the various categories. Third, the lattice can be conditioned on more than one categorical variable, which allows for the discovery of more complicated relationships in the data.

> **Exercise 2.3** Using the 2007 airline data, create a lattice plot of elapsed time of flight histograms by airline. Hint: Start by rerunning the first four lines of code from Example 2.5. Then use the `histogram()` function from the R `lattice` package, where you will be plotting the `ActualElapsedTime` variable conditioned on the `UC` variable using the syntax
>
> ```
> histogram(~data07$ActualElapsedTime | UC)
> ```

2.2.5 Box Plots

Box plots are useful for depicting the distributions of continuous data. They do so by displaying summary statistics of the data, including the median and the lower and upper hinges (or first and third quartiles in some statistical software programs). As a result, box plots only require one dimension – unlike histograms, which need two dimensions. Because the box plot is based on summary statistics, some information is lost, but box plots can still be very informative, particularly when comparing the distributions of two or more sets of data.

As shown in Figure 2.11, to construct a box plot, first calculate the median and the hinges of a set of data. A box is then plotted that connects the upper and lower hinges (roughly the 25th and 75th percentiles), and a line is added inside the box to show the median. At each end of the box, *whiskers* are added by extending lines from the box that are 1.5 times the interquartile range (IQR). These lines are

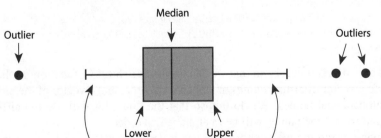

Figure 2.11 An illustration of how a box plot is constructed from summary statistics. Box plots are useful for depicting the distribution of continuous data in one dimension.

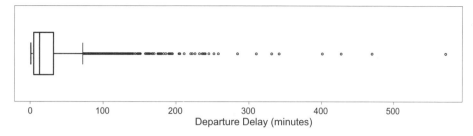

Figure 2.12 Example box plot: departure delay for 2007 flights originating in Monterey, California, for flights that were delayed.

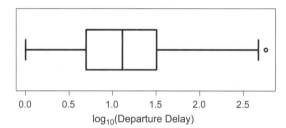

Figure 2.13 Box plot of the log (base 10) transformed departure delays for 2007 flights originating in Monterey, California, for flights that were delayed.

then truncated back to the last point contained within the line. Each whisker thus terminates at an actual data point, which means the whiskers will likely be of different lengths. Finally, observations that fall outside of the whiskers are indicated by dots and are designated outliers.

A box plot displays a lot of information: a measure of central tendency, the median; measures of how variable the data are as indicated by the width of the box (which is or is close to the IQR) and the length of the whiskers; and potentially unusual points, the outliers. Note that how box plots are drawn can vary by software package. For example, when drawing the whiskers some packages allow the user to choose multiples other than 1.5 times the IQR and some extend the whiskers all the way to the minimum and maximum values in the data.

Figure 2.12 shows a horizontal box plot of the departure delay for 2007 flights originating in Monterey, California (for flights that were actually delayed). Focusing on the box, we see that most delayed flights were delayed for a relatively short period of time – roughly a half-hour or less. However, the long right tail shows that some flights were delayed for very long times, with the longest delay being greater than 500 minutes (or more than 8 hours).

As with histograms, box plots can also be distorted and hard to read when the data are extremely skewed. Figure 2.12 is still reasonably readable, but a box plot equivalent to the histogram in Figure 2.9(a) would be equivalently distorted and thus difficult to read. In such cases, just like with histograms, transforming the data can make the plot easier to read. For example, Figure 2.13 is a box plot of the log-transformed 2007 departure delay data for Monterey flights. Compare Figure 2.13 to Figure 2.12.

The log base 10 transformation is particularly useful because it is easy to mentally convert back to the natural units: zero on the log base 10 scale is 1 on the natural scale, 1 is 10 on the natural scale, 2 is 100 on the natural scale, etc. Just think of integers in the log base 10 scale as counting the zeros (with a 1 in front) on the natural scale.

■ **Example 2.6 — R code for the box plots in Figures 2.12 and 2.13.** Using the DepDelay variable in the 2007 data, subset to only the Monterey flights using the Origin variable, the code below generates the box plots in Figures 2.12 and 2.13.

```
boxplot(data07$DepDelay[data07$Origin=="MRY" & data07$DepDelay > 0],
        xlab="Departure Delay (minutes)",horizontal=TRUE)
boxplot(log10(data07$DepDelay[data07$Origin=="MRY" & data07$DepDelay > 0]),
        xlab=expression(paste(log[10],"(Departure Delay)",sep="")),horizontal=TRUE)
```

The `horizontal=TRUE` option plots the box plots horizontally; the default is to plot them vertically. The use of the `expression()` function with the `xlab` option allows for writing mathematical expressions on the axis labels. ■ ■ ■

A useful way to compare observations grouped by a categorical variable is via side-by-side box plots where, similar to lattice plots, a separate box plot is created for each category. Thus, side-by-side box plots require a continuous and a categorical measure on each observation; for example, departure delay and airline. Also like lattice plots, one of the reasons side-by-side box plots are powerful is that all the box plots are graphed on the same scale, which facilitates comparisons between the categories.

For example, Figure 2.14 is a side-by-side box plot of the $\log_{10}$-transformed departure delays for 2007 by airline. The figure clearly shows which airlines had better departure delay performance (meaning their departure delay distributions tend toward smaller delays) – for example, Hawaiian – and which airlines had worse departure delay performance (meaning their departure delay distributions tend toward larger delays) – for example, Mesa.

■ **Example 2.7 — Creating side-by-side box plots in R.** Using the `UC` vector created for the lattice plot in Example 2.5, the code below generates the plot in Figure 2.14.

```
par(mar = c(9,5,4,4))
boxplot(log10(data07$DepDelay[data07$DepDelay>0])~UC[data07$DepDelay>0],
ylab=expression(paste(log[10],"(Departure Delay)",sep="")),las=2,xlab="")
```

First note that we're using the same `boxplot()` function as we did in Example 2.6. The difference here is in the arguments, particularly the syntax

```
log10(data07$DepDelay[data07$DepDelay>0]) ~ UC[data07$DepDelay>0]
```

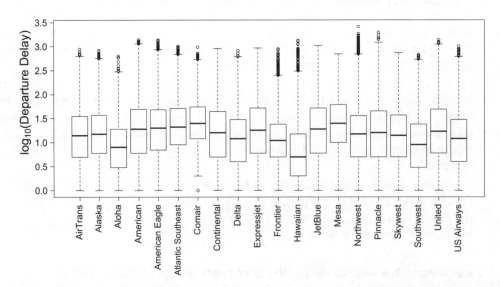

Figure 2.14 Example side-by-side box plot: plot of the $\log_{10}$-transformed departure delays for 2007 by airline.

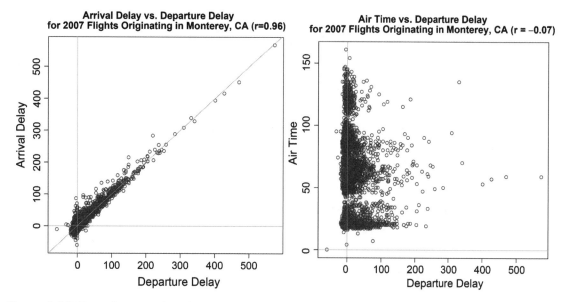

Figure 2.15 Example scatterplots: for all 2007 flights originating from the Monterey, California airport, arrival delay (in minutes) versus departure delay (in minutes) in (a) and flight air time (in minutes) versus departure delay in (b).

This tells the `boxplot()` function to plot the $\log_{10}$-transformed departure delay values (for those observations with positive departure delays) as a function of the UC vector (again, only for those UC values corresponding to flights with positive departure delays).[2] That is, this syntax tells R to create a separate box plot for each unique value in the UC vector; note that this only makes sense if the UC vector is categorical.

Second, the `las=2` argument tells the `boxplot()` function to rotate the labels (the airline names) 90 degrees so that they are readable. (Run the code without this argument and see what you get.) However, because some of the names are long, we also need to increase the lower margin of the plot so that the names all fit. That is done in the first line, where the `par()` function controls lots of graphical parameters. The option `mar` is for adjusting the margins, where the vector sets the margins starting at the bottom and working clockwise around the plot. Note that the margins remain at the specified setting until either the `par()` function is rerun with the `mar` option or the plotting window is closed.   ■ ■ ■

Exercise 2.4 Using the 2007 airline data, create a box plot of elapsed time of flight for all flights. Then create side-by-side box plots of elapsed time of flight by airline (using the UC variable).

2.2.6 Scatterplots

A scatterplot is a graph of one continuous variable versus another, with a dot for each ordered pair in the data. Scatterplots are useful for summarizing data and they are particularly effective at showing whether there is a relationship between two variables, where such a relationship would show up as a pattern in the plot. Anscombe's data in Section 2.1 is the canonical example of how scatterplots can reveal patterns that might otherwise go unnoticed.

Figure 2.15 shows two examples of scatterplots, where (a) is a scatterplot of arrival delay versus departure delay for all the 2007 flights originating from the Monterey, California airport. What the

---

[2]The $\sim$ symbol means "as a function of." We will use this notation more in later chapters.

scatterplot shows, perhaps not surprisingly, is that there is a high correlation ($r = 0.96$) between the two variables where, presumably, departure delays cause arrival delays. The plot also shows that there are some flights with little or no departure delay that do have arrival delays, probably because of weather and air traffic congestion encountered en route. However, the plot makes it quite clear that the largest driver of arrival delay is departure delay.

The scatterplot in Figure 2.15(b) is another example, plotting flight air time versus departure delay for all of the 2007 flights originating from the Monterey airport. Here we see that there are clusters of what look like four groupings of flights that take roughly the same air time. The scatterplot shows that these air times are essentially (and perhaps unsurprisingly) uncorrelated with departure delay ($r = -0.07$). That is, the plot shows that, once a flight is airborne, its flight time is not affected by the length of departure delay.

■ **Example 2.8 — R code for the scatterplots in Figure 2.15.** Using the DepDelay, ArrDelay, and AirTime vectors from the 2007 data, the code below generates the scatterplots in Figure 2.15.

```
plot( data07$DepDelay[data07$Origin=="MRY"] ,
      data07$ArrDelay[data07$Origin=="MRY"] ,
      xlab="Departure Delay",ylab="Arrival Delay" ,
      main="Arrival Delay vs. Departure Delay\n
      for 2007 Flights Originating in Monterey, CA (r=0.96)" , col="blue")
   lines(c(-500,3000),c(0,0),col="gray")
   lines(c(0,0),c(-500,3000),col="gray")
   lines(c(-500,4000),c(-500,4000),col="gray")
plot( data07$DepDelay[data07$Origin=="MRY"] ,
      data07$AirTime[data07$Origin=="MRY"] ,
      xlab="Departure Delay",ylab="Air Time" ,
      main="Air Time vs. Departure Delay\n
      for 2007 Flights Originating in Monterey, CA (r=-0.07)" , col="blue")
   lines(c(-500,3000),c(0,0),col="gray")
   lines(c(0,0),c(-500,3000),col="gray")
```

While the code looks complicated, the basic syntax for a scatterplot of vectors x and y is plot(x, y). The lines() function overlays a line on the plot, where the line connects the two points $\{x_1, y_1\}$ and $\{x_2, y_2\}$ via the syntax lines(c(x_1, x_2),c(y_1, y_2)).   ■ ■ ■

Now, we might want to check on the hypothesis that, in fact, there are four groups of flight times. Table 2.1 shows that flights from Monterey went to 11 different destinations in 2007 (where the vast majority went to Los Angeles or San Francisco).

Digging a bit deeper, the side-by-side box plots in Figure 2.16 show that there are either five or six groups of similar flight times. For example, Figure 2.16 shows that the flight times to Las Vegas (LAS) and Los Angeles (LAX) are similar, as are the flight times to Long Beach (LGB) and Ontario (ONT) airports, and as are the flight times to Phoenix (PHX) and Salt Lake City (SLC). Flight times to Denver (DEN) and San Francisco (SFO) are clearly different from the other destinations, though if we look a bit more closely we see that sometimes it takes as long to fly to Phoenix as to go to Denver, and San Diego (SAN) flight times are somewhere in between LAS/LAX and LGB/ONT times.

Table 2.1 Counts of all 2007 flights originating from Monterey, California, by destination airport.

Destination	DEN	FAT	LAS	LAX	LGB	ONT	PHX	PIH	SAN	SFO	SLC
# Flights	364	1	140	3,513	101	413	714	1	464	2,475	582

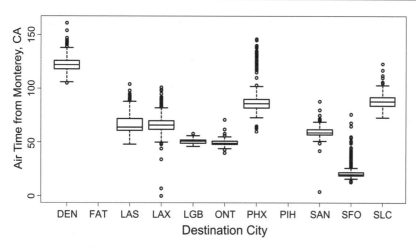

Figure 2.16 Side-by-side box plots of flight air time (in minutes) by destination airport for all the 2007 flights originating from the Monterey, California airport.

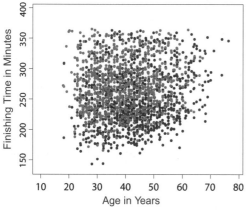

Figure 2.17 A scatterplot of age versus finishing time for the Napa marathon, with red points for women and blue points for men.

Exercise 2.5 Using the 2007 airline data, create a scatterplot of elapsed time of flight versus flight air time (`AirTime`) for all the flights. Now only plot those flights that originate from Dallas Love Field airport (`data07$Origin=="DAL"`).

Scatter Plots with Grouping Variables

Often there is a third categorical variable, and we would like to illustrate this on a scatterplot by using a different color or plotting character (or both). We illustrate this with the marathon data, where the two continuous variables are age and finishing time and the categorical variable is gender.

■ **Example 2.9 — Marathon data.** Here we plot the finishing time (on the vertical axis) and the age of the runner (on the horizontal axis), with the color of the symbol representing the gender of the runner (male or female). This can be done in R by first plotting the points for men, then adding the points for women with the `points()` command. Here we use blue for men and red for women. The R code to do this is:

```
ageF  <- napa$Age[napa$Gender == "F"]
timeF <- napa$Hours[napa$Gender == "F"]*60
plot( ageM , timeM , pch=19 , col="blue" , xlab="Age in Years" ,
      ylab="Finishing Time in Minutes" , xlim=c(10,80) , ylim=c(100,400) )
points( ageF , timeF , pch=19 , col="red" )
```

Here the variables ageM and timeM give the age and finishing time for males; ageF and timeF give the same for females. We have specified the *x* and *y* axes to go from 10 to 80 and 100 to 400, respectively.

The resulting plot is shown in Figure 2.17. From this we can see that the age distributions are about the same, but the times for women are a bit higher than those for men. We could find the correlations separately for men, women, and combined by typing the following:

```
cor( ageM , timeM )
cor( ageF , timeF )
cor( napa$Age, napa$Hours*60 )
```

which yields the output

```
[1] 0.1672689
[1] 0.1018639
[1] 0.08664575
```

Note that the correlation in the combined group is the smallest. We will revisit this phenomenon in Chapter 13 when we study Simpson's paradox. ∎∎∎

Scatter Plots for Multiple Continuous Variables

When there are more than two continuous variables, a *scatterplot matrix* is useful for displaying the data. A scatterplot matrix uses a lattice-like format to show scatterplots for all pairs of variables. For example, Figure 2.18 shows a scatterplot matrix of AirTime, Distance, ArrDelay, and DepDelay data for the 2007 Monterey flights.

To read the plot, note that the diagonal identifies the variables associated with the axes of each of the individual scatterplots. Thus, for example, ArrDelay is the variable on the *y*-axis of all the plots in the top row and it is the variable on the *x*-axis of all the plots in the left column. So, for example, the top left scatterplot shows ArrDelay versus DepDelay – note that it matches the left scatterplot in Figure 2.15. Similarly, the plot in the third row of the second column is a scatterplot of AirTime versus DepDelay and it matches the scatterplot in Figure 2.15(b). Also note how each pair of variables is plotted twice, but with the axes reversed. In fact, all the plots above the diagonal have a matching reversed plot below the diagonal, so it is really only necessary to examine half of the plots in the matrix. In fact, a scatterplot matrix of k variables has $k(k-1)/2$ unique plots.

Example 2.10 shows the code for generating the scatterplot matrix in Figure 2.18. Be careful when using the pairs() function; if too many variables are involved, each of the plots will be too small.

∎ **Example 2.10 — R code for the scatterplot matrix in Figure 2.18.** The pairs() function run on a matrix or data frame with more than two continuous variables produces a scatterplot matrix. In the code below, it was run on the 14th, 15th, 16th, and 19th columns of the data07 data frame, which correspond to the AirTime, ArrDelay, DepDelay, and Distance variables, and only for those rows in which the Origin variable is equal to "MRY."

```
pairs( data07[data07$Origin=="MRY",c(15,16,14,19)],col="blue" )
```
∎∎∎

2.3 Graphics for Longitudinal Data

Graphics for longitudinal data, that is data collected on multiple subjects across time, are in many ways similar to those for cross-sectional data. (See Section 1.4.1 for additional discussion about types of data.) A key difference, however, is that time is an important variable that must be displayed in some useful and informative way. For many longitudinal graphics, that often means that the variable plotted on the *x*-axis is time.

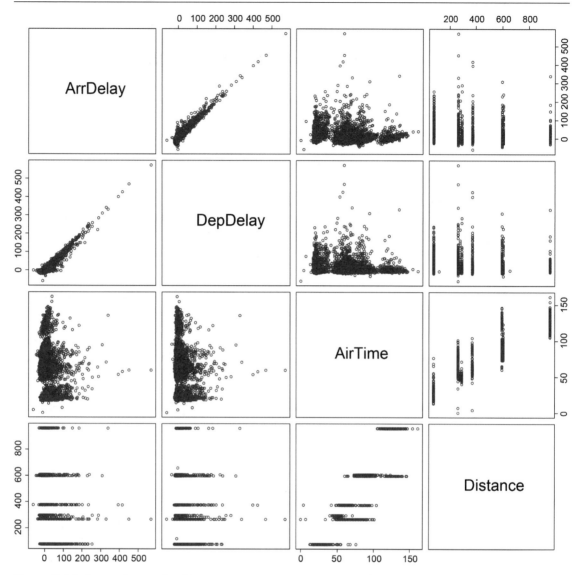

Figure 2.18 Scatterplot matrix of `ArrDelay`, `DepDelay`, `AirTime`, and `Distance` variables for all 2007 flights originating from Monterey, California.

2.3.1 Time Series Plots

Time series are data collected or measured repeatedly over time, where the term "time series" is often used as a synonym for longitudinal data. However, time series *plots* display longitudinal data with respect to time. By convention, the data are plotted with the magnitude of the observation on the *y*-axis and time, in the appropriate units, on the *x*-axis. Time series plots are useful for assessing whether there are trends in the data, such as a regular increase in the number of flights departing from an airport, perhaps resulting from population growth in the region served by the airport, or whether there are cycles in the data, perhaps as a result of weekly, seasonal, or other influences.

Figure 2.19 is a time series plot of the total number of daily flights out of LAX, JFK, and MRY. The *x*-axis contains the sequential days of the year (referred to as the *Julian date*, where 1 corresponds to January 1 and December 31 is either 365 or 366, depending on whether or not it is a leap year). The *y*-axis is the total number of flights on each day of the year.

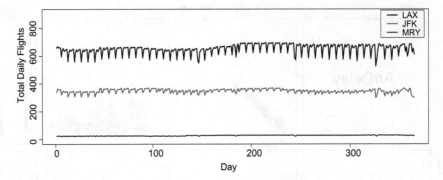

Figure 2.19 Time series plot of the total number of daily flights out of LAX, JFK, and MRY in 2007. The *Julian date* is just the days of the year sequentially numbered, where 1 corresponds to January 1, and December 31 is either 365 or 366, depending on whether it is a leap year or not.

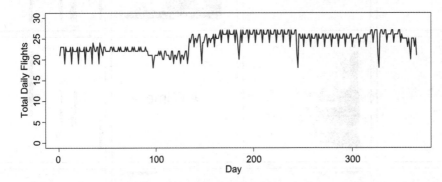

Figure 2.20 Time series plot of the total number of daily flights out of MRY in 2007, where the *y*-axis scale now makes the weekly cycle much more apparent.

It is a convention in time series and other longitudinal plots to display the data using lines that connect what otherwise would be discrete points on the plot. That is, with these data for each day and each airport there is a daily count. While we could have plotted these data as points in a scatterplot-like display, much like the plots in Section 2.2.6, by using lines to connect the data it helps make the time trends visually clear and it also helps to communicate to the reader that this is a time series plot and not a scatterplot. Note, for example, how the lines in Figure 2.19 show a clear weekly cycle in the number of flights, where the regular dips in the number of flights occur on Fridays. We also see a very slight increase in the number of flights in the summer, at least in LAX, and less regularity in the weekly pattern toward the end of the year during the holidays.

Figure 2.20 shows only the Monterey data, where the *y*-axis scale now makes the weekly cycle much more apparent for these data. Also, the discrete nature of the data is more obvious in this plot, as well as what may be some seasonal effects. For example, we see a slight increase in the number of summer flights, and the major holidays are visible as downward spikes: the day before Memorial Day, the day before Labor Day, Thanksgiving and the day after, Christmas, and New Years Eve (corresponding to Julian dates 148, 244, 325, 326, 359, and 365).

2.3.2 Repeated Cross-Sectional Plots

Another way to present time series data is via a series of repeated plots, where each plot represents a different (but sequential) time period. For example, box plots repeated over time can be very informative. Figure 2.21 shows this approach for the LAX departure delay data, where each box plot is based on a month of data.

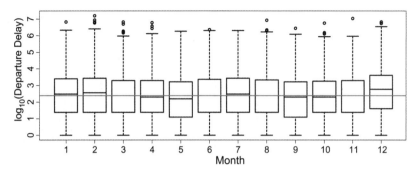

Figure 2.21 Repeated box plots by month of departure delay (for those flights that were delayed) for 2007 LAX data.

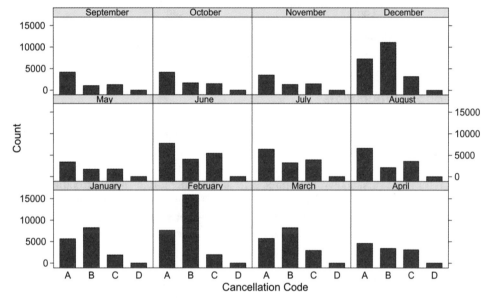

Figure 2.22 Lattice plot of cancellation codes by month for all of the flights canceled in 2007.

Remember that the horizontal line in the middle of a box plot is the median, and in Figure 2.21 the horizontal line across the entire plot corresponds to the median departure delay for the entire year. Comparing these, we see that there was a cycle in median departure delays with greater median delays in the winter months (December, January, and February) and lower median delays in the spring and fall. Now, note that the y-axis on this plot is on a log base 10 scale, so actual differences are larger than they visually appear in the plot. For example, the median departure delay for all 2007 flights was 11 minutes, while in December it was 16 minutes (i.e., $10^{1.20412} = 16$).

Lattice plots can also be used to display repeated cross-sectional plots if the conditioning variable corresponds to time. For example, Figure 2.22 is a lattice plot of the cancellation codes for all the flights canceled in 2007. While this type of plot is not as useful as those in Figures 2.19–2.21 for showing trends, it does show that there are very different cancellation patterns and numbers of flight cancellations throughout the year. For example, we see that in December, January, February, and March, weather-related cancellations are the most frequent cause (particularly in February 2007). However, throughout the rest of the year, carrier-related cancellations are the most frequent cause. Also, the plot clearly shows that there were more cancellations in the winter months (followed by the summer months) than in other times of the year.

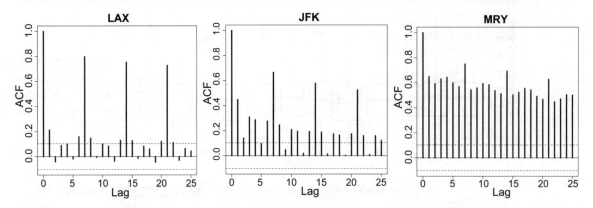

Figure 2.23 Autocorrelation plots for the number of daily flights from LAX, JFK, and MRY for 2007 (see Figures 2.19 and 2.20).

2.3.3 Autocorrelation Plots

The lag k autocorrelation is the correlation between values of the same time series separated by k time units. The autocorrelation plot, a plot of the lag k autocorrelation against k, can be used for checking whether a time series has trends or cycles. See Section 16.3 for a complete description of autocorrelation.

Figure 2.23 shows the autocorrelation plots for the number of daily flights from LAX, JFK, and MRY for 2007 (first shown in Figures 2.19 and 2.20). "Large" autocorrelations are those with bars that fall outside the two dotted lines (where the "ACF" label on the y-axis label stands for *autocorrelation function*). Those inside the dotted lines can be considered close enough to zero that they can be ignored.[3] Note that at lag 0 we have the correlation of a variable with itself, which is always 1.

So, the large autocorrelation values in the LAX plot at lags 7, 14, and 21 indicate that there is a strong weekly cycle in the data (which we also observed in Figure 2.19). The JFK plot has similarly large autocorrelation values at lags 7, 14, and 21, also indicative of a strong weekly cycle. In addition, that plot shows decreasing autocorrelation values at other lag values, which indicates that the JFK data also have a weak trend. Finally, the MRY data show both large autocorrelation values at lags 7, 14, and 21, and generally large but decreasing values at all of the other lags, which indicates a weekly cycle and other trends in the data. These were evident in Figure 2.20, where we saw a number of different patterns throughout the year.

2.4 Chapter Summary

In this chapter we have learned about many different methods for visualizing data, from traditional statistical graphics for cross-sectional and longitudinal data, to more recent types of visualizations for text and geospatial data. As with the descriptive statistics in Chapter 1, the purpose of visualization is to summarize data and convey information, though in this case via plots of the data. The goal of data visualization is to facilitate efficient and effective data insight and exploration. Just as with descriptive statistics, the goal is *not* to find a visualization to demonstrate a preconceived theory or story. Rather, the visualizations should provide clear and objective insights into the data where, as with descriptive statistics, good visualizations should let the data speak for itself.

We've only scratched the surface of data visualization. There are a multitude of books about statistical graphics, data visualization, and R graphics. For those who would like to dig a bit deeper, the following are good references from which to start.

---

[3]That is, the autocorrelations within the dotted lines are "not statistically different from zero." We'll learn more about this in later chapters on hypothesis testing.

- For general guidance on design principles for good quantitative graphics, see the following texts.
  - *The Visual Display of Quantitative Information* and *Envisioning Information*, both by E.R. Tufte, are the canonical guides to principles of good quantitative graphics. Also see *Beautiful Evidence* and *Visual Explanations: Images and Quantities, Evidence and Narrative* by Tufte.
  - *Visualizing Data* and *The Elements of Graphing Data*, by W.S. Cleveland are the original and definitive treatments of classical statistical graphics.
  - *The Wall Street Journal Guide to Information Graphics: The Dos and Don'ts of Presenting Data, Facts, and Figures*, by D.M. Wong provides a business news take on graphical display.
- To learn more about using R to create good visualizations, see the following texts.
  - *R Graphics*, by P. Murell.
  - *R Graphics Cookbook*, by W. Chang.
  - *ggplot2: Elegant Graphics for Data Analysis*, by H. Wickham.
- To learn about interactive graphics, see the books listed below. Most of these discuss dynamic graphics in the context of a specific software system, but the principles and ideas generalize.
  - *Visual Statistics: Seeing Data with Dynamic Interactive Graphics*, by F.W. Young, P.M. Valero-Mora, and M. Friendly.
  - *Dynamic Graphics for Statistics*, edited by W.S. Cleveland and M.E. McGill.
  - *Interactive and Dynamic Graphics for Data Analysis: With R and GGobi*, by D. Cook and D.F. Swayne.

2.5 Problems

Note: *For all plots and graphs in these problems, turn in a plot with informatively labeled axes and be sure to appropriately format, label, and title your plots. A good data visualization should communicate exactly what is being plotted and, to the greatest extent possible, the graph should be easy to understand and interpret.*

Problem 2.1 In your own words, describe how and why data visualization can be useful.

Problem 2.2 Find three examples of good data visualizations in popular media or on the internet. Discuss what the visualizations show and explain why you consider them good visualizations.

Problem 2.3 Find three examples of poor data visualizations in popular media or on the internet. Discuss what the visualizations are intended to show and explain why they are poor visualizations. Then, describe an alternative way to visualize your examples and justify why your visualization is better.

Problem 2.4 In R, plot the data from Problem 1.15 as a histogram and a bar chart. Compare your plots, both to each other and to the summary statistics from Problems 1.15, 1.18, and 1.20. Which one do you prefer for visualizing these data and why?

Problem 2.5 In R, plot the data from Problem 1.15 using a box plot. Identify the parts of the box plot and label them with the appropriate numerical values taken from the summary statistics calculated in Problems 1.15, 1.18, and 1.20. (If you want to label your plot within R, the `arrows` and `text()` functions will be useful.) Comparing the box plot to the histogram of Problem 2.4, which do you prefer for visualizing these data and why?

Problem 2.6 In R, plot the data from Problem 1.22 using a scatterplot. Compare the plot to the correlation calculated in Problem 1.22. Do you learn anything more about the data from the plot?

Problem 2.7 The moving average of size m of a set of longitudinal data $x_1, x_2, \ldots, x_n$ is the set of averages

$$\bar{x}_t = \sum_{i=t-m+1}^{t} x_i, \ t = m, m+1, \ldots, n.$$

Calculate the moving average with $m = 5$ for the following sequence of data:

13, 26, 11, 20, 21, 25, 15, 13, 21, 24, 31, 21, 22, 14,
22, 22, 30, 17, 23, 16, 20, 32, 21, 31, 24, 24, 20, 18,
19, 17, 15, 15, 15, 31, 25, 20, 13, 18, 11, 18, 10, 12

Use R to plot this longitudinal data set along with the moving average. What do you learn from the plot?

Problem 2.8 Using the definition of the moving average from the previous problem, compute the moving average with $m = 3$ for the following longitudinal data set:

0.74, 0.91, 0.10, −0.46, −1.13, −1.17, −0.03, −0.67,
−0.04, −0.58, 1.09, 0.72, 0.26, 0.49, −0.33, −0.60,
−0.62, −0.60, −0.01, 0.08, 0.90

Use R to plot this longitudinal data set along with the moving average. What do you learn from the plot?

Problem 2.9 Using R, plot a histogram and a box plot of the 2007 aircraft flight arrival delay data (contained in the `ArrDelay` variable). In conjunction with the descriptive statistics calculated in Problem 1.23, discuss what you have learned from the plots and the descriptive statistics.

Problem 2.10 Using R, plot a histogram and a box plot of the 2007 aircraft flight times (contained in the `AirTime` variable). In conjunction with the descriptive statistics calculated in Problem 1.24, discuss what you have learned from the plots and the descriptive statistics.

Problem 2.11 Using R, plot the 2007 flight arrival delay (the `ArrDelay` variable) versus the aircraft flight time (the `AirTime` variable) in a scatterplot. Compare the plot to the correlation calculated in Problem 1.25 and describe your results.

Problem 2.12 Using R, plot the 2007 aircraft flight time (the `AirTime` variable) versus the distance flown (the `Distance` variable) in a scatterplot. Compare the plot to the correlation calculated in Problem 1.26 and describe your results.

Problem 2.13 Using R, graphically display the counts from the table created in Problem 1.27 appropriately in a bar chart.

Problem 2.14 Using R, graphically display the data from the table created in Problem 1.28 appropriately in a bar chart.

Problem 2.15 As described in Problem 1.29, the `women` data set comes with the base package in R. It contains the heights and weights of 15 women. (Type `women` to view the data and `?women` to get some information about the data.) Graphically summarize the data using the appropriate plots and then discuss what the plots show.

Problem 2.16 As described in Problem 1.30, the `faithful` data set installed with the R base package has data on 272 eruptions of Old Faithful. The data set contains the length of each eruption (in minutes) and the time between eruptions, also in minutes. (Type `faithful` to view the data and `?faithful` for more information about the data.) Graphically summarize the data using the appropriate plots and then discuss what the plots show.

Problem 2.17 As described in Problem 1.31, the `cars` data set installed with the R base package has data on 50 observations of car speed (in miles per hour) and distance to stop (in feet). Type `cars` to view the data and `?cars` for more information about the data. Graph the data appropriately and describe what you learn from the plot.

Problem 2.18 As described in Problem 1.32, the `mtcars` data set installed with the R base package has data on 32 cars from the early 1970s. Type `cars` to view the data and `?cars` for more information about the data. Graphically summarize each of the variables in the data appropriately using one or more plots. Using a scatterplot matrix, determine which of the other 10 variables is most positively correlated with miles per gallon (*mpg*) and which is most negatively correlated. (Hint: The `plot()` function run on the whole data frame will produce a scatterplot matrix on all the variables in the data frame – though note that with 10 variables each individual scatterplot will probably be too small to view.) Using the scatterplot matrix and other plots, what can you say about the data?

Problem 2.19 A word cloud visualizes text data. The size and hue of a word is proportional to the number of times it appears. Word clouds can be created using the R `wordcloud` package.

(a) Install the `wordcloud` package. Type

```
?wordcloud::wordcloud
```

to learn about the `wordcloud` function in the `wordcloud` package.

(b) Using a text file of your choice, create a word cloud of it using the `wordcloud` package.

(c) Describe whether and how the word cloud is an effective visualization of the text.

Problem 2.20 Consider the time series defined in the R array

```
x = c( 15, 21, 20, 19, 25, 22, 25,
       27, 31, 30, 32, 26, 27, 22,
       25, 20, 24, 28, 28, 34)
```

(a) Plot the data in a time series graph.
(b) Execute the R commands `x[2:20]` and `x[1:19]`. Explain the results.
(c) Explain what `cor( x[2:20] , x[1:19] )` does.
(d) How would you compute the lag 1 autocorrelation?
(e) How would you compute the lag 2 autocorrelation?

Problem 2.21 The closing prices for Netflix stock over the first 20 trading days in 2020 are shown below.

01–02	329.81	01–16	338.62
01–03	325.90	01–17	339.67
01–06	335.83	01–21	338.11
01–07	330.75	01–22	326.00
01–08	339.26	01–23	349.60
01–09	335.66	01–24	353.16
01–10	329.05	01–27	342.88
01–13	338.92	01–28	348.52
01–14	338.69	01–29	343.16
01–15	339.07	01–30	347.74

We will study this data set more thoroughly in Chapter 16.

(a) Plot the time series.
(b) Use the method suggested in parts (d) and (e) of Problem 2.20 to estimate the autocorrelations for lag 1 through 3.

3 — Basic Probability

3.1 Introduction

To do statistics you must first be able to "speak probability." In this chapter we are going to concentrate on the basic ideas of probability. In probability, the mechanism that generates outcomes is assumed known and the problems focus on calculating the chance of observing particular types or sets of outcomes. Classical problems include flipping "fair" coins (where fair means that on one flip of the coin the chance it comes up heads is equal to the chance it comes up tails) and "fair" dice (where fair now means the chance of landing on any side of the die is equal to that of landing on any other side). The mechanism is, of course, the fair coin or the fair die.

Statistics, on the other hand, starts with the outcomes (data) and asks questions about the generating mechanism. Typical statistics questions would be: "Given that I flipped a coin 10 times and observed 10 heads, do I believe that the coin is fair?" and "Given that I rolled a pair of six-sided dice 11 times and observed the sequence $\{2, 3, 4, \ldots, 12\}$, do I believe that the dice are fair?" These are rather simple examples, but they form the foundation for more sophisticated statistical questions like, "Given a poll of 1,500 randomly selected voters, what is the chance that candidate X will win the election?" Or, "After a clinical trial with 500 people, is drug Y effective at curing a particular disease?" Or, "Given an A/B test of two website designs, which design more effectively induces visitors to click through to an advertiser's content?"

At this point, it may be worthwhile to think through the example of 10 heads on 10 tosses. What does your intuition say about the fairness of the coin? You would probably say "No, I don't believe the coin is fair." If pressed for a reason you might say "getting 10 heads on 10 tosses is very unlikely; it is so unlikely that I don't think it could occur by chance." This is very nearly correct. Getting 10 heads on 10 tosses is very unlikely *if the coin is fair*. If the coin is biased, for example a novelty "two-headed" coin, then 10 out of 10 is not at all unlikely.

3.2 Events and Sample Spaces

A *sample space*, often denoted by the capital letter S, is a list of all possible outcomes that might be observed, perhaps from an experiment or other form of data collection. A *sample point* is one particular outcome in the sample space. For example, the sample space for one flip of a two-sided coin is the set consisting of sample points "heads" and "tails." Mathematically, $S = \{\text{heads}, \text{tails}\}$, or more succinctly,

$S = \{H, T\}$, where "H" stands for the coin coming up heads and "T" stands for the coin coming up tails. As another example, for the roll of one six-sided die, S would be the set of integers between one and six, inclusive, so $S = \{1, 2, 3, 4, 5, 6\}$. A third example might be the lifetime of a machine selected at random. Here the observation is a continuous measure that can take on only non-negative values, so for this example the sample space is the non-negative part of the real line: $S = [0, \infty)$.

For reasons of mathematical simplicity, in this chapter we will focus on discrete sample spaces.

> **Definition 3.2.1 — Discrete Sample Space.** A discrete sample space has either a finite or a countably infinite number of distinct sample points.[1]

In the previous examples, $S = \{H, T\}$ and $S = \{1, 2, 3, 4, 5, 6\}$ are discrete sample spaces, as is $S = \{1, 2, 3, \ldots\}$, but $S = [0, \infty)$ is not.

Events are subsets of sample spaces.

> **Definition 3.2.2 — Event.** In a discrete sample space, an event is any collection (i.e., subset) of sample points in S.

In the die example, we might define event A as "roll an odd number," $A = \{1, 3, 5\}$, and event B as "roll a number less than 6," $B = \{1, 2, 3, 4, 5\}$. Here A is a *subset* of B (denoted $A \subset B$) because all of the points in A are also contained in B. Both A and B are events because $A \subset S$ and $B \subset S$. The symbol $\in$ denotes "is an element of" where here, for example, $2 \in B$. The *null* or *empty set* is the set consisting of no points and is denoted $\emptyset$. The null set is a subset of every set. So, in the die example, $\emptyset \subset A \subset B \subset S$.

> **Exercise 3.1** Write out the sample space corresponding to simultaneously flipping two coins. If you win $2 when both coins come up heads and you lose $1 otherwise, write out the sample points that correspond to the event "after flipping both coins, lose $1."
>
> *Answer*: $S = \{(H_1, H_2), (H_1, T_2), (T_1, H_2), (T_1, T_2)\}$, where (H_1, T_2) means the first coin comes up heads and the second comes up tails, for example.
>
> Let A denote the event "after flipping both coins, lose $1." Then $A = \{(H_1, T_2), (T_1, H_2), (T_1, T_2)\}$.

> **Definition 3.2.3 — Union and Intersection.** The *union* of sets A and B is
>
> $$A \cup B = \{\text{all elements that are in set } A \text{ or set } B\}$$
>
> and the *intersection* is
>
> $$A \cap B = \{\text{all elements that are in both set } A \text{ and set } B\}.$$
>
> The ideas of union and intersection carry over to a collection of sets:
>
> $$A_1 \cup A_2 \cup \cdots \cup A_n = \{\text{all elements that are in at least one set } A_i\}$$
>
> and
>
> $$A_1 \cap A_2 \cap \cdots \cap A_n = \{\text{all elements that are in every set } A_i\}.$$

---

[1] A set is countably infinite if it has infinitely many elements in it, but the elements can be put in a list. For example the set $A = 1, 2, 3, \ldots$ is infinite, but the elements are in a list. By contrast, the interval $[0, 1]$ has infinitely many elements but they cannot be put in a list; this set is not countably infinite.

For example, if $S = 2, 3, \ldots, 12$ is the sample space, and if $A_1 = \{2, 3\}$, $A_2 = \{2, 4, 6, 8, 10, 12\}$, and $A_3 = \{7, 11\}$, then

$$
\begin{aligned}
A_1 \cup A_2 &= \{2, 3, 4, 6, 8, 10, 12\} \\
A_1 \cup A_3 &= \{2, 3, 7, 11\} \\
A_1 \cap A_2 &= \{2\} \\
A_1 \cap A_3 &= \emptyset \\
A_1 \cup A_2 \cup A_3 &= \{2, 3, 4, 6, 7, 8, 10, 11, 12\}.
\end{aligned}
$$

We will often use the word "or" instead of "$\cup$" and "and" instead of "$\cap$," especially when we describe an event in words rather than in set notation. This should seem natural because of the way these operations were defined.

> **Definition 3.2.4 — Disjoint or Mutually Exclusive Sets.** Sets A and B are said to be *disjoint* or *mutually exclusive* if $A \cap B = \emptyset$; that is, A and B contain no elements in common. Equivalently, we could say that it is impossible for A and B to occur at the same time. A collection of sets $A_1, A_2, \ldots, A_n$ is said to be *pairwise disjoint* if every pair of sets, say A_i and A_j for $i \neq j$ selected from the collection are disjoint. We will also use the term *pairwise mutually exclusive*.

> **Definition 3.2.5 — Complement of an Event.** For an event A, the *complement* is defined to be the set of all elements in the sample space that are not in A. The complement of A is denoted $\bar{A}$.

3.2.1 Probability Axioms

A *probability measure* on the sample space S is a function Pr from subsets of S to the real numbers. The notation $\Pr(A)$ is spoken "the probability event A occurs," and mathematically Pr maps every event in S to a real number in the interval $[0, 1]$ according to the following axioms.

> **Definition 3.2.6 — Probability Axioms.** For a function Pr to be a probability measure, it must meet the following axioms:
>
> Axiom 1: For each event $A \in S$, $\Pr(A) \geq 0$.
>
> Axiom 2: $\Pr(S) = 1$.
>
> Axiom 3: If $A_1, A_2, A_3, \ldots$ is a sequence of pairwise disjoint events in S, then
>
> $$\Pr(\text{event } A_1 \text{ occurs or event } A_2 \text{ occurs or event } A_3 \text{ occurs or } \cdots) = \sum_{i=1}^{\infty} \Pr(A_i).$$

The first axiom says that the probability of each event in the sample space occurring must be non-negative. The second axiom says that the total probability in the sample space is equal to 1. Taken together, Axioms 1 and 2 mean that for any event A, $0 \leq \Pr(A) \leq 1$. The third axiom says that the probability of a set of events that do not have any sample points in common is the sum of the probabilities of the individual events. We'll discuss this axiom in more detail in the next section.

Let's apply the axioms to the fair coin and die examples. For the coin, with $S = \{H, T\}$, because it's fair and from Axiom 1 we know that $\Pr(\text{heads}) = \Pr(\text{tails}) \geq 0$. Then, from Axioms 2 and 3 we can infer that

$$\Pr(S) = \Pr(\text{flip heads or flip tails}) = \Pr(\text{heads}) + \Pr(\text{tails}) = 1,$$

where Axiom 3 applies since it's not possible to flip a coin and have it come up both heads and tails simultaneously. Combining these two facts, we conclude that $\Pr(\text{heads}) = \Pr(\text{tails}) = 1/2$.

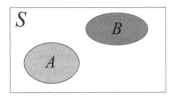

Figure 3.1 A simple Venn diagram representing two *disjoint* events, A and B.

Similarly, for a fair die with $S = \{1,2,3,4,5,6\}$, from Axiom 1 we know that $\Pr(\text{roll } i) \geq 0$, for $i = 1,2,3,4,5,6$. And, from Axioms 2 and 3 we can infer that

$$
\begin{aligned}
\Pr(S) &= \Pr(\text{roll 1 or roll 2 or roll 3 or roll 4 or roll 5 or roll 6}) \\
&= \Pr(\text{roll 1}) + \Pr(\text{roll 2}) + \Pr(\text{roll 3}) + \Pr(\text{roll 4}) + \Pr(\text{roll 5}) + \Pr(\text{roll 6}) = 1.
\end{aligned}
$$

Thus, because the die is fair, it must be that $\Pr(\text{roll } i) = 1/6$, where $i = 1,2,3,4,5,6$.

3.2.2 Union of Events

Axiom 3 defines a rule for calculating the probability of the union of disjoint events. *Union* means *together*, so the union of events A and B consists of all the sample points that are either in event A or in event B (or in both).

Using this notation, we can now write Axiom 3 more succinctly as follows. If $A_1, A_2, A_3, \ldots$ is a sequence of pairwise disjoint events in S, then

$$
\Pr(A_1 \cup A_2 \cup A_3 \cup \cdots) = \sum_{i=1}^{\infty} \Pr(A_i).
$$

We can visualize Axiom 3 using a *Venn diagram*. For example, consider two disjoint events A and B in sample space S. Figure 3.1 shows a Venn diagram where the box represents the entire sample space and the circles are the events A and B. The box has a total area of 1, corresponding to the probability that something in the sample space must happen (Axiom 2), and the size of the circles correspond to $\Pr(A)$ and $\Pr(B)$. Since the events are disjoint, the circles do not overlap and, because they do not overlap, the probability of $A \cup B$ is simply the probability that A occurs plus the probability that B occurs. That is, for disjoint events, Axiom 3 says

$$
\Pr(A \cup B) = \Pr(A) + \Pr(B).
$$

We can now use the axioms to calculate various probabilities.

■ **Example 3.1** Given a single roll of a fair six-sided die, what is the probability of rolling either a 1 or a 2?

Solution: Since the event "roll a 1" is disjoint from the event "roll a 2" (they can't both happen at the same time):

$$
\Pr(\text{roll 1 or roll 2}) = \Pr(\text{roll 1} \cup \text{roll 2}) = \Pr(\text{roll 1}) + \Pr(\text{roll 2}) = 1/6 + 1/6 = 1/3.
$$

The last line follows because the die is assumed to be fair.   ■ ■ ■

> **Exercise 3.2** Returning to Exercise 3.1, if the two coins are fair then all the events in the sample space are equally likely. Given this, what is the probability you will lose \$1 on one simultaneous flip of the coins?
>
> *Answer*: The probability is 0.75.

■ **Example 3.2** On a single roll of a fair six-sided die, what is the probability of rolling an even number?

Solution:

$$
\begin{aligned}
\text{Pr(roll an even number)} &= \text{Pr(roll 2 or roll 4 or roll 6)} \\
&= \text{Pr(roll 2} \cup \text{roll 4} \cup \text{roll 6)} \\
&= \text{Pr(roll 2)} + \text{Pr(roll 4)} + \text{Pr(roll 6)} \\
&= 1/6 + 1/6 + 1/6 \\
&= 1/2.
\end{aligned}
$$

Again, the last line follows because the die is fair. For a *biased* die where

$$
\text{Pr(roll } i) = 0.15, \quad i = 1,2,3,4,5 \qquad \text{and} \qquad \text{Pr(roll 6)} = 0.25,
$$

we would have

$$
\text{Pr(roll an even number)} = \text{Pr(roll 2)} + \text{Pr(roll 4)} + \text{Pr(roll 6)} = 0.15 + 0.15 + 0.25 = 0.55.
$$

■ ■ ■

> **Exercise 3.3** Demonstrate that the biased coin described in Example 3.2 has a legitimate probability distribution. That is, show that it meets Probability Axioms 1 and 2.

There is a more general rule for events that may or may not be disjoint. The Venn diagram of Figure 3.2(a) illustrates the union of two events, A and B, that are not disjoint. The event "A and B" is the *intersection* of the two events, denoted $A \cap B$, and is depicted by the dark-shaded region in the Venn diagram of Figure 3.2(b). A sample point can now be in four different regions of the sample space: A, B, "A and B," and "neither A nor B." The additive law of probability is the general rule for calculating $A \cup B$. For a proof of this theorem, see Problem 3.49.

> **Theorem 3.2.1 — Additive Law of Probability.** For any two events A and B,
>
> $$\text{Pr}(A \cup B) = \text{Pr}(A) + \text{Pr}(B) - \text{Pr}(A \cap B).$$

The additive law of probability says that, in general, the probability of the union of two events is the sum of their individual probabilities minus the probability of their intersection. The way to think about this is to see that the sum of the two events' individual probabilities counts the intersection area twice, and that area must be subtracted to make everything add up correctly.

Note that the previous rule for disjoint events is really a special case of the general rule. When events are disjoint, the probability of their intersection is zero (i.e., $\text{Pr}(A \cap B) = 0$), so the expression in Theorem 3.2.1 simply reduces to $\text{Pr}(A \cup B) = \text{Pr}(A) + \text{Pr}(B)$.

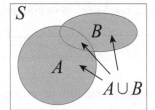 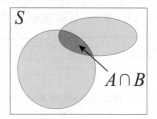

Figure 3.2 (a) The shaded area in the Venn diagram is the union of two nondisjoint events, A and B: $A \cup B$. (b) The dark area in the Venn diagram is the intersection of A and B: $A \cap B$.

■ **Example 3.3** Consider a standard deck of playing cards.[2] What is the probability of picking a face card or a card with the suit of clubs?

Solution: Let event A be "pick a face card" and event B be "pick any club card." If it's a fair deck, so every card is equally likely, then via the additive law $\Pr(A) = 12/52$ since there are 12 face cards (king, queen, and jack from each of the four suits) out of 52 cards, $\Pr(B) = 13/52$ since there are 13 club cards out of 52 cards, and $\Pr(A \cap B) = 3/52$ since there are 3 cards that are both clubs and face cards (jack of clubs, queen of clubs, and king of clubs) out of the 52 cards. Then:

$$\Pr(A \cup B) \;=\; \Pr(A) + \Pr(B) - \Pr(A \cap B) \;=\; 12/52 + 13/52 - 3/52 \;=\; 22/52.$$

Of course, this matches what we would get via enumerating the sample space and counting: there are 13 clubs cards and nine other non-clubs face cards for a total of 22 cards out of 52.   ■■■

The additive law of probability can be generalized to the union of more than two events. For example, for any three events A, B, and C,

$$\Pr(A \cup B \cup C) \;=\; \Pr(A) + \Pr(B) + \Pr(C) - \Pr(A \cap B) - \Pr(A \cap C) - \Pr(B \cap C) + \Pr(A \cap B \cap C).$$

3.2.3 Intersection of Independent Events

A useful rule deals with the probability of the intersection of two independent events. Intuitively, two events are *independent* if knowing the outcome for one of the events provides no information about the outcome of the other event. For example, in two independent flips of a fair coin, knowing that a head occurred on the first flip provides no information about what will happen on the second flip. Probabilistically, the condition that two events are independent is defined as follows.

> **Definition 3.2.7 — Independence.** Two events A and B are independent if
>
> $$\Pr(A \cap B) \;=\; \Pr(A) \times \Pr(B).$$

Thus, under the condition of independence, the probability that the event "A and B" occurs (that is, the shaded region of the diagram in Figure 3.2(b): $A \cap B$) can be calculated as the probability that event A occurs times the probability that event B occurs.

■ **Example 3.4** Consider two people each independently flipping fair coins. We know the coin flips are independent because the people are in separate rooms and they can't communicate or see what the other is doing. What is the probability they both flip heads?

Solution: Since the coins are fair, $\Pr(H_i) = \Pr(T_i) = 1/2$, $i = 1, 2$, where, for example, H_i is the event that person i flipped a head. If person 1's flip is completely independent of person 2's, then the probability they both flip heads is:

$$\Pr(\text{both people flip heads}) \;=\; \Pr(H_1 \text{ and } H_2) \;=\; \Pr(H_1 \cap H_2) \;=\; \Pr(H_1) \times \Pr(H_2) \;=\; \frac{1}{2} \times \frac{1}{2} \;=\; \frac{1}{4}.$$

■■■

---

[2] A standard deck of 52 cards consists of 13 cards from each of four suits: clubs (♣), diamonds (♢), spades (♠), and hearts (♡). The ranks are ace (highest), king, queen, jack, and then the numbers 10 down to 2 (lowest). Diamonds and hearts are red and clubs and spades are black. King, queen, and jack are considered "face cards." The whole list of 52 cards is:

Ace (A)♣, king (K)♣, queen (Q)♣, jack (J) ♣, 10♣, 9♣, 8♣, 7♣, 6♣, 5♣, 4♣, 3♣, 2♣
Ace (A)♢, king (K)♢, queen (Q)♢, jack (J) ♢, 10♢, 9♢, 8♢, 7♢, 6♢, 5♢, 4♢, 3♢, 2♢
Ace (A)♠, king (K)♠, queen (Q)♠, jack (J) ♠, 10♠, 9♠, 8♠, 7♠, 6♠, 5♠, 4♠, 3♠, 2♠
Ace (A)♡, king (K)♡, queen (Q)♡, jack (J) ♡, 10♡, 9♡, 8♡, 7♡, 6♡, 5♡, 4♡, 3♡, 2♡

Definition 3.2.7 generalizes to more than two events and can be quite useful for calculating the probability of the intersection of multiple independent events. For example, if events A, B, and C are independent, then

$$\text{Pr}(A \cap B \cap C) = \text{Pr}(A) \times \text{Pr}(B) \times \text{Pr}(C).$$

■ **Example 3.5** What is the probability of failing to get a 4 on any of three (independent) rolls of a fair die?

Solution: If the rolls are independent, then the probability of not getting a 4 on three rolls of the die is the probability of not getting a 4 on the first roll of the die *and* not getting a 4 on the second roll of the die *and* not getting a 4 on the third roll of the die. The probability of failing to get a 4 on one roll is the same as rolling 1 or 2 or 3 or 5 or 6. From Axiom 3 we know that

$$
\begin{aligned}
\text{Pr}(\text{roll 1 or 2 or 3 or 5 or 6}) &= \text{Pr}(\text{roll 1} \cup \text{roll 2} \cup \text{roll 3} \cup \text{roll 5} \cup \text{roll 6}) \\
&= \text{Pr}(\text{roll 1}) + \text{Pr}(\text{roll 2}) + \text{Pr}(\text{roll 3}) + \text{Pr}(\text{roll 5}) + \text{Pr}(\text{roll 6}) \\
&= \frac{1}{6} + \frac{1}{6} + \frac{1}{6} + \frac{1}{6} + \frac{1}{6} \\
&= \frac{5}{6}.
\end{aligned}
$$

Thus,

$$
\begin{aligned}
\text{Pr}(\text{fail to roll a 4 three times}) \\
&= \text{Pr}(\text{not get a 4 on roll 1 and not get a 4 on roll 2 and not get a 4 on roll 3}) \\
&= \text{Pr}(\text{not get a 4 on roll 1} \cap \text{not get a 4 on roll 3} \cap \text{not get a 4 on roll 3}) \\
&= \text{Pr}(\text{not roll a 4}) \times \text{Pr}(\text{not roll a 4}) \times \text{Pr}(\text{not roll a 4}) \\
&= \frac{5}{6} \times \frac{5}{6} \times \frac{5}{6} \\
&= \frac{125}{216}.
\end{aligned}
$$

■ ■ ■

So, how do you tell when the question is about the union or the intersection of events? The key word for union is *or*, as in the union of events A and B comprises all sample points that are in either A or B. The key word for intersection is *and*, as in the intersection of events A and B comprises all sample points that are in A and B.

3.2.4   Complementary Events

Using the axioms, additional useful rules for calculating probabilities can be derived. For example, consider an event A for which it is easy to calculate $\text{Pr}(A)$, but perhaps what is of interest is the probability that anything other than A happens. In other words, the event of interest is "not A," which is denoted $\overline{A}$ for the complement of A. Rather than calculating $\text{Pr}(\overline{A})$ directly we can apply the complement rule to compute $\text{Pr}(A)$ and then take $\text{Pr}(\overline{A}) = 1 - \text{Pr}(A)$.

Theorem 3.2.2 — Complement Rule. For any event A and its complement $\overline{A}$,

$$\text{Pr}(\overline{A}) = 1 - \text{Pr}(A).$$

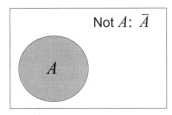

Figure 3.3 A simple Venn diagram representing an event A and its complement $\overline{A}$.

The complement rule follows directly from Axioms 1 and 2, which follow from the algebra of sets where, by construction, $A \cup \overline{A} = S$ and $A \cap \overline{A} = \emptyset$. These properties are visually evident in the Venn diagram of Figure 3.3, where the box represents S and the circle is the event A. As before, the box has a total probability of 1 and the size of the circle corresponds to $\Pr(A)$, so it should be intuitively clear that $\Pr(\overline{A}) + \Pr(A) = 1$. More formally, given that $A \cup \overline{A} = S$, we apply the probability operator to both sides of the equation to get

$$\Pr(A \cup \overline{A}) = \Pr(S),$$

and thus by Probability Axioms 2 and 3 we have

$$\Pr(A) + \Pr(\overline{A}) = 1.$$

Rearranging, it follows that $\Pr(\overline{A}) = 1 - \Pr(A)$. A simple example illustrates the utility of this rule.

■ **Example 3.6** With a six-sided fair die, what is the probability of not rolling a 4 on one roll?

Solution: Let A denote the event "roll a 4 on a single roll of a fair six-sided die," where $\Pr(A) = 1/6$. From the complement rule, it follows that

$$\Pr(\overline{A}) = 1 - \Pr(A) = 1 - \frac{1}{6} = \frac{5}{6}.$$

This matches the result we obtained in the first part of Example 3.5.   ■ ■ ■

■ **Example 3.7** What is the probability of getting a 4 on at least one of three rolls of a fair die? This is a problem that is more complicated to calculate directly since there are $6^3 = 216$ events in the sample space: $S = \{(1,1,1),(1,1,2),(1,1,3),\ldots,(6,6,6)\}$.

Solution: Recognizing that the complement to the desired probability is the probability of getting no fours on all three rolls of the die simplifies the problem. On one roll, the probability of not getting a 4 is 5/6 and from Example 3.5 we know not getting a 4 on three rolls is $(5/6)^3$. So the probability of getting a 4 on at least one of three rolls is $1 - (5/6)^3 \approx 0.42$.   ■ ■ ■

Exercise 3.4 Consider a new coin flipping game where, on each round of the game, three people each independently flip one fair coin. If they all flip heads, each person gets $10, but otherwise they each have to pay $1. What is the probability the players each lose $1 on one round?

Answer: The probability is 0.875.

3.2.5 Conditional Probability

If two events are *dependent*, then knowing the outcome of one event *does* provide information about the probability of the other event. The notation is $\Pr(A \mid B)$, which is read "the probability of A given B," meaning the probability that event A will occur given that event B has occurred.

> **Definition 3.2.8 — Conditional Probability.** Assuming $\Pr(B) \neq 0$, the probability of A given B is defined as
>
> $$\Pr(A \mid B) = \frac{\Pr(A \cap B)}{\Pr(B)}.$$

The idea behind this definition is that, given event B occurred, the relevant sample space becomes the sample points in B rather than S and the relevant probability measure is now over B. What Definition 3.2.8 is doing is re-normalizing the probabilities from the entire sample space S to B, since it is now known that the event must be one of those in B. If A and B are independent then

$$\Pr(A \mid B) = \frac{\Pr(A \cap B)}{\Pr(B)} = \frac{\Pr(A) \times \Pr(B)}{\Pr(B)} = \Pr(A).$$

Thus, if A and B are independent, then $\Pr(A \mid B) = \Pr(A)$ and, equivalently, $\Pr(B \mid A) = \Pr(B)$. The interpretation is that, if A and B are independent, then knowing whether B occurred provides no additional information about $\Pr(A)$.

■ **Example 3.8** Consider two fair dice rolled simultaneously so that their outcomes are independent. Given that the total of the two rolls is 4, what is the probability that the first die came up a 2?

Solution: Let's start by calculating the probability of rolling a total of 4. Let A be the event that the total of the two dice is 4: $A = \{(1,3),(3,1),(2,2)\}$, where $(1,3)$ means the first die came up 1 and the second came up 3. Write out the sample space if this is not obvious. Then,

$$
\begin{aligned}
\Pr(A) &= \Pr[(1,3) \text{ or } (3,1) \text{ or } (2,2)] \\
&= \Pr[(1,3) \cup (3,1) \cup (2,2)] \\
&= \Pr[(1,3)] + \Pr[(3,1)] + \Pr[(2,2)] \\
&= (1/6 \times 1/6) + (1/6 \times 1/6) + (1/6 \times 1/6) \\
&= 3/36.
\end{aligned}
$$

The result is just an application of the additive law of probability to independent events.

Let B be the event that the first die shows a 2. Then $\Pr(A \cap B) = 1/36$, since there is only one way to get a total of 4 with the first die showing a 2 out of $6^2 = 36$ total outcomes, and as we just calculated $\Pr(A) = 3/36$. So,

$$\Pr(B \mid A) = \frac{1/36}{3/36} = \frac{1}{3}.$$

To check this result, note that the event A consists of only three events from S: $A = \{(1,3),(3,1),(2,2)\}$. Given that A has occurred, the relevant probability calculation is then based on only these three possible outcomes. Since there is only one event out of three in A where the first roll is a 2, the conditional probability is $1/3$.   ■ ■ ■

The expression for conditional probability in Definition 3.2.8 can be turned around to define a more general way to calculate the intersection of two events. This more general rule for calculating $A \cap B$ is the multiplicative law of probability. See Problem 3.51.

> **Theorem 3.2.3 — Multiplicative Law of Probability.** Let A and B be events, not necessarily independent, and suppose $\Pr(B) \neq 0$. Then
>
> $$\Pr(A \cap B) = \Pr(B)\Pr(A \mid B).$$
>
> Because the intersection operator "$\cap$" is symmetric (i.e., $A \cap B = B \cap A$), we can also express this theorem as
>
> $$\Pr(A \cap B) = \Pr(A)\Pr(B \mid A).$$

■ **Example 3.9** We know that 9.2% of the visitors to a particular web page click on a certain advertiser's link. We also know that, of those who click on the ad link, 13.5% buy something from the advertiser's site. What is the proportion of web page visitors who both click on the advertiser's link and buy something from the advertiser?

Solution: Let event B denote the event that a web page visitor clicks on the advertiser's link, so $\Pr(B) = 0.092$, and let event A denote the event that, of those who click on the ad link, 13.5% buy something from the advertiser's site, so $\Pr(A \mid B) = 0.135$. Then via the multiplicative law of probability, it follows that:

$$\Pr(A \cap B) = \Pr(A \mid B)\Pr(B) = 0.135 \times 0.092 = 0.01242.$$

That is, just over 1% of the visitors to the web page click on the advertiser's link and then subsequently buy something. ■■■

■ **Example 3.10** A stockroom contains six computers that look identical. However, unknown to the stockroom personnel, four are good and two are defective. Two will be drawn from the stockroom at random. What is the probability that two good computers are selected?

Solution: Let G_1 and G_2 denote the events of selecting a good computer on the first and second draws, respectively. From the multiplicative law of probability, it follows that:

$$\Pr(G_1 \cap G_2) = \Pr(G_1)\Pr(G_2 \mid G_1) = \frac{4}{6} \times \frac{3}{5} = \frac{6}{15}.$$

That is, assuming each computer is equally likely to be drawn, on the first draw there are four good computers out of six total, so the probability a good computer is drawn is 4/6. On the second draw, if we *condition* on having picked a good computer on the first draw, then on the second draw there are now only three good ones left out of the five remaining computers. So, on the second draw the probability of picking a good computer *given a good computer was picked on the first draw*, $\Pr(G_2 \mid G_1)$, is 3/5.

If the preceding explanation is not obvious, write out all the possible unique pairs of good and bad computers that can occur and then count how many of them consist of two good ones. ■■■

3.3 Calculating Probabilities

To this point, we've been using probability to describe the random behavior of coins and dice and, while these may seem a bit trivial, their use in probability problems dates back to the Renaissance with the study of gambling. In fact, in the 1600s Blaise Pascal and Pierre de Fermat did the first mathematical treatment of probability by studying dice. Today, probability is widely studied using very sophisticated mathematics. However, strip away all the mathematics and at its core, probability is a means for formally quantifying one's belief in the occurrence of events. For example, when we say a coin is "fair" we're saying we believe that the occurrence of heads on the next flip is as likely as tails. But, with another coin or under other circumstances, we might say that $\Pr(\text{heads}) = 0.9$. In this case, we're saying we believe that the chance of observing a heads is 9 times more likely than observing a tails.

There are two main philosophical approaches to probability:

- **Objective probability**. The idea of objective probability is that probability describes some objective, typically physical, process. The most common type of objective probability is what is called *frequentist*, where the idea is that probability describes the long-run relative frequency of occurrences from the process. For example, with our fair coin, $\Pr(\text{heads}) = 1/2$ comes from the fact that if the coin was flipped a *very* large number of times then the number of heads should be half of the total number of flips.

- **Subjective probability**. The idea of subjective probability is that probability describes an individual's degree of belief that each outcome will occur. The most common type of subjective probability is *Bayesian*, which provides a mathematically formal method for combining expert information with experimental data to produce probabilities. For example, with our coin, a Bayesian probabilist first specifies what is called a *prior* that describes his or her belief in the probability of the coin coming up heads, and this probability is subsequently updated by collecting some data by flipping the coin.

Both of these philosophical approaches carry over into statistics, where there are frequentist statistical methods and Bayesian statistical methods. In this book, we will mostly take the frequentist viewpoint, mainly because it is the most direct way to learn about probability and introductory statistics. However, it is important to note that both approaches make important contributions to the generation of knowledge and, depending on the particular problem, one approach may be better suited to solving the problem than the other. We will discuss Bayesian statistics in Chapter 15. Data scientists interested in broadening their education should take a complete Bayesian statistics course to fully round-out their statistical skills.

3.3.1 Sample Point Method

Under the assumption that every sample point in S is equally likely to occur, the probability that some event A occurs is defined as the number of sample points that are in event A divided by the total number of sample points in S. Using the notation $n(A)$ to mean "the number of sample points in event A" and $n(S)$ to mean "the total number of sample points in the sample space," then

$$\Pr(A) = \frac{\text{number of outcomes in } A}{\text{total number of outcomes in } S} = \frac{n(A)}{n(S)}.$$

This is often referred to as the *sample point method* for determining probabilities.

For example, for a "fair" coin, the probability of getting heads is $1/2$, since $n(H) = 1$ and $n(S) = 2$. On the other hand, the probability of getting heads on a two-headed coin is 1 (i.e., the coin is guaranteed to come up heads), since $n(H) = 2$. Of course, as before, this is an idealized model of coin flipping since the sample point "the coin lands on its side" is not in the sample space.

■ **Example 3.11** Consider a standard 52-card deck. Each card is a sample point in S and if the deck is fair then each has probability $1/52$ of being randomly picked from the deck. What is the probability of choosing an ace from the deck?

Solution: Define event A as randomly choosing an ace from the deck. Then, since there are four aces in a deck of cards, $n(A) = 4$ and thus

$$\Pr(A) = \frac{n(A)}{n(S)} = \frac{4}{52} = \frac{1}{13}. \qquad \blacksquare\blacksquare\blacksquare$$

Of course, this is basically what we've been doing throughout Section 3.2, but now we've formalized the idea. You may be wondering what happens if the individual sample points in S are *not* equally likely. For example, what if the coin is not fair? What do we do then? We'll address this in the next two chapters when we introduce various *probability distributions* that will allow us to generalize and account for these types of problems.

For now we will assume that all sample points in S are equally likely. And, while that sounds quite restrictive, it can still allow us to solve quite a number of relevant probability problems *as long as we can count the number of sample points*. When we're only dealing with coins and dice this is usually pretty straightforward, but it need not be so. Let's return to the airline data from Chapters 1 and 2 to estimate some probabilities.

■ **Example 3.12** The airlines data has information on flight delays and cancellations. With it, we can ask questions such as: "What is the probability that a randomly selected domestic US flight from 2007 experienced a delay?" and "What is the probability that the flight was canceled?"

Solution: To answer these questions, note that the sample space is all US domestic flights in 2007, where we can calculate $n(S)$ from the data:

```
data07 <- read.csv("2007.csv")
dim(data07)

[1] 7453215       29
```

Thus, $n(S) = 7,453,215$. We can similarly count the number of flights that were delayed and the number that were canceled:

```
table(data07$DepDelay > 0)
table(data07$CancellationCode != "")

  FALSE    TRUE
4147653 3144814

  FALSE    TRUE
7292466  160749
```

So, if we define event A as "a randomly selected 2007 domestic US flight experienced a delay" and event B as "the flight was canceled," then $n(A) = 3,144,814$ and $n(B) = 160,749$. This gives

$$\Pr(A) = \frac{n(A)}{n(S)} = \frac{3,144,814}{7,453,215} = 0.422$$

and

$$\Pr(B) = \frac{n(B)}{n(S)} = \frac{160,749}{7,453,215} = 0.022.$$

Thus, in 2007, a random flight had a 42.2% chance of being delayed (though, as we saw in Chapters 1 and 2, most of those delays were quite small) and a 2.2% chance of being canceled.

Now, rather than doing all the counting and division manually, there's an easier way to do these calculations in R using the mean of indicator variables. For example:

```
mean(as.numeric(data07$CancellationCode != ""))
```

which yields

[1] 0.02156774

Of course, this is a nationwide probability that is not particularly relevant to individual flights in the sense that the probabilities likely vary by location, airline, time of year, etc.. For example, here's what the Monterey, CA airport looked like for flight delays in 2007 by month of the year:

	Jan	Feb	Mar	Apr	May	Jun	Jul	Aug	Sep	Oct	Nov	Dec
$n(A)$:	228	250	213	208	203	240	302	286	211	272	242	284
$n(S)$:	668	602	674	617	697	748	780	797	739	758	721	758
$\Pr(A)$:	0.34	0.42	0.32	0.34	0.29	0.32	0.39	0.36	0.29	0.36	0.34	0.37

In the table, we see that the probability of delay was (slightly) higher in summer and winter than in the rest of the year and, overall, it's slightly lower than the national probability.   ■ ■ ■

3.3.2 Counting Sample Points

In this section, we'll look at some useful counting techniques that, when combined with what we've learned so far, will allow us to calculate the probabilities of some fairly complicated events.

> **Definition 3.3.1 — mn-Rule.** Let event A have m sample points, $A = \{a_1, a_2, \ldots, a_m\}$, and let event B have n sample points, $B = \{b_1, b_2, \ldots, b_n\}$. Then there are $m \times n$ pairs of sample points: (a_i, b_j), $i = 1, 2, \ldots, m$, $j = 1, 2, \ldots, n$.

To see why the *mn*-rule is true, consider the table shown in Figure 3.4. Here it should be clear that there are $m \times n$ pairs of sample points: (a_i, b_j), $i = 1, 2, \ldots, m$, $j = 1, 2, \ldots, n$.

■ **Example 3.13** How many outcomes are possible when rolling two six-sided dice?

Solution: For each die there are six possible outcomes. Therefore there are $6 \times 6 = 36$ possible pairs of sample points: $(1, 1), (1, 2), (1, 3), \ldots, (6, 5), (6, 6)$.   ■ ■ ■

■ **Example 3.14** Imagine randomly sampling 10 people, say out of your data science class, and recording their birthdays. Ignoring leap years, how many "10-tuples" are in the sample space? That is, how many different sets of 10 birthdays are possible?

Solution: We solve this problem with repeated application of the *mn*-rule. For the first two people chosen, there are $365 \times 365 = 365^2$ pairs of birthdays. For three people, there are $365^2 \times 365 = 365^3$ 3-tuples; for four people there are $365^3 \times 365 = 365^4$ 4-tuples. By now the pattern should be apparent. For 10 people, there are 365^{10} different sets of 10 birthdays, which is approximately 4.2×10^{25}.   ■ ■ ■

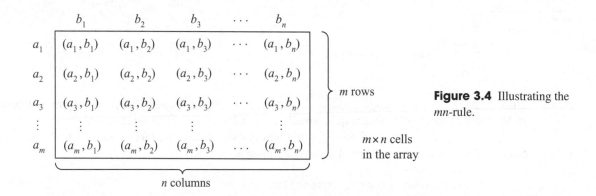

Figure 3.4 Illustrating the *mn*-rule.

Exercise 3.5 A new internet start-up company allows you to buy customized smartphones. They have 10 phone models that each come in 8 colors with 5 different calling/data plans and you get to choose one of 3 different protective cases. How many different customized phones are there?

Answer: 1,200.

Factorials

For a group of distinguishable objects of size n, where n is an integer, there are

$$n! = n \times (n-1) \times (n-2) \times \cdots \times 2 \times 1$$

ways that the group can be ordered. The notation $n!$ is spoken "n factorial."

Definition 3.3.2 — Factorial. The factorial of a non-negative integer n, denoted by $n!$, is the product of all the positive integers less than or equal to n. By convention, $0! = 1$.

Intuitively, it is easy to derive this result by imagining sequentially selecting an ordering. On the first draw there are n items that can be selected. On the second draw, once the first item is chosen, there are $n-1$ items left to be chosen. On the third draw there are $n-2$ left to be chosen, and so on down to the last draw for which there is only one item left. Hence, the total number of ways any ordering can occur is simply their product, or $n!$.

■ **Example 3.15** How many different ways can the letters A, B, C, D be arranged?

Solution: The answer is $4! = 24$. To see this most fundamentally, write out all the possible sequences:

ABCD	BCDA	CDAB	DABC
ABDC	BCAD	CDBA	DACB
ACBD	BDAC	CABD	DBAC
ACDB	BDCA	CADB	DBCA
ADBC	BACD	CBAD	DCAB
ADCB	BADC	CBDA	DCBA

There are exactly 24 unique orderings of the four letters. Note that the characteristic that the objects must be distinguishable is important. If some or all of the objects are indistinguishable from one another then the number of orderings is reduced. ■ ■ ■

In R, the `factorial()` function is handy for calculating factorials. For example, `factorial(4)` yields 24, whereas `factorial(26)` yields `4.032915e+26`, an extraordinarily large number.

The number of different ways of ordering (or sequentially selecting) k objects out of n distinguishable objects, $k \leq n$, is $n \times (n-1) \times \cdots \times (n-k+1)$. That is, if we think of this as randomly sequentially selecting the objects, there are n objects to select from first, followed by $n-1$ second, on down to $n-k+1$ on the last selection. The total number of ways to select (or order) the objects is their product.

A compact way of writing the product $n \times (n-1) \times \cdots \times (n-k+1)$ is $n!/(n-k)!$. If this isn't obvious, write out the ratio in terms of their products and then cancel out the quantities that are in both the numerator and denominator to see it.

Definition 3.3.3 — Permutation. An ordering of k distinct objects out of n is called a *permutation* and the total number of permutations is denoted $P(n,k)$, where

$$P(n,k) = \frac{n!}{(n-k)!}.$$

Note that a factorial is a special permutation where we are selecting n out of n objects:

$$P(n,n) \;=\; \frac{n!}{(n-n)!} \;=\; \frac{n!}{0!} \;=\; \frac{n!}{1} \;=\; n!.$$

■ **Example 3.16** How many ways can two letters out of A, B, C, D be ordered?

Solution: This is a permutation question since the letters are distinct and we're interested in the number of orderings of two letters. Thus, we have

$$P(4,2) \;=\; \frac{4!}{(4-2)!} \;=\; \frac{4 \times 3 \times \cancel{2} \times \cancel{1}}{\cancel{2} \times \cancel{1}} \;=\; 4 \times 3 \;=\; 12.$$

To verify, here are all the ordered two-letter possibilities: (A,B), (A,C), (A,D), (B,A), (B,C), (B,D), (C,A), (C,B), (C,D), (D,A), (D,B), (D,C).   ■ ■ ■

■ **Example 3.17** In horse racing, the trifecta is a type of bet where, to win, you have to pick the first-, second-, and third-place horses precisely in order. In a race with nine horses that are exactly the same, so that it's equally likely that any of the three finish in first, second, and third place, what is the probability of winning the trifecta?

Solution: Because order matters, the number of ways three of the nine horses can come in first, second, and third is

$$P(9,3) \;=\; \frac{9!}{(9-3)!} \;=\; 504.$$

Since there is only one way to win the bet, the probability of winning the trifecta is 1/504 or about 0.002. Of course, in the real world, not all horses are equal, so with information about each horse's capability it is possible to improve on these odds. But a trifecta is still very hard to win.   ■ ■ ■

Next, let's do a famous problem that requires permutations. The "birthday problem" has a solution that runs counter to our intuition.

■ **Example 3.18** Find the smallest number of randomly selected people such that there is at least a 50% probability that at least two have the same birthday. For simplicity, disregard leap years and seasonal or weekday variation in the distribution of birthdays. That is, assume that when picking a random person there are 365 possible birthdays and they are all equally likely.

Solution: To solve this problem, we will use the sample point method. Now, from Example 3.14, we know that if we pick n people, there are 365^n n-tuples of birthdays. This is $n(S_n)$, the total number of sample points in the sample space if n people are selected.

If we were to solve this directly, we would need to find $n(A_n)$, where the event A_n is the number of n-tuples with at least one matching birthday. This is hard to solve, so instead we will use the complement rule and compute $n(\overline{A}_n)$, where the event $\overline{A}_n$ is the number of n-tuples with no matching birthdays.

Now, it turns out that $n(\overline{A}_n) = P(365,n)$. This result is easy to see if you think about the problem in terms of picking people sequentially. The first person chosen can have a birthday on any of 365 days, but the second person can only have one of 364 days since he or she cannot match the birthday of the first person. The third person can only have one of 363 days since he or she cannot match the birthday of the first or second person. And on it goes up to the nth person, who can only have one of $365 - n + 1$ days since he or she cannot match the birthday of any of the previous $n - 1$ people. From this, it follows that

$$n(\overline{A}_n) \;=\; 365 \times 364 \times \cdots \times (365 - n + 2) \times (365 - n + 1) \;=\; 365!/n! \;=\; P(365,n).$$

Thus,

$$\Pr(\overline{A}_n) \;=\; \frac{n(\overline{A}_n)}{n(S_n)} \;=\; \frac{P(365,n)}{365^n},$$

and if you try various values of n you will find that $\Pr(\overline{A}_{23}) = 0.493$, so that $\Pr(A_{23}) = 1 - 0.493 = 0.507$. That is, given all the assumptions in the problem statement, with $n = 23$ people there is slightly more than a 50% chance that at least two people will have a matching birthday.   ■ ■ ■

> **Exercise 3.6** How big does n have to be to have a 95% chance that at least two people will have a matching birthday?
>
> *Answer:* $n = 47$.

The number of different ways of selecting k objects out of n objects *when the order of selection does not matter*, $k \leq n$, is called a *combination*. For example, imagine tossing a fair coin three times and recording the number of heads and tails. Coin toss results of $H_1 H_2 T_3$, $H_1 T_2 H_3$, and $T_1 H_2 H_3$ (where the subscripts denote the sequential coin-flip order) are all the same event (so long as we are just recording the number of heads), since they all consist of two heads and one tail and we don't care what order the heads and tails occurred in.

> **Definition 3.3.4 — Combination.** The unordered selection of k objects out of n is called a *combination* and the total number of combinations is denoted $C(n,k)$ or equivalently $\binom{n}{k}$, where
>
> $$P(n,k) \equiv \binom{n}{k} \;=\; \frac{n!}{(n-k)!\,k!}.$$

So $C(n,k)$ gives the possible number of combinations of n objects taken k at a time without regard to the order in which the items were selected. To see why $\binom{n}{k}$ is equal to $P(n,k) \times k!$ we can take an indirect approach and use the *mn*-rule. How many ways can we order k units out of a set of n? We know that the answer is $P(n,k)$. We could also look at this problem in another way, with two stages: first, select k objects out of n where order doesn't matter, then order these k objects. The first can be done in $\binom{n}{k}$ ways and the second in $k!$ ways. Since we've counted the same quantity in two different ways, the two must be equal. Thus,

$$P(n,k) \;=\; \binom{n}{k} \times k!$$

or equivalently

$$\binom{n}{k} \;=\; \frac{P(n,k)}{k!} \;=\; \frac{n!}{k!(n-k)!}.$$

■ **Example 3.19** Returning to Example 3.16, how many combinations of two letters can be selected out of A, B, C, D?

Solution: $C(4,2) = 4!/(2! \times 2!) = 6$. This result follows from the solution to Example 3.16, where we wrote out the twelve permutations of two letters. Note that each pair of letters occurs twice, so if ordering doesn't matter then there are six unique pairs.   ■ ■ ■

The previous example is relatively easy to calculate by hand by writing out all the possible combinations or permutations. However, as n and k get moderately large, the number of combinations and permutations can get very large. For example, suppose we have a class of 30 students and wish to know how many different basketball teams of five students could be formed. The list of possible combinations of five students out of the 30 in the class is huge:

$$\binom{30}{5} = \frac{30!}{25!5!} = \frac{30 \times 29 \times 28 \times 27 \times 26}{5 \times 4 \times 3 \times 2} = \frac{17,100,720}{120} = 142,506.$$

■ **Example 3.20** You're preparing for an exam by studying a list of 10 problems, where the exam will be a random selection of 5 of these 10 problems. You can solve 8 of the 10 problems. What is the probability that you will be able to solve all five problems that will be on the exam?

Solution: To solve this problem, we must first determine the number of ways that 5 problems can be selected from the 10 possible problems. It is $\binom{10}{5}$, and this is the number of sample points in the sample space: $n(S)$. We also need to know how many ways that 5 problems can be selected from the 8 you know how to solve. It is $\binom{8}{5}$, and this is the number of sample points in event A, $n(A)$, which is the event that you will be able to solve all 5 problems on the exam. Then,

$$\Pr(A) = \frac{n(A)}{n(S)} = \frac{\binom{8}{5}}{\binom{10}{5}} = \frac{\frac{8!}{5!3!}}{\frac{10!}{5!5!}} = \frac{8!5!5!}{10!5!3!} = \frac{5 \times 4}{10 \times 9} = \frac{20}{90} \approx 0.222. \qquad \blacksquare\,\blacksquare\,\blacksquare$$

In R, the `factorial()` function can be used to calculate combinations directly from the formula or the `choose()` function can do it directly.

■ **Example 3.21 — Using R to calculate combinations.** Compute (a) $\binom{4}{2}$, and (b) $\binom{10}{5}$. Use the result to recompute the answer to Example 3.20.

Solution: We can compute these directly or use R's choose function:

```
factorial(4)/(factorial(2)*factorial(4-2))
choose(4,2)
factorial(10)/(factorial(5)*factorial(10-5))
choose(10,5)
choose(8,5)/choose(10,5)
```

which yields

```
[1] 6
[1] 6
[1] 252
[1] 252
[1] 0.2222222
```
                                                                                     $\blacksquare\,\blacksquare\,\blacksquare$

3.3.3 Combining Events

Thinking about events as sets of sample points allows us to use set algebra to solve for combinations of events and, from them, probabilities. Indeed, that is essentially what we were doing in Section 3.2 (e.g., in the derivation of the complement rule). The idea is to express the event (set) of interest in terms of other sets for which you know their probabilities. Then, perhaps after some set algebra, you can find the probability of the event of interest.

In terms of set algebra, union and intersection are associative and commutative (just like addition and multiplication in arithmetic) and the relation "subset" is reflexive, antisymmetric, and transitive (just like, for example, the arithmetic relation "less than or equal").

Two sets of laws from set algebra are useful for calculating the probability of complex events. They are the *distributive laws for sets* and *DeMorgan's laws*.

Theorem 3.3.1 — Distributive Laws for Sets. For any sets A, B, and C,

$$A \cap (B \cup C) = (A \cap B) \cup (A \cap C)$$

and

$$A \cup (B \cap C) = (A \cup B) \cap (A \cup C).$$

With set algebra, just like with regular algebra, you first calculate the quantities within parentheses. The distributive laws for sets say that you can carry the intersection or union operation through an expression using the opposite operation. DeMorgan's laws connect the complements of unions or intersections of sets (events) to the intersection or union of the complements of the individual sets (events).

Theorem 3.3.2 — DeMorgan's Laws. For any two sets A and B,

$$\overline{(A \cap B)} = \overline{A} \cup \overline{B}$$

and

$$\overline{(A \cup B)} = \overline{A} \cap \overline{B}.$$

Exercise 3.7 Draw Venn diagrams for the expressions on the left and right sides of DeMorgan's laws to show that they are, in fact, equal. Do it for two cases: (1) when A and B are disjoint and (2) when A and B are not disjoint.

3.4 Bringing It All Together

Important extensions of the multiplicative law are the law of total probability and Bayes' Theorem, both of which can be invaluable for computing conditional probabilities, particularly conditional probabilities that might otherwise be very difficult to calculate directly. Let's start with the law of total probability and then develop and explain Bayes' Theorem.

3.4.1 Law of Total Probability

Consider Figure 3.5, showing a sample space is composed of four disjoint events: $S = B_1 \cup B_2 \cup B_3 \cup B_4$ where, from Probability Axioms 2 and 3, it follows that $\Pr(B_1) + \Pr(B_2) + \Pr(B_3) + \Pr(B_4) = 1$. In addition, there is another event A, which is not disjoint from some or all of events B_1, B_2, B_3, and B_4 and we are interested in determining $\Pr(A)$.

To calculate $\Pr(A)$, we start by finding an expression for A, where if we know $A \cap B_i$ for $i = 1, \ldots, 4$ then

$$A = (A \cap B_1) \cup (A \cap B_2) \cup (A \cap B_3) \cup (A \cap B_4).$$

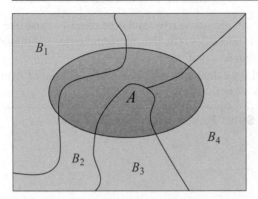

Figure 3.5 A Venn diagram showing $S = B_1 \cup B_2 \cup B_3 \cup B_4$ and an event A that intersects the other B_i events.

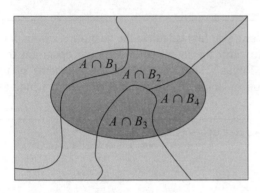

Figure 3.6 Given $A \cap B_i$ for $i = 1, \ldots, 4$ then $A = (A \cap B_1) \cup (A \cap B_2) \cup (A \cap B_3) \cup (A \cap B_4)$.

See Figure 3.6. Then, because the B_i are all disjoint, so are all the $A \cap B_i$, and thus from Axiom 3 it follows that

$$\Pr(A) \;=\; \Pr(A \cap B_1) + \Pr(A \cap B_2) + \Pr(A \cap B_3) + \Pr(A \cap B_4).$$

In other words, if x is an element in A, then it must be in one and only one of the B_i. Combine this result with the fact that $\Pr(A \cap B_i) = \Pr(B_i)\Pr(A \mid B_i)$, via the multiplicative law, and we have the law of total probability.

Theorem 3.4.1 — Law of Total Probability. For $B_1, B_2, \ldots, B_n$ where $\displaystyle\bigcup_{i=1}^{n} B_i = S$ and $B_i \cap B_j = \emptyset$ for $i \neq j$, and with $\Pr(B_i) > 0$ for all i, then for any event A,

$$\Pr(A) \;=\; \sum_{i=1}^{n} \Pr(A \mid B_i)\Pr(B_i).$$

Proof. The law of total probability is straightforward to prove using the distributive law for sets:

$$\Pr(A) \;=\; \Pr(A \cap S) \;=\; \Pr\!\left(A \cap \left(\bigcup_{i=1}^{n} B_i \right) \right) \;=\; \Pr\!\left(\bigcup_{i=1}^{n} (A \cap B_i) \right).$$

Now, because the B_i are disjoint, then the $A \cap B_i$ must also be disjoint, and thus by Axiom 3,

$$\Pr\!\left(\bigcup_{i=1}^{n} (A \cap B_i) \right) \;=\; \sum_{i=1}^{n} \Pr(A \cap B_i) \;=\; \sum_{i=1}^{n} \Pr(A \mid B_i)\Pr(B_i),$$

where the last equality follows from the multiplicative law of probability.   ■

The law of total probability is useful in situations where it is not obvious how to calculate $\Pr(A)$ directly but in which $\Pr(A \mid B_i)$ and $\Pr(B_i)$ are more straightforward to calculate or are known.

■ **Example 3.22** A company that manufactures electric cars has two factories. Factory 1 is the older of the two and makes 75% of the cars. Factory 2 is newer than factory 1 and makes the remaining 25% of the cars. Of the cars that factory 1 produces, 4% require some sort of corrective work, while only 1% of factory 2 cars require rework. If we choose a car at random, what is the probability that it will require rework?

Solution: Let the event F_i denote a car produced by factory i, $i = 1, 2$. Let R denote the event that the car requires rework. Then, we're given that for a randomly selected car, $\Pr(F_1) = 0.75$ and $\Pr(F_2) = 0.25$. We're also given that $\Pr(R \mid F_1) = 0.04$ while $\Pr(R \mid F_2) = 0.01$. Therefore, via the law of total probability, we have

$$\Pr(R) = \Pr(R \mid F_1)\Pr(F_1) + \Pr(R \mid F_2)\Pr(F_2) = 0.04 \times 0.75 + 0.01 \times 0.25 = 0.0325.$$

Thus, if we choose a car at random, the probability that it will require rework is 3.25%.   ■ ■ ■

Figure 3.7 graphically depicts the problem and solution to Example 3.22. In it, we see all the information we were given in tree form. Working from left to right, the first two *branches* depict the fact that a random car could have come from one of two factories and thus there are two initial branches. Each of the branches is labeled with the probability of traversing down that branch which, in this problem, corresponds to the probability a car is built by factory 1 or factory 2.

Each of these branches then splits into two further branches, depending on whether a car requires rework. And, as before, each of the branches is labeled with the appropriate probability. Note that these probabilities are now conditional probabilities where the condition is that we first traversed down the main branch. To fully label the tree requires recognizing that for any sets A and B for which $\Pr(B) \geq 0$,

$$\Pr(A \mid B) + \Pr(\overline{A} \mid B) = 1.$$

This is straightforward to prove – just apply the definition of conditional probability to the above expression and solve.

The *leaves* of the tree are the values that result from multiplying the probabilities along each branch pathway and, via the multiplicative law, they are the intersection of the events to which the branches

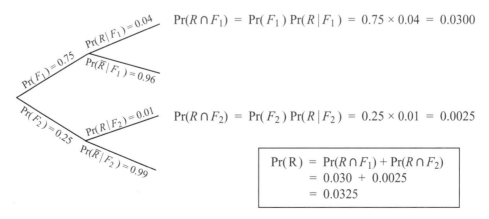

Figure 3.7 Using a tree to solve Example 3.22.

correspond. Here we only label and solve for the two leaves relevant to this problem, $\Pr(R \cap F_1)$ and $\Pr(R \cap F_2)$, which we then sum to get the desired answer.

3.4.2 Bayes' Theorem

Bayes' Theorem is named after the Reverend Thomas Bayes (1701–1761), who first derived it as a way to update probabilities. The field of Bayesian statistics that we discussed briefly in Section 3.3 (and will discuss more thoroughly in Chapter 15) carries his name as a result. Bayes' Theorem is also useful for inverting conditional probabilities. For example, it is useful for calculating $\Pr(B_i \mid A)$ for some event B_i when the only $\Pr(B_i)$s and the $\Pr(A \mid B_i)$s are known.

Unfortunately, it is not as easy to draw pictures for Bayes' Theorem as it was for the law of total probability, so we will simply state it as a formula. Notice, however, that the denominator is obtained from the law of total probability, which can be illustrated by a tree diagram.

> **Theorem 3.4.2 — Bayes Theorem.** Let A and $B_1, \ldots, B_n$ be events where the B_i are disjoint, $\bigcup_{i=1}^{n} B_i = S$, and $\Pr(B_i) > 0$ for all i. Then
>
> $$\Pr(B_j \mid A) = \frac{\Pr(A \mid B_j)\Pr(B_j)}{\sum_{i=1}^{n}\Pr(A \mid B_i)\Pr(B_i)}.$$

Proof. This theorem follows directly from the law of total probability and the multiplication rule substituted into the definition for conditional probability:

$$\Pr(B_j \mid A) = \frac{\Pr(B_j \cap A)}{\Pr(A)} = \frac{\Pr(A \mid B_j)\Pr(B_j)}{\sum_{i=1}^{n}\Pr(A \mid B_i)\Pr(B_i)}. \qquad \blacksquare$$

Let's illustrate the use of Bayes' Theorem using a classic example of determining the probability of getting a *true positive* result from a medical test, where a true positive result is one in which the test outcome correctly matches the existence of the condition being tested for. That is, for a person with a particular disease, a true positive result means the test correctly indicated the presence of the disease. In comparison, a *false positive* occurs when a test indicates the presence of a disease in a person that actually does not have the disease. The probability that a person randomly selected from the population[3] tests positive given that they have the disease is called the *sensitivity* of the test. The probability that a person randomly selected from the population tests negative given that they do not have the disease is called the *specificity*. The unconditional probability that a person randomly selected from the population has the disease is called the *base rate*.

■ **Example 3.23** Consider a medical test whose sensitivity is 0.95 and whose specificity is 0.99. The base rate is 0.005. Find

(a) the probability that a randomly selected person tests positive, and

(b) the probability that a person has the disease given a positive test.

Solution: See Figure 3.8. Let D be the event that a person has the disease and let E be the event that his or her test is positive. We're given $\Pr(E \mid D) = 0.95$, $\Pr(E \mid \overline{D}) = 0.01$, and $\Pr(D) = 0.005$. We want to find $\Pr(D \mid E)$. Using the law of total probability,

$$\Pr(E) = \Pr(E \mid D)\Pr(D) + \Pr(E \mid \overline{D})\Pr(\overline{D}).$$

---

[3]The "selected from the population" is important because in the absence of a test result, the probability of having the disease is equal to the base rate. If a person shows symptoms of the disease, then before the test the probability of having the disease is probably much higher than the base rate.

The only quantity we are missing is $\Pr(\overline{D})$, and this is easily obtained using the complement rule: $\Pr(\overline{D}) = 1 - \Pr(D) = 1 - 0.005 = 0.995$. The probability of testing positive is thus

$$\Pr(E) = \Pr(E \mid D)\Pr(D) + \Pr(E \mid \overline{D})\Pr(\overline{D}) = (0.95)(0.005) + (0.01)(0.995) = 0.0147.$$

Thus,

$$\Pr(D \mid E) = \frac{\Pr(E \mid D)\Pr(D)}{\Pr(E)} = \frac{(0.95)(0.005)}{0.0147} = 0.32.$$

Thus, for this particular test with this particular disease prevalence, there is only a 32% chance that a person who tests positive is actually sick. This is a counterintuitive result, given how seemingly "accurate" the test is. What's going on here? Well, the disease is really quite rare in the population, so a positive result is more likely to be a false positive than a true positive.   ■■■

Next, we computed the probability of having the disease given a positive test. This is called the *positive predictive value (PPV)*. In mathematical language,

$$\text{PPV} = \Pr(\text{having the disease} \mid \text{positive test}).$$

Similarly, the *negative predictive value (NPV)* is

$$\text{NPV} = \Pr(\text{not having the disease} \mid \text{negative test}).$$

In the next example, let's see what effect the base rate of the disease in the population has on the PPV.

■ **Example 3.24** Let's revisit Example 3.23. What is the probability that a person has the disease given that he or she has a positive test if 10% of the population has the disease?

Solution: See Figure 3.8. As before, let D be the event that a person has the disease and let E be the event that his or her test is positive. Then,

$$\Pr(D \mid E) = \frac{\Pr(E \mid D)\Pr(D)}{\Pr(E \mid D)\Pr(D) + \Pr(E \mid \overline{D})\Pr(\overline{D})} = \frac{(0.95)(0.1)}{(0.95)(0.1) + (0.01)(0.9)} = 0.91.$$

With a 10% prevalence, the test performs much better. The lesson: With a very rare disease a test has to be extremely accurate or most of the positive test results will be false positives.   ■■■

Exercise 3.8 For the problem in Example 3.24, what is the probability a person does not have the disease given he or she has a negative test?

Answer: $\Pr(\overline{D} \mid \overline{E}) = 0.99$.

Let's wrap up the chapter with an example of using Bayes' Theorem to update probabilities.

■ **Example 3.25** Suppose that a plane is missing and has an equal chance of going down in any one of three regions. If the plane is actually located in region i, the probability that a search for it in that region is successful is $1 - \alpha_i$ for $0 \le \alpha_i \le 1$, $i = 1, 2, 3$. What is the probability that the plane is in region 1 given a search of region 1 was unsuccessful?

Solution: We'll begin by defining some notation. Let L_i denote that the plane is located in region i and let S_i denote a successful search of region i. Thus, $\overline{L}_i$ denotes that the plane is not located in region i and $\overline{S}_i$ denotes an unsuccessful search of region i. Then, $P(L_i) = 1/3$, $P(S_i|L_i) = 1 - \alpha_i$ for $i = 1, 2, 3$, and we

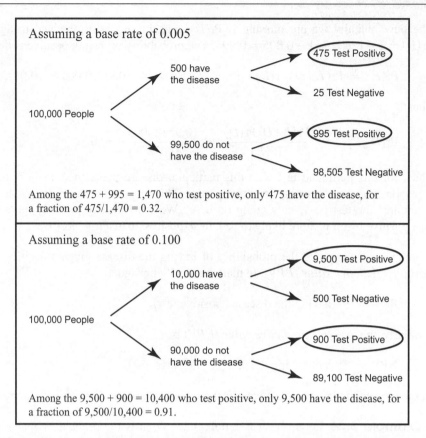

Figure 3.8 Illustrating the positive predictive value (PPV) for base rates of 0.005 and 0.100.

need to find $P(L_1|\overline{S}_1)$. Let's start by using the definition of conditional probability and then appropriately express the numerator and denominator in terms of quantities that we know or can determine:

$$P(L_1|\overline{S}_1) = \frac{P(L_1 \cap \overline{S}_1)}{P(\overline{S}_1)} = \frac{P(L_1)P(\overline{S}_1|L_1)}{P(L_1)P(\overline{S}_1|L_1) + P(L_2)P(\overline{S}_1|L_2) + P(L_3)P(\overline{S}_1|L_3)}.$$

To determine the values of the various pieces in the equation:

(a) We're given that $P(L_i) = 1/3, i = 1, 2, 3$.

(b) Also, given that $P(S_i|L_i) = 1 - \alpha_i$, so $P(\overline{S}_1|L_1) = 1 - P(S_1|L_1) = 1 - (1 - \alpha_1) = \alpha_1$.

(c) What remains is $P(\overline{S}_1|L_2)$ and $P(\overline{S}_1|L_3)$. However, if we just convert the notation into what it means, we can determine their values directly. That is, $P(\overline{S}_1|L_2)$ is the probability that a search of region 1 is unsuccessful given that the plane is actually located in region 2. This probability is 1 because a search of region 1 must be unsuccessful if the plane is not there. Similarly, $P(\overline{S}_1|L_3) = 1$.

Plugging these values into the equation, we have:

$$P(L_1|\overline{S}_1) = \frac{\alpha_1/3}{\alpha_1/3 + 1/3 + 1/3} = \frac{\alpha_1}{\alpha_1 + 2}.$$

What does this mean? Let's illustrate with a specific example. Imagine that $\alpha_1 = \alpha_2 = \alpha_3 = 0.1$, so that the probability of an unsuccessful search in any of the three regions, given the plane is in the region being searched, is 0.1 (and thus the probability of detecting it is 0.9). Prior to searching, we know that the probability the plane is in region 1 is 1/3. However, after unsuccessfully searching region 1, we update the probability that the plane is in region 1 to $P(L_1|\bar{S}_1) = 0.1/(0.1+2) = 0.04762$.

Now that we've searched region 1 once and didn't find the plane, our updated probability that the plane is there decreases from 1/3 to about 1/20. Note that the probability is not zero. There's still some chance the plane is there and we just missed it in the search. However, given that we had a 90% chance of finding it, our assessment of the probability that it's there is now much lower.   ■ ■ ■

3.5 Chapter Summary

In this chapter we've learned about the basic rules of probability. As discussed in the introduction, the point is not to turn you into a probabilist – though an appreciation for and ability to apply probability can be useful in many situations – but rather to establish a requisite foundation of knowledge so that you can understand the statistical methods yet to come.

Thus, in this chapter we've learned about the axioms of probability, events, and sample spaces, and about how to think about more complicated types of events based on unions, intersections, and complements. We also learned about ways to count sample points, such as with the *mn*-rule, factorials, combinations, and permutations. We then applied these techniques to do various probability calculations, ultimately building up to the law of total probability and Bayes' Theorem.

In the next chapter we're going to build on this material and learn about a very important concept: the random variable. The utility of random variables is that they provide something of a shortcut to doing useful probability calculations in a variety of real-world situations. Making the connection between the basic probability concepts and principles of this chapter to the more abstract notion of random variables introduced in Chapter 4 (and further developed in Chapters 5 and 6) is critical, as the statistical methods we will later learn about all follow from the idea of the random variable.

3.6 Problems

Problem 3.1 The repair of a broken cell phone is either completed on time or it is late and the repair is either done satisfactorily or unsatisfactorily. What is the sample space for a cell phone repair?

Problem 3.2 What is the sample space for a two-sided coin that is tossed two times? What is the sample space for the same coin that is tossed three times?

Problem 3.3 Consider a family with two smartphones. Let a denote a smartphone that runs the Android operating system, o denotes a smartphone running iOS, and w denotes a phone with a windows operating system. Then the sample space is

$$S = \{aa, ao, aw, oo, ow, ww\}.$$

Let event A denote the event that neither phone has an Android operating system, let event B denote both phones run the iOS operating system, and let event C denote at least one phone runs the Windows operating system. List the elements of $A, B, C, A \cap B, A \cup B, A \cap C, A \cup C, B \cap C, B \cup C,$ and $C \cap \overline{B}$.

Problem 3.4 Let A and B be two events. Mathematically express the following events using unions, intersections, and complements.

(a) Both events occur.
(b) Neither event occurs.
(c) At least one event occurs.
(d) Exactly one event occurs.

Problem 3.5 For nondisjoint sets A, B, and C, use Venn diagrams to show that the following are true:

(a) $A = (A \cap B) \cup (A \cap \overline{B})$
(b) $A \cap C = (A \cap B \cap C) \cup (A \cap \overline{B} \cap C)$
(c) If $B \subset C$ then $C = B \cup (C \cap \overline{B})$

Problem 3.6 An experiment has three possible outcomes, A, B, and C, only one of which can occur at a time. If outcome B is twice as likely as outcome A and if outcome C is three times as likely as outcome A, what are

(a) $\Pr(A)$?
(b) $\Pr(B)$?
(c) $\Pr(C)$?

Problem 3.7 A sample space consists of five sample points: $S = \{s_1, s_2, s_3, s_4, s_5\}$. If $\Pr(\{s_1\}) = \Pr(\{s_2\}) = 0.1$, $\Pr(\{s_3\}) = 0.3$, and $2 \times \Pr(\{s_4\}) = \Pr(\{s_5\})$, find $\Pr(\{s_4\})$ and $\Pr(\{s_5\})$.

Problem 3.8 If birthdays are equally likely to fall on any day throughout the year (which is not true, but assume it is for the purposes of this problem), for a non-leap year, what is the probability that a person chosen at random has a birthday in January, February, ..., November, and December?

Problem 3.9 For two events A and B, $\Pr(A) = 0.5$, $\Pr(B) = 0.4$, and $\Pr(A \cup B) = 0.7$.

(a) What is $\Pr(A \cap B)$? Show your result in a Venn diagram.
(b) If instead $\Pr(A \cup B) = 0.9$, how does that change your Venn diagram?

Problem 3.10 For two events X and Y, $\Pr(X) = 0.2$, $\Pr(Y) = 0.3$, and $\Pr(X \cap Y) = 0.1$.

(a) What is $\Pr(X \cup Y)$? Show your result in a Venn diagram.
(b) If instead $\Pr(X \cap Y) = 0.2$, how does that change your Venn diagram?

Problem 3.11 For two events A and B, all you know is that $\Pr(A) = 0.6$ and $\Pr(B) = 0.7$. What can you say about $\Pr(A \cup B)$ and $\Pr(A \cap B)$?

Problem 3.12 For two events C and D, all you know is that $\Pr(C) = 0.3$ and $\Pr(C \cup D) = 0.7$.

(a) What can you say about $\Pr(D)$ and $\Pr(C \cap D)$?
(b) If you're now told that C and D are disjoint, what can you say about $\Pr(D)$ and $\Pr(C \cap D)$?

Problem 3.13 What is the probability that at least one of the following events occurs on a single throw of two fair six-sided dice?

- Event A: The dice total 5.
- Event B: The dice total 6.
- Event C: The dice total 7.

Problem 3.14 What is the probability that at least one of the following events occurs on a single throw of two fair *four*-sided dice?

- Event A: The dice total 5.
- Event B: The dice total 6.
- Event C: The dice total 7.

Problem 3.15 In an experiment, two *biased* coins are tossed, where $\Pr(\text{"flip a heads"}) = 0.6$, and the sides facing up are recorded.

(a) Write out the sample space for this experiment.
(b) Assuming the coins are independent, what are the probabilities for each of the points in the sample space?
(c) Let A denote the event that at least one tail is observed and let B denote the event that exactly one tail is observed. List the sample points contained in A and B.

(d) Find $\Pr(A), \Pr(B), \Pr(A \cap B), \Pr(A \cup B)$, and $\Pr(\overline{A} \cap B)$.

Problem 3.16 A single fair six-sided die is rolled twice. Let A be the event of getting a total of 8 on both rolls. Let B be the event that a 5 comes up on the first roll. What are:

(a) $\Pr(A \mid B)$?
(b) $\Pr(A \cup B)$?

Problem 3.17 Two fair six-sided dice are rolled simultaneously. Let A be the event of getting a total of 8 on the roll. Let B be the event that a 5 is rolled on a single die out of the pair. What are:

(a) $\Pr(A \mid B)$?
(b) $\Pr(A \cup B)$?

Problem 3.18 A card is drawn at random from a standard deck of playing cards. What is the probability that at least one of the following two events will occur:

(a) Event A: a heart is drawn.
(b) Event B: a card that is *not* a face card (jack, queen, or king) is drawn.

Problem 3.19 How many ordered pairs can be selected that are composed of one from each of the following two groups of letters: $\{a,b,c,d,e,f,g\}$ and $\{u,v,w,x,y,z\}$?

Problem 3.20 How many ordered pairs can be selected that are composed of one from each of the following two groups: $\{\mathbf{A},\mathbf{K},\mathbf{Q},\mathbf{J},\mathbf{10},\mathbf{9},\mathbf{8},\mathbf{7},\mathbf{6},\mathbf{5},\mathbf{4},\mathbf{3},\mathbf{2}\}$ and $\{\heartsuit,\clubsuit,\spadesuit,\diamondsuit\}$?

Problem 3.21 A sommelier is asked to rank three wines from best to worst (no ties allowed). How many ways can she rank the three wines?

Problem 3.22 A menu has six appetizers, four soups, eight main courses, and five desserts. How many ways can a four-course meal be chosen from this menu? How about a three-course meal, where the diner only chooses one of either an appetizer or soup for the first course?

Problem 3.23 For $n = 12$ different donuts, calculate the number of groups of 5 donuts that can be selected from the 12 when the order of selection is not important.

Problem 3.24 A shipment of 20 printers are received of which 15 are laser printers and 5 are inkjet models. If 10 of these are selected at random to be quality checked, how many different ways can the sample be chosen so that exactly 5 laser printers and 5 inkjet printers are inspected? Now, if only 6 are selected, how many ways can the sample be chosen so that exactly 3 laser printers and 3 inkjet printers are inspected?

Problem 3.25 The manager of a new data analytics company has to assign 15 data scientists to three projects in groups of sizes 10, 3, and 2. How many ways can the data scientists be assigned to these three groups if each person is in one and only one group?

Problem 3.26 A card is drawn at random from a standard 52-card deck. What are the following conditional probabilities?

(a) $\Pr(\mathbf{A}\heartsuit \mid$ card from a red suit)
(b) $\Pr(\mathbf{A}\heartsuit \mid$ card from a black suit)
(c) $\Pr(\text{club} \mid$ card from a black suit)

Problem 3.27 A fair six-sided die is rolled. What are the following conditional probabilities?

(a) $\Pr(5 \mid$ a prime number is rolled)
(b) $\Pr(\text{a prime number is rolled} \mid 5)$
(c) $\Pr(\text{number is less than 3} \mid$ a prime number)
(d) $\Pr(\text{a prime number} \mid$ number is less than 3)

Problem 3.28 A poker hand consists of five cards drawn randomly from a standard 52-card deck.

(a) How many possible poker hands are there?
(b) How many hands are there that consist entirely of cards from the diamond suit?
(c) How many hands contain all four aces?
(d) How many hands contain cards with four cards of the same rank (i.e., with the same number or picture)?

Problem 3.29 Continuing with Problem 3.28:

(a) What is the probability of getting a hand that contains all four aces?
(b) What is the probability of getting a hand with four cards of the same rank (i.e., with the same number or picture)?
(c) What is the probability of getting a hand with only cards from the diamond suit (a diamond flush)?

Problem 3.30 For two events A and B, let $\Pr(A) = 0.5, \Pr(B) = 0.2$, and $\Pr(A \cap B) = 0.1$.

(a) Are A and B independent? Why or why not?
(b) Find $\Pr(A \mid B)$ and $\Pr(A \cap B \mid A \cup B)$.

Problem 3.31 For a person randomly selected from all men in the United States, let event A be that he is over 6 feet tall and event B be that he is a professional basketball player. Which is larger: $\Pr(A|B)$ or $\Pr(B|A)$? Explain your answer.

Problem 3.32 For two events A and B, let $\Pr(A) = 0.5, \Pr(B) = 0.3$, and $\Pr(A \cap B) = 0.1$. Find the following quantities:

(a) $\Pr(A \mid A \cup B)$

(b) $\Pr(A \cup B \mid A \cap B)$

Problem 3.33 Consider three disjoint events B_1, B_2, and B_3 where $B_1 \cup B_2 \cup B_3 = S$ and $\Pr(B_1) = 1/5$, $\Pr(B_2) = \Pr(B_3) = 2/5$. Within the sample space S there is another event A for which we know $\Pr(A \mid B_1) = 1/4$, $\Pr(A \mid B_2) = 1/4$, and $\Pr(A \mid B_3) = 1/2$. Find $\Pr(A)$.

Problem 3.34 In college basketball, when a foul occurs and certain conditions hold,[4] the fouled player gets to go to the free-throw line and shoots what is called a "one and one." This works as follows. The player takes the first shot; if the player is successful on the first shot, then a second shot is taken. If the player misses the first shot, the attempt is over and no additional shots are attempted. Under no circumstances does the player ever get to attempt more than two shots. Each successful shot counts one point.

(a) In terms of points, what are the possible outcomes?

(b) Assuming that the player has a 0.7 probability of making any shot, and successive shots are independent, find the probability for each of the possible outcomes in part (a).

Problem 3.35 Returning to the 2007 airlines data, let event A be that a 2007 domestic US flight was canceled, let event B be that the flight was scheduled to leave Dulles airport (airport code IAD), and let event C be that the flight was in February. From Example 3.12 we know that for a random flight $\Pr(A) = 0.02157$. Find:

(a) $\Pr(A \mid B)$

(b) $\Pr(A \mid C)$

(c) $\Pr(A \mid B \cap C)$

Problem 3.36 The table below summarizes the 2007 airlines data in terms of the fraction of flights that were delayed or not by whether the flight was on a weekend or weekday.

	Delayed (A)	Not Delayed ($\overline{A}$)	Total
Weekday (B)	0.417	0.318	0.735
Weekend ($\overline{B}$)	0.152	0.113	0.265
Total	0.569	0.431	1.000

For a random 2007 flight, where event A is the flight was delayed and event B is the flight was on a weekday, what are the probabilities of the following events?

[4]When the number of fouls for a team in a half is between seven and nine (inclusive), their opponents will attempt a one-and-one on any foul on the opponent except (1) when the fouled player was attempting a shot, or (2) when the player commits a player control foul.

(a) $\Pr(A \cap B)$ and $\Pr(\overline{A} \cap \overline{B})$

(b) $\Pr(A \mid B)$ and $\Pr(\overline{A} \mid \overline{B})$

(c) Are events A and B independent?

Problem 3.37 Consider an aircraft detection system consisting of n radar sets that operate independently. They each have a probability of detecting an aircraft coming into the airport's airspace (a 30 mile radius around the airport) of 0.9.

(a) If $n = 2$, what is the probability at least one radar set detects an arriving aircraft 30 miles away?

(b) How large must n be to achieve a probability of detecting an aircraft coming into the airport's airspace of 0.9999?

Problem 3.38 Customers can buy one of four products, each having its own web page with a "buy" link. When they click on the link, they are redirected to a common web page containing a registration and payment form. Once there, the customer either buys the desired product (generically labeled 1, 2, 3, and 4) or they fail to complete and a sale is lost. Let event A_i be that a customer is on product i's web page and let event B be the event that the customer buys the product. For the purposes of this problem, assume that each potential customer visits at most one product page and so he or she buys at most one product. For the probabilities shown in the table below, find the probability that a customer buys a product.

Product (i)	$\Pr(B \mid A_i)$	$\Pr(A_i)$
1	0.72	0.14
2	0.90	0.04
3	0.51	0.11
4	0.87	0.02

Problem 3.39 Consider a cache of historical Twitter data where each tweet can be assigned to a particular region of the United States: west, midwest, northeast, and south. For all the data from one week in 2014, let event A denote that a randomly selected tweet contained the word "Obama" and let event B_i be the event that a randomly selected tweet originated from region i. For the probabilities shown in the table below, find the probability that a randomly selected tweet contains the word "Obama."

i	Region	$\Pr(A \mid B_i)$	$\Pr(B_i)$
1	West	0.015	0.35
2	Midwest	0.003	0.20
3	Northeast	0.009	0.31
4	South	0.011	0.14

Problem 3.40 Returning to Problem 3.38, if a random purchase is selected, find the probability that

it was item 1 that was purchased. Similarly, for a randomly selected purchase, find the probability of purchase for item 2, item 3, and item 4. What can you say about these four probabilities?

Problem 3.41 Returning to Problem 3.39, for a randomly selected tweet that contains the word "Obama," what is the probability that it came from the west? Similarly, for a randomly selected tweet that contains the word "Obama," find the probability that it came from the midwest, from the northeast, and from the south. What can you say about the four probabilities?

Problem 3.42 Consider a new type of commercial unmanned aerial vehicle (UAV) that has been outfitted with a transponder so that if it crashes it can easily be found and reused. Other older UAVs do not have transponders. Eighty percent of all UAVs are recovered and, of those recovered, 75% have a transponder. Further, of those not recovered, 90% do not have a transponder. Denote recovery as $R+$ and failure to recover as $R-$. Denote having a transponder as $T+$ and not having a transponder as $T-$. Find:

(a) $\Pr(T+)$
(b) The probability of not recovering a UAV given that it has a transponder.

Problem 3.43 Consider a diagnostic test designed to determine whether a piece of equipment is faulty. Let S denote that the item is not operating correctly (i.e., it's "sick") and $\overline{S}$ denote that it is operating correctly. Also, let D denote that the diagnostic test signals that the equipment is not operating correctly (perhaps correctly and perhaps not) and $\overline{D}$ that the diagnostic test does not signal. One percent of the equipment fleet is faulty, so $\Pr(S) = 0.01$. This diagnostic test

has the following characteristics: $\Pr(D \mid S) = 0.9$ and $\Pr(\overline{D} \mid \overline{S}) = 0.9$. Find $\Pr(S \mid D)$.

Problem 3.44 A company that manufactures high-speed wireless routers tests the routers before shipping them to suppliers. Let $D+$ be the event that a router is defective and $D-$ be the event that the router is not defective. Let $T+$ be the event that a test indicates that the router is defective and $T-$ be the event that the test says the router is not defective. For the router and associated test: $\Pr(D+) = 0.001$, $\Pr(T+ \mid D+) = 0.99$, and $\Pr(T+ \mid D-) = 0.001$. Find the probability that a randomly selected router actually is defective given the test says it is defective: $\Pr(D+ \mid T+)$.

Problem 3.45 Suppose that the test for a disease has sensitivity 0.90 and specificity 0.999. The base rate for the disease is 0.02. Find:

(a) The probability that someone selected at random from the population tests positive.
(b) The probability that the person has the disease given a positive test.
(c) The probability that the person does not have the disease given a negative test.

Problem 3.46 Repeat Problem 3.45 for the case where the sensitivity is 0.80, the specificity is 0.98, and the base rate is 0.003.

Problem 3.47 If A and B are disjoint events, what can you say about $\Pr(A \cup B)$ and $\Pr(A \cap B)$?

Problem 3.48 Prove that for any sets A and B, for which $\Pr(B) > 0$, $\Pr(A \mid B) + \Pr(\overline{A} \mid B) = 1$.

Problem 3.49 Prove Theorem 3.2.1.

Problem 3.50 Prove DeMorgan's laws.

Problem 3.51 Prove Theorem 3.2.3.

4 — Random Variables

4.1 Introduction

In Chapter 3 we learned how to do basic probability calculations and even put them to use solving some fairly complicated probability problems. In this chapter and the next two, we generalize how we do probability calculations, where we will transition from working with sets and events to working with random variables.

> **Definition 4.1.1 — Random Variable.** A *random variable* is the assignment of a number to each outcome in the sample space for some random phenomenon. Random variables are denoted by capital letters (e.g., X or Y), whereas possible outcomes are denoted with lower case letters (e.g., x or y).

A *random variable* is thus a variable whose value is subject to chance. The values a random variable could assume might be the possible outcomes of an experiment that has not yet been performed. Or, it might be the possible outcomes of a past experiment whose value is not yet known. The connection between Chapter 3 and this one is that here we will assign *numbers* to the sample points in a sample space S.

4.2 Discrete Random Variables

To begin, let's define what it means for a random variable to be discrete.

> **Definition 4.2.1 — Discrete Random Variable.** A random variable is discrete if it can only take on a finite or countably infinite number of values.

4.2.1 Probability Mass Function

As we just learned, a random variable is a variable whose value varies from one experiment to the next. That is, its value is not fixed but instead the possible values the random variable can take on follows a probability distribution, which for discrete random variables is called a *probability mass function* (PMF).

(a)   (b)

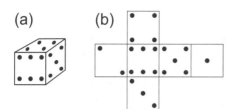

Figure 4.1 A fair die, with a 3D rendering (a) and an "exploded" cube (b).

> **Definition 4.2.2 — Probability Mass Function.** The probability mass function (PMF) for a discrete random variable X is a function that gives the probability $\Pr(X = x)$ of each value that the random variable can assume. We use the shorthand $p_X(x)$ for $\Pr(X = x)$, where the notation $p_X(x)$ means the probability at x for random variable X. When the random variable is understood, we will simply write $p(x)$. We will also use other lower case letters, such as f, to denote the PMF.

■ **Example 4.1** Consider a random variable X that represents the outcome of a fair die. Prior to rolling the die, X can take on any one of six values: $X \in \{1, 2, 3, 4, 5, 6\}$. Find the PMF.

Solution: See Figure 4.1 for a depiction of a fair die. Because the die is assumed to be fair, each of the six outcomes is equally likely. Since the probabilities of the outcomes must sum to 1, each must have probability $1/6$ and the PMF is

$$\Pr(X = x) \equiv p(x) = \begin{cases} 1/6, & \text{if } x = 1 \\ 1/6, & \text{if } x = 2 \\ 1/6, & \text{if } x = 3 \\ 1/6, & \text{if } x = 4 \\ 1/6, & \text{if } x = 5 \\ 1/6, & \text{if } x = 6. \end{cases}$$

We can write this more succinctly by

$$p(x) = \begin{cases} \frac{1}{6}, & \text{if } x = 1, 2, \ldots, 6 \\ 0, & \text{if otherwise.} \end{cases}$$

Here the points in the sample space and the values that the random variable X can take on *are exactly the same*. This PMF is referred to as a *discrete uniform distribution* since the probability mass is uniformly distributed over the possible values that X can assume. We can now write probability statements like $\Pr(X = 3) = 1/6$ to represent the statement "the probability of rolling a 3 with a fair, six-sided die is one-sixth." We can also write this more compactly as $p(3) = 1/6$.   ■ ■ ■

Notation is important here. Capital letters mean random variables; lower case letters mean possible values of these random variables. Thus, when we write $X = 1$ we mean the event that the random variable equals 1, and $X = 17$ means the event that the random variable X equals 17. In general, $X = x$ means the event that random variable equals the value x, where x is some number. Thus, $\Pr(X = x)$ is the general mathematical statement "the probability that random variable X takes on value x." This is just what we mean by the PMF. If the context makes it clear which random variable we are referring to, we will just write $p(x)$. We will use this convention throughout the text, so it's important to spend some time getting used to it.

Figure 4.2 is a plot of the PMF for Example 4.1. Here we see that there is a "mass" of probability at each discrete value that the random variable X can assume, which is denoted by the dot with the line underneath. PMFs must satisfy the probability axioms of Chapter 3.

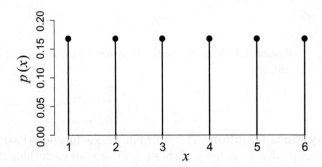

Figure 4.2 Graphical representation of the PMF for random variable X in Example 4.1.

Theorem 4.2.1 — Properties of a PMF. PMFs must satisfy the following two properties:

1. $0 \leq p(x) \leq 1$ for all x.
2. $\displaystyle\sum_{\text{all } x} p(x) = 1$.

That is, the probability that a random variable X takes on any value x must be between 0 and 1, inclusive. Also, if we sum up $\Pr(X = x)$ over all the values x that the random variable can take on, the probabilities must sum to 1. These properties follow from the probability axioms and apply to random variables just like they applied to events and sample spaces in Chapter 3.

Exercise 4.1 Show that the PMF in Exercise 4.1 satisfies the criteria of Theorem 4.2.1.

Let's look at an example that is not a simple one-to-one mapping from a sample space to the values of a random variable.

■ **Example 4.2** Let the random variable Y represent whether we roll an even number on one roll of a fair die, where $Y = 1$ denotes the event "roll an even number" and $Y = 0$ denotes the event "roll an odd number." Find the PMF of Y.

Solution: From Example 3.2 we know that

$$\Pr(Y = 1) \equiv \Pr(\text{roll an even number}) = \frac{1}{2},$$

and so, by the complement rule (Theorem 3.2.2), $\Pr(Y = 0) = 1 - 1/2 = 1/2$. The figure below shows the mapping from the original sample space S to the values that the random variable Y can assume:

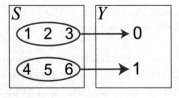

We can now write the PMF succinctly as

$$p(y) = \Pr(Y = y) = \begin{cases} 1/2, & \text{for } y = 0 \\ 1/2, & \text{for } y = 1. \end{cases}$$

■ ■ ■

Exercise 4.2 Let X represent the outcome of a fair four-sided die, $X \in \{1, 2, 3, 4\}$. Write out the equation for the PMF for X, similar to the PMF in Example 4.1, and then graphically depict the PMF, similar to the plot in Figure 4.2.

The distribution in Example 4.2 is a specific example of a more general type of distribution: the *Bernoulli distribution*.[1] A random variable X that follows a Bernoulli distribution can only take on one of two values, 0 and 1. The Bernoulli distribution has one *parameter*, p, which typically denotes $\Pr(X = 1)$. In Example 4.2, $p = 1/2$.

> **Definition 4.2.3 — Bernoulli Random Variable.** Let X denote a discrete random variable that can take on one of two possible outcomes: 0 or 1. Then for p between 0 and 1, the PMF is
>
> $$\Pr(X = x) = \begin{cases} 1 - p, & x = 0 \\ p, & x = 1. \end{cases}$$

Notice that Definition 4.2.3 describes a *family* of distributions, where a specific distribution results when we choose a particular value of the parameter p. Many of the distributions in this chapter, and all of the distributions in Chapters 5 and 6, will have one or more parameters.

> **Definition 4.2.4 — Parameter.** When referring to probability distributions, a *parameter* is a variable in the PMF that determines the specific form of the distribution.

Thus, we would say that the random variable Y in Example 4.2 is a Bernoulli random variable, that is, a random variable that follows the Bernoulli distribution, with parameter $p = 1/2$.

■ **Example 4.3** Let the random variable X denote the outcome of one coin flip of a possibly biased coin, where $X = 1$ denotes that the coin came up heads and $X = 0$ denotes that it came up tails. Then, let $\Pr(X = 1) = p$ and $\Pr(X = 0) = 1 - p$, so that X has a Bernoulli distribution with parameter p. Find and graph the PMF for X for an arbitrary p.

Solution: We can write

$$\Pr(X = x) = \begin{cases} 1 - p, & x = 0 \\ p, & x = 1. \end{cases}$$

Figure 4.3 illustrates the PMF for the case when $p > 1/2$.   ■ ■ ■

If in Example 4.3 we let $p = 1/2$ then we have specified that the coin is fair. For any other p the coin is biased. In the most extreme cases, if $p = 1$ then the coin will always be a head, and, if $p = 0$ then the coin will always be a tail.

Note how, in Example 4.3, the Bernoulli distribution gives us a convenient *model* for a coin or, for that matter, any other probabilistic process that only has two outcomes. Depending on our choice of the parameter p we can make the model fit the behavior of any coin – as long as we rule out the chance that the coin can land on its side. If, when writing the PMF, we want to be explicit that only two outcomes are possible, we can also write the PMF as:

$$\Pr(X = x) = \begin{cases} 1 - p, & \text{for } x = 0 \\ p, & \text{for } x = 1 \\ 0, & \text{otherwise.} \end{cases}$$

---

[1]The Bernoulli distribution is named after Swiss scientist Jacob Bernoulli (1655–1705), who was a professor of mathematics at the University of Basel. His *Ars Conjectandi*, published in 1713, derived the expected winnings for various games of chance.

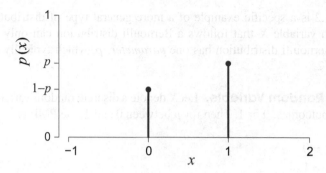

Figure 4.3 Probability mass function for a Bernoulli random variable X with $p > 1 - p$.

4.2.2 Cumulative Distribution Function

In addition to the PMF, discrete random variables have cumulative distribution functions.

> **Definition 4.2.5 — Cumulative Distribution Function.** For random variable X, the *cumulative distribution function* (CDF) is
>
> $$F_X(x) = \Pr(X \le x), \quad -\infty < x < \infty.$$

The CDF is a function that, for every value x on the real line, gives the cumulative probability that random variable X is less than or equal to x. As with the PMF, if the random variable associated with the CDF is clear, then we will use the simpler notation $F(x)$.

For a discrete random variable, the CDF is calculated as

$$F(x) = \Pr(X \le x) = \sum_{z \le x} \Pr(X = z).$$

■ **Example 4.4** Find the CDF for random variable X (the outcome of a fair six-sided die) with the PMF from Example 4.1.

Solution: To find the CDF, we need to find $F(x) = \Pr(X \le x)$ for all values of x, $-\infty < x < \infty$. While this initially sounds daunting, note that because X is a discrete random variable, $F(x)$ is a step function that only changes at those points where $p(x) > 0$. Thus, for example, because $p(x) = 0$ for all $x < 1$, $F(x) = 0$ for all $x < 1$. However, $F(x) = 1/6$ for $1 \le x < 2$, because, for any x in the interval $[1, 2)$,

$$F(x) = \Pr(X \le x) = \sum_{z \le x} \Pr(X = z) = \Pr(X = 1) = 1/6.$$

Similarly, for any x in the interval $[2, 3)$,

$$F(x) = \Pr(X \le x) = \sum_{z \le x} \Pr(X = z) = \Pr(X = 1) + \Pr(X = 2) = \frac{1}{6} + \frac{1}{6} = \frac{1}{3}.$$

Thus, working this all the way through for all $-\infty < x < \infty$, we have

$$F(x) = \begin{cases} 0, & \text{for } x < 1 \\ 1/6, & \text{for } 1 \le x < 2 \\ 2/6, & \text{for } 2 \le x < 3 \\ 3/6, & \text{for } 3 \le x < 4 \\ 4/6, & \text{for } 4 \le x < 5 \\ 5/6, & \text{for } 5 \le x < 6 \\ 1, & \text{for } x \ge 6. \end{cases}$$

Figure 4.4 graphically depicts the CDF.   ■ ■ ■

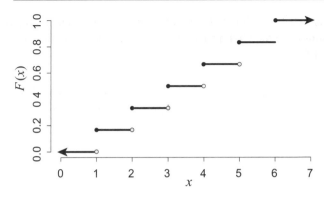

Figure 4.4 Cumulative distribution function derived in Example 4.4 for random variable X from Example 4.1.

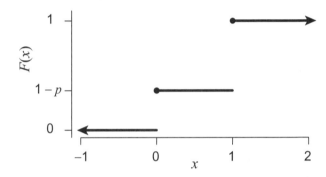

Figure 4.5 Cumulative distribution function for a Bernoulli random variable where the "step" at $F(x) = p$ could be anywhere between 0 and 1, depending on the specific value of p.

Regardless of the fact that the PMF for a discrete random variable is nonzero at no more than a countable number of points on the real line, the CDF is defined for all values of the real line. That is, for any value x, it is always possible to compute the value $\Pr(X \le x)$. Furthermore, note that $F(x)$ is 0 for every x less than the smallest value that random variable X can take on. Similarly, $F(x)$ is always 1 for any x greater than or equal to the largest value that random variable X can take on.

■ **Example 4.5** Find the CDF for a Bernoulli random variable.

Solution: Using logic similar to that in Example 4.4, we have that $F(x) = 0$ for $x < 0$, while for $x \in [0, 1)$,

$$F(x) \;=\; \Pr(X \le x) \;=\; \Pr(X = 0) \;=\; p,$$

and for $x \ge 1$,

$$F(x) \;=\; \Pr(X \le x) \;=\; \Pr(X = 0) + \Pr(X = 1) \;=\; 1 - p + p \;=\; 1.$$

Thus, the CDF for a Bernoulli random variable is

$$F(x) \equiv \Pr(X \le x) \;=\; \begin{cases} 0, & \text{for } x < 0 \\ 1 - p, & \text{for } 0 \le x < 1 \\ 1, & \text{for } x \ge 1. \end{cases}$$

The plot of the CDF is shown in Figure 4.5 where, of course, the "step" of $1 - p$ that occurs at $x = 0$ could be anywhere from 0 and 1, depending on the value of p. ■■■

Exercise 4.3 Continuing Exercise 4.2, let X denote the outcome of a fair four-sided die. Write out the equation for the CDF for X, similar to the CDF in Example 4.4, and then graphically depict the CDF, similar to the plot in Figure 4.4.

The CDF gives the probability in the "left tail" (also referred to as the "lower tail" in R) of the PMF up through x. We can apply the complement rule to get the probability in the right tail: $\Pr(X > x)$. Returning to the fair die example, the probability of rolling something greater than a 4 is:

$$\Pr(X > 4) = 1 - \Pr(X \le 4) = 1 - F(4),$$

where, in general,

$$\Pr(X > x) = 1 - \Pr(X \le x) = 1 - F(x).$$

However, note that again by the complement rule,

$$\Pr(X \ge x) = 1 - \Pr(X < x),$$

where, in general, $1 - \Pr(X < x) \neq 1 - F(x)$. For example, returning to the fair die, if we were asked what the probability is of rolling a 4 or greater, we have

$$\Pr(X \ge 4) = 1 - \Pr(X < 4) = 1 - \Pr(X \le 3) = 1 - F(3).$$

The key take-away here is that, when working with the CDF of discrete random variables, be aware of whether a value of x is included in the calculation. In particular, when applying the complement rule to solve for a probability, pay close attention to the endpoints of the interval and whether you are working with a strict inequality or not.

4.2.3 Expected Value

The expected value of a random variable can be thought of as the average of a very large number (infinite, really) of observations of the random variable. We denote the expected value of a random variable X as $\mathbb{E}(X)$. Calculation of the expected value of X for discrete random variables is straightforward: it is the sum of the products of each possible outcome times the probability of the outcome.

Definition 4.2.6 — Expected Value of a Discrete Random Variable. If the discrete random variable X can take on outcomes $x_1, x_2, x_3, \ldots$ with probabilities $p(x_i) = \Pr(X = x_i)$, $i = 1, 2, \ldots$, then the *expected value* of X, denoted $\mathbb{E}(X)$, is

$$\mathbb{E}(X) = \sum_{i=1}^{\infty} x_i \Pr(X = x_i).$$

The expected value is also called the *mean* of the random variable, or the mean of the distribution. The Greek letter μ is often used to denote $\mathbb{E}(X)$.

While it is useful intuition to think about the expected value as taking the average of a large number of observations, calculating an expected value does not require a sample of data like the calculation for the sample mean does. Instead, applying Definition 4.2.6 to a discrete random variable results in an expression that is a function of the distribution's parameters.

■ **Example 4.6** Find the expected value of a Bernoulli random variable.

Solution: Returning to Definition 4.2.3, the expected value of random variable X that has a Bernoulli distribution is

$$\mathbb{E}(X) \equiv \mu = \sum_{x=0}^{1} x \Pr(X=x) = 0 \times (1-p) + 1 \times p = p.$$

So, the expected value of a Bernoulli random variable is simply $\mathbb{E}(X) = p$.   ■ ■ ■

Thus, for a fair coin, if we let 0 denote tails and 1 denote heads, then $\mathbb{E}(X) = 1/2$. This should make sense when thinking about the expected value as the long run fraction of times the coin comes up heads. If the coin is fair, we'd expect to get heads in half of the flips. If the probability were p (not necessarily $1/2$) then in a large number of trials we would expect the proportion of heads to be about p. For example, in $1,000,000$ tosses we would expect about $1,000,000p$ heads and $1,000,000(1-p)$ tails. This would average out to

$$\frac{1}{1,000,000}(1,000,000p) = p,$$

which is the expected value of X.

■ **Example 4.7** What is the expected value for one roll of a fair die?

Solution: A fair die has the PMF shown in Example 4.1. So, the expected value is simply the sum of each possible outcome (1, 2, 3, 4, 5, and 6) times its probability of occurrence which, in the case of a fair die, is always 1/6. That is:

$$\mathbb{E}(X) = \sum_{x=1}^{6} x \Pr(X=x) = \sum_{x=1}^{6}\left(x \times \frac{1}{6}\right) = \frac{1}{6}+\frac{2}{6}+\frac{3}{6}+\frac{4}{6}+\frac{5}{6}+\frac{6}{6} = \frac{21}{6}.$$

The result: $\mathbb{E}(X) = 20/6 = 3.5$, which can be interpreted this way: If a fair die were rolled repeatedly, then the resulting long run average of all those rolls would be 3.5. Note how the expected value of X does not have to be an integer – or even one of the possible outcomes.   ■ ■ ■

We can verify the result from Example 4.7 via simulation in R. To do this, we'll use the `sample()` function to simulate draws from a discrete distribution with equal probabilities of $\frac{1}{6}$ across the integers 1 through 6. The first argument to the `sample()` function is the set from which we would like to sample; here that is the set of numbers 1, 2, 3, 4, 5, 6, which we can represent in R by `1:6`. The second argument tells R how many we would like to sample. The default is to sample *without* replacement. For example, if we would like to select two numbers from 1 through 6 we could type `sample(1:6,2)` and we would get two numbers selected from `1:6`. We might get 5 and 2, or 1 and 3, but we could not get the same number twice, such as 4 and 4, because the default is to sample without replacement. In our case we'd like to sample 1,000,000 values so we have to sample *with* replacement. Thus we need to tell R in the third argument that we should sample with replacement:

```
x <- sample(1:6,1000000,replace=TRUE )
tab <- table(x)
barplot( tab/sum(tab) , col="blue" , ylim=c(0,0.20) ,
        xlab="Number on Die" , ylab="Proportion" )
```

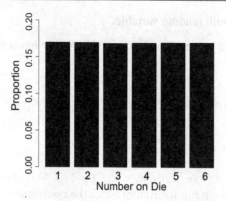

Figure 4.6 A bar chart showing the distribution of 10 million simulated rolls of a fair die.

Figure 4.6 is the result of running the above code, and it looks very much like the PMF in Figure 4.2 in the sense that we see that the rolls are roughly evenly divided between the integers 1 though 6. Though not readily visible in the figure, for this simulation the precise fraction of times the simulated die rolled a 1 through 6 are:

Outcome:	1	2	3	4	5	6
Fraction:	0.166194	0.167191	0.165785	0.166769	0.166399	0.167662

All of these are quite close to $1/6 = 0.1\overline{6}$, where the fact that they don't match precisely is because of the randomness inherent in the simulation. The average over these 1,000,000 simulations can be calculated in R by typing `mean(x)`, which yields

```
mean(x)
```

```
[1] 3.499929
```

The result is not precisely 3.5 because $\mathbb{E}(X) = 3.5$ is a theoretical result predicated on doing an unlimited number of rolls.

> **Exercise 4.4** Continuing Exercises 4.2 and 4.3, calculate the expected value of X, the outcome of a fair four-sided die. Verify your result by simulating a four-sided die in R.
>
> *Answer*: $\mathbb{E}(X) = 2.5$.

Let's return to the biased die of Example 3.2.

■ **Example 4.8** What is the expected value for the biased die of Example 3.2?

Solution: The biased die has PMF:

$$p(y) = \begin{cases} 0.15, & \text{for } x = 1 \\ 0.15, & \text{for } x = 2 \\ 0.15, & \text{for } x = 3 \\ 0.15, & \text{for } x = 4 \\ 0.15, & \text{for } x = 5 \\ 0.25, & \text{for } x = 6. \end{cases}$$

Then,

$$\mathbb{E}(X) = \sum_{x=1}^{6} x p(x) = 1 \times 0.15 + 2 \times 0.15 + 3 \times 0.15 + 4 \times 0.15 + 5 \times 0.15 + 6 \times 0.25 = 3.75.$$

So, a die that is biased to give more sixes than a fair die also has a larger expected value.   ■■■

We next define the expectation of a function of a random variable.

> **Definition 4.2.7 — Expected Value of a Function of a Discrete Random Variable.** If X is a discrete random variable with PMF $p(x)$ and g is a function of X, then
>
> $$\mathbb{E}[g(X)] = \sum_x g(x)p(x).$$

■ **Example 4.9** Compute $\mathbb{E}(X^2)$ for the outcome of one roll of a fair six-sided die.

Solution:

$$E(X^2) = \sum_{x=1}^{6} x^2 p(x) = 1^2\frac{1}{6} + 2^2\frac{1}{6} + 3^2\frac{1}{6} + 4^2\frac{1}{6} + 5^2\frac{1}{6} + 6^2\frac{1}{6} = \frac{91}{6}.$$

■ ■ ■

4.2.4 Variance and Standard Deviation

The variability of a random variable is measured by its variance and standard deviation, which are defined below.

> **Definition 4.2.8 — Variance of a Random Variable.** The *variance* of random variable X is
>
> $$\mathbb{V}(X) \equiv \mathbb{E}\left([X - \mathbb{E}(X)]^2\right) = \mathbb{E}\left([X - \mu]^2\right).$$

The shorthand σ_X^2, or just σ^2 if it is clear which random variable we're talking about, is often used to denote $\mathbb{V}(X)$. The *standard deviation* σ is the square root of the variance.

The units for the standard deviation σ are the same as the units for the original data. For example, if X is measured in US dollars, then μ and σ both have units of US dollars. The variance of X would have units of dollars^2.

For a discrete random variable, we calculate the variance as

$$\mathbb{V}(X) = \mathbb{E}\left([X - \mathbb{E}(X)]^2\right) = \sum_{i=1}^{\infty} (x_i - \mu)^2 p(x_i).$$

So, the variance is just the expected squared distance of a random variable from its mean. The larger the variance, the more one should expect to see observations far from the expected value. The smaller the variance, the more such observations are likely to be closer to the expected value and thus closer together. Thus, the variance measures the "spread-out-ness" of the probability.

There is a shortcut formula for the variance that is similar to the shortcut in Chapter 2 for the sample variance.

> **Theorem 4.2.2 — Variance Shortcut Formula.** For a random variable X, an expression that is mathematically equivalent to Definition 4.2.8 is
>
> $$\mathbb{V}(X) = \mathbb{E}(X^2) - [\mathbb{E}(X)]^2.$$

We will defer a demonstration of this until Section 4.4, where we cover properties of expectations. As with the sample variance shortcut formula, the shortcut formula for the variance of random variables is sometimes easier to use for calculating the variance.

■ **Example 4.10** Returning to Definition 4.2.3 and Example 4.6, calculate the variance of X, a Bernoulli random variable.

Solution: The variance of the Bernoulli random variable X in Definition 4.2.3 can be calculated using the shortcut formula as follows. First, we have

$$\mathbb{E}(X^2) = \sum_{x=0}^{1} x^2 p(x) = 0^2 \times (1-p) + 1^2 \times p = p.$$

Then, using the shortcut formula and the fact that we know $\mathbb{E}(X) = p$ from Example 4.6, we have

$$\mathbb{V}(X) = \mathbb{E}(X^2) - [\mathbb{E}(X)]^2 = p - p^2 = p(1-p).$$ ■ ■ ■

■ **Example 4.11** Demonstrate that, for a Bernoulli random variable, Definition 4.2.8 for the variance gives the same result as the shortcut formula in Example 4.10.

Solution: We know $\mu \equiv \mathbb{E}(X) = p$, so

$$\mathbb{V}(X) = \sum_{x=0}^{1} (x-p)^2 p(x) = (0-p)^2 \times (1-p) + (1-p)^2 \times p = p^2(1-p) + (1-p)^2 p = p(1-p).$$

■ ■ ■

■ **Example 4.12** Returning to Examples 4.1 and 4.7, calculate the variance of the outcome of one roll of a fair die.

Solution: In Example 4.9 we found that $\mathbb{E}(X^2) = 91/6$. Then, using the result from Example 4.7 that $\mathbb{E}(X) = 21/6$,

$$\mathbb{V}(X) = \mathbb{E}(X^2) - [\mathbb{E}(X)]^2 = 91/6 - (21/6)^2 = \frac{546}{36} - \frac{441}{36} = \frac{105}{36} \approx 2.917.$$ ■ ■ ■

■ **Example 4.13** Calculate the variance and standard deviation for the outcome of the biased die from Example 4.8.

Solution: As in the previous examples, we begin by calculating

$$\begin{aligned} \mathbb{E}(X^2) &= \sum_{x=1}^{6} x^2 p(x) \\ &= 1^2 \times 0.15 + 2^2 \times 0.15 + 3^2 \times 0.15 + 4^2 \times 0.15 + 5^2 \times 0.15 + 6^2 \times 0.25 \\ &\approx 16.11. \end{aligned}$$

Then, from Example 4.8 that showed $\mathbb{E}(X) = 3.75$, we have

$$\mathbb{V}(X) = \mathbb{E}(X^2) - [\mathbb{E}(X)]^2 \approx 16.11 - (3.75)^2 \approx 2.0475.$$

Thus, $\sigma = \sqrt{2.0475} \approx 1.431.$ ■ ■ ■

Exercise 4.5 Continuing Exercises 4.2–4.4, calculate the variance and standard deviation of X, the outcome of a fair four-sided die. Verify your result by simulating a four-sided die in R.

Answer: $\mathbb{V}(X) = 1.25$ and $\sigma = 1.12$.

4.3 Continuous Random Variables

In Chapter 3 and the first half of this chapter, we focused on discrete probability and discrete random variables. As we have seen, these are largely based on counting sample points. In contrast, as we will see in this section, continuous random variables are based on the notion of continuous measures.

> **Definition 4.3.1 — Continuous Random Variable.** A continuous random variable is one that can take on values in one or more intervals of the real line.

The idea is that measurements of things such as time, distance, weight, and volume are continuous, and when such measurements are random they must be represented by random variables. Because the measurements are continuous, the mathematics shifts from being based on summation to using integration. However, the intuition we have developed thus far will carry over.

4.3.1 Probability Density Function

Probabilities for discrete random variables are found by summing the PMF over the appropriate outcomes. For continuous random variables, summing becomes *integrating*.

> **Definition 4.3.2 — Probability Density Function.** The *probability density function* (PDF) for a continuous random variable X is a non-negative function $f_X(x)$ defined on the real line having the property that for any set A of real numbers,[2]
>
> $$\Pr(X \in A) = \int_A f_X(x)dx.$$

The notation $f_X(x)$ means that the PDF f of a random variable X is evaluated at the point x. When the corresponding random variable is clear, the notation can be abbreviated to $f(x)$. Also, the notation $\int_A$ indicates the integral over the set A, which is usually an interval. Definition 4.3.2 says

$$\Pr\left(X \in [a,b]\right) = \Pr\left(a \le X \le b\right) = \int_a^b f(x)\,dx.$$

Since the definite integral of a nonnegative function (such as a PDF) is the area under the graph of $y = f(x)$, we see that *probability equals area under the PDF*. The *uniform distribution* is a good place to start to demonstrate the concept of continuous distributions and their associated calculations.

> **Definition 4.3.3 — Uniform Distribution.** The PDF for a uniformly distributed random variable X on the interval $[\theta_1, \theta_2]$ is
>
> $$f(x) = \begin{cases} \dfrac{1}{\theta_2 - \theta_1}, & \text{for } x \in [\theta_1, \theta_2] \text{ with } \theta_2 > \theta_1 \\ 0, & \text{otherwise.} \end{cases}$$
>
> *Parameters*: The uniform distribution has two parameters, θ_1 and θ_2. For the *standard uniform distribution*, $\theta_1 = 0$ and $\theta_2 = 1$.

---

[2]Technically, we cannot take the integral for *every* possible set. The set A must be *measurable*. See a text on measure theory to study the meaning of this term.

Notation: $X \sim \text{UNIF}(\theta_1, \theta_2)$, which stands for "The random variable X has a uniform distribution on the interval $[\theta_1, \theta_2]$."

■ **Example 4.14** Let $X \sim \text{UNIF}(\theta_1, \theta_2)$. Find $\Pr(X \in [a, b])$, where $\theta_1 \le a < b \le \theta_2$.

Solution: From Definition 4.3.2, the probability that $X \in [a, b]$ where $\theta_1 \le a < b \le \theta_2$ is

$$\Pr\left(X \in [a, b]\right) = \int_a^b f(x)\, dx = \int_a^b \left(\frac{1}{\theta_2 - \theta_1}\right) dx = \left[\frac{1}{\theta_2 - \theta_1} x\right]_a^b = \frac{b - a}{\theta_2 - \theta_1}.$$

The bracket notation with subscripts and superscripts, used in the second last equality, is defined this way:

$$\left[g(x)\right]_a^b = g(b) - g(a).$$

If, in particular, X has a standard uniform distribution, then $\Pr(X \in [0, 0.5]) = \frac{0.5 - 0}{1 - 0} = 0.5$.   ■ ■ ■

The uniform PDF is the continuous analog of the discrete uniform PMF that we saw for one roll of a fair die in Example 4.1. A graph of the density function is shown in Figure 4.7. The interpretation of this density is that every interval of equal length that lies between θ_1 and θ_2 is equally likely, while nothing outside of the interval $[\theta_1, \theta_2]$ can occur.

■ **Example 4.15** For a continuous random variable X with PDF

$$f(x) = \begin{cases} -\dfrac{x}{4}, & -2 \le x \le 0 \\[2mm] \dfrac{x}{4}, & 0 \le x \le 2 \\[2mm] 0, & \text{otherwise,} \end{cases}$$

sketch the PDF and calculate $\Pr(-0.5 \le X \le 0.5)$.

Solution: The PDF is plotted in Figure 4.8, where the probability $\Pr(-0.5 \le X \le 0.5)$ is shown in gray shading as the area under the PDF between -0.5 and 0.5. The desired probability is

$$\begin{aligned} \Pr(-0.5 \le X \le 0.5) &= \int_{-0.5}^{0.5} f(x)\, dx \\ &= \int_{-0.5}^{0} f(x)\, dx + \int_{0}^{0.5} f(x)\, dx \\ &= \int_{-0.5}^{0} \left(-\frac{x}{4}\right) dx + \int_{0}^{0.5} \left(\frac{x}{4}\right) dx \end{aligned}$$

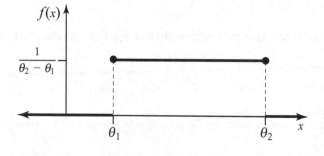

Figure 4.7 The PDF $f(x)$ for $X \sim \text{UNIF}(\theta_1, \theta_2)$.

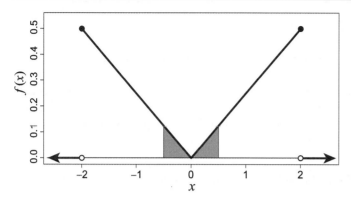

Figure 4.8 For Example 4.15, the area under $f(x)$ over the range $[-0.5, 0.5]$ is $\Pr(-0.5 < X < 0.5)$.

$$= \left[-\frac{x^2}{8}\right]_{-0.5}^{0} + \left[\frac{x^2}{8}\right]_{0}^{0.5}$$

$$= 0 - \left(-\frac{(-0.5)^2}{8}\right) + \frac{(0.5)^2}{8} - 0$$

$$= \frac{1}{16}.$$

∎∎∎

Exercise 4.6 Find $\Pr(1 \leq X \leq 4)$ and $\Pr(X \geq 6)$ for the continuous random variable X with PDF

$$f(x) = \begin{cases} \dfrac{x^3}{5,000}(10-x), & 0 \leq x \leq 10 \\ 0, & \text{otherwise.} \end{cases}$$

Answer: $\Pr(1 \leq X \leq 4) \approx 0.087$ and $\Pr(X \geq 6) \approx 0.663$.

The probability that X is in the set A is the area under the PDF over the range of A. For example, Figure 4.8 illustrates the $\Pr(-0.5 \leq X \leq 0.5)$ for Example 4.15, where the gray is the area under the PDF. In this case we could have determined the probability using simple geometry, because we know that the area of a triangle is $\frac{1}{2} \times$ base $\times$ height. That is, since $f(-0.5) = -(-0.5/4) = 1/8$,

$$\Pr(-0.5 \leq X \leq 0) = \frac{1/8 \times 1/2}{2} = \frac{1}{32}.$$

Because of the symmetry of the PDF around $x = 0$, $\Pr(-0.5 \leq X \leq 0) = \Pr(0 \leq X \leq 0.5)$. Thus,

$$\Pr(-0.5 \leq X \leq 0.5) = 2 \times \Pr(-0.5 \leq X \leq 0) = \frac{1}{16}.$$

While the PDF is the continuous analog of the PMF, there are some important differences. The most important is that $f(x)$ is *not* the probability that random variable X takes on the value x. This is because for any continuous random variable X and any value x, $\Pr(X = x) = 0$. This follows from Definition 4.3.2:

$$\Pr(X = x) = \int_{x}^{x} f(z)\,dz = 0.$$

This bears repeating and emphasizing: For a continuous random variable

$$\boxed{f(x) \neq \Pr(X = x).}$$

Because $\Pr(X = x) = 0$ for any x, it also follows that

$$\Pr(a \leq X \leq b) = \Pr(a < X \leq b) = \Pr(a \leq X < b) = \Pr(a < X < b).$$

That is, the inclusion or exclusion of the interval endpoints does not affect the probability that a continuous random variable is in the interval. This property does not hold for discrete random variables.

Just like discrete random variables, PDFs must satisfy the properties of the probability axioms of Chapter 3, which results in the following theorem.

> **Theorem 4.3.1 — Properties of a PDF.** Probability density functions must satisfy the following two properties:
> 1. $f(x) \geq 0$ for $-\infty < x < \infty$.
> 2. $\int_{-\infty}^{\infty} f(x)\,dx = 1$.

Note that the two criteria in Theorem 4.3.1 imply that $0 \leq \Pr(X \in A) \leq 1$ for any set A. That is, for any interval A,

$$\int_A f(x)dx \leq \int_{-\infty}^{\infty} f(x)dx = 1.$$

Also, since $f(x) \geq 0$, for any set A it follows that $\int_A f(x)dx \geq 0$.

■ **Example 4.16** Verify that the PDF for the uniform distribution of Definition 4.3.3 meets the requirements in Theorem 4.3.1.

Solution: By the definition of $f(x)$ for the UNIF(θ_1, θ_2) distribution, we see that $f(x) \geq 0$ for all x. For the second criterion:

$$
\begin{aligned}
\int_{-\infty}^{\infty} f(x)\,dx &= \int_{-\infty}^{\theta_1} f(x)\,dx + \int_{\theta_1}^{\theta_2} f(x)\,dx + \int_{\theta_2}^{\infty} f(x)\,dx \\
&= \int_{-\infty}^{\theta_1} 0\,dx + \int_{\theta_1}^{\theta_2} \left(\frac{1}{\theta_2 - \theta_1} \right) dx + \int_{\theta_2}^{\infty} 0\,dx \\
&= 0 + \left(\frac{1}{\theta_2 - \theta_1} \right) \int_{\theta_1}^{\theta_2} dx + 0 \\
&= \frac{\theta_2 - \theta_1}{\theta_2 - \theta_1} \\
&= 1.
\end{aligned}
$$

■ ■ ■

> **Exercise 4.7** Verify that $f(x)$ in Exercise 4.6 meets the conditions of Theorem 4.3.1.

■ **Example 4.17** The length X of a call coming into a help desk is randomly distributed over the interval 0 to 600 seconds (10 minutes). Thus, $X \sim$ UNIF$(0, 600)$. What is:

(a) the probability a call takes more than 7 minutes to resolve?
(b) the probability a call is completed in 25 seconds or less?

Solution: Here we're given that $X \sim \text{UNIF}(\theta_1 = 0, \theta_2 = 600)$. Thus,

$$f(x) = \begin{cases} \dfrac{1}{600 - 0}, & 0 \leq x \leq 600 \\ 0, & \text{otherwise.} \end{cases}$$

Then, the probability that a call takes more than 7 minutes = 420 seconds to resolve is

$$\Pr(X \geq 420) = \int_{420}^{600} \frac{1}{600} \, dx = \left[\frac{x}{600}\right]_{420}^{600} = \frac{600 - 420}{600} = \frac{180}{600} = \frac{3}{10},$$

and, the probability a call is completed in 25 seconds or less is

$$\Pr(X \leq 25) = \int_{0}^{25} \frac{1}{600} \, dx = \left[\frac{x}{600}\right]_{0}^{25} = \frac{25 - 0}{600} = \frac{25}{600} = \frac{1}{24}. \qquad \blacksquare\blacksquare\blacksquare$$

4.3.2 Cumulative Distribution Function

The CDF for a continuous random variable is defined in exactly the same way as that for a discrete random variable. The only difference is in the calculation, where for a continuous random variable we calculate the CDF as an integral.

> **Definition 4.3.4 — Cumulative Distribution Function of a Continuous Random Variable.** The CDF of a continuous random variable X having PDF $f(x)$ is
>
> $$F(x) = \Pr(X \leq x) = \int_{-\infty}^{x} f(z) \, dz.$$

That is, for a continuous random variable, the cumulative probability that random variable X is less than or equal to value x is the area under the PDF from $-\infty$ to x. From this, it also follows using the first fundamental theorem of calculus (Varberg *et al.*, 2007) that

$$F'(x) = \frac{d}{dx} \int_{-\infty}^{x} f(z) \, dz = f(x), \qquad (4.1)$$

whenever the derivative exists. In other words, the PDF is the derivative of the CDF.

■ **Example 4.18** Find the CDF for the random variable with PDF

$$f(x) = \begin{cases} \dfrac{4 - 3x^2}{3}, & 0 \leq x \leq 1 \\ 0, & \text{otherwise.} \end{cases}$$

Solution: First, since $f(x) = 0$ for $x < 0$, we know that for $x < 0$

$$F(x) = \int_{-\infty}^{x} f(z) dz = \int_{-\infty}^{x} 0 \, dz = 0.$$

Now, for $0 \leq x \leq 1$, we have

$$F(x) = \int_{-\infty}^{x} f(z) \, dz = \int_{-\infty}^{0} 0 \, dz + \int_{0}^{x} \frac{4 - 3z^2}{3} \, dz = \left[\frac{4z}{3} - \frac{z^3}{3}\right]_{0}^{x} = \frac{x(4 - x^2)}{3}.$$

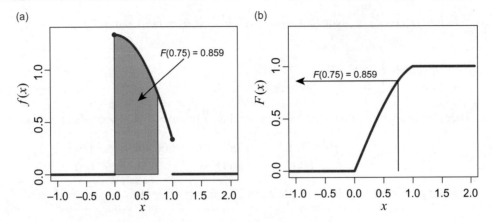

Figure 4.9 Plots of $f(x)$ and $F(x)$ for Example 4.18, both showing $\Pr(-\infty < x \le 0.75)$.

Finally, for $x > 1$, $F(X) = 1$, so the CDF is

$$F(x) = \begin{cases} 0, & x < 0 \\ \dfrac{x(4 - x^2)}{3}, & 0 \le x \le 1 \\ 1, & x > 1. \end{cases}$$

    ■ ■ ■

Figure 4.9 plots the PDF and CDF for Example 4.18, where here we see the correspondence between the PDF (a) and CDF (b). In the graph of the PDF, the shaded area under the curve corresponds to $\Pr(-\infty < x \le 0.75)$. In the graph of the CDF, we can read the value of $F(0.75) = \Pr(-\infty < x \le 0.75)$ right off the graph.

Exercise 4.8 Find the CDF for the PDF in Exercise 4.6. Plot the PDF and CDF.

The CDF can be used to find probabilities for bounded intervals, as shown in the following theorem.

Theorem 4.3.2 — Finding Probabilities Using the CDF. If X is a continuous random variable and $a < b$, then

$$\Pr(X \in [a,b]) = \Pr(a \le X \le b) = F(b) - F(a).$$

Proof. Let $A = (-\infty, a)$ and $B = [a, b]$. Since A and B are disjoint and $A \cup B = (-\infty, b]$, we can write

$$\Pr(X \le b) = \Pr(A \cup B) = \Pr(A) + \Pr(B) = \Pr(X < a) + \Pr(a \le X \le b).$$

Because X is a continuous random variable, $\Pr(X < a) = \Pr(X \le a)$. The above can be written as

$$\Pr(X \le b) = \Pr(A) + \Pr(B) = \Pr(X \le a) + \Pr(a \le X \le b),$$

which can be written in terms of the CDF $F(x)$ as

$$F(b) = F(a) + \Pr(a \le X \le b).$$

The result follows by subtracting $F(a)$ from both sides.

    ■

For a continuous random variable X, the probability that X equals any particular value is zero, so we can conclude that

$$\Pr(a \le X \le b) = \Pr(a < X \le b) = \Pr(a \le X < b) = \Pr(a < X < b) = F(b) - F(a).$$

See Problem 4.68 for a situation that you can apply to discrete random variables.

■ **Example 4.19** Let X have the following PDF:

$$f(x) = \begin{cases} c(2-x), & 0 \le x \le 2 \\ 0, & \text{otherwise.} \end{cases}$$

(a) Determine the constant c so that X is a proper PDF.
(b) Using $f(x)$, find $F(x)$ and graph $f(x)$ and $F(x)$.
(c) Use $F(x)$ to calculate $\Pr(1 \le X \le 2)$.
(d) Use geometry and the graph of $f(x)$ to check the previous solution.

Solution: Because $f(x)$ is a PDF, it must be that $\int_{-\infty}^{\infty} f(x)dx = 1$. Thus,

$$1 = \int_{-\infty}^{\infty} f(x)\, dx = \int_0^2 c(2-x)dx = c\int_0^2 (2-x)dx = c\left[2x - \frac{x^2}{2}\right]_0^2 = 2c,$$

and therefore $c = 1/2$. Thus,

$$f(x) = \begin{cases} 1 - \dfrac{x}{2}, & 0 \le x \le 2 \\ 0, & \text{otherwise.} \end{cases}$$

Now, because $f(x)$ is nonzero only on the interval $0 < x < 2$, we know that $F(x) = 0$ for $x < 0$. Similarly, because we know that $f(x)$ integrates to 1 over the interval $0 \le x \le 2$, we also know that $F(x) = 1$ for $x \ge 2$. So, all that remains is to find $F(x)$ over the interval $0 \le x \le 2$:

$$F(x) = \int_{-\infty}^{x} f(z)\, dz = \int_0^x \left(1 - \frac{z}{2}\right)\, dz = \left[z - \frac{z^2}{4}\right]_0^x = x - \frac{x^2}{4}.$$

Thus, the CDF is

$$F(x) = \begin{cases} 0, & x < 0 \\ x - \dfrac{x^2}{4}, & 0 \le x < 2 \\ 1, & x \ge 2. \end{cases}$$

Plots of $f(x)$ and $F(x)$ are shown in Figure 4.10. Using the CDF to compute $\Pr(1 \le X \le 2)$ gives

$$\begin{aligned}
\Pr(1 \le X \le 2) &= \Pr(X \le 2) - \Pr(X < 1) \\
&= F(2) - F(1) \qquad [\text{X is continuous, so } F(1) = \Pr(X \le 1) = \Pr(X < 1)] \\
&= \left(2 - \frac{2^2}{4}\right) - \left(1 - \frac{1^2}{4}\right) \\
&= \frac{1}{4},
\end{aligned}$$

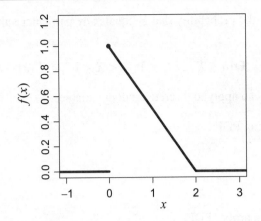 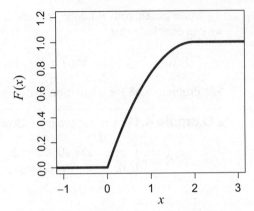

Figure 4.10 Plots of $f(x)$ and $F(x)$ for Example 4.19.

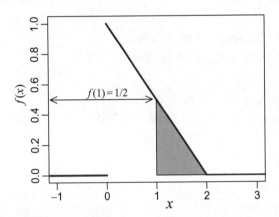

Figure 4.11 From Example 4.19, $\Pr(1 \leq X \leq 2)$ is the area under the PDF over the range $1 \leq x \leq 2$.

where the equality in the first line follows because we are working with a continuous random variable and so $\Pr(X < 1) = \Pr(X \leq 1)$.

Finally, to find $\Pr(1 \leq X \leq 2)$ using the PDF, we need to calculate the area under the PDF over the region $[1, 2]$. As shown in Figure 4.11, we can do the calculation using simple geometry (because the region under the curve is just a triangle). The length of the side on the horizontal axis is $2 - 1 = 1$ and the vertical side of the triangle has height $f(1) = 1 - 1/2 = 1/2$. Therefore, the area under the curve, that is, the shaded triangle in the figure, is $1 \times 1/2 \times 1/2 = 1/4$. ∎∎∎

R can be used to calculate the CDF for the uniform (and many other) distributions. For $X \sim$ UNIF(θ_1, θ_2), you can calculate $F(x)$ using the `punif()` function. The first argument to `punif()` is the value of x at which we'd like to evaluate the CDF. The second and third arguments are θ_1 and θ_2, respectively. This is illustrated in the next example.

■ **Example 4.20** Use R to find $\Pr(X \leq 25)$ and $\Pr(X \geq 420)$ from Example 4.17.

Solution: Given $\Pr(X \leq 25) = F(25)$:

```
punif(25,0,600)
[1] 0.04166667
```

Now, $\Pr(X \geq 420) = 1 - \Pr(X < 420) = 1 - \Pr(X \leq 420) = 1 - F(420)$:

```
1-punif(420,0,600)
[1] 0.3
```
∎∎∎

4.3.3 Expected Value

The definition of the expected value of a continuous random variable is similar to that of a discrete random variable except the sum is replaced by an integral.

> **Definition 4.3.5 — Expected Value of a Continuous Random Variable.** The expected value of continuous random variable X with PDF $f(x)$ is
>
> $$\mathbb{E}(X) = \int_{-\infty}^{\infty} x f(x)\, dx.$$

As with discrete random variables, the Greek letter μ is also often used to denote $\mathbb{E}(X)$.

■ **Example 4.21** Find the expected value of X, if $X \sim \mathrm{UNIF}(\theta_1, \theta_2)$.

Solution: The expected value of a uniform random variable is

$$\mathbb{E}(X) = \int_{-\infty}^{\infty} x f(x) dx = \frac{1}{\theta_2 - \theta_1} \int_{\theta_1}^{\theta_2} x\, dx = \frac{1}{\theta_2 - \theta_1} \cdot \frac{x^2}{2}\Big|_{\theta_1}^{\theta_2} = \frac{\theta_2^2 - \theta_1^2}{2(\theta_2 - \theta_1)} = \frac{\theta_2 + \theta_1}{2}.$$

This result should seem intuitive. The mean of a random variable that is uniformly distributed from θ_1 to θ_2 is the midpoint, $(\theta_1 + \theta_2)/2$, of the interval. In particular, if X has a standard uniform distribution, $X \sim \mathrm{UNIF}(0,1)$, then $\mathbb{E}(X) = 1/2$. ■■■

We can check this calculation via an R simulation. The function `runif(n,`θ_1,θ_2`)` will generate n random observations from a uniform distribution on the interval $[\theta_1, \theta_2]$. Running `mean(runif(10000000,0,1))` yields 0.5001523.

■ **Example 4.22** Find the expected value of random variable X with density function

$$f(x) = \begin{cases} 1 - \dfrac{x}{2}, & 0 \leq x \leq 2 \\ 0, & \text{otherwise.} \end{cases}$$

Solution:

$$\mathbb{E}(X) = \int_{-\infty}^{\infty} x f(x)\, dx = \int_0^2 x\left(1 - \frac{x}{2}\right) dx = \left[\frac{x^2}{2} - \frac{x^3}{6}\right]_0^2 = \frac{4}{2} - \frac{8}{6} = \frac{2}{3}.$$ ■■■

> **Exercise 4.9** Calculate the expected value of X in Exercise 4.6.
>
> *Answer:* $\mathbb{E}(X) = 6.6\overline{6}$.

4.3.4 Variance and Standard Deviation

The variability of a continuous random variable is measured by its variance, and the variance is defined the same way as for discrete random variables (Definition 4.2.8), except the expectation is done with an integral rather than a sum:

$$\mathbb{V}(X) = \mathbb{E}\left((X - \mu)^2\right) = \int_{-\infty}^{\infty} (x - \mu)^2 f(x)\, dx.$$

As before, the shorthand σ^2 is often used to denote $\mathbb{V}(X)$, and, as with discrete random variables, the standard deviation is the square root of the variance: $\sigma = \sqrt{\mathbb{V}(X)}$.

The variance shortcut formula (Definition 4.2.2) also directly applies to continuous random variables. Again, the only difference is in how it's calculated:

$$\mathbb{V}(X) = \mathbb{E}(X^2) - (\mathbb{E}(X))^2 = \int_{-\infty}^{\infty} x^2 f(x)\,dx - \left[\int_{-\infty}^{\infty} x f(x)\,dx\right]^2.$$

■ **Example 4.23** Returning to Example 4.21, find the variance of $X \sim \text{UNIF}(\theta_1, \theta_2)$.

Solution: Using the variance shortcut formula, we start by calculating $\mathbb{E}(X^2)$:

$$\mathbb{E}(X^2) = \int_{-\infty}^{\infty} x^2 f(x)\,dx = \frac{1}{\theta_2 - \theta_1} \int_{\theta_1}^{\theta_2} x^2\,dx = \frac{1}{\theta_2 - \theta_1}\left[\frac{x^3}{3}\right]_{\theta_1}^{\theta_2} = \frac{\theta_2{}^3 - \theta_1{}^3}{3(\theta_2 - \theta_1)},$$

and so the variance is

$$\begin{aligned}
\mathbb{V}(X) = \mathbb{E}(X^2) - [\mathbb{E}(X)]^2 &= \frac{\theta_2{}^3 - \theta_1{}^3}{3(\theta_2 - \theta_1)} - \left(\frac{\theta_2 + \theta_1}{2}\right)^2 \\
&= \frac{(\theta_2 - \theta_1)(\theta_2^2 + \theta_2\theta_1 + \theta_1^2)}{3(\theta_2 - \theta_1)} - \frac{\theta_2^2 + 2\theta_2\theta_1 + \theta_1^2}{4} \\
&= \frac{4\theta_2^2 + 4\theta_2\theta_1 + 4\theta_1^2}{12} - \frac{3\theta_2^2 + 6\theta_2\theta_1 + 3\theta_1^2}{12} \\
&= \frac{\theta_2^2 - 2\theta_2\theta_1 + \theta_1^2}{12} \\
&= \frac{(\theta_2 - \theta_1)^2}{12}.
\end{aligned}$$

■ ■ ■

■ **Example 4.24** Returning to Example 4.22, find the variance of random variable X with PDF

$$f(x) = \begin{cases} 1 - \dfrac{x}{2}, & 0 \le x \le 2 \\ 0, & \text{otherwise.} \end{cases}$$

Solution: As usual, we'll use the shortcut formula, which means we first need to calculate $E(X^2)$:

$$\mathbb{E}(X^2) = \int_{-\infty}^{\infty} x^2 f(x)\,dx = \int_0^2 x^2\left(1 - \frac{x}{2}\right)dx = \int_0^2 \left(x^2 - \frac{x^3}{2}\right)dx = \left[\frac{x^3}{3} - \frac{x^4}{8}\right]_0^2 = \frac{8}{3} - \frac{16}{8} = \frac{2}{3}.$$

Thus,

$$\mathbb{V}(X) = \mathbb{E}(X^2) - [\mathbb{E}(X)]^2 = \frac{2}{3} - \left(\frac{2}{3}\right)^2 = \frac{2}{9}.$$

■ ■ ■

Exercise 4.10 Calculate the variance and standard deviation of X in Exercise 4.6.

Answer: $\mathbb{V}(X) \approx 3.175$ and $\sigma \approx 1.782$.

4.4 Expected Value and Variance Properties

The expected value has a number of properties that we will make repeated use of (and, in a couple of cases, have already used). The properties in this section and the next apply to both discrete and continuous random variables. The first property is that the expected value of a constant is just the constant.

> **Theorem 4.4.1 — Expected Value of a Constant.** Let X be a random variable either with PMF $p(x)$ or PDF $f(x)$ and let c be a constant. Then $\mathbb{E}(c) = c$.

Proof. Suppose X is a discrete random variable; then,

$$\mathbb{E}(c) = \sum_{i=1}^{\infty} c \Pr(X = x_i) = c \sum_{i=1}^{\infty} \Pr(X = x_i) = c \times 1 = c,$$

where, from Theorem 4.2.1, $\sum_{i=1}^{\infty} \Pr(X = x_i) = 1$.

If X is a continuous random variable, then the proof proceeds similarly:

$$\mathbb{E}(c) = \int_{-\infty}^{\infty} c f(x)\, dx = c \int_{-\infty}^{\infty} f(x)\, dx = c \times 1 = c,$$

where, from Theorem 4.3.1, $\int_{-\infty}^{\infty} f(x)dx = 1$. ∎

The proofs for the discrete and continuous cases are the same, except that sums are used in the discrete case and integrals are used in the continuous case. Since sums and integrals obey similar linear properties (such as factoring constants), the proofs will be similar for the two cases. Usually, we will present a proof for one kind of random variable.

The second property is that the expected value of a constant times a random variable X is just the constant times the expected value of X.

> **Theorem 4.4.2 — Expected Value of a Constant Times a Random Variable.** Let X be a random variable with PMF $p(x)$ or PDF $f(x)$ and let c be a constant. Then $\mathbb{E}(cX) = c\mathbb{E}(X)$.

Proof. As with the proof for Theorem 4.4.1, the proof of this property follows directly from the application of the definition of expected value. Here we will only prove it for the discrete case:

$$\mathbb{E}(cX) = \sum_{i=1}^{\infty} c x_i \Pr(X = x_i) = \sum_{i=1}^{\infty} c x_i\, p(x_i) = c \sum_{i=1}^{\infty} x_i\, p(x_i) = c\mathbb{E}(X). \qquad \blacksquare$$

> **Definition 4.4.1 — Support of a Random Variable.** The *support* of a random variable X is the set of x for which the PMF (if X is discrete) or the PDF (if X is continuous) is positive. We can think of the support as the set of possible values that the random variable can take on.

Often the values x_i of the discrete random variable X will themselves be integers. When this occurs, we will usually write the sum using the index x; for example, $\sum_{x=1}^{\infty} p(x)$ if the support of X is $\{1, 2, \ldots\}$.

The third property says that the expected value of the sum of functions of a random variable is equal to the sum of the expected values of the functions of the random variable.

> **Theorem 4.4.3 — Expected Value of the Sum of Functions of a Random Variable.** Let X be a random variable either with PMF $p(x)$ or PDF $f(x)$, and let $g_1, \ldots, g_k$ be k functions of X.

Then,

$$\mathbb{E}\left(\sum_{i=1}^{k} g_i(X)\right) = \sum_{i=1}^{k} \mathbb{E}\left(g_i(X)\right).$$

Proof. For the discrete case:

$$\mathbb{E}\left(\sum_{i=1}^{k} g_i(X)\right) = \sum_{j=1}^{\infty}\left(\sum_{i=1}^{k} g_i(x_j)\right)p(x_j) = \sum_{j=1}^{\infty}\sum_{i=1}^{k} g_i(x_j)p(x_j) = \sum_{i=1}^{k}\sum_{j=1}^{\infty} g_i(x_j)p(x_j) = \sum_{i=1}^{k} \mathbb{E}\left(g_i(X)\right).$$

Switching the order of summation is justified because the infinite series converges absolutely. For the continuous case, the result is also true but it's a bit more mathematically complicated as we have to justify switching summation and integration operations. ∎

Note that, in general, $\mathbb{E}(g(X)) \neq g(\mathbb{E}(X))$, so switching the order of the function and expectation symbols is *incorrect*. The fourth property says that the expected value of a sum of random variables is equal to the sum of their expected values. We leave the proof as Problems 4.59 and 4.60.

> **Theorem 4.4.4 — Expected Value of a Sum of Random Variables.** Let $X_1, X_2, \ldots X_k$ be random variables, either all with the same PMF or PDF or perhaps each with a different PMF or PDF. Then,
>
> $$\mathbb{E}\left(\sum_{i=1}^{k} X_i\right) = \sum_{i=1}^{k} \mathbb{E}(X_i).$$

All of these theorems are examples of the fact that expectation is a linear operator. Working with sums of random variables is usually straightforward.

Next, we prove the shortcut formula for the variance from Section 4.2.

> **Theorem 4.4.5 — Shortcut Formula for the Variance of a Random Variable.** For any random variable X with mean μ and finite variance, the variance can be computed as
>
> $$\mathbb{V}(X) = \mathbb{E}(X^2) - \mu^2.$$

Proof.

$$\mathbb{V}(X) = \mathbb{E}\left[(X-\mu)^2\right] = \mathbb{E}\left(X^2 - 2X\mu + \mu^2\right) = \mathbb{E}\left(X^2\right) - 2\mu^2 + \mu^2 = \mathbb{E}\left(X^2\right) - \mu^2. \quad ∎$$

The variance has a number of additional properties we will make use of, including the following theorems. The first property is that the variance of a constant is zero.

> **Theorem 4.4.6 — Variance of a Constant.** Let c be a constant. Then $\mathbb{V}(c) = 0$.

Proof. The proof for Theorem 4.4.6 follows directly from Definition 4.2.2. If X is a random variable that equals c with probability 1, then

$$\mathbb{V}(X) = \mathbb{V}(c) = \mathbb{E}(c^2) - [\mathbb{E}(c)]^2 = c^2 - c^2 = 0. \quad ∎$$

The second property is that the variance of a constant times a random variable X is just the constant squared times the variance of X.

Theorem 4.4.7 — Variance of a Constant Times a Random Variable. Let X be a random variable and let c be a constant. Then $\mathbb{V}(cX) = c^2 \mathbb{V}(X)$.

Proof. For any random variable X and constant c,

$$
\begin{aligned}
\mathbb{V}(cX) &= \mathbb{E}(c^2 X^2) - [\mathbb{E}(cX)]^2 \\
&= c^2 \mathbb{E}(X^2) - [c\mathbb{E}(X)]^2 \\
&= c^2 \left(\mathbb{E}(X^2) - \mathbb{E}(X)^2 \right) \\
&= c^2 \mathbb{V}(X).
\end{aligned}
$$
∎

The third property puts Theorems 4.4.6 and 4.4.7 together.

Theorem 4.4.8 — Variance of aX+b. Let X be a random variable and let a and b be constants. Then $\mathbb{V}(aX + b) = a^2 \mathbb{V}(X)$.

Proof. For any random variable X and constants a and b,

$$
\begin{aligned}
\mathbb{V}(aX + b) &= \mathbb{E}\left[(aX + b)^2\right] - \left[\mathbb{E}(aX + b)\right]^2 \\
&= \mathbb{E}\left[a^2 X^2 + 2abX + b^2\right] - \left[a\mathbb{E}(X) + b\right]^2 \\
&= \mathbb{E}(a^2 X^2) + \mathbb{E}(2abX) + \mathbb{E}(b^2) - \left[a^2 \mathbb{E}(X)^2 + 2ab\mathbb{E}(X) + b^2\right] \\
&= a^2 \mathbb{E}(X^2) + 2ab\mathbb{E}(X) + b^2 - a^2 \mathbb{E}(X)^2 - 2ab\mathbb{E}(X) - b^2 \\
&= a^2 \left[\mathbb{E}(X^2) - \mathbb{E}(X)^2\right] \\
&= a^2 \mathbb{V}(X).
\end{aligned}
$$
∎

4.5　Joint Distributions for Discrete Random Variables

So far in this chapter we have talked about the probability distribution (discrete or continuous) of one random variable at a time. We now look at how we can describe the behavior of two or more random variables at a time. To make these ideas more concrete, let's look at a simple example.

■ **Example 4.25**　Suppose a large population consists of 40% Democrats, 30% Republicans, and 30% Independents. A sample of size 3 is taken. Let X denote the number of Democrats in the sample and Y the number of Republicans in the sample. Find all possible combinations of X and Y and their corresponding probabilities.

Solution: The outcomes x and y of the random variables X and Y must satisfy $x \geq 0$, $y \geq 0$, and $x + y \leq 3$. Of course, x and y must be integers. The possible outcomes are then the ordered pairs

```
(0,0)    (0,1)    (0,2)    (0,3)
(1,0)    (1,1)    (1,2)
(2,0)    (2,1)
(3,0)
```

Table 4.1 All possible samples of size 3 from a large population consisting of 40% Democrats, 30% Republicans, and 30% Independents.

Outcome	X	Y	Prob	Outcome	X	Y	Prob
III	0	0	$0.3^3 = 0.027$	RRR	0	3	$0.3^3 = 0.027$
IID	1	0	$0.3^2 \times 0.4 = 0.036$	DDR	2	1	$0.4^2 \times 0.3 = 0.048$
IDI	1	0	$0.3^2 \times 0.4 = 0.036$	DRD	2	1	$0.4^2 \times 0.3 = 0.048$
DII	1	0	$0.3^2 \times 0.4 = 0.036$	RDD	2	1	$0.4^2 \times 0.3 = 0.048$
DDI	2	0	$0.4^2 \times 0.3 = 0.048$	RRD	1	2	$0.3^2 \times 0.4 = 0.036$
DID	2	0	$0.4^2 \times 0.3 = 0.048$	RDR	1	2	$0.3^2 \times 0.4 = 0.036$
IDD	2	0	$0.4^2 \times 0.3 = 0.048$	DRR	1	2	$0.3^2 \times 0.4 = 0.036$
DDD	3	0	$0.4^3 = 0.064$	IDR	1	1	$0.3 \times 0.4 \times 0.3 = 0.036$
IIR	0	1	$0.3^2 \times 0.3 = 0.027$	IRD	1	1	$0.3 \times 0.4 \times 0.3 = 0.036$
IRI	0	1	$0.3^2 \times 0.3 = 0.027$	DIR	1	1	$0.3 \times 0.4 \times 0.3 = 0.036$
RII	0	1	$0.3^2 \times 0.3 = 0.027$	DRI	1	1	$0.3 \times 0.4 \times 0.3 = 0.036$
RRI	0	2	$0.3^2 \times 0.3 = 0.027$	RDI	1	1	$0.3 \times 0.4 \times 0.3 = 0.036$
RIR	0	2	$0.3^2 \times 0.3 = 0.027$	RID	1	1	$0.3 \times 0.4 \times 0.3 = 0.036$
IRR	0	2	$0.3^2 \times 0.3 = 0.027$				

Table 4.2 Joint PMF for X and Y.

		\(y\) 0	1	2	3	
	0	0.027	0.081	0.081	0.027	0.216
x	1	0.108	0.216	0.108	0	0.432
	2	0.144	0.144	0	0	0.288
	3	0.064	0	0	0	0.064
		0.343	0.441	0.189	0.027	1.000

The set of all possible ordered collections of three is shown in Table 4.1, along with the corresponding probabilities. The set of pairs (x,y) with the corresponding probabilities are shown in Table 4.2. Note that many combinations are impossible, as indicated by a probability of 0. It is not possible, for example, to get $x = 3$ Democrats and at the same time $y = 1$ Republican in the sample of size 3. ∎ ∎ ∎

Definition 4.5.1 — Joint PMF. The *joint probability mass function* (joint PMF) for two random variables X and Y is

$$f(x,y) = \Pr\left(X = x, Y = y\right).$$

The comma in the above probability is interpreted as "and." For a set of n random variables, $X_1, X_2, \ldots, X_n$, the *joint PMF* is

$$f(x_1, x_2, \ldots, x_n) = \Pr\left(X_1 = x_1, X_2 = x_2, \ldots, X_n = x_n\right).$$

Table 4.2 shows the *joint PMF* for the pair (X, Y) in Example 4.25. If we know the joint PMF for X and Y we can find the distribution of either one of them by themselves. The distribution of either X or Y by itself is called the marginal distribution.

Definition 4.5.2 — Marginal PMF. Suppose X and Y have joint PMF $f(x,y)$. The *marginal PMFs* are

$$f_1(x) = \Pr\left(X = x\right)$$
$$f_2(y) = \Pr\left(Y = y\right).$$

The marginal PMFs can be determined from the joint PMF by summing across all possible values of the other variable. For example, in the sample of size 3 in Example 4.25, the marginal PMF of X can be determined as follows:

$$\Pr(X=0) = \Pr(X=0,Y=0) + \Pr(X=0,Y=1) + \Pr(X=0,Y=2) + \Pr(X=0,Y=3)$$
$$= f(0,0) + f(0,1) + f(0,2) + f(0,3)$$
$$\Pr(X=1) = \Pr(X=1,Y=0) + \Pr(X=1,Y=1) + \Pr(X=1,Y=2) + \Pr(X=1,Y=3)$$
$$= f(1,0) + f(1,1) + f(1,2) + f(1,3)$$

etc.

In general, the marginal PMFs are

$$f_1(x) = \sum_{\text{all } y} f(x,y) \tag{4.2}$$

$$f_2(y) = \sum_{\text{all } x} f(x,y). \tag{4.3}$$

Expected values for functions of (X,Y) can be obtained by summing over all possible joint outcomes.

Definition 4.5.3 — Expected Value of a Function of Two Random Variables. Suppose (X,Y) have joint PMF $f(x,y)$, and suppose $g(X,Y)$ is some function of the random variables X and Y. The expected value of $g(X,Y)$ is

$$\mathbb{E}(g(X,Y)) = \sum_{\text{all } x} \sum_{\text{all } y} g(x,y) f(x,y).$$

■ **Example 4.26**   For the random variables X and Y in Example 4.25, find

(a) the marginal PMF for X,

(b) the marginal PMF for Y,

(c) $\mathbb{E}(X)$,

(d) $\mathbb{E}(Y)$, and

(e) $\mathbb{E}(XY)$.

Solution: The marginal PMF for X is obtained by summing across the rows in Table 4.2. This has been done by writing the row sums in the right margin of Table 4.2. This gives the marginal PMF of X as

x	0	1	2	3
$f_1(x)$	0.216	0.432	0.288	0.064

Similarly, the marginal PMF for Y is

y	0	1	2	3
$f_2(y)$	0.343	0.441	0.189	0.027

For the situation in Example 4.25, the expected values of X and Y could be obtained using the marginals calculated above; however, there is an easier way. If you reread the problem statement, ignoring the definition of Y, you'll notice that X is the number of successes (selecting a Democrat) on three independent trials with probability of success 0.40 on each trial. Thus, $X \sim \text{BIN}(3, 0.4)$. Similar reasoning leads to $Y \sim \text{BIN}(3, 0.3)$. Thus,

$$\mathbb{E}(X) = 3 \times 0.4 = 1.2 \qquad \text{and} \qquad \mathbb{E}(Y) = 3 \times 0.3 = 0.9.$$

The expected value of XY is obtained by summing xy times the PMF across all possible outcomes:

$$
\begin{aligned}
E(XY) &= \sum_{x=0}^{3} \sum_{y=0}^{3} xy f(x, y) \\
&= 0{\times}0{\times}0.027 + 0{\times}1{\times}0.081 + 0{\times}2{\times}0.081 + 0{\times}3{\times}0.027 \\
&\quad + 1{\times}0{\times}0.108 + 1{\times}1{\times}0.216 + 1{\times}2{\times}0.108 + 1{\times}3{\times}0 \\
&\quad + 2{\times}0{\times}0.144 + 2{\times}1{\times}0.144 + 2{\times}2{\times}0 + 2{\times}3{\times}0 \\
&\quad + 3{\times}0{\times}0.064 + 3{\times}1{\times}0 + 3{\times}2{\times}0 + 3{\times}3{\times}0 \\
&= 1{\times}1{\times}0.216 + 1{\times}2{\times}0.108 + 2{\times}1{\times}0.144 \\
&= 0.720.
\end{aligned}
$$

∎

Exercise 4.11 Compute the expectations of X and Y by summing

$$\mathbb{E}(X) = \sum_{x=0}^{3} x f_1(x) \qquad \text{and} \qquad \mathbb{E}(Y) = \sum_{y=0}^{3} y f_2(y),$$

and verify that you get the same result as in Example 4.26.

Definition 4.5.4 — Covariance. The *covariance* between two random variables X and Y is

$$\text{cov}(X, Y) = \mathbb{E}\Big[(X - \mu_X)(Y - \mu_Y)\Big],$$

where μ_X and μ_Y are the expected values of X and Y, respectively.

Don't confuse the covariance of two random variables with the covariance between a set of ordered pairs $(x_1, y_1), (x_2, y_2), \dots, (x_n, y_n)$ given in Definition 1.5.10. Definition 4.5.4 says to take the product of $X - \mu_X$ and $Y - \mu_Y$ and then take its expectation. The sample covariance in Definition 1.5.10 can be thought of as the estimate of the covariance between the random variables X and Y if the sample data $(x, y_1), (x_2, y_2), \dots, (x_n, y_n)$ is considered a random sample from the joint distribution of X and Y. An equivalent formula for the covariance, which is easier to apply in practice, is given in the next theorem.

Theorem 4.5.1 — Shortcut Formula for Covariance. Suppose X and Y are discrete random variables with joint PMF $f(x, y)$. Let $\mu_X = \mathbb{E}(X)$ and $\mu_Y = \mathbb{E}(Y)$. Then the covariance is

$$\text{cov}(X, Y) = \mathbb{E}(XY) - \mu_X \mu_Y.$$

Proof.

$$
\begin{aligned}
\text{cov}(X, Y) &= \mathbb{E}\big[(X - \mu_X)(Y - \mu_Y)\big] \\
&= \mathbb{E}\big[XY - \mu_X Y - \mu_Y X + \mu_X \mu_Y\big]
\end{aligned}
$$

$$= \mathbb{E}(XY) - \mu_X \mathbb{E}(Y) - \mu_Y \mathbb{E}(X) + \mu_X \mu_Y$$
$$= \mathbb{E}(XY) - \mu_X \mu_Y - \mu_Y \mu_X + \mu_X \mu_Y$$
$$= \mathbb{E}(XY) - \mu_X \mu_Y.$$ ∎

The covariance is related to what is perhaps a more familiar term, the correlation.

Definition 4.5.5 — Correlation. The *correlation* between random variables X and Y is

$$\mathrm{corr}(X,Y) = \frac{\mathrm{cov}(X,Y)}{\sigma_X \sigma_Y},$$

where σ_X and σ_Y are the standard deviations of the random variables X and Y, respectively. The correlation is often written ρ_{XY} or simply ρ.

The correlation is a dimensionless quantity, in the sense that it has no units, and it is constrained to be between -1 and 1.

■ **Example 4.27** For the random variables in Example 4.25, compute the covariance and the correlation.

Solution: In Example 4.26 we did most of the work needed to get the covariance. If we apply Theorem 4.5.1, we obtain

$$\mathrm{cov}(X,Y) = \mathbb{E}(XY) - \mu_X \mu_Y = 0.72 - 1.2 \times 0.9 = -0.36.$$

Since the marginals are binomial, the standard deviations are

$$\sigma_X = \sqrt{3 \times 0.4 \times (1 - 0.4)} = \sqrt{0.72} \qquad \text{and} \qquad \sigma_Y = \sqrt{3 \times 0.3 \times (1 - 0.3)} = \sqrt{0.63}.$$

The correlation is therefore

$$\rho = \frac{-0.36}{\sqrt{0.72}\sqrt{0.63}} \approx -0.535.$$ ■ ■ ■

Notice that the covariance in this last example is negative. This is to be expected since, if the number of Democrats in the sample is large (say two or three), then the number of Republicans would tend to be small, and vice versa.

Definition 4.5.6 — Independent Discrete Random Variables. We say that discrete random variables X and Y are *independent* if the joint PMF factors into the product of the marginals:

$$f(x,y) = f_1(x) f_2(y), \qquad \text{for all } x, y.$$

Theorem 4.5.2 — Expected Value of Product of Functions of Independent Random Variables. If X and Y are independent random variables, and $g(X)$ and $h(Y)$ are functions of X and Y, respectively, then

$$\mathbb{E}\big(g(X)h(Y)\big) = \mathbb{E}\big(g(X)\big)\mathbb{E}\big(h(Y)\big).$$

Proof. We will prove this for discrete random variables.

$$
\begin{aligned}
\mathbb{E}\big(g(X)h(Y)\big) &= \sum_{\text{all } x}\sum_{\text{all } y} g(x)h(y)f(x,y) \\
&= \sum_{\text{all } x}\sum_{\text{all } y} g(x)h(y)f_1(x)f_2(y) \qquad \text{[because } X \text{ and } Y \text{ are independent]} \\
&= \sum_{\text{all } x} g(x)f_1(x)\left(\sum_{\text{all } y} h(y)f_2(y)\right) \\
&= \sum_{\text{all } x} g(x)f_1(x)\,\mathbb{E}\big(h(Y)\big) \\
&= \mathbb{E}\big(h(Y)\big)\sum_{\text{all } x} g(x)f_1(x) \\
&= \mathbb{E}\big(g(X)\big)\,\mathbb{E}\big(h(Y)\big).
\end{aligned}
$$

■

This theorem can be generalized to n random variables.

Theorem 4.5.3 — Generalization of Theorem 4.5.2. If $X_1, X_2, \ldots, X_n$ are independent random variables, and $g_1(X_1), g_2(X_2), \ldots, g_n(X_n)$ are functions of $X_1, X_2, \ldots, X_n$, then

$$
\mathbb{E}\big(g_1(X_1)\times g_2(X_2)\times\cdots\times g_n(X_n)\big) = \mathbb{E}\big(g_1(X_1)\big)\times\mathbb{E}\big(g_2(X_2)\big)\times\cdots\times\mathbb{E}\big(g_n(X_n)\big).
$$

An important special case of Theorem 4.5.2 occurs when $g(X) = X$ and $h(Y) = Y$:

$$
\mathbb{E}(XY) = \mathbb{E}(X)\mathbb{E}(Y).
$$

When X and Y are independent, the covariance is zero. This is because

$$
\text{cov}(X,Y) = \mathbb{E}(XY) - \mathbb{E}(X)\mathbb{E}(Y) = \mathbb{E}(X)\mathbb{E}(Y) - \mathbb{E}(X)\mathbb{E}(Y) = 0. \tag{4.4}
$$

Be careful, though; the converse is not true. It is possible for the covariance to be zero while the random variables are not independent. See Problem 4.49.

With these results in hand, we can give some formulas for variances and covariances of sums of random variables.

Theorem 4.5.4 — Properties of Variance and Covariance. For any random variables X, Y, and Z:

(a) $\mathbb{V}(X+Y) = \mathbb{V}(X) + \mathbb{V}(Y) + 2\text{cov}(X,Y)$.

(b) If X and Y are independent, then $\mathbb{V}(X+Y) = \mathbb{V}(X) + \mathbb{V}(Y)$.

(c) $\text{cov}(aX,Y) = a\,\text{cov}(X,Y)$.

(d) $\text{cov}(X,aY) = a\,\text{cov}(X,Y)$.

(e) $\text{cov}(X+Y,Z) = \text{cov}(X,Z) + \text{cov}(Y,Z)$.

(f) $\text{cov}(X,Y+Z) = \text{cov}(X,Y) + \text{cov}(X,Z)$.

Proof. We will prove only the result in (a) and leave the rest as exercises. See Problem 4.45. Starting with the definition of the variance, we can write

$$\mathbb{V}(X+Y) = \mathbb{E}\left(\left((X+Y)-\mathbb{E}(X+Y)\right)^2\right)$$
$$= \mathbb{E}\left(\left((X-\mu_X)+(Y-\mu_Y)\right)^2\right)$$
$$= \mathbb{E}\left((X-\mu_X)^2\right)+\mathbb{E}\left((Y-\mu_Y)^2\right)+2\mathbb{E}\left((X-\mu_X)(Y-\mu_Y)\right)$$
$$= \mathbb{V}(X)+\mathbb{V}(Y)+2\text{cov}(X,Y). \blacksquare$$

Theorem 4.5.4 can be generalized to the case of an arbitrary number of random variables. For a proof of this theorem, see Problem 4.46.

Theorem 4.5.5 — More Properties of Variance and Covariance. Suppose $X_1, X_2, \ldots, X_n$ and $Y_1, Y_2, \ldots, Y_m$ are random variables. Then

1. $\mathbb{V}\left(\sum_{i=1}^{n} X_i\right) = \sum_{i=1}^{n}\mathbb{V}(X_i)+2\sum_{i=1}^{n}\sum_{j=1}^{i-1}\text{cov}(X_i, X_j).$

2. If $X_1, X_2, \ldots, X_n$ are independent, then $\mathbb{V}\left(\sum_{i=1}^{n} X_i\right) = \sum_{i=1}^{n}\mathbb{V}(X_i).$

3. $\text{cov}\left(\sum_{i=1}^{n} X_i, \sum_{j=1}^{m} Y_i\right) = \sum_{i=1}^{n}\sum_{j=1}^{m}\text{cov}(X_i, Y_j).$

4.6 Conditional Distributions for Discrete Random Variables

While the joint PMF for two random variable X and Y gives the probability of observing a particular pair of outcomes, say $X = x$ and $Y = y$, the conditional PMF of one random variable given the other involves a conditional probability.

Definition 4.6.1 — Conditional PMF. Suppose (X, Y) have joint PMF $f(x, y)$. The conditional PMF of Y given $X = x$, assuming $\Pr(X = x) \neq 0$, is

$$f(y|x) = \frac{\Pr(X = x, Y = y)}{\Pr(X = x)} = \frac{f(x, y)}{f_1(x)}. \tag{4.5}$$

The conditional PMF of X given $Y = y$, assuming $\Pr(Y = y) \neq 0$, is

$$f(x|y) = \frac{\Pr(X = x, Y = y)}{\Pr(Y = y)} = \frac{f(x, y)}{f_2(y)}. \tag{4.6}$$

If there is ever ambiguity about which random variables we're referring to, we will write $f_{Y|x}(y|x)$.

■ **Example 4.28** For the joint PMF described in Example 4.25 and Table 4.2, find

(a) $f(y|x = 0)$,
(b) $f(y|x = 3)$.
(c) Explain, in the context of this problem why the answer to part (b) is obvious.

Solution: For part (a),

$$f(0|x=0) = \frac{f(0,0)}{f_1(0)} = \frac{0.027}{0.216} = \frac{27}{216} = \frac{1}{8}$$

$$f(1|x=0) = \frac{f(0,1)}{f_1(0)} = \frac{0.081}{0.216} = \frac{81}{216} = \frac{3}{8}$$

$$f(2|x=0) = \frac{f(0,2)}{f_1(0)} = \frac{0.081}{0.216} = \frac{81}{216} = \frac{3}{8}$$

$$f(3|x=0) = \frac{f(0,3)}{f_1(0)} = \frac{0.027}{0.216} = \frac{27}{216} = \frac{1}{8}.$$

Note that for part (b),

$$f(0|x=3) = \frac{f(3,0)}{f_1(3)} = \frac{0.064}{0.064} = 1$$

$$f(1|x=3) = \frac{f(3,1)}{f_1(3)} = \frac{0.000}{0.064} = 0$$

$$f(2|x=3) = \frac{f(3,2)}{f_1(3)} = \frac{0.000}{0.064} = 0$$

$$f(3|x=3) = \frac{f(3,3)}{f_1(3)} = \frac{0.000}{0.064} = 0.$$

If there are three Democrats in a sample of size 3 from the population, there must be zero Republicans and zero Independents. Thus, $f(0|3) = 1$ and $f(y|3) = 0$ for all $y \geq 1$. ∎ ∎ ∎

Exercise 4.12 Repeat Example 4.28, interchanging the roles of X and Y.

The concept of conditional probability is related to independence.

Theorem 4.6.1 — Condition for Random Variables Being Independent. Suppose X and Y have joint PDF $f(x,y)$. Then X and Y are independent if and only if $f(y|x) = f_2(y)$.

Proof. First, suppose X and Y are independent. Then,

$$f(x,y) = f_1(x)f_2(y).$$

If we multiply both sides of (4.5) by $f_1(x)$, we see that

$$f(x,y) = f_1(x)f(y|x).$$

Equating these two expressions for $f(x,y)$ yields

$$f_1(x)f_2(y) = f_1(x)f(y|x),$$

which implies, as long as $f_1(x) \neq 0$, that

$$f(y|x) = f_2(y).$$

To prove the other direction of the "if and only if" statement, suppose that $f(y|x) = f_2(y)$. Then,

$$f(x,y) = f_1(x)f(y|x) = f_1(x)f_2(y),$$

so X and Y are independent. ∎

Exercise 4.13 Suppose X and Y are random variables with PMF $f(x,y)$. Suppose that $f_1(x_0) > 0$ and $f_2(y_0) > 0$, but $f(x_0,y_0) = 0$. Can X and Y be independent?

Answer: No.

We have stated and proved many of the results for joint PMFs and conditional PMFs for just two random variables. The results carry over to higher dimensions in a natural way. For example, if random variables $X_1, X_2, \ldots, X_n$ have joint PMF $f(x_1, x_2, \ldots, x_n)$, then the marginal of X_1 is

$$f_1(x_1) = \sum_{\text{all } x_2} \sum_{\text{all } x_3} \cdots \sum_{\text{all } x_n} f(x_1, x_2, \ldots, x_n).$$

Similar definitions apply to getting the marginals $f_2(x_2), \ldots, f_n(x_n)$. In other words, when we want the marginal for one of the random variables, we sum across all possible values of all other variables. If we want the (joint) marginal distribution for two random variables, we sum across all possible values of all other variables. For example, if we want the joint PMF of X_1 and X_n, we would have to evaluate

$$f_{1n}(x_1, x_n) = \sum_{\text{all } x_2} \sum_{\text{all } x_3} \cdots \sum_{\text{all } x_{n-1}} f(x_1, x_2, \ldots, x_n).$$

We say that $X_1, X_2, \ldots, X_n$ are independent if

$$f(x_1, x_2, \ldots, x_n) = f_1(x_1) f_2(x_2) \cdots f_n(x_n).$$

Conditional PMFs are defined in a similar fashion. For example, the conditional PMF of X_1 given all the others (i.e. $X_2, X_3, \ldots, X_n$) is

$$f(x_1 | x_2, x_3, \ldots, x_n) = \frac{f(x_1, x_2, \ldots, x_n)}{f(x_2, \ldots, x_n)}.$$

4.7 Joint Distributions for Continuous Random Variables

We introduced joint PMFs in Section 4.5. When the random variables are both continuous we have the analogous concept of the joint PDF.

Definition 4.7.1 — Joint PDF. Suppose X_1 and X_2 are continuous random variables. The joint PDF is the function $f(x_1, x_2)$ with the property that probabilities involving X_1 and X_2 can be found by integrating the PDF over the region whose probability we'd like to find:

$$\Pr\big((X_1, X_2) \in A\big) = \iint_A f(x_1, x_2) \, dx_2 \, dx_1.$$

The probability of (X_1, X_2) being in region A is the volume underneath the joint PDF $f(x_1, x_2)$ over region A.

The double integral in Definition 4.7.1 can often be evaluated by writing it as an iterated integral. The marginal distributions of X_1 and X_2 are defined in an analogous way to how marginals were defined for discrete random variables.

Definition 4.7.2 — Marginal PDF. If (X_1, X_2) have joint PDF $f(x_1, x_2)$, then the marginal of X_1 is

$$f_1(x_1) = \int_{\text{all } x_2} f(x_1, x_2) \, dx_2$$

and the marginal of X_2 is

$$f_2(x_2) = \int\limits_{\text{all } x_1} f(x_1, x_2)\, dx_1.$$

You may notice that many of the definitions and theorems from Section 4.5 involve sums, whereas in this section we use integrals. For example, to evaluate probabilities involving two or more continuous random variables or to obtain a marginal distribution, we have to select the appropriate limits of integration, whereas for discrete random variables we have to select the appropriate values over which to sum. The concept of independent continuous random variables and the expectation for functions of two continuous random variables carries over from Definitions 4.5.3, 4.5.4, and 4.5.6, and Theorem 4.5.1 with sums replaced by integrals.

Definition 4.7.3 — Independent Continuous Random Variables. Random variables X and Y with joint PDF $f(x,y)$ are independent provided

$$f(x,y) = f_1(x)f_2(y), \qquad \text{for all } x,y.$$

Definition 4.7.4 — Expected Value of a Function of Two Continuous Random Variables. The expected value of a function $g(X,Y)$ of two random variables X and Y with joint PDF $f(x,y)$ is

$$\mathbb{E}\big(g(X,Y)\big) = \iint\limits_{\text{all } x \text{ and } y} f(x,y)\, dy\, dx.$$

Definition 4.7.5 — Covariance for Continuous Random Variables. The *covariance* between two random variables X and Y is

$$\text{cov}(X,Y) = \mathbb{E}\Big[(X - \mu_X)(Y - \mu_Y)\Big]$$

where μ_X and μ_Y are the expected values of X and Y, respectively.

Theorem 4.7.1 — Shortcut for Covariance for Continuous Random Variables.
Suppose X and Y are continuous random variables with joint PDF $f(x,y)$. Then the covariance is

$$\text{cov}(X,Y) = \mathbb{E}(XY) - \mu_X \mu_Y.$$

The proof of this theorem is identical to that in Theorem 4.5.1, except the expectations are integrals in this case, rather than sums as in the proof of Theorem 4.5.1.

Theorem 4.7.2 — Expected Value Property for Independent Continuous Random Variables. If X and Y are independent continuous random variables, and $g(X)$ and $h(Y)$ are functions of X and Y, respectively, then

$$\mathbb{E}\big(g(X)h(Y)\big) = \mathbb{E}\big(g(X)\big)\mathbb{E}\big(h(Y)\big).$$

Once again, the proof for the continuous case is similar to the proof of the analogous result (Theorem 4.5.2) for discrete random variables, except sums are replaced by integrals. This theorem can be generalized to n random variables.

Theorem 4.7.3 — Generalization of Theorem 4.7.2. If $X_1, X_2, \ldots, X_n$ are independent continuous random variables, and $g_1(X_1), g_2(X_2), \ldots, g_n(X_n)$ are functions of $X_1, X_2, \ldots, X_n$, then

$$\mathbb{E}\big(g_1(X_1) \times g_2(X_2) \times \cdots \times g_n(X_n)\big) = \mathbb{E}\big(g_1(X_1)\big) \times \mathbb{E}\big(g_2(X_2)\big) \times \cdots \times \mathbb{E}\big(g_n(X_n)\big).$$

■ **Example 4.29** Suppose X and Y have the following joint PDF:

$$f(x,y) = \begin{cases} k, & 0 \le x \le 10 \text{ and } 0 \le y \le 10 \\ 0, & \text{otherwise.} \end{cases}$$

(a) Find k so that this is a valid joint PDF.
(b) Find $\Pr(0 \le X \le 5, 0 \le Y \le 5)$.
(c) Find $\Pr(X < Y)$.
(d) Find $\Pr(X^2 + Y^2 \le 25)$.
(e) Find the marginal $f_1(x)$.

Solution:

(a) The PDF must satisfy

$$\int_0^{10} \int_0^{10} f(x,y) \, dy \, dx = 1,$$

which means

$$1 = \int_0^{10} \int_0^{10} k \, dy \, dx = \int_0^{10} \Big[ky \Big]_0^{10} dx = \int_0^{10} 10k \, dx = \Big[10kx \Big]_0^{10} = 100k.$$

Thus, $k = 1/100$.

(b) To evaluate this probability, we integrate over the region $0 \le x \le 5, 0 \le y \le 5$:

$$\Pr(0 \le X \le 5, 0 \le Y \le 5) = \int_0^5 \int_0^5 \frac{1}{100} \, dy \, dx$$
$$= \int_0^5 \left[\frac{1}{100} y \right]_0^5 dx$$
$$= \int_0^5 \frac{5}{100} \, dx$$
$$= \left[\frac{5}{100} x \right]_0^5$$
$$= \frac{1}{4}.$$

Since the PDF is flat, or uniform, across the support $\{(x,y): 0 \le x \le 10, \ 0 \le y \le 10\}$, we can evaluate the double integral by taking the base area ($5 \times 5 = 25$) times the height ($1/100$):[3]

$$\Pr(0 \le x \le 5, 0 \le y \le 5) = \underbrace{(5 \times 5)}_{\text{base area}} \times \underbrace{\left(\frac{1}{100} \right)}_{\text{height}} = \frac{25}{100} = \frac{1}{4}.$$

---

[3]This trick of base area times height works only when the PDF is flat across its support.

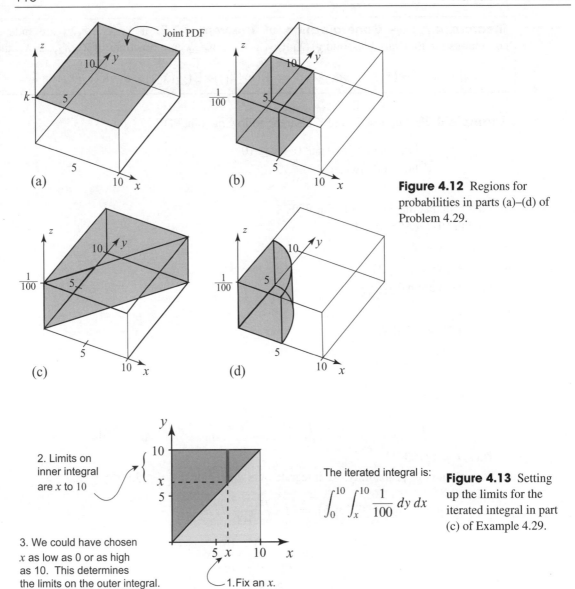

Figure 4.12 Regions for probabilities in parts (a)–(d) of Problem 4.29.

2. Limits on inner integral are x to 10

3. We could have chosen x as low as 0 or as high as 10. This determines the limits on the outer integral.

1. Fix an x.

The iterated integral is:

$$\int_0^{10} \int_x^{10} \frac{1}{100}\, dy\, dx$$

Figure 4.13 Setting up the limits for the iterated integral in part (c) of Example 4.29.

(c) We can evaluate this probability using double integrals:

$$\Pr\left(X < Y\right) = \iint\limits_A \frac{1}{100}\, dy\, dx, \tag{4.7}$$

where A is the region $\{(x,y) : 0 \leq x < y \leq 10\}$. Figure 4.13 shows how we can determine the limits for the iterated integral. To determine these, work your way inside out. For a fixed value of the outer variable, x, we allow the inner variable y to go from a low of x to a high of 10; this gives the limits on the inner integral. Finally, ask yourself: What is the *least* value of x we could choose and still be in the support? And, what is the *greatest* value of x we could choose and still be in the support? The answers are that 0 is the least we could choose for x and 10 is the greatest we could choose for x. This gives the limits on the outer integral. Continuing from (4.7),

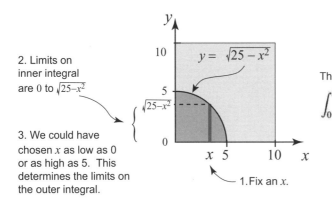

2. Limits on inner integral are 0 to $\sqrt{25-x^2}$

3. We could have chosen x as low as 0 or as high as 5. This determines the limits on the outer integral.

$y = \sqrt{25 - x^2}$

1. Fix an x.

The iterated integral is:

$$\int_0^5 \int_0^{\sqrt{25-x^2}} \frac{1}{100}\, dy\, dx$$

Figure 4.14 Setting up the limits for the iterated integral in part (d) of Example 4.29.

$$
\begin{aligned}
\Pr\left(X < Y\right) &= \int_0^{10} \int_x^{10} \frac{1}{100}\, dy\, dx \\
&= \int_0^{10} \left[\frac{1}{100}\, y\right]_x^{10} dx \\
&= \int_0^{10} \left(\frac{10}{100} - \frac{x}{100}\right) dx \\
&= \left[\frac{10x}{100} - \frac{x^2}{200}\right]_0^{10} \\
&= \frac{1}{2}.
\end{aligned}
$$

We could also obtain the volume under the (constant) PDF by taking the base area of S times the height. In this case, S is a triangle, so the base area is $\frac{1}{2}\times 10 \times 10 = 50$ and the height is $\frac{1}{100}$. Thus, the desired probability is $50 \times \frac{1}{100} = \frac{1}{2}$, which agrees with our answer obtained using double integrals.

(d) The values of x and y satisfy $x^2 + y^2 \le 25$, but the support for our PDF is $\{(x,y) : 0 \le x \le 10,\ 0 \le y \le 10\}$. The random ordered pair (X, Y) must be in the quarter of the circle $x^2 + y^2 \le 25$ that is in the first quadrant. This probability is most easily done using the base area times height method. The base area is $\frac{1}{4}\pi 5^2 = \frac{25\pi}{4}$. The height of the PDF is $\frac{1}{100}$. Thus,

$$
\Pr(X^2 + Y^2 \le 25) = \underbrace{\left(\frac{25\pi}{4}\right)}_{\text{base area}} \times \underbrace{\left(\frac{1}{100}\right)}_{\text{height}} = \frac{25\pi}{400} = \frac{\pi}{16} \approx 0.196.
$$

We could set up and evaluate an iterated integral (see Figure 4.14), but this leads to a rather obscure integral formula:

$$
\begin{aligned}
\Pr(X^2 + Y^2 \le 25) &= \int_0^5 \int_0^{\sqrt{25-x^2}} \frac{1}{100}\, dy\, dx \\
&= \frac{1}{100} \int_0^5 \sqrt{5^2 - x^2}\, dx \\
&= \frac{1}{100} \left[\frac{x}{2}\sqrt{5^2 - x^2} + \frac{5^2}{2}\tan^{-1}\frac{x}{\sqrt{5^2 - x^2}}\right]_0^5
\end{aligned}
$$

$$= \frac{1}{100}\left[\left(0 + \frac{5^2}{2}\lim_{x\to 5^-}\tan^{-1}\frac{x}{\sqrt{5^2-x^2}}\right) - \left(0 + \frac{5^2}{2}\tan^{-1}\frac{0}{\sqrt{5^2-0^2}}\right)\right]$$

$$= \frac{1}{100}\frac{25}{2}\frac{\pi}{2}$$

$$= \frac{\pi}{16}.$$

The limit is needed on the third from the last line because $\tan^{-1}\left(x/\sqrt{5^2-x^2}\right)$ is undefined when $x = 5$. The limit is

$$\lim_{x\to 5^-}\tan^{-1}\frac{x}{\sqrt{5^2-x^2}} = \lim_{z\to\infty}\tan^{-1}z = \frac{\pi}{2}.$$

The answer of $\pi/16 \approx 0.1963$ seems reasonable; the region in part (d) is slightly smaller than the region in part (b), where the probability was equal to $1/4$.

(e) The marginal for X is found by integrating over all possible y for each x. If $0 \leq x \leq 10$, then

$$f_1(x) = \int_0^{10}\frac{1}{100}\,dy = \left[\frac{y}{100}\right]_0^{10} = \frac{1}{10}, \qquad 0 \leq x \leq 10.$$

For all other values of x, the PDF is 0. Thus, the marginal of X is the UNIF(0,10) distribution. ∎

We could do parts (b), (c), and (d) of Example 4.29 by taking the area of the desired region times the height of the PDF (because the PDF is constant over the support). We caution the reader that this only works when the PDF is flat across its entire support. For the next example, the PDF is not flat over its support and we need to use integration.

■ **Example 4.30** Suppose X and Y have joint PDF

$$f(x,y) = \begin{cases} \dfrac{1}{4}e^{-x/2}, & 0 \leq y \leq x \\ 0, & \text{otherwise.} \end{cases}$$

Find:

(a) $\Pr(Y \leq 3)$,
(b) the marginal of X,
(c) the marginal of Y,
(d) $\text{cov}(X,Y)$.

Solution:

(a) We must take the integral over the region where $y \leq 3$ and x is unspecified. The region of integration is shown as the darker shading in Figure 4.15(b). We must be careful to integrate over the region satisfied by the inequality $y \leq 3$ *and* the support of the random variables X and Y. In this case it is easier to fix y and integrate over x first.[4] This leads to

---

[4]If we were to fix on x first, then the limits for the inner integral will depend on whether x is less than 3 or greater than or equal to 3. This would involve having to write two separate double integrals to get the desired probability.

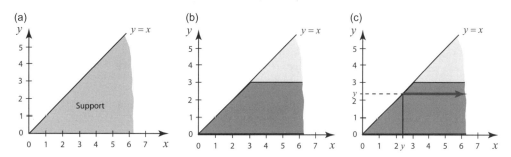

Figure 4.15 Setting up the limits for the integral in Example 4.30 part (a).

$$\Pr(Y \leq 3) = \int_0^3 \int_y^\infty \frac{1}{4} e^{-x/2} \, dx \, dy$$

$$= \int_0^3 \left[-\frac{1}{2} e^{-x/2} \right]_y^\infty dy$$

$$= \int_0^3 \left[0 + \frac{1}{2} e^{-y/2} \right] dy$$

$$= \int_0^3 \frac{1}{2} e^{-y/2} \, dy$$

$$= \left[-e^{-y/2} \right]_0^3$$

$$= 1 - e^{-3/2}$$

$$\approx 0.777.$$

(b) To get the marginal for X we fix x and integrate over y. For $x > 0$,

$$f_1(x) = \int_0^x \frac{1}{4} e^{-x/2} \, dy = \frac{1}{4} e^{-x/2} \int_0^x 1 \, dy = \frac{1}{4} e^{-x/2} [y]_0^x = \frac{x}{4} e^{-x/2}, \qquad x > 0.$$

(c) To get the marginal for Y, we fix y and integrate over x. For $y > 0$,

$$f_2(y) = \int_y^\infty \frac{1}{4} e^{-x/2} \, dx = \frac{1}{4} \left[-2e^{-x/2} \right]_y^\infty = \frac{1}{4} \left[2e^{-y/2} \right] = \frac{1}{2} e^{-y/2}, \quad y > 0.$$

(d) To get $\mathrm{cov}(X, Y)$, we begin with the expected value for Y, which can be written as

$$\mathbb{E}(Y) = \int_0^\infty y \frac{1}{2} e^{-y/2} \, dy = \frac{1}{2} \int_0^\infty y e^{-y/2} \, dy = \frac{1}{2} \int_0^\infty \underbrace{y}_{u} \underbrace{e^{-y/2} \, dy}_{dv}.$$

This integral can be evaluated using integration by parts. Let $u = y$ and $dv = e^{-y/2} \, dy$ so that $du = dy$ and $v = -2e^{-y/2}$. The limits for u are 0 to ∞. Thus,

$$\mathbb{E}(Y) = \frac{1}{2} \left([uv]_0^\infty - \int_0^\infty v \, du \right) = \frac{1}{2} \left(\left[y(-2e^{-y/2}) \right]_0^\infty - \int_0^\infty (-2e^{-y/2}) \, dy \right).$$

Note that in the case of an infinite upper limit, the bracket notation is really a limit:

$$\mathbb{E}(Y) = \frac{1}{2}\left(\left[y(-2e^{-y/2})\right]_0^\infty + \int_0^\infty 2e^{-y/2}\,dy\right)$$

$$= \frac{1}{2}\left(\lim_{y\to\infty}\left(-2ye^{-y/2}\right) - \left(-2\times 0e^{-0/2}\right) + \left[-4e^{-y/2}\right]_0^\infty\right)$$

$$= \frac{1}{2}\left((0-0) + \left(\lim_{y\to\infty}\left(-4e^{-y/2}\right) - (-4e^{-0/2})\right)\right)$$

$$= \frac{1}{2}(0+0-(-4))$$

$$= 2.$$

See Problem 4.58 for an explanation of how this limit is evaluated. Now let's consider $\mathbb{E}(X)$:

$$\mathbb{E}(X) = \int_0^\infty x\frac{x}{4}e^{-x/2}\,dx = \frac{1}{4}\int_0^\infty x^2 e^{-x/2}\,dx. \tag{4.8}$$

This integral can be evaluated using integration by parts twice (see Problem 4.58), or by using the gamma function:

$$\Gamma(a) = \int_0^\infty x^{a-1}e^{-x}dx.$$

It can be shown that if a is a positive integer, then

$$\Gamma(a) = (a-1)!$$

If we make the substitution $u = x/2$ in (4.8) we obtain

$$\mathbb{E}(X) = \frac{1}{4}\int_0^\infty (2u)^2 e^{-u}2\,du = 2\int_0^\infty u^{3-1}e^{-u}\,du = 2\Gamma(3) = 4.$$

Finally, let's evaluate $\mathbb{E}(XY)$:

$$\mathbb{E}(XY) = \int_0^\infty \int_0^x xy\frac{1}{4}e^{-x/2}\,dy\,dx$$

$$= \frac{1}{4}\int_0^\infty xe^{-x/2}\int_0^x y\,dy\,dx$$

$$= \frac{1}{4}\int_0^\infty xe^{-x/2}\left[\frac{y^2}{2}\right]_0^x\,dx$$

$$= \frac{1}{8}\int_0^\infty x^3 e^{-x/2}\,dx. \tag{4.9}$$

This integral can be evaluated using integration by parts three times (see Problem 4.58), or we can manipulate it so that it becomes a gamma function. Making the substitution $u = x/2$, with $du = \frac{1}{2}dx$ and $dx = 2\,du$, yields

$$\mathbb{E}(XY) = \frac{1}{8}\int_0^\infty (2u)^3 e^{-u}\,(2\,du) = \frac{1}{8}2^4\int_0^\infty u^{4-1}e^{-u}\,du = \frac{16\Gamma(4)}{8} = 2\times 3! = 12. \tag{4.10}$$

The covariance is therefore

$$\mathrm{cov}(X,Y) = \mathbb{E}(XY) - \mathbb{E}(X)\mathbb{E}(Y) = 12 - 4\times 2 = 4.$$ ∎∎∎

4.8 **Conditional Distributions for Continuous Random Variables**

The definitions of conditional distributions for continuous random variables are similar to that given in Section 4.6.

> **Definition 4.8.1 — Conditional PDF.** Suppose (X,Y) have joint PDF $f(x,y)$. The conditional PDF of Y given $X = x$, assuming $f_1(x) \neq 0$, is
>
> $$f(y|x) = \frac{f(x,y)}{f_1(x)}. \tag{4.11}$$
>
> The conditional PDF of X given $Y = y$, assuming $f_2(y) \neq 0$, is
>
> $$f(x|y) = \frac{f(x,y)}{f_2(y)}. \tag{4.12}$$

■ **Example 4.31** For the random variables X and Y in Example 4.30, find

(a) $f(y|x)$,
(b) $f(x|y)$.

(a) For a fixed x satisfying $x > 0$,

$$f(y|x) = \frac{f(x,y)}{f_1(x)} = \frac{\frac{1}{4}e^{-x/2}}{\frac{x}{4}e^{-x/2}} = \frac{1}{x}, \qquad 0 < y < x.$$

Thus,

$$f(y|x) = \begin{cases} \dfrac{1}{x}, & 0 \leq y \leq x \\ 0, & \text{otherwise.} \end{cases}$$

While it is always important to keep track of the support for marginal and joint distributions, it is particularly important for conditional distributions. We usually think of the variable being conditioned on (x in this case) to be *fixed*. The "variable" in this part of the example is y. Thus, conditioned on a particular value x for the random variable X, we see that Y is uniformly distributed from 0 to x. We write this as $Y|(X=x) \sim \text{UNIF}(0,x)$.

(b) Now, we fix on a value for y satisfying $y > 0$:

$$f(x|y) = \frac{\frac{1}{4}e^{-x/2}}{\frac{1}{2}e^{-y/2}} = \left(\frac{1}{2}e^{y/2}\right)e^{-x/2}, \qquad x \geq y.$$

Thus,

$$f(x|y) = \begin{cases} \left(\dfrac{1}{2}e^{y/2}\right)e^{-x/2}, & x \geq y \\ 0, & x < y. \end{cases} \qquad \blacksquare\blacksquare\blacksquare$$

> **Exercise 4.14** Verify that both conditional distributions in Example 4.31 integrate to 1.

We can define conditional expectations in a natural way.

Definition 4.8.2 — Conditional Expectation. Suppose X and Y are continuous random variables with joint PDF $f(x,y)$ and conditional PDFs $f(y|x)$ and $f(x|y)$. The conditional expectation of Y given $X = x$ is

$$\mathbb{E}(Y|X=x) = \int_{\text{all } y \text{ for given } x} y f(y|x) \, dy.$$

The conditional expectation of X given $Y = y$ is

$$\mathbb{E}(X|Y=y) = \int_{\text{all } x \text{ for given } y} x f(x|y) \, dx.$$

If there is no ambiguity we will write $\mathbb{E}(Y|x)$ and $\mathbb{E}(X|y)$.

Definitions for conditional expectations of functions of X or Y are given in the following.

Definition 4.8.3 — Conditional Expectation of a Function of a Random Variable. For continuous random variables X and Y with joint PDF $f(x,y)$ and conditional PDFs $f(y|x)$ and $f(x|y)$, the conditional expectation of $g(Y)$ given $X = x$ is

$$\mathbb{E}(g(Y)|X=x) = \int_{\text{all } y \text{ for given } x} g(y) f(y|x) \, dy,$$

and the conditional expectation of $h(X)$ given $Y = y$ is

$$\mathbb{E}(h(X)|Y=y) = \int_{\text{all } x \text{ for given } y} h(x) f(x|y) \, dx.$$

If there is no ambiguity we will write $\mathbb{E}(g(Y)|x)$ and $\mathbb{E}(h(X)|y)$.
   The conditional variances are defined as

$$\mathbb{V}(Y|x) = \mathbb{E}\left((Y - \mathbb{E}(Y|x))^2 | x\right)$$

and

$$\mathbb{V}(X|y) = \mathbb{E}\left((X - \mathbb{E}(X|y))^2 | y\right).$$

Exercise 4.15 Verify that

$$\mathbb{V}(Y|x) = \mathbb{E}(Y^2|x) - (\mathbb{E}(Y|x))^2 \qquad \text{and} \qquad \mathbb{V}(X|y) = \mathbb{E}(X^2|y) - (\mathbb{E}(X|y))^2.$$

■ **Example 4.32** Consider the random variables X and Y from Example 4.30 that have joint PDF

$$f(x,y) = \begin{cases} \frac{1}{4}e^{-x/2}, & 0 \le y \le x \\ 0, & \text{otherwise.} \end{cases}$$

Find:

(a) $\mathbb{E}(Y|x=3)$,
(b) $\mathbb{E}(Y|x)$,

(c) $\mathbb{E}(X|y)$,

(d) $\mathbb{E}(Y^2|x)$,

(e) $\mathbb{V}(Y|x)$.

Solution:

(a) Here we are conditioning on $X = 3$, so

$$\mathbb{E}(Y|X = 3) = \int_0^3 y\frac{1}{3}\,dy = \frac{1}{3}\left[\frac{1}{2}y^2\right]_0^3 = \frac{9}{6} = \frac{3}{2}.$$

(b) Next, we condition on an arbitrary x and integrate:

$$\mathbb{E}(Y|x) = \int_0^x y\frac{1}{x}\,dy = \frac{1}{x}\left[\frac{1}{2}y^2\right]_0^x = \frac{x^2}{2x} = \frac{x}{2}.$$

Note that we could have obtained the answer to part (b) by observing that $Y|(X = x) \sim \text{UNIF}(0,x)$ and the expected value of a $\text{UNIF}(0,x)$ random variable is $x/2$.

(c) Next, condition on the value of y and apply integration by parts:

$$\mathbb{E}(X|y) = \int_y^\infty x\,f(x|y)\,dx$$

$$= \left(\frac{1}{2}e^{y/2}\right)\int_y^\infty x\,e^{-x/2}\,dx$$

$$= \left(\frac{1}{2}e^{y/2}\right)\left(\left[-2xe^{-x/2}\right]_y^\infty + \int_y^\infty 2\,e^{-x/2}dx\right)$$

$$= \left(\frac{1}{2}e^{y/2}\right)\left(2ye^{-y/2} + \left[-4e^{-x/2}\right]_y^\infty\right)$$

$$= \left(\frac{1}{2}e^{y/2}\right)\left(2ye^{-y/2} + 4e^{-y/2}\right)$$

$$= y+2.$$

(d) $\mathbb{E}(Y^2|x) = \int_0^x y^2\frac{1}{x}\,dy = \frac{1}{x}\left[\frac{1}{3}y^3\right]_0^x = \frac{1}{3x}(x^2) = \frac{x^2}{3}, \qquad x > 0.$

(e) $\mathbb{V}(Y) = \mathbb{E}(Y^2|x) - \left[\mathbb{E}(Y|x)\right]^2 = \frac{x^2}{3} - \frac{x^2}{4} = \frac{x^2}{12} \qquad x > 0.$

Figure 4.16 shows graphs of $y = \mathbb{E}(Y|x)$ (a) and $x = E(X|y) = y+2$ (b). These are both *lines*, but they are not the same line.

■ ■ ■

4.9 Conditioning on a Random Variable

It is instructive to note that in Example 4.32, $\mathbb{E}(Y|X = 3)$ is a *number*; it was $3/2$ in that example. The expression $\mathbb{E}(Y|x)$ is a function of x, $\mathbb{E}(Y|x) = x/2$ in Example 4.32. What sort of object do you think we would obtain if we conditioned on a random variable, say X? In Example 4.32, if we substituted the random variable X for the fixed number x, we would obtain $\mathbb{E}(Y|X) = X/2$. The right side of this is $X/2$, which is a random variable! Thus, when we condition on a random variable, we obtain a

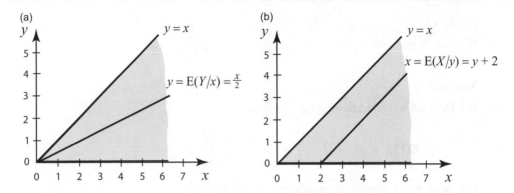

Figure 4.16 Graphs of the conditional expectation functions for Example 4.32.

random variable. This applies to both jointly continuous and jointly discrete random variables. If fact, the concept applies to one continuous and one discrete random variable.

If $\mathbb{E}(Y|X)$ is a random variable, then it makes sense to talk about its expectation $\mathbb{E}\big(\mathbb{E}(Y|X)\big)$ and its variance $\mathbb{V}\big(\mathbb{E}(Y|X)\big)$. One of the most remarkable theorems in statistics is the following, which gives a relationship between these quantities and the unconditional mean and variance.

> **Theorem 4.9.1 — Expected Value and Variance of Conditional Expectation.** If X and Y are random variables with joint PDF or PMF $f(x,y)$, and conditional PDF or PMF $f(y|x)$, then
> 1. $\mathbb{E}(Y) = \mathbb{E}\big(\mathbb{E}(Y|X)\big)$, and
> 2. $\mathbb{V}(Y) = \mathbb{V}\big(\mathbb{E}(Y|X)\big) + \mathbb{E}\big(\mathbb{V}(Y|X)\big)$.

Proof. We will give a proof for continuous random variables. For the first part,

$$
\begin{aligned}
\mathbb{E}(Y) &= \int_{-\infty}^{\infty} \int_{-\infty}^{\infty} y\, f(x,y)\, dy\, dx \\
&= \int_{-\infty}^{\infty} \int_{-\infty}^{\infty} y\, f(x)\, f(y|x)\, dy\, dx \\
&= \int_{-\infty}^{\infty} f(x) \int_{-\infty}^{\infty} y\, f(y|x)\, dy\, dx \\
&= \int_{-\infty}^{\infty} f(x) \underbrace{\mathbb{E}(Y|x)}_{\text{this is a function of } x \text{ only}} dx \\
&= \mathbb{E}\big(\mathbb{E}(Y|X)\big).
\end{aligned}
$$

The proof for the variance is also straightforward, but a little messier. We begin with the conditional variance of Y given X:

$$
\mathbb{V}(Y|X) = \mathbb{E}\big(Y^2|X\big) - \big[\mathbb{E}(Y|X)\big]^2. \tag{4.13}
$$

If we take the expectation over X on each side of (4.13) we obtain

$$
\mathbb{E}\big(\mathbb{V}(Y|X)\big) = \mathbb{E}\big(\mathbb{E}(Y^2|X)\big) - \mathbb{E}\big(\big[\mathbb{E}(Y|X)\big]^2\big). \tag{4.14}
$$

By the definition of conditional variance,

$$\mathbb{V}\big(\mathbb{E}(Y|X)\big) = \mathbb{E}\Big([\mathbb{E}(Y|X)]^2\Big) - \Big[\mathbb{E}\big(\mathbb{E}(Y|X)\big)\Big]^2. \tag{4.15}$$

Now let's add the expressions for $\mathbb{E}\big(\mathbb{V}(Y|X)\big)$ and $\mathbb{V}\big(\mathbb{E}(Y|X)\big)$ in (4.14) and (4.15):

$$\begin{aligned}
\mathbb{E}(\mathbb{V}(Y|X)) + \mathbb{V}(\mathbb{E}(Y|X)) &= \mathbb{E}\Big(\mathbb{E}(Y^2|X)\Big) - \mathbb{E}\Big([\mathbb{E}(Y|X)]^2\Big) + \mathbb{E}\Big([\mathbb{E}(Y|X)]^2\Big) - \Big[\mathbb{E}\big(\mathbb{E}(Y|X)\big)\Big]^2 \\
&= \mathbb{E}\Big(\mathbb{E}(Y^2|X)\Big) - \Big[\mathbb{E}\big(\mathbb{E}(Y|X)\big)\Big]^2 \\
&= \mathbb{E}(Y^2) - [\mathbb{E}(Y)]^2 \\
&= \mathbb{V}(Y). \quad\blacksquare
\end{aligned}$$

The proof of Theorem 4.9.1 illuminates which variables we are taking the expected value of. For example, in the inner expectation $\mathbb{E}(Y|X)$, the expectation is taken over Y, while in the outer expectation $\mathbb{E}\big(\mathbb{E}(Y|X)\big)$ we are taking the expectation over X. Theorem 4.9.1 applies to two continuous random variables, two discrete random variables, and one of each.

■ **Example 4.33** Suppose that in a manufacturing process the defect rate P varies from day to day according to a uniform distribution on the interval $(0,1)$. At the end of the day, two units from the process are selected and tested independently. Let Y denote the number of defectives in this sample of size 2. Find $\mathbb{E}(Y)$ and $\mathbb{V}(Y)$.

Solution: Possible values for Y are 0, 1, and 2. Conditioned on $P = p$, these have probabilities

$$\begin{aligned}
\Pr(Y=2|p) &= \Pr(\text{both items are defective}) = p^2 \\
\Pr(Y=0|p) &= \Pr(\text{both items are nondefective}) = (1-p)^2 \\
\Pr(Y=1|p) &= 1 - \Pr(Y=0|p) - \Pr(Y=2|p) = 1 - p^2 - (1-p)^2 = 2p(1-p).
\end{aligned}$$

Still conditioned on $P = p$, we have

$$\begin{aligned}
\mathbb{E}(Y|p) &= 0 \times \Pr(Y=0|p) + 1 \times \Pr(Y=1|p) + 2 \times \Pr(Y=2|p) \\
&= 0 \times (1-p)^2 + 1 \times 2p(1-p) + 2 \times p^2 \\
&= 2p
\end{aligned}$$

and

$$\begin{aligned}
\mathbb{E}(Y^2|p) &= 0^2 \times \Pr(Y=0|p) + 1^2 \times \Pr(Y=1|p) + 2^2 \times \Pr(Y=2|p) \\
&= 0 \times (1-p)^2 + 1 \times 2p(1-p) + 4 \times p^2 \\
&= 2p + 2p^2.
\end{aligned}$$

The conditional variance is therefore

$$\mathbb{V}(Y|p) = \mathbb{E}(Y^2|p) - [\mathbb{E}(Y|p)]^2 = 2p + 2p^2 - (2p)^2 = 2p(1-p).$$

Applying Theorem 4.9.1 gives

$$\mathbb{E}(Y) = \mathbb{E}\big(\mathbb{E}(Y|P)\big) = \mathbb{E}(2P) = 2\int_0^1 p \times 1\, dp = 2 \times \frac{1}{2} = 1$$

and

$$\mathbb{V}(Y) = \mathbb{V}\big(\mathbb{E}(Y|P)\big) + \mathbb{E}\big(\mathbb{V}(Y|P)\big) = \mathbb{V}(2P) + \mathbb{E}\big(2P(1-P)\big) = 4\mathbb{V}(P) + 2\mathbb{E}(P) - 2\mathbb{E}(P^2).$$

Since $\mathbb{E}(P^2) = \mathbb{V}(P) + \big[\mathbb{E}(P)\big]^2$, we can write the variance as

$$\mathbb{V}(Y) = 4\mathbb{V}(P) + 2\mathbb{E}(P) - 2\mathbb{E}(P^2) = \frac{4}{12} + 2\times\frac{1}{2} - 2\Big(\mathbb{V}(P) + \big[\mathbb{E}(P)\big]^2\Big) = \frac{2}{3}.$$

Here we have used the fact that $\mathbb{V}(P) = 1/12$ for the uniform distribution on $(0,1)$. ∎

4.10 Chapter Summary

In this chapter we've learned about random variables, both discrete and continuous, as well as their associated distributions: mass, density, and cumulative. We also learned how to calculate the expected value, variance, and standard deviation for random variables, quantities that are useful for characterizing the location and variability of a random variable. In later chapters we will use the sample mean and sample variance introduced in Chapter 1 to infer the values of the expected value and variance of a random variable from data. This is the connection between the probability we are learning about now and statistics that we will learn about later.

To illustrate the idea of a discrete random variable, we studied a particular type: the Bernoulli random variable. We also saw how the Bernoulli distribution provides a useful model for real-world phenomena where there are only two outcomes. Similarly, to illustrate the concept of a continuous random variable, we learned about another particular type, the (continuous) uniform random variable, where we also saw how the uniform distribution can be applied to solve various probability problems. In the next two chapters we will learn about many more discrete and continuous distributions, each of which is a useful probability model for various types of real-world phenomena.

Finally, in this chapter we learned about various theoretical properties of the expected value and variance for both discrete and continuous random variables.

4.11 Problems

Problem 4.1 Let X have PMF

$$p(x) = \begin{cases} 0.2, & x = 1 \\ 0.3, & x = 2 \\ 0.5, & x = 4 \\ 0, & \text{otherwise.} \end{cases}$$

Show that the PMF satisfies the properties of Theorem 4.2.1.

Problem 4.2 Let X have discrete uniform PMF on the values $x \in \{-1, 0, 1\}$. Write out the equation for the PMF and show that it satisfies the properties of Theorem 4.2.1.

Problem 4.3 Let X have a Bernoulli PMF with $p = \Pr(X = 1) = 0.7$. Write out the equation for the PMF and show that it satisfies the properties of Theorem 4.2.1.

Problem 4.4 Let X denote the sum of two fair, six-sided dice. Specify the domain that X can take on and then write out the equation for the PMF and show that it satisfies the properties of Theorem 4.2.1.

Problem 4.5 For the situation described in Problem 4.4, find the probability that $X = 7$.

Problem 4.6 Let X have the PMF in Problem 4.2. Find the following.

(a) $\Pr(X < -1)$ and $\Pr(X \le -1)$
(b) $\Pr(X > -1)$ and $\Pr(X \ge -1)$
(c) $\Pr(X > 0)$ and $\Pr(X \ge 0)$

Problem 4.7 Let X have the PMF in Problem 4.3. Find the following.

(a) $\Pr(X < 0)$ and $\Pr(X \le 0)$
(b) $\Pr(X > 1)$ and $\Pr(X \ge 1)$

Problem 4.8 Calculate the CDF from the PDF in Problem 4.2. Write out the expression for $F(x)$ and plot the PMF and CDF.

Problem 4.9 Let X have the PMF in Problem 4.4. Find the following.

(a) $\Pr(X < 2)$ and $\Pr(X \le 2)$
(b) $\Pr(X > 2)$ and $\Pr(X \ge 2)$
(c) $\Pr(X > 4)$ and $\Pr(X \ge 4)$

Problem 4.10 Calculate the CDF from the PMF in Problem 4.4. Write out the expression for $F(x)$ and plot the PMF and CDF.

Problem 4.11 Using the CDF from Problem 4.10, find the probabilities specified in Problem 4.9.

Problem 4.12 Using the CDF plotted in Problem 4.8, find the probabilities specified in Problem 4.6.

Problem 4.13 Calculate the CDF from the PMF in Problem 4.3. Write out the expression for $F(x)$ and plot the PMF and CDF.

Problem 4.14 Using the CDF plotted in Problem 4.13, find the probabilities specified in Problem 4.7.

Problem 4.15 Calculate the expected value, variance, and standard deviation of the random variable X with the PMF in Problem 4.1.

Problem 4.16 Calculate the expected value, variance, and standard deviation of the random variable X with the PMF in Problem 4.2.

Problem 4.17 Calculate the expected value, variance, and standard deviation of the random variable X with the PMF in Problem 4.3.

Problem 4.18 Calculate the expected value, variance, and standard deviation of the random variable X with the PMF in Problem 4.4.

Problem 4.19 In R, the `rbinom(n,1,p)` function will generate n observations from a Bernoulli distribution with probability of success p: $x_1, x_2, \ldots, x_n$. By generating random observations from the Bernoulli distribution in R, for some $0 < p < 1$ of your choice, empirically illustrate that, as $n \to \infty$, $\bar{x} \to \mathbb{E}(X) = p$.

Of course, you cannot really let the sample size n go to infinity, but you can let it get really large, and so what you are empirically showing is that as the sample size n gets really large, the sample mean gets really close to p, the expected value of a Bernoulli distribution. Summarize your results with a line graph of n versus $\bar{x}$.

Problem 4.20 Returning to Problem 4.19, note that the plot is for only one sequence of $(n, \bar{x})$ pairs. If you repeat the experiment, you will get a different sequence, so there is the possibility that the sequence you observed just happened to look like it converged. So, another way to convincingly demonstrate that the sample mean converges is to show that the variance of the difference between the sample means and $\mathbb{E}(X) = p$ gets vanishingly small as n gets big. To do this, for each value of n, generate m random sets of data and for each set of data calculate the sample mean, $\bar{x}_i, i = 1, \ldots, m$. Now, using the `var()` function in R, calculate the sample variance of $\bar{x}_i - p, i = 1, \ldots, m$ and show that as n gets large the variance gets very small. Summarize your results with a line graph of n versus the sample variance. To do these calculations, create a function that calculates the desired variance and embed that in a loop.

Problem 4.21 Let X have PDF

$$f(x) = \begin{cases} 1 - x/2, & 0 \le x \le 2 \\ 0, & \text{otherwise.} \end{cases}$$

Sketch the PDF and show that the PDF satisfies the properties of Theorem 4.3.1.

Problem 4.22 Let X have a uniform PDF on the interval $[-1, 1]$. Write out the PDF and show that it satisfies the properties of Theorem 4.3.1.

Problem 4.23 Let X have PDF

$$f(x) = \begin{cases} \dfrac{1}{2}(1 - x^3), & -1 \le x \le 1 \\ 0, & \text{otherwise.} \end{cases}$$

Sketch the PDF and show that the PDF satisfies the properties of Theorem 4.3.1.

Problem 4.24 Let X have the PDF in Problem 4.21. Find the following:

(a) $\Pr(X > 0)$ and $\Pr(X \ge 0)$
(b) $\Pr(X > 1)$ and $\Pr(X \ge 1)$

Problem 4.25 Let X have the PDF in Problem 4.22. Find:

(a) $\Pr(X < 0)$ and $\Pr(X \le 0)$
(b) $\Pr(X > -1)$ and $\Pr(X \ge -1)$
(c) $\Pr(X > 1/2)$ and $\Pr(X = 1/2)$

Problem 4.26 Let X have the PDF in Problem 4.23. Find:

(a) $\Pr(X < 0)$ and $\Pr(X \le 0)$
(b) $\Pr(X > 1/2)$ and $\Pr(X = 1/2)$

Problem 4.27 Calculate the CDF from the PDF in Problem 4.21. Write out the expression for $F(x)$ and plot the PDF and CDF.

Problem 4.28 Calculate the CDF from the PDF in Problem 4.22. Write out the expression for $F(x)$ and plot the PDF and CDF.

Problem 4.29 Calculate the CDF from the PDF in Problem 4.23. Write out the expression for $F(x)$ and plot the PDF and CDF.

Problem 4.30 Suppose the random variable X takes on the values $0, 1, 2, \ldots, 100$, each with probability $1/101$. Write an expression for the CDF for all real numbers. (*Hint:* You may need to use the floor function $\lfloor \cdot \rfloor$.)

Problem 4.31 Calculate the expected value, variance, and standard deviation of the random variable X with the PDF in Problem 4.22.

Problem 4.32 Calculate the expected value, variance, and standard deviation of the random variable X with the PDF in Problem 4.23.

Problem 4.33 Using the CDF from Problem 4.27, find the probabilities in Problem 4.24.

Problem 4.34 Using the CDF plotted from Problem 4.28, find the probabilities in Problem 4.25.

Problem 4.35 Using the CDF plotted in Problem 4.29, find the probabilities specified in Problem 4.26.

Problem 4.36 In R, the `runif(n)` function will generate n observations from a standard uniform distribution: $x_1, x_2, \ldots, x_n$. By generating random observations from the uniform distribution on $[0,1]$, empirically illustrate that as $n \to \infty$, $\bar{x} \to \mathbb{E}(X) = 1/2$. Of course, you cannot really let the sample size n go to infinity, but you can let it get really large, and so what you are empirically showing is that as the sample size n gets really large, the sample mean gets really close to $1/2$, the expected value of a standard uniform distribution. Summarize your results with a line graph of n versus $\bar{x}$.

Problem 4.37 Note that in Problem 4.36 the plot is for only one sequence of $(n, \bar{x})$ pairs. If you repeated the experiment, you would have gotten a different sequence, so there is the possibility that the sequence you observed just happened to look like it converged. Another way to demonstrate that the sample mean converges is to show that the variance of the difference between the sample means and $\mathbb{E}(X) = 1/2$ gets vanishingly small as n gets big. To do this, for each value of n, generate m random sets of data and for each set of data calculate the sample mean, $\bar{x}_i, i = 1, \ldots, m$. Now, using the `var()` function in R, calculate the sample variance of $\bar{x}_i - 1/2, i = 1, \ldots, m$ and show that as n gets large the variance gets very small. Summarize your results with a line graph of n versus the sample variance. To do these calculations, it will most likely be helpful to create a function that calculates the desired variance and embed that in a loop.

Problem 4.38 For a *continuous* random variable X and constant c, prove that $\mathbb{E}(cX) = c\mathbb{E}(X)$.

Problem 4.39 For a continuous random variable X and constants a and b, prove that $\mathbb{E}(aX + b) = a\mathbb{E}(X) + b$.

Problem 4.40 For a continuous random variable X and for functions g and h, what can you say about $\mathbb{E}(g(X) + h(X))$? What if there are three functions, g_1, g_2, g_3? How does this problem relate to Theorem 4.4.3?

Problem 4.41 Suppose X has a distribution with mean $\mu = 10$ and standard deviation $\sigma = 2$. Find the following:

(a) $\mathbb{E}(5X+3)$
(b) $\mathbb{V}(X+10)$
(c) $\mathbb{V}(10X)$
(d) $\mathbb{V}(1-2X)$

Problem 4.42 Suppose X has the PMF given in Problem 4.1. Find the following:

(a) $\mathbb{E}(X-1)$
(b) $\mathbb{V}(X-1)$
(c) $\mathbb{V}(\frac{1}{2}x)$
(d) $\mathbb{V}(1-X)$

Problem 4.43 Suppose that X and Y are random variables with $\mathbb{E}(X)=12$ and $\mathbb{E}(Y)=8$.

(a) Find $\mathbb{E}(X-Y)$.
(b) Find $\mathbb{E}(5X-6Y)$.
(c) Is it possible to determine $\mathbb{E}(X^2)$ with the given information? Explain.

Problem 4.44 Suppose that X and Y are random variables with $\mathbb{E}(X)=0$, $\mathbb{E}(Y)=0$, $\mathbb{V}(X)=1$, and $\mathbb{V}(Y)=4$.

(a) Find $\mathbb{E}(1-X-Y)$.
(b) Find $\mathbb{E}(5X-6Y)$.
(c) Is it possible to determine $\mathbb{E}(XY)$ with the given information? Explain.

Problem 4.45 Prove parts (b) through (f) of Theorem 4.5.4.

Problem 4.46 Prove all three parts of Theorem 4.5.5 for the case of $n=3$ and $m=3$.

Problem 4.47 Suppose X and Y have the joint PMF given in Table 4.3.

Table 4.3 Joint PMF for Problem 4.47.

		y		
		-1	0	2
	-1	0.0	0.1	0.0
x	0	0.1	0.6	0.1
	1	0.0	0.1	0.0

(a) Find the marginal PMF of X.
(b) Find the marginal PMF of Y.
(c) Find $\mathbb{E}(X)$.
(d) Find $\mathbb{V}(X)$.
(e) Find $\mathbb{E}(Y)$.
(f) Find $\mathbb{V}(Y)$.
(g) Find $\text{cov}(X,Y)$.
(h) Find ρ_{XY}.
(i) Are X and Y independent?

Problem 4.48 Suppose X and Y have the joint PMF given in Table 4.4.

Table 4.4 Joint PMF for Problem 4.48.

		y			
		0	1	2	3
	0	0.04	0.04	0.04	0.04
x	1	0.20	0.10	0.10	0.04
	2	0.24	0.10	0.06	0

(a) Find the marginal PMF of X.
(b) Find the marginal PMF of Y.
(c) Find $\mathbb{E}(X)$.
(d) Find $\mathbb{V}(X)$.
(e) Find $\mathbb{E}(Y)$.
(f) Find $\mathbb{V}(Y)$.
(g) Before you do any calculations, do you believe the covariance is positive or negative?
(h) Find $\text{cov}(X,Y)$.
(i) Find ρ_{XY}.
(j) Are X and Y independent?

Problem 4.49 Suppose that X_1 and X_2 are discrete random variables taking on the ordered pairs $(-1,1)$, $(0,0)$, and $(1,1)$, each with probability $1/3$. The support consists of the three dots in the following figure. Find $\text{cov}(X_1,X_2)$ and ρ_{XY}. Are these random variables independent?

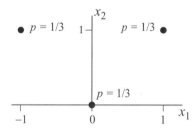

Problem 4.50 Given the results of Equation (4.4) and Problem 4.49, what can you say about the two random variables X and Y? Specifically, what can you say about the following two statements? Are they equivalent? Does one imply the other?

(a) Random variables X and Y are independent.
(b) The correlation between the random variables X and Y is 0.

Problem 4.51 Suppose that X and Y have the joint PDF

$$f(x,y) = \begin{cases} k, & 0 \le x \le y \le 1 \\ 0, & \text{otherwise.} \end{cases}$$

(a) Find k.

(b) Find $\Pr(X < Y)$.

(c) Find $\Pr\left(X < \frac{1}{2}\right)$.

(d) Find the marginal of X.

(e) Find the marginal of Y.

(f) Are X and Y independent?

Problem 4.52 Suppose that X and Y have the joint PDF

$$f(x,y) = \begin{cases} k(x+y), & 0 \le x \le 1; 0 \le y \le 1 \\ 0, & \text{otherwise.} \end{cases}$$

(a) Find k.

(b) Find $\Pr(X < Y)$.

(c) Find $\Pr\left(X < \frac{1}{2}\right)$.

(d) Find the marginal of X.

(e) Find the marginal of Y.

(f) Are X and Y independent?

Problem 4.53 Suppose that X and Y have the joint PDF

$$f(x,y) = \begin{cases} k(x+y), & 0 \le x \le y \le 1 \\ 0, & \text{otherwise.} \end{cases}$$

(a) Find k.

(b) Find $\Pr(X < Y)$.

(c) Find $\Pr\left(X < \frac{1}{2}\right)$.

(d) Find the marginal of X.

(e) Find the marginal of Y.

(f) Find $\mathbb{E}(X)$.

(g) Find $\mathbb{E}(Y)$.

(h) Find $\mathbb{E}(XY)$.

(i) Find $\text{cov}(X,Y)$.

Problem 4.54 Suppose that X and Y have the joint PDF

$$f(x,y) = \begin{cases} ke^{-x^2}, & 0 \le y \le x < \infty \\ 0, & \text{otherwise.} \end{cases}$$

(a) Find k.

(b) Find $\Pr(X < 2)$.

(c) Find the marginal of X.

(d) Are X and Y independent? Explain.

Problem 4.55 Prove a result analogous to Theorem 4.6.1 that says X and Y are independent if and only if $f(x|y) = f_1(x)$.

Problem 4.56 Suppose X and Y are continuous random variables with joint PDF $f(x,y)$ and marginals $f_1(x)$ and $f_2(y)$. If (x_0, y_0) is an ordered pair for which $f_1(x_0) > 0$ and $f_2(y_0) > 0$, but $f(x_0, y_0) = 0$, could X and Y be independent? Use the reasoning in this problem to infer that the random variables in Problem 4.53 are not independent.

Problem 4.57 In Example 4.30 we asserted that

$$\lim_{y \to \infty} y e^{-y} = 0.$$

(a) Prove this result using L'Hopital's rule.

(b) Prove that

$$\lim_{y \to \infty} y^2 e^{-y} = 0.$$

(c) What do you think is the value of

$$\lim_{y \to \infty} y^n e^{-y}?$$

Problem 4.58 Refer to the integrals in Example 4.30.

(a) Perform integration by parts twice to evaluate the integral in Equation (4.8).

(b) Perform integration by parts three times to evaluate the integral in Equation (4.9).

Problem 4.59 Suppose that the random variables in Theorem 4.4.4, $X_1, X_2, \ldots X_k$, are discrete with PMF $f(x_1, x_2, \ldots, x_k)$. Show that

$$\mathbb{E}\left(\sum_{i=1}^{k} X_i\right) = \sum_{i=1}^{k} \mathbb{E}(X_i)$$

(a) for $k = 2$

(b) for an arbitrary k.

Problem 4.60 Suppose that the random variables in Theorem 4.4.4, $X_1, X_2, \ldots X_k$, are continuous with PDF $f(x_1, x_2, \ldots, x_k)$. Show that

$$\mathbb{E}\left(\sum_{i=1}^{k} X_i\right) = \sum_{i=1}^{k} \mathbb{E}(X_i)$$

(a) for $k = 2$

(b) for an arbitrary k.

Problem 4.61 Consider the joint PMF given in Problem 4.47 and Table 4.3.

(a) Find the conditional PMF of Y given $x = -1$.

(b) Find the conditional PMF of Y given $x = 0$.

(c) Find the conditional PMF of Y given $x = 1$.

Problem 4.62 Consider the joint PMF given in Problem 4.48 and Table 4.4.

(a) Find the conditional PMF of X given $y = 0$.

(b) Find the conditional PMF of X given $y = 1$.

(c) Find the conditional PMF of X given $y = 2$.

(d) Find the conditional PMF of X given $y = 3$.

Problem 4.63 Consider the joint PDF in Problem 4.52.

(a) Find $f(y|x)$.

(b) Find $\mathbb{E}(Y|x)$.

(c) Find $f(x|y)$.

(d) Find $\mathbb{E}(X|y)$.

Problem 4.64 Consider the joint PDF in Problem 4.53.

(a) Find $f(y|x)$.

(b) Find $\mathbb{E}(Y|x)$.

(c) Find $f(x|y)$.

(d) Find $\mathbb{E}(X|y)$.

Problem 4.65 Suppose X_1, X_2, X_3, and X_4 have the joint PMF $f(x_1, x_2, x_3, x_4)$. Give a definition for the following:

(a) the marginal of X_3,

(b) the joint marginal of X_1, X_2, and X_4,

(c) the joint marginal of X_2 and X_4,

(d) the joint marginal of X_1 and X_3,

(e) the conditional $f(x_1, x_2, x_3|x_4)$,

(f) the conditional $f(x_3|x_1, x_2, x_4)$,

(g) the conditional $f(x_1|x_3)$. (*Hint:* First define the joint marginal of (X_1, X_3).)

Problem 4.66 Suppose X_1, X_2, X_3, and X_4 have the joint PDF $f(x_1, x_2, x_3, x_4)$. Give a definition for the quantities in Problem 4.65 parts (a)–(g).

Problem 4.67 Suppose that a manufacturing process employs one of three production workers, who have varying rates of producing defective units. Worker 1 produces 5% defective, worker 2 produces 8% defective, and worker 3 produces 20% defective. Each day one of the three workers is assigned (at random with equal probability) to the process. At the end of the day, three items are selected and tested. Let Y denote the number of defectives in this sample of three, and let D denote the defect rate for the selected worker.

(a) Find the PMF of Y.

(b) Find $\mathbb{E}(Y|D = d)$.

(c) Find $\mathbb{V}(Y|D = d)$.

(d) Find $\mathbb{E}(Y)$.

(e) Find $\mathbb{V}(Y)$.

Problem 4.68 Show that for any random variable X (discrete or continuous) with CDF $F(x)$,

$$\Pr(a < X \le b) = F(b) - F(a).$$

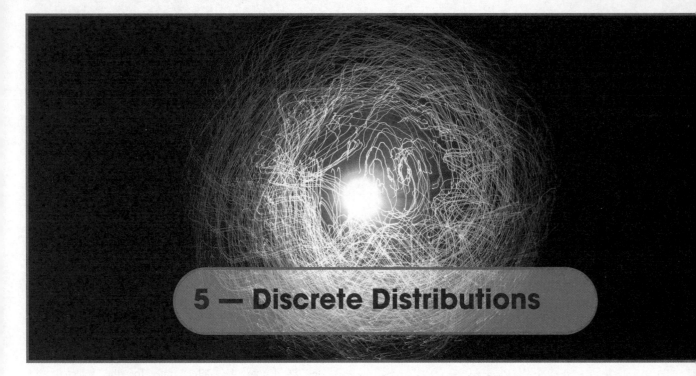

5 — Discrete Distributions

5.1 Introduction

In Chapter 3 we learned about the fundamental ideas of probability, and in Chapter 4 we generalized the notion of probability from working with sets to working with random variables and distributions. In many ways, random variables and their associated distributions can simplify probability calculations and, appropriately applied, are useful models for real-world phenomena. There are certain situations that occur in practice that deserve special attention. In these cases, the random variable will have a distribution that has a known and well-studied form. The key is to recognize the random mechanism that led to our random variable and identify the distribution. For example, a clinical trial of 100 people, each of whom is either cured or not cured, is the same mechanism as tossing 100 times a (possibly) biased coin. Recognizing that these two situations are the same is important in identifying the random mechanism. Often, we would be interested in the number cured, or the number of heads. Thus, the same distribution would be applied to the two problems.

We've already developed one discrete distribution; in Chapter 4 we developed the Bernoulli distribution to model binary discrete events, where it allowed us to model phenomena like flipping a coin. In this chapter we will learn about six more discrete distributions: (1) binomial, (2) geometric, (3) negative binomial, (4) hypergeometric, (5) Poisson, and (6) multinomial. There are many more, but these will suffice for our purposes.

As with the Bernoulli distribution, these new discrete distributions are useful for modeling various types of processes and phenomena that are encountered in the real world. The beauty of these distributions, once you know them, is that you can skip all the derivation work and simply apply them to solving the problem at hand.

5.2 Binomial Distribution

We begin with the idea of Bernoulli trials, which plays a central role in the development of the binomial distribution.

Definition 5.2.1 — Bernoulli Trials. *Bernoulli trials* have the following five characteristics:

1. The experiment consists of exactly *n* identical trials.

2. Each trial can have only one of two outcomes, typically called "success" or "failure."
3. The probability of success (p) is constant for every trial.
4. The trials are independent.
5. The random variable of interest (X) is the number of successes out of the n trials.

For example, imagine flipping a coin n times, where the random variable X is the number of heads observed out of the n flips. Here we define "success" as flipping a head and "failure" as flipping a tail. As long as the probability of getting a head is constant for each flip, with probability p, and trials are independent, then the number of successes X has a binomial distribution.

Definition 5.2.2 — Binomial Distribution. The probability mass function (PMF) of a binomial random variable X with parameters n and p is

$$\Pr(X = x) \;=\; p(x) \;=\; \binom{n}{x} p^x (1-p)^{n-x}, \quad x = 0, 1, 2, \ldots, n,$$

where $\binom{n}{x} = n!/(n-x)!x!$ is the number of ways x objects can be chosen from n objects. Note that $0!$ is defined to be 1.

Parameters and support: The number of trials, n, is a positive, nonzero integer and p is the probability of "success," $0 \le p \le 1$. Except for the trivial cases of $p = 0$ and $p = 1$, the PMF is nonzero on the integers from 0 to n, inclusive: $x = 0, 1, 2, \ldots, n$.

Notation: $X \sim \mathrm{BIN}(n,p)$, which stands for "The random variable X has a binomial distribution with parameters n and p."

Where does the PMF for the binomial distribution come from? Imagine a sequence of n trials, where each trial can only have one of two outcomes ("S" or "F"). Here's one possible sequence:

$$\underbrace{FFFSFSSFFF\ldots SF}_{n \text{ trials}}.$$

Within this sequence, there are x successes and $n - x$ failures. Rearranging the order of the outcomes, we can display the results as:

$$\underbrace{SSSSSS\ldots SSS}_{x \text{ successes}} \underbrace{FFFFF\ldots FF}_{n-x \text{ failures}}.$$

Now, if each of the trials is independent of the others and has probability p of a success and probability $1 - p$ of a failure, then by the multiplicative rule, the probability of this sequence of outcomes is:

$$\underbrace{p \times p \times p \cdots p \times p}_{x \text{ terms}} \times \underbrace{(1-p) \times (1-p) \cdots (1-p) \times (1-p)}_{n-x \text{ terms}} \;=\; p^x(1-p)^{n-x}.$$

But the above sequence is just one way that we can observe x successes and $n - x$ failures in a sequence of n trials. Since there are $\binom{n}{x}$ ways to observe *where* among the n trials the x successes occur, the probability of observing x successes out of n trials is

$$\Pr(X = x) \;=\; \binom{n}{x} p^x (1-p)^{n-x}, \qquad x = 0, 1, 2, \ldots, n.$$

Note that for $n = 1$ the binomial distribution reduces to the Bernoulli distribution:

$$\Pr(X = x) \;=\; \binom{1}{x} p^x (1-p)^{1-x}, \qquad x = 0, 1.$$

■ **Example 5.1** Imagine that a fair coin is flipped four times, each flip being independent of the others. What is the probability of obtaining two heads and two tails on the four flips?

Solution: Let X denote the number of heads that occur on four flips of the coin. Because the flips (trials) are identical and independent with only two outcomes, heads (success) and tails (failure), with constant probability of heads ($p = 1/2$), X has a binomial distribution with parameters $n = 4$ and $p = 1/2$. By Definition 5.2.2,

$$\Pr(X = 2) = \binom{4}{2}\left(\frac{1}{2}\right)^2\left(\frac{1}{2}\right)^{4-2} = \frac{4\cdot3\cdot2\cdot1}{2\cdot1\cdot2\cdot1}\left(\frac{1}{2}\right)^4 = \frac{3}{8}.$$

To verify this, write out all the possible sequences of four flips of a fair coin:

Sequence	# Heads	Sequence	# Heads	Sequence	# Heads	Sequence	# Heads
TTTT	0	HTTT	1	THHT	2	HTHH	3
TTTH	1	TTHH	2	HTHT	2	HHTH	3
TTHT	1	THTH	2	HHTT	2	HHHT	3
THTT	1	HTTH	2	THHH	3	HHHH	4

There are $2\times2\times2\times2 = 16$ possible sets of outcomes, six of which have two heads and two tails. ■■■

> **Exercise 5.1** Continuing with Example 5.1, what is the probability of getting three heads and one tail in four flips of the coin? And, what is the probability of getting all heads on the four flips?
>
> *Answers*: $\Pr(X = 3) = 1/4$, $\Pr(X = 4) = 1/16$.
>
> Without doing any further calculations, can you guess what the probability is of getting all tails?

To get an idea of what the binomial distribution looks like, Figure 5.1 shows a variety of binomial PMFs for $n = 10, 20$ and $p = 0.1, 0.5, 0.8$. Here we see that the x-axis ranges from 0 to n, where $p(x) > 0$ for the integer values. However, we also see that for some values of x, $p(x)$ is very small. We also see that for $p = 0.5$ the distributions are symmetric, but for other values of p it is skewed. For example, consider the plot (a) for a binomial experiment with $n = 10$ and $p = 0.1$. The PMF shows that you are very likely under these conditions to observe 0, 1, or 2 successes and very unlikely to observe five or more successes. This should make sense, given that on any particular trial you only have a 10% chance of a success.

By comparison, plot (e) shows that for a binomial experiment with $n = 10$ and $p = 0.8$ you are very unlikely to see four or fewer successes. This too should make sense since on each trial you have an 80% chance of getting a success. At the most extreme, the chance of seeing no successes is $0.2^{10} = 0.0000001024$ – quite small and thus very unlikely.

R is handy for calculating probabilities from binomial (and many other) distributions. For a binomial random variable with n trials and probability of success p, you can calculate the values of the PMF and cumulative distribution function (CDF) at x using the following syntax:

$p(x) \equiv \Pr(X = x)$: `dbinom(x, n, p)`
$F(x) \equiv \Pr(X \le x)$: `pbinom(x, n, p)`

For the binomial distribution, the CDF is

$$F(x) = \begin{cases} 0, & x < 0 \\ \sum_{z=0}^{x}\binom{n}{z}p^z(1-p)^{n-z}, & x = 0, 1, \ldots, n \\ 1, & x > n. \end{cases}$$

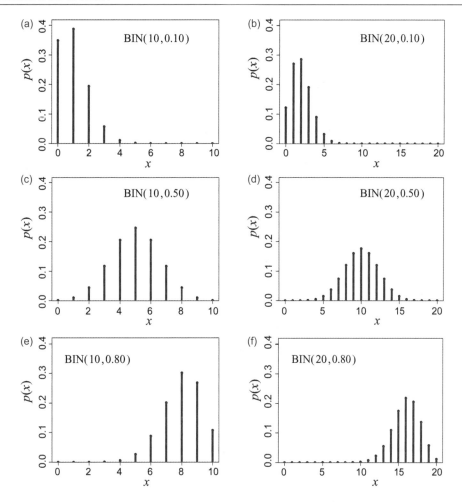

Figure 5.1 Illustration of six binomial PMFs: $n = 10$ (first column), $n = 20$ (second column), and $p = 0.10$, $p = 0.50$, and $p = 0.80$ for the three rows.

This defines F only for integral values of x, but of course $F(x)$ is defined for all real numbers x. We can write $F(x)$ for all x using the floor function, written as $\lfloor x \rfloor$, that truncates x to the nearest integer less than or equal to x. The CDF is mostly flat, but takes jumps at each possible outcome $x = 0, 1, \ldots, n$. The height of the jump at x is equal to the probability of getting exactly x successes. For example, if $X \sim \mathrm{BIN}(4, \frac{1}{2})$, then

$$F(3) = \Pr(X \le 3) = \frac{1}{16} + \frac{4}{16} + \frac{6}{16} + \frac{4}{16} = \frac{15}{16}.$$

But also,

$$F(3.5) = F(3.9) = F(3.9999) = \frac{15}{16}.$$

The CDF doesn't take its last jump until $x = 4$, when it jumps from $15/16$ to $\Pr(X \le 4) = 1$. In general, the CDF of the binomial distribution can then be written for all x as

$$F(x) = \begin{cases} 0, & x < 0 \\ \sum_{z=0}^{\lfloor x \rfloor} \binom{n}{z} p^z (1-p)^{n-z}, & 0 \le x \le n \\ 1, & x > n. \end{cases}$$

■ **Example 5.2** Use R to find $p(2)$ from Example 5.1, as well as $F(2)$.

Solution: The R code

```
dbinom(2,4,0.5)
pbinom(2,4,0.5)
dbinom(0,4,0.5)+dbinom(1,4,0.5)+dbinom(2,4,0.5)
```

yields

```
[1] 0.375
[1] 0.6875
[1] 0.6875
```

■ ■ ■

We can also use R to generate random observations from a binomial distribution with n trials and probability p of success using the `rbinom()` function. For example, `rbinom(48,100,0.5)` performs 48 simulations of the number of heads observed on 100 flips of a fair coin and yields

```
 [1] 55 52 50 44 41 43 53 48 45 48 52 45 46 45 49 46 47 57 54 48 53 51 44 53
[25] 52 40 50 41 51 44 55 55 48 50 47 47 51 52 59 54 43 42 47 44 45 46 42 53
```

Here we see that in the 48 flips the most extreme observations are 40 heads and 59 heads.

> **Exercise 5.2** Use R to check your solutions to Exercise 5.1.

■ **Example 5.3** Let X denote the number of heads (successes) out of 100 flips (trials) of a fair coin. What are the probabilities associated with the following events:
  (a) Observe exactly 40 heads: $\Pr(X = 40) = p(40)$,
  (b) Observe exactly 59 heads: $\Pr(X = 59) = p(59)$,
  (c) Observe 40 or fewer heads: $\Pr(X \le 40) = F(40)$,
  (d) Observe at least 59 heads: $\Pr(X \ge 59) = 1 - F(58)$,
where, of course, the equivalence in the last part comes from the complement rule.

Solution: We can obtain these probabilities using the `dbinom()` and `pbinom()` functions in R:

```
dbinom(40,100,0.5)
dbinom(59,100,0.5)
pbinom(40,100,0.5)
1-pbinom(58,100,0.5)
pbinom(58,100,0.5,lower.tail=FALSE)
```

This produces

```
[1] 0.01084387
[1] 0.01586907
[1] 0.02844397
[1] 0.04431304
[1] 0.04431304
```

■ ■ ■

Exercise 5.3 Let $X \sim \text{BIN}(1000, 0.17)$. Use R to find $\Pr(X = 170)$, $\Pr(X \leq 170)$, and $\Pr(150 \leq X \leq 190)$.

Answers: $\Pr(X = 170) \approx 0.0336$, $\Pr(X \leq 170) \approx 0.5205$, and $\Pr(150 \leq X \leq 190) \approx 0.9562$.

Theorem 5.2.1 — Expected Value and Variance of a Binomial Random Variable. The expected value of a binomial random variable is $\mathbb{E}(X) = np$ and the variance is $\mathbb{V}(X) = np(1-p)$.

Proof. We will start by proving the expected value result. By Definitions 4.2.6 and 5.2.2 we have

$$
\begin{aligned}
\mathbb{E}(X) &= \sum_{x=0}^{n} x \binom{n}{x} p^x (1-p)^{n-x} \\
&= \sum_{x=1}^{n} x \frac{n!}{x!(n-x)!} p^x (1-p)^{n-x} \\
&= \sum_{x=1}^{n} \frac{n!}{(x-1)!(n-x)!} p^x (1-p)^{n-x},
\end{aligned}
$$

where the second line follows from the first because the first term in the sum is zero and the third line follows by canceling terms.

Now, note that the last expression still looks much like a binomial and here we'll apply a technique that is often useful for solving these types of problems: We're going to manipulate the terms in the expression until we get the sum of a binomial PMF over its entire range of values (which we know is equal to 1). That is, we're first going to factor np out of each term and then do a substitution of variables, $z = x - 1$:

$$
\begin{aligned}
\mathbb{E}(X) &= np \sum_{x=1}^{n} \frac{(n-1)!}{(x-1)!(n-x)!} p^{x-1} (1-p)^{n-x} \\
&= np \sum_{z=0}^{n-1} \frac{(n-1)!}{z!(n-1-z)!} p^z (1-p)^{n-1-z} \\
&= np \underbrace{\sum_{z=0}^{n-1} \binom{n-1}{z} p^z (1-p)^{(n-1)-z}}_{\text{sum is } 1}.
\end{aligned}
$$

In the last line, the sum is over all possible outcomes of a $\text{BIN}(n-1, p)$ distribution; thus, the sum is 1. Therefore, $\mathbb{E}(X) = np \times 1 = np$.

Now, for the variance, one might be tempted to evaluate $\mathbb{E}(X^2)$ following an approach similar to the one we just used for the expected value, in order to use the variance shortcut from Definition 4.2.2. Unfortunately, directly evaluating $\mathbb{E}(X^2)$ won't work, but notice that

$$
\mathbb{E}(X^2) = \mathbb{E}(X(X-1)) + \mathbb{E}(X).
$$

So, let's evaluate $\mathbb{E}(X(X-1))$, from which we can then calculate $\mathbb{E}(X^2)$:

$$
\begin{aligned}
\mathbb{E}\Big(X(X-1)\Big) &= \sum_{x=0}^{n} x(x-1) \binom{n}{x} p^x (1-p)^{n-x} \\
&= \sum_{x=2}^{n} \frac{x(x-1)n!}{x!(n-x)!} p^x (1-p)^{n-x}
\end{aligned}
$$

$$= \sum_{x=2}^{n} \frac{n!}{(x-2)!(n-x)!} p^x (1-p)^{n-x}$$

$$= n(n-1)p^2 \sum_{x=2}^{n} \frac{(n-2)!}{(x-2)!(n-x)!} p^{x-2}(1-p)^{n-x}$$

$$= n(n-1)p^2 \sum_{z=0}^{n-2} \frac{(n-2)!}{z!(n-z-2)!} p^z (1-p)^{n-z-2}$$

$$= n(n-1)p^2 \underbrace{\sum_{z=0}^{n-2} \binom{n-2}{z} p^z (1-p)^{(n-2)-z}}_{\text{sum is } 1}$$

$$= n(n-1)p^2,$$

where the change in summation indices in the second line follows because the first two terms in the summation are zero; the third line follows by canceling the two terms in the numerator of the fraction with the first two terms of the factorial in the denominator; in the fourth line we pull the n, $(n-1)$, and p^2 factors out in front of the summation; and in the fifth line we apply a change of variables ($z = x - 2$), which gives us in the six line another binomial summed over the range of its support, so it yields a sum that is 1. The variance is therefore

$$\mathbb{V}(X) = \mathbb{E}(X^2) - [\mathbb{E}(X)]^2$$
$$= \mathbb{E}(X(X-1)) + \mathbb{E}(X) - [\mathbb{E}(X)]^2$$
$$= n(n-1)p^2 + np - [np]^2$$
$$= n^2p^2 - np^2 + np - n^2p^2$$
$$= np(1-p).$$

The formula for the expected value should be intuitive. The product np gives the proportion p of the sample size n. In a sequence of n trials, each with probability p of success, wouldn't we expect a fraction p of the n trials to be successes? For example, if we toss a fair coin 100 times, we would expect on average that half, that is, $100 \times \frac{1}{2} = 50$, of them will be heads. The formula for the variance is less intuitive, but two special cases can be described, and in these two cases the results match our intuition. If $p = 0$, then a success is impossible and every sequence of n trials will result in $X = 0$ successes. Since $X = 0$ is the only possible outcome, X will never vary, so its variance will be 0. This matches the formula $\mathbb{V}(X) = np(1-p) = n \times 0 \times (1-0) = 0$. The other special case involves $p = 1$. Here the event is *sure* so every trial will result in a success, and every sequence of n trials will result in n successes. Thus, $X = n$ *always*! As a result, the variance will be 0 and the formula agrees: $\mathbb{V}(X) = np(1-p) = n \times 1 \times (1-1) = 0$.

■ **Example 5.4** Returning to Example 5.1, what are the expected value, variance, and standard deviation of X?

Solution: $\mathbb{E}(X) = np = 4 \times 1/2 = 2$, so if this experiment were conducted a large number of times, on average, two heads will be observed. The variance is $\mathbb{V}(X) = np(1-p) = 4 \times 1/2 \times 1/2 = 1$. So, the variance is 1 and thus $\sigma = 1$. ■■■

Exercise 5.4 For $X \sim \mathrm{BIN}(1000, 0.17)$ in Exercise 5.3, what is the expected value, variance, and standard deviation of X?

Answers: $\mathbb{E}(X) = 170$, $\mathbb{V}(X) = 141.1$, and so $\sigma_X \approx 11.879$.

■ **Example 5.5** Suppose that a computer virus has infected 1% of all computers in a company. The IT department randomly selects 20 computers and tests for the virus.

(a) What is the expected number of infected computers they will find?
(b) What is the standard deviation?
(c) What is the chance they will find no infected computers?
(d) What is the chance they will find exactly one infected computer?
(e) What is the chance they will find two or more infected computers?

Solution: The expected number of computers with the virus is $E(X) = np = 20 \times 0.01 = 0.2$. The standard deviation is $\sqrt{\mathbb{V}(X)} = \sqrt{np(1-p)} = \sqrt{20 \times 0.01 \times 0.99} \approx 0.445$. The R code

```
dbinom(0,20,0.01)
dbinom(1,20,0.01)
1-dbinom(0,20,0.01)-dbinom(1,20,0.01)
```

gives

```
[1] 0.8179069
[1] 0.1652337
[1] 0.01685934
```

There is an 81.8% chance they will not find any infected computers out of 20 tested; there's a 16.5% chance of finding exactly one infected computer and a 1.7% chance of finding two or more infected computers. The percentages computed above add up to 100%, as they should, since there are no other possible outcomes.   ■ ■ ■

■ **Example 5.6** Returning to the aircraft delay data introduced in Chapter 1, we find that the proportion of flights at the Monterey airport that were delayed was 0.29. Assuming the probability of delay is constant for every flight and flights are independent, for a random day with $n = 25$ flights and probability $p = 0.29$ that a flight will be delayed:[1]

(a) What is the expected number of delayed flights on a day with 25 flights?
(b) What is the probability that at least one flight will be delayed that day?
(c) What is the probability that the number of delayed flights will be between 4 and 11, inclusive?
(c) How do the results of the model compare to the actual data from May 2007?

Solution: For $X \sim \mathrm{BIN}(25, 0.29)$, the expected number of delayed flights is

$$\mathbb{E}(X) = np = 25 \times 0.29 = 7.25.$$

The probability that at least one flight will be delayed is found by

```
1-dbinom(0,25,0.29)
```

---

[1]These assumptions are unlikely to be perfectly met in practice since the probability of delay could vary by time of day, airline, and many other factors. Similarly, it's unlikely that flight delays are independent. Nonetheless, the binomial can be used as an *approximate* model for delays.

which gives

    [1] 0.9998088

and $\Pr(4 \leq X \leq 11) = \Pr(X \leq 11) - \Pr(X < 4) = F(11) - F(3)$ is

    pbinom(11,25,0.29)-pbinom(3,25,0.29)

    [1] 0.9238509

According to this model, on an average day in May 2007 we could expect slightly more than 7 flights out of 25 to be delayed; the chance that at least one flight will be delayed is nearly a certainty; and there's a better than 90% chance that the number of delays will be between 4 and 11 (inclusive). What do the actual data look like? The table below shows the number of delayed flights on the 10 days in May 2007 with exactly 25 flights:

Day:	14	16	17	18	20	21	23	25	30	31
# delayed flights:	9	5	9	5	9	7	4	8	7	12

So, the actual average number of delayed flights on these 10 days was 7.5, which is fairly close to the model's expected value, there were no days with no delayed flights, and 90% of the observed values were between 4 and 11 (inclusive). At least for these particular data, the binomial looks to be a reasonable model.   ■ ■ ■

In some situations we are given information (or can easily deduce information) about a conditional distribution. We then would like to determine an unconditional distribution. The next example illustrates this.

■ **Example 5.7** Suppose that a manufacturing process is highly dependent on the ambient humidity. On days when the relative humidity exceeds 40%, the defect rate is 10%. On days when the relative humidity is 40% or below, the defect rate is 5%. Half of the time, the humidity exceeds 40% and half of the time it is 40% or less. At the end of the day, we select $n = 10$ items for inspection. Let Y denote the number of defective items observed in this sample of 10 items. Find the PMF for Y.

Solution: It might seem as if Y has a binomial distribution, since Y is the number of "successes" (i.e., getting a defective item) on a fixed number of trials. This would be incorrect, however. The humidity must factor into the distribution. Let's define X to be

$$X = \begin{cases} 0, & \text{if humidity is 40\% or less} \\ 1, & \text{if humidity exceeds 40\%} \end{cases}$$

What we *can* say is this:

$$Y \,|\, (X = 0) \sim \text{BIN}(10, 0.05) \qquad \text{and} \qquad Y \,|\, (X = 1) \sim \text{BIN}(10, 0.10).$$

The joint PMF can then be given separately for $x = 0$ (low humidity) and $x = 1$ (high humidity):

$$\Pr(X = 0, Y = y) = \Pr(X = 0)\Pr(Y = y | X = 0) = \frac{1}{2}\binom{10}{y} 0.05^y (1 - 0.05)^{10-y}$$

$$\Pr(X = 1, Y = y) = \Pr(X = 1)\Pr(Y = y | X = 1) = \frac{1}{2}\binom{10}{y} 0.10^y (1 - 0.10)^{10-y}$$

for $y = 0, 1, \ldots, 10$. The marginal PMF for Y can then be found by summing across all values for x:

$$
\begin{aligned}
f_2(y) &= \Pr(Y = y) \\
&= \Pr(X = 0, Y = y) + \Pr(X = 1, Y = y) \\
&= \frac{1}{2}\binom{10}{y} 0.05^y (1 - 0.05)^{10-y} + \frac{1}{2}\binom{10}{y} 0.10^y (1 - 0.10)^{10-y}, \qquad y = 0, 1, \ldots, 10.
\end{aligned}
$$

We can compute this in R by running

```
fy = 0.5*dbinom(0:10,10,0.05) + 0.5*dbinom(0:10,10,0.10)
print(fy)
sum(fy)
```

which yields

```
> print(fy)
[1] 4.737077e-01 3.512726e-01 1.341725e-01 3.393534e-02 6.062535e-03
[6] 7.744850e-04 7.022680e-05 4.414189e-06 1.830432e-07 4.509277e-09
[11] 5.004883e-11
> sum(fy)
[1] 1
```

The 11 components of the array `fy` give the probabilities for $y = 0$ through $y = 10$. Note that the probabilities sum to 1, as they must. As we mentioned at the beginning, this is not a binomial distribution.
∎ ∎ ∎

5.3 Geometric Distribution

For our next discrete distribution, just like with the binomial, imagine a situation with a sequence of independent trials, where there are only two outcomes (success and failure), and the trials have constant probability of success p. However, unlike with the binomial, we're now interested in the number of trials X until we observe the first success. The number of trials is therefore *not* fixed – we just keep going, trial after trial, until we get a success. The random variable X is defined to be the number of trials needed to get one success. The distribution of X is called the geometric distribution.

> **Definition 5.3.1 — Geometric Experiment.** A *geometric experiment* has the following five characteristics:
>
> 1. The experiment consists of a sequence of identical trials.
> 2. Each trial can have only one of two outcomes, typically called "success" or "failure."
> 3. The probability of success p is constant for every trial.
> 4. The trials are independent.
> 5. The random variable X is the number of trials until the first success is observed.
>
> For any experiment or phenomenon that has these characteristics, X has a geometric distribution.

If the first success occurs on trial x, then the first $x - 1$ trials must be failures. Our sequence of S and F must therefore look like this:

$$
\underbrace{FFF \ldots FF}_{x-1 \text{ failures}} S.
$$

If each of the trials is independent of the others, and the probability is $1 - p$ for an F outcome and p of the final S outcome, then by the multiplicative rule the probability of this sequence of outcomes yields

$$\underbrace{(1-p) \times (1-p) \times (1-p) \cdots (1-p) \times (1-p)}_{x-1 \text{ times}} \times p = (1-p)^{x-1}p.$$

The PMF of the geometric distribution is given in the next definition.

> **Definition 5.3.2 — Geometric Distribution.** The PMF of a geometric random variable X with parameter p is
>
> $$\Pr(X = x) \equiv p(x) = (1-p)^{x-1}p, \quad x = 1, 2, 3, \ldots.$$
>
> *Parameters and support*: The parameter p is the probability of "success," $0 \le p \le 1$. The PMF is nonzero on the positive integers: $x = 1, 2, 3, \ldots$.
>
> *Notation*: $X \sim \text{GEO}(p)$, which stands for "The random variable X has a geometric distribution with parameter p."

For $X \sim \text{GEO}(p)$, you can calculate the values of the PMF and CDF at x in R, as well as generate random observations from a geometric distribution, using the following syntax:

$p(x) \equiv \Pr(X = x)$: `dgeom(x-1,p)`
$F(x) \equiv \Pr(X \le x)$: `pgeom(x-1,p)`
Generate n random variates: `rgeom(n,p)`

The reason for the $x - 1$ in the above expression is that R defines a geometric random variable as *the number of failures before the first success*. That is, it does not count the final success. So, to make the R functions consistent with Definition 5.3.2, the first argument in the `dgeom()` and `pgeom()` functions must be $x - 1$.

The CDF of the geometric distribution is

$$F(x) = \begin{cases} \sum_{z=1}^{\lfloor x \rfloor} (1-p)^{z-1}p, & x \ge 1 \\ 0 & x < 1, \end{cases}$$

where $\lfloor x \rfloor$ is the floor function.

Figure 5.2 shows a variety of geometric PMFs for $p = 0.1, 0.3, 0.6, 0.9$. Here we see that for small values of p (the probability of success) x can take on large values. This should make sense since the smaller p is, the more likely you will see many failures before the first success. On the other hand, for larger values of p, x is more likely to be small. This too should make sense according to the same logic. If p is large then it is relatively easy to observe a success and thus it will most likely occur after only a few failures. The most extreme case is when $p = 1$, then you are guaranteed to observe a success on the first flip and so X will always take on the value 1.

■ **Example 5.8** Imagine that a fair coin is sequentially flipped, each flip being independent of the others. What is the probability that the first head is observed on flip x of the coin, $x = 1, 2, 3, \ldots$?

Solution: Let X denote the number of coin flips until the first heads is observed. Because the flips (trials) are identical and independent with only two outcomes, heads (success) and tails (failure), with constant probability of heads ($p = 1/2$), X has a geometric distribution with parameter $p = 1/2$. By Definition 5.3.2,

$$\Pr(X = x) \equiv p(x) = \left(1 - \frac{1}{2}\right)^{x-1} \frac{1}{2} = \left(\frac{1}{2}\right)^x.$$

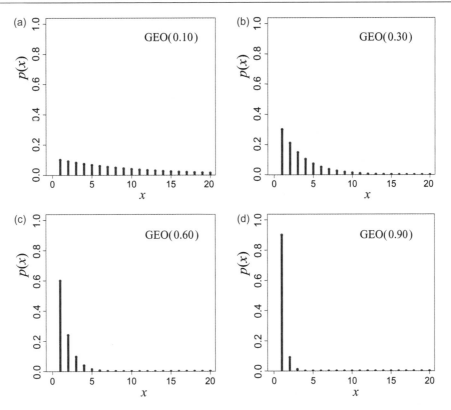

Figure 5.2 Illustration of four geometric PMFs: $p = 0.10$ (a), $p = 0.30$ (b), $p = 0.60$ (c) and $p = 0.9$ (d).

Note that the set of possible values for x is $\{1, 2, 3, \ldots\}$, a set that has infinitely many members. This is because there is no upper bound on how long it might take to get one success. The table below gives some values of $p(x)$ for various x:

x	1	2	3	4	5	$\cdots$
$p(x)$	0.5	0.25	0.125	0.0625	0.03125	$\cdots$

So, it's pretty hard to flip a coin and get five or more tails in a row before flipping the first heads:

$$\Pr(X \geq 6) = 1 - p(1) - p(2) - p(3) - p(4) - p(5) = 1 - 0.5 - 0.5^2 - 0.5^3 - 0.5^4 - 0.5^5 = 0.03125.$$

That's about three chances in 100.   ■ ■ ■

> **Exercise 5.5** Consider a biased coin, where the probability of flipping a head is 0.4. Letting X denote the number of flips until the first heads is observed, what is the probability of getting five or more tails in a row before you observe the first heads? (Calculate your answer manually and then check it using R.)
>
> *Answer*: $\Pr(X \geq 6) \approx 0.07776$.

■ **Example 5.9** Thirty percent of applicants for a data scientist position at a social media company have advanced R programming skills. The applicants, selected randomly from the pool, are interviewed sequentially. Find the probability that the first applicant with advanced R programming skills is found on or before the fourth interview.

Solution: Because the interview process is sequential and we're looking for the occurrence of the first "success," which is having an interview with a candidate who has advanced R programming skills, this is a geometric distribution problem. Thus, we have that $p = \mathrm{Pr}(\text{"success"}) = P(\text{"interview an applicant with advanced R programming skills"}) = 0.3$, and if we define X as the number of applicants interviewed up to and including the first applicant with advanced R programming skills, then we want to find $P(X \leq 4)$. This is

$$
\begin{aligned}
P(X \leq 4) &= p(1) + p(2) + p(3) + p(4) \\
&= q^0 p + q^1 p + q^2 p + q^3 p \\
&= 0.3 + 0.7 \times 0.3 + 0.7^2 \times 0.3 + 0.7^3 \times 0.3 \\
&= 0.7599.
\end{aligned}
$$

We see that the result matches the result from R:

```
pgeom(4-1,0.3)
```

```
[1] 0.7599
```

∎

Theorem 5.3.1 — Expected Value and Variance of a Geometric Random Variable.
The expected value of a geometric random variable is $\mathbb{E}(X) = 1/p$ and the variance is $\mathbb{V}(X) = (1-p)/p^2$.

Proof. We start by proving the expected value result. By Definitions 4.2.6 and 5.3.2, we have that

$$
\mathbb{E}(X) = \sum_{x=1}^{\infty} x(1-p)^{x-1} p = p \sum_{x=1}^{\infty} x(1-p)^{x-1} = p \sum_{x=1}^{\infty} x q^{x-1},
$$

where we've substituted $q = 1-p$. Note that for $x \geq 1$,

$$
\frac{d}{dq}(q^x) = x q^{x-1},
$$

so that we have

$$
\sum_{x=1}^{\infty} x q^{x-1} = \sum_{x=1}^{\infty} \frac{d}{dq}(q^x) = \frac{d}{dq}\left(\sum_{x=1}^{\infty} q^x\right),
$$

where interchanging the order of summation and differentiation can be justified here. Substituting the latter expression into the expected value gives

$$
\mathbb{E}(X) = p \sum_{x=1}^{\infty} x q^{x-1} = p \frac{d}{dq}\left(\sum_{x=1}^{\infty} q^x\right).
$$

Because the sum in the last expression is a geometric series,

$$
\sum_{x=1}^{\infty} q^x = \frac{q}{1-q}.
$$

Therefore,

$$\mathbb{E}(X) = p\frac{d}{dq}\left(\frac{q}{1-q}\right) = p\frac{(1-q)1-q(-1)}{(1-q)^2} = p\frac{1}{(1-q)^2} = \frac{p}{p^2} = \frac{1}{p}.$$

To derive the expression for the variance, not surprisingly, we will first evaluate $\mathbb{E}(X^2)$ and then use the shortcut formula. So,

$$\mathbb{E}(X^2) = \sum_{x=1}^{\infty} x^2(1-p)^{x-1}p = p\sum_{x=1}^{\infty} x^2(1-p)^{x-1} = p\sum_{x=1}^{\infty} x^2 q^{x-1}.$$

If we apply similar reasoning to that above, where we found $\sum_{i=1}^{\infty} xq^{x-1}$, we find that

$$\sum_{x=1}^{\infty} x^2 q^{x-1} = \frac{1+q}{(1-q)^3}.$$

Thus,

$$\mathbb{E}(X^2) = \frac{p(1+q)}{(1-q)^3} = \frac{p(1+1-p)}{p^3} = \frac{2-p}{p^2},$$

and so

$$\mathbb{V}(X) = E(X^2) - [E(x)]^2 = \frac{2-p}{p^2} - \left(\frac{1}{p}\right)^2 = \frac{2-p-1}{p^2} = \frac{1-p}{p^2}. \qquad \blacksquare$$

■ **Example 5.10** Returning to Example 5.9, find the expected number of applicants that will have to be interviewed in order to find the first one with advanced R programming skills. What is the standard deviation?

Solution: Since we know X has a geometric distribution, we know that $E(X) = 1/p$. Therefore, $E(X) = 1/0.3 \approx 3.33$. Hence, on average, the social media company will have to interview slightly more than three people to find the first one with advanced R programming skills. The standard deviation is $\sigma_X = \sqrt{(1-p)/p^2} = \sqrt{0.7/0.3^2} \approx 7.78$. Note that the expected value in this case is not an integer, nor does it necessarily have to be, because it represents a long-run average. Thus, while on any particular interview cycle there has to be an integer outcome – the company cannot interview a fraction of a person – the average over a large number of interview cycles can be a noninteger. ■ ■ ■

> **Exercise 5.6** Returning to Exercise 5.5, with $X \sim \text{GEO}(0.4)$, what is the expected value, variance, and standard deviation of X?
>
> *Answer:* $\mathbb{E}(X) = 2.5$, $\mathbb{V}(X) = 3.75$, and $\sigma_X \approx 1.94$.

■ **Example 5.11** Returning to the aircraft delay data from Example 5.6, where we estimated the probability of a delayed flight at the Monterey airport in May 2007 to be $p = 0.29$:

(a) What is the probability distribution for the number of flights until the first delayed flight of the day?

(b) What is the expected number of flights until the first delayed flight?

Solution: Let X be the first delayed flight of the day, where we will assume that probability of delay p is constant for every flight and flights are independent. The PMF for X is

$$p(x) = 0.29(1 - 0.29)^{x-1}, \qquad x = 1, 2, \ldots$$

which for the first 10 possible values is computed to be

x	1	2	3	4	5	6	7	8	9	10	$\cdots$
$p(x)$	0.290	0.206	0.146	0.104	0.074	0.052	0.037	0.026	0.019	0.013	$\cdots$

The expected value of the number of flights until the first delay is

$$\mathbb{E}(X) = \frac{1}{p} = \frac{1}{0.29} \approx 3.45. \qquad \blacksquare\blacksquare\blacksquare$$

5.4 Negative Binomial

With the geometric, the outcome of interest is the number of observations until the first success. However, sometimes the outcome of interest is the number of observations until the rth success, $r \geq 1$. So, as with the geometric distribution, imagine a situation with a sequence of independent trials, where there are only two outcomes (success and failure), and the trials have constant probability of success p. However, unlike with the geometric, we're now interested in the number of trials X until we observe the rth success. Then the random variable X has a negative binomial distribution.

> **Definition 5.4.1 — Negative Binomial Experiment.** A *negative binomial experiment* has the following five characteristics:
>
> 1. The experiment consists of a sequence of identical trials.
> 2. Each trial can have only one of two outcomes, typically called "success" or "failure."
> 3. The probability of success p is constant for every trial.
> 4. The trials are independent.
> 5. The random variable X is the number of trials until the rth success is observed.
>
> For any experiment or phenomenon that has these characteristics, X has a negative binomial distribution and, for $x = r, r+1, r+2, \ldots$, the PMF of X, $p(x)$, is calculated according to Definition 5.4.2.

The derivation of the negative binomial proceeds in a way similar to those for the binomial and the geometric distributions. To begin, consider a sequence of x independent trials, where we will think of x as fixed, and within which there are r successes ($r \leq x$). Furthermore, the last trial is a success (since, with the negative binomial, we will stop as soon as we have observed the rth success). Let's define this as event A; here is one possible event A sequence:

$$\underbrace{SSFSFFSFSSSFF\ldots SFF}_{x-1\text{ trials with } r-1 \text{ successes}}\ S.$$

Note that, within this or any equivalent sequence, in the first $x-1$ trials there must be exactly $r-1$ successes. Let's define this as event B,

$$B = \{\text{the first } x-1 \text{ trials have } r-1 \text{ successes}\},$$

and let's define event C as

$$C = \{\text{trial } x \text{ is a success}\}.$$

Then, because the trials are independent, event C is independent of event B and $A = B \cap C$ so that by the multiplicative rule $\Pr(A) = \Pr(B) \times \Pr(C)$. Now, it should be evident that the probability of event B can be computed using the binomial distribution, so

$$\Pr(B) = \binom{x-1}{r-1} p^{r-1} (1-p)^{x-r},$$

and by definition $\Pr(C) = p$. Thus, for $x = r,\ r+1,\ r+2,\ \ldots$

$$\Pr(X = x) = \Pr(A) = \Pr(B) \times \Pr(C) = \binom{x-1}{r-1} p^{r-1} (1-p)^{x-r} \times p = \binom{x-1}{r-1} p^{r} (1-p)^{x-r}.$$

> **Definition 5.4.2 — Negative Binomial Distribution.** The PMF of a negative binomial random variable X with parameters r and p is
>
> $$\Pr(X = x) = \binom{x-1}{r-1} p^{r} (1-p)^{x-r}, \qquad x = r, r+1, r+2, \ldots.$$
>
> *Parameters and support*: The number of successes, r, is a positive integer and p is the probability of "success," $0 \le p \le 1$. The PMF is nonzero on the integers $x = r, r+1, r+2, \ldots$. Notice that there is no upper bound on the possible values for X.
>
> *Notation*: $X \sim \mathrm{NB}(r,p)$, which stands for "The random variable X has a negative binomial distribution with parameters r and p."

Figure 5.3 shows a variety of negative binomial PMFs, using Definition 5.4.2, for select values of r and p.

For $X \sim \mathrm{NB}(r,p)$, you can calculate using R the values of the PMF and CDF at x, as well as generate random observations from a negative binomial distribution, using the following syntax:

$p(x) \equiv \Pr(X = x)$: `dnbinom(x - r, r, p)`
$F(x) \equiv \Pr(X \le x)$: `pnbinom(x - r, r, p)`
Generate n random variates: `rnbinom(n, x - r, r, p)`

The reason for the $x - r$ in the above expressions is that R defines a negative binomial random variable as *the number of failures before the rth success*. It does not count the r successes. So, to make the R functions consistent with Definition 5.4.2, the first argument in `dnbinom()` and `pnbinom()` is $x - r$.

■ **Example 5.12** Consider an experiment that involves simultaneously rolling two six-sided fair dice. If the two dice match then you win and otherwise you lose. What is the probability of getting the second win on the third roll?

Solution: Let X denote the number of trials until you win twice. Then $X \sim \mathrm{NB}(2, 1/6)$, and

$$\Pr(X = 3) = \binom{3-1}{2-1} p^{2}(1-p)^{3-2} = \frac{2!}{1!1!}\left(\frac{1}{6}\right)^{2}\left(\frac{5}{6}\right) \approx 0.046.$$

Here, $p = \Pr(\text{"success"}) = \Pr(\text{"the two dice match"}) = 6/36 = 1/6$. So there's roughly a 1-in-20 chance of the second match of the two dice occurring on the third roll. ■ ■ ■

> **Exercise 5.7** Returning to the biased coin of Exercise 5.5, where $p = 0.4$, what is the probability of flipping exactly five tails before observing two heads?
>
> *Answer*: $\Pr(X = 7) \approx 0.075$.

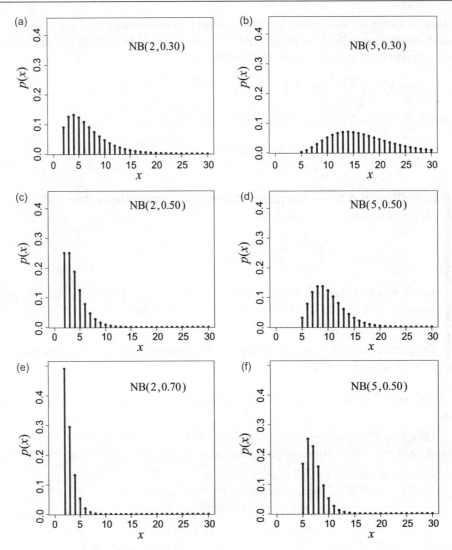

Figure 5.3 Illustrative negative binomial PMFs, using the PMF from Definition 5.4.2, $r = 2, 5$ and $p = 0.30,$ $0.50, 0.70.$

■ **Example 5.13** Employees of a firm that manufactures fluorescent light bulbs are being tested for the presence of mercury in their system. Within this population, 5% of the employees have positive indications of mercury in their system and the firm is asked to randomly select three employees with positive indications for further testing.

(a) Suppose we test people sequentially until we find three with indications of mercury. What is the probability that exactly 10 employees will have to be tested in order to find 3 with positive indications?

(b) Suppose now we test a fixed number n of people. How large must n be in order to have at least a 50% chance of finding three employees with positive indications?

Solution: (a) This is a negative binomial distribution problem because we are running independent trials with two possible outcomes (mercury or no mercury) and a constant probability of having indications of mercury. Define the random variable X as the number of trials required to find the third employee who tests positive. Hence, we have that $p = \text{Pr}(\text{"success"}) = \text{Pr}(\text{"employee tests positive"}) = 0.05$, with $r = 3$, and we want to find $\text{Pr}(X = 10)$. Then,

$$\text{Pr}(X = 10) = \binom{10-1}{r-1} p^r q^{10-r} = \binom{9}{2}(0.05)^3(0.95)^7 \approx 0.0031.$$

Using R we find that

```
dnbinom(10-3,3,0.05)
[1] 0.003142518
```

So, it's pretty unlikely that the third employee with positive indications will be found as the tenth person tested.

The second question is useful as a planning exercise. It is asking us to find the smallest value of n so that $\text{Pr}(X \leq 3) \geq 0.5$ and this is most easily solved in R. By trying different values of n we find:

```
pnbinom(53-3,3,0.05)
[1] 0.4981841
pnbinom(54-3,3,0.05)
[1] 0.5107752
```

Thus, we will need to plan on testing 54 employees in order to have at least a 50% chance of finding three employees with positive indications.

We can also solve for the desired quantile of the distribution directly in R without having to manually hunt for it using the pnbinom() function. You must specify r, the probability of success p, and the quantile q. Then, for $X \sim \text{NB}(r,p)$, R returns the smallest x such that $\text{Pr}(X \leq x) \geq q$. In this problem,

```
qnbinom(0.5,3,0.05)
[1] 51
```

Thus, we will need to plan on observing 51 employees without positive indications of mercury to observe 3 with indications, which means we will need to test $51 + 3 = 54$ employees in order to have a 50% chance of finding 3 employees with positive indications.   ■ ■ ■

Theorem 5.4.1 — Expected Value and Variance of a Negative Binomial Random Variable. The expected value of a negative binomial random variable is $\mathbb{E}(X) = r/p$ and the variance is $\mathbb{V}(X) = r(1-p)/p^2$.

The proof of Theorem 5.4.1 is left as an exercise (see Problem 5.44).

Exercise 5.8 What is the expected number of people and the standard deviation of the number of people who need to be tested in Example 5.13?

Answer: $\mathbb{E}(X) = 60$; $\sigma_X \approx 33.76$.

5.5 Hypergeometric Distribution

Consider a situation in which there is a finite population N of items. Each of the items is one of two types that we'll generically call "red" and "black," where there are r red items in the population and $N - r$ black items in the population. Out of the population, you will sample n of the items. To make this more concrete, imagine a large bowl filled with r red and $N - r$ black marbles, into which you stick your hand and (without looking) grab n random marbles. The random variable of interest is X, the number of red items (marbles) in the sample of size n. Under these conditions, X has a hypergeometric distribution.

You might have noticed that the hypergeometric distribution sounds a lot like the binomial distribution. If we think of selecting a red item as a success and selecting a black item as a failure, then X is the number of successes out of n items (i.e., trials). However, the difference is that, because the population is finite, successive trials are not independent. For example, imagine sequentially selecting the n marbles one by one. On the first draw, the probability of selecting a red marble is r/N but, if you do select a red marble on the first draw, then the probability of selecting a red on the second draw is now $(r-1)/(N-1)$. That is, the act of selecting a marble in the first draw inevitably changes the probabilities in the second and successive draws. So, the binomial distribution does not apply under these conditions (though, when N is large and r and $N - r$ are not too small, the binomial distribution can be used to approximate hypergeometric probabilities).

> **Definition 5.5.1 — Hypergeometric Distribution.** The PMF of a hypergeometric random variable X with parameters r, N, and n is
>
> $$\Pr(X = x) = \frac{\binom{r}{x}\binom{N-r}{n-x}}{\binom{N}{n}}, \qquad x = \max\left(0, n - (N-r)\right), \ldots, \min\left(r, n\right).$$
>
> *Parameters and support*: N is the population size and n is the sample size, so $n \leq N$. The parameter r is the number of "red" elements in the population, so $r \leq N$ and thus there are $N - r$ "black" elements. The formulas for the support (i.e., the set of numbers for which the PMF is nonzero) is a bit complicated, and it is explained below. The PMF is nonzero on the integers $\max(0, n - (N - r)), \ldots, \min(r, n)$.
>
> *Notation*: $X \sim \text{HYPER}(r, N, n)$, which stands for "The random variable X has a hypergeometric distribution with parameters r, N, and n."

To derive the hypergeometric PMF, we begin by noting that there are $\binom{N}{n}$ possible sets of n items out of the population of N items. Assuming each set of n items is equally likely, each has probability $1/\binom{N}{n}$. We then apply the sample point method of Section 3.3.1 to calculate the probability that a sample of size n will contain r red items. To do so, define A as the event "a sample of size n contains exactly x red items." Then,

$$\Pr(A) = \frac{\text{number of outcomes in } A}{\text{total number of outcomes in } S} = \frac{n(A)}{n(S)},$$

where $N(S) = \binom{N}{n}$ and it remains to find $n(A)$, which is the number of sample points in S that contain x red items and $n - x$ black items. Applying the mn-rule of Section 3.3.2, first note that the number of ways of choosing x red items out of the total of r red items is $\binom{r}{x}$ and the number of ways to choose the remaining $n - x$ black items out of the total of $N - r$ black items is $\binom{N-r}{n-x}$. Thus,

$$n(A) = \binom{r}{x}\binom{N-r}{n-x},$$

and we've derived the hypergeometric PMF:

$$\Pr(X = x) \;=\; \Pr(A) \;=\; \frac{\binom{r}{x}\binom{N-r}{n-x}}{\binom{N}{n}}, \qquad x \le r \text{ and } n - x \le N - r.$$

Now let's address the support. Suppose r red balls and $N - r$ black balls make up the population. A sample of n balls is selected. The number of red balls in this sample of n certainly cannot exceed n, but it also cannot exceed the number of red balls in the population. Thus, the largest possible value for X is $\min(r,n)$. For example, if $N = 100$, $r = 5$, and $n = 10$, then X cannot exceed 5 and it cannot exceed 10; thus, X must be less than or equal to $\min(r,n) = 5$. Conversely, the number of black balls selected in the sample, $n - x$, cannot exceed the number of black balls in the population, which is $N - r$. Thus, $n - x \le N - r$, which is equivalent to $x \ge n - (N - r)$. It's also true that the number x of red balls must be at least 0. Thus, the smallest value that X can take on is $\max(0, n - (N - r))$. For example, if $N = 100$, $r = 95$, and $n = 10$, then X could be as small as $\max(0, 10 - (100 - 95) = \max(0,5) = 5$ or as large as $\min(10,95) = 10$. The possible outcomes for X are therefore $x = 5, 6, \ldots, 10$.

> **Definition 5.5.2 — Hypergeometric Experiment.** As a result of the preceding derivation, a *hypergeometric experiment* has the following four characteristics:
>
> 1. Each item in a finite population of N items can have only one of two characteristics, generically called "red" and "black."
> 2. A sample of size n is drawn randomly from the population.
> 3. Every sample of n items out of the N items has an equal chance of being selected.
> 4. The random variable X is the number of red items in the sample of n.
>
> For any experiment or phenomenon that has these characteristics, X has a hypergeometric distribution and, for $x = 0, 1, 2, \ldots, n$, the PMF of X, $p(x)$, is calculated according to Definition 5.5.1.

Figure 5.4 shows a variety of hypergeometric PMFs, using Definition 5.4.2, for select values of r and N (with n fixed at 20).

For $X \sim \text{HYPER}(r, N, n)$, you can calculate the values for the PMF and CDF in R, as well as generate random observations from a hypergeometric distribution, using the following syntax:

$p(x) \equiv \Pr(X = x)$: `dhyper`$(x, r, N - r, n)$
$F(x) \equiv \Pr(X \le x)$: `phyper`$(x, r, N - r, n)$
Generate m random variates: `rhyper`$(m, r, N - r, n)$

Note that the R function arguments are the number of "red" items (r), the number of "black" items ($N - r$), and the sample size (n).

■ **Example 5.14** Consider a bowl that contains 10 marbles: 5 are green, 2 are blue, and 3 are red. Three marbles are drawn from the bowl, one at a time without replacement. Find the probability that all three marbles will be green.

Solution: This is a hypergeometric distribution problem. We have a finite population of marbles: $N = 10$. We also know that we're sampling $n = 3$ marbles with $r = 5$ green marbles (where we don't care about the specific colors of the other marbles; they're just non-green). Let X be the number of green marbles drawn from the bowl of 10 marbles. Then,

$$\Pr(X = 3) \;=\; \frac{\binom{5}{3}\binom{5}{0}}{\binom{10}{3}} \;=\; \frac{5 \times 4 \times 3}{10 \times 9 \times 8} \;=\; \frac{1}{12} \;\approx\; 0.0833.$$

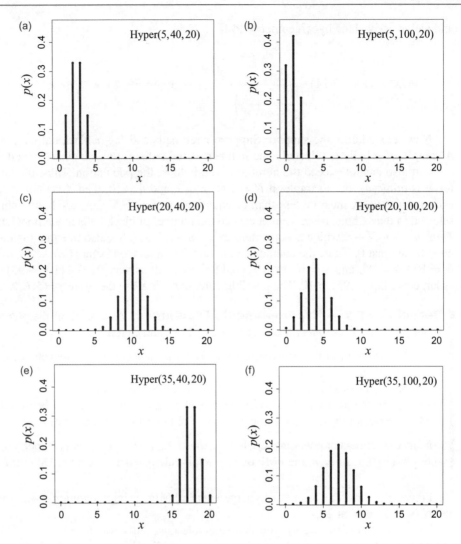

Figure 5.4 Illustrative hypergeometric PMFs, using the PMF from Definition 5.5.1, for $r = 5, 20, 35$, $N = 40, 100$, and $n = 20$.

Evaluating this in R gives

```
dhyper(3,5,10-5,3)
[1] 0.08333333
```

Exercise 5.9 For the bowl in Example 5.14, manually calculate the probability that (a) two of the three marbles selected will be green, and (b) that none of the marbles selected will be green.

Answer: $\Pr(X = 2) \approx 0.417$; $\Pr(X = 0) \approx 0.083$.

■ **Example 5.15** Suppose that a data analytics group in a company consists of 10 data scientists: 5 have a bachelor's degree, 2 have a master's degree, and 3 have a PhD. Three of the 10 are selected at random for a special project. What is the probability that all three have only a bachelor's degree?

Solution: This is exactly the same problem as in Example 5.14, where green = "bachelor's degree," blue = "master's degree," and red = "PhD." Choosing three with only a bachelor's degree is equivalent to choosing three green balls. The probability is therefore the same, namely $1/12$. ▪▪▪

Note that we often describe distributions using everyday phenomena like tossing coins, choosing balls from a bowl, etc. Often, a bit of critical thinking is needed to match the problem at hand with the distribution described using these everyday items. Once this is done, the correct distribution can be determined and we are well on our way to solving the problem.

Theorem 5.5.1 — Expected Value and Variance of a Hypergeometric Random Variable. The expected value of a hypergeometric random variable is

$$\mathbb{E}(X) = \frac{nr}{N} = n\left(\frac{r}{N}\right)$$

and the variance is

$$\mathbb{V}(X) = n\left(\frac{r}{N}\right)\left(\frac{N-r}{N}\right)\left(\frac{N-n}{N-1}\right).$$

The proof of Theorem 5.5.1 is left as an exercise (see Problem 5.45).

The formula for the expected value should be intuitive: $\frac{r}{N}$ is the fraction of red balls in the population and n is the sample size. We would expect a fraction $\frac{r}{N}$ of the n balls sampled to be red. For example, if there are $n = 1,000$ balls, 250 of which are red, then we would expect about one-fourth $\left(\frac{250}{1,000}\right)$ of the $n = 100$ balls selected, that is, $\frac{250}{100} \times 100 = 25$, to be red. Also, notice that the variance is related to the variance of the binomial. (Recall that if $Y \sim \mathrm{BIN}(n,p)$ then $V(Y) = np(1-p)$.) The variance for the hypergeometric is

$$\mathbb{V}(X) = \underbrace{n}_{\text{sample size}} \times \underbrace{\frac{r}{N}}_{\text{fraction of } S \text{ in pop.}} \times \underbrace{\left(1 - \frac{r}{N}\right)}_{\text{fraction of } F \text{ in pop.}} \times \underbrace{\frac{N-n}{N-1}}_{\text{finite pop. correction}}.$$

Here we have described the red balls as "successes" and the black balls as "failures." The last expression in this product, called the *finite population correction*, is always less than 1, and is close to 1 if the sample size is small relative to the population size. The finite population correction often occurs in problems where we sample from a finite population.

▪ **Example 5.16** Returning to Example 5.14, what is the expected number of green marbles drawn? What are the variance and standard deviation?

Solution: $\mathbb{E}(X) = nr/N = 3 \times 5/10 = 1.5$. For the variance:

$$\mathbb{V}(X) = 3\left(\frac{5}{10}\right)\left(\frac{5}{10}\right)\left(\frac{7}{9}\right) \approx 0.583.$$

Thus, $\sigma_X \approx \sqrt{0.583} \approx 0.765$. ▪▪▪

▪ **Example 5.17** A jury of 12 people have been selected from a pool of 40 potential jurors, 26 of whom are white and the other 14 are minorities. (The fraction of minorities in the jury pool reflects the make-up of the community.) The jury of 12 was supposedly selected at random from the jury pool, but it contained only two minority jurors. Is there reason to doubt the claim that the jury of 12 was randomly selected from the jury pool of 40?

Solution: This is a hypergeometric distribution problem with $N = 40$ potential jurors, of which $r = 14$ are potential minority jurors, and from which $n = 12$ jurors were selected. Let X be the number of minority jurors selected from the pool of potential jurors. We want to find $\Pr(X \leq 2)$.

Why do we want to find $\Pr(X \leq 2)$ and not $\Pr(X = 2)$? The logic is as follows. We want to determine how unusual the outcome is that we observed. "Unusual" is the probability of the outcome we observed *or something even more extreme.* Thus, in this case, because we're concerned about the lack of diversity on the jury, we want to know how likely it is that out of the potential jury pool we would randomly get two *or fewer* minority jurors: $P(X \leq 2)$. Then,

$$\Pr(X \leq 2) = \Pr(X = 0) + \Pr(X = 1) + \Pr(X = 2) = \frac{\binom{14}{0}\binom{26}{12}}{\binom{40}{12}} + \frac{\binom{14}{1}\binom{26}{11}}{\binom{40}{12}} + \frac{\binom{14}{2}\binom{26}{10}}{\binom{40}{12}} = 0.1076.$$

Using R to evaluate these probabilities gives

```
phyper(2,14,40-14,12)
[1] 0.1076083
```

Do these results give us reason to doubt the randomness of the jury selection? With the given jury pool, it's not that unlikely that there would be only two or fewer minority jurors in 12 randomly selected jurors. That is, there's slightly higher than a 1-in-10 chance of this occurring.

In later chapters you'll learn that a typical criterion for assessing whether something more than just randomness accounts for the observed outcome is if the probability of observing that outcome or something more extreme is less than 0.05. Under this criterion there is only a 1-in-20 or less chance that the observed outcome (or something more extreme) could have occurred due to chance, and so when the probability is less than 0.05 it is often reasonable to conclude that the observed outcome is due to something more than naturally occurring randomness. ■ ■ ■

> **Exercise 5.10** For Example 5.17, find the expected value and the standard deviation for the number of minority jurors selected.
>
> *Answer:* $\mathbb{E}(X) = 4.2$; $\sigma_X \approx 1.38$.

5.6 Poisson Distribution

The *Poisson distribution* is often used to model the number of events occurring in an interval of time or space, where the events are "rare," meaning they have a very low probability of occurrence in a small interval. To apply, the events must occur at a constant rate and independently. The distribution is named after its discoverer, French mathematician Siméon Denis Poisson (1781–1840). Examples of types of events that can have a Poisson distribution include the number of:

- deaths from horse kicks in the Prussian army;
- typographical errors on a page of printed text;
- website hits per minute;
- radioactive particles that decay in a given period of time;
- number of oak trees with oak wilt disease in a 100 square meter area; and
- traffic accidents at a given intersection in one month.

In fact, the number of deaths from horse kicks in the Prussian army was the very first application of the Poisson distribution by Ladislaus Bortkiewicz (1868–1931). Note how all of these examples involve counting the number of events either in a region of space or over an interval of time. A key assumption is that only one event can occur in an (infinitesimally) small region of space or interval of time. That is, two or more events cannot occur in *exactly* the same location or point in time.

> **Definition 5.6.1 — Poisson Distribution.** The PMF for the Poisson distribution, with parameter λ, is
>
> $$\Pr(X = x) \ = \ \frac{\lambda^x e^{-\lambda}}{x!}, \qquad x = 0, 1, 2, 3, \ldots,$$
>
> where e is the base of the natural logarithm ($e = 2.71828\ldots$).
>
> *Parameter and support*: The parameter λ can be thought of as the average number of events per unit interval, so it must be positive. The PMF is nonzero on the integers $x = 0, 1, 2, 3, \ldots$.
>
> *Notation*: $X \sim \text{POIS}(\lambda)$, which stands for "The random variable X has a Poisson distribution with parameter λ."

Unlike the previous discrete distributions, we can't derive the Poisson from the probability principles developed in Chapters 3 and 4. However, we can show that it is a special limiting form of the binomial distribution. The idea is as follows.

Thinking about the traffic accidents example, imagine splitting the time interval up into n subintervals, where each is so small that at most one accident could occur in any subinterval. The subinterval could be an hour or, if that's too large, maybe it's a minute or a second. Then,

$$\Pr(x \text{ accidents occur in a subinterval of time}) \ = \ \begin{cases} 1 - p, & x = 0 \\ p, & x = 1 \\ 0, & x \geq 2. \end{cases}$$

If we can assume that what happens in one interval of time is independent of what happens in any other interval (that doesn't intersect the first one), then this starts to look like a binomial. However, p depends on n (since larger n means smaller intervals which means smaller p). For any finite choice of n, we cannot *guarantee* that the probability of two or more accidents is zero.

What we can do, though, is look at what happens to the binomial distribution when we let $n \to \infty$. We are going to do this in a special way, where as n gets large, p gets small so that their product is the constant $\lambda = np$. Taking the limit of the binomial PDF as $n \to \infty$, we have:

$$\lim_{n \to \infty} \binom{n}{x} p^x (1-p)^{n-x} \ = \ \lim_{n \to \infty} \frac{n(n-1)\cdots(n-x+1)}{x!} \left(\frac{\lambda}{n}\right)^x \left(1 - \frac{\lambda}{n}\right)^{n-x}$$

$$= \ \lim_{n \to \infty} \frac{\lambda^x}{x!} \left(1 - \frac{\lambda}{n}\right)^n \left(1 - \frac{\lambda}{n}\right)^{-x} \frac{n(n-1)\cdots(n-x+1)}{n^x}$$

$$= \ \frac{\lambda^x}{x!} \lim_{n \to \infty} \left(1 - \frac{\lambda}{n}\right)^n \left(1 - \frac{\lambda}{n}\right)^{-x} \left(1 - \frac{1}{n}\right) \left(1 - \frac{2}{n}\right) \cdots \left(1 - \frac{x-1}{n}\right).$$

Note that $\lim_{n \to \infty} \left(1 - \frac{\lambda}{n}\right)^n = e^{-\lambda}$ and all other terms to the right of it have a limit of 1. Thus,

$$\lim_{n \to \infty} \binom{n}{x} p^x (1-p)^{n-x} \ = \ \frac{e^{-\lambda} \lambda^x}{x!}.$$

So, the Poisson distribution is a limiting form of the binomial distribution when we let the number of subintervals or subregions go to infinity in those situations where the probability of falling in the subinterval or subregion goes to zero but their product is $\lambda = np$.

> **Definition 5.6.2 — Poisson Experiment.** A *Poisson experiment* has the following four characteristics:
>
> 1. We observe the occurrence of events in time or space. In any small time or space interval we either observe or don't observe an event.
> 2. The rate of events that occur per unit of space or time is constant.
> 3. The probability an event will occur is proportional to the interval size.
> 4. The probability that more than one event occurs in an infinitesimally small unit of space or time is zero.
>
> For any experiment or phenomenon that has these characteristics, X has a Poisson distribution and, for $x = 0, 1, 2, \ldots$, the PMF of X, $p(x)$, is calculated according to Definition 5.6.1.

■ **Example 5.18** Let random variable X have a Poisson distribution with $\lambda = 2$. Find the following probabilities:

(a) $P(X = 4)$,

(b) $P(X \geq 4)$,

(c) $P(X < 4)$,

(d) $P(X \geq 4 | X \geq 2)$.

Solution: For parts (a)–(c) we can directly apply Definition 5.6.1:

(a) $P(X = 4) = \dfrac{\lambda^x e^{-\lambda}}{x!} = \dfrac{2^4 e^{-2}}{4!} \approx 0.09.$

(b) $P(X \geq 4) = 1 - P(X \leq 3) = 1 - \dfrac{2^0 e^{-2}}{0!} - \dfrac{2^1 e^{-2}}{1!} - \dfrac{2^2 e^{-2}}{2!} - \dfrac{2^3 e^{-2}}{3!} \approx 1 - 0.857 = 0.143.$

(c) $P(X < 4) = 1 - P(X \geq 4) \approx 1 - 0.143 \approx 0.857.$

For part (d), we proceed by applying the definition of conditional probability (see Definition 3.2.8), where here we have

$$\Pr(X \geq 4 | X \geq 2) = \frac{\Pr(X \geq 4 \cap X \geq 2)}{\Pr(X \geq 2)} = \frac{\Pr(X \geq 4)}{\Pr(X \geq 2)} = \frac{\Pr(X \geq 4)}{1 - \Pr(X \leq 1)}.$$

We have the numerator from part (b) and

$$P(X \leq 1) = \Pr(X = 0) + \Pr(X = 1) = \frac{2^0 e^{-2}}{0!} + \frac{2^1 e^{-2}}{1!} = e^{-2} + 2e^{-2} \approx 0.406,$$

so that

$$P(X \geq 4 | X \geq 2) = \frac{P(X \geq 4)}{1 - P(X \leq 1)} \approx \frac{0.143}{1 - 0.406} \approx 0.241. \qquad ■■■$$

For $X \sim \text{POIS}(\lambda)$, you can calculate the values of the PMF and CDF at x in R, as well as generate random observations from a Poisson distribution, using the following syntax:

$p(x) \equiv \Pr(X = x)$: dpois(x, λ)

$F(x) \equiv \Pr(X \leq x)$: ppois$(x, \lambda)$

Generate n random variates: rpois(n, λ)

■ **Example 5.19** Use R to confirm the results of Example 5.18.

Solution: In R, the commands

```
dpois(4,2)
ppois(3,2,lower.tail=FALSE)
ppois(3,2)
ppois(3,2,lower.tail=FALSE)/ppois(1,2,lower.tail=FALSE)
```

yield

```
[1] 0.09022352
[2] 0.1428765
[3] 0.8571235
[4] 0.2405353
```
■■■

We might be curious about whether the Poisson distribution is, in fact, a proper probability distribution. Of course, to determine this, we need to verify that the PDF meets the two conditions

$$p(x) = \frac{\lambda^x e^{-\lambda}}{x!} \geq 0 \qquad \text{and} \qquad \sum_{x=0}^{\infty} p(x) = 1 \tag{5.1}$$

for any $\lambda > 0, x = 0, 1, 2, 3, \ldots$.

It follows that $p(x) \geq 0$ because with $\lambda > 0$ and $x = 0, 1, 2, 3, \ldots$ it's clear that $\lambda^x \geq 0$, $e^{-\lambda} > 0$, and $x! > 0$. Since $\sum_{x=0}^{\infty} \lambda^x / x!$ is the power series for e^λ, we have

$$\sum_{x=0}^{\infty} p(x) = e^{-\lambda} \underbrace{\sum_{x=0}^{\infty} \frac{\lambda^x}{x!}}_{\text{power series for } e^\lambda} = e^{-\lambda} \times e^\lambda = e^0 = 1.$$

Figure 5.5 shows some illustrative Poisson PMFs for select values of λ.

■ **Example 5.20** Consider a computer chip manufacturing operation where the average number of defects is two per silicon wafer. If there are more than four defects then the silicon wafer has to be scrapped. Assuming the number of defects per wafer follows a Poisson distribution, what is the probability that a randomly selected wafer must be scrapped?

Solution: Let X be the number of defects on a wafer. Using R to find $\Pr(X > 4) = 1 - \Pr(X \leq 3)$,

```
1 - ppois(3,2)
```

```
[1] 0.1428765
```

Thus, roughly one out of every seven wafers must be scrapped.   ■■■

Exercise 5.11 Returning to Example 5.20, what is the probability that there are no defects on a randomly selected wafer? One defect? Two defects? Three defects? Four defects? First, manually calculate the probabilities and then check your solutions in R.

Answers: $\Pr(X = 0) \approx 0.135$; $\Pr(X = 1) \approx 0.271$; $\Pr(X = 2) \approx 0.271$; $\Pr(X = 3) \approx 0.180$; $\Pr(X = 4) \approx 0.090$.

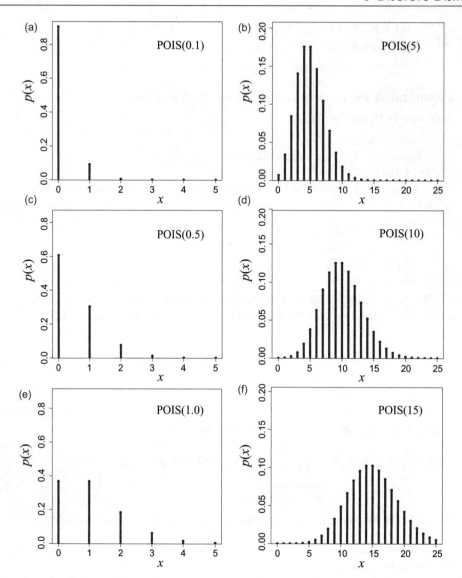

Figure 5.5 Illustrative Poisson PMFs for $\lambda = 0.1, 0.5, 1, 5, 10, 15$. (Note the change in y-axis scale between the left and right columns of PMFs.)

Theorem 5.6.1 — Expected Value and Variance of a Poisson Random Variable. The expected value of a Poisson random variable is $\mathbb{E}(X) = \lambda$ and the variance is $\mathbb{V}(X) = \lambda$.

Proof. Let's start with the expected value result. By Definitions 4.2.6 and 5.6.1 we have

$$
\begin{aligned}
\mathbb{E}(X) &= \sum_{x=0}^{\infty} x \frac{\lambda^x e^{-\lambda}}{x!} \\
&= \sum_{x=1}^{\infty} x \frac{\lambda^x e^{-\lambda}}{x!} \qquad\qquad \text{[because the first term of the sum is 0]}
\end{aligned}
$$

$$= \lambda \sum_{x=1}^{\infty} \frac{\lambda^{x-1} e^{-\lambda}}{(x-1)!} \qquad \text{[because } x/x! = 1/(x-1)! \text{ and } \lambda^x = \lambda\,\lambda^{x-1}\text{]}$$

$$= \lambda \underbrace{\sum_{z=0}^{\infty} \frac{\lambda^z e^{-\lambda}}{z!}}_{\text{sum is } 1} \qquad \text{[shifting the index by 1 with } z = x-1\text{]}$$

$$= \lambda.$$

To evaluate the variance, we will first find $\mathbb{E}(X^2)$ and then use the variance shortcut formula:

$$\mathbb{E}(X^2) = \sum_{x=0}^{\infty} x^2 \frac{\lambda^x e^{-\lambda}}{x!}$$

$$= \lambda \sum_{x=1}^{\infty} \frac{x \lambda^{x-1} e^{-\lambda}}{(x-1)!}$$

$$= \lambda \sum_{z=0}^{\infty} \frac{(z+1) \lambda^z e^{-\lambda}}{z!}$$

$$= \lambda \left[\sum_{z=0}^{\infty} \frac{z \lambda^z e^{-\lambda}}{z!} + \sum_{z=0}^{\infty} \frac{\lambda^z e^{-\lambda}}{z!} \right]$$

$$= \lambda(\lambda+1),$$

using reasoning similar to what was used to evaluate $\mathbb{E}(X)$. Finally, putting these two results together, we have

$$\mathbb{V}(X) = \mathbb{E}(X^2) - [\mathbb{E}(X)]^2 = \lambda(\lambda+1) - \lambda^2 = \lambda. \qquad \blacksquare$$

■ Example 5.21 Customers arrive at an ATM according to a Poisson distribution at an average rate of seven per hour (i.e., $\lambda = 7$). Find:

(a) the probability that at least one customer arrives during a given hour;
(b) the probability that exactly five customers arrive during a given hour;
(c) the expected number of customers arriving in a given hour;
(d) the standard deviation of the number of customers arriving in a given hour.

Solution: Given $X \sim \text{POIS}(\lambda = 7)$, we have:

(a) $\Pr(X \geq 1) = 1 - \Pr(Y = 0) = 1 - \dfrac{7^0 e^{-7}}{0!} \approx 0.999$;

(b) $\Pr(X = 5) = \dfrac{7^5 e^{-7}}{5!} \approx 0.128$;

(c) $\mathbb{E}(X) = \lambda = 7$; $\sigma_X = \sqrt{7} \approx 2.65$.

The R code

```
1-ppois(0,7)
dpois(5,7)
```

gives

```
[1] 0.9990881
[1] 0.1277167
```

■ ■ ■

Given that the Poisson distribution is a limiting form of the binomial distribution, you might ask what the difference is between binomial and Poisson random variables. Remember, with the binomial distribution an event is a success or failure in a sequence of n independent trials. For example, we might be interested in the number of infected computers out of a sample of n computers. Note that for the binomial, $\mathbb{E}(X) \geq \mathbb{V}(X)$ since $np \geq np(1-p)$ with equality only occurring in the trivial case when $p = 0$. In comparison, for the Poisson distribution, events occur on a per unit basis (per unit time, per unit area, etc.). As we've just seen, for the Poisson distribution the expected value is always equal to the variance: $\mathbb{E}(X) = \mathbb{V}(X) = \lambda$.

5.7 Multinomial Distribution

In Example 4.25 we described a situation where a large population consisted of 40% Democrats, 30% Republicans, and 30% Independents. We took a sample of size $n = 3$ and defined X to be the number of Democrats in the sample and Y to be the number of Republicans in the sample; we then obtained the joint PMF for X and Y. We could have also defined Z to be the number of Independents in the sample, but because everyone in the population belongs to exactly one of the sets of Democrats, Republicans, and Independents, we must have $X + Y + Z = n$. Thus, there are really only *two* random variables because Z could be obtained from $Z = n - X - Y$. This is an example of a general situation that leads to the *multinomial distribution*.

The multinomial distribution can be thought of as a generalization of the binomial. For Bernoulli trials, each trial must result in one of the two outcomes we called success (S) and failure (F). The probability of a success on any trial is fixed at p. For a multinomial process, each trial can result in one of k outcomes. See Figure 5.6.

Definition 5.7.1 — Multinomial Experiment. A *multinomial experiment* has the following five characteristics:

1. The experiment consists of exactly n identical trials.
2. Each trial can have one of k outcomes, that we call "bin 1," "bin 2," etc.
3. The set of probabilities $[p_1, p_2, \ldots, p_k]$, where p_i is the probability of falling in bin i, is constant for every trial. Note that we require $\sum_{i=1}^{k} p_i = 1$.
4. The trials are independent.
5. The random variables of interest are $X_1, X_2, \ldots, X_n$, where X_i is the number of times outcome i occurred out of the n trials.

For any experiment or phenomenon that has these characteristics, the vector of the number of times each possible outcome occurred, $[X_1, X_2, \ldots, X_n]$, has a *multinomial distribution*.

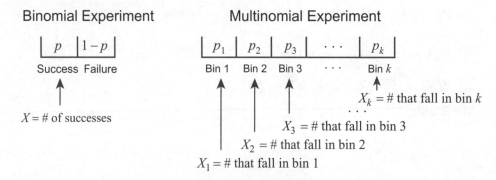

Figure 5.6 Comparison of binomial and multinomial experiments.

Returning to Example 4.25, let's investigate the counting argument that will lead us to the multinomial PMF. Table 4.1 shows all possible samples of size 3 from a large population consisting of 40% Democrats, 30% Republicans, and 30% Independents.[2] Let $[X_1, X_2, X_3]$ denote the vector containing the numbers of Democrats, Republicans, and Independents. (In Example 4.25 we considered only the first two of these and called them X and Y.) Let's focus on those outcomes that lead to exactly two Democrats, one Republican, and no Independents – that is, $X_1 = 2, X_2 = 1, X_3 = 0$. There are three of these, as shown in the upper half of the second column of Table 4.1: DDR, DRD, and RDD, each with probability $0.4^2 \times 0.3 \times 0.3^0 = 0.048$. Let's analyze why there are exactly three ways to get $X_1 = 2, X_2 = 1, X_3 = 0$. Imagine three slots __ __ __ that are to be filled with two Ds, one R, and zero Is. Task 1 is to choose where the two Ds go; this can be done in $\binom{3}{2} = 3$ ways. Task 2 is to choose where the one R goes. Regardless of the outcome of Task 1, there will be one slot remaining in which we must insert one R. This task can be done in $\binom{1}{1} = 1$ way. In Task 3 we must place the zero Is. (Since there are no slots remaining, this isn't a problem.) This can be done in $\binom{0}{0} = 1$ ways. Thus, the number of ways we can do the three tasks together is

$$\binom{3}{2} \times \binom{1}{1} \times \binom{0}{0} = 3 \times 1 \times 1 = 3.$$

Thus, since there are three ways to obtain $X_1 = 2, X_2 = 1, X_3 = 0$ and each has probability 0.048,

$$\Pr\left(X_1 = 2, X_2 = 1, X_3 = 0\right) = \binom{3}{2} \times \binom{1}{1} \times \binom{0}{0} 0.4^2 \times 0.3^1 \times 0.3^0 \approx 0.144.$$

In general, if we have n trials with k bins, every outcome satisfying $X_1 = x_1, X_2 = x_2, \ldots, X_k = x_k$ has probability $p_1^{x_1} p_2^{x_2} \cdots p_k^{x_k}$. We will assume that $\sum_{i=1}^{k} x_i = n$; otherwise, the probability of observing $X_1 = x_1, X_2 = x_2, \ldots, X_k = x_k$ is zero. We can determine the number of different ways $X_1 = x_1, X_2 = x_2, \ldots, X_k = x_k$ can occur by

$$\binom{n}{x_1} \times \binom{n - x_1}{x_2} \times \binom{n - x_1 - x_2}{x_3} \times \cdots \times \binom{n - x_1 - x_2 - \cdots - x_{k-1}}{x_k}.$$

If we write each of these binomial coefficients in terms of factorials, a lot of simplification occurs:

$$\frac{n!}{x_1!(n-x_1)!} \times \frac{(n-x_1)!}{x_2!(n-x_1-x_2)!} \times \frac{(n-x_1-x_2)!}{x_3!(n-x_1-x_2-x_3)!} \times \cdots \times \frac{(n-x_1-x_2-\cdots-x_{k-1})!}{x_k!(n-x_1-x_2-\cdots-x_k)!} = \frac{n!}{x_1!x_2!\cdots x_k!}.$$

Putting this all together, we see that

$$\Pr\left(X_1 = x_1, X_2 = x_2, \ldots, X_k = x_k\right) = \begin{cases} \dfrac{n!}{x_1!x_2!\cdots x_k!} p_1^{x_1} p_2^{x_2} \cdots p_k^{x_k}, & \text{if } x_i \geq 0 \text{ and } \sum_{i=1}^{k} x_i = n \\ 0, & \text{otherwise.} \end{cases}$$

Definition 5.7.2 — Multinomial Distribution. The PMF of a multinomial random vector $[X_1, X_2, \ldots, X_n]$ with parameters n and $[p_1, p_2, \ldots, p_k]$ is

$$\Pr(X_1 = x_1, X_2 = x_2, \ldots, X_k = x_k) = p(x_1, x_2, \ldots, x_n) = \frac{n!}{x_1!x_2!\cdots x_k!} p_1^{x_1} p_2^{x_2} \cdots p_k^{x_k}.$$

Parameters and support: The number of trials, n, is a positive, nonzero integer and $p_1, p_2, \ldots, p_k$ are

---

[2]The "large" population is actually assumed to be infinite. With an infinite population, successive draws are independent and the probabilities of bin membership are constant.

the probabilities of the k outcomes. The support is the set of all k-tuples of integers $[x_1, x_2, \ldots, x_n]$ that satisfy $x_i \geq 0$, $i = 1, 2, \ldots, k$, and $\sum_{i=1}^{k} x_i = n$.

Notation: $[X_1, X_2, \ldots, X_k] \sim \text{MULT}\left(n, [p_1, p_2, \ldots, p_k]\right)$.

It can be shown mathematically that the marginal distribution of X_i is $\text{BIN}(n, p_i)$, but this is easy to see directly. The random variable X_i is the number of times outcome i occurred in a sequence of n multinomial trials. Imagine we have just two outcomes on any trial: outcome i and *not* outcome i. Call it a success if we observe outcome i and a failure if we observe anything else. Then X_i is the number of successes on n trials, so $X_i \sim \text{BIN}(n, p_i)$. Some other properties of the multinomial distribution are given in the following theorem.

> **Theorem 5.7.1 — Properties of the Multinomial Distribution.** Suppose $X_1, X_2, , \ldots, X_k \sim$ $\text{MULT}(n, [p_1, p_2, \ldots, p_k])$. Then,
> (a) $\mathbb{E}(X_i) = np_i$,
> (b) $\mathbb{V}(X_i) = np_i(1 - p_i)$,
> (c) $\text{cov}(X_r, X_s) = -np_r p_s$.

Proof. The proofs for (a) and (b) follow directly from the observation that the marginal distribution for X_i is $\text{BIN}(n, p_i)$. For the third part, let r and s be distinct outcomes or bins, and define

$$U_i = \begin{cases} 1, & \text{if trial } i \text{ results in outcome } r \\ 0, & \text{otherwise} \end{cases} \quad \text{and} \quad V_i = \begin{cases} 1, & \text{if trial } i \text{ results in outcome } s \\ 0, & \text{otherwise.} \end{cases}$$

Since X_r is the number of times outcome r occurs on the n trials, we can obtain X_r by adding a 1 each time outcome r occurs ($U_i = 1$). Similarly, X_s is obtained by adding a 1 for each time outcome s occurs ($V_i = 1$). In other words,

$$X_r = \sum_{i=1}^{n} U_i \quad \text{and} \quad X_s = \sum_{i=1}^{n} V_i .$$

Thus,

$$\text{cov}(X_r, X_s) = \text{cov}\left(\sum_{i=1}^{n} U_i, \sum_{j=1}^{n} V_j\right) = \sum_{i=1}^{n} \text{cov}(U_i, V_i) + \sum\sum_{i \neq j} \text{cov}(U_i, V_j).$$

Note that when $i = j$, we have

$$\text{cov}(U_i, V_j) = \text{cov}(U_i, V_i) = \mathbb{E}(U_i V_i) - \mathbb{E}(U_i)\mathbb{E}(V_i).$$

Since the joint PMF of (U_i, V_i) can be written as

		u_i	
		0	1
v_i	0	$1 - p_r - p_s$	p_r
	1	p_s	0

we can write

$$\mathbb{E}(U_i V_i) = 0 \times 0 \times (1 - p_r - p_s) + 0 \times 1 \times p_r + 0 \times 1 \times p_s + 1 \times 1 \times 0 = 0.$$

Thus,

$$\text{cov}(U_i,V_i) \;=\; \mathbb{E}(U_iV_i) - \mathbb{E}(U_i)\mathbb{E}(V_i) \;=\; 0 - p_ip_j \;=\; -p_ip_j.$$

Also, because successive trials are independent, we know that U_i and V_j are independent so long as $i \neq j$. Thus,

$$\sum_{i\neq j}\sum \text{cov}(U_i,V_j) \;=\; \sum_{i\neq j}\sum 0 \;=\; 0.$$

The covariance between X_r and X_s is thus

$$\text{cov}(X_r,X_s) \;=\; \text{cov}\left(\sum_{i=1}^{n}U_i, \sum_{j=1}^{n}V_j\right) \;=\; \sum_{i=1}^{n}\text{cov}(U_i,V_i) + \sum_{i\neq j}\sum \text{cov}(U_i,V_j) \;=\; -np_rp_s + 0 \;=\; -np_rp_s. \quad\blacksquare$$

The negative covariance should be expected. If the number of bin 1 outcomes is high, there are fewer opportunities to have bin 2 outcomes. Thus, when X_1 is on the high side, we would expect the value for X_2 to be on the low side.

There are two R functions that can be used for the multinomial distribution:
- $\texttt{dmultinom}(x,n,p)$, which gives the PMF for the vector x of length k, the number of trials n, and with probabilities contained in the vector p.
- $\texttt{rmultinom}(N,n,p)$, which simulates N random vectors (each of size k) from the MULT(n,p) where p is a vector of probabilities that sum to 1.

■ **Example 5.22** Suppose that a polygraph test has three possible outcomes: (1) deceptive, (2) not deceptive, and (3) inconclusive. The probabilities for these three outcomes for someone who is telling the truth are $p_1 = 0.1$, $p_2 = 0.7$, and $p_3 = 0.2$. Suppose 100 people are screened with a polygraph test. Let X_1 be the number who are found to be deceptive, X_2 the number who are found to be not deceptive, and X_3 the number whose test is inconclusive. Assuming that all are telling the truth, find

(a) the probability that the tests result in $X_1 = 5$, $X_2 = 80$, and $X_3 = 15$;

(b) the probability that at most five people are found to be deceptive; and

(c) the covariance between all pairs of the three random variables.

Solution: The probability for (a) is

$$\Pr(X_1=5, X_2=80, X_3=15) \;=\; \frac{100!}{5!\,80!\,15!}\,0.1^5 \times 0.7^{80} \times 0.2^{15} \;\approx\; 0.001104.$$

This result was obtained in R using the command $\texttt{dmultinom(c(5,80,15),100,c(0.1,0.7,0.2))}$.

To tackle part (b), note that the marginal of X_1 is BIN$(100,0.10)$, so $\Pr(X_1 \le 5)$ can be obtained using the command $\texttt{pbinom(5,100,0.10)}$, which yields 0.05757. The covariances are

$$\text{cov}(X_1,X_2) = -np_1p_2 = -100 \times 0.1 \times 0.7 = -7$$
$$\text{cov}(X_1,X_3) = -np_1p_3 = -100 \times 0.1 \times 0.2 = -2$$
$$\text{cov}(X_2,X_3) = -np_2p_3 = -100 \times 0.7 \times 0.2 = -14.$$
■ ■ ■

Exercise 5.12 Output from a manufacturing process is classified as conforming (90%), rework (8%), or scrap (2%). If a sample of size $n = 5$ is selected from the output of this process, what is the probability that three are conforming, one is rework, and one is scrap?

5.8 Chapter Summary

In this chapter we've learned about six specific discrete distributions: the binomial, geometric, negative binomial, hypergeometric, Poisson, and multinomial. Along with the Bernoulli distribution from Chapter 4, you now have seven discrete distributions available. As the examples show, these have wide applicability to real-world problems. Table 5.1 is a summary of the six univariate discrete distributions, their PMFs, expected values, and variances.

Table 5.2 summarizes the R functions for calculating the probability at x, $p(x)$, and the cumulative probability at x, $F(x)$, for the six distributions. These functions are useful for calculating the probabilities associated with the various distributions. Why learn all of these distributions? First, as the examples and exercises should have illustrated, these new discrete distributions are useful for modeling real-world problems and phenomena. Furthermore, without them, if you faced a problem like one of those in this chapter, you would have had to derive the distribution "from scratch" using the tools you learned in Chapters 3 and 4. But now that you know them, you can skip all the derivation work and simply apply them appropriately to solve the problem at hand.

Second, and perhaps most importantly from the perspective of statistics, in order to do statistical analysis one must first master probability. That is, as we discussed in the introductory section of Chapter 3, to do statistics you must first be able to "speak probability." In some sense, what we are going to do in later chapters is "reverse engineer" probability problems where, for example, if we have data that were

Table 5.1 Summary of the six univariate discrete distributions we have learned thus far, their PMFs, expected values, and variances. For the PMFs, the table lists the support and the restrictions on the distributional parameter(s), where of course for the Bernoulli, binomial, geometric, and negative binomial we have $0 \le p \le 1$.

Distribution	Probability mass function $\Pr(X = x)$	Expected value $\mathbb{E}(X)$	Variance $\mathbb{V}(X)$
Bernoulli	$p(x) = \begin{cases} 1-p, & x=0 \\ p, & x=1 \end{cases}$	p	$p(1-p)$
Binomial $\mathrm{BIN}(n,p)$	$p(x) = \binom{n}{x}p^x(1-p)^{n-x},$ $x = 0, 1, 2, \ldots, n$	np	$np(1-p)$
Geometric $\mathrm{GEO}(p)$	$p(x) = (1-p)^{x-1}p,$ $x = 1, 2, 3, \ldots$	$\dfrac{1}{p}$	$\dfrac{1-p}{p^2}$
Hypergeometric HYPER	$p(x) = \dfrac{\binom{r}{x}\binom{N-r}{n-x}}{\binom{N}{n}}, r \le N,$ $x = \max\left(0, n-(N-r)\right), \ldots, \min\left(r,n\right)$	$\dfrac{nr}{N}$	$n\,\dfrac{r}{N}\dfrac{N-r}{N}\dfrac{N-n}{N-1}$
Negative binomial $\mathrm{NB}(r,p)$	$p(x) = \binom{x-1}{r-1}p^r(1-p)^{x-r},$ $x = r, r+1, r+2, \ldots$	$\dfrac{r}{p}$	$\dfrac{r(1-p)}{p^2}$
Poisson $\mathrm{POIS}(\lambda)$	$p(x) = \dfrac{\lambda^x e^{-\lambda}}{x!}, \lambda > 0,$ $x = 0, 1, 2, 3, \ldots$	λ	λ

Table 5.2 Summary of R functions for calculating $p(x)$ and $F(x)$.

Distribution	Probability mass $p(x) = \Pr(X = x)$	Cumulative probability $F(x) = \Pr(X \leq x)$
Bernoulli	$\texttt{dbinom}(x, 1, p)$	$\texttt{pbinom}(x, 1, p)$
Binomial	$\texttt{dbinom}(x, n, p)$	$\texttt{pbinom}(x, n, p)$
Geometric	$\texttt{dgeom}(x - 1, p)$	$\texttt{pgeom}(x - 1, p)$
Hypergeometric	$\texttt{dhyper}(x, r, N - r, n)$	$\texttt{phyper}(x, r, N - r, n)$
Negative binomial	$\texttt{dnbinom}(x - r, r, p)$	$\texttt{pnbinom}(x - r, r, p)$
Poisson	$\texttt{dpois}(x, \lambda)$	$\texttt{ppois}(x, \lambda)$
Multinomial	$\texttt{dmultinom(x,n,p)}$	NA

generated from a process that follows the binomial distribution, then we can use those data to estimate p, the probability of success, or perhaps the expected value, μ. In order to do this, we must first master the probabilistic details of the binomial distribution.

5.9 Problems

Problem 5.1 Describe a real-world phenomenon (unrelated to cards, coins, and dice) that is a binomial experiment. Explicitly describe how it meets the five criteria in Definition 5.2.1.

Problem 5.2 Describe a real-world phenomenon (unrelated to cards, coins, and dice) that is a geometric experiment. Explicitly describe how it meets the five criteria in Definition 5.3.1.

Problem 5.3 A colleague describes an experiment his internet start-up company conducted in which they recruited 100 volunteers to test a web advertisement. Each person was independently shown the page on a computer screen and, unbeknownst to the participant, whether they clicked on the ad or not was recorded. The outcome of the experiment was the total number of participants who clicked on the advertisement. Is this a binomial experiment? Why or why not?

Problem 5.4 A fair coin is flipped $n = 10$ times. Let X be the number of heads observed. Construct a table of X versus $p(x) = \Pr(X = x)$ for $x = 0, 1, \ldots, 10$.

(a) What value or values of x have the highest probability of occurring?

(b) What value or values of x have the lowest probability of occurring?

(c) Let Y be the number of heads observed on a biased coin, flipped $n = 10$ times, where the probability of heads is $p = 0.3$. Construct a table of Y versus $p(y) = \Pr(Y = y)$ for $y = 0, 1, \ldots, 10$.

(d) What value or values of y have the highest probability of occurring?

(e) What value or values of y have the lowest probability of occurring?

Problem 5.5 Suppose $X \sim \mathrm{BIN}(100, 0.8)$. Find the following.

(a) $\Pr(X = 80)$ (b) $\Pr(X \leq 80)$
(c) $\Pr(X < 80)$ (d) $\mu_X = \mathbb{E}(X)$
(e) $\sigma_X^2 = \mathbb{V}(X)$
(f) $\Pr(\mu_X - 2\sigma_X < X < \mu_X + 2\sigma_X)$

Problem 5.6 Suppose $Y \sim \mathrm{BIN}(100, 0.2)$. Find the following.

(a) $\Pr(Y = 20)$ (b) $\Pr(Y \leq 20)$
(c) $\Pr(Y < 20)$ (d) $\mu_Y = \mathbb{E}(Y)$
(e) $\sigma_Y^2 = \mathbb{V}(Y)$
(f) $\Pr(\mu_Y - 2\sigma_Y < Y < \mu_Y + 2\sigma_Y)$

Problem 5.7 Suppose $X \sim \mathrm{BIN}(n, p)$ and $Y \sim \mathrm{BIN}(n, 1-p)$. Note that n and p are the same in the definitions of both X and Y.

(a) What is the relationship between $\Pr(X = x)$ and $\Pr(Y = n - x)$?

(b) How are $\mathbb{E}(X)$ and $\mathbb{E}(Y)$ related?

(c) How are $\mathbb{V}(X)$ and $\mathbb{V}(Y)$ related?

Problem 5.8 Returning to Problem 5.4, where $X \sim \mathrm{BIN}(10, 1/2)$ and $Y \sim \mathrm{BIN}(10, 0.3)$, find the following probabilities.

(a) $\Pr(X < 5)$ and $\Pr(Y < 5)$
(b) $\Pr(X > 5)$ and $\Pr(Y > 5)$
(c) $\Pr\big(\mathbb{E}(X) - 1 \leq X \leq \mathbb{E}(X) + 1\big)$
(d) $\Pr\big(\mathbb{E}(X) - 2 \leq X \leq \mathbb{E}(X) + 2\big)$
(e) $\Pr\big(\mathbb{E}(X) - \sigma_X \leq X \leq \mathbb{E}(X) + \sigma_X\big)$
(f) $\Pr\big(\mathbb{E}(Y) - 1 \leq Y \leq \mathbb{E}(Y) + 1\big)$
(g) $\Pr\big(\mathbb{E}(Y) - 2 \leq Y \leq \mathbb{E}(Y) + 2\big)$
(h) $\Pr\big(\mathbb{E}(Y) - \sigma_Y \leq Y \leq \mathbb{E}(Y) + \sigma_Y\big)$

Problem 5.9 Consider a situation where you're testing a new challenge in an electronic game and you are trying to get some insight into how many tries it will take a player to pass the challenge. When players get to a certain point in the game, they're presented with the challenge. If they pass it, they go on to the next level of the game. If they fail, the game resets to the beginning of the challenge and they have to try it again, where each try is independent of all past tries. This process repeats until the player passes. Is this a geometric experiment? Why or why not?

Problem 5.10 In a variant of the scenario in Problem 5.9, instead of just one challenge, you are considering presenting the player with a sequence of successively harder challenges where, to move on to the next level, he or she must pass three of them. If they fail any one, they have to start all over and this process repeats until the player passes three challenges. If the probability of passing any one of these is 0.7, and successive tries are independent, what is the probability of moving to the next level?

Problem 5.11 Among the volunteers who have agreed to donate blood at a clinic, 90% have Rh factor present in their blood. Suppose that 24 people are selected. Find the probability that at least 20 of them have Rh factor present.

Problem 5.12 Continuation of Problem 5.11. Suppose now that 20 people with Rh factor are needed. How large must a sample be so that the probability is at least 0.95 that we have at least 20 people with Rh factor. (*Hint*: Trial and error for the sample size n.)

Problem 5.13 The proportions for each blood type for the United States are:

O	A	B	AB
0.44	0.42	0.10	0.04

Suppose $n = 12$ people are selected at random from this population. Let X denote the number with type O blood. Find the probability that $X = 6$.

Problem 5.14 For the distribution in Problem 5.13 and with a sample of $n = 120$, find the probability that exactly 60 of those selected have type O blood.

Problem 5.15 For the distribution in Problem 5.13 and with a sample of $n = 120$, find the probability that 60 or fewer of those selected have type O blood.

Problem 5.16 *Plantar fasciitis* occurs when there are tears to a ligament near the heel bone. Although this usually occurs in one foot, it is possible to affect both feet. Suppose that n people with planar fasciitis in both feet are recruited for a clinical trial of two shoe inserts. Each person receives insert 1 on one foot and insert 2 on the other; the selection of which foot receives which insert is done at random. After one week, each participant is asked which worked better. (The participants did not know which inserts they received on which foot, and neither did their physicians. This is called a *double blind* study.) Let X denote the number among the n participants who chose insert 1 over insert 2 (ties were not allowed). Assume that the two inserts are equally effective in treating pain due to *plantar fasciitis*.

(a) What is the distribution of X?
(b) If $n = 10$, find the probability that exactly eight chose insert 1.
(c) Again assuming $n = 10$, find the probability that eight or more preferred insert 1 over insert 2.

Problem 5.17 For the situation described in Problem 5.16, 100 people are sampled and given insert 1 on one foot and insert 2 on the other.

(a) What is the probability that the number of subjects who prefer insert 1 is 90 or more, assuming that in reality the two inserts are equally effective?
(b) If you observed that 90 out of 100 people preferred insert 1, would you believe that the two inserts are equally effective? Explain your reasoning.

Problem 5.18 Suppose $X \sim$ GEO(0.2). Find the following.

(a) $\Pr(X = 5)$ (b) $\Pr(X \geq 5)$
(c) $\Pr(X < 10)$ (d) $\mathbb{E}(X)$
(e) $\mathbb{V}(X)$

Problem 5.19 Suppose $X \sim$ GEO(0.9). Find the following.

(a) $\Pr(X = 5)$ (b) $\Pr(X \geq 5)$
(c) $\Pr(X \leq 1)$ (d) $\mathbb{E}(X)$
(e) $\mathbb{V}(X)$

Problem 5.20 Suppose $X \sim$ NB(3,0.1) Find the following.

(a) $\Pr(X \leq 10)$ (b) $\Pr(X > 20)$
(c) $\mathbb{E}(X)$ (d) $\mathbb{V}(X)$

Problem 5.21 A "claw machine" is an arcade game in which a three-prong claw can be maneuvered in an attempt to pick up a prize from the inside of a clear box. Most of the time, the claw catches nothing and the user has wasted the money used to buy a chance. Suppose the probability is 0.10 of successfully clawing a prize and the cost to play is $0.25 for each attempt. Successive tries are independent.

(a) If I play the game 10 times, what is the probability that I win at least one prize?
(b) Suppose I play the game until I win a prize. What is the probability that it takes me at least 10 tries?
(c) Suppose I play the game until I win a prize. What is the probability that it takes 10 or fewer tries?
(d) What is the expected number of tries needed to win a prize?
(e) What is the expected cost to win a prize?
(f) Suppose that I have already played the game 10 times without winning a prize. What is the probability that I will win a prize within the next 10 tries?

Problem 5.22 In Problem 5.21, what is the standard deviation of the cost that must be expended in order to win one prize?

Problem 5.23 Building on Problem 5.11, suppose we test people for Rh factor, and we continue testing (one person after another) until we get 20 with the Rh factor. Let Y denote the number of people tested until we reach the required threshold of 20. Find the probability that Y is less than or equal to 24.

Problem 5.24 Building on Problem 5.21, suppose that a claw machine has 15 prizes in it, and the probability is 0.10 that a single try will result in winning a prize. Suppose further that we would like to empty the machine by winning all 15 prizes. If we continue to play until all 15 prizes are won, what is the probability that it takes at least 100 attempts?

Problem 5.25 Type O negative (denoted O−) blood is a universal donor in the sense that any other person can receive this type. It is estimated that about 7% of the US population has O− blood. Suppose we would like to obtain O− blood from five people, and we select people (at random) until we obtain five with type O− blood. Let X denote the number of people we have to test until we find five such people. You may nccd R to do parts (b) and (c).

(a) What is the distribution of X?

(b) What is the probability that we must test 10 or fewer people?

(c) What is the probability that at least 20 people will have to be tested?

Problem 5.26 Suppose X is the number of independent trials with success probability p needed to get r successes. Then $X \sim \text{NB}(r,p)$. Suppose we define Y to be the number of failures before we obtain the rth success. (This is the definition of the negative binomial distribution that R uses.)

(a) What is the support of Y?

(b) What is the support of X?

(c) Explain why $Y = X - r$.

(d) Find the PMF for Y.

Problem 5.27 A manufacturer makes a batch of 200 units each day. Suppose on a given day the number of defective units produced is 20. Let X denote the number of defectives in a sample of 25.

(a) What are the possible values for X?

(b) Find $\Pr(X = 3)$.

(c) Find $\Pr(X \le 5)$.

(d) Find $\mathbb{E}(X)$.

(e) Find $\mathbb{V}(X)$.

Problem 5.28 A manufacturer makes a batch of 2,000 units each day. Suppose on a given day the number of defective units produced is 200. Let X denote the number of defectives in a sample of 25.

(a) What are the possible values for X?

(b) Find $\Pr(X = 30)$.

(c) Find $\Pr(X \le 50)$.

(d) Find $\mathbb{E}(X)$.

(e) Find $\mathbb{V}(X)$.

Problem 5.29 A manufacturer makes many thousands of units each day. Suppose on a given day the number of defective units produced is 10% of the total produce for the day. Let X denote the number of defectives in a sample of size 25. Assume that successive units being defective (or not) are independent, and the probability of an item being defective is constant at 0.10.

(a) What are the possible values for X?

(b) Find $\Pr(X = 3)$.

(c) Find $\Pr(X \le 5)$.

(d) Find $\mathbb{E}(X)$.

(e) Find $\mathbb{V}(X)$.

Problem 5.30 Compare your answers from Problems 5.28 and 5.29. What does this mean in practice?

Problem 5.31 Suppose $X \sim \text{POIS}(5)$. Find the following.

(a) $\Pr(X = 0)$

(b) $\Pr(X \le 5)$

(c) $\Pr(X > 10)$

(d) $\mathbb{V}(X)$

Problem 5.32 Suppose $X \sim \text{POIS}(15)$. Find the following.

(a) $\Pr(X = 0)$

(b) $\Pr(X \le 15)$

(c) $\Pr(X > 30)$

(d) $\mathbb{V}(X)$

Problem 5.33 Suppose $X \sim \text{BIN}(100, 0.05)$ and $Y \sim \text{POIS}(5)$. Plot the PMFs for both X and Y. How similar are they?

Problem 5.34 Suppose $X \sim \text{BIN}(1000, 0.05)$ and $Y \sim \text{POIS}(50)$. Plot the PMFs for both X and Y. How similar are they?

Problem 5.35 Pertussis, also called whooping cough, is a somewhat rare disease that, unfortunately, has increased in recent years. Suppose that the probability that a person has pertussis at a particular point in time is about 0.000002. Missouri has about 6,000,000 residents. If each person has probability 0.000002 of having pertussis, then the number X who have pertussis has a binomial distribution with $n = 6,000,000$ and $p = 0.000002$. Find the following.

(a) $\mathbb{E}(X)$

(b) $\mathbb{V}(X)$

(c) The probability that exactly 12 people in Missouri have pertussis at a given time.

(d) The probability that 20 or fewer Missourians have pertussis at a given time.

Problem 5.36 We saw that the Poisson distribution is an approximation to the binomial when n is large and p is small.

(a) Explain why the situation in Problem 5.35 fits this assumption.

(b) Using the Poisson distribution, approximate the probability of having exactly 12 Missourians with pertussis.

(c) Again, assuming the Poisson distribution, approximate the probability that 20 or fewer Missourians have pertussis.

(d) Compare your answers to those of Problem 5.35.

Problem 5.37 Suppose $X \sim \text{POIS}(\lambda)$ and let $f(x)$ be the PMF for X. Show that

$$\frac{f(x+1)}{f(x)} = \frac{\lambda}{x+1}.$$

Problem 5.38 Write the result of Problem 5.37 as

$$f(x+1) = \frac{\lambda}{x+1} f(x).$$

(a) For what values of x is $f(x+1) > f(x)$?
(b) For what value of x is $f(x)$ greatest?

Problem 5.39 Consider the distribution of blood types given in Problem 5.13. From a sample of size $n = 6$ selected from this population, let X_1 denote the number with type O, X_2 the number with type A, X_3 the number with type B, and X_4 the number with type AB.

(a) Explain why $X_1 + X_2 + X_3 + X_4 = 6$ with probability 1.
(b) Find $\Pr(X_1 = 3, X_2 = 3, X_3 = 0, X_4 = 0)$.
(c) Find $\Pr(X_1 = 2, X_2 = 3, X_3 = 1, X_4 = 0)$.
(d) Find $\Pr(X_1 = 2, X_2 = 2, X_3 = 1, X_4 = 0)$.

Problem 5.40 Suppose that in a population, 40% of people have never had COVID, 35% of people have had it once, and 25% have had it more than once. Suppose we take a sample of size $n = 20$. Let X_1, X_2, X_3 denote the number that have had COVID zero times, exactly one time, and more than one time, respectively. Find the following.

(a) $\Pr(X_1 = 12, X_2 = 6, X_3 = 2)$
(b) $\Pr(X_1 = 12, X_2 = 7, X_3 = 2)$
(c) $\Pr(X_1 = 8, X_2 = 6, X_3 = 6)$

Problem 5.41 Continuing with Problem 5.39, find the following.

(a) $\Pr(X_1 = 3)$
(b) $\mathbb{E}(X_2)$

(c) $\mathbb{V}(X_2)$
(d) $\text{cov}(X_1, X_2)$
(e) $\text{cov}(X_1, X_4)$

Problem 5.42 Continuing with Problem 5.40, find the following.

(a) $\Pr(X_2 = 9)$
(b) $\mathbb{E}(X_3)$
(c) $\mathbb{V}(X_1)$
(d) $\text{cov}(X_1, X_2)$
(e) $\text{cov}(X_1, X_3)$

Problem 5.43 Suppose we roll a fair six-sided die and observe the outcome N. We then select N cards from a standard 52-card deck (which contains 13 hearts and 39 non-hearts). Let M denote the number of hearts selected. Find $\mathbb{E}(M)$ and $\mathbb{V}(M)$.

Problem 5.44 Prove Theorem 5.4.1. (*Hints*: For the expected value part, note that

$$x\binom{x-1}{r-1} = r\binom{x}{r}.$$

For the variance, use an approach similar to the binomial variance proof by first finding $\mathbb{E}\big(X(X+1)\big)$.)

Problem 5.45 Prove Theorem 5.5.1. (*Hints*: For the expected value and variance proofs, note that

$$\binom{n}{k} = \frac{n}{k}\binom{n-1}{k-1}.$$

Also for the variance proof, start from Definition 4.2.8 applied to a hypergeometric random variable, complete the square, and work through the algebra.)

6 — Continuous Distributions

6.1 Introduction

In Chapter 5 we learned about a number of discrete distributions. In this chapter we focus on continuous distributions, which are useful as models of various real-world events. By the end of this chapter you will know nine continuous and eight discrete distributions. There are many more continuous distributions, but these nine will suffice for our purposes. These continuous distributions are useful for modeling various types of processes and phenomena that are encountered in the real world. These particular distributions are an important part of the statistical methods that we will learn about starting in Chapter 7. Finally, it is worth mentioning that these models are just approximations to reality. A famous statistician[1] said "all models are wrong, but some are useful." For example, no random phenomenon yields data that come from a distribution that is exactly normal (i.e., bell-shaped), but enough are close enough to make the normal distribution a useful model.

6.2 Uniform Distribution

Suppose I get off of a plane and walk to the ground transportation area to catch a shuttle bus to the parking lot where I parked my car. I know that the buses run every 20 minutes, but when I arrive I don't know how long it's been since the last bus came by. The amount of time that I must wait is a continuous random variable that takes on a value between 0 (if I'm really lucky and I arrive a moment before the bus leaves) and 20 (if I'm really unlucky and I arrive the moment the bus pulls away). Any value between 0 and 20 is possible, including nonintegers such as 17.931 and π. The random variable X that represents my waiting time is therefore continuous with support $[0, 20]$.

In real life, though, measurements of such phenomena are always rounded to some degree of accuracy; in this case, they may be rounded to the nearest second. When this rounding is taken into account, we see that X is really a discrete random variable. In cases like this, where a continuous variable is rounded, we will often treat the outcome as a continuous random variable, but we must be aware that this is just an approximation.

---

[1]George E. P. Box (1919–2013). The longer quote from Box and Draper, (1987) is this: "Essentially, all models are wrong, but some are useful. However, the approximate nature of the model must always be borne in mind."

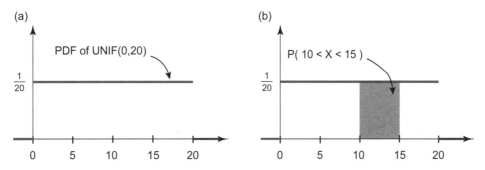

Figure 6.1 (a) Uniform distribution over $[0, 20]$. The PDF is equal to the nonzero constant $c = 1/20$ over this interval and zero otherwise. The same PDF is shown in (b) and the area that corresponds to $\Pr(10 \le X \le 15)$ is shaded.

We saw in Section 4.3 that for continuous random variables, probability is computed as area under the PDF. If I arrive at a random time and I have no idea at all how long ago the last bus came, it would seem reasonable that any interval of a fixed width is equally likely. In other words, the probability that my wait is 0–3 minutes, or 1–4 minutes, or any interval of length 3 minutes, should have the same probability. Thus, it seems reasonable to assume that the PDF is *flat* over the interval $[0, 20]$; that is,

$$f(x) = \begin{cases} c, & 0 \le x \le 20 \\ 0, & \text{otherwise.} \end{cases}$$

The graph in Figure 6.1(a) shows the PDF being equal to the value c between 0 and 20, and equal to 0 everywhere else. The value c must be chosen so the PDF integrates to 1. Since the PDF is flat, the area of interest is a rectangle, so no calculus is need to find the area. Since the base of the rectangle is 20, the height must be $\frac{1}{20}$ so that the area under the curve is 1. Thus,

$$f(x) = \begin{cases} \dfrac{1}{20}, & 0 \le x \le 20 \\ 0, & \text{otherwise.} \end{cases}$$

If we wanted $P(10 \le X \le 15)$, we could get it as the area under the PDF between 10 and 15; see Figure 6.1(b). Since this is a rectangle, we can find the area without calculus:

$$P(10 \le X \le 15) = \frac{1}{20}(15 - 10) = \frac{1}{4}.$$

Of course, you could use calculus to find the area under the PDF by integration:

$$P(10 \le X \le 15) = \int_{10}^{15} \frac{1}{20}\, dx = \left[\frac{1}{20}x \right]_{10}^{15} = \frac{1}{20}15 - \frac{1}{20}10 = \frac{1}{4},$$

which is the same answer as before. This is an example of the uniform distribution.

Definition 6.2.1 — Uniform Distribution. The random variable X has a uniform distribution over the interval $[a, b]$ if the PDF is of the form

$$f(x) = \begin{cases} \dfrac{1}{b - a}, & a \le x \le b \\ 0, & \text{otherwise.} \end{cases}$$

Parameters and support: The parameters are a and b, and the support is the interval $[a, b]$.

Notation: The notation is $X \sim \text{UNIF}(a, b)$.

Exercise 6.1 Suppose $Y \sim \text{UNIF}(50,100)$. (a) Sketch the PDF. (b) Find $P(60 \leq Y \leq 90)$. (c) Find $P(80 \leq Y \leq 120)$. (Be careful with the support.)

Answer: $\frac{3}{5}$, $\frac{2}{5}$.

Next we determine the mean and variance of the uniform distribution.

Theorem 6.2.1 — Expected Value and Variance of the Uniform Distribution. If $X \sim \text{UNIF}(a,b)$, then

$$\mathbb{E}(X) = \frac{a+b}{2} \qquad \text{and} \qquad \mathbb{V}(X) = \frac{(b-a)^2}{12}.$$

Proof.

$$
\begin{aligned}
\mathbb{E}(X) &= \int_{-\infty}^{\infty} x\, f(x)\, dx \\
&= \int_{-\infty}^{a} x \cdot 0\, dx + \int_{a}^{b} x\, \frac{1}{b-a}\, dx + \int_{b}^{\infty} x\, 0\, dx \\
&= 0 + \frac{1}{b-a} \left[\frac{x^2}{2} \right]_a^b + 0 \\
&= \frac{1}{2(b-a)} (b^2 - a^2) \\
&= \frac{a+b}{2}.
\end{aligned}
$$

To get the variance, we begin with $E(X^2)$:

$$
\begin{aligned}
\mathbb{E}(X^2) &= \int_{-\infty}^{\infty} x^2\, f(x)\, dx \\
&= \int_{a}^{b} x^2\, \frac{1}{b-a}\, dx \qquad \text{[The integrals over } (-\infty, a) \text{ and } (b, \infty) \text{ are equal to 0.]} \\
&= \frac{1}{b-a} \left[\frac{x^3}{3} \right]_a^b \\
&= \frac{1}{3(b-a)} (b^3 - a^3) \\
&= \frac{1}{3(b-a)} (b-a)(b^2 + ab + a^2) \\
&= \frac{b^2 + ab + a^2}{3}.
\end{aligned}
$$

Thus,

$$\mathbb{V}(X) = \mathbb{E}(X^2) - [\mathbb{E}(X)]^2 = \frac{b^2 + ab + a^2}{3} - \left(\frac{a+b}{2} \right)^2 = \frac{b^2 - 2ab + a^2}{12} = \frac{(b-a)^2}{12}. \qquad \blacksquare$$

■ **Example 6.1** For the waiting time for the shuttle bus, find the expected value, variance, and standard deviation.

Solution: We have $X \sim \mathrm{UNIF}(0,20)$, so $a = 0$ and $b = 20$. Thus,

$$\mathbb{E}(X) \;=\; \frac{0+20}{2} = 10$$

$$\mathbb{V}(X) \;=\; \frac{(20-0)^2}{12} = \frac{400}{12} = \frac{100}{3} \approx 33.33.$$

The standard deviation is thus $\sqrt{100/3} \approx 5.7735$.   ∎∎∎

Note that the expected value is the midpoint of the support interval. My expected wait time is therefore 10 minutes, the midpoint of $[0,20]$; this shouldn't be surprising.

> **Exercise 6.2** If $W \sim \mathrm{UNIF}(-100,100)$, find $E(W)$ and $V(W)$.
>
> *Answer*: $0,\ 14{,}333/4 \approx 3{,}333.3$.

6.3 Exponential Distribution

The *exponential distribution* is often a useful model for the waiting time between events. For example, we might want to model the time until the next call comes into a call center, or the time until a machine breaks down. In these examples, the waiting time is unknown and so it is often appropriate to think of the wait time as a random variable having an exponential distribution.

> **Definition 6.3.1 — Exponential Distribution.** The PDF of an exponential random variable X with parameter λ is
>
> $$f(x) = \begin{cases} \lambda e^{-\lambda x}, & 0 \le x < \infty \\ 0, & -\infty < x < 0. \end{cases} \tag{6.1}$$
>
> *Parameters and support*: The Greek letter λ is the rate of events per unit of time, where it must be that $\lambda > 0$; $x \ge 0$.
>
> *Notation*: $X \sim \mathrm{EXP}(\lambda)$, which stands for "The random variable X has an exponential distribution with parameter λ."

> **Theorem 6.3.1 — The Function in (6.1) is a Valid PDF.** The distribution in Definition 6.3.1 is a proper PDF since $f(x) \ge 0$ for all x and $\int_{-\infty}^{\infty} f(x)\,dx = 1$.

Proof. Since by definition, $\lambda > 0$ and e^x is always greater than or equal to 0, it follows that $f(x) \ge 0$ for $-\infty < x < \infty$. The integral over $(-\infty, \infty)$ is

$$\int_{-\infty}^{\infty} f(x)\,dx = \int_{0}^{\infty} \lambda e^{-\lambda x}\,dx = -e^{-\lambda x}\Big|_{0}^{\infty} = 0 - (-1) = 1. \qquad \blacksquare$$

> **Theorem 6.3.2 — CDF for the Exponential Distribution.** The CDF for the $\mathrm{EXP}(\lambda)$ distribution is
>
> $$F(x) = \Pr(X \le x) = \begin{cases} 1 - e^{-\lambda x}, & x \ge 0 \\ 0, & x < 0. \end{cases}$$

Proof. For $x < 0$, $\Pr(X \le x) = 0$. For $x > 0$, the result follows from integrating the PDF:

$$F(x) = \int_{-\infty}^{x} f(z)dz = \int_{0}^{x} \lambda e^{-\lambda z}dz = -e^{-\lambda z}\Big|_{0}^{x} = 1 - e^{-\lambda x}. \qquad \blacksquare$$

The exponential distribution is sometimes used to model the time to failure of some equipment (or person). For any random variable X used to model time to failure, the conditional probability of failure in the small interval $t < X \le t + \Delta t$ conditioned on survival to the beginning of the interval is a function of t. The limit of this function is called the *hazard function*, which can be expressed in terms of the PDF and CDF of X.

Definition 6.3.2 — Hazard Function. If X is a continuous random variable with support $[0, \infty)$, then the hazard function is

$$h(t) = \lim_{\Delta t \to 0} \frac{P(t < X \le t + \Delta t | X > t)}{\Delta t}.$$

Theorem 6.3.3 — Hazard Function in Terms of the PDF and CDF. If X is a continuous random variable with support $[0, \infty)$, PDF $f(t)$, and CDF $F(t)$, then the hazard function is

$$h(t) = \frac{f(t)}{1 - F(t)}.$$

Proof.

$$
\begin{aligned}
h(t) &= \lim_{\Delta t \to 0} \frac{\Pr(t < X \le t + \Delta t | X > t)}{\Delta t} \\
&= \lim_{\Delta t \to 0} \frac{\Pr(t < X \le t + \Delta t \text{ and } X > t)}{\Delta t \, P(X > t)} \\
&= \lim_{\Delta t \to 0} \frac{\Pr(t < X \le t + \Delta t)}{\Delta t \, (1 - F(t))} \\
&= \lim_{\Delta t \to 0} \frac{F(t + \Delta t) - F(t)}{\Delta t} \frac{1}{1 - F(t)} \\
&= F'(t) \frac{1}{1 - F(t)} \\
&= \frac{f(t)}{1 - F(t)}. \qquad \blacksquare
\end{aligned}
$$

The hazard function measures how likely an imminent failure is, given survival up to the current time. If this probability increases as the item ages, then we can say that the item is wearing out. For the exponential distribution, the hazard is just the constant λ, as we demonstrate in the next theorem.

Theorem 6.3.4 — Hazard Function for Exponential Distribution. The hazard function for the EXP(λ) distribution is $h(t) = \lambda$.

Proof.

$$h(t) = \frac{f(t)}{1 - F(t)} = \frac{\lambda e^{-\lambda t}}{1 - (1 - e^{-\lambda t})} = \lambda. \qquad \blacksquare$$

There is a connection between the exponential and the Poisson distributions. If the Poisson is an appropriate model for the number of events per interval of time, then the time between events has an exponential distribution. Conversely, if the time between events is exponentially distributed and the events are independent, then the number of events per unit of time has a Poisson distribution.

■ **Example 6.2** Suppose that accidents on a hazardous stretch of road can be well modeled by an exponential distribution and they occur at the rate of two per week (i.e., $\lambda = 2$). If an accident has just occurred, what is the probability that the time until the next accident is one week or less?

Solution: Let X denote the time between accidents, and we want to know $\Pr(X \leq 1)$. Then,

$$\Pr(X \leq 1) = \int_0^1 2e^{-2x}\, dx = -e^{-2x}\Big|_0^1 = 1 - e^{-2} \approx 0.865.$$

Thus, there is an 86.5% chance that the next accident will occur in one week or less. Conversely, there is an 13.5% chance that the next accident will not occur for more than a week. ■ ■ ■

■ **Example 6.3** Returning to Example 6.2, what's the probability that the time to the next accident is one day or less?

Solution: As before, let X denote the time between accidents; what we now want to know is the cumulative probability $\Pr(X \leq 1/7)$. Then,

$$\Pr(X \leq 1/7) = 1 - e^{-2 \times 1/7} \approx 0.249.$$

So, there is an almost 25% chance that the next accident will occur within a day. ■ ■ ■

Theorem 6.3.5 — Expected Value and Variance of an Exponential Random Variable.
For $X \sim \text{EXP}(\lambda)$, $\mathbb{E}(X) = 1/\lambda$, and $\mathbb{V}(X) = 1/\lambda^2$.

Proof. To find $\mathbb{E}(X)$ we can apply integration by parts:

$$\mathbb{E}(X) = \int_{-\infty}^{\infty} x f(x)\, dx = \int_0^{\infty} x \lambda e^{-\lambda x} dx = -x e^{-\lambda x}\Big|_0^{\infty} + \int_0^{\infty} e^{-\lambda x} dx = (0 - 0) + \left[-\frac{1}{\lambda} e^{-\lambda x} \right]_0^{\infty} = \frac{1}{\lambda}.$$

For the variance, we'll take the usual approach and first find $\mathbb{E}(X^2)$:

$$\begin{aligned}
\mathbb{E}(X^2) &= \int_0^{\infty} x^2 \lambda e^{-\lambda x} dx \\
&= \left[-x^2 e^{-\lambda x} \right]_0^{\infty} + \int_0^{\infty} 2x e^{-\lambda x} dx \\
&= (0 - 0) + \left[-\frac{2}{\lambda} x e^{-\lambda x} \right]_0^{\infty} + \frac{2}{\lambda} \int_0^{\infty} e^{-\lambda x} dx \\
&= (0 - 0) + (0 - 0) + \frac{2}{\lambda} \left[-\frac{1}{\lambda} e^{-\lambda x} \right]_0^{\infty} \\
&= \frac{2}{\lambda^2},
\end{aligned}$$

where the third and fourth lines follow by using integration by parts. Thus, the variance is

$$\mathbb{V}(X) = \mathbb{E}(X^2) - [\mathbb{E}(X)]^2 = \frac{2}{\lambda^2} - \left(\frac{1}{\lambda}\right)^2 = \frac{1}{\lambda^2}.$$ ∎

■ **Example 6.4** Again returning to Example 6.2, what's the expected time between accidents? The standard deviation of the time between accidents?

Solution: From Theorem 6.3.5, for X the time between accidents with $\lambda = 2$, we have that

$$\mathbb{E}(X) = 1/\lambda = 1/2 = 0.5.$$

That is, the expected time between accidents is one-half of a week. The standard deviation is

$$\sigma_X = \sqrt{1/\lambda^2} = 1/\lambda = 1/2 = 0.5.$$ ■ ■ ■

Exercise 6.3 Between midnight and 3 a.m., the rate of cars arriving at a bridge toll booth is 10 per hour. What is the probability that, after the next arrival, the booth attendant will not see another car for at least 15 minutes? What is the mean time between arrivals? The variance?

Answers: $\Pr(X > 1/4) \approx 0.082$, $\mathbb{E}(X) = 0.1$ hour or 6 minutes, $\mathbb{V}(X) = 0.01$.

Similar to the notion of sample quantiles (see Section 1.5), the pth quantile for a distribution is defined as follows.

Definition 6.3.3 — Quantile. For a random variable X with continuous distribution F, the pth *quantile* ϕ_p is the smallest value of ϕ_p such that $F(\phi_p) = \Pr(X \le \phi_p) = p$. If random variable X has a discrete distribution, the quantile ϕ_p is the smallest value for which $\Pr(X \le \phi_p) \ge p$.

For an exponential random variable with rate parameter λ, you can calculate the values of the PDF and CDF at x using the following R functions:

$f(x)$: dexp(x, λ)

$F(x)$: pexp(x, λ)

Though we haven't discussed them much to this point, for every distribution R has two additional functions: one for calculating the pth *quantile*, ϕ_p, of a distribution, and another for generating random draws from the distribution. For the EXP(λ) distribution they are:

ϕ_p: qexp(p, λ)

Generate n random variates: rexp(n, λ)

The quantile is that value on the horizontal axis of the plot of a CDF corresponding to a given cumulative probability p. As shown in Figure 6.2, for a continuous CDF F, ϕ_p is the value on the x-axis that corresponds to the value p on the vertical axis. Think of it as reading the graph in reverse of how you normally would: Find the desired value of p on the vertical axis, trace that across to the function, and then down to the value of $x = \phi_p$ on the horizontal axis.

For continuous random variables whose CDFs have discontinuities, difficulties arise because there may not be one unique value ϕ_p that corresponds to p. In these cases, according to Definition 6.3.3, ϕ_p is defined as the *smallest* value such that $F(\phi_p) \ge p$. Things get even more complicated for discrete random variables, because the CDF is a step function and with such functions it is often the case that for a given value of p there is no value of ϕ_p such that $F(\phi_p) = p$. Thus, for discrete random variables, ϕ_p is the smallest value such that $F(\phi_p) \ge p$. See Figure 6.3.

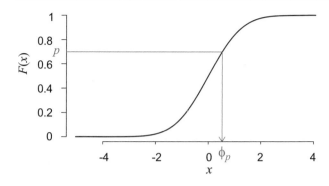

Figure 6.2 For a continuous random variable, a quantile is that value on the horizontal axis of the CDF corresponding to a given cumulative probability p.

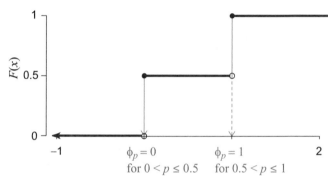

Figure 6.3 For a discrete random variable, a quantile is the smallest value such that $F(\phi_p) \geq p$. In this plot of a Bernoulli random variable, $\phi_p = 0$ for $0 \leq p \leq 0.5$ and $\phi_p = 1$ for $0.5 < p \leq 1$.

■ **Example 6.5** Use R to find $F(1)$ and $F(1/7)$ from Examples 6.2 and 6.3.

Solution: The following R code

```
pexp(1,2)          # F(1)
pexp(1/7,2)        # F(1/7)
```

yields

```
[1] 0.8646647
[1] 0.2485227
```
■ ■ ■

Exercise 6.4 In R, for $X \sim \text{EXP}(\lambda = 10)$:

(a) Calculate $f(0.25)$ and $F(0.25)$.

(b) Find the quartiles of the distribution. That is, find the 0.25, 0.5, and 0.75 quantiles.

Answers: $f(0.25) \approx 0.821$; $F(0.25) \approx 0.918$, $\phi_{0.25} \approx 0.029$; $\phi_{0.5} \approx 0.069$, $\phi_{0.75} = 0.139$.

Not all wait times are exponentially distributed. For example, Figure 6.4 is a histogram of the 2007 airline carrier delays. Overlaid is an exponential distribution with λ equal to the inverse of the sample mean of the carrier delays. If the carrier delays were truly exponential, the red curve would closely follow the histogram. Instead, what we see in Figure 6.4 is that the bars that correspond to delays less than 25 minutes or so are too large, while those corresponding to delays between 25 and about 100 minutes are too small. This is evidence that, while the histogram might have an "exponential-like" shape, the underlying phenomenon is not exponential. This should not be too surprising, however, because airline delays are caused by a number of factors that could be unique to a particular flight or that may affect the whole airport (such as extreme weather).

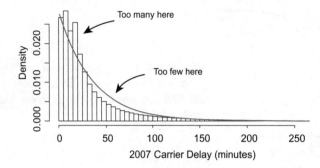

Figure 6.4 A histogram of the 2007 carrier delays (less than 250 minutes), with an exponential density function overlaid in red, showing that carrier delays do not follow an exponential distribution.

Exercise 6.5 Using the airline data, visually assess whether for some specific airport (your choice), `CarrierDelay` looks to be exponentially distributed. For the same airport, assess whether one of the other types of airline delay (`DepartureDelay`, `WeatherDelay`, `NASDelay`, `SecurityDelay`, or `LateAircraftDelay`) is exponentially distributed.

Exponentially distributed random variables have an important property referred to as the memoryless property.

Theorem 6.3.6 — Memoryless Property of Exponential Random Variables. For any exponentially distributed random variable X, $\Pr(X > t_1 + t_2 | X > t_1) = \Pr(X > t_2)$ where $t_1 > 0$ and $t_2 > 0$.

Proof. To prove Theorem 6.3.6, let's start by appropriately re-expressing the conditional probability,

$$\Pr(X > t_1 + t_2 | X > t_1) = \frac{\Pr(X > t_1 + t_2 \text{ and } X > t_1)}{\Pr(X > t_1)} = \frac{\Pr(X > t_1 + t_2)}{\Pr(X > t_1)}.$$

The second equality follows because the event $X > t_1$ is contained within $X > t_1 + t_2$. Now,

$$\Pr(X > t_1 + t_2) = 1 - \Pr(X \leq t_1 + t_2) = 1 - \left(1 - e^{-\lambda(t_1 + t_2)}\right) = e^{-\lambda(t_1 + t_2)}.$$

Similarly, since $\Pr(X > t_1) = e^{-\lambda t_1}$, we can write

$$\Pr(X > t_1 + t_2 | X > t_1) = \frac{e^{-\lambda(t_1 + t_2)}}{e^{-\lambda t_1}} = \frac{e^{-\lambda t_1} e^{-\lambda t_2}}{e^{-\lambda t_1}} = e^{-\lambda t_2} = \Pr(X > t_2). \quad \blacksquare$$

If we think about this in terms of waiting times, what the memoryless property says is that the probability of waiting for length of time $t_1 + t_2$, given you've already been waiting for t_1 time periods, is precisely the same as if you didn't know you had already been waiting for t_1 time periods and just had to wait for the remaining t_2 time periods. That is, an exponentially distributed random variable has no memory of how long it has been waiting (probabilistically speaking).

The exponential distribution is often parameterized differently from that given in Definition 6.3.1.

Definition 6.3.4 — Alternative Exponential Distribution. The PDF of an exponential random variable X with mean θ is

$$f(x) = \begin{cases} \dfrac{1}{\theta} e^{-x/\theta}, & 0 \leq x < \infty \\ 0, & -\infty < x < 0. \end{cases} \tag{6.2}$$

Parameters and support: The Greek letter θ is the expected value, where it must be that $\theta > 0; x \geq 0$.

Notation: $X \sim \text{EXP1}(\theta)$, which stands for "The random variable X has an exponential distribution with mean θ."

The parameters of the two parameterizations of the exponential distribution are related by $\theta = 1/\lambda$. Parameterized in terms of λ, the mean and variance are $1/\lambda$ and $1/\lambda^2$, and the hazard is λ, whereas if the exponential distribution is parameterized in terms of θ, the mean, variance, and hazard are θ, θ^2, and $1/\theta$, respectively. The exponential distribution is often used as a model for lifetimes, and is used by those in the medical sciences (where a lifetime is the time to death or time to recovery of a patient) and in engineering (where a lifetime is the time until a piece of equipment no longer functions). Although not universal, the $\text{EXP}(\lambda)$ parameterization is often used in medical sciences, whereas the $\text{EXP1}(\theta)$ parameterization is usually used in engineering.

■ **Example 6.6** The operational lifetime of an electronic component for a computer follows an exponential distribution with mean five years. Given that the component has successfully operated for five years, what is the probability that it continues to operate for at least another two years?

Solution: Let X denote the operational lifetime of the component. Then we know that $\mathbb{E}(X) = 5$, so $\lambda = 0.2$. The question is asking $\Pr(X > 7 | X > 5)$. Via the memoryless property of the exponential distribution, we know that $\Pr(X > 7 | X > 5) = \Pr(X > 2)$. Using the R command

```
pexp(2,0.2,lower.tail=FALSE)
```

we find that this probability is 0.67032. So, there is slightly more than a 67% chance that the component, which has been operating for the past five years will continue to operate for at least another two years.
■ ■ ■

In the next example we use the results from Chapter 4 regarding the expectation of the conditional expectation.

■ **Example 6.7** Suppose an insurance company insures homes for damage. For a given house, the number of claims in a given year is a (discrete) random variable K having a $\text{POIS}(0.2)$ distribution. For claim i the amount Y_i paid by the insurance company has an EXP1 distribution with a mean of $\$5,000$. The claim amount is independent of the number of claims made during the year. Find:

(a) the expected amount paid by the insurance company for all claims during the year;
(b) the standard deviation of the amount paid by the company for all claims during the year.

Solution: Let K be the number of claims during the year and let $Y_1, Y_2, \ldots, Y_K$ denote the amounts of these K claims. (Note that we have a random number of random variables.) Conditioned on $K = k$, the total amount W paid by the insurance company in the given year is $W = Y_1 + Y_2 + \cdots + Y_k$. Since the claim amounts are i.i.d.,

$$\mathbb{E}(W | K = k) = \mathbb{E}(Y_1 + Y_2 + \cdots + Y_k) = k \mathbb{E}(Y_1) = k \times \$5,000 = \$5,000k.$$

If we condition on a random number K of claims,

$$\mathbb{E}(W | K) = \$5,000 K$$

Thus, using Theorem 4.9.1,

$$\mathbb{E}(W) = \mathbb{E}\Big(\mathbb{E}(W | K)\Big) = \mathbb{E}(\$5,000 K) = \$5,000 \mathbb{E}(K) = \$5,000 \times 0.2 = \$1,000.$$

To get the variance of the yearly amount paid, first note that $Y_i|k \sim \text{EXP1}(5,000)$ so

$$\mathbb{V}(W|k) = \mathbb{V}(Y_1 + Y_2 + \cdots + Y_k) = k\mathbb{V}(Y_1) = (\$5,000)^2 k.$$

Then apply the second part of Theorem 4.9.1:

$$\mathbb{V}(W) = \mathbb{V}(\$5,000K) + \mathbb{E}\big((\$5,000)^2 K\big) = (\$5,000)^2\mathbb{V}(K) + (\$5,000)^2\mathbb{E}(K).$$

Since the mean and variance of K are both 0.2,

$$\mathbb{V}(W) = (\$5,000)^2 \times 0.2 + (\$5,000)^2 \times 0.2 = 0.4 \times (\$5,000)^2 = (\$)^2 10,000,000.$$

The units on the variance are "dollars squared." The standard deviation, measured in dollars, is

$$\sqrt{\mathbb{V}(W)} = \sqrt{(\$)^2 10,000,000} \approx \$3,162.28.$$

■■■

6.4 Normal Distribution

The *normal distribution* is an important distribution in statistics – perhaps the *most* important distribution. Many natural physical phenomena and statistics closely follow a normal distribution. This happens for a very good reason that will be described more fully in Section 8.5 when we discuss the *Central Limit Theorem* (CLT). Briefly, what the CLT says is that sums and averages of (i.i.d.) random variables tend toward the normal distribution. The more random variables that are summed or averaged, the closer the distribution of the sum or average is to a normal distribution.

> **Definition 6.4.1 — Normal Distribution.** The PDF of a normally distributed random variable X with parameters μ and σ is
>
> $$f(x) = \frac{1}{\sqrt{2\pi}\,\sigma} \exp\left(-\frac{(x-\mu)^2}{2\sigma^2}\right), \qquad -\infty < x < \infty.$$

The normal distribution is also often called the *Gaussian distribution*, in honor of Carl Friedrich Gauss (1777–1855), a famous German mathematician. Note that the name "normal" does not imply that random variables that follow another distribution are abnormal – they're simply not normally distributed.

Parameters and support: As we will see, the Greek letter μ is the mean of the distribution, with $-\infty < \mu < \infty$, and σ is the standard deviation, $\sigma > 0$. The support is $-\infty < x < \infty$.

Notation: $X \sim N(\mu, \sigma^2)$, which stands for "The random variable X has a normal distribution with mean μ and variance σ^2."

Theorem 6.4.1 — Expected Value and Variance of a Normally Distributed Random Variable. For $X \sim N(\mu, \sigma^2)$, $\mathbb{E}(X) = \mu$ and $\mathbb{V}(X) = \sigma^2$.

What Theorem 6.4.1 says is that the mean of a normally distributed random variable is the parameter μ and the variance of a normally distributed random variable is the parameter σ^2. From this, it follows that the standard deviation of a normally distributed random variable is σ. Although we can't prove this until Section 6.9, after we've introduced a few more tools, we can run some simulations to see that sample means and variances are close to the true values of μ and σ^2.

■ **Example 6.8** Generate one million random variates from the $N(10, 10)$ distribution and calculate the sample mean and sample standard deviation.

Solution: To simulate one million random numbers and compute the sample mean and variance, we can run the R code

```
some.data - rnorm(1000000,10,sqrt(10))
mean(some.data)
sd(some.data)
```

which produces the output

```
[1] 10.00064
[1] 3.163256
```

What we see is that the sample mean and standard deviation are very close to the true mean $\mu = 10$ and standard deviation $\sigma = \sqrt{10} \approx 3.162278$. They do not match precisely because there is randomness in the sample. If you reproduce this example on your computer, you will get slightly different values for your sample mean and standard deviation because you will have generated a different set of observations.   ■ ■ ■

The normal distribution with mean $\mu = 0$ and variance $\sigma^2 = 1$ is called the *standard normal distribution*. Often we use the letter Z to represent a standard normal random variable. In Figure 6.5 we see why the normal distribution is frequently referred to as the "bell curve," though it is important to know that it is not the only density function that looks like this. Soon we will learn about the t distribution, which looks very similar. Note that the normal distribution is symmetric about μ, meaning its PDF to the left of μ is the mirror image of the PDF to the right of μ. That is, if f is the normal density function, then $f(\mu + \delta) = f(\mu - \delta)$ for all $\delta \geq 0$. This is evident in Figure 6.5.

The normal distribution also has "thin tails," meaning that the curve drops quite sharply as you move away from the mean. For example, 99.7% of the probability of a normal distribution is within plus or minus three standard deviations of the mean. This is evident in Figure 6.5, where the density function $f(x)$ is nearly zero for $x > 3$ and $x < -3$. This property of thin tails is in particular reference to the t distribution described in Section 6.6.2.

Figure 6.6 compares the density functions for the standard normal with two other normal densities, one with mean $\mu = -2$ and the other with mean $\mu = +1$ (and both with variance $\sigma^2 = 1$). What Figure 6.6 shows is that changing the mean of the distribution simply shifts it along the x-axis. Figure 6.7 then compares the standard normal with two other normal densities, one with variance $\sigma^2 = 2$ and the other with variance $\sigma^2 = 3$ (and both with mean $\mu = 0$). Figure 6.6 shows that changing the variance of the distribution changes the spread of the probability, with larger σ^2 leading to greater dispersion in the probability.

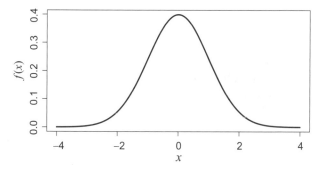

Figure 6.5 The PDF for a standard normal distribution.

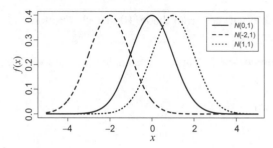

Figure 6.6 Comparison of the standard normal PDF versus two alternative normal PDFs with means -2 and $+1$ (and variance 1).

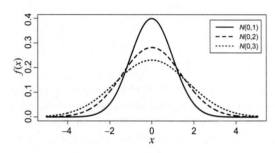

Figure 6.7 Comparison of the standard normal PDF versus two alternative normal PDFs with variances 2 and 3 (and mean 0).

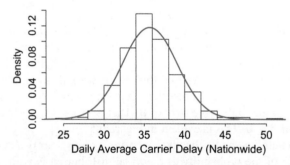

Figure 6.8 A histogram of the 2007 *average daily* carrier delays with the equivalent normal density function (setting $\mu = \bar{x}$ and $\sigma^2 = s^2$) overlaid in red. The plot shows that the average daily carrier delays seems to follow a normal distribution.

We saw in Figure 6.4 that 2007 airline carrier delays are not exponentially distributed. However, Figure 6.8 shows that the nationwide daily average carrier delay looks reasonably *normally* distributed. In the figure, the histogram shows the empirical distribution of 365 daily average carrier delays calculated from the data; the red line is the normal density fitted by setting the distributional parameters μ to $\bar{x} = 35.67$ minutes and σ^2 to $s^2 = 11.47$.

The normal PDF does not have an antiderivative that can be expressed in terms of elementary functions.[2] Fortunately, we can simply use R for the task. For a normally distributed random variable with parameters μ and σ^2, you can calculate the values of the PDF and CDF at x, as well as the pth quantile, ϕ_p, and you can generate random observations, using the following R functions:

$f(x)$: dnorm(x, μ, σ)

$F(x)$: pnorm(x, μ, σ)

ϕ_p: qnorm(p, μ, σ)

Generate n random variates: rnorm(n, μ, σ)

Note that the default value for μ is 0 and for σ is 1, so R syntax such as rnorm(n) will generate n standard normal variates. Also note that the argument for all of the R functions is σ and not σ^2.

---

[2]Elementary functions include polynomial, rational, trigonometric, logarithmic, and exponential functions, their inverses, and any composition of these. Basically, these are the functions that you study in algebra and precalculus.

■ **Example 6.9** For $X \sim N(0,1)$, use R to find $\Pr(Z \leq 0)$ and $\Pr(Z > 3)$. Next, for $X \sim N(2,3)$, use R to find $\Pr(X \leq 0)$ and $\Pr(X > 3)$.

Solution: The R code

```
pnorm(0)
pnorm(3,lower.tail=FALSE)
pnorm(0,2,sqrt(3))
1-pnorm(3,2,sqrt(3))
```

produces

```
[1] 0.5
[1] 0.001349898
[1] 0.1241065
[1] 0.2818514
```

Note how the use of the `lower.tail=FALSE` gives the complementary probability, just as we can use $\Pr(X > x) = 1 - \Pr(X \leq x)$. ■ ■ ■

Exercise 6.6 For $X \sim N(-4,7)$, use R to find $\Pr(X \leq -3)$ and $\Pr(X > 1)$.

Answers: $F(-3) \approx 0.647$, $1 - F(1) \approx 0.029$.

■ **Example 6.10** For $Z \sim N(0,1)$, use R to find the 0.025 and 0.975 quantiles. Next, find the same quantiles for $X \sim N(2,3)$.

Solution: The R code

```
qnorm(0.025)
qnorm(0.975)
qnorm(0.025,2,sqrt(3))
qnorm(0.975,2,sqrt(3))
```

produces

```
[1] -1.959964
[1] 1.959964
[1] -1.394757
[1] 5.394757
```

■ ■ ■

Exercise 6.7 For $X \sim N(-4,7)$, use R to find the first and third quartiles.

Answers: $\phi_{0.25} \approx -5.785$, $\phi_{0.75} \approx -2.215$.

6.4.1 Standardizing

It is conventional to use the letter Z to represent a random variable from a standard normal distribution. The symbol $\Phi(z)$ is used to represent the cumulative probability that a random variable Z that has a standard normal distribution is less than or equal to some number z: $\Phi(z) = \Pr(Z \leq z)$ In other words, Φ is the CDF of the standard normal distribution.

Definition 6.4.2 — Standardizing. An observation X is *standardized* by subtracting the mean and dividing by the standard deviation: $Z = (X - \mu)/\sigma$.

The pth quantile of the standard normal distribution is that value ϕ_p that satisfies $\Pr(Z \leq \phi_p) = p$, or using the CDF Φ, that value that satisfies $\Phi(\phi_p) = p$. Often, we will need an *upper* quantile of the normal distribution. If α is some (usually small) number between 0 and 1, the value that makes the right tail area of the standard normal distribution equal to α is denoted z_α. We will use this notation for other distributions as well.

For normally distributed random variables, the following theorem says that the standardized random variable has a standard normal distribution. We will prove this in Section 6.9.

Theorem 6.4.2 — Standardized Normally Distributed Random Variables. For $X \sim N(\mu, \sigma^2)$, the standardized random variable $Z = (X - \mu)/\sigma \sim N(0,1)$.

Theorem 6.4.2 can be useful for calculating various probabilities. For example, for $X \sim N(\mu, \sigma^2)$,

$$\Pr(X \leq x) = \Pr(X - \mu < x - \mu) = \Pr\left(\frac{X-\mu}{\sigma} \leq \frac{x-\mu}{\sigma}\right) = \Pr\left(Z \leq \frac{x-\mu}{\sigma}\right) = \Phi\left(\frac{x-\mu}{\sigma}\right).$$

Similarly, for $X \sim N(\mu, \sigma^2)$,

$$\Pr(a < X < b) = \Pr\left(\frac{a-\mu}{\sigma} < Z < \frac{b-\mu}{\sigma}\right) = \Phi\left(\frac{b-\mu}{\sigma}\right) - \Phi\left(\frac{a-\mu}{\sigma}\right).$$

To demonstrate the equivalence empirically in R, let $X \sim N(5,9)$. Then notice that the following calculations for $\Pr(X \leq 7)$ give the same result:

```
pnorm(7,5,3)
pnorm((7-5)/3,0,1)
```

which produces

```
[1] 0.7475075
[1] 0.7475075
```

Standardization is useful because it allows us to relate any normal distribution to the standard normal. We can also go the other direction: If $Z \sim N(0,1)$ then $X = \mu + Z \times \sigma$ has a $N(\mu, \sigma^2)$ distribution.

Now, given that we can use R to calculate probabilities for any normal distribution, you might be wondering why standardization is necessary. One answer is that before the existence of software such as R, it was necessary to look up probabilities for normal distributions in tables. Yet, with standardization, only one table for the standard normal distribution was necessary since the probabilities for any other normal distribution could be found using the same table. Table 6.1 is an (abbreviated) example of such a table, giving $\Pr(-z \leq Z \leq z)$.

All normal distributions are essentially the same if one thinks about distance in terms of standard deviations (rather than the natural units of the problem). For example, in Table 6.1, the bold values are both the probability that a random variable Z with a standard normal distribution is within ± 1, ± 2, and ± 3 from 0 and, as Theorem 6.4.4 states, the probability that a random variable X with a $N(\mu, \sigma^2)$ distribution is within ± 1, ± 2, and ± 3 standard deviations (σ) from μ.

Table 6.1 The probability that a random variable with a standard normal distribution falls between $-z$ and z: $\Pr(-z \le Z \le z)$. The bold numbers show the probabilities for falling with one, two, and three standard deviations of the mean.

z	$\Pr(-z \le Z \le z)$	z	$\Pr(-z \le Z \le z)$
0.100	0.0797	1.400	0.8385
0.200	0.1585	1.500	0.8664
0.300	0.2358	1.600	0.8904
0.400	0.3108	1.645	0.9000
0.500	0.3829	1.700	0.9109
0.600	0.4515	1.800	0.9281
0.700	0.5161	1.900	0.9426
0.800	0.5763	1.960	0.9500
0.900	0.6319	**2.000**	**0.9545**
1.000	**0.6827**	2.500	0.9876
1.100	0.7287	2.576	0.9900
1.200	0.7699	**3.000**	**0.9973**
1.300	0.8064	4.000	0.9999

Theorem 6.4.3 — A Special Probability. For $X \sim N(\mu, \sigma^2)$, $Z \sim N(0,1)$, and $k > 0$,

$$\Pr(|X - \mu| \le k\sigma) = \Pr(-k \le Z \le k).$$

Proof. The proof follows directly from Definition 6.4.2 and Theorem 6.4.2. That is, for $X \sim N(\mu, \sigma^2)$, $Z \sim N(0,1)$, and $k > 0$,

$$\Pr(|X - \mu| \le k\sigma) = \Pr(-k\sigma \le X - \mu \le k\sigma)$$

$$= \Pr\left(\frac{-k\sigma}{\sigma} \le \frac{X - \mu}{\sigma} \le \frac{k\sigma}{\sigma} \right)$$

$$= \Pr(-k \le Z \le k),$$

where $Z \sim N(0,1)$.   ∎

■ **Example 6.11** What is the probability that a normally distributed random variable is within 2.5 standard deviations of the mean?

Solution: From Table 6.1, $\Pr(|X - \mu| \le 2.5\sigma) = 0.9876$, and we can verify the result in R with the command `pnorm(2.5)-pnorm(-2.5)` which gives   ■ ■ ■

```
[1] 0.9875807
```

Although we solved this problem for the *standard* normal distribution, it is true for any normal distribution. For example, if $X \sim N(2,4)$, then

```
pnorm(2+2.5*2,2,2)-pnorm(2-2.5*2,2,2)
```

yields the same result.

Exercise 6.8 What is the probability that for $X \sim N(\mu, \sigma^2)$, X is more than three standard deviations away from the mean?

Answer: $\Pr(X < \mu - 3\sigma \text{ or } X > \mu + 3\sigma) \approx 0.0027$.

Theorem 6.4.3 and Table 6.1 imply that, for a normally distributed random variable:

- the probability an observation is within one standard deviation of the mean is 68%;[3]
- the probability an observation is within two standard deviations of the mean is 95%; and
- the probability an observation is within three standard deviations of the mean is 99.7%.

For nonnormal but reasonably symmetric distributions, these probabilities are referred to as the *empirical rule*. By Theorem 6.4.4, the empirical rule is a useful rule of thumb that provides a rough estimate of the probability an observation is within one, two, or three standard deviations of the mean.

> **Theorem 6.4.4 — Empirical Rule.** For X that has a nonnormal but reasonably symmetric distribution:
> $$\Pr(|X - \mu| \leq \sigma) \approx 0.68,$$
> $$\Pr(|X - \mu| \leq 2\sigma) \approx 0.95, \text{ and}$$
> $$\Pr(|X - \mu| \leq 3\sigma) \approx 0.997.$$

The mean average daily delay in the flight data from Chapter 1 was $\bar{x} = 35.7$ minutes, with a sample standard deviation of $s = 3.4$ minutes. So, out of the 365 daily carrier delay averages, 70.4% fall in the interval $[\bar{x}-s, \bar{x}+s]$, 95.6% fall in the interval $[\bar{x}-2s, \bar{x}+2s]$, and 99.2% fall in the interval $[\bar{x}-3s, \bar{x}+3s]$. All are quite close to the empirical rule, as expected, since the distribution of average carrier delay shown in Figure 6.8 looks quite normal.

On the other hand, returning to the flight data from Chapter 1, the mean carrier delay was $\bar{x} = 35.4$ minutes with a sample standard deviation of $s = 53.5$ minutes. So, out of the 814,922 carrier delays in 2007, 91.4% fall in the interval $[\bar{x}-s, \bar{x}+s]$, 96.6% fall in the interval $[\bar{x}-2s, \bar{x}+2s]$, and 98.4% fall in the interval $[\bar{x}-3s, \bar{x}+3s]$. Here we see that, because the distribution of carrier delays shown in Figure 6.4 is neither normal nor symmetric, the empirical rule does not work nearly as well, particularly at plus or minus one standard deviation around the mean.

> **Exercise 6.9** Using the `ActualElapsedTime` variable in the 2007 airline data, find the fraction of the 7,275,288 flights within one-, two-, and three-sample standard deviations of the sample mean.
>
> *Answers*: 76.4% of 2007 flight times are in the interval $[\bar{x}-s, \bar{x}+s]$, 92.5% are in the interval $[\bar{x}-2s, \bar{x}+2s]$, and 96.1% are in the interval $[\bar{x}-3s, \bar{x}+3s]$.

6.4.2 Bivariate and Multivariate Normal Distributions

The bivariate normal distribution is a *joint distribution* of two variables, X_1 and X_2. Probabilities that (X_1, X_2) falls in some region R can be found by taking the double integral

$$\Pr\left((X_1, X_2) \in R\right) = \iint_R f(x_1, x_2) \, dx_2 \, dx_1,$$

where $f(x_1, x_2)$ is the joint PDF of the bivariate normal, given in Definition 6.4.3. This is the two-dimensional generalization of the univariate normal distribution we have studied so far.

---

[3]Note that "percent" means "per hundred," so 68 percent means 68 per 100, a ratio that equals 0.68. We sometimes talk about probabilities in percentage terms, but this should be interpreted as the percentage divided by 100, which will always be between 0 and 1.

Definition 6.4.3 — Bivariate Normal Distribution. The bivariate normal PDF has five parameters: μ_1, μ_2, σ_1^2, σ_2^2, and σ_{12}:

$$f(x_1,x_2) = \frac{1}{2\pi\sqrt{\sigma_1^2\sigma_2^2 - \sigma_{12}^2}} \exp\left(-\frac{1}{2}\begin{bmatrix} x_1 - \mu_1 & x_2 - \mu_2 \end{bmatrix} \begin{bmatrix} \sigma_1^2 & \sigma_{12} \\ \sigma_{12} & \sigma_2^2 \end{bmatrix}^{-1} \begin{bmatrix} x_1 - \mu_1 \\ x_2 - \mu_2 \end{bmatrix}\right).$$

This can be written in the form

$$f(x_1,x_2) = \frac{1}{2\pi\sigma_1\sigma_2\sqrt{1-\rho^2}} \exp\left(-\frac{\frac{(x_1-\mu_1)^2}{\sigma_1^2} - \frac{2\rho(x_1-\mu_1)(x_2-\mu_2)}{\sigma_1\sigma_2} + \frac{(x_2-\mu_2)^2}{\sigma_2^2}}{2(1-\rho^2)}\right), \tag{6.3}$$

where $\rho = \sigma_{12}/\sigma_1\sigma_2$.

Parameters and support: The means of X_1 and X_2 are μ_1 and μ_2, which can be any real numbers $-\infty < \mu_1, \mu_2 < \infty$. The variances of X_1 and X_2 are σ_1^2 and σ_2^2, which must be positive. Finally, the correlation ρ between the two variables must satisfy $-1 \leq \rho \leq 1$. If $\rho = 0$, then X_1 and X_2 are independent.[4] The density is defined over the entire real plane: $-\infty < x_1, x_2 < \infty$.

Notation: If X_1 and X_2 have the bivariate distribution with PDF (6.3), then we write

$$\begin{bmatrix} X_1 \\ X_2 \end{bmatrix} \sim \text{BVN}\left(\mu_1, \mu_2, \sigma_1^2, \sigma_2^2, \rho\right) \qquad \text{or} \qquad \begin{bmatrix} X_1 \\ X_2 \end{bmatrix} \sim N_2\left(\begin{bmatrix} \mu_1 \\ \mu_2 \end{bmatrix}, \Sigma\right)$$

where

$$\Sigma = \begin{bmatrix} \sigma_1^2 & \rho\sigma_1\sigma_2 \\ \rho\sigma_1\sigma_2 & \sigma_2^2 \end{bmatrix}$$

is the *covariance matrix*, that is, the matrix whose diagonal elements are the variances, and the off-diagonal element σ_{ij} is the covariance between X_i and X_j.

The following R code produces an interactive plot using the `plotly` package. After you run this plot, type either `fig1` or `fig2` at the prompt. When the plot you chose is rendered, you can click and drag to see the surface from different positions. Moving the mouse over the contour plot shows you the (x,y,z) coordinate of a point on the surface.

```
install.packages(c("mnormt","plotly"))
library( mnormt )
library(plotly)
mu1 = 0
mu2 = 0
mu = c(mu1,mu2)
sigma1 = 2
sigma2 = 2
rho = 0.7
sigma12 = sigma1*sigma2*rho
```

---

[4]It is always true that if X and Y are independent, then $\rho = 0$. The converse (that $\rho = 0$ implies independence) is not always true. There are examples of random variables X and Y for which $\rho = 0$ but X and Y are not independent (see Problem 4.49). However, if the joint distribution is a bivariate normal distribution with parameters $\mu_1, \mu_2, \sigma_1, \sigma_2$, and ρ, and if $\rho = 0$, then X and Y are independent. See Problem 6.27.

(a)

(b)

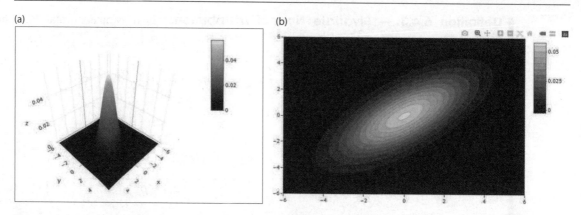

Figure 6.9 Result of running `plotly` to produce a surface plot (with `fig1`) or a contour plot (with `fig2`) using the R code in this section. Clicking and dragging on the surface plot in (a) allows you to see the surface from various angles. Mousing over the contour plot (b) shows you the (x, y, z) position of a point on the plot.

```
sigma = matrix( c( sigma1^2 , sigma12 , sigma12 , sigma2^2 ) , nrow=2 )
x = seq( mu1 - 3*sigma1, mu1 + 3*sigma1 , length=100 )
y = seq( mu2 - 3*sigma2, mu2 + 3*sigma2 , length=100 )
f = function(x, y) dmnorm(cbind(x, y), mu, sigma)
z = outer(x, y, f)
fig <- plot_ly( x=x , y=y , z=z )
fig1 = fig %>% add_surface()
fig2 = fig %>% add_contour
##  At the R prompt in the R console, type  fig1  or  fig2
```

Figure 6.9 shows the surface plot and the contour plot produced with the code above. We invite the reader to experiment with different values of `mu1`, `mu2`, `sigma1`, `sigma2`, and `rho` (the correlation) to see the effect of each of these parameters on the distribution.

The concepts of the CDF and quantiles still apply to the bivariate normal distribution (as well as any other multivariate distribution), though they become more complicated. For example, the CDF F for a bivariate distribution is defined as $F(x_1, x_2) = \Pr(X_1 \le x_1, X_2 \le x_2)$. That is, it is the probability that $X_1 \le x_1$ and $X_2 \le x_2$.

The bivariate normal can be further generalized as the *multivariate normal distribution*. Let $\mathbf{X}$ denote a k-dimensional vector: $\mathbf{X} = [X_1, X_2, \ldots, X_k]'$. Then the multivariate normal distribution in k dimensions is defined as follows.

Definition 6.4.4 — Multivariate Normal Distribution. The multivariate normal probability density function is:

$$f(x_1, \ldots, x_k) = \frac{1}{(2\pi)^{k/2} \mid \Sigma \mid^{1/2}} \exp\left(-\frac{1}{2} (\mathbf{x} - \boldsymbol{\mu})' \Sigma^{-1} (\mathbf{x} - \boldsymbol{\mu}) \right),$$

where $\mid \Sigma \mid$ denotes the determinant of Σ.

Parameters and support: The mean vector is $\boldsymbol{\mu}$ and the variance–covariance matrix is Σ. The density is defined over the entire k-dimensional space: $-\infty < x_1, x_2, \ldots, x_k < \infty$.

Notation: $\mathbf{X} \sim N_k(\boldsymbol{\mu}, \Sigma)$ denotes that $\mathbf{X}$ has a multivariate normal distribution with mean vector $\boldsymbol{\mu}$ and covariance matrix Σ.

For $k = 2$, Definition 6.4.4 is equivalent to Definition 6.4.3 with $\mu = \left[\mu_1, \mu_2\right]'$ and

$$\Sigma = \left[\begin{array}{cc} \sigma_1^2 & \rho\sigma_1\sigma_2 \\ \rho\sigma_2\sigma_1 & \sigma_2^2 \end{array} \right].$$

Similarly, for $k = 1$, Definition 6.4.4 reduces to the univariate normal distribution with the PDF shown in Definition 6.4.1, where Σ is a 1×1 matrix (i.e., it's a real number).

For a multivariate normally distributed random variable in $k > 1$ dimensions, the `mvtnorm` package is useful for PDF, CDF, quantile, and random variate generation calculations:

$f(x)$: `dmvnorm(`x`, mean=`μ`, sigma=`Σ`)`

$F(x)$: `pmvnorm(`x`, mean=`μ`, sigma=`Σ`)`

ϕ_p: `qmvnorm(`p`, mean=`μ`, sigma=`Σ`)`

Generate n random variates: `rmvnorm(`n`, mean=`μ`, sigma=`Σ`)`

Here, μ is a k-dimensional mean vector and Σ is a $k \times k$ variance–covariance matrix of the multivariate normal distribution. In the `dmvnorm()` function, x is the k-dimensional vector at which to calculate the PDF or CDF value. In the `pmvnorm()` function, $x = [x_1, \ldots, x_k]'$ is the k-dimensional vector at which to calculate the cumulative probability $F(x)$. The `qmvnorm()` function computes the vector $x = [x, x, \ldots, x]$, with all components equal, such that $\Pr(X_1 \leq x, X_2 \leq x, \ldots, X_k \leq x) = p$.

6.5 Gamma and Weibull Distributions

The gamma and Weibull distributions are two-parameter families of continuous probability distributions that are often used to model lifetimes.

6.5.1 Gamma Distribution

We begin with the gamma distribution. It is defined on the positive real line and is skewed to the right. It is widely used to model physical quantities that only take on positive values.

> **Definition 6.5.1 — Gamma Distribution.** The PDF for random variable X having a gamma distribution with parameters α and β is
>
> $$f(x) = \frac{x^{\alpha-1}}{\beta^\alpha \Gamma(\alpha)} e^{-x/\beta}, \qquad x > 0,$$
>
> where the gamma function, $\Gamma()$, is defined as
>
> $$\Gamma(t) = \int_0^\infty x^{t-1} e^{-x} dx. \tag{6.4}$$
>
> *Parameters and support*: The "shape" parameter is α, $\alpha > 0$, and the "scale" parameter is β, $\beta > 0$. The support is $x > 0$.
>
> *Notation*: $X \sim \text{GAM}(\alpha, \beta)$, which stands for "The random variable X has a gamma distribution with parameters α and β."

The gamma function is built into R as `gamma()`. As shown in Figure 6.10, a wide variety of PDF shapes are possible with the gamma distribution. In fact, the χ^2 and exponential distributions are special cases of the gamma distribution:
- The exponential distribution is the gamma distribution with $\alpha = 1$ and $\beta = 1/\lambda$.

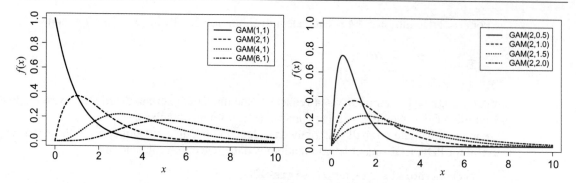

Figure 6.10 Some gamma distribution PDFs for various values of α and β.

- The chi-squared distribution with ν degrees of freedom for a positive integer ν (covered in Section 6.6.1) is the gamma distribution with parameters $\alpha = \nu/2$ and $\beta = 2$.

Theorem 6.5.1 — Mean and Variance of the Gamma Distribution. If X has a gamma distribution with parameters α and β, then $\mathbb{E}(X) = \alpha\beta$ and $\mathbb{V}(X) = \alpha\beta^2$.

Proof.

$$
\begin{aligned}
\mathbb{E}(X) &= \int_{-\infty}^{\infty} x f(x) dx \\
&= \int_0^{\infty} x \left(\frac{x^{\alpha-1} e^{-x/\beta}}{\beta^\alpha \Gamma(\alpha)} \right) dx \\
&= \frac{\beta^{\alpha+1} \Gamma(\alpha+1)}{\beta^\alpha \Gamma(\alpha)} \underbrace{\int_0^{\infty} \frac{x^{(\alpha+1)-1}}{\beta^{\alpha+1} \Gamma(\alpha+1)} e^{-x/\beta} dx}_{\text{integral is 1 because the integrand is the GAM}(\alpha,\beta)\text{ PDF}} \\
&= \frac{\beta\left(\alpha\Gamma(\alpha)\right)}{\Gamma(\alpha)} \\
&= \alpha\beta.
\end{aligned}
$$

The fifth line follows from the fourth because $\Gamma(\alpha+1) = \alpha\Gamma(\alpha)$ for any positive number α. See Problem 6.38.

Now, for the variance, we start by finding $\mathbb{E}(X^2)$ using the same reasoning as above. Problem 6.41 asks you to show that $\mathbb{E}(X^2) = \alpha(\alpha+1)\beta^2$. The variance is therefore

$$
\mathbb{V}(X) = \mathbb{E}(X^2) - \mathbb{E}(X)^2 = \alpha(\alpha+1)\beta^2 - (\alpha\beta)^2 = \alpha^2\beta^2 + \alpha\beta^2 - \alpha^2\beta^2 = \alpha\beta^2. \qquad\blacksquare
$$

■ **Example 6.12** Let's return to the 2007 airline data and fit a gamma distribution to the flight times. To do so, we'll peek ahead to Chapter 9 and use the "method of moments." The idea is to estimate the distributional parameters by equating the mean and variance of the estimated distribution and the mean and variance from the sample. In Example 9.14 we show that the estimators $\hat{\alpha}$ and $\hat{\beta}$ are[5]

$$
\hat{\alpha} = \frac{\bar{x}^2}{\frac{n-1}{n} S^2} \qquad \text{and} \qquad \hat{\beta} = \frac{\frac{n-1}{n} S^2}{\bar{x}}.
$$

---

[5]Here we employ the convention that estimators for parameters are indicated with a "hat" over the parameter; thus, $\hat{\alpha}$ indicates an estimate for α based on a sample from the distribution. More on this in Chapter 9.

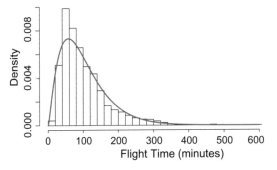

Figure 6.11 A gamma distribution fit to the 2007 airlines `AirTime` data with $\hat{\alpha} = 2.27$ and $\hat{\beta} = 45.30$.

Plot the fitted gamma distribution along with a histogram of the 2007 flight times.

Solution: Figure 6.11 shows the histogram of 2007 flight times (the `AirTime` data) with the fitted gamma distribution with $\hat{\alpha} = 2.27$ and $\hat{\beta} = 45.30$ overlaid. We see that the gamma PDF matches the overall shape of the data, though there are some deviations where the gamma underestimates the number of flights with flight times between 40 and 100 minutes and it overestimates the number of flights with flight times between 160 and 220 minutes or so.   ∎∎∎

The R syntax for the gamma distribution PDF, CDF, quantile, and random variate generation calculations is:

$f(x)$: `dgamma(x, α, 1/β)`

$F(x)$: `pgamma(x, α, 1/β)`

ϕ_p: `qgamma(p, α, 1/β)`

Generate n random variates: `rgamma(n, α, 1/β)`

> **Exercise 6.10** For $X \sim \mathrm{GAM}(2.27, 45.3)$, as shown in Figure 6.11, find:
>
> (a) $f(60)$,
>
> (b) $\Pr(X \le 100)$, $\Pr(X > 300)$, and
>
> (c) the median.
>
> *Answers:* $f(60) \approx 0.0073$; $\Pr(X \le 100) \approx 0.572$, $\Pr(X > 300) \approx 0.015$; $\phi_{0.5} \approx 88.19$.

The gamma distribution is also used as a model for lifetimes. Its hazard function is

$$h(t) = \frac{f(t)}{1 - F(t)} = \frac{\dfrac{x^{\alpha-1}}{\beta^{\alpha}\Gamma(\alpha)}\exp(-x/\beta)}{\displaystyle\int_x^{\infty}\frac{u^{\alpha-1}}{\beta^{\alpha}\Gamma(\alpha)}\exp(-u/\beta)\,du} = \frac{x^{\alpha-1}\exp(-x/\beta)}{\displaystyle\int_x^{\infty}u^{\alpha-1}\exp(-u/\beta)\,du}.$$

Let's make the substitution $y = u/\beta$ in the integral in the denominator. Then $dy = (1/\beta)\,du$. If $u = x$ then $y = x/\beta$, and if $u \to \infty$ then $y \to \infty$. We can therefore write the denominator as

$$\int_x^{\infty} u^{\alpha-1}\exp(-u/\beta)\,du = \int_{x/\beta}^{\infty}(\beta y)^{\alpha-1}\exp(-y)(\beta\,dy) = \beta^{\alpha}\int_{x/\beta}^{\infty}y^{\alpha-1}\exp(-y)\,dy = \beta^{\alpha}\Gamma\left(\alpha, \frac{x}{\beta}\right).$$

Here, $\Gamma(\alpha, z) = \int_z^{\infty} y^{\alpha-1}\exp(-y)\,dy$ is the upper incomplete gamma function. The R package `expint` includes the function `gammainc()` that can be used to evaluate the incomplete gamma function.

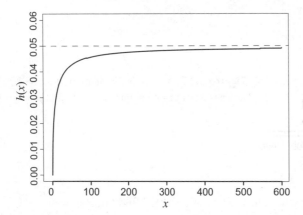

Figure 6.12 Hazard function for the GAM(1.5,20) distribution.

■ **Example 6.13** Sketch the hazard function for the GAM$(1.5, 100)$ distribution.

Solution: The following R code accomplishes this task:

```
install.packages("expint")
alpha = 1.5
beta = 20
library(expint)
x = seq( 0 , 600 , 1 )
y = dgamma( x , shape=alpha , scale=beta ) /
                    ( 1 - pgamma( x , shape=alpha , scale=beta ) )
y1 = x^(alpha-1) * exp(-x/beta) / ((beta^alpha)*gammainc(alpha,x/beta))
plot( x , y1 , type="l" , ylim=c(0,1.2/beta), ylab="h(x)")
abline( h=1/beta, lty=2)
```

The variable y is the hazard computed using the "d" and "p" functions for the gamma distribution. The variable y1 uses the incomplete gamma function `incgamma`. Plotting x against y or y1 gives exactly the same graph, which is shown in Figure 6.12. Note how the hazard begins at 0 when $x = 0$ and increases quickly at first, then levels off at 0.05.   ■ ■ ■

The phenomenon in Figure 6.12 is typical for the gamma distribution. The asymptotic value is

$$\lim_{x\to\infty} \frac{\dfrac{x^{\alpha-1}}{\beta^\alpha\,\Gamma(\alpha)\,e^{x/\beta}}}{\displaystyle\int_x^\infty \dfrac{u^{\alpha-1}}{\beta^\alpha\,\Gamma(\alpha)\,e^{u/\beta}}\,du} = \frac{1}{\beta}. \tag{6.5}$$

See Problem 6.42 for a demonstration of this.

6.5.2 Weibull Distribution

The Weibull distribution is defined below.

Definition 6.5.2 — Weibull Distribution. The PDF for the random variable X having a Weibull distribution with parameters α and β is

$$f(x) = \frac{\beta}{\theta}\left(\frac{x}{\theta}\right)^{\beta-1}\exp\left(-\left(\frac{x}{\theta}\right)^\beta\right), \qquad x > 0,$$

where, as before, the gamma function is $\Gamma(t) = \int_0^\infty x^{t-1} e^{-x} dx$.

Parameters and support: The "shape" parameter is β, $\beta > 0$, and the "scale" parameter is θ, $\theta > 0$. The support is $x > 0$.

Notation: $X \sim \mathrm{WEIB}(\theta, \beta)$, which stands for "The random variable X has a Weibull distribution with parameters θ and β."

The support for the Weibull is $(0, \infty)$, so it is often used to model random variables that are constrained to be positive, such as lifetimes, impurities, etc. The Weibull can take on a number of possible shapes, as shown in Figure 6.13. The PDFs shown in Figure 6.13(a) have inherently different shapes. When $\beta = 0.8$ the PDF approaches ∞ as $x \to 0^+$; when $\beta = 1$, the PDF satisfies $f(0) = 1/\theta = 0.025$; and when $\beta > 1$, the PDF begins at 0, increases to a maximum, then decreases, approaching 0 as $x \to \infty$. The graph in (b) holds β fixed while θ varies. Note how all of these PDFs in Figure 6.13(b) have the same basic shape: They start at 0, reach a maximum, then decrease and approach 0. Varying the scale parameter θ changes the scaling of the distribution, with larger values of θ leading to "stretched out" PDFs.

The CDF of the Weibull has a closed-form expression:

$$F(x) = \int_0^x \frac{\beta}{\theta} \left(\frac{u}{\theta}\right)^{\beta-1} \exp\left(-\left(\frac{u}{\theta}\right)^\beta\right) du = 1 - \exp\left(-\left(\frac{x}{\theta}\right)^\beta\right), \qquad x > 0. \qquad (6.6)$$

The survival function for a random variable X is defined as follows.

Definition 6.5.3 — Survival Function. Let X be a random variable with PDF $f(x)$ and CDF $F(x)$. The *survival function* $S(x)$ is defined to be the probability that the random variable *exceeds* x:

$$S(x) = \Pr(X > x) = 1 - \Pr(X \le x) = 1 - F(x).$$

For a random variable that represents the lifetime of a person, animal, or item, $S(x)$ is the probability of survival past time x.

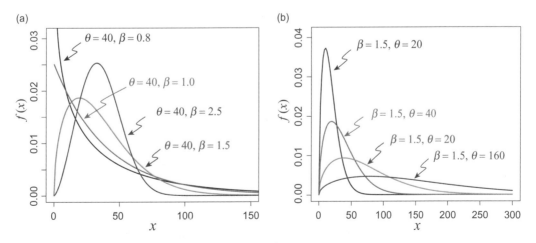

Figure 6.13 Various shapes for the Weibull PDF. In (a), the scale parameter θ is fixed at 40 while β varies. In (b), the shape parameter β is fixed at 1.5 while θ varies.

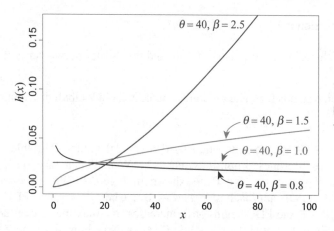

Figure 6.14 Hazard function for varying values of β.

The survival function for the Weibull also has a closed-form solution:

$$S(x) = 1 - \left[1 - \exp\left(-\left(\frac{x}{\theta}\right)^\beta \right) \right] = \exp\left(-\left(\frac{x}{\theta}\right)^\beta \right), \qquad x > 0.$$

Since both the PDF and survival function have closed-form expressions, so does the hazard function:

$$h(x) = \frac{\frac{\beta}{\theta} \left(\frac{x}{\theta}\right)^{\beta-1} \exp\left(-\left(\frac{x}{\theta}\right)^\beta \right)}{\exp\left(-\left(\frac{x}{\theta}\right)^\beta \right)} = \frac{\beta}{\theta} \left(\frac{x}{\theta}\right)^{\beta-1}, \qquad x > 0. \tag{6.7}$$

The shape of the hazard function depends on the parameter β. When $0 < \beta < 1$, the hazard function is decreasing because the exponent on x in (6.7), which is $(\beta - 1)$, is negative. When $\beta = 1$, the exponent is 0, so the hazard is constant. For $1 < \beta < 2$, the exponent is between 0 and 1, so the hazard is increasing but concave down. Finally, for $\beta > 2$, the exponent is greater than 1, so the hazard is increasing and concave up. These situations are illustrated in Figure 6.14.

The mean and variance of the Weibull distribution are given in the following theorem.

> **Theorem 6.5.2 — Mean and Variance of the Weibull Distribution.** Let $X \sim \text{WEIB}(\theta, \beta)$. Then
>
> $$\mathbb{E}(X) = \theta\, \Gamma\left(1 + \frac{1}{\beta} \right) \tag{6.8}$$
>
> $$\mathbb{V}(X) = \theta^2 \left[\Gamma\left(1 + \frac{2}{\beta} \right) - \Gamma^2\left(1 + \frac{1}{\beta} \right) \right]. \tag{6.9}$$

Proof. We can evaluate $\mathbb{E}(X)$ by integrating:

$$\mathbb{E}(X) = \int_0^\infty x\, \frac{\beta}{\theta} \left(\frac{x}{\theta}\right)^{\beta-1} \exp\left(-\left(\frac{x}{\theta}\right)^\beta \right) dx = \frac{\beta}{\theta^\beta} \int_0^\infty x^\beta \exp\left(-\left(\frac{x}{\theta}\right)^\beta \right) dx.$$

Following the change of variable $u = (x/\theta)^\beta$, with $x = \theta\, u^{1/\beta}$ and $dx = (\theta/\beta) u^{(1/\beta)-1}$, we can write

$$\mathbb{E}(X) = \frac{\beta}{\theta^\beta} \int_0^\infty \left(\theta\, u^{1/\beta} \right)^\beta e^{-u} \frac{\theta}{\beta} u^{(1/\beta)-1}\, du = \theta \int_0^\infty u^{(1+1/\beta)-1} e^{-u}\, du = \theta\, \Gamma\left(1 + \frac{1}{\beta} \right).$$

We leave the proof of the variance as a problem. See Problem 6.47.   ∎

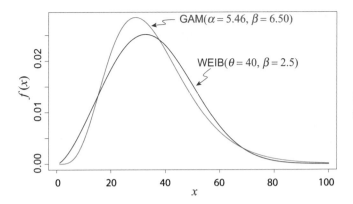

Figure 6.15 Gamma and Weibull PDFs for Example 6.14.

■ **Example 6.14** Consider the following distributions that might be used to model the lifetimes of a mechanical system:

1. GAM(5.46,6.50),

2. WEIB(20,2.5).

Compare these distributions as possible models for the lifetimes.

Solution: The following R code computes the mean, variance, 0.05 quantile, and the probability of surviving past time 80.

```
q = 0.05
x0 = 80
alpha = 5.46;  beta = 6.50
E1 = alpha*beta
V1 = alpha*beta^2
p1 = pgamma(x0,shape=alpha,scale=beta,lower.tail=FALSE)
q1 = qgamma(q,shape=alpha,scale=beta)
c(E1,V1,p1,q1)
theta = 40;  beta = 2.5
E2 = theta*gamma(1+1/beta)
V2 = theta^2 * ( gamma(1+2/beta) - (gamma(1+1/beta))^2 )
p2 = pweibull(x0,shape=beta,scale=theta,lower.tail=FALSE)
q2 = qweibull(q,shape=beta,scale=theta)
c(E2,V2,p2,q2)
```

The output from this code is

```
[1] 3.549000e+01 2.306850e+02 9.963911e-03 1.470107e+01
[1] 3.549055e+01 2.306347e+02 3.493489e-03 1.219226e+01
```

We see that the means and variances are nearly the same:

$$\mu_1 = 3.549000 \quad \text{vs.} \quad \mu_2 = 3.549055$$
$$\sigma_1^2 = 230.685 \quad \text{vs.} \quad \sigma_2^2 = 230.6347.$$

The 0.05 quantiles differ somewhat, 14.70 vs. 12.19. There is a fairly big discrepancy in the survival probability for 80 time units. For the gamma distribution, the survival probability is 0.00996, while for the Weibull distribution it is 0.00349. The two PDFs are shown in 6.15.   ■ ■ ■

The gamma and Weibull distributions have similar shapes, and they can agree near the center of the distributions, but have different properties in the tails. As we saw in the last example, the quantiles and tail probabilities can be quite different. Unfortunately, in life testing the tails are often of primary interest. This makes the problem of model selection challenging.

6.6 Distributions Related to the Normal

In this section we will learn about three distributions related to the normal distribution: the chi-square (or χ^2) distribution, the t distribution, and the F distribution. In later chapters these distributions, along with the normal distribution, will have large roles to play in various statistical methods.

6.6.1 Chi-square (χ^2) Distribution

The χ^2 distribution is defined below.

> **Definition 6.6.1 — χ^2 Distribution.** The PDF of a random variable X that follows a χ^2 distribution with ν degrees of freedom is
>
> $$f(x) = \frac{x^{\nu/2-1}}{2^{\nu/2}\Gamma(\nu/2)} e^{-x/2}, \qquad x > 0, \tag{6.10}$$
>
> where the gamma function is defined as before.
>
> *Parameters and support*: The Greek letter ν is the degrees of freedom, $\nu > 0$. The density is defined over the positive real line, $x > 0$.
>
> *Notation*: $X \sim \chi^2(\nu)$, which stands for "The random variable X has a chi-squared distribution with ν degrees of freedom."

A close inspection of (6.10) indicates that if we take a gamma distribution with parameters $\alpha = \nu/2$ and $\beta = 2$, we obtain the $\chi^2(\nu)$ distribution.

The following two results relate the normal distribution to the chi-square distribution. The proofs require techniques that we haven't introduced yet, so the proofs are deferred to later in this chapter.

> **Theorem 6.6.1 — The Square of a Standard Normal Has a $\chi^2(1)$ Distribution.** If $Z \sim N(0,1)$, then the random variable $X = Z^2$ has a $\chi^2(1)$ distribution.

> **Theorem 6.6.2 — Distribution of the Sum of Independent χ^2 Random Variables.** If $X_1 \sim \chi^2(\nu_1)$, $X_2 \sim \chi^2(\nu_2)$, ..., $X_n \sim \chi^2(\nu_n)$ are independent random variables, then
>
> $$\sum_{i=1}^{n} X_i \sim \chi^2\left(\sum_{i=1}^{n} \nu_i\right).$$
>
> In other words, the sum of independent χ^2 random variables has a χ^2 distribution whose degrees of freedom is the sum of the degrees of freedom of the Xs.

Theorem 6.6.2 allows the degrees of freedom of the Xs to differ. If the degrees of freedom are all the same, then the random variables $X_1, X_2, \ldots, X_n$ are *identically distributed* as well as being independent. This concept of being independent and identically distributed is important in statistics.

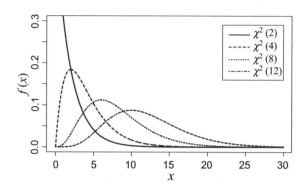

Figure 6.16 Some χ^2, or chi-square, PDFs for various degrees of freedom.

> **Definition 6.6.2 — Independent and Identically Distributed (i.i.d.).** If the random variables $X_1, X_2, \ldots, X_n$ are independent and they all have the same distribution, then we say they are *independent and identically distributed*, which we abbreviate i.i.d. We write this as $X_1, X_2, \ldots, X_n \sim$ i.i.d. NAME, where NAME refers to the name of the distribution, for example $N(\mu, \sigma^2)$ or $WEIB(\theta, \beta)$.

When in Theorem 6.6.2 the random variables are i.i.d. we have the following theorem.

Theorem 6.6.3 — Distribution of the Sum of Squares of Independent $N(0,1)$ Random Variables. If $Z_1, Z_2, \ldots, Z_n \sim$ i.i.d. $N(0,1)$, then

$$Z_1^2 + Z_2^2 + \cdots + Z_n^2 \sim \chi^2(n).$$

Because the $\chi^2(\nu)$ distribution is just a gamma distribution with $\alpha = \nu/2$ and $\beta = 2$, the mean and variance of the $\chi^2(\nu)$ are

$$\mathbb{E}(X) = \alpha\beta = \frac{\nu}{2}2 = \nu \qquad \text{and} \qquad \mathbb{V}(X) = \alpha\beta^2 = \frac{\nu}{2}2^2 = 2\nu.$$

As shown in Figure 6.16, when the number of degrees of freedom is small the χ^2 distribution is noticeably asymmetric. As the number of degrees of freedom increases, the resulting distribution becomes more symmetric.

The R syntax for the $\chi^2(\nu)$ distribution PDF, CDF, quantile, and random variate generation are

$f(x)$: dchisq(x, ν)

$F(x)$: pchisq(x, ν)

ϕ_p: qchisq(p, ν)

Generate n random variates: rchisq(n, ν)

■ **Example 6.15** As with the t-test, in later chapters you will learn about a statistical test called the chi-squared test which involves the use of various quantiles from the χ^2 distribution, often the 0.975 and 0.995 quantiles. Using R, find these quantiles for a χ^2 distribution with $\nu = 20$ degrees of freedom.

Solution: The R code

```
qchisq(0.975,20)
qchisq(0.995,20)
```

yields

```
[1] 34.16961
[1] 39.99685
```

■ ■ ■

The notation for the $1 - \alpha$ quantile of the $\chi^2(\nu)$ distribution is $\chi^2_\alpha(\nu)$. Notice that the probability in the subscript is the *right* tail area, not the cumulative probability to the left as in the definition of the CDF. With this notation, the answers to the previous example are $\chi^2_{0.025}(20) \approx 34.16961$ and $\chi^2_{0.005}(20) \approx 39.99685$.

> **Exercise 6.11** Using R, find the 0.975 and 0.995 quantiles for a χ^2 distribution with $\nu = 8$ degrees of freedom.
>
> *Answers*: $\phi_{0.975} \approx \chi^2_{0.025}(8) \approx 17.53$, $\phi_{0.995} \approx \chi^2_{0.005}(8) \approx 21.95$.

6.6.2 *t* Distribution

We know from Theorem 6.4.2 that if X has a $N(\mu, \sigma^2)$ distribution then $Z = (X - \mu)/\sigma$ has a standard normal distribution. However, it is often the case that the population standard deviation is not known and must be estimated from data. Under such circumstances, the sample variance S is used in the standardization calculation in place of the population variance σ, $T = (X - \mu)/S$, and then the random variable T has a *t distribution* with $\nu = n - 1$ "degrees of freedom" (where n is the number of observations used in the calculation of the sample variance).

> **Definition 6.6.3 — *t* distribution.** Suppose $Z \sim N(0,1)$ and $V \sim \chi^2(\nu)$ are independent random variables. Define the random variable T to be
>
> $$T = \frac{Z}{\sqrt{V/\nu}}. \tag{6.11}$$
>
> Then T has a *t* distribution with ν degrees of freedom. The PDF of a random variable T that follows a *t* distribution with ν degrees of freedom is
>
> $$f(x) = \frac{1}{\sqrt{\nu}B(\frac{1}{2}, \frac{\nu}{2})} \left(1 + \frac{x^2}{\nu}\right)^{-(\nu+1)/2}, \qquad -\infty < x < \infty,$$
>
> where B is the beta function, $B(\alpha, \beta) = \Gamma(\alpha)\Gamma(\beta)/\Gamma(\alpha + \beta)$.
>
> *Parameters and support*: The Greek letter ν is the degrees of freedom, $\nu > 0$. The density is defined over the entire real line, $-\infty < x < \infty$.
>
> *Notation*: $X \sim t(\nu)$, which stands for "The random variable X has a *t* distribution with ν degrees of freedom."

The *t* distribution is symmetric and bell-shaped, like the normal distribution, but has "heavier tails," meaning that it is more likely to produce values that fall farther from its mean (compared to the normal distribution). When standardizing, the heavier tails account for the extra uncertainty that is introduced into the standardized value because the variance has been estimated.

Figure 6.17 shows four different *t* distributions, corresponding to four different degrees of freedom, compared to a standard normal distribution. When looking at the tails of the curves in Figure 6.17, the curve with the largest tails (corresponding to "$t(2)$") is a *t* distribution with 2 degrees of freedom; the curve with the next largest tails (corresponding to "$t(3)$") is a *t* distribution with 3 degrees of freedom,

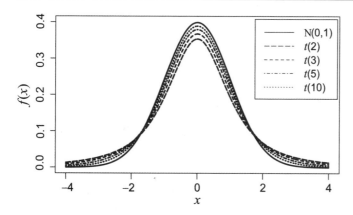

Figure 6.17 Examples of the t distribution for 2, 3, 5, and 10 degrees of freedom compared to the standard normal distribution.

followed by a t distribution with 5 degrees of freedom; and the curve with the smallest tails is a t distribution with 10 degrees of freedom. What this graph shows is that the fewer degrees of freedom, the heavier the tails. Conversely, the more degrees of freedom, the closer the t distribution gets to the normal. As the degrees of freedom approaches $+\infty$, the t distribution approaches the normal distribution.

In Section 10.4 we will show that when we standardize a normally distributed variable, but use an estimate of the standard deviation (instead of the population deviation), we obtain a statistic of the form (6.11).

The R syntax for the t distribution PDF, CDF, quantile, and random variate generation is:

$f(x)$: dt(x, v)

$F(x)$: pt(x, v)

ϕ_p: qt(p, v)

Generate n random variates: rt(n, v)

■ **Example 6.16** In later chapters, you will learn about a statistical test called the t-test, which involves the use of various quantiles from the t distribution, often the 0.975 and 0.995 quantiles. Using R, find these quantiles for a t distribution with $v = 10$ degrees of freedom.

Solution: The R code

```
qt(0.975,10)
qt(0.995,10)
```

produces

```
[1] 2.228139
[1] 3.169273
```

We can write these quantiles using the "right tail" notation we introduced for the χ^2 distribution: $t_{0.025}(10) \approx 2.228139$ and $t_{0.005}(10) \approx 3.169273$. ■ ■ ■

■ **Example 6.17** Compare the quantiles for the t distribution found in Example 6.16 to the same quantiles from a t distribution with $v = 3$ degrees of freedom and the standard normal distribution.

Solution: The R code

```
qt(0.975,3)
qt(0.995,3)
qnorm(0.975)
qnorm(0.995)
```

provides the output

```
[1] 3.182446
[1] 5.840909
[1] 1.959964
[1] 2.575829
```

We see that if the degrees of freedom is small, the t distribution's PDF differs quite a bit from the normal.   ∎ ∎ ∎

Exercise 6.12 Using R, find the 0.975 and 0.995 quantiles for a t distribution with $v = 13$ degrees of freedom.

Answers: $\phi_{0.975} \approx 2.16$, $\phi_{0.995} \approx 3.01$.

∎ **Example 6.18** Now, let's use R to empirically demonstrate that the t distribution does arise when using the sample standard deviation to standardize a normally distributed random variable. To do this, we'll generate repeated random samples of size $n = 4$ from a standard normal distribution from which we'll standardize one of the observations using the distributional mean ($\mu = 0$) and sample standard deviation calculated on the remaining three observations. We'll do this many times and then plot a histogram of the standardized values and overlay it with a t distribution with $v = n - 1 = 3 - 1 = 2$ degrees of freedom.

Solution: Here's the R code:

```
std.vals <- vector()
for(i in 1:100000) {  std.vals <- c(std.vals,(rnorm(1)-0)/sd(rnorm(3))) }
hist(std.vals,freq=FALSE,main="",breaks=5000,xlim=c(-6,6),ylim=c(0,0.4),
  xlab="x",ylab="f(x)")
  lines(seq(-7,7,0.01),dt(seq(-7,7,0.01),2),col="blue",lwd=2)
  lines(seq(-7,7,0.01),dnorm(seq(-7,7,0.01)),col="red",lwd=2)
```

The results are shown in Figure 6.18, where the histogram is the empirical density for 100,000 randomly generated observations standardized using the sample standard deviation, the blue curve is the PDF for a t distribution with $v = 2$ degrees of freedom, and the red curve is the PDF for a standard normal distribution. The t distribution clearly fits the data, while the normal does not.   ∎ ∎ ∎

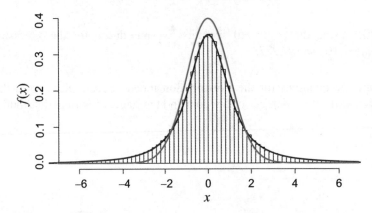

Figure 6.18 Results from Example 6.18. The histogram is the empirical density for 100,000 randomly generated observations standardized using the sample standard deviation. The empirical distribution follows the PDF for a t distribution with $v = 2$ degrees of freedom (blue curve) and not the PDF for the standard normal distribution (red curve).

Although we will not prove this here, the $t(v)$ distribution has mean $\mu = 0$ provided $v > 1$ and variance $\sigma^2 = v/(v-2)$ provided $v > 2$. The mean of the $t(1)$ distribution does not exist because the integrals $\int_0^\infty t\, f(t)\, dt$ and $\int_{-\infty}^0 t\, f(t)\, dt$ both diverge. Similar integrals required to get the variance diverge when $v \leq 2$.

6.6.3 *F* Distribution

The *F distribution* is related to the χ^2 distribution. Let Y_1 and Y_2 be independent χ^2 random variables with v_1 and v_2 degrees of freedom respectively. Then the random variable

$$X = \frac{Y_1/v_1}{Y_2/v_2} \tag{6.12}$$

has an F distribution with parameters v_1 and v_2.

> **Definition 6.6.4 — *F* Distribution.** The PDF of a random variable X that follows an F distribution with parameters v_1 and v_2 is
>
> $$f(x) = \frac{1}{B\left(\frac{v_1}{2}, \frac{v_2}{2}\right)} \frac{x^{(v_1 - 1)/2}}{(1+x)^{(v_1+v_2)/2}}, \qquad x > 0,$$
>
> where B is the beta function, $B(\alpha, \beta) = \Gamma(\alpha)\Gamma(\beta)/\Gamma(\alpha + \beta)$.
>
> *Parameters and support*: $v_1 > 0$ and $v_2 > 0$; $x \geq 0$.
>
> *Notation*: $X \sim F(v_1, v_2)$, which stands for "The random variable X has an F distribution with parameters v_1 and v_2." The parameters v_1 and v_2 are called the *numerator* and *denominator degrees of freedom*, respectively. Note that the order of v_1 and v_2 is important.

As shown in Figure 6.19, a wide variety of PDF shapes is possible with the F distribution. The R syntax for the F distribution PDF, CDF, quantile, and random variate generation is:

$f(x)$: $\texttt{df}(x, v_1, v_2)$
$F(x)$: $\texttt{pf}(x, v_1, v_2)$
ϕ_p: $\texttt{qf}(p, v_1, v_2)$
Generate n random variates: $\texttt{rf}(n, v_1, v_2)$

■ **Example 6.19** In later chapters you will learn about a statistical test called the F test which involves the use of various quantiles from the F distribution, often the 0.975 and 0.995 quantiles. Using R, find these quantiles for an F distribution with $v_1 = 12$ and $v_1 = 4$ degrees of freedom.

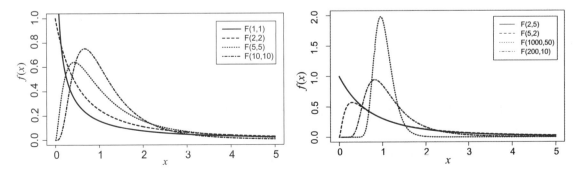

Figure 6.19 Some F distribution PDFs for various degrees of freedom.

Solution: The R code

```
qf(0.975,12,4)
qf(0.995,12,4)
```

gives

```
[1] 8.751159
[1] 20.70469
```

These can be written using the notation $F_{0.025}(12,4) \approx 8.751159$ and $F_{0.005}(12,4) \approx 20.70469$. ▪▪▪

Exercise 6.13 Using R, find the 0.975 and 0.995 quantiles for an F distribution with $v_1 = 3$ and $v_2 = 5$ degrees of freedom.

Answers: $\phi_{0.975} = F_{0.025}(12,4) \approx 7.76$, $\phi_{0.995} = F_{0.005}(12,4) \approx 16.53$.

6.7 Beta Distribution

As with the gamma distribution, the beta distribution is also is a two-parameter family of continuous probability distributions. It is defined on the interval $(0,1)$ and, depending on the parameter choices, can be symmetric or skewed either to the right or left. It is often used to model proportions, such as the fraction of time a machine is in repair or the fraction of time your internet service is down.

Definition 6.7.1 — Beta Distribution. The PDF for random variable X having a beta distribution with parameters α and β is

$$f(x) = \frac{1}{B(\alpha,\beta)} x^{\alpha-1}(1-x)^{\beta-1}, \qquad 0 < x < 1,$$

where B is the beta function, defined as $B(\alpha,\beta) = \Gamma(\alpha)\Gamma(\beta)/\Gamma(\alpha+\beta)$.

Parameters and support: $\alpha > 0$, $\beta > 0$; $0 < x < 1$.

Notation: $X \sim \text{BETA}(\alpha,\beta)$, which stands for "The random variable X has a beta distribution with parameters α and β."

▪ **Example 6.20** Let $X \sim \text{BETA}(4,2)$. Find $f(0.3)$ and $F(0.3)$.

Solution: First, the density height is

$$f(0.3) = \frac{0.3^3(1-0.3)^1}{B(4,2)} = \frac{\Gamma(6)}{\Gamma(4)\Gamma(2)}(0.027 \times 0.7) = \frac{5!}{3!1!}(0.0189) \approx 0.378,$$

where for any positive integer α, $\Gamma(\alpha) = (\alpha-1)!$. Now, for the cumulative probability:

$$F(0.3) = \int_0^{0.3} \frac{x^3(1-x)^1}{B(4,2)}\, dx = \frac{5!}{3!1!}\int_0^{0.3}(x^3-x^4)\,dx = 20\left(\frac{x^4}{4}\Big|_0^{0.3} - \frac{x^5}{5}\Big|_0^{0.3}\right) \approx 0.03078.$$

▪▪▪

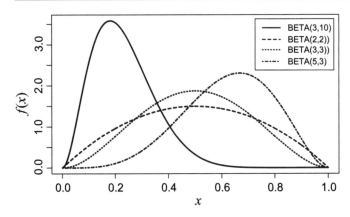

Figure 6.20 Beta distribution PDFs for various values of α and β.

As shown in Figure 6.20, a wide variety of PDF shapes is possible with the beta distribution.

Theorem 6.7.1 — Expected Value and Variance of a Beta-Distributed Random Variable. If X has a beta distribution with parameters α and β, then

$$\mathbb{E}(X) = \frac{\alpha}{\alpha + \beta} \qquad \text{and} \qquad \mathbb{V}(X) = \frac{\alpha\beta}{(\alpha+\beta)^2(\alpha+\beta+1)}.$$

Proof.

$$\mathbb{E}(X) = \int_{-\infty}^{\infty} x\, f(x)\, dx$$

$$= \int_0^1 x \left(\frac{x^{\alpha-1}(1-x)^{\beta-1}}{B(\alpha,\beta)} \right) dx$$

$$= \frac{1}{B(\alpha,\beta)} \int_0^1 x^{(\alpha+1)-1}(1-x)^{\beta-1}\, dx$$

$$= \frac{B(\alpha+1,\beta)}{B(\alpha,\beta)} \underbrace{\int_0^1 \frac{1}{B(\alpha+1,\beta)} x^{(\alpha+1)-1}(1-x)^{\beta-1}\, dx}_{\text{integral is 1 because the integrand is the BETA}(\alpha+1,\beta)\text{ distribution}}$$

$$= \frac{\Gamma(\alpha+\beta)}{\Gamma(\alpha)\Gamma(\beta)} \times \frac{\Gamma(\alpha+1)\Gamma(\beta)}{\Gamma(\alpha+\beta+1)}$$

$$= \frac{\Gamma(\alpha+\beta)}{\Gamma(\alpha)\Gamma(\beta)} \times \frac{\alpha\Gamma(\alpha)\Gamma(\beta)}{(\alpha+\beta)\Gamma(\alpha+\beta)}$$

$$= \frac{\alpha}{\alpha+\beta}.$$

The proof for $\mathbb{V}(X)$ begins with the computation of $\mathbb{E}(X)$; this is done using reasoning similar to that above. The variance is then obtained using $\mathbb{V}(X) = \mathbb{E}(X^2) - \left[\mathbb{E}(X)\right]^2$. See Problem 6.68. ∎

■ **Example 6.21** Returning to Example 6.20, what is the expected value and standard deviation of $X \sim \text{BETA}(4,2)$?

Solution:

$$\mathbb{E}(X) = \frac{\alpha}{\alpha+\beta} = \frac{4}{6} = \frac{2}{3}.$$

$$\mathbb{V}(X) = \frac{\alpha\beta}{(\alpha+\beta)^2(\alpha+\beta+1)} = \frac{8}{36\times7} = \frac{8}{252} = \frac{2}{63}.$$

So, $\sigma_X = \sqrt{2/63} \approx 0.178$.   ■ ■ ■

The R syntax for the beta distribution PDF, CDF, quantile, and random variate generation is:

$f(x)$: dbeta(x,α,β)

$F(x)$: pbeta(x,α,β)

ϕ_p: qbeta(p,α,β)

Generate n random variates: rbeta(n,α,β)

■ **Example 6.22** Check the answers to Example 6.20 with R.

Solution: The code

```
dbeta(0.3,4,2)
pbeta(0.3,4,2)
```

yields

```
[1] 0.378
[1] 0.03078
```

■ ■ ■

Exercise 6.14 Suppose $X \sim \text{BETA}(3,10)$. Find:

(a) $f(0.2)$,

(b) $\Pr(0.2 \le X \le 0.8)$, and

(c) the 95th percentile.

Answers: $f(0.2) \approx 3.54$; $\Pr(0.2 \le X \le 0.8) \approx 0.56$; $\phi_{0.95} \approx 0.438$.

There are many BETA distributions with the same mean. For example, the BETA$(4,1)$, BETA$(8,2)$, and BETA$(40,10)$ have the same mean, namely $4/(4+1) = 0.8$. The variances for these three distributions are different, however. Note that the variance of the BETA(α,β) distribution can be written as

$$\mathbb{V}(X) = \frac{\alpha\beta}{(\alpha+\beta)^2(\alpha+\beta+1)} = \frac{\mu(1-\mu)}{\alpha+\beta+1},$$

where $\mu = \alpha/(\alpha+\beta)$ (see Problem 6.69). Thus, as α and β become larger, so long as $\alpha/(\alpha+\beta)$ remains constant, the variance decreases. The variances of the three BETA distributions mentioned above are

$$\mathbb{V}(X_1) = \frac{0.8\times(1-0.8)}{4+1+1} \approx 0.02667$$

$$\mathbb{V}(X_2) = \frac{0.8\times(1-0.8)}{8+2+1} \approx 0.01455$$

$$\mathbb{V}(X_3) = \frac{0.8\times(1-0.8)}{40+10+1} \approx 0.00314.$$

6.8 **Transformations**

A common question in probability is this: Given a random variable X having a known distribution, what is the distribution of a new random variable Y which is a function of X? There are several approaches for answering this question. We will cover one of them, called the CDF technique, in this section. We make use of the notation that the subscript on a CDF or PDF indicates the random variable having the given CDF or PDF. For example, $F_U(u)$ is the CDF of U, while $F_X(x)$ is the CDF of X. It will also be important to adhere to the notation that uses capital letters for random variables and lower case letters for possible values of these random variables.

The CDF technique can be described as follows:
1. Let X denote the random variable whose distribution is known and let $Y = g(X)$ be a new random variable whose distribution is desired. Begin with the CDF of the new random variable:

$$F_Y(y) = \Pr(Y \le y).$$

2. Substitute $g(X)$ for Y:

$$F_Y(y) = \Pr(Y \le y) = \Pr(g(X) \le y).$$

3. Manipulate the event $g(X) \le y$ to get this in terms of X. This is the step that varies from problem to problem.
4. Simplify the result to get the CDF of $Y = g(X)$.
5. If the PDF of Y is desired, then evaluate

$$f_Y(y) = F_Y'(y).$$

We begin with a simple example.

■ **Example 6.23** Suppose $U \sim \text{UNIF}(0,1)$ and define $Y = 4X$. What is the distribution of Y?

Solution: We know that the CDF of U is

$$F_U(u) = \begin{cases} 0, & u < 0 \\ u, & 0 \le u \le 1 \\ 1, & u > 1. \end{cases}$$

The CDF technique involves directly finding the CDF of the new random variable. In this case,

$$\begin{aligned} F_X(x) &= \Pr(X \le x) \\ &= \Pr(4U \le x) \\ &= \Pr(U \le x/4). \end{aligned} \tag{6.13}$$

The smallest that X can be is $4 \times 0 = 0$, which occurs when $U = 0$ (the smallest possible value for U), and the largest that X can be is $4 \times 1 = 4$, which occurs when $U = 1$ (the largest possible value for U). Suppose, for now, that x is between 0 and 4. If this is the case, then $x/4$ is between 0 and 1, and we know from (6.13) that for values in this interval, the CDF is equal to $x/4$. (Remember, we're getting $\Pr(U \le x/4)$.) Thus,

$$F_X(x) = \frac{x}{4}, \quad 0 \le x \le 4.$$

If x were to be less than 0, then $\Pr(X \leq x) = 0$; similarly, if x were to be greater than 4, then $\Pr(X \leq x) = 1$. Thus,

$$F_X(x) = \Pr(U \leq x/4) = \begin{cases} 0, & x < 0 \\ x/4 & 0 \leq x \leq 4 \\ 1 & x > 4. \end{cases}$$

If we differentiate this CDF we will obtain the PDF of X:

$$f(x) = F'(x) = \begin{cases} \frac{1}{4}, & 0 \leq x \leq 4 \\ 0, & \text{otherwise.} \end{cases}$$

This is the PDF of the UNIF$(0,4)$ distribution. Thus, if $U \sim$ UNIF$(0,1)$, then the distribution of $X = 4U$ is UNIF$(0,4)$. ∎∎∎

This result seems intuitive, and you probably could have seen it without going through the details of the CDF techniques. The next example is less intuitive.

■ **Example 6.24** Suppose $U \sim$ UNIF$(0,1)$. Let $X = -\log U$. What is the distribution of X?

Solution: The CDF technique begins with the direct evaluation of the CDF $F_X(x)$:

$$\begin{aligned} F_X(x) &= \Pr(X \leq x) \\ &= \Pr(-\log U \leq x) \\ &= \Pr(\log U \geq -x) \\ &= \Pr\left(U \geq e^{-x}\right) \\ &= 1 - \Pr\left(U \leq e^{-x}\right) \\ &= 1 - F_U\left(e^{-x}\right). \end{aligned}$$

Notice that the least x can be is 0, and this occurs when $u = 1$, its *largest* possible value. Also note that as $u \to 0^+$, $\log u \to -\infty$. Thus, as $u \to 0^+$, $-\log u \to \infty$. The random variable X must therefore be between 0 and ∞. For x between 0 and ∞, e^{-x} is between 0 and 1. For values u between 0 and 1 (such as e^{-x}) we know that $F_U(u) = u$. Thus,

$$F_X(x) = 1 - F_U\left(e^{-x}\right) = 1 - e^{-x}, \quad x > 0.$$

The PDF of X can be found by differentiating the CDF:

$$f_X(x) = F'_X(x) = \begin{cases} e^{-x}, & x > 0 \\ 0, & \text{otherwise.} \end{cases}$$

This is the PDF of the EXP(1) distribution. Thus, $X \sim$ EXP(1). ∎∎∎

We can now prove Theorem 6.6.1 that says if $Z \sim N(0,1)$, then $X = Z^2 \sim \chi^2(1)$.

Proof. We can prove this theorem using the CDF technique. For $x > 0$, we can write

$$
\begin{aligned}
F_X(x) &= \Pr(X \le x) \\
&= \Pr\left(Z^2 \le x\right) \\
&= \Pr\left(-\sqrt{x} \le Z \le \sqrt{x}\right) \\
&= \int_{-\sqrt{x}}^{\sqrt{x}} \frac{1}{\sqrt{2\pi}} e^{-z^2/2}\, dz \\
&= 2\int_{0}^{\sqrt{x}} \frac{1}{\sqrt{2\pi}} e^{-z^2/2}\, dz.
\end{aligned}
$$

At first glance this doesn't appear very helpful because it still involves an integral. Moreover, there is no analytic way to get an antiderivative (in terms of elementary functions) for the integrand. If we want to get the PDF of X, we could simply take the derivative. In this case we use the first fundamental theorem of calculus (Varberg *et al.*, 2007) that says

$$
\frac{d}{dw} \int_{a}^{w} h(y)\, dy = h(w).
$$

Since the upper limit in our integral is a function of x, namely $\sqrt{x}$, we will need to apply the chain rule in addition to the fundamental theorem of calculus:

$$
\begin{aligned}
f_X(x) &= \frac{d}{dx} 2\int_{0}^{\sqrt{x}} \frac{1}{\sqrt{2\pi}} e^{-z^2/2}\, dz \\
&= 2\frac{1}{\sqrt{2\pi}} \exp\left(-(\sqrt{x})^2/2\right) \times \underbrace{\left(\frac{d}{dx}\sqrt{x}\right)}_{\text{chain rule}} \\
&= \sqrt{\frac{2}{\pi}} \exp\left(-x/2\right) \times \left(\frac{1}{2}x^{-1/2}\right) \\
&= \frac{x^{1/2-1}}{2^{1/2}\sqrt{\pi}} \exp\left(-x/2\right), \qquad x > 0.
\end{aligned}
$$

This is the PDF of the $\chi^2(1)$ distribution.[6] This shows that if $Z \sim N(0,1)$, then $X = Z^2 \sim \chi^2(1)$. ∎

6.8.1 Simulating from Distributions

Computers don't simulate random numbers. They generate what are called *pseudo* random numbers. The idea is to start with a number and then apply some function to it to get another number. This process repeats itself for as many random numbers as needed. The beginning number is called the *seed*, and once it is selected all subsequent numbers are determined. (This is why we say that the numbers are not truly random.) Usually, software packages like R begin by simulating UNIF(0,1) random variables. It is usually possible to transform these pseudo random numbers to obtain pseudo random numbers from some other distribution. The transformation from uniform to some other distribution can be determined with the help of the following theorem.

---

[6]If you compare this PDF with the PDF of the $\chi^2(1)$ distribution given in (6.10) you will see that the factors that depend on x match up exactly when $\nu = 1$. If this occurs, then the rest of the PDF, involving the constants, must also match up, since the PDF must integrate to 1. The $2^{1/2}$ matches fine, but in (6.10) with $\nu - 1$ we have the factor $\Gamma(1/2)$, whereas in the expression here we have $\sqrt{\pi}$. Thus, in the process of finding the PDF of $X = Z^2$, we've managed to show that $\Gamma(1/2) = \sqrt{\pi}$.

> **Theorem 6.8.1 — Distribution of the Inverse CDF of a Uniform Random Variable.**
> Suppose F is a continuous CDF with inverse function F^{-1} and suppose $U \sim \text{UNIF}(0,1)$. Then the random variable $F^{-1}(U)$ has CDF F.

Proof. The proof uses the CDF technique described previously in this section. Since F is an increasing function,

$$P(X \le x) = \Pr(F^{-1}(U) \le x) = \Pr\left(U \le F(X)\right) = \int_0^{F(x)} \underbrace{\frac{1}{1-0}}_{\text{PDF of UNIF}(0,1)} du = \left[u\right]_0^{F(x)} = F(x).$$

Thus, the random variable X has CDF F.   ∎

Given this theorem, the algorithm for simulating from a distribution with PDF F is simple:

Simulating from a Distribution with CDF $F(x)$
  1. Simulate u from the UNIF$(0,1)$ distribution.
  2. Set $x = F^{-1}(u)$.

■ **Example 6.25** Explain how to simulate from the EXP(1) distribution.

Solution: The EXP(1) distribution has CDF

$$F(x) = 1 - \exp(-x), \quad x > 0.$$

To find the inverse function, set $u = F(x)$ and solve for x:

$$
\begin{aligned}
u &= F(x) \\
u &= 1 - \exp(-x) \\
\exp(-x) &= 1 - u \\
\log \exp(-x) &= \log(1 - u) \\
x &= -\log(1 - u).
\end{aligned}
$$

The rule is then
  1. Simulate $U \sim \text{UNIF}(0,1)$
  2. Set $X = -\log(1 - U)$.   ■■■

■ **Example 6.26** Give an algorithm for simulating from the distribution with PDF

$$
f(x) = \begin{cases} \dfrac{1}{x^2}, & x > 1 \\ 0, & \text{otherwise.} \end{cases}
$$

Solution: First find the CDF:

$$F(x) = \int_1^x \frac{1}{y^2}\, dy = \left[-\frac{1}{y}\right]_1^x = 1 - \frac{1}{x}, \quad x > 0.$$

Next, we must find the inverse function:

$$u = 1 - \frac{1}{x}$$

$$\frac{1}{x} = 1 - u$$

$$x = \frac{1}{1-u}.$$

The rule is therefore

1. Simulate $U \sim \text{UNIF}(0,1)$.

2. Set $X = \dfrac{1}{1-U}$. ■ ■ ■

If we know the marginal PDF of X and the conditional PDF of $Y|(X = x)$, we can find the joint PDF using

$$f(x,y) = f_1(x) f(y|x).$$

This way of thinking provides us with a way to simulate from a joint distribution, given one marginal and the other conditional PDF.

■ **Example 6.27** Explain how you could simulate from the joint distribution in Example 4.30.

Solution: We can simulate from the distribution of X, which has marginal distribution GAM(4,2). The `rgamma()` function can be used for this task. Then, conditioned on a value x for X, we can simulate from the conditional distribution of Y given x, which is the UNIF$(0,x)$ distribution. ■ ■ ■

6.9 Moment Generating Functions

We begin with the definition of the moment generating function.

> **Definition 6.9.1 — Moment Generating Function (MGF) for a Continuous Random Variable.** Suppose X is a continuous random variable with PDF $f(x)$. The MGF $M_X(t)$ of X is defined to be the expectation
>
> $$M_X(t) = \mathbb{E}\big(e^{tX}\big) = \int_{-\infty}^{\infty} e^{tx} f(x)\, dx.$$
>
> The domain for $M_X(t)$ is the set of all t for which this expectation exists and is finite. The expectation certainly exists when $t = 0$ (see Problem 6.79), and in many cases it exists in an open interval containing $t = 0$.

We will explain why such a function is called the moment generating function and why such a function is useful. But first, let's determine the MGF for a few distributions.

■ **Example 6.28** Find the MGF of the EXP1(θ) distribution.

Solution: Let $X \sim \text{EXP1}(\theta)$. Using the definition, we can write[7]

$$M_X(t) = \mathbb{E}\big(e^{tX}\big) = \int_0^{\infty} e^{tx} \frac{1}{\theta} e^{-x/\theta}\, dx = \frac{1}{\theta} \int_0^{\infty} \exp\left(-x\left(\frac{1}{\theta} - t\right)\right) dx.$$

---

[7]Keep in mind that t is a *constant* as far as the integral goes. In this example, the PDF is a function of x, so factors involving t only can be factored out of the integral, but those involving x or both x and t cannot.

This integral exists for $t < 1/\theta$ and is equal to

$$M_X(t) = \frac{1}{\theta}\left[-\frac{1}{\frac{1}{\theta}-t}\exp\left(-x\left(\frac{1}{\theta}-t\right)\right)\right]_0^{\infty} = \frac{1}{\theta}\left[-0+\frac{\theta}{1-\theta t}e^0\right] = \frac{1}{1-\theta t}.$$

This MGF exists for $t < 1/\theta$. ∎

■ **Example 6.29** Find the MGF of the $\text{GAM}(\alpha,\beta)$ distribution.

Solution: We will use the trick of manipulating the constants, similar to what was done in the proof of Theorem 6.2.1. The MGF for a random variable X having a $\text{GAM}(\alpha,\beta)$ distribution is

$$M_X(t) = \int_0^{\infty} e^{tx} \frac{x^{\alpha-1}}{\beta^{\alpha}\Gamma(\alpha)} e^{-x/\beta}\, dx$$

$$= \frac{1}{\beta^{\alpha}\Gamma(\alpha)}\int_0^{\infty} x^{\alpha-1}\exp\left(-x\left(\frac{1}{\beta}-t\right)\right)dx$$

$$= \frac{1}{\beta^{\alpha}\Gamma(\alpha)}\int_0^{\infty} x^{\alpha-1}\exp\left(-x\Big/\left(\frac{\beta}{1-\beta t}\right)\right)dx$$

$$= \frac{1}{\beta^{\alpha}\Gamma(\alpha)}\left(\frac{\beta}{1-\beta t}\right)^{\alpha}\Gamma(\alpha)\underbrace{\int_0^{\infty}\frac{x^{\alpha-1}}{\left(\frac{\beta}{1-\beta t}\right)^{\alpha}\Gamma(\alpha)}\exp\left(-x\Big/\left(\frac{\beta}{1-\beta t}\right)\right)dx}_{\text{integral is 1 because the integrand is the }\text{GAM}\left(\alpha,\frac{\beta}{1-\beta t}\right)\text{ PDF}}$$

$$= \frac{\beta^{\alpha}}{\beta^{\alpha}(1-\beta t)^{\alpha}}$$

$$= (1-\beta t)^{-\alpha}.$$

The integral on line four is equal to 1 because the integrand is a gamma PDF; but this is only true if $\beta/(1-\beta t) > 0$, which is equivalent to $t < 1/\beta$. Thus, the MGF exists for $t < 1/\beta$. ∎

■ **Example 6.30** Find the MGF of the $N(\mu,\sigma^2)$ distribution.

Solution: Let $X \sim N(\mu,\sigma^2)$. We will apply two tricks to obtain the MGF. First, we will complete the square in the exponential function, and then we will manipulate the integrand so that it is the integral of a valid PDF (as in Example 6.29):

$$M_X(t) = \int_{-\infty}^{\infty} e^{tx}\frac{1}{\sqrt{2\pi}\,\sigma}\exp\left(-\frac{(x-\mu)^2}{2\sigma^2}\right)dx$$

$$= \int_{-\infty}^{\infty} \frac{1}{\sqrt{2\pi}\,\sigma}\exp\left(\frac{2\sigma^2 tx}{2\sigma^2}-\frac{(x-\mu)^2}{2\sigma^2}\right)dx$$

$$= \int_{-\infty}^{\infty} \frac{1}{\sqrt{2\pi}\,\sigma}\exp\left(-\frac{1}{2\sigma^2}\left[x^2-2(\mu+\sigma^2 t)x+\mu^2\right]\right)dx$$

$$= \int_{-\infty}^{\infty} \frac{1}{\sqrt{2\pi}\,\sigma}\exp\left(-\frac{1}{2\sigma^2}\left[\left(x-(\mu+\sigma^2 t)\right)^2-2\mu\sigma^2 t-\sigma^4 t^2\right]\right)dx$$

$$= \exp\left(-\frac{1}{2\sigma^2}\left(-2\sigma^2\mu t - \sigma^4 t^2\right)\right) \underbrace{\int_{-\infty}^{\infty} \frac{1}{\sqrt{2\pi}\,\sigma} \exp\left(-\frac{1}{2\sigma^2}\left(x - (\mu + \sigma^2 t)\right)^2\right) dx}_{\text{integrand is the } N(\mu + \sigma^2 t, \sigma^2) \text{ PDF so integral is 1.}}$$

$$= \exp\left(\mu t + \frac{\sigma^2 t^2}{2}\right).$$

This MGF exists for all t.　　　　　　　　　　　　　　　　　　　　　　　■ ■ ■

The MGF has two main uses. First, it can be used to find the moments of a random variable. If X is a random variable (continuous or discrete), the kth *population moment* (also called the kth moment about the origin, or the uncorrected kth moment) is

$$\mu_k' = \mathbb{E}(X^k).$$

The next theorem shows how moments of a random variable are related to its MGF.

Theorem 6.9.1 — Obtaining Moments from the MFG. Let X be a random variable with PDF $f(x)$ and MGF $M_X(t)$. Suppose that $M_X(t)$ is twice differentiable on an open interval containing 0. Then,

$$\mu_1' = M_X'(0)$$
$$\mu_2' = M_X''(0).$$

Proof. The derivative of the MGF $M_X(t)$ is

$$M_X'(t) = \frac{d}{dt}\int_{-\infty}^{\infty} e^{tx} f(x)\,dx = \int_{-\infty}^{\infty} \frac{\partial}{\partial t} e^{tx} f(x)\,dx = \int_{-\infty}^{\infty} x e^{tx} f(x)\,dx. \qquad (6.14)$$

We have moved the derivative inside the integral; justifying this requires mathematical techniques beyond the scope of this book, so we omit it here. Since the integrand is a function of both t and x we must take the partial derivative in the integrand. Evaluating the MGF at $t = 0$ gives

$$M_X'(0) = \int_{-\infty}^{\infty} x e^{0 \cdot x} f(x)\,dx = \int_{-\infty}^{\infty} x f(x)\,dx = \mathbb{E}(X).$$

Picking up from (6.14), we can obtain the second derivative:

$$M_X''(t) = \frac{d}{dt}\int_{-\infty}^{\infty} x e^{tx} f(x)\,dx = \int_{-\infty}^{\infty} x \frac{\partial}{\partial t} e^{tx} f(x)\,dx = \int_{-\infty}^{\infty} x^2 e^{tx} f(x)\,dx.$$

Substituting $t = 0$ gives

$$M_X''(0) = \int_{-\infty}^{\infty} x^2 e^{0 \cdot x} f(x)\,dx = \int_{-\infty}^{\infty} x^2 f(x)\,dx = \mathbb{E}(X^2). \qquad \blacksquare$$

We could continue this process for higher-order moments and higher-order derivatives, but the first two moments will suffice. The variance of a random variable can then be obtained using MGFs:

$$\mathbb{V}(X) = \mathbb{E}(X^2) - \left[\mathbb{E}(X)\right]^2 = M_X''(0) - \left[M_X'(0)\right]^2.$$

In Section 6.4.1 we stated, but didn't prove, that the mean and variance of the $N(\mu, \sigma^2)$ distribution are μ and σ^2, respectively. We can now prove this result using MGFs.

■ **Example 6.31** Suppose $X \sim N(\mu, \sigma^2)$. Use MGFs to show that $\mathbb{E}(X) = \mu$ and $\mathbb{V}(X) = \sigma^2$.

Solution: The MGF for the normal distribution is $M_X(t) = \exp\left(\mu t + \sigma^2 t^2/2\right)$. Since

$$M_X'(t) = \left(\mu + \sigma^2 t\right) \exp\left(\mu t + \sigma^2 t^2/2\right), \tag{6.15}$$

we can write

$$\mathbb{E}(X) = M_X'(0) = \left(\mu + \sigma^2 \times 0\right) \exp\left(\mu \times 0 + \sigma^2 \times 0^2/2\right) = \mu.$$

The second derivative requires applying the product rule to (6.15):

$$M_X''(t) = (\mu + \sigma^2 t)^2 \exp\left(\mu t + \sigma^2 t^2/2\right) + \exp\left(\mu t + \sigma^2 t^2/2\right)\sigma^2.$$

Substituting 0 for t yields

$$\mathbb{E}(X^2) = M_X''(0) = (\mu + \sigma^2 \times 0)^2 \exp\left(\mu \times 0 + \sigma^2 \times 0^2/2\right) + \sigma^2 \exp\left(\mu \times 0 + \sigma^2 0^2/2\right) = \mu^2 + \sigma^2.$$

Thus,

$$\mathbb{V}(X) = \mathbb{E}(X^2) - \left[\mathbb{E}(X)\right]^2 = M_X''(0) - \left[M_X'(0)\right]^2 = \mu^2 + \sigma^2 - \mu^2 = \sigma^2. \quad ■■■$$

Another use of MGFs is to find the distribution of a sum of independent random variables. We begin by stating two important theorems. The first may seem obvious, but is deceptively hard to prove.

Theorem 6.9.2 — Uniqueness Theorem for MGFs. Suppose X and Y are random variables with respective MGFs $M_X(t)$ and $M_Y(t)$ that are continuous on a open interval containing 0. If $M_X(t) = M_Y(t)$, then X and Y have the same distribution.

Theorem 6.9.3 — MGF of the Sum of Independent Random Variables. If $X_1, X_2, \ldots, X_n$ are independent random variables with respective MGFs $M_1(t), M_2(t), \ldots, M_n(t)$, then the random variable $S = X_1 + X_2 + \cdots + X_n$ has MGF

$$M_S(t) = M_1(t) \times M_2(t) \times \cdots \times M_n(t).$$

Proof.

$$M_S(t) = \mathbb{E}\left(e^{tS}\right) = \mathbb{E}\left(e^{t(X_1 + X_2 + \cdots + X_n)}\right) = \mathbb{E}\left(e^{tX_1} e^{tX_2} \cdots e^{tX_n}\right).$$

Because the X_i are independent, we can write the above as

$$M_S(t) = \mathbb{E}\left(e^{tX_1}\right) \times \mathbb{E}\left(e^{tX_2}\right) \times \cdots \times \mathbb{E}\left(e^{tX_n}\right) = M_1(t) \times M_2(t) \times \cdots \times M_n(t). \quad ■$$

Theorem 6.9.3 can be used to prove the following theorem, which is related to Theorem 6.4.2.

Theorem 6.9.4 — Distribution of $aX+b$ if $X \sim N(\mu, \sigma^2)$. If $X \sim N(\mu, \sigma^2)$, then $W = aX + b \sim N(a\mu + b, a^2\sigma^2)$.

Proof. The MGF for W is

$$M_W(t) = \mathbb{E}(e^{tW}) = \mathbb{E}(e^{t(aX+b)}) = \mathbb{E}(e^{taX}e^{tb}) = e^{tb}\mathbb{E}(e^{(ta)X}) = e^{tb}M_X(ta).$$

We showed in Example 6.31 that $M_X(t) = \exp(\mu t + \sigma^2 t^2/2)$. We can continue with the expression for $M_W(t)$ to obtain

$$M_W(t) = e^{tb}M_X(ta) = \exp(tb)\exp\left(\mu ta + \frac{\sigma^2(ta)^2}{2}\right) = \exp\left((a\mu+b)t + \frac{(a^2\sigma^2)t^2}{2}\right).$$

This is the MGF of the $N(a\mu+b, a^2\sigma^2)$ distribution. Thus, $W = aX+b \sim N(a\mu+b, a^2\sigma^2)$. ∎

We can now prove Theorem 6.4.2, which says that if $X \sim N(\mu, \sigma^2)$, then $Z = (X - \mu)/\sigma \sim N(0,1)$.

Proof. In Theorem 6.9.4, take $a = 1/\sigma$ and $b = -\mu/\sigma$. Then,

$$\frac{1}{\sigma}X - \frac{\mu}{\sigma} = \frac{X-\mu}{\sigma} \sim N\left(\frac{1}{\sigma}\mu - \frac{\mu}{\sigma}, \frac{1}{\sigma^2}\sigma^2\right).$$

This simplifies to the $N(0,1)$ distribution. ∎

We will also use Theorem 6.9.3 to show that the sum of normal random variables is normally distributed.

Theorem 6.9.5 — Distribution of the Sum of Independent Normal Random Variables.
Suppose $X_i \sim N(\mu_i, \sigma_i^2)$ are independent. Then

$$\sum_{i=1}^n X_i \sim N\left(\sum_{i=1}^n \mu_i, \sum_{i=1}^n \sigma_i^2\right).$$

Proof. Let $S = \sum_{i=1}^n X_i$. The MGF for each X_i is

$$M_i(t) = \exp\left(\mu_i t + \frac{\sigma_i^2 t^2}{2}\right).$$

By Theorem 6.9.3, the MGF of S is

$$M_S(t) = \prod_{i=1}^n M_i(t) = \prod_{i=1}^n \exp(\mu_i t + \sigma_i^2 t^2/2) = \exp\left(t\sum_{i=1}^n \mu_i + \left(\sum_{i=1}^n \sigma_i^2\right)t^2/2\right).$$

This is the MGF for the $N(\sum_{i=1}^n \mu_i, \sum_{i=1}^n \sigma_i^2)$ distribution. By the uniqueness theorem (Theorem 6.9.2) we conclude that

$$S \sim N\left(\sum_{i=1}^n \mu_i, \sum_{i=1}^n \sigma_i^2\right).$$ ∎

■ **Example 6.32** Prove Theorem 6.6.2. If $X_1 \sim \chi^2(\nu_1)$, $X_2 \sim \chi^2(\nu_2)$, ..., $X_n \sim \chi^2(\nu_n)$ are independent random variables, then

$$S = \sum_{i=1}^{n} X_i \sim \chi^2\left(\sum_{i=1}^{n} \nu_i\right).$$

Solution: Since we are looking for the distribution of the sum of independent random variables, we consider the MGF of the sum. Recall that the $\chi^2(\nu)$ distribution is just the GAM($\alpha = \nu/2, \beta = 2$) distribution. Thus, using the result from Example 6.29, we conclude that the MGF of X_i is

$$M_i(t) = (1-2t)^{-\nu_i/2}.$$

Applying Theorem 6.9.3, we have

$$M_S(t) = \prod_{i=1}^{n} (1-2t)^{-\nu_i/2} = (1-2t)^{-\sum_{i=1}^{n} \nu_i/2}.$$

This is just the MGF for the $\chi^2\left(\sum_{i=1}^{n} \nu_i\right)$ distribution. By the uniqueness theorem (Theorem 6.9.2) we conclude that $S \sim \chi^2\left(\sum_{i=1}^{n} \nu_i\right).$ ■ ■ ■

So far we have defined the MGF only for continuous random variables, but the same idea applies to discrete random variables.

Definition 6.9.2 — MGF for Discrete Random Variables. Suppose X is a discrete random variable with PMF $f(x)$. The MGF $M_X(t)$ of X is defined to be the expectation

$$M_X(t) = \mathbb{E}(e^{tX}) = \sum_{\text{all } x} e^{tx} f(x)\, dx.$$

■ **Example 6.33** Find the MGF of the GEO(p) distribution.
Solution: Let $X \sim$ GEO(p). Then

$$M_X(t) = \sum_{x=1}^{\infty} e^{tx}(1-p)^{x-1}p = \sum_{x=1}^{\infty} \left[e^t(1-p)\right]^x (1-p)^{-1}p = \frac{p}{1-p}\sum_{x=1}^{\infty} \left[e^t(1-p)\right]^x.$$

The series in this last expression is the sum of a geometric sequence, so it converges to[8]

$$\sum_{x=1}^{\infty} \left[e^t(1-p)\right]^x = \frac{e^t(1-p)}{1-e^t(1-p)}, \qquad \left|e^t(1-p)\right| < 1.$$

The MGF is therefore

$$M_X(t) = \frac{p}{1-p}\sum_{x=1}^{\infty} \left[e^t(1-p)\right]^x = \frac{p}{1-p}\frac{e^t(1-p)}{1-e^t(1-p)} = \frac{pe^t}{1-(1-p)e^t}.$$

The MGF exists for $\left|e^t(1-p)\right| < 1$, which is equivalent to $t < -\log(1-p)$. Since p is between 0 and 1, $-\log(1-p)$ is always positive. Thus, for any parameter p, the MGF exists on the interval $(-\infty, -\log(1-p))$, which is an open interval containing 0. ■ ■ ■

[8]Recall that $1+a+a^2+a^3+\cdots = 1/(1-a)$ for $|a| < 1$. If the first term isn't 1, the series is

$$b+b \times a+b \times a^2+b \times a^3+\cdots = \frac{b}{1-a}, \qquad |a| < 1.$$

We are applying this latter result with $b = e^t(1-p)$ and $a = e^t(1-p)$.

6.10 **Quantile–Quantile Plots**

Visually assessing whether histograms like those in Figures 6.4 and 6.8 follow some theoretical distribution is difficult at best and fraught with the potential for error. For example, we see in Figure 6.8 that the bar centered at $x = 35$ seems a bit too tall. Is this sufficient to conclude that the data do not follow a normal distribution?

A better way to visually assess distributional fit is through the use of *quantile–quantile plots* (often called "QQ plots for short). A one-sample Q-Q plot graphs an ordered set of data versus the theoretical quantiles from a specified distribution. That is, for a sample of size n and theoretical distribution F, the QQ plot is a scatterplot of the pairs $\left(F^{-1}\left((i-1/2)/n\right), x_{(i)}\right)$, $i = 1, \ldots, n$, where $x_{(i)}$ is the ith ordered observation. If the data come from the hypothesized distribution, then $F^{-1}\left((i-1/2)/n\right) \approx x_{(i)}$ and thus on a scatterplot the points corresponding to the n pairs should fall close to a straight line.

In R, the function qqnorm() does the calculation automatically for the normal distribution. For example, the following syntax plots the normal Q-Q plot that corresponds to the histogram of Figure 6.8:

```
data07$MonthDay <- data07$Month+data07$DayofMonth/100
avgs <- by(data07$CarrierDelay[data07$CarrierDelay>0],
         data07$MonthDay[data07$CarrierDelay>0],mean)
qqnorm(avgs)
qqline(avgs)
```

The first line creates a unique month–day variable that is used in the second line to calculate the daily carrier delay averages using the by() function. Note here that we're calculating the average of only those days in which the carrier delay is greater than zero since when it is equal to zero there is no delay and when it is negative the flight left early. The result, which is now the nationwide average daily carrier delay, is assigned to the avgs variable from which the normal Q-Q plot is plotted. Finally, a line running through the first and third quartiles (i.e., the 25th and 75th percentiles) is added to the plot using the qqline() function . The output is shown in Figure 6.21. The plot largely follows a straight line, indicating that the normal distribution is a fairly good fit to the data.

However, there is also a slight curve visible for the larger quantiles, suggesting a slight skew or some large outliers that may not be entirely consistent with the normal model. But that also should not be too surprising since a normal model suggests that carrier delays are simply random and one day in the year is just like any other day. Yet we know that the National Air System is affected by weather, holidays, and other phenomena. Indeed, just a cursory investigation into the 2007 data shows that of the 39 days

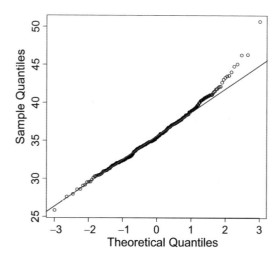

Figure 6.21 Normal QQ plot of average daily carrier delays with a qqline() overlaid, showing that there is some deviation from the normal distribution in the right tail.

with average carrier delays over 40 minutes, many are associated with major storms in 2007 and, in fact, the day with the highest average carrier delay is Christmas Eve.

So, while it seems that the normal distribution fits most of the data well, including carrier delays around $x = 35$ in Figure 6.8, the right tail of the empirical distribution contains a number of observations a bit larger than what one would expect to observe under the normal model. Yet it's still not too far off. For example, if we were to assume $X \sim (35.7, 3.4^2)$, using the sample mean ($\bar{x} = 35.7$) and standard deviation ($s = 3.4$) from the data, then the probability a carrier delay is greater than 40 is computed using `1-pnorm(40,35.7,3.4)`, which gives

```
[1] 0.1029884
```

So, we would expect to see $365 \times 0.103 = 37.6$ days with delays greater than 40 minutes and we actually observed 39 such days. That's actually quite close.

Returning to the t distribution, which looks very much like a normal distribution, it is often impossible to tell just by looking at a histogram whether a set of data with a "bell shape" has a normal distribution or a t distribution. It is much easier to see departures from normality with a normal Q-Q plot.

For example, the plot in Figure 6.22(a) is a histogram of 100,000 observations from a t distribution with $\nu = 10$ degrees of freedom. With the equivalent normal curve overlaid in red, the data look like they fit the normal distribution. However, the right normal Q-Q plot clearly shows that the "tails" of the empirical distribution are much larger than what they would be if the observations had come from a normal distribution.

Here's the R code for the plots:

```
data <- rt(100000,10)
hist(data,freq=FALSE,breaks=100,xlab="x")
   lines(seq(-6,6,0.01),dnorm(seq(-6,6,0.01),mean(data),sd(data)),lwd=2,
   col="red")
qqnorm(data)
   qqline(data,col="blue")
```

You may be wondering how to tell when the points on a QQ plot deviate too far from the line. After all, in Figure 6.21 there was some deviation but not enough to negate the normal distribution, while in Figure 6.22 there was more than enough deviation to conclude that the normal distribution did not fit the data. In later chapters we will develop formal methods, such as goodness-of-fit tests, that we can use to

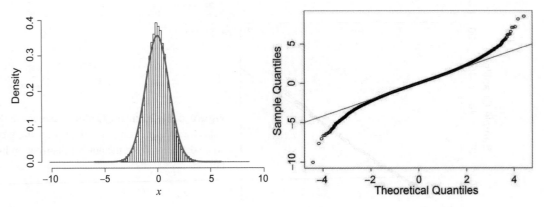

Figure 6.22 (a) Histogram of 100,000 observations from a t distribution with $\nu = 10$ degrees of freedom with the estimated normal density. (b) Normal QQ plot for the same data. With this histogram it is hard to discern that the data depart from normality, but the QQ plot makes it clear.

test the fit of data to a distribution. For now we will make the determination subjectively, where, as you gain experience looking at the QQ plots, you will get better at interpreting them correctly.

There are two things to keep in mind when looking at QQ plots. The first is that there will always be some deviation from the line, particularly with the smaller and larger observations in the data. The second is that how closely the points fall on the line depends on sample size. Smaller samples are also inherently noisy and that will show up as greater deviation from the line, while larger samples will be much less noisy and the points should hew fairly closely to the line.

Thus, one reason we can confidently say that the data in Figure 6.22 are not normally distributed is that with a sample size of 100,000 the points should fall quite close to the line. Another reason is that the blue line in the plot shows that even with 100,000 observations it would be very rare to observe points more than 5 standard deviations from the mean, yet in the data we see one point that is 10 standard deviations below the mean and many points more than 5 standard deviations away from the mean.

The probability of being five or more standard deviations below the mean for a normal distribution is very small: in R, pnorm(-5) yields $2.866516e\text{-}07$. With 100,000 observations we would on average expect to see $100,000 \times 2.866516 \times 10^{-07} \approx 0.0287$ observations (i.e., less than 1) five or more standard deviations below the mean. Yet, in the data in Figure 6.22 there are 36 such observations.

One way to develop your intuition is to create some normal QQ plots using simulated data that you know are normally distributed. Do it for a variety of sample sizes to get some idea of how far away from the line the observations fall and, in so doing, you will go a long way toward developing your intuition about the size of deviations that are to be expected. (See Problem 6.90.)

To create a Q-Q plot for distributions other than the normal takes a bit more work. For example, returning to the data shown in Figure 6.4, the following code generates a QQ plot for the exponential distribution:

```
n <- length(data07$CarrierDelay[data07$CarrierDelay>0])
lambda.est <- 1/mean(data07$CarrierDelay[data07$CarrierDelay>0])
theoretical.quantiles <- qexp((1:n-1/2)/n,lambda.est)
sample.quantiles <- sort(data07$CarrierDelay[data07$CarrierDelay>0])
plot(theoretical.quantiles, sample.quantiles,xlab="Theoretical Quantiles",
    ylab="Sample Quantiles")
```

The first line simply assigns to n the number of observations in the data. The second line estimates λ from the data using $\hat{\lambda} = 1/\bar{x}$, and then the third line calculates the theoretical quantiles for n observations from an exponential distribution with $\hat{\lambda}$. The fourth line defines the sample quantiles as the ordered data. Finally, the last line plots the theoretical quantiles versus the sample quantiles and appropriately annotates the plot.

Adding a line through the first and third quartiles also takes more work, as shown in the following R code:

```
line.slope <- (quantile(sample.quantiles,0.75) -
            quantile(sample.quantiles,0.25))/
              (qexp(0.75*(n-1/2)/n,lambda.est)-
            qexp(0.25*(n-1/2)/n,lambda.est))
line.intercept <- quantile(sample.quantiles,0.75)-line.slope*
            qexp(0.75*(n-1/2)/n,lambda.est)
abline(line.intercept,line.slope,col="blue",lwd=2)
```

Here, the first line calculates the slope of the line passing through the first and third quartiles, while the second line calculates the associated intercept. Then the third line of code adds the line to the QQ plot. The result is shown in Figure 6.23, where we can immediately see that the data do not follow an exponential distribution.

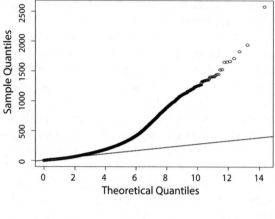

Figure 6.23 Exponential QQ plot of 2007 carrier delay data with line overlaid, clearly showing a lack of linearity, indicating that the data do not follow an exponential distribution.

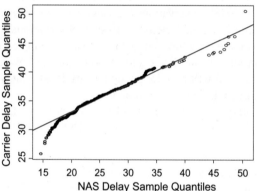 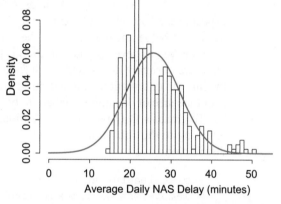

Figure 6.24 (a) Two-sample QQ plot of the average daily NAS delay vs. the average daily carrier delay. There are some significant differences in the left tails of the distributions (because the dots deviate quite a bit from the line). (b) Histogram of the average daily NAS delay with the estimated normal density in red, where we see that the left tail is truncated at about 15 minutes.

We have used Q-Q plots to compare a set of data to a theoretical distribution for the purpose of assessing whether it is reasonable to assume that the data follow that distribution. It is possible to use Q-Q plots to compare two sets of data to assess whether they have the same underlying distribution. Such a comparison does not answer the question of which distribution may have generated the data, but rather whether the two sets of data come from the same family of distributions (normal, gamma, etc.). In R, the function to use is qqplot(), where the first and second arguments are vectors corresponding to the two sets of data. What qqplot() does is compare the sample quantiles of the first set of data against the sample quantiles of the second set of data. If the two data sets are the same size, this amounts to a scatterplot of the ordered pairs of data (though note that the two sets of data do not have to be of the same size to use the qqplot() function).

To illustrate with the airline data, imagine we are interested in determining whether the average daily NAS delay has the same distribution as the average daily carrier delay, where we already know that the average daily carrier delay is reasonably normally distributed, and we might hypothesize that the average daily NAS delay might also be so distributed. The plot in Figure 6.24(a) shows the two-sample QQ plot of the average daily NAS delay versus the average daily carrier delay. The plot shows there are differences in the left tails of the distributions because the dots deviate quite a bit from the line in the lower left quadrant of the plot. The plot in Figure 6.24(b) shows why; we can see that the left tail of the average daily NAS delay distribution is truncated at around 15 minutes.

Table 6.2 Summary of the nine continuous distributions we have learned thus far, their PDFs, expected values, and variances. The table lists the range of x values for which the PDF is nonzero and the restrictions, if any, on the distributional parameter(s).

Distribution	Density	Parameters	Support
Beta BETA(a,b)	$f(x) = \dfrac{\Gamma(\alpha+\beta)}{\Gamma(\alpha)\Gamma(\beta)} x^{\alpha-1}(1-x)^{\beta-1}$	$\alpha > 0, \beta > 0$	$0 < x < 1$
Chi-squared $\chi^2(\nu)$	$f(x) = \dfrac{1}{2^{\nu/2}\Gamma(\frac{\nu}{2})} x^{\nu/2-1} \exp(-x/2)$	$\nu > 0$	$x > 0$
Exponential EXP(λ)	$f(x) = \lambda e^{-\lambda x}$	$\lambda > 0$	$x \geq 0$
Alt exp EXP1(θ)	$f(x) = \dfrac{1}{\theta} e^{-x/\theta}$	$\theta > 0$	$x \geq 0$
F $F(\nu_1, \nu_2)$	$f(x) = \dfrac{\Gamma((\nu_1+\nu_2)/2)}{\Gamma(\nu_1/2)\Gamma(\nu_2/2)} \dfrac{x^{(\nu_1-1)/2}}{(1+x)^{(\nu_1+\nu_2)/2}}$	$\nu_1, \nu_2 > 0$	$x \geq 0$
Gamma GAM(α,β)	$f(x) = \dfrac{x^{\alpha-1}}{\beta^\alpha \Gamma(\alpha)} e^{-x/\beta}$	$\alpha > 0, \beta > 0$	$x > 0$
Normal $N(\mu,\sigma^2)$	$f(x) = \dfrac{1}{\sqrt{2\pi}\sigma} \exp\left(-\dfrac{(x-\mu)^2}{2\sigma^2}\right)$	$-\infty < \mu < \infty,\ \sigma > 0$	$-\infty < x < \infty$
t $t(\nu)$	$f(x) = \dfrac{\Gamma((1+\nu)/2)}{\sqrt{\nu}\Gamma(1/2)\Gamma(\nu/2)} \left(1+\dfrac{x^2}{\nu}\right)^{-(\nu+1)/2}$	$\nu > 0$	$-\infty < x < \infty$
Uniform UNIF(θ_1,θ_2)	$f(x) = \dfrac{1}{\theta_2 - \theta_1}$	$-\infty < \theta_1 < \theta_2 < \infty$	$\theta_1 \leq x \leq \theta_2$
Weibull WEIB(θ,β)	$f(x) = \dfrac{\beta}{\theta}\left(\dfrac{x}{\theta}\right)^{\beta-1} \exp\left(-\left(\dfrac{x}{\theta}\right)^\beta\right)$	$\alpha > 0,\ \beta > 0$	$x > 0$

Table 6.3 Summary of R functions for continuous distribution PDF, CDF, quantile, and random variate generation calculations. The multivariate normal functions require the `mvtnorm` package.

Distribution	$f(x)$	$F(x)$	ϕ_p	n variates
Beta	$\texttt{dbeta}(x,\alpha,\beta)$	$\texttt{pbeta}(x,\alpha,\beta)$	$\texttt{qbeta}(p,\alpha,\beta)$	$\texttt{rbeta}(n,\alpha,\beta)$
Chi-squared	$\texttt{dchisq}(x,\nu)$	$\texttt{pchisq}(x,\nu)$	$\texttt{qchisq}(p,\nu)$	$\texttt{rchisq}(n,\nu)$
Exponential	$\texttt{dexp}(x,\lambda)$	$\texttt{pexp}(x,\lambda)$	$\texttt{qexp}(p,\lambda)$	$\texttt{rexp}(n,\lambda)$
F	$\texttt{df}(x,\nu_1,\nu_2)$	$\texttt{pf}(x,\nu_1,\nu_2)$	$\texttt{qf}(p,\nu_1,\nu_2)$	$\texttt{rf}(n,\nu_1,\nu_2)$
Gamma	$\texttt{dgamma}(x,\alpha,1/\beta)$	$\texttt{pgamma}(x,\alpha,1/\beta)$	$\texttt{qgamma}(p,\alpha,1/\beta)$	$\texttt{rgamma}(n,\alpha,1/\beta)$
Mult. norm.	$\texttt{dmvnorm}(x,\mu,\Sigma)$	$\texttt{pmvnorm}(x,\mu,\Sigma)$	$\texttt{qmvnorm}(p,\mu,\Sigma)$	$\texttt{rmvnorm}(n,\mu,\Sigma)$
Normal	$\texttt{dnorm}(x,\mu,\sigma)$	$\texttt{pnorm}(x,\mu,\sigma)$	$\texttt{qnorm}(p,\mu,\sigma)$	$\texttt{rnorm}(n,\mu,\sigma)$
t	$\texttt{dt}(x,\nu)$	$\texttt{pt}(x,\nu)$	$\texttt{qt}(p,\nu)$	$\texttt{rt}(n,\nu)$
Uniform	$\texttt{dunif}(x,\theta_1,\theta_2)$	$\texttt{punif}(x,\theta_1,\theta_2)$	$\texttt{qunif}(p,\theta_1,\theta_2)$	$\texttt{runif}(n,\theta_1,\theta_2)$
Weibull	$\texttt{dweibull}(x,\beta,\theta)$	$\texttt{pweibull}(x,\beta,\theta)$	$\texttt{qweibull}(p,\beta,\theta)$	$\texttt{rweibull}(n,\beta,\theta)$

6.11 Chapter Summary

In this chapter we've learned about nine specific continuous distributions: the exponential, normal, multivariate normal, chi-squared, F, t, gamma, Weibull, and beta. Along with the uniform distribution from Chapter 4, you now have ten continuous distributions available to you which have wide applicability to real-world phenomena and statistical inference. These distributions can be used to model and analyze myriad physical phenomena. For example, using Q-Q plots we have just shown that the average daily carrier delays can be modeled by a normal distribution.

Table 6.2 is a summary of the continuous distributions: their PMFs, expected values, and variances. Table 6.3 summarizes the R functions for the PDF, CDF, quantile, and random variate generation calculations. As discussed in this chapter, these functions are useful for calculating the probabilities associated with the various distributions. Software such as R can be very handy.

We considered the problem of finding the distribution of a function of a random variable having a known distribution. This shows us how many of the distributions are related, and it provides a method for simulating random numbers from a distribution. One technique for finding the distribution of a function of a random variable is the CDF technique. The other involves the MGF, which is particularly useful in determining the distribution of a sum of independent random variables.

6.12 Problems

Problem 6.1 Suppose $U \sim \text{UNIF}(0,20)$. Find the following.
(a) $\Pr\left(U \leq 10\right)$ (b) $\Pr\left(5 \leq U \leq 12\right)$
(c) $\Pr(U = 10)$ (d) The CDF of U
(e) $\mathbb{E}(U)$ (f) $\mathbb{V}(U)$

Problem 6.2 Suppose $U \sim \text{UNIF}(100,200)$. Find the following.
(a) $\Pr\left(U \geq 50\right)$ (b) $\Pr\left(50 \leq U \leq 120\right)$
(c) $\Pr\left(U = 150\right)$ (d) The CDF of U
(e) $\mathbb{E}(U)$ (f) $\mathbb{V}(U)$

Problem 6.3 Suppose $U \sim \text{UNIF}(-5,5)$. Find the following.
(a) $\Pr\left(U \leq 0\right)$ (b) $\Pr\left(-a \leq U \leq a\right)$
(c) The CDF of U (d) $\mathbb{E}(U)$
(e) $\mathbb{V}(U)$

Problem 6.4 Suppose $U \sim \text{UNIF}(0,\theta)$. Find the following.
(a) $\Pr\left(U = \theta/2\right)$ (b) $\Pr\left(U < \theta/2\right)$
(c) The CDF of U (d) $\mathbb{E}(U)$
(e) $\mathbb{V}(U)$ (f) $\mathbb{E}(U^2)$

Problem 6.5 Suppose that the shuttle buses from the airport terminal to the long-term parking lots run every 12 minutes. If I arrive at a random point in time, my waiting time is therefore a $\text{UNIF}(0,12)$ random variable X.

(a) What is the probability that I wait at least 10 minutes?
(b) What is the probability that I wait for 2 or fewer minutes?
(c) What is my expected waiting time?

Problem 6.6 Continuation of Problem 6.5. Suppose that in a given month I make three trips through the airport. What is the probability that when I take the shuttle from the terminal to the parking lot, I must wait at least 10 minutes every time?

Problem 6.7 Suppose $X \sim \text{EXP}(0.25)$. Find the following:
(a) $\Pr\left(X > 4\right)$ (b) $\Pr(X = 4)$
(c) The CDF of X (d) $\Pr(X > 4 \,|\, X > 2)$
(e) $\mathbb{V}(X)$ (f) The hazard function

Problem 6.8 Suppose $X \sim \text{EXP1}(12)$. Find the following.
(a) $\Pr\left(X > 12\right)$ (b) $\Pr\left(5 \leq X \leq 12\right)$
(c) $\Pr(X = 12)$ (d) The CDF of X
(e) The hazard function

Problem 6.9 Suppose $X \sim \text{EXP1}(800)$. Find the following.

(a) $\Pr\left(X > 800\right)$
(b) $\Pr(X = 800)$
(c) The CDF of X
(d) $\Pr(X > 1,000 \,|\, X > 800)$
(e) What is the hazard function for X?

Problem 6.10 Suppose $X \sim \text{EXP}(0.00125)$. Find the following.

(a) $\Pr\left(X > 800\right)$
(b) $\Pr(X = 800)$
(c) The CDF of X
(d) $\Pr(X > 1,600 \,|\, X > 800)$

Problem 6.11 Suppose that a car's headlight has a lifetime (in hours) that has an $\text{EXP1}(2,000)$ distribution.

(a) What is the probability that a new light bulb will last for at least 100 hours?
(b) What is the probability that a 1,000-hour-old working light bulb will last for at least 100 additional hours?
(c) What is the probability that a 2,000-hour-old working light bulb will last for at least 100 additional hours?
(d) The manufacturer suggests replacing *both* headlights when one burns out. If the lifetime distribution is exponentially distributed, is this good advice?

Problem 6.12 Suppose that florescent light bulbs have a lifetime (in hours) that has an $\text{EXP1}(2,000)$ distribution. Suppose a room has two such light bulbs and that the bulbs act independently.

(a) What is the probability the first bulb lasts for 2,000 or more hours?
(b) What is the probability the second bulb lasts for 2,000 or more hours?
(c) What is the probability that both bulbs last for 2,000 or more hours?
(d) When a bulb (either one) burns out, we must call the maintenance crew to replace it. What is the probability we must call the maintenance crew within the first 1,000 hours?
(e) Suppose we do not call the maintenance crew when a bulb burns out. What is the probability that at the 3,000-hour mark we are in the dark because both bulbs have burned out?

Problem 6.13 Repeat all parts of Problem 6.12 using three bulbs instead of two.

Problem 6.14 Repeat all parts of Problem 6.12 using n bulbs instead of two.

Problem 6.15 Suppose $Z \sim N(0,1)$. Find the following.

(a) $\Pr(Z > 0)$

(b) $\Pr(Z = 0)$

(c) $\Pr(-2 < Z < 2)$

(d) $\Pr(-1.96 \leq Z \leq 1.96)$

(e) $\Pr(1 \leq Z \leq 3)$

Problem 6.16 Suppose $Z \sim N(0,1)$. Find the following.

(a) $\Pr(Z < 0)$

(b) $\Pr(-1.64 < Z < 1.64)$

(c) $\Pr(-3 \leq Z \leq 1)$

(d) Find c such that $\Pr(Z > c) = 0.02$. Approximate c to two decimal places.

Problem 6.17 Suppose $X \sim N(50,4)$. Find the following.

(a) $\Pr(X > 52)$

(b) $\Pr(46 < X < 54)$

(c) Find c such that $\Pr(X < c) = 0.05$.

Problem 6.18 Suppose that in a population, the height of adult men is approximately normal with mean $\mu = 70$ inches and standard deviation $\sigma = 4$ inches. One man is selected from the population and his height is determined. Let X denote this random variable. Find the following.

(a) $\Pr(X \geq 72)$

(b) $\Pr(64 < X < 76)$

(c) 5% of men in this population are shorter than what height?

Problem 6.19 A manufacturing process produces capacitors designed to produce 100 VDC (volts of direct current). Capacitors produced by this company are normally distributed with a mean of 100.7 and a standard deviation of 1.8. Products whose capacitance is between 95 and 105 are considered acceptable; others are unacceptable and discarded.

(a) What is the probability that an item produced by this process must be discarded?

(b) What fraction of the output from this process must be discarded?

Problem 6.20 Find the following.

(a) $z_{0.5}$ (b) $z_{0.025}$

(c) $z_{0.975}$ (d) $z_{0.75}$

Problem 6.21 Find the following.

(a) $z_{0.05}$ (b) $z_{0.95}$

(c) $z_{0.9}$ (d) $z_{0.99}$

Problem 6.22 Explain why $z_{1-\alpha} = -z_\alpha$.

Problem 6.23 For random variable Z with a standard normal distribution, prove that $\mathbb{E}(Z) = 0$.

Problem 6.24 For random variable Z with a standard normal distribution, prove that $\mathbb{V}(Z) = 1$.

Problem 6.25 For random variable Z with a standard normal distribution, find $\mathbb{E}(Z^3)$.

Problem 6.26 If $Z \sim N(0,1)$ what do you think $\mathbb{E}(Z^n)$ is when n is odd? Construct an argument to justify your assertion.

Problem 6.27 Show that if $(X_1, X_2) \sim N_2(\mu_1, \mu_2, \sigma_1, \sigma_2, 0)$ then the joint PDF factors into the product of the marginal PDFs, and therefore X_1 and X_2 are independent.

Problem 6.28 Write out and simplify the PDF for the $N_2(0,0,1,1,\rho)$ distribution.

Problem 6.29 Write out and simplify the PDF for the multivariate normal distribution with

$$\mu = \begin{bmatrix} 0 \\ 0 \\ 0 \end{bmatrix} \quad \text{and} \quad \Sigma = \begin{bmatrix} 1 & \rho & \rho \\ \rho & 1 & \rho \\ \rho & \rho & 1 \end{bmatrix}.$$

Problem 6.30 For the bivariate normal distribution in Problem 6.28 with $\rho = 0.5$, use R to find the following.

(a) $\Pr(X_1 \leq 0, X_2 \leq 0)$

(b) $\Pr(X_1 \leq 1.96, X_2 \leq 1.96)$

Problem 6.31 For the multivariate normal distribution in Problem 6.29 with $\rho = 0.5$, use R to find the following.

(a) $\Pr(X_1 \leq 0, X_2 \leq 0, X_3 \leq 0)$

(b) $\Pr(X_1 \leq 1.96, X_2 \leq 1.96, X_3 \leq 0)$

Problem 6.32 Repeat Problem 6.30 for $\rho = 0.8$, $0.9, 0.95, 0.99$. What do you think happens to this probability as $\rho \to 1^-$? Explain.

Problem 6.33 What is the PDF for the $GAM(1, \theta)$ distribution? What is another name for this distribution?

Problem 6.34 What is the PDF for the $WEIB(\theta, 1)$ distribution? What is another name for this distribution?

Problem 6.35 Suppose $X \sim GAM(2,100)$. Use R's built-in functions to find the following.

(a) $\Pr(X \leq 100)$ (b) $\Pr(X < 100)$

(c) $\Pr(X > 500)$ (d) $\mathbb{E}(X)$

(e) $\mathbb{V}(X)$ (f) The 95th percentile

Problem 6.36 Suppose $X \sim \text{GAM}(0.5, 1000)$. Use R's built-in functions to find the following.

(a) $\Pr(X \leq 1000)$ (b) $\Pr(X = 1000)$
(c) $\Pr(X > 2000)$ (d) $\mathbb{E}(X)$
(e) $\mathbb{V}(X)$ (f) The 90th percentile

Problem 6.37 For the gamma function $\Gamma()$ defined in Equation (6.4), find the following.

(a) $\Gamma(1)$

(b) $\Gamma(2)$ (*Hint:* Use integration by parts.)

(c) $\Gamma(3)$

Problem 6.38 For the gamma function $\Gamma()$ defined in (6.4), show that for $a > 1$, $\Gamma(a) = (a-1)\Gamma(a-1)$. Use this result to show that if a is a positive integer, then $\Gamma(a) = (a-1)!$.

Problem 6.39 The lifetime in hours X of an electrical component has a $\text{GAM}(2, 1000)$ distribution. A larger system is made up of four of these components, and the larger system fails if any one of the smaller components fails. Such a system is called a *series* system. What is the probability that the larger system lasts 500 or more hours?

Problem 6.40 Repeat Problem 6.39 but now assume that the system fails whenever all four of the components fail. Such a system is called a *parallel* system.

Problem 6.41 For the $\text{GAM}(\alpha, \beta)$ distribution, find $\mathbb{E}(X^2)$ using the technique in the proof of Theorem 6.5.1.

Problem 6.42 Show that the hazard function for the gamma distribution, given in (6.5), satisfies

$$\lim_{x \to \infty} h(x) = \frac{1}{\beta}.$$

Hint: Use L'Hopital's rule.

Problem 6.43 Suppose $X \sim \text{WEIB}(100, 2)$. Find the following.

(a) $\Pr(X \leq 100)$ (b) $\Pr(X < 100)$
(c) $\Pr(X > 500)$ (d) $\mathbb{E}(X)$
(e) $\mathbb{V}(X)$ (f) The 95th percentile

Problem 6.44 Suppose $X \sim \text{WEIB}(200, 0.8)$. Find the following.

(a) $\Pr(X \leq 200)$ (b) $\Pr(X < 200)$
(c) $\mathbb{E}(X)$ (d) $\Pr(X = 500)$
(e) $\mathbb{V}(X)$ (f) The 99th percentile

Problem 6.45 Suppose $X \sim \text{WEIB}(\theta, \beta)$. Find a formula for the quantile q of X.

Problem 6.46 Suppose $X \sim \text{WEIB}(\theta, \beta)$. Show that

$$\mathbb{E}(X^2) = \theta^2 \Gamma\left(1 + \frac{2}{\beta}\right).$$

Problem 6.47 Use the result of Problem 6.46 to show that the variance of the Weibull distribution is

$$\mathbb{V}(X) = \theta^2 \left[\Gamma\left(1 + \frac{2}{\beta}\right) - \Gamma^2\left(1 + \frac{1}{\beta}\right)\right],$$

as given in Equation (6.9).

Problem 6.48 Suppose $X \sim \text{WEIB}(\theta, \beta)$. Find a formula for $\mathbb{E}(X^p)$, where p is a positive integer.

Problem 6.49 Suppose that $X \sim \chi^2(5)$. Find the following.

(a) $\mathbb{E}(X)$ (b) $\mathbb{V}(X)$
(c) $\Pr(X > 5)$ (d) $\Pr(X = 5)$
(e) $\Pr(X \leq 1)$

Problem 6.50 Suppose that $X \sim \chi^2(25)$. Find the following.

(a) $\mathbb{E}(X)$ (b) $\mathbb{V}(X)$
(c) $\Pr(X > 25)$ (d) $\Pr(X \leq 5)$

Problem 6.51 Find the following.

(a) $\chi^2_{0.05}(1)$ (b) $\chi^2_{0.05}(10)$
(c) $\chi^2_{0.05}(100)$ (d) $\chi^2_{0.50}(10)$

Problem 6.52 Find the following

(a) $\chi^2_{0.50}(1)$ (b) $\chi^2_{0.50}(10)$
(c) $\chi^2_{0.50}(50)$ (d) $\chi^2_{0.50}(100)$

Problem 6.53 Suppose $T \sim t(10)$. Find the following.

(a) $\Pr(T > 2)$ (b) $\Pr(T < -2)$
(c) $\mathbb{E}(T)$ (d) The median of T

Problem 6.54 Suppose $T \sim t(50)$. Find the following.

(a) $\Pr(T > 1)$

(b) $\Pr(T < -1)$

(c) The 50th percentile of T

Problem 6.55 Find the following.

(a) $t_{0.025}(10)$ (b) $t_{0.025}(100)$
(c) $t_{0.025}(1000)$ (d) $t_{0.975}(10)$

Problem 6.56 Find the following.

(a) $t_{0.01}(10)$ (b) $t_{0.01}(100)$
(c) $t_{0.99}(10)$ (d) $t_{0.99}(100)$

Problem 6.57 Repeat the simulation in Example 6.18 with $n = 10, 20, 30,$ and 40. What do you conclude from these results?

Problem 6.58 Explain why $t_{1-\alpha}(v) = -t_\alpha(v)$.

Problem 6.59 Is it true that

$$\chi^2_{1-\alpha}(v) = -\chi^2_\alpha(v)?$$

Explain.

Problem 6.60 Is it true that

$$F_{1-\alpha}(v_1, v_2) = -F_\alpha(v_1, v_2)?$$

Explain.

Problem 6.61 Suppose $W \sim F(3, 20)$. Find the following.

(a) $\Pr(W < 1.5)$ (b) $\Pr(W = 1.5)$
(c) $\Pr(W > 4)$ (d) $\Pr(W \geq 4)$

Problem 6.62 Find the following.

(a) $F_{0.01}(3, 20)$ (b) $F_{0.05}(3, 20)$
(c) $F_{0.50}(3, 20)$ (d) $F_{0.95}(3, 20)$

Problem 6.63 Using a simulation approach like that in Example 6.18, empirically demonstrate that the sum of the squared values of n observations drawn from a standard normal distribution has a χ^2 distribution with n degrees of freedom.

Problem 6.64 Using a simulation approach like that in Example 6.18, empirically demonstrate that if Y_1 and Y_2 are independent χ^2 random variables with v_1 and v_2 degrees of freedom, respectively, then the random variable $X = \frac{Y_1/v_1}{Y_2/v_2}$ has an F distribution with parameters v_1 and v_2.

Problem 6.65 Suppose $X \sim \text{BETA}(a, b)$ and let $b_\alpha(a, b)$ denote that value for which $\Pr\left(X > b_\alpha(a, b)\right) = \alpha$. Find the following.

(a) $b_{0.05}(5, 20)$ (b) $b_{0.95}(20, 5)$
(c) $b_{0.50}(8, 8)$ (d) $b_{0.10}(1, 1)$

Problem 6.66 Suppose $X \sim \text{BETA}(1, 9)$. Find the following.

(a) $\Pr(X > 0.5)$ (b) $\Pr(X = 0.5)$
(c) $\Pr(0.1 < X < 0.2)$ (d) $\mathbb{E}(X)$
(e) $\mathbb{V}(X)$

Problem 6.67 Suppose that a manufacturing process is quite variable from day to day. Some days most of the units produced are good, while other days many units are bad and must be scrapped. The distribution of the daily defective rate P, measured as a proportion, has a $\text{BETA}(1, 4)$ distribution.

(a) Plot the PDF of P.

(b) Assuming the defective rates are i.i.d. across days, are the majority of units defective or nondefective? Explain.

(c) Find the probability that on a given day the defective rate exceeds 0.20.

(d) Find $\mathbb{E}(P)$.

(e) Find $\mathbb{V}(P)$.

Problem 6.68 Returning to Theorem 6.7.1, prove that if $X \sim \text{BETA}(\alpha, \beta)$, then

$$\mathbb{V}(X) = \frac{\alpha\beta}{(\alpha+\beta)^2(\alpha+\beta+1)}.$$

Problem 6.69 Show that for the $\text{BETA}(\alpha, \beta)$, the variance can be written as

$$\mathbb{V}(X) = \frac{\mu(1-\mu)}{\alpha+\beta+1},$$

where $\mu = \alpha/(\alpha+\beta)$.

Problem 6.70 Find the mode (i.e., the value that maximizes the PDF) of the $\text{BETA}(\alpha, \beta)$ distribution.

Problem 6.71 Suppose $U \sim \text{UNIF}(0, \theta)$. Define $X = U^2$. Find the PDF of X.

Problem 6.72 Suppose $U \sim \text{UNIF}(-1, 1)$. Define $Y = U^2$. Find the PDF of Y.

Problem 6.73 Suppose $U \sim \text{UNIF}(0, 1)$.

(a) Explain why the distribution of $X = 1 - U$ is $\text{UNIF}(0, 1)$.

(b) Example 6.24 showed that $-\log U \sim \text{EXP}(1)$; but in Example 6.24 we showed that $-\log(1 - U) \sim \text{EXP}(1)$. Use part (a) to explain why both methods work to simulate from the $\text{EXP}(1)$ distribution.

(c) Which method do you think is used in practice? Why?

Problem 6.74 Suppose $X \sim \text{WEIB}(\theta, \beta)$. Let $Y = X^\beta$. Find the PDF of Y. What is the name and parameter(s) of this distribution.

Problem 6.75 Suppose $X \sim \text{WEIB}(\theta, \beta)$. Let $W = (X/\theta)^\beta$. Find the PDF of W. What are the name and parameter(s) of this distribution.

Problem 6.76 Give an algorithm for generating pseudo random numbers from the $\text{UNIF}(-1, 1)$ distribution.

Problem 6.77 Give an algorithm for generating pseudo random numbers from the $\text{WEIB}(\theta, \beta)$ distribution.

Problem 6.78 Give an algorithm for generating pseudo random numbers from the distribution with PDF

$$f(x) = \begin{cases} x, & 0 \leq x \leq 1 \\ 2 - x, & 1 < x \leq 2 \\ 0, & \text{otherwise.} \end{cases}$$

Problem 6.79 Suppose $M_X(t)$ is the MGF for a random variable X. What is $M_X(0)$?

Problem 6.80 Suppose $X_1 \sim N(\mu_1, \sigma_1^2)$ and $X_2 \sim N(\mu_2, \sigma_2^2)$ are independent. Use MGFs to show that $c_1 X_1 + c_2 X_2$ has a normal distribution for constants c_1 and c_2.

Problem 6.81 Find the MGF of the $\text{UNIF}(a, b)$ distribution.

Problem 6.82 Use the MGF for the gamma distribution in Example 6.29 to find the mean and variance of the gamma distribution.

Problem 6.83 Suppose $X_1, X_2, \ldots, X_n \sim$ i.i.d. $\mathrm{EXP1}(\theta)$. Use the MGF to find the distribution of $\sum_{i=1}^{n} X_i$.

Problem 6.84 Find the MGF of the $\mathrm{POIS}(\lambda)$ distribution.

Problem 6.85 Use the result of Problem 6.84 to derive the mean and variance of the Poisson distribution.

Problem 6.86 Suppose $X_1, X_2, \ldots, X_n \sim$ i.i.d. $\mathrm{POIS}(\lambda)$. Use the MGF to find the distribution of $\sum_{i=1}^{n} X_i$.

Problem 6.87 Let $X_1, X_2, X_3, X_4 \sim$ i.i.d. $\mathrm{GEO}(p)$. Find the MGF of $S = \sum_{i=1}^{4} X_i$.

Problem 6.88 Consider the $\mathrm{NB}(4, p)$ distribution.

(a) Find the MGF.

(b) What is the distribution of S from Problem 6.87?

Problem 6.89 Construct a normal Q-Q plot for the data in Table 6.4 on the tensile strength of aluminum. Is the normal distribution a reasonable model?

Table 6.4 Data for Problem 6.89

Unit	Tensile strength	Unit	Tensile strength
1	46,200	6	47,500
2	48,100	7	49,100
3	40,300	8	41,900
4	43,100	9	43,800
5	44,000	10	44,500

Problem 6.90 To develop your intuition with normal QQ plots, create a series of normal Q-Q plots using simulated standard normal data of sizes $n = 10, 25, 50, 100, 1000$, and 10000. What do you observe as the sample size gets large? Now, repeat but with data from a t distribution with $\nu = 5$ degrees of freedom. At what sample size can you start to confidently conclude that the simulated data do not come from a normal distribution?

Problem 6.91 Consider the data in the file `liver_patient.csv`, which is available on the book's website. Construct normal QQ plots for the following variables.

(a) `Total_Bilirubin`

(b) `Direct_Bilirubin`

(c) `Alkaline_Phosphotase`

(d) `Alamine_Aminotransferase`

Which of these variables can be modeled by a normal distribution?

Problem 6.92 Repeat Problem 6.91 for men and women separately.

Problem 6.93 Using the 2007 airlines data and R, recreate the plots in Figure 6.24.

Problem 6.94 Suppose that in a manufacturing process the daily defect rate varies according to a $\mathrm{BETA}(1, 9)$ distribution. Let D denote the defect rate on a given day and let Y denote the number of defective items in a sample of size $n = 5$.

(a) Find $\mathbb{E}(Y|D)$.

(b) Find $\mathbb{E}\big(\mathbb{E}(Y|D)\big)$.

(c) Find $\mathbb{V}(Y)$.

(d) Find the marginal distribution of Y.

7 — About Data and Data Collection

7.1 Introduction

In statistics, we are often interested in some characteristics of a population. Maybe we are interested in the mean of some measurable characteristic, or maybe we are interested in the proportion of the population that have some property. In all but the simplest cases, the population is so large that it is impossible, or at least impractical, to take the measurement on *every* item in the population. We therefore have to settle on taking a *sample* and measuring those units selected for this sample. What much of the rest of this book is about is making an inference regarding population characteristics (called parameters) based on the results in a sample. This is called *inferential statistics*. We summarize some of these terms in the following definitions:

> **Definition 7.1.1 — Population, Parameter, Sample, Statistic.** The *population* is the set of people or objects that we would like to learn something about.
>
> A *population parameter*, or simply a parameter, is some characteristic of the population that we would like to know.
>
> A *sample* is a subset of the population selected at random for the purpose of representing the whole population.
>
> A *statistic* is some function of the observed data from the sample.

Figure 7.1 shows the relationships between these concepts. Populations are generally too large to measure each person or object, so we must rely on a sample selected, usually at random, from the population. Then we compute numerical summaries, called statistics, from the sample. We use these to say something about the parameters – for example, we might use the sample mean as an estimate of the population mean. (More about estimation in Chapter 9.) Finally, we make inference to the population. The term *statistics* is used in two senses. Here we have used the term in a narrow and technical sense to mean summaries computed from a sample. Statistics is also the term used to describe the entire field of using data to make inference. Context will usually determine which kind of "statistics" we are talking about.

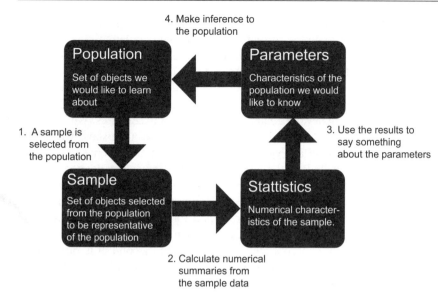

Figure 7.1 Steps in making inference from sample to population.

If a population is infinite and can be described by a continuous probability distribution, in the sense that the random variable defined as the measurement of a subject or object selected at random has that distribution, then two important properties are these:

- successive selections (random variables) have the same distribution, since they are selected from the same population; and

- successive selections are independent, since the selection of one item from an assumed infinite population does not affect the probability distribution of the next one chosen.

This idea leads to the following definition.

Definition 7.1.2 — Random Sample. If random variables $X_1, X_2, \ldots, X_n$ are *independent and identically distributed*, abbreviated *i.i.d.*, with PDF or PMF $f(x)$, then these random variables are said to be a *random sample* from that distribution.

Of course, no population is infinite, and any assumed probability distribution is likely a simplifying assumption. There are ways to account for finite populations, and these are needed if the sample makes up a significant portion of the population. For many problems, though, these assumptions are reasonable.

It is important to consider the ideas of population/sample and parameter/statistic. This chapter will deal with the issue of collecting sample data that are in some way characteristic of the population. Chapters 8 and 9 deal with the related issue of using statistics computed from samples to infer something about the population parameters.

When doing data science, it is not enough to focus on the analytical methods. At least as important are the data. Statistical analysis, however sophisticated, of weak or poor data is akin to building a skyscraper on a shaky foundation. In Chapters 1 and 2 we learned about numerical and graphical methods for describing data. These descriptive statistical methods can be used with data from an entire population or with data just from a sample of the population. The goal is to appropriately summarize data, whether from a sample or a population, so that we can better understand them.

If all we are doing is using descriptive statistics to characterize the sample itself when working with sample data, simply applying such descriptive statistics to the data is fine. If we want to use the sample for making inference from the sample to the (unobserved) population, then we have to know how the sample was selected. That is, when the purpose shifts from simply summarizing the sample of data to using the sample to characterize the population, then we must also incorporate information about how the sample was collected in order to do the correct inference to the population.

Furthermore, depending on the question one is trying to answer or the experiment one is trying to conduct, there are better and worse ways to sample. Thus, if possible, we want to design how the sampling is conducted so that we maximize the information in the data that are collected.

In this chapter we'll start by learning about the scientific method, which is a formal approach for advancing knowledge that has been developed and refined over many centuries. We'll then compare and contrast the scientific method to the "data science cycle" in order to understand what can and cannot be concluded from a typical statistical analysis. We will then learn about some important concepts in statistics and data analysis and conclude with methods for sampling when collecting data.

The notion of sampling to obtain data may seem foreign or even unnecessary in the current environment, in which data seem to be being collected from everyone on everything. However, there are a number of important reasons to have an appreciation for sampling. First, many of the standard statistical methods assume particular types of sampling, so when using them it is important to understand those assumptions to ensure you are appropriately applying them. Second, it is rarely the case – even today – that one has data on an entire population. Rather, most data, even when you have it in enormous quantities, misses parts of the population and have arisen according to some process that may make parts of the data more or less accurate or appropriate to use.

Thus, any good analysis *always* begins with a careful consideration of the data: How and from where did it originate? Who or what might be excluded from the data set? Which variables seem to contain good data and which seem problematic? How strong or robust or generalizable are the results derived from these data? Can I conclude that there is a causal connection? Are there important caveats to my findings?

7.2 Data and the Scientific Method

As we discussed in Chapter 1, statistics is the study of how to extract useful information from data (see Definition 1.1.2). That is, the purpose of statistics is to learn about some phenomenon of interest by analyzing data. However, in any data analysis, we need to guard against the possibility of false discovery. That is, if we take any data set, particularly a large one, and just fish through it for correlations, some of the observed correlations will be spurious, by which we mean that the correlation is just an artifact of that particular data set that does not exist in the real world.

For example, imagine that you are given the data from an examination of 1,000 eighth-grade students who each took a 100-question multiple-choice math test. Your goal is to identify those students who have demonstrated a proficiency in mathematics and, to do so, you use a cutoff of 70%: Those students who score above 70% are classified as proficient. However, it is possible to misclassify students as proficient that perhaps just got lucky by guessing and, similarly, it's possible to misclassify students who really are proficient but who made some sort of error, say in marking their responses on the answer sheet.

The point of this example is to illustrate that in any data analysis it is important to minimize the "discovery" of spurious (i.e., false) results. To do this, analyses must be firmly grounded, and there is no firmer ground than the scientific method.

> **Definition 7.2.1 — Scientific Method.** The *scientific method* is the systematic pursuit of knowledge based on observation, measurement, and experimentation that includes the formulation and empirical testing of hypotheses. A key tenet of the scientific method is that results of an experiment can and should be confirmed via replication.

The scientific method is a formal approach for investigating phenomena in the pursuit of new knowledge. For a method to be called scientific it must be based on empirical evidence *and* formal principles of reasoning. Furthermore, the results of any scientific inquiry should be confirmed via some form of replication of the results. For example, returning to the eighth-grade students, one way to identify

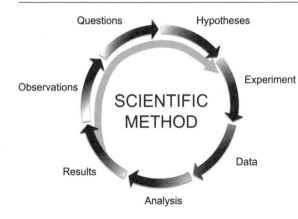

Figure 7.2 Steps in the scientific method, including reproducing the results via replication.

those students who just got lucky by guessing is to give the students a second test. Those who do well on both tests are much more likely to actually be proficient in math, since it's unlikely that a student will just get lucky twice in a row. Figure 7.2 illustrates the scientific method. Briefly, each of the steps is described below.

- **Observation**: The scientific method begins with some observation about the phenomenon of interest. By "observation" we do not mean data, but rather we mean that a scientist or researcher has an insight into the phenomenon. For example, perhaps based on experience or prior knowledge, a researcher might observe that eighth-grade student mathematics proficiency seems to be related to where students live.

- **Questions**: Based on this observation, the researcher would begin by asking questions. For example, with the eighth-grade exam, the researcher might have questions about how the exam is administered, what subjects are on the exam, etc. Here the researcher seeks to understand the phenomenon by asking questions and doing background research.

- **Hypotheses**: As a result of the information obtained in the background research, the scientist then poses some hypotheses that seek to explain the observed phenomenon *and* that can be formally, empirically tested to determine whether the hypotheses are true or false. For example, with the eighth-grade mathematics exam, the researcher might pose hypotheses such as (a) "Eighth-grade mathematics proficiency is higher for students who live in urban areas as compared to students who live in rural areas" and (b) "Eighth-grade mathematics proficiency is higher for students who come from families with higher incomes as compared to students who come from families with lower incomes."

- **Experiment**: The hypotheses are then tested by conducting an appropriately designed, rigorous experiment.[1] In our example, this might consist of giving a mathematics exam to a representative sample of eighth-grade students while simultaneously collecting home address and family income information on each of the students.

- **Data**: The experiment then results in data where, for the example, these data would be test score results and associated demographic information.

- **Analysis**: The data are then analyzed using appropriate statistical techniques to formally test the hypotheses. The hypotheses were posed *prior* to conducting the experiment and analyzing the data. To first look at the data and then pose hypotheses that fit the observed data is cheating, and doing so is likely to result in spurious findings.

---

[1]In this section we use "experiment" to mean something slightly different than in Section 7.3. Here, experiment means a way to collect data using some study design. In Section 7.3 we distinguish study designs that allow the researcher to impose treatments, called *experiments*, and those that do not, called *observational studies*.

- **Results**: From the analyses come the results, which are the determination of whether the hypotheses that were proposed are true or false. That is, the results are the determinations of whether the data collected from the experiment support or refute the original hypotheses.

- **Reproduce**: Finally, an important part of the scientific method is reproducing the results. Typically this involves other scientists conducting independent experiments to see whether the original results can be replicated. Replication is important because it helps ensure that the significant results from the original results are not spurious. It also ensures that the results of the original experiment are not unique to either the particular data or the researcher, and they can help determine whether the original results can be generalized.

It is difficult to overstate the importance of replication in the scientific method. In Figure 7.2, replication is the gray arrow, where the idea is to independently repeat the experiment to see whether the original results hold up to scrutiny. For example, returning to the eighth-grade math proficiency experiment, imagine that the original researcher's experiment was conducted using California students and it found that proficiency is associated with income and not with urban/rural households. Then a new researcher might assess whether the experimental methodology is sound and repeat the experiment to see whether the conclusions continue to hold. Another researcher might repeat the experiment using students from Texas to see whether the California results generalize to Texas. From these and similar replications of the experiment, we gain confidence in (or perhaps refute) the original results by learning about how robust the original results are.

A failure to understand the importance of replication in medical studies is one major reason why the popular media frequently reports conflicting medical news. Indeed, some say that, for a variety of reasons, the claims of a majority of published medical studies turn out to be false (see Ioannidis, 2005). Today wine is good for you and tomorrow it's bad. An aspirin a day helps reduce the risk of cancer – or does it? At issue is that news organizations latch onto the latest medical or other study results and report them without waiting to see whether the conclusions can be replicated. The outcome is that the news contains an unnecessarily high rate of spurious results and we get cynical about modern medicine because it seems like the research community keeps reversing itself.

Data science does not always fit neatly into the scientific method, often because the data scientist has little control over data collection, perhaps using convenient data that are readily available. One of the frequently touted benefits of data science is the extraction of useful information from data that already exist, which speeds up the time to discovery. Thus, the data science cycle, illustrated in Figure 7.3, is something of a subset of the scientific method:

- **Questions**: In a manner similar to the scientific method, the data science cycle starts with questions of interest. These may be business-related questions or they may be scientific questions,

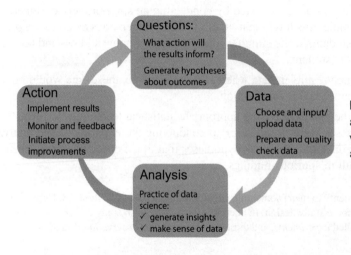

Figure 7.3 The data science cycle is an abridgment of the scientific method, where it is usually focused on extracting actionable information from existing data.

but they are questions to which the statistician or data scientist would like to find answers. Note that the questions may ultimately be formulated in terms of formal hypotheses to be tested or they may be more informally expressed.

- **Data**: Given the questions, the statistician or data scientist then seeks to identify relevant data that typically already exist but that are relevant for answering, or at least informing, the questions. Because the data scientist is working with existing data, he or she is usually not involved in the data collection and, in fact, may not even have definitive knowledge of how the data were collected. Furthermore, the data may not be the result of any sort of formal experimentation; they may just be data that were collected as part of some business or administrative process.

- **Analysis**: With the data in hand, the data scientist then analyzes the data to try to answer the questions. The analysis may range from the purely descriptive, using the techniques we learned about in Chapters 1 and 2, to more formal analyses that we will cover later.

- **Action**: Ultimately, the results of the analysis will inform and some action will be taken. In a business context, the action might be some sort of product or process improvement; in a scientific context, the results may feed into the observation step in the scientific method.

As we just described, within the data science cycle the data scientist is often working with existing *administrative data*, by which we mean data that were likely collected for some purpose other than answering the data scientist's question or questions.

> **Definition 7.2.2 — Administrative Data.** Administrative data is information collected primarily for administrative (not research) purposes.

Sometimes administrative data are available on an entire population, which can render the need to conduct inferential statistics moot and so simple descriptive statistics are sufficient for the analysis. On the other hand, administrative data can be analytically and methodologically challenging because they have generally been collected for purposes other than to answer the question of interest, and data may be missing or inaccurate depending on how the data set was created, collected, maintained, and updated. Finally, because the researcher likely has no control over how the data were collected, the conclusions that can be drawn are usually limited. Administrative data are one form of *observational* data – that is, data not collected as a result of a formal experiment.

7.3   Experimental vs. Observational Data

Experimental data are data that result from conducting a formal experiment in which the researchers impose treatments on the subjects or objects. Such experiments are often characterized by the random assignment of subjects to conditions. The goal is usually to establish cause-and-effect by demonstrating that a particular outcome occurs for a given condition.

A specific type of experiment is the *clinical trial*, in which people are randomly assigned to either get a medical treatment or a placebo. The randomization is necessary to avoid inadvertent bias, and such trials are often "double blind," meaning that neither the patient nor the diagnosing physician knows whether a patient receives the treatment or the placebo. Once data are collected, the differences between what are called the "treatment" and "control" groups are examined to see whether the treatment works as desired.

What randomization is designed to do is to avoid bias in the results. Without random assignment, it might be that sicker people are more likely to be assigned to the treatment group. But if this happens, then it's possible that an effective treatment will be determined to be ineffective because the treatment group was sicker than the control group and thus perhaps had worse outcomes even with an effective treatment. This is an example of *confounding*, where the effect of the treatment is confounded with (i.e., cannot be separated from) the effects of initial level of sickness.

So, we might be tempted to then assign equally sick people to each group to try to control for the initial level of sickness. But then perhaps the treatment group is systematically assigned older people or perhaps poorer people or by any number of other visible or invisible characteristics of the population. Any of these might also bias the results as well and, of course, we don't know in advance what might affect the outcome of the experiment. (If we did, we wouldn't need to do the experiment.) The only way to ensure that the two groups are not systematically biased according to some characteristic or trait that might confound the outcome is to randomize assignment.

This is also the reason that clinical trials are double blind. If they were not, then individuals who know they have received the placebo may have different behaviors than those who get the treatment. Similarly, if the physician treating a patient knows whether the patient is receiving the treatment or the placebo, then he or she might treat the patient differently. In these cases, the outcome of the experiment could be confounded with patient or doctor behavior that again might mask the true effect of the treatment.

The key characteristics of experiments are:

- the existence of a control group which is used to characterize the baseline state;
- the randomization of the treatment to subjects is controlled by the scientist; and
- the experiment should be carefully designed and controlled, say by blinding and other such protocols, in order to minimize the chance of confounding the experimental results with other factors.

Since the researcher controls and manipulates the treatment, observing the effect of that manipulation on the outcome is what allows for the determination of whether the treatment *causes* an outcome. For example, in clinical trials, it's the fact that the drug is or is not administered to the people in the trial that allows the researcher to determine whether the drug causes people to get better.

Observational data, in contrast to experimental data, arise when it is impractical, unethical, or perhaps too expensive to conduct an experiment or to apply random assignment of subjects to conditions. Instead of the scientist actively controlling the treatment, he or she only observes data that are already in existence, or that naturally arise via some process that the researcher does not control, where the mechanism by which the treatment was "assigned" may be unknown.

In observational studies it is difficult to ascribe whether any observed difference is due to the treatment or some other confounding factor or factors. At issue is that when two variables A and B are found to be correlated in observational data, the correlation could be because A causes B, or because B causes A, or perhaps because some third factor C causes both A and B.

For example, it turns out that ice cream sales and the number of people who die by drowning is positively correlated. Does this mean that eating ice cream causes someone to have an increased risk of drowning? Clearly the causality cannot go the other way: Someone who has drowned cannot eat more ice cream. However, as it turns out, the correlation is caused by a third factor: seasonal temperatures. When it's hot out, more people swim, so the number of people who drown goes up because there are more people swimming in the water; at the same time, more people are eating ice cream because it's hot. Variables, such as the season, that may affect both of the variables under consideration (ice cream sales and drownings) are called *lurking variables*. Sometimes, as in the ice cream and drowning example, a lurking variable can be determined. In other cases, the lurking variables can be quite subtle. We will study lurking variables more deeply in Section 13.8.

■ **Example 7.1** There seems to be a relationship between dementia and levels of vitamin B12. Those with low levels of vitamin B12 have a higher risk of dementia. However, adding a daily supplement of B12 seems to have no effect on the risk of dementia. How can we resolve this paradox?

Solution: There must be at least one lurking variable here. Something else (e.g., diet, exercise, genetics) may affect both B12 levels and the risk of dementia. Adding B12 supplements does not affect these other factors, so B12 supplements have little or no effect on the risk of dementia.   ■ ■ ■

■ **Example 7.2** Does proximity to overhead power lines cause leukemia in children? A study published in 2016 in the *British Journal of Cancer* studied this question. The study was conducted in California and involved 5,788 children with leukemia and 3,308 control subjects. For each subject, the distance to overhead power lines was determined. The study compared the distances for the leukemia cases and the control cases. Is this an experiment or an observational study?

Solution: This is an observational study because the investigators did not impose any treatment on the subjects. ■ ■ ■

■ **Example 7.3** From a group of 100 laboratory mice, researchers selected 50 and subjected them to close proximity to a high-voltage power line for 30 days. The other 50 mice were held out as controls. All 100 mice were then tested for leukemia and the disease rates were computed for the mice subjected to the power line and the controls. Is this an experiment or an observational study?

Solution: This is an experiment because the investigators imposed a treatment by placing some mice in close proximity to a power line. ■ ■ ■

The necessity of working with observational data is well illustrated by the question of whether smoking causes cancer. As early as the 1930s, some doctors suspected there was a connection between smoking and lung cancer. However, because of ethical and other practical considerations, conducting an experiment to establish cause and effect was impossible. (It's neither practical nor ethical to force people to smoke, particularly when there is a suspicion that smoking is harmful.) Thus, the medical and epidemiological studies had to be based on observational data, which were based on comparing various sets of data on smokers and nonsmokers.

However, because all the studies were observational, it took many decades and many studies to convincingly establish a connection between smoking and lung cancer. For example, the British Minister of Health told Parliament in 1953, "The Standing Advisory Committee on Cancer and Radiology and the Medical Research Council both advised me that the relationship [between smoking and lung cancer] is not necessarily causal" (White, 1990). Furthermore, for a long time the cigarette manufacturers were able to effectively deflect lawsuits by pointing out that cause and effect could not be established scientifically with observational data. They, along with some scientists, pointed to a number of other possible causal factors, including environmental chemicals, industrial pollution, and even genetics (where the idea was that individuals with certain genes were both more likely to smoke and to get lung cancer). Indeed, it was not until 2000 that the first large lawsuit was won against a tobacco company in the United States, and that was in part because documents came to light showing that cigarette companies knew of the addictive nature of tobacco.

■ **Example 7.4** It has been argued that smoking does not cause lung cancer. Rather, there are other lurking variables that affect the propensity to smoke and to acquire lung cancer. However, it has been observed that stopping smoking reduces the risk of lung cancer. Does this prove that smoking causes lung cancer?

Solution: This does not prove that smoking causes lung cancer, but it does provide additional evidence of it. Compare this with Example 7.1, where there was a relationship between dementia and vitamin B12, but the intervention (adding supplements) had *no* effect. In the smoking/lung cancer example, the intervention (stop smoking) does have an effect on the risk of lung cancer. It may still be the case that other lurking variables simultaneously affect the propensity to smoke, the ability to quit smoking, and the chance of getting lung cancer. This complicated structure makes it less likely to believe that such lurking variables exist. ■ ■ ■

Because the statistician's data are often observational, establishing causation is difficult. Instead, the data scientist must usually be content with establishing associations (i.e., correlations), not causation.

Conducting experiments is sometimes possible and the data scientist, when given the opportunity, should favor experiment-based analyses to observational data-based analyses.

7.3.1 Convenience vs. Probability Sampling

Related to the question of experimental vs. observational data is the issue of how the data are collected. *Convenience sampling* occurs when the researcher does not control how the data are obtained. For example, pop-up surveys on websites are a form of convenience sampling when the survey is offered to every visitor to the site and each visitor decides whether to take the survey. In comparison, in scientific surveys the researcher uses probability to draw a random sample of the population so that the representativeness of the sample (in comparison to the population) is known.

To illustrate how convenience sampling can result in data that lead to incorrect conclusions, during the Monica Lewinsky scandal MSNBC found that 73% of 200,000 respondents to their "Live Vote" poll said President Bill Clinton should leave office. An NBC–*Wall Street Journal* scientifically conducted poll found that 34% of 2,005 survey respondents said the president should leave office (with margin of error of 2.2%).[2] Hence, the Live Vote poll overstated the sentiment in the general public by more than double. The issue, of course, is that those who felt President Clinton should leave office participated in the Live Vote poll at higher rates than the rest of the population.

By contrast, in *probability sampling* (which is also commonly referred to as random sampling), the researcher uses randomization to sample from the population. For example, in the NBC–*Wall Street Journal* poll, the 2,005 survey respondents were randomly chosen from the entire population in such a way that the pollsters could ensure that the resulting survey results were representative of the entire US population. Thus, just as in experiments the scientist or researcher must control the mechanism by which the treatment is assigned, in other sorts of data collection the researcher must also control how the data are sampled from the population. In the absence of such control, with convenience samples, the resulting data can end up misrepresenting the underlying population.

An advantage of convenience samples is that they often require much less time and effort to collect, and are less costly to generate. Indeed, administrative and observational data are often the result of convenience sampling. But because the sampling mechanism is not known, they generally do not support statistical inference. However, convenience sampling can be useful to researchers in other ways. For example, early in the course of research, responses from a convenience sample might be useful in developing research hypotheses. Responses from convenience samples might also be useful for identifying issues, defining ranges of alternatives, or collecting other sorts of noninferential data.

■ **Example 7.5** Investigators are interested in the relationship between diet and dementia in older adults. They approach the management at a retirement community about interviewing residents regarding their diet and level of dementia. Is this a probability sample or a convenience sample?

Solution: This is a convenience sample because all participants are (conveniently) located at one place. It would be difficult to generalize the results of this study.   ■ ■ ■

For data scientists, the main point is that when working with convenience data collected via some unknown mechanism, such as observational or administrative data, the generalizability of the conclusions from the data is limited.

7.4 Accuracy vs. Precision

Thus far we've covered two very important concepts that circumscribe what one can conclude from any particular analysis: experimental versus observational data and probability versus convenience samples. In this section we talk about a number of additional concepts.

---

[2] www.nbcnews.com/id/3704453/ns/about/t/live-votes-msnbccom/. More on "margin of error" later.

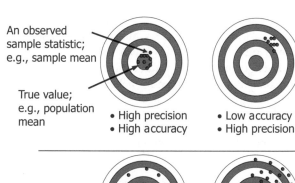

- High precision   - Low accuracy
- High accuracy   - High precision

- High accuracy   - Low accuracy
- Low precision   - Low precision

Figure 7.4 An illustrative comparison of accuracy versus precision.

If a sample is systematically not representative of the larger population in some way, the analysis can be *biased*. Bias means that the estimates resulting from the sample may not correctly measure the desired characteristic in the population. For example, results from a survey conducted over the internet about personal computer usage is not likely to reflect the general population because only those with access to a computer can take the survey.

The issue with bias is that no matter how big the sample, the estimated quantity is not going to reflect the true value if the entire population were observed. For example, the estimate of average computer usage from the internet survey is going to overestimate the average usage in the entire US population. That is, it will be biased upward. Even if you took a larger sample of internet users, the results would still be biased upwards.

Variance, on the other hand, is simply a measure of the variability of the estimate of the population parameter. It is a reflection of the reality that a different sample would very likely result in a different sample statistic and it gives you some idea of the precision of your estimates. That is, from the variance you can calculate the standard deviation of the data, as we have done, and quantify your uncertainty.

The targets in Figure 7.4 show the difference between bias and variance. Imagine for the sake of illustration that these are the results for various contestants in an Olympic rifle competition. The goal, of course, is to hit the bullseye consistently.

In the upper left target we see a contestant with low variance and low bias: The hits are tightly clustered right in or around the bullseye. This reflects an experienced shooter with a properly calibrated rifle. Compare these results to the upper right target, where the next contestant also has low variance but high bias. This could possibly be due to a misaligned rifle sight – that is, there is something systematically wrong with the rifle. Thus, the shooter is experienced but their shots are systematically missing the bullseye.

In the lower left we see a contestant with low bias but high variance. This could be a novice shooter with a good rifle. Hence, their shots are correctly centered on the target, but they are not very consistent in their shooting. Finally, in the lower right, we see a contestant with both high variance and high bias. This pattern might occur with a novice shooter with a bad rifle – they can neither shoot consistently nor sight accurately.

The analogy is that the target bullseye is the population parameter that we wish to estimate, such as the population mean, and the dots are the sample statistics. With low bias and low variance, the sample statistics are consistently close to the population parameter. But with high bias the sample statistics consistently miss the population parameter.

Randomization is used to minimize the chance of bias. The idea is that by randomly choosing potential survey respondents you tend to get a little bit of everything in the sample. However, even with randomization, surveys can encounter other kinds of bias when they are studying people. This is because people might be inclined to over- or understate certain things, particularly with sensitive questions (particularly questions about income and various types of behavior).

■ **Example 7.6** A manufacturer is given a bar determined by a government agency to be 10 cm in length. The bar is measured eight times using a hand micrometer and eight times with a laser distance measurer. The results are shown below:

Micrometer:	9.997	9.990	9.987	9.991	9.983	9.981	9.994	9.979
Laser:	9.999	10.001	10.004	9.997	9.998	10.000	10.001	9.998

Are these measurement systems accurate? Which are precise?

Solution: The laser measurement system is accurate because the values are centered around 10 cm, the exact length of the bar. The micrometer is not accurate because these values are not centered at 10 cm; they are all below the known length of the bar.

In terms of precision, the laser measurer is more precise because the variability is considerably less than the variability of the micrometer. The standard deviation for the micrometer is 0.00639, while the standard deviation for the laser measurer is 0.00225.   ■ ■ ■

7.5 Types of Random Samples

Sampling can be grouped into two broad categories: probability-based (also loosely called "random") and convenience sampling. Here we concentrate much of our discussion on probability-based surveying – simple random sampling in particular – because the probability selection mechanism allows for valid statistical inference about the entire population. A list of all subjects or units is called the *sampling frame*. If a sampling frame exists, the random sample is usually selected from it.

A sample is a probability sample when the probability with which every respondent was sampled is known. The sampling probabilities are not necessarily equal. Often the population can be enumerated in some fashion (the sampling frame), from which the potential respondents are probabilistically selected. This enumeration may be an actual enumeration, in the form of some type of complete list of the population, or it may be implied, such as when a multistage sampling scheme is used and only the members of selected primary sampling units are actually enumerated.

Types of probability samples include the following:

- *Simple random sampling* means that for any n, every sample of size n has an equal chance of being drawn.

- *Stratified random sampling* is useful when the population is composed of a number of homogeneous groups. In these cases, it can be either practically or statistically advantageous (or both) to first stratify the population into the homogeneous groups and then randomly draw samples from each group. Mathematically speaking, the population of N units is first divided into k mutually exclusive groups or *strata* of sizes $N_1, N_2, \ldots, N_k$. Then simple random samples are drawn independently from each stratum.

- *Cluster sampling* is useful when the sampling units naturally occur in groups or clusters. For example, sometimes to survey individuals it is useful to use the household as the sampling unit. The household is the cluster and in this scheme a sample of households are first selected and then individuals are randomly selected from within each household.

▪ Example 7.7 One hundred people buy a raffle ticket for a chance to win one of three prizes, which are $10, $20, and $50 gift certificates. The administrators of the raffle select the third-place winner ($10 gift certificate), then the second-place winner, and finally the first-place winner. Can the group of three winners be considered a random sample from the population of 100 ticket holders?

Solution: The probability of selecting any particular group of three is $1/\binom{100}{3}$. Thus, every group of three ticket holders has the same chance of being selected. This is a simple random sample. ▪ ▪ ▪

▪ Example 7.8 The university is interested in the opinion of undergraduate students on the availability of campus parking. They select at random 20 first-year, 20 second-year, 20 third-year, and 20 fourth-year students to participate in focus groups. Is this sample of 80 a simple random sample?

Solution: No. Not every group of 80 students has the same chance of being selected. For example, a group of 80 first-year students has no chance of being selected. This is a stratified random sample. ▪ ▪ ▪

▪ Example 7.9 Business transactions are conducted by phone and are recorded and stored on cassette tapes. After hundreds of tapes are accumulated, the business practices are questioned. Each cassette includes somewhere between 5 and 20 conversations between the company and a potential customer. Investigators would like to inspect the population of all conversations to determine what proportion mention a particular business practice, but it is impractical to listen to all of the conversations on all of the tapes, so some kind of random sampling is needed. Simple random sampling is not feasible, because inevitably this would call for listening to one transaction on one tape, two on another tape, none on the third tape, and so on. From a practical point of view, once you load a tape into a player, you might as well listen to all of the conversations on the tape. This suggests the following sampling plan: Take a simple random sample of cassette tapes, and on each of the selected tapes listen to all of the transactions. What kind of sampling plan is this?

Solution: This is a cluster sample. Cassette tapes are the clusters and a simple random sample of these is selected. Once the clusters are selected, all units in the cluster are measured. (Here "measured" means that someone listened to all conversations and made a determination of whether the questionable business practice was used.) ▪ ▪ ▪

7.5.1 Sources of Bias

Bias can creep into data in many different ways. Here we briefly describe some of the more common sources of bias.

- *Frame coverage bias* occurs when the frame misses some important part of the population. For example, a mail survey based on address lists will miss the homeless population. Telephone surveys will miss those who do not have a phone.
- *Selection bias* results when some units in the population are more likely to be selected than others. For example, those with strong opinions tend to answer the survey more than those without strong opinions. This can also occur if survey participation depends on the respondents having access to particular equipment.
- *Nonresponse bias* occurs if those who refuse to answer the survey are somehow systematically different that those who do answer the survey. For example, in telephone surveys those with higher incomes often have voice mail to screen calls and are more likely not to be at home. Hence, the resulting survey responses are likely to be biased toward those with lower incomes.
- *Sensitivity bias* occurs with sensitive questions that people either refuse to answer or perhaps choose to answer untruthfully due to embarrassment and/or privacy concerns.

- *Reporting bias* occurs because results that are statistically insignificant often don't get reported. Nearly every study you hear about in the popular press links something to cancer. That's because those studies that fail to find a link simply do not get reported! Reporting bias is sometimes called file drawer bias, as those studies that do not find a statistically significant result often end up in a file drawer, never to see the light of day.

- *Size bias* occurs when larger units have a greater chance of being selected than smaller units. For example, suppose you want to conduct a random sample of store customers. You decide to select every tenth customer coming into the store. If some of the customers visit much more frequently than others, then your sample could suffer from size bias, as those who visit frequently are more likely to be selected into the sample than those who do not.

■ **Example 7.10** Researchers are interested in feedback from patients regarding the care they get from their primary care physician. The community uses an online medical charting system where patients have access to their records, appointments, test results, etc. Patients can also communicate with their physician using the system. Patients who do not enroll in this service communicate using traditional methods such as in-person visits and the telephone. The researchers send a questionnaire to a random sample of those patients who have accounts in the medical charting system. What kind of bias is likely to occur in this example?

Solution: This is coverage bias. The population consists of all patients in the region, but sampling is done on a smaller group, those with accounts. It is likely that younger patients, who are on average more computer-savvy than older patients, will be overrepresented in the sample.   ■ ■ ■

■ **Example 7.11** Researchers develop a questionnaire to assess the level of stress on nurses in a region during the COVID pandemic. The questionnaire has two parts. The first asks for demographic information, while the second asks about their stress during the pandemic. About 10% of the respondents fill out the demographic part but skip the second part. The subsequent analysis is done only on those who completed both parts. What kind of bias is this?

Solution: This is nonresponse bias. If those who completed the questionnaire differ in a substantial way from those who didn't, the resulting sample may not be representative of the population.   ■ ■ ■

■ **Example 7.12** On December 1, 1969, a lottery was held at the Selective Service headquarters in Washington, DC. The purpose was to determine the order that young men would be drafted into military service. The lottery was based on the birthday of males born between 1944 and 1950. This is often called the 1970 draft lottery because it applied to those who would be drafted in 1970. The lottery was conducted as follows. Each of the 366 possible birthdays (there were two leap years between 1944 and 1950, so February 29 had to be included) was typed out on a piece of paper and put in a small capsule. Once all the January birthdays were completed, they were dumped in a bin. Then the February birthdays were put into capsules and placed in the same bin and mixed. Then the March birthdays, etc., all the way to the December birthdays. Then, on national television, they selected the first capsule; it contained the date September 14. The second capsule selected corresponded to April 24. This continued until there was just one capsule remaining, June 8. This determined the order in which people were drafted. The full results of the 1970 draft lottery are in the file 1970DraftLottery.csv, available on the book's website. Is there evidence of a kind of bias? If so, what kind of bias is it?

Solution: Plots of the draft number (vertical axis) and birth date (horizontal axis) are shown in Figure 7.5. The plots show a clear decrease in the lottery number across the birth date or birth month. This means that those born late in the year were more likely to be drafted than those born earlier in the year. This seems to be an example of selection bias.

In subsequent years a double randomization scheme was used. There were two bins. One contained, as before, capsules for each possible birth date. The other was for the sequence of numbers 1 through 365

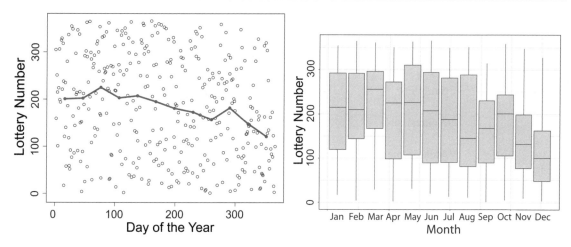

Figure 7.5 Results of the 1970 draft lottery. (a) The lottery number selected for each possible birth date, along with a line in red showing the monthly means. (b) Box plots aggregated by month.

or 366. Under the new system, a birth date was selected and a random sequence number was selected. The birth date selected was then given the random sequence number selected. For example, when July 31 was selected, the number 339 was selected. The birth date July 31 was then given sequence number 339 regardless of when July 31 was selected. ■■■

7.6 Types of Error

The primary purpose of a survey is to gather information about a population. However, even when a survey is conducted as a census, the results are affected by several sources of error. A good survey design seeks to reduce all types of errors, not only the sampling error arising from surveying only part of the population.

Survey error is commonly characterized in terms of the precision of statistical estimates. Yet characterizing survey error only in terms of variability and response rates misses other ways in which error can enter into a survey process. Table 7.1 lists the four general categories of survey error as presented and defined by Groves and Lyberg (2010) as part of the "Total Survey Error" approach. This section presents a basic treatment of these sources of error (see Groves and Lyberg (2010) if you are interested in exploring this topic in greater detail).

Errors of coverage occur when some part of the population of interest cannot become part of the sample. To be precise, Groves and Lyberg (2010) specify four different populations:

1. The "population of inference" is the population that the researcher ultimately intends to draw conclusions about.
2. The "target population" is the population of inference less various groups that the researcher has chosen to disregard.
3. The "frame population" is that portion of the target population that can be enumerated via a sampling frame.
4. The "survey sample" are those members of the sampling frame who were chosen to be surveyed. Coverage error is then defined as the difference between the statistics calculated on the frame population and the target population. The two most common approaches to reducing coverage error are: (a) obtaining as complete a sampling frame as possible, and (b) *post-stratifying* to weight the survey sample to match the population of inference on some key characteristics. In some cases, it is also possible to employ a frameless sampling strategy that, when properly designed, may allow every member of the target population a chance to be sampled.

Table 7.1 Sources of survey error according to Groves and Lyberg (2010).

Source of error	Definition
Coverage	"...from the failure to give any chance of sample selection to some persons in the population."
Sampling	"...from heterogeneity on the survey measure among persons in the population."
Nonresponse	"...from the failure to collect data on all persons in the sample."
Measurement	"...from inaccuracies in responses recorded on the survey instruments. These arise from: (a) effects of interviewers on the respondents' answers to survey questions; (b) error due to respondents, from the inability to answer questions, lack of requisite effort to obtain the correct answer, or other psychological factors; (c) error due to the weakness in the wording of survey questionnaires; and, (d) error due to effects of the mode of data collection, the use of face to face or telephone communications."

Sampling error arises when a subset of the target population is surveyed yet inference is made about the whole population. Assuming no difference between the population of inference and the target population, the sampling error is simply a quantification of the uncertainty in the sample statistic. This uncertainty can be decomposed into a variance component and a bias component. Groves and Lyberg (2010) state that variance characterizes the variability in the sample. As we have already discussed, bias is the systematic difference between the sample statistic and the actual population parameter of interest.

The precision of statistical estimates from probabilistic sampling mechanisms can be improved with larger sample sizes, which can be achieved either by selecting a larger sample of potential respondents, minimizing their nonresponse via various mechanisms, or through a combination of both.

Nonresponse errors occur when data are not collected on either entire respondents (*unit nonresponse*) or individual survey questions (*item nonresponse*). Groves and Lyberg (2010) state that "Nonresponse is an error of nonobservation." The response rate, which is the ratio of the number of respondents to the number sampled, is often taken as a measure of goodness. Higher response rates are taken to imply lower nonresponse bias, where the bias occurs when those that do not respond to the survey itself or a particular survey item systematically differ in their responses.

Measurement error arises when the survey response differs from the "true" response. For example, respondents may not answer sensitive questions honestly, or respondents may make errors in answering questions or misinterpret questions. These measurement errors may be mitigated or exacerbated by the mode of data collection. This is discussed more fully in the next section.

7.7 Historical Gaffes in Data Collection

Literary Digest Poll: A famous example of a survey that reached exactly the wrong conclusion as a result of sampling frame bias is the *Literary Digest* poll in the 1936 presidential election. For their survey, the Literary Digest assembled a sampling frame from telephone numbers and club membership lists. While using telephone numbers today might result in a fairly representative sample of the population, in 1936 only one in four households had a telephone. Those households that did were wealthier. Compounding this was the addition of club membership lists to the sampling frame, which only further skewed the frame toward people with higher incomes. Clearly, this sampling frame was biased.

Figure 7.6 Harry Truman holds up a copy of the *Chicago Daily Tribune* that mistakenly announced that Thomas Dewey had won the presidential election. © Associated Press. Photograph taken by Byron Rollins on November 4, 1948 in Saint Louis, MO.

In fielding their survey, the *Literary Digest* sent surveys to about 10 million people. However, only about one in four returned the survey, a clear case of nonresponse bias and perhaps even selection bias, since those who returned the survey likely felt more strongly about the election than those who did not.

From the poll, the *Literary Digest* predicted that Landon would beat Roosevelt 57% to 43%. Since you have probably never heard of Landon, you can guess the *Literary Digest* was wrong. In fact, FDR beat Landon by 62% to 38%. This was the largest error ever made by a major poll. Unlike the Bush/Gore election, for example, where the difference between the percentage of voters for each candidate was miniscule, the *Literary Digest* was off by almost 20 percentage points for each candidate! The Gallup organization, however, called it right even though they used significantly less data. The *Literary Digest* went out of business shortly thereafter.

Gallup Poll: But even Gallup, a pioneer in modern survey methods, didn't always get it right. The photograph of Harry S. Truman holding up a paper with the headline "DEWEY DEFEATS TRUMAN" is a famous one reflecting another miscue in the prediction of presidential races.

In this case, Gallup used a sampling technique called quota sampling. The basic idea is that each pollster is given a particular set of quotas of types of people to interview based on demographics. For example, interviewer "X" is to interview 13 people. Of these, 7 should be men and 6 should be women. Of the men, get 3 under 40 and 4 over 40, etc. The idea was to get the sample to match the population for various demographics in an effort to make it representative.

That seems like a good idea. So what went wrong? What happened was that survey interviewers chose to interview Republicans more often than Democrats. This could have resulted because Republicans were easier to locate or because, consciously or subconsciously, the pollsters simply preferred to interview them. This is a good example of why it is important to select the sample via a truly random mechanism. If left up to people, it is impossible to really draw a random sample.

7.8 Chapter Summary

Inferential statistics, the subject of most of the remainder of this book, is about saying something about a population, given the results of a sample taken from the population. This chapter focuses on the process of collecting data from a population. We learn about the world through the scientific method, which involves steps of posing questions and hypotheses, collecting and analyzing data, and looking at the results.

We distinguish observational studies and experiments. In experiments, the investigators have the ability to impose treatments. In observational studies, this aspect is not present, and we simply observe what is there. Making causal inferences is difficult for observational data.

A measurement process has two characteristics: precision and accuracy. Precision refers to the ability of the process to produce consistent (though possibly biased) results. A measurement process is accurate if, on average, there is no bias in the resulting measurements. We looked at several ways of collecting samples, including simple random sampling, stratified sampling, and cluster sampling.

7.9 Problems

For Problems 7.1–7.6, identify whether each underlined term is a population, parameter, sample, or statistic.

Problem 7.1 The Youth Risk Behavior Surveillance System (YRBSS) is administered by the CDC and it monitors health status and behaviors among <u>youth in the United States</u>. In 2019, the CDC estimated through the YRBSS that <u>21.7%</u> of youth have used marijuana in the 30 days prior to the survey. This 21.7% is an estimate of the <u>proportion of youth in the United States who have used marijuana in the past 30 days</u>.[3]

Problem 7.2 A small company that employs <u>600 employees</u> (550 union members and 50 managers) would like to learn about the employees' opinion regarding the new health insurance plan. Specifically, the company would like to know what <u>proportion of employees approve of the new plan</u>. From a sample of 50 selected at random, <u>74%</u> approve.

Problem 7.3 Rasmussen Reports[4] tracks the presidential approval rating (a percentage) on a daily basis. The tracking poll estimates the job approval among <u>likely voters in the United States</u>. In early 2023, <u>48% of 1,500 people selected and interviewed</u> approved of the president's job performance.

Problem 7.4 The National Health and Nutrition Examination Survey, described in Section 10.7.4, examines <u>5,000</u> people annually. The website cdc.gov/nchs/nhanes/about_nhanes.htm says "The sample for the survey is selected to represent the <u>U.S. population of all ages</u>." One of the NHANES surveys estimated that <u>2.9%</u> of the population has diabetes.

Problem 7.5 Based on census and voting records, it is known that <u>67%</u> of <u>adults in the United States</u> voted in the 2020 general election. A survey that asked adults in face-to-face interviews whether they voted in the 2020 election reported that <u>72%</u> said they had voted.

Problem 7.6 A company with <u>3,000 employees</u> is interested in the commuting patterns among its employees, specifically the <u>mean commuting distance</u>. Suppose that <u>200 employees</u> are selected and interviewed. The average commuting distance reported in those selected was <u>12.2</u> miles.

Problem 7.7 Consider the situation described in Problem 7.2. The company's employees consist of 550 union members and 50 managers.

(a) If we took an SRS of 50 employees, how many managers would we expect to see in the sample?

(b) If we took an SRS of 200 employees, how many managers would we expect to see in the sample?

(c) Suppose that instead of selecting an SRS of 200 employees, we selected 160 union members and 40 managers. What kind of sample is this?

(d) In the sample described in part (c), are managers overrepresented or underrepresented in the sample of 200?

Problem 7.8 Consider the situation described in Problem 7.6. The 3,000 employees consist of 2,000 women and 1,000 men. Describe how a stratified random sample could be selected from the population of employees.

Problem 7.9 Analyze the data from the 1971 draft lottery as in Example 7.12 to see whether there was any trend in the birth dates. The data are contained in the file 1971DraftLottery.csv.

Problem 7.10 Repeat Problem 7.9 for the results of the 1975 draft lottery, which are contained in the file 1975DraftLottery.csv.

Problem 7.11 Suppose that a government agency would like to conduct a survey of US households on the effects of COVID. Investigators would be sent to each residence selected to conduct in-person interviews. Explain why a simple random sample is not physically practical in this situation. What kind of sampling plan might be more appropriate?

For Problems 7.12–7.19, say whether the bias is frame coverage bias, selection bias, nonresponse bias, sensitivity bias, reporting bias, or size bias.

Problem 7.12 Researchers would like to know the average number of children in families from a community. They select a random sample of people and ask how many children were in their family when they were growing up; that is, how many siblings the respondent had plus themselves.

Problem 7.13 Survey panels involve preselected individuals who are paid to answer questionnaires. It was found that in a particular state, 18% of the population is 65 or older. In the survey panel, only 5% of the people are 65 or older.

Problem 7.14 A survey is designed to gauge the opinion of teachers on the use of artificial intelligence software with the capacity to write a brief essay. The survey is long and takes nearly an hour to complete. When the results were analyzed, the investigators found that many respondents stopped the questionnaire when they realized it was too long. Those teachers who taught computer classes tended to finish the questionnaire, so the resulting data set contained responses of more computer teachers than any other subject.

---

[3] www.cdc.gov/healthyyouth/data/yrbs
[4] rasmussenreports.com

Problem 7.15 A researcher is studying the relationship between herbal tea and the incidence of coronary artery disease. In a literature review, he has found that four studies concluded a relationship and one that concluded no relationship. The researcher is, however, aware of at least two studies that were conducted but never published that concluded no relationship.

Problem 7.16 On a survey of financial knowledge, researchers ask whether the respondent has ever cheated on their income taxes. They were surprised that only 2% report that they have cheated.

Problem 7.17 A survey after a presidential debate asks viewers of a news network to go online and select who they believe won the debate.

Problem 7.18 Truffles are a rare kind of mushroom found growing wild in Europe. Researchers want to study the chemical composition of these fungi and search a forest for specimens. (Truffles have resisted large-scale cultivation.) They find several large truffles, but only a few small ones.

Problem 7.19 Researchers ask families with young children whether they have exposed their child to classical music. Children for whom the answer was yes tended to do better, on average, in school when compared to those who answered no.

For Problems 7.20–7.25, say whether the error is coverage error, sampling error, nonresponse error, or measurement error.

Problem 7.20 Suppose that the health department in a given county conducts COVID tests and sends them to a local lab for analysis and determination of whether the person tests positive. On a given day, 60 tests were conducted but the lab was only able to analyze 30 tests, of which 6 were positive. The lab reports to the county that 6 tests were positive but fails to mention that half the tests were not completed. The county reports the positivity rate as $6/60 = 0.10$.

Problem 7.21 Sampling inspection of a manufacturing product involved testing $n = 20$ units out of a batch of $N = 200$ units. In the sample, it was found that 3 of the 20 (15%) were defective. This led to the inspection

of all 200 units, where it was found that only 10 out of 200 (5%) were defective.

Problem 7.22 Scales are used to measure the weight of produce that is packaged and sold on a per pound basis. Suppose the worker fails to account for the packaging that holds the produce in assessing the weight and therefore the price of the package. The weight of the packaging is called the "tare" weight. Every package of produce is therefore measured on the high side by exactly the tare weight.

Problem 7.23 Survey panels involve preselected individuals who are paid to answer questionnaires. It was found that in a particular state 17% of the population is 65 or older. In the survey panel only 6% of the people are 65 or older. A researcher considering using such a panel is concerned that the older demographic will not be represented in the sample.

Problem 7.24 On a survey of financial knowledge among the population of US adults, researchers reported that 28% of respondents failed to answer the question about their race. Researchers are concerned that such a large proportion of the sample did not answer this question.

Problem 7.25 The capacitance is to be measured on each item produced by a manufacturing process. Products whose capacitance is between 95 and 105 are considered acceptable. A new employee boots up the measuring device incorrectly and obtains results of 0 for each item. These readings of 0 are entered into the system.

Problem 7.26 Consider the situation described in Problem 7.22. Suppose that when there is no packaging, the scale measures a standard 0.1 kg bar consistently at 0.1 kg. When the scale is used to weigh produce and ignores the tare weight, is the measuring process precise? Is it accurate?

Problem 7.27 One way to measure a patient's pulse is to count heart beats in a 10-second interval and multiply by 6 to get beats per minute. Another way would be to count heart beats in a full minute. Which method is more accurate? Which is more precise?

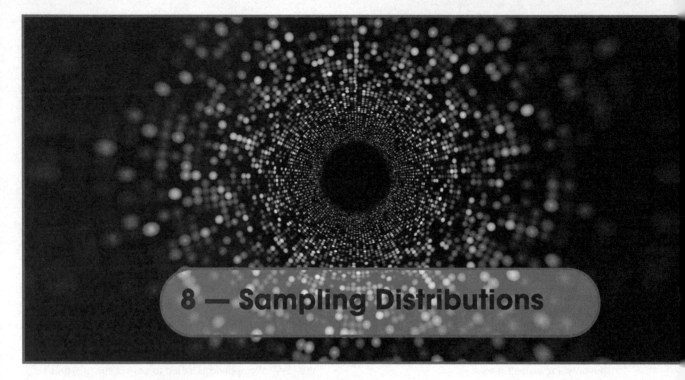

8 — Sampling Distributions

8.1 Introduction

Sampling joke: "If you don't believe in random sampling, the next time you have a blood test, tell the doctor to take it all." At the beginning of Chapter 7 we introduced the ideas of population vs. sample and parameter vs. statistic. We build on this in the current chapter. The key concept in this chapter is that if we were to take different samples from a distribution and compute some statistic, such as the sample mean, then we would get different results. Thus, for example, repeated samples will result in different sample means. The probability distribution of these sample means is what we mean by the *sampling distribution*; thus, we have to view the sample mean $\overline{X}$ as a random variable with some distribution. While this may seem obvious, it is essential to recognize this because otherwise the terminology may sound odd and contradictory.

8.2 Linear Combinations of Random Variables

The objective of this chapter is to give the sampling distributions of many of the common statistics. We begin by reviewing some of the results from Sections 5.4 and 5.5. These results are given as Theorems 4.4.3 and 4.4.7, and are restated here for convenience.

(a) If $X_1, X_2, \ldots X_n$ are random variables each with finite expectation, then

$$\mathbb{E}\left(\sum_{i=1}^{n} X_i\right) = \sum_{i=1}^{n} \mathbb{E}(X_i). \tag{8.1}$$

(b) If X is a random variable with finite variance, then

$$\mathbb{V}(X + c) = \mathbb{V}(X). \tag{8.2}$$

(c) If X is a random variable with finite variance, then

$$\mathbb{V}(cX) = c^2 \mathbb{V}(X). \tag{8.3}$$

(d) If $X_1, X_2, \ldots X_n$ are random variables each with finite variance, then

$$\mathbb{V}\left(\sum_{i=1}^{n} X_i\right) = \sum_{i=1}^{n} \mathbb{V}(X_i) + 2 \sum_{i<j} \mathrm{cov}(X_i, X_j). \tag{8.4}$$

(d) If $X_1, X_2, \ldots X_n$ are independent random variables each with finite variance, then

$$\mathbb{V}\left(\sum_{i=1}^{n} X_i\right) = \sum_{i=1}^{n} \mathbb{V}(X_i). \tag{8.5}$$

Definition 8.2.1 — Linear Combination. Suppose $X_1, X_2, \ldots X_n$ are random variables, and $c_1, c_2, \ldots, c_n$ are numbers. Then an expression of the form

$$c_1 X_1 + c_2 X_2 + \cdots + c_n X_n \tag{8.6}$$

is said to be a *linear combination* of the random variables $X_1, X_2, \ldots X_n$.

■ **Example 8.1** Suppose X_1, X_2 are independent random variables each with mean 10 and variance 4. Find the following:

(a) $\mathbb{E}(X_1 + X_2)$
(b) $\mathbb{E}(X_1 - X_2)$
(c) $\mathbb{V}(X_1 + X_2)$
(d) $\mathbb{V}(X_1 - X_2)$
(e) $\mathbb{E}(3X_1 - 2X_2)$
(f) $\mathbb{V}(3X_1 - 2X_2)$

Solution: This problem is asking for the mean or variance of a linear combination of the random variables X_1 and X_2. We can use the results summarized above.

For the first part, the linear combination is $1 \times X_1 + 1 \times X_2$, so

$$\mathbb{E}(X_1 + X_2) = \mathbb{E}(1 \times X_1 + 1 \times X_2) = \mathbb{E}(1 \times X_1) + \mathbb{E}(1 \times X_2) = 1 \times \mathbb{E}(X_1) + 1 \times \mathbb{E}(X_2)$$
$$= \mathbb{E}(X_1) + \mathbb{E}(X_2) = 10 + 10 = 20.$$

The variance of the sum of the two independent random variables is

$$\mathbb{V}(X_1 + X_2) = \mathbb{V}(1 \times X_1 + 1 \times X_2) = \mathbb{V}(1 \times X_1) + \mathbb{V}(1 \times X_2) = 1^2 \times \mathbb{V}(X_1) + 1^2 \times \mathbb{V}(X_2)$$
$$= \mathbb{V}(X_1) + \mathbb{V}(X_2) = 4 + 4 = 8.$$

The next two parts look at the expected value and variance of the difference $X_1 - X_2$ and are handled similarly:

$$\mathbb{E}(X_1 - X_2) = \mathbb{E}(1 \times X_1 - 1 \times X_2) = \mathbb{E}(1 \times X_1) - \mathbb{E}(1 \times X_2) = 1 \times \mathbb{E}(X_1) - 1 \times \mathbb{E}(X_2)$$
$$= \mathbb{E}(X_1) - \mathbb{E}(X_2) = 10 - 10 = 0$$

and

$$\mathbb{V}(X_1 - X_2) = \mathbb{V}(1 \times X_1 - 1 \times X_2) = \mathbb{V}(1 \times X_1) + \mathbb{V}((-1) \times X_2)$$
$$= 1^2 \times \mathbb{V}(X_1) + (-1)^2 \times \mathbb{V}(X_2) = \mathbb{V}(X_1) + \mathbb{V}(X_2) = 4 + 4 = 8.$$

Note that the variance of the difference $X_1 - X_2$ is the *sum* of the variances. When the (-1) term factors out of the variance we square it to get a positive 1.

The last two are handled using the rules for expectations and variance stated above:

$$\mathbb{E}(3X_1 - 2X_2) = \mathbb{E}(3 \times X_1 - 2 \times X_2) = \mathbb{E}(3 \times X_1) + \mathbb{E}(-2 \times X_2) = 3 \times \mathbb{E}(X_1) - 2 \times \mathbb{E}(X_2)$$
$$= 3\mathbb{E}(X_1) - 2\mathbb{E}(X_2) = 30 - 20 = 10$$

and

$$\mathbb{V}(3X_1 - 2X_2) = \mathbb{V}(3 \times X_1 - 2 \times X_2) = \mathbb{V}(3 \times X_1) + \mathbb{V}(-2 \times X_2) = 3^2 \times \mathbb{V}(X_1) + (-2)^2 \times \mathbb{V}(X_2)$$
$$= 9\mathbb{V}(X_1) + 4\mathbb{V}(X_2) = 36 + 16 = 52.$$

∎∎∎

The results of this last example suggest that means and variances of a linear combination of independent random variables can be obtained using the formulas from Chapter 5.

Theorem 8.2.1 — Expected Value and Variance of a Linear Combination of Independent Random Variables. If $X_1, X_2, \ldots, X_n$ are independent random variables, then

$$\mathbb{E}\left(\sum_{i=1}^{n} c_i X_i\right) = \sum_{i=1}^{n} c_i \mathbb{E}(X_i) \tag{8.7}$$

and

$$\mathbb{V}\left(\sum_{i=1}^{n} c_i X_i\right) = \sum_{i=1}^{n} c_i^2 \mathbb{V}(X_i). \tag{8.8}$$

These theorems could have been used to find the means and variances of the random variables in the previous exercise, but with the solutions given we are able to see how the rules for means and variances of linear combinations work. The assumptions in the previous theorem can be relaxed to require only that the random variables be uncorrelated. (Recall that independent random variables are uncorrelated, but not necessarily vice versa.)

If X_1 and X_2 are normally distributed, then whether they are independent or not, the linear combination

$$c_1 X_1 + c_2 X_2 \tag{8.9}$$

is also normally distributed. The proof involves MGFs (see Problem 6.80). This idea generalizes to the following theorem.

Theorem 8.2.2 If $X_i \sim N(\mu_i, \sigma_i^2)$, and if $X_1, X_2, \ldots, X_n$ are independent, then

$$\sum_{i=1}^{n} c_i X_i \sim N\left(\sum_{i=1}^{n} c_i \mu_i, \sum_{i=1}^{n} c_i^2 \sigma_i^2\right). \tag{8.10}$$

The result is simpler if the random variables are independent and have the same distribution (i.e., they are i.i.d.), as the next theorem indicates.

> **Theorem 8.2.3 — Normality of a Linear Combination of Normal Independent Random Variables.** If $X_1, X_2, \ldots, X_n \sim$ i.i.d. $N(\mu, \sigma^2)$, and if $c_1, c_2, \ldots, c_n$ are constants, then
>
> $$\sum_{i=1}^{n} c_i X_i \sim N \left(\mu \sum_{i=1}^{n} c_i, \sigma^2 \sum_{i=1}^{n} c_i^2 \right). \tag{8.11}$$

■ **Example 8.2** Heights of adult men in the United States are approximately normally distributed with mean 70 inches and standard deviation 4 inches. For women, the mean is 64 inches and the standard deviation is 3 inches. If one man and one woman are selected at random, find the distribution of the difference in height (man's height minus woman's height). Also, find the probability that the woman is taller than the man.

Solution: Let Y denote the height of a man and X denote the height of a woman. We are interested in the distribution of the linear combination $D = Y - X$, which can be written as

$$D = Y - X = 1Y + (-1)X.$$

The expected value is therefore

$$\mathbb{E}(D) = 1\mathbb{E}(Y) + (-1)\mathbb{E}(X) = 70 - 64 = 6,$$

and the variance is

$$\mathbb{V}(D) = 1^2\mathbb{V}(Y) + (-1)^2\mathbb{V}(X) = 4^2 + 3^2 = 25.$$

The distribution of the difference is then $N(6, 25)$. The probability that the woman is taller than the man is equal to the probability that the difference is less than zero. Thus,

$$\Pr(\text{woman is taller than man}) = \Pr(D < 0) = \Pr\left(\frac{D-6}{\sqrt{25}} < \frac{0-6}{\sqrt{25}}\right) = \Pr(Z < -1.2) \approx 0.1151.$$

■ ■ ■

■ **Example 8.3** Continuing with the same information from the previous example, suppose that four women and four men are selected at random. Find the probability that the mean height of the four women is greater than the mean height of the four men.

Solution: Let Y_1, Y_2, Y_3, Y_4 denote the heights of the four men, and X_1, X_2, X_3, X_4 the heights of the four women. We are then asking for the probability that

$$D = \overline{Y} - \overline{X} = \frac{X_1 + X_2 + X_3 + X_4}{4} - \frac{Y_1 + Y_2 + Y_3 + Y_4}{4} < 0. \tag{8.12}$$

The above difference can be written as a linear combination of the eight random variables as

$$D = \frac{1}{4}Y_1 + \frac{1}{4}Y_2 + \frac{1}{4}Y_3 + \frac{1}{4}Y_4 - \frac{1}{4}X_1 - \frac{1}{4}X_2 - \frac{1}{4}X_3 - \frac{1}{4}X_4. \tag{8.13}$$

Theorem 8.2.2 can be applied to obtain the expected value

$$\begin{aligned}
\mathbb{E}(D) &= \frac{1}{4} \times 70 + \frac{1}{4} \times 70 + \frac{1}{4} \times 70 + \frac{1}{4} \times 70 - \frac{1}{4} \times 64 - \frac{1}{4} \times 64 - \frac{1}{4} \times 64 - \frac{1}{4} \times 64 \\
&= 4 \times \frac{70}{4} - 4 \times \frac{64}{4} \\
&= 6.
\end{aligned}$$

The variance is

$$\mathbb{V}(D) = \left(\frac{1}{4}\right)^2 \times 4^2 + \left(\frac{1}{4}\right)^2 \times 4^2 + \left(\frac{1}{4}\right)^2 \times 4^2 + \left(\frac{1}{4}\right)^2 \times 4^2$$

$$+ \left(-\frac{1}{4}\right)^2 \times 3^2 + \left(-\frac{1}{4}\right)^2 \times 3^2 + \left(-\frac{1}{4}\right)^2 \times 3^2 + \left(-\frac{1}{4}\right)^2 \times 3^2$$

$$= \frac{4}{16} \times 16 + \frac{4}{16} \times 9$$

$$= 6.25.$$

Once the mean and variance are obtained, we can find the probability using properties of the normal distribution:

$$\Pr(D < 0) = \Pr\left(\frac{D-6}{\sqrt{6.25}} < \frac{0-6}{\sqrt{6.25}}\right) = \Pr(Z < -2.40) \approx 0.0082.$$

Notice how much less likely it is for the differences in averages to be less than zero than for differences of a single observation each.   ■ ■ ■

■ **Example 8.4** Redo Example 8.3 by first finding the distribution of the means $\overline{Y}$ and $\overline{X}$ and then combining the result.

Solution: The random variable $\overline{Y}$ can be written as

$$\overline{Y} = \frac{1}{4}Y_1 + \frac{1}{4}Y_2 + \frac{1}{4}Y_3 + \frac{1}{4}Y_4,$$

so the mean is

$$\mathbb{E}(\overline{Y}) = \mathbb{E}\left(\frac{1}{4}Y_1 + \frac{1}{4}Y_2 + \frac{1}{4}Y_3 + \frac{1}{4}Y_4\right) = 4 \times \frac{70}{4} = 70$$

and the variance is

$$\mathbb{V}(\overline{Y}) = \mathbb{V}\left(\frac{1}{4}Y_1 + \frac{1}{4}Y_2 + \frac{1}{4}Y_3 + \frac{1}{4}Y_4\right) = 4 \times \frac{16}{4^2} = 4.$$

It should not be surprising that the mean of $\overline{Y}$ is equal to 70, the same as the mean in the population. In other words, the expected value of an average of four from a population is equal to the population mean. The result for the variance may be a bit surprising. It says that the variance of the mean of four observations is equal to one-fourth the variance of a single observations ($\sigma^2 = 4^2 = 16$ in the population, whereas, the variance of the mean of four is just 4). Thus,

$$\overline{Y} \sim N\left(70, \frac{16}{4}\right).$$

Similar reasoning would lead us to the result that

$$\overline{Y} \sim N\left(64, \frac{9}{4}\right).$$

We can now combine these results and use Theorem 8.2.3 to determine that

$$\mathbb{E}(\overline{Y} - \overline{X}) = 70 - 64 = 6$$

and

$$\mathbb{V}(\overline{Y} - \overline{X}) = \frac{16}{4} + \frac{9}{4} = \frac{25}{4} = 6.25.$$

Thus, the difference has distribution $D = \overline{Y} - \overline{X} \sim N(6, 6.25)$, so we can say

$$\Pr(D < 0) = \Pr\left(\frac{D-6}{\sqrt{6.25}} < \frac{0-6}{\sqrt{6.25}}\right) \approx 0.0082,$$

giving the same answer as before.   ■■■

8.3 Sampling Distributions for Sums and Means

If two people were to take a random sample from the same distribution, and each computes the sample mean, it is almost certain[1] that the sample means will differ. This is easy to see, but also easy to overlook. It makes sense, therefore to think about repeatedly sampling from a population (or, as we will usually think of it, sampling from a probability distribution). We could ask the question, "What is the distribution of some sample statistic if we take repeated samples from the same distribution?"

We can see this property using simulated samples generated in R. The `rnorm()` function can be used to simulate from a normal distribution. Building on the examples from the previous section, we can select a sample of size $n = 25$ from the $N(70, 4^2)$ distribution with the command

```
rnorm( 25 , mean=70 , sd=4 )    # set.seed(12347)
```

The output from R is

```
 [1] 74.08026 73.21952 62.29362 66.02102 71.44194 71.01013 69.27269
 [8] 62.00181 61.59456 60.63125 73.32265 62.92451 67.01523 65.59731
[15] 62.31772 72.96401 74.63966 67.01532 67.98544 66.20838 66.09311
[22] 62.15484 74.34240 68.08690 70.51037
```

If we were to take the sample mean, variance, and standard deviation of this sample of size 25, we might obtain

$$\bar{x} = 67.71, \qquad s^2 = 20.949, \text{ and} \qquad s = 4.577. \tag{8.14}$$

Remember, there is chance involved in selecting the random sample, so we would not expect the sample mean $\bar{x} = 67.71$ to be equal to the population mean $\mu = 70$. Here they differ by more than a full two inches. If we were to take a second sample of 100, and again compute the sample mean, variance, and standard deviation, we would obtain

$$\bar{x} = 68.55, \qquad s^2 = 9.383, \text{ and} \qquad s = 3.063. \tag{8.15}$$

A third time might yield

$$\bar{x} = 70.12, \qquad s^2 = 11.865, \text{ and} \qquad s = 3.445. \tag{8.16}$$

You can see that repeated samples give different sample means, as well as sample variances and standard deviations.

---

[1] The probability is 1 that the sample means will differ if the data are taken from a continuous probability distribution.

We could easily have R take 10,000 samples, each of size $n = 25$, using the following code:

```
xbar = c()
sampvar = c()
sampsd = c()
for (i in 1:10000)
{
  x = rnorm( 25 , mean=70 , sd=4 )
  xbar = c(xbar,mean(x))
  sampvar = c(sampvar,var(x))
  sampsd = c(sampsd,sd(x))
}
```

With this code, each iteration of the loop generates a new sample of size $n = 25$ from the $N(70, 4^2)$ distribution. It computes the sample mean and concatenates it to the xbar array; then it does the same for the sample variance s^2 and the sample standard deviation s. Figure 8.1 shows histograms of these 10,000 sample statistics $(\bar{x}, s^2, s)$. Figure 8.1(a) for the sample mean seems to be centered at 70 and is fairly symmetric. The histogram for the sample variance (b) is centered around 16 and has a distinct right-skewness.

These plots illustrate the concept of the *sampling distribution*, which is the distribution of the sample statistic across many samples. If we take the perspective that the data have yet to be collected, then the observations in the sample are random variables: $X_1, X_2, \ldots, X_n$. In terms of these random variables, each sample statistic is a function of the random variables in this set. For example,

(a)
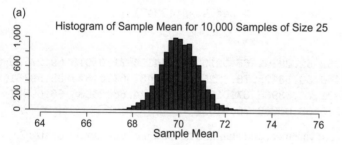

(b)
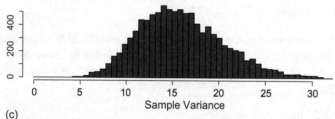

Figure 8.1 Sampling distributions of the sample mean (a), sample variance (b), and sample standard deviation (c) for 10,000 samples of size 25 from the normal distribution with mean 70 and variance 16.

(c)
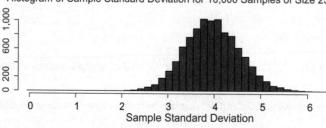

$$\overline{X} = \frac{1}{n}\sum_{i=1}^{n}X_i \tag{8.17}$$

and

$$S^2 = \frac{1}{n-1}\sum_{i=1}^{n}(X_i - \overline{X})^2. \tag{8.18}$$

Note the use of capital letters above. Capital letters represent random variables, while lower case letters represent numbers. Hence, $\overline{X}$ denotes the average of a set of random variables. Because the random variables are random, so is $\overline{X}$, and thus $\overline{X}$ has a PDF, which is referred to as the sampling distribution of $\overline{X}$. In contrast, $\overline{y}$ is a sample average, which is a number, and hence cannot have a distribution.

A function of a set of random variables is itself a random variable with its own distribution, referred to as the *sampling distribution*. The next two theorems give some results on the sampling distribution for the sum of i.i.d. random variables.

Theorem 8.3.1 — Expected Value and Variance of a Sum of i.i.d. Random Variables.
If $X_1, X_2, \ldots, X_n$ are i.i.d. from some distribution (not necessarily a normal distribution) with mean μ and variance σ^2, then

$$\mathbb{E}\left(\sum_{i=1}^{n}X_i\right) = n\mu \tag{8.19}$$

and

$$\mathbb{V}\left(\sum_{i=1}^{n}X_i\right) = n\sigma^2. \tag{8.20}$$

Proof. The proof is simple and relies on the expected value and variance for the sum of independent random variables:

$$\mathbb{E}\left(\sum_{i=1}^{n}X_i\right) = \sum_{i=1}^{n}\mathbb{E}(X_i) = \sum_{i=1}^{n}\mu = n\mu.$$

Similarly,

$$\mathbb{V}\left(\sum_{i=1}^{n}X_i\right) = \sum_{i=1}^{n}\mathbb{V}(X_i) = \sum_{i=1}^{n}\sigma^2 = n\sigma^2. \qquad\blacksquare$$

Theorem 8.3.2 — Normality of a Sum of i.i.d. Normal Random Variables. If $X_1, X_2, \ldots, X_n \sim$ i.i.d. $N(\mu, \sigma^2)$, then

$$\sum_{i=1}^{n}X_i \sim N(n\mu, n\sigma^2). \tag{8.21}$$

Proof. The mean and variance are the same as in the previous theorem. The only added assumption is that the sample comes from the normal distribution. As we mentioned in the previous section, a linear combination of normal random variables has a normal distribution, so the sum has a $N(n\mu, n\sigma^2)$ distribution. $\qquad\blacksquare$

The results for the sample mean $\overline{X}$ follow directly.

> **Theorem 8.3.3 — Expected Value and Variance of the Sample Mean.** If $X_1, X_2, \ldots, X_n$
> are i.i.d. from some distribution (not necessarily a normal distribution) with mean μ and variance σ^2,
> then
>
> $$\mathbb{E}\left(\overline{X}\right) = \mu \qquad \text{and} \qquad \mathbb{V}\left(\overline{X}\right) = \frac{\sigma^2}{n}. \qquad (8.22)$$
>
> If, in addition, the random variables are normally distributed, then
>
> $$\overline{X} \sim N\left(\mu, \frac{\sigma^2}{n}\right). \qquad (8.23)$$

Proof. Begin by assuming that $X_1, X_2, \ldots, X_n$ are i.i.d. from some distribution (not necessarily a normal distribution) with mean μ and variance σ^2. Then,

$$\mathbb{E}(\overline{X}) = \mathbb{E}\left(\frac{1}{n}\sum_{i=1}^{n}X_i\right) = \frac{1}{n}\,\mathbb{E}\left(\sum_{i=1}^{n}X_i\right) = \frac{1}{n}\,n\mu = \mu$$

and

$$\mathbb{V}(\overline{X}) = \mathbb{V}\left(\frac{1}{n}\sum_{i=1}^{n}X_i\right) = \frac{1}{n^2}\sum_{i=1}^{n}\mathbb{V}(X_i) = \frac{1}{n^2}\,n\sigma^2 = \frac{\sigma^2}{n}.$$

If we assume that the sample is from a normal distribution, then the sum, and hence also the mean, is normally distributed. ∎

■ **Example 8.5** Suppose that a sample of size $n = 9$ is selected from the population of adult women. (Recall that the population of heights of adult women is approximately normal with mean 64 inches and standard deviation 3 inches.) Find the probability that the average height in this sample exceeds 65 inches. Repeat this example for sample sizes of $n = 25$ and $n = 100$.

Solution: Let $X_1, X_2, \ldots, X_n$ denote the observed heights of the $n = 9$ women selected. Then by Theorem 8.3.3, $\overline{X} \sim N\left(64, 3^2/9\right)$. The desired probability is then

$$\Pr(\overline{X} > 65) = \Pr\left(\frac{\overline{X} - 64}{\sqrt{3^2/9}} > \frac{65 - 64}{\sqrt{3^2/9}}\right) = \Pr(Z > 1) = 0.1587.$$

Thus, there is about a 16% chance that the sample mean $\overline{X}$ will exceed the true mean $\mu = 64$ by more than one inch.

If the sample size is $n = 25$ instead, then the probability is

$$\Pr(\overline{X} > 65) = \Pr\left(\frac{\overline{X} - 64}{\sqrt{3^2/25}} > \frac{65 - 64}{\sqrt{3^2/25}}\right) = \Pr(Z > 1.667) = 0.0478.$$

This is only about a 5% chance.

Finally, the probability that the average height of $n = 100$ women exceeds 65 inches is

$$\Pr(\overline{X} > 65) = \Pr\left(\frac{\overline{X} - 64}{\sqrt{3^2/100}} > \frac{65 - 64}{\sqrt{3^2/100}}\right) = \Pr(Z > 3.333) = 0.0004.$$

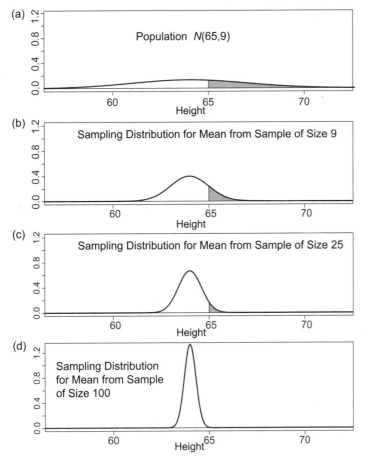

Figure 8.2 Original population of heights (a) is normal with mean 64 inches and standard deviation 3 inches. Sampling distribution of the sample mean for samples of size $n=9, 25, 100$.

This is a very small chance, less than 0.1%, that the sample mean $\overline{X}$ from a sample of size $n = 100$ exceeds 65 inches.

Notice how the probability of exceeding 65 inches, which is one inch above the mean, gets smaller as the sample size gets larger. This is because the variance decreases as n increases. (Notice the n in the formula for $\mathbb{V}(\overline{X})$ is in the denominator.) The standard deviation for the sample mean is $\sigma/\sqrt{n}$. Figure 8.2 shows the sampling distributions for samples of size $n = 1, 9, 25, 100$. ∎ ∎ ∎

Exercise 8.1 For the population of heights of adult women, find the probability that the mean from a sample of size $n = 9$, 25, and 100 lies between 63 and 65 inches.

Exercise 8.2 For the population of heights of adult men, which is $N(70, 16)$, find the probability that the mean from a sample of size $n = 9$, 25, 100, and 400 lies between 69.9 and 70.1 inches.

8.4 Sampling Distribution for the Sample Variance

The sample mean $\overline{X}$ is not the only statistic that we might compute from a sample. There are many other important statistics, such as the sample variance,

$$S^2 = \frac{1}{n-1} \sum_{i=1}^{n} (X_i - \overline{X})^2, \tag{8.24}$$

or its square root, the standard deviation S. There are others as well, such as the median, the range (largest − smallest), and the interquartile range (third quartile − first quartile). Here we investigate the sample variance.

Keep in mind that our perspective is *before* we take the sample, so the observations, and hence the statistics $\overline{X}$ and S^2, are themselves random variables. Let's begin by selecting a random sample from a normal distribution with mean 70 and standard deviation 4, and compute the sample variance.

From Figure 8.1(b) we can see that the sampling distribution of S^2 is not going to be normal. The true distribution is related to the χ^2 distribution, a special case of the gamma distribution discussed in Section 7.6.

Theorem 8.4.1 — Sampling Distribution of $(n-1)S^2/\sigma^2$. Suppose $X_1, X_2, \ldots, X_n$ are i.i.d. $N(\mu, \sigma^2)$ and let S^2 be the sample variance defined in (8.24). Then,

$$\frac{(n-1)S^2}{\sigma^2} \sim \chi^2(n-1). \tag{8.25}$$

Proof. A rigorous proof involves techniques that we have not introduced, so we can only sketch the proof here. The first result we need is that if $Z \sim N(0,1)$, then $Z^2 \sim \chi^2(1)$. The second result we need is that the sum of independent χ^2 random variables has a χ^2 distribution whose degrees of freedom is equal to the sum of the degrees of freedom of the random variables that make up the sum. The third result we need is that if we take a normal random variable, subtract the mean and divide by the standard deviation, we have a standard normal distribution – that is, a normal distribution with mean 0 and variance 1. Thus, for each i,

$$\frac{X_i - \mu}{\sigma} \sim N(0,1). \tag{8.26}$$

By the first result we know that

$$\left(\frac{X_i - \mu}{\sigma}\right)^2 \sim \chi^2(1) \tag{8.27}$$

and, using the result on the sum of independent χ^2 random variables, we conclude that

$$\left(\frac{X_1 - \mu}{\sigma}\right)^2 + \left(\frac{X_2 - \mu}{\sigma}\right)^2 + \cdots + \left(\frac{X_n - \mu}{\sigma}\right)^2 = \sum_{i=1}^{n} \left(\frac{X_i - \mu}{\sigma}\right)^2$$

$$= \frac{n\left(\frac{1}{n}\sum_{i=1}^{n}(X_i - \mu)^2\right)}{\sigma^2} \sim \chi^2(n).$$

The term in parentheses in the numerator of the last line looks suspiciously like the sample variance, but it is not. It involves subtracting the true mean μ, whereas the sample variance in (8.24) involves subtracting the sample mean $\overline{X}$. If we replace the true mean with the estimated mean, namely the sample mean $\overline{X}$, then we have imposed the restriction that

$$\sum_{i=1}^{n}(X_i - \overline{X}) = 0, \tag{8.28}$$

whereas there is no such restriction on

$$\sum_{i=1}^{n}(X_i - \mu). \tag{8.29}$$

This forces a reduction in the degrees of freedom from n to $n-1$; the details of this are rather technical. See Bain and Engelhardt (1992) or Wackerly *et al.* (2014) for a rigorous demonstration of this. Thus,

$$\frac{(n-1)S^2}{\sigma^2} \sim \chi^2(n-1). \tag{8.30}$$

■ **Example 8.6** Suppose we take a sample of size $n = 25$ from the population of adult males and measure their heights. Recall that this distribution is normal with mean 70 inches and standard deviation 4 inches. Find the probability that the sample variance computed from this sample will exceed 20.

Solution: Let S^2 denote the sample variance computed from this sample. We would like to know $\Pr(S^2 > 20)$. If we begin with this and manipulate it so that it involves $(n-1)S^2/\sigma^2$ we can obtain the desired probability as follows:

$$\Pr\left(S^2 > 20\right) = \Pr\left(\frac{(25-1)S^2}{4^2} > \frac{(25-1)\times 20}{4^2}\right) = \Pr\left(Q > 30\right),$$

where $Q \sim \chi^2(25-1)$. We can use R to approximate this probability using the command

```
1 - pchisq(30,24)
```

This yields

$$\Pr(S^2 > 20) = 0.1848. \tag{8.31}$$

Note that in the above R code, the function `pchisq()` gives the left tail area, whereas we want the right tail area; thus, we must subtract the computed value from 1.   ■ ■ ■

8.5 The Central Limit Theorem

The *Central Limit Theorem* (CLT) says that the distribution of means of random variables tends toward a normal distribution as the number of random variables gets large, even if the random variables are not themselves normally distributed.

> **Theorem 8.5.1 — Central Limit Theorem.** Let $X_1, X_2, \ldots, X_n$ be i.i.d. from a distribution having mean μ and variance σ^2, and let $\overline{X}_n$ denote the sample mean from this sample of size n. Then, for every x,
>
> $$\lim_{n\to\infty} \Pr\left(\frac{\overline{X}_n - \mu}{\sigma/\sqrt{n}} \le x\right) = \Phi(x), \qquad -\infty < x < \infty,$$
>
> where Φ is the CDF of the standard normal distribution.

Proof. We give a sketch of the proof. We call this a "sketch" because we use a number of results that we haven't (and won't) prove. For example, we will use MGFs, which don't always exist. Second, we will prove that the MGFs of the standardized random variables $Z_n = (\overline{X}_n - \mu)/(\sigma/\sqrt{n})$ converge to the MGF of the normal distribution. To make this rigorous, we would have to show that convergence in the space of MGFs implies convergence of the CDFs.

We will need several results in our proof. The first is this. If X is a random variable with MGF $M_X(t)$, then the MGF of cX is related to the MGF $M_X(t)$ by

$$M_{cX}(t) = M_X(ct). \tag{8.32}$$

The second result we need is the formula for the Taylor series with remainder. If f is a function that has derivatives of order one and two on an interval I containing a, then for every x in I there exists a c between a and x such that

$$f(x) = f(a) + \frac{f'(a)}{1!}(x-a) + \frac{f''(c)}{2!}(x-a)^2.$$

If $a = 0$, then a Taylor series is often called a Maclaurin series. For example, the Maclaurin series with remainder for the exponential function is

$$e^x = e^0 + \frac{e^0}{1!}(x-0) + \frac{e^c}{2!}(x-0)^2 = 1 + \frac{x}{1!} + \frac{e^c x^2}{2!} \qquad \text{[for some } c \text{ between 0 and } x\text{]}.$$

The third result we need is a limit from calculus. A well-known result is that

$$\lim_{n \to \infty} \left(1 + \frac{1}{n}\right)^n = e,$$

which can be generalized to

$$\lim_{n \to \infty} \left(1 + \frac{A}{n}\right)^n = e^A.$$

A further generalization is this. If A_n is a sequence of real numbers that converges to A, then

$$\lim_{n \to \infty} \left(1 + \frac{A_n}{n}\right)^n = e^A.$$

We can now proceed with the proof of the CLT and write S_n as

$$S_n = \frac{\overline{X} - \mu}{\sigma/\sqrt{n}} = \frac{\sum_{i=1}^{n}(X_i - \mu)}{\sqrt{n}\,\sigma} = \sum_{i=1}^{n} \frac{X_i - \mu}{\sqrt{n}\,\sigma}. \tag{8.33}$$

The MGF of S_n can be written as

$$M_{S_n}(t) = \mathbb{E}\left(e^{tS_n}\right)$$

$$= \mathbb{E}\left(\exp\left(t\sum_{i=1}^{n} \frac{X_i - \mu}{\sqrt{n}\,\sigma}\right)\right) \qquad \text{[using (8.33)]}$$

$$= \prod_{i=1}^{n} \mathbb{E}\left(\exp\left(\frac{t}{\sqrt{n}\,\sigma}(X_i - \mu)\right)\right) \qquad \text{[because the } X_i \text{ are independent]}$$

$$= \prod_{i=1}^{n} M_{X_i - \mu}\left(\frac{t}{\sqrt{n}\,\sigma}\right) \qquad \text{[using (8.32)]}$$

$$= \left[M_{X_1 - \mu}\left(\frac{t}{\sqrt{n}\,\sigma}\right)\right]^n \qquad \text{[because all } X_i \text{ have the same MGF].}$$

$$\text{(8.34)}$$

At this point, we write the Maclaurin series with remainder for $M_{X_1 - \mu}$:

$$M_{X_1 - \mu}(t) \;=\; M_{X_1 - \mu}(0) + \frac{M'_{X_1 - \mu}(0)}{1!}t + \frac{M''_{X_1 - \mu}(c)}{2!}t^2 \qquad \text{[for some } c \text{ between 0 and } t\text{].}$$

By Theorem 6.9.1, $M'_Y(0) = \mathbb{E}(Y)$ for any random variable Y. Thus,

$$M'_{X_1 - \mu}(0) \;=\; \mathbb{E}(X_1 - \mu) \;=\; 0.$$

Additionally, since $M(0) = 1$ for every MGF (see Problem 6.79), we can write

$$M_{X_1 - \mu}(t) \;=\; 1 + \frac{0}{1!}t + \frac{M''_{X_1 - \mu}(c)}{2!}t^2 \;=\; 1 + \frac{M''_{X_1 - \mu}(c)}{2!}t^2 \qquad \text{[for some } c \text{ between 0 and } x\text{].}$$

Using these results and picking up from the last line of (8.34),

$$M_{S_n}(t) \;=\; \left[M_{X_1 - \mu}\left(\frac{t}{\sqrt{n}\,\sigma}\right)\right]^n = \left[1 + \frac{M''_{X_1 - \mu}(c_n)}{2}\left(\frac{t}{\sqrt{n}\,\sigma}\right)^2\right]^n = \left[1 + \frac{M''_{X_1 - \mu}(c_n)}{2}\frac{t^2}{n\sigma^2}\right]^n, \quad \text{(8.35)}$$

for some c_n between 0 and $t/(\sqrt{n}\,\sigma)$. Note that because $t/(\sqrt{n}\,\sigma) \to 0$ as $n \to \infty$, and because c_n is "sandwiched" between 0 and $t/(\sqrt{n}\,\sigma)$, it must be the case that

$$\lim_{n \to \infty} M''_{X_1 - \mu}(c_n) \;=\; M''_{X_1 - \mu}(0) \;=\; \mathbb{E}\left((X_1 - \mu)^2\right) \;=\; \mathbb{V}(X) \;=\; \sigma^2.$$

This implies that

$$\lim_{n \to \infty} \frac{M''_{X_1 - \mu}(c_n)}{\sigma^2} \;=\; 1.$$

Now, take the limit as $n \to \infty$ on both sides of (8.35):

$$\lim_{n \to \infty} M_{S_n}(t) = \lim_{n \to \infty}\left[1 + \frac{M''_{X_1 - \mu}(c_n)}{2}\frac{t^2}{n\sigma^2}\right]^n = \lim_{n \to \infty}\left[1 + \frac{\left(M''_{X_1 - \mu}(c_n)/\sigma^2\right)\times t^2/2}{n}\right]^n = \exp\left(t^2/2\right).$$

This is the MGF of the $N(0,1)$ distribution. What we have shown is that the MGF of $S_n = (\overline{X}_n - \mu)/(\sigma/\sqrt{n})$ converges to the MGF of the $N(0,1)$ distribution. As we stated at the beginning of the proof, this is sufficient to conclude that the CDF of S_n converges to the CDF of the $N(0,1)$ distribution. ∎

The CLT uses a more general concept called convergence in distribution.

Definition 8.5.1 — Convergence in Distribution. If $W_1, W_2, \ldots$ is a sequence of random variables such that

$$\lim_{n \to \infty} \Pr\left(W_n \leq x\right) \;=\; F(x),$$

then we say that W_n *converges in distribution* to the distribution with CDF $F(x)$ and we write $W_n \overset{d}{\to} F$, although sometimes we will write the name of the distribution in place of F.

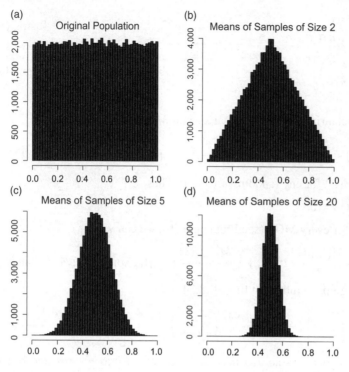

Figure 8.3 An illustration of the CLT applied to sums of uniformly distributed random variables. (a) A histogram resulting from 100,000 draws from a UNIF(0,1) distribution. (b) 100,000 means of two draws from the same distribution. (c) plots 100,000 means of five UNIF(0,1) random variables, and (d) is 100,000 means of 20.

We can restate the CLT by saying that if $X_1, X_2, \ldots, X_i, \ldots, X_n \sim$ i.i.d. from a distribution with mean μ and variance σ^2, and if $\overline{X}_n$ is the sample mean, then

$$S_n = \frac{\overline{X}_n - \mu}{\sigma/\sqrt{n}} \xrightarrow{d} N(0,1).$$

The CLT applies both to discrete and continuous random variables. Figure 8.3 illustrates the CLT for data with a uniform distribution. Figure 8.3(a) is a histogram resulting from 100,000 draws from a $U(0,1)$ distribution. Figure 8.3(b) is 100,000 means of two draws from the same distribution. Figure 8.3(c) plots 100,000 means of five $U(0,1)$ random variables, and Figure 8.3(d) is 100,000 means of 20. Notice two things. First, the variability gets smaller as the sample size gets larger. Second, the distribution of the sample mean tends toward the normal distribution as the sample size gets large.

■ **Example 8.7** Suppose the number of customer arrivals in one hour has a Poisson distribution with mean $\lambda = 6$. Use the CLT to approximate (a) the probability that the average number of arrivals in a 48-hour period is eight or greater, and (b) the probability that the sum of the numbers of arrivals in these 48 one-hour periods exceeds 300.

Solution: Let X_i denote the number of arrivals in hour i; then $X_1, X_2, \ldots, X_n \sim$ i.i.d. POIS(6). The standard deviation of X_i is $\sigma = \sqrt{6}$. Applying the CLT, we have

$$\Pr\left(\overline{X} \geq 8\right) = \Pr\left(\frac{\overline{X} - 6}{\sqrt{6}/\sqrt{48}} \geq \frac{8-6}{\sqrt{6}/\sqrt{48}}\right) \approx \Pr\left(Z \geq 5.66\right) \approx 0.$$

It is therefore unlikely that we would see an average of 8 or more per hour over a two-day period. If such an event did occur, we would doubt that the mean of the Poisson distribution is really 6.

Part (b) asks for a probability involving the sum, but this can easily be transformed to a problem involving the mean:

$$\Pr\left(\sum_{i=1}^{48} X_i > 300\right) = \Pr\left(\frac{1}{48}\sum_{i=1}^{48} X_i > \frac{1}{48} 300\right)$$

$$= \Pr\left(\frac{\overline{X} - 6}{\sqrt{6}/\sqrt{48}} > \frac{300/48 - 6}{\sqrt{6}/\sqrt{48}}\right)$$

$$\approx \Pr\left(Z > 0.7071\right)$$

$$\approx 0.2398.$$

∎∎∎

In Section 6.6.3 we introduced the F distribution, which is the distribution of

$$X = \frac{Y_1/\nu_2}{Y_2/\nu_2},$$

where Y_1 and Y_2 are independent χ^2 random variables with ν_1 and ν_2 degrees of freedom, respectively. We can use this result to find probabilities involving the ratios of sample variances under certain assumptions.

■ **Example 8.8** Suppose $X_1, X_2, \ldots, X_9 \sim$ i.i.d. $N(\mu_1, \sigma^2)$ and $Y_1, Y_2, \ldots, Y_9 \sim$ i.i.d. $N(\mu_2, \sigma^2)$ are independent samples from two normal populations. Note that the population means may differ, but the variances are the same. What is the probability that the sample variance from the first sample is more than double the sample variance from the second sample?

Solution: Let

$$S_1^2 = \frac{1}{9-1}\sum_{i=1}^{9}(X_i - \overline{X})^2$$

$$S_2^2 = \frac{1}{9-1}\sum_{i=1}^{9}(Y_i - \overline{Y})^2.$$

The ratio of sample variances can be written as

$$F = \frac{S_1^2}{S_2^2} = \frac{(n-1)S_1^2/\sigma^2}{(n-1)S_2^2/\sigma^2} = \frac{\left((9-1)S_1^2/\sigma^2\right)/(9-1)}{\left((9-1)S_2^2/\sigma^2\right)/(9-1)} = \frac{V_1/(9-1)}{V_2/(9-1)},$$

where V_1 and V_2 are independent χ^2 random variables with $9 - 1 = 8$ degrees of freedom. Thus, the ratio F satisfies $F \sim F(8,8)$ so we can write

$$\Pr\left(S_1^2 > 2S_2^2\right) = \Pr\left(\frac{S_1^2}{S_2^2} > 2\right) = \Pr\left(\frac{V_1/8}{V_2/8} > 2\right) = \texttt{pf(2,8,8,lower.tail=FALSE)}$$

The last line yields 0.1732968 when this command is executed in R.

∎∎∎

8.6 Normal Approximation to the Binomial

The CLT from the last section applies to a sample from any distribution, as long as the mean and variance are finite. In particular, it applies to sampling from the Bernoulli or binomial distributions. If we have a sequence of Bernoulli trials, each resulting in a success or failure, and if we let X_i be 1 if the ith trial is a success and 0 if it is a failure, then

$$Y = \sum_{i=1}^{n} X_i \sim \text{BIN}(n, p),$$ (8.36)

where p is the probability of success on any trial. Each X_i has a Bernoulli (or $\text{BIN}(1,p)$) distribution, which certainly has a finite mean and variance. Thus we can conclude from the CLT that for a large number n of trials, the binomial distribution is approximately normally distributed with mean

$$\mu = np \tag{8.37}$$

and variance

$$\sigma^2 = np(1-p). \tag{8.38}$$

Of course, the larger the number of trials, the better will be the normal approximation. The probability of success also affects the accuracy of the approximation. When p is near $\frac{1}{2}$, the binomial is roughly symmetric and the normal approximation works well. When p is near 0 or 1, the binomial is highly skewed (to the right if p is near 0, and to the left if p is near 1), so the normal approximation does not work so well. As a rough rule of thumb, the normal approximation will work satisfactorily if $np > 5$ and $n(1-p) > 5$.

■ **Example 8.9** A crossover clinical trial is designed to give each participant both drugs being tested, but at different times. Usually, for each participant, one drug is selected at random and given first. The response is recorded, and then after a fixed period of time, called the "washout," the other drug is given. The response is then recorded and comparisons are made between the outcomes of the two treatments. These crossover designs can only be run for short-term regimens. They are generally not applicable to long-term issues such as diabetes and cancer. Suppose that a crossover trial is run, and each participant is asked at the end of the trial which drug worked better at relieving pain. Each participant must reply "Drug A" or "Drug B" (ties are not allowed). Suppose that Drug A is a new treatment and Drug B is a placebo. Out of 200 participants, 120 of them chose Drug A as providing more pain relief. Assume that the drugs are equally effective, so that the probability of a person preferring Drug A over Drug B is $\frac{1}{2}$. Under this assumption, what is the probability that 120 or more people would prefer Drug A over Drug B.

Solution: Let Y denote the number out of 200 who prefer Drug A. Then under the assumptions given above, $Y \sim \text{BIN}(200, 0.5)$. The exact probability that $Y \geq 120$ is

$$\Pr(Y \geq 120) = \Pr(Y = 120) + \Pr(Y = 121) + \cdots + \Pr(Y = 200)$$
$$= \binom{200}{120}(0.5)^{120}(1-0.5)^{200-120} + \cdots + \binom{200}{200}(0.5)^{200}(1-0.5)^{200-200}.$$

The last expression is a formidable sum to evaluate (unless you have software such as R). Fortunately, the normal approximation works well here, because np and $n(1-p)$ are each 100, and are well over 5. We can therefore write

$$\Pr(Y \geq 120) = \Pr\left(\frac{Y-np}{\sqrt{np(1-p)}} \geq \frac{120-np}{\sqrt{np(1-p)}}\right)$$
$$= \Pr\left(\frac{Y-200\times 0.5}{\sqrt{200\times 0.5\times(1-0.5)}} \geq \frac{120-200\times 0.5}{\sqrt{200\times 0.5\times(1-0.5)}}\right)$$
$$\approx \Pr(Z \geq 2.828)$$
$$\approx 0.0023.$$

The last lines can be computed in R using the code

```
z = (120 - 200*0.5) / sqrt( 200*0.5*(1-0.5) )
pnorm( z , 0 , 1 , lower.tail = FALSE )
```

which gives 0.0023. Thus, there is a very small probability that 120 or more people will prefer Drug A. Remember, that this is calculated under the assumption that the drugs are equally effective. Think for a moment what this means. If the drugs are equally effective, we've observed a fairly unlikely event (one with probability 0.0023). Doesn't this make you doubt the assumption that the drugs are equally effective? This is basically the logic behind hypothesis testing, which we will cover in Chapters 12 and 13.   ■ ■ ■

The exact probability can be computed from R using the following code:

```
pbinom( 120 , 200 , 0.5 , lower.tail=FALSE )
```

The result for this probability is 0.0018, which is a bit off from the approximation in the previous example. We will investigate this discrepancy in the next example. For very large sample sizes, the exact method can be very slow. There are also cases where we must use the analytic result in a subsequent formula. In this case the normal approximation is needed.

Suppose $Y \sim \text{BIN}(20, 0.4)$ The probability distribution for Y is shown in Figure 8.4(a). The height of each bar is the probability of a particular y. Since the width of each bar is 1 (extending from $y - \frac{1}{2}$ to $y + \frac{1}{2}$) the area of a bar is equal to the height of the bar. If we wanted to obtain the probability that

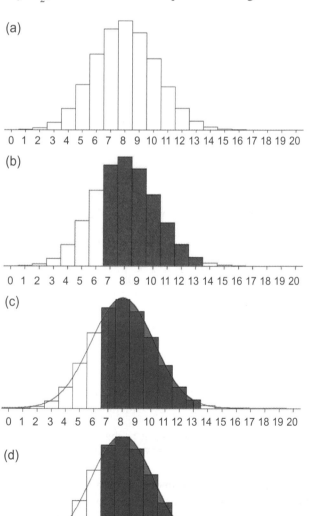

Figure 8.4 Using the normal approximation to the binomial. Part (a) shows the exact binomial distribution BIN(20,0.4); (b) shows the bars corresponding to $\Pr(7 \leq X \leq 13)$; (c) shows the normal approximation with the area shaded from 8 to 14; and (d) shows the continuity correction, with the area under the normal curve shaded from 7.5 to 14.5.

$7 \leq Y \leq 13$, then we would want the sum of the areas of the bars that cover 7 through 13, inclusive; these are the dark blue bars in Figure 8.4(b). If we were to use the normal approximation to the binomial we would compute the red shaded area in Figure 8.4(c). Look closely and you will see that this shaded area is under the smooth normal curve, starting at 7 and ending at 13. Compare (b) and (c) in this figure, and you will see that the blue shading begins at 6.5 in the second graph (midway between 6 and 7) and extends to 13.5 (midway between 13 and 14). By contrast, the area under the normal curve goes from 7.0 to 13.0. Might the approximation be better if we took the area under the normal curve between 6.5 and 13.5? Usually, the answer is yes. This adjustment is called the *continuity correction*.

■ **Example 8.10** Use the continuity correction to evaluate the probabilities:
   (a) $\Pr(7 \leq Y \leq 13)$, where $Y \sim \text{BIN}(20, 0.40)$;
   (b) $\Pr(Y \geq 120)$, where $Y \sim \text{BIN}(200, 0.50)$.

Solution: For the first problem, we can write

$$\Pr(7 \leq Y \leq 13) = \Pr(6.5 \leq Y \leq 13.5)$$

$$= \Pr\left(\frac{6.5 - 20 \times 0.4}{\sqrt{20 \times 0.4(1 - 0.4)}} \leq \frac{Y - 20 \times 0.4}{\sqrt{20 \times 0.4(1 - 0.4)}} \leq \frac{13.5 - 20 \times 0.4}{\sqrt{20 \times 0.4(1 - 0.4)}} \right)$$

$$\approx \Pr(-0.6847 \leq Z \leq 2.5104)$$

$$\approx 0.7472.$$

If we were to not use the continuity correction, we would have a different approximation:

$$\Pr(7 \leq Y \leq 13) = \Pr\left(\frac{7 - 20 \times 0.4}{\sqrt{20 \times 0.4(1 - 0.4)}} \leq \frac{Y - 20 \times 0.4}{\sqrt{20 \times 0.4(1 - 0.4)}} \leq \frac{13 - 20 \times 0.4}{\sqrt{20 \times 0.4(1 - 0.4)}} \right)$$

$$\approx \Pr(-0.4564 \leq Z \leq 2.2822)$$

$$\approx 0.6647.$$

The exact value, found using the code `sum(dbinom(7:13,20,0.4))` in R, is 0.7435. Using the continuity correction gives a much more accurate approximation to the probability.

The second part is handled similarly, but in this case $Y \geq 120$ becomes $Y \geq 119.5$:

$$\Pr(Y \geq 120) = \Pr(Y \geq 119.5)$$

$$= \Pr\left(\frac{Y - 200 \times 0.5}{\sqrt{200 \times 0.5(1 - 0.5)}} \geq \frac{119.5 - 200 \times 0.5}{\sqrt{200 \times 0.5(1 - 0.5)}} \right)$$

$$\approx \Pr(Z \geq 2.7577)$$

$$\approx 0.0029.$$

Without the continuity correction we approximated the probability to be 0.0023; with the continuity correction we obtained 0.0029. The exact value, using `sum(dbinom(120:200,200,0.5))`, yields 0.0028; once again, the continuity correction yields a closer approximation.   ■ ■ ■

The rough rule of thumb for using the normal approximation to the binomial is that $np > 5$ and $n(1 - p) > 5$. This is because the binomial distribution tends to be fairly symmetric, though not perfectly so, when p is near $\frac{1}{2}$ and n is large. Figure 8.5 shows binomial distributions for $p = 0.4$ (the four graphs in (a)) and for $p = 0.1$ (the four graphs in (b)). For each group of four plots, the probability histograms for $n = 10, 20, 40$, and 80 are shown. For all eight plots, the normal curve is shown. When $p = 0.4$, the normal approximation seems to work well even when n is 10 or 20. By contrast, when $p = 0.1$, the normal curve does not seem to fit well until n gets to be around 80.

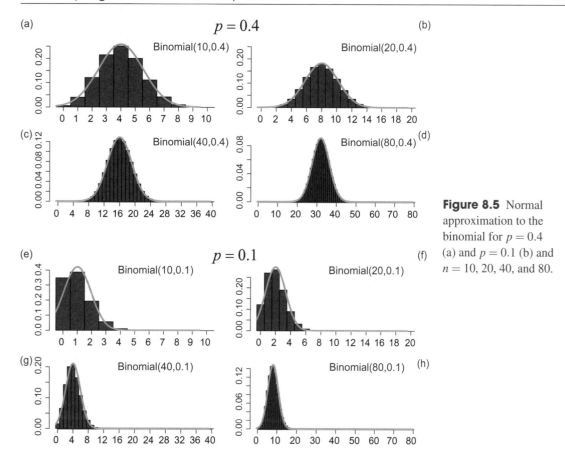

Figure 8.5 Normal approximation to the binomial for $p = 0.4$ (a) and $p = 0.1$ (b) and $n = 10, 20, 40,$ and 80.

8.7 Sampling Distributions for Proportions

Suppose we take a sample of size n from a large population and record whether each subject or unit in the sample has (or doesn't have) some property. If we assume a virtually infinite population, this is equivalent to performing n Bernoulli trials where the probability of success p is equal to the proportion of the population having that property. Under these conditions, the number X of successes (i.e., the number in the sample having the desired property) has a binomial distribution: $X \sim \mathrm{BIN}(n, p)$. While p is the proportion of population units with some property, we let $\hat{p}$ be the fraction of *sample* units with the property.[2] Thus,

$$\hat{p} = \frac{X}{n}.$$

In this section we look for the sampling distribution of $\hat{p}$. The exact sampling distribution can be obtained directly from the binomial PMF:

$$P\left(\hat{p} = \frac{a}{n}\right) = \binom{n}{a} p^a (1-p)^{n-1}, \qquad a = 0, 1, \ldots, n.$$

If the sample size is large, we can use the normal approximation to the binomial, described in Section 8.6. This says that

$$X \stackrel{\mathrm{approx}}{\sim} N(np, np(1-p)).$$

---

[2] The use of the "hat" notation was first introduced in Chapter 1 to show that the sample mean $\hat{\mu}$ is an estimate of the population mean μ. In the same way here, $\hat{p}$ is an estimate of the population proportion p. We will explain this further in the next chapter.

Since $\hat{p}$ is just a constant times X, we can say that its distribution is approximately normal, with

$$\mathbb{E}(\hat{p}) = \mathbb{E}\left(\frac{X}{n}\right) = \frac{1}{n}\mathbb{E}(X) = \frac{1}{n}np = p \tag{8.39}$$

and

$$\mathbb{V}(\hat{p}) = \mathbb{V}\left(\frac{X}{n}\right) = \frac{1}{n^2}\mathbb{E}(X) = \frac{1}{n^2}np(1-p) = \frac{p(1-p)}{n}. \tag{8.40}$$

Thus,

$$\hat{p} \stackrel{approx}{\sim} N\left(p, \frac{p(1-p)}{n}\right). \tag{8.41}$$

This equation gives us the sampling distribution for the proportion of successes in a sample of size n where the probability of successes is p.

■ **Example 8.11**   Suppose that in a city, the proportion of those vaccinated against COVID is 0.65. What is the probability that in a sample of size 600 the number of those who have been vaccinated exceeds 0.70?

Solution: We can solve this in two ways: by converting the problem back to the binomial distribution, and by using the normal approximation given in (8.41).[3] We'll start with the latter. Let X denote the number of vaccinated persons in the sample of 600 and let $\hat{p} = X/n$. Using (8.41), we can write

$$\Pr(\hat{p} > 0.70) = \Pr\left(\frac{\hat{p} - 600 \times 0.65}{\sqrt{600 \times 0.65 \times 0.35}} > \frac{0.70 - 0.65}{\sqrt{0.65 \times 0.35/600}}\right)$$

$$\approx \Pr(Z > 2.567763)$$

$$\approx 0.0051.$$

The computation on the last line was done using the `pnorm()` function in R.

If we transform to the binomial random variable we can apply the continuity correction and write

$$\Pr(\hat{p} > 0.70) = \Pr\left(\frac{X}{600} > 0.7\right) = \Pr(X > 420) = \Pr(X > 420.5)$$

$$= \Pr\left(\frac{X - 600 \times 0.65}{\sqrt{600 \times 0.65 \times 0.35}} > \frac{420.5 - 600 \times 0.65}{\sqrt{600 \times 0.65 \times 0.35}}\right) \approx \Pr(Z > 2.61) \approx 0.0045.$$

The answers differ because the former used the continuity correction and the latter didn't. As we discussed previously, the continuity correction usually gives the more accurate answer.   ■ ■ ■

---

[3]We could, of course, use the continuity correction on the distribution of $\hat{p}$, but this is a little tricky. We have to look at the neighboring possible values for $\hat{p}$ and take the midpoint. Here, the possible outcomes for $\hat{p}$ include 419/600, 420/600, and 421/600. We could use the midpoint of 420/600 and 421/600 and find the probability of exceeding that. It's usually easier to just transform back to the binomial and work from there. See Problem 8.36.

8.8 Tchebysheff's Theorem and the Law of Large Numbers

Tchebysheff's Theorem provides a probabilistic bound on the distance a random variable can be from its expected value. In particular, it provides a conservative lower bound for the probability that a random variable falls in the interval $\mu \pm k\sigma$ for any $k > 0$. It is particularly useful because it applies to any random variable regardless of its distribution (so long as the mean and variance exist) and it can be used to prove the weak law of large numbers later in this section.

> **Theorem 8.8.1 — Tchebysheff's Theorem.** Let X be a random variable with finite mean μ and variance σ^2. Then, for any constant $k > 0$,
>
> $$\Pr(|X - \mu| \geq k\sigma) \leq \frac{1}{k^2}$$
>
> or, equivalently,
>
> $$\Pr(|X - \mu| < k\sigma) \geq 1 - \frac{1}{k^2}.$$

Proof. Note that the theorem statement places no restrictions on the distribution of the random variable other than that its mean and variance must be finite. Hence, it applies to both discrete and continuous random variables.

To prove the result, we begin with the definition of the variance for a continuous random variable. (The proof in the discrete case proceeds in an analogous way and is left as an exercise.)

$$
\begin{aligned}
\text{Var}(X) &= \int_{-\infty}^{\infty} (x - \mu)^2 f(x)\, dx \\
&= \int_{-\infty}^{\mu - k\sigma} (x - \mu)^2 f(x)\, dx + \int_{\mu - k\sigma}^{\mu + k\sigma} (x - \mu)^2 f(x)\, dx + \int_{\mu + k\sigma}^{\infty} (x - \mu)^2 f(x)\, dx \\
&\geq \int_{-\infty}^{\mu - k\sigma} (x - \mu)^2 f(x)\, dx + \int_{\mu + k\sigma}^{\infty} (x - \mu)^2 f(x)\, dx \\
&\geq k^2 \sigma^2 \int_{-\infty}^{\mu - k\sigma} f(x)\, dx + k^2 \sigma^2 \int_{\mu + k\sigma}^{\infty} f(x)\, dx.
\end{aligned}
$$

The second line follows from the first as simply a restatement of the integral. The third line follows from the second because the middle integral, which must be nonnegative, is dropped. The fourth line follows from the third because if $x < \mu - k\sigma$, then $x - \mu < -k\sigma$; but $x - \mu < 0$ and $-k\sigma < 0$ imply $(x - \mu)^2 \geq k^2\sigma^2$. Similarly, if $x > \mu + k\sigma$, then $(x - \mu)^2 \geq k^2\sigma^2$. This implies

$$
\begin{aligned}
\text{Var}(X) &\geq k^2\sigma^2 \left[\int_{-\infty}^{\mu - k\sigma} f(x)\, dx + \int_{\mu + k\sigma}^{\infty} f(x)\, dx \right] \\
&= k^2\sigma^2 \left[\Pr(X \leq \mu - k\sigma) + \Pr(X \geq \mu + k\sigma) \right] \\
&= k^2\sigma^2 \Pr(|X - \mu| \geq k\sigma).
\end{aligned}
$$

Thus, all that's left is to substitute σ^2 for $\text{Var}(X)$ and simplify:

$$\sigma^2 \geq k^2\sigma^2 \left[\Pr(|X - \mu| \geq k\sigma) \right], \qquad \text{or equivalently,} \qquad \Pr(|X - \mu| \geq k\sigma) \leq \frac{1}{k^2}.$$

Finally, the other bound, $\Pr(|X - \mu| < k\sigma) \geq 1 - 1/k^2$, follows from the complement rule. That is, we know that

$$\Pr(|X - \mu| < k\sigma) + \Pr(|X - \mu| \geq k\sigma) = 1,$$

so if $\Pr(|X - \mu| \geq k\sigma) \leq 1/k^2$ then it must be that $\Pr(|X - \mu| < k\sigma) \geq 1 - 1/k^2$. ∎

Tchebysheff's Theorem can be used to prove the law of large numbers, which says that the sample mean will get close to the true population mean as the sample size gets large. This rather vague statement is made more precise in the next theorem.

> **Theorem 8.8.2 — The Weak Law of Large Numbers.** Suppose $X_1, X_2, \ldots, X_n$ are i.i.d. from a distribution with mean μ and variance σ^2. Let $\overline{X}$ denote the mean of this sample of size n. Then for every $\varepsilon > 0$,
>
> $$\lim_{n \to \infty} \Pr\left(|\overline{X} - \mu| < \varepsilon\right) = 1. \tag{8.42}$$

Proof. We know from Theorem 8.3.3 that

$$\mathbb{E}(\overline{X}) = \mu \qquad \text{and} \qquad \mathbb{V}(\overline{X}) = \frac{\sigma^2}{n}.$$

Thus, by Tchebysheff's Theorem applied to the distribution of $\overline{X}$,

$$\Pr\left(|\overline{X} - \mu| < k\frac{\sigma}{\sqrt{n}}\right) \geq 1 - \frac{1}{k^2}.$$

Choose k so that $k\sigma/\sqrt{n} = \varepsilon$, that is, $k = \varepsilon\sqrt{n}/\sigma$. Thus,

$$\Pr\left(|\overline{X} - \mu| < k\frac{\sigma}{\sqrt{n}}\right) = \Pr\left(|\overline{X} - \mu| < \varepsilon\right) \geq 1 - \frac{\sigma^2}{n\varepsilon^2}.$$

As $n \to \infty$, the right-hand side goes to 1. Therefore,

$$\lim_{n \to \infty} \Pr\left(|\overline{X} - \mu| < \varepsilon\right) = 1. \qquad ∎$$

In our statement of the weak law of large numbers, we assumed that the population variance is finite. Although this simplifies the proof, it is not a necessary condition. If the population mean is finite, but the population variance is infinite, the weak law of large numbers still holds, but the proof is much more difficult. The weak law of large numbers applies to the sample mean from any distribution so long as the mean is finite. In particular, it applies to the binomial distribution, or more specifically, the Bernoulli distribution. If $X_1, X_2, \ldots, X_n \sim$ i.i.d. Bern(p), then

$$Y_n = \sum_{i=1}^n X_i \sim \text{BIN}(n, p).$$

The weak law of large numbers applies to the mean

$$\frac{Y_n}{n} = \frac{1}{n}\sum_{i=1}^n X_i = \text{fraction of times a success occurred in } n \text{ trials}$$

to say that

$$\lim_{n\to\infty} \Pr\left(\left|\frac{Y_n}{n} - p\right| < \varepsilon\right) = 1.$$

In other words, the fraction of successes will, in the long run, equal the probability of a success.

We must be careful to interpret what the weak law of large numbers says and doesn't say. It is really a *limit theorem*, so it only says something about the limit of probabilities. It says nothing about the occurrence in a finite sequence of trials.

■ **Example 8.12** Suppose we toss a fair coin repeatedly, with a head being considered a success. Suppose that the first 10 tosses result in the outcome

   SSSFSSSFSS.

What is the probability that the eleventh toss is an *S*?

Solution: If the coin is really fair, and if tosses are really independent (as required by the definition of Bernoulli trials) then the probability of a success on the eleventh toss is $\frac{1}{2}$, just as it is on any other toss. It doesn't matter that 8 of the first 10 tosses were successes.   ■ ■ ■

A common misconception of the weak law of large numbers is that subsequent trials tend to balance out earlier trials. For example, many people will see 8 of 10 successes and claim that if successes and tails are to "balance out" the next toss is more likely to be a tail.

■ **Example 8.13** Which of these sequences of Bernoulli trials is more likely if $p = \Pr(S) = 1/2$?

   SSSSSS   or   *FFSFSSS.*

Solution: The first sequence of six straight successes seems nonrandom and quite unlikely. The second looks much more random and not so unlikely. Actually, though, the first sequence has probability $(1/2)^6 = 1/64$ while (because the trials are independent, each with $p = 1/2$) the second has probability $(1/2)^7 = 1/128$. Thus, the first sequence, *SSSSSS*, is more likely.   ■ ■ ■

The weak law of large numbers is related to the concept of *convergence in probability*.

Definition 8.8.1 — Convergence in Probability. The infinite sequence of random variables $Y_1, Y_2, \ldots$ *converges in probability* to the constant θ if for every $\varepsilon > 0$

$$\lim_{n\to\infty} \Pr(|Y_n - \theta| < \varepsilon) = 1.$$

If the sequence $Y_1, Y_2, \ldots$ converges in probability to θ, then we write

$$Y_n \xrightarrow{P} \theta.$$

By the weak law of large numbers we can conclude that the sample mean from a population with finite mean and variance converges in probability to the mean of the population, which we would write as

$$\overline{X} \xrightarrow{P} \mu.$$

For the special case of binomial random variables, we have

$$\frac{Y_n}{n} \xrightarrow{P} p.$$

While these are true, there are other situations where random variables converge in probability to some constant. This will be an important property in the next chapter.

8.9 **Chapter Summary**

The key idea from this chapter is that statistics, such as the sample mean or sample variance, are themselves random variables. As such, they have a probability distribution called the *sampling distribution*.

Linear combinations of normal random variables have a normal distribution, and from this we concluded that the mean of a sample from a normal distribution has a normal distribution. The sample variance $S^2 = \sum_{i=1}^{n}(X_i - \overline{X})^2/(n-1)$ is another statistic computed from the random variables $X_1, X_2, \ldots, X_n$ that comprise a sample. The random variable $(n-1)S^2/\sigma^2$ has a $\chi^2(n-1)$ distribution. The CLT goes one step further and says that the mean computed from a sample from a distribution with finite mean and variance has an approximate normal distribution when n is large.

The weak law of large numbers asserts that the mean of a sample will converge in probability to the true mean. This means that as the sample size gets large (i.e., without bound) the probability of $\overline{X}$ being within ε of the true mean μ will converge to 1 for every $\varepsilon > 0$.

8.10 Problems

Problem 8.1 Suppose that X and Y are independent random variables with means $\mu_X = 80$ and $\mu_Y = 72$, and variances $\sigma_X^2 = 16$ and $\sigma_Y^2 = 25$. Find the following.

(a) $\mathbb{E}(X+Y)$
(b) $\mathbb{V}(X+Y)$
(c) $\mathbb{E}(X-Y)$
(d) $\mathbb{V}(X-Y)$
(e) $\mathbb{E}(3X-4Y)$

Problem 8.2 Suppose that X and Y are independent random variables with means $\mu_X = 0.150$ and $\mu_Y = 0.180$, and standard deviations $\sigma_X = 0.005$ and $\sigma_Y = 0.004$. Find the following.

(a) $\mathbb{E}(X+Y)$
(b) $\mathbb{V}(X+Y)$
(c) $\mathbb{E}(X-Y)$
(d) $\mathbb{V}(X-Y)$
(e) $\mathbb{E}(100X - 50Y)$

Problem 8.3 Dowel pins are made to fit into a hole in order to attach one part to another. The diameter of the dowel pin is normally distributed with a mean of 1.18 cm and a standard deviation of 0.02 cm. The diameter of the hole is normally distributed with mean of 1.23 cm and a standard deviation of 0.02 cm. Let X and Y denote the diameters of the drilled holes and the dowel pin, respectively, as shown in Figure 8.6. The dowel pin will fit provided its diameter is less than the diameter of the hole. What is the probability that the dowel pin will fit if we select one dowel pin and one part at random?

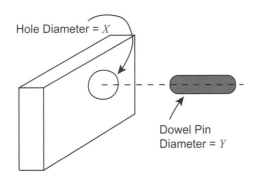

Figure 8.6 Situation described in Problem 8.3.

Problem 8.4 In a given population, the height of adult men is normally distributed with a mean of 69 inches and a standard deviation of 4 inches. The height of adult women is normally distributed with mean 63 inches and standard deviation 3 inches. Suppose one man and one woman are chosen at random. What is the probability that the woman is taller than the man?

Problem 8.5 Suppose that in the situation described in Problem 8.4 random samples of n men and n women are selected independently from the populations. What is the probability that the average height of the women is greater than the average height of the men if

(a) $n = 2$?
(b) $n = 4$?
(c) $n = 6$?

Problem 8.6 Suppose that in Problem 8.3 production workers repeatedly select one dowel pin and one unit with a hole. Suppose that 100 pairs of dowel pins and units with holes are selected. Let W denote the number of times out of 100 that the dowel pin fits into the hole. Find the following.

(a) $\Pr(W > 90)$
(b) $\mathbb{E}(W)$
(c) $\mathbb{V}(W)$

Problem 8.7 Suppose $X_1, X_2, \ldots, X_n \sim$ i.i.d. from a distribution with mean $\mu = 30$ and variance $\sigma^2 = 4$. Suppose that n is even. Find the following.

(a) $\mathbb{E}\left(\sum_{i=1}^{n} X_i\right)$

(b) $\mathbb{V}\left(\sum_{i=1}^{n} X_i\right)$

(c) $\mathbb{E}\left(X_1 - X_2 + X_3 - X_4 + \cdots - X_n\right)$
(d) $\mathbb{V}\left(X_1 - X_2 + X_3 - X_4 + \cdots - X_n\right)$
(e) $\mathbb{E}\left(\overline{X}\right)$, where $\overline{X} = (1/n)\sum_{i=1}^{n} X_i$
(f) $\mathbb{V}\left(\overline{X}\right)$

Problem 8.8 Suppose $X_1, X_2, \ldots, X_n \sim$ i.i.d. from a distribution with mean $\mu = 0.6$ and variance $\sigma^2 = 0.01$. Suppose that n is even. Find the following.

(a) $\mathbb{E}\left(\sum_{i=1}^{n} X_i\right)$

(b) $\mathbb{V}\left(\sum_{i=1}^{n} X_i\right)$

(c) $\mathbb{E}\left(X_1 - X_2 + X_3 - X_4 + \cdots - X_n\right)$
(d) $\mathbb{V}\left(X_1 - X_2 + X_3 - X_4 + \cdots - X_n\right)$
(e) $\mathbb{E}\left(\overline{X}\right)$, where $\overline{X} = (1/n)\sum_{i=1}^{n} X_i$
(f) $\mathbb{V}\left(\overline{X}\right)$

Problem 8.9 Based on Problem 8.7, make the additional assumption that the distribution is normal. What can you say about the distribution each of the following?

(a) $\sum_{i=1}^{n} X_i$

(b) $\overline{X}$

(c) How do these answers differ from your answers to Problem 8.7?

Problem 8.10 Suppose $X_1, X_2, \ldots, X_n \sim$ i.i.d. $N(\mu, 1)$. What is the probability that the sample variance exceeds 2 when

(a) $n = 10$?

(b) $n = 20$?

(c) $n = 30$?

(d) Explain why the answers to the above questions are independent of the value of μ.

Problem 8.11 Suppose $X_1, X_2, \ldots, X_n \sim$ i.i.d. $N(\mu, 10)$. What is the probability that the sample variance exceeds 20 when

(a) $n = 10$?

(b) $n = 20$?

(c) $n = 30$?

Problem 8.12 Suppose $X_1, X_2, \ldots, X_{25} \sim$ i.i.d. $N(\mu_1, \sigma^2)$ and $Y_1, Y_2, \ldots, Y_{25} \sim$ i.i.d. $N(\mu_2, \sigma^2)$ are independent samples from two normal populations. Find the probability that the sample variance from the first sample is more than c times the sample variance from the second sample in the following cases.

(a) $c = 1$

(b) $c = 2$

(c) $c = 4$

(d) $c = 6$

Problem 8.13 Suppose $X_1, X_2, \ldots, X_{20} \sim$ i.i.d. $N(\mu_1, \sigma^2)$ and $Y_1, Y_2, \ldots, Y_{10} \sim$ i.i.d. $N(\mu_2, \sigma^2)$ are independent samples from two normal populations.

(a) Find the probability that the sample variance from the first sample is more than double the sample variance from the second sample.

(b) Find the probability that the sample variance from the second sample is more than double the sample variance from the first sample.

Problem 8.14 Suppose $X_1, X_2, \ldots, X_n \sim$ i.i.d. $N(\mu_1, \sigma^2)$ and $Y_1, Y_2, \ldots, Y_n \sim$ i.i.d. $N(\mu_2, \sigma^2)$ are independent samples from two normal populations. Find the probability that the sample variance from the first sample is more than double the sample variance from the second sample in the following cases.

(a) $n = 10$

(b) $n = 20$

(c) $n = 40$

(d) $n = 80$

Problem 8.15 Suppose $X_1, X_2, \ldots, X_n \sim$ i.i.d. $N(\mu_1, \sigma^2)$ and $Y_1, Y_2, \ldots, Y_n \sim$ i.i.d. $N(\mu_2, \sigma^2)$ are independent samples from two normal populations. Find the probability that the sample standard deviation from the first sample is less than half the sample standard deviation from the second sample in the following cases.

(a) $n = 10$

(b) $n = 20$

(c) $n = 40$

(d) $n = 80$

Problem 8.16 If $V_1 \sim \chi^2(\nu_1)$ and $V_2 \sim \chi^2(\nu_2)$ are independent χ^2 random variables, it can be shown that $V_1 + V_2 \sim \chi^2(\nu_1 + \nu_2)$. Suppose $X_1, X_2, \ldots, X_{n_1} \sim$ i.i.d. $N(\mu_1, \sigma^2)$ and $Y_1, Y_2, \ldots, Y_{n_2} \sim$ i.i.d. $N(\mu_2, \sigma^2)$ are independent samples from two normal populations. What is the distribution of

$$\frac{\sum_{i=1}^{n_1}(X_i - \overline{X})^2 + \sum_{i=1}^{n_2}(Y_i - \overline{Y})^2}{\sigma^2} ?$$

Problem 8.17 Use R to simulate 1,000 samples each of size $n = 10$ from the $N(10, 1)$ distribution. Simulate a second set of 1,000 samples, also of size 10 from the $N(20, 1)$ distribution. For each sample, compute the sample variance.

(a) Plot a histogram of the 1,000 ratios of S_1^2 (computed from the first set of 1,000 simulated samples) and S_2^2 (computed from the second set of 1,000 simulated samples).

(b) What fraction of the ratios computed from part (a) exceed 2?

(c) What is the probability that $S_1^2 > 2S_2^2$?

Problem 8.18 Suppose $X_1, X_2, \ldots, X_{50} \sim$ i.i.d. $\text{EXP1}(\theta = 100)$. Approximate the probability that $\overline{X} > c$ for the following.

(a) $c = 100$

(b) $c = 120$

(c) $c = 140$

(d) $c = 160$

Problem 8.19 Repeat Problem 8.18 for a sample of $n = 10$.

Problem 8.20 Suppose $X_1, X_2, \ldots, X_{50} \sim$ i.i.d. $\text{GAM}(\alpha = 2, \beta = 100)$. Use the CLT to approximate the probability that $\overline{X} > c$ for the following.

(a) $c = 200$

(b) $c = 220$

(c) $c = 240$

(d) $c = 260$

Problem 8.21 Repeat Problem 8.20 for a sample of $n = 10$.

Problem 8.22 Suppose $X_1, X_2, \ldots, X_{100} \sim$ i.i.d. POIS(5). Use the CLT to approximate the probability that

(a) $\overline{X} > 5.20$

(b) $\displaystyle\sum_{i=1}^{100} > 520$

(c) $\overline{X} > 5.40$

Problem 8.23 It can be shown that if $X_1, X_2, \ldots, X_n$ are independent Poisson random variables, where $X_i \sim$ POIS(λ_i), then

$$\sum_{i=1}^{n} X_i \sim \text{POIS}\left(\sum_{i=1}^{n} \lambda_i\right).$$

(a) Use this fact to find the distribution of $\sum_{i=1}^{100} X_i$ for the random variables introduced in Problem 8.22.

(b) Use the result of part (a) along with the ppois() function in R to find the exact probability in part (b) of Problem 8.22.

Problem 8.24 Plot the PMF for the POIS(500) distribution over the domain $x = 400, 401, \ldots, 600$. Comment on whether this distribution looks like the normal.

Problem 8.25 Simulate $N = 1,000$ samples each of $n = 50$ from the distributions below. For each sample, compute the sample mean $\overline{y}$ for each sample and compute

$$z = \frac{\overline{y} - \mu}{\sigma/\sqrt{n}},$$

where μ and σ are the mean and standard deviation of the distribution. Then plot a histogram of the sample means. You may use the following code as a guide:

```
N = 1000
n = 50
mean.y = 20                      #---
sd.y = 20                        #---
y = rexp( n*N , rate=1/mean.y )  ####
y.mat = matrix( y , ncol=n )
ybar = apply( y.mat , 1 , mean )
z = ( ybar - mean.y ) / (sd.y/sqrt(n))
h1 = hist(z)
binw = h1$breaks[2] - h1$breaks[1]
hist( z , ylim=c(0,1.4*max(h1$counts)) )
x = seq(-3,3,0.1)
lines( x , N*binw*dnorm(x) , col="red" )
```

Comment on whether the histograms appear to be close to normal. Here are the distributions.

(a) EXP1(20)

(b) GAM(2, 10)

(c) GAM(0.5, 40)

(d) $t(20)$ (*Hint:* the variance of the t distribution is $v/(v-2)$.)

(e) $t(3)$

You will have to specify in the R code the distributions, including parameters (the line marked with ####), and (on the lines marked with #--) the population mean and standard deviation for each distribution.

Problem 8.26 Repeat Problem 8.25 for sample sizes of $n = 10$.

Problem 8.27 Repeat Problem 8.25 with $n = 50$ for the following distributions.

(a) $t(3)$

(b) $t(2.5)$; the t distribution is allowed to have noninteger degrees of freedom

(c) $t(2.1)$

(d) What happens with the $t(2)$ distribution?

Problem 8.28 Use the result from Problem 8.23 and the CLT to explain why, for large λ, the POIS(λ) PMF looks like the normal distribution.

Problem 8.29 It is known that 6% of visitors to a website will purchase a product. Suppose that on a given day the number of visitors is 600.

(a) Use the pbinom() command to approximate the probability that the number of sales exceeds 40.

(b) Use the normal approximation to the binomial *without* the continuity correction to approximate the probability that sales exceeds 40.

(c) Use the normal approximation to the binomial *with* the continuity correction to approximate the probability that sales exceeds 40.

Problem 8.30 It is known that 6% of visitors to a website will purchase a product. Suppose that on a given day the number of visitors is 6,000.

(a) Use the pbinom() command to approximate the probability that the number of sales exceeds 400.

(b) Use the normal approximation to the binomial *without* the continuity correction to approximate the probability that the number of sales exceeds 400.

(c) Use the normal approximation to the binomial *with* the continuity correction to approximate the probability that sales exceeds 400.

Problem 8.31 It is known that 20% of visitors to a website will purchase a product. Suppose that on a given day the number of visitors is 600.

(a) Use the `pbinom()` command to approximate the probability that the number of sales exceeds 140.

(b) Use the normal approximation to the binomial *without* the continuity correction to approximate the probability that the number of sales exceeds 140.

(c) Use the normal approximation to the binomial *with* the continuity correction to approximate the probability that sales exceeds 140.

Problem 8.32 Suppose that an internet merchant has tested two website designs and found that Design A leads to a purchase rate of 10% and Design B leads to a purchase rate of 14%. Suppose that the next 1,200 visitors are sent to Designs A and B alternately (i.e., the first visitor is assigned to Design A, the second to Design B, the third to Design A, etc.). With this scheme, 600 visitors are shown Design A and another 600 are shown Design B. Use the normal approximation to find the probability that the number out of those shown Design A who purchase a product exceeds the number out of those shown Design B.

Problem 8.33 Consider a crossover clinical trial like that described in Example 8.9. Suppose 1,000 patients participate. Let X denote the number who prefer Treatment A over Treatment B. (Remember, ties are not allowed.) Assuming the treatments are equally effective (i.e., the probabilities of preferring a treatment is the same for both treatments), use the normal approximation to the binomial to find the probability that

(a) X is 510 or more;

(b) X is 530 or more;

(c) X is 550 or more;

(d) the proportion of those who prefer Treatment A is 0.55 or higher.

Problem 8.34 Consider Problem 8.31 where 20% of visitors to a website purchase a product. What is the probability that in a sample of 600 visitors, the proportion of those who purchase a product is 0.15 or below.

Problem 8.35 Consider a clinical trial, similar to that described in Problem 8.33, except this trial is designed so that 500 people are given Treatment A and 1,000 are given Treatment B. Assuming the treatments are equally effective (i.e., the probabilities of being cured is the same for both treatments), approximate the probability that the proportion cured in the sample of 500 given Treatment A is 0.10 greater than the proportion cured in the sample of 1,000 given Treatment B.

Problem 8.36 Explain how you could apply the continuity correction when applying the normal approximation to a sample proportion.

Problem 8.37 For the EXP1(100) distribution, find the probability that one observation X from this distribution is within two standard deviations of the mean, that is, $\Pr(|X - \mu| < 2\sigma)$ where μ and σ are the mean and standard deviation, respectively, of the EXP1(100) distribution. Compare this with the bound in Tchebysheff's Theorem.

Problem 8.38 For the UNIF$(-1, 1)$ distribution, find the probability that one observation X from this distribution is within two standard deviations of the mean, that is, $\Pr(|X - \mu| < 2\sigma)$ where μ and σ are the mean and standard deviation, respectively, of the UNIF$(-1, 1)$ distribution. Compare this with the bound in Tchebysheff's Theorem.

Problem 8.39 Suppose $X_1 \sim N(\mu_1, \sigma_1^2)$ and $X_2 \sim N(\mu_2, \sigma_2^2)$. Show that $c_1 X_1 + c_2 X_2 \sim N(c_1 \mu_1 + c_2 \mu_2, c_1^2 \sigma_1^2 + c_2^2 \sigma_2^2)$. (*Hint*: There are three parts: (1) show that the linear combination $c_1 X_1 + c_2 X_2$ is normally distributed (use MGFs); (2) the mean is $c_1 \mu_1 + c_2 \mu_2$; and (3) the variance is $c_1^2 \sigma_1^2 + c_2^2$.)

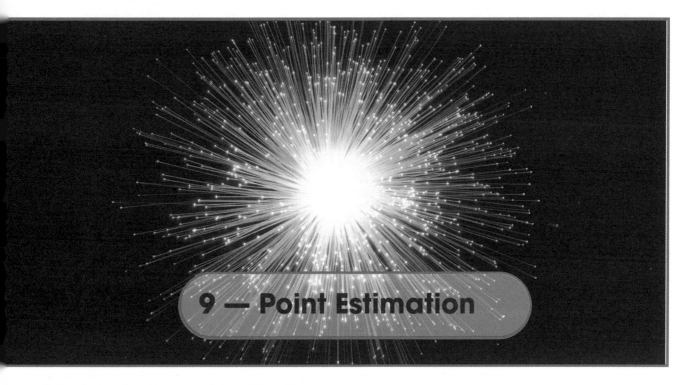

9 — Point Estimation

9.1 Introduction and Intuitive Estimators

We began the last chapter by reviewing the terms population, parameter, sample, and statistic. Parameters are numerical characteristics of a population that we would like to know, but since the population is nearly always too large to make a measurement on every unit, we often rely on a sample from the population. We let $X_1, X_2, \ldots, X_n$ represent the observations in a sample. Assuming an infinite population, this implies that these random variables are *independent and identically distributed*, abbreviated i.i.d. A statistic (used in the narrow, technical sense) is some function of these random variables. For example, the population mean μ may be of interest, and the statistic we compute could be the sample mean $\overline{X}$. We would then use $\overline{X}$ as an estimator for μ. It is important to recognize that these are different. Indeed, without knowing the difference, the rest of this chapter, and much of the rest of the book, will make little sense.

> **Definition 9.1.1 — Estimator of a Parameter.** If θ is an unknown parameter, and $s(X_1, X_2, \ldots, X_n)$ is some function of a random sample $X_1, X_2, \ldots, X_n$ that is used as our approximation as to the value of the parameter, then $s(X_1, X_2, \ldots, X_n)$ is said to be the *estimator* of the parameter θ. We denote an estimator by putting a hat or caret (^) above the name of the parameter θ and write $\hat{\theta}$. Note that θ is a fixed (nonrandom) variable, whereas $\hat{\theta}$ is a random variable.

There are several cases where a statistic we could use as an estimator is intuitive, and we give several examples of these. There are more complicated situations where an estimator is not so obvious. In Sections 9.3 and 9.4 we discuss two general approaches to finding estimates. If there is no obvious and intuitive estimator for a parameter, these general approaches can be applied. In many cases, these general approaches yield the intuitive estimators.

■ **Example 9.1** Suppose $X_1, X_2, \ldots, X_n$ is a random sample from the $N(\mu, \sigma^2)$ distribution. Find intuitive estimators for μ and σ.

Solution: The most obvious estimator of the population mean μ is the sample mean $\overline{X}$, so we write

$$\hat{\mu} = \overline{X}.$$

The variance is trickier. Since the population variance is $\sigma^2 = \mathbb{E}\left[(X-\mu)^2\right]$ we might think that the average squared deviation from μ, that is $(1/n)\sum_{i=1}^{n}(X_i-\mu)^2$, would be a good estimator. The problem, however, is that this depends on a parameter, namely μ, that we don't know. Estimators are only allowed to depend on the observed random variables, not on any unknown parameters. The obvious thing to do is to replace μ with an estimator $\overline{X}$ to obtain

$$\hat{\sigma}_0^2 = \frac{1}{n}\sum_{i=1}^{n}(X_i-\overline{X})^2.$$

This is a reasonable estimator and it is related to what we have called the sample variance,

$$S^2 = \frac{1}{n-1}\sum_{i=1}^{n}(X_i-\overline{X})^2.$$

They are related by

$$\hat{\sigma}_0^2 = \frac{n-1}{n}S^2.$$

Both of these would be reasonable estimators. In the next section we will give some criteria that can be used as a basis for comparing estimators. ∎∎∎

■ **Example 9.2** Suppose that $X \sim \text{BIN}(n,p)$. Find an intuitive estimator for p.

Solution: The value of p is the probability of getting a success on one Bernoulli trial. The random variable X is the number of successes on n trials. The fraction of successes on the n trials seems like an intuitive estimator for p. This gives $\hat{p} = X/n$. ∎∎∎

■ **Example 9.3** Suppose $X_1, X_2, \ldots X_n$ are observed lifetimes and are assumed to follow an exponential distribution with mean θ. Find an intuitive estimator for θ.

Solution: The mean of the $\text{EXP1}(\theta)$ distribution is θ, so it makes sense to use the sample mean $\overline{X}$ to estimate the population mean. Thus, $\hat{\theta} = \overline{X}$. ∎∎∎

Exercise 9.1 Suppose $X_1, X_2, \ldots X_{31}$ are the daily numbers of patients who arrive at the emergency department at a community hospital in the month of July. It is assumed that this is a random sample from the Poisson distribution $\text{POIS}(\lambda)$. Find an intuitive estimator for λ.

Answer: $\hat{\lambda} = \overline{X}$.

In this section we have defined what an estimator is. A related term is *estimate*. It sounds like *estimate* and *estimator* should be synonymous, but they are not. There is an important distinction. An estimator is a function of the random variables that make up a sample, and as such, an estimator is a random variable. The point of reference here is *before* actually taking a sample. Once a sample is taken, and actual numbers are substituted for the random variables, what we have is an *estimate* of the parameter. Think of the *estimator* as the *rule* that tells us what to do with the observations; think of the *estimate* as the result of this rule when the actual data are used.

■ **Example 9.4** Suppose, as in Exercise 9.1, that we observe daily counts $X_1, X_2, \ldots, X_{31}$ on the number of patients arriving at the emergency department. Suppose that the observed data for 31 days are

20	21	16	18	25	22	14	19	13	25	23	24	19	29	19	15
15	21	22	21	31	22	24	16	17	28	25	26	24	16	17	

Distinguish between an estimator for λ and an estimate for λ.

Solution: The estimator for λ is the sample mean,

$$\hat{\lambda} = \frac{1}{31} \sum_{i=1}^{31} X_i.$$

This is the rule for what to do with the data to form an estimate. The perspective is *before* data are collected. Before data are collected, we say we should take the average and use it to estimate λ.

The estimate for λ using this particular random sample of data is

$$\hat{\lambda} = \frac{1}{31} \sum_{i=1}^{31} x_i = \frac{1}{31}(20 + 21 + 16 + \cdots + 16 + 17) \approx 20.87.$$

The perspective here is *after* we have collected and recorded the data.   ■■■

An estimator is always a random variable; an estimate is always a number. We will, however, use the "hat" notation, for example $\hat{\lambda}$, to indicate both. The context will determine whether $\hat{\lambda}$ refers to an estimator or an estimate.

9.2 Estimation Criteria

We saw in Example 9.1 from the previous section that there can be more than one reasonable estimator for a parameter. There, the sample variance S^2 and $\hat{\sigma}^2 = \frac{1}{n} \sum_{i=1}^{n} (X_i - \overline{X})^2$ both seem intuitive and reasonable. In this section we define criteria that can be used to compare various estimators.

9.2.1 Unbiased Estimators

Let's begin with the concept of *bias*.

> **Definition 9.2.1 — Unbiased Estimator.** Let $\hat{\theta}$ be some estimator (a random variable) of the parameter θ (a fixed but unknown number that is a property of the population). We say that $\hat{\theta}$ is an *unbiased estimator* if
>
> $$\mathbb{E}(\hat{\theta}) = \theta. \tag{9.1}$$
>
> In other words, an estimator is unbiased if its expectation is equal to the parameter it estimates. An estimator is biased if it is not unbiased. The bias of an estimator $\hat{\theta}$ is defined to be
>
> $$\text{Bias}(\hat{\theta}) = \mathbb{E}(\hat{\theta}) - \theta.$$

Let's take a look at some familiar estimators and see whether they are unbiased.

■ **Example 9.5** Suppose we observe n Bernoulli trials with probability θ of a success on any trial. Let X denote the number of successes. The usual estimator for θ is $\hat{\theta} = X/n$. Is $\hat{\theta}$ unbiased?

Solution: The expected value of the estimator is

$$\mathbb{E}\left(\hat{\theta}\right) = \mathbb{E}\left(\frac{X}{n}\right) = \frac{1}{n}\mathbb{E}(X) = \frac{1}{n}(n\theta) = \theta, \tag{9.2}$$

so the estimator $\hat{\theta}$ is unbiased.   ■■■

■ **Example 9.6** Suppose now that we have a sample $X_1, X_2, \ldots, X_n$ from the $N(\mu, \sigma^2)$ distribution. The usual estimators for μ and σ^2 are

$$\hat{\mu} = \overline{X} = \frac{1}{n} \sum_{i=1}^{n} X_i$$

$$\hat{\sigma}^2 = S^2 = \frac{1}{n-1} \sum_{i=1}^{n} (X_i - \overline{X})^2.$$

Are these estimators unbiased?

Solution: Determining whether $\hat{\mu}$ is unbiased is rather straightforward; from Theorem 8.3.1 we know that

$$\mathbb{E}(\hat{\mu}) = \mathbb{E}(\overline{X}) = \mu,$$

so we know that $\hat{\mu} = \overline{X}$ is unbiased. Now, recall from Theorem 8.4.1 that

$$\frac{(n-1)S^2}{\sigma^2} \sim \chi^2(n-1). \tag{9.3}$$

Since the expected value of a chi-square random variable is equal to its degrees of freedom, we can take the expected value on both sides of this last result to obtain successively

$$\mathbb{E}\left(\frac{(n-1)S^2}{\sigma^2}\right) = n-1$$

$$\frac{n-1}{\sigma^2} \mathbb{E}(S^2) = n-1$$

$$\mathbb{E}(S^2) = \sigma^2. \tag{9.4}$$

Thus, S^2 is also unbiased. This also means the estimator $\hat{\sigma}_0^2$ is biased and the bias is $-\sigma^2/n$. This is the justification usually given for why the divisor is $n-1$, rather than n, in the formula for the sample variance. ■■■

■ **Example 9.7** Suppose we have a sample $X_1, X_2, \ldots, X_n$ from the $N(\mu, \sigma^2)$ distribution. Is $S = \sqrt{\frac{1}{n-1} \sum_{i=1}^{n} (X_i - \overline{X})^2}$ an unbiased estimator of σ?

Solution: To see whether the estimator is unbiased we take the expectation and see whether it is σ:

$$\mathbb{E}(S) = \mathbb{E}\left(\frac{\sigma}{\sqrt{n-1}} \sqrt{\frac{(n-1)S^2}{\sigma^2}}\right) = \frac{\sigma}{\sqrt{n-1}} \mathbb{E}(\sqrt{Y}),$$

where the random variable Y has a $\chi^2(n-1)$ distribution. This expectation can be written as

$$\mathbb{E}(S) = \frac{\sigma}{\sqrt{n-1}} \mathbb{E}(\sqrt{Y}) = \frac{\sigma}{\sqrt{n-1}} \int_0^\infty \sqrt{y} \, \frac{y^{(n-1)/2-1}}{2^{(n-1)/2} \Gamma(\frac{n-1}{2})} \exp(-y/2) \, dy.$$

This is a formidable integral, but it can be evaluated exactly using a computer algebra system such as *Mathematica*™. This integral is found to be

$$\mathbb{E}(S) = \frac{\sigma}{\sqrt{n-1}} \frac{\sqrt{2}\Gamma(n/2)}{\Gamma((n-1)/2)}.$$

This is never equal to σ, and it can be shown that it is always less than σ. It is, however, true that

$$\lim_{n\to\infty} \mathbb{E}(S) = \lim_{n\to\infty} \frac{\sigma}{\sqrt{n-1}} \frac{\sqrt{2}\Gamma(n/2)}{\Gamma((n-1)/2)} = \sigma.$$

Thus, the estimator $\hat{\sigma}$ is biased, and the bias $\mathbb{E}(S) - \sigma$ is always negative. In other words, the sample standard deviation tends to underestimate the population standard deviation just a little bit, and the amount of this bias tends to 0 as the sample size tends to infinity.   ■■■

It may seem odd that S^2 is an unbiased estimator of σ^2, but S is a biased estimate of σ. This occurs because the relationship between S^2 and S is nonlinear. If we have two parameters along with two estimators that are linearly related, say $\theta_0 = 2\theta$ and $\hat{\theta}_0 = 2\hat{\theta}$, if one is unbiased then the other is unbiased. This is because of some of the properties of expectations; for example,

$$\mathbb{E}(\hat{\theta}) = \mathbb{E}(2\hat{\theta}) = 2\mathbb{E}(\hat{\theta}) = 2\theta.$$

Nonlinear functions do not share the property. For a sample from the normal distribution,

$$\mathbb{E}\left(\sqrt{S^2}\right) \neq \sqrt{\mathbb{E}(S^2)}.$$

Thus, except for some odd circumstances that rarely arise in applications, nonlinear functions of unbiased estimators are biased.

Exercise 9.2 Refer to Example 9.4, where we had a sample from the Poisson distribution with unknown mean λ. (a) Is $\overline{X}$ an unbiased estimator of λ? (b) Is $\sqrt{\overline{X}}$ an unbiased estimator of the population standard deviation?

Answer (a) Yes; (b) No.

■ **Example 9.8** Suppose we have a sample $X_1, X_2, \ldots, X_n$ from the $N(\mu, \sigma^2)$ distribution. Is

$$\hat{\phi} = \frac{1}{n}\sum_{i=1}^{n} X_i^2$$

an unbiased estimator of the population parameter $\phi = E(X^2)$?

Solution: We can write

$$\mathbb{E}(\hat{\phi}) = \mathbb{E}\left(\frac{1}{n}\sum_{i=1}^{n} X_i^2\right) = \frac{1}{n}\sum_{i=1}^{n} \mathbb{E}\left(X_i^2\right) = \frac{1}{n}\sum_{i=1}^{n} \phi = \frac{1}{n}\times n\phi = \phi$$

and conclude that $\hat{\phi} = \sum_{i=1}^{n} X_i^2/n$ is an unbiased estimator of ϕ.   ■■■

9.2.2 Consistent Estimators

In Definition 8.8.1 we introduced the concept of convergence in probability. We restate this here but in the context where the random variables are estimators and n is the sample size on which the estimator is based.

Definition 9.2.2 — Consistent Estimator. The infinite sequence of estimators $\hat{\theta}_1, \hat{\theta}_2, \ldots$ *converges in probability* to the parameter θ if for every $\varepsilon > 0$,

$$\lim_{n \to \infty} \Pr\left(\left|\hat{\theta}_n - \theta\right| < \varepsilon\right) = 1, \tag{9.5}$$

in which case we write $\hat{\theta}_n \overset{p}{\to} \theta$ and say that $\hat{\theta}_n$ is a *consistent estimator* of θ.

The weak law of large numbers, Theorem 8.8.2, guarantees that a sample average will be a consistent estimator of the population mean μ. We restate this theorem here in the context of estimators for μ.

Theorem 9.2.1 — $\overline{X}$ is a consistent estimator of the population mean μ. Suppose $X_1, X_2, \ldots, X_n$ are i.i.d. from a distribution with mean μ and variance σ^2. Let $\hat{\mu} = \overline{X}$ denote the mean of this sample of size n, which serves as the estimator for μ. Then for every $\varepsilon > 0$,

$$\lim_{n \to \infty} \Pr\left(\left|\overline{X} - \mu\right| < \varepsilon\right) = 1. \tag{9.6}$$

In other words, $\hat{\mu}$ is a consistent estimator of μ.

Roughly speaking, being a consistent estimator guarantees that the estimator can be arbitrarily close to the true parameter value with high probability as long as we take a large enough sample. This property holds in most realistic situations.

■ **Example 9.9** Explain why the estimators in the following cases are consistent.
   (a) $X_1, X_2, \ldots, X_n \sim$ i.i.d. $\mathrm{EXP1}(\theta)$ and $\hat{\theta} = \overline{X}$
   (b) $X_1, X_2, \ldots, X_n \sim$ i.i.d. $\mathrm{POISSON}(\lambda)$ and $\hat{\lambda} = \overline{X}$
   (c) $X \sim \mathrm{BIN}(n, \theta)$ and $\hat{\theta} = X/n$

Solution: In the first two cases, the parameter (θ in the case of the exponential distribution and λ in the case of the Poisson distribution) is equal to the population mean. By the weak law of large numbers, the sample mean $\overline{X}$ converges to the population mean, so both estimators are consistent. In the third case, we could think of the random variable X, the number of successes in n Bernoulli trials, as being the sum of $Y_1, Y_2, \ldots, Y_n$, where

$$Y_i = \begin{cases} 1, & \text{if the } i\text{th trial leads to a success} \\ 0, & \text{if the } i\text{th trial leads to a failure.} \end{cases}$$

Thus, $X = \sum_{i=1}^{n} Y_i$ and $\hat{\theta} = X/n = \overline{Y}$, so the estimator $\hat{\theta}$ is equal to the mean of a sample of size n from the $\mathrm{BIN}(1, \theta)$ distribution. By the weak law of large numbers, $\hat{\theta}$ is a consistent estimator.　　　■ ■ ■

The following theorem is useful in establishing that other unbiased estimators are consistent.

Theorem 9.2.2 — Condition that Guarantees a Consistent Estimator. Let $\hat{\theta}_n$ denote an estimator of a parameter θ based on a sample of size n. If, for every n, the estimator $\hat{\theta}_n$ is unbiased, and if

$$\lim_{n \to \infty} \mathbb{V}(\hat{\theta}_n) = 0,$$

then $\hat{\theta}_n$ is a consistent estimator.

The result of this theorem should be fairly intuitive. If the variance of an estimator goes to zero, then the distribution should shrink down to the point where it becomes a point mass. If the estimator is also unbiased, this point mass should equal the parameter being estimated.

■ **Example 9.10** Suppose again that we have a sample $X_1, X_2, \ldots, X_n$ from the $N(\mu, \sigma^2)$ distribution. Is S_n^2 a consistent estimator for σ^2?

Solution: We've established that S_n^2 is an unbiased estimator of σ^2. We must therefore establish the other part of Theorem 9.2.2 that requires the variance to converge to zero. Since $(n-1)S_n^2/\sigma^2 \sim \chi^2(n-1)$, we can write

$$
\begin{aligned}
\lim_{n\to\infty} \mathbb{V}(S_n^2) &= \lim_{n\to\infty} \mathbb{V}\left(\frac{\sigma^2}{n-1} \frac{(n-1)S_n^2}{\sigma^2} \right) \\
&= \lim_{n\to\infty} \left(\frac{\sigma^2}{n-1} \right)^2 \mathbb{V}\left(\frac{(n-1)S_n^2}{\sigma^2} \right) \\
&= \lim_{n\to\infty} \left(\frac{\sigma^2}{n-1} \right)^2 2(n-1) \\
&= \lim_{n\to\infty} \frac{2\sigma^4}{n-1} \\
&= 0.
\end{aligned}
$$

Thus, S_n^2 is a consistent estimator of σ^2. ■■■

The following theorem regarding convergence in probability is useful in establishing the consistency of estimators. We will use it in one example in this section, and again in the next section.

Theorem 9.2.3 — Properties of Convergence in Probability. If $X_n \xrightarrow{\text{P}} c$ and $Y_n \xrightarrow{\text{P}} d$, then
1. $(X_n + Y_n) \xrightarrow{\text{P}} (c+d)$;
2. $(X_n \times Y_n) \xrightarrow{\text{P}} c \times d$;
3. $(X_n/Y_n) \xrightarrow{\text{P}} c/d$ so long as $d \neq 0$.
4. If $g()$ is a real-valued function that is continuous at c, then $g(X_n) \xrightarrow{\text{P}} g(c)$.

Theorem 9.2.3 applies if one of the sequences of random variables, say Y_n, is a sequence of constants.

■ **Example 9.11** Suppose once again that we have a sample $X_1, X_2, \ldots, X_n$ from the $N(\mu, \sigma^2)$ distribution. Is the sample standard deviation S_n a consistent estimator of σ?

Solution: Let $\hat{\sigma}_n = S_n$, the sample standard deviation based on a sample of size n. Since S_n is a biased estimator of σ, we can't apply Theorem 9.2.2; we need to be a little more clever. Let's define $g(y) = \sqrt{y}$ and write

$$
S_n = \sqrt{S_n^2} = g(S_n^2).
$$

Since g is a continuous function, and $S_n^2 \xrightarrow{\text{P}} \sigma^2$, we can apply part 4 of Theorem 9.2.3 to conclude

$$
S_n = g(S_n^2) \xrightarrow{\text{P}} g(\sigma^2) = \sigma.
$$

Thus, S_n is a consistent estimator of σ. ■■■

9.3 Method of Moments

When intuitive methods fail, a systematic method is needed to find good point estimates of parameters. In this section we describe the *method of moments* (MOM) and in the next we describe the *method of maximum likelihood.* The method of moments is quite intuitive, but it has limitations. First, it applies only to cases where we have a random sample from some distribution. This excludes important cases such as regression, where we estimate the relationships among variables. Despite being an intuitive approach to parameter estimation, method of moments estimators are usually not very efficient, especially in more complicated situations.

We introduced the idea of moments of a distribution (or, more precisely, the moments of a random variable having that distribution) in Section 6.9. The *k*th *sample moment* is

$$m'_k = \frac{1}{n} \sum_{i=1}^{n} X_i^k.$$

Note that the population moment is a parameter because it is a characteristic of the population or distribution. On the other hand, the sample moment is a statistic because it is a characteristic of the sample. If there are *p* unknown parameters, the method of moments can be described simply as

> Equate the first *p* population moments with the corresponding sample moments and solve for the parameters.

In other words, select as estimates for the parameters those values for which the population and sample moments agree. For a one-parameter distribution, this is particularly simple: Take the estimate of the population mean to be the sample mean.

■ **Example 9.12** Suppose we observe *n* Bernoulli trials with unknown probability θ of success. The observed variable is the number of successes, which of course has a $\text{BIN}(n, \theta)$ distribution (when viewed before any data are collected). Find the MOM of θ.

Solution: Here we observe just a single random variable, X, so the sample moment is simply

$$\frac{1}{1} \sum_{i=1}^{1} X = X.$$

The first population moment is the expected value

$$\mathbb{E}(X) = np.$$

Equating these yields

$$X = np,$$

which has the solution

$$\hat{p} = \frac{X}{n}.$$

On the last line we use a hat above the parameter *p* to indicate that it is an estimate of *p*, not the parameter *p* itself as is convention.   ■ ■ ■

■ **Example 9.13** Suppose $X_1, X_2, \ldots, X_n \sim$ i.i.d. $N(\mu, \sigma^2)$. Find the MOM estimators for μ and σ^2.

Solution: The first population moment of the normal distribution is $\mu'_1 = \mu$ and the first sample moment is $m'_1 = \bar{x}$. Equating these gives the estimator for μ to be

$$\hat{\mu} = \overline{X}.$$

The second population moment of the normal distribution is related to the variance. Recall that the variance is

$$\sigma^2 = E[(X-\mu)^2] = \mathbb{E}(X^2) - \mu^2,$$

so

$$\mathbb{E}(X^2) = \mu_2' = \sigma^2 + \mu^2.$$

The second sample moment is

$$m_2' = \frac{1}{n}\sum_{i=1}^{n} X_i^2.$$

Equating these gives

$$\sigma^2 + \mu^2 = \frac{1}{n}\sum_{i=1}^{n} X_i^2.$$

We can now substitute the estimate for μ to obtain

$$\hat{\sigma}^2_{\text{MOM}} = \frac{1}{n}\sum_{i=1}^{n} X_i^2 - \hat{\mu}^2 = \frac{1}{n}\sum_{i=1}^{n} X_i^2 - \overline{X}^2 = \frac{1}{n}\sum_{i=1}^{n}(X_i - \overline{X})^2. \qquad \blacksquare\blacksquare\blacksquare$$

The MOM estimator for the variance is not quite the sample variance, because the divisor is n rather than the usual $n-1$. For large n there is very little difference between the two since they are related by

$$\hat{\sigma}^2_{\text{MOM}} = \frac{n-1}{n}\left(\frac{1}{n-1}\sum_{i=1}^{n}(X_i - \overline{X})^2\right) = \frac{n-1}{n}S^2.$$

Thus, for a two-parameter distribution the MOM estimates are obtained by matching the population mean and variance with the sample mean and variance, so long as the sample variance is interpreted as above, with a divisor of n rather than $n-1$.

■ **Example 9.14** Suppose we place electrical items on test, and we observe how many weeks until they fail. If the observed lifetimes are 12, 17, 40, 71, and if we assume the gamma distribution $\text{GAM}(\alpha,\beta)$, find the MOMs of the parameters and estimate the mean lifetime for the population.

Solution: The population moments are obtained from Theorem 6.5.1:

$$\mu_1' = \alpha\beta$$
$$\mu_2' = \mathbb{V}(X) + \mathbb{E}(X)^2 = \alpha\beta^2 + \alpha^2\beta^2.$$

The sample moments are

$$m_1' = \bar{x}$$
$$m_2' = \frac{1}{n}\sum_{i=1}^{n} x_i^2.$$

Equating these yields the two equations in two unknowns, specified by

$$\alpha\beta = \bar{x}$$

$$\alpha\beta^2 + \alpha^2\beta^2 = \frac{1}{n}\sum_{i=1}^{n} x_i^2.$$

The first equation can be solved for α in terms of β to be

$$\alpha = \frac{\bar{x}}{\beta},$$

which can then be substituted into the second equation to yield

$$\frac{\bar{x}}{\beta}\beta^2 + \frac{\bar{x}^2}{\beta^2}\beta^2 = \frac{1}{n}\sum_{i=1}^{n} x_i^2.$$

After some simplification and a bit of rearranging, we find that

$$\beta = \frac{\frac{1}{n}\left(\sum_{i=1}^{n} x_i^2 - n\bar{x}^2\right)}{\bar{x}}.$$

If we define

$$S_0^2 = \frac{1}{n}\sum_{i=1}^{n}(x_i - \bar{x})^2 = \frac{n-1}{n}S^2,$$

where S^2 is the usual sample variance, then we can write the MOM estimator for β as

$$\hat{\beta} = \frac{S_0^2}{\bar{x}}.$$

Substituting this into the equation for α yields

$$\hat{\alpha} = \frac{\bar{x}}{\hat{\beta}} = \frac{\bar{x}^2}{S_0^2}.$$

The following R code can be used to compute the MOM estimates for the given data:

```
x = c( 12 , 17 , 40 , 71 )
m1 = mean( x )
m2 = sum( x^2 ) / length(x)
alphahat = m1^2 / ( m2 - m1^2 )
betahat = ( m2 - m1^2 ) / m1
print( c( alphahat , betahat ) )
print( alphahat*betahat )
```

which yields

```
[1]  2.25391 15.52857
[1] 35
```

The parameter estimates are then

$$\hat{\alpha} = 2.25391 \quad \text{and} \quad \hat{\beta} = 15.52857.$$

The estimated mean life is

$$\hat{\mu} = \hat{\alpha}\hat{\beta} = 35.$$

■ ■ ■

■ **Example 9.15** Are the MOM estimators from Example 9.14 consistent?

Solution: By the weak law of large numbers,

$$m_1' = \overline{X}_n = \frac{1}{n}\sum_{i=1}^{n} X_i \xrightarrow{\text{P}} \mu = \alpha\beta.$$

By applying the fourth part of Theorem 9.2.3, we see that

$$m_2' = \frac{1}{n}\sum_{i=1}^{n} X_i^2 \xrightarrow{\text{P}} \mathbb{E}(X^2) = \alpha\beta^2 + \alpha^2\beta^2.$$

The MOM estimators of α and β are

$$\hat{\alpha} = \frac{m_2' - (m_1')^2}{m_1'}$$

and

$$\hat{\beta} = \frac{m_1'}{m_2' - (m_1')^2}.$$

We can now apply the various parts of Theorem 9.2.3 to conclude that

$$\hat{\alpha} = \frac{(m_1')^2}{m_2' - (m_1')^2} \xrightarrow{\text{P}} = \frac{(\mu_1')^2}{\mu_2' - (\mu_1')^2} = \frac{(\alpha\beta)^2}{(\alpha\beta^2 + \alpha^2\beta^2) - (\alpha\beta)^2} = \alpha.$$

Also,

$$\hat{\beta} = \frac{m_2' - (m_1')^2}{m_1'} \xrightarrow{\text{P}} \frac{\mu_2' - (\mu_1')^2}{\mu_1'} = \frac{(\alpha\beta^2 + \alpha^2\beta^2) - (\alpha\beta)^2}{\alpha\beta} = \beta.$$

Both estimators converge in probability to the parameters they estimate. Thus, both estimators are consistent. ■ ■ ■

It can be shown that under some reasonable conditions, the MOM estimators will be consistent. Consistency, however, is the least we expect from an estimator. All it says is that if the sample size is sufficiently large, the estimator will tend to be "close" to the parameter it estimates. In small sample sizes the MOM estimators are not that efficient. In the next section we present a method that (1) works more generally (i.e., it is not restricted to the i.i.d. case), and (2) is more efficient in the sense of providing estimators that tend to be closer to the parameter they estimate.

> **Exercise 9.3** Suppose $X_1, X_2, \ldots, X_n \sim$ i.i.d. POIS(λ). Find the MOM estimator for λ. Is this estimator unbiased? Is it consistent?

9.4 Maximum Likelihood

The method of maximum likelihood is a general method for estimating parameters that can be applied in nearly any circumstance. We begin by recalling the definition of the joint PDF or PMF for all of the observed random variables. Often, but not always, our observations are assumed to be independent and to come from some distribution with unknown parameters. For example, suppose $X_1, X_2, \ldots, X_n \sim$ i.i.d.

with PDF or PMF $f(x|\theta)$. Here, θ could be a vector of parameters or it could be a scalar; for now we assume it is a scalar. The joint PDF or PMF of all observations[1] $\boldsymbol{X} = [X_1, X_2, \ldots, X_n]$ is therefore

$$f(\boldsymbol{x}|\theta) = \prod_{i=1}^{n} f(x_i|\theta). \tag{9.7}$$

In developing this result we have assumed that θ is some *fixed* but *unknown* parameter (a number) and the variables are $x_1, x_2, \ldots, x_n$.

Let's reorient our thinking and look at the variables in (9.7) differently. Let's assume that $x_1, x_2, \ldots, x_n$ have already been observed, and are therefore known numbers, and let's think of θ as the variable. The likelihood is then defined to be

$$L(\theta|\boldsymbol{x}) = f(\boldsymbol{x}|\theta). \tag{9.8}$$

The likelihood is therefore the same function as the joint PDF of the observed values; it's just that our perspective is different. In the likelihood function, we think of the x_is as fixed so the likelihood is a function of the parameter θ.

The principle of maximum likelihood is then this:

Select as the estimator for θ the value of θ that maximizes the likelihood function.

The estimator obtained this way is called the maximum likelihood estimator (MLE).

■ **Example 9.16** Consider the setup of Example 9.12 where we observe n Bernoulli trials with unknown probability θ of success. The number of successes has a $\text{BIN}(n, \theta)$ distribution (when viewed before any data are collected). Find the MLE of θ.

Solution: The PMF of X is

$$f(x|\theta) = \binom{n}{x} \theta^x (1-\theta)^{n-x}, \quad x = 0, 1, \ldots, n. \tag{9.9}$$

The likelihood function is therefore

$$L(\theta|x) = \binom{n}{x} \theta^x (1-\theta)^{n-x}, \qquad 0 \le \theta \le 1. \tag{9.10}$$

Finding the maximum of a function is a classic problem in calculus. We proceed by taking the derivative with respect to θ, setting the result equal to zero, and solving. The derivative requires the use of the product rule and the chain rule and yields

$$\begin{aligned}
\frac{dL}{d\theta} &= \binom{n}{x} \left[\theta^x (n-x)(1-\theta)^{n-x-1}(-1) + x\theta^{x-1}(1-\theta)^{n-x} \right] \\
&= \binom{n}{x} \theta^{x-1}(1-\theta)^{n-x-1} \left[-(n-x)\theta + x(1-\theta) \right] \\
&= \binom{n}{x} \theta^{x-1}(1-\theta)^{n-x-1} \left[-n\theta + x \right].
\end{aligned}$$

---

[1]Throughout the rest of the book, we use boldface to indicate a vector. For example, the vector of random variables will be indicated $\boldsymbol{X} = [X_1, X_2, \ldots, X_n]$ and the vector of observed values will be indicated $\boldsymbol{x} = [x_1, x_2, \ldots, x_n]$.

We then set this result equal to zero and solve for θ:

$$\binom{n}{x} \theta^{x-1}(1-\theta)^{n-x-1}[-n\theta + x] = 0.$$

We can divide both sides by $\binom{n}{x} \theta^{x-1}(1-\theta)^{n-x-1}$ since this can never be zero. This yields

$$x - n\theta = 0,$$

which has the solution

$$\hat{\theta} = \frac{x}{n}.$$

We leave it as an exercise to show that $d^2L/d\theta^2|_{\hat{\theta}} < 0$, assuring that we have found a maximum. In this case, the MLE of the probability of success for Bernoulli trials is equal to the proportion of successes that were observed, which is also the intuitive estimator in Example 9.2. It is also the MOM estimator from Example 9.12. ∎∎∎

Look carefully at the notation. The final estimator is written $\hat{\theta}$, with the "hat" above θ indicating that it is an estimate. We typically carry out the algebra writing simply θ and put the "hat" on it only at the end.

Having a zero derivative does not always imply that a function achieves a maximum. The function could actually achieve a minimum (consider, for example, the function $f(\theta) = \theta^2$) or it could achieve neither (consider the function $f(\theta) = \theta^3$). We can apply the second derivative test to demonstrate that the derivative of the likelihood function yields a solution that maximizes a function. Recall that if the second derivative of a function evaluated at a stationary point (a point where the derivative is zero) is negative, then a maximum is guaranteed. Showing that the likelihood function is maximized is easier when we work with the log likelihood, which we describe next.

The required calculus and algebra can often be simplified if we work with the (natural) logarithm of the likelihood function, denoted $\ell(\theta|x)$. The log function is strictly increasing, which means that $a < b$ if and only if $\log a < \log b$. Thus, if $\hat{\theta}$ is the MLE of θ, then it maximizes L, which means

$$L(\hat{\theta}|x) > L(\theta|x)$$

for any other value of θ. By the increasing property of the logarithm function stated above, this implies that

$$\log L(\hat{\theta}|x) > \log L(\theta|x).$$

Thus, we can maximize either $L(\theta|x)$ or $\ell(\theta|X) = \log L(\theta|x)$, whichever is easier. Usually, maximizing the logarithm is easier. Figure 9.1 illustrates the likelihood and log likelihood functions for the case where $n = 100$ and $x = 72$; the maximum occurs at $\hat{\theta} = 0.72$. The curves look different, but they both reach their maximum at the value $\hat{\theta} = x/n = 0.72$. From now on, we will usually maximize the log likelihood function, rather than the likelihood function.

∎ **Example 9.17** For the setup described in Example 9.16, find the MLE of θ by maximizing the log likelihood function.

Solution: The log likelihood function is

$$\ell(\theta|x) = \log \binom{n}{x} + x\log\theta + (n-x)\log(1-\theta).$$

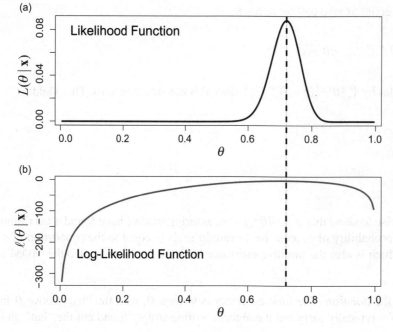

Figure 9.1 Likelihood function (a) and log likelihood function (b) for the binomial example. Both functions yield a maximum at the same value of θ, which in this case is $\hat{\theta} = 0.72$.

Differentiating with respect to θ yields

$$\frac{d\ell}{d\theta} = \frac{x}{\theta} + \frac{n-x}{1-\theta}(-1).$$

Setting this equal to zero yields

$$\frac{x}{\theta} - \frac{n-x}{1-\theta} = 0,$$

which has the solution

$$\hat{\theta} = \frac{x}{n}.$$

This is the same solution as before. The second derivative of the log likelihood function is

$$\frac{d^2\ell}{d\theta^2} = -\frac{x}{\theta^2} - \frac{n-x}{(1-\theta)^2},$$

which is always negative, implying that we have achieved a maximum.   ∎∎∎

The method of maximum likelihood finds the value of the unknown parameter that is most likely, in the sense of having the highest likelihood or log likelihood. Usually, very large or very small values of the parameter make the likelihood function (or the log likelihood function) very small. Often an inspection of the likelihood function will indicate that $L(\theta) \to 0$ as $\theta \to \infty$ or $\theta \to -\infty$. If this is the case for a nonnegative function (which the likelihood function is), and if there is a single stationary point, then the stationary point yields a maximum.

∎ **Example 9.18** Suppose, as in Example 9.1, we observe daily counts of the number of patients presenting at an emergency department at a hospital with influenza-like illness (ILI); that is, they have symptoms of the flu, but have not been tested. For the daily counts $X_1, X_2, \ldots, X_n$, what is a reasonable distribution for these data? Find the MLE of the parameter(s) for this distribution.

Solution: As we discussed in Section 5.6, the Poisson is often a good model for the number of observed counts in a fixed period of time. The PMF for the Poisson is

$$f(x|\lambda) = \frac{\lambda^x \exp(-\lambda)}{x!}, \qquad x = 0, 1, \ldots.$$

If we make the assumption that the patient counts for each day have a POI(λ) distribution, then the likelihood function is

$$L(\lambda|\boldsymbol{x}) = \prod_{i=1}^{n} \frac{\lambda^{x_i} \exp(-\lambda)}{x_i!} = \frac{\lambda^{\sum_{i=1}^{n} x_i} \exp(-n\lambda)}{\prod_{i=1}^{n} x_i!},$$

from which we obtain the log likelihood function

$$\ell(\lambda|\boldsymbol{x}) = \left(\sum_{i=1}^{n} x_i \right) \log \lambda - n\lambda - \sum_{i=1}^{n} \log x_i!.$$

If we differentiate with respect to λ and set the result equal to zero, we obtain

$$\frac{d\ell}{d\lambda} = \frac{\sum_{i=1}^{n} x_i}{\lambda} - n = 0,$$

which yields the solution

$$\hat{\lambda} = \frac{\sum_{i=1}^{n} x_i}{n} = \bar{x}.$$

Thus, the estimate for λ is the sample mean $\bar{x}$. This shouldn't be surprising. The mean of the Poisson distribution is λ, so to estimate the mean of a distribution we use the mean of the sample. We leave it to the reader to take the second derivative of the log likelihood and show that it is negative. ■■■

Perhaps the most important distribution in statistics is the normal, which is a two-parameter distribution. When there is more than a single parameter we take the partial derivatives[2] with respect to each parameter and set the result equal to zero. Thus, if there are k parameters, there will be k equations in k unknowns. In the next example we find the MLEs of the parameters of the normal distribution.

■ **Example 9.19** Suppose $X_1, X_2, \ldots, X_n \sim$ i.i.d. $N(\mu, \sigma^2)$. Find the MLEs for μ and σ^2.

Solution: We begin by letting $\tau = \sigma^2$ since it is a bit awkward to take a derivative with respect to σ^2 (it almost looks like a second derivative with the 2 in the exponent). The likelihood function is

$$L(\mu, \tau|\boldsymbol{x}) = \prod_{i=1}^{n} \frac{1}{\sqrt{2\pi\tau}} \exp\left((x_i - \mu)^2/2\tau\right) = \frac{1}{(2\pi)^{n/2} \tau^{n/2}} \exp\left(\sum_{i=1}^{n} (x_i - \mu)^2/2\tau\right)$$

[2]The partial derivative of a function $f(x,y)$ with respect to x is obtained by treating y as a constant and applying the usual rules for derivatives. Similarly, for the partial with respect to y we hold x constant. The notation for partial derivatives uses the "partial" symbol ∂ instead of the d that is used for a single variable.

and so the log likelihood function is

$$\ell(\mu, \tau | \boldsymbol{x}) = -\frac{n}{2} \log 2\pi - \frac{n}{2} \log \tau - \frac{1}{2\tau} \sum_{i=1}^{n} (x_i - \mu)^2.$$

Taking partial derivatives yields

$$\frac{\partial \ell}{\partial \mu} = 0 - \frac{1}{2\tau} \sum_{i=1}^{n} 2(x_i - \mu)(-1)$$

$$\frac{\partial \ell}{\partial \tau} = -\frac{n}{2\tau} + \frac{1}{2\tau^2} \sum_{i=1}^{n} (x_i - \mu)^2.$$

Setting the first partial derivative equal to zero yields

$$-2 \sum_{i=1}^{n} (x_i - \mu) = 0$$

$$\sum_{i=1}^{n} x_i - n\mu = 0$$

$$\mu = \bar{x}.$$

Thus, the first equation can be solved for the parameter μ.[3] The equation obtained by setting the partial derivative with respect to τ equal to zero can be written as

$$-\frac{n}{2\tau} + \frac{1}{2\tau^2} \sum_{i=1}^{n} (x_i - \mu)^2 = 0$$

$$\frac{2\tau^2}{n} \left[-\frac{n}{2\tau} + \frac{1}{2\tau^2} \sum_{i=1}^{n} (x_i - \mu)^2 \right] = \frac{2\tau^2}{n} \times 0$$

$$\tau = \frac{1}{n} \sum_{i=1}^{n} (x_i - \mu)^2.$$

Substituting the solution from the first equation gives the MLEs for μ and $\sigma^2 = \tau$:

$$\hat{\mu} = \bar{x}$$

$$\hat{\sigma}^2 = \hat{\tau} = \frac{1}{n} \sum_{i=1}^{n} (x_i - \hat{\mu})^2.$$

The result for $\hat{\mu}$ is as expected, but the result for $\hat{\sigma}$ is a bit surprising since the divisor is n rather than the $n-1$ that is found in the sample variance:

$$S^2 = \frac{1}{n-1} \sum_{i=1}^{n} (X_i - \hat{\mu})^2.$$

For large values of n the difference will be negligible. The two estimators of σ^2 are related by

$$S^2 = \frac{n}{n-1} \hat{\sigma}^2.$$

---

[3]It is rather unusual that we are able to solve one equation for a variable. Usually, we obtain two equations in two unknowns and we have to solve one equation for one variable and substitute this into the other equation.

Recall the difference between an estimator and an estimate. An estimator is the function of random variables that we use to estimate a parameter; an estimator is therefore a *random variable*. Once we collect data and substitute these values into the formula for the estimator, we end up with a number that estimates the parameter; an estimate is a *number*, not a random variable.

The method of maximum likelihood is well suited when we would like to estimate some function of the model parameters. The following result says that if we want to obtain the MLE of some function, say $g(\boldsymbol{\theta})$, we simply find the MLE of $\boldsymbol{\theta}$ and substitute it into the function g. This is called the *invariance property* of the MLE.

Theorem 9.4.1 — Invariance Property of the MLE. If $g(\boldsymbol{\theta})$ is a function of $\boldsymbol{\theta}$ and $\hat{\boldsymbol{\theta}}$ is the MLE of $\boldsymbol{\theta}$, then the MLE of $g(\boldsymbol{\theta})$ is $g(\hat{\boldsymbol{\theta}})$.

■ **Example 9.20** For the case of a sample from a normal distribution with unknown mean and unknown variance, find the MLEs of (a) the standard deviation σ and (b) the parameter $\phi = \mathbb{E}(X^2)$.

Solution: (a) Since $\sigma = \sqrt{\sigma^2}$, the invariance property implies that the MLE of σ is

$$\hat{\sigma} = \sqrt{\hat{\sigma}^2} = \sqrt{\frac{1}{n}\sum_{i=1}^{n}(X_i - \bar{x})^2}.$$

(b) The parameter ϕ can be written as

$$\phi = \mathbb{E}(X^2) = \mathbb{V}(X) + \mathbb{E}(X) = \sigma^2 + \mu^2.$$

The MLE of ϕ is therefore

$$\hat{\phi} = \hat{\sigma}^2 + \hat{\mu}^2 = \frac{1}{n}\sum_{i=1}^{n}(x_i - \bar{x})^2 - \bar{x}^2 = \frac{1}{n}\sum_{i=1}^{n}x_i^2.$$

■■■

9.5 Approximating MLEs

For all of the examples in the previous section we were able to use algebra to solve for the parameters once we took the first derivatives. This led us to formulas for the MLEs. We are not always this lucky. Often, the equations obtained by equating the first derivatives to zero are nonlinear and cannot be solved using ordinary algebra. For these cases, we must rely on a numerical approximation to the solution.

We begin with an interesting, but rather old,[4] example from genetics.

■ **Example 9.21** Genetic theory suggests that an organism will fall into one of four categories, which have probabilities

$$\frac{1}{2} + \frac{\theta}{4}, \quad \frac{1}{4}(1 - \theta), \quad \frac{1}{4}(1 - \theta), \quad \frac{1}{4}\theta.$$

---

[4]This example was first published by Fisher and Balmukand (1928).

Note that the probabilities sum to 1, as they must if each organism is to be put in one and only one category. A collection of 197 organisms was categorized, and the numbers that fell into these four categories were

$$125, \quad 18, \quad 20, \quad 34.$$

Estimate θ.

Solution: There is no obvious and intuitive way to proceed in this example, so we look to the method of maximum likelihood. Let $[X_1, X_2, X_3, X_4]$ be the numbers of animals out of the $n = 197$ that fall into each category. Then $[X_1, X_2, X_3, X_4]$ has the multinomial distribution

$$\text{MULT}\left(n, \left[\frac{2+\theta}{4}, \frac{1-\theta}{4}, \frac{1-\theta}{4}, \frac{\theta}{4}\right]\right)$$

Given data $x = [x_1, x_2, x_3, x_4]$, the likelihood function is

$$L(\theta|x) = \binom{n}{x_1\, x_2\, x_3\, x_4}\left(\frac{2+\theta}{4}\right)^{x_1}\left(\frac{1-\theta}{4}\right)^{x_2}\left(\frac{1-\theta}{4}\right)^{x_3}\left(\frac{\theta}{4}\right)^{x_4}.$$

The log likelihood function is therefore

$$\ell(\theta|x) = c + x_1\log\left(\frac{2+\theta}{4}\right) + x_2\log\left(\frac{1-\theta}{4}\right) + x_3\log\left(\frac{1-\theta}{4}\right) + x_4\log\left(\frac{\theta}{4}\right),$$

where the constant c is equal to the logarithm of the multinomial coefficient, which does not depend on the parameter θ. Differentiating with respect to θ is straightforward; setting this derivative equal to zero leads to

$$\ell'(\theta|x) = \frac{x_1}{\frac{2+\theta}{4}}\left(\frac{1}{4}\right) + \frac{x_2}{\frac{1-\theta}{4}}\left(-\frac{1}{4}\right) + \frac{x_3}{\frac{1-\theta}{4}}\left(-\frac{1}{4}\right) + \frac{x_4}{\frac{\theta}{4}}\left(\frac{1}{4}\right) = 0$$

$$\frac{x_1}{2+\theta} - \frac{x_2+x_3}{1-\theta} + \frac{x_4}{\theta} = 0.$$

Figure 9.2(a) shows the log likelihood function $\ell(\theta|x)$. Figure 9.2(b) shows the derivative $\ell'(\theta|x)$, suggesting that the solution of $\ell'(\theta|x) = 0$ is slightly to the right of 0.6. Figure 9.2(c) gives a "zoomed in" plot of $\ell'(\theta|x)$ from which we can approximate the solution to be 0.627. Thus, $\hat{\theta} \approx 0.627$.

If a more accurate approximation to the MLE is required, we could use the optim() function in R. The first argument to optim() is an initial guess[5] to the solution of $\ell'(\theta|x) = 0$. The second argument is the name of the function; in our case we want to solve the equation that sets the log likelihood function equal to zero. The third argument, which is optional, allows us to specify the method to be used to find the optimum. Here we have chosen the method BFGS, which is related to Newton's method, but doesn't require specification of the derivative. The R function optim finds a *minimum*, not a *maximum*. If we are searching for a maximum, we can apply optim() to the negative of the function we want to maximize. Type ?optim to see more details.

In the following R code we define the negative of the log likelihood function as 11, call the function optim() and print the estimated solution opt$par:

---

[5]Most optimization algorithms are iterative, meaning they start with an initial guess and successively get better estimates of the optimum until some convergence criterion is reached.

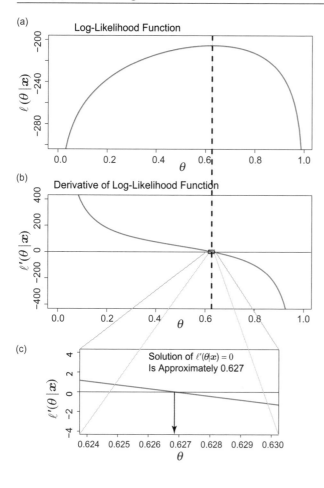

Figure 9.2 Log likelihood (a) for the multinomial genetics example, and derivative of log likelihood (b). The MLE occurs at the point where the $\ell(\theta|x)$ reaches a maximum, which is the same as the point where $\ell'(\theta|x) = 0$.

```
x = c(125, 18, 20, 34)
ll = function(theta) {
        z = x[1]*log(0.5+0.25*theta) + (x[2]+x[3])*log(0.25*(1-theta)) + x[4]*log(theta/4)
        return( - z )
}
opt = optim( 0.5 , ll, method = "BFGS" )
opt$par
```

This yields the output

```
[1] 0.626821
```

which is a more accurate estimate of the MLE.   ∎ ∎ ∎

In multiparameter distributions is it usually difficult or impossible to obtain closed-form expressions for the MLEs. Sometimes we are able to solve one equation for one variable, only to be left with one nonlinear equation in the other variable. This is what occurs for the Weibull distribution.

∎ **Example 9.22** Reconsider the problem we described in Example 9.14, where we placed electrical items on test and observed how many weeks until failure. The observed lifetimes are 12, 17, 40, 71. This time, let's assume the Weibull distribution $\mathrm{WEIB}(\theta, \beta)$. Find the MLEs of the parameters and estimate the mean lifetime for the population.

Solution: The likelihood function for an arbitrary sample of size n is

$$L(\theta, \beta | \mathbf{t}) = \prod_{i=1}^{n} \left(\frac{\beta}{\theta} \right) \left(\frac{t_i}{\theta} \right)^{\beta-1} \exp\left(- \left(\frac{t_i}{\theta} \right)^{\beta} \right) = \frac{\beta^n}{\theta^{n\beta}} \left(\prod_{i=1}^{n} t_i \right)^{\beta-1} \exp\left(-\sum_{i=1}^{n} \left(\frac{t_i}{\theta} \right)^{\beta} \right),$$

where $\mathbf{t} = [12, 17, 40, 71]$ is the vector of observed lifetimes. The log likelihood function is thus

$$\ell(\theta, \beta | \mathbf{t}) = n\log\beta - n\beta\log\theta + (\beta - 1)\sum_{i=1}^{n} \log t_i - \sum_{i=1}^{n} \left(\frac{t_i}{\theta} \right)^{\beta}.$$

If we differentiate with respect to θ and β, and set the results equal to zero, we obtain

$$\frac{\partial \ell}{\partial \theta} = -\frac{n\beta}{\theta} + \sum_{i=1}^{n} \beta\theta^{-\beta-1}t_i^{\beta} = 0$$

$$\frac{\partial \ell}{\partial \beta} = \frac{n}{\beta} - n\log\theta + \sum_{i=1}^{n} \log t_i - \sum_{i=1}^{n} \left(\frac{t_i}{\theta} \right)^{\beta} \log\frac{t_i}{\theta} = 0.$$

(9.11)

The first equation can be solved for θ in terms of β, yielding

$$\theta = \left(\frac{\sum_{i=1}^{n} t_i^{\beta}}{n} \right)^{1/\beta}.$$

(9.12)

If we were to substitute this result into the second equation, we would obtain (after much messy algebra omitted here)

$$\beta = \frac{n \sum_{i=1}^{n} t_i^{\beta}}{n \sum_{i=1}^{n} t_i^{\beta} \log t_i - \left(\sum_{i=1}^{n} \log t_i \right) \left(\sum_{i=1}^{n} t_i^{\beta} \right)}.$$

(9.13)

This is one equation in one variable, but it is nonlinear and it is not possible to find an analytic solution for β. Figure 9.3 shows the graph of the right side of (9.13) along with the graph of $y = \beta$. The MLE of β is the β coordinate of the intersection, which is approximately $\hat{\beta} = 1.5783$. Once the MLE of β is determined, the MLE of θ can be found from (9.12). This is found to be

$$\hat{\theta} = \left(\frac{\sum_{i=1}^{n} t_i^{\hat{\beta}}}{n} \right)^{1/\hat{\beta}} = 39.25.$$

A contour plot of the likelihood function, as a function of both β and θ, is shown in Figure 9.4. The dot inside the highest level curve is the estimate of the MLE.

The mean of the $\text{WEIB}(\theta, \beta)$ distribution is $\theta \times \Gamma(1 + 1/\beta)$, so using the invariance property of the MLE from Theorem 9.4.1, we find that the estimate of the mean lifetime is

$$\hat{\mu} = \hat{\theta} \times \Gamma\left(1 + \frac{1}{\hat{\beta}} \right) = 39.25 \times \Gamma\left(1 + \frac{1}{1.5783} \right) \approx 35.24.$$

■■■

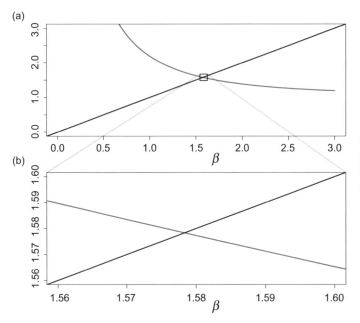

Figure 9.3 The right side of (9.13) (red) plotted against the line $y = \beta$. The value of β at the point of intersection is the MLE. This is approximately $\hat{\beta} = 1.5783$.

There are many software packages that provide estimates for the parameters of the Weibull distribution. In many cases, the software will find the MLEs in a manner similar to that described above. In other software packages, the logarithms of the failure times are considered. If a lifetime has a Weibull distribution, then the logarithm has an extreme value distribution, for which it is a bit easier to find the MLEs. This leads to different parameterizations, which means that an understanding of the exact model fit is needed, along with some intuition about how to convert into the parameterization we are working with. In R, we can use the survival package to produce estimates. The code below fits a Weibull distribution to the data from the previous example.

```
library(survival)
t = c( 12 , 17 , 40 , 71 )
n = length( t )
WeibullFit = survreg( Surv(t) ~ 1 , dist="weibull" )
summary( WeibullFit )
```

The survreg() function fits a regression model, but in our case there are no predictors. The Surv() function converts the vector t into a survival object.[6] We therefore specify the model as Surv(t) ~ 1, meaning there is nothing in the regression model except the intercept. The resulting output is shown below.

```
Call:
survreg(formula = Surv(t) ~ 1, dist = "weibull")
             Value Std. Error     z       p
(Intercept)  3.670      0.336 10.94 <2e-16
Log(scale)  -0.456      0.394 -1.16  0.25

Scale= 0.634
```

---

[6]We have ignored one important aspect of life testing models, namely censoring. Censoring occurs when the life test is terminated before all units (people) have failed (died). Information about whether an item failed or was censored is usually contained in a censoring variable, which is also part of the survival object.

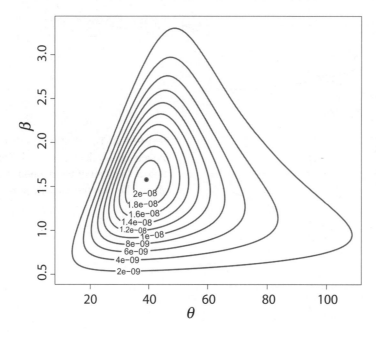

Figure 9.4 Weibull likelihood function in terms of both β and θ for the data in Example 9.22.

```
Weibull distribution
Loglik(model)= -17.7    Loglik(intercept only)= -17.7
Number of Newton-Raphson Iterations: 6
n= 4
```

The estimates don't seem to match the ones we calculated above. This is because R analyzes the logarithms of the failure time and gives the estimates in terms of this model. The parameter estimates can be obtained from R's output as follows:

$$\hat{\beta} = \frac{1}{(\text{Scale})} = \frac{1}{0.634} \approx 1.577$$

$$\hat{\theta} = \exp((\text{Intercept})) = \exp(-3.670) \approx 39.25.$$

Both of these estimates are very close to the ones we obtained.

Exercise 9.4 Suppose we run an experiment by repeatedly testing people until we find someone with type O blood. If it takes X trials until we obtain the first success, find the MLE of the proportion of the population with type O blood.

9.6 Sufficiency

Is it possible to summarize the information from an entire data set in just a few statistics? For example, if you have a sample from the normal distribution with unknown mean and unknown variance, does the sample mean and sample variance contain all of the information the sample has to offer? It would seem like the answer is "yes" because the population has only two parameters, one that describes the center (the mean) and one that describes the variability (the variance), and the sample mean and variance describe, respectively, the sample center and variability. To be a bit more precise, does knowing each and every data value $X_1, X_2, \ldots, X_n$ give us any *additional* information if we already know $\overline{X}$ and S^2?

> **Definition 9.6.1 — Sufficient Statistics.** Suppose $X_1, X_2, \ldots, X_n$ is a random sample from some distribution with PDF or PMF $f(x|\theta)$ where θ could be a vector of parameters. Let $\boldsymbol{X} = [X_1, X_2, \ldots, X_n]$ and suppose $T(\boldsymbol{X})$ is a possibly vector-valued statistic that is a function of the random sample. If knowing the values of all data points $\boldsymbol{X}$ provides no additional information beyond knowing $T(\boldsymbol{X})$, then $T(\boldsymbol{X})$ is said to be a vector of *sufficient statistics*.
>
> This idea can be made more precise by saying that the conditional PDF of $\boldsymbol{X}$ given $T(\boldsymbol{X})$,
>
> $$f(x_1, x_2, \ldots, x_n | T(\boldsymbol{x})),$$
>
> does not depend on the parameter θ. If the conditional PDF f does not depend on θ then the full data set cannot give any additional information that is not already contained in the sufficient statistic.

In a practical situation, this definition is difficult to test. For example, if we were given a distribution, such as the Poisson, and asked for a sufficient statistic, it would be difficult to proceed. Fortunately there is an easy test, given in the theorem below, about the structure of the likelihood function.

Theorem 9.6.1 — Factorization Theorem. Suppose $\boldsymbol{X} = [X_1, X_2, \ldots, X_n]$ is a random sample from some distribution with PDF $f(x|\theta)$. If the likelihood function can be factored as

$$L(\boldsymbol{\theta}|\boldsymbol{x}) = \underbrace{h(\boldsymbol{x})}_{\text{data only}} \times \underbrace{g(\boldsymbol{\theta}, T(\boldsymbol{x}))}_{\text{parameter } \theta \text{ and data only through } T(\boldsymbol{x})} \tag{9.14}$$

then the function $T(\boldsymbol{x})$ is a vector of sufficient statistics for the parameter vector $\boldsymbol{\theta}$.

Let's give a few examples of how the factorization theorem is used.

■ **Example 9.23** Suppose $X_1, X_2, \ldots, X_n \sim$ i.i.d. POIS(λ). Find a sufficient statistic for λ.

Solution: The likelihood function for the Poisson distribution was found in Example 9.18 to be

$$L(\lambda|\boldsymbol{x}) = \frac{\lambda^{\sum_{i=1}^{n} x_i} \exp(-n\lambda)}{\prod_{i=1}^{n} x_i!} = \underbrace{\left(\prod_{i=1}^{n} x_i!\right)}_{\text{function of data}} \times \underbrace{\exp(-n\lambda)\lambda^{\sum_{i=1}^{n} x_i}}_{\text{function of } \lambda \text{ and } \sum_{i=1}^{n} x_i}.$$

Thus, $T = \sum_{i=1}^{n} X_i$ is a sufficient statistic.[7]   ■ ■ ■

The sufficient statistic is never unique. Any constant multiple, for example, is also sufficient. Here, we might suggest that the sample mean $\overline{X}$ is sufficient, because the second part of the factorization could be written as

$$\underbrace{\exp(-n\lambda)\lambda^{n\overline{x}}}_{\text{function of } \lambda \text{ and } \bar{x}}.$$

It shouldn't be surprising that the sum of the observations or the sample mean should be a sufficient statistic. After all, the Poisson is a one-parameter distribution whose parameter equals the population mean and the sufficient statistic is a constant multiple of the sample mean.

---

[7]Note that at the end we write the sufficient statistic as depending on the random variables $X_1, X_2, \ldots, X_n$, rather than the observed values $x_1, x_2, \ldots, x_n$. This is a subtle point, but agrees with our thinking that statistics are random variables that are computed from the random variables in a sample.

■ **Example 9.24** Find the sufficient statistics for a sample $X_1, X_2, \ldots, X_n$ from the $N(\mu, \sigma^2)$ distribution.

Solution: The likelihood function is

$$
\begin{aligned}
L(\mu, \sigma^2 | x) &= \frac{1}{(2\pi)^{n/2}\sigma^n} \exp\left(-\frac{1}{2\sigma^2}\sum_{i=1}^{n}(x_i - \mu)^2\right) \\
&= \frac{1}{(2\pi)^{n/2}\sigma^n} \times \exp\left(-\frac{1}{2\sigma^2}\left(\sum_{i=1}^{n}x_i^2 - 2\mu\sum_{i=1}^{n}x_i + \mu^2\right)\right).
\end{aligned}
$$

In the last line we could think of $(2\pi)^{-n/2}$ as being the "data only" part of the likelihood's factorization. It, of course, doesn't depend on the data; when we say that this part depends only on the data, what we really mean is that it is independent of the unknown parameters. The second part is a function of the parameters μ and σ^2 and the statistics are $\sum_{i=1}^{n}x_i^2$ and $\sum_{i=1}^{n}x_i$. Thus, the statistics

$$
\begin{aligned}
T_1(\boldsymbol{X}) &= \sum_{i=1}^{n}X_i \\
T_2(\boldsymbol{X}) &= \sum_{i=1}^{n}X_i^2
\end{aligned}
$$

are sufficient.   ■ ■ ■

A couple of important notes are needed regarding this example. First, while it might be tempting to say that $\sum_{i=1}^{n}X_i^2$ is sufficient for σ^2, and $\sum_{i=1}^{n}X_i$ is sufficient for μ, we cannot say this. The definition of *sufficient* says that a vector of statistics can be sufficient for a vector of parameters, but a one-to-one match up is not possible. Many books use the terms sufficient statistic to refer to a one-dimensional parameter and statistic, and *jointly sufficient statistics* for higher dimensions. In this example, $\sum_{i=1}^{n}X_i$ and $\sum_{i=1}^{n}X_i^2$ would be called *jointly sufficient statistics* for $[\mu, \sigma^2]$. Second, the second part of the factorization could be written in a different way; for example,

$$
\begin{aligned}
L(\mu, \sigma^2 | x) &= \frac{1}{(2\pi)^{n/2}\sigma^n} \exp\left(-\frac{1}{2\sigma^2}\sum_{i=1}^{n}(x_i - \mu)^2\right) \\
&= \frac{1}{(2\pi)^{n/2}\sigma^n} \exp\left(-\frac{1}{2\sigma^2}\sum_{i=1}^{n}(x_i - \bar{x} + \bar{x} - \mu)^2\right) \\
&= \frac{1}{(2\pi)^{n/2}\sigma^n} \exp\left(-\frac{1}{2\sigma^2}\left(\sum_{i=1}^{n}(x_i - \bar{x})^2 + n(\bar{x} - \mu)^2\right)\right) \\
&= \frac{1}{(2\pi)^{n/2}\sigma^n} \exp\left(-\frac{1}{2\sigma^2}\left((n-1)s^2 + n(\bar{x} - \mu)^2\right)\right),
\end{aligned}
$$

and conclude that $\bar{X}$ and S^2 are sufficient statistics. This seems even more reasonable since we can now say that the sample mean and sample variance are sufficient for the population mean μ and population variance σ^2.

Exercise 9.5 Suppose we have a sample $X_1, X_2, \ldots, X_n$ from the GAMMA(α, β) distribution which has PDF

$$f(x|\alpha,\beta) \;=\; \frac{x^{\alpha-1}}{\beta^{\alpha}\Gamma(\alpha)}\exp(-x/\beta), \quad x > 0.$$

Find sufficient statistics for α and β.

Answer: $T_1(\boldsymbol{X}) = \sum\limits_{i=1}^{n} X_i, \quad T_2(\boldsymbol{X}) = \prod\limits_{i=1}^{n} X_i.$

The idea behind the definition of sufficient statistics is that knowing the sufficient statistics is enough (i.e., sufficient) to know all of the information in the sample. To make this connection between the definition of sufficient statistics and the factorization theorem a bit clearer, we state the likelihood principle, which many, but not all, statisticians believe.[8]

Definition 9.6.2 — Likelihood Principle. The *likelihood principle* states that all information from the sample is contained in the likelihood function.

If you accept the likelihood principle, and if sufficient statistics exist, then the likelihood function has a factor that depends only on the data and a factor in which the parameters and sufficient statistics are mixed. The "data only" part serves to scale the likelihood, which has no effect on the MLEs of the parameters. The shape of the likelihood function (as a function of the parameters) depends only on the factor that combines the parameters and the sufficient statistics. Thus, knowing the sufficient statistics is enough to know the shape of the likelihood (but not its scaling, which is affected by the "data only" part).

In particular, the MLEs are functions of the sufficient statistics. To see why, let's take the log of the likelihood in its factorized form:

$$\ell(\boldsymbol{\theta}|\boldsymbol{x}) \;=\; \log L(\boldsymbol{\theta}|\boldsymbol{x}) \;=\; \log h(\boldsymbol{x}) + \log g(\boldsymbol{\theta}, T(\boldsymbol{x})).$$

When we take the derivative with respect to any component of $\boldsymbol{\theta}$, the first term will be zero because the function h does not depend on any parameters. The derivatives of the second part will be functions of the sufficient statistics. Thus, we can conclude that the MLEs will depend on the data only through the sufficient statistics. If we know the values of the sufficient statistics, we can compute the MLEs without knowing all of the data values in the sample.

Finally, we must point out that when we say $T(\boldsymbol{x})$ is a sufficient statistic, we are assuming that the model is correct. If it is not, and some other model is the correct one, then $T(\boldsymbol{x})$ may not be sufficient.

■ **Example 9.25** Bilirubin is a substance found in blood and its level can be a sign of liver function. Normal levels are between 0.3 and 1.2 milligrams per deciliter. The file `liver_patient.csv` from the book's website contains data from a study done in India on 583 patients. Suppose your colleague assumes that the 583 bilirubin measurements came from a random sample from the normal distribution with mean μ and σ^2, both of which are unknown. Your colleague computes the sample mean and variance to be

$$\bar{x} \;=\; 3.299$$
$$s^2 \;=\; 38.558$$

---

[8]See Section 15.4 for a situation which calls into question the likelihood principle.

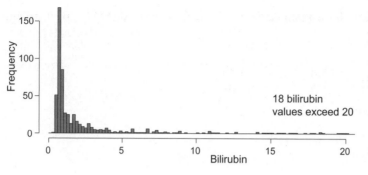

Figure 9.5 Histogram of bilirubin values. Note that 18 of the 583 are off to the right of 20. This distribution is highly skewed to the right, indicating that the normal distribution is not a reasonable model.

and claims we can dispense with the raw data and proceed with just the sample mean and variance. Is this justifiable?

Solution: This is only justifiable if the normal distribution is the appropriate model. A histogram for the bilirubin values, shown in Figure 9.5, clearly shows that the normal distribution is not an appropriate model, because the values are highly skewed to the right. Further investigation reveals that the data come from men and women aged 4 to 90, and some have liver disease and some don't. The assumption that the data come from a single population is highly dubious.

■ ■ ■

9.7 Chapter Summary

As we've said several times so far in this book, statistics is about saying something about a population, given a sample taken from it. In this chapter we looked at the problem of *point estimation*, that is, giving a single number as our estimate of the population parameter. We looked at criteria for comparing estimators, including bias, variance, and consistency (i.e., the property that the estimator converges to the true population parameter as the sample size gets large).

After looking at intuitive methods of estimation (e.g., estimating the population mean using the sample mean), we then looked at general methods of estimation, including the method of moments and maximum likelihood. Because the method of maximum likelihood can lead to equations that cannot be solved analytically we must sometimes *approximate* values that maximize the likelihood function. We describe some methods for performing this approximation.

We conclude the chapter with a discussion of *sufficiency*. It is sometimes possible to summarize the information from a sample in just a few statistics, called sufficient statistics. This is an important topic in statistics and data science because it is possible to store just a few statistics, rather than the entire data set (which can be quite large).

9.8 Problems

Problem 9.1 Suppose we conduct a sequence of independent trials where each trial results in a success or failure. We repeat this until we observe the first success. Let X denote the number of trials needed to get this one success. Give an intuitive estimator for the probability p of a success.

Problem 9.2 Suppose we conduct a sequence of independent trials where each trial results in a success or failure. We repeat this until we observe the rth success. Let X denote the number of trials needed to get r successes. Give an intuitive estimator for the probability p of a success.

Problem 9.3 Suppose you take a sample of 100 student athletes at your school and find that the sample mean height is 71 inches and the sample standard deviation is 4 inches. Assume the population of heights is normally distributed. Find intuitive estimators for

(a) the population mean μ;
(b) the population variance σ^2;
(c) the proportion of athletes whose height exceeds 72 inches.

Problem 9.4 Suppose you have a sample $X_1, X_2, \ldots, X_n \sim$ i.i.d. GAM$(2,\beta)$. What does $\overline{X}$ estimate? Can you use this to obtain an estimator for β?

Problem 9.5 Suppose after returning home from a cross-country flight I walk to the ground transportation part of the airport to wait for the shuttle bus to take me back to the parking lot where I left my car. The buses run every θ minutes, but I don't know what θ is. Since I arrive at a random point in time, it is reasonable to assume that my waiting time is UNIF$(0,\theta)$. Suppose I make n trips and I have waiting times $X_1, X_2, \ldots, X_n \sim$ i.i.d. UNIF$(0,\theta)$. Show that $\hat{\theta} = 2\overline{X}$ is an unbiased estimator of θ.

Problem 9.6 What is the variance of the estimator $\hat{\theta}$ from Problem 9.5?

Problem 9.7 Consider the situation described in Problem 9.5. It can be shown (see Problem 9.35) that if $X_1, X_2, \ldots, X_n \sim$ i.i.d. UNIF$(0,\theta)$, then the PDF of $Y = \max(X_1, \ldots, X_n)$ is

$$f_n(x) = \begin{cases} \frac{n}{\theta}\left(\frac{x}{\theta}\right)^{n-1}, & 0 < x < \theta \\ 0, & \text{otherwise.} \end{cases}$$

(a) Find $\mathbb{E}(Y)$.
(b) Find $\mathbb{V}(Y)$.
(c) Show that $\tilde{\theta} = \big((n+1)/n\big)Y$ is an unbiased estimator for θ.
(d) Find $\mathbb{V}(\tilde{\theta})$.

(e) Which estimator is better: the $\hat{\theta}$ from Problem 9.5 or the $\tilde{\theta}$ from this problem? Explain your reasoning.

Problem 9.8 Is the sample mean $\overline{X}$ always an unbiased estimator of the population mean μ? Justify your answer.

Problem 9.9 Suppose $X_1, X_2, \ldots, X_n \sim$ i.i.d. $N(\mu, \sigma^2)$. Let $\overline{X}$ denote the sample mean and S^2 the sample variance, $S^2 = \sum_{i=1}^n (X_i - \overline{X})^2/(n-1)$. Is S a consistent estimator of σ?

Problem 9.10 Suppose $X_1, X_2, \ldots, X_n \sim$ i.i.d. $N(\mu, \sigma^2)$. Let $\overline{X}$ denote the sample mean. Let $\tilde{\sigma}^2 = \sum_{i=1}^n (X_i - \overline{X})^2/(n+1)$ be an estimator for σ^2.

(a) Is $\tilde{\sigma}^2$ a consistent estimator of σ^2?
(b) Is $\tilde{\sigma}$ a consistent estimator of σ?

Problem 9.11 Let $X_1, X_2, \ldots, X_n \sim$ i.i.d. UNIF$(0,\theta)$. We gave the PDF of $Y = \max(X_1, X_2, \ldots, X_n)$ in Problem 9.7. Is Y a consistent estimator of θ?

Problem 9.12 Let $X_1, X_2, \ldots, X_n \sim$ i.i.d. UNIF$(0,\theta)$. Is $2\overline{X}$ a consistent estimator of θ?

Problem 9.13 Consider the situation in Problem 9.1, but suppose we repeatedly perform the independent trials until we get a success. Let $X_1, X_2, \ldots, X_n$ denote the numbers of trials needed to get the first success on these n sequences of trials. Is the estimator $\hat{p} = 1/\overline{X}$ a consistent estimator of p?

Problem 9.14 Suppose X has PDF

$$f(x) = \frac{\alpha}{\theta}\left(\frac{x}{\theta}\right)^{\alpha-1}, \qquad 0 < x < \theta.$$

(a) Find the CDF $F(x)$.
(b) Find $\mathbb{E}(X)$.
(c) Find $\mathbb{V}(X)$.
(d) For what value of α does this reduce to the UNIF$(0,\theta)$ distribution?
(e) Suppose that α is known and we are trying to estimate θ. Consider

$$Y_n = \max(X_1, X_2, \ldots, X_n).$$

Is Y_n an unbiased estimator of θ?
(f) Find the multiple c (which may depend on n and α) such that $\tilde{\theta} = cY_n$ is an unbiased estimator of θ.
(g) Find the variance of the unbiased estimator $\tilde{\theta}$.

Problem 9.15 Consider the situation described in Problem 9.14, except now let's consider a multiple of $\overline{X}$ as an estimator for θ. Assume α is known.

(a) Find $\mathbb{E}(\overline{X})$.

(b) Find the multiple c (which may depend on n and α) such that $\hat{\theta} = c\overline{X}$ is an unbiased estimator of θ.

(c) Find the variance of the unbiased estimator $\hat{\theta}$.

(d) Compare the variances of $\hat{\theta}$ and $\tilde{\theta}$ from Problem 9.14. Which estimator would you prefer?

Problem 9.16 Suppose $X \sim \text{BIN}(n, p)$. Suppose we want to estimate the variance of X, which is $\sigma^2 = np(1-p)$. If we were to substitute the estimator $\hat{p} = X/n$ for p we would have $\hat{\sigma}^2 = (X/n)(1 - X/n)$.

(a) Show that this is a biased estimator of σ^2.

(b) Starting with $\mathbb{E}(\hat{\sigma}^2)$ from part (a), modify this result to obtain an unbiased estimator for σ^2.

Problem 9.17 Suppose $X_1, X_2, \ldots, X_n \sim$ i.i.d. EXP (λ).

(a) Find the MOM estimator for λ.

(b) Find the MLE for λ.

(c) Are the estimators from parts (a) and (b) unbiased?

(d) Are the estimators from parts (a) and (b) consistent?

Problem 9.18 Suppose $X_1, X_2, \ldots, X_n \sim$ i.i.d. EXP1 (θ).

(a) Find the MOM estimator for θ.

(b) Find the MLE for θ.

(c) Are the estimators from parts (a) and (b) unbiased?

(d) Are the estimators from parts (a) and (b) consistent?

Problem 9.19 Consider the data in Table 9.1 given by Wang (2000) on the failure times of an electrical component. Assuming an EXP1 distribution, find the MLE and the MOM estimator of θ for this data set.

Table 9.1 Data for Problem 9.19.

i	t_i	i	t_i
1	5	10	165
2	11	11	196
3	21	12	224
4	31	13	245
5	46	14	293
6	75	15	321
7	98	16	330
8	122	17	350
9	145	18	420

Problem 9.20 Suppose $X_1, X_2, \ldots, X_n \sim$ i.i.d. GAM $(2, \beta)$.

(a) Find the MOM estimator for β.

(b) Find the MLE for β.

(c) Are the estimators from parts (a) and (b) unbiased?

(d) Are the estimators from parts (a) and (b) consistent?

Problem 9.21 Suppose $X_1, X_2, \ldots, X_n \sim$ i.i.d. UNIF $(0, \theta)$.

(a) Find the MOM estimator for θ.

(b) Is the estimator from part (a) unbiased?

(c) Is the estimator from part (b) consistent?

Problem 9.22 Suppose $X_1, X_2, \ldots, X_n \sim$ i.i.d. BETA (α, β). Find the MOM estimators for α and β.

Problem 9.23 Suppose $X_1, X_2, \ldots, X_n \sim$ i.i.d. GEO (p).

(a) Find the MOM estimator for p.

(b) Find the MLE for p.

(c) Are the estimators from parts (a) and (b) consistent? Justify your conclusion. (*Hint*: You will need to use Theorem 9.2.3 part (d).)

Problem 9.24 Bartholomew (1957) gave the following data on the lifetimes in months of 10 mechanical items:

2	119	51	77	33	27	14	24	4	37

Assume an EXP1(θ) distribution for these outcomes.

(a) Estimate the mean θ of the exponential distribution.

(b) Use the exponential distribution and the estimate from part (a) to estimate the probability that an item from this population will last for 120 months (10 years).

Problem 9.25 Consider the data given in Problem 9.24, except now let's assume the distribution of lifetimes is GAM(α, β).

(a) Assuming the GAM(α, β) distribution for the lifetimes, find the likelihood and log likelihood functions.

(b) Find $\partial \ell / \partial \beta$.

(c) Set the result from part (b) equal to 0 and solve for β in terms of α.

(d) Find $\partial \ell / \partial \alpha$. (*Hint*: The expression $\Gamma'(\alpha)/\Gamma(\alpha)$ is a commonly encountered function in mathematics and statistics. It is called the *digamma* function and is usually denoted

$$\psi(\alpha) = \frac{\Gamma'(\alpha)}{\Gamma(\alpha)}.$$

In R, you can compute ψ with the built-in function `digamma()`.)

(e) Substitute your answer from part (c) into the equation obtained by setting $\partial\ell/\partial\alpha$ equal to 0. You should have one equation in one unknown (α).

(f) Construct an array of α values from 0.5 to 2.0 in increments of 0.01. Evaluate the left side of the equation from part (e) at each point in the grid and plot this result over the array of α. Approximate where this function equals zero. You may have to zoom in on the plot to see where the graph of the function crosses the α axis.

(g) Your answer to part (f) is your approximation to the MLE for α. Substitute this value into your result from part (c).

(h) Estimate the probability that an item from this population will last for 120 months.

Problem 9.26 Consider the discrete random variable X having PMF

$$f(x) = \frac{\Gamma(x+\phi)}{\Gamma(x+1)\Gamma(\phi)}\left(\frac{\mu}{\mu+\phi}\right)^x\left(\frac{\phi}{\mu+\phi}\right)^\phi, \; x=0,1,\ldots$$
$$(9.15)$$

(a) Show that the distribution whose PMF is given in (9.15) reduces to the negative binomial distribution described in Problem 5.26 when $r = \phi$ and $\mu = \phi/(\mu+\phi)$.

(b) Show that as $\phi \to \infty$ the PMF in (9.15) converges to the POIS(μ) PMF for every x.

(c) It can be shown that the mean and variance of the PMF in (9.15) are $\mathbb{E}(X) = \mu$ and $\mathbb{V}(X) = \mu + \mu^2/\phi$. Use these results to find the MOM estimators for μ and ϕ if we have a sample of size n from the distribution with PMF given in (9.15).

(d) Show that the MLE of μ is $\overline{X}$.

(e) Approximate the MLE of ϕ.

Problem 9.27 Suppose $X_1, X_2, \ldots, X_n \sim$ i.i.d. BIN($1,p$). Find a sufficient statistic for p.

Problem 9.28 Suppose $X_1, X_2, \ldots, X_n \sim$ i.i.d. GEO(p). Find a sufficient statistic for p.

Problem 9.29 Suppose $X_1, X_2, \ldots, X_n \sim$ i.i.d. NB(r,p) where r is known. Find a sufficient statistic for p.

Problem 9.30 Suppose $X_1, X_2, \ldots, X_n \sim$ i.i.d. EXP(λ). Find a sufficient statistic for λ.

Problem 9.31 Suppose $X_1, X_2, \ldots, X_n \sim$ i.i.d. EXP1(θ). Find a sufficient statistic for θ.

Problem 9.32 Suppose $X_1, X_2, \ldots, X_n \sim$ i.i.d. with PDF

$$f(x) = \frac{x}{\theta^2}\exp(-x/\theta), \quad x > 0.$$

Find a sufficient statistic for θ.

Problem 9.33 Suppose $X_1, X_2, \ldots, X_n \sim$ i.i.d. with PDF

$$f(x) = \frac{x^2}{2\theta^3}\exp(-x/\theta), \quad x > 0.$$

Find a sufficient statistic for θ.

Problem 9.34 Suppose $X_1, X_2, \ldots, X_n \sim$ i.i.d. with arbitrary (continuous) CDF $F(x)$. Let

$$Y_n = \max(X_1, X_2, \ldots, X_n).$$

(a) Explain why

$$\Pr(Y_n \le y) = \Pr(X_1 \le y \cap X_2 \le y \cap \cdots \cap X_n \le y).$$

(b) Use the independence of $X_1, X_2, \ldots, X_n$ to write $\Pr(Y_n \le y)$ in terms of the CDF of the Xs. The result is the CDF of Y_n.

(c) Find the PDF of Y_n. (*Hint*: Differentiate the CDF).

Problem 9.35 Suppose $X_1, X_2, \ldots, X_n \sim$ i.i.d. UNIF($0,\theta$). Use the result of Problem 9.34 to find the CDF of $Y_n = \max(X_1, \ldots, X_n)$.

Problem 9.36 Suppose $X_1, X_2, \ldots, X_n \sim$ i.i.d. with the PDF given in Problem 9.14. Find the CDF of $Y_n = \max(X_1, \ldots, X_n)$.

10 — Confidence Intervals

10.1 Introduction

The problem of statistical inference can be described as follows. There is a population and we would like to know certain aspects of the units that make up the population. For example, we might want to know what proportion have a certain property, or what the mean value (of some measure) of all units in the population is. The population is too large to sample in its entirety, so we rely on information from a sample taken from the population. The properties of the population that we don't know (but would like to) are called *parameters*. The summaries of the data we have in the sample are called *statistics*. We would like to use the statistics to say something about the parameters.

The last chapter dealt with point estimation, that is, the process of finding a single number to use as our best guess for the value of an unknown population parameter. In this chapter we aim to give a measure of uncertainty to a point estimate. For a parameter, we would like to find an interval we believe includes or covers the true (but unknown!) value of the population parameter. Of course, the method for finding this interval will depend on the distributional assumptions we make and which parameter we are addressing. This interval, called a *confidence interval*, should cover all "plausible" values of the parameters. We will give a precise definition of what we mean by plausible in the next section.

We begin this chapter with a few examples where intuition is sufficient to tell us whether a value is plausible or not. In general, though, intuition is not enough. This is why we spend this chapter covering confidence intervals in a variety of circumstances. The form of the confidence interval will depend on the distribution of various statistics computed from the sample, so we will rely heavily on the distributional results stated in Chapter 8.

■ **Example 10.1** Visitors to a website are given the option to place an order or not. Some people in your group claim that with a new website design, half of the visitors will place an order. Being skeptical, you suggest a study where 10,000 visitors are shown the new design. Of the 10,000 visitors 3,000 place an order. Is 0.5 a plausible value for the probability that a visitor will place an order? Is 0.3 a plausible value for the probability? How about 0.32?

Solution: Intuitively, if the true value, say θ, were equal to 0.5, we would likely get a proportion much closer to 0.5 than the $3,000/10,000 = 0.3$ we observed. What we observed based on a large sample, $n = 10,000$, seems quite inconsistent with a true probability of 0.5. On the other hand, the data seem completely consistent with $\theta = 0.3$. Thus, using just our intuition, we would say that 0.3 should be in a confidence interval for θ but 0.5 should not be in the interval. Our intuition fails when we address the issue of whether 0.32 is a plausible value. This case requires a little closer scrutiny.   ■ ■ ■

As you know, this example describes a sequence of Bernoulli trials, so the number of "successes" has a binomial distribution. To be a bit more precise, observing $x = 3,000$ and $\hat{\theta} = 0.3$ is inconsistent with $\theta = 0.5$, but completely consistent with $\theta = 0.3$.

■ **Example 10.2** Suppose, in the previous example, we had chosen a sample of $n = 10$, rather than $n = 10,000$. Now, in addition, suppose that three visitors placed an order. Are $0.5, 0.3$, and 0.32 plausible values for θ?

Solution: If $\theta = 0.50$, this is like tossing 10 fair coins. Our intuition says getting 3 heads on 10 tosses isn't that unusual. (Try it a few times!) If the true probability were $\theta = 0.3$ or $\theta = 0.32$, getting four heads (or visitors to the website) doesn't seem that unusual. Thus, all of these values should be in the confidence interval.   ■ ■ ■

We see from these last two examples that the confidence interval should depend on how many successes we observe *and* on how large the sample size is.

10.2 Basic Properties

What we are seeking is an interval that we believe covers the true value of an unknown population parameter. The endpoints of the interval depend on the data, so if our perspective is *before* data are collected, then the endpoints are random. Thus, a confidence interval is a random interval, and we'd like it to contain the true parameter with high probability, say $1 - \alpha$, where α is small.

> **Definition 10.2.1 — Confidence Interval.** A *confidence interval* is a random interval that will contain the true value of a parameter with probability $1 - \alpha$. The value $1 - \alpha$ is called the *confidence level* of the interval.

Our starting point is therefore

$$\Pr(L < \theta < U) = 1 - \alpha. \tag{10.1}$$

It is important to keep straight which parts of (10.1) are random and which are fixed. The parameter θ is a fixed number, even though we don't know what it is.[1] The endpoints L and U of the confidence are random variables, which will vary depending on the data that are collected.

We begin with a simple example to illustrate the method we use to determine a confidence interval, how we compute the interval, and how we interpret it. The example is too simple, really, because it assumes that the variance is known (which it never is).

■ **Example 10.3** Suppose $X_1, X_2, \ldots, X_n \sim N(\mu, 100)$. Note that we are assuming that the mean is unknown, but the variance is known to be $\sigma^2 = 100$. Find a formula for a confidence interval for the parameter μ.

---

[1] In Bayesian statistics, which we cover in Chapter 15, the uncertainty in parameters is expressed in probabilistic terms. Parameters are therefore treated as random variables. In this chapter, however, we follow the classical approach to statistics where parameters are fixed (not random!) but unknown values.

Solution: Equation (10.1) is where we'd like to end up, except with μ in the middle instead of θ. Our starting point depends on the problem we're addressing. In our case, we have a normal distribution so we know that

$$Z = \frac{\overline{X} - \mu}{\sigma/\sqrt{n}} = \frac{\overline{X} - \mu}{10/\sqrt{n}} \sim N(0,1).$$

We can use R (or a table of the normal distribution) to look up the quantiles of the normal distribution. For example, if we want a 95% interval, we would select $\alpha = 0.05$ and $\alpha/2 = 0.025$. We then run

```
alpha = 0.05
zalpha = qnorm( 1 - alpha/2 , 0 , 1 )
print( zalpha )
```

which yields

```
[1] 1.959964
```

giving $z_{0.025} = 1.96$, approximately. (Recall that z_γ is that value of z that makes the right tail area $\Pr(Z > z)$ equal to γ. See Figure 10.1.) Thus, and this is the key idea,

$$\Pr\left(-z_{\alpha/2} < Z < z_{\alpha/2}\right) = 1 - \alpha,$$

which in our case is

$$\Pr\left(-1.96 < \frac{\overline{X} - \mu}{10/\sqrt{n}} < 1.96\right) = 0.95.$$

With a little algebra, we can put μ by itself in the middle. This yields the general expression

$$\Pr\left(\overline{X} - z_{\alpha/2}\frac{\sigma}{\sqrt{n}} < \mu < \overline{X} + z_{\alpha/2}\frac{\sigma}{\sqrt{n}}\right) = 1 - \alpha,$$

or in the specific case

$$\Pr\left(\overline{X} - 1.96\frac{10}{\sqrt{n}} < \mu < \overline{X} + 1.96\frac{10}{\sqrt{n}}\right) = 0.95.$$

Once again, it is important to keep in mind what is random inside the above probability: It is the first and third pieces to the inequality. The confidence interval is composed of the values of μ that satisfy

$$\underbrace{\overline{X} - z_{\alpha/2}\frac{10}{\sqrt{n}}}_{\text{random}} < \underbrace{\mu}_{\text{fixed}} < \underbrace{\overline{X} + z_{\alpha/2}\frac{10}{\sqrt{n}}}_{\text{random}}.$$

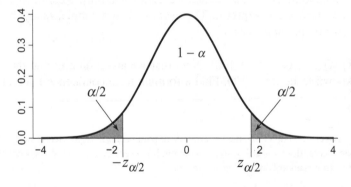

Figure 10.1 $z_{\alpha/2}$ is that value that makes the right tail area under the standard normal density function equal to $\alpha/2$.

Confidence intervals are often written in parentheses/bracket notation.[2] The 95% confidence interval for a sample of data of size n from a population following a normal distribution is thus

$$\left(\overline{X} - 1.96\frac{10}{\sqrt{n}}, \overline{X} + 1.96\frac{10}{\sqrt{n}} \right). \tag{10.2}$$

■ ■ ■

Because this is a random interval, in the sense that the endpoints of the interval are random, the interpretation is rather subtle, and confidence intervals are often misinterpreted. If we want a $100(1-\alpha)\%$ confidence interval for the mean of a population where the variance is known, we would interpret the confidence interval in 10.2 as follows:

> The probability is $1 - \alpha$ that a confidence interval calculated using the method in (10.2) contains the true parameter.

The above statement is true *before* we collect and analyze data.

■ **Example 10.4** Suppose we take a sample of size $n = 5$ from the $N(\mu, 100)$ distribution, and the sample consists of the values

$$105, \quad 78, \quad 121, \quad 90, \quad 95.$$

Find a 95% confidence interval for μ and interpret it.

Solution: Here $\alpha = 0.05$, so $\alpha/2 = 0.025$ and $z_{\alpha/2} = 1.96$. A little bit of R coding gives us the confidence interval:

```
x = c(105, 78, 121, 90, 95 )
n = length( x )
sigma = 10
xbar = mean(x)
alpha = 0.05
z = qnorm( 1-alpha/2 , 0 , 1 )
L = xbar - z*sigma/sqrt(n)
U = xbar + z*sigma/sqrt(n)
print( c(L,U) )
```

which gives

```
[1]   89.03477 106.56523
```

The confidence interval is therefore the interval $(89.03, 106.57)$. The interpretation of this interval is this: The probability that an interval calculated in this way includes the true value of the population mean μ is 0.95. Another way to say this is to say that in the long run, 95% of the time an interval calculated in this way will include the true value of μ.

■ ■ ■

The confidence interval is one of the most widely misinterpreted ideas in statistics. The viewpoint for making a probability statement must be *before* data are collected, not *after*. Once data are collected, there is no randomness remaining, so the probability statement no longer makes sense. Here we discuss two of the most common misunderstandings and try to explain why they are wrong. Let's take the previous example to provide context for why these explanations are wrong:

---

[2]In this notation, parentheses indicate that an endpoint is not in the interval and square brackets indicate that the endpoint is included. For example, $(0,1)$ consists of those numbers x such that $0 < x < 1$, whereas $(0,1]$ consists of those satisfying $0 < x \leq 1$.

⊘ $\Pr(89.03 < \mu < 106.57) = 0.95$.   As we have been emphasizing throughout the book, we treat parameters as fixed, nonrandom numbers.[3] Just because we don't know μ, doesn't mean it's random. The numbers 89.03 and 106.57 are clearly not random. So, what's left to be random inside the probability? Either the interval $(89.03, 106.57)$ contains the true value of μ or it doesn't, but we don't know which.

⊘ 95% of the population lies in the interval $(89.03, 106.57)$.   The confidence interval is for the *mean of the population*, not the set of all data values from the population. The sample size n appears as a square root in the denominator; thus, the larger n is, the narrower the confidence interval. With enough data, a confidence interval can be very narrow, indicating that we are quite confident what the mean is. The individual data values may be far from the mean, so a confidence interval for the mean will often include only a small fraction of the population.

If the probabilistic interpretation is incorrect, how should we interpret the confidence interval? As stated at the end of the previous example, we could say that the probability is 0.95 that an interval calculated in this way will include the true mean when viewed before data are collected. The *confidence* is in the method, not the particular interval we end up with. Equivalently, we could say that in the long run, 95% of the confidence intervals will include μ.

So, once we collect the data, why should we believe that the interval contains the parameter μ? If we were to calculate confidence intervals all day long (a rather dull job to be sure) we would expect 95% of them to contain μ and 5% to miss μ. Of course, we never know which case we have, since in the real world we only get to take one sample and get one confidence interval. If you repeatedly performed a task, and 95% of the time you get outcome A and 5% of the time you get outcome B, when you perform the task the next time you expect to see outcome A and you would be a bit surprised if outcome B occurs. Thus, we believe that μ is in our 95% confidence interval because the method works 95% of the time.

Imagine the following illustrative, albeit contrived, situation. You have a bowl of small red and white balls. Most, say 95%, are white, and the remaining 5% are red. Think of selecting a ball as computing a confidence interval, where a white ball means your confidence interval includes the true parameter, and a red ball means your confidence interval misses the true parameter. Now imagine you get to reach in and select one ball without being able to see its color. You have the ball in your hand (Figure 10.2), but you don't know its color. Is it white or red? If you were forced to say, you would probably say the ball is white. Think carefully about why. You probably think that most of the time when you select a ball it will be white, because, after all, 95% of the balls are white. At this point, there is no randomness because the ball has already been selected; there is, however, uncertainty about the ball's color. The ball in your hand is either white or red, you just don't know which. When you compute a confidence interval, the interval you end up with either includes or misses the true value of the parameter. You just don't know which. You know that in the long run, 95% of the intervals you compute will include the true parameter. In real life, though, you only get to take one sample.

As for the second common misinterpretation of the confidence interval, we should point out that there are a number of types of intervals. If we want an interval that contains most of the population with high confidence, we want a *tolerance interval*. If you do much data analysis, it is important to know the

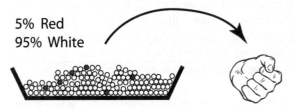

5% Red
95% White

Figure 10.2 Picking a ball from a bowl. Suppose the ball contains 95% white balls and 5% red balls. White is analogous to a confidence interval containing the true parameter, whereas red is analogous to a confidence interval that misses the true parameter.

---

[3]The Bayesian approach works differently. We will cover this in Chapter 15.

types of intervals that arise in practice. Prediction intervals, another type of interval that is useful when we want to predict future observations, are covered in Chapter 14.

■ **Example 10.5** Consider the same data as in the previous example, where we took a sample of size $n = 5$ from the $N(\mu, 100)$ distribution, and obtained the data

105, 78, 121, 90, 95.

Find 90%, 95%, and 99% confidence intervals for μ and discuss the effect of the confidence level $1 - \alpha$ on the position and length of the intervals.

Solution: For the first part $\alpha = 0.10$, so $z_{\alpha/2} = 1.645$; everything else in the formula remains the same. The 90% confidence interval is therefore

$$\left(97.8 - 1.645\frac{10}{\sqrt{5}}, 97.8 + 1.645\frac{10}{\sqrt{5}}\right) = (90.44, 105.16).$$

For the 95% and 99% intervals, only the value for $z_{\alpha/2}$ changes. The only change to the code is to replace `alpha = 0.10` with `alpha = 0.05` and `alpha = 0.01`, respectively. The three intervals are

90%:	(90.44, 105.16),
95%:	(89.03, 106.57),
99%:	(86.28 109.32).

As the confidence level increases toward 1, the confidence interval gets wider. If we want to be *really* sure that our interval contains μ we have to hedge and give a wide interval. If we are willing to be a little less confident, for example with a 90% interval rather than a 99% interval, then a narrower interval will do the job. ■■■

Exercise 10.1 Suppose that a sample of size n is taken from the $N(\mu, 100)$ distribution. Suppose also that the sample mean is the same as in Example 10.3, namely $\bar{x} = 97.8$. Find 95% confidence intervals for the cases where $n = 10, 20, 40$, and discuss how the width depends on n.

Answer: (91.60, 104.00); (93.42, 102.18); (94.70, 100.90)

We close this section by pointing out again that the problem that we addressed in this section is completely unrealistic. We've assumed that the mean of the population is unknown, but the variance of the population is known. Usually, if we don't know the mean, we don't know the variance either. This example does, however, allow us to gain some insight into other confidence intervals. It is like a stepping stone to the more realistic, and messier, problems that follow. If you can understand the concepts for this simple example, you will be able to better follow the rest of this chapter.

10.3 Large Sample Confidence Intervals

The CLT was stated in Chapter 8. It said that if we take a sample from a distribution with mean μ and variance σ^2, then for large n the distribution of the sample mean $\overline{X}$ is approximately normal with mean μ (the same as the population mean) and variance σ^2/n. This can be the starting point for obtaining a confidence interval for the mean of a population.

For example, if we conduct a sequence of n Bernoulli trials where the probability of success on any trial is θ, and let Y_i be 1 if the ith trial is a success and 0 otherwise, then

$$\overline{Y} = \frac{1}{n}\sum_{i=1}^{n}Y_i = \frac{\text{number of successes}}{n} = \hat{\theta}.$$

Thus we could think of our usual estimator $\hat{\theta}$ as being the sample mean of the Y_i, $i = 1, 2, \ldots, n$. Thus, the CLT applies and allows us to say that for large n

$$\frac{\overline{Y} - \theta}{\sqrt{\theta(1-\theta)/n}} \overset{\text{approx}}{\sim} N(0,1). \tag{10.3}$$

This is *almost* the result we need to get a confidence interval for θ. We say "almost" because we would like for θ to appear just once, preferably in the numerator of (10.3) so that we can perform some algebra and obtain θ in the middle of the inequality and data on the outside parts. (As is, we would have θ on both of the outside parts.) We can resolve this issue with the following theorem.

Theorem 10.3.1 — A Condition for Asymptotic Normality. If Y_n is a sequence of random variables that converges in distribution to the standard normal, and if U_n is another sequence of random variables that converges in probability to 1, then

$$\frac{Z_n}{U_n} \overset{d}{\to} N(0,1).$$

We won't attempt to prove this here, but the result is rather intuitive. In the case of the binomial distribution described above, we know that $\hat{\theta} = \overline{Y}$ is a consistent estimator of θ; that is,

$$\hat{\theta} \overset{P}{\to} \theta.$$

Thus, by Theorem 9.2.3,

$$\hat{\theta}(1-\hat{\theta}) \overset{P}{\to} \theta(1-\theta)$$

and so

$$\sqrt{\frac{\hat{\theta}(1-\hat{\theta})}{\theta(1-\theta)}} \overset{P}{\to} 1.$$

Thus,

$$\frac{\hat{\theta} - \theta}{\sqrt{\hat{\theta}(1-\hat{\theta})/n}} = \underbrace{\frac{\hat{\theta} - \theta}{\sqrt{\theta(1-\theta)/n}}}_{\text{converges in dist to } N(0,1)} \Bigg/ \underbrace{\frac{\sqrt{\hat{\theta}(1-\hat{\theta})}}{\sqrt{\theta(1-\theta)}}}_{\text{converges in prob to } 1}.$$

Finally, we conclude that

$$\frac{\hat{\theta} - \theta}{\sqrt{\hat{\theta}(1-\hat{\theta})/n}} \overset{d}{\to} N(0,1).$$

This is an important result, because it tells us that we can substitute the estimator $\hat{\theta}$ for θ in the denominator of the formula from the CLT, and the standard normal distribution is still the right approximate distribution. This is a subtle point, but it is quite powerful and useful.

We can therefore obtain a confidence interval for θ by starting with

$$\Pr\left(-z_{\alpha/2} < \frac{\hat{\theta} - \theta}{\sqrt{\hat{\theta}(1-\hat{\theta})/n}} < z_{\alpha/2}\right) = 1 - \alpha.$$

This can be manipulated to put θ in the middle and the random parts (the ones that involve $\hat{\theta}$) on the outside:

$$\Pr\left(\hat{\theta} - \frac{z_{\alpha/2}}{\sqrt{\hat{\theta}(1-\hat{\theta})/n}} \;<\; \theta \;<\; \hat{\theta} + \frac{z_{\alpha/2}}{\sqrt{\hat{\theta}(1-\hat{\theta})/n}}\right) \;=\; 1 - \alpha. \tag{10.4}$$

This gives us the confidence interval

$$\left(\hat{\theta} - \frac{z_{\alpha/2}}{\sqrt{\hat{\theta}(1-\hat{\theta})/n}} \,,\, \hat{\theta} + \frac{z_{\alpha/2}}{\sqrt{\hat{\theta}(1-\hat{\theta})/n}}\right). \tag{10.5}$$

Many, but not all, of the confidence intervals you run across will be of the form

$$\hat{\theta} \pm z_{\alpha/2} \times \text{something}.$$

This "something" is usually called the standard error.

> **Definition 10.3.1 — Standard Error of an Estimator.** Let θ be a parameter and $\hat{\theta}$ be an estimator. The *standard error of the estimator* $\hat{\theta}$ is the standard deviation of $\hat{\theta}$. Recall that estimators are random variables, so it makes sense to talk of their expectation, which is important for the concept of unbiasedness, as well as the variance or standard deviation. We abbreviate the standard error of the estimator $\hat{\theta}$ by s.e.$(\hat{\theta})$.

The term standard deviation is used in this definition in the same sense as when we defined the standard deviation of a random variable:

$$\text{s.e.}(\hat{\theta}) \;=\; \sqrt{\mathbb{V}(\hat{\theta})}.$$

We will often talk about the standard deviation of a random variable and of an estimator (which is a random variable). We will also talk about the standard error of an estimator, but we will never talk about the standard error of a random variable, unless the random variable is an estimator of a population parameter. In nearly all realistic cases, s.e.$(\hat{\theta})$ will depend on parameters that we don't know. For example, for the case of estimating the probability θ in a sequence of Bernoulli trials, we found that

$$\text{s.e.}(\hat{\theta}) \;=\; \sqrt{\frac{\theta(1-\theta)}{n}}, \tag{10.6}$$

which depends on θ. Since θ is unknown, this won't do. When we use an estimator for the parameter (in place of the parameter itself) in the standard error, the argument given above suggests that the normal approximation is still valid, so long as $\hat{\theta}$ is a consistent estimator.

We will distinguish between the standard error of the estimator $\hat{\theta}$, which is

$$\text{s.e.}(\hat{\theta}) \;=\; \sqrt{\frac{\theta(1-\theta)}{n}},$$

and the *estimated standard error*,

$$\widehat{\text{s.e.}}(\hat{\theta}) \;=\; \sqrt{\frac{\hat{\theta}(1-\hat{\theta})}{n}},$$

which is based on using a consistent estimator for any unknown parameters.

■ **Example 10.6** Consider the data on police stops in Denver. We can read this from the data file `Denver_Police_Pedestrian_Stops_and_Vehicle_Stops.csv`, available from the book's website. What is the estimate of the probability that an arrest is made on a vehicle stop in Precinct 111? Find a 95% confidence interval for this parameter.

Solution: The code below uses the pipe operator `%>%`, so we must first install and load the `tidyverse` library. After filtering for the precinct being number 111 and the stop being a vehicle stop, we create a logical variable `arrestMade` which is true if the call disposition contains the term "Arrest Made". There were 26,357 vehicle stops in precinct 111, and 2,529 of them led to an arrest being made. We have here a sequence of Bernoulli trials with $n = 26,357$ and $x = 2,529$.

```
Denver = read.csv("Denver_Police_Pedestrian_Stops_and_Vehicle_Stops.csv")
library( tidyverse )
Precinct111 = Denver %>%
    filter( PRECINCT_ID == "111" & PROBLEM == "Vehicle Stop" ) %>%
    group_by( CALL_DISPOSITION )
arrestMade = grepl( "Arrest Made" , Precinct111$CALL_DISPOSITION )
n111 = length( arrestMade )
theta111hat = sum( arrestMade ) / n111
est_se = sqrt( theta111hat*(1-theta111hat)/n111 )
print( c( theta111hat-1.96*est_se , theta111hat+1.96*est_se ) )
```

This produces

```
[1] 0.09239600 0.09950748
```

The point estimate for the probability that a vehicle stop leads to an arrest is $\hat{\theta} = \frac{2,529}{26,357} = 0.0960$. The sample size is plenty large for the normal approximation to the binomial to work well. The formula for the confidence interval in (10.5) is applied in the last two lines of code. The confidence interval for the probability of a vehicle stop leading to an arrest in precinct 111 is $(0.0924, 0.0995)$. This is a very narrow confidence interval, indicating we have much information about the magnitude of the probability. With a sample of size $n = 26,357$, a narrow interval is to be expected.   ■ ■ ■

This example brings up a subtle point about statistical inference. One might argue that we have the entirety of police stops, not a sample from some population. There is, of course, some merit for this assertion. If you think about this a little deeper, you should realize that there is always some uncertainty about whether a stop will lead to an arrest. We are making an assumption about this probability (for example, that it is constant across time), which may not hold and should certainly be investigated. We might envision our population as all stops that might be made under conditions as they exist today. This is rather nebulous, but such assumptions are often made, even if they are not admitted. For example, criminologists may compare the arrest rates among precincts. We could rerun the analysis from this previous example on precinct 113 to obtain a point estimate of $\hat{\theta} = \frac{1,229}{14,018} = 0.0877$ with a 95% confidence interval of $(0.0830, 0.0924)$. The confidence intervals for precincts 111 and 113 are nearly nonoverlapping, with the lower endpoint of the precinct 111 interval equaling the upper endpoint of the precinct 113 interval. This suggests that the arrest rates are different for the two precincts.[4] This kind of comparison is done all the time, even on data that can be thought of as the whole population. Many statisticians think of a *process* that produces the outcomes we observe. Inference is then made to the process. Confidence intervals in situations like this make sense only if we envision a larger population consisting of all stops that might be made in these precincts; in other words, we think of all possible outcomes that the process might produce.

---

[4]We will see better ways to approach this problem later in this chapter and in the next chapter.

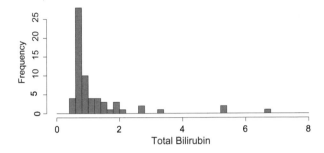

Figure 10.3 Bilirubin levels for men aged 20–49, with normal liver function.

The next example uses continuous measurements and asks for a confidence interval for the population mean. In the solution, we'll develop the formula for a confidence interval for μ.

■ **Example 10.7** We found in Example 9.25 that the liver data set was made up of a heterogeneous group of people: men and women aged 4 to 90, with and without liver disease. Let's consider a subset of the 583 people, namely men between the ages of 20 and 49 with normal liver function. This reduces the sample size to 64, but it creates a much more homogeneous group. Find a 95% confidence interval for the population mean bilirubin reading for men aged 20 to 49 with normal bilirubin. Interpret the interval.

Solution: Figure 10.3 shows a histogram of the bilirubin levels for the 64 men in the sample. This is clearly a nonnormal distribution since the data are skewed to the right. The CLT (Theorem 8.3.1), however, says that the sample mean from a population with mean μ and variance σ^2 is approximately normally distributed with mean μ and variance σ^2/n, even if the population is not normally distributed. In other words,

$$\frac{\overline{X} - \mu}{\sigma/\sqrt{n}} \overset{\text{approx}}{\sim} N(0,1).$$

Here we run into the same problem as in the previous example, where the standard error of the estimator depends on an unknown parameter. In the binomial example, the unknown parameter was θ in the expression $\sqrt{\hat{\theta}(1-\hat{\theta})/n}$, while here it is σ in the expression $\sigma/\sqrt{n}$. The same reasoning, applies, however: The sample standard deviation S is a consistent estimator of σ so $S/\sigma \overset{p}{\to} 1$. Thus, we can use the estimate S for the standard deviation in place of the true (but unknown) standard deviation σ and the approximate normal distribution still holds. That is,

$$\frac{\overline{X} - \mu}{S/\sqrt{n}} \overset{\text{approx}}{\sim} N(0,1).$$

This allows us to write

$$\Pr\left(-z_{\alpha/2} < \frac{\overline{X} - \mu}{S/\sqrt{n}} < z_{\alpha/2}\right) \approx 1 - \alpha,$$

which can be manipulated to yield the confidence interval for μ:

$$\left(\overline{X} - z_{\alpha/2}\frac{S}{\sqrt{n}}, \overline{X} + z_{\alpha/2}\frac{S}{\sqrt{n}}\right). \tag{10.7}$$

We can implement this formula using R; the code is shown below. We first read in the data frame from a csv file, then define the variables of interest. The line that defines `bilirubinM` selects only those subjects who are men aged 20–49 who have normal liver function (group=2).

```
liver = read.csv("liver_patient.csv")
gender = liver$Gender
bilirubin = liver$Total_Bilirubin
age = liver$Age
group = liver$Dataset
bilirubinM = bilirubin[ gender=="Male" & age > 19 & age < 50 & group==2 ]
n = length(bilirubinM)
xbar = mean( bilirubinM )
stdev = sd( bilirubinM )
print( c( xbar-1.96*stdev/sqrt(n) , xbar+1.96*stdev/sqrt(n) ) )
```

This gives the confidence interval as

    [1] 1.024396 1.591229

The confidence interval for the population mean μ is therefore $(1.024, 1.591)$. In the long run, 95% of the time an interval calculated in this way will include the true value of the mean. ∎

10.4 Small Sample Confidence Intervals

When sample sizes are small, the CLT cannot be applied. In cases like this, we have to work a little harder to find a confidence interval for a population parameter. What made the method from the previous section work was that for the case of data taken from a population with mean μ and variance σ^2,

$$Z = \frac{\overline{X} - \mu}{S/\sqrt{n}} \overset{\text{approx}}{\sim} N(0,1).$$

From this we were able to look up in a table – the normal table in this case – the number $z_{\alpha/2}$ such that

$$\Pr\left(-z_{\alpha/2} < \frac{\overline{X} - \mu}{S/\sqrt{n}} < z_{\alpha/2}\right) = 1 - \alpha. \tag{10.8}$$

The key is that we knew the approximate distribution of Z. This leads to the concept of a *pivotal quantity* which we define below.

Definition 10.4.1 — Pivotal Quantity. Suppose $X_1, X_2, \ldots, X_n \sim$ i.i.d. from some distribution with PDF $f(x|\theta)$ where θ is an unknown, possibly vector-valued, parameter. Let $Q(\boldsymbol{X}, \theta)$ be a function of the data $\boldsymbol{X} = [X_1, X_2, \ldots, X_n]$ and of the parameter θ. (Since Q is a function of the random sample, it is a random variable.) We say that $Q(\boldsymbol{X}, \theta)$ is a *pivotal quantity* if its distribution does not depend on any parameters that we don't know.

Definition 10.4.2 — Approximate Pivotal Quantity. If $Q(\boldsymbol{X}, \theta)$ is a function of the data $\boldsymbol{X} = [X_1, X_2, \ldots, X_n]$ and of the parameter θ, and if the distribution of Q is approximated by a distribution that does not depend on any unknown parameters (for example, Q might have an approximate normal distribution when the sample size n is large), then we say that Q is an *approximate pivotal quantity*.

We pointed out in Section 10.2 that assuming a known variance when the mean is unknown is a rather unrealistic problem, but it gave us insight into how to proceed in the general case, which we address now. Here, we make the more realistic assumption that $X_1, X_2, \ldots, X_n \sim$ i.i.d. $N(\mu, \sigma^2)$ where both μ and σ^2 are unknown. The formula in (10.5) does us no good, because it involves σ which we do not know. The most natural thing to do is to replace the population parameter σ with an estimate, say the sample standard deviation

$$S = \sqrt{\frac{1}{n-1}\sum_{i=1}^{n}(X_i - \overline{X})^2}.$$

The statistic that gets the ball rolling would then become

$$T = \frac{\overline{X} - \mu}{S/\sqrt{n}}.$$

Unfortunately, this statistic no longer has a normal distribution. Recall from Definition 6.6.3 in Chapter 6 that the t distribution was defined to be the distribution of the ratio

$$T = \frac{Z}{\sqrt{V/v}}, \tag{10.9}$$

where $Z \sim N(0,1)$ and $V \sim \chi^2(v)$ are independent random variables. For the case where we have a sample of size n from the $N(\mu, \sigma^2)$ distribution, we know that if H_0 is true, then

$$\frac{\overline{X} - \mu}{\sigma/\sqrt{n}} \sim N(0,1),$$

$$\frac{(n-1)S^2}{\sigma^2} \sim \chi^2(n-1).$$

We can then write the T statistic from above as

$$T = \frac{\overline{X} - \mu}{S/\sqrt{n}} = \frac{\dfrac{\overline{X} - \mu_0}{\sigma/\sqrt{n}}}{\dfrac{S/\sqrt{n}}{\sigma/\sqrt{n}}} = \frac{\dfrac{\overline{X} - \mu_0}{\sigma/\sqrt{n}}}{\sqrt{\dfrac{(n-1)S^2/\sigma^2}{n-1}}} = \frac{Z}{\sqrt{V/v}},$$

where $Z = (\overline{X} - \mu_0)/(\sigma/\sqrt{n})$, $V = (n-1)S^2/\sigma^2$, and $v = n-1$. Thus, the statistic T has a t distribution with $n-1$ degrees of freedom. We indicate by writing $T \sim t(n-1)$. Since this t distribution does not depend on any unknown parameters, T is a pivotal quantity.

The $t(v)$ distribution has PDF

$$f(t|v) = \frac{\Gamma\left(\frac{v+1}{2}\right)}{\sqrt{v\pi}\,\Gamma\left(\frac{v}{2}\right)}\left(1 + \frac{x^2}{v}\right)^{-(v+1)/2}, \qquad -\infty < t < \infty,$$

although we will never need to use this. Even though it doesn't look like the PDF for the normal, the shape of the t distribution does look much like the normal. As the degrees of freedom $v \to \infty$, the t distribution approaches the normal distribution. The t distribution is tabulated in many books, but its percentiles can be obtained easily using R.

In general, once a pivotal quantity and its distribution are identified, we can (in theory, at least) determine percentiles of that distribution and then make a probabilistic statement like the following:

$$\Pr(\text{lower quantile} < \text{pivotal quantity} < \text{lower quantile}) = 1 - \alpha,$$

where the lower quantile and upper quantile represent the values which leave $\alpha/2$ in the left and right tails, respectively, of the distribution for the pivotal quantity. The trick is then to use algebra to manipulate the inequalities inside the probability to get the unknown parameter in the middle, and the information from the random sample on the outside. We saw how we manipulated the Z statistic to leave μ in the middle and $\overline{X} \pm \sigma/\sqrt{n}$ as the endpoints. Applying this method yields

$$\Pr\left(-t_{\alpha/2}(n-1) < \frac{\overline{X} - \mu}{S/\sqrt{n}} < t_{\alpha/2}(n-1) \right), \tag{10.10}$$

and we proceed with virtually the same steps as in Section 10.2, except (1) use S in place of σ, and (2) use $t_{\alpha/2}(n-1)$ in place of $z_{\alpha/2}$. This yields the confidence interval

$$\left(\overline{X} - t_{\alpha/2}\frac{S}{\sqrt{n}}, \overline{X} + t_{\alpha/2}\frac{S}{\sqrt{n}} \right). \tag{10.11}$$

■ **Example 10.8** The file `Diabetes.csv` contains health information about 768 diabetic women who were Pima people (an American Indian tribe, mostly from southern Arizona and New Mexico, also known as Akimel O'odham). Let's consider only those women who were 40 or older and had 10 or more pregnancies. (We also filtered out those observations for which `BloodPressure` was less than 50, because these are likely to be errors.) The R code below loads and filters the data, plots a histogram so we can assess the normality assumption, and computes the confidence interval.

```
diabetes = read.csv( "Diabetes.csv" )
diabetes0 = diabetes %>%
    filter( Pregnancies >= 10 & Age >= 40 & BloodPressure > 50 )
DBP = diabetes0$BloodPressure
hist( DBP , breaks=seq(50,170,5) , xlab="Diastolic Blood Pressure" )
n = length( DBP )
xbar = mean( DBP )
stdev = sd( DBP )
t025 = qt( 0.975 , n-1 )
print( c( xbar - t025*stdev/sqrt(n) , xbar + t025*stdev/sqrt(n) ))
```

This produces

```
[1] 75.40948 82.94949
```

The value of $t_{\alpha/2}$ is computed using the `qt()` function in R and is found to be $t_{\alpha/2} = 2.024$. Note that we have defined t_γ to be that number that makes the *right* tail area equal to γ, whereas R uses the left tail area by default. If we want the number that makes the right tail area equal to 0.025, we find the value that makes the left tail area equal to 0.975.

The confidence interval for the mean diastolic blood pressure (the lower number in a blood pressure reading) is found to be $(75.41, 82.95)$. The histogram in Figure 10.4 is slightly skewed to the right,

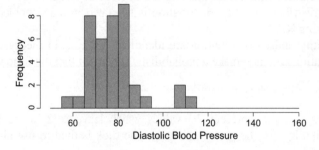

Figure 10.4 Histogram of diastolic blood pressure (DBP) for 39 diabetic women aged 40 or greater who have had 10 or more pregnancies.

indicating that the normal distribution may not be the appropriate distribution. Fortunately, the method for obtaining a confidence interval for the mean of a population is robust with respect to the assumption of a normal distribution. (We saw the term *robust* in Section 1.5 in connection with the use of the median rather than the mean. When we say that a method is robust, we mean that it works well even when the assumptions are violated.)   ■ ■ ■

If we are sampling from a normal distribution, we can obtain a confidence interval for the population variance by using the result that

$$\frac{(n-1)S^2}{\sigma^2} \sim \chi^2(n-1),$$ (10.12)

where S is the usual sample standard deviation. The result in (10.12) is a pivotal quantity because its distribution, $\chi^2(n-1)$, does not depend on any unknown parameters. To get a confidence interval for the population variance σ^2 we can do as before and put the pivotal quantity in the middle of an inequality and look up the endpoints for the appropriate distribution, χ^2 in this case. The χ^2 distribution is not symmetric like the normal and the t distributions, so we must look up both endpoints separately. The point that makes the upper tail of the $\chi^2(v)$ distribution have area γ is denoted $\chi^2_\gamma(v)$. If we want the point that makes the left tail area equal to γ, we must make the right tail area equal to $1 - \gamma$. (We can't just take the negative of the upper tail value; as we mentioned, the χ^2 distribution is not symmetric.) See Figure 10.5 for an illustration of how the upper and lower values are determined. The derivation begins with the probability statement

$$\Pr\left(\chi^2_{1-\alpha/2} < \frac{(n-1)S^2}{\sigma^2} < \chi^2_{\alpha/2}\right) = 1 - \alpha.$$

This leads to the confidence interval for σ^2:

$$\left(\frac{(n-1)S^2}{\chi^2_{1-\alpha/2}} < \sigma^2 < \frac{(n-1)S^2}{\chi^2_{\alpha/2}}\right).$$ (10.13)

We leave the details of the derivation as an exercise.

■ **Example 10.9** For the data in the previous example, find a 99% confidence interval for the population variance.

Solution: We can look up the quantiles of the $\chi^2(38)$ using R:

```
qchisq(0.005,38,lower.tail=TRUE)
qchisq(0.995,38,lower.tail=TRUE)
```

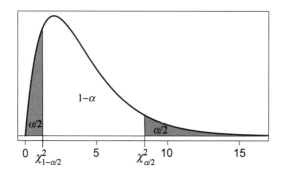

Figure 10.5 Determining lower and upper quantiles of the χ^2 distribution. Note, the χ^2 distribution is not symmetric, so both quantiles must be obtained from a table or from software.

which gives

```
[1] 19.28891
[2] 64.18141
```

Using code similar to that given in the preceding example, we find that the sample variance is

```
var( DBP )
```

which gives

```
[1] 135.2564
```

The confidence interval for σ^2 is thus

$$\left(\frac{(39-1) \times 135.2564}{64.18141} , \frac{(39-1) \times 135.2564}{19.28891} \right) = \left(206.526, 687.189 \right).$$

■ ■ ■

If we wanted a confidence interval for the population standard deviation σ, we would simply take the square root of the endpoints in the confidence interval for σ^2. See Problem 10.29. This would yield the 99% confidence interval $(14.37, 26.21)$.

> **Exercise 10.2** If the systolic blood pressures on 10 persons selected at random from some population yielded
>
> 120,   155,   104,   95,   128,   115,   104,   99,   142,   131,
>
> find 95% confidence intervals for the population mean and standard deviation.
>
> *Answer*: $(105.2089, 133.3911)$ and $(13.54898, 35.96087)$.

10.5 Confidence Intervals for Differences

Confidence intervals for individual parameters can give us a good idea about the magnitude of some quantity. Often, however, we are more interested in comparing one group with another group. For example, we might ask whether the arrest rates from vehicle stops are the same in two different precincts. Or, we might ask whether systolic blood pressure is the same in women who have had several pregnancies compared to those who have not. Questions such as these, which involve population parameters, can be answered using confidence intervals for differences. In the former case, we would define θ_1 and θ_2 to be the arrest rates in precincts 1 and 2 and we would want a confidence interval for the new parameter $\theta_1 - \theta_2$. The population in this case would be the *process* that generates vehicle stops in these precincts, whether it be in the past or future. (Admittedly, this is rather vague, but we usually want to infer something about a process that generated our observed data.) In the case of diastolic blood pressure, we would define μ_1 and μ_2 to be the mean systolic blood pressure in groups 1 and 2 and we would want a confidence interval for $\mu_1 - \mu_2$.

Once we have derived confidence intervals for parameters, obtaining confidence intervals for differences is easy. We use the rules for expectations and variances of differences. Recall that for independent random variables X and Y,

$$\mathbb{E}(X - Y) = \mathbb{E}(X) - \mathbb{E}(Y)$$

and

$$\mathbb{V}(X - Y) = \mathbb{V}(X) + \mathbb{V}(Y).$$

10.5.1 Confidence Intervals for Differences of Proportions

Let's apply these rules to the problem of differences of population proportions. If we collect data from each of two groups, and count the number of successes (i.e., units with the property that we are studying), we would define X to be the number of successes out of the n_1 units selected in group 1, and Y to be the number of successes out of the n_2 units selected in group 2. Under these assumptions

$$X \sim \text{BIN}(n_1, \theta_1) \qquad \text{and} \qquad Y \sim \text{BIN}(n_2, \theta_2),$$

where θ_1 and θ_2 are the true (unknown) population proportions in populations 1 and 2. Estimates for θ_1 and θ_2 are then

$$\hat{\theta}_1 = \frac{X}{n_1} \qquad \text{and} \qquad \hat{\theta}_2 = \frac{Y}{n_2}. \tag{10.14}$$

Using the result from (8.41), these estimators have variances

$$\mathbb{V}(\hat{\theta}_1) = \frac{\theta_1(1-\theta_1)}{n_1} \qquad \text{and} \qquad \mathbb{V}(\hat{\theta}_2) = \frac{\theta_2(1-\theta_2)}{n_2}. \tag{10.15}$$

These variances can be estimated by

$$\widehat{\mathbb{V}(\hat{\theta}_1)} = \frac{\hat{\theta}_1(1-\hat{\theta}_1)}{n_1} \qquad \text{and} \qquad \widehat{\mathbb{V}(\hat{\theta}_2)} = \frac{\hat{\theta}_2(1-\hat{\theta}_2)}{n_2}. \tag{10.16}$$

Since the samples are independent, the statistics $\hat{\theta}_1$ and $\hat{\theta}_2$ are independent, so we can write

$$\mathbb{V}(\hat{\theta}_1 - \hat{\theta}_2) = \mathbb{V}(\hat{\theta}_1) + \mathbb{V}(\hat{\theta}_2) = \frac{\theta_1(1-\theta_1)}{n_1} + \frac{\theta_2(1-\theta_2)}{n_2}.$$

The *estimated* variance for the difference in sample proportions is

$$\widehat{\mathbb{V}}\left(\hat{\theta}_1 - \hat{\theta}_2\right) = \frac{\hat{\theta}_1(1-\hat{\theta}_1)}{n_1} + \frac{\hat{\theta}_2(1-\hat{\theta}_2)}{n_2}. \tag{10.17}$$

Thus, for large n_1 and n_2

$$\frac{\hat{\theta}_i - \theta_i}{\sqrt{\dfrac{\hat{\theta}_i(1-\hat{\theta}_i)}{n_i}}} \overset{\text{approx}}{\sim} N(0,1).$$

The above rules for the mean and variance of a difference imply that

$$\frac{(\hat{\theta}_1 - \hat{\theta}_2) - (\theta_1 - \theta_2)}{\sqrt{\dfrac{\hat{\theta}_1(1-\hat{\theta}_1)}{n_1} + \dfrac{\hat{\theta}_2(1-\hat{\theta}_2)}{n_2}}} \overset{\text{approx}}{\sim} N(0,1). \tag{10.18}$$

The fraction on the left side of the above equation is seen to be an approximate pivotal quantity, because its approximate distribution doesn't depend on any unknown parameters. To derive a $100 \times (1-\alpha)\%$ confidence interval for $\theta_1 - \theta_2$, we would then look up the percentiles $\pm z_{\alpha/2}$ from the normal distribution, put the pivotal quantity between these percentiles, and do some algebra to obtain $\theta_1 - \theta_2$ in the middle, and the random endpoints of the confidence interval on the ends. (Keep in mind that $\hat{\theta}_1$ and

$\hat{\theta}_2$ are random variables, so the endpoints of the confidence interval, which are functions of $\hat{\theta}_1$ and $\hat{\theta}_2$, are random variables.) The resulting confidence interval is then

$$\left((\hat{\theta}_1 - \hat{\theta}_2) - z_{\alpha/2}\sqrt{\frac{\hat{\theta}_1(1-\hat{\theta}_1)}{n_1} + \frac{\hat{\theta}_2(1-\hat{\theta}_2)}{n_2}}, (\hat{\theta}_1 - \hat{\theta}_2) + z_{\alpha/2}\sqrt{\frac{\hat{\theta}_1(1-\hat{\theta}_1)}{n_1} + \frac{\hat{\theta}_2(1-\hat{\theta}_2)}{n_2}} \right). \quad (10.19)$$

The derivation of this result is left as an exercise.

If we are interested in whether the parameters θ_1 and θ_2 are equal, we would inspect the confidence interval to see whether it included zero. If it does include zero, then zero is a plausible value for $\theta_1 - \theta_2$ and the parameters may be the same. (It doesn't say they *are* the same, though.) If the confidence interval does not include zero, then zero is not a plausible value, so we have evidence that the parameters differ. We can never be certain that the parameters differ, but the confidence interval will provide information about how much they differ.

■ **Example 10.10** Find 99% confidence intervals for the arrest rates for vehicle stops in both precincts 111 and 113, and give a 99% confidence interval for the difference.

Solution: In precinct 111 there were 26,357 vehicle stops with 2,529 arrests, and in precinct 113 there were 14,018 vehicle stops with 1,229 arrests. The estimated parameters for the process of vehicle stops in these two precincts are

$$\hat{\theta}_1 = \frac{2,529}{26,357} \approx 0.0960 \quad \text{and} \quad \hat{\theta}_2 = \frac{1,229}{14,018} \approx 0.0877.$$

For 99% confidence intervals, the z value will be $z_{0.01/2} = 2.576$. With this change, the code from Example 10.6 can be edited to yield the following 99% confidence intervals:

$$(0.0913, 0.1006) \qquad \text{for } \theta_1$$
$$(0.0815, 0.0938) \qquad \text{for } \theta_2.$$

Figure 10.6 shows these intervals graphically. Notice that there is some overlap in the intervals, suggesting that the rates could be equal for the two arrest processes. This is misleading, however, because points in the overlap are at the right or left end of the confidence interval. Combinations of these values of θ_1 and θ_2 are slightly inconsistent with the observed data, and taken together could be quite inconsistent. A better approach is to look at the confidence interval for the difference.

To get the two individual confidence intervals we apply the method from Section 10.3 and for the confidence interval for the difference we apply equation (10.19). The code to do this is:

```
Denver = read.csv("Denver_Police_Pedestrian_Stops_and_Vehicle_Stops.csv")
library( tidyverse )
z = qnorm( 0.005 , 0 , 1 , lower.tail = FALSE )   ## Giving a 99% CI
```

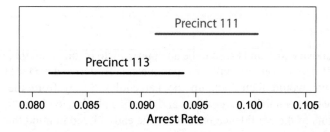

Figure 10.6 Comparison of 99% confidence intervals for the arrest rates during vehicle stops for precincts 111 and 113. Notice a slight overlap in the intervals.

```
Precinct111  =  Denver  %>%
        filter( PRECINCT_ID == "111" & PROBLEM == "Vehicle Stop" ) %>%
        group_by( CALL_DISPOSITION )
arrestMade111 = grepl( "Arrest Made" , Precinct111$CALL_DISPOSITION)
n111 = length( arrestMade111 )
theta111hat = sum( arrestMade111 ) / n111
est_se111 = sqrt( theta111hat*(1-theta111hat)/n111 )
print( c( theta111hat-z*est_se111 , theta111hat+z*est_se111 ) )
Precinct113  =  Denver  %>%
        filter( PRECINCT_ID == "113" & PROBLEM == "Vehicle Stop" ) %>%
        group_by( CALL_DISPOSITION )
arrestMade113 = grepl( "Arrest Made" , Precinct113$CALL_DISPOSITION)
n113 = length( arrestMade113 )
theta113hat = sum( arrestMade113 ) / n113
est_se113 = sqrt( theta113hat*(1-theta113hat)/n113 )
print( c( theta113hat-z*est_se113 , theta113hat+z*est_se113 ) )
est_se_diff = sqrt( est_se111^2 + est_se113^2 )
print( c( theta111hat-theta113hat - z*est_se_diff ,
        theta111hat-theta113hat + z*est_se_diff ))
```

which produces the 99% confidence intervals for θ_1, θ_2, and $\theta_1 - \theta_2$, respectively:

```
[1]  0.09127879 0.10062469
[1]  0.08152006 0.09382593
[1]  0.000552489 0.016005006
```

The confidence interval for the difference $\theta_1 - \theta_2$ is the interval $(0.00055, 0.01601)$, which does not include zero. Since the entire confidence interval lies on positive values for the difference, we can conclude that there is evidence that the arrest rate θ_1 in precinct 111 is higher than the arrest rate θ_2 in precinct 113. ∎∎∎

Notice how we phrased the conclusion. We said "There is evidence that the arrest rate in precinct 111 is higher than in precinct 113"; we don't say outright that "the arrest rate in precinct 111 is higher than in precinct 113." In statistics there is always uncertainty in the inferences we make, which requires us to say "there is evidence that ..."

10.5.2 Confidence Intervals for Differences in Means

Suppose we have samples from each of two populations where the normal distribution is a reasonable assumption. In other words, we have

$$X_1, X_2, \ldots, X_{n_1} \quad \sim \quad \text{i.i.d. } N(\mu_1, \sigma_1^2)$$
$$Y_1, Y_2, \ldots, Y_{n_2} \quad \sim \quad \text{i.i.d. } N(\mu_2, \sigma_2^2).$$

In addition, we assume that the two samples are independent of one another. We are interested in the difference in means $\mu_1 - \mu_2$. There are two options that one may follow in this problem, depending on whether it is reasonable to assume that the two population variances, σ_1^2 and σ_2^2, are equal. The classical approach assumes they are equal, whereas Welch's method does not. We begin with the classical approach.

If we assume equal population variances, the above assumptions become

$$X_1, X_2, \ldots, X_{n_1} \quad \sim \quad \text{i.i.d. } N(\mu_1, \sigma^2)$$

$$Y_1, Y_2, \ldots, Y_{n_2} \quad \sim \quad \text{i.i.d. } N(\mu_2, \sigma^2).$$

There are three parameters to be estimated: μ_1, μ_2, and σ^2. The estimators for μ_1 and μ_2 are obvious:

$$\hat{\mu}_1 = \overline{X} \quad \text{and} \quad \hat{\mu}_2 = \overline{Y}.$$

These are the MLEs and are unbiased estimates of μ_1 and μ_2, respectively. The parameter σ^2 raises a conundrum. For the case of the means, sample 1 contains no information about μ_2, and sample 2 contains no information about μ_1. *Both* samples, however, contain information about σ^2. (See Problems 10.48 and 10.49.) The estimators

$$\hat{\sigma}^2_{\text{samp 1}} = \frac{1}{n_1 - 1} \sum_{i=1}^{n_1} (X_i - \overline{X})^2 = S_1^2$$

and

$$\hat{\sigma}^2_{\text{samp 2}} = \frac{1}{n_2 - 1} \sum_{i=1}^{n_2} (Y_i - \overline{Y})^2 = S_2^2$$

are both unbiased, so in theory either could then be used to estimate the common variance σ^2. It seems more reasonable, however, to combine them to create a single estimate using data from both samples. The optimal way to combine the sample variances S_1^2 and S_2^2 is the weighted average

$$S_p^2 = \frac{(n_1 - 1)S_1^2 + (n_2 - 1)S_2^2}{n_1 + n_2 - 2} = \frac{\sum_{i=1}^{n_1} (X_i - \overline{X})^2 + \sum_{i=1}^{n_2} (Y_i - \overline{Y})^2}{n_1 + n_2 - 2}.$$

This is called the *pooled estimate of the variance* and it is an unbiased estimator of σ^2. The quantity

$$\frac{(\overline{X} - \overline{Y}) - (\mu_1 - \mu_2)}{S_p \sqrt{\frac{1}{n_1} + \frac{1}{n_2}}} \tag{10.20}$$

is a pivotal quantity; it has a t distribution with $n_1 + n_2 - 2$ degrees of freedom. This allows us to look up quantiles of the pivotal quantity, put the pivotal quantity between these quantiles, and rearrange to get the unknown difference of parameters in the middle. This leads to

$$\left((\overline{X} - \overline{Y}) - t_{\alpha/2} S_p \sqrt{\frac{1}{n_1} + \frac{1}{n_2}} , \ (\overline{X} - \overline{Y}) + t_{\alpha/2} S_p \sqrt{\frac{1}{n_1} + \frac{1}{n_2}} \right). \tag{10.21}$$

■ **Example 10.11** Consider the data set on diabetes from Example 10.8. Let's compare the diastolic blood pressure measurement for the two populations of (1) women who are 40 or over and have had 10 or more pregnancies, and (2) women who are 40 or over and have had fewer than 10 pregnancies. Find confidence intervals for each of the two population means and a confidence interval for the difference in means between these two populations.

Solution: We have a sample of size $n_1 = 39$ from the first population (age 40 or over with 10 or more pregnancies) and a sample of size $n_2 = 161$ from the second population (age 40 or over with fewer than 10 pregnancies). The sample statistics from these two samples can be computed with the following code:

```
library( tidyverse )
diabetes = read.csv( "Diabetes.csv" )
diabetes1  =  diabetes  %>%
    filter( Pregnancies >= 10 & Age >= 40 & BloodPressure > 50 )
DBP1 = diabetes1$BloodPressure
n1 = length( DBP1 )
xbar1 = mean( DBP1 )
stdev1 = sd( DBP1 )
t025_1 = qt( 0.975 , n1-1 )
print( c( xbar1-t025_1*stdev1/sqrt(n1) , xbar1+t025_1*stdev1/sqrt(n1) ))
diabetes2  =  diabetes  %>%
    filter( Pregnancies < 10 & Age >= 40 & BloodPressure > 50 )
DBP2 = diabetes2$BloodPressure
n2 = length( DBP2 )
xbar2 = mean( DBP2 )
stdev2 = sd( DBP2 )
t025_2 = qt( 0.975 , n2-1 )
print( c( xbar2-t025_2*stdev2/sqrt(n2) , xbar2+t025_2*stdev2/sqrt(n2) ))
spooled = sqrt( ((n1-1)*stdev1^2 + (n2-1)*stdev2^2) / (n1+n2-2) )
t025_diff = qt( 0.975 , n1+n2-2 )
print( c( xbar1-xbar2 - t025_diff*spooled*sqrt((1/n1)+(1/n2)) ,
          xbar1-xbar2 + t025_diff*spooled*sqrt((1/n1)+(1/n2)) ) )
```

which yields confidence intervals for μ_1, μ_2, and $\mu_1 - \mu_2$, respectively:

```
[1]  75.40948 82.94949
[1]  76.35857 79.69112
[1]  -2.677796  4.987081
```

The confidence intervals for the means μ_1 and μ_2 are $(75.409, 82.949)$ and $(76.359, 79.691)$, respectively. The confidence interval for the difference in means is $(-2.678, 4.987)$, which covers zero. Thus, zero is a plausible value for the true difference in population means, $\mu_1 - \mu_2$. We have no evidence that the difference in the population means is zero. Another way of saying this is that there is no evidence that the population means differ.   ■ ■ ■

Once again, notice the phrasing of the conclusion. "There is no evidence that the population means differ." We are not saying that the population means are the same; we just don't have evidence that they differ.

In the previous discussion we assumed that the population variances were equal. This is an assumption that may not hold in all cases. When such an assumption is untenable, we can apply a procedure known as Welch's method, or Satterthwaite's method (or the Welch–Sattherwaite method). In the method described previously, where we assumed the variances were the same, we used data from both samples to estimate the common σ^2. The result was the pooled variance, S_p^2, which led to the pivotal quantity in (10.20). If we were to use separate estimates for the variances in the two populations, σ_1^2 and σ_2^2, then the statistic would become

$$\frac{(\overline{X} - \overline{Y}) - (\mu_1 - \mu_2)}{\sqrt{S_1^2/n_1 + S_2^2/n_2}}, \tag{10.22}$$

where S_1^2 and S_2^2 are the sample variances from the two independent samples. In this case, the quantity in (10.22) is not a pivotal quantity, but is an approximate pivotal quantity because its distribution is approximately a t distribution where the degrees of freedom parameter is

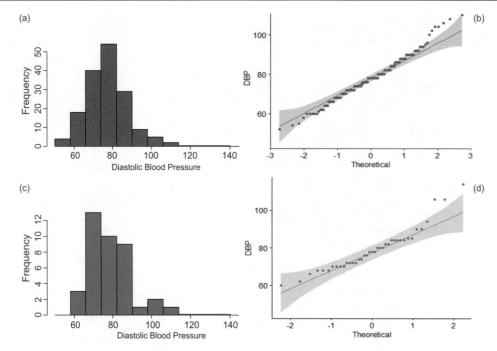

Figure 10.7 Histograms and QQ plots for diastolic blood pressure among those with fewer than 10 and 10 or more pregnancies.

$$v = \frac{\left(\dfrac{S_1^2}{n_1} + \dfrac{S_2^2}{n_2}\right)^2}{\dfrac{S_1^4}{n_1^2(n_1-1)} + \dfrac{S_2^4}{n_2^2(n_2-1)}}. \tag{10.23}$$

The degrees of freedom for the t distribution needn't be an integer, and the expression in (10.23) for the degrees of freedom for Welch's t test is rarely an integer.

■ **Example 10.12** Suppose in Example 10.11 we are unwilling to assume the variances were equal, and instead we applied Welch's approach. Find a 95% confidence interval for the difference in means and compare it with the result from Example 10.11.

Solution: The information from Example 10.11 can be summarized as follows:

$$n_1 = 39 \qquad n_2 = 161$$
$$\bar{x}_1 = 79.17949 \qquad \bar{x}_1 = 78.02484$$
$$s_1 = 11.62998 \qquad s_2 = 10.70569.$$

The degrees of freedom parameter for Welch's procedure is

$$v = \frac{\left(\dfrac{S_1^2}{n_1} + \dfrac{S_2^2}{n_2}\right)^2}{\dfrac{S_1^4}{n_1^2(n_1-1)} + \dfrac{S_2^4}{n_2^2(n_2-1)}} = \frac{\left(\dfrac{11.62998^2}{39} + \dfrac{10.70569^2}{161}\right)^2}{\dfrac{11.62998^4}{39^2(39-1)} + \dfrac{10.70569^4}{161^2(161-1)}} \approx 37.735.$$

Notice that the degrees of freedom parameter is not an integer as it was in all previous examples. If you are using a table of the t distribution, this would cause a problem, but if you use R to compute the quantiles of the t distribution it is not a problem at all. The R function qt() that computes quantiles of the t distributions can accept a noninteger for the degrees of freedom. After running the R code from the previous example to load the data and define the sample statistics, we can compute the degrees of freedom parameter and the margin of error quite easily:

```
### Welch's Method
df.diff.WS = ( stdev1^2/n1 + stdev^2/n2)^2 /
             ( stdev1^4/(n1^2*(n1-1)) + stdev2^4/(n2^2*(n2-1)) )
t025.diff.WS = qt( 0.975 , df.diff.WS )
print( df.diff.WS )
print( c(stdev1^2,stdev2^2) )
print( c(spooled^2) )
MOE.WS = t025.diff.WS * sqrt( stdev1^2/n1 + stdev2^2/n2 )
print( c( xbar1-xbar2-MOE.WS , xbar1-xbar2+MOE.WS ) )
```

This yields

```
[1] 58.05874
[1] 135.2564 114.6119
[1] 118.574
[1] -2.937785  5.247070
```

The confidence interval for the difference in means is $(-2.938, 5.247)$. This is a bit wider than the interval from before, when we assumed the variances were equal. This is mainly due to the t quantile, which previously was based on $n_1 + n_2 - 2 = 198$ degrees of freedom. Here, the degrees of freedom parameter is much lower, 37.735. Consequently the t quantile is slightly larger when using Welch's approach ($t_{0.025}(37.753) = 2.025$) than when using the classical approach ($t_{0.025}(198) = 1.972$). The larger t quantile leads to a somewhat wider confidence interval. ∎∎∎

With Welch's approach, the t distribution for the approximate pivotal quantity in (10.22) is usually a good approximation. There is a simpler version of Welch's method for which the degrees of freedom is easy to calculate and gives nearly the same result. The simplified version is to take the degrees of freedom to be

$$df = \min(n_1 - 1, n_2 - 1).$$

For the previous example, this would lead to

$$df = \min(n_1 - 1, n_2 - 1) = \min(39 - 1, 161 - 1) = 38.$$

The difference between using 37.753 degrees of freedom and 38 degrees of freedom is very slight, since $t_{0.025}(37.753) = 2.024861$ and $t_{0.025}(38) = 2.024394$. The R code to perform the simple version of Welch's method is:

```
df.simple = min( n1-1 , n2-1 )
t025.diff.WS.simple = qt( 0.975 , df.simple )
MOE.WS.simple = t025.diff.WS.simple * sqrt( stdev1^2/n1 + stdev2^2/n2 )
print( c( xbar1-xbar2-MOE.WS.simple , xbar1-xbar2+MOE.WS.simple ) )
```

which produces the output

```
[1] -2.984235  5.293520
```

With the simple version, the confidence interval is $(-2.984, 5.294)$, nearly identical to the interval calculated before.

In this section we assumed that we had samples from each of two completely separate populations, and we assumed that the samples were collected independently. The procedure discussed here is often called the *two-sample confidence interval for the difference of two means*, or sometimes the *independent samples confidence interval for the difference of two means*.

10.5.3 Confidence Interval for Paired Data

Rather than having independent samples, we sometimes collect data in a way that pairs a certain kind of measurement. For example, we might collect a before and after measurement on each subject. Or we might measure a manufactured part with two different measuring devices. In cases like this the samples are clearly not independent, so the method of the previous section does not apply.

When we have data such as these, we are not really asking about the difference of the means, but rather the mean of the differences. If we subtract one variable from the other, we obtain a difference that often has a clear interpretation. For example, the *after* measurement on a subject may tend to be larger than the *before* measurement; or measuring device A may yield smaller measurements than device B even when measuring the same parts. In cases like this, we should think of the *differences* as the observed data. We would like to know the mean of the distribution of differences.

Suppose our observed data are of the form:

i	Group 1	Group 2	Difference
1	$X_1 \sim N(\mu_1, \sigma_1^2)$	$Y_1 \sim N(\mu_2, \sigma_2^2)$	$D_1 = Y_1 - X_1 \sim N(\mu_d, \sigma_d^2)$
2	$X_2 \sim N(\mu_1, \sigma_1^2)$	$Y_2 \sim N(\mu_2, \sigma_2^2)$	$D_2 = Y_2 - X_2 \sim N(\mu_d, \sigma_d^2)$
$\vdots$	$\vdots$	$\vdots$	$\vdots$
n	$X_n \sim N(\mu_1, \sigma_1^2)$	$Y_n \sim N(\mu_2, \sigma_2^2)$	$D_n = Y_n - X_n \sim N(\mu_d, \sigma_d^2)$

Here, $\mu_d = \mu_2 - \mu_1$ and $\sigma_d^2 = \sigma_1^2 + \sigma_2^2 - 2\text{cov}(X_i, Y_i)$. When we analyze paired data like this, the focus is often on the population mean of the differences D_i. This is equal to

$$\mu_d = \mathbb{E}(D_i) = \mathbb{E}(Y_i - X_i) = \mathbb{E}(Y_i) - \mathbb{E}(X_i) = \mu_2 - \mu_1. \tag{10.24}$$

Once differences are taken, the problem becomes a one-sample problem where we want a confidence interval for the mean of a population (the population of *differences*). This was the topic of Section 10.4. The confidence interval for μ_d is therefore

$$\left(\overline{D} - t_{\alpha/2}\frac{S_d}{\sqrt{n}}, \ \overline{D} + t_{\alpha/2}\frac{S_d}{\sqrt{n}}\right). \tag{10.25}$$

Here, S_d is the sample standard deviation of the differences, and the degrees of freedom for the t value is $n-1$. Note that n is the number of *pairs* of data values, or equivalently the number of differences.

■ **Example 10.13** Consider a study done in the 1960s about the relationship between cholesterol and a coronary event, such as a heart attack. Here we focus only on the cholesterol variables. In this study, 20 subjects were studied from 1952 to 1962, with cholesterol levels being taken at the beginning and end of this time period. A study like this is called a *prospective cohort study*, since a fixed cohort of subjects was followed forward in time (or prospectively). Such data are called longitudinal data.

The data for the 20 subjects are shown in Table 10.1 and are also available in the file DixonMassey Cholesterol1952-1962.csv. The difference is calculated as the 1962 value minus the 1952 value, so that negative values indicate a *decrease* in the cholesterol level (Figure 10.8). Since interest is focused on the differences, we can ignore the original cholesterol and focus on the D_i.

Table 10.1 Cholesterol study from 1952 to 1962.

i	Chol. in 1952	Chol. in 1962	Difference
1	240	209	−31
2	243	209	−34
3	250	173	−77
4	254	165	−89
5	264	239	−25
6	279	270	−9
7	284	274	−10
8	285	254	−31
9	290	223	−67
10	298	209	−89
11	302	219	−83
12	310	281	−29
13	312	251	−61
14	315	208	−107
15	322	227	−95
16	337	269	−68
17	348	299	−49
18	384	238	−146
19	386	285	−101
20	520	325	−195

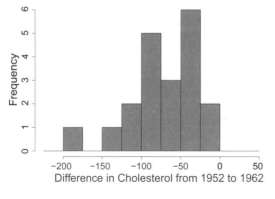

Figure 10.8 Histogram of *differences* in cholesterol from 1952 to 1962 for 20 subjects.

The following code loads the data, then defines the needed variables chol52, chol62, and diff. The next few lines determine the number of pairs along with the sample mean and standard deviation. The confidence interval for μ_d is computed and printed on the last line of code. Note that n in this case is 20 because that is the number of pairs. (Even though there are 40 data values, there are only 20 *pairs* of data values.)

```
chol = read.csv("DixonMasseyCholesterol1952-1962.csv")
chol52 = chol$chol52
chol62 = chol$chol62
diff = chol62 - chol52
n = length( diff )
t = qt( 0.025 , n-1 , lower.tail = FALSE )
dbar = mean( diff )
sddiff = sd( diff )
print( c( dbar - t*sddiff/sqrt(n) , dbar + t*sddiff/sqrt(n) ) )
```

This gives us the confidence interval

```
[1] -91.59243 -48.00757
```

We see that the confidence interval for the mean of the difference $\mu_d = \mu_2 - \mu_1$ is $(-91.59, -48.01)$. This interval excludes 0, suggesting that cholesterol levels dropped from 1952 to 1962.   ■ ■ ■

10.6 Determining the Sample Size

Before data are collected, we often need to make a decision about how large of a sample will be required to estimate the population parameter of interest. In the formulas for the confidence intervals derived above, the sample size n usually appears in the denominator and inside a square root. For example, the confidence interval for the probability θ in Bernoulli trials is

$$\left(\hat{\theta} - z_{\alpha/2}\sqrt{\frac{\hat{\theta}(1-\hat{\theta})}{n}}, \hat{\theta} + z_{\alpha/2}\sqrt{\frac{\hat{\theta}(1-\hat{\theta})}{n}} \right).$$

The margin of error (MOE) is the quantity that is subtracted from and added to the point estimate $\hat{\theta}$, which, for the binomial case, is

$$\text{MOE} = z_{\alpha/2}\sqrt{\frac{\hat{\theta}(1-\hat{\theta})}{n}}.$$

Suppose we would like to take a sufficiently large sample so that the MOE is equal to some predetermined value, say $\text{MOE} = E$. If we set this equal to the margin of error, we get

$$E = z_{\alpha/2}\sqrt{\frac{\hat{\theta}(1-\hat{\theta})}{n}}.$$

Squaring both sides gives

$$E^2 = z_{\alpha/2}^2 \frac{\hat{\theta}(1-\hat{\theta})}{n},$$

which can be solved for n:

$$n = z_{\alpha/2}^2 \frac{\hat{\theta}(1-\hat{\theta})}{E^2}. \tag{10.26}$$

There is a "catch" at this point. In order to find the sample size for a confidence interval for θ, we need to know what the estimate $\hat{\theta}$ will be. Before we take a sample size, we don't know what that will be. Knowing the true value of θ would work too, but we don't know that either. We are determining the sample size required to estimate the parameter θ so we can't use either $\hat{\theta}$ or θ in the formula. There are two ways around this conundrum. One is to note that the parameter in the MOE appears through the quantity

$$q(\theta) = \theta(1-\theta).$$

If we were to graph $q(\theta)$ against θ, we obtain the downward opening parabola as shown in Figure 10.9. This function seems to reach a maximum at $\theta = \frac{1}{2}$ and the maximum is $\frac{1}{2}(1-\frac{1}{2}) = \frac{1}{4}$. (See Problem 10.53 at the end of the chapter to demonstrate this.) Using $\theta = \frac{1}{2}$ makes the MOE as large as it could

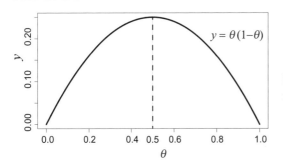

Figure 10.9 The graph of $q(\theta) = \theta(1 - \theta)$ reaches a maximum of $1/4$ when $\theta = 1/2$.

possibly be, so this approach leads to a wider margin of error. Statisticians would call this a *conservative* confidence interval because the estimated margin of error is likely on the high side. That is, once the data are collected, if $\hat{\theta} < 1/2$, then the margin of error will be smaller than E.

A second approach would be to use some preliminary estimate of θ, possibly from previous related studies. This value of θ could be used in place of θ in the computation of the MOE and the required sample size n. If θ is close to $\frac{1}{2}$, then the two approaches will be similar. If θ is near 0 or 1, then there may be a large difference between the two.

■ **Example 10.14** Polling organizations periodically take samples and ask those selected whether they approve of the job the president is doing. These are the job approval ratings that you hear about. How large a sample must be taken so that the margin of error for a 95% confidence interval for the proportion of those who approve of the president's job performance is 0.03?

Solution: Often the president's job approval is near 50%, so let's use the conservative value of $\hat{\theta} = 1/2$ in the formula for the required sample size. Then,

$$n = z_{\alpha/2}^2 \frac{\hat{\theta}(1 - \hat{\theta})}{E^2} = 1.96^2 \frac{1/4}{0.03^2} \approx 1{,}067.1.$$

We would ordinarily round up to $n = 1{,}068$.   ■ ■ ■

■ **Example 10.15** In the previous example, suppose that of the 1,068 people interviewed, 485 approved of the president's job performance. Find the margin of error and compare it with the value of MOE $= 0.03$ specified previously.

Solution: The point estimate for the approval proportion is $\hat{\theta} = \frac{485}{1{,}068} \approx 0.4541$ and the MOE is

$$\text{MOE} = 1.96 \sqrt{\frac{0.4541(1 - 0.4541)}{1{,}068}} \approx 0.02986,$$

which is less than but very close to the desired MOE of 0.03.   ■ ■ ■

Exercise 10.3 Suppose we had information beforehand, for example, a previous week's estimate, that the true value of θ was about 0.45. What would be the required sample size in this case?

Answer: $n = 1{,}057$ after rounding up.

Next, let's consider the problem of a confidence interval for the mean of a normal (or approximately normal) distribution. For the more realistic case where we do not know the population variance, the confidence interval uses the t distribution. Recall that if $X_1, X_2, \ldots, X_n \sim$ i.i.d. $N(\mu, \sigma^2)$, then a confidence interval for μ is

$$\left(\overline{X} - t_{\alpha/2} \frac{S}{\sqrt{n}},\ \overline{X} + t_{\alpha/2} \frac{S}{\sqrt{n}} \right).$$

The margin of error in this case is

$$\text{MOE} = t_{\alpha/2} \frac{S}{\sqrt{n}}.$$

If we have a desired MOE of E, then we could set $E = t_{\alpha/2} S / \sqrt{n}$ and solve for n to get

$$n = \frac{t_{\alpha/2}^2 S^2}{E^2}.$$

Here the catch is that we would need to know the value of the sample variance S^2, or, even better, the true value of σ^2. The only way forward here is to use a previously determined estimate of σ^2, call it σ_0^2, and use it in place of S^2 to yield

$$n = \frac{t_{\alpha/2}^2 \sigma_0^2}{E^2}. \tag{10.27}$$

Since the confidence interval is of the form $\left(\overline{X} - E, \overline{X} + E \right)$, the width of the resulting confidence interval is then $2E$. Thus, half of the width of the confidence interval is E. For this reason, the margin of error E is often called the *half-width* of the confidence interval.

There is one additional catch here. We haven't really solved for n because the value $t_{\alpha/2}$ depends on the degrees of freedom, $n-1$, which of course depends on n. The t distribution has nearly the same percentiles as the normal if the degrees of freedom is large. For example, with 50 degrees of freedom, $t_{0.025} = 2.01$, compared to $z_{0.025} = 1.96$ for the normal. In cases where the needed sample size is 50 or higher, using $t_{\alpha/2} \approx 2$ will be adequate.

A few things about (10.26) and (10.27) are worth pointing out. First, the margin of error E appears in the denominator of the formula for the required sample size. Thus, a narrower interval will require a larger sample size, and a wider interval can be obtained with a smaller sample size. This should match our intuition. A narrow confidence interval says a lot about the unknown population parameter, and so a lot of information – that is a large sample – is necessary. Second, if we want the confidence level $1 - \alpha$ to be higher, then α will have to be smaller, which means $t_{\alpha/2}$ must be larger. This, in turn, leads to a wider confidence interval.

■ **Example 10.16** Suppose we want to take a sample of residents near a town with a primary lead smelter. Suppose that in normal populations the mean blood lead level (BLL) is about 3.4 micrograms per deciliter (mg/dl), with a standard deviation of about 0.8. We would like to estimate the mean with a 95% margin of error equal to 0.1. (In other words, we want the 95% confidence interval to be of the form $\overline{x} \pm 0.1$.) How large must the sample be?

Solution: We will assume that the standard deviation is about 0.8, as it is assumed to be in normal populations. Using the approximation $t_{\alpha/2} \approx 2$, we have from (10.27) that the required sample size is

$$n = \frac{t_{\alpha/2}^2 S^2}{E^2} = \frac{2^2 \times 0.8^2}{0.1^2} = 256.$$

■ ■ ■

What if our original estimate for σ is off? The answer is predictable. If our prior estimate of σ is too high, and the sample standard deviation is smaller than we thought, then the confidence interval will be narrower than we initially specified (a good thing). On the other hand, if our prior estimate of σ is too low and the sample standard deviation is higher than we thought, then the confidence interval will be wider than we specified (a bad thing).

If we are interested in confidence intervals for differences, the procedure is similar. For the difference in proportions we proceed in a similar manner. The margin of error for the difference is

$$\text{MOE for } (\hat{\theta}_1 - \hat{\theta}_2) = z_{\alpha/2}\sqrt{\frac{\hat{\theta}_1(1 - \hat{\theta}_1)}{n_1} + \frac{\hat{\theta}_2(1 - \hat{\theta}_2)}{n_2}}. \tag{10.28}$$

We would set this MOE equal to our desired value E and solve for the sample sizes n_1 and n_2. The usual conundrum of not knowing θ_1 and θ_2 is present here as it was in Section 10.5, so the solution should be familiar. We could either use $\hat{\theta}_i = \frac{1}{2}$ and create a conservative confidence interval, or we could go with initial estimates for θ_1 and θ_2 and use these in place of $\hat{\theta}_1$ and $\hat{\theta}_2$. Unless θ_1 or θ_2 are close to 0 or 1, the two methods are likely to give similar results. Another problem that didn't come up previously is that we now have two unknowns: n_1 and n_2. Without an additional constraint, there will be a number of solutions to (10.28). Usually we would want equal sample sizes in the two groups, so that $n_1 = n_2 = n$, leaving us with one equation in one unknown:

$$E = z_{\alpha/2}\sqrt{\frac{\hat{\theta}_1(1 - \hat{\theta}_1)}{n} + \frac{\hat{\theta}_2(1 - \hat{\theta}_2)}{n}}. \tag{10.29}$$

If we take the first approach and go with conservative confidence intervals, we would take $\hat{\theta}_1 = \hat{\theta}_2 = \frac{1}{2}$, yielding

$$E = z_{\alpha/2}\sqrt{\frac{\frac{1}{2}(1 - \frac{1}{2})}{n} + \frac{\frac{1}{2}(1 - \frac{1}{2})}{n}} = z_{\alpha/2}\sqrt{\frac{1}{2n}},$$

which has the solution

$$n = \frac{z_{\alpha/2}^2}{2E^2}. \tag{10.30}$$

■ **Example 10.17** Suppose you work for a hospital and your team would like to design an intervention (e.g., home checkups) that will reduce the chance of a hospital readmission within 30 days. Your experiment is designed so that n discharged patients are assigned to the intervention group, and another n are assigned to receive no treatment. The latter is the *placebo* or *control* group. You are interested in the proportions θ_1 and θ_2 of patients readmitted within 30 days, so you would like a confidence interval for $\theta_1 - \theta_2$. In the past, readmission rates at this hospital have been about 8%, for a proportion of 0.08. How large must the sample be for each group so that the 95% confidence interval for $\theta_1 - \theta_2$ has a margin of error equal to 0.04? Apply both the methods of (1) being conservative by taking $\theta_1 = \theta_2 = \frac{1}{2}$ and (2) using the prior estimate of 0.08.

Solution: Our desired margin of error is $E = 0.04$, so applying (10.30) yields

$$n = \frac{z_{\alpha/2}^2}{2E^2} = \frac{1.96^2}{2 \times 0.04^2} = 1,200.5.$$

Rounding up, we would have $n_1 = n_2 = 1,201$. Altogether, 2,402 patients would be needed.

If we were to use the prior estimate of 0.08 instead of the conservative 0.50, the margin of error from (10.28) would become

$$z_{\alpha/2}\sqrt{\frac{0.08 \times 0.92}{n} + \frac{0.08 \times 0.92}{n}} = z_{\alpha/2}\sqrt{\frac{0.1472}{n}}.$$

Setting this equal to 0.04 and solving for n gives

$$n = \frac{1.96^2 \times 0.1472}{0.04^2} = 353.4.$$

Thus, we would need 354 in each group, for a total of 708 discharged patients. In this case, there is a substantial difference between the two approaches. This is due to the estimated proportion being close to 0, so $\theta(1-\theta) = 0.08 \times 0.92 = 0.0736$, which is quite a bit below the maximum of 0.25 (see Problem 10.53). ∎ ∎ ∎

Exercise 10.4 Suppose the experiment from the previous example is run with 1,201 patients in each group. The results were that 72 out of the 1,201 patients in the intervention group were readmitted, and 108 of the patients in the control group were readmitted. Find a 95% confidence interval for the difference in proportions.

Answer: $(-0.0510, -0.0090)$.

Exercise 10.5 Suppose we want to find a confidence interval for the difference in means between two normally distributed populations. Given a preliminary estimate σ_0^2 for the common variance, find a formula for the required sample size n for each population. Assume the sample size is large enough so that the t distribution percentiles are nearly the same as the normal percentiles.

Answer: $n = 2\sigma^2 z_{\alpha/2}^2/E^2$.

10.7 Confidence Intervals from Complex Survey Data

10.7.1 Sampling from a Finite Population

So far we have equated i.i.d. random variables with a random sample. This is reasonable if the population is very large relative to the sample size. If the population is not so large, then the independence assumption fails. For example, if we take a sample of size 20 from a population of size 100 and if we measure a categorical variable, calling the outcomes success and failure, then successive draws are *dependent*. Suppose that of the 100 in the population there are 60 with the characteristic being measured (successes) and 40 without (failures). The probability that the first trial is a success is therefore 0.60. If the first one selected is a success, then (assuming we are sampling without replacement) the probability that the second one is a success is $59/99 \approx 0.596$. The probability distribution for the number of successes in 20 trials has a hypergeometric distribution, not a binomial distribution. The two distributions are shown in Figure 10.10. A careful examination indicates that the hypergeometric probability is slightly more concentrated around the mean, which is $20 \times 0.6 = 12$ for both distributions.

Using results from Chapter 5 we can say that if we sample 20 items from a virtually infinite population, the distribution of the number of successes is $\mathrm{BIN}(20, 0.6)$, which has mean and variance

$$\mu_{\mathrm{Bin}} = np = 20 \times 0.6 = 12 \quad \text{and} \quad \sigma_{\mathrm{Bin}}^2 = np(1-p) = 20 \times 0.6 \times (1-0.6) = 4.8.$$

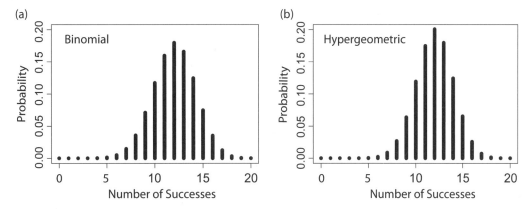

Figure 10.10 Comparison of binomial and hypergeometric PMFs

On the other hand, if the population size is just 100, then the mean and variance are

$$\mu_{\text{Hyp}} = 20\frac{60}{100} = 12 \quad \text{and} \quad \sigma^2_{\text{Hyp}} = 20 \times \frac{60}{100} \times \left(1 - \frac{60}{100}\right) \times \left(\frac{100 - 20}{100}\right) = 3.84.$$

The hypergeometric has a smaller variance than the binomial.

   If we sample without replacement from a finite population, where the variable of interest is binary (consisting of successes and failures), the finite nature of the population must be reflected in the confidence interval. The standard error of the estimate of the proportion of successes is

$$\text{s.e.}(\hat{\theta}) = \sqrt{\frac{\theta(1 - \theta)}{n - 1}\left(1 - \frac{n}{N}\right)},$$

where n is the sample size, N is the population size, and θ is the true population proportion of successes. Of course, θ is unknown so it must be estimated. The estimated standard error is therefore

$$\widehat{\text{s.e.}}(\hat{\theta}) = \sqrt{\frac{\hat{\theta}(1 - \hat{\theta})}{n - 1}\left(1 - \frac{n}{N}\right)},$$

which leads to the confidence interval

$$\left(\hat{\theta} - z_{\alpha/2}\sqrt{\frac{\hat{\theta}(1 - \hat{\theta})}{n - 1}\left(1 - \frac{n}{N}\right)} \, , \, \hat{\theta} + z_{\alpha/2}\sqrt{\frac{\hat{\theta}(1 - \hat{\theta})}{n - 1}\left(1 - \frac{n}{N}\right)}\right). \tag{10.31}$$

The factor $1 - n/N$ is called the finite population correction.

■ **Example 10.18** Consider the presidential job approval poll described in Example 10.14. Suppose a sample of 1,200 is selected from the population of about 160 million registered voters. In the sample, suppose 550 approve of the president's job performance. Estimate the proportion of the population who approve of the president's job performance and give a 90% confidence interval.

Solution: The point estimate is simply the fraction in the sample who approve; thus,

$$\hat{\theta} = \frac{550}{1,200} \approx 0.4583.$$

Applying the result from (10.31) yields

$$\hat{\theta} \pm z_{0.05} \sqrt{\frac{\hat{\theta}(1-\hat{\theta})}{n-1}\left(1-\frac{n}{N}\right)} = 0.4583 \pm 1.645\sqrt{\frac{0.4583(1-0.4583)}{n-1}\left(1-\frac{n}{N}\right)}$$

$$= 0.4583 \pm 0.0237.$$

The confidence interval is therefore $(0.4347, 0.4820)$ for the population proportion who approve of the president's job performance. Note that in this case the finite population correction factor is

$$1 - \frac{n}{N} = 0.9999925,$$

which is very close to 1. We would get nearly identical results if we assumed an infinite population. ∎ ∎ ∎

∎ **Example 10.19** Suppose a university selects a sample of 1,200 students from the population of all 16,000 students, and the question is whether the person approves of the job of the university's chancellor. Suppose 550 say they approve.

Solution: Note that all the numbers are the same as in the previous example except the population size. The confidence interval is found to be $(0.4356, 0.4811)$, which is just a bit narrower than the interval from the previous example. This is because the sample size makes up a much larger percentage of the population, 7.5% here compared to 0.00075% previously.   ∎ ∎ ∎

In the preceding discussion we were interested in a population proportion. If we are interested in a population mean rather than a proportion, a similar method applies when we sample from a finite population. Note that the normal distribution cannot apply to a finite population, since the normal is a continuous distribution, and if we sample from a finite population the possible outcomes are discrete. The population may be symmetric and bell-shaped, but it is not possible to be normally distributed. Because of this, we must necessarily rely on the CLT. The variance of the sample mean can be shown to be

$$\mathbb{V}(\overline{Y}) = \frac{\sigma^2}{n}\left(1-\frac{n}{N}\right).$$

The estimated standard error is obtained by using the unbiased estimator S^2 in place of σ^2 and taking the square root; that is,

$$\widehat{\text{s.e.}}(\overline{Y}) = \frac{S}{\sqrt{n}}\sqrt{1-\frac{n}{N}}.$$

This leads directly to the following confidence interval for the population mean when taking a sample of size n from a population of size N:

$$\left(\overline{Y} - t_{\alpha/2}\frac{S}{\sqrt{n}}\sqrt{1-\frac{n}{N}}, \overline{Y} + t_{\alpha/2}\frac{S}{\sqrt{n}}\sqrt{1-\frac{n}{N}}\right). \tag{10.32}$$

Exercise 10.6 Suppose a simple random sample of 100 university graduates selected from the population of 2,000 graduates yielded a mean student debt of \$40,200, with standard deviation of \$14,900. Find a 95% confidence interval for the mean student debt level in the population of all 2,000 students.

Answer: $(\$37,243.52, \$43,156.48)$

10.7.2 Stratified Random Samples

In Section 7.5 we discussed sampling strategies beyond the usual simple random sample. Here we focus on stratified sampling, where we take a simple random sample from each of the strata that make up the population. The sampling method must be taken into account when making point and interval estimates for parameters. The next example should make it clear why this is so.

■ **Example 10.20** Suppose a company has 3,000 employees: 1,000 men and 2,000 women. They select a stratified sample of 100 men and 100 women. One of the questions they ask in the survey is "Do you approve of the new policy for a leave of absence?" The results are summarized in the table below. Find a point estimate for the population proportion who would answer "yes."

	Do you approve of the new policy?		
	Yes	No	Total
Men	60	40	100
Women	50	50	100
Total	110	90	200

Solution: First, we must realize that this is not a simple random sample, so the sampling plan, a stratified random sample in this case, must be taken into account. The naive (and incorrect) solution, which ignores the stratification and just assumes that the sample is a simple random sample, would give us 110 successes in 200 trials, yielding

$$\hat{\theta}_{naive} = \frac{110}{200} = 0.55.$$

The naive confidence interval is therefore

$$\hat{\theta}_{naive} \pm z_{\alpha/2}\sqrt{\frac{\hat{\theta}_{naive}(1-\hat{\theta}_{naive})}{n-1}\left(1-\frac{n}{N}\right)} = 0.55 \pm \sqrt{\frac{0.55 \times (1-0.55)}{200-1}\left(1-\frac{20}{300}\right)}$$
$$= 0.55 \pm 0.0668.$$

A closer look at the data shows that men approve of the policy at a higher rate than women. Men approve 60% to 40%, whereas for women it is a 50–50 split. But there are more women in the population than men, even though the sampling design called for 100 of each to be selected. We can see that 10% of men were selected (100 out of 1,000), whereas only 5% of women were selected (100 out of 2,000). We would say that men were *oversampled* and women were *undersampled* because of the deviations of the proportion selected from the proportion in the population. Each man in the sample represented 10 men, whereas each woman in the sample represented 20 women. The sample of women must therefore be given more weight. It seems intuitive that the correct estimate for the population proportion who favor the policy should be closer to the 50% number for women than the 60% for men, because there are more women in the population and samples of equal size were taken.

In order to take into account the sampling design, let's look at men and women separately. Rather than trying to estimate the *proportion* in each group who favor the policy, let's try to estimate the *total number* in each group who favor the policy. If 50% of women in the sample approve of the policy, we would estimate that half of the 2,000 women in the company, that is, 1,000 women, would favor the policy. On the other hand, if 60% of men in the sample favor the policy, then we would estimate that 600 of the 1,000 men (60% of the men) would be in favor. The total number of people in favor would be estimated to be $600 + 1,000 = 1,600$. Since there are 3,000 people in the population, the estimate for the proportion who approve is $\hat{\theta} = 1,600/3,000 \approx 0.5333$. As expected, the estimate is closer to the estimate from the women's number (50%) than the men's number (60%). ■ ■ ■

The more general case of stratification would have k strata. For $i = 1, 2, \ldots, k$, let n_i be the sample size for stratum i, let N_i be the number of subjects (or sampling units) in stratum i, and let y_i be the number of observed successes[5] from stratum i. Also, let $\hat{\theta}_i = y_i/n_i$ denote the estimate of the proportion of successes in stratum i. The general formula for estimating a proportion from a stratified sample is

$$\hat{\theta}_{\text{strat}} = \frac{N_1\hat{\theta}_1 + N_2\hat{\theta}_2 + \cdots + N_k\hat{\theta}_k}{N_1 + N_2 + \cdots + N_k}.$$

The numerator can be seen as an estimate of the total number of successes in the population, and the denominator is the total population size across all k strata. The above formula can be written in a slightly different way as

$$\hat{\theta}_{\text{strat}} = \frac{N_1}{N}\hat{\theta}_1 + \frac{N_2}{N}\hat{\theta}_2 + \cdots + \frac{N_k}{N}\hat{\theta}_k,$$

where $N = N_1 + N_2 + \cdots + N_k$ is the population size. This shows that the estimate of the population proportion of successes is a weighted average of the within-stratum estimates.

Let τ denote the total number of successes in the population. The estimate for τ can therefore be written as

$$
\begin{aligned}
\hat{\tau} &= (\text{est. of total from stratum 1}) + (\text{est. of total from stratum 2}) + \cdots + (\text{est. of total from stratum } k) \\
&= N_1\frac{y_1}{n_1} + N_2\frac{y_2}{n_2} + \cdots + N_k\frac{y_k}{n_k} \\
&= \frac{N_1}{n_1}y_1 + \frac{N_2}{n_2}y_2 + \frac{N_k}{n_k}y_k.
\end{aligned}
$$

This representation gives the estimate of the total number of successes as a weighted average of the observed data $y_1, y_2, \ldots, y_k$. The weight w_i gives the number of subjects or units in the population that are represented by each member contained in the sample. For example, in Example 10.20,

$$\hat{\tau} = \frac{N_1}{n_1}y_1 + \frac{N_2}{n_2}y_2 = \frac{1,000}{100}60 + \frac{2,000}{100}50 = 10 \times 60 + 20 \times 50 = 1,600.$$

Let's dig deeper into the data and define x_{ij} to be 1 if the jth person from stratum i is a success and 0 otherwise. Then $y_i = \sum_{j=1}^{n_i} x_{ij}$. We can write the estimate of the total as the weighted sum of all of the x_i:

$$
\begin{aligned}
\hat{\tau} &= \frac{N_1}{n_1}(x_{11} + x_{12} + \cdots + x_{1n_1}) + \frac{N_2}{n_2}(x_{21} + x_{22} + \cdots + x_{2n_2}) + \cdots + \frac{N_k}{n_k}(x_{k1} + x_{k2} + \cdots + x_{kn_k}) \\
&= \underbrace{\frac{N_1}{n_1}x_{11} + \cdots + \frac{N_1}{n_1}x_{kn_1}}_{\text{all } n_1 \text{ weights are } \frac{N_1}{n_1}} + \underbrace{\frac{N_2}{n_2}x_{21} + \cdots + \frac{N_2}{n_2}x_{kn_2}}_{\text{all } n_2 \text{ weights are } \frac{N_2}{n_2}} + \cdots + \underbrace{\frac{N_k}{n_k}x_{11} + \cdots + \frac{N_k}{n_k}x_{kn_k}}_{\text{all } n_k \text{ weights are } \frac{N_k}{n_k}}. \quad (10.33)
\end{aligned}
$$

Note that the weights on the individual data points x_{ij} in (10.33) sum to the population size N. This may not appear obvious, but because the weights are constant within each stratum, we have

[5]By "success" we mean selecting a person with the property that we are studying, which is "favoring the policy."

$$\sum_{i=1}^{k} w_i = \underbrace{\frac{N_1}{n_1} + \frac{N_1}{n_1} + \cdots + \frac{N_1}{n_1}}_{n_1 \text{ times}} + \underbrace{\frac{N_2}{n_2} + \frac{N_2}{n_2} + \cdots + \frac{N_2}{n_2}}_{n_2 \text{ times}} + \cdots + \underbrace{\frac{N_k}{n_k} + \frac{N_k}{n_k} + \cdots + \frac{N_k}{n_k}}_{n_k \text{ times}}$$

$$= n_1 \frac{N_1}{n_1} + n_2 \frac{N_2}{n_2} + \cdots + n_k \frac{N_k}{n_k}$$

$$= N_1 + N_2 + \cdots + N_k$$

$$= N.$$

Thus, the weight associated with each individual in the sample is equal to N_i/n_i, where i is the stratum that contains that individual. This is equal to the number of individuals who are represented by each individual in the sample. The reciprocal of the weight,

$$\pi_{ij} = \frac{1}{w_{ij}} = \frac{n_i}{N_i},$$

is equal to the probability that individual j in stratum i would be selected in the sample. These ideas of sample weights are important when we use software to estimate parameters in complex surveys, as we describe below.

In the previous example, we could write the estimate of the proportion of the population who favor the policy as the weighted average of the observed proportions for men and women:

$$\hat{\theta} = \frac{1,000}{3,000} \times 0.60 + \frac{2,000}{3,000} \times 0.50 \approx 0.5333.$$

The sampling weights are equal to N_i/n_i, which for men is $w_{1j} = N_1/n_1 = 1,000/100 = 10$ and for women is $w_{2j} = N_2/n_2 = 2,000/100 = 20$. Thus, as we described earlier, each man in the sample represents 10 men in the population, and each woman in the sample represents 20 women in the population. The sum of the weights for all those men and women in the sample is

$$\sum_{i=1}^{2} \sum_{j=1}^{n_i} w_{ij} = \underbrace{10 + 10 + \cdots + 10}_{100 \text{ times}} + \underbrace{20 + 20 + \cdots + 20}_{100 \text{ times}} = 100 \times 10 + 100 \times 20 = 3,000.$$

Sampling weights also play a role when we want to estimate the mean of some population. For example, suppose that in Example 10.20, we want to know the average commute distance in miles for all 3,000 employees (1,000 men and 2,000 women) from our stratified sample of 100 men and 100 women. Let μ_1 and μ_2 be the population means that are estimated by the respective sample means $\overline{X}$ and $\overline{Y}$. If we were to obtain estimates of $\bar{x} = 20$ and $\bar{y} = 12$ then the total commute times for men and women are estimated to be

$$\hat{\tau}_1 = \text{estimate of total for men} \times 1,000 = 20 \times 1,000 = 20,000$$

and

$$\hat{\tau}_2 = \text{estimate of total for women} \times 2,000 = 12 \times 2,000 = 24,000.$$

Thus, the estimate for the total among all 3,000 employees is the sum of these two estimated totals:

$$\hat{\tau} = \hat{\tau}_1 + \hat{\tau}_2 = 20,000 + 24,000 = 44,000.$$

The estimate for the population mean commute distance is therefore

$$\hat{\mu} = \frac{\hat{\tau}_1 + \hat{\tau}_2}{1,000 + 2,000} = \frac{20,000 + 24,000}{3,000} \approx 14.67 \text{ miles.} \tag{10.34}$$

This is closer to the women's estimated mean commute distance (12 miles) than to the men's (20 miles).

Given this example, it should be clear how to proceed for the general case. Let's get the notation down first. Suppose we take a stratified random sample from a population with k strata where the number of units in stratum i is N_i. We take a sample of size n_i from stratum i. The mean and total of all N_i units in stratum i are μ_i and τ_i; these are unknown population parameters. Let $N = N_1 + N_2 + \cdots + N_k$ denote the number of units in the population. The overall population total and mean are

$$\tau = \sum_{i=1}^{k} \text{total in stratum } i = \sum_{i=1}^{k} \tau_i \qquad \text{and} \qquad \mu = \frac{\tau}{N}.$$

Let x_{ij} denote the measurement on the jth unit in stratum i and let

$$\bar{x}_i = \frac{\sum_{j=1}^{n_i} x_{ij}}{n_i}$$

denote the sample mean from stratum i. We would estimate the stratum mean μ_i by

$$\hat{\mu}_i = \bar{x}_i.$$

The estimate of the total in stratum i is therefore

$$\hat{\tau}_i = N_i \bar{x}_i.$$

The estimate of the total from the entire population is

$$\hat{\tau} = \sum_{i=1}^{k} \hat{\tau}_i = \sum_{i=1}^{k} N_i \bar{x}_i = \sum_{i=1}^{k} \sum_{j=1}^{n_i} \frac{N_i}{n_i} x_{ij}.$$

The quantity $w_i = N_i/n_i$ is called the *sampling weight*. The sampling weight for unit j in stratum i is the coefficient of x_{ij} in the estimate for the total. The sum of the weights is

$$\sum_{i=1}^{k} \sum_{j=1}^{n_i} w_{ij} = \sum_{i=1}^{k} \sum_{j=1}^{n_i} \frac{N_i}{n_i} = \sum_{i=1}^{k} n_i \frac{N_i}{n_i} = \sum_{i=1}^{k} N_i = N.$$

The sum of the weights is equal to the population size. Note that weights are assigned only to those units that were selected in the stratified sample. Also, the probability that unit j from stratum i is selected is $\pi_{ij} = n_i/N_i$ because we sample from every stratum, and sampling is done without replacement. The reciprocal of the selection probability, $1/\pi_{ij}$ is equal to the sampling weight w_i. The estimate of the population mean is therefore

$$\hat{\mu} = \frac{\text{estimate of total across all strata}}{\text{total number of units in population}} = \frac{\sum_{i=1}^{k} \hat{\tau}_i}{N}.$$

For the example discussed above we were interested in the total and mean commute distance for the population of 3,000 employees. Here, the sample weight for each of the 100 men in the sample is $w_{1j} = N_1/n_1 = 1,000/100 = 10$ and the sample weight for each of the 100 women in the sample is $w_{ij} = N_2/n_2 = 2,000/100 = 20$. There are 100 men, each with a sampling weight of 10, and 100 women each with a sampling weight of 20, so the sum of the weights is $10 \times 100 + 20 \times 100 = 3,000$, which is the population size. Another way of looking at the sampling weights is this: Each man in the sample represents 10 men, and each woman in the sample represents 20 women.

The estimated standard error for the estimated mean can be shown to be[6]

$$\mathrm{MOE}(\hat{\mu}) = z_{\alpha/2} \sqrt{\sum_{i=1}^{k} \left(1 - \frac{n_i}{N_i} \right) \left(\frac{N_i}{N} \right)^2 \frac{s_i^2}{n_i}}. \tag{10.35}$$

The margin of error can be used to obtain a confidence interval for the mean of a population given data from a stratified random sample.

■ **Example 10.21** Suppose that in the stratified random sample of the company with 3,000 employees, the sample standard deviations were $s_1 = 8$ and $s_2 = 6$. Find a 95% confidence interval for the mean commute distance.

Solution: The point estimate for the mean, $\hat{\mu} = 14.67$, was determined in (10.34). The margin of error can be found from (10.35):

$$\begin{aligned}
\mathrm{MOE} &= z_{\alpha/2} \sqrt{\sum_{i=1}^{k} \left(1 - \frac{n_i}{N_i} \right) \left(\frac{N_i}{N} \right)^2 \frac{s_i^2}{n_i}} \\
&= 1.96 \sqrt{\left(1 - \frac{n_1}{N_1} \right) \left(\frac{N_1}{N} \right) \frac{s_1^2}{n_1} + \left(1 - \frac{n_2}{N_2} \right) \left(\frac{N_2}{N} \right) \frac{s_2^2}{n_2}} \\
&= 0.9109.
\end{aligned}$$

The R code to perform these calculations is:

```
n1 = 100;          n2 = 100
N1 = 1000;         N2 = 2000
N = N1 + N2
xbar1 = 20;        xbar2 = 12
s1 = 8;            s2 = 6
z = 1.96;          muhat = (20000+24000)/3000
MOE = z * sqrt( (1-n1/N1)*(N1/N)^2*s1^2/n1 + (1-n2/N2)*(N2/N)^2*s2^2/n2 )
print( muhat )
print( c( muhat - MOE , muhat + MOE ))
```

which gives

```
[1] 14.66667
[1] 13.75574 15.57759
```

The 95% confidence interval for the mean commute distance is therefore $(13.76, 15.58)$ miles.   ■ ■ ■

---

[6]A demonstration of this is beyond the scope of this book; see Lohr (2010).

10.7.3 Cluster Sampling

Another type of sampling plan is *cluster sampling*. Consider the situation described in Example 7.9, where conversations were recorded on cassette tapes, with multiple conversations per tape. The sampling plan was to select a random sample of tapes and then listen to all of the conversations on the selected tapes. This is the concept behind *one-stage cluster sampling*. Suppose the population is divided into many small clusters. The clusters may be households, city blocks, cassette tapes, or some other sort of group. One-stage cluster sampling then involves two steps:

1. Select a simple random sample of the clusters.
2. Inspect all of the units in the sampled clusters.

The clusters are often homogeneous. People in the same household are often in the same family and therefore in the same socioeconomic group. Also, calls on the same cassette are likely to be made by the same business associate. This homogeneity must be taken into account when we construct the margin of error. Suppose we take a cluster sample of 100 households with 1–8 members, totaling 400 people. Compare this with a simple random sample of 400 people. Both samples contain 400 people, but the cluster sample contains less information. This is because the people in a particular cluster are likely to have similar opinions or experiences compared to those in the simple random sample. Cluster samples are not more efficient (statistically), but they are easier to conduct. In fact, a simple random sample of 100 people might be as costly as a cluster sample of size 400 because of the efficiency of collecting data from all members of a household once that household is selected.

Two-stage cluster samples are similar, but involve selecting a simple random sample of all of the outcomes in each selected cluster. Therefore, two-stage cluster sampling involves these three steps:

1. Select a simple random sample of the clusters.
2. Within each selected cluster, take a simple random sample of units.
3. Inspect all of the selected units.

The methods of inference for cluster sampling are messy, although not much more difficult to perform. Here we will emphasize the larger picture that one-stage and two-stage cluster samples are easier to conduct and less costly to perform, but the homogeneity within clusters must be accounted for. There are really two sources of variability in cluster sampling: one is the *within-cluster* variability, and the other is the *cluster-to-cluster* variability. If the within-cluster variability is small, cluster sampling is less efficient in estimating the population parameters. In the extreme case where all the units in a cluster are identical (i.e., no within-cluster variability), inspecting one element of the cluster is equivalent to inspecting *all* elements of the cluster. Thus, the information in the whole cluster will be the same as the information in a single unit, so in this case cluster sampling is not a good strategy.

Stratified and cluster samples look to be similar, and people often confuse the two. There are some fundamental differences between the two methods:

- Usually, the number of strata is small, whereas the number of clusters is large.

- Usually, the number of units in each stratum is large, whereas the number of units in each cluster is small.

- Clusters are often close together in some physical sense (same household, cassette tape, etc.), whereas units in strata are intermingled (e.g., male and female employees).

- We select a simple random sample from *each* stratum, whereas only a few clusters are selected and many clusters are not inspected at all.

Figure 10.11 illustrates the ideas of stratified random sampling (red, on the left), one-stage cluster sampling (blue, in the middle), and two-stage cluster sampling (green, on the right).

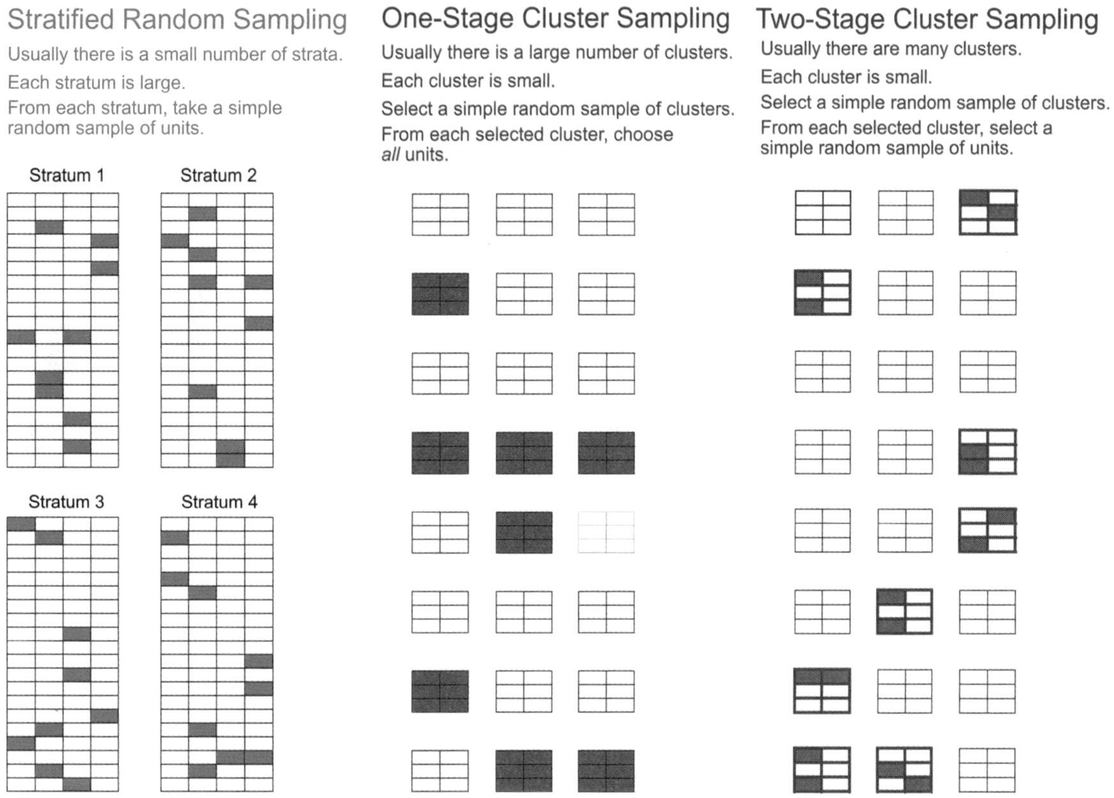

Stratified Random Sampling
Usually there is a small number of strata.
Each stratum is large.
From each stratum, take a simple random sample of units.

One-Stage Cluster Sampling
Usually there is a large number of clusters.
Each cluster is small.
Select a simple random sample of clusters.
From each selected cluster, choose *all* units.

Two-Stage Cluster Sampling
Usually there are many clusters.
Each cluster is small.
Select a simple random sample of clusters.
From each selected cluster, select a simple random sample of units.

Stratum 1   Stratum 2

Stratum 3   Stratum 4

Figure 10.11 Differences between stratified and cluster sampling.

10.7.4 Secondary Data Sources

In general, the research paradigm can be described as follows:
1. Identify a population of interest.
2. Frame a research question about the population.
3. Take a sample from the population and make the necessary measurements.
4. Analyze the data and answer the research question.

The most expensive part of this process is number 3, sampling and measuring units from the population. There are, however, a number of surveys collected by government agencies on various aspects of health, crime, etc.

Primary data are data collected by a research team for the specific purpose of addressing their research question. *Secondary data*, by contrast, are "[data] collected by someone else for some other purpose" (Boslaugh, 2007). Data can be both primary data for one purpose and research team, and secondary data for another purpose and research team. If research team A collects data to answer their research question and publishes their results, along with their data (on the journal's website), and later research team B downloads the data to analyze for their own research question, then the data are primary for research team A and secondary for research team B.[7]

The Centers for Disease Control and Prevention (CDC) conducts the National Health and Nutrition Examination Survey (NHANES), a study of health and nutritional status of adults and children living

---

[7]Many journals now require researchers to make available their data and computer code as a condition of publication. This is done in the interest of *reproducibility*; this means that a reader who downloads the same data and code should be able to get exactly the same results. *Replicability* refers to separate research teams repeating the same experiment from scratch, and reaching the same conclusions.

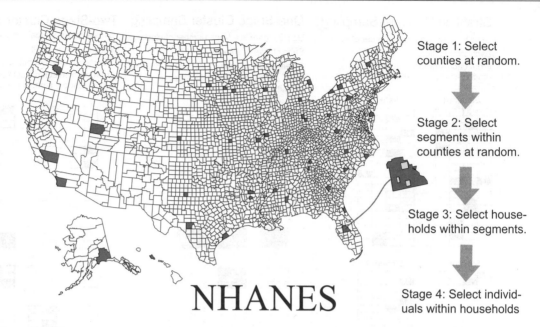

Stage 1: Select counties at random.

Stage 2: Select segments within counties at random.

Stage 3: Select house-holds within segments.

Stage 4: Select individ-uals within households

NHANES

Figure 10.12 Illustration of NHANES sampling procedure, which involves four stages.

in the United States. The study involves both interviews and physical exams of about 5,000 people per year. The CDC describes the NHANES sampling plan as follows:[8]

> "Stage 1: Primary sampling units (PSUs) are selected. These are mostly single counties or, in a few cases, groups of contiguous counties with probability proportional to a measure of size (PPS).
>
> Stage 2: The PSUs are divided up into segments (generally city blocks or their equivalent). As with each PSU, sample segments are selected with PPS.
>
> Stage 3: Households within each segment are listed, and a sample is randomly drawn. In geographic areas where the proportion of age, ethnic, or income groups selected for oversampling is high, the probability of selection for those groups is greater than in other areas.
>
> Stage 4: Individuals are chosen to participate in NHANES from a list of all persons residing in selected households. Individuals are drawn at random within designated age-sex-race/ethnicity screening subdomains. On average, 1.6 persons are selected per household."

The sampling procedure is illustrated in Figure 10.12. NHANES oversamples some groups so that separate inferences can be made for each group. Recall that the margin of error for estimating a mean mainly depends mostly on the sample size, not on the population size. If there were no oversampling, the proportion of minorities in the sample would be roughly proportional to their presence in the population, but these numbers would be fairly small, leading to wide confidence intervals.

Because it is a complex survey design, involving both stratification and clustering, special care must be taken when analyzing the NHANES data. The usual techniques of inference, including the confidence interval formulas, assume a simple random sample; since NHANES involves a complex survey design, these formulas are not applicable. Software must account for the stratification and clustering in the design. We will discuss how to use the R package survey for complex surveys later in this chapter.

---

[8]See www.cdc.gov/nchs/tutorials/nhanes/SurveyDesign/SampleDesign/Info1.htm.

The Behavioral Risk Factor Surveillance System (BRFSS, often pronounced "burfuss") is also conducted by the CDC. They say:

> BRFSS's objective is to collect uniform state-specific data on health risk behaviors, chronic diseases and conditions, access to health care, and use of preventive health services related to the leading causes of death and disability in the United States.[9]

The survey began in 1984 with data from 15 states. It now includes all 50 states plus the District of Columbia and three US territories. It involves 400,000 interviews of adults every year. Like NHANES, BRFSS is also a complex survey that involves stratification and clustering. The survey designers warn researchers that "Potential bias resulting from selection probabilities and noncoverage among segments of the population can be reduced through weighting."[10]

The Youth Risk Behavior Surveillance System (YRBSS),[11] also run by the CDC, collects data on health-related risk behaviors, including sexual behavior, alcohol and drug use, diet, tobacco, and physical activity. It is also a complex survey that uses stratification and clustering. The National Crime Victimization Survey (NCVS)[12] is conducted each year by the Bureau of Justice Statistics, which is part of the US Department of Justice. About 160,000 people are interviewed each year regarding criminal victimization. Data on age, sex, race, marital status, and income are collected, along with whether they have been the victim of a crime. Data on the nature of the crime, such as the time, the place, and the use of weapons, are also collected. The NCVS involves stratification and clustering, which must be accounted for when performing the data analysis.

Besides the four mentioned above, there are a number of other sources for secondary data. Most involve a complex survey design and care is necessary to analyze the data properly. See Boslaugh (2007) for more details on finding these surveys and using the available data.

There are a number of advantages to using these secondary data sources. The most obvious is cost. Many government sources are free, but others involve monetary cost. Even when there is a cost involved, a secondary source may still be cheaper. When you account for materials (e.g., tablets, computers), transportation, collection time of workers, data entry, etc., the cost of collecting your own data can be high. Also, many of the secondary data sources are compiled by government agencies that take data across the entire United States; doing this to obtain primary data would be a formidable challenge for individual research teams. Finally, these secondary data sources are usually compiled by governmental agencies that have experts who can assign the correct sampling weights and cluster variables. The major drawback to using secondary data is that you're stuck with the questions asked in the survey, which may not be directly applicable to your research question.

10.7.5 Software for Analyzing Data from Complex Surveys

Complex surveys, that is those with stratification and/or multistage clustering, call for special software so that the design can be taken into account. As we pointed out earlier, this is necessary to avoid potentially serious sources of bias. The required computations can be done in R using the `survey` package, which we illustrate below. Other software packages can also do the required calculations. With SAS software, there are a number of procedures that can be used, depending on the type of analysis. There are procedures for analyzing frequencies and means, as well as procedures for doing regression and logistic regression. The next example describes how to use R's survey package to analyze data from the YBRSS.

---

[9] `www.cdc.gov/brfss/annual_data/2017/pdf/overview-2017-508.eps`
[10] `www.cdc.gov/brfss/annual_data/2018/pdf/Complex-Smple-Weights-Prep-Module-Data-Analysis-2018-508.pdf`
[11] `www.cdc.gov/healthyyouth/data/yrbs/index.htm`
[12] `www.census.gov/programs-surveys/ncvs.html`

■ **Example 10.22** Data from the YBRSS from 2007 and 2017 are available in the files YRBS_2007.csv and YRBS_2017.csv. Use the height and weight data to compute the body mass index (BMI) for each respondent in the survey using the formula

$$\text{BMI} = \frac{\text{weight in kilograms}}{\text{height in meters squared}}.$$

The stratum, cluster, and sampling weights are given in the variables strata, psu, and weight, respectively. Use these to obtain point and interval estimates for the mean BMI level in the population in 2007 and 2017. Use results to find a confidence interval for the difference $\mu_{2017} - \mu_{2007}$.

Solution: The following code performs the required tasks. We begin by loading the survey package. If this package has not been installed on your computer, you must first install it; after that it should be able to be loaded with the library(survey) command. The next line loads the data for 2007. We define the yrbs2007$BMI2007 variable using the formula for how BMI is a function of height and weight. After that, we define a new data frame, yrbs, that consists of the important variables, and we omit those records that contain one or more missing values.

```
library(survey)
yrbs2007 = read.csv( file="YRBS_2007.csv" , header=TRUE  )
yrbs2007$BMI2007 = (yrbs2007$HowMuchDoYouWeighWithoutShoesInKG)/
                   (yrbs2007$HowTallAreYouWithoutShoesInMeters)^2
yrbs = data.frame( BMI=yrbs2007$BMI2007 , psu=yrbs2007$psu ,
                   strata=yrbs2007$stratum , weight=yrbs2007$weight )
yrbs_na = na.omit(yrbs)
design = svydesign( id = ~ psu , weight = ~ weight ,
                    strata = ~ strata, data = yrbs_na )
svymean( ~ BMI , design )
```

which yields

```
        mean     SE
BMI 23.411 0.1012
```

Nearly identical code (with 2017 replacing 2007) can be used to obtain the population mean BMI for 2017. The svydesign function is a function in the survey package that defines the design of the survey. The needed input includes the primary sampling unit (psu) that defines the cluster, the sampling weights, the stratum, and the name of the data frame. Finally, the svymean function computes the estimated mean and its standard error. There are five lines for each of 2007 and 2017. The estimates of the two population means (the mean BMI in 2007 and the mean BMI in 2017) along with their standard errors, are

$$\hat{\mu}_1 = 23.411, \quad \widehat{s.e.}_1 = 0.1012 \quad \text{[for 2007]}$$
$$\hat{\mu}_2 = 23.589, \quad \widehat{s.e.}_1 = 0.0880 \quad \text{[for 2017]}.$$

These estimates are not simply the sample means from the respective years (i.e., 2007 or 2017). Because YRBSS oversamples some groups and undersamples others, the sample weights must be used to estimate the mean parameters. The interval estimates for the two years' mean BMI levels are

$$(23.411 - 1.96 \times 0.1012, 23.411 + 1.96 \times 0.1012) = (23.21265, 23.60935) \quad \text{[for 2007]}$$
$$(23.589 - 1.96 \times 0.0880, 23.589 + 1.96 \times 0.0880) = (23.41652, 23.76148) \quad \text{[for 2017]}.$$

The estimate of the difference of means is

$$\hat{\mu}_2 - \hat{\mu}_1 = 23.589 - 23.411 = 0.178,$$

so the BMI seems to have increased over this decade. A confidence interval for the difference in means $(\mu_2 - \mu_1)$ can be easily computed because the samples are independent. The variance for the difference in means is

$$\mathbb{V}(\hat{\mu}_2 - \hat{\mu}_1) = \mathbb{V}(\hat{\mu}_2) + \mathbb{V}(\hat{\mu}_1).$$

The estimated standard error for the estimate of the difference of means is therefore

$$\widehat{s.e.}(\hat{\mu}_2 - \hat{\mu}_1) = \sqrt{\hat{\mathbb{V}}(\hat{\mu}_2) + \hat{\mathbb{V}}(\hat{\mu}_1)} = \sqrt{\widehat{s.e.}^2(\hat{\mu}_2) + \widehat{s.e.}^2(\hat{\mu}_1)} = \sqrt{0.1012^2 + 0.0880^2} = 0.1341.$$

A confidence interval for the difference in means is thus

$$\left((23.589 - 23.411) - 1.96 \times 0.1341, (23.589 - 23.411) + 1.96 \times 0.1341\right) = \left(-0.085, 0.441\right).$$

Thus, the estimated mean BMI increased slightly from 23.411 to 23.589, but since the confidence interval $(-0.085, 0.441)$ includes zero, zero is a plausible value for the population difference. ■ ■ ■

10.8 Chapter Summary

While the previous chapter focused on getting point estimates for parameters, we've focused in this chapter on getting confidence intervals for parameters. A confidence interval is a random interval (when viewed *before* data are collected) that contains the true value of the parameter with a given probability, usually close to 1. Once data are collected we would, for example, interpret a 95% confidence interval for a parameter as an interval computed by a method that includes the true parameter 95% of the time. The confidence is really in the method used to compute the interval.

When a pivotal quantity (i.e., a function of the unknown parameter and the data whose distribution does not depend on the unknown parameter) exists for a parameter, we can obtain quantiles $q_{1-\alpha/2}$ and $q_{\alpha/2}$ of the distribution of this pivotal quantity and use them to obtain a confidence interval. There are a number of situations where such a pivotal quantity exists. If we sample from a normal distribution with unknown mean μ and unknown variance σ^2, we can use the pivotal quantity $t = (\overline{X} - \mu)/(S/\sqrt{n})$ to obtain a confidence interval for μ. Similarly, the pivotal quantity $(n-1)S^2/\sigma^2$ can be used to obtain a confidence interval for σ^2. If data are collected from a design involving stratification or clustering, special care must be taken to obtain proper estimates of the standard error.

10.9 Problems

Problem 10.1 Explain why a confidence interval has random endpoints and therefore the interval itself is random.

Problem 10.2 Explain why the probability that a confidence interval contains the true parameter equals $1 - \alpha$ only when viewed *before* we have collected and analyzed the data.

Problem 10.3 You collect a sample of size $n = 100$ for which the sample mean is $\bar{x} = 3.76$. Calculate a 95% confidence interval for μ if the population has a $N(\mu, \sigma^2 = 9)$ distribution. Explain what your confidence interval means.

Problem 10.4 Calculate a 90% confidence interval for the data in Problem 10.3.

Problem 10.5 Calculate a 95% confidence interval for μ for the following data, 28.67, 15.91, 6.07, 16.99, 27.53, 4.58, 15.48, 13.90, −2.98, 23.18, which comes from a $N(\mu, \sigma^2 = 100)$ distribution. Explain your confidence interval.

Problem 10.6 Calculate a 98% confidence interval for the data in Problem 10.5.

Problem 10.7 Calculate a 95% confidence interval for μ for the following data, −2.5, −4.0, −6.5, −5.8, −6.2, −7.9, which comes from a $N(\mu, \sigma^2 = 20)$ distribution. Explain your confidence interval.

Problem 10.8 Calculate a 92% confidence interval for the data in Problem 10.7.

Problem 10.9 Why is the interval in Problem 10.7 wider than the one in Problem 10.8? Does this make sense? Justify your answer.

Problem 10.10 You collect a sample of size $n = 10,000$ with sample mean $\bar{x} = 0.537$ and $s = 0.129$. Because n is so large, you can use a large sample confidence interval calculation assuming

$$\Pr\left(\bar{X} - 1.96\frac{S}{\sqrt{n}} < \mu < \bar{X} + 1.96\frac{S}{\sqrt{n}}\right) \approx 1 - \alpha.$$

Using this approximation, calculate a 95% confidence interval for μ of the population.

Problem 10.11 Calculate the small sample confidence interval for Problem 10.10 using the appropriate t distribution critical value. Why was the approximate solution in Problem 10.10 so close to this solution?

Problem 10.12 Calculate the 90% confidence interval for the data in Problem 10.10.

Problem 10.13 To estimate the "click-through" rate for a website banner advertisement, you display the page to 9,481 customers. Of them, 393 click through and the rest do not. Calculate a 95% large sample confidence interval for the fraction of customers who click on the banner advertisement.

Problem 10.14 Calculate an approximate 99% large sample confidence interval for Problem 10.13. Explain why this interval is wider than the interval in Problem 10.13.

Problem 10.15 Calculate a small sample 95% confidence interval for μ for the data in Problem 10.5, which are assumed to come from a population having a normal distribution with an unknown variance. Explain why this interval is wider than the one calculated in Problem 10.5.

Problem 10.16 Calculate a small sample 98% confidence interval for the data in Problem 10.15.

Problem 10.17 Calculate a small sample 95% confidence interval for μ for the data in Problem 10.7, which comes from a population that has a normal distribution with unknown variance. Explain your confidence interval. Explain why this interval is different from the one calculated in Problem 10.7.

Problem 10.18 Calculate a small sample 92% confidence interval for the data in Problem 10.17.

Problem 10.19 Calculate a 95% confidence interval for the difference in arrest rates between precincts 111 and 113 for vehicle stops given in Example 10.10.

Problem 10.20 In a survey taken in February 2021,[13] 336 out of 550 men and 406 out of 612 women surveyed said they strongly approved or somewhat approved of the CDC's handling of the coronavirus pandemic. Find a 95% confidence interval for the difference in the population proportions of men and women who would approve of the CDC's handling.

Problem 10.21 Suppose that in a test of two website designs, visitors to the site were directed randomly to one of the two designs, labeled A and B. Of the 10,219 visitors to design A, 512 clicked through to the next page, while 578 of the 9,658 visitors to design B clicked through. Find a 90% confidence interval for the difference of the population proportions that would click through.

Problem 10.22 Calculate an approximate 90% large sample confidence interval for the difference in proportions using the data in Problem 10.13. Explain why this interval is narrower than the interval in Problem 10.13.

---

[13]See www.dataforprogress.org

Problem 10.23 Two formulations of aluminum were used to create bars that were tested for tensile strength. The strengths for six bars of each type are given below:

Type 1: 46.2, 48.1, 40.3, 43.1, 44.0, 47.4
Type 2: 40.1, 41.9, 43.3, 43.0, 39.9, 40.2

Assuming equal population variances, compute a 95% confidence interval for the difference in means between the populations of type 1 and type 2 bars.

Problem 10.24 Repeat Problem 10.23 without assuming equal variances.

Problem 10.25 Continuation of Problem 10.23. Four specimens of a third type of bar were also tested, yielding the following strengths:

Type 3: 38.0, 36.9, 38.7, 37.3

(a) Find a 95% confidence interval for the difference in strengths between type 1 and type 3 bars.
(b) Find a 95% confidence interval for the difference in strengths between type 2 and type 3 bars.
(c) Interpret the confidence intervals from Problem 10.23 and parts (a) and (b) above. Are we 95% confident that *all* three intervals simultaneously contain the true mean strength differences? (We will take up this issue in Chapter 20.)

Problem 10.26 Repeat Problem 10.25 without assuming equal variances.

Problem 10.27 Consider the aluminum strength data given in Problems 10.23 and 10.25. Find confidence intervals for the variances for each of the three types of aluminum.

Problem 10.28 To reduce road noise when asphalt is used for pavement, rubber is often added to the asphalt mixture. An important characteristic of asphalt is the stabilized viscosity. Fifteen measurements of this variable for 15 asphalt batches are shown below.

3193, 3124, 3153, 3145, 3093,
3466, 3355, 2979, 3182, 3227,
3256, 3332, 3204, 3282, 3170

Assume these data come from a normal distribution with unknown mean μ and unknown variance σ^2.

(a) Find a 95% confidence interval for the population mean μ.
(b) Find a 95% confidence interval for the population variance σ^2.

Problem 10.29 Let $X_1, X_2, \ldots, X_n$ be a sample from a $N(\mu, \sigma^2)$ distribution. Suppose we want a confidence interval for the standard deviation σ. Explain why we can first find a confidence interval for σ^2 and then take the square roots of the endpoints.

Problem 10.30 Use the method from Problem 10.29 to find a 95% confidence interval for the standard deviation σ for the data in Problem 10.28.

Problem 10.31 Suppose $X_1, X_2, \ldots, X_m \sim$ i.i.d. $N(\mu_1, \sigma_1^2)$ and $Y_1, Y_2, \ldots Y_n \sim$ i.i.d. $N(\mu_2, \sigma_2^2)$. Let S_1^2 and S_2^2 be the respective sample variances.

(a) What is the distribution of

$$\frac{S_1^2}{S_2^2} \frac{\sigma_2^2}{\sigma_1^2} \, ?$$

(b) Is the statistic in part (a) a pivotal quantity?

Problem 10.32 Use the result of Problem 10.31 to obtain a 95% confidence interval for the ratio of variances σ_2^2/σ_1^2 for the type 1 and type 2 aluminum strength data in Problem 10.23.

Problem 10.33 Suppose that $X_1, X_2, \ldots, X_n \sim$ i.i.d. EXP1(θ) which has PDF

$$f(x|\theta) = \begin{cases} \dfrac{1}{\theta} \exp(-x/\theta), & \text{if } x > 0 \\ 0, & \text{otherwise.} \end{cases}$$

(a) It can be shown that

$$Y = \frac{2\sum_{i=1}^n X_i}{\theta} \sim \chi^2(2n).$$

Is Y a pivotal quantity? Explain.

(b) Suppose we are testing a new drug to fight COVID. Our outcome is the recovery time, that is, the number of days it takes until a patient is symptom free. We give n patients the new drug and record their recovery time, X_i. We assume that $X_1, X_2, \ldots, X_n$ i.i.d. EXP1(θ). Develop a confidence interval for θ. Explain your reasoning.

(c) Suppose that in a test of $n = 300$ patients, the average recovery time is $\bar{x} = 10.5$ days for the patients given the new drug. Find a confidence interval for the population mean recovery time. State your results in easily understood terms.

Problem 10.34 Suppose $X_1, X_2, \ldots, X_m \sim$ i.i.d. EXP1(θ_1) and $Y_1, Y_2, \ldots Y_n \sim$ i.i.d. EXP1(θ_2). Let $\overline{X}$ and $\overline{Y}$ denote the sample means.

(a) Using the result from Problem 10.33, explain why

$$W = \frac{\overline{X}}{\overline{Y}} \frac{\theta_2}{\theta_1} \sim F(2m, 2n).$$

(b) Is W a pivotal quantity?

(c) Use the results from parts (a) and (b) to derive a confidence interval for the ratio of population means θ_2/θ_1.

Problem 10.35 Suppose that a new drug to fight COVID symptoms is being investigated. Twenty subjects with COVID are assigned to one of two groups. Ten are given the new drug and 10 are given a placebo. The average time in days until symptoms are gone are 10.5 for the new drug, and 14.0 for the placebo. Assume that the time to being symptom-free has an exponential distribution (EXP1) with mean θ_1 for the new drug and θ_2 for the placebo. Find a 95% confidence interval for the ratio of population means θ_2/θ_1.

Problem 10.36 Suppose that engineers are working on a new design for a toaster. Fifty items are produced with the old design and another 50 are produced with a new design. Assume exponential distributions $\text{EXP1}(\theta_1)$ and $\text{EXP1}(\theta_2)$ for the lifetimes of the toasters in the old and new design groups. The sample mean lifetime for the old design was 62,400 cycles, while the sample mean lifetime for the new design was 87,700. Find a 90% confidence interval for the ratio of population means.

Problem 10.37 Suppose that $X_1, X_2, \ldots, X_n \sim$ i.i.d. $N(\mu, 1)$.

(a) What is the 95% confidence interval for μ.

(b) How large must n be so that the confidence interval for μ has a half-width of 0.05?

Problem 10.38 Suppose that $X_1, X_2, \ldots, X_n \sim$ i.i.d. $N(\mu, 25)$.

(a) What is the 95% confidence interval for μ.

(b) How large must n be so that the confidence interval for μ has a half-width of 0.05?

Problem 10.39 Suppose that $X_1, X_2, \ldots, X_n \sim$ i.i.d. $N(\mu, \sigma^2)$, where σ^2 is unknown but believed to be around 0.01. How large should n be so that the margin of error for a 90% confidence interval for μ is 0.008?

Problem 10.40 Repeat Problem 10.39 for a margin of error of 0.002.

Problem 10.41 Suppose that investigators are interested in the fraction of a population in a particular county who are reluctant to get the COVID vaccine. How large must the sample size be so that the MOE for a 95% confidence interval is equal to 0.04? (Assume that no initial estimate of this fraction is available.)

Problem 10.42 Repeat Problem 10.41 with the prior assumption that the proportion is believed to be about 0.30.

Problem 10.43 Traffic engineers would like to know what fraction of motorists exceed the posted speed limit. They set up a radar device that determines the speed of every tenth car. How large must the sample size be so that the half-width of the confidence for the proportion of speeding motorists is 0.06?

Problem 10.44 Consider the situation described in Problem 10.43, except this time the radar device is connected to an electronic sign that flashes red and blue lights whenever a speeding car is detected. Also, the speed of every car (not every tenth car) is determined. Discuss the issues of bias involved here in estimating the fraction of speeding drivers.

Problem 10.45 Suppose $X_1, X_2, \ldots, X_m \sim$ i.i.d. $N(\mu_1, \sigma_1^2)$ and $Y_1, Y_2, \ldots, Y_n \sim$ i.i.d. $N(\mu_2, \sigma_2^2)$. The variances are unknown, but are assumed equal and roughly equal to 100. We want to obtain a 95% confidence interval for $\mu_1 - \mu_2$. Assuming equal sample sizes ($n_1 = n_2$), how large must n_1 and n_2 be so that MOE is equal to 4?

Problem 10.46 A university has a faculty of 700 professors spread across the ranks assistant professor (200), associate professor (200), and professor (300). To learn about the faculty's opinion regarding a new tenure policy, the administration selected a stratified sample of 96 assistant professors, 65 associate professors, and 69 professors. The results are shown in Table 10.2.

Table 10.2 Data for Problem 10.46.

Do you approve of the new tenure policy?			
	Yes	No	Total
Assistant professor	45	51	96
Associate professor	42	23	65
Professor	58	11	69

(a) Find a point estimate for the proportion of the faculty who approve of the new tenure policy.

(b) Find a 95% confidence interval for the population proportion in part (a).

(c) Suppose that instead of selecting a stratified random sample, the university sent the survey to *all* faculty, and 96 assistant professors, 65 associate professors, and 69 professors responded. How would you estimate the population proportion?

(d) Discuss the possible bias in your point estimate from part (c).

Problem 10.47 As part of the same survey described in Problem 10.46, the administration also asked "How many papers have you published in the last five years?" The results are summarized in Table 10.3. Assuming that the sample was really a stratified sample, give a point estimate and a confidence interval for the population mean of the number of papers published in the previous five years.

Table 10.3 Data for Problem 10.47.

How many papers have you published in the last five years?			
	n	$\bar{x}$	s^2
Assistant professor	96	8.4	2.0
Associate professor	75	7.5	3.6
Professor	79	11.2	5.8

Problem 10.48 For the case of independent samples from a normal distribution discussed in Section 10.5.2, show that the MLEs of the parameters are

$$\hat{\mu}_1 = \bar{X}$$
$$\hat{\mu}_2 = \bar{Y}$$
$$\hat{\sigma}^2 = \frac{1}{n_1 + n_2}\left(\sum_{i=1}^{n_1}(X_i - \bar{X})^2 + \sum_{i=1}^{n_2}(Y_i - \bar{Y})^2\right).$$

Problem 10.49 Consider the case of independent samples from a normal distribution discussed in Section 10.5.2. Let $\hat{\sigma}^2$ be the estimator for the common variance defined by

$$\hat{\sigma}^2 = \frac{1}{n_1 + n_2 - 2}\left(\sum_{i=1}^{n_1}(X_i - \bar{X})^2 + \sum_{i=1}^{n_2}(Y_i - \bar{Y})^2\right).$$

(a) What is the distribution of

$$\frac{(n_1 + n_2 - 2)\hat{\sigma}^2}{\sigma^2}?$$

(*Hints:* For a sample of size n from a normal distribution, $(n-1)S^2/\sigma^2 \sim \chi^2(n-1)$. Also, the sum of independent χ^2 random variables has a χ^2

distribution whose degrees of freedom is the sum of the degrees of freedom of the random variables in the sum.)

(b) Use this result to obtain a formula for a confidence interval for the common variance from independent samples from two normal distributions.

Problem 10.50 Use the result from Problem 10.49 to obtain a 95% confidence interval for the (assumed) common variance for the type 1 and type 2 aluminum strengths given in Problem 10.23.

Problem 10.51 Generalize the result from Problem 10.49 to the case of three independent samples.

Problem 10.52 Use the result from Problem 10.51 to find a 90% confidence interval for the (assumed) constant variance for the three types of aluminum whose strength data are given in Problems 10.23 and 10.25.

Problem 10.53 Use calculus to demonstrate that the function $q(\theta) = \theta(1 - \theta)$ reaches a maximum of $\frac{1}{4}$ at $\theta = \frac{1}{2}$.

Problem 10.54 Derive the confidence interval in equation (10.19).

Problem 10.55 Look up information on the Behavioral Risk Factor Surveillance System (BRFSS) at www.cdc.gov/brfss.

(a) In what sense is BRFSS a stratified sample?

(b) What groups are undersampled? Oversampled?

(c) How does the BRFSS design use clustering?

Problem 10.56 Repeat the questions in Problem 10.55 for the Youth Behavioral Risk Surveillance System (YRBSS). Information can be found at www.cdc.gov/healthyyouth.

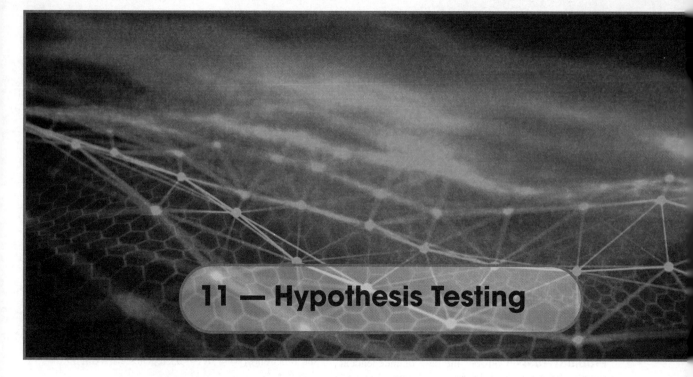

11 — Hypothesis Testing

11.1 Introduction

The last two chapters have covered the basic concepts of estimation. In Chapter 9 we studied the problem of giving a single number to estimate a parameter. In Chapter 10 we looked at ways to give an interval that we believe will include the true parameter. In many applications, we want to ask some very specific questions about the parameter(s). For example, is the mean of a normal distribution equal to 100, or is there evidence that it is not? Is the probability of some random trial equal to $\frac{1}{2}$, or is there evidence that it is not? Is the mean lifetime of a person using a new drug the same as the lifetime with a standard drug?

Let's consider the situation described in Example 8.9 about a clinical trial that is designed to give each participant both drugs, but at different times. Such a trial is called a crossover trial, because each participant "crosses" from Drug A to Drug B (or vice versa); the choice as to which comes first is determined randomly. A judgment must be made about which drug works better. For now, ties are not allowed. Often, one drug is new and is being compared to a standard drug, or possibly no drug at all. If the control is "no drug" then participants are usually given a pill that looks identical to the drug being tested, but has no active ingredients; this is called a *placebo*. Of interest is the probability that for a person chosen at random the new drug works better than the old. Call this parameter θ. In addition to estimating θ, we would like to know whether there is evidence that $\theta > \frac{1}{2}$, because this would tell us that Drug A (the new drug) works better than the placebo.

To further illustrate the situation, suppose we recruited 200 participants who took both Drug A (the new drug) and Drug B (the placebo). Let X denote the number out of the 200 for whom Drug A worked better than Drug B. If the drugs really work the same, then $X \sim \text{BIN}(200, \frac{1}{2})$. Suppose that for 101 people Drug A worked better and for the other 99 people Drug B worked better. This leads to $X = 101$. One might say that our estimate for θ is $\hat{\theta} = \frac{101}{200} = 0.505$, which is greater than $\frac{1}{2}$, so Drug A works better, right? Another, playing the devil's advocate, might say this:

> Wait a minute, 101 is only a little larger than what we expect, which is 100, and 101 could easily have occurred by chance even if the drugs work the same.

Note carefully what the quote says, because this contains the essential idea of hypothesis testing. The quote makes the assumption that the two drugs work the same, which is equivalent to assuming that the true parameter θ is equal to $\frac{1}{2}$. It also says that 101 is fairly likely if $\theta = \frac{1}{2}$. According to this reasoning, we would say that there really is no evidence that θ differs from $\frac{1}{2}$. The probability of observing 101 or more people for whom Drug A worked better than Drug B, given that the two drugs worked the same, can be obtained from the binomial distribution, or from the normal approximation to the binomial. If we use the binomial directly, the desired probability is

$$\Pr\left(X \geq 101 \,\middle|\, \theta = \frac{1}{2}\right) = 1 - \Pr\left(X \leq 100 \,\middle|\, \theta = \frac{1}{2}\right).$$

(The binomial distribution is discrete, so a number strictly between 100 and 101 is impossible; thus, the complement of $X \geq 101$ is $X \leq 100$.) We can use the binomial distribution and a little R code to approximate this:

```
1 - pbinom(100,200,0.5)
```

which gives

```
[1] 0.4718258
```

Obtaining 101 successes or more is not that unlikely *if* $\theta = \frac{1}{2}$.

Let's look at the same situation with different data. Suppose now that for 150 out of the 200 people, Drug A worked better than Drug B. Here our estimate is $\hat{\theta} = \frac{150}{200} = 0.75$. Could this happen if the true parameter is $\theta = \frac{1}{2}$? This would mean the split is 150 for Drug A and 50 for Drug B. Our intuition should say that this is unlikely if the drugs work the same. We can work this out as before:

$$\Pr\left(X \geq 150 \,\middle|\, \theta = \frac{1}{2}\right) = 1 - \Pr\left(X \leq 149 \,\middle|\, \theta = \frac{1}{2}\right).$$

Here is the R code to work this out:

```
1 - pbinom(149,200,0.5)
```

which produces

```
[1] 4.196643e-13
```

The probability of observing $X \geq 150$ is approximately $4.2 \times 10^{-13} = 0.00000000000042$ or about one chance in 2.4 trillion. The argument of the devil's advocate would now have to go something like this:

> It is possible (with probability 4.2×10^{-13}) to obtain 150 or more people for whom Drug A works better than Drug B, so we haven't any evidence that Drug A works better.

This argument now sounds absurd. While it is certainly possible to get $X \geq 150$, it is extremely unlikely; with a probability of 0.00000000000042 we might say that it is nearly impossible. Thus, if we observe $X = 150$ out of 200 trials, we would say that there *is* evidence that Drug A works better than Drug B because what we observed ($X = 150$) is unlikely to have occurred by chance if the two drugs work the same.

The two situations described above, where $X = 101$ and $X = 150$, lead to the obvious conclusions that there is no evidence that Drug A works better when $X = 101$, and there is strong evidence that Drug A works better when $X = 150$. You can imagine some intermediate cases where it isn't so obvious. For example, what if we observed $X = 120$; this is a split of 120 to 80 in terms of which drug worked better. It is not obvious how likely such an outcome is. This is why we studied probability so thoroughly in the first half of this book, and in particular why we studied the binomial distribution in Section 5.2. The exact probability that we would get 120 or more can be computed with similar R code as before:

```
1 - pbinom(119,200,0.5)
```

which gives the output

```
[1] 0.002842578
```

The probability is about 0.0028, a small number, but not nearly as small as the 4.2×10^{-13} we observed for $X = 150$. In Chapter 8 we used the normal approximation to the binomial and got an estimated probability of 0.0023, fairly close to our result above. The devil's advocate might then say

> Getting 120 or more people for whom Drug A worked better is possible if the two drugs work the same, so there is no evidence that Drug A works better.

This argument is not as absurd as before, when Drug A worked better for 150 people. Still though, 0.0028 is pretty small, so we might be inclined to say that the observed result of $X = 120$ is unlikely to have occurred by chance if the two drugs work the same (i.e., $\theta = \frac{1}{2}$). If we don't believe this outcome occurred by chance, then we would have to conclude that $\theta > \frac{1}{2}$ and that Drug A does work better.

If we observed $X = 120$ out of the 200 trials, then one of two things must be true:

1. Drug A works better, that is, $\theta > \frac{1}{2}$; or
2. the two drugs worked the same and we observed an unlikely event ($X = 120$, which has a probability of 0.0028).

The smaller the probability in item 2 is, the less likely we are to believe that the drugs are the same, and the more we will believe that Drug A works better.

The above discussion encapsulates the key ideas of hypothesis testing. You can probably see that there should be some cutoff to decide which statement we will conclude. The next section covers the terminology and concepts of hypothesis testing. Subsequent sections in this chapter cover some particular types of hypothesis tests.

11.2 Elements of a Statistical Test

Let's begin with some definitions.

> **Definition 11.2.1 — Hypothesis.** A *hypothesis* is a statement about one or more parameters. The *null hypothesis* is usually a specific statement about the parameters and is the statement against which we seek evidence. The *alternative hypothesis* is the statement that we conclude if we decide against the null hypothesis.
>
> The null hypothesis is denoted H_0 and the alternative hypothesis is denoted H_1.

In ordinary English, the word "null" means "nothing," so the null hypothesis is like a "nothing hypothesis." It often is a statement about parameters (or the population itself, if you prefer) that amounts to nothing. For example, two treatments may yield the *same* mean (no difference), the suicide rate today is the *same* as it has been in the past, or two random variables have *no* correlation, etc. In all of these cases, the null hypothesis is saying "there is nothing going on."

In the previous section we described the crossover experiment where we assigned 200 people to receive both Drugs A and B sequentially, with the selection of the first drug being done at random. We observed X, the number of times Drug A (the new drug) worked better than Drug B (the placebo), which, if the drugs really act the same, has a binomial distribution with $n = 200$ and $\theta = \frac{1}{2}$. In this problem we were looking to see whether there was evidence that the drugs are not the same. The null hypothesis here is therefore H_0: $\theta = \frac{1}{2}$. The alternative is H_1: $\theta > \frac{1}{2}$.

Often, but not always, the null and alternative hypotheses taken together cover the entire parameter space, which in this case is the interval $[0, 1]$ since θ is a probability. In this example, however, the two hypotheses together cover only the interval $[\frac{1}{2}, 1]$. The reason for this lies in the nature of drug testing. If the new drug works better than a placebo (i.e., $\theta > \frac{1}{2}$), then it is effective in treating the condition. If either $\theta = \frac{1}{2}$ or $\theta < \frac{1}{2}$, then the drug is not effective. In the former case it works the same as a placebo, and in the latter case it works worse than the placebo; either way, the drug is not effective. In this book, we choose to make the null hypothesis as specific as possible and choose H_0: $\theta = \frac{1}{2}$ rather than H_0: $\theta \leq \frac{1}{2}$. Be aware that some books will take the alternative to be H_1: $\theta \leq \frac{1}{2}$.

The main idea behind hypothesis testing is this:

Assume H_0 is true until we get evidence that it is false. $\qquad$ (11.1)

What we need to pin down is what we mean by "evidence that H_0 is false." Continuing with our crossover study, the decision must be based on the observed data X. Recall from the first section of this chapter that if X is large, we have evidence that the new drug is effective. We found that the chance of getting $X \geq 150$ is very small, telling us that this result is inconsistent with θ being equal to $\frac{1}{2}$. On the other hand $X = 101$ seemed to be quite consistent with H_0 being true, since the probability of getting $X \geq 101$ was 0.4718. We should probably choose our cutoff somewhere between 101 and 150. In order to make such a choice, we must have a criterion for making a decision.

Let's first distinguish between the reality (or ground truth) of which hypothesis is true and what our decision will be. Since there are two possibilities for each, there are four situations that could arise as shown in Table 11.1.

Table 11.1 Table showing Type I and Type II errors in the terminology of deciding H_0 or H_1.

		Reality	
		H_0 is true	H_0 is false
Decision	H_0	No error	Type II error
	H_1	Type I error	No error

Two of the boxes lead to no error at all: decide H_0 when H_0 is true, and decide H_1 when H_1 is true. The other two are errors:

1. Reject H_0 when H_0 is true (a "Type I error").
2. Do not reject H_0 when H_0 is false (a "Type II error").

The standard approach in hypothesis testing is to determine the cutoff point for our decision so that the probability of making a Type I error is some small predetermined number. Suppose we would like to make the probability of committing a Type I error to be about 0.05. This means that the cutoff point x_c must be selected so that

$$0.05 = \Pr(X \geq x_c | H_0 \text{ is true}).$$

We already know that $X = 101$ is too small since it makes the probability equal to 0.4718, and $X = 150$ is too large since it makes the probability 0.00000000000042. We could use the pbinom() function in R to zero in on the right value. In the code below we create a variable xc consisting of all integers from 100 to 120, inclusive. We then calculate the probability of X being xc or greater. (Note, we have to find the probability using the complement rule: $\Pr(X \geq x_c | H_0 \text{ is true}) = 1 - \Pr(X \leq x_c - 1 | H_0 \text{ is true})$.)

```
xc = 100:120
pr = 1 - pbinom( xc-1 , 200 , 0.5 )
print( cbind( xc , pr ) )
```

Part of the output is reproduced below:

```
           xc           pr
   [1,]  100  0.528174240
   [2,]  101  0.471825760
   ....................
   [12,] 111  0.068683325
   [13,] 112  0.051819519
   [14,] 113  0.038418816
   [15,] 114  0.027982870
   ....................
```

Because X is a discrete random variable we are unable to choose x_c so that $\Pr(X > x_c | H_0 \text{ is true})$ precisely equals 0.05. We can go a little higher, and choose $x_c = 112$, making the probability 0.051819519, or we could choose $x = 113$, making the probability 0.038418816. Usually we choose x_c so that the probability is at most 0.05; thus we choose $x_c = 113$.

The decision rule is simple. If the number of people X for whom Drug A works better than Drug B is 113 or greater, decide H_1, and if X is 112 or less, decide on H_0.

Because the logic of hypothesis testing assumes the null hypothesis until there is evidence to disbelieve it, we usually do not talk of "deciding on H_0" or "deciding on H_a". Usually, we talk of rejecting or failing to reject H_0. If we decide on H_a we say we reject H_0, and if we decide on H_0 we say we fail to reject H_0. Table 11.2 is a revision of Table 11.1 using the terminology of rejecting or failing to reject the null hypothesis.

Table 11.2 Type I and Type II errors in the terminology of rejecting and failing to reject H_0.

		Reality	
		H_0 is true	H_0 is false
Decision	Fail to reject H_0	No error	Type II error
	Reject H_0	Type I error	No error

The preceding discussion has introduced several concepts to which we now assign some terminology.

Definition 11.2.2 — Test Statistic, Rejection Region, Level of Significance. The *test statistic* is some function of the data and possibly the specific numbers in the null hypothesis that is used to make the decision about whether to reject H_0 or fail to reject H_0.

The *rejection region*, also called the *critical region*, is the set of values of the test statistic that lead us to reject H_0.

The probability that we choose for the probability of rejecting H_0 when it is true, usually denoted α, is called the *size* of the test, or the *level of significance* of the test.

Rejecting H_0 with a small α is a stronger conclusion than rejecting with a larger value. Choosing α is like setting the bar. If we set the probability α of making a Type I error to be large, say $\alpha = 0.2$, we are saying that we're not too concerned about rejecting H_0 when it is true. This makes it pretty easy to reject H_0. On the other hand, if we set α to be a very small number, say $\alpha = 0.001$, we're saying we want to make the probability of committing a Type I error very low. This is like setting the bar high for rejecting H_0. If we set the bar high and still reject H_0, then we can reach a strong conclusion.

In the previous example the test statistic was the data value X itself. In more complicated cases it may depend on a larger set of data and on what specific values the parameters take on in the null hypothesis. The critical region was $X \geq 113$, and the size of the test was $\alpha = 0.038418816$.

■ **Example 11.1** Suppose that a hospital emergency department sees patients with influenza-like illness (ILI) at an average rate of six per day, except when there is a flu outbreak, in which case the rate is higher. As for many count variables like this, the number of patients has a Poisson distribution, in this case with a mean of $\lambda = 6$ when there is no outbreak. Each day, the hospital records the number of ILI cases and tests whether there is evidence of an outbreak. Given data from a single day, say X cases of ILI, design a test of size $\alpha = 0.01$ to see whether there is evidence of an outbreak.

Solution: The null hypothesis here is H_0: $\lambda = 6$, with the alternative being H_1: $\lambda > 6$. The test statistic is the single value $X \sim \text{POIS}(\lambda)$. We would reject H_0 if X is sufficiently large. Thus, we need to find the value x_c such that $\Pr(X \geq x_c | H_0$ is true) is roughly (slightly smaller than if necessary) $\alpha = 0.01$. We can do this with R using similar logic and code as given previously for the binomial situation:

```
xc = 6:15
pr = 1 - ppois( xc-1 , 6 )
print( cbind( xc , pr ) )
```

which yields

```
           xc          pr
 [1,]   6 0.554320359
 [2,]   7 0.393697218
 [3,]   8 0.256020240
 [4,]   9 0.152762506
 [5,]  10 0.083924017
 [6,]  11 0.042620924
 [7,]  12 0.020091964
 [8,]  13 0.008827484
 [9,]  14 0.003628493
[10,]  15 0.001400354
```

We see that we cannot make the size of the test exactly equal to the desired value of $\alpha = 0.01$; the closest we can come and stay below 0.01 is to make the critical region $X \geq 13$, which yields $\alpha = 0.008827484$. Thus, if we observe 13 or more ILI cases then we would reject the null hypothesis of no outbreak and say that there is evidence of an outbreak. If we get 12 or fewer ILI cases, we have no evidence of an outbreak. ■ ■ ■

Note how we carefully worded the conclusion when $X \leq 12$. We said there is no evidence of an outbreak; we did not say there is no outbreak. This latter statement is too bold. The reason for this lies in the logic of hypothesis testing. In (11.1) we pointed out that we assume the null hypothesis is true, and then look for evidence (in the form of having observed an unlikely event) to reject it. You can't get evidence that H_0 is true by assuming it is true. This is why it is preferable to say "fail to reject H_0" rather than "accept H_0," although some books do talk about accepting the null hypothesis.

■ **Example 11.2** Suppose that one week's worth of data from the hospital gave the number of ILI patients per day as

    7   5   9   9   12   12   8

Set up a hypothesis test of H_0: $\lambda = 6$ against the alternative that $\lambda > 6$ with $\alpha = 0.01$.

Solution: A reasonable test statistic here is

$$S = \sum_{i=1}^{7} X_i.$$

The description of the Poisson distribution in Section 5.6 said that the number of events (ILI cases in this example) in an interval of length T with a rate of λ has a Poisson distribution with mean $T\lambda$. (See Problem 6.86.) In the present case, this means that

$$S = \sum_{i=1}^{7} X_i \sim \text{POIS}(7 \times 6 = 42).$$

We can then use the POIS(42) distribution to find the rejection region:

```
xc = 55:60
pr = 1 - ppois( xc-1 , 42 )
print( cbind( xc , pr ) )
```

which gives

```
        xc         pr
[1,]    55 0.030882973
[2,]    56 0.022279733
[3,]    57 0.015827304
[4,]    58 0.011072882
[5,]    59 0.007630024
[6,]    60 0.005179177
```

The critical region is therefore $S \geq 59$ and the size of the test is $\alpha = 0.007630024$. With the data given here, the sum over the seven days is 62, which is in the critical region since it is greater than or equal to 59. Thus, there is evidence that the true mean of the Poisson is greater than 6. ∎∎∎

There are cases where we would like to test a null hypothesis $H_0\colon \theta = \theta_0$ for some particular θ_0 of interest against an alternative that θ is not equal to $H_1\colon \theta \neq \theta_0$. Such tests are said to be *two-sided*, in contrast to the examples in this section where the tests were *one-sided*. We will learn more about two-sided tests in the remainder of this chapter.

11.3 Power

We defined α to be the probability that we reject H_0 given that it is true. But what about the probability of rejecting H_0 when H_0 isn't true (i.e., H_1 is true)? This gets to the idea of *power*, which we will define shortly. Recall that the null hypothesis is usually specific in the sense that it pins down a parameter to be equal to some particular number. The alternative hypothesis usually involves an interval, such as $\theta > \frac{1}{2}$ in the binomial example, or $\lambda > 6$ in the Poisson example. When H_0 is false, the probability of rejecting the null hypothesis will depend on which value (and there are many of them) under the alternative hypothesis is the true parameter value. For example, we might ask "what is the probability of rejecting the null hypothesis $H_0\colon \theta = \frac{1}{2}$ in the binomial case in favor of the alternative hypothesis $H_1\colon \theta > \frac{1}{2}$ when $\theta = 0.6$." This is like asking "if the true value of the probability is $\theta = 0.6$, what is the probability of (correctly) rejecting H_0?" We could ask the same thing for $\theta = 0.7$, or for any other value of $\theta > \frac{1}{2}$.

> **Definition 11.3.1 — Power.** Let θ be a parameter of some distribution, and θ_0 be some particular value for θ that we would like to test. The probability of rejecting $H_0\colon \theta = \theta_0$ in favor of some alternative hypothesis H_a is called the *power* of the test. Specifically,
>
> $$\text{power} = \pi(\theta) = \Pr(\text{reject } H_0 | \theta \text{ is the true value of the parameter}). \qquad (11.2)$$
>
> Thus, the power is a function of the unknown parameter.

In the next example we compute the power function for the binomial example that arose in the crossover trial from Section 11.1.

■ **Example 11.3** For the crossover trial described in Section 11.2, where 200 people were given both drugs and H_0 is rejected for $x \geq 113$, find the power function and graph it for θ in the interval $[0, 1]$.

Solution: The power function is

$$\pi(\theta) \;=\; \Pr(X \geq 113 | \theta) \;=\; \sum_{x=113}^{200} \Pr(X = x | \theta) \;=\; 1 - \sum_{x=0}^{112} \Pr(X = x | \theta).$$

This last expression can be evaluated in R using the `pbinom()` command. For example, if $\theta = 0.6$, which is one of the possible values for θ under the alternative hypothesis, we can compute the power $\pi(0.6) = 1$ - `pbinom(112,200,0.6)` which yields 0.8603357. We can define `thetaVec` to be a vector of values from 0 to 1 in increments of 0.01, and then compute the power for each. The result can then be plotted to give a graph of the power function, as shown in Figure 11.1.

```
thetaVec = seq( from=0 , to=1 , by=0.01 )
power = 1 - pbinom( 112 , 200 , thetaVec )
plot( thetaVec , power, type="l" , col="red" , lwd=2 ,
      xlab="True value of theta" , ylab="Power" )
print( cbind( thetaVec , power ) )
```

Part of the output is shown below:

```
            thetaVec          power
     [1,]       0.00 0.000000e+00
     [2,]       0.01 0.000000e+00
. . . . . . . . . . . . . . . . . . . . . . . . . .
    [49,]       0.48 9.759854e-03
    [50,]       0.49 2.007261e-02
    [51,]       0.50 3.841882e-02
    [52,]       0.51 6.856811e-02
    [53,]       0.52 1.143543e-01

. . . . . . . . . . . . . . . . . . . . . . . . . .
    [60,]       0.59 7.858898e-01
    [61,]       0.60 8.603357e-01
    [62,]       0.61 9.151410e-01
    [63,]       0.62 9.521730e-01
    [64,]       0.63 9.750961e-01
    [65,]       0.64 9.880671e-01
    [66,]       0.65 9.947598e-01

. . . . . . . . . . . . . . . . . . . . . . . . . .
   [100,]       0.99 1.000000e+00
   [101,]       1.00 1.000000e+00
```

■ ■ ■

Notice in Figure 11.1 the power curve goes through the point $(0.5, 0.0384)$. The x-coordinate is the null value for the probability of having the new drug work better, $\theta = 0.5$, and the y-coordinate is the size of the test, $\alpha = 0.0384$. This is not coincidental. We determined the rejection region so that when H_0 is true the probability of rejecting H_0 would be α. In general, the point (θ_0, α) will be on the power curve. Notice also that the power function is strictly increasing. This means that the larger θ is, the more likely we are to reject the null hypothesis H_0: $\theta = \frac{1}{2}$ in favor of H_1: $\theta > \frac{1}{2}$. This seems intuitive. When θ is large, the probability of a success on a single trial is high, so we would expect to see many

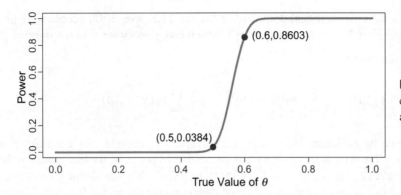

Figure 11.1 Power curve for crossover design with $n = 200$ and rejection region $x \geq 113$.

successes, leading to rejection of H_0. If θ is very close to the null value $\frac{1}{2}$, then we would expect nearly half of all trials to be successes, making it unlikely that we reject H_0. Finally, notice that the power curve approaches 1 as $\theta \to 1$ and it approaches 0 as $\theta \to 0$. This behavior is typical of one-sided tests. In Section 11.5.2 we will discuss the power curve for two-sided hypotheses.

The concept of power is closely related to the problem of sample size selection. In Chapter 10 the criterion for selecting the sample size was the width of the interval. For example, we asked how large the sample must be so that a 95% confidence interval has a fixed predetermined width. Larger sample sizes lead to narrower intervals, so we equated the desired margin of error with the actual margin of error (which depends on the sample size n) and solved for n. In the context of hypothesis testing, the sample size is determined so that the power at a fixed value of the alternative hypothesis is equal to a predetermined value. For the last example, we might ask what sample size is necessary to achieve a power of 0.85 when the true value of θ is $\theta = 0.6$. When we computed and graphed the power we found that the power was 0.8603. Thus, a sample of size 200 is sufficient to make the power be at least as large as the desired 0.85. If a power of 0.85 was desired, we could have gotten away with a slightly smaller sample size. If, instead, we wanted the power to be 0.95 for the alternative $\theta = 0.6$, then a larger sample size would be necessary.

11.4 *P*-values

To many, the abrupt cutoff in the rejection region seems arbitrary. Consider the crossover trial we have been studying where 200 people were assigned both drugs. For each person it was determined which drug worked better. For the test of H_0: $\theta = \frac{1}{2}$ against the one-sided alternative H_1: $\theta > \frac{1}{2}$ the critical region was $X \geq 113$. Thus, had we observed $X = 113$ we would have rejected H_0, but had we observed just one fewer success, in which case $X = 112$, we would have failed to reject. Is there really that much of a difference between observing 112 versus 113 successes? Should our decision come down to the outcome for a single person, whose outcome determines the conclusion of the study? The approach described so far in this chapter can be described as follows:

1. Determine the appropriate null and alternative hypotheses.
2. Choose a value of α.
3. Determine the rejection region.
4. From the observed data, compute the test statistic.
5. If the test statistic is in the rejection region, then reject H_0; otherwise, fail to reject H_0.

Some would argue that step 2, the selection of α, is arbitrary. To avoid this, many researchers prefer the use of *P*-values.

> **Definition 11.4.1 — P-value.** The *P-value* is the probability of getting a test statistic as extreme or more extreme than what was actually observed given that H_0 is true. Here, more extreme means getting a test statistic further away from the cutoff point that determines the rejection region. The smaller the *P*-value, the stronger the evidence against the null hypothesis H_0.

The next example illustrates this concept.

■ **Example 11.4** For the crossover trial with 200 people, find the *P*-value if (a) $X = 101$, (b) $X = 113$, (c) $X = 120$, and (d) $X = 150$, and assess the strength of evidence against H_0: $\theta = \frac{1}{2}$.

Solution: All of the required probabilities have been calculated in Section 11.1, but they were not called *P*-values at the time. We compute the *P*-values as right tail areas using the complement rule,

$$\Pr\left(X \geq x \mid \theta = \frac{1}{2}\right) = 1 - \Pr\left(X \leq x - 1 \mid \theta = \frac{1}{2}\right).$$

The R code to do all four parts is shown below:

```
PV101 = 1 - pbinom( 101-1 , 200 , 0.5 )
PV113 = 1 - pbinom( 113-1 , 200 , 0.5 )
PV120 = 1 - pbinom( 120-1 , 200 , 0.5 )
PV150 = 1 - pbinom( 150-1 , 200 , 0.5 )
print( c(PV101,PV113,PV120,PV150) )
```

which produces the output

```
[1] 4.718258e-01 3.841882e-02 2.842578e-03 4.196643e-13
```

With a *P*-value of 0.4718, getting $X = 101$ provides virtually no evidence against H_0. A *P*-value of 0.0384 associated with getting 113 successes is marginal evidence against H_0. A *P*-value of 0.00284 corresponding to $X = 120$ is very small, and provides fairly strong evidence against H_0. Finally, the *P*-value of 4.19×10^{-13} is extremely small, providing overwhelming evidence against H_0.   ■ ■ ■

Remember, the *P*-value is calculated under the assumption that the null hypothesis is true. The following logic may help your interpretation of the *P*-value. Imagine you observed a very small *P*-value, such as the 4.19×10^{-13} observed when $X = 150$. After seeing this result you must believe one of two things:

1. H_0 is true and you observed a very unlikely event, one with probability 4.19×10^{-13}; or
2. H_0 is false.

The smaller the probability in the first part, the less tenable H_0 seems to be. Unless the *P*-value equals zero, you are allowed to continue to believe that H_0 is true; you just have to admit that we observed an event or something more extreme with probability equal to the *P*-value. For the case of $X = 150$, if you still want to believe that the two drugs have equal effectiveness, you must admit that we observed an event with probability 4.19×10^{-13}.

11.5 Testing the Mean: Variance Known

11.5.1 Hypothesis Tests for the Mean from a Population with Known Variance

Suppose now that we observe a sample from a normal distribution: $X_1, X_2, \ldots, X_n \sim N(\mu, \sigma^2)$. For now, we will assume that σ^2 is known.[1] This problem is a good starting point because many of the calculations

[1] We made a similar assumption at the beginning of the previous chapter, where we obtained a confidence interval for the mean of a normal distribution. This is usually an unreasonable assumption to make, because when we don't know the mean, ordinarily we wouldn't know the variance either.

can be done exactly, so it is a good stepping stone to the more realistic problem where the variance σ^2 is unknown. We are interested in testing the null hypothesis H_0: $\mu = \mu_0$, where μ_0 is a fixed number that is determined by the context of the problem. The alternative hypothesis can be one of the following three:

$$H_1: \mu \neq \mu_0 \qquad \text{or} \qquad H_1: \mu < \mu_0 \qquad \text{or} \qquad H_1: \mu > \mu_0.$$

The type of problem addressed will determine which alternative is appropriate.

If we make the key assumption that H_0: $\mu = \mu_0$ is true, then

$$Z = \frac{\overline{X} - \mu_0}{\sigma/\sqrt{n}} \sim N(0,1). \tag{11.3}$$

Using the terminology of the previous chapter, we would say that Z is a pivotal quantity assuming H_0 is true. This is a strong assumption, and we must keep in the back of our minds that it has a $N(0,1)$ distribution only if the null hypothesis is true.

Suppose for now that the alternative is two-sided: H_1: $\mu \neq \mu_0$. In this case, if we take a sample $X_1, X_2, \ldots, X_n \sim N(\mu, \sigma^2)$ and observe an $\overline{X}$ that is much greater or much smaller than the hypothesized μ_0, we would believe this is evidence against the null hypothesis H_0: $\mu = \mu_0$. But getting an $\overline{X}$ way above or way below μ_0 is equivalent to getting a value of Z in (11.3) that is way above or way below zero. We can determine the cutoff values like we did previously with the binomial or Poisson distributions by setting

$$\alpha = \text{Pr(Reject the null hypothesis when it is true)}.$$

We will take a rejection region that is symmetric about zero, so we reject if $Z > z_{\alpha/2}$ or $Z < -z_{\alpha/2}$. See Figure 11.2.

■ **Example 11.5** Suppose we have a sample of BMI levels for 200 adults. We assume the BMI levels are normally distributed with mean μ, which is unknown, and a standard deviation of 5.0. Suppose the sample mean of the 200 adults was $\overline{x} = 22.9$. An ideal BMI is 22. Test whether the population BMI is equal to this ideal value of 22 against the alternative that it is not 22. Assume $\alpha = 0.05$.

Solution: Here the null hypothesis is H_0: $\mu = 22$ and the alternative is H_1: $\mu \neq 22$. The test statistic is

$$z = \frac{\overline{x} - \mu_0}{\sigma/\sqrt{n}} = \frac{22.9 - 22.0}{5/\sqrt{200}} = 2.546.$$

The rejection region would be

$$z < -1.96 \quad \text{or} \quad z > 1.96.$$

Since our test statistic $z = 2.546$ exceeds 1.96 we would reject H_0. There is evidence that the population mean is not the ideal value of 22.   ■ ■ ■

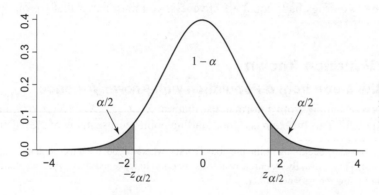

Figure 11.2 Determining the rejection region for the test on the mean of a normal distribution.

In Definition 11.4.1 we defined the P-value as the probability of getting a test statistic as extreme or more extreme than what was actually observed given that H_0 is true. When the alternative is two-sided, as in the previous example, the concept of "as extreme or more extreme" must be carefully interpreted. If we obtained a z statistic of 2.546, as in the previous example, then -2.546 should be considered just as extreme. Note that 22.9 is 0.9 BMI units *above* the hypothesized value of 22. Had we observed a sample mean of 21.1, which is 0.9 BMI units *below* the hypothesized value, then the observed test statistic would be

$$z = \frac{21.1 - 22.0}{5/\sqrt{200}} = -2.546.$$

Thus, 2.546 and -2.546 are equally extreme. Also, any sample mean above 22.9 would be more extreme than the observed value, and any sample mean below 21.1 would be more extreme. In general, if we observe a test statistic equal to z, then for the two-sided alternative, the P-value is

$$P = \Pr(Z < -z) + \Pr(Z > z) = 2 \times (\text{area in one tail})$$

because of the symmetry of the normal distribution.

■ **Example 11.6** What is the P-value for the situation in the previous example?

Solution: The P-value is

$$P = \Pr(Z < -2.546) + \Pr(Z > 2.546) = 2 \times \Pr(Z > 2.546) = 0.0109.$$

R can be used to approximate this P-value:

```
z0 = (22.9-22)/(5/sqrt(200))
PV = 2*min(pnorm(z0),1-pnorm(z0))
print(PV)
```

This gives

```
[1] 0.0109095
```

A P-value this small is evidence against H_0.   ■ ■ ■

This last example suggests that the P-value could provide more information than the result of the hypothesis test. In Example 11.5 we rejected the null hypothesis that $\mu = 22$ at the level $\alpha = 0.05$. Would we have rejected H_0: $\mu = 22$ with a level of significance of $\alpha = 0.04$? Since $z = 2.546$ is greater than $z_{0.04/2} = 2.054$, we would reject. What if $\alpha = 0.02$? In this case, $z_{0.02/2} = 2.326$, so yes, we would again reject. You might notice that we are getting close to the observed value of 2.546. Let's now choose 0.01; this leads to $z_{0.01/2} = 2.576$ and we would *fail to reject* because our observed z statistic of $z = 2.546$ does not exceed $z_{0.01/2} = 2.576$. Had we chosen $\alpha = 0.0109095$ (notice that this is exactly the P-value), then we would be right on the fence about rejecting H_0 or failing to reject H_0. Figure 11.3 illustrates this. With this reasoning we can see that the following is an alternative definition of the P-value.

Definition 11.5.1 — P-value – Alternative Definition. The P-value is the smallest value of α that leads to rejection of the null hypothesis. A technical note is needed here. Since the rejection region *excludes* the cutoff, if our z statistic were actually equal to the cutoff we would fail to reject H_0. Because of this, the minimum doesn't exist. (Instead, we would define the P-value as the *least upper bound*, also known as the *infimum*.)

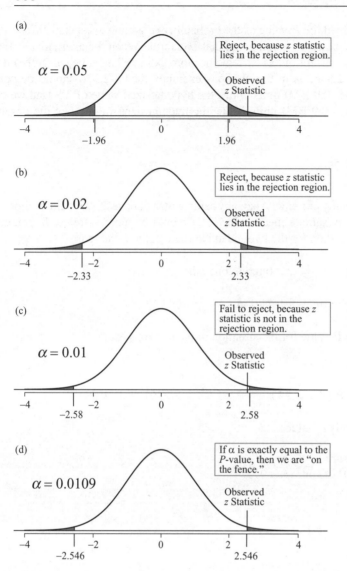

(a)

$\alpha = 0.05$

Reject, because z statistic lies in the rejection region.

Observed z Statistic

−4   −1.96   0   1.96   4

(b)

$\alpha = 0.02$

Reject, because z statistic lies in the rejection region.

Observed z Statistic

−4   −2 −2.33   0   2 2.33   4

(c)

$\alpha = 0.01$

Fail to reject, because z statistic is not in the rejection region.

Observed z Statistic

−4   −2 −2.58   0   2 2.58   4

(d)

$\alpha = 0.0109$

If α is exactly equal to the P-value, then we are "on the fence."

Observed z Statistic

−4   −2 −2.546   0   2 2.546   4

Figure 11.3 Illustration of P-value as the smallest α leading to rejection of the null hypothesis.

11.5.2 Power

Let's next consider the power for the hypothesis test described earlier in this section. Suppose we have H_0: $\mu = \mu_0$ against the two-sided alternative H_1: $\mu \neq \mu_0$. The decision would be to reject H_0 if $z < -z_{\alpha/2}$ or $z > z_{\alpha/2}$. If H_0 is true, the probability of rejection is α, but what if H_0 is false? The probability of rejecting H_0 when it is false is the power, which we defined in Definition 11.3.1. Suppose that the true, but unknown, value for the population mean is μ. It is important to keep straight the difference between μ and μ_0: μ is the true population mean, whereas μ_0 is the number we hypothesize μ to be. For the situation here, the power is

$$
\begin{aligned}
\pi(\mu) &= \Pr(\text{Reject } H_0 | \mu) \\
&= \Pr(Z < -z_{\alpha/2}) + \Pr(Z > z_{\alpha/2}) \\
&= \Pr\left(\frac{\overline{X} - \mu_0}{\sigma/\sqrt{n}} < -z_{\alpha/2}\right) + \Pr\left(\frac{\overline{X} - \mu_0}{\sigma/\sqrt{n}} > z_{\alpha/2}\right).
\end{aligned}
$$

This last line is equal to α only if the null hypothesis $H_0: \mu = \mu_0$ is true. If H_0 is false, then μ does not equal μ_0, so

$$\frac{\overline{X} - \mu_0}{\sigma/\sqrt{n}} \tag{11.4}$$

does not have a standard normal distribution. The wrong mean is subtracted in the numerator. We have to undo the standardization and then restandardize using the correct mean:

$$\pi(\mu) = \Pr\left(\overline{X} < \mu_0 - z_{\alpha/2}\frac{\sigma}{\sqrt{n}}\right) + \Pr\left(\overline{X} > \mu_0 + z_{\alpha/2}\frac{\sigma}{\sqrt{n}}\right)$$

$$= \Pr\left(\frac{\overline{X}-\mu}{\sigma/\sqrt{n}} < \frac{\mu_0-\mu-z_{\alpha/2}\sigma/\sqrt{n}}{\sigma/\sqrt{n}}\right) + \Pr\left(\frac{\overline{X}-\mu}{\sigma/\sqrt{n}} > \frac{\mu_0-\mu+z_{\alpha/2}\sigma/\sqrt{n}}{\sigma/\sqrt{n}}\right).$$

Here we have taken $\overline{X}$ and subtracted the correct mean, μ, and divided by the correct standard deviation, $\sigma/\sqrt{n}$, so it has a standard normal distribution. Further simplification gives

$$\pi(\mu) = \Pr\left(Z < \frac{\mu_0-\mu}{\sigma/\sqrt{n}} - z_{\alpha/2}\right) + \Pr\left(Z > \frac{\mu_0-\mu}{\sigma/\sqrt{n}} + z_{\alpha/2}\right)$$

$$= \Phi\left(\frac{\mu_0-\mu}{\sigma/\sqrt{n}} - z_{\alpha/2}\right) + 1 - \Phi\left(\frac{\mu_0-\mu}{\sigma/\sqrt{n}} + z_{\alpha/2}\right). \tag{11.5}$$

Here Φ is the CDF of the standard normal distribution. To calculate the power for a particular situation, we could either use the normal table or use R to obtain the normal probabilities.

■ **Example 11.7** Consider the diabetes data from Example 10.11 for women who have had 10 or more pregnancies and are at least 40 years of age. Suppose the standard deviation of diastolic blood pressure (DBP) is known to be $\sigma = 12$.

(a) Test using $\alpha = 0.05$ whether the mean DBP is 80 against the alternative that it is not; that is, $H_0: \mu = 80$ and $H_1: \mu \neq 80$. A DBP of 80 is considered normal.

(b) Find the P-value for this result.

(c) Find the power of this test when the true mean is 75.

(d) Sketch the power function for μ between 70 and 90.

Solution: For the first part, the rejection region is $z < -1.96$ or $z > 1.96$. The sample mean of the $n = 39$ cases is $\bar{x} = 79.18$, so the test statistic is

$$z = \frac{\bar{x}-\mu_0}{\sigma/\sqrt{n}} = \frac{79.18-80}{12/\sqrt{39}} = -0.43.$$

Since this z statistic satisfies neither $z < -1.96$ nor $z > 1.96$, we fail to reject H_0. Here "as extreme or more extreme" means having a z statistic less than -0.43 or greater than 0.43, so the P-value is

$$P = \Pr(Z < -0.43) + P(Z > 0.43) = 2\Pr(Z < -0.43) = 2 \times 0.3336 = 0.6672.$$

The large P-value suggests that there is little evidence that the mean DBP differs from the normal value of 80.

The power when $\mu = 75$ can be computed from (11.5):

$$\pi(90) = \Phi\left(\frac{80-75}{12/\sqrt{39}} - 1.96\right) + 1 - \Phi\left(\frac{80-75}{12/\sqrt{39}} + 1.96\right) \approx 0.7396.$$

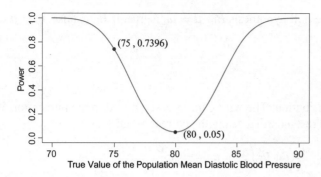

Figure 11.4 Power curve for Example 11.7. The dot at the bottom of the curve tells us that the power when H_0: $\mu = 80$ is true, the probability of rejecting is α. The dot at the top left of the curve shows that the power is 0.7396 when $\mu = 75$.

The power curve can be plotted using the `pnorm()` command in R:

```
n = 39
alpha = 0.05
z = - qnorm(alpha/2)
mu = seq( 70 , 90 , 0.25 )
pwr = pnorm( (80-mu)/(12/sqrt(39)) - 1.96 ) +
        1 - pnorm( (80-mu)/(12/sqrt(39)) + 1.96 )
plot( mu , pwr , type="l" , col="red" , lwd=2 ,
        xlab="True value of the population mean diastolic blood pressure" ,
        ylab="Power" , ylim=c(0,1) )
```

The power curve is shown in Figure 11.4.   ■ ■ ■

Note a few things from Figure 11.4. First, the point $(80, 0.05)$ lies on the power curve. This is expected since the rejection region was chosen to make the probability of rejecting equal to $\alpha = 0.05$ when $\mu = 80$. Second, this point is the global minimum of the power function. In other words, the probability of rejecting H_0 is least when H_0 is true. Third, the power function approaches 1 as μ moves away from 80 in either direction. Fourth, the power function is symmetric about the vertical line at 80. This implies that the power when $\mu = 85$ is the same as the power when $\mu = 75$ since both are 5 units away from the null value of $\mu = 80$.

As we mentioned previously, the power is closely related to the problem of determining the required sample size. Often, investigators have a particular alternative hypothesis in mind and would like to find the sample size that would make the power at that value close to 1, such as 0.90. The next example addresses this issue.

■ **Example 11.8** Investigators are planning a study to test whether the population mean diastolic blood pressure is equal to 80 against the alternative that it differs from 80. The size of the test is to be $\alpha = 0.05$, and the standard deviation is known to be $\sigma = 12$.

1. The investigators would like the power to be 0.90 when the true population mean is $\mu = 75$. Is a sample size of $n = 39$, as in the previous example, sufficient to attain this power?
2. What sample size is needed to achieve a power of 0.90 when $\mu = 75$?

Solution: The power at $\mu = 75$ was found in the previous example to be 0.7396, which is below the desired 0.90. A larger sample is therefore needed.

Figure 11.5 shows the power functions of the test of H_0: $\mu = 80$ vs. H_1: $\mu \neq 80$ for samples of size $n = 39$, 50, 60, and 70. For every sample size, the power curve passes through the point $(80, 0.05)$. (In general, the power curve will pass through (μ_0, α).) At all other points, the power curve is higher for larger sample sizes. The R code below defines a function called `power()` that takes as arguments mu, mu0, `sigma`, `alpha`, and n, and returns the power for those values. We define nn to be the vector of

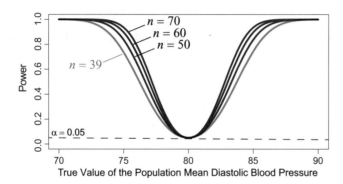

Figure 11.5 Power for test of $H_0 : \mu = 80$ vs. $H_1 : \mu \neq 80$ for samples of size $n = 39$, 50, 60, and 70, if we assume σ is known to be equal to 12.

values from 39 through 70, inclusive, and then compute the power by calling `pwr = power( 75 , 80, 12 , 0.05 , nn )`. To make the results easily readable, we bind the two vectors, as column vectors, to create a matrix with two columns, and display the result.

```
power = function(mu,mu0,sigma,alpha,n)
{
    z = - qnorm(alpha/2)
    pnorm( (mu0-mu)/(sigma/sqrt(n)) - z ) +
    1 - pnorm( (mu0-mu)/(sigma/sqrt(n)) + z )
}
nn = 39:70
pwr = power( 75 , 80 , 12 , 0.05 , nn )
cbind( nn , pwr )
```

Some rows of the output have been omitted to save space.

```
          nn        pwr
 [1,] 39 0.7396044
 [2,] 40 0.7502492
 . . . . . . . . . . . . . . .
[12,] 50 0.8380110
[13,] 51 0.8450979
 . . . . . . . . . . . . . . .
[21,] 59 0.8926073
[22,] 60 0.8975158
[23,] 61 0.9022203
[24,] 62 0.9067282
 . . . . . . . . . . . . . . .
[31,] 69 0.9333390
[32,] 70 0.9365100
```

From the above results, we see that the power is just below 0.90 when $n = 60$ and just above 0.90 when $n = 61$. The choice of $n = 61$ yields a power of 0.9022203, so we would have to take a sample of size 61. ∎∎∎

We have now discussed both one-sided and two-sided tests for the mean of a normal distribution with known variance. The context of the problem will determine what the alternative hypothesis is.

Hypothesis tests for the mean of a normal distribution: variance known

In all cases the test statistic is

$$z = \frac{\bar{x} - \mu_0}{\sigma/\sqrt{n}}$$

H_0: $\mu = \mu_0$ vs. H_1: $\mu \neq \mu_0$	Reject H_0 if $z < -z_{\alpha/2}$ or $z > z_{\alpha/2}$.
H_0: $\mu = \mu_0$ vs. H_1: $\mu < \mu_0$	Reject H_0 if $z < -z_\alpha$.
H_0: $\mu = \mu_0$ vs. H_1: $\mu > \mu_0$	Reject H_0 if $z > z_\alpha$.

11.6 Testing the Mean: Variance Unknown

The situation where we know the variance, discussed thoroughly in the previous section, is unrealistic, but the math worked out nicely, and because of this it gave us insight into how we might proceed in the case where we don't know the variance. The variance-known case was like a stepping stone to the more realistic case where σ^2 is unknown. To make things more precise, let's assume that we have a sample $X_1, X_2, \ldots, X_n \sim$ i.i.d. $N(\mu, \sigma^2)$, but now μ and σ^2 are both unknown. We wish to test the hypotheses H_0: $\mu = \mu_0$ against the alternative H_1: $\mu \neq \mu_0$. Previously, we used the test statistic

$$Z = \frac{\overline{X} - \mu_0}{\sigma/\sqrt{n}}. \tag{11.6}$$

If we don't know σ, using this test statistic creates a problem. The most obvious solution is to use an estimate of σ in place of σ. Recall from Chapter 9 that

$$\hat{\sigma}^2 = S^2 = \frac{1}{n-1} \sum_{i=1}^{n} (X_i - \overline{X})^2$$

is an unbiased estimate of σ^2. We normally take the square root to estimate σ:

$$\hat{\sigma} = S = \sqrt{\frac{1}{n-1} \sum_{i=1}^{n} (X_i - \overline{X})^2},$$

although this is a biased estimator of σ. If we make this substitution in (11.6), we usually call the result T rather than Z, and write

$$T = \frac{\overline{X} - \mu_0}{S/\sqrt{n}}. \tag{11.7}$$

In Chapter 10 we explained why this has a t distribution with $n-1$ degrees of freedom. If the null hypothesis H_0: $\mu = \mu_0$ is true, then the statistic becomes

$$T = \frac{\overline{X} - \mu_0}{S/\sqrt{n}} \sim t(n-1).$$

The t distribution is tabulated in many statistics books, and R has built-in functions that give the PDF, CDF, and quantiles; the respective R functions are `dt()`, `pt()`, and `qt()`. The rules for testing the mean of a normal distribution are analogous to that for the known-variance case, with the exceptions that (1) we use the sample standard deviation S in place of σ, and (2) we look up the cutoff values from the t distribution, not the standard normal distribution. Here are the rules for testing the mean of a normal distribution with unknown variance.

Hypothesis tests for the mean of a normal distribution: variance unknown

In all cases the test statistic is

$$t = \frac{\bar{x} - \mu_0}{s/\sqrt{n}}$$

H_0: $\mu = \mu_0$ vs. H_1: $\mu \neq \mu_0$ Reject H_0 if $t < -t_{\alpha/2}(n-1)$ or $t > t_{\alpha/2}(n-1)$.

H_0: $\mu = \mu_0$ vs. H_1: $\mu < \mu_0$ Reject H_0 if $t < -t_{\alpha}(n-1)$.

H_0: $\mu = \mu_0$ vs. H_1: $\mu > \mu_0$ Reject H_0 if $t > t_{\alpha}(n-1)$.

Let's now take another look at Example 11.7, except this time we do not assume a known variance.

■ **Example 11.9** Consider the diabetes data from Example 11.7 for women who have had 10 or more pregnancies and are at least 40 years of age.

(a) Test using $\alpha = 0.05$ the hypotheses H_0: $\mu = 80$ and H_1: $\mu \neq 80$.

(b) Find the P-value for this result.

Solution: Using R, we find the sample mean and standard deviation to be

$$\bar{x} = 79.17949 \qquad \text{and} \qquad s = 11.62998.$$

The t statistic is therefore

$$t = \frac{79.17949 - 80}{11.62998/\sqrt{39}} = -0.44.$$

The quantile of the t distribution can be obtained from R and is $t_{0.025} = 2.02$. The rule would be to reject H_0 if $t < -2.02$ or $t > 2.02$. The R code to perform all of these calculations is:

```
library( tidyverse )
diabetes = read.csv( "Diabetes.csv" )
diabetes1  =  diabetes  %>%
   filter( Pregnancies >= 10 & Age >= 40 & BloodPressure > 50 )
DBP1 = diabetes1$BloodPressure
n1 = length( DBP1 )
xbar1 = mean( DBP1 )
stdev1 = sd( DBP1 )
t025_1 = qt( 0.975 , n1-1 )
tStat = (xbar1 - 80)/(stdev1/sqrt(n1))
alpha = 0.05
t_alpha_2 = qt( 0.975 , n1-1 )
print( tStat )
print( t_alpha_2 )
```

which yields

```
[1] -0.4405942
[1] 2.024394
```

Since $t = -0.44$ is not less than $-t_{0.025} = -2.02$, nor is it greater than $t_{0.025} = 2.02$, we fail to reject H_0. The P-value is equal to the sum of the tail areas of the t distribution to the left of $t = -0.44$ and to the right of $t = 0.44$. This is found to be $P = 0.6620$ using the code below:

```
PValue = pt(-abs(tStat),n1-1 ) + 1 - pt(abs(tStat),n1-1)
print( PValue )
```

yielding

```
[1] 0.6620032
```

Note that the conclusions are basically the same as in Example 11.9. In both cases, the decision was to fail to reject H_0: $\mu = 80$. Here the P-value is equal to 0.6620; previously, the P-value was 0.6672. When the sample size is fairly large, say over 30, the two methods will yield similar results. Because the assumption of a known variance is usually suspect, we should treat the t-test as the correct approach. ∎

When the null hypothesis is false, and μ is equal to some value under the alternative, say μ_1, which differs from μ_0, computing the power is a bit more difficult than in the case where σ^2 is known. In the known-variance case, we were able to compute the power by unstandardizing and restandardizing the statistic $\overline{X}$. For the variance-unknown case, if $\mu \neq \mu_0$, then the statistic

$$T = \frac{\overline{X} - \mu_0}{S/\sqrt{n}} \tag{11.8}$$

does not have the usual t distribution. In Chapter 6 and again in Chapter 10 we defined the t distribution to be the distribution of the ratio

$$T = \frac{Z}{\sqrt{V/v}}, \tag{11.9}$$

where $Z \sim N(0,1)$ and $V \sim \chi^2(v)$ are independent random variables. If we alter the assumptions so that the numerator is $X = Z + \lambda$, while keeping the assumptions that $V \sim \chi^2(v)$ and that Z and V are independent random variables, then the result in (11.9) has a noncentral t distribution with $n-1$ degrees of freedom and noncentrality parameter λ. (Many books call the noncentrality parameter δ. Because R uses delta to denote the difference in the means, not the standardized difference between the means, we reserve δ for that meaning and use λ here to represent the noncentrality parameter.) We indicate this by writing

$$T = \frac{X}{\sqrt{V/v}} = \frac{Z+\lambda}{\sqrt{V/v}} \sim t(n-1,\lambda). \tag{11.10}$$

When μ is any value, we can still write

$$T = \frac{\overline{X} - \mu_0}{S/\sqrt{n}} = \frac{\dfrac{\overline{X} - \mu_0}{\sigma/\sqrt{n}}}{\sqrt{\dfrac{(n-1)S^2/\sigma^2}{n-1}}} = \frac{\dfrac{\overline{X} - \mu}{\sigma/\sqrt{n}} + \dfrac{\mu - \mu_0}{\sigma/\sqrt{n}}}{\sqrt{\dfrac{(n-1)S^2/\sigma^2}{n-1}}} = \frac{Z+\lambda}{\sqrt{V/v}},$$

where

$$\lambda = \frac{\mu - \mu_0}{\sigma/\sqrt{n}}.$$

Here, though, the expression in the numerator, $Z + \lambda$, has a $N(\lambda, 1)$ distribution, not a $N(0,1)$ distribution. Thus,

$$T = \frac{\overline{X} - \mu_0}{S/\sqrt{n}} \sim t(n-1, \lambda).$$

Note that if $\mu = \mu_0$, in other words, if H_0 is true, then the noncentral t distribution becomes the usual t distribution. To emphasize this we sometimes call our usual t distribution the *central t* distribution.

Fortunately, the R functions dt(), pt(), and qt() can be used to evaluate the PDF, CDF, and quantiles, respectively, of the noncentral t distribution. For example, to get the CDF at a particular value x of the noncentral t distribution with $n-1$ degrees of freedom and noncentrality parameter λ, we would type pt(x , n-1 , lambda). The only difference in R between the noncentral t distribution and the central t distribution is the additional argument for the noncentrality parameter.

We can now write the power for the two-sided t test of H_0: $\mu = \mu_0$ versus H_1: $\mu \neq \mu_0$ as

$$\pi(\mu) = \Pr\left(T < -t_{\alpha/2}\right) + \Pr\left(T > t_{\alpha/2}\right)$$
$$= \text{pt(-t.alpha.2,n-1,lambda) + 1 - pt(t.alpha.2,n-1,lambda)}$$

In the last line we have used R notation where t.alpha.2 is the $1 - \alpha/2$ quantile of the central $t(n-1)$ distribution and lambda is the noncentrality parameter $(\mu - \mu_0)/(\sigma/\sqrt{n})$.

We could use the R function pt() for the noncentral t distribution to obtain the power, but fortunately R has a built-in function that gives the power function for the t test. The documentation for R describes the function as follows:

```
power.t.test(n = NULL, delta = NULL, sd = 1, sig.level = 0.05,
             power = NULL, type = c("two.sample", "one.sample", "paired"),
             alternative = c("two.sided", "one.sided"),
             strict = FALSE, tol = .Machine$double.eps^0.25)
```

This definition gives the arguments with names. If we use these names we don't need to remember the order of the arguments. Note that three of the default arguments are NULL; for these we must specify two of the three and the other one will be determined. For example, if we specify n and delta but not power, then the power will be computed as a function of n and delta. R uses delta to represent the difference $\mu - \mu_0$ between the true mean and the hypothesized mean. The noncentrality parameter $\lambda = (\mu - \mu_0)/(\sigma/\sqrt{n})$ is a distinct, but related quantity.

If in power.t.test we specify the sample size, n, and the difference in means, delta, then it will compute the power for this combination. The output from this function is a list containing the information for this analysis of power. For example, if we say

```
n = 39
stdev = 12
mu = 75
delta =  mu - 80
power.t.test( n=n , delta=delta , sd = stdev , sig.level=0.05 ,
              type="one.sample" , alternative="two.sided" , strict=TRUE )
```

then the output would be

```
        One-sample t test power calculation
              n = 39
          delta = 5
             sd = 12
      sig.level = 0.05
          power = 0.7176593
    alternative = two.sided
```

from which we can see that the power for testing H_0: $\mu = 80$ versus the alternative H_1: $\mu \neq 80$ when $n = 39$, and $\mu = 75$ is equal to 0.7176593. Note that R took the absolute value of δ. This yields the same power because the power function is symmetric about μ_0 so $\mu - \mu_0 = -5$ has the same power as $\mu_0 - \mu = 5$. We can send the function power.t.test a vector of values for delta, the difference in means, and it will return a vector of values for the power. The code

```
n = 39;      stdev = 12
mu = seq(70,90,5)
delta =  mu - 80
pwr = power.t.test( n=n , delta=delta , sd = stdev , sig.level=0.05 ,
           type="one.sample" , alternative="two.sided" , strict=TRUE )
print(pwr)
```

returns

```
        One-sample t test power calculation
              n = 39
          delta = 10, 5, 0, 5, 10
             sd = 12
      sig.level = 0.05
          power = 0.9990610, 0.7176593, 0.0500000, 0.7176593, 0.9990610
    alternative = two.sided
```

The structure of the object pwr is gotten by using the structure function str():

```
str( pwr )
List of 8
 $ n          : num 39
 $ delta      : num [1:5] 10 5 0 5 10
 $ sd         : num 12
 $ sig.level  : num 0.05
 $ power      : num [1:5] 0.999 0.718 0.05 0.718 0.999
 $ alternative: chr "two.sided"
 $ note       : NULL
 $ method     : chr "One-sample t test power calculation"
 - attr(*, "class")= chr "power.htest"
```

Thus, to obtain the values of the power for the elements of the vector delta, we can say

```
power39 = pwr$power
print( power39 )
```

which gives

```
[1] 0.9990610 0.7176593 0.0500000 0.7176593 0.9990610
```

We could instead send `power.t.test` a scalar for `delta` and a vector for `n` and it will return the power at this particular difference in means for each value of `n`.

■ **Example 11.10** Consider the diabetes data from Example 11.9 for women who have had 10 or more pregnancies and are at least 40 years of age. Now suppose that the t test will be applied, rather than the z test. Suppose the standard deviation is believed to be $\sigma = 12$.

(a) Find the power of this test when the true mean is 75.

(b) Sketch the power function for μ between 70 and 90.

(c) How large would the sample size have to be so that the power at $\mu = 75$ is at least 0.90.

Solution: The power was obtained above to be 0.7176593. The power curve for $n = 50$ can be created with the R commands

```
n = 50
stdev = 12
mu = seq(70,90,0.1)
delta =  mu - 80
pwr = power.t.test( n=n , delta=delta , sd = stdev , sig.level=0.05 ,
                type="one.sample" , alternative="two.sided" , strict=TRUE )
plot( mu , pwr$power , type="l" , col="black" , lwd=2 , ylim=c(0,1) ,
      xlab="True Value of Mean" , ylab="Power")
```

You could compute the power curves for other sample sizes n by changing the first line that defines n. The power curves for $n = 39, 50, 60$, and 70 are shown in Figure 11.6.

To obtain the required sample size to achieve a power of 0.90 or higher at $\mu = 75$, we can send to `power.t.test` a vector of values for n, much like we did for the z test discussed previously. After a little trial and error regarding the value of n, we could narrow it down to the low 60s. The following code computes the power for $n = 61$ through $n = 65$:

```
power.t.test( n=61:65 , delta=-5 , sd = 12 , sig.level=0.05 ,
              type="one.sample" , alternative="two.sided" , strict=TRUE )
```

which produces the output

```
One-sample t test power calculation
        n = 61, 62, 63, 64, 65
    delta = 5
       sd = 12
```

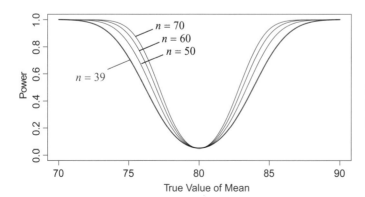

Figure 11.6 Power curves for t test of H_0: $\mu = 80$ versus H_1: $\mu \neq 80$ for $n = 39$, 50, 60, and 70.

```
        sig.level = 0.05
            power = 0.8928437, 0.8977435, 0.9024397, 0.9069393, 0.9112494
      alternative = two.sided
```

From this we see that the power for $n = 62$ falls just short of 0.90, whereas $n = 63$ yields a power of 0.9024397, just over the desired 0.90. The required sample size is thus $n = 63$. ∎∎∎

You might have caught the seeming contradiction in the setup to this example. We assumed that the population standard deviation was about 12, but we used the t test which assumes the standard deviation is unknown. It seems like if we knew σ we would use a z test. What gives? If asked the question "What test do you plan to use after you collect the data?" the answer would be "We would use the t test because we don't know σ." But if this is all we know, we don't know enough to answer the question regarding how large a sample is required. If the true standard deviation were $\sigma = 1.0$, then a fairly small sample will do the job. If $\sigma = 20$, then a much larger sample is needed. We must know some ballpark value for σ before we can determine the required sample size. In the previous example, the correct interpretation would be this: *If* the true value of the standard deviation is $\sigma = 12$, then the sample size would have to be $n = 63$ if we are to have a power of 0.90 when the true mean is $\mu = 75$. Always think of the power for the t test as a "what if" question. Here we would say "What if the true standard deviation were $\sigma = 12$. How large must the sample be?"

11.7 Testing a Proportion

We can test hypotheses regarding the proportion in a population using either the binomial distribution or the normal approximation to the binomial. If the sample size is small – that is, one for which $n\theta < 5$ or $n(1 - \theta) < 5$ – then we should use the binomial distribution.

■ **Example 11.11** Suppose we run a crossover study as in Example 8.9 with $n = 12$ patients and that Drug A worked better for 11 of the 12 patients. Is this evidence that Drug A is more effective than Drug B? Test using $\alpha = 0.05$ and give the P-value.

Solution: Since we are interested in whether Drug A is *better* than Drug B, we are testing $H_0: \theta = \frac{1}{2}$ against the alternative that $H_0: \theta > \frac{1}{2}$. We will use $X =$ number of cases where Drug A worked better than Drug B. The following R code computes the right tail area for each possible x, for $x = 0, 1, \ldots, 11, 12$.

```
x = 0:12
p.right.tail = 1 - pbinom(x-1,12,0.5)
print( cbind( x , round( p.right.tail , 3) ) )
```

Part of the output is:

```
[1,]   0 1.000
[2,]   1 1.000
[3,]   2 0.997
. . . . . . . .
[8,]   7 0.387
[9,]   8 0.194
[10,]  9 0.073
[11,] 10 0.019
[12,] 11 0.003
[13,] 12 0.000
```

We can see that no choice of a cutoff value x_c makes the tail probability (i.e., the probability of rejecting H_0 given that it is true) equal to 0.05. The largest value of α that we could use and stay below 0.05 is

$\alpha = 0.019$, which corresponds to $x = 10$. Thus a test of H_0: $\theta = \frac{1}{2}$ against H_1: $\theta > \frac{1}{2}$ of size $\alpha = 0.019$ is to reject H_0 if $x \geq 10$. Since $x = 11$ in this case, we would reject H_0 and conclude that there is evidence that Drug A works.

The *P*-value can be calculated as

$$P = \Pr\left(X \geq 11 | \theta = \frac{1}{2}\right) = 1 - \texttt{pbinom(11-1,12,0.5)} = 0.003.$$

A small *P*-value like this is evidence that H_0 is false.   ▪▪▪

Exercise 11.1 Repeat Example 11.11 for the case where 13 of 15 patients preferred Drug A.

When the sample size is large, we can use the normal approximation to the binomial described in Section 8.6.

▪ Example 11.12 Suppose that in a crossover study with 300 patients, 162 preferred Drug A when compared to the placebo, Drug B. Is this evidence that Drug A is more effective than the placebo at the $\alpha = 0.05$ level?

Solution: We observe $X =$ number of people out of 300 who prefer Drug A. We are testing H_0: $\theta = \frac{1}{2}$ against H_1: $\theta > \frac{1}{2}$. Let's compute the *P*-value and assess the evidence against H_0. Under H_0, $X \sim$ BIN$(300, 0.5)$, so

$$X \overset{\text{approx}}{\sim} N\left(300 \times 0.5, 300 \times 0.5 \times (1 - 0.5)\right) = N(150, 75).$$

Figure 11.7 shows the binomial and normal approximation for the random variable X under the assumption that H_0 is true. We can see that the normal approximation should work well here.

Using the continuity correction for the normal approximation to the binomial, we find the *P*-value to be

$$P = \Pr(X \geq 162) = \Pr(X \geq 161.5) = \Pr\left(\frac{X - 150}{\sqrt{75}} \geq \frac{161.5 - 150}{\sqrt{75}}\right) \approx \Pr(Z > 1.328) \approx 0.092.$$

Since $P > \alpha = 0.05$, we fail to reject H_0. There is not sufficient evidence to say that Drug A is more effective than the placebo.   ▪▪▪

Exercise 11.2 Use R's `pbinom()` function to compute the *P*-value exactly for the data in Example 11.12.

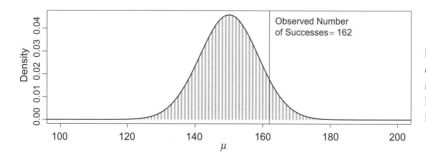

Figure 11.7 Binomial distribution and the normal approximation for the data in Example 11.12, assuming H_0 is true.

Exercise 11.3 Suppose that in the crossover trial described in Example 11.12, 187 of the 300 preferred Drug A. Is this evidence that Drug A is more effective than the placebo, Drug B?

We can determine the power along with the sample size needed to achieve a desired power using the normal approximation to the binomial, provided the sample size is sufficiently large.

■ **Example 11.13** Suppose it is known that 60% of people respond favorably to an allergy drug. A new drug in development is being tested to see whether it is more effective in treating allergies. Let p denote the proportion of people who would respond favorably to the new drug. We wish to test H_0: $p = 0.60$ against H_1: $p > 0.60$ with a sample of size 400 using $\alpha = 0.05$. Find and sketch the power curve.

Solution: Let X denote the number of people out of the 400 who respond favorably. We will use the normal approximation to the binomial and base our test on the test statistic

$$Z = \frac{X - 400 \times 0.6}{\sqrt{400 \times 0.6 \times (1 - 0.6)}} = \frac{X - 240}{9.798}.$$

The decision rule would be to reject H_0: $p = 0.6$ if the observed test statistic satisfies $z > z_{0.05} = 1.645$. The power function is

$$\begin{aligned}
\pi(p) &= \Pr\left(Z > 1.645\right) \\
&= \Pr\left(\frac{X - 240}{9.798} > 1.645\right) \\
&= \Pr\left(X > 256.12\right) \\
&= \Pr\left(\frac{X - 400p}{\sqrt{400p(1-p)}} > \frac{256.12 - 400p}{\sqrt{400p(1-p)}}\right) \\
&= 1 - \texttt{pnorm((256.12-400*p)/sqrt(400*p*(1-p)))}
\end{aligned}$$

A plot of the power curve is shown in Figure 11.8.   ■ ■ ■

Exercise 11.4 Suppose we would like the power for the problem described in Example 11.13 to be 0.90 when the true proportion is $p = 0.7$. Would we need a larger sample? Or might a smaller sample suffice?

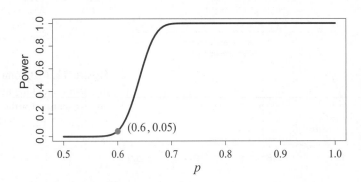

Figure 11.8 Power function for Example 11.13.

11.8 **Testing the Variance**

In the previous sections of this chapter we have seen that if a pivotal quantity (see Definition 10.4.1) exists for a parameter, it can often be used to obtain a test statistic. We saw that when we have normal data from a distribution with known standard deviation σ,

$$Z = \frac{\overline{X} - \mu_0}{\sigma/\sqrt{n}}$$

is a pivotal quantity, having a standard normal distribution, that can be used to test hypotheses regarding the population mean μ. When σ is unknown, we saw that

$$T = \frac{\overline{X} - \mu_0}{S/\sqrt{n}}$$

is a pivotal quantity, having a $t(n-1)$ distribution, that can be used to test hypotheses regarding the population mean μ. When we want to test whether a proportion is equal to some particular value, we can (when the sample size is sufficiently large) use the approximate pivotal quantity

$$Z = \frac{X - np}{\sqrt{np(1-p)}},$$

which has an approximate standard normal distribution when n is large.

Suppose now that we have data $X_1, X_2, \ldots, X_n \sim$ i.i.d. $N(\mu, \sigma^2)$ and we want to test H_0: $\sigma^2 = \sigma_0^2$ against one of the alternative hypotheses

$$H_1\colon \sigma_2 \neq \sigma_0^2 \qquad \text{or} \qquad H_1\colon \sigma_2 < \sigma_0^2 \qquad \text{or} \qquad H_1\colon \sigma_2 > \sigma_0^2.$$

A pivotal quantity for the sample variance S^2 is

$$\frac{(n-1)S^2}{\sigma^2} \sim \chi^2(n-1),$$

which can be used to test whether the variance is equal to some particular value σ_0^2. Our test statistic is simply

$$W = \frac{(n-1)S^2}{\sigma_0^2}. \tag{11.11}$$

Depending on the alternative hypothesis, we would reject H_0 for small values of W, large values of W, or either small or large values of W.

■ **Example 11.14** Montgomery (2020b) describes a manufacturing process for silicon wafers. An important characteristic of the process is the resistivity after an epitaxial layer is deposited in the deposition process. It was found that the normal distribution was a good fit to the logarithm of resistivity. It is important in many manufacturing processes to minimize the process variability as measured by the variance or standard deviation. The process mean, variance, and standard deviation of the log resistivity were found to be 5.444, 0.300, and 0.0893, respectively. Suppose that after investing in newer equipment a sample of size $n = 12$ yielded the following values for log resistivity:

5.332	5.398	5.756	5.458	5.470	5.787
5.536	5.191	5.307	5.355	5.689	5.516

Is there evidence that the new equipment yields a smaller variance than before? Compute and interpret the P-value.

Solution: The sample mean and variance are found to be $\bar{x} = 5.483$ and $s^2 = 0.03424754$. We are testing H_0: $\sigma^2 = 0.0893$ against the alternative H_1: $\sigma^2 < 0.0893$. This is a left-tailed test because we are interested in whether the process variance got smaller after installing the new equipment. The test statistic from (11.11) is then

$$w = \frac{(12-1) \times 0.03424754}{0.0893} = 4.217711.$$

Since our alternative is H_1: $\sigma^2 < 0.0893$ we reject for *small* values of w. The P-value is thus

$$P = \Pr\left(W \le 4.217711\right) = \texttt{pchisq(4.217711 , 12-1)} = 0.0368899.$$

This small P-value provides evidence that the process variance has indeed been reduced. ∎

> **Exercise 11.5** Repeat Example 11.14 for the following values of log resistivity:
>
> 5.305   5.448   5.368   5.465   5.587   5.518   5.514   5.421   5.464

11.9 Likelihood Ratio Tests

So far, our construction of a rejection region (or critical region) for testing a pair of hypotheses has been intuitive. We selected what we believed to be a reasonable test statistic and then we tried to figure out the distribution of the statistic, or some function of it. Often this test statistic was a function of a sufficient statistic. For example, when testing the mean of a normal distribution (H_0: $\mu = \mu_0$) based on a sample of size n from a normal distribution, we reasoned that the test should be based on the sample mean $\overline{X}$. Standardizing gave us $(\overline{X} - \mu_0)/(\sigma/\sqrt{n})$, but unfortunately this depends on the parameter σ, which is unknown. If we substituted the estimate S for σ (a very intuitive thing to do) we obtained $(\overline{X} - \mu_0)/(S/\sqrt{n})$, which turned out to have a $t(n-1)$ distribution. We then based our test on that.

There are times, however, when our intuition fails. Some problems are so complicated that no sufficient statistic, and therefore no obvious test statistic, exists. We need some systematic way to construct a test when there is no clear choice for a test statistic. One approach is the likelihood ratio test.[2] Before we define the likelihood ratio we need to introduce some terminology. Let θ denote an unknown parameter or a vector of unknown parameters. We will let Ω denote the entire parameter space. Our null hypothesis is written as H_0: $\theta \in \Omega_0$, where Ω_0 is a subset of Ω. The alternative hypothesis is H_1: $\theta \notin \Omega_0$. The likelihood ratio is then defined to be

$$\lambda = \frac{\max\limits_{\Omega_0} L(\theta|\text{data})}{\max\limits_{\Omega} L(\theta|\text{data})}. \tag{11.12}$$

The likelihood ratio, which describes how much better H_1 explains the given data than H_0 does, leads to a rejection region of the form

$$\frac{\max\limits_{\Omega_0} L(\theta|\text{data})}{\max\limits_{\Omega} L(\theta|\text{data})} < k, \tag{11.13}$$

---

[2] Another approach will be described in Section 17.1.

where k is some constant determined so that the size of the test is the desired α. The direction of this inequality should be intuitive: If the maximum of the likelihood under the null hypothesis is considerably less than the overall maximum of the likelihood, then we tend to disbelieve H_0. Note, too, that λ will *always* be less than or equal to 1. The maximum in the denominator is taken over a larger set (Ω) than the maximum in the numerator (Ω_0), so the denominator can never be smaller than the numerator.

The likelihood ratio test gives us only the *form* of the test; that is, it gives us a general rule for rejecting the null hypothesis. It doesn't tell us what the rejection region should be. This can sometimes be done once the statistic λ is simplified. When simplification of λ does not lead to an obvious test statistic we can approximate the distribution of $-2\log\lambda$ by a χ^2 distribution. This is described later in this section.

Many of the intuitive test statistics that we have developed so far are actually likelihood ratio tests. One case where this happens is described in the next example.

■ **Example 11.15** Suppose $X_1, X_2, \ldots X_n \sim$ i.i.d. $N(\mu, \sigma^2)$. We would like to test H_0: $\mu = 0$ against the alternative H_1: $\mu \neq 0$. Derive the likelihood ratio test.

Solution: The parameter space is $\Omega = \{(\mu, \sigma) : \sigma > 0\}$ and the parameter space under H_0 is $\Omega = \{(\mu, \sigma) : \sigma > 0, \mu = 0\}$. These are illustrated in Figure 11.9, where the vertical gray line indicates where in the (μ, σ) space H_0 is true. The likelihood function for a fixed $x = (x_1, x_2, \ldots, x_n)$ is

$$L(\mu, \sigma | x) = \prod_{i=1}^{n} \frac{1}{\sqrt{2\pi}\,\sigma} \exp\left(-\frac{1}{2\sigma^2}(x_i - \mu)^2\right) = (2\pi)^{-n/2}\sigma^{-n} \exp\left(-\sum_{i=1}^{n}(x_i - \mu)^2\right).$$

In Example 9.19 we showed that the MLEs for μ and σ^2 are

$$\hat{\mu} = \bar{x} \qquad \text{and} \qquad \hat{\sigma}^2 = \frac{1}{n}\sum_{i=1}^{n}(x_i - \bar{x})^2.$$

When H_0: $\mu = 0$ is true, the likelihood function is

$$L(0, \sigma | x) = \prod_{i=1}^{n} \frac{1}{\sqrt{2\pi}\,\sigma} \exp\left(-\frac{1}{2\sigma^2}(x_i - 0)^2\right) = (2\pi)^{-n/2}\sigma^{-n} \exp\left(-\frac{1}{2\sigma^2}\sum_{i=1}^{n}x_i^2\right). \quad (11.14)$$

It is not difficult to show (see Problem 11.53) that the maximum of this likelihood function occurs at

$$\hat{\sigma}_0^2 = \frac{1}{n}\sum_{i=1}^{n}x_i^2,$$

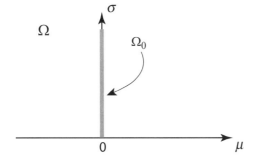

Figure 11.9 Parameter spaces Ω and Ω_0. The parameter space with no restrictions, ω, is the upper half-plane; the parameter space under H_0 is the positive σ axis.

the subscript 0 indicating that this is the estimator for σ^2 under the null hypothesis. The likelihood ratio is therefore

$$
\begin{aligned}
\lambda &= \frac{\max\limits_{\Omega_0} L(\mu,\sigma)}{\max\limits_{\Omega} L(\mu,\sigma)} \\[2mm]
&= \frac{L(0,\hat{\sigma}_0)}{L(\hat{\mu},\hat{\sigma})} \\[2mm]
&= \frac{(2\pi)^{-n/2}\hat{\sigma}_0^{-n}\exp\left(-\frac{1}{2\hat{\sigma}_0^2}\sum_{i=1}^{n}x_i^2\right)}{(2\pi)^{-n/2}\hat{\sigma}^{-n}\exp\left(-\frac{1}{2\hat{\sigma}^2}\sum_{i=1}^{n}(x_i-\mu)^2\right)} \\[2mm]
&= \frac{\hat{\sigma}_0^{-n}\exp\left(-\frac{1}{2\hat{\sigma}_0^2}n\hat{\sigma}_0^2\right)}{\hat{\sigma}^{-n}\exp\left(-\frac{1}{2\hat{\sigma}^2}n\hat{\sigma}^2\right)} \\[2mm]
&= \frac{\hat{\sigma}_0^{-n}\exp(-n/2)}{\hat{\sigma}^{-n}\exp(-n/2)} \\[2mm]
&= \left(\frac{\hat{\sigma}_0^2}{\hat{\sigma}^2}\right)^{-n/2} \\[2mm]
&= \left(\frac{\frac{1}{n}\sum_{i=1}^{n}x_i^2}{\frac{1}{n}\sum_{i=1}^{n}(x_i-\bar{x})^2}\right)^{-n/2} \\[2mm]
&= \left(\frac{\sum_{i=1}^{n}(x_i-\bar{x})^2+n\bar{x}^2}{\sum_{i=1}^{n}(x_i-\bar{x})^2}\right)^{-n/2} \quad [\text{because }\sum_{i=1}^{n}(x_i-\bar{x})^2=\sum_{i=1}^{n}x_i^2-n\bar{x}^2] \\[2mm]
&= \left(\frac{(n-1)S^2+n\bar{x}^2}{(n-1)S^2}\right)^{-n/2} \quad [\text{where }S^2=\sum_{i=1}^{n}(x_i-\bar{x})^2/(n-1)] \\[2mm]
&= \left(1+\frac{n}{n-1}\frac{\bar{x}^2}{S^2}\right)^{-n/2}. \tag{11.15}
\end{aligned}
$$

The likelihood ratio test provides us with a test statistic and a direction – that is, reject for large (or small or both) values of the test statistic. It does not provide us with the appropriate rejection region. From (11.13) we see that the general rule is to reject H_0 if λ is sufficiently *small*. Since the general rule is to reject H_0 for $\lambda < k$ (without specifying exactly what constitutes "small"), we will simplify the expression for λ and rename the constant or the right side. For example, when we raise both sides to the $n/2$ power, as in the second line below, we will simply rename $k^{n/2}$ to be the constant k_1. Note that for positive numbers A and B, $A^{-1} < B^{-1}$ if and only if $A > B$. We can now write the rejection region as

$$\lambda \; < \; k$$

$$\left(1 + \frac{n}{n-1} \frac{\bar{x}^2}{S^2}\right)^{-n/2} < \; k$$

$$\left(1 + \frac{n}{n-1} \frac{\bar{x}^2}{S^2}\right)^{-1} < \; k_1 \qquad \text{[raising both sides to the } 2/n \text{ power]}$$

$$1 + \frac{n}{n-1} \frac{\bar{x}^2}{S^2} > \; k_2 \quad \text{[for positive numbers } A \text{ and } B, A^{-1} < B^{-1} \text{ if and only if } A > B]$$

$$\frac{n}{n-1} \frac{\bar{x}^2}{S^2} > \; k_3$$

$$\frac{\bar{x}^2}{S^2} > \; k_4$$

$$\left(\frac{\bar{x} - 0}{S/\sqrt{n}}\right)^2 > \; k_5.$$

The rule is therefore to reject H_0 if

$$\left(\frac{\bar{x} - 0}{S/\sqrt{n}}\right)^2 > \; k_5$$

where the statistic inside the parentheses on the left side is just the usual t statistic T for testing H_0: $\mu = 0$ against H_1: $\mu \neq 0$. The inequality

$$T^2 > k_5$$

is equivalent to

$$T < -\sqrt{k_5} \quad \text{or} \quad T > \sqrt{k_5}.$$

The value of k_5 is then chosen so that the probability of rejecting H_0 when it is true is α; that is, $\sqrt{k_5} = t_{\alpha/2}(n-1)$ since $T \sim t(n-1)$ when H_0 is true. Thus, the likelihood ratio is exactly equivalent to the one-sample t-test. ∎∎∎

That was a lot of work to find out that the likelihood ratio test is simply the one-sample t-test. For many of the familiar tests that you have seen, or will see in the next chapter, the likelihood ratio test reduces to the familiar test. (Example 12.3 derives the likelihood ratio test for the case where we test the equality of two means.)

The real use of the likelihood ratio test is for situations where an intuitive or familiar procedure is not available. In cases like this, we usually cannot determine the exact distribution of any function of the likelihood ratio λ. Instead we rely on the large sample approximation given in the next theorem.

Theorem 11.9.1 — Approximate Distribution of −2 logλ. Suppose $X_1, X_2, \ldots, X_n$ are random variables with joint PDF $f(x_1, x_2, \ldots, x_n | \theta)$. Let $L(\theta | x_1, x_2, \ldots, x_n) = f(x_1, x_2, \ldots, x_n | \theta)$ denote the likelihood function. Suppose that the parameter vector θ has r components and we want to test H_0: $\theta \in \Omega_0$ where Ω_0 is an r_0-dimensional subset of Ω. Then, for large n,

$$-2 \log \lambda \; \overset{\text{approx}}{\sim} \; \chi^2(r - r_0).$$

Thus, in a complicated hypothesis-testing situation where there is no obvious test statistic, we can apply the likelihood ratio test and, if the sample is sufficiently large, we can determine the rejection region using the χ^2 distribution. The rule, based on this approximation, is to reject H_0 if

$$-2\log\lambda > \chi_\alpha^2(r - r_0).$$

■ **Example 11.16** Suppose that the lifetimes of 23 items placed on test are

48,   102,   51,   6,   96,   9,   72,   15,   48,   65,   7,   29,
23,   27,   40,   71,   15,   39,   21,   10,   3,   92,   17

Assume the Weibull distribution WEIB(θ, β) for these observations. When $\beta = 1$, the Weibull distribution reduces to the exponential distribution, a much simpler model. Test H_0: $\beta = 1$ against the alternative H_1: $\beta \neq 1$ using $\alpha = 0.05$.

Solution: There is no obvious test statistic here, so we apply the likelihood ratio test. The likelihood (not assuming H_0 is true) is

$$L(\theta, \beta | t_1, t_2, \ldots, t_n) = \frac{\beta^n}{\theta^{n\beta}} \left(\prod_{i=1}^{n} t_i \right)^{\beta-1} \exp\left(-\sum_{i=1}^{n} \left(\frac{t_i}{\theta} \right)^\beta \right)$$

and the log likelihood is

$$\ell(\theta, \beta | t_1, t_2, \ldots, t_n) = n\log\beta - n\beta\log\theta + (\beta - 1)\log\prod_{i=1}^{n} t_i - \sum_{i=1}^{n} \left(\frac{t_i}{\theta} \right)^\beta.$$

Closed-form solutions do not exist for the MLEs, so an approximate method like that described in Example 9.22 must be used. We leave it as an exercise (see Problem 11.54) to find that the MLEs are approximately

$$\hat{\beta} \approx 1.287 \qquad \text{and} \qquad \hat{\theta} \approx 42.59.$$

The maximum of the likelihood under no restrictions is therefore

$$\max_{\Omega} L(\hat{\theta}, \hat{\beta} | t_1, t_2, \ldots, t_n) = L(42.59, 1.287 | t_1, t_2, \ldots, t_n) \approx 5.778 \times 10^{-47}.$$

In the contour plot of the likelihood function shown in in Figure 11.10, we can see where the likelihood function is maximized and we can see that the maximum exceeds the highest shown contour, which is 5×10^{-47}.

We must now find the maximum of the likelihood function over the space Ω_0, that is, the space over which H_0 is true. Since the null hypothesis is H_0: $\beta = 1$, the likelihood function under this constraint is

$$L(\theta, 1 | t_1, t_2, \ldots, t_n) = \frac{1^n}{\theta^{n \times 1}} \left(\prod_{i=1}^{n} t_i \right)^{1-1} \exp\left(-\sum_{i=1}^{n} \left(\frac{t_i}{\theta} \right)^1 \right) = \frac{1}{\theta} \exp\left(\frac{1}{\theta} \sum_{i=1}^{n} t_i \right).$$

This is simply the likelihood function for the exponential distribution. The value of θ that maximizes this function (see Problem 9.18) is

$$\hat{\theta}_0 = \frac{1}{n} \sum_{i=1}^{n} t_i = \bar{t} \approx 39.39.$$

The maximum of the likelihood function over Ω_0 is therefore

$$\max_{\Omega_0} L(\theta, \beta | t_1, t_2, \ldots, t_n) = L(\bar{t}, 1 | t_1, t_2, \ldots, t_n) \approx 2.075 \times 10^{-47}.$$

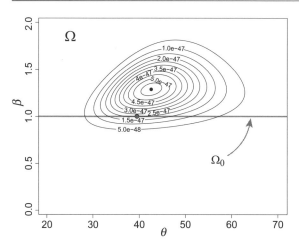

Figure 11.10 Weibull likelihood function in terms of both β and θ for data in Example 11.16. The region Ω_0 that specifies the null hypothesis is the thick gray line.

The horizontal gray line in Figure 11.10 is the parameter space constrained to satisfy H_0. The maximum along this line occurs at $(\bar{t}, 1)$, which is approximately $(39.39, 1)$, and the maximum is 2.075×10^{-47}.

The likelihood ratio is therefore

$$
\lambda = \frac{\max_{\Omega_0} L(\theta, \beta | t_1, t_2, \ldots, t_n)}{\max_{\Omega} L(\theta, \beta | t_1, t_2, \ldots, t_n)}
$$

$$
\approx \frac{L(39.39, 1 | t_1, t_2, \ldots, t_n)}{L(42.59, 1.287 | t_1, t_2, \ldots, t_n)}
$$

$$
\approx \frac{2.075 \times 10^{-47}}{5.778 \times 10^{-47}}
$$

$$
\approx 0.359.
$$

The distribution of $-2 \log \lambda$ is approximately χ^2 with degrees of freedom

$$
\text{df} = \dim(\Omega) - \dim(\Omega_0) = 2 - 1 = 1.
$$

For our data, we have

$$
-2 \log \lambda \approx -2 \log 0.359 \approx 2.05.
$$

With one degree of freedom we would reject for $-2 \log \lambda > \chi^2_{0.05}(1) \approx 3.84$. Since 2.05 does not exceed 3.84, we fail to reject H_0: $\beta = 1$. There is no evidence that the parameter β differs from 1. ∎ ∎ ∎

Exercise 11.6 The generalized gamma distribution contains both the Weibull and the gamma distributions as special cases. The generalized gamma PDF is

$$
f(t | \alpha, \beta, \theta) = \frac{\beta}{\theta \, \Gamma(\alpha)} \left(\frac{t}{\theta} \right)^{\alpha \beta - 1} \exp \left[-\left(\frac{t}{\theta} \right)^{\beta} \right], \qquad t > 0.
$$

Suppose we have a sample of size n from this distribution, where n is sufficiently large for the χ^2 approximation to the distribution of $-2 \log \lambda$. If we wanted to test H_0: $\beta = \alpha = 1$ against the alternative H_1: $\beta \neq 1$ or $\alpha \neq 1$, what would be the degrees of freedom for the χ^2 test?

11.10 Chapter Summary

Hypothesis testing is one of the most commonly applied methods in scientific inquiry. We suppose a null hypothesis H_0 (often a hypothesis of no effect) is true and we use data to assess whether there is evidence against it. The alternative hypothesis H_1 is the one we conclude if we find sufficient evidence against H_0. We base a hypothesis test on a test statistic T. We choose the rejection region (i.e., the set of values that will lead us to rejecting H_0) so that the probability of rejecting H_0 when it is true is some predetermined, usually small, number α.

Rejecting H_0 when it is true is called a Type I error, and accepting the null hypothesis when it is false is a Type II error. In these terms, the probability of a Type I error is equal to α. There are many situations where we can construct an exact test, meaning that we can determine a test statistic with a known distribution and find the critical region so that the probability of rejecting H_0 when it is true is α. Tests involving the mean of a normal distribution can be based on a T statistic. Tests involving the variance of a normal distribution can be based on the pivotal quantity $(n-1)S^2/\sigma^2$, which has a $\chi^2(n-1)$ distribution. More complicated situations often rely on large sample normal approximations to the test statistic.

Power is defined to be the probability of rejecting H_0 given specific values of the parameters, often the "non-null" values. Questions of how large a sample is needed are often based on the power for a particular alternative. For example, we may ask how large a sample is needed to give a power of 0.90 for a specific alternative.

When no obvious test statistic exists, we can often derive a test statistic using the likelihood ratio approach. When the distribution of the resulting test statistic is unknown or intractable, we can apply the large-sample approximation that $-2\log\lambda \overset{\text{approx}}{\sim} \chi^2$ where the degrees of freedom is the difference between the dimension of the entire parameter space and the dimension of the parameter space when it is restricted to the null hypothesis.

11.11 Problems

Problem 11.1 Suppose that 2,000 people are randomly assigned to receive either a vaccine or a placebo. Among the 1,000 in the placebo group, 45 develop the disease within six months. Among the 1,000 in the vaccine group, none develop the disease. Based on your intuition gained from the probability covered so far, do you believe the vaccine is effective? Explain.

Problem 11.2 Suppose that 2,000 people are randomly assigned to receive either a vaccine or a placebo. Among the 1,000 in the placebo group, 25 develop the disease within six months. Among the 1,000 in the vaccine group, 23 develop the disease. Based on your intuition gained from the probability covered so far, do you believe the vaccine is effective? Explain.

Problem 11.3 Suppose I tell the golf pro at the local driving range that I can drive a golf ball as far as the professional. To test this claim, we agree to make five drives each and we measure the length of the drive in yards. Here are the data:

Pro: 305, 295, 270, 310, 325
Me: 225, 190, 175, 205, 95

Based on your intuition, do you believe my claim that I can drive as far as a pro?

Problem 11.4 Suppose that a woman claims that she can tell from tasting whether a cup of tea was prepared with tea poured first and then the milk, or vice versa. Is this claim valid? (This example is the one alluded to in the title of the book by Salsburg, 2001). To test this claim, the woman was given eight glasses of tea, four of which were prepared by adding milk to a cup of tea and four prepared by adding tea to a small amount of milk. The lady was able to identify all eight cups correctly. Do you believe she has the ability to distinguish tea-first and milk-first cups of tea? What if the woman was given 20 cups of tea, with half prepared each way, and correctly identifies all cups?

Problem 11.5 In Problems 11.1–11.4, identify the null and alternative hypotheses.

Problem 11.6 In Problems 11.1–11.4, identify the test statistic and its distribution.

Problem 11.7 A county's population is equally divided between an eastern half and a western half. Jurors should be selected at random from across the county. Suppose that in a sample of 100 jurors, 30 are from the eastern half and 70 are from the western half. Design a test to see whether there is evidence that juror selection is not random. What is your decision for this data?

Problem 11.8 Consider the crossover clinical trial described in Example 8.9. In such a trial, subjects receive both treatments and say which worked better. Suppose $n = 50$ participants are enrolled and must declare at the end of the study whether Drug A or Drug B worked better. Design a two-sided test to test whether the drugs are equally effective. If 27 people said Drug A was more effective and 23 people said Drug B was more effective, what is your decision?

Problem 11.9 Consider the situation described in Example 11.1. Design a test of size $\alpha = 0.05$.

Problem 11.10 Again, consider the situation from Example 11.1, except suppose that the rate is 8 per day. Design a test of size $\alpha = 0.05$.

Problem 11.11 Suppose $X_1, X_2, \ldots, X_9 \sim$ i.i.d. $N(\mu, 1)$.

(a) What is the rejection region for a size $\alpha = 0.05$ test for $H_0\colon \mu = 10$ with alternative $H_1\colon \mu < 10$?

(b) What is the power when $\mu = 11$?

Problem 11.12 Suppose $X_1, X_2, \ldots, X_9 \sim$ i.i.d. $N(\mu, 1)$.

(a) What is the rejection region for a size $\alpha = 0.05$ test for $H_0\colon \mu = 10$ with alternative $H_1\colon \mu \neq 10$?

(b) What is the power when $\mu = 11$?

Problem 11.13 Consider again the crossover design described in Problem 11.8. Suppose now that $n = 10$.

(a) Design a two-sided test for this situation with α as large as possible but $\alpha \leq 0.05$.

(b) Compute the exact power for $p = 0.3, 0.4, 0.5, 0.6$, and 0.7, where p is the probability that a subject selected from the population prefers Drug A over Drug B.

Problem 11.14 Figure 11.11 shows the power curve for testing $H_0\colon \mu = 0$. Using this figure, answer the following questions and explain your answer:

(a) Is the alternative $\mu < 0$, $\mu > 0$, or $\mu \neq 0$?

(b) What is the size α for the test?

(c) It is believed that there will be an effect of about 1 unit. In other words, they believe that the true

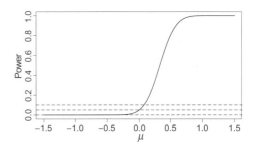

Figure 11.11 Power curve for Problem 11.14.

value of μ will be about 1, and they want to be confident that the null hypothesis will be rejected if this is the case. Is the sample size large enough, or is a larger sample size needed?

Problem 11.15 Figure 11.12 shows the power curve for testing H_0: $\mu = 0$. Using this figure, answer the following questions and explain your answer:

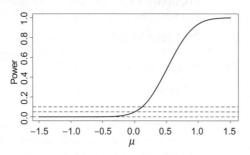

Figure 11.12 Power curve for Problem 11.15

(a) Is the alternative $\mu < 0$, $\mu > 0$, or $\mu \neq 0$?
(b) What is the size α for the test?
(c) Is the sample size for this problem larger than the sample size for Problem 11.14?

Problem 11.16 Figure 11.13 shows the power curve for testing H_0: $\mu = 0$. Using this figure, answer the following questions and explain your answer:

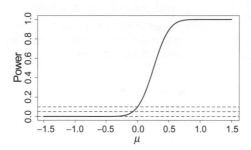

Figure 11.13 Power curve for Problem 11.16.

(a) Is the alternative $\mu < 0$, $\mu > 0$, or $\mu \neq 0$?
(b) What is the size α for the test?

Problem 11.17 Figure 11.14 shows the power curve for testing H_0: $\mu = 0$. Using this figure, answer the following questions and explain your answer:

(a) Is the alternative $\mu < 0$, $\mu > 0$, or $\mu \neq 0$?
(b) What is the size α for the test?

Problem 11.18 Figure 11.15 shows the power curve for testing H_0: $\mu = 0$. Using this figure, answer the following questions and explain your answer:

(a) Is the alternative $\mu < 0$, $\mu > 0$, or $\mu \neq 0$?

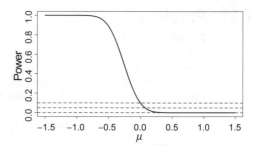

Figure 11.14 Power curve for Problem 11.17.

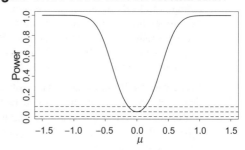

Figure 11.15 Power curve for Problem 11.18.

(b) What is the size α for the test?

Problem 11.19 Figure 11.16 shows the power curve for testing H_0: $\mu = 0$. Using this figure, answer the following questions and explain your answer:

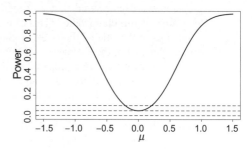

Figure 11.16 Power curve for Problem 11.19.

(a) Is the alternative $\mu < 0$, $\mu > 0$, or $\mu \neq 0$?
(b) What is the size α for the test?
(c) Is the sample size for this problem larger than the sample size for Problem 11.18?

Problem 11.20 Once again, consider the crossover design described in Problem 11.8. Suppose that n subjects participate and at the end of the study they must indicate whether Drug A or Drug B worked better. (No ties are allowed.) Suppose that Drug A is a new drug and is being compared to a standard of care which is Drug B. Investigators are performing a one-sided test where they may reject the null hypothesis that the drugs are the same in favor of the alternative that Drug A is better. Suppose that all n subjects select Drug A. Find the exact P-value if

(a) $n = 5$;
(b) $n = 10$;
(c) $n = 20$.

Problem 11.21 Repeat Problem 11.20 with $n = 80$, but this time using the normal approximation to the binomial.

Problem 11.22 Repeat Problem 11.20 if the test is two-sided and

(a) all subjects prefer Drug A;
(b) all subjects prefer Drug B.

Problem 11.23 Consider the lady tasting tea from Problem 11.4. The lady is presented with eight cups of tea, four prepared with tea first and four prepared with milk first. The null hypothesis is that she is guessing which four were prepared with milk first. Suppose the decision rule is to reject that she is guessing (in favor of the alternative that she can distinguish the two) if she identifies all four milk-first cups correctly. If she gets all four milk-first cups correct, what is the P-value? (*Hint*: If she is guessing, then the number of milk-first cups she selects has a hypergeometric distribution.)

Problem 11.24 Repeat Problem 11.23 if n cups of tea are presented, with $n/2$ being milk-first and the other $n/2$ being tea-first. Here n is an even number. Suppose the lady selects all of the milk-first cups correctly. What is the P-value if

(a) $n = 10$;
(b) $n = 12$;
(c) $n = 16$.

Problem 11.25 In Problem 6.19 we considered a manufacturing process that makes capacitors designed to produce 100 VDC (volts of direct current). Capacitors produced by this company are normally distributed with a mean of 100.5 and a standard deviation of 2.0. Occasionally something happens to the process to change the mean from $\mu_0 = 100.5$ to some other value μ_1; the standard deviation remains constant at 2.0. At the end of each day, the company selects 10 capacitors and tests them to check for stability of the manufacturing process. On one particular day, the measured output in VDC was

99.2 100.1 96.3 97.8 95.0
99.2 100.5 99.5 96.2 97.8

(a) Test, using $\alpha = 0.05$, whether the process mean has remained constant at $\mu_0 = 100.5$ against the alternative that it has changed in either direction. Assume that the standard deviation is known to be $\sigma = 2.0$.

(b) What is the P-value?

(c) This process of selecting 10 items to be tested is continued every day for the 250 work days in a given year. Suppose that the process remains stable at $\mu = \mu_0 = 100.5$ throughout the entire period. How many times would you expect that we would (incorrectly) reject the null hypothesis over this 250-day period. In the quality control literature, incorrectly rejecting the null hypothesis is called a *false alarm*.

(d) Again, supposing that the process remains stable at $\mu = \mu_0 = 100.5$ throughout the entire period, what is the distribution of the number of working days until the first false alarm?

(e) Once again, supposing that the process remains stable at $\mu = \mu_0 = 100.5$ throughout the entire period, what is the expected value of the number of days until a false alarm occurs?

Problem 11.26 A manufacturing process for engine pistons produces pistons whose diameters are normally distributed with a mean of $\mu_0 = 74.001$ and $\sigma = 0.0103$ when the process is stable (Montgomery, 2020b). A sample of eight pistons selected at random from the process yields a sample mean of $\bar{x} = 74.067$.

(a) Test, using $\alpha = 0.05$, whether the process mean has remained constant at $\mu_0 = 74.001$ against the alternative that it has changed in either direction. Assume that the process standard deviation is known to be $\sigma = 0.0103$.

(b) What is the P-value?

(c) This process of selecting eight items to be tested is continued every day for the 250 work days in a given year. Suppose that the process remains stable at $\mu = \mu_0 = 74.001$ throughout the entire period. How many times would you expect that we would (incorrectly) reject the null hypothesis over this 250-day period. In the quality control literature, incorrectly rejecting the null hypothesis is called a *false alarm*.

(d) Again, supposing that the process remains stable at $\mu = \mu_0 = 74.001$ throughout the entire period, what is the distribution of the number of working days until the first false alarm?

(e) Once again, supposing that the process remains stable at $\mu = \mu_0 = 74.001$ throughout the entire period, what is the expected value of the number of days until a false alarm occurs?

Problem 11.27 For the situation described in Problem 11.25, find and graph the power function for μ between 95 and 105 in increments of 0.05.

Problem 11.28 For the situation described in Problem 11.26, find and graph the power function for μ between 73.9 and 74.1 in increments of 0.001.

Problem 11.29 Repeat Problem 11.25 using $\alpha = 0.005$. Compare the false alarm rates from part (c). Discuss the implications of varying α. (*Hint*: There is an inherent trade-off between a small and large α. Do you see what this trade-off is?)

Problem 11.30 Repeat Problem 11.26 using $\alpha = 0.001$. Compare the false alarm rates from part (c). Discuss the trade-offs involved in varying α.

Problem 11.31 Consider the situation and the data described in Problem 10.28. The ideal stabilized viscosity is 3,200. Test, using α, whether the true mean is 3,200 against the alternative that it differs from 3,200. What is the P-value?

Problem 11.32 It is known that there are health risks due to grief and stress after the death of a spouse. One of the consequences could be unplanned weight loss. Suppose that in a study of 58 persons who lost a spouse, the mean weight loss over one year was 2.03 lb, with a sample standard deviation of 8.13 lb. Let μ denote the population mean weight loss over one year for persons losing a spouse. (The issue of weight loss after the death of a spouse was studied in Shahar *et al.* (2001), although the design was different than that described here.)

(a) Test the null hypothesis H_0: $\mu = 0$ against the alternative H_1: $\mu \neq 0$ using $\alpha = 0.05$.
(b) What is the P-value?
(c) Comment on the design of the study. Can you think of a more effective design?

Problem 11.33 Repeat Problem 11.32 for the alternative H_1: $\mu > 0$. Do your conclusions differ from those obtained in Problem 11.32?

Problem 11.34 Repeat Problem 11.32 if the sample mean and standard deviation are $\bar{x} = 2.2$ and $s = 4.2$.

Problem 11.35 Suppose you are designing a follow-up study to that described in Problem 11.32. Initially, you select a sample size of $n = 50$. Using the sample standard deviation given in Problem 11.32 as a rough estimate of the population standard deviation, compute the power of a one-sided test with alternative H_1: $\mu > 0$ for μ running from 0 to 4 in increments of 0.05.

Problem 11.36 Repeat Problem 11.35 for a sample size of 80.

Problem 11.37 For the situation described in Problem 11.35, how large must the sample size be so that the power is 0.90 when the true mean weight loss is 2.0 lb?

Problem 11.38 Consider again the situation described in 11.35. Suppose someone argues that the variability in the population to be studied in the follow-up experiment is larger, say $\sigma \approx 10$ rather than $\sigma \approx 8.12$. Repeat Problems 11.35, 11.36, and 11.37 with this revised σ.

Problem 11.39 The Environmental Protection Agency sets a level of 15 μg/L for lead in drinking water. When there is evidence that the level exceeds this value, the EPA takes action. Since there is substantial measurement error in the process of determining the lead concentration, a water sample is selected and tested 10 times, yielding the results (in μg/L).

10.7 12.4 10.3 15.2 12.7
10.4 13.0 13.5 13.2 11.4

Assume that these represent a random sample from the $N(\mu, \sigma^2)$ distribution, where μ is the true mean lead level in the water tank. Suppose we want to test the null hypothesis H_0: $\mu = 15$ against the alternative H_1: $\mu > 15$.

(a) Does the alternative hypothesis say that the water meets the standard?
(b) Describe in plain words what Type I and Type II errors are in this context.
(c) Using $\alpha = 0.01$, test the null and alternative hypotheses.
(d) What is the P-value?

Problem 11.40 Repeat Problem 11.39, except formulate the hypotheses as H_0: $\mu = 15$ and H_1: $\mu < 15$.

Problem 11.41 For the data in 11.39, find a 99% confidence interval for μ.

Problem 11.42 In Problems 11.39, 11.40, and 11.41, we took three approaches for addressing the question of water safety. Discuss the appropriateness of these approaches if we want to know whether the water meets the 15 μg/L standard.

Problem 11.43 Consider the situation described in Problem 11.39, but with the hypotheses formulated as in Problem 11.40: H_0: $\mu = 15$ and H_1: $\mu < 15$. Use $\alpha = 0.01$. Suppose the standard deviation of the measurement error is thought to be $\sigma \approx 1.5$.

(a) What is the power for a sample size of $n = 10$ when the true value of μ is 14?
(b) Plot the power function for μ between 5 and 20.

Problem 11.44 Consider again the situation described in Problem 11.40. How large must the sample size be if the power is to equal 0.90 for $\mu = 14$?

Problem 11.45 Suppose that in a study of 200 people whose spouse had died, 56% lost weight in the subsequent year. Let p denote the population proportion of those who lost weight in the year after the death of a spouse. Test using $\alpha = 0.05$ the hypotheses H_0: $p = 0.5$ against H_1: $p > 0.5$.

Problem 11.46 Repeat Problem 11.45 using the two-sided alternative H_1: $p \neq 0.5$.

Problem 11.47 A school bond issue requires approval from 60% of voters who cast a vote. In a survey of 200 likely voters, 71% favored the bond issue. If p is the population proportion who approve of the bond issue, test H_0: $p = 0.6$ against the alternative H_1: $p > 0.6$ using $\alpha = 0.05$.

Problem 11.48 For the situation described in Problem 11.47, find a 95% confidence interval for p.

Problem 11.49 Consider again the situation from Problem 11.39. Based on past history, the engineers who run the water treatment plant have observed that the standard deviation of the measurement error is 1.5. In an effort to reduce the variability of the measurement process, the engineers purchased more sophisticated equipment to measure the lead content. They took a sample of water and tested it 20 times. They observed $\bar{x} = 8.5$ and $s = 0.85$.

(a) Test the hypotheses H_0: $\sigma^2 = 1.5^2$ against the alternative H_1: $\sigma^2 \neq 1.5^2$. Use $\alpha = 0.05$.

(b) Find a 95% confidence interval for σ^2.

Problem 11.50 Repeat part (a) of Problem 11.49 using the one-sided alternative H_1: $\sigma^2 < 1.5^2$.

Problem 11.51 Consider the manufacturing system for engine pistons described in Problem 11.26. When this process was operating in a stable fashion, the mean was 74.001 and the standard deviation was 0.0103. A sample of size $n = 5$ yielded

   73.988   74.005   73.984   74.034   74.008

(a) Using data taken from this set of five, test H_0: $\mu = 74.001$ against H_1: $\mu \neq 74.001$ with $\alpha = 0.05$.

(b) What is the P-value?

(c) Again, using this set of five data points, test H_0: $\sigma^2 = 0.0103^2$ against H_1: $\sigma^2 \neq 0.0103^2$.

(d) What is the P-value from the test in part (c)?

Problem 11.52 It has been documented that researchers tend to *underestimate* the standard deviation when providing information for a power analysis. For example, if the true standard deviation is $\sigma = 2.0$, the researchers may say $\sigma = 1.5$.

(a) Discuss the implications of this regarding the computed power for a fixed value of the parameter under the alternative hypothesis.

(b) Discuss the implications of this regarding the determination of the required sample size for a fixed value of the parameter under the alternative hypothesis.

Problem 11.53 Show that the likelihood function in (11.14) is maximized when $\hat{\sigma}_0 = \sum_{i=1}^{n} x_i^2 / n$.

Problem 11.54 For the data in Example 11.16, show that the MLEs are approximately $\hat{\beta} \approx 1.287$ and $\hat{\theta} \approx 42.59$.

Problem 11.55 Suppose $X_1, X_2, \ldots, X_n \sim$ i.i.d. $N(\mu, \sigma^2)$. Derive the likelihood ratio test for testing H_0: $\mu = \mu_0$ vs. H_1: $\mu \neq \mu_0$, where μ_0 is a fixed number.

Problem 11.56 Suppose $X_1, X_2, \ldots, X_n \sim$ i.i.d. $N(\mu, \sigma^2)$. Derive the likelihood ratio test for testing H_0: $\sigma^2 = 1$ vs. H_1: $\sigma^2 \neq 1$.

Problem 11.57 Suppose the data in Example 11.16 constitute a random sample from the $GAM(\alpha, \beta)$ distribution.

(a) Derive the likelihood ratio test for testing H_0: $\alpha = 1$ vs. H_1: $\alpha \neq 1$.

(b) Why might we want to test H_0: $\alpha = 1$?

12 — Hypothesis Tests for Two or More Populations

12.1 Introduction

A common problem in statistics is to compare groups. Does a new drug work better at reducing the time of hospitalization from COVID? Which pop-up ad generates a higher click-rate? Which type of metal – aluminum, brass, or stainless steel – will produce the most reliable product? Usually, the question involves either the mean response or the proportion of responses. The question always involves the parameters (not the sample statistics). Thus, for example, we might test whether μ_1, the mean hospitalization time for COVID patients given a new drug, is equal to μ_2, the mean hospitalization time for COVID patients given a standard drug.[1] The hypotheses would then be H_0: $\mu_1 = \mu_2$ against H_1: $\mu < \mu_2$ (or possibly H_1: $\mu_1 \neq \mu_2$). Other problems may compare more than two means; for example, the experiment regarding which metal to use to increase a product's reliability would compare three means.

In this chapter we look at comparing groups. We mostly compare means, but we also address comparing proportions and variances. The design of an experiment or observational study often dictates which methods to use to test the appropriate hypotheses.

12.2 Testing Two Independent Samples

12.2.1 Comparing Two Means

One of the most common problems in statistics is to test whether the means of the two populations are equal. (We can also have a one-sided test where the alternative hypothesis is that one is greater than the other.) Let's introduce some notation to make the ideas more concrete. Let

$$X_1, X_2, \ldots, X_{n_1} \sim \text{ i.i.d. from some distribution with mean } \mu_1 \text{ and variance } \sigma_1^2$$

$$Y_1, Y_2, \ldots, Y_{n_2} \sim \text{ i.i.d. from some distribution with mean } \mu_2 \text{ and variance } \sigma_2^2.$$

We further assume that the two samples are independent of one another.

---

[1]Clinical trials like this will often compare a new drug with a placebo, but for serious diseases it would be unethical to withhold treatment from a person by giving a knowingly inferior treatment (like a placebo) when other treatments exist. Experiments with a serious disease will often compare a new drug with the best existing treatment.

Under these conditions, can we construct a test statistic whose distribution is known (at least approximately)? It seems reasonable to base our estimates of μ_1 and μ_2 on the sample means, $\overline{X}$ and $\overline{Y}$, respectively. By the CLT (Theorem 8.3.1), both $\overline{X}$ and $\overline{Y}$ are approximately normally distributed (assuming that both populations have finite variances). Thus, the difference $\overline{X} - \overline{Y}$ is approximately normally distributed. The expected value of $\overline{X} - \overline{Y}$ is

$$\mathbb{E}(\overline{X} - \overline{Y}) = \mathbb{E}(\overline{X}) - \mathbb{E}(\overline{Y}) = \mu_1 - \mu_2,$$

and its variance is

$$\mathbb{V}(\overline{X} - \overline{Y}) = \mathbb{V}(\overline{X}) + \mathbb{V}(\overline{Y}) - 2\mathrm{cov}(\overline{X}, \overline{Y}) = \frac{\sigma_1^2}{n_1} + \frac{\sigma_2^2}{n_2}.$$

Because the samples are independent, $\mathrm{cov}(\overline{X}, \overline{Y}) = 0$. Standardizing (subtracting the mean and dividing by the standard deviation) yields

$$Z = \frac{(\overline{X} - \overline{Y}) - (\mu_1 - \mu_2)}{\sqrt{\dfrac{\sigma_1^2}{n_1} + \dfrac{\sigma_2^2}{n_2}}} \overset{\text{approx}}{\sim} N(0, 1).$$

The quality of this approximation depends on the magnitude of n_1 and n_2. Technically, the distribution of Z converges to $N(0, 1)$ as both $n_1 \to \infty$ and $n_2 \to \infty$. The quality also depends on the distribution of the samples $X_1, X_2, \ldots, X_{n_1}$ and $Y_1, Y_2, \ldots, Y_{n_2}$. If these distributions are close to normally distributed, then the approximation will be better than if the distributions were distinctly nonnormal.

In practice, we do not know the values of σ_1^2 and σ_2^2. The usual practice is to use the estimates

$$\hat{\sigma}_1^2 = S_1^2 = \frac{1}{n_1 - 1} \sum_{i=1}^{n_1} (X_i - \overline{X})^2$$

$$\hat{\sigma}_2^2 = S_2^2 = \frac{1}{n_2 - 1} \sum_{i=1}^{n_2} (Y_i - \overline{Y})^2$$

in place of σ_1^2 and σ_2^2. This yields a different test statistic:

$$T = \frac{(\overline{X} - \overline{Y}) - (\mu_1 - \mu_2)}{\sqrt{\dfrac{S_1^2}{n_1} + \dfrac{S_2^2}{n_2}}} \overset{\text{approx}}{\sim} N(0, 1).$$

For large n_1 and n_2, the approximate distribution is still $N(0, 1)$ even with estimators used in place of the parameters.[2] The justification for calling this statistic T will come shortly. The decision rule would then be to reject H_0: $\mu_1 = \mu_2$ in favor of H_1: $\mu_1 \neq \mu_2$ if

$$T < -z_{\alpha/2} \quad \text{or} \quad T > z_{\alpha/2}.$$

If the alternative were H_1: $\mu_1 < \mu_2$, then the rejection region would be

$$T < -z_\alpha.$$

---

[2] A better approximation will come later in this chapter.

A similar result holds for the alternative $H_1: \mu_1 > \mu_2$. This is a useful result if both n_1 and n_2 are reasonably large. But what if we have small samples?

Let's consider what the distribution of T is if we assume a normal distribution and if we assume that both populations have the *same* variance. Specifically, we assume

$$X_1, X_2, \ldots, X_{n_1} \sim \text{ i.i.d. } N(\mu_1, \sigma^2)$$

$$Y_1, Y_2, \ldots, Y_{n_2} \sim \text{ i.i.d. } N(\mu_2, \sigma^2).$$

We still assume that the two samples are independent of one another. These assumptions lead to an interesting phenomenon. The Xs give us information about μ_1 and the Ys give us information about μ_2, but *both* the Xs and Ys give us information about σ^2. This is because the parameter σ^2 is assumed to be common across both populations. We should therefore use information from both the Xs and Ys to estimate σ^2. We know from (9.4) that

$$\mathbb{E}(S_1^2) = \sigma^2$$

and

$$\mathbb{E}(S_2^2) = \sigma^2,$$

so both S_1^2 and S_2^2 are unbiased, but both estimators use partial information from the two samples. A weighted average of S_1^2 and S_2^2 can be used as an estimator for σ^2. Specifically,

$$\hat{\sigma}^2 = S_p^2 = \frac{(n_1 - 1)S_1^2 + (n_2 - 1)S_2^2}{n_1 + n_2 - 2}. \tag{12.1}$$

This is called the *pooled* estimate of the variance. We next prove two important results regarding S_p^2.

Theorem 12.2.1 — Properties of the Pooled Variance. Under the assumptions above,

$$\mathbb{E}(S_p^2) = \sigma^2 \qquad \text{and} \qquad \frac{(n_1 + n_2 - 2)S_p^2}{\sigma^2} \sim \chi^2(n_1 + n_2 - 2).$$

Proof. To prove the first part, we can write

$$\mathbb{E}(S_p^2) = \mathbb{E}\left(\frac{(n_1 - 1)S_1^2 + (n_2 - 1)S_2^2}{n_1 + n_2 - 2}\right)$$

$$= \frac{1}{n_1 + n_2 - 2}\, \mathbb{E}\left((n_1 - 1)S_1^2 + (n_2 - 1)S_2^2\right)$$

$$= \frac{1}{n_1 + n_2 - 2}\left((n_1 - 1)\mathbb{E}(S_1^2) + (n_2 - 1)\mathbb{E}(S_2^2)\right)$$

$$= \frac{1}{n_1 + n_2 - 2}\left((n_1 - 1)\sigma^2 + (n_2 - 1)\sigma^2\right)$$

$$= \frac{\sigma^2}{n_1 + n_2 - 2}\left(n_1 - 1 + n_2 - 1\right)$$

$$= \sigma^2.$$

Thus, S_p^2 is an unbiased estimator of the common variance σ^2. We know that

$$\frac{(n_1-1)S_1^2}{\sigma^2} \sim \chi^2(n_1-1) \quad \text{and} \quad \frac{(n_2-1)S_2^2}{\sigma^2} \sim \chi^2(n_2-1).$$

Furthermore, $(n_1-1)S_1^2/\sigma^2$ and $(n_2-1)S_2^2/\sigma^2$ are independent because the Xs are independent of the Ys. Also, the sum of independent χ^2 random variables has a χ^2 distribution and the degrees of freedom add (see Example 6.32). That is, if $U \sim \chi^2(v_1)$, $V \sim \chi^2(v_2)$, and U and V are independent, then $U + V \sim \chi^2(v_1 + v_2)$. Thus,

$$\frac{(n_1-1)S_1^2}{\sigma^2} + \frac{(n_2-1)S_2^2}{\sigma^2} \sim \chi^2(n_1-1+n_2-1).$$

This can be written as

$$\frac{\dfrac{(n_1-1)S_1^2 + (n_2-1)S_2^2}{n_1+n_2-2}}{\dfrac{\sigma^2}{n_1+n_2-2}} \sim \chi^2(n_1+n_2-2),$$

which implies

$$\frac{(n_1+n_2-2)S_p^2}{\sigma^2} \sim \chi^2(n_1+n_2-2). \qquad\blacksquare$$

The t distribution was defined in Definition 6.6.3 as the distribution of the random variable

$$T = \frac{Z}{\sqrt{V/v}},$$

where $Z \sim N(0,1)$ and $V \sim \chi^2(v)$ are independent random variables. We can take

$$Z = \frac{(\overline{X}-\overline{Y})-(\mu_1-\mu_2)}{\sqrt{\dfrac{\sigma^2}{n_1}+\dfrac{\sigma^2}{n_2}}} = \frac{(\overline{X}-\overline{Y})-(\mu_1-\mu_2)}{\sigma\sqrt{\dfrac{1}{n_1}+\dfrac{1}{n_2}}}$$

and

$$V = \frac{(n_1+n_2-2)S_p^2}{\sigma^2}.$$

Thus,

$$T = \frac{\dfrac{(\overline{X}-\overline{Y})-(\mu_1-\mu_2)}{\sqrt{\dfrac{\sigma^2}{n_1}+\dfrac{\sigma^2}{n_2}}}}{\sqrt{\dfrac{(n_1+n_2-2)S_p^2/\sigma^2}{n_1+n_2-2}}} \sim t(n_1+n_2-2), \qquad (12.2)$$

because the numerator is a standard normal random variable and the denominator is the square root of a $\chi^2(\nu)$ random variable divided by its degrees of freedom. The T statistic in (12.2) can be written as

$$T = \frac{\dfrac{(\overline{X} - \overline{Y}) - (\mu_1 - \mu_2)}{\sigma\sqrt{\frac{1}{n_1} + \frac{1}{n_2}}}}{\dfrac{S_p}{\sigma}\sqrt{\dfrac{n_1 + n_2 - 2}{n_1 + n_2 - 2}}} = \frac{(\overline{X} - \overline{Y}) - (\mu_1 - \mu_2)}{S_p\sqrt{\frac{1}{n_1} + \frac{1}{n_2}}}.$$

If the null hypothesis H_0: $\mu_1 = \mu_2$ is true, then the above statistic becomes

$$T = \frac{\overline{X} - \overline{Y}}{S_p\sqrt{\dfrac{1}{n_1} + \dfrac{1}{n_2}}} \sim t(n_1 + n_2 - 2). \tag{12.3}$$

For the two-sided alternative H_1: $\mu_1 \neq \mu_2$, the rule is to reject H_0: if $T < -t_{\alpha/2}$ or $T > t_{\alpha/2}$.

■ **Example 12.1** The file `FirstYearGPA.csv` (Cannon *et al.*, 2018) contains data on 219 students who entered college. Variables include standardized test scores and some demographic data on students. The response is the student's first-year GPA in college. We will analyze these data thoroughly in Chapter 14, but for now let's look at the effect of being a first-generation college student. This is defined to be a student whose parents or grandparents did not attend college. Is the first-year GPA affected by being a first-generation college student (coded as 1 in the `FirstGen` variable)?

Solution: The following code reads the file and extracts the needed variables from the data frame:

```
first = read.csv( "FirstYearGPA.csv" , header=TRUE )
FirstGen = first$FirstGen
GPA  = first$GPA
```

We will first perform the computations directly in R using (12.3). First we extract separate vectors for the first-generation college students and the non-first-generation college students:

```
GPA0 = GPA[ FirstGen == 0 ]
GPA1 = GPA[ FirstGen == 1 ]
```

Then we determine the lengths of the two vectors, along with the two sample means and the pooled variance:

```
n1 = length( GPA0 )
n2 = length( GPA1 )
s2pooled = ( (n1-1)*var(GPA0) + (n2-1)*var(GPA1) ) / (n1+n2-2)
GPAbar0 = mean( GPA0 )
GPAbar1 = mean( GPA1 )
print( c(GPAbar0,GPAbar1) )
```

We can see from R's output that the sample mean GPA from the non-first-generation college student group is 3.122268 and the sample mean of the first-generation college student group is 2.893600. The non-first-generation college student group had a higher GPA than the first-generation college student group. The difference is not that great though. What a hypothesis test does is to determine whether the observed difference can be explained by chance. The t statistic is

```
T = ( GPAbar0 - GPAbar1 ) / sqrt(s2pooled*(1/n1+1/n2))
```

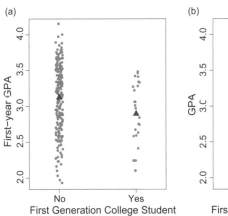
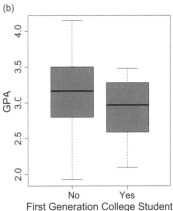

Figure 12.1 First-year GPA for non-first-generation college students and first-generation college students. The triangles indicate the mean for the group.

This works out to $t = 2.335333$. The critical values are determined using the `qt()` command in R:

```
qt( 0.025 , n1+n2-2 , lower.tail=FALSE )
```

This yields $t_{0.025}(n_1 + n_2 - 2) = 1.971$. The decision rule is to reject H_0 if

$$t < -t_{0.025} = -1.971 \quad \text{or} \quad t > t_{0.025} = 1.971.$$

Since $t = 2.335$, we reject H_0 and conclude that there is a difference in the population means for the non-first-generation college student group and the first-generation college student group. The P-value for this test can be computed in R using the command

```
2*min( c( pt( -T , n1+n2-2 ) , pt( T , n1+n2-2 ) ) )
```

This may seem like a convoluted way to get the P-value, but in general we don't know whether the T statistic is positive or negative. Here it is positive, so the P-value is `2*pt( -T , n1+n2-2 )`, but had it been negative it would have been `2*pt( T , n1+n2-2 )`. The use of the `min()` function covers both cases. In this case the P-value is $P = 0.0204$.

A dot plot of the raw data, along with a box plot for the two groups, is shown in Figure 12.1. The blue triangles in Figure 12.1(a) indicate the position of the mean GPA. We can see that the mean for non-first-generation college students is above the mean for first-generation college students. Figure 12.1(b) shows the side-by-side box plots. Even though we reject the hypothesis of equal means, we can see from the figures that there is considerable overlap in the two samples; many first-generation college students had higher scores than many of the students who were not first-generation college students.

R has built-in functions to perform t-tests like this. We can use the `t.test` command to do the required computations. We send the two vectors containing the data for the two groups. If we assume that the variances are the same, we would also send the optional argument `var.equal = TRUE`; the default is `FALSE`. In our case, we would enter the command

```
t.test( GPA0 , GPA1 , var.equal = TRUE )
```

The output from this command is shown below:

```
        Two Sample t-test

data:  GPA0 and GPA1
t = 2.3353, df = 217, p-value = 0.02044
alternative hypothesis: true difference in means is not equal to 0
```

```
95 percent confidence interval:
0.03567853 0.42165756
sample estimates:
mean of x mean of y
3.122268  2.893600
```

The numeric values obtained using t.test agree with those we did previously when we performed the computations from scratch using R.   ■ ■ ■

The plots in Figure 12.1 might seem to indicate that the variability in the non-first-generation group is greater than the variability in the first-generation group. This can be deceiving because there were considerably more in the former group than in the latter. The box plots indicate that the interquartile ranges are about the same for the two groups, but there are fewer values in the extremes. When one sample size is smaller than the other, we would expect fewer observations in the extremes. The standard deviations of the two groups are

$$s_{\text{non-first-gen}} = 0.4637245 \qquad \text{and } s_{\text{first-gen}} = 0.4365001,$$

which are nearly the same. Here, the assumption of equal population variances seems reasonable.[3]

When it is unreasonable to assume that the variances in the two groups are unequal, we can use an approximation due to Welch or Satterthwaite. Using this method, our statistic becomes

$$T = \frac{(\overline{X} - \overline{Y}) - (\mu_1 - \mu_2)}{\sqrt{\frac{S_1^2}{n_1} + \frac{S_2^2}{n_2}}} \overset{\text{approx}}{\sim} t(\nu), \tag{12.4}$$

where

$$\nu = \frac{\left(\frac{s_1^2}{n_1} + \frac{s_2^2}{n_2}\right)^2}{\frac{(s_1^2/n_1)^2}{n_1 - 1} + \frac{(s_2^2/n_2)^2}{n_2 - 1}}. \tag{12.5}$$

Unfortunately, the distribution of this test statistic is not a t distribution, but in most situations it is well approximated by the $t(\nu)$ distribution. We will call the method based on (12.4) and (12.5) Welch's method or the Welch–Satterthwaite method.

Another important fact to keep in mind is that the value of ν will almost never reduce to an integer, so tables of the t distribution cannot be used directly. There are several options here. One could simply round to the nearest integer and look up that value in the t table. Another option is to use tables and do a linear interpolation in the table. For example, if we want $t_{0.01}(27.5)$ we could look up $t_{0.01}(27)$ and get 2.473, and then look up $t_{0.01}(28)$ and get 2.467. Since 27.5 is midway between 27 and 28, the appropriate quantile for the $t(27.5)$ could be approximated by the midpoint of 2.473 and 2.467, which works out to 2.470. A third option is to use R to obtain the needed quantiles. The appropriate commands in R, namely dt(), pt(), and qt(), can be used with noninteger degrees of freedom.

The good news is that these procedures lead to nearly the same conclusions for most problems. Since we are assuming access to and familiarity with R, we will use R to get the needed information from the t distribution.

---

[3]Note that we are assuming equal *population* variances. In practice, the *sample* variances will almost always differ by chance, even when the population variances are equal.

■ **Example 12.2** Bilirubin measures liver function, with higher levels indicating the possibility of poor liver function. Consider the data set of liver patients in `liver_patient.csv`. While this data set gives information about a number of variables, we will focus on the difference between the bilirubin levels in women and men. Specifically, test whether the mean bilirubin level is the same in populations of female and male liver patients. Do not assume equal population variances and apply the Welch–Satterthwaite method.

Solution: The following code reads in the required data, extracts the needed variables from the data frame, and does the computations for the Welch–Satterthwaite method. It then prints all of the needed information, including the *P*-value.

```
liver = read.csv("liver_patient.csv")
names(liver)      #### to see the variable names

gender = liver$Gender
sex = as.numeric( gender == "Male" )
bilirubin = liver$Total_Bilirubin
bilirubinM = bilirubin[ gender == "Male" ]
bilirubinF = bilirubin[ gender == "Female" ]

n1 = length( bilirubinM )
n2 = length( bilirubinF )
s2M = var(bilirubinM)
s2F = var(bilirubinF)
bilirubinM.mean = mean( bilirubinM )
bilirubinF.mean = mean( bilirubinF )
T = ( bilirubinM.mean - bilirubinF.mean  ) / sqrt(s2M/n1+s2F/n2)
nu = ( s2M/n1 + s2F/n2 )^2 / ( (s2M/n1)^2/(n1-1) + (s2F/n2)^2/(n2-1) )

print( c(bilirubinM.mean,bilirubinF.mean) )
print( c(sd(bilirubinM),sd(bilirubinF) ) )
print( nu )
print( T )
print( qt( 0.025 , nu , lower.tail=FALSE ) )
PV = 2*min( c( pt( -T , nu ) , pt( T , nu ) ) )
print(PV)
```

The output from this code is:

```
[1] 3.613152 2.322535
[1] 6.558631 4.863417
[1] 319.3899
[1] 2.511335
[1] 1.967419
[1] 0.0125208
```

From this, we can see that the *t* statistic is 2.511 and the degrees of freedom for the Welch's approximation is 319.39. R has no trouble giving $t_{0.025}(319.39)$ for the noninteger degrees of freedom: $t_{0.025}(319.39) = 1.967$. Note how close this is to $z_{0.025}$, the normal quantile; this is because the degrees of freedom is so large. Since $T > t_{0.025}(319.39)$, we reject the null hypothesis of equal population means for the bilirubin variable for females and males.

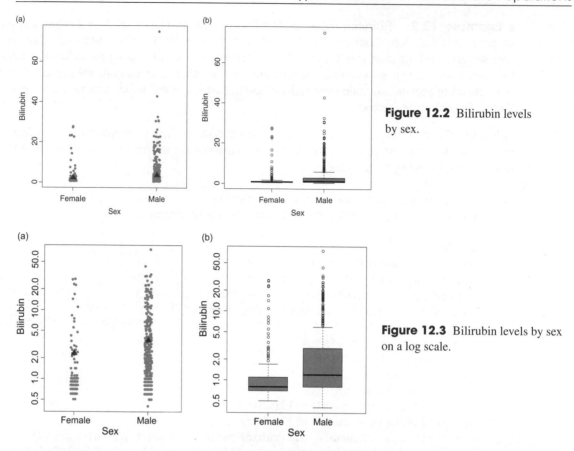

Figure 12.2 Bilirubin levels by sex.

Figure 12.3 Bilirubin levels by sex on a log scale.

It's always a good idea to plot data because plots will often uncover aspects of the data that are not apparent in a straightforward analysis. Figure 12.2 shows the dot plot (with the sex variable jittered) for women and men. From this we can see that the distributions of bilirubin are skewed to the high side, indicating a substantial deviation from the normal distribution. ∎∎∎

In Example 12.2, we saw that the response variable (bilirubin) was skewed to the high (or right) side. When this occurs it is often advisable to transform the data by taking logs or square roots (or some other transformation) to achieve something closer to a normal distribution. Figure 12.2 shows the same plot as Figure 12.2, except the vertical axis (bilirubin) is on a log scale. It appears that the data are certainly less skewed, but there still seems to be some skewness to the high side.

We introduced the likelihood ratio test in Section 11.9 and showed that the likelihood ratio test for testing the value of the mean of a normal distribution is the usual one-sample t-test.

To illustrate this terminology, suppose we have independent samples from two normal distributions and we would like to test whether the means are equal. That is, we have $X_1, X_2, \ldots, X_{n_1} \sim$ i.i.d. $N(\mu_1, \sigma^2)$ and $Y_1, Y_2, \ldots Y_{n_2} \sim$ i.i.d. $N(\mu_2, \sigma^2)$. (Notice that we have assumed equal variances.) Here the parameter space is $\Omega = \{(\mu_1, \mu_2, \sigma) | \sigma > 0\}$ and Ω_0 is the subset of this three-dimensional space for which $\mu_1 = \mu_2$. The restriction $\mu_1 = \mu_2$ determines a *plane* in the (μ_1, μ_2, σ) space. The situation is illustrated in Figure 12.4. The null and alternative hypotheses can then be written as

$$H_0: (\mu_1, \mu_2, \sigma) \in \Omega_0 \qquad \text{versus} \qquad H_1: (\mu_1, \mu_2, \sigma) \notin \Omega_0.$$

The likelihood ratio test for testing whether two means (from normal populations with common variance) are equal also leads to a familiar test.

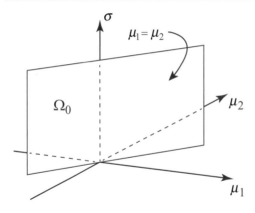

Figure 12.4 The "null" region Ω_0 defined by the restriction $\mu_1 = \mu_2$ in the (μ_1, μ_2, σ) space.

■ **Example 12.3** Suppose we have $X_1, X_2, \ldots, X_{n_1} \sim$ i.i.d. $N(\mu_1, \sigma^2)$ and $Y_1, Y_2, \ldots Y_{n_2} \sim$ i.i.d. $N(\mu_2, \sigma^2)$. Find the likelihood ratio test for testing $H_0: \mu_1 = \mu_2$ against the alternative $H_1: \mu_1 \neq \mu_2$.

Solution: The likelihood function (not assuming H_0 is true) is

$$L(\mu_1, \mu_2, \sigma) = \prod_{i=1}^{n_1} \frac{1}{\sqrt{2\pi}\,\sigma} \exp\left(-\frac{1}{2\sigma^2}(x_i - \mu_1)^2\right) \times \prod_{i=1}^{n_2} \frac{1}{\sqrt{2\pi}\,\sigma} \exp\left(-\frac{1}{2\sigma^2}(y_i - \mu_2)^2\right)$$

$$= (2\pi)^{-(n_1+n_2)/2}\, \sigma^{-n_1-n_2} \exp\left(-\frac{1}{2\sigma^2}\left[\sum_{i=1}^{n_1}(x_i - \mu_1)^2 + \sum_{i=1}^{n_2}(y_i - \mu_2)^2\right]\right).$$

We leave it as an exercise (Problem 12.31) to show that the MLEs are

$$\hat{\mu}_1 = \bar{x} \tag{12.6}$$

$$\hat{\mu}_2 = \bar{y} \tag{12.7}$$

$$\hat{\sigma}^2 = \frac{1}{n_1 + n_2}\left(\sum_{i=1}^{n_1}(x_i - \bar{x})^2 + \sum_{i=1}^{n_2}(y_i - \bar{y})^2\right). \tag{12.8}$$

The maximum of the likelihood function occurs at the MLEs of the parameters. Thus,

$$\max_{\Omega} L(\mu_1, \mu_2, \sigma) = L(\hat{\mu}_1, \hat{\mu}_2, \hat{\sigma}^2)$$

$$= (2\pi)^{-(n_1+n_2)/2}\, \hat{\sigma}^{-n_1-n_2} \exp\left(-\frac{1}{2\hat{\sigma}^2}\left[\sum_{i=1}^{n_1}(x_i - \bar{x})^2 + \sum_{i=1}^{n_2}(y_i - \bar{y})^2\right]\right)$$

$$= (2\pi)^{-(n_1+n_2)/2}\, \hat{\sigma}^{-n_1-n_2} \exp\left(-\frac{1}{2}\frac{(n_1+n_2)\hat{\sigma}^2}{\hat{\sigma}^2}\right)$$

$$= (2\pi)^{-(n_1+n_2)/2}\, \hat{\sigma}^{-n_1-n_2} \exp\left(-\frac{n_1+n_2}{2}\right).$$

When $H_0: \mu_1 = \mu_2$ is true – that is $(\mu_1, \mu_2, \sigma) \in \Omega_0$ – the two samples really constitute a single sample from the $N(\mu, \sigma^2)$ distribution. The MLEs, assuming H_0 is true, are therefore

$$\hat{\mu} = \frac{1}{n_1 + n_2}\left(\sum_{i=1}^{n_1} x_i + \sum_{i=1}^{n_2} y_i\right) = \frac{n_1\bar{x} + n_2\bar{y}}{n_1 + n_2} \tag{12.9}$$

$$\hat{\sigma}_0^2 = \frac{1}{n_1 + n_2} \left(\sum_{i=1}^{n_1} (x_i - \hat{\mu})^2 + \sum_{i=1}^{n_2} (y_i - \hat{\mu})^2 \right). \tag{12.10}$$

Thus, when restricted to Ω_0, the maximum of the likelihood in the numerator of (11.12) is

$$\max_{\Omega_0} L(\mu_1, \mu_2, \sigma) = L(\hat{\mu}, \hat{\mu}, \hat{\sigma}_0^2)$$

$$= (2\pi)^{-(n_1+n_2)/2} \hat{\sigma}_0^{-n_1-n_2} \exp\left(-\frac{1}{2\hat{\sigma}_0^2} \left[\sum_{i=1}^{n_1} (x_i - \hat{\mu})^2 + \sum_{i=1}^{n_2} (y_i - \hat{\mu})^2 \right] \right)$$

$$= (2\pi)^{-(n_1+n_2)/2} \hat{\sigma}_0^{-n_1-n_2} \exp\left(-\frac{1}{2} \frac{(n_1 + n_2)\hat{\sigma}_0^2}{\hat{\sigma}_0^2} \right)$$

$$= (2\pi)^{-(n_1+n_2)/2} \hat{\sigma}_0^{-n_1-n_2} \exp\left(-\frac{(n_1 + n_2)}{2} \right).$$

The likelihood ratio is therefore

$$\lambda = \frac{\max\limits_{\Omega_0} L(\mu_1, \mu_2, \sigma)}{\max\limits_{\Omega} L(\mu_1, \mu_2, \sigma)}$$

$$= \frac{(2\pi)^{-(n_1+n_2)/2} \hat{\sigma}_0^{-n_1-n_2} \exp\left(-\frac{n_1+n_2}{2} \right)}{(2\pi)^{-(n_1+n_2)/2} \hat{\sigma}^{-n_1-n_2} \exp\left(-\frac{n_1+n_2}{2} \right)}$$

$$= \left(\frac{\hat{\sigma}_0^2}{\hat{\sigma}^2} \right)^{-(n_1+n_2)/2}. \tag{12.11}$$

Expanding the first summation in the estimator $\hat{\sigma}_0^2$ from (12.10) yields

$$\sum_{i=1}^{n_1} (x_i - \hat{\mu})^2 = \sum_{i=1}^{n_1} (x_i - \bar{x} + \bar{x} - \hat{\mu})^2$$

$$= \sum_{i=1}^{n_1} (x_i - \bar{x})^2 + \sum_{i=1}^{n_1} (\bar{x} - \hat{\mu})^2 + 2 \sum_{i=1}^{n_1} (x_i - \bar{x})(\bar{x} - \hat{\mu})$$

$$= \sum_{i=1}^{n_1} (x_i - \bar{x})^2 + n_1 (\bar{x} - \hat{\mu})^2 + 2(\bar{x} - \hat{\mu}) \underbrace{\sum_{i=1}^{n_1} (x_i - \bar{x})}_{\text{sum is } 0}$$

$$= (n_1 - 1)S_1^2 + n_1 (\bar{x} - \hat{\mu})^2.$$

Similarly,

$$\sum_{i=1}^{n_2} (y_i - \hat{\mu})^2 = (n_2 - 1)S_2^2 + n_2 (\hat{\mu} - \bar{y})^2.$$

Here, S_1^2 and S_2^2 are the respective sample variances. Thus, the estimate of σ^2 under H_0 is

$$\hat{\sigma}_0^2 = \frac{1}{n_1 + n_2} \left((n_1 - 1)S_1^2 + n_1 (\bar{x} - \hat{\mu})^2 + (n_2 - 1)S_2^2 + n_2 (\bar{y} - \hat{\mu})^2 \right). \tag{12.12}$$

Problem 12.32 asks you to show that

$$\bar{x} - \hat{\mu} = \frac{n_2}{n_1 + n_2}(\bar{x} - \bar{y}) \tag{12.13}$$

$$\bar{y} - \hat{\mu} = -\frac{n_1}{n_1 + n_2}(\bar{x} - \bar{y}). \tag{12.14}$$

Applying these results and continuing from (12.12) gives

$$\hat{\sigma}_0^2 = \frac{1}{n_1 + n_2}\left((n_1 - 1)S_1^2 + n_1\left(\frac{n_2}{n_1 + n_2}(\bar{x} - \bar{y})\right)^2 + (n_2 - 1)S_2^2 + n_2\left(\frac{n_1}{n_1 + n_2}(\bar{x} - \bar{y})\right)^2\right)$$

$$= \frac{1}{n_1 + n_2}\left((n_1 - 1)S_1^2 + (n_2 - 1)S_2^2 + \frac{n_1 n_2^2}{(n_1 + n_2)^2}(\bar{x} - \bar{y})^2 + \frac{n_1^2 n_2}{(n_1 + n_2)^2}(\bar{x} - \bar{y})^2\right)$$

$$= \frac{1}{n_1 + n_2}\left((n_1 - 1)S_1^2 + (n_2 - 1)S_2^2 + \frac{n_1 n_2}{(n_1 + n_2)^2}(\bar{x} - \bar{y})^2(n_2 + n_1)\right)$$

$$= \frac{1}{n_1 + n_2}\left((n_1 - 1)S_1^2 + (n_2 - 1)S_2^2 + \frac{n_1 n_2}{n_1 + n_2}(\bar{x} - \bar{y})^2\right). \tag{12.15}$$

Combining (12.15) and (12.8) and substituting these into (12.11) yields

$$\lambda = \left(\frac{\hat{\sigma}_0^2}{\hat{\sigma}^2}\right)^{-(n_1 + n_2)/2}$$

$$= \left(\frac{\frac{1}{n_1 + n_2}\left((n_1 - 1)S_1^2 + (n_2 - 1)S_2^2 + \frac{n_1 n_2}{n_1 + n_2}(\bar{x} - \bar{y})^2\right)}{\frac{1}{n_1 + n_2}\left(\sum_{i=1}^{n_1}(x_i - \bar{x})^2 + \sum_{i=1}^{n_2}(y_i - \bar{y})^2\right)}\right)^{-(n_1 + n_2)/2}$$

$$= \left(\frac{(n_1 - 1)S_1^2 + (n_2 - 1)S_2^2 + \frac{n_1 n_2}{n_1 + n_2}(\bar{x} - \bar{y})^2}{(n_1 - 1)S_1^2 + (n_2 - 1)S_2^2}\right)^{-(n_1 + n_2)/2}$$

$$= \left(\frac{W + \frac{n_1 n_2}{n_1 + n_2}(\bar{x} - \bar{y})^2}{W}\right)^{-(n_1 + n_2)/2}$$

$$= \left(1 + \frac{n_1 n_2}{n_1 + n_2}(\bar{x} - \bar{y})^2 \Big/ W\right)^{-(n_1 + n_2)/2},$$

where

$$W = (n_1 - 1)S_1^2 + (n_2 - 1)S_2^2 = (n_1 + n_2 - 2)S_p^2$$

and S_p^2 is the pooled estimate of the variance given in (12.1). The likelihood ratio test gives a test statistic and a direction; it does not provide us with the critical region. The general rule is to reject H_0 if λ is sufficiently *small* (see (11.13)). The likelihood ratio test of H_0: $\mu_1 = \mu_2$ leads to rejecting H_0 when

$$\lambda < k$$

$$\left(1 + \frac{n_1 n_2}{n_1 + n_2}(\bar{x} - \bar{y})^2 \Big/ W\right)^{-(n_1+n_2)/2} < k$$

$$1 + \frac{n_1 n_2}{n_1 + n_2}(\bar{x} - \bar{y})^2 \Big/ W > k_1$$

$$\frac{n_1 n_2}{n_1 + n_2} \frac{(\bar{x} - \bar{y})^2}{W} > k_2$$

$$\frac{(\bar{x} - \bar{y})^2}{W} > k_3$$

$$\frac{(\bar{x} - \bar{y})^2}{S_p^2} > k_4$$

$$\left(\frac{\bar{x} - \bar{y}}{S_p \sqrt{1/n_1 + 1/n_2}}\right)^2 > k_5$$

$$T^2 > k_5.$$

Here, T is the usual t statistic for testing the equality of two means given in (12.3). Rejecting for $T^2 > k_5$ is equivalent to rejecting when

$$T > \sqrt{k_5} \quad \text{or} \quad T < -\sqrt{k_5}.$$

We would then have to determine k_5 so that the probability of rejecting H_0 when it is true is α; thus, since $T \sim t(n_1 + n_2 - 2)$, we have $k_5 = t_{\alpha/2}$. This is exactly the two-sample t-test derived earlier in this section. Thus, the likelihood ratio test for testing the equality of means from normal distributions with common variance is the usual two-sample t-test. ∎

12.2.2 Comparing Two Proportions

Suppose now that we have independent samples from two populations, where each subject or unit either has or does not have some property. Let θ_1 and θ_2 denote the population proportions in the two populations. If we define X_i to be the number of subjects or units in the sample from population i with some property, then if the populations are large (so that successive draws from each population are independent), then

$$X_1 \sim \text{BIN}(n_1, \theta_1) \quad \text{and} \quad X_2 \sim \text{BIN}(n_2, \theta_2).$$

Interest is focused on comparing the two proportions θ_1 and θ_2; specifically, whether $\theta_1 = \theta_2$. We will test $H_0: \theta_1 = \theta_2$ against the alternative $H_0: \theta_1 \neq \theta_2$, although one-sided tests are easily constructed. If the null hypothesis is true, then we are in a situation like the one in the previous section where both samples contributed information to estimate σ^2; in that case, we pooled the information to get a better estimator. Here, when H_0 is true, we have two sources of information for the common proportion $\theta = \theta_1 = \theta_2$. Again, we pool the information from both samples. Under H_0, it's as if we have one large sample of size $n_1 + n_2$ from a population where the proportion of the quantity of interest is θ. In this case, the pooled estimate is

$$\hat{\theta}_p = \frac{X_1 + X_2}{n_1 + n_2}.$$

The variance of the estimator of the difference in estimated proportions is

$$\mathbb{V}(\hat{\theta}_1 - \hat{\theta}_2) = \mathbb{V}(\hat{\theta}_1) + \mathbb{V}(\hat{\theta}_2) = \frac{\theta_1(1 - \theta_1)}{n_1} + \frac{\theta_2(1 - \theta_2)}{n_2},$$

which, under H_0, can be estimated by

$$\hat{\mathbb{V}}(\hat{\theta}_1 - \hat{\theta}_2) = \frac{\hat{\theta}_p(1 - \hat{\theta}_p)}{n_1} + \frac{\hat{\theta}_p(1 - \hat{\theta}_p)}{n_2} = \hat{\theta}_p(1 - \hat{\theta}_p)\left(\frac{1}{n_1} + \frac{1}{n_2}\right).$$

Thus, if H_0 is true, we can standardize the difference in estimated proportions:

$$Z = \frac{\hat{\theta}_1 - \hat{\theta}_2}{\sqrt{\hat{\theta}_p(1 - \hat{\theta}_p)\left(\frac{1}{n_1} + \frac{1}{n_2}\right)}}. \tag{12.16}$$

If n_1 and n_2 are both large, then $Z \overset{\text{approx}}{\sim} N(0,1)$.

■ **Example 12.4** Data for Progress (2021) conducts various surveys, including those involving job approval for both individual office holders and organizations. One such survey was conducted on February 2, 2021 and asked about the CDC's handling of the coronavirus pandemic. The survey included 1,162 people – 550 men and 612 women – and the response choices were: strongly approve, somewhat approve, somewhat disapprove, strongly disapprove, and not sure. The results are shown in Table 12.1. These percentages have been converted into raw counts in Table 12.2.

For this example, we'll focus on the proportions who approve of the CDC's handling by summing the strongly approve and somewhat approve counts. Thus, out of the 550 men, $116 + 220 = 336$ approved and $118 + 74 + 22 = 214$ did not approve of the CDC's handling of the coronavirus pandemic. (In the

Table 12.1 Sample proportions of men and women who approve of the CDC's handling of the coronavirus pandemic.

Handling of coronavirus: CDC	Overall (%)	Male (%)	Female (%)
Strongly approve	23.93	21.18	26.49
Somewhat approve	39.89	39.93	39.86
Somewhat disapprove	18.11	21.42	15.02
Strongly disapprove	12.26	13.43	11.17
Not sure	5.81	4.03	7.46
n	1,162	550	612

Table 12.2 Counts of men and women who approve of the CDC's handling of the coronavirus pandemic.

Handling of coronavirus: CDC	Total	Male	Female
Strongly approve	278	116	162
Somewhat approve	464	220	244
Somewhat disapprove	210	118	92
Strongly disapprove	142	74	68
Not sure	68	22	46
n	1,162	550	612

latter group, either they disapproved or were not sure.) Among the 612 women, $162 + 244 = 406$ approved and $92 + 68 + 46 = 206$ did not. Labeling men as group 1 and women as group 2, we can write

$$\hat{\theta}_1 = \frac{336}{550} = 0.6109 \quad \text{and} \quad \hat{\theta}_2 = \frac{406}{612} = 0.6634.$$

We can see that in the sample, women had a higher approval of the CDC than men: 66.34% compared to 61.09% for men. But is there really a difference in the population of men and women? Or is the observed difference (66.34% vs. 61.09%) just due to random noise? This is the question that the hypothesis test addresses. The pooled estimate of the common θ (under H_0) is

$$\hat{\theta} = \frac{742}{1,162} = 0.6386.$$

The z statistic is therefore

$$z = \frac{\dfrac{336}{550} - \dfrac{406}{612}}{\sqrt{\dfrac{742}{1,162}\left(1 - \dfrac{742}{1,162}\right)\left(\dfrac{1}{550} + \dfrac{1}{612}\right)}} \approx -1.860.$$

With a significance level of $\alpha = 0.05$, the rule would be to reject H_0: $\theta_1 = \theta_2$ if $z < -1.96$ or $z > 1.96$. Since our test statistic is -1.86, we fail to reject H_0. The P-value for this test is 0.0629. ■ ■ ■

■ **Example 12.5** Many ecommerce companies use experiments to see which on-screen presentation leads to further "clicks" because users who peruse more deeply are more likely to purchase a product. This is called *A/B testing*, or *split testing*. Suppose that in a designed experiment visitors to an ecommerce site are randomly shown screen A or screen B. Among the 9,581 who are shown screen A, 1,198 clicked further, whereas among the 9,611 who were shown screen B, 1,101 clicked further. Test whether the population proportions of those who "click through" are the same for screens A and B. Use $\alpha = 0.05$.

Solution: If we let θ_1 and θ_2 denote the probabilities of clicking deeper into the website for those visitors shown screen A and B, respectively, then we are testing H_0: $\theta_1 = \theta_2$ against H_1: $\theta_1 \neq \theta_2$. For this data set, we have

$$\hat{\theta}_1 = \frac{1,198}{9,581} = 0.1250$$

$$\hat{\theta}_2 = \frac{1,101}{9,611} = 0.1146$$

$$\hat{\theta} = \frac{1,198 + 1,101}{9,581 + 9,611} = 0.1198.$$

The test statistic is therefore

$$z = \frac{\hat{\theta}_1 - \hat{\theta}_2}{\sqrt{\hat{\theta}(1 - \hat{\theta})(1/n_1 + 1/n_2)}} = \frac{0.1250 - 0.1146}{\sqrt{0.1198(1 - 0.1198)(1/9581 + 1/9611)}} = 2.236.$$

For $\alpha = 0.05$, the rule is to reject H_0 if $z < -1.96$ or $z > 1.96$, so in this case we reject H_0. There is evidence that the click rates for the two screens are different. The P-value for this test is

$$2P(Z > 2.236189) = 0.025.$$ ■ ■ ■

The fairly small difference in click rates, 12.50% vs. 11.46%, was statistically significant in Example 12.5 because with such large samples, nearly 10,000 in this example, the estimated standard error is small, making small differences significant. It is often preferable to report a confidence interval for the difference in population proportions in addition to the results of the hypothesis test.

■ **Example 12.6**   Find 95% confidence intervals for the difference between proportions in Examples 12.4 and 12.5.

Solution: The formula for the confidence interval for the difference of two proportions was given in (10.19). For the CDC approval data, where we compare the proportions of men and women who approve of the CDC's handling of the coronavirus,

$$\left((\hat{\theta}_1 - \hat{\theta}_2) - z_{\alpha/2} \sqrt{\frac{\hat{\theta}_1(1-\hat{\theta}_1)}{n_1} + \frac{\hat{\theta}_2(1-\hat{\theta}_2)}{n_2}} \, , \, (\hat{\theta}_1 - \hat{\theta}_2) + z_{\alpha/2} \sqrt{\frac{\hat{\theta}_1(1-\hat{\theta}_1)}{n_1} + \frac{\hat{\theta}_2(1-\hat{\theta}_2)}{n_2}} \right)$$

$$\left(0.6109 - 0.6634 - 1.96\sqrt{0.0007971} \, , \, 0.6109 - 0.6634 + 1.96\sqrt{0.0007971} \right)$$

$$\left(-0.1078 \, , \, 0.0028 \right).$$

For the A/B testing data of Example 12.5:

$$\left(0.1250 - 0.1146 - 1.96\sqrt{0.00002197} \, , \, 0.1250 - 0.1146 + 1.96\sqrt{0.00002197} \right)$$

$$\left(0.0013 \, , \, 0.0197 \right).$$

We can see that the confidence interval for the difference of proportions in the A/B testing example is quite narrow because of the large sample sizes   ■ ■ ■

For many testing procedures, the null hypothesis of equality of parameters (means or proportions) will be rejected if and only if the 95% confidence interval for the difference excludes 0.[4] This is not true for the test statistic given in (12.16) and the confidence interval given in (10.19). The advantage of using the pooled estimate for the common value of θ under H_0 is that it uses information a little more efficiently and therefore has higher power. A disadvantage is that it is not consistent with the confidence interval method.

Some books forgo pooling to estimate the common parameter θ when testing the equality of proportions and instead use the statistic

$$Z_1 = \frac{(\hat{\theta}_1 - \hat{\theta}_2) - (\theta_1 - \theta_2)}{\sqrt{\dfrac{\hat{\theta}_1(1-\hat{\theta}_1)}{n_1} + \dfrac{\hat{\theta}_2(1-\hat{\theta}_2)}{n_2}}} \overset{\text{approx}}{\sim} N(0,1). \tag{12.17}$$

The advantage of using this test statistic instead of (10.19) is that it is consistent with the confidence interval method, in the sense that we reject H_0: $\theta_1 = \theta_2$ if and only if 0 is not contained in the confidence interval. Usually, the results of (12.16) and (12.17) are similar.

---

[4]This should make intuitive sense. If 0 is contained in the confidence interval for the difference, then 0 is a plausible value for the difference. In other words, the parameters could be the same. On the other hand, if 0 is not contained in the confidence interval for the difference, then 0 is not a plausible value for the difference, so we have evidence against the null hypothesis of equality.

■ **Example 12.7**   Apply the test statistics in (12.16) and (12.17) to both the CDC approval (Example 12.4) and A/B testing (Example 12.5).

Solution: For the CDC approval data:

$$z_1 = \frac{\dfrac{336}{550} - \dfrac{406}{612}}{\sqrt{\dfrac{1}{550}\left[\dfrac{336}{550}\left(1 - \dfrac{336}{550}\right)\right] + \dfrac{1}{612}\left[\dfrac{406}{612}\left(1 - \dfrac{406}{612}\right)\right]}} \approx -1.859218.$$

Using the pooled estimate for the common θ, we found for Example 12.4 that $z = -1.859545$.
    For the A/B testing example in Example 12.5, the statistic is

$$z_1 = \frac{\dfrac{1,198}{9,581} - \dfrac{1,101}{9,611}}{\sqrt{\dfrac{1}{9,581}\left[\dfrac{1,198}{9,581}\left(1 - \dfrac{1,198}{9,581}\right)\right] + \dfrac{1}{9,611}\left[\dfrac{1,101}{9,611}\left(1 - \dfrac{1,101}{9,611}\right)\right]}} \approx 2.236348.$$

The test statistics from using the two Z statistics are nearly identical in both cases: -1.859545 vs. -1.859218 for the CDC approval data, and 2.236189 vs. 2.236348 for the A/B testing data.[5] In most cases, it makes little difference which test statistic is used.   ■ ■ ■

12.2.3   Comparing Variances

There are situations where we wish to compare population variances rather than population means or proportions. We will make the assumptions that both populations are normal and they are independent of one another. Thus, we have independent samples

$$X_1, X_2, \ldots, X_{n_1} \sim \text{ i.i.d. } N(\mu_1, \sigma_1^2)$$

$$Y_1, Y_2, \ldots, Y_{n_2} \sim \text{ i.i.d. } N(\mu_2, \sigma_2^2).$$

We want to test H_0: $\sigma_1^2 = \sigma_2^2$ against the alternative H_1: $\sigma_1^2 \neq \sigma_2^2$, although one-sided tests are possible. Rather than looking at the *difference* of the estimated variances, we look at the ratio

$$\frac{S_1^2}{S_2^2},$$

where S_1^2 and S_2^2 are the sample variances of the Xs and Ys, respectively. Recall from (6.12) that the F distribution is defined to be the distribution of the ratio of independent χ^2 random variables, each divided by its degrees of freedom. If U_1 and U_2 are independent χ^2 random variables with ν_1 and ν_2 degrees of freedom, respectively, then the random variable

$$F = \frac{U_1/\nu_1}{U_2/\nu_2} \tag{12.18}$$

has an F distribution with parameters ν_1 numerator degrees of freedom, and ν_2 denominator degrees of freedom. Recall also that $(n_1 - 1)S_1^2/\sigma_1^2 \sim \chi^2(\nu_1 - 1)$ and $(n_2 - 1)S_2^2/\sigma_2^2 \sim \chi^2(\nu_2 - 1)$; moreover, since S_1^2 and S_2^2 are based on independent samples, they are independent. Thus,

---

[5]In making these comparisons we reworked the test statistics in Examples 12.4 and 12.5, keeping more significant digits.

$$\frac{\dfrac{(n_1-1)S_1^2}{\sigma_1^2}/(n_1-1)}{\dfrac{(n_2-1)S_2^2}{\sigma_2^2}/(n_2-1)} \sim F(n_1-1, n_2-1),$$

but this reduces to

$$\frac{S_1^2/\sigma_1^2}{S_2^2/\sigma_2^2} \sim F(n_1-1, n_2-1).$$

Furthermore, if the null hypothesis of equal variances is true, then this result becomes

$$F = \frac{S_1^2}{S_2^2} \sim F(n_1-1, n_2-1).$$

This gives us a test statistic for testing H_0: $\sigma_1^2 = \sigma_2^2$ against H_1: $\sigma_1^2 \neq \sigma_2^2$. The rule is to reject H_0 if

$$F < F_{1-\alpha/2}(n_1-1, n_2-1) \qquad \text{or} \qquad F > F_{\alpha/2}(n_1-1, n_2-1),$$

where $F_{\alpha/2}(n_1-1, n_2-1)$ and $F_{1-\alpha/2}(n_1-1, n_2-1)$ are the values that leave $\alpha/2$ and $1-\alpha/2$ in the tails of the $F(n_1-1, n_2-1)$ distribution. Remember, the F distribution is not symmetric, so you need to look up the lower and upper quantiles separately, which can be done with the R function `qf()`.

■ **Example 12.8**   In Example 12.1 we compared the mean first-year GPA of first-generation college students against the mean of non-first-generation college students and found that there is a statistically significant difference between these means. The dot plot in Figure 12.1(a) suggested that the variances may differ. This is difficult to ascertain, though, because the sample sizes are vastly different, with 194 non-first-generation college students and 25 first-generation college students.

Solution: The F-statistic for testing H_0: $\sigma_1^2 = \sigma_2^2$ against a two-sided alternative is

$$F = \frac{S_1^2}{S_2^2} = \frac{0.1905}{0.2150} = 0.8860.$$

The sample variance of the first-generation college group is actually the larger of the two. The different sample sizes, 194 vs. 25, make it appear that the non-first-generation college student group has the larger variance. We can determine the quantiles of the $F(n_1-1, n_2-1) = F(193, 24)$ distribution using R's `qf()` function:

```
n1 = 194
n2 = 25
f025 = qf(0.025,n1-1,n2-1,lower.tail=FALSE)
f975 = qf(0.975,n1-1,n2-1,lower.tail=FALSE)
```

Using our mathematical notation, this yields

$$F_{0.025}(193, 24) = 1.982 \qquad \text{and} \qquad F_{0.975}(193, 24) = 0.583.$$

These quantiles are shown in Figure 12.5. Our observed F ratio, 0.886, does not lead to rejection of H_0. There is no evidence that the variances differ.   ■ ■ ■

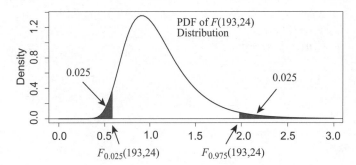

Figure 12.5 PDF of the F distribution with $n1 - 1 = 193$ numerator degrees of freedom and $n_2 - 1 = 23$ denominator degrees of freedom.

The F-test for equality of variances is often used as a preliminary step when interest focuses on the comparisons of the means. The exact t-test from (12.3) assumes equal variances, so it is tempting to first apply the F-test for the equality of variances, and if this is not rejected, to then use the t-test from (12.3). If the null hypothesis of equality of variances is rejected, then use Welch's t-test from (12.4). A few cautions should be mentioned. First, failing to reject H_0: $\sigma_1^2 = \sigma_2^2$ is not evidence that the variances are the same.[6] Second, the exact t-test and Welch's t-test yield nearly identical P-values in practice. If you suspect that the variances may differ, just use Welch's t-test.

12.3 Testing Paired Samples

The preceding section assumed that the two samples were independent of one another. Under this assumption, all of the random variables $X_1, X_2, \ldots, X_{n_1}, Y_1, Y_2, \ldots, Y_{n_2}$ are independent. In particular, X_i and Y_i are independent. There are situations for which the data are naturally paired. For example, X_i might be the response of patient i when given treatment A, and Y_i is the response by the *same* patient when given treatment B. Since we are measuring the response on the *same* person (even if it is the response from two different treatments), then X_i and Y_i are dependent. For example, a patient may be given cream A to treat a skin rash on the left leg, and cream B to treat a skin rash on the right leg. The response would be the extent to which the rash was reduced. If X_i and Y_i represent the amount of reduction for patient i on the left and right legs, respectively, then X_i and Y_i are likely to be dependent random variables. The data are naturally *paired* because X_i and Y_i are outcomes from the same person. In cases like this the samples are clearly not independent, so the method of the previous section does not apply.

If we want to test whether the two treatments work the same against the alternative that they work differently, we are really testing whether the mean of the differences

$$D_1 = X_1 - Y_1, \ \ D_2 = X_2 - Y_2, \ \ \ldots, \ \ D_n = X_n - Y_n$$

is equal to zero. The alternative is that the mean of the differences is not zero.

When we have data such as these, we are not really asking about the difference of the means, but rather the mean of the differences. If we subtract one variable from the other, we obtain a difference that often has a clear interpretation. For example, the *after* measurement on a subject may tend to be larger than the *before* measurement; or measuring device A may yield smaller measurements than device B even when measuring the same parts. In cases like this, we should think of the *differences* as the observed data. What we would like to know is the size of the mean of the distribution of differences. Once we take the differences, the actual data values X_i and Y_i are no longer needed.

Whether data are from independent samples (as in the previous section) or paired (as in this section) is dependent on the design of the study. Thus, once a study is run it is not a matter of selecting a method

---

[6]There is no evidence that they differ, which is not the same thing.

to analyze the data; the design dictates what kind of analysis should be run. The data structure is as shown below:

i	Group 1	Group 2	Difference
1	$X_1 \sim N(\mu_1,\sigma_1^2)$	$Y_1 \sim N(\mu_2,\sigma_2^2)$	$D_1 = Y_1 - X_1 \sim N(\mu_d,\sigma_d^2)$
2	$X_2 \sim N(\mu_1,\sigma_1^2)$	$Y_2 \sim N(\mu_2,\sigma_2^2)$	$D_2 = Y_2 - X_2 \sim N(\mu_d,\sigma_d^2)$
$\vdots$	$\vdots$	$\vdots$	$\vdots$
n	$X_n \sim N(\mu_1,\sigma_1^2)$	$Y_n \sim N(\mu_2,\sigma_2^2)$	$D_n = Y_n - X_n \sim N(\mu_d,\sigma_d^2)$

Here, $\mu_d = \mu_2 - \mu_1$ and $\sigma_d^2 = \sigma_2^2 + \sigma_2^2 - 2\text{cov}(X_i,Y_i)$. When we analyze paired data like these, we often want to test whether the population mean of the differences is equal to zero. Once differences are taken, the problem becomes a one-sample test. We want to test H_0: $\mu_d = 0$ against a two-sided, or possibly a one-sided, alternative. This one-sample test was discussed in Section 11.6. The test statistic is then

$$t = \frac{\overline{d} - 0}{s_d/\sqrt{n}}, \tag{12.19}$$

where $\overline{d}$ is the sample mean of the differences, s_d is the sample standard deviation of the differences, and the degrees of freedom for the t value is $n-1$. Note that n is the number of *pairs* of data values, or equivalently the number of differences. Because the data are paired, we must have the same number of Xs as Ys. (This need not be the case for independent samples.)

■ **Example 12.9** Let's consider the study described in Example 10.13 on the cholesterol change for 20 subjects from 1952 to 1962. This study was a prospective cohort study, since a fixed cohort of subjects was followed prospectively. Such data are called longitudinal data. Since the difference is calculated as the 1962 value minus the 1952 value, negative values indicate a *decrease* in the cholesterol level. The data are shown in Table 10.1. Test whether there is evidence that the cholesterol levels of the population from which these 20 subjects were selected has a mean of zero.

Solution: The sample mean and standard deviation of the differences are found from Example 10.13 to be

$$\overline{d} = -69.8 \quad \text{and} \quad s_d = 46.5636.$$

To test H_0: $\mu_d = 0$ against H_1: $\mu_d \neq 0$, the test statistic from (12.19) is

$$t = \frac{\overline{d} - 0}{s_d/\sqrt{n}} = -6.704.$$

The rule to reject H_0: $\mu_d = 0$ is

$$t < -t_{0.025}(19) = -2.093 \quad \text{or} \quad t > t_{0.025}(19) = 2.093.$$

We found these t-values using t = qt(0.025 , n-1 , lower.tail=FALSE). Since our observed t statistic is well below the threshold of -2.093, we reject the null hypothesis. The P-value for this test is

$$P = 2\Pr(T < -6.704) = 2.08 \times 10^{-6}.$$

This is strong evidence that the mean cholesterol difference has changed. This is consistent with the confidence interval from Example 10.13, which we calculated to be $(-91.59, -48.01)$. ■ ■ ■

In the last example the data are naturally paired by person; in other words, we have the "before" and "after" cholesterol measurement *on each person*. There are times where the experiment can be designed using the method that gives us the more precise method for estimating the difference in means: $\mu_1 - \mu_2$ in the case of the independent samples t-test, and μ_d in the case of the paired t-test. The next example examines which is the more efficient design.

■ **Example 12.10** Suppose that a water treatment facility is being planned to take water from a river and process it for use by the local community. The designers must select the position of the intake valve in the river. It is suspected that impurities in the river may depend on whether we take water at the surface or near the bottom. The investigators would like to design an experiment to collect water samples at the bottom and at the surface to see whether the impurity levels differ. Two plans are suggested:

1. Select n positions along the river and from each collect a sample from the bottom. Also, select n positions along the river and collect a sample from the surface. This is the *independent samples design*.
2. Select n positions along the river and at each position collect a sample from the bottom and the surface. This is the *paired samples design* because at each location we are taking a pair of measurements.

Under what circumstances is the paired samples design more efficient?

Solution: We must first make some assumptions about the distribution of the data. Let's assume that the surface impurity measurements are normally distributed with a mean of μ_1 and a variance of σ_1^2. Similarly, we'll assume that the bottom measurements are normally distributed with a mean of μ_2 and a variance of σ_2^2. As always, a transformation, such as the log or square root, is sometimes necessary to achieve something close to a normal distribution. Further, we will assume that the correlation between the surface and the bottom measurements taken at the same location is ρ. Thus,

$$X_1, X_2, \ldots, X_n \sim N(\mu_1, \sigma_1^2)$$
$$Y_1, Y_2, \ldots, Y_n \sim N(\mu_2, \sigma_2^2)$$
$$\text{cov}(X_i, Y_i) = \rho \sigma_1 \sigma_2 \quad \text{if } X_i \text{ and } Y_i \text{ are collected at the same location.}$$

Second, we must decide on some criterion to compare the two methods. For this, we will use the variance of the estimated difference between the mean impurity level at the surface and the mean impurity at the bottom. Since we'd like to estimate the difference with as much precision as possible, we prefer the method with the smaller variance of the estimated difference.

With the assumptions laid out, and the criterion specified, we can compute the variance of the estimator for the difference in means. For the independent samples design,

$$\mathbb{V}(\overline{X} - \overline{Y}) = \mathbb{V}(\overline{X}) + \mathbb{V}(\overline{Y}) - 2\text{cov}(\overline{X}, \overline{Y}) = \mathbb{V}(\overline{X}) + \mathbb{V}(\overline{Y}) = \frac{\sigma_1^2}{n} + \frac{\sigma_2^2}{n} = \frac{\sigma_1^2 + \sigma_2^2}{n}. \quad (12.20)$$

For the paired samples design, define $D_i = X_i - Y_i$. Then,

$$\mathbb{V}(\overline{D}) = \mathbb{V}\left(\frac{1}{n}\sum_{i=1}^{n} D_i\right)$$
$$= \frac{1}{n^2}\sum_{i=1}^{n} \mathbb{V}(D_i)$$
$$= \frac{1}{n^2}\sum_{i=1}^{n} \left[\mathbb{V}(X_i) + \mathbb{V}(Y_i) - 2\text{cov}(X_i, Y_i)\right]$$

$$= \frac{1}{n^2} \sum_{i=1}^{n} \left[\sigma_1^2 + \sigma_2^2 - 2\rho\,\sigma_1\sigma_2 \right]$$

$$= \frac{1}{n^2}\, n \left[\sigma_1^2 + \sigma_2^2 - 2\rho\,\sigma_1\sigma_2 \right]$$

$$= \frac{\sigma_1^2 + \sigma_2^2 - 2\rho\,\sigma_1\sigma_2}{n}. \tag{12.21}$$

We can now compare (12.20) and (12.21). The paired samples design is better whenever

$$\mathbb{V}(\overline{D}) < \mathbb{V}(\overline{X} - \overline{Y})$$

$$\frac{\sigma_1^2 + \sigma_2^2 - 2\rho\,\sigma_1\sigma_2}{n} < \frac{\sigma_1^2 + \sigma_2^2}{n}$$

$$\sigma_1^2 + \sigma_2^2 - 2\rho\,\sigma_1\sigma_2 < \sigma_1^2 + \sigma_2^2$$

$$-2\rho\,\sigma_1\sigma_2 < 0$$

$$\rho > 0.$$

Thus, whenever $\rho > 0$, the paired samples design yields an estimator of the difference between the surface and the bottom that has a smaller variance than that obtained from the independent samples design. Ordinarily, we would expect a positive correlation between the surface and bottom measurements when taken at the same location. If our selected location is a place on the river where impurities are high, then we would expect high measurements at both the surface and the bottom, and conversely for places in the river where impurities are low. The paired samples design is more efficient whenever $\rho > 0$, which is most of the time. ∎∎∎

■ **Example 12.11** Assume the paired-samples design for the water intake problem described in the previous example, and that this experiment yielded the data in Table 12.3. Determine whether a transformation of the data is needed and apply the paired t-test to the (possibly) transformed data.

Solution: The $n = 40$ pairs of data are shown in Table 12.3; also included are the differences between the surface and bottom impurity levels. Histograms for the surface and bottom measurements are shown in the top two plots of Figure 12.6(a). The surface measurements are highly skewed to the right. The bottom measurements are also skewed to the right, but not as much as the surface measurements. We will try the log and square root transformations on the data to see which distribution most closely resembles a normal distribution (no transformation, log transformation, or square root transformation). Remember, it is the distribution of the *difference* that matters for paired data like this. Figure 12.6(b) shows a scatterplot of the original measurements and a histogram of the differences $X_i - Y_i$ on the right. The histogram is nearly symmetric and centered just to the left of -0.02. Figure 12.6(c) shows the scatterplot of $\log X_i$ and $\log Y_i$ and a histogram of the differences. Figure 12.6(d) does the same for the square root transformation. Looking at the three histograms, the one that seems to be most symmetric is the one with the original data.[7]

The following R code reads in the data and computes the mean and standard deviation of the differences:

```
water.intake = read.csv( "WaterIntake.csv" )
n = nrow( water.intake )
dbar = mean( water.intake$Difference )
sd.diff = sd( water.intake$Difference )
```

---

[7]The distribution of the differences is fairly symmetric, even though both of the histograms for the data are skewed to the right.

Table 12.3 River intake impurities using paired samples design with data taken at 40 locations.

Location	Surface	Bottom	Difference	Location	Surface	Bottom	Difference
1	0.033	0.085	−0.052	21	0.096	0.099	−0.003
2	0.021	0.049	−0.028	22	0.032	0.078	−0.046
3	0.055	0.106	−0.051	23	0.078	0.076	0.002
4	0.043	0.034	0.009	24	0.076	0.143	−0.067
5	0.104	0.113	−0.009	25	0.043	0.071	−0.028
6	0.046	0.054	−0.008	26	0.024	0.044	−0.020
7	0.015	0.042	−0.027	27	0.074	0.091	−0.017
8	0.025	0.042	−0.017	28	0.055	0.098	−0.043
9	0.132	0.172	−0.040	29	0.017	0.053	−0.036
10	0.118	0.107	0.011	30	0.073	0.095	−0.022
11	0.042	0.054	−0.012	31	0.031	0.060	−0.029
12	0.029	0.041	−0.012	32	0.062	0.078	−0.016
13	0.037	0.06	−0.023	33	0.019	0.024	−0.005
14	0.028	0.061	−0.033	34	0.014	0.043	−0.029
15	0.029	0.093	−0.064	35	0.053	0.093	−0.040
16	0.032	0.045	−0.013	36	0.022	0.040	−0.018
17	0.069	0.083	−0.014	37	0.025	0.054	−0.029
18	0.028	0.066	−0.038	38	0.057	0.068	−0.011
19	0.027	0.048	−0.021	39	0.021	0.031	−0.010
20	0.024	0.076	−0.052	40	0.020	0.044	−0.024

For this example, we wish to test H_0: $\mu_d = 0$ against the alternative H_1: $\mu_d \neq 0$. The following code computes the t-statistic and the t quantile and prints the important variables.

```
t = ( dbar - 0 ) / (sd.diff/sqrt(n) )
t.quantile = qt( 0.025 , n-1 , lower.tail=FALSE )
print( c(n,dbar,sd.diff,t,t.quantile) )
```

The output from this print command is

```
40.00000000 -0.02462500  0.01825698 -8.53055529  2.02269092
```

The important quantities are therefore

$$n = 40$$

$$\overline{d} = -0.02462500$$

$$s_d = 0.01825698$$

$$t = \frac{\overline{d} - 0}{s_d \sqrt{n}} = \frac{-0.02462500 - 0}{0.01825698/\sqrt{40}} = -8.53$$

$$t_{0.025}(39) = 2.023 \quad \text{and} \quad -t_{0.025}(39) = -2.023.$$

Since our observed t statistic is below $-t_{0.025}(39) = -2.023$, we reject H_0. A confidence interval for the mean of the differences is computed with the following R code:

```
LCL = dbar - t.quantile*sd.diff/sqrt(n)
UCL = dbar + t.quantile*sd.diff/sqrt(n)
print( c( LCL , UCL ) )
```

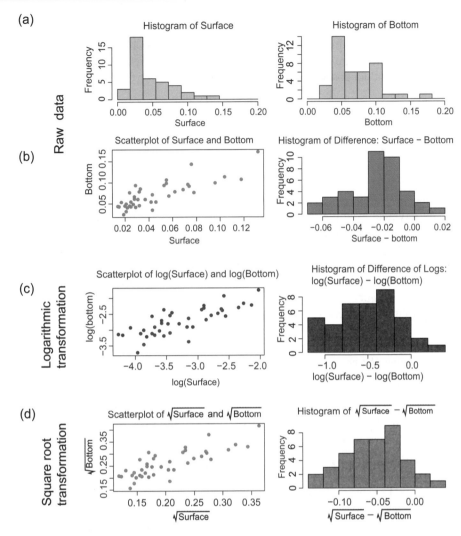

Figure 12.6 Paired data for water intake example.

This results in the 95% confidence interval $(-0.0305, -0.0188)$, and indicates that we can be confident that the impurities are higher at the bottom than at the surface. We would reach similar conclusions if we analyzed the log of impurities or the square root of impurities. If an intake is to be constructed, it should be placed near the surface to minimize the impurities. ∎ ∎ ∎

Exercise 12.1 Apply the paired t-test to log and square root transformed data from the previous example.

12.4 Single-Factor Analysis of Variance

In previous sections we have compared the means (or variances or proportions) of two populations. Often, we want to compare more than two groups. Here we suppose that there are k groups and we would like to compare the means of these k groups.[8] Let

---

[8]Often the groups will be the various treatments in an experiment.

Y_{ij} = response of subject j within group i; $j = 1, 2, \ldots, n_i$; $i = 1, 2, \ldots, k$.

The sample sizes within each group, $n_1, n_2, \ldots, n_k$, are possibly different, although they can be the same. (Many experiments are designed to have $n_1 = n_2 = \cdots = n_k$.) We will make the following assumptions about these random observations from the k groups:

$$
\begin{aligned}
Y_{11}, Y_{12}, \ldots, Y_{1n_1} &\sim N(\mu_1, \sigma^2) \\
Y_{21}, Y_{22}, \ldots, Y_{2n_2} &\sim N(\mu_2, \sigma^2) \\
\vdots \qquad \vdots \quad \vdots & \\
Y_{k1}, Y_{k2}, \ldots, Y_{kn_k} &\sim N(\mu_k, \sigma^2).
\end{aligned}
\tag{12.22}
$$

Note that the variance is assumed to be the same across all k groups.

As a typical example of this situation, consider the testing of metal bars made from an aluminum alloy. One property of a metal bar is the tensile strength, which is the maximum strain put on the bar without causing irreversible deformations. Aluminum alloys are made mostly of aluminum, but can contain other elements such as copper, magnesium, tin, zinc, and others. Thus, various kinds of aluminum bars can be made by combining different levels of these metals. Suppose we have access to three types of aluminum alloy bars and we would like to know whether there are differences in their tensile strengths. We could select a number of bars of each type and test them to determine the tensile strength of each. Suppose we do this for six units on two of the alloys and four units on the other and obtain the data shown in Table 12.4. We would like to test the null hypothesis H_0: $\mu_1 = \mu_2 = \mu_3$ against the alternative H_1: not all μ_i are the same. In general, we wish to test

$$H_0\colon \mu_1 = \mu_2 = \cdots = \mu_k \qquad \text{vs.} \qquad H_1\colon \text{not all } \mu_i \text{ are the same.}$$

The alternative hypothesis does not say that all of the μ_i are different, only that they are not all the same. For example, if $\mu_1 = \mu_2 = 40,000$ while $\mu_3 = 38,000$, then H_0 is false and therefore H_1 is true, even though the μ_i are not all distinct.

Table 12.4 Tensile strengths of specimens for each of three types of aluminum.

Type of aluminum	Tensile strength	Type of aluminum	Tensile strength	Type of aluminum	Tensile strength
1	46,200	2	40,100	3	38000
1	48,100	2	41,900	3	36,900
1	40,300	2	43,300	3	38,700
1	43,100	2	43,000	3	37,300
1	44,000	2	39,900		
1	47,400	2	40,200		

The point estimators for the μ_i are simply the sample means for each group:

$$\hat{\mu}_1 = \frac{1}{n_1} \sum_{j=1}^{n_1} y_{1j} = \frac{1}{n_1} y_{\cdot 1} = \bar{y}_{1\cdot}$$

$$\hat{\mu}_2 = \frac{1}{n_2} \sum_{j=1}^{n_2} y_{2j} = \frac{1}{n_2} y_{\cdot 2} = \bar{y}_{2\cdot}$$

$$\vdots \ \vdots \qquad\qquad \vdots \qquad \vdots$$

$$\hat{\mu}_k = \frac{1}{n_k} \sum_{j=1}^{n_k} y_{kj} = \frac{1}{n_2} y_{\cdot k} = \bar{y}_{k\cdot}.$$

Here we have made use of the "dot" notation:

$$y_{i\cdot} = \sum_{j=1}^{n_i} y_{ij} \qquad \text{and} \qquad \bar{y}_{i\cdot} = \frac{1}{n_i} \sum_{j=1}^{n_i} y_{ij}.$$

Thus, the dot without a bar indicates a *sum* across all observations in a group, while the dot with a bar indicates *averaging* within the group.

Information about the parameter σ^2 is contained in all of the k groups. It should be clear from the definition of the random variables given in (12.22) that

$$\hat{\sigma}_1^2 = \frac{1}{n_1 - 1} \sum_{j=1}^{n_1} (Y_{1j} - \bar{Y}_{1\cdot})^2$$

$$\hat{\sigma}_2^2 = \frac{1}{n_2 - 1} \sum_{j=1}^{n_2} (Y_{2j} - \bar{Y}_{2\cdot})^2$$

$$\vdots$$

$$\hat{\sigma}_k^2 = \frac{1}{n_k - 1} \sum_{j=1}^{n_k} (Y_{kj} - \bar{Y}_{k\cdot})^2$$

are all unbiased estimators for the common value of σ^2. Moreover,

$$\frac{(n_i - 1)\hat{\sigma}_i^2}{\sigma^2} \sim \chi^2(n_i - 1)$$

and $\hat{\sigma}_1^2$, $\hat{\sigma}_2^2$, ..., and $\hat{\sigma}_k^2$ are independent. Since the sum of independent χ^2 random variables is a chi-square distribution whose degrees of freedom is equal to the sum of the degrees of freedom of the terms in the sum,

$$\frac{(n_1 - 1)\hat{\sigma}_1^2}{\sigma^2} + \frac{(n_2 - 1)\hat{\sigma}_2^2}{\sigma^2} + \cdots + \frac{(n_k - 1)\hat{\sigma}_k^2}{\sigma^2} \sim \chi^2(n_1 - 1 + n_2 - 1 + \cdots + n_k - 1).$$

Thus,

$$\frac{(n_1 - 1)\hat{\sigma}_1^2 + (n_2 - 1)\hat{\sigma}_2^2 + \cdots + (n_k - 1)\hat{\sigma}_k^2}{\sigma^2} \sim \chi^2(N - k),$$

where

$$N = \sum_{i=1}^{k} n_i.$$

Thus, if we combine the k estimators by taking a weighted average, with weights equal to the degrees of freedom, we obtain the estimator

$$\hat{\sigma}^2 = \frac{(n_1 - 1)\hat{\sigma}_1^2 + (n_2 - 1)\hat{\sigma}_2^2 + \cdots + (n_k - 1)\hat{\sigma}_k^2}{(n_1 - 1) + (n_2 - 1) + \cdots + (n_k - 1)}$$

$$= \frac{(n_1 - 1)\hat{\sigma}_1^2 + (n_2 - 1)\hat{\sigma}_2^2 + \cdots + (n_k - 1)\hat{\sigma}_k^2}{N - k}$$

$$= \frac{\sum_{j=1}^{n_1}(Y_{1j}-\overline{Y}_1)^2 + \sum_{j=1}^{n_2}(Y_{2j}-\overline{Y}_2)^2 + \cdots + \sum_{j=1}^{n_k}(Y_{kj}-\overline{Y}_k)^2}{N-k}$$

$$= \frac{\sum_{i=1}^{k}\sum_{j=1}^{n_i}(Y_{ij}-\overline{Y}_i)^2}{N-k}. \tag{12.23}$$

The estimator $\hat{\sigma}^2$ from (12.23) therefore satisfies

$$\frac{(N-k)\hat{\sigma}^2}{\sigma^2} \sim \chi^2(N-k). \tag{12.24}$$

From this, we can say that

$$\mathbb{E}\left(\frac{(N-k)\hat{\sigma}^2}{\sigma^2}\right) = N-k,$$

from which we conclude that

$$\mathbb{E}(\hat{\sigma}^2) = \sigma^2.$$

Thus, $\hat{\sigma}^2$ is an unbiased estimator for σ^2.

Exercise 12.2 Show that the MLE of σ^2 is

$$\hat{\sigma}^2_{\text{MLE}} = \frac{\sum_{i=1}^{k}\sum_{j=1}^{n_i}(Y_{ij}-\overline{Y}_i)^2}{N}.$$

Our unbiased estimator is therefore related to the MLE by

$$\hat{\sigma}^2 = \frac{N}{N-k}\,\sigma^2_{\text{MLE}}.$$

A test of $H_0: \mu_1 = \mu_2 = \cdots = \mu_k$ is based on the decomposition of the sum of squared deviations about the overall mean:

$$\overline{\overline{y}} = \frac{1}{N}\sum_{i=1}^{k}\sum_{j=1}^{n_i} y_{ij}.$$

In this section we are dealing with just a single factor that has k levels. Let's call that factor A. Then, we define the following sums of squares.

Definition 12.4.1 — Sums of Squares. The *total sum of squares*, denoted SST, is

$$\text{SST} = \sum_{i=1}^{k}\sum_{j=1}^{n_i}\left(y_{ij}-\overline{\overline{y}}\right)^2.$$

If we were to assume that H_0 is true, then all the population means are the same. Remember, too, that we've assumed that all k variances are the same. Thus, under H_0, we essentially have a sample of size N from a normal distribution; thus, the variance is estimated as $\text{SST}/(N-1)$.

The *sum of squares due to error*, denoted SSE, is

$$\text{SSE} = \sum_{i=1}^{k} \sum_{j=1}^{n_i} (y_{ij} - \bar{y}_{i\cdot})^2.$$

This is the numerator in the unbiased estimator for σ^2 from (12.23). The deviations between y_{ij} and the within-group mean $\bar{y}_{i\cdot}$ is called the *residual*, so SSE might be more appropriately called the sum of squared residuals, but the term SSE is more commonly used. (Also, we will use SSR to mean something else, namely, the sum of squares due to regression, in Chapter 14.)

The *sum of squares due to factor A* is

$$\text{SSA} = \sum_{i=1}^{k} \sum_{j=1}^{n_i} (\bar{y}_{i\cdot} - \bar{\bar{y}})^2 = \sum_{i=1}^{k} n_i (\bar{y}_{i\cdot} - \bar{\bar{y}})^2.$$

SSA measures the variability of the sample means for factor A: $\bar{y}_{1\cdot}, \bar{y}_{2\cdot}, \ldots, \bar{y}_{k\cdot}$.

The sums of squares SSE and SSA are often referred to as the *within-group* and the *between-group* sums of squares. This distinction of *within* and *between* is one of the central ideas in the analysis of variance.

The following theorem gives the decomposition of these sums of squares and forms the basis for the test statistic to test H_0: $\mu_1 = \mu_2 = \cdots = \mu_k$.

Theorem 12.4.1 — Sum of Squares Decomposition.   The sums of squares defined above satisfy

$$\text{SST} = \text{SSA} + \text{SSE}.$$

Proof. We begin with the left side, and within the squared term we subtract and add $\bar{y}_{i\cdot}$:

$$\begin{aligned}
\text{SST} &= \sum_{i=1}^{k} \sum_{j=1}^{n_i} (y_{ij} - \bar{\bar{y}})^2 \\
&= \sum_{i=1}^{k} \sum_{j=1}^{n_i} ((y_{ij} - \bar{y}_{i\cdot}) + (\bar{y}_{i\cdot} - \bar{\bar{y}}))^2 \\
&= \sum_{i=1}^{k} \sum_{j=1}^{n_i} (y_{ij} - \bar{y}_{i\cdot})^2 + \sum_{i=1}^{k} \sum_{j=1}^{n_i} (\bar{y}_{i\cdot} - \bar{\bar{y}})^2 + 2\sum_{i=1}^{k} \sum_{j=1}^{n_i} (y_{ij} - \bar{y}_{i\cdot})(\bar{y}_{i\cdot} - \bar{\bar{y}}) \\
&= \text{SSE} + \text{SSA} + 2\, CP,
\end{aligned}$$

where CP indicates the sum of the cross product term, which can be written as

$$\begin{aligned}
CP &= \sum_{i=1}^{k} \sum_{j=1}^{n_i} (y_{ij} - \bar{y}_{i\cdot})(\bar{y}_{i\cdot} - \bar{\bar{y}}) \\
&= \sum_{i=1}^{k} (\bar{y}_{i\cdot} - \bar{\bar{y}}) \sum_{j=1}^{n_i} (y_{ij} - \bar{y}_{i\cdot}) \\
&= \sum_{i=1}^{k} (\bar{y}_{i\cdot} - \bar{\bar{y}}) \left(\sum_{j=1}^{n_i} y_{ij} - \sum_{j=1}^{n_i} \bar{y}_{i\cdot} \right)
\end{aligned}$$

$$= \sum_{i=1}^{k} \left(\bar{y}_{i\cdot} - \bar{\bar{y}} \right) \left(n\bar{y}_{i\cdot} - n\bar{y}_{i\cdot} \right)$$

$$= \sum_{i=1}^{k} \left(\bar{y}_{i\cdot} - \bar{\bar{y}} \right) \times 0$$

$$= 0. \tag{12.25}$$

Thus, $\mathrm{SST} = \mathrm{SSA} + \mathrm{SSE}$. ∎

For each sum of squares, we define the mean squared errors as follows:

$$\mathrm{MSE} = \frac{\mathrm{SSE}}{N-k}$$

$$\mathrm{MST} = \frac{\mathrm{SST}}{N-1}$$

$$\mathrm{MSA} = \frac{\mathrm{SSA}}{k-1}.$$

The analysis of variance (ANOVA) table is often used to summarize the information from a study where we compare the means across k groups. The ANOVA table for the one-factor design is given in Table 12.5.

Table 12.5 ANOVA table for one-factor design.

Source	Sum of squares	df	Mean square	F	P
Factor A	$\mathrm{SSA} = \sum_{i=1}^{k} \sum_{j=1}^{n_i} \left(\bar{y}_{i\cdot} - \bar{\bar{y}} \right)^2$	$k-1$	$\dfrac{\mathrm{SSA}}{k-1}$	$F = \dfrac{\mathrm{MSA}}{\mathrm{MSE}}$	
Error	$\mathrm{SSE} = \sum_{i=1}^{k} \sum_{j=1}^{n_i} \left(y_{ij} - \bar{y}_{i\cdot} \right)^2$	$N-k$	$\dfrac{\mathrm{SSE}}{N-k}$		
Total	$\mathrm{SST} = \sum_{i=1}^{k} \sum_{j=1}^{n_i} \left(y_{ij} - \bar{\bar{y}} \right)^2$	$N-1$	$\dfrac{\mathrm{SST}}{N-1}$		

By Theorem 12.4.1, the sums of squares SSA and SSE add to give the total sum of squares. It's clear from the table that the degrees of freedom (df) for factor A and for error add to give the degrees of freedom for the total sum of squares. For all three rows of the ANOVA table, the mean square is the sum of squares divided by the degrees of freedom. We will discuss the last two columns, F and P, shortly.

The next theorem gives us a result we can use to obtain a test statistic for testing H_0: $\mu_1 = \mu_2 = \cdots = \mu_k$.

Theorem 12.4.2 — Distributional Result for SSE. Under the distributional assumptions in (12.22):

1. $\dfrac{\mathrm{SSE}}{\sigma^2} = \dfrac{(N-k)\hat{\sigma}^2}{\sigma^2} = \dfrac{(N-k)\mathrm{MSE}}{\sigma^2} \sim \chi^2(N-k)$.

2. $\dfrac{\mathrm{SSA}}{\sigma^2} = \dfrac{(k-1)\mathrm{MSA}}{\sigma^2} \sim \chi^2(k-1)$ provided H_0 is true.

3. $\dfrac{\text{SST}}{\sigma^2} = \dfrac{(N-1)\text{MST}}{\sigma^2} \sim \chi^2(N-1)$ provided H_0 is true.

4. SSA and SSE are independent.

Proof. The first result was justified in the discussion preceding (12.24). If H_0 is true, then we can think of our k samples as being one big sample from the $N(\mu, \sigma^2)$ where μ is the common mean. In this case we can apply (8.4.1) to establish the third result. The independence of SSA and SSE is beyond our scope, so we will not demonstrate this here. By Theorem 12.4.1,

$$\text{SST} = \text{SSA} + \text{SSE}$$
$$(N-1)\text{MSE} = (k-1)\text{MSA} + (N-k)\text{MSE}$$
$$\underbrace{\frac{(N-1)\text{MSE}}{\sigma^2}}_{\chi^2(N-1)} = \underbrace{\frac{(k-1)\text{MSA}}{\sigma^2}}_{?} + \underbrace{\frac{(N-k)\text{MSE}}{\sigma^2}}_{\chi^2(N-k)}. \tag{12.26}$$

If H_0 is true, we have then established that two of the three terms in the above formula have chi-square distributions. We showed in Example 6.32 that if $W_1 \sim \chi^2(v_1)$ and $W_2 \sim \chi^2(v_2)$ are independent, then $W_1 + W_2 \sim \chi^2(v_1 + v_2)$. A related result is that if $W_1 = W_2 + W_3$ where $W_1 \sim \chi^2(v_1)$ and $W_3 \sim \chi^2(v_3)$, with $v_1 > v_3$ and if W_2 and W_3 are independent, then $W_2 \sim \chi^2(v_1 - v_3)$. (See Problem 12.33.) This is exactly the result we need to conclude from (12.26) that

$$\frac{(k-1)\text{MSA}}{\sigma^2} \sim \chi^2(k-1) \quad \text{provided } H_0 \text{ is true.} \qquad \blacksquare$$

Theorem 12.4.2 allows us to give the following interpretation.

> The degrees of freedom for a sum of squares is the divisor required to make the sum of squares an unbiased estimator of σ^2 given that H_0 is true.

Note that the mean squared error $\text{MSE} = \text{SSE}/(N-k)$ is an unbiased estimator of σ^2 regardless of whether H_0 is true or false. For the other two, SST and SSA, the null hypothesis must be true in order for the corresponding MSE (the sum of squares divided by the degrees of freedom) to be an unbiased estimator.

Theorem 12.4.3 — F Statistic. Under the assumptions given in (12.22),

$$F = \frac{\text{MSA}}{\text{MSE}} \sim F(k-1, N-k).$$

Proof. We can write this F statistic as

$$F = \frac{\text{MSA}}{\text{MSE}} = \frac{(\text{SSA}/\sigma^2)/(k-1)}{(\text{SSE}/\sigma^2)/(N-k)}.$$

By Theorem 12.4.2, SSA/σ^2 has a $\chi^2(k-1)$ distribution and SSE/σ^2 has a $\chi^2(N-k)$ distribution; moreover, the sums of squares in the numerator and denominator are independent. Thus, by the definition of the F distribution given in Definition 6.6.4, we conclude that $F \sim F(k-1, N-k)$. $\qquad \blacksquare$

This result gives us a test statistic that we can use to test H_0: $\mu_1 = \mu_2 = \cdots = \mu_k$. Deviations from H_0 in any direction will tend to inflate the numerator in the F statistic, namely, SSA. Thus, the F-test for testing the equality of k means rejects H_0 for *large* values of F.

Theorem 12.4.4 — Expected Mean Square. Under the assumptions in (12.22) the expected value of the mean square due to factor A is

$$\mathbb{E}(\text{MSA}) = \sigma^2 + \frac{1}{k-1} \sum_{i=1}^{k} n_i(\mu_i - \overline{\mu})^2, \tag{12.27}$$

where

$$\overline{\mu} = \frac{1}{N} \sum_{i=1}^{k} n_i \mu_i.$$

Proof. The proof is long and messy, but not conceptually difficult. Several times in the proof we will need the result that for any random variable, $\mathbb{V}(X) = \mathbb{E}(X^2) - [\mathbb{E}(X)^2]$; if $\mathbb{E}(X) = 0$, then $\mathbb{E}(X^2) = \mathbb{V}(X)$. The model in (12.22) can be written as $Y_{ij} = \mu_i + \varepsilon_{ij}$ where the ε_{ij} are i.i.d. $N(0, \sigma^2)$. Let's first write $\overline{Y}_{i\cdot}$ and $\overline{\overline{Y}}$ in terms of the μ_i and ε_i:

$$\overline{Y}_{i\cdot} = \frac{1}{n_i} \sum_{j=1}^{n_i} Y_{ij} = \frac{1}{n_i} \sum_{j=1}^{n_i} (\mu_i + \varepsilon_{ij}) = \mu_i + \overline{\varepsilon}_{i\cdot}, \tag{12.28}$$

where

$$\overline{\varepsilon}_{i\cdot} = \frac{1}{n_i} \sum_{j=1}^{n_i} \varepsilon_{ij}.$$

Similarly, we can write

$$\overline{\overline{Y}} = \frac{1}{N} \sum_{i=1}^{k} \sum_{j=1}^{n_i} Y_{ij} = \frac{1}{N} \sum_{i=1}^{k} \sum_{j=1}^{n_i} (\mu_i + \varepsilon_{ij}) = \frac{1}{N} \sum_{i=1}^{k} n_i \mu_i + \overline{\overline{\varepsilon}} = \overline{\mu} + \overline{\overline{\varepsilon}}$$

where

$$\overline{\overline{\varepsilon}} = \frac{1}{N} \sum_{i=1}^{k} \sum_{j=1}^{n_i} \varepsilon_{ij}.$$

Thus,

$$\overline{\overline{\varepsilon}} \sim N\left(0, \frac{\sigma^2}{N}\right) \quad \text{and} \quad \overline{\varepsilon}_i \sim N\left(0, \frac{\sigma^2}{n_i}\right).$$

Since both expected values are equal to 0,

$$\mathbb{E}\left(\overline{\overline{\varepsilon}}^2\right) = \mathbb{V}\left(\overline{\overline{\varepsilon}}^2\right) = \frac{\sigma^2}{N} \quad \text{and} \quad \mathbb{E}\left(\overline{\varepsilon}_{i\cdot}^2\right) = \mathbb{V}\left(\overline{\varepsilon}_{i\cdot}\right) = \frac{\sigma^2}{n_i}.$$

With these definitions out of the way, we can proceed directly with the MSA:

$$\text{MSA} = \frac{1}{k-1} \sum_{i=1}^{k} n_i \left(\overline{Y}_{i\cdot} - \overline{\overline{Y}}\right)^2$$

$$= \frac{1}{k-1} \sum_{i=1}^{k} n_i \left(\mu_i + \overline{\varepsilon}_{i\cdot} - \overline{\mu} - \overline{\overline{\varepsilon}} \right)^2$$

$$= \frac{1}{k-1} \sum_{i=1}^{k} n_i \left((\mu_i - \overline{\mu}) + (\overline{\varepsilon}_{i\cdot} - \overline{\overline{\varepsilon}}) \right)^2$$

$$= \underbrace{\frac{1}{k-1} \sum_{i=1}^{k} n_i (\mu_i - \overline{\mu})^2}_{\textcircled{1}} + \underbrace{\frac{1}{k-1} \sum_{i=1}^{k} 2n_i (\mu_i - \overline{\mu}) \left(\overline{\varepsilon}_{i\cdot} - \overline{\overline{\varepsilon}} \right)}_{\textcircled{2}} + \underbrace{\frac{1}{k-1} \sum_{i=1}^{k} n_i \left(\overline{\varepsilon}_{i\cdot} - \overline{\overline{\varepsilon}} \right)^2}_{\textcircled{3}}.$$

$$\text{(12.29)}$$

The expected value of the MSA is then equal to the sum of the expected values of the three terms above. Let's take these one at a time, keeping in mind that μ_i are fixed, and the ε_{ij}, $\overline{\varepsilon}_{i\cdot}$, and $\overline{\overline{\varepsilon}}$ are random:

$$\mathbb{E}\left(\textcircled{1} \right) = \mathbb{E}\left(\frac{1}{k-1} \sum_{i=1}^{k} n_i (\mu_i - \overline{\mu})^2 \right) = \frac{1}{k-1} \sum_{i=1}^{k} n_i (\mu_i - \overline{\mu})^2 ; \qquad \text{(12.30)}$$

$$\mathbb{E}\left(\textcircled{2} \right) = \mathbb{E}\left(\frac{1}{k-1} \sum_{i=1}^{k} 2n_i (\mu_i - \overline{\mu}) \left(\overline{\varepsilon}_{i\cdot} - \overline{\overline{\varepsilon}} \right) \right)$$

$$= \frac{1}{k-1} \sum_{i=1}^{k} 2n_i (\mu_i - \overline{\mu}) \mathbb{E}\left(\overline{\varepsilon}_{i\cdot} - \overline{\overline{\varepsilon}} \right)$$

$$= 0; \qquad \text{(12.31)}$$

and

$$\mathbb{E}\left(\textcircled{3} \right) = \mathbb{E}\left(\frac{1}{k-1} \sum_{i=1}^{k} n_i \left(\overline{\varepsilon}_{i\cdot} - \overline{\overline{\varepsilon}} \right)^2 \right)$$

$$= \frac{1}{k-1} \mathbb{E}\left(\sum_{i=1}^{k} n_i \left(\overline{\varepsilon}_{i\cdot}^2 - 2\overline{\varepsilon}_{i\cdot} + \overline{\overline{\varepsilon}}^2 \right) \right)$$

$$= \frac{1}{k-1} \mathbb{E}\left(\sum_{i=1}^{k} n_i \overline{\varepsilon}_{i\cdot}^2 - 2\overline{\overline{\varepsilon}} \sum_{i=1}^{k} n_i \overline{\varepsilon}_{i\cdot} + \overline{\overline{\varepsilon}}^2 \sum_{i=1}^{k} n_i \right)$$

$$= \frac{1}{k-1} \left[\sum_{i=1}^{k} n_i \mathbb{E}\left(\overline{\varepsilon}_{i\cdot}^2 \right) - N \mathbb{E}\left(\overline{\overline{\varepsilon}}^2 \right) \right]$$

$$= \frac{1}{k-1} \left[\sum_{i=1}^{k} n_i \mathbb{V}\left(\overline{\varepsilon}_{i\cdot} \right) - N \mathbb{V}\left(\overline{\overline{\varepsilon}} \right) \right]$$

$$= \frac{1}{k-1} \left[\sum_{i=1}^{k} n_i \frac{\sigma^2}{n_i} - N \frac{\sigma^2}{N} \right]$$

$$= \frac{1}{k-1} \left[k\sigma^2 - \sigma^2 \right]$$

$$= \sigma^2. \qquad \text{(12.32)}$$

We can now combine the results from (12.30), (12.31), and (12.32) into the expression for MSA from (12.29):

$$\mathbb{E}(\text{MSA}) = \mathbb{E}\left(\textcircled{1}\right) + \mathbb{E}\left(\textcircled{2}\right) + \mathbb{E}\left(\textcircled{3}\right)$$

$$= \frac{1}{k-1} \sum_{i=1}^{k} n_i (\mu_i - \overline{\mu})^2 + 0 + \sigma^2$$

$$= \sigma^2 + \frac{1}{k-1} \sum_{i=1}^{k} n_i (\mu_i - \overline{\mu})^2. \qquad \blacksquare$$

This theorem says that if H_0: $\mu_1 = \mu_2 = \cdots = \mu_k$ is true, then

$$\mathbb{E}(\text{MSA}) = \sigma^2 + 0 = \sigma^2,$$

because when all the μ_i are the same they all equal $\overline{\mu}$, making the sum of squares on the right side of (12.27) equal to 0. Recall that the F statistic is

$$F = \frac{\text{MSA}}{\text{MSE}},$$

so when H_0 is true, both the numerator and denominator have expected values that are equal to σ^2. When H_0 is false, the numerator has an expected value that is greater than σ^2. Thus, the F statistic will tend to be larger than 1 when H_0 is false, suggesting that the test for equality of means should reject H_0 when F is sufficiently *large*. Since $F \sim F(k-1, N-k)$ when H_0 is true, we would reject whenever

$$F = \frac{\text{MSA}}{\text{MSE}} > F_\alpha(k-1, N-k). \qquad (12.33)$$

■ **Example 12.12**   For the aluminum data given in Table 12.4, test whether all three means are equal.

Solution: We read in the data frame and define the necessary variables with the following code:

```
alum = read.csv("Aluminum.csv")
TypeOfAluminum = alum$TypeOfAluminum
TensileStrength = alum$TensileStrength
y1 = TensileStrength[ TypeOfAluminum == 1 ]
y2 = TensileStrength[ TypeOfAluminum == 2 ]
y3 = TensileStrength[ TypeOfAluminum == 3 ]
k = 3
n1 = length(y1)
n2 = length(y2)
n3 = length(y3)
N = n1 + n2 + n3
y1bar = mean(y1)
y2bar = mean(y2)
y3bar = mean(y3)
ybarbar = sum( c(y1,y2,y3) )/N
s1sq = var(y1)
s2sq = var(y2)
s3sq = var(y3)
```

The pooled estimate of the variance is found as follows:

```
sSquared = ( (n1-1)*s1sq + (n2-1)*s2sq + (n3-1)*s3sq ) / (N - 3)
```

We can obtain these values with some print statements:

```
print( c(y1bar,y2bar,y3bar) )
print( c(s1sq,s2sq,s3sq) )
print( sSquared )
```

which yields

```
[1]  44850 41400 37725
[1]  8675000.0 2360000.0  629166.7
[1]  4389423
```

Thus,

$$n_1 = 6, \qquad n_2 = 6, \qquad n_3 = 4$$

$$\bar{y}_1 = 44,850, \qquad \bar{y}_2 = 41,400, \qquad \bar{y}_3 = 37,725$$

$$s_1^2 = 8,675,000, \qquad s_2^2 = 2,360,000, \qquad s_3^2 = 629,167$$

$$s^2 = 4,389,423.$$

The sums of squares and the mean squares can be computed from this code:

```
SST = (N-1)*var( c(y1,y2,y3) )
MST = SST/(N-1)
SSE = sum( (y1-y1bar)^2 ) + sum( (y2-y2bar)^2 ) + sum( (y3-y3bar)^2 )
MSE = SSE/(N-k)
SSA = sum( c(n1,n2,n3)*(c(y1bar,y2bar,y3bar)-ybarbar)^2 )
MSA = SSA/(k-1)
```

The results can be obtained using some print statements:

```
print( SST )
print( SSA )
print( SSE )
print( SSA + SSE )
```

which yields

```
[1]  180250000
[1]  123187500
[1]  57062500
[1]  180250000
```

Notice that $SSA + SSE = SST$, as it must. The ANOVA table is given in Table 12.6.

The P-value is obtained using the `pf()` function in R:

```
F = MSA/MSE
pf( F , 2 , 13 , lower.tail=FALSE )
```

yielding P-value $= 0.0005663578$. Thus, there is strong evidence that the mean tensile strengths of the three types of aluminum are not the same.

There are built-in R functions that can be used to construct the ANOVA table. The above was done so you can see what calculations are needed to obtain the ANOVA table. Here is the code:

```
alum.lm = lm( TensileStrength ~ as.factor( TypeOfAluminum ) )
summary( alum.lm )
alum.anova = anova( alum.lm )
print( alum.anova )
```

The linear model command lm() is used to fit the model. The predictor, or independent, variable TypeOfAluminum must be a factor variable; otherwise R will run a linear regression (described in Chapter 14). The output from summary(alum.lm) is

```
Call:
lm(formula = TensileStrength ~ as.factor(TypeOfAluminum))

Residuals:
Min    1Q Median    3Q    Max
-4550  -1225   -75  1412   3250

Coefficients:
Estimate Std. Error t value Pr(>|t|)
(Intercept)                 44850.0    855.3  52.437  < 2e-16 ***
as.factor(TypeOfAluminum)2  -3450.0   1209.6  -2.852 0.013600 *
as.factor(TypeOfAluminum)3  -7125.0   1352.4  -5.268 0.000152 ***
---
Signif. codes:  0 '***' 0.001 '**' 0.01 '*' 0.05 '.' 0.1 ' ' 1

Residual standard error: 2095 on 13 degrees of freedom
Multiple R-squared:  0.6834, Adjusted R-squared:  0.6347
F-statistic: 14.03 on 2 and 13 DF,  p-value: 0.0005664
```

The output below Coefficients gives the average deviations of the TypeOfAluminum2 and TypeOfAluminum3 from the baseline TypeOfAluminum1. It also gives t statistics for testing whether the means at levels 2 and 3 differ from that at level 1.

The ANOVA table can be obtained using the anova() command. This function takes as its argument the linear model object obtained from the lm() function. The output from anova(alum.lm) is

```
Analysis of Variance Table

Response: TensileStrength
Df    Sum Sq  Mean Sq F value    Pr(>F)
as.factor(TypeOfAluminum)  2 123187500 61593750   14.032 0.0005664 ***
Residuals                 13  57062500  4389423
---
Signif. codes:  0 '***' 0.001 '**' 0.01 '*' 0.05 '.' 0.1 ' ' 1
```

Table 12.6 ANOVA table for the data in Example 12.12.

Source	Sum of squares	Degrees of freedom	Mean square	F	P
Factor A	123,187,500	2	61,593,750	14.032	0.00057
Error	57,062,500	13	4,389,423		
Total	180,250,000	15	12,016,667		

R does not print out the "Total" row of the ANOVA table as we have done. Of course, the Total row can be obtained by adding the sums of squares and the degrees of freedom columns.

The F statistic for testing whether all three means are the same is $F = 14.032$ with 2 and 13 degrees of freedom. The P-value for this test is the probability of observing an $F(2,13)$ statistic greater than 14.032, which is found using `pf(14.032,2,13,lower.tail=FALSE)` to be 0.000566; thus, there is strong evidence that the means are not all the same. ∎ ∎ ∎

The one-factor ANOVA model can be written in an alternative way. This model is equivalent to that given in (12.22). In this model we assume

$$Y_{ij} = \mu + \tau_i + \varepsilon_{ij}, \qquad \varepsilon_{ij} \sim \text{i.i.d. } N(0, \sigma^2), \qquad i = 1, 2, \ldots, k; \; j = 1, 2, \ldots, n_i, \tag{12.34}$$

where

$$\sum_{i=1}^{k} n_i \tau_i = 0.$$

If we look at the models in (12.22) and (12.34) we can see that $\mu_i = \mu + \tau_i$, or equivalently, $\tau_i = \mu_i - \mu$. Thus, the τ_i can be thought of as a deviation from some overall mean. Some group or treatment means will be higher, $\tau_i > 0$, and some will be lower, $\tau_i < 0$. The null hypothesis $H_0: \mu_1 = \mu_2 = \cdots = \mu_k$ is equivalent to testing $H_0: \tau_1 = \tau_2 = \cdots = \tau_k = 0$. The test statistic for the latter is the same F statistic as for the former, and the rejection rule is the same as given in (12.33).

Posing the model as in (12.34) has several advantages. First, as described below, it allows us to partition the observed values into various components. This is particularly helpful when there is more than one factor. Second, when we compute the power for the ANOVA test we will find in Section 12.7 that the power depends on the sum of τ_i^2. Third, it makes for a simpler description of the model when we generalize the one-factor model to two or more factors.

We can use the method of maximum likelihood to estimate the treatment effects τ_i along with the overall mean μ and the error variance σ^2. If we assume that

$$Y_{ij} \overset{\text{indep}}{\sim} N(\mu + \tau_i, \sigma^2),$$

then the likelihood is

$$L(\mu, \tau_1, \ldots, \tau_k, \sigma^2) = \prod_{i=1}^{k} \prod_{j=1}^{n_i} \frac{1}{\sqrt{2\pi}\sigma} \exp\left(-(y_{ij} - \mu - \tau_i)^2 / 2\sigma^2\right)$$

$$= (2\pi)^{-N/2} \sigma^{-N/2} \exp\left(-\sum_{i=1}^{k} \sum_{j=1}^{n_i} (y_{ij} - \mu - \tau_i)^2 / 2\sigma^2\right).$$

Taking the logarithm gives the log-likelihood function

$$\ell(\mu, \tau_1, \ldots, \tau_k, \sigma^2) = -\frac{N}{2} \log(2\pi) - \frac{N}{2} \log \sigma - \sum_{i=1}^{k} \sum_{j=1}^{n_i} (y_{ij} - \mu - \tau_i)^2 / 2\sigma^2.$$

Taking the derivative with respect to μ and setting the result equal to 0 gives

$$\frac{\partial \ell}{\partial \mu} = -\frac{1}{2\sigma^2} \sum_{i=1}^{k} \sum_{j=1}^{n_i} 2(y_{ij} - \mu - \tau_i)(-1) = \frac{1}{\sigma^2} \sum_{i=1}^{k} \sum_{j=1}^{n_i} (y_{ij} - \mu - \tau_i) = 0. \tag{12.35}$$

Taking the derivative with respect to τ_I $(I = 1, 2, \ldots, k)$ gives

$$\frac{\partial \ell}{\partial \tau_I} = -\frac{1}{2\sigma^2} \sum_{j=1}^{n_I} 2(y_{Ij} - \mu - \tau_I)(-1) = \frac{1}{\sigma^2} \sum_{j=1}^{n_I} (y_{Ij} - \mu - \tau_I) = 0. \tag{12.36}$$

When we differentiate with respect to τ_i, only the $i = I$ term in the outer sum depends on I; all other terms are independent of τ_i and therefore the derivative is 0. Finally, differentiating with respect to σ gives

$$\frac{\partial \ell}{\partial \sigma} = -\frac{N}{2}\frac{1}{\sigma} + \frac{2}{2\sigma^3} \sum_{i=1}^{k} \sum_{j=1}^{n_I} (y_{ij} - \mu - \tau_i)^2 = 0. \tag{12.37}$$

If we simplify (12.35) and (12.36), we obtain the following system of equations:

$$\frac{1}{\sigma^2} \sum_{i=1}^{k} \sum_{j=1}^{n_i} (y_{ij} - \mu - \tau_i) = 0$$

$$\frac{1}{\sigma^2} \sum_{j=1}^{n_I} (y_{Ij} - \mu - \tau_I) = 0.$$

Expanding the sums gives the system of equations:

$$N\bar{\bar{y}} - N\mu - \sum_{i=1}^{k} n_i \tau_i = 0$$

$$\sum_{j=1}^{n_I} y_{Ij} - n_I\mu - n_I\tau_i = 0.$$

If we write this out for each I, we see that the estimators $\hat{\mu}$, $\hat{\tau}_1, \ldots \hat{\tau}_k$ must satisfy

$$\begin{aligned}
N\bar{\bar{y}} &= N\hat{\mu} + n_1\hat{\tau}_1 + n_2\hat{\tau}_2 + \cdots + n_k\hat{\tau}_k \\
n_1\bar{y}_{1.} &= n_1\hat{\mu} + n_1\hat{\tau}_1 \\
n_2\bar{y}_{2.} &= n_2\hat{\mu} \qquad\quad + n_2\hat{\tau}_2 \\
&\vdots \qquad\qquad\qquad \vdots \\
n_k\bar{y}_{k.} &= n_k\hat{\mu} \qquad\qquad\qquad\quad + n_k\hat{\tau}_k.
\end{aligned} \tag{12.38}$$

If we add together the second through the last equations, we obtain precisely the first equation; thus, this system does not have a unique solution unless we impose an additional constraint. Here we impose the restriction

$$\sum_{i=1}^{k} n_i \hat{\tau}_i = 0.$$

With this restriction, the estimators become

$$\begin{aligned}
\hat{\mu} &= \bar{\bar{y}} \\
\hat{\tau}_i &= \bar{y}_{i.} - \bar{\bar{y}}.
\end{aligned} \tag{12.39}$$

With this result, the estimated mean for treatment i is

$$\hat{\mu}_i = \hat{\mu} + \hat{\tau}_i = \hat{\mu} + (\bar{y}_{i.} - \hat{\mu}) = \bar{y}_{i.}.$$

As expected, the estimated mean for treatment i is the mean in the sample from treatment i.

Exercise 12.3 Beginning with (12.37), show that the MLE of σ^2 is

$$\hat{\sigma}^2 = \frac{1}{N} \sum_{i=1}^{k} \sum_{j=1}^{n_i} (y_{ij} - \bar{y}_{i.})^2.$$

The population mean within each treatment could also be thought of as the predicted value for a new observation from that treatment; we call this predicted value $\hat{Y}_{i.}$. The observed value Y_{ij} can be decomposed as

$$y_{ij} = \underbrace{\bar{\bar{y}}}_{\text{overall mean}} + \underbrace{\bar{y}_{i.} - \bar{\bar{y}}}_{\hat{\tau}_i = \text{treatment effects}} + \underbrace{y_{ij} - \hat{y}_{ij}}_{r_{ij} = \text{residual}}$$

Rearranging gives

$$\underbrace{y_{ij} - \bar{\bar{y}}}_{\text{deviation from overall mean}} = \underbrace{\bar{y}_{i.} - \bar{\bar{y}}}_{\hat{\tau}_i = \text{treatment effects}} + \underbrace{y_{ij} - \hat{y}_{ij}}_{r_{ij} = \text{residual}}.$$

Theorem 12.4.1 states that the sum of the squared terms on the left side equals the sum of squared terms on the right side. Thus,

$$\underbrace{\sum_{i=1}^{n} \sum_{j=1}^{n_i} (y_{ij} - \bar{\bar{y}})^2}_{\text{sum of squared deviations from overall mean}} = \underbrace{\sum_{i=1}^{n} \sum_{j=1}^{n_i} (\bar{y}_{i.} - \bar{\bar{y}})^2}_{\text{sum of squared treatment effects}} + \underbrace{\sum_{i=1}^{n} \sum_{j=1}^{n_i} (y_{ij} - \hat{y}_{ij})^2}_{\text{sum of squared residuals}}.$$

In other words, SST = SSA + SSE, as claimed by Theorem 12.4.1.

■ **Example 12.13** Illustrate the partition of the original observations for the aluminum data from Example 12.12.

Solution: We can decompose the original observations as follows:

$$\begin{bmatrix} 46,200 & 40,100 & 38,000 \\ 48,100 & 41,900 & 36,900 \\ 40,300 & 43,300 & 38,700 \\ 43,100 & 43,000 & 37,300 \\ 44,000 & 39,900 \\ 47,400 & 40,200 \end{bmatrix} = \begin{bmatrix} 41,775 & 41,775 & 41,775 \\ 41,775 & 41,775 & 41,775 \\ 41,775 & 41,775 & 41,775 \\ 41,775 & 41,775 & 41,775 \\ 41,775 & 41,775 \\ 41,775 & 41,775 \end{bmatrix} + \begin{bmatrix} 3,075 & -375 & -4,050 \\ 3,075 & -375 & -4,050 \\ 3,075 & -375 & -4,050 \\ 3,075 & -375 & -4,050 \\ 3,075 & -375 \\ 3,075 & -375 \end{bmatrix} + \begin{bmatrix} 1,350 & -1,300 & 275 \\ 3,250 & 500 & -825 \\ -4,550 & 1,900 & 975 \\ -1,750 & 1,600 & -425 \\ -850 & -1,500 \\ 2,550 & -1,200 \end{bmatrix}.$$

If we sum the squares of all the entries in the last two matrices, we get

$$6 \times 3,075^2 + 6(-375)^2 + 4(-4,050)^2 = 123,187,500 = \text{SSA}$$

and

$$1,350^2 + 3,250^2 + (-4,550)^2 + \cdots + (-425)^2 = 57,062,500 = \text{SSE}.$$

The sum of squares total (SST) could be obtained by subtracting the first matrix on the right from the matrix on the left. Summing the squares of these values gives the SST. ■ ■ ■

We can obtain confidence intervals for τ_i or μ_i, or differences of parameters such as $\tau_{i_1} - \tau_{i_2} = \mu_{i_1} - \mu_{i_2}$. This procedure should be familiar. We know that

$$\frac{\hat{\mu}_i - \mu_i}{\sigma/\sqrt{n_i}} \sim N(0,1).$$

Because we have assumed a constant variance across all groups or treatments, we can use the pooled estimate of the variance $\text{SSE}/(n-k)$. Since

$$\frac{(n-k)\text{MSE}}{\sigma^2} \sim \chi^2(n-k),$$

we can assert that

$$\frac{\hat{\mu}_i - \mu_i}{\sqrt{\text{MSE}}/\sqrt{n_i}} \sim t(n-k).$$

Details of showing that this has a t distribution are left as an exercise.

A $100 \times (1-\alpha)\%$ confidence interval for μ_i is

$$\left(\hat{\mu}_i - t_{\alpha/2} \frac{\sqrt{\text{MSE}}}{\sqrt{n_i}} \,,\, \hat{\mu}_i + t_{\alpha/2} \frac{\sqrt{\text{MSE}}}{\sqrt{n_i}} \right) = \left(\bar{y}_{i\cdot} - t_{\alpha/2} \frac{\sqrt{\text{MSE}}}{\sqrt{n_i}} \,,\, \bar{y}_{i\cdot} + t_{\alpha/2} \frac{\sqrt{\text{MSE}}}{\sqrt{n_i}} \right).$$

Note that differences in the τs are equal to differences in the means μ_i; for example,

$$\tau_{i_1} - \tau_{i_2} = (\mu_{i_1} - \mu) - (\mu_{i_2} - \mu) = \mu_{i_1} - \mu_{i_2}.$$

The needed distributional result to obtain a confidence interval for the difference in means (or effects) is given in the next exercise.

Exercise 12.4 Show that

$$\frac{(\hat{\mu}_{i_1} - \hat{\mu}_{i_2}) - (\mu_{i_1} - \mu_{i_2})}{\sqrt{\text{MSE}} \sqrt{\dfrac{1}{n_{i_1}} + \dfrac{1}{n_{i_2}}}} \sim t(n-k).$$

Using this result, we can obtain the confidence interval for $\mu_{i_1} - \mu_{i_1}$, or equivalently, for $\tau_{i_1} - \tau_{i_1}$:

$$\left((\bar{y}_{i_1} - \bar{y}_{i_2}) - t_{\alpha/2} \sqrt{\text{MSE}\left(\frac{1}{n_{i_1}} + \frac{1}{n_{i_2}} \right)} \,,\, (\bar{y}_{i_1} - \bar{y}_{i_2}) + t_{\alpha/2} \sqrt{\text{MSE}\left(\frac{1}{n_{i_1}} + \frac{1}{n_{i_2}} \right)} \right). \quad (12.40)$$

The trouble with confidence intervals such as this is that we often want to make comparisons of the differences of all pairs of treatment effects. If, for example, we have $k = 4$ treatments, we will have $\binom{4}{2} = 6$ paired comparisons to make. Assuming $\alpha = 0.05$, we would be 95% confident that each of these intervals would contain the true difference. Often, we would like to be 95% confident that *all* intervals contain their true values. If k were equal to 12, then there would be $\binom{12}{2} = 66$ confidence intervals. If each one were a 95% interval, we'd have to admit it is likely that some of these intervals may not contain the true difference.

We would often like to make a statement like:

We are 95% confident that all of these intervals contain all of the parameter differences.

It should be clear that we would have to be more confident (than 95%) for each confidence interval in order to be 95% confident for the entire ensemble of confidence intervals. This issue of *multiple comparisons* is covered in Chapter 20.

12.5 **Two-Factor ANOVA**

Suppose that an experiment is designed with two factors, that is, predictor variables that can be set by the investigators. The two factors are assumed to have a and b levels, respectively. Let

Y_{ijk} = outcome of subject k for factor combination (i, j); i.e., factor A at level i and factor B at level j.

A combination of factor levels is called a *treatment*. If there are a levels for factor A and b levels for factor B, then there are $a \times b$ treatments.

One model we might assume about the data is that

$$Y_{ijk} = \mu + \tau_i + \eta_j + (\tau\eta)_{ij} + \varepsilon_{ijk}, \qquad i = 1,\ldots,a; \; j = 1,\ldots,b; \; k = 1,\ldots,n_{ij}, \tag{12.41}$$

where $\varepsilon_{ijk} \overset{\text{indep}}{\sim} N(0, \sigma^2)$ We will assume that the same numbers of subjects are assigned to each of the ab treatments; that is, $n_{11} = \cdots = n_{ab} = n$. Such a design is said to be *balanced*; otherwise it is *unbalanced*. R is capable of handling the unbalanced case, but the notation and concepts are easier to see for a balanced design. The symbol $(\tau\eta)_{ij}$ must be thought of as a single entity (not a product of τ and η). This is a type of interaction between factors A and B. The observation Y_{ijk} is then assumed to be equal to an overall mean plus an effect due to level i for factor A, plus an effect due to level j for factor B, plus an effect due to the interaction between factors A and B. The restrictions on the parameters are

$$\sum_{i=1}^{a} \tau_i = 0$$

$$\sum_{j=1}^{b} \eta_j = 0$$

$$\sum_{i=1}^{a} (\tau\eta)_{ij} = 0 \qquad \text{for each } j = 1,\ldots,b$$

$$\sum_{j=1}^{b} (\tau\eta)_{ij} = 0 \qquad \text{for each } i = 1,\ldots,a.$$

With these restrictions, the model in (12.41) allows for a different mean for each of the ab treatments.

Using a method similar to that described in Section 12.4, parameter estimates can be obtained and are found to be

$$\begin{aligned}
\hat{\tau}_i &= \bar{y}_{i\cdot\cdot} - \bar{y}_{\cdots} \\
\hat{\eta}_j &= \bar{y}_{\cdot j\cdot} - \bar{y}_{\cdots} \\
\widehat{(\tau\eta)}_{ij} &= \bar{y}_{ij\cdot} - \bar{y}_{i\cdot\cdot} - \bar{y}_{\cdot j\cdot} + \bar{y}_{\cdots} \,.
\end{aligned} \tag{12.42}$$

Here we have used the "dot" notation, which is ubiquitous in describing ANOVA. If a variable contains a bar above it, then a dot in the subscript indicates averaging over all of the "dotted" subscripts. Thus,

$$\bar{y}_{\cdots} = \frac{1}{abn} \sum_{i=1}^{a} \sum_{j=1}^{b} \sum_{k=1}^{n} y_{ijk}$$

$$\bar{y}_{i\cdot\cdot} = \frac{1}{bn} \sum_{j=1}^{b} \sum_{k=1}^{n} y_{ijk}$$

Table 12.7 ANOVA table for two factors.

Source	Sum of squares	Degrees of freedom	Mean square	F	P
Factor A	SSA	$a-1$	$\text{MSA} = \dfrac{\text{SSA}}{a-1}$	$F = \dfrac{\text{MSA}}{\text{MSE}}$	
Factor B	SSB	$b-1$	$\text{MSB} = \dfrac{\text{SSB}}{b-1}$	$F = \dfrac{\text{MSB}}{\text{MSE}}$	
AB interaction	SSAB	$(a-1)(b-1)$	$\text{MSAB} = \dfrac{\text{SSAB}}{(a-1)(b-1)}$	$F = \dfrac{\text{MSAB}}{\text{MSE}}$	
Error	SSE	$(n-1)ab$	$\text{MSE} = \dfrac{\text{SSA}}{a-1}$		
Total	SST	$nab-1$	$\dfrac{\text{SST}}{(n-1)ab}$		

$$\bar{y}_{.j.} = \frac{1}{an}\sum_{i=1}^{a}\sum_{k=1}^{n} y_{ijk}\,.$$

If the bar is omitted, the dots indicate summing (not averaging) across all dotted subscripts; for example,

$$y_{...} = \sum_{i=1}^{a}\sum_{j=1}^{b}\sum_{k=1}^{n} y_{ijk}\,.$$

With this notation, the estimated mean outcome for treatment combination (i,j) is $\bar{y}_{ij.}$. The residual, defined to be the difference between the observed and the estimated mean for a given treatment, is

$$r_{ijk} = y_{ijk} - \bar{y}_{ij.}$$

The ANOVA table for a two-factor model with interaction is shown in Table 12.7.

Exercise 12.5 Verify that the degrees of freedom for factors A and B, for the AB interaction, and for error sum to the degrees of freedom for the total sum of squares.

The effect of either factor or the interaction can be tested with an F-test by taking, for example,

$$F = \frac{\dfrac{\text{SSA}}{a-1}}{\dfrac{\text{SSE}}{(n-1)ab}} = \frac{\text{MSA}}{\text{MSE}}.$$

Under the assumption of no effect due to factor A, this has an $F(a-1,(n-1)ab)$ distribution. The null hypothesis of no effect due to factor A is rejected for large values of F. Similar tests can be constructed for factor B or the interaction.

■ **Example 12.14** In an article in the *Journal of the American Medical Association*, Thomas *et al.* (2021) describe a clinical trial to test the efficacy of using high doses of zinc and/or ascorbic acid (vitamin C) on the recovery time for COVID-19. They studied a number of outcome variables, including the time to 50% reduction in symptoms. They assigned volunteers to the clinical trial into one of four categories:

Table 12.8 Summary of the clinical trial on whether high doses of zinc or ascorbic acid reduce the time to 50% reduction in symptoms. Times are in days.

		Zinc	
		No	Yes
Ascorbic acid	No	$n = 50$ $\bar{y} = 6.7$ $s = 4.4$	$n = 58$ $\bar{y} = 5.9$ $s = 4.9$
	Yes	$n = 48$ $\bar{y} = 5.5$ $s = 3.7$	$n = 58$ $\bar{y} = 5.5$ $s = 3.4$

1. standard of care (no ascorbic acid, no zinc);
2. ascorbic acid only;
3. zinc only;
4. ascorbic acid and zinc.

Describe the factors and treatments for this experiment.

Solution: We can think of this as a two-factor experiment with factors ascorbic acid and zinc. There are two levels of ascorbic acid (none and high dose) and two levels of zinc (none and high dose).

There were a total of 214 patients who completed the trial. Table 12.8 summarizes the data on days until symptoms were reduced by 50%; the table shows the number of subjects in each treatment, along with the mean and standard deviation of days until 50% symptom reduction. ∎∎∎

■ **Example 12.15** The data file COVID-Zinc-AscorbicAcid.csv contains a smaller data set designed to mimic the data from the *JAMA* article described in the previous exercise. Here the sample sizes are assumed to be 5 in each treatment (compared to the roughly 50 in each treatment in the original study). The following R code can be used to read in the 20 data values and define the needed variables:

```
JAMA = read.csv( "COVID-Zinc-AscorbicAcid.csv" )
DaysUntil50PctReduction = JAMA$DaysUntil50PctReduction
AscorbicAcid = JAMA$AscorbicAcid
Zinc = JAMA$Zinc
```

The variables are

AscorbicAcid = Subject took ascorbic acid, coded as -1 for "No" and 1 for "Yes"
Zinc = Subject took zinc, coded as -1 for "No" and 1 for "Yes".

The data are summarized in Table 12.9. Estimate the model parameters and illustrate the partitioning of the observations. Finally, construct the ANOVA table and test whether either factor has an effect.

Solution: The estimates of model parameters are

$$\hat{\mu} = \bar{y}_{...} = 5.895$$

$$\hat{\tau}_1 = \bar{y}_{1..} - \bar{y}_{...} = 6.3 - 5.895 = 0.405$$

$$\hat{\tau}_2 = \bar{y}_{2..} - \bar{y}_{...} = 5.49 - 5.895 = -0.405$$

$$\hat{\eta}_1 = \bar{y}_{.1.} - \bar{y}_{...} = 6.1 - 5.895 = 0.205$$

$$\hat{\eta}_2 = \bar{y}_{.2.} - \bar{y}_{...} = 5.69 - 5.895 = -0.205$$

Table 12.9 Summary of hypothetical clinical trial on whether high doses of zinc or ascorbic acid reduce the time to 50% reduction in symptoms. Sample sizes are five for each of the four treatments. Data are in days.

		Zinc	
		No	Yes
Ascorbic acid	No	$n = 5$ $\bar{y} = 6.70$ $s = 4.40$	$n = 5$ $\bar{y} = 5.90$ $s = 4.88$
	Yes	$n = 5$ $\bar{y} = 5.50$ $s = 3.70$	$n = 5$ $\bar{y} = 5.48$ $s = 3.38$

$$\widehat{(\tau\eta)}_{11} = \bar{y}_{11.} - \bar{y}_{1..} - \bar{y}_{.1.} + \bar{y}_{...} = 6.7 - 6.3 - 6.1 + 5.895 = 0.195$$

$$\widehat{(\tau\eta)}_{12} = \bar{y}_{12.} - \bar{y}_{1..} - \bar{y}_{.2.} + \bar{y}_{...} = 5.9 - 6.3 - 5.69 + 5.895 = -0.195$$

$$\widehat{(\tau\eta)}_{21} = \bar{y}_{21.} - \bar{y}_{2..} - \bar{y}_{.1.} + \bar{y}_{...} = 5.5 - 5.49 - 6.1 + 5.895 = -0.195$$

$$\widehat{(\tau\eta)}_{22} = \bar{y}_{22.} - \bar{y}_{2..} - \bar{y}_{.2.} + \bar{y}_{...} = 5.48 - 5.49 - 5.69 + 5.895 = 0.195.$$

Each observation can be written as the overall mean plus an effect due to ascorbic acid, plus an effect due to zinc, plus an interaction effect, plus the residual:

$$y_{ijk} = \hat{\mu} + \hat{\tau}_i + \hat{\eta}_j + \widehat{(\tau\eta)}_{ij} + r_{ijk}.$$

This can be illustrated in the following matrix sum. Note that the observations are arranged in a 2×2 array for the four distinct treatments:

$$
\begin{bmatrix}
\text{AA = No} & \text{AA = No} \\
& \\
\text{Zinc = No} & \text{Zinc = Yes} \\
& \\
& \\
& \\
& \\
\text{AA = Yes} & \text{AA = Yes} \\
& \\
\text{Zinc = No} & \text{Zinc = Yes} \\
& \\
& \\
& \\
&
\end{bmatrix}
=
\begin{bmatrix}
2.6 & 13.3 \\
4.4 & 8.0 \\
14.1 & 0.8 \\
6.0 & 3.2 \\
6.4 & 4.2 \\
10.8 & 9.5 \\
5.8 & 6.4 \\
6.0 & 0.5 \\
4.4 & 6.9 \\
0.5 & 4.1
\end{bmatrix}
$$

$$
=
\begin{bmatrix}
5.895 & 5.895 \\
5.895 & 5.895 \\
5.895 & 5.895 \\
5.895 & 5.895 \\
5.895 & 5.895 \\
5.895 & 5.895 \\
5.895 & 5.895 \\
5.895 & 5.895 \\
5.895 & 5.895 \\
5.895 & 5.895
\end{bmatrix}
+
\begin{bmatrix}
0.405 & 0.405 \\
0.405 & 0.405 \\
0.405 & 0.405 \\
0.405 & 0.405 \\
0.405 & 0.405 \\
-0.405 & -0.405 \\
-0.405 & -0.405 \\
-0.405 & -0.405 \\
-0.405 & -0.405 \\
-0.405 & -0.405
\end{bmatrix}
+
\begin{bmatrix}
0.205 & -0.205 \\
0.205 & -0.205 \\
0.205 & -0.205 \\
0.205 & -0.205 \\
0.205 & -0.205 \\
0.205 & -0.205 \\
0.205 & -0.205 \\
0.205 & -0.205 \\
0.205 & -0.205 \\
0.205 & -0.205
\end{bmatrix}
+
\begin{bmatrix}
0.195 & -0.195 \\
0.195 & -0.195 \\
0.195 & -0.195 \\
0.195 & -0.195 \\
0.195 & -0.195 \\
-0.195 & 0.195 \\
-0.195 & 0.195 \\
-0.195 & 0.195 \\
-0.195 & 0.195 \\
-0.195 & 0.195
\end{bmatrix}
+
\begin{bmatrix}
-4.1 & 7.4 \\
-2.3 & 2.1 \\
7.4 & -5.1 \\
-0.7 & -2.7 \\
-0.3 & -1.7 \\
5.3 & 4.02 \\
0.3 & 0.92 \\
0.5 & -4.98 \\
-1.1 & 1.42 \\
-5.0 & -1.38
\end{bmatrix}.
$$

The total sum of squares is then the sum of the squared deviations of the original observations from the overall mean. The sums of squares due to factor A (ascorbic acid), factor B (zinc), or the AB interaction can be obtained by summing the squares of the entries in the appropriate matrix in the above partition. This yields

$$\text{SSA} = 10 \times 0.405^2 + 10 \times (-0.405)^2 = 3.2805$$

$$\text{SSB} = 10 \times 0.205^2 + 10 \times (-0.205)^2 = 0.8405$$

$$\text{SSAB} = 10 \times 0.195^2 + 10 \times (-0.195)^2 = 0.7605.$$

The sum of squares total (SST) is found to be

$$\text{SST} = \sum_{i=1}^{a} \sum_{j=1}^{b} \sum_{k=1}^{n} \left(y_{ijk} - \bar{y}_{...}\right)^2 = 278.0495,$$

and finally, the sum of squares due to error (SSE) is

$$\sum_{i=1}^{a} \sum_{j=1}^{b} \sum_{k=1}^{n} (y_{ijk} - \bar{y}_{ij.})^2 = 273.1680.$$

The results are summarized in the ANOVA table below:

Source	Sum of squares	Degrees of freedom	Mean square	F	P
Factor A: ascorbic acid	3.2805	1	3.2805	0.1921	0.6670
Factor B: zinc	0.8405	1	0.8405	0.0492	0.8272
Interaction: ascorbic acid:zinc	0.7605	1	0.7605	0.0445	0.8355
Error	273.1680	16	17.0730		
Total	278.0495	19	14.6342		

First, we test whether there is an interaction between ascorbic acid and zinc. From the ANOVA table we see that the F statistic is 0.0445, with a P-value of 0.8355. This provides no evidence (from this hypothetical study) that there is an interaction between ascorbic acid and zinc. The F statistic for testing whether ascorbic acid has an effect on the mean number of days until 50% reduction in symptoms is $F = 0.1921$, yielding a P-value of 0.6670; this provides no evidence that taking ascorbic acid affects mean time to 50% symptom reduction. Similarly, the test for whether zinc affects the outcome results in a P-value of 0.8272; again, providing no evidence of an effect.

The actual study considered several outcomes besides average time to 50% reduction in symptoms. The report published in the *Journal of the American Medical Association* indicates that the trial was stopped early for futility. The study found no effects due to ascorbic acid or zinc. You can verify the previous calculations in R by running the code

```
summary( aov( DaysUntil50PctReduction
        ~AscorbicAcid+Zinc+AscorbicAcid*Zinc ) )
```

■ ■ ■

If the interaction terms $(\tau\eta)_{ij}$ are assumed to be zero, then the model in (12.41) is said to be *additive*. This means that factors A and B have effects on the outcome that are independent of the level of the other variables. For a balanced design we can write

$$\text{SST} = \sum_{i=1}^{a} \sum_{j=1}^{b} \sum_{k=1}^{n} \left(y_{ijk} - \bar{y}_{...}\right)^2$$

$$= \sum_{i=1}^{a} \sum_{j=1}^{b} \sum_{k=1}^{n} \left((\bar{y}_{i..} - \bar{y}_{...}) + (\bar{y}_{.j.} - \bar{y}_{...}) + (\bar{y}_{ij.} - \bar{y}_{i..} - \bar{y}_{.j.} + \bar{y}_{...}) + (y_{ijk} - \bar{y}_{ij.}) \right)^2$$

$$= \sum_{i=1}^{a} \sum_{j=1}^{b} \sum_{k=1}^{n} (\bar{y}_{i..} - \bar{y}_{...})^2 + \sum_{i=1}^{a} \sum_{j=1}^{b} \sum_{k=1}^{n} (\bar{y}_{.j.} - \bar{y}_{...})^2 + \sum_{i=1}^{a} \sum_{j=1}^{b} \sum_{k=1}^{n} (\bar{y}_{ij.} - \bar{y}_{i..} - \bar{y}_{.j.} + \bar{y}_{...})^2$$

$$+ \sum_{i=1}^{a} \sum_{j=1}^{b} \sum_{k=1}^{n} (y_{ijk} - \bar{y}_{ij.})^2 + \text{sum of all 3 cross-product terms}$$

$$= \text{SSA} + \text{SSB} + \text{SSAB} + \text{SSE}. \tag{12.43}$$

Showing that each of the three cross product sums is equal to 0 is a long and tedious exercise. In principle, it is no different than showing that the cross product sum in (12.25) is equal to zero. This procedure must be done for each of the three cross product sums, so we omit it here.

This decomposition of the sums of squares requires that the number of observations in each treatment be the same. Otherwise, there is no guarantee that $\text{SST} = \text{SSA} + \text{SSB} + \text{SSE}$.

12.6   Other Designs for Experiments

An *experiment* is a study in which the investigators have the control to impose treatments on the subjects or units. An *observational study* is one where the investigators cannot impose treatments, and instead simply observe the responses or outcomes along with one or more predictor variables. Clinical trials involve the imposition of some treatment, usually with one of the treatments being a control or placebo. Often, *covariates*, such as age, gender, race, zip code, etc., are recorded on each subject to help determine the effect of the treatment. We cover this in greater detail in Chapter 14.

The models we have considered here do not involve any covariates and simply compare various treatments. Such designs occur more often in engineering and science, since the subjects or units on which the experiment is conducted can be much more homogeneous than in the health, medical, or social sciences. Aluminum bars made with the same alloys are much more uniform than people. Thus, the models we describe here are more often used in the physical sciences.

One obvious generalization of the method from Section 12.5 is to higher dimensions. For example, if there are three factors, A, B, and C, with a, b, and c levels, respectively, and if each treatment contains n observations, then the ANOVA table looks like that shown in Table 12.10.

> **Exercise 12.6** Verify that the degrees of freedom for error and for the three factors and all interactions add to the "total" degrees of freedom.

Designs that make runs at all possible combinations of factors are called *factorial* designs. These designs can be run for any number of factors, although if the number of factors is large, then the number of combinations can become too large. In principle, factorial designs can be run for any number of factors. For example, the model for three factors would be

$$y_{ijk\ell} = \mu + \tau_i + \eta_j + \omega_k + (\tau\eta)_{ij} + (\tau\omega)_{ik} + (\eta\omega)_{jk} + (\tau\eta\omega)_{ijk} + \varepsilon_{ijk\ell},$$

where $\varepsilon_{ijk\ell} \overset{\text{indep}}{\sim} N(0, \sigma^2)$. Often, the higher-order interaction terms, such as the three-factor interaction $(\tau\eta\omega)_{ijk}$, are assumed to be zero. It is generally recommended that if an interaction term is included in a model, then all lower-level interactions and main effects contained in that interaction should be included; this is called the *hierarchy principle*. For example, if we include the AB interaction in the model, then we should also include both the A and B main effects. Special care must be taken in the unbalanced case. See Montgomery (2020a) and Cobb (1998) for more complete descriptions for the analysis of factorial designs.

Table 12.10 ANOVA table for a model with three factors and all interactions.

Source	Sum of squares	Degrees of freedom	Mean square	F
Factor A	SSA	$a-1$	$\text{MSA} = \dfrac{\text{SSA}}{a-1}$	$F = \dfrac{\text{MSA}}{\text{MSE}}$
Factor B	SSB	$b-1$	$\text{MSB} = \dfrac{\text{SSB}}{b-1}$	$F = \dfrac{\text{MSB}}{\text{MSE}}$
Factor C	SSC	$c-1$	$\text{MSC} = \dfrac{\text{SSC}}{c-1}$	$F = \dfrac{\text{MSC}}{\text{MSE}}$
AB interaction	SSAB	$(a-1)(b-1)$	$\text{MSAB} = \dfrac{\text{SSAB}}{(a-1)(b-1)}$	$F = \dfrac{\text{MSAB}}{\text{MSE}}$
AC interaction	SSAC	$(a-1)(c-1)$	$\text{MSAC} = \dfrac{\text{SSAC}}{(a-1)(c-1)}$	$F = \dfrac{\text{MSAC}}{\text{MSE}}$
BC interaction	SSBC	$(b-1)(c-1)$	$\text{MSBC} = \dfrac{\text{SSBC}}{(b-1)(c-1)}$	$F = \dfrac{\text{MSBC}}{\text{MSE}}$
ABC interaction	SSABC	$(a-1)(b-1)(c-1)$	$\text{MSABC} = \dfrac{\text{SSABC}}{(a-1)(b-1)(c-1)}$	$F = \dfrac{\text{MSABC}}{\text{MSE}}$
Error	SSE	$(n-1)abc$	$\text{MSE} = \dfrac{\text{SSA}}{(n-1)abc}$	
Total	SST	$nabc-1$	$\dfrac{\text{SST}}{(n-1)ab}$	

12.6.1 Two-Level Factorial Designs

When the number of factors is large, we often set the factors to just two levels each. If there are k factors, this results in 2^k treatments. If we make n runs at each treatment combination, the design will require $n2^k$ runs.

■ **Example 12.16** An experiment is conducted to help determine which characteristics of a drill bit affect the time until the bit breaks. This example mimics a real example described by Wu and Hamada (2000). Here we focus on just 5 factors instead of the 11 in the original experiment. The data presented here are hypothetical. The five factors are characteristics of the drill bit itself: type of carbide, length, thickness, taper, and moment of inertia. For this example, we will label these factors as A, B, C, D, and E. Two drill bits were produced at all 32 possible combinations of the low and high for the five factors. The response variable is the logarithm of the number of cycles to failure.[9] Table 12.11 shows the design, that is the runs for which each factor is at the low ($-$) or high ($+$) level. The responses, along with columns for all two-factor, three-factor, four-factor, and five-factor interactions, are shown in Table 12.12. Estimate the main and interaction effects.

Solution: It is always good practice to plot the data before performing any analysis. Figure 12.7 shows the number of cycles to failure plotted against each of the five factors. This figure indicates that the distribution of the response is skewed to the high side, with a few drill bits lasting a very long time. Figure 12.8 shows what happens when we take the (natural) logarithm of the number of cycles. Here the distribution looks more symmetric. The lines in these plots run from the mean at the low level of the factor to the mean at the high level. Both plots suggest that factors B and C may have an effect on the response.

---

[9]The distribution of a variable such as cycles to failure is often highly skewed to the right, since some drill bits will last a long, long time, while others will fail quickly.

Table 12.11 Design of the drill bit experiment. There are five factors, A, B, C, D, and E (each at two levels), and thus $2^5 = 32$ different treatments. Two drill bits were produced at each treatment.

Runs	A	B	C	D	E	Run	A	B	C	D	E
1-2	−	−	−	−	−	33-34	−	−	−	−	+
3-4	+	−	−	−	−	35-36	+	−	−	−	+
5-6	−	+	−	−	−	37-38	−	+	−	−	+
7-8	+	+	−	−	−	39-40	+	+	−	−	+
9-10	−	−	+	−	−	41-42	−	−	+	−	+
11-12	+	−	+	−	−	43-44	+	−	+	−	+
13-14	−	+	+	−	−	45-46	−	+	+	−	+
15-16	+	+	+	−	−	47-48	+	+	+	−	+
17-18	−	−	−	+	−	49-50	−	−	−	+	+
19-20	+	−	−	+	−	51-52	+	−	−	+	+
21-22	−	+	−	+	−	53-54	−	+	−	+	+
23-24	+	+	−	+	−	55-56	+	+	−	+	+
25-26	−	−	+	+	−	57-58	−	−	+	+	+
27-28	+	−	+	+	−	59-60	+	−	+	+	+
29-30	−	+	+	+	−	61-62	−	+	+	+	+
31-32	+	+	+	+	−	63-64	+	+	+	+	+

For balanced two-level factorial designs like this, the *factor effect* is often defined to be the difference of the mean at the high level and the mean at the low level for that factor. This terminology is at odds with our previous meaning of factor effect – for example, in the notation $\hat{\tau}_i = \bar{y}_{i\cdot\cdot} - \hat{\mu}$. For this example, we will use the new terminology that the estimated effect due to a factor is the difference between the means at the high and low levels. The estimated factor effects will be labeled $\hat{\psi}_A$, $\hat{\psi}_B$, etc. We then have

$$\hat{\psi}_A = 3.9198980 - 4.1077260 = -0.1878281$$

$$\hat{\psi}_B = 4.3890953 - 3.6385278 = 0.7505675$$

$$\hat{\psi}_C = 3.6318447 - 4.3957784 = -0.7639337$$

$$\hat{\psi}_D = 4.1314501 - 3.8961730 = 0.2352771$$

$$\hat{\psi}_E = 4.2491295 - 3.7784936 = 0.4706358.$$

These estimated factor effects are obtained by taking the average at the high level minus the average at the low level. Equivalently, we could have obtained these by taking the "dot product" of the factor column (consisting of −1s and +1s) with the response variable log(Cycles) in this case, and dividing by half the number of runs (32 in this case).[10] For example, the estimated factor effect for A is

$$[-1,-1,1,1,-1,-1,\ldots,1] \cdot [3.1355, 4.1589, 5.8171, 3.0910, 3.4012, 3.4340, \ldots, 3.8918]/32$$

$$= \Big((-1)(3.135) + (-1)(4.159) + (1)(5.817) + (1)(3.091) + (-1)(3.401)$$

$$+ (-1)(3.434) + \cdots + (1)(3.892)\Big)\Big/32$$

$$= -0.1878.$$

---

[10]Recall that the dot product of two vectors $\boldsymbol{a} \cdot \boldsymbol{b} = [a_1, a_2, \ldots, a_n] \cdot [b_1, b_2, \ldots, b_n] = a_1 b_1 + a_2 b_2 + \cdots a_n b_n$. Thus the dot product of two vectors is the scalar obtained by adding the products of corresponding components.

Table 12.12: Factors and interactions for 2^5 design, along with the responses: Cycles and logarithm of Cycles.

A	B	C	D	E	AB	AC	AD	AE	BC	BD	BE	CD	CE	DE	ABC	ABD	ABE	ACD	ACE	ADE	BCD	BCE	BDE	CDE	ABCD	ABCE	ABDE	ACDE	BCDE	ABCDE	Cycles	log(Cycles)
−1	−1	−1	−1	−1	+1	+1	+1	+1	+1	+1	+1	+1	+1	+1	−1	−1	−1	−1	−1	−1	−1	−1	−1	−1	+1	+1	+1	+1	+1	−1	23	3.135
−1	−1	−1	−1	−1	+1	+1	+1	+1	+1	+1	+1	+1	+1	+1	−1	−1	−1	−1	−1	−1	−1	−1	−1	−1	+1	+1	+1	+1	+1	−1	64	4.159
+1	−1	−1	−1	−1	−1	−1	−1	−1	+1	+1	+1	+1	+1	+1	+1	+1	+1	+1	+1	+1	−1	−1	−1	−1	−1	−1	−1	−1	+1	+1	336	5.817
+1	−1	−1	−1	−1	−1	−1	−1	−1	+1	+1	+1	+1	+1	+1	+1	+1	+1	+1	+1	+1	−1	−1	−1	−1	−1	−1	−1	−1	+1	+1	22	3.091
−1	+1	−1	−1	−1	−1	+1	+1	+1	−1	−1	−1	+1	+1	+1	+1	+1	+1	−1	−1	−1	+1	+1	+1	−1	−1	−1	−1	+1	−1	+1	30	3.401
−1	+1	−1	−1	−1	−1	+1	+1	+1	−1	−1	−1	+1	+1	+1	+1	+1	+1	−1	−1	−1	+1	+1	+1	−1	−1	−1	−1	+1	−1	+1	31	3.434
+1	+1	−1	−1	−1	+1	−1	−1	−1	−1	−1	−1	+1	+1	+1	−1	−1	−1	+1	+1	+1	+1	+1	+1	−1	+1	+1	+1	−1	−1	−1	30	3.401
+1	+1	−1	−1	−1	+1	−1	−1	−1	−1	−1	−1	+1	+1	+1	−1	−1	−1	+1	+1	+1	+1	+1	+1	−1	+1	+1	+1	−1	−1	−1	30	3.401
−1	−1	+1	−1	−1	+1	−1	+1	+1	−1	+1	+1	−1	−1	+1	+1	−1	−1	+1	+1	−1	+1	+1	−1	+1	−1	−1	+1	−1	−1	+1	34	3.526
−1	−1	+1	−1	−1	+1	−1	+1	+1	−1	+1	+1	−1	−1	+1	+1	−1	−1	+1	+1	−1	+1	+1	−1	+1	−1	−1	+1	−1	−1	+1	28	3.332
+1	−1	+1	−1	−1	−1	+1	−1	−1	−1	+1	+1	−1	−1	+1	−1	+1	+1	−1	−1	+1	+1	+1	−1	+1	+1	+1	−1	+1	−1	−1	79	4.369
+1	−1	+1	−1	−1	−1	+1	−1	−1	−1	+1	+1	−1	−1	+1	−1	+1	+1	−1	−1	+1	+1	+1	−1	+1	+1	+1	−1	+1	−1	−1	52	3.951
−1	+1	+1	−1	−1	−1	−1	+1	+1	+1	−1	−1	−1	−1	+1	−1	+1	+1	+1	+1	−1	−1	−1	+1	+1	+1	+1	−1	−1	+1	−1	17	2.833
−1	+1	+1	−1	−1	−1	−1	+1	+1	+1	−1	−1	−1	−1	+1	−1	+1	+1	+1	+1	−1	−1	−1	+1	+1	+1	+1	−1	−1	+1	−1	24	3.178
+1	+1	+1	−1	−1	+1	+1	−1	−1	+1	−1	−1	−1	−1	+1	+1	−1	−1	−1	−1	+1	−1	−1	+1	+1	−1	−1	+1	+1	+1	+1	97	4.575
+1	+1	+1	−1	−1	+1	+1	−1	−1	+1	−1	−1	−1	−1	+1	+1	−1	−1	−1	−1	+1	−1	−1	+1	+1	−1	−1	+1	+1	+1	+1	41	3.714
−1	−1	−1	+1	−1	+1	+1	−1	+1	+1	−1	+1	−1	+1	−1	−1	+1	−1	+1	−1	+1	+1	−1	+1	+1	−1	+1	−1	−1	−1	+1	35	3.555
−1	−1	−1	+1	−1	+1	+1	−1	+1	+1	−1	+1	−1	+1	−1	−1	+1	−1	+1	−1	+1	+1	−1	+1	+1	−1	+1	−1	−1	−1	+1	27	3.296
+1	−1	−1	+1	−1	−1	−1	+1	−1	+1	−1	+1	−1	+1	−1	+1	−1	+1	−1	+1	−1	+1	−1	+1	+1	+1	−1	+1	+1	−1	−1	59	4.078
+1	−1	−1	+1	−1	−1	−1	+1	−1	+1	−1	+1	−1	+1	−1	+1	−1	+1	−1	+1	−1	+1	−1	+1	+1	+1	−1	+1	+1	−1	−1	1,242	7.124
−1	+1	−1	+1	−1	−1	+1	−1	+1	−1	+1	−1	−1	+1	−1	+1	−1	+1	+1	−1	+1	−1	+1	−1	+1	+1	−1	+1	−1	+1	−1	25	3.219
−1	+1	−1	+1	−1	−1	+1	−1	+1	−1	+1	−1	−1	+1	−1	+1	−1	+1	+1	−1	+1	−1	+1	−1	+1	+1	−1	+1	−1	+1	−1	18	2.89
+1	+1	−1	+1	−1	+1	−1	+1	−1	−1	+1	−1	−1	+1	−1	−1	+1	−1	−1	+1	−1	−1	+1	−1	+1	−1	+1	−1	+1	+1	+1	32	3.466
+1	+1	−1	+1	−1	+1	−1	+1	−1	−1	+1	−1	−1	+1	−1	−1	+1	−1	−1	+1	−1	−1	+1	−1	+1	−1	+1	−1	+1	+1	+1	63	4.143
−1	−1	+1	+1	−1	+1	−1	−1	+1	−1	−1	+1	+1	−1	−1	+1	+1	−1	−1	+1	+1	−1	+1	+1	−1	+1	−1	−1	+1	+1	−1	31	3.434
−1	−1	+1	+1	−1	+1	−1	−1	+1	−1	−1	+1	+1	−1	−1	+1	+1	−1	−1	+1	+1	−1	+1	+1	−1	+1	−1	−1	+1	+1	−1	21	3.045
+1	−1	+1	+1	−1	−1	+1	+1	−1	−1	−1	+1	+1	−1	−1	−1	−1	+1	+1	−1	−1	−1	+1	+1	−1	−1	+1	+1	−1	+1	+1	206	5.328
+1	−1	+1	+1	−1	−1	+1	+1	−1	−1	−1	+1	+1	−1	−1	−1	−1	+1	+1	−1	−1	−1	+1	+1	−1	−1	+1	+1	−1	+1	+1	51	3.932
−1	+1	+1	+1	−1	−1	−1	−1	+1	+1	+1	−1	+1	−1	−1	−1	−1	+1	−1	+1	+1	+1	−1	−1	−1	−1	+1	+1	+1	−1	+1	23	3.135
−1	+1	+1	+1	−1	−1	−1	−1	+1	+1	+1	−1	+1	−1	−1	−1	−1	+1	−1	+1	+1	+1	−1	−1	−1	−1	+1	+1	+1	−1	+1	16	2.773
+1	+1	+1	+1	−1	+1	+1	+1	−1	+1	+1	−1	+1	−1	−1	+1	+1	−1	+1	−1	−1	+1	−1	−1	−1	+1	−1	−1	−1	−1	−1	111	4.71
+1	+1	+1	+1	−1	+1	+1	+1	−1	+1	+1	−1	+1	−1	−1	+1	+1	−1	+1	−1	−1	+1	−1	−1	−1	+1	−1	−1	−1	−1	−1	32	3.466
−1	−1	−1	−1	+1	+1	+1	+1	−1	+1	+1	−1	+1	−1	−1	−1	−1	+1	−1	+1	+1	−1	+1	+1	+1	+1	−1	−1	−1	−1	+1	55	4.007
−1	−1	−1	−1	+1	+1	+1	+1	−1	+1	+1	−1	+1	−1	−1	−1	−1	+1	−1	+1	+1	−1	+1	+1	+1	+1	−1	−1	−1	−1	+1	66	4.19
+1	−1	−1	−1	+1	−1	−1	−1	+1	+1	+1	−1	+1	−1	−1	+1	+1	−1	+1	−1	−1	−1	+1	+1	+1	−1	+1	+1	+1	−1	−1	59	4.078
+1	−1	−1	−1	+1	−1	−1	−1	+1	+1	+1	−1	+1	−1	−1	+1	+1	−1	+1	−1	−1	−1	+1	+1	+1	−1	+1	+1	+1	−1	−1	87	4.466
−1	+1	−1	−1	+1	−1	+1	+1	−1	−1	−1	+1	+1	−1	−1	+1	+1	−1	−1	+1	+1	+1	−1	−1	+1	−1	+1	+1	−1	+1	−1	14	2.639
−1	+1	−1	−1	+1	−1	+1	+1	−1	−1	−1	+1	+1	−1	−1	+1	+1	−1	−1	+1	+1	+1	−1	−1	+1	−1	+1	+1	−1	+1	−1	18	2.89
+1	+1	−1	−1	+1	+1	−1	−1	+1	−1	−1	+1	+1	−1	−1	−1	−1	+1	+1	−1	−1	+1	−1	−1	+1	+1	−1	−1	+1	+1	+1	66	4.19
+1	+1	−1	−1	+1	+1	−1	−1	+1	−1	−1	+1	+1	−1	−1	−1	−1	+1	+1	−1	−1	+1	−1	−1	+1	+1	−1	−1	+1	+1	+1	57	4.043
−1	−1	+1	−1	+1	+1	−1	+1	−1	−1	+1	−1	−1	+1	−1	+1	−1	+1	+1	−1	+1	+1	−1	+1	−1	−1	+1	−1	+1	+1	−1	395	5.979
−1	−1	+1	−1	+1	+1	−1	+1	−1	−1	+1	−1	−1	+1	−1	+1	−1	+1	+1	−1	+1	+1	−1	+1	−1	−1	+1	−1	+1	+1	−1	42	3.738
+1	−1	+1	−1	+1	−1	+1	−1	+1	−1	+1	−1	−1	+1	−1	−1	+1	−1	−1	+1	−1	+1	−1	+1	−1	+1	−1	+1	−1	+1	+1	115	4.745
+1	−1	+1	−1	+1	−1	+1	−1	+1	−1	+1	−1	−1	+1	−1	−1	+1	−1	−1	+1	−1	+1	−1	+1	−1	+1	−1	+1	−1	+1	+1	196	5.278
−1	+1	+1	−1	+1	−1	−1	+1	−1	+1	−1	+1	−1	+1	−1	−1	+1	−1	+1	−1	+1	−1	+1	−1	−1	+1	−1	+1	+1	−1	+1	21	3.045
−1	+1	+1	−1	+1	−1	−1	+1	−1	+1	−1	+1	−1	+1	−1	−1	+1	−1	+1	−1	+1	−1	+1	−1	−1	+1	−1	+1	+1	−1	+1	21	3.045
+1	+1	+1	−1	+1	+1	+1	−1	+1	+1	−1	+1	−1	+1	−1	+1	−1	+1	−1	+1	−1	−1	+1	−1	−1	−1	+1	−1	−1	−1	−1	35	3.555
+1	+1	+1	−1	+1	+1	+1	−1	+1	+1	−1	+1	−1	+1	−1	+1	−1	+1	−1	+1	−1	−1	+1	−1	−1	−1	+1	−1	−1	−1	−1	32	3.466
−1	−1	−1	+1	+1	+1	+1	−1	−1	+1	−1	−1	−1	−1	+1	−1	+1	+1	+1	+1	−1	+1	+1	−1	−1	−1	−1	+1	+1	+1	−1	64	4.159
−1	−1	−1	+1	+1	+1	+1	−1	−1	+1	−1	−1	−1	−1	+1	−1	+1	+1	+1	+1	−1	+1	+1	−1	−1	−1	−1	+1	+1	+1	−1	66	4.19
+1	−1	−1	+1	+1	−1	−1	+1	+1	+1	−1	−1	−1	−1	+1	+1	−1	−1	−1	−1	+1	+1	+1	−1	−1	+1	+1	−1	−1	+1	+1	51	3.932
+1	−1	−1	+1	+1	−1	−1	+1	+1	+1	−1	−1	−1	−1	+1	+1	−1	−1	−1	−1	+1	+1	+1	−1	−1	+1	+1	−1	−1	+1	+1	148	4.997
−1	+1	−1	+1	+1	−1	+1	−1	−1	−1	+1	+1	−1	−1	+1	+1	−1	−1	+1	+1	−1	−1	−1	+1	−1	+1	+1	−1	+1	−1	+1	20	2.996
−1	+1	−1	+1	+1	−1	+1	−1	−1	−1	+1	+1	−1	−1	+1	+1	−1	−1	+1	+1	−1	−1	−1	+1	−1	+1	+1	−1	+1	−1	+1	21	3.045
+1	+1	−1	+1	+1	+1	−1	+1	+1	−1	+1	+1	−1	−1	+1	−1	+1	+1	−1	−1	+1	−1	−1	+1	−1	−1	−1	+1	−1	−1	−1	74	4.304
+1	+1	−1	+1	+1	+1	−1	+1	+1	−1	+1	+1	−1	−1	+1	−1	+1	+1	−1	−1	+1	−1	−1	+1	−1	−1	−1	+1	−1	−1	−1	140	4.942
−1	−1	+1	+1	+1	+1	−1	−1	−1	−1	−1	−1	+1	+1	+1	+1	+1	+1	−1	−1	−1	−1	−1	−1	+1	+1	+1	+1	−1	−1	+1	479	6.172
−1	−1	+1	+1	+1	+1	−1	−1	−1	−1	−1	−1	+1	+1	+1	+1	+1	+1	−1	−1	−1	−1	−1	−1	+1	+1	+1	+1	−1	−1	+1	53	3.97
+1	−1	+1	+1	+1	−1	+1	+1	+1	−1	−1	−1	+1	+1	+1	−1	−1	−1	+1	+1	+1	−1	−1	−1	+1	−1	−1	−1	+1	−1	−1	1,230	7.115
+1	−1	+1	+1	+1	−1	+1	+1	+1	−1	−1	−1	+1	+1	+1	−1	−1	−1	+1	+1	+1	−1	−1	−1	+1	−1	−1	−1	+1	−1	−1	88	4.477
−1	+1	+1	+1	+1	−1	−1	−1	−1	+1	+1	+1	+1	+1	+1	−1	−1	−1	−1	−1	−1	+1	+1	+1	+1	−1	−1	−1	−1	+1	−1	63	4.143
−1	+1	+1	+1	+1	−1	−1	−1	−1	+1	+1	+1	+1	+1	+1	−1	−1	−1	−1	−1	−1	+1	+1	+1	+1	−1	−1	−1	−1	+1	−1	358	5.881
+1	+1	+1	+1	+1	+1	+1	+1	+1	+1	+1	+1	+1	+1	+1	+1	+1	+1	+1	+1	+1	+1	+1	+1	+1	+1	+1	+1	+1	+1	+1	82	4.407
+1	+1	+1	+1	+1	+1	+1	+1	+1	+1	+1	+1	+1	+1	+1	+1	+1	+1	+1	+1	+1	+1	+1	+1	+1	+1	+1	+1	+1	+1	+1	49	3.892

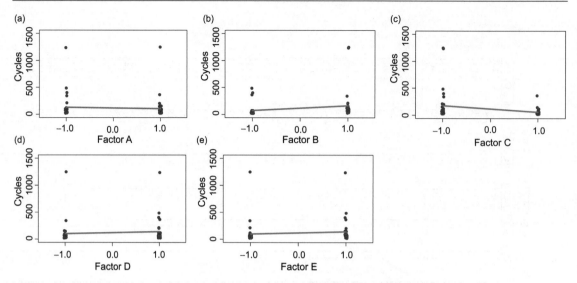

Figure 12.7 Scatterplots of the number of cycles to failure against each of the five factors.

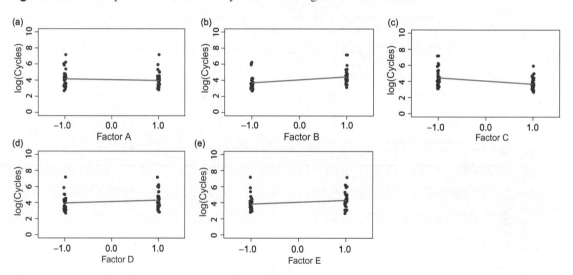

Figure 12.8 Scatterplots of log(cycles) to failure against each of the five factors.

In general, the effect is estimated by

$$\text{dot product of column with } -1\text{s and } 1\text{s with data column} \Big/ \left(n\,2^{k-1} \right),$$

where k is the number of factors and n is the number of replicates. In our example here, we have $k = 5$ and $n = 2$, so the divisor is $\left(n\,2^{k-1} \right) = 2 \times 2^4 = 32$.

Interaction effects can be estimated by a similar procedure. For example, we can estimate the AB interaction by taking the dot product of the column labeled AB in Table 12.12 with the response column:

$$[1, 1, -1, -1, -1, -1, \ldots, 1] \cdot [3.135, 4.159, 5.817, 3.091, 3.401, 3.434, \ldots, 3.892]/32$$

$$= \Big((1)(3.135) + (1)(4.159) + (-1)(5.817) + (-1)(3.091) + (-1)(3.401)$$

$$+ (-1)(3.434) + \cdots + (1)(3.892) \Big) \Big/ 32$$

Table 12.13 All estimated effects (main effects and all interactions) from the drill bit experiment using the log of cycles to failure.

Main effects	Second-order interactions	Third-order interactions	Fourth-order interactions	Fifth-order interaction
$\hat{\psi}_A = -0.1878$	$\hat{\psi}_{AB} = -0.0425$	$\hat{\psi}_{ABC} = -0.1616$	$\hat{\psi}_{ABCD} = -0.1845$	$\hat{\psi}_{ABCDE} = -0.0424$
$\hat{\psi}_B = 0.7506$	$\hat{\psi}_{AC} = 0.1992$	$\hat{\psi}_{ABD} = -0.1651$	$\hat{\psi}_{ABCE} = -0.1966$	
$\hat{\psi}_C = -0.7639$	$\hat{\psi}_{AD} = -0.4331$	$\hat{\psi}_{ABE} = 0.1317$	$\hat{\psi}_{ABDE} - 0.0754$	
$\hat{\psi}_D = 0.2353$	$\hat{\psi}_{AE} = 0.0031$	$\hat{\psi}_{ACD} = 0.2979$	$\hat{\psi}_{ACDE} = 0.2368$	
$\hat{\psi}_E = 0.4706$	$\hat{\psi}_{BC} = -0.0552$	$\hat{\psi}_{ACE} = 0.2260$	$\hat{\psi}_{BCDE} = -0.3853$	
	$\hat{\psi}_{BD} = -0.1411$	$\hat{\psi}_{ADE} = -0.0589$		
	$\hat{\psi}_{BE} = -0.2631$	$\hat{\psi}_{BCD} = -0.0854$		
	$\hat{\psi}_{CD} = -0.0221$	$\hat{\psi}_{BCE} = 0.2072$		
	$\hat{\psi}_{CE} = -0.1743$	$\hat{\psi}_{BDE} = -0.2259$		
	$\hat{\psi}_{DE} = 0.3797$	$\hat{\psi}_{CDE} = -0.2949$		

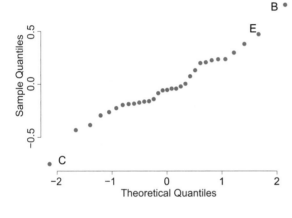

Figure 12.9 Normal probability plot for all 31 estimated main and interaction effects for the drill bit experiment. The main effects for factors B, C, and E are noted on the plot.

$$= -0.1878.$$

The complete set of estimated effects (main effects and all interaction effects) is shown in Table 12.13. We see that the largest effects are the main effects due to factors B and C. Factor B has a positive effect, meaning that the responses, number of cycles to failure, is larger when B = +1. Factor C has a negative effect. The next largest effect is for factor E, which seems to have a positive effect.

If none of the factors or interactions have an effect on the outcome, then it can be shown that the estimated main and interaction effects are independent and normally distributed with a constant mean. Thus, one way we can infer which factors or interactions have a nonzero effect on the outcome is to construct a normal probability plot (a QQ plot). The factors or interactions that "stick out" are the ones that have a true effect. Alternatively, an ANOVA table can be constructed as described below. The normal QQ plot for the drill bit experiment is shown in Figure 12.9.

■ ■ ■

An ANOVA table can be constructed for a 2^k design. The sums of squares for a factor can be shown to be

$$\text{SS} = n2^{k-2} (\text{estimated effect})^2.$$

The total sum of squares can be computed in the usual way:

$$\text{SST} = \sum_{i_1=1}^{2} \sum_{i_2=1}^{2} \cdots \sum_{i_k=1}^{2} \left(y_{i_1 i_2 \cdots i_k} - \bar{y} \right)^2,$$

where $\bar{y}$ is the mean of all $n2^k$ observations. Finally, the sum of squares due to error (SSE) can be obtained by subtracting the sum of the sum of squares for each factor and interaction from SST:

$$\text{SSE} = \text{SST} - \sum_{\text{all main and interaction effects}} \left(\text{SS due to factor or interaction}\right).$$

Because each factor has only two levels, the degrees of freedom for each main or interaction effect is 1. The degrees of freedom for SST is $n2^k - 1$, and because the degrees of freedom add, we conclude that the degrees of freedom for error is

$$\text{df}_{\text{Error}} = \left(n2^k - 1\right) - \text{number of main and interaction effects} = \left(n2^k - 1\right) - \left(2^k - 1\right) = (n-1)2^k.$$

Another way to think about df_{Error} is this: For each treatment there are n observations, giving $n-1$ degrees of freedom for estimating the error variance σ^2, and since there are 2^k treatments, there are $2^k(n-1)$ degrees of freedom for error.

■ **Example 12.17**   For the drill bit experiment in Example 12.16, construct an ANOVA table and infer which factors or interactions have a nonzero effect on the number of cycles to failure.

Solution: The ANOVA table is shown in Table 12.14. From this we can see that factors B and C have a statistically significant effect on the response, with P-values of 0.0006 and 0.0005, respectively. Factor E has a P-value of 0.0237, which would lead us to reject the null hypothesis of no effect if $\alpha = 0.05$.[11] The AD interaction also has a small P-value (0.0364), suggesting that factors A and D may interact.

■ ■ ■

For a balanced design such as in the previous examples, if we assume that higher-order interactions are zero, we can take them out of the model. The corresponding sums of squares and the degrees of freedom for the sums of squares get absorbed in SSE and df_{Error}. This is illustrated in the next example.

■ **Example 12.18**   Assume that all three-factor interactions and higher are zero for the drill bit experiment and adjust the ANOVA table accordingly.

Solution: Table 12.15 shows the ANOVA table with all third-order and higher interactions absorbed into the error term.

Note that the MSE changed very little, from 0.6288 to 0.6483. This means that the F statistics changed very little. The degrees of freedom used to compute the P-value (shown in the last column) for the F-test changed from $v_1 = 1$, $v_2 = 32$ to $v_1 = 1$, $v_2 = 48$.

■ ■ ■

12.6.2 Fractional Factorial Designs

With five factors each at two levels, as in the previous examples, there are $2^5 = 32$ different treatments. Such a factorial design is usually called a 2^5 design. This means that an unreplicated factorial experiment (i.e., an experiment where each treatment has exactly one run) must contain 32 runs. Is it possible to estimate main effects and some interactions by running a fraction of the 32 treatments? It turns out that the answer is "yes." We can often estimate main effects and sometimes interactions with just half as many runs. Sometimes we can run one-fourth, one-eighth, etc., of the full design. These are called *fractional factorial designs*.

For example, if we have five factors each at two levels, we can construct a one-half fraction of the full design by taking the first four factors, A, B, C, and D, and constructing a full 2^4 design. The levels of

[11]Table 12.14 has 31 F statistics, and therefore 31 P-values. We should account for the fact that we are performing multiple hypothesis tests. Chapter 20 takes up this issue.

Table 12.14 ANOVA table for the drill bit experiment showing sums of squares for all main and interaction effects, along with SSE and SST.

Source	SS	df	MS	F	P
A	0.5642	1	0.5642	0.8972	0.3506
B	9.0143	1	9.0143	14.3352	0.0006
C	9.3353	1	9.3353	14.8457	0.0005
D	0.8862	1	0.8862	1.4093	0.2439
E	3.5462	1	3.5462	5.6394	0.0237
AB	0.0291	1	0.0291	0.0463	0.8310
AC	0.6354	1	0.6354	1.0105	0.3223
AD	3.0003	1	3.0003	4.7712	0.0364
AE	0.0001	1	0.0001	0.0002	0.9879
BC	0.0486	1	0.0486	0.0772	0.7829
BD	0.3188	1	0.3188	0.5070	0.4816
BE	1.1085	1	1.1085	1.7629	0.1937
CD	0.0077	1	0.0077	0.0123	0.9125
CE	0.4860	1	0.4860	0.7728	0.3859
DE	2.3062	1	2.3062	3.6675	0.0645
ABC	0.4172	1	0.4172	0.6634	0.4214
ABD	0.4358	1	0.4358	0.6930	0.4113
ABE	0.2779	1	0.2779	0.4419	0.5110
ACD	1.4212	1	1.4212	2.2600	0.1426
ACE	0.8179	1	0.8179	1.3007	0.2626
ADE	0.0556	1	0.0556	0.0885	0.7680
BCD	0.1165	1	0.1165	0.1853	0.6697
BCE	0.6862	1	0.6862	1.0913	0.3040
BDE	0.8170	1	0.8170	1.2992	0.2628
CDE	1.3915	1	1.3915	2.2129	0.1467
ABCD	0.5452	1	0.5452	0.8670	0.3588
ABCE	0.6180	1	0.6180	0.9828	0.3290
ABDE	0.0910	1	0.0910	0.1447	0.7062
ACDE	0.8975	1	0.8975	1.4273	0.2410
BCDE	2.3766	1	2.3766	3.7795	0.0607
ABCDE	0.0288	1	0.0288	0.0458	0.8320
Error	20.1223	32	0.6288		
Total	62.4029	63	0.9905		

the factor E are defined to be the product ABCD. By product, we mean the product of -1 when the level is $-$ and $+1$ when the level is $+$; thus, $(-) \times (-) = (+)$, $(-) \times (+) = (-)$, etc. This idea is presented in Table 12.16. This design is called a half-fraction of a 2^5 design, and denoted 2^{5-1}. It requires just 16 runs, compared to the 32 required by the full factorial design.

Of course, if we define E = ABCD then the estimate of the four-factor interaction ABCD will exactly equal the estimate of the E main effect. In this case, we say that E is *confounded* with ABCD. Since higher-order interactions are often zero or negligible, we can safely assume that the dot product of column E and the response column is estimating the E main effect.

With the 2^{5-1} design defined by E = ABCD, there is also confounding between the other four-factor interactions and a main effect. For example, if we were to multiply columns B, C, D, and E, we would obtain exactly column A. You can see this directly by performing the 16 multiplications in Table 12.16.

Table 12.15 ANOVA table for the drill bit experiment assuming all three-factor and higher interactions are zero.

Source	SS	df	MS	F	P
A	0.5642	1	0.5642	0.8703	0.3555
B	9.0143	1	9.0143	13.9055	0.0005
C	9.3353	1	9.3353	14.4008	0.0004
D	0.8862	1	0.8862	1.3670	0.2481
E	3.5462	1	3.5462	5.4703	0.0236
AB	0.0291	1	0.0291	0.0449	0.8331
AC	0.6354	1	0.6354	0.9802	0.3271
AD	3.0003	1	3.0003	4.6282	0.0365
AE	0.0001	1	0.0001	0.0002	0.9880
BC	0.0486	1	0.0486	0.0749	0.7855
BD	0.3188	1	0.3188	0.4918	0.4865
BE	1.1085	1	1.1085	1.7101	0.1972
CD	0.0077	1	0.0077	0.0119	0.9135
CE	0.4860	1	0.4860	0.7497	0.3909
DE	2.3062	1	2.3062	3.5576	0.0653
Error	31.1161	48	0.6483		
Total	62.4029	63	0.9905		

Table 12.16 One-half fraction of the 2^5 design, called a 2^{5-1} design, where E = ABCD.

Run	A	B	C	D	E = ABCD
1	−	−	−	−	+
2	+	−	−	−	−
3	−	+	−	−	−
4	+	+	−	−	+
5	−	−	+	−	−
6	+	−	+	−	+
7	−	+	+	−	+
8	+	+	+	−	−
9	−	−	−	+	−
10	+	−	−	+	+
11	−	+	−	+	+
12	+	+	−	+	−
13	−	−	+	+	+
14	+	−	+	+	−
15	−	+	+	+	−
16	+	+	+	+	+

You can also see it by noting that multiplying columns is a commutative and associative operation. In other words, A(BC) is the same as (CA)B. Also, the product of a column with itself is equal to a column of all +s. Such a column is denoted I (for identity).[12] This is because I serves as a multiplicative identity; for example IA = AI = A.

[12]When the number of factors is large, we usually label them A, B, C, D, E, F, G, H, J, K, etc. Note the missing I, which is reserved for the identity column, i.e., the column of all +s.

Using this shorthand notation for columns, we can multiply both sides of the defining relationship $E = ABCD$ by BCD to get

$$E = ABCD$$
$$(BCD)E = (BCD)(ABCD)$$
$$BCDE = BBCCDDA$$
$$BCDE = IIIA$$
$$A = BCDE.$$

Thus, the A main effect is confounded with the BCDE interaction.

Exercise 12.7 What four-factor interactions are confounded with the B, C, and D main effects in the 2^{5-1} design?

Exercise 12.8 Construct a 2^{4-1} design.

12.6.3 Block Designs

Another set of useful designs consists of those that contain a *blocking* variable. By this we mean a variable that is not so much of interest in the context of the problem, but one that may affect the response variable. For example, the response may depend on the day that the run was made. This would occur if the mechanics of shutting down and restarting equipment, including measurement equipment, may affect the data obtained. It may also be the case that the experimental units must be produced on one of four different machines. If there is machine-to-machine variability, the resulting data may be affected. In these two examples we are not particularly interested in the day-to-day effect, or the machine-to-machine effect, but we must admit that these effects may be present and should therefore be accounted for in the model.

Here, we will describe one particular block design, the situation where there is a single factor with m levels and one blocking variable with b levels. Suppose that there is one observation for each treatment/block combination. For example, if the blocking variable is days, with the experiment being conducted over 10 days, and if there are 4 treatments, then there are 40 treatment–day combinations. With one run per treatment–day, there were a total of 40 runs. Let's define

$$Y_{ij} = \text{outcome from treatment } i \text{ in block } j.$$

The randomized block design for the 4 treatments with 10 blocks is shown in Table 12.17.

Table 12.17 Randomized block design with 4 treatments and 10 blocks, and 1 observation per treatment–block combination.

Block	Treatment			
	1	2	3	4
1	y_{11}	y_{21}	y_{31}	y_{41}
2	y_{12}	y_{22}	y_{32}	y_{42}
$\vdots$	$\vdots$	$\vdots$	$\vdots$	$\vdots$
10	$y_{1,10}$	$y_{2,10}$	$y_{3,10}$	$y_{4,10}$

The model for a randomized block design with one run per treatment–block combination is

$$Y_{ij} = \mu + \tau_i + \beta_j + \varepsilon_{ij}, \qquad \varepsilon_{ij} \sim N(0, \sigma^2), \tag{12.44}$$

where

$$\sum_{i=1}^{m} m\tau_i = 0 \qquad \text{and} \qquad \sum_{j=1}^{b} \beta_j = 0.$$

This is essentially the model for the two-factor design, except there is no interaction term. Usually in a block design we assume no interaction between treatments and blocks; in other words, the effect of a treatment is the same within each block. Using the dot notation introduced earlier, we define

$$\overline{Y}_{i\cdot} = \frac{1}{b} \sum_{j=1}^{b} Y_{ij} = \text{mean within treatment } i \tag{12.45}$$

$$\overline{Y}_{\cdot j} = \frac{1}{m} \sum_{i=1}^{m} Y_{ij} = \text{mean within block } j. \tag{12.46}$$

$$\overline{Y}_{\cdot\cdot} = \sum_{i=1}^{m} \sum_{j=1}^{b} Y_{ij}. \tag{12.47}$$

We can then determine that

$$\mathbb{E}\left(\overline{Y}_{i\cdot}\right) = \frac{1}{b} \sum_{j=1}^{b} \mathbb{E}\left(Y_{ij}\right) = \frac{1}{b} \sum_{j=1}^{b} \mathbb{E}\left(\mu + \tau_i + \beta_j + \varepsilon_{ij}\right) = \mu + \tau_i$$

and

$$\mathbb{V}\left(\overline{Y}_{i\cdot}\right) = \frac{1}{b^2} \sum_{j=1}^{b} \mathbb{V}\left(Y_{ij}\right) = \frac{1}{b^2} \sum_{j=1}^{b} \mathbb{V}\left(\mu + \tau_i + \beta_j + \varepsilon_{ij}\right) = \frac{1}{b^2} \sum_{j=1}^{b} \sigma^2 = \frac{\sigma^2}{b}.$$

Exercise 12.9 Similar to the above derivations, show that

$$\mathbb{E}\left(\overline{Y}_{\cdot j}\right) = \mu + \beta_j$$

$$\mathbb{V}\left(\overline{Y}_{\cdot j}\right) = \frac{\sigma^2}{m}.$$

The sums of squares for the randomized block design are then

$$\text{SST} = \text{sum of squares total} = \sum_{i=1}^{m} \sum_{j=1}^{b} \left(y_{ij} - \overline{y}_{\cdot\cdot}\right)^2$$

$$\text{SSBlocks} = \text{sum of squares due to blocks} = m \sum_{j=1}^{b} \left(\overline{y}_{\cdot j} - \overline{y}_{\cdot\cdot}\right)^2$$

$$\text{SSTreatments} = \text{sum of squares due to treatments} = b \sum_{i=1}^{m} \left(\overline{y}_{i\cdot} - \overline{y}_{\cdot\cdot}\right)^2$$

Table 12.18 ANOVA table for the randomized block design.

Source	Sum of squares	df	MS	F	P-value
Treatments	SSTreatments	$m-1$	$\dfrac{\text{SSTreatments}}{m-1}$	$\dfrac{\text{SSTreatments}}{\text{SSE}}$	
Blocks	SSBlocks	$b-1$	$\dfrac{\text{SSBlocks}}{b-1}$	$\dfrac{\text{SSBlocks}}{\text{SSE}}$	
Error	SSE	$n-b-m+1$	$\dfrac{\text{SSE}}{n-b-m+1}$		
Total	SST	$n-1$	$\dfrac{\text{SST}}{n-1}$		

$$\text{SSE} = \text{sum of squares due to error} = \text{SST} - \text{SSBlocks} - \text{SSTreatments}.$$

The ANOVA table for the randomized block design is therefore as shown in Table 12.18. Usually the main question is whether the treatment means are equal. Thus, we would like to test H_0: $\tau_1 = \cdots = \tau_m = 0$ against the alternative that at least one of the τ_i is not zero.[13] A test of H_0 can then be based on the test statistic

$$F = \frac{\text{SSTreatments}/(m-1)}{\text{SSE}/(n-b-k+1)} = \frac{\text{MSTreatments}}{\text{MSE}}.$$

The distribution of F under H_0 is $F(m-1, n-b-m+1)$. We would reject H_0 for

$$F > F_\alpha(m-1, n-b-m+1).$$

We can also test whether there is a block effect. The test is based on

$$F = \frac{\text{SSBlocks}/(b-1)}{\text{SSE}/(n-b-k+1)} = \frac{\text{MSBlocks}}{\text{MSE}}.$$

The reference distribution is now $F(b-1, n-b-m+1)$ and we reject for sufficiently large values of F. Usually, we recognize that block effects are likely present and testing whether they are present is not very informative. Blocking is usually done to control for the effect of the outcome and to make inferences more precise. For confidence intervals for differences in treatment means, this means narrower intervals. For hypothesis tests regarding differences between treatment means, this means more powerful tests.

It is possible to have a randomized block design with multiple runs in each treatment–block combination. For this case, the notation generalizes easily; we just need an additional index to indicate the observation within the treatment–block. It is also possible to have fewer runs in a block than there are treatments. This would occur, for example, if we had four treatments and wanted to block by day, but only two units could be produced per day. In this case we have what is called an *incomplete block design*. Table 12.19 gives an example of a design with 5 factors and 10 blocks where each block can contain just 2 runs. In this table we see that each pair of treatments occurs the same number of times (once in this case). In other words, 1 and 2 occur together in a block once (in block 1); 1 and 3 occur together in a block once (in block 2), etc. A design in which every pair of treatments occur in a block the same number of times is called a *balanced incomplete block design*. The design in Table 12.19 is a balanced incomplete design.

---

[13]If one of the τs is nonzero, then at least one of the other τ_is must be nonzero, because the sum of them must be 0.

Table 12.19 Randomized block design with $m = 5$ treatments, $b = 10$ blocks, and $s = 2$ runs in each block.

Block	Treatment 1	2	3	4	5
1	y_{11}	y_{21}			
2	y_{12}		y_{32}		
3	y_{13}			y_{43}	
4	y_{14}				y_{54}
5		y_{25}	y_{35}		
6		y_{26}		y_{46}	
7		y_{27}			y_{57}
8			y_{38}	y_{48}	
9			y_{39}		y_{59}
10				$y_{4,10}$	$y_{5,10}$

Table 12.20 Balanced incomplete block design with $m = 7$ treatments, $s = 3$ runs per block, and just 7 blocks.

Run	Treatment 1	2	3	4	5	6	7
1	y_{11}	y_{21}		y_{41}			
2		y_{22}	y_{32}		y_{52}		
3			y_{33}	y_{43}		y_{63}	
4				y_{44}	y_{54}		y_{74}
5	y_{15}				y_{55}	y_{65}	
6		y_{26}				y_{66}	y_{76}
7	y_{17}		y_{37}				y_{77}

It may be clear that you can always construct a balanced incomplete block design with $\binom{m}{s}$ blocks, where m is the number of treatments and s is the number of runs that are possible in each block. For example, the balanced incomplete block design in Table 12.19 has $m = 5$ and $s = 2$ so $\binom{5}{2} = 10$. Each of the 10 pairs of treatments occurs once in the design. It is possible to have *fewer* blocks than $\binom{m}{s}$. For example, suppose we have a design with $m = 7$ treatments and a block size of only $s = 3$ (i.e., only three runs are possible per block). In this case $\binom{7}{3} = 35$, so a balanced incomplete block design certainly exists for a design with 35 blocks. In this case, however, a smaller design exists. Table 12.20 shows a balanced incomplete block design with just seven blocks.

12.6.4 Some Experimental Design Principles

Experimental design is a broad and deep subject and many books have been written on the topic. In this section we have only been able to scratch the surface of the field. See Montgomery (2020a), Cobb (1998), and Wu and Hamada (2000) for a more thorough treatment.

We finish this section with some broad recommendations regarding the conduct of designing experiments.

1. **Randomization** Experimental designs are usually stated in a systematic fashion, but this is not the order in which designs should be run. For example, if the outcome is the life of drill bits, as in Table 12.16, then we would want to test the 16 drill bits in a random order. This helps to assure us that no systematic factor can wreak havoc. If, for example, there was a trend that the testing

of the drill bit became less severe over time, possibly because the material being drilled becomes softer over time, then the drill bits tested first would tend to have shorter lifetimes. If the design were run in the order presented in Table 12.16, then all of the units with D= − will be run before those with D= +. If units tested earlier tended to have shorter lifetimes, then we would likely see an effect due to D, even if no such effect exists. In general, randomization of units to treatments helps assure there is a balance in the units across all treatments.

2. **Blocking** Blocking means grouping similar units together (into what are called *blocks*). If we believe the blocking variable may affect the outcome, we can assign treatments within each of the levels of the blocking variable. For example, if we are testing a COVID-19 therapy, we might suspect that age is a predictor for whether the therapy is successful. It is likely to be more successful for younger patients than those who are older. We might then divide our patients into age groups (the blocks); for example, under 40, 40–65, and over 65. Then, within each of these three blocks, would assign patients to receive the new therapy or to receive the standard of care.[14] This way, we guarantee that both the new therapy and the standard-of-care are tested on groups of people with similar age distributions. Randomization alone may not provide this assurance. We saw the idea of blocking in Section 12.3. The paired t-test uses blocks of size two; within each block (pair) one unit is assigned to one treatment, while the remaining unit is assigned to the second treatment.

3. **Crossing** Crossing refers to the simultaneous varying of more than one factor. The alternative is to try to hold everything constant except one factor and then vary that factor. This is called the one-factor-at-a-time (OFAT) approach. If there exists an interaction between any of the factors, then OFAT can lead to spurious conclusions. For example, suppose that the life of a drill bet depends on two factors, A and B, and the true means for each of the four treatments are as shown in the table below:

		Factor B	
		−	+
Factor	−	500	400
A	+	300	700

Normally, we never know the true mean responses. That's why we run the experiment: to learn something about what the mean response is for each treatment! Suppose we had elected to hold everything constant except factor A. If we set B= −, then a first experiment would likely show that factor A has a negative effect; higher responses are generated when A= − than when A=+. Since we're trying to *maximize* the lifetime of the drill bit, we would choose A= −. Then, in a second experiment we would hold A constant at A= − and vary B, again, finding that there is a negative effect. When A= − the best level for factor B is B= −. We would find that the optimum setting is then A=− and B= −. We would never discover that the combination A= + and B= +, with a mean outcome of 700, is actually a better setting. We missed this because we varied one factor at a time and missed the obvious interaction between A and B. Had we run a factorial experiment we could have detected the interaction and determined the true optimum.

4. **Replication** Repeating, or replicating, a design allows us to obtain better estimates of σ^2, which in turn increases the chance that we will detect a factor effect when one truly exists. This involves the power for ANOVA tests, which is discussed in Section 12.7. A single replication in a full factorial model with all interaction terms leaves no degrees of freedom to estimate σ^2. It is often reasonable to assume that higher-order interactions are zero, and to gain some degrees of freedom

---

[14]For a serious disease like COVID, use of a placebo would be unethical. The alternative to the new therapy should be the current best treatment, called the "standard of care."

to estimate σ^2 by combining the sums of squares for the interactions that are assumed negligible. *Pure error* is a term used to refer to a way of estimating σ^2 directly, without having to assume some terms in a model are zero. For example, in the drill bit experiment of Example 12.16, we had two replicates at each of 32 treatments. Since two observations yield one degree of freedom for estimating σ^2, we have 32 degrees of freedom for error, even if we keep all possible interactions; this is called pure error. Replication gives us the ability to estimate σ^2 using pure error.

12.7 Power

In Definition 11.3.1 of Section 11.3 we defined the power to be the probability of rejecting the null hypothesis given some particular values for the parameters. These values of the parameters could be those that make H_0 true; in this case the power is the probability of rejecting H_0 given that it is true, which is equal to α. Usually when we talk of power, we mean the probability of rejecting H_0 for some values of the parameters other than those that make the null hypothesis true.

Before getting into the specifics of determining power for the various tests in this chapter, let's give definitions of some of the important noncentral distributions.

Definition 12.7.1 — Noncentral t Distribution. Suppose that $X \sim N(\lambda, 1)$ and $V \sim \chi^2(v)$. Suppose further that X and V are independent random variables. Then,

$$T = \frac{X}{\sqrt{V/v}} \tag{12.48}$$

has a noncentral t distribution with $n-1$ degrees of freedom and noncentrality parameter λ. We write

$$T = \frac{X}{\sqrt{V/v}} \sim t(n-1, \lambda).$$

If in the above notation the function t has two arguments, then the *noncentral t* distribution is meant, where the first argument is the degrees of freedom and the second argument is the noncentrality parameter. Note that when $\lambda = 0$, then the numerator is $N(0,1)$ and the denominator is $\chi^2(v)$, making the ratio $Z/\sqrt{V/v}$ have a t distribution, which we sometimes call the *central t* distribution. Thus, $t(v, 0) = t(v)$.

Definition 12.7.2 — Noncentral χ^2 Distribution. Suppose $X_1, X_2, \ldots, X_r$ are independent random variables with $X_i \sim N(\mu_i, 1)$. The random variable

$$Y = X_1^2 + X_2^2 + \cdots + X_r^2 \tag{12.49}$$

has a noncentral χ^2 distribution with r degrees of freedom and noncentrality parameter

$$\lambda = \sum_{i=1}^{r} \mu_i^2.$$

We write this as

$$Y \sim \chi^2(r, \lambda).$$

With one argument, $\chi^2(r)$ means the usual χ^2 distribution with r degrees of freedom; with two arguments, $\chi^2(r, \lambda)$ represents the *noncentral* χ^2 distribution with r degrees of freedom and noncentrality parameter λ. If $\mu_i = 0$ for all i, then the noncentral χ^2 distribution reduces to the usual χ^2 distribution.

Definition 12.7.3 — Noncentral F Distribution. Suppose $Y_1 \sim \chi^2(\nu_1, \lambda)$, $Y_2 \sim \chi^2(\nu_2)$, and Y_1 and Y_2 are independent. Then the random variable

$$F = \frac{Y_1/\nu_1}{Y_2/\nu_2} \tag{12.50}$$

has a *noncentral F distribution* with ν_1 numerator degrees of freedom, ν_2 denominator degrees of freedom, and noncentrality parameter λ. We indicate this distribution by writing

$$F = \frac{Y_1/\nu_1}{Y_2/\nu_2} \sim F(\nu_1, \nu_2, \lambda).$$

If $\lambda = 0$, then the noncentral F distribution reduces to the usual F distribution defined in Definition 6.6.4.

We needn't give the PDF for these distributions because they are not needed in what follows. The R functions for the t, χ^2, and F distribution also work for the corresponding noncentral distributions. For the noncentral t distribution, these functions and their arguments are

$$\text{dt}(\ \underbrace{\text{x}}_{\text{argument}}\ ,\ \underbrace{\text{df}}_{\text{degrees of freedom}}\ ,\ \underbrace{\text{ncp}}_{\text{noncentrality parameter } \lambda}\ ) = \text{Density of noncentral } t \text{ distribution}$$

$$\text{pt}(\ \underbrace{\text{x}}_{\text{argument}}\ ,\ \underbrace{\text{df}}_{\text{degrees of freedom}}\ ,\ \underbrace{\text{ncp}}_{\text{noncentrality parameter } \lambda}\ ) = \text{CDF of noncentral } t \text{ distribution}$$

$$\text{qt}(\ \underbrace{\text{q}}_{\text{argument}}\ ,\ \underbrace{\text{df}}_{\text{degrees of freedom}}\ ,\ \underbrace{\text{ncp}}_{\text{noncentrality parameter } \lambda}\ ) = \text{Quantile function of noncentral } t \text{ distribution}$$

$$\text{rt}(\ \underbrace{\text{n}}_{\text{argument}}\ ,\ \underbrace{\text{df}}_{\text{degrees of freedom}}\ ,\ \underbrace{\text{ncp}}_{\text{noncentrality parameter } \lambda}\ ) = n \text{ simulated random values from } t \text{ distribution}$$

Similar functions exist for the noncentral χ^2 distribution (dchisq, pchisq, qchisq, and rchisq) and the noncentral F distribution (df, pf, qf, and rf).

12.7.1 Power for Two-Sample t-Test

Let's first consider the power for the two-sample problem where we wish to test the equality of means μ_1 and μ_2. Suppose we have $X_1, X_2, \ldots, X_{n_1} \sim$ i.i.d. $N(\mu_1, \sigma^2)$ and $Y_1, Y_2, \ldots, Y_{n_2} \sim N(\mu_2, \sigma^2)$. Here we are assuming the equality of variances in the two groups. The t statistic for testing H_0: $\mu_1 = \mu_2$ against H_1: $\mu_1 \neq \mu_2$ is

$$T = \frac{\overline{X} - \overline{Y}}{S_p \sqrt{\dfrac{1}{n_1} + \dfrac{1}{n_2}}}. \tag{12.51}$$

When H_0 is true, $T \sim t(n_1 + n_2 - 2)$. Let's investigate its distribution when H_0 may be false. The T statistic can be written as

$$T = \frac{\overline{X} - \overline{Y}}{S_p\sqrt{\dfrac{1}{n_1} + \dfrac{1}{n_2}}} = \frac{\dfrac{\overline{X} - \overline{Y}}{\sigma\sqrt{\frac{1}{n_1} + \frac{1}{n_2}}}}{\dfrac{S_p\sqrt{\frac{1}{n_1} + \frac{1}{n_2}}}{\sigma\sqrt{\frac{1}{n_1} + \frac{1}{n_2}}}} = \frac{U}{\sqrt{\dfrac{S_p^2}{\sigma^2}}} = \frac{U}{\sqrt{\dfrac{(n_1 + n_2 - 2)S_p^2/\sigma^2}{n_1 + n_2 - 2}}} = \frac{U}{\sqrt{V/(n_1 + n_2 - 2)}}.$$

Here, S_p^2 is the pooled variance and $V = (n_1 + n_2 - 2)S_p^2/\sigma^2 \sim \chi^2(n_1 + n_2 - 2)$. Also,

$$U = \frac{\overline{X} - \overline{Y}}{S_p\sqrt{1/n_1 + 1/n_2}},$$

which is normally distributed with mean

$$\mathbb{E}(U) = \mathbb{E}\left(\frac{\overline{X} - \overline{Y}}{\sigma\sqrt{\frac{1}{n_1} + \frac{1}{n_2}}}\right) = \frac{1}{\sigma\sqrt{\frac{1}{n_1} + \frac{1}{n_2}}}\,\mathbb{E}(\overline{X} - \overline{Y}) = \frac{\mu_1 - \mu_2}{\sigma\sqrt{\frac{1}{n_1} + \frac{1}{n_2}}}$$

and variance

$$\mathbb{V}(U) = \mathbb{V}\left(\frac{\overline{X} - \overline{Y}}{\sigma\sqrt{\frac{1}{n_1} + \frac{1}{n_2}}}\right) = \left(\frac{1}{\sigma\sqrt{\frac{1}{n_1} + \frac{1}{n_2}}}\right)^2 \mathbb{V}(\overline{X} - \overline{Y}) = \frac{\mathbb{V}(\overline{X}) + \mathbb{V}(\overline{Y})}{\sigma^2\left(\frac{1}{n_1} + \frac{1}{n_2}\right)} = \frac{\frac{\sigma^2}{n_1} + \frac{\sigma^2}{n_2}}{\sigma^2\left(\frac{1}{n_1} + \frac{1}{n_2}\right)} = 1.$$

Thus, the t statistic from (12.51) has a noncentral t distribution with $n_1 + n_2 - 2$ degrees of freedom, and noncentrality parameter

$$\lambda = \frac{\mu_1 - \mu_2}{\sigma\sqrt{\dfrac{1}{n_1} + \dfrac{1}{n_2}}}.$$

The null hypothesis would then be rejected if $T < -t_{\alpha/2}(n_1 + n_2 - 2)$ or $T > -t_{\alpha/2}(n_1 + n_2 - 2)$.

We can then derive the power function for the two-sample t-test. This function depends only on the value of the noncentrality parameter λ and the sample sizes n_1 and n_2:

$$\pi(\lambda) = \Pr(\text{Reject } H_0 | \lambda)$$

$$= \Pr\left(T < -t_{\alpha/2}(n_1 + n_2 - 2)\right) + \Pr\left(T > t_{\alpha/2}(n_1 + n_2 - 2)\right)$$

$$= \Pr\left(\frac{\overline{X} - \overline{Y}}{S_p\sqrt{\dfrac{1}{n_1} + \dfrac{1}{n_2}}} < -t_{\alpha/2}(n_1 + n_2 - 2)\right)$$

$$+ \Pr\left(\frac{\overline{X} - \overline{Y}}{S_p\sqrt{\dfrac{1}{n_1} + \dfrac{1}{n_2}}} > t_{\alpha/2}(n_1 + n_2 - 2)\right)$$

$$= \text{pt}\left(-t_{\alpha/2}(n_1 + n_2 - 2)\,,\ n_1 + n_2 - 2\,,\ (\mu_1 - \mu_2)/(\sigma\sqrt{(1/n_1) + 1/n_2})\right)$$

$$+ 1 - \text{pt}\left(t_{\alpha/2}(n_1 + n_2 - 2)\,,\ n_1 + n_2 - 2\,,\ (\mu_1 - \mu_2)/(\sigma\sqrt{(1/n_1) + 1/n_2})\right).$$

$$(12.52)$$

In order to determine the power for a two-sample t-test we should know the true values for the individual population means, μ_1 and μ_2 (or at least their difference $\mu_1 - \mu_2$), the assumed constant variance σ^2, and the size α of the test. Power for the two-sample t-test can be done in R using the pt() function as described in the next example.

■ **Example 12.19** Suppose we would like to design a study similar to that described in Example 12.1. We would like to test whether GPAs of first-generation and non-first-generation college students have the same population means. The alternative hypothesis is that the population means differ. Assume that $\alpha = 0.05$, $\sigma = 0.45$, and the sample sizes will be the same for the two groups. Find and graph the power function for $\delta = \mu_1 - \mu_2 = -0.3, -0.29, \ldots, 0.29, 0.30$, and for $n_1 = n_2 = 20, 40, 60, 80$.

Solution: The R code needed to obtain the power is:

```
sigma = 0.45
n1 = 20
n2 = 20
delta = seq( -0.30 , 0.30 , 0.01 )
t.crit = qt( 0.025 , n1+n2-2 , lower.tail=FALSE )
lambda = delta / (sigma*sqrt(1/n1+1/n2))
pwr20 = pt(-t.crit,n1+n2-2,lambda) + 1 - pt(t.crit,n1+n2-2,lambda)
```

The first few lines define the constants $\sigma = 0.45$ and $n_1 = n_2 = 20$. In the fourth line we define a vector delta to be a sequence from -0.30 to 0.30 in increments of 0.01. The next line computes the t critical value $t_{\alpha/2}(n_1 + n_2 - 1)$. The variable lambda is then the vector of noncentrality parameters. (Since delta is a vector, so is lambda.) Finally, pwr20 is the vector containing the power for the two-sample t-test. The formula for pwr20 coded in the last line is that given in (12.52). Similar code (not shown here) can be constructed for sample sizes of $n = n_1 = n_2 = 40, 60$, and 80. Plots of the power as a function of δ are shown in Figure 12.10 for the various sample sizes. A few things to note about the power curves: (1) the curves are symmetric about $\delta = 0$, which is to be expected because the alternative is two-sided; (2) $\pi(0) = 0.05 = \alpha$, also to be expected since when $\delta = 0$ the null hypothesis is true, and the probability of rejecting H_0 when it is true is α; and (3) as $\delta \to \pm\infty$, the power goes to 1. ■ ■ ■

> **Exercise 12.10** Derive the power function for the one-sided alternative $\mu_1 > \mu_2$. Plot the power function for the same set of δ values as in Example 12.19.

Power is closely related to sample size. Often we would like to determine the sample size needed to achieve a given power for a particular alternative. The next example illustrates this.

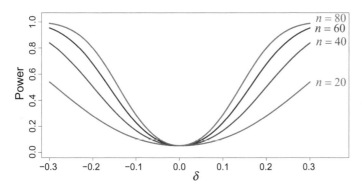

Figure 12.10 Power for a two-sample t-test with sample sizes of $n = 20, 40, 60$, and 80 for each group.

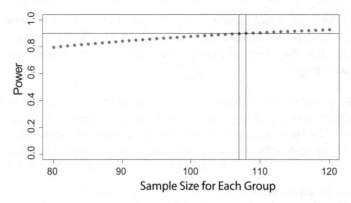

Figure 12.11 Power as a function of sample size. Vertical lines are at 107 and 108.

■ **Example 12.20**   Consider the situation described in Example 12.19. Suppose we believe that the difference in GPA will be about -0.20 and the standard deviation is $\sigma = 0.45$. How large a sample size is needed to achieve a power of 0.90?

Solution: We can see from Figure 12.10 that a sample size of $n = 80$ in each group is insufficient, since it gives a power of approximately 0.80. Thus, a slightly larger sample is needed. In the following R code we fix on a value for $\delta = \mu_1 - \mu_2$, namely $\delta = -0.20$, and create a vector for the sample size n in each group. This time, the degrees of freedom argument to pt() is a vector, while the noncentrality parameter is a scalar, the opposite of what occurred in Example 12.19.

```
n = 80:120
n1 = n
n2 = n
sigma = 0.45
delta = -0.2
t.crit = qt( 0.025 , n1+n2-2 , lower.tail=FALSE )
lambda = delta / (sigma*sqrt(1/n1+1/n2))
pwr = pt(-t.crit,n1+n2-2,lambda ) + 1 - pt(t.crit,n1+n2-2,lambda)
cbind( n , pwr )
```

Part of the output from the cbind() statement at the end is as follows:

```
[27,]  106 0.8963111
[28,]  107 0.8990410
[29,]  108 0.9017056
[30,]  109 0.9043062
[31,]  110 0.9068443
```

From this we can see that a sample size of $n = 107$ is not quite large enough, yielding a power of 0.899, while $n = 108$ is sufficient, yielding a power of 0.902. We would need $n = 108$ subjects in each group, for a total of $n_1 + n_2 = 216$ subjects.

Figure 12.11 shows the power of the two-sample t-test as a function of the sample size n. As expected, the function is increasing and surpasses 0.90 when $n = 108$.   ■ ■ ■

12.7.2   Power for One-Factor ANOVA

In the case of one-factor ANOVA, the null hypothesis is H_0: $\mu_1 = \mu_2 = \cdots = \mu_k$. There are infinitely many values of the vector $\mu = [\mu_1, \mu_2, \ldots, \mu_k]$ that make H_0 true. For example, the vectors $[0, 0, \ldots, 0]$ and $[17, 17, \ldots, 17]$ both make the null hypothesis true. There are also many vectors $[\mu_1, \mu_2, \ldots, \mu_k]$ that make the null hypothesis false; for example, if $k = 4$, then when $\boldsymbol{\mu} = [0,0,0,1]$ and $[1,1,0,0]$. Thus, finding the power of an F-test for a one-factor ANOVA is a bit more complicated than finding the power for a two-sample t-test.

Recall that the assumptions behind one-factor ANOVA are

$$Y_{ij} = \text{response of subject } j \text{ within group } i \stackrel{\text{indep}}{\sim} N(\mu_i, \sigma^2); \; j = 1, 2, \ldots, n_i; \; i = 1, 2, \ldots, k. \quad (12.53)$$

The usual F statistic for testing whether $\mu_1 = \mu_2 = \cdots = \mu_k$ is

$$F = \frac{\text{SSA}/(k-1)}{\text{SSE}/(N-k)}.$$

Under H_0, $(k-1)\text{SSA}/\sigma^2$ and $(N-k)\text{SSE}$ have independent χ^2 distributions with $k-1$ and $N-k$ degrees of freedom. In fact, the denominator has a $\chi^2(N-k)$ distribution regardless of whether H_0 is true or false. However, when H_0 is false, the numerator has a noncentral χ^2 distribution, as stated in the next theorem.

> **Theorem 12.7.1 — Distributional Result for MS Due to Factor A.** Under the assumptions in (12.53),
>
> $$\frac{(k-1)\text{MSA}}{\sigma^2} \sim \chi^2(k-1, \lambda),$$
>
> where
>
> $$\lambda = \frac{1}{\sigma^2} \sum_{i=1}^{k} n_i \tau_i^2 \quad \text{and} \quad \tau_i = \mu - \mu_i.$$

The proof of this theorem is beyond the scope of this book, so we omit it here (see Larsen and Marx, 2005).

> **Theorem 12.7.2 — F Statistic for One-Factor ANOVA.** Under the assumptions in (12.53),
>
> $$F = \frac{\text{MSA}}{\text{MSE}} = \frac{\text{SSA}/(k-1)}{\text{SSE}/(N-k)} \sim F(k-1, N-k, \lambda),$$
>
> where the noncentrality parameter λ is as stated in the previous theorem.

Proof. By Theorem 12.4.2, SSA and SSE are independent. The F statistic can then be written as

$$F = \frac{\frac{\text{SSA}}{\sigma^2}/(k-1)}{\frac{\text{SSE}}{\sigma^2}/(N-k)} = \frac{\text{a } \chi^2(k-1,\lambda)/(k-1) \text{ random variable}}{\text{a } \chi^2(N-k)/(N-k) \text{ random variable}} \sim F(k-1, N-k, \lambda),$$

using the definition of the noncentral F distribution from Definition 12.7.3 and Theorem 12.7.2. ∎

The noncentral F distribution can then be used to compute power for the one-factor ANOVA situation. We need to provide the sample sizes, the error variance σ^2, the size α of the test, and a set of assumed values for the population means. For given values of n_i, σ^2, and α, the power depends only on the noncentrality parameter λ. The power is

$$\pi(\lambda) = P(F > F_\alpha(k-1, N-k)) = \text{pf}(k-1, N-k, \lambda), \tag{12.54}$$

where pf is R's CDF function for the noncentral F distribution.

■ **Example 12.21** Suppose we want to test three levels of a cholesterol-reducing drug along with a placebo. We plan to assign equal numbers of subjects to the $k = 4$ treatments. We believe that the three levels of the drug will have the following effects on change in cholesterol (positive numbers indicate cholesterol reduction):

$$\mu_1 = 0 \quad \text{[Placebo]} \qquad \text{and} \qquad \mu_2 = \mu_3 = \mu_4 = 12 \text{ mg/dL}.$$

We believe $\sigma \approx 20$, and we will use a significance level of $\alpha = 0.05$. How large must n be so that the power for detecting[15] this difference is 0.95?

Solution: The following code can be used to compute the power for a fixed sample size. Once we have a rough estimate of the sample size needed to achieve a power of 0.95, we can run this code with an array for n.

```
k = 4
mu = c( 0 , 12 , 12 , 12 )
mubar = mean(mu)
n.each = 20
n = rep( n.each , k )
N = sum(n)
tau = mu - mubar
sigma = 20
dfn = k - 1
dfd = sum(n) - 4
lambda = sum( n*tau^2 ) / sigma^2
F.crit = qf(0.05,k-1,N-k,lower.tail=FALSE)

pwr = pf( F.crit , k-1 , N-k , lambda , lower.tail=FALSE )
print( c( n.each , sigma, lambda , pwr ) )
```

This code produces the output

```
[1] 20.0000000 20.0000000  5.4000000  0.4509211
```

so we see that for sample sizes of 20 in each group, the power is about 0.45. Clearly, a substantially larger sample size is needed. This does, however, give us a starting point in our search for the required sample size. We can then run the following code to determine the power for each value of n over a grid; the grid here is n running from 21 to 80 in increments of 1.

```
npts = 60
n = rep( NA , npts )
pwr = rep( NA , npts )
for (i in 1:npts)
{
```

---

[15] By "detecting" we mean rejecting the null hypothesis of equal means.

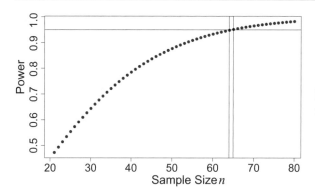

Figure 12.12 Power for the ANOVA F-test as a function of sample size n.

```
  n.each = i + 20
  n[i] = n.each
  N = k*n[i]
  dfn = k - 1
  dfd = N - 4
  lambda[i] = sum( n[i]*tau^2 ) / sigma^2
  F.crit = qf(0.05,k-1,N-k,lower.tail=FALSE)
  pwr[i] = pf( F.crit , k-1 , N-k , lambda[i] , lower.tail=FALSE )
}
```

If we use the cbind(n,pwr) command, part of the resulting output is as follows

```
[43,] 63 0.9446579
[44,] 64 0.9480984
[45,] 65 0.9513429
[46,] 66 0.9544013
[47,] 67 0.9572827
[48,] 68 0.9599961
```

Thus, we see that $n = 65$ is the smallest sample size that achieves a power of 0.95. Figure 12.12 shows how the power depends on the sample size n. The vertical lines are at $n = 64$ and $n = 65$, the values that yield a power just below or just above the desired 0.95. Thus, the experiment will need $n = 65$ subjects at each of the four levels, for a total of 260 subjects. ■ ■ ■

Exercise 12.11 Explain why the power from the previous example is the same for the situation where $\mu_1 = \mu_2 = \mu_3 = 12$ and $\mu_4 = 0$.

12.8 Chapter Summary

This chapter focused on the development of hypothesis tests for testing parameters from two or more populations. For example, we might compare the means of two populations given a sample from each of the populations. If we assume that the two populations have equal variances, the statistic $T = (\overline{X} - \overline{Y})/\left(S_p\sqrt{\frac{1}{n_1} + \frac{1}{n_2}}\right)$ leads to an exact test that is based on a t distribution. If we are reluctant to assume that variances are equal, then we can apply the Welch–Satterthwaite method, which is not exact, but the test statistic has a distribution that is very close to the t distribution. Tests for the equality of variances from two normal populations can be based on the ratio of sample variances S_1^2/S_2^2.

If we want to compare the equality of k means from k populations we can apply the analysis of variance (ANOVA) and compare our test statistic to an F distribution. More complicated designs, such as two-factor experiments or k-factor experiments, can also be handled using ANOVA with an F-test.

Power can be computed for many of the scenarios described here. For example, if we want to compare means from two populations, the test statistic T has a noncentral t distribution. This distribution can be handled by R with the usual built-in functions for the t distribution if we provide the noncentrality parameter. When we compare k means, the test statistic has a noncentral F distribution when H_0 is false. We can use R's built-in function for the F distribution to handle the noncentral case and determine the power. When we plan a study, we often want to determine the sample size that will give us the desired power for a predetermined alternative hypothesis.

12.9 Problems

Problem 12.1 A clinical trial for an eye drop used to treat dry eyes involved assigning a number of people to each of two treatments: the new drug and a vehicle.[16] The outcome was the inferior corneal staining score (ICSS). After 12 weeks, the difference from the baseline ICSS measurement taken at the beginning of the study was determined. The results of one such study are shown below:

Treatment	Vehicle	NewDrug
Sample Size	360	358
Sample Mean	-22.8	-35.3
Sample SD	28.6	28.4

Apply the z-test to test $H_0: \mu_{\text{Vehicle}} = \mu_{\text{New}}$ versus $H_0: \mu_{\text{Vehicle}} > \mu_{\text{New}}$. Use $\alpha = 0.05$ and approximate the P-value.

Problem 12.2 For the data in the previous problem, apply the two-sample t-test assuming equal variances. Do the conclusions differ?

Problem 12.3 For the data in Problem 12.1, apply the Welch–Satterthwaite method.

Problem 12.4 For the data in Problem 12.1, is it possible to assess the normality assumption of the responses?

Problem 12.5 Two machines are used for filling plastic bottles with a net volume of 16.0 ounces. The fill volume can be assumed to be normally distributed. A member of the engineering staff suspects that both machines fill to the same mean net volume, whether or not this volume is 16.0 ounces. A random sample of 10 bottles is taken from the output of each machine. The data are as follows:

Machine 1		Machine 2	
16.03	16.01	16.02	16.03
16.04	15.96	15.97	16.04
16.05	15.98	15.96	16.02
16.05	16.02	16.01	16.01
16.02	15.99	15.99	16.00

(a) What are your conclusions about the mean fill volumes? Use $\alpha = 0.05$. What is the P-value?

(b) Construct a 95% confidence interval on the difference in means. Provide a practical interpretation of this interval.

---

[16] A vehicle consists of the eye drop solution minus the active ingredient. In some sense, a vehicle is like a placebo, although the use of *some* eye drop, even one without the active ingredient being studied, can improve the severity of symptoms.

Problem 12.6 A polymer is manufactured in a batch chemical process. Viscosity measurements are typically made on each batch, and long experience with the process has indicated that viscosity is normally distributed. Fifteen batch viscosity measurements are given as follows:

724, 718, 776, 760, 745, 759,
795, 756, 742, 740, 761, 749,
739, 747, 742

A process change that involves switching the type of catalyst used in the process is made. Following the process change, eight batch viscosity measurements are taken:

735, 775, 729, 755, 783, 760,
738, 780

Assume that process variability is unaffected by the catalyst change.

(a) Formulate and test whether the population means are equal against a two-sided alternative using $\alpha = 0.10$. What are your conclusions?

(b) Find the P-value.

(c) Construct a 90% confidence interval on the difference in mean batch viscosity resulting from the process change.

(d) Compare the results of parts (a) and (c) and discuss your findings.

Problem 12.7 Consider the viscosity data in Problem 12.6. Is it reasonable to conclude that the variability in viscosity measurements is the same for both types of catalyst?

Problem 12.8 An article in the magazine *Industrial Engineer* (September 2012) reported on a study of potential sources of injury to equine veterinarians conducted at a university veterinary hospital. Forces on the hand were measured for several common activities that veterinarians engage in when examining or treating horses. The authors considered the forces on the hands for two tasks, lifting and using ultrasound. Assume that both sample sizes are 6, the sample mean force for lifting was 6.0 pounds with standard deviation 1.5 pounds, and the sample mean force for using ultrasound was 6.2 pounds with standard deviation 0.3 pounds (data read from graphs in the article). Is there evidence to conclude that the two activities result in significantly different mean forces on the hands?

Problem 12.9 Consider the hypothesis test $H_0: \mu_1 = \mu_2$ against $H_0: \mu_1 \neq \mu_2$. Suppose that sample sizes are $n_1 = 15$ and $n_2 = 15$, and that the sample means are 4.7 and 7.8, respectively. The sample standard deviations are 4 and 6.25, respectively. Assume that the variances are equal and that the data are drawn from normal distributions. Use $\alpha = 0.05$.

(a) Test the hypothesis and find the *P*-value.
(b) Explain how the test could be conducted with a confidence interval.
(c) What is the power of the test in part (a) for a true difference in means of 3? Take the value of σ^2 to be the pooled estimate from part (a).
(d) Assume that sample sizes are equal. What sample size should be used to obtain $\beta = 0.05$ if the true difference in means is -2? Assume that $\alpha = 0.05$.

Problem 12.10 Two suppliers manufacture a plastic gear used in a laser printer. The impact strength of these gears measured in foot-pounds is an important characteristic. A random sample of 10 gears from supplier 1 results in sample average impact strength of 290 and sample standard deviation of 12, and another random sample of 16 gears from the second supplier results in sample average impact strength of 321 and sample standard deviation of 22.

(a) Is there evidence to support the claim that supplier 2 provides gears with higher mean impact strength? Use $\alpha = 0.05$, and assume that both populations are normally distributed but the variances are not equal. What is the *P*-value?
(b) Do the data support the claim that the mean impact strength of gears from supplier 2 is at least 25 foot-pounds higher than that of supplier 1? Make the same assumptions as in part (a).
(c) Construct a 95% confidence interval for the difference in mean impact strength, and explain how this interval could be used to answer the question posed regarding supplier-to-supplier differences.

Problem 12.11 Consider the impact strength of gears studied in Problem 12.10. Is the assumption of unequal variances supported by the sample data?

Problem 12.12 An article in *Radio Engineering and Electronic Physics* [1984, Vol. 29(3), pp. 63–66] investigated the behavior of a stochastic generator in the presence of external noise. The number of periods was measured in a sample of 100 trains for each of two different levels of noise voltage, 100 and 150 mV. For 100 mV, the sample mean number of periods in a train was 7.9 with $s = 2.6$. For 150 mV, the sample mean was 6.9 with $s = 2.4$.

(a) It was originally suspected that raising noise voltage would reduce the mean number of periods. Do the data support this claim? Use $\alpha = 0.01$ and assume that each population is normally distributed, and the two population variances are equal. What is the *P*-value?

(b) Calculate a confidence interval to answer the question in part (a).

Problem 12.13 Is the assumption of equal variances in Problem 12.12 supported by the sample data?

Problem 12.14 The distance in yards traveled by a golf ball is tested by hitting the ball with Iron Byron, a mechanical golfer developed by the United States Golf Association with a swing that is said to emulate the distance hit by the legendary champion Byron Nelson. Ten randomly selected balls of two different brands are tested and the distance measured. The data follow:

Brand 1: 275, 286, 287, 271, 283, 271, 279, 275, 263, 267
Brand 2: 258, 244, 260, 265, 273, 281, 271, 270, 263, 268

(a) Is there evidence that distance is approximately normally distributed? Is an assumption of equal variances justified?
(b) Test the hypothesis that both brands of ball have equal mean distance. Use $\alpha = 0.05$. What is the *P*-value?
(c) Construct a 95% two-sided confidence interval for the mean difference in distance for the two brands of golf balls.
(d) What is the power of the statistical test in part (b) to detect a true difference in mean distance of 5 yards?
(e) What sample size would be required to detect a true difference in mean distance of 3 yards with power of approximately 0.75?

Problem 12.15 An electrical engineer must design a circuit to deliver the maximum amount of current to a display tube to achieve sufficient image brightness. Within her allowable design constraints, she has developed two candidate circuits and tests prototypes of each. The resulting data (in microamperes) are

Circuit 1: 251, 255, 258, 257, 250, 251, 254, 250
Circuit 2: 250, 253, 249, 256, 259, 252, 260, 251

(a) Develop and test an appropriate set of hypotheses for this problem.
(b) Test whether the variances are equal.
(c) Construct and interpret a 96% confidence interval for the difference in means.

Problem 12.16 The manager of a fleet of automobiles is testing two brands of radial tires and assigns one tire of each brand at random to the two rear wheels of eight cars and runs the cars until the tires wear out. The data (in kilometers) are in Table 12.21.

Table 12.21 Data for Problem 12.16.

Car	Brand 1	Brand 2
1	36,925	34,318
2	45,300	42,280
3	36,240	35,500
4	32,100	31,950
5	37,210	38,015
6	48,360	47,800
7	38,200	37,810
8	33,500	33,215

Find a 99% confidence interval on the difference in mean life. Which brand would you prefer based on this calculation?

Problem 12.17 A computer scientist is investigating the usefulness of two design languages to improve programming tasks. Twelve programmers who are familiar with both languages are asked to code a standard function in both languages and the time (in minutes) is recorded. The data are in Table 12.22.

Table 12.22 Data for Problem 12.17.

Programmer	Time Design language 1	Design language 2
1	17	18
2	16	14
3	21	19
4	14	11
5	18	23
6	24	21
7	16	10
8	14	13
9	21	19
10	23	24
11	13	15
12	18	20

(a) Is the assumption that the difference in coding time is normally distributed reasonable?

(b) Find a 95% confidence interval for the mean difference in coding times.

(c) Is there any indication that one design language is preferable?

Problem 12.18 A random sample of 1,000 adult residents of Maricopa County, Arizona indicated that 685 were in favor of increasing the highway speed limit to 75 mph, and another sample of 800 adult residents of Pima County indicated that 467 were in favor of the increased speed limit.

(a) Do these data indicate that there is a difference in the support for increasing the speed limit for the residents of the two counties? Use $\alpha = 0.05$. What is the P-value?

(b) Construct a 95% confidence interval on the difference in the two proportions.

Problem 12.19 Medical researchers conducted an experiment to judge the efficacy of surgery on men diagnosed with prostate cancer. They randomly assigned about half (347) of the 695 men in the study to have surgery, and 18 of them eventually died of prostate cancer compared with 31 of the 348 who did not have surgery. Is there any evidence to suggest that the surgery lowered the proportion of those who died of prostate cancer? Answer this question by testing appropriate hypotheses.

Problem 12.20 Consider the ANOVA table:

Source	df	SS	MS	F	P-value
Factor	?	117.4	39.1	?	?
Error	16	396.8	?		
Total	19	?			

(a) How many levels of the factor were used in this experiment?

(b) How many replicates did the experimenter use?

(c) Fill in the missing information in the ANOVA table. What are your conclusions about this experiment?

Problem 12.21 Montgomery (2020a) described an experiment in which the tensile strength of a synthetic fiber was of interest to the manufacturer. It is suspected that strength is related to the percentage of cotton in the fiber. Five levels of cotton percentage were used, and five replicates were run in random order, resulting in the data in Table 12.23.

(a) Does cotton percentage affect breaking strength? Draw comparative box plots and perform an analysis of variance. Use $\alpha = 0.05$.

(b) Plot average tensile strength against cotton percentage and interpret the results.

(c) Analyze the residuals and comment on model adequacy.

Table 12.23 Data for Problem 12.21.

Cotton pct.	Tensile strength				
15	7	7	15	11	9
20	12	17	12	18	18
25	14	18	18	19	19
30	19	25	22	19	23
35	7	10	11	15	11

Problem 12.22 Serafini *et al.* (2003) described an experiment on the effect of consuming chocolate on cardiovascular health. The experiment used three types of chocolates: 100 g of dark chocolate, 100 g of dark chocolate with 200 ml of full-fat milk, and 200 g of milk chocolate. Twelve subjects participated in the experiment. On different days, a subject consumed one of the chocolate-factor levels, and one hour later total antioxidant capacity of that person's blood plasma was measured in an assay. Data similar to that in the article are given in Table 12.24.

Table 12.24 Data for Problem 12.22.

Dark choc.	Dark choc. + milk choc.	Milk choc.
118.8	105.4	102.1
122.6	101.1	105.8
115.6	102.7	99.6
113.6	97.1	102.7
119.5	101.9	98.8
115.9	98.9	100.9
115.8	100.0	102.8
115.1	99.8	98.7
116.9	102.6	64.7
115.4	100.9	97.8
115.6	104.5	99.7
107.9	93.5	98.6

(a) Construct comparative box plots and study the data. What visual impression do you have from examining these plots?

(b) Analyze the experimental data using an ANOVA. If $\alpha = 0.05$, what conclusions would you draw? What would you conclude if $\alpha = 0.01$?

(c) Is there evidence that the dark chocolate increases the mean antioxidant capacity of the subjects' blood plasma more than milk chocolate?

(d) Analyze the residuals from this experiment.

Problem 12.23 An article in *Industrial Quality Control* (1956, pp. 5–8) describes an experiment to investigate the effect of two factors (glass type and phosphor type) on the brightness of a television tube. The response variable measured is the current (in microamps) necessary to obtain a specified brightness level. The data are shown in Table 12.25.

(a) State the hypotheses of interest in this experiment.

(b) Test the hypotheses in part (a) and draw conclusions using ANOVA with $\alpha = 0.05$.

(c) Analyze the residuals from this experiment.

Problem 12.24 Patton *et al.* (1986) described a study on the effects of two variables – polysilicon

Table 12.25 Data for Problem 12.23.

Glass type	Phosphor type 1	2	3
1	280	300	290
	290	310	285
	285	295	290
2	230	260	220
	234	240	225
	240	235	230

doping (PS dop) and anneal conditions (time and temperature) – on the base current of a bipolar transistor. The data from this experiment are given in Table 12.26.

Table 12.26 Data for Problem 12.24

PS dop	Anneal (temperature/time) 900 60	900 180	950 60	1,000 15	1,000 30
1×10^{20}	4.40	8.30	10.15	10.29	11.01
	4.60	8.89	10.20	10.30	10.58
2×10^{20}	3.20	7.81	9.38	10.19	10.81
	3.50	7.75	10.02	10.10	10.60

(a) Is there any evidence to support the claim that either polysilicon doping level or anneal conditions affect base current? Do these variables interact? Use $\alpha = 0.05$.

(b) Graphically analyze the interaction.

(c) Analyze the residuals from this experiment.

Problem 12.25 Soylak *et al.* (2005) described a 2^3 factorial design (with two replicates) to find lead level using flame atomic absorption spectrometry (FAAS). The data are given in Table 12.27. ST is shaking time and RC is reagent concentration.

(a) Calculate the main effect estimates. Which effects appear to be large?

(b) Conduct an analysis of variance to confirm your findings for part (a).

(c) Analyze the residuals from this experiment. Are there any problems with model adequacy?

Problem 12.26 Hare (1988) described a factorial experiment to study filling variability of dry soup mix packages. The factors are A = number of mixing ports through which the vegetable oil was added (1, 2), B =

Table 12.27 Data for Problem 12.25.

	Factors			Lead recovery %	
Run	ST	pH	RC	Rep 1	Rep 2
1	−	−	−	39.8	42.1
2	+	−	−	51.3	48.0
3	−	+	−	57.9	58.1
4	+	+	−	78.9	85.9
5	−	−	+	78.9	84.2
6	+	−	+	84.2	84.2
7	−	+	+	94.4	90.9
8	+	+	+	94.7	105.3

temperature surrounding the mixer (cooled, ambient), C = mixing time (60, 80 s), D = batch weight (1,500, 2,000 lb), and E = number of days of delay between mixing and packaging (1, 7). Between 125 and 150 packages of soup were sampled over an 8-hour period for each run in the design, and the standard deviation of package weight was used as the response variable. The design and resulting data are given in Table 12.28.

Table 12.28 Data for Problem 12.26.

Std order	A	B	C	D	D	y Std dev.
1	−	−	−	−	−	1.13
2	+	−	−	−	+	1.25
3	−	+	−	−	+	0.97
4	+	+	−	−	−	1.70
5	−	−	+	−	+	1.47
6	+	−	+	−	−	1.28
7	−	+	+	−	−	1.18
8	+	+	+	−	+	0.98
9	−	−	−	+	+	0.78
10	+	−	−	+	−	1.36
11	−	+	−	+	−	1.85
12	+	+	−	+	+	0.62
13	−	−	+	+	−	1.09
14	+	−	+	+	+	1.10
15	−	+	+	+	+	0.76
16	+	+	+	+	−	2.10

(a) What is the generator for this design?

(b) Estimate the factor main effects. Which effects are large?

(c) Conduct an appropriate analysis of variance and draw conclusions about this filling process.

Problem 12.27 Theobald (1981) described an experiment in which a shape measurement was determined for several different nozzle types at different levels of jet efflux velocity. Interest in this experiment focuses primarily on nozzle type, and velocity is a nuisance

factor. The data are in Table 12.29. Does nozzle type affect shape measurement? Compare the nozzles with box plots and the analysis of variance.

Table 12.29 Data for Problem 12.27.

Nozzle type	Jet efflux velocity (m/s)					
	11.73	14.37	16.56	20.43	23.46	28.74
1	0.78	0.80	0.81	0.75	0.77	0.78
2	0.85	0.85	0.92	0.86	0.81	0.83
3	0.93	0.92	0.95	0.89	0.89	0.83
4	1.14	0.97	0.98	0.88	0.86	0.83
5	0.97	0.86	0.78	0.76	0.76	0.76

Problem 12.28 Lamberton *et al.* (1976) described a field test for detecting the presence of arsenic in urine samples. The test has been proposed for use among forestry workers because of the increasing use of organic arsenics in that industry. The experiment compared the test as performed by both a trainee and an experienced trainer to an analysis at a remote laboratory. Four subjects were selected for testing and are considered as blocks. The response variable is arsenic content (in ppm) in the subject's urine. The data are given in Table 12.30. Is there any difference in the arsenic test procedure?

Table 12.30 Data for Problem 12.28.

	Subject			
Test	1	2	3	4
Trainee	0.05	0.05	0.04	0.15
Trainer	0.05	0.05	0.04	0.17
Lab	0.04	0.04	0.03	0.10

Problem 12.29 An article published in the *Proceedings of the National Academy of Sciences* (Talalay *et al.*, 2007) studied the effect of broccoli sprout extracts on sunburn. A compound was produced and applied to volunteer subjects and the amount of sunburn experienced was measured. Comment on the following designs.

(a) Apply the compound to n subjects and observe the results.

(b) Divide the available volunteers into two groups. One group receives the experimental compound while the other receives a placebo (a compound created with no active ingredients).

(c) For each subject, apply the experimental compound to one side of the body and the placebo to the other side.

Problem 12.30 The design of the study in Problem 12.29 was actually the matched-pairs design described in part (c). The outcome was the mean reduction in erythema (a measure of the intensity of the sunburn). The sample size was $n = 6$, with each subject being his of her own matched pair. The values of the mean reduction of erythema were

38.3,   23.8,   8.37,   78.1,   53.9,   23.5

Test whether this broccoli sprout extract is helpful in preventing sunburn. Use $\alpha = 0.01$.

Problem 12.31 Derive the MLEs in equations (12.6), (12.7), and (12.8).

Problem 12.32 Derive the equalities in (12.13) and (12.14).

Problem 12.33 Suppose that $W_1 = W_2 + W_3$ and that $W_1 \sim \chi^2(v_1)$ and $W_3 \sim \chi^2(v_3)$, where $v_1 > v_3$. Suppose further that W_2 and W_3 are independent. Use MGFs to show that $W_2 \sim \chi^2(v_1 - v_3)$.

Problem 12.34 Consider the situation described in Problem 12.16, except assume that we would like to compare six brands of tire. Since a car has only four wheels, we can only test four brands at a time. Develop a design in which we could test six tire brands.

Problem 12.35 Polymer clay is hardened by placing the formed piece (such as a sculpture) in an oven for a fixed period of time. We would like to test eight polymer clay formulations, but our oven can hold only three pieces at a time. Develop a design in which we could test all eight formulations.

Problem 12.36 Suppose we would like to test the means of two populations. We will select a sample of size $n = 10$ units from each population (20 items altogether). It is believed that the standard deviation in each population is about $\sigma = 12.5$. It is also believed that the means in the two populations will be about $\mu_1 = 80$ and $\mu_2 = 90$. The alternative hypothesis is two-sided and the significance level is $\alpha = 0.05$.

(a) What is the power for this test?

(b) How large must n (the sample size for each sample) be so that the power is 0.95?

Problem 12.37 Repeat Problem 12.36, assuming a significance level of $\alpha = 0.01$ and the one-sided alternative H_1: $\mu_2 > \mu_1$.

Problem 12.38 Suppose that in Problem 12.36 the researchers were overly optimistic in saying that $\sigma = 12.5$; it is actually $\sigma = 20$. Answer both parts of Problem 12.36 assuming $\sigma = 20$. Describe the effect of misspecifying σ.

Problem 12.39 Suppose we are designing an experiment with $k = 3$ treatments: one placebo, a new drug at a low dose, and the new drug at a high dose. The outcome is change in kidney function, which is measured by the glomerular filtration rate (GFR). Normal levels are 60 or above. We have available 60 patients whose GFR level is near 40, which indicates kidney disease. The outcome variable in the experiment comparing the three treatments is the increase in kidney function. It is believed that $\mu_1 = 0$, and $\mu_2 = \mu_3 = 5$. The standard deviation of the increase in kidney function is believed to be $\sigma = 4$.

(a) What is the power if the significance level is $\alpha = 0.05$?

(b) How large must be the sample size for each treatment so that the power is 0.9?

13 — Hypothesis Tests for Categorical Data

13.1 Introduction

Often we look at the relationships between categorical variables, such as which hospital a patient is admitted to, or whether a person has diabetes, pre-diabetes, or no diabetes at all. These variables can be nominal (like the hospital) or ordinal (like diabetes, pre-diabetes, or no diabetes). In many cases we want to know something about how these variables are related. Tests regarding the relationships between variables are usually carried out using χ^2 tests. For such tests, the test statistic has an approximate χ^2 distribution.

Usually, this leads to the idea of a contingency table – that is, a table of the frequency or count of the number of units in each possible combination of the variables. For example, a study of hospitals in the Netherlands looked at prophylactic use of antibiotics to prevent a hospital-acquired urinary tract infection (UTI). (See Reintjes *et al.*, 2000.) In an observational study, the investigators found that the counts for the patients who were given/not given antibiotics and those who did/did not acquire a UTI are as follows:

		UTI	
		Yes	No
Prophylactic	Yes	42	1,237
Antibiotic	No	104	2,136

This is an example of a contingency table.

We see that among the $42 + 1,237 = 1,279$ who received antibiotics, 42 developed a UTI; this is 3.28%. On the other hand, among the $104 + 2,136 = 2,240$ who did not receive antibiotics, 104 developed a UTI; this is 4.64%. Thus, a lower percentage developed a UTI in the antibiotics group than in the no-antibiotics group. Two questions arise here. First, is this difference too big to be explained by chance (in other words, is the difference statistically significant)? And second, what does this say about the use of antibiotics for preventing UTIs? (Remember, this was an observational study, so there may be other variables lurking.)

Most of this chapter deals with these questions about contingency tables. We begin the chapter in Section 13.2 with the concept of testing whether a sample comes from a particular distribution. This leads to the use of an approximate χ^2 test. The χ^2 test from contingency tables can then be seen as a generalization of the test statistic for testing whether discrete data come from a particular distribution.

13.2   Goodness-of-Fit Tests

Suppose we have data and we would like to know whether they came from some particular completely specified distribution. For example, suppose that we use a computer to simulate n uniform random variables from 0 to 1.[1] Then we expect about one-tenth of them $(n/10)$ to be in each of the intervals $[0.0, 0.1]$, $(0.1, 0.2]$, $\ldots, (0.9.1.0]$. Let

O_i = observed frequency in interval i; i.e., the interval from $(i-1)/10$ to $i/10$, $i = 1, 2, \ldots, 10$

E_i = expected frequency in interval i.

When the observed frequencies differ markedly from the expected frequencies, we should see evidence that the assumption of a UNIF(0,1) distribution is not tenable. Our null hypothesis here is that the data came from the UNIF$(0,1)$ distribution.

■ **Example 13.1**   Use R's uniform random number generator to simulate 100 observations from UNIF$(0,1)$. Identify the observed and expected frequencies and make an informal comparison.

Solution: We can simulate the 100 observations using the R code

```
set.seed(123)
n = 100
u = runif(n,0,1)
```

The set.seed(123) sets the random number seed to a fixed number, 123 in this case. Once this number is set, the simulation can be repeated and we will obtain the same sequence of random numbers. We can use R's cut() function to determine in which interval, among $[0.0, 0.1]$, $(0.1, 0.2]$, $\ldots, (0.9.1.0]$, each observation falls. The summary() command then determines how many of the 100 observations fall in each interval.

```
cut.u = cut( u , breaks=seq(0,1,0.1) )
summary.u = summary( cut.u )
print( summary.u )
```

The result of the last statement is

(0,0.1]	(0.1,0.2]	(0.2,0.3]	(0.3,0.4]	(0.4,0.5]	...	(0.8,0.9]	(0.9,1]
7	12	11	9	14	...	11	7

Thus, $O_1 = 7$, $O_2 = 12, \ldots, O_{10} = 7$. Before taking the sample, the number of simulated values that fall in a particular interval, such as the first one $(0, 0.1)$, has a binomial distribution with $n = 100$ and

$$p = \int_0^{0.1} 1 \, du = 0.1.$$

Thus, the expected frequency for each interval is $np = 100 \times 0.1 = 10$.

Figure 13.1 shows a histogram. The observed frequencies (i.e., counts) are roughly, though not precisely, flat over the interval $[0, 1]$. The expected frequency is 10 for each interval. In this simulated data set, none of the intervals had an observed frequency of exactly 10.   ■ ■ ■

---

[1] Systems such as R can simulate from almost any distribution, but at the heart they generate numbers from a uniform distribution and then transform these in some way to get simulations from a desired distribution. For example, if $U \sim \text{UNIF}(0,1)$, then $Y = -\log U$ has an EXP(1) distribution. See Section 6.9. Nowadays, random number generators from well-established systems, like R, are generally very good at simulating from a desired distribution, such as the uniform. In the "olden days" before these trusted systems were developed, users who used simulation had to test out random number generators. One of the qualities of a good random number generator was that results mimicked the desired distribution. This is the spirit of the example that follows.

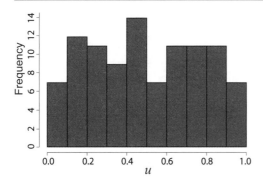

Figure 13.1 Histogram for 100 simulated values from the UNIF$(0,1)$ distribution.

In order to test H_0 that the data come from a UNIF$(0,1)$ distribution against the alternative H_1 that data do not come from a UNIF$(0,1)$ distribution, we must develop a test statistic and determine its distribution, or an approximation to its distribution. Let's focus for a moment on the first interval $(0,0.1)$. The exact distribution for the frequency in this cell is BIN$(100,0.1)$, although for large n the distribution is approximately normal with mean np and variance $np(1-p)$. If the number of cells k is large, then each p_i is small, so the variance for the number in cell i is approximately $np(1-p_i) \approx np_i = E_i$. Using this approximation for cell 1, we have,

$$\frac{O_1 - E_1}{\sqrt{E_1}} \sim N(0,1). \tag{13.1}$$

Since the square of a standard normal is $\chi^2(1)$, and the sum of independent χ^2 random variables has a χ^2 distribution with the sum of the degrees of freedom for the terms, we might jump to the conclusion that if we add up terms, such as those in (13.1), we would get that

$$\sum_{i=1}^{k} \left(\frac{O_i - E_i}{\sqrt{E_i}} \right)^2 = \sum_{i=1}^{k} \frac{(O_i - E_i)^2}{E_i}$$

has an approximate $\chi^2(k)$. This conclusion, though, is based on some faulty reasoning. First, the variance is not exactly equal to np; it is $np(1-p)$. Second, since $\sum_{i=1}^{k} O_i = n$ the $(O_i - E_i)^2/E_i$ terms are not independent. In this case, the faulty reasoning led to an incorrect conclusion. The true distributional result is that $\sum_{i=1}^{k} \frac{(O_i - E_i)^2}{E_i}$ is approximately $\chi^2(k-1)$. This is a difficult result to show in general, so we omit it here. We will, however, show it for the case of $k = 2$ cells. The proof will demonstrate the importance of taking into account the restriction that $p_1 + p_2 = 1$.

Theorem 13.2.1 — Approximate Distribution of $\sum_{i=1}^{k}(O_i - E_i^2)/E_i$. If n observations are independently placed into k cells where the probability of each object falling in cell i is p_i, where $p_1 + p_2 + \cdots + p_k = 1$, and if we let O_i denote the number falling in cell i and $E_i = np_i$, then

$$X^2 = \sum_{i=1}^{k} \frac{(O_i - E_i)^2}{E_i} \overset{\text{approx}}{\sim} \chi^2(k-1). \tag{13.2}$$

Proof. We will prove this only for the case of $k = 2$:

$$X^2 = \sum_{i=1}^{2} \frac{(O_i - E_i)^2}{E_i}$$

$$= \frac{(O_1 - E_1)^2}{E_1} + \frac{(O_2 - E_2)^2}{E_2}$$

$$= \frac{(O_1 - np_1)^2}{np_1} + \frac{((n - O_1) - n(1 - p_1))^2}{n(1 - p_1)}$$

$$= \frac{(O_1 - np_1)^2}{np_1} + \frac{(np_1 - O_1)^2}{n(1 - p_1)}$$

$$= (O_1 - np_1)^2 \left(\frac{1}{np_1} + \frac{1}{n(1 - p_1)} \right)$$

$$= \frac{(O_1 - np_1)^2}{np_1(1 - p_1)}$$

$$= \left(\frac{O_1 - np_1}{\sqrt{np_1(1 - p_1)}} \right)^2 .$$

Note that $O_i \sim \mathrm{BIN}(n, p_1)$ so $\mathbb{E}(O_i) = np_1$ and $\mathbb{V}(O_i) = np_1(1 - p_1)$, so using the normal approximation to the binomial, the last expression above is the square of an approximate standard normal random variable, so $X^2 \overset{\mathrm{approx}}{\sim} \chi^2(1)$.   ∎

In this proof, we see how the restriction $p_2 = 1 - p_1$ is used. In general, the restriction that $p_1 + p_2 + \cdots + p_k = 1$ causes the degrees of freedom for the approximate χ^2 to be the number of cells reduced by one.

■ **Example 13.2** For the data in Example 13.1, apply the χ^2 goodness-of-fit test to test whether R's uniform random number generator actually produces UNIF(0,1) outcomes.

Solution: Our null hypothesis is H_0: the data come from a UNIF(0,1) distribution, with the alternative being H_1: the data do not come from a UNIF(0,1) distribution. Once we run the code to simulate 100 random observations[2] we can compute the X^2 statistic as follows:

```
sum( (summary.u - n/10)^2 / (n/10) )
```

The output from this line is 5.2. With 10 cells (intervals) we have $10 - 1 = 9$ degrees of freedom. Since χ^2 tests are right-tailed tests, we can determine the rejection to be $X^2 > \chi^2_\alpha$. With $\alpha = 0.05$, we can find χ^2_α by the command

```
qchisq( 0.05 , 9 , lower.tail=FALSE )
```

yielding $\chi^2_\alpha = 16.9$. Since our observed test statistic $X^2 = 5.2$ does not exceed 16.9, we have no evidence to reject the null hypothesis that the data have a UNIF(0,1) distribution.   ■ ■ ■

In the previous example the distribution was completely specified – that is, there were no unknown parameters. Often, the distribution family is specified, but the parameter(s) of the distribution are unknown. For example, we may want to test whether the Poisson distribution is the distribution of a count variable, but the mean of the Poisson distribution is unknown. In this case, we first estimate the parameter(s); then, using the estimated parameters, we compute the X^2 statistic from (13.2) for a finite number of intervals. When there is no upper bound on the hypothesized distribution, we take the last category to be unbounded on the right. The degrees of freedom for the χ^2 distribution is then k (the number of cells) minus one (because of the restriction that $p_1 + p_2 + \cdots + p_k = 1$) minus an additional one for each parameter that must be estimated. If one parameter is estimated, then the approximate distribution is $\chi^2(k - 2)$.

---

[2]Since we set the seed at the same value as before, namely `set.seed(123)`, we will get exactly the same sequence of random numbers.

■ **Example 13.3** Electronic integrated circuits are produced from thin wafers that are cut from some material. The wafers produced sometimes have tiny flaws on them that make part of the wafer unusable. Suppose we produce 100 wafers and for each we determine the number of flaws. A summary of the outcomes is:

Number of flaws	0	1	2	3	4	5 or more
Observed frequency	42	22	13	8	6	9

Test whether these data came from a Poisson distribution.

Solution: Since we don't know the mean of the Poisson, we first estimate it. Getting the MLE is a little tricky here because we don't know the number of observations that were equal to 5 or 6 or ...; we only know that there were nine observations that were 5 or more. We will need to use the multinomial distribution where the categories are: 0 defects, 1 defect, 2 defects, 3 defects, 4 defects, and 5 or more defects. The corresponding probabilities are

$$p_0 = \frac{\lambda^0 e^{-\lambda}}{0!} = e^{-\lambda}$$

$$p_1 = \frac{\lambda^1 e^{-\lambda}}{1!} = \lambda e^{-\lambda}$$

$$p_2 = \frac{\lambda^2 e^{-\lambda}}{2!} = \frac{\lambda^2 e^{-\lambda}}{2}$$

$$p_3 = \frac{\lambda^3 e^{-\lambda}}{3!} = \frac{\lambda^3 e^{-\lambda}}{6}$$

$$p_4 = \frac{\lambda^4 e^{-\lambda}}{4!} = \frac{\lambda^4 e^{-\lambda}}{24}$$

$$p_5 = 1 - e^{-\lambda} - \lambda e^{-\lambda} - \frac{\lambda^2 e^{-\lambda}}{2} - \frac{\lambda^3 e^{-\lambda}}{6} - \frac{\lambda^4 e^{-\lambda}}{24}.$$

The following R code computes and plots the log likelihood

```
log.likelihood = function(lambda,O)
{
  p0 = exp(-lambda)
  p1 = lambda*exp(-lambda)
  p2 = lambda^2*exp(-lambda)/2
  p3 = lambda^3*exp(-lambda)/6
  p4 = lambda^4*exp(-lambda)/24
  p5 = 1 - p0 - p1 - p2 - p3 - p4
  z = O[1]*log(p0) + O[2]*log(p1) + O[3]*log(p2) + O[4]*log(p3) +
      O[5]*log(p4) + O[6]*log(p5)
  return(z)
}
```

Note that we have defined p0, p1, p2, etc., to be the probabilities for observing 0, 1, 2, etc. defects on a wafer, but O[1], O[2], O[3], etc. to denote the actual number of defects, so there is a bit of a disconnect in the notation. Ideally, we would use O[0] to represent the observed frequency for wafers with zero defects, but R does not allow 0 as an index.[3]

---
[3]Some languages, such as C and C++, use 0 to index the "first" element of an array, but R does not allow this.

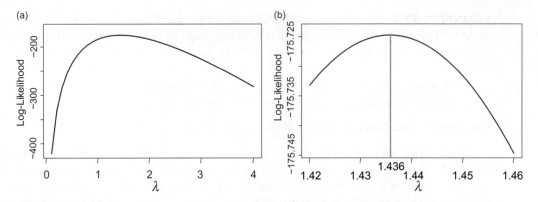

Figure 13.2 Log-likelihood function for data in Example 13.3. The graph in (b) shows a "zoomed in" version of the log likelihood function where the maximum occurs near $\hat{\lambda} = 1.436$.

Figure 13.2(a) shows graphs of the log likelihood function over the interval (0,4), and Figure 13.2(b) shows this over (1.420,1.460). By zooming in a little more we find that the maximum occurs at approximately 1.436. The MLE is therefore $\hat{\lambda} = 1.436$.

The estimated probabilities for cells 0, 1, 2, 3, 4, and 5, along with the corresponding expected frequencies, are then found using the R code

```
lambdahat = 1.436
p0hat = exp(-lambdahat)
p1hat = lambdahat*exp(-lambdahat)
p2hat = lambdahat^2*exp(-lambdahat)/2
p3hat = lambdahat^3*exp(-lambdahat)/6
p4hat = lambdahat^4*exp(-lambdahat)/24
p5hat = 1 - p0hat - p1hat - p2hat - p3hat - p4hat
phat.vec = c(p0hat,p1hat,p2hat,p3hat,p4hat,p5hat)
print( phat.vec )
E = 100*phat.vec
print( phat.vec )
print( E )
```

This results in the following output:

```
[1] 0.23787737 0.34159190 0.24526298 0.11739922 0.04214632 0.01572221
[1] 23.787737 34.159190 24.526298 11.739922  4.214632  1.572221
```

We summarize the observed and expected frequencies in Table 13.1. The same data are summarized in Figure 13.3, where we see that the Poisson distribution would predict fewer observations at the endpoints (0 defects and 5+ defects). Thus, there seems to be more variability than the Poisson distribution can provide. Recall that the mean and variance are the same for the Poisson distribution. Here we see the phenomenon of *overdispersion*, where the variance is greater than the mean. But is this discrepancy more than would be expected by chance? In other words, are the results statistically significant?

Table 13.1 also shows the terms that make up the X^2 statistic. We see the large discrepancies at 0 defects and at 5+ defects. The sum is

$$X^2 = \sum_{i=0}^{5} \frac{(O_i - E_i)^2}{E_i} \approx 63.728.$$

Table 13.1 Observed and expected frequencies for wafer data.

Number of defects	Observed frequency	Expected frequency	$\dfrac{(O-E)^2}{E}$
0	42	23.787737	13.944
1	22	34.159190	4.328
2	13	24.526298	5.417
3	8	11.739922	1.191
4	6	4.214632	0.756
5+	9	1.572221	35.092
Sum	100	100	63.728

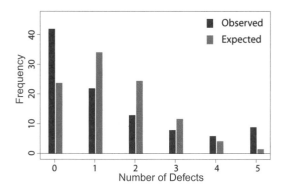

Figure 13.3 Observed frequencies (blue) and estimated frequencies (red) assuming a Poisson distribution.

The distribution of the test statistic X^2 is $\chi^2(6-1-1) = \chi^2(4)$. We subtract 1 because the sum of the O_i must be equal to 100, and we must subtract an additional 1 because we estimated one parameter in the distribution. We can find the critical value using qchisq(0.05,4,lower.tail=FALSE). The result is 9.488, so we would reject the null hypothesis of a Poisson distribution if $X^2 > 9.488$. Our observed X^2 statistic of 63.728 is well beyond 9.488. The P-value is obtained from the R command pchisq(63.728,4,lower.tail=FALSE), which works out to be 4.8×10^{-13}. There is strong evidence that the data do not come from a Poisson distribution. ∎∎∎

13.3 Contingency Tables: Testing Independence

In this section we will look again at a goodness-of-fit test, but this time there is some additional structure to the cells. We will consider a table giving the cross tabs of two discrete (nominal or ordered categorical) variables. Suppose one variable (the row variable) has r categories, and the other (the column variable) has c categories. We assume that each subject or unit falls in exactly one of the rc cells. The question we address is whether the two variables are independent, against the alternative that they are not independent.

Data for Progress[4] is an organization that routinely conducts surveys across time. Recently, one of their emphases has been on the COVID pandemic and the government response. One question they asked was whether the person approves of the CDC's handling of the COVID pandemic. The results from February 2, 2021 are given in Table 13.2 (which is reproduced here from Table 12.1). We analyzed these data in Example 12.4 by comparing the proportions of men and women who approved (strongly or somewhat) versus those who did not approve (strongly or somewhat disapproved, or not sure). In other

---

[4]See www.dataforprogress.org. They describe themselves as "a multidisciplinary group of experts using state-of-the-art techniques in data science to support progressive activists and causes."

Table 13.2 Probabilities for the contingency table of the CDC's handling of the coronavirus pandemic.

Handling of coronavirus: CDC	Male	Female	Total
Strongly approve	116	162	278
Somewhat approve	220	244	464
Somewhat disapprove	118	92	210
Strongly disapprove	74	68	142
Not sure	22	46	68
Total	550	612	1,162

words, we collapsed some of the categories so that there were only two possible responses: approve and not approve. Here we will take a different approach and keep all five responses.

We will test whether sex (male versus female) is independent of the opinion of approval/non-approval of the CDC's handling of the pandemic. To make the problem more precise, let's introduce some notation:

p_{ij} = probability of falling in cell (i, j); that is, the row variable is i and the column variable is j

$p_{i.}$ = probability that the row variable is category i

$p_{.j}$ = probability that the column variable is category j.

If the row variable and column variable are independent, then

$$p_{ij} = p_{i.}p_{.j}, \quad \text{for all } i, j.$$

Similarly, we define

O_{ij} = observed frequency in cell (i, j)

r_i = observed frequency in row i

c_j = observed frequency in column j.

The notation is illustrated for the CDC approval data in Tables 13.3 and 13.4.

Table 13.3 Probabilities for all categories for those who approve of the CDC's handling of the coronavirus pandemic.

Handling of coronavirus: CDC	Male	Female	Total
Strongly approve	p_{11}	p_{12}	$p_{1.}$
Somewhat approve	p_{21}	p_{22}	$p_{2.}$
Somewhat disapprove	p_{31}	p_{32}	$p_{3.}$
Strongly disapprove	p_{41}	p_{42}	$p_{4.}$
Not sure	p_{51}	p_{52}	$p_{5.}$
Total	$p_{.1}$	$p_{.2}$	n

Table 13.4 Counts of men and women who approve of the CDC's handling of the coronavirus pandemic.

Handling of coronavirus: CDC	Male	Female	Total
Strongly approve	O_{11}	O_{12}	r_1
Somewhat approve	O_{21}	O_{22}	r_2
Somewhat disapprove	O_{31}	O_{32}	r_3
Strongly disapprove	O_{41}	O_{42}	r_4
Not sure	O_{51}	O_{52}	r_5
Total	c_1	c_2	n

In a realistic problem, we don't know $p_{ij}, p_{i.}$, and $p_{.j}$, so these must be estimated. The MLE of the probability of some event is just the fraction of times that event occurred in a random sample. Applying this rule successively to the individual cells, the row margins, and the column margins gives

$$\hat{p}_{ij} = \frac{O_{ij}}{n}$$

$$\hat{p}_{i.} = \frac{r_i}{n}$$

$$\hat{p}_{.j} = \frac{c_j}{n}.$$

The hypotheses we are testing are

H_0: the row and column variable are independent   vs. H_1: the row and column variable

are not independent.

Under the assumption that H_0 is true, $p_{ij} = p_{i.} \, p_{.j}$, so our estimate for the expected count in cell (i, j) is

$$E_{ij} = \widehat{\mathbb{E}(O_{ij})} = n\hat{p}_{i.} \times \hat{p}_{.j} = n \left(\frac{r_i}{n}\right) \left(\frac{c_j}{n}\right) = \frac{r_i c_j}{n}. \tag{13.3}$$

This allows us to obtain an expected count for cell (i, j) under the assumption that H_0 is true. The χ^2 test proceeds by computing (just as in Section 13.2) the sum of the observed minus expected squared over expected:

$$X^2 = \sum_{i=1}^{r} \sum_{j=1}^{c} \frac{(O_{ij} - E_{ij})^2}{E_{ij}}. \tag{13.4}$$

This result has an approximate χ^2 distribution. The only issue left to settle is the degrees of freedom. As a general principle we begin with rc degrees of freedom (because that is how many cells there are) and subtract 1 for each parameter estimated. We really have to estimate only the marginal probabilities $p_{1.}, p_{2.}, \ldots, p_{r.}$ and $p_{.1}, p_{.2}, \ldots, p_{.r}$. However, each of these sets of probabilities must add to 1, so there are really $(r-1)$ of the row probabilities to estimate, and similarly, $(c-1)$ of the column probabilities to estimate. There is also the restriction that all of the probabilities must sum to 1. Thus, the degrees of freedom for this χ^2 test is

$$\text{df} = rc - (r-1) - (c-1) - 1 = rc - r - c + 1 = (r-1)(c-1).$$

We would reject H_0 for $X^2 > \chi_\alpha^2((r-1)(c-1))$.

■ **Example 13.4** For the data in Table 13.2, test whether sex and opinion on the CDC handling of the COVID pandemic are independent.

Solution: The expected frequencies obtained from (13.3) are shown in parentheses in Table 13.5, along with the observed frequencies. The test statistic is therefore

$$X^2 = \frac{(116 - 131.58)^2}{131.58} + \frac{(162 - 146.42)^2}{146.42} + \cdots + \frac{(46 - 35.81)^2}{35.81} = 17.54.$$

We used the following R code to compute the expected frequencies and the sum of the observed minus expected squared over expected:

Table 13.5 Survey results on the CDC handling of the COVID pandemic: observed and expected (in parentheses) frequencies

Handling of coronavirus: CDC	Male	Female	Total
Strongly approve	116 (131.58)	162 (146.42)	278
Somewhat approve	220 (219.62)	244 (244.38)	464
Somewhat disapprove	118 (99.40)	92 (110.60)	210
Strongly disapprove	74 (67.21)	68 (74.79)	142
Not sure	22 (32.19)	46 (35.81)	68
Total	550	612	1,162

```
O = matrix( c( 116 , 162 ,
               220 , 244 ,
               118 ,  92 ,
                74 ,  68 ,
                22 ,  46 ) ,
            nrow=5 , byrow=TRUE )
c.margin = apply( O , 2 , sum )
r.margin = apply( O , 1 , sum )
E = round( outer( r.margin , c.margin )/1162 , 2 )
sum( (O-E)^2/E )
```

The degrees of freedom is $(5-1)(2-1)=4$. With $\alpha=0.05$, the rejection region is $X^2 > 9.49$. Since $17.54 > 9.49$, we reject the null hypothesis of independence. Thus, sex and the opinion of the CDC's handling of the coronavirus pandemic are not independent. The P-value for this test is computed from

```
pchisq( 17.54 , 4 , lower.tail=FALSE )
```

which gives the P-value as 0.00152; this is strong evidence of nonindependence. ∎∎∎

If the contingency table is contained in a matrix or a data frame, then the χ^2 can be done in R using the chisq.test() command. In this case, once we define the observation matrix O, we can enter

```
chisq.test( O )
```

This yields the output

```
        Pearson's Chi-squared test

data:  O
X-squared = 17.538, df = 4, p-value = 0.001519
```

In the example described above, the data have already been summarized into a matrix. The raw data set consists of $n=1,162$ rows, each corresponding to one person. For example, we could have the data frame whose first few rows are

```
         HandlingByCDC Gender
1       Strongly Approve Female
2 Somewhat Disapprove Female
3      Somewhat Approve   Male
4              Not Sure   Male
5       Strongly Approve   Male
```

If this is the case, we can first apply the `table()` command to create a two-dimensional table, on which we can then apply the `chisq.test()` function. Usually, we would want the same ordering of the levels as in the original table. The default in R is to present them in alphabetical order, so `"Not Sure"` would come first. If we want to assure the same order, we can redefine the variables in a data frame using the `factor()` command and explicitly spell out the labels in the order we want. For example, if the data frame is called `tall` and has columns `HandlingByCDC` and `Gender`, we could enter the following to redefine the variables as factor variables with the desired ordering of factor levels:

```
read.csv(tall)
tall$Gender = factor( tall$Gender , levels=c( "Male" , "Female" ) )
tall$HandlingByCDC = factor( tall$HandlingByCDC ,
levels = c("Strongly Approve" , "Somewhat Approve" ,
         "Somewhat Disapprove" , "Strongly Disapprove" , "Not Sure" ) )
```

We could then apply the `table()` command

```
table( tall )
```

to obtain

```
                     Gender
HandlingByCDC       Male Female
Strongly Approve     116    162
Somewhat Approve     220    244
Somewhat Disapprove  118     92
Strongly Disapprove   74     68
Not Sure              22     46
```

Finally, we would apply the `chisq.test()` command to obtain a summary of the χ^2 test:

```
chisq.test( table( tall ) )
```

We can visually present the result of a contingency table using a stacked bar chart. For example, Figure 13.4 shows the results of stacking the `HandlingByCDC` variable for the levels of gender. We can

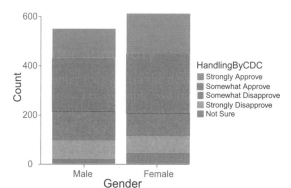

Figure 13.4 Stacked bar chart showing various opinions regarding the CDC's handling of the coronavirus pandemic by gender.

see the biggest discrepancy is that a much higher proportion of women than men responded "Strongly Agree".

This figure was produced using the `ggplot2` library and the following commands:

```
library(ggplot2)
ggplot(tall) +
  aes(x = Gender, fill = HandlingByCDC) +
  geom_bar()
```

We could produce a similar plot by interchanging the roles of `Gender` and `HandlingByCDC`, as given in Figure 13.5.

The stacked bar chart shows the totals as heights of individual bars. If one group has more observations than the other, then the total bar heights will be different, making assessments of the relative proportions difficult. The *mosaic plot* is an alternative to the stacked bar chart. The size of a group is handled by adjusting the width and the height of the bars. The combined height of all the bars is the same for both males and females. The width of the bars is adjusted to indicate the numbers of males and females.

The code to create the mosaic plot in Figure 13.6 is

```
mosaicplot( t(table(tall)) ,  las=1 , main="" ,
            xlab = "Gender",
            ylab = "CDC Handling of COVID" ,
            color = c("blue","purple","magenta","red","gray") )
```

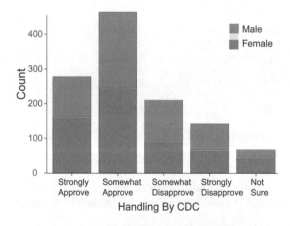

Figure 13.5 Stacked bar chart showing breakdown by gender for each level of `HandlingByCDC`.

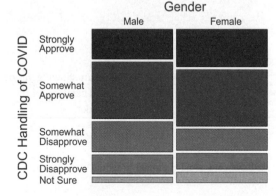

Figure 13.6 Mosaic chart showing breakdown by gender for each level of `HandlingByCDC`.

Here, $t(0)$ gives the transpose of the observation matrix O. The las=1 command adjusts the orientation of the text so that it is horizontal; otherwise the labels for CDC Handling of COVID would be vertical, and several labels would run together. In Figure 13.6 we can see that the Female bars are slightly wider because a few more females (612) were included in the survey than males (550).

The distribution of the X^2 statistic is approximately $\chi^2\big((r-1)(c-1)\big)$. This is really an asymptotic result, with the distribution of X^2 converging to the $\chi^2\big((r-1)(c-1)\big)$ distribution as $n \to \infty$. As a rule, the larger the n, the better the approximation. The advice often given about whether the χ^2 approximation can be adequate is this: (1) no cell should have an expected frequency less than 1; and (2) no more than 20% of the cells should have an observed value that is less than 5. Others say the expected frequency should be at least five for all cells. Of course, there is no magical cutoff over which the χ^2 approximation is good. The χ^2 approximation gets progressively better as n gets larger. In Example 13.5 the conditions are clearly met. We will address what to do when these conditions are not met in Sections 13.5 and 13.6.

Finally, we point out again that the χ^2 test assumes that both variables are random, and each subject or unit will fall into exactly one of the rc cells in the table. This assumption does not hold for all contingency tables. In the next section we address one case where this assumption fails.

13.4 Contingency Tables: Homogeneity

Suppose now that the sampling strategy was to select a fixed number of women and men, and then for each subject (male or female) we ask the person's opinion on some question. In this case, it is not true that the two variables are random. The response to the question is random, but gender is not. The margins for gender have been fixed in advance. This phenomenon where one set of marginal totals is fixed is common. For example, in clinical trials it is common to assign people at random to one of two or more treatments. Often, the researchers attempt to assign equal numbers to each treatment, but because some people drop out of the study, it is rare to see clinical trials with exactly the same numbers for each group.

There have been a number of clinical trials for treatments of COVID-19. The one below is hypothetical, but it is similar to many of the trials that have been conducted. Suppose that subjects are assigned to one of two treatments: an experimental drug and a placebo.[5] Suppose that 650 people in the experimental group and 647 in the placebo group complete the study. These margins are now considered fixed. The outcome was whether the person was (1) not hospitalized and survived, (2) hospitalized but survived, and (3) died. We assume that the data are as shown in Table 13.6.

Since the number of people assigned to each treatment is no longer random, it makes no sense to test for independence. Instead, we test whether the probability of being in each of the outcome groups is the same across all treatments. Table 13.7 shows the probabilities for the clinical trial with two treatments and three possible outcomes. The row totals are assumed to be fixed. In this case, the null hypothesis is

H_0: $p_{11} = p_{21}$, $p_{12} = p_{22}$, $p_{31} = p_{32}$.

Table 13.6 Results of hypothetical clinical trial testing for an experimental drug against a placebo.

	Outcome			
	Not hospitalized	Hospitalized	Died	Total
Experimental drug	572	67	11	650
Placebo	543	81	23	647
Total	1,115	148	34	1,297

[5]Often those who receive the placebo will also receive the standard care, that is, the best available treatment.

Table 13.7 Probabilities of a subject being in each column, given the treatment (experimental drug or placebo). Note that the row sums are 1; the column sums need not be 1.

	Outcome			
	Not hospitalized	Hospitalized	Died	Total
Experimental drug	p_{11}	p_{12}	p_{13}	$p_{11} + p_{12} + p_{13} = 1$
Placebo	p_{21}	p_{22}	p_{23}	$p_{21} + p_{22} + p_{23} = 1$

In other words, the probability that someone falls in the not-hospitalized group is the same for both the experimental drug and the placebo, and the probability of being in the hospitalized group is the same for both groups, and the probability of being in the group who died is the same for both groups. In general, the null hypothesis is

$$H_0 \colon p_{1j} = p_{2j} = \cdots = p_{rj}, \text{ for all } j.$$

We are testing whether the probabilities are the same (homogeneous) across all treatments. Such a test is called a test of *homogeneity*.

Given r_1, the row 1 total, the number of patients who fall in the not-hospitalized group has a $\mathrm{BIN}(r_1, p_{11})$ distribution, so the expected count is $r_1 p_{11}$. In general, given the row total r_i, the expected count in cell (i, j) is $r_i p_{ij}$, which under H_0 is estimated by

$$E_{ij} = \widehat{\mathbb{E}(O_{ij})} = r_i \hat{p}_{ij} = r_i \frac{c_j}{n} = \frac{r_i c_j}{n}.$$

Under H_0, $p_{1j} = p_{2j} = \cdots = p_{rj}$, so we would estimate any one of these by the column total, divided by the total across all cells in the contingency table. The expected frequencies are the *same* as in the test of independence described in Section 13.3, and the test statistic is also the same:

$$X^2 = \sum_{i=1}^{r} \sum_{j=1}^{c} \frac{(O_{ij} - E_{ij})^2}{E_{ij}}. \tag{13.5}$$

The degrees of freedom is also the same as in the case of a test for independence. There are rc probabilities in all. There are r constraints on the row sums (since the row totals are fixed). In addition we must estimate $c - 1$ column probabilities. The degrees of freedom is therefore

$$\mathrm{df} = rc - r - (c - 1) = rc - r - c + 1 = (r - 1)(c - 1).$$

The mechanics for the χ^2 test for independence (Section 13.3) and the test for homogeneity (this section) are identical. By this we mean that the test statistics X^2 are computed using the same formula, and the approximate distribution of the test statistic is the same $\chi^2\big((r-1)(c-1)\big)$. Thus, R has a single function chisq.test() that performs the test. When interpreting the results, though, it is important to keep straight whether we are testing the independence of two random variables or the homogeneity of probabilities across groups.

■ **Example 13.5** For the hypothetical clinical trial whose outcomes are given in Table 13.6, test whether the probabilities for the various outcomes are the same for the experimental drug and for the placebo.

Solution: Table 13.8 shows the expected frequencies for the clinical trial data. If we apply the chisq.test() function in R, we obtain the following output:

```
              Pearson's Chi-squared test

    data:  O
    X-squared = 6.307, df = 2, p-value = 0.0427
```

Table 13.8 Observed and expected results of hypothetical clinical trial testing for an experimental drug against a placebo.

	Outcome			
	Not hospitalized	Hospitalized	Died	Total
Experimental drug	572	67	11	650
	(623.71)	(82.79)	(19.02)	
Placebo	543	81	23	647
	(620.83)	(82.41)	(18.93)	
Total	1,115	148	34	1,297

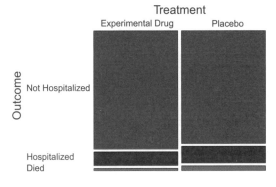

Figure 13.7 Mosaic plot for outcomes (not hospitalized, hospitalized, died) for the experimental drug and the placebo.

The P-value for this test is 0.0427, so there is some evidence that the outcome probabilities are not the same for the placebo and the experimental drug. A mosaic plot for these data, Figure 13.7, shows that more people died in the placebo group, and also more people in the experimental drug group were not hospitalized.

∎ ∎ ∎

13.5 Fisher's Exact Test

The tests of independence and homogeneity described above use the X^2 statistic. The distribution of X^2 converges to the χ^2 distribution as the sample size n in the case of a test of independence, or the fixed row (or column) totals go to infinity. For small sample sizes the χ^2 approximation may not be very good. For a 2×2 contingency, we can apply the Fisher exact test (Fisher, 1935). He describes the experiment this way:

> A lady declares that by tasting a cup of tea made with milk she can discriminate whether the milk or the tea infusion was first added to the cup. We will consider the problem of designing an experiment by means of which this assertion can be tested ... Our experiment consists in mixing eight cups of tea, four in one way and four in the other, and presenting them to the subject for judgment in a random order ... Her task is to divide the 8 cups into two sets of 4, agreeing, if possible, with the treatments received.

There are four cups made with tea poured first and four with milk poured first. The lady selects the four for which she believes the tea was poured first; the other four are the ones where she believes the milk was poured first. The results of the experiment can be summarized in a 2×2 table, as illustrated in Table 13.9, where the lady correctly identified all "tea-first" cups.

Note that the number of cups in the $(1, 1)$ cell (tea poured first and lady identified as tea poured first) is sufficient to determine all other numbers in the table. The row totals are fixed at four because four cups

Table 13.9 The lady tasting tea selects all four "tea-first" cups correctly.

| | Lady's guess | | |
	Tea	Milk	Total
Tea	4	0	4
Milk	0	4	4
Total	4	4	8

Table 13.10 All five possible 2 × 2 tables for the tea-tasting experiment.

| | Lady's guess | | |
	Tea	Milk	Total
Tea	4	0	4
Milk	0	4	4
Total	4	4	8
	$X = 4$		

| | Lady's guess | | |
	Tea	Milk	Total
Tea	3	1	4
Milk	1	3	4
Total	4	4	8
	$X = 3$		

| | Lady's guess | | |
	Tea	Milk	Total
Tea	2	2	4
Milk	2	2	4
Total	4	4	8
	$X = 2$		

| | Lady's guess | | |
	Tea	Milk	Total
Tea	1	3	4
Milk	3	1	4
Total	4	4	8
	$X = 1$		

| | Lady's guess | | |
	Tea	Milk	Total
Tea	0	4	4
Milk	4	0	4
Total	4	4	8
	$X = 0$		

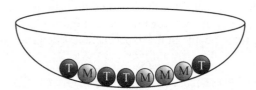

Figure 13.8 Bowl with eight balls, each corresponding to one cup of tea. Those marked with "T" are those with tea poured first, and those marked with "M" are those with milk poured first.

were prepared with each method. The column totals are fixed because the lady must select exactly four tea-first cups. Let's define the random variable X to be the number of tea-first cups correctly identified – that is, the (1,1) entry in the table. The null hypothesis is that the lady has no ability to discriminate between tea-first and milk-first, and therefore is just guessing. There are then five possible tables. We list these in Table 13.10.

The obvious test statistic is X, the number of tea-first cups identified. We must then determine the correct distribution for X. We can envision the problem as one with a bowl containing eight balls, each one corresponding to one cup of tea prepared in a specific way (Figure 13.8). Four of the balls are marked with "T," indicating tea first, and the other four are marked with "M," indicating milk first. Correctly identifying all four cups correctly is like selecting four balls, all of which are marked "T." The probability of this is

$$\Pr(X = 4) = \Pr(\text{selecting all four "T" balls in four draws}) = \frac{\binom{4}{4} \times \binom{4}{0}}{\binom{8}{4}} = \frac{1}{70}.$$

The numerator is the number of ways of selecting four "T" balls and zero "M" balls; the denominator is the number of ways of selecting four balls out of eight. Since this works out to 1/70, selecting all four correctly is a rather unlikely event if the lady is guessing. In this case we might agree that she is able to tell the difference.

But what if she identifies only three correctly? The table would now look like the second one in Table 13.10. How likely is this event?

$$\Pr(X = 3) = \frac{\binom{4}{3} \times \binom{4}{1}}{\binom{8}{4}} = \frac{16}{70}.$$

We see that the distribution of X is the hypergeometric distribution, described in Section 5.5. Thus,

$$\Pr(X = 2) = \frac{\binom{4}{2} \times \binom{4}{2}}{\binom{8}{4}} = \frac{36}{70},$$

$$\Pr(X = 1) = \frac{\binom{4}{1} \times \binom{4}{3}}{\binom{8}{4}} = \frac{16}{70},$$

and

$$\Pr(X = 0) = \frac{\binom{4}{0} \times \binom{4}{4}}{\binom{8}{4}} = \frac{1}{70}.$$

The PMF for X is therefore

x	0	1	2	3	4
$P(X = x)$	$\frac{1}{70}$	$\frac{16}{70}$	$\frac{36}{70}$	$\frac{16}{70}$	$\frac{1}{70}$

We see that getting all four correct is rather unlikely under the null hypothesis, but getting exactly three correct is not so unlikely. If the lady had guessed three correctly, our P-value would be $\frac{1}{70} + \frac{16}{70} \approx 0.2429$, which would not lead to rejection of H_0. Thus, only $X = 4$, that is, correctly identifying all four tea-first cups, would lead us to conclude that the lady can discriminate between the two methods.

What we have described here is really a one-sided test. We look for departures from guessing in only one direction – the direction of being able to *correctly* identify the cups. What would we think if we observed $X = 0$? In this case, the woman would select all tea-first cups and identify them as milk-first, or vice versa. In this case we might infer that the woman has some ability to discriminate, but she has the rule backward. Some practical testing problems are one-sided and some are two-sided. This is a subject matter decision, not a statistical decision. Had this been a two-sided test where we observed $X = 0$ or $X = 4$, our P-value would have been

$$P = \frac{1}{70} + \frac{1}{70} = \frac{2}{70} \approx 0.029,$$

which would lead us to reject the null hypothesis at $\alpha = 0.05$.[6] For the case where the row (and column) totals are the same, four in this example, the two-sided P-value is simply double the one-sided P-value. When the row totals differ, the problem is a little more complicated, and we address this later in this section.

The one-sided Fisher's exact test for a 2×2 table is conducted as follows:

1. Create a 2×2 contingency table from the raw data.
2. Find the probability distribution for X, the count for the upper left corner of the table.
3. Determine what values of X are more extreme than that observed.
4. Compute the P-value by summing the probabilities of all outcomes for X that are as extreme or more extreme than that observed.

R has a built-in function for the Fisher exact test called `fisher.test()`. We must provide a matrix or a data frame for the contingency table. For a one-sided test we use the option `alternative="greater"`. For the case where the lady selected three cups correctly, the R code would be:

```
O = matrix( c( 3 , 1 ,
               1 , 3 ) , nrow=2 , ncol=2 , byrow=TRUE )
fisher.test( O , alternative="greater" )
```

The output from the `fisher.test()` is

```
            Fisher's Exact Test for Count Data

data:  O
p-value = 0.2429
alternative hypothesis: true odds ratio is greater than 1
95 percent confidence interval:
   0.3135693        Inf
sample estimates:
odds ratio
      6.408309
```

The P-value from `fisher.test()` agrees with the value we calculated directly.

■ **Example 13.6** Suppose that the lady had been presented with 16 cups, 8 of which were prepared with tea first and 8 with milk first. Suppose she selects $X = 7$ of the tea-first cups correctly. Is this evidence that she can tell the tea-first cups from the milk-first cups? What is the P-value?

Solution: In this case it is as if we were selecting eight balls from the bowl containing eight marked "T" and eight marked "M" (Figure 13.9). The PMF for the number of correct tea-first selections is

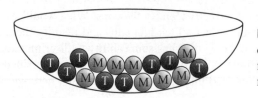

Figure 13.9 Bowl with 16 balls, each corresponding to one cup of tea. Those marked with "T" are those with tea poured first, and those marked with "M" are those with milk poured first.

---

[6]Many practitioners recommend against the use of selecting α beforehand; instead, they suggest letting the P-value stand as evidence against the null hypothesis. Smaller values of the P-value give us stronger evidence against the null.

$$\Pr(X = x) \;=\; \frac{\binom{8}{x} \times \binom{8}{8-x}}{\binom{16}{8}}, \qquad x = 0, 1, 2, 3, 4, 5, 6, 7, 8.$$

The PMF for the statistic X is computed using the R code

```
pmfX = dhyper( 0:8 , 8 , 8 , 8 )
sum( pmfX )
round(pmfX,4)
```

The arguments to `dhyper()` are the outcomes for which we would like the probabilities (0 through 8), the number of balls of one type ("T," which is 8), the number of balls of the other type ("M," which is also 8), and the number of balls selected from the bowl (8). The output is

```
0.0001 0.0050 0.0609 0.2437 0.3807 0.2437 0.0609 0.0050 0.0001
```

Since we observed $X = 7$, the P-value is $0.0050 + 0.0001 = 0.0051$, giving us fairly strong evidence that the lady can indeed distinguish the tea-first and the milk-first cups. ∎ ∎ ∎

If the alternative hypothesis were two-sided, we would have to consider $X = 0$ and $X = 1$ to be as extreme or more extreme. Because of the symmetry in the problem, eight of each kind and eight being selected from the bowl, we would compute the P-value as

$$\begin{aligned} P\text{-value} \;&=\; \Pr(X = 0) + \Pr(X = 1) + \Pr(X = 7) + \Pr(X = 8) \\ &=\; 0.0001 + 0.0050 + 0.0050 + 0.0001 \;=\; 0.0102. \end{aligned}$$

In other words, we would simply double the one-sided P-value.

The problem is more complicated when there is no symmetry. Suppose, for example, that there are 20 cups, 12 prepared with tea first and 8 prepared with milk first. Twelve cups (or balls, using the bowl/ball analogy) are then selected by the lady to be the 12 prepared with tea first (see Figure 13.10). Now the distribution of X, the number of correct tea-first cups, is not symmetric. The PMF is

$$\Pr(X = x) \;=\; \frac{\binom{12}{x} \times \binom{8}{12-x}}{\binom{20}{12}}, \qquad x = 4, 5, \ldots, 12.$$

The fewest number of correct tea-first cups that the lady could get is 4, since she must choose 12 cups; even if all 8 of the milk-first cups are selected, the remaining four must be tea-first cups. The PMF for X is then found to be

4	5	6	7	8	9	10	11	12
0.0039	0.0503	0.2054	0.3521	0.2751	0.0978	0.0147	0.0008	0.0000

Figure 13.10 Bowl with 20 balls, each corresponding to one cup of tea. Those marked with "T" are those with tea poured first, and those marked with "M" are those with milk poured first.

If we were doing a one-sided test, and observed $X = 10$ correct tea-first cups, then our P-value would be P-value $= 0.0147 + 0.0008 + 0.0000 = 0.0155$. But what would be our P-value if we had a two-sided alternative? What does "as extreme or more extreme" mean in a distribution like this that is not symmetric? There are various was to interpret this, but the one that R's `fisher.test()` uses is this: any outcome with a smaller probability than the one observed is "as extreme or more extreme." Thus, if we observed $X = 10$, we would interpret the outcomes $X = 4$ to be "as extreme or more extreme than $X = 8$, but no other outcome in the left tail would be so considered. Thus, the P-value is

$$P = \Pr(X = 4) + \Pr(X = 10) + \Pr(X = 11) + \Pr(X = 12)$$

$$\approx 0.0039 + 0.0147 + 0.0008 + 0.0000 = 0.0194.$$

If we use `fisher.test()` on the matrix with $X = 10$ using the commands

```
O = matrix( c( 10 , 2 ,
               2 , 6 ) , nrow=2 , ncol=2 , byrow=TRUE )
fisher.test( O )
```

we obtain the output

```
           Fisher's Exact Test for Count Data

data:  O
p-value = 0.01937
alternative hypothesis: true odds ratio is not equal to 1
95 percent confidence interval:
   1.191683 228.451626
sample estimates:
odds ratio
   12.4628
```

Fisher's exact test can be applied to any 2×2 contingency table. It is often applied when the counts are uniformly low.

■ **Example 13.7** *Clostridioides difficile* (*C. difficile*) is a bacterial infection of the colon that often causes abdominal pain and watery diarrhea, and can lead to serious damage to the colon. Van Nood *et al.* (2013) describe a study involving two treatments for *C. difficile*. One is to use the antibiotic vancomycin; the other is to transplant healthy feces from a donor. The results are summarized in Table 13.11. Test whether the two treatments are equally effective in curing *C. difficile*.

Solution: The following R commands define the table in matrix form and then apply `fisher.test()`:

```
C.difficile = matrix( c(  13 , 3 ,
                          4 , 9 ) , nrow=2 , ncol=2 , byrow=TRUE )
fisher.test( C.difficile )
```

Table 13.11 Clinical trial comparing the antibiotic vancomycin and a fecal transplant for treating *C. difficile*.

		Cured		
		Yes	No	Total
Treatment	Fecal transplant	13	3	16
	Vancomycin	4	9	13
	Total	17	12	29

This produces the output:

```
              Fisher's Exact Test for Count Data

data:  C.difficile
p-value = 0.00953
alternative hypothesis: true odds ratio is not equal to 1
95 percent confidence interval:
   1.373866 78.811505
sample estimates:
odds ratio
   8.848725
```

The P-value is therefore 0.00953, providing strong evidence that the effectiveness is not the same for the two treatments. A one-sided test would yield a P-value of 0.008401. ∎ ∎ ∎

Exercise 13.1 Verify that the one-sided test yields a P-value of 0.008401.

13.6 **The Continuity Correction and Simulation**

When the sample sizes are not large enough to justify the χ^2 approximation, there are two additional techniques to consider: the continuity correction and the use of simulation.

For 2×2 contingency tables, Yates (1934) suggested the *continuity correction*, whereby we subtract 0.5 from the absolute difference $|O_{ij} - E_{ij}|$ before squaring. With this correction, the test statistic is

$$X^2 = \sum_{i=1}^{2} \sum_{j=1}^{2} \frac{\left(|O_{ij} - E_{ij}| - 0.5\right)^2}{E_{ij}}.$$

This is only applicable for 2×2 tables and it reduces the value of X^2 compared to the usual X^2 statistic; thus, the P-value is larger. In R, the default is to apply the continuity correction for 2×2 tables. If the user does not want to apply the continuity correction, the argument `correction=FALSE` must be used.

■ **Example 13.8** Apply Yates' continuity correction to the fecal transplant data from Example 13.7.

Solution: Table 13.12 shows the observed and expected (in parentheses) counts. If we apply the continuity correction, the X^2 statistic is

$$X^2 = \frac{(|4-7.6207|-0.5)^2}{7.6207} + \frac{(|9-5.3793|-0.5)^2}{5.3793} + \frac{(|13-9.3793|-0.5)^2}{9.3793} + \frac{(|3-6.6207|-0.5)^2}{6.6207}$$
$$= 5.5976.$$

Table 13.12 Fecal transplant data showing observed counts and expected counts in parentheses.

		Cured Yes	Cured No	Total
	Fecal transplant	13	3	16
		(9.3793)	(6.6207)	
Treatment	Vancomycin	4	9	13
		(7.6207)	(5.3793)	
	Total	17	12	29

This can be computed automatically in R with the code chisq.test(C.difficile,correct=TRUE), which yields the output

```
        Pearson's Chi-squared test with Yates' continuity correction

data:  C.difficile
X-squared = 5.5976, df = 1, p-value = 0.01799
```

The command chisq.test(C.difficile) gives the same output because the default setting for 2×2 tables is to use the continuity correction. The command chisq.test(C.difficile, correct=FALSE) yields the output

```
        Pearson's Chi-squared test

data:  C.difficile
X-squared = 7.535, df = 1, p-value = 0.006051
```

The P-value is considerably larger with the continuity correction than without, although in this case they both give results that would be significant at the $\alpha = 0.05$ level.   ∎∎∎

Many believe that the continuity correction overcorrects. This is a deeper question than we can address here, but you should know about the continuity correction because it is the default method in R. For contingency tables that are larger than 2×2, adding the argument correct has no effect on the output. For these larger tables, R never uses the continuity correction.

A second method for testing independence or homogeneity is simulation, which can be applied to $r \times c$ tables (for any r and c). This can be done with small to moderate sample sizes where we don't trust the asymptotic χ^2 approximation.

For this case we can use simulation to approximate the P-value. The idea is to simulate the table entries from the marginal totals assuming either independence or homogeneity, and calculate the X^2 statistic. Specifically, the algorithm is this:

1. Compute the statistic

$$X^2 = \sum_{i=1}^{r} \sum_{j=1}^{c} \frac{(O_{ij} - E_{ij})^2}{E_{ij}}$$

using data from the observed table.

2. From the (assumed fixed) margins r_i and c_j, simulate the table entries O_{ij} under the supposition of independence or homogeneity.
3. For the simulated table, compute the X^2 statistic.
4. Repeat steps 2 and 3 B times, keeping track of the X^2 statistic on each simulation.
5. The estimated P-value is the proportion of times the simulated X^2 exceeds the X^2 statistic computed from the observed table.

R's chisq.test() function includes an optional argument for the user to specify that the P-value should be computed using this simulation method. To simulate the P-value, add simulate.pvalue=TRUE to the argument list for chisq.test(). The default number of simulations is 2,000, but this can be changed with the optional argument B=.

∎ **Example 13.9** Suppose we conduct a clinical trial on a COVID-19 experimental drug therapy and compare it against a placebo, like the one described in Section 13.4, except the sample sizes are much smaller. The data are shown in Table 13.13. Use the method of simulation to estimate the P-value to

Table 13.13 Results of hypothetical clinical trial testing for an experimental drug against a placebo.

	Outcome			
	Not hospitalized	Hospitalized	Died	Total
Experimental drug	47	4	1	52
Placebo	41	12	3	56
Total	88	16	4	108

test for homogeneity of probabilities across the three possible outcomes: not hospitalized, hospitalized, and died.

Solution: We can perform 100,000 simulations of the *P*-value using the following code:

```
tab = matrix( c( 47 , 4 , 1 ,
                 41 , 12 , 3 ) , nrow=2 , byrow=TRUE )
chisq.test( tab , simulate.p.value = TRUE , B = 100000 )
```

This produces the output

```
Pearson's Chi-squared test with simulated p-value (based on 1e+05
replicates)

data:  tab
X-squared = 5.2682, df = NA, p-value = 0.05951
```

The code `chisq.test( tab )` gives a *P*-value of 0.07178 along with a warning that `Chi-squared approximation may be incorrect.` ■ ■ ■

13.7 McNemar's Test

In Chapter 12 we introduced block designs. In Section 12.6.3 we defined a blocking variable as one that is not of particular interest in the context of the problem, but one that may affect the response variable. The paired *t*-test is an example of the use of a blocking variable, where in each block one subject or unit is assigned to one treatment and the other subject or unit is assigned to the other. When we have blocking with categorical variables, we can apply McNemar's test.

To illustrate the method, let's consider a specific example. Johnson and Johnson (1972) undertook a study to see whether Hodgkin's lymphoma was associated with having previously had a tonsillectomy. They collected data on Hodgkin's patients and paired them with a healthy sibling. The sibling pair constitutes a block in this example. The study involved 85 Hodgkin's lymphoma patients and their siblings, for a total of 170 people in the study. Data for such a study would have the structure shown below:

Pair	Hodgkin's patient	Healthy sibling
1	No tonsillectomy	No tonsillectomy
2	Tonsillectomy	No tonsillectomy
3	No tonsillectomy	No tonsillectomy
⋮	⋮	⋮
85	No tonsillectomy	Tonsillectomy

Table 13.14 Contingency table for McNemar's test on the Hodgkin's lymphoma and tonsillectomy study.

		Healthy sibling	
		No tonsillectomy	Tonsillectomy
Hodgkin's	No tonsillectomy	37	7
patient	Tonsillectomy	15	26

Table 13.15 General case for McNemar's test on the Hodgkin's lymphoma and tonsillectomy data.

		Healthy sibling	
		No tonsillectomy	Tonsillectomy
Hodgkin's	No tonsillectomy	a	b
patient	Tonsillectomy	c	d

The data set for the Hodgkin's lymphoma study is summarized in Table 13.14. The pairs of siblings in the (1,1) cell represent siblings both of whom did not have previous tonsillectomies. These tell us nothing about whether Hodgkin's lymphoma and having had one's tonsils removed are related. These pairs are said to be *concordant* since the siblings agree. The same is true for the pairs of siblings in the (2,2) cell; these are also concordant. The information lies in the (1,2) and (2,1) cells, where one of the siblings had a tonsillectomy and the other did not. These outcomes are *discordant*.

The general case is illustrated in Table 13.15. After removing the concordant pairs, which give us no information about the relationship between Hodgkin's lymphoma and tonsillectomy, we are left with the entries b and c. We condition on the sum $b+c$ and focus attention on b. Let $n = b+c$. We now look at the conditional distribution of the count in cell (1,2) given n. For now, let's think of B as being random. Define

$$p_{b|n} = \Pr\left(\text{a discordant pair falls in cell (1,2)}\right)$$

$$p_{c|n} = \Pr\left(\text{a discordant pair falls in cell (2,1)}\right)$$

$$= 1 - \Pr\left(\text{a discordant pair falls in cell (1,2)}\right).$$

The null hypothesis is

$$H_0: p_{b|n} = p_{c|n}.$$

Since $p_{b|n} + p_{c|n} = 1$, this is equivalent to testing

$$H_0: p_{b|n} = \frac{1}{2}.$$

One-sided and two-sided alternatives are possible.
If H_0 is true, then

$$B|n \sim \text{BIN}\left(n, \frac{1}{2}\right). \tag{13.6}$$

Then,

$$\mathbb{E}(B|n) = \frac{n}{2} \quad \text{and} \quad \mathbb{V}(B|n) = n\frac{1}{2}\left(1-\frac{1}{2}\right) = \frac{n}{4}.$$

If n is large, we can use the normal approximation to the binomial to obtain

$$B \mid n \overset{\text{approx}}{\sim} N\left(\frac{n}{2}, \frac{n}{4}\right).$$

Let C denote the random number of sibling pairs that fall in cell (2,1). Since we are conditioning on $B + C = n$, we have $C = n - B$. Thus, the difference between the two discordant counts, $B - C$, satisfies

$$\mathbb{E}(B - C \mid n) = \mathbb{E}(B - (n - B) \mid n) = \mathbb{E}(2B - n \mid n) = 2\frac{n}{2} - n = 0, \tag{13.7}$$

and

$$\mathbb{V}(B - C \mid n) = \mathbb{V}(2B - n \mid n) = 2^2 \mathbb{V}(B \mid n) = 4n\frac{1}{2}\left(1 - \frac{1}{2}\right) = n. \tag{13.8}$$

Thus,

$$B \mid n \overset{\text{approx. under } H_0}{\sim} N(0, n), \tag{13.9}$$

so

$$\frac{B - C}{\sqrt{n}} \Big| n \overset{\text{approx. under } H_0}{\sim} N(0, 1). \tag{13.10}$$

One version of McNemar's test statistic is

$$Z = \frac{B - C}{\sqrt{B + C}} \overset{\text{approx. under } H_0}{\sim} N(0, 1). \tag{13.11}$$

This test statistic can be used to test either a one-sided or a two-sided alternative when the sample sizes are large. When the alternative is two-sided, McNemar's test is usually given as

$$Z^2 = \left(\frac{B - C}{\sqrt{B + C}}\right)^2 = \frac{(B - C)^2}{B + C} \overset{\text{approx. under } H_0}{\sim} \chi^2(1). \tag{13.12}$$

This result follows from the fact that the square of a standard normal random variable has a $\chi^2(1)$ distribution.

R has a built-in function for performing the McNemar test. The function is `mcnemar.test()`. The only required argument is a matrix containing the contingency table.

■ **Example 13.10** Apply the χ^2 version of the McNemar test to the Hodgkin's lymphoma data.

Solution: For this data set $b = 7$ and $c = 15$. Thus,

$$Z^2 = \frac{(7 - 15)^2}{7 + 15} = 2.909.$$

The 0.95 quantile of the $\chi^2(1)$ distribution is $\chi^2_{0.05}(1) = 3.84$. Since $Z^2 = 2.909$ does not exceed 3.84, we do not have evidence that the conditional probabilities of having had a tonsillectomy differ.

R can perform the necessary calculations using the following code:

```
counts = matrix( c( 37 , 7 ,
                    15 , 26 ) , nrow=2 , ncol=2 , byrow=TRUE )
mcnemar.test( counts , correct=FALSE )
```

The `correct=FALSE` argument tells R not to use the continuity correction. We will discuss this correction shortly. The output is

```
McNemar's Chi-squared test
```

```
data: counts
McNemar's chi-squared = 2.9091, df = 1, p-value = 0.08808
```

Of course, the results from R agree with our calculations from scratch. ■ ■ ■

The sample sizes in the previous problem are fairly small (7 and 15), so the asymptotic $\chi^2(1)$ approximation from (13.12) or the asymptotic $\chi^2(1)$ distribution from (13.11) are questionable. One approach for small samples is to use the *continuity correction*, whereby

$$Z^2 = \frac{(|B-C|-1)^2}{B+C} \overset{\text{approx. under } H_0}{\sim} \chi^2(1). \tag{13.13}$$

In R, the default method is to use the continuity correction. This is why we had to say `correct=FALSE` in Example 13.10.

A second approach is to get the *P*-value using the exact conditional distribution for B given n, which is $\text{BIN}\left(n, \frac{1}{2}\right)$ as given in (13.6). If $b \geq c$, then the one-sided *P*-value is

$$\text{exact one-sided } P \text{ value} = \sum_{i=b}^{n} \binom{n}{i} \left(\frac{1}{2}\right)^i \left(1-\frac{1}{2}\right)^{n-i}. \tag{13.14}$$

If $c \geq b$, then the one-sided *P*-value is

$$\text{exact one-sided } P \text{ value} = \sum_{i=c}^{n} \binom{n}{i} \left(\frac{1}{2}\right)^i \left(1-\frac{1}{2}\right)^{n-i}.$$

For a two-sided alternative, the one-sided *P*-value should be doubled.

Exercise 13.2   For the Hodgkin's lymphoma data set, find the *P*-value using the continuity correction and also the exact *P*-value from (13.14).

McNemar's test can be applied whenever we have blocked categorical data. For example, we might test a political ad for a particular candidate. We would ask each person whether they had a favorable or unfavorable opinion of the candidate before being shown the ad. Then the same question is asked after seeing the ad. The data might look like this:

		Before watching ad	
		Favorable	Unfavorable
After	Favorable	a	b
watching ad	Unfavorable	c	d

In this case each person serves as his or her own control.

As another example, we could test a cream for treating tinea pedis, commonly known as athlete's foot. The experiment could be designed to assign one foot at random to receive the cream and the other to receive a placebo. Data would be summarized as:

		Foot treated with cream	
		Cured	Not cured
Foot treated	Cured	a	b
with placebo	Not cured	c	d

Again, each person is his or her own control. The cells that provide information are left-foot-cured/right-foot-not-cured, and right-foot-cured/left-foot-not-cured.

13.8 Higher-Dimensional Tables and Simpson's Paradox

With data from an observational study there are rarely just two variables being considered. There are often several, and when we look at only two at a time we can be led to unjustified conclusions. When we deal with randomized experiments these other variables will usually be distributed roughly evenly over the treatments, so these effects should balance out across the subjects.

To make these concepts clearer, consider this example. You need to have a serious operation, and you have to decide between two surgeons: Surgeon A and Surgeon B. For this particular procedure Surgeon A has a 70% survival rate while Surgeon B has an 80% survival rate. The counts for the numbers of surgeries and the outcome are:

	Survived	Died	Total
Surgeon A	140	60	200
Surgeon B	160	40	200

The decision may seem clear: Surgeon B has the higher survival rate (80% vs. 70%). In fact, there is a statistically significant difference between the survival rates. The output from R's chisq.test() without the continuity correction is

```
        Pearson's Chi-squared test

data:  Surgery
X-squared = 5.3333, df = 1, p-value = 0.02092
```

The data in the above table are from an observational study, not a designed experiment. The patients who went to Drs. A and B chose which surgeon to perform the surgery, and the surgeons chose which patients to accept. There was no random assignment of patients to surgeons like there would be in a designed experiment. On further investigation, you find that Surgeon A is a recognized expert in this type of operation and accepts patients who are at high risk.[7] Surgeon B rarely accepts high-risk patients and usually refers them to another surgeon, like Surgeon A. When data are broken down by risk group, the tables are as follows:

	Low risk				High risk		
	Survived	Died	Total		Survived	Died	Total
Surgeon A	40	6	46	Surgeon A	100	54	154
Surgeon B	150	26	176	Surgeon B	10	14	24
Total	190	32	222	Total	110	68	178

Now we see that for those at low risk, 40 of 46 of Surgeon A's patients survived (87.0%), whereas 150 of 176 of Surgeon B's patients survived (85.2%). If we applied the χ^2 test of homogeneity to just the low-risk group without the continuity correction, we find that $X^2 = 0.088$, yielding a P-value of 0.7662. Among those at high risk, 100 of 154 of Surgeon A's patients survived (64.9%) and 10 of 24 of Surgeon B's patients survived (41.7%). The X^2 statistic for the high-risk group is 4.76, with a P-value of 0.0291.

For those at low risk, Surgeon A seems better (87.0% vs. 85.2%), although the difference is not statistically significant. For the high-risk group, Surgeon A is again better (64.9% vs. 41.7%), a result that *is* statistically significant. The table that does not separate low- vs. high-risk patients suggests that Surgeon B is better.

Thus, ignoring my condition (low vs. high risk), I would select Surgeon B. But, if I'm at low risk, I would select Surgeon A; if I'm at high risk, I would also select Surgeon A. This is the heart of *Simpson's paradox*: We get different conclusions if we do or don't consider the third variable, the risk level.

---

[7]There are many indices for measuring whether a patient is at high risk. For example, cardiac surgeons often use the Parsonnet score. This score can range from 0 for someone with very little risk to over 60 for those at high risk.

Data that involve only the two variables of interest (surgeon and survival in this example) are said to be *aggregated*, meaning that the data set has been formed by combining several unrelated parts. In this case the parts are the patients at both low and high risk, as well as any other separating variable that we have not accounted for. There could be many other variables that affect survival: age, gender, race, and socioeconomic status, among others. These variables are called *lurking variables*. Data that have been pulled apart using a third (and possibly additional) variable are said to be *disaggregated*. Simpson's paradox is that we reach different conclusions if we look at aggregated data and disaggregated data.

In the example of the surgeons, the lurking variable risk-group is related to *both* the selection of the surgeon and the survival. A 2×2 table of surgeon vs. risk group is:

	Low risk	High risk	Total
Surgeon A	46	154	200
Surgeon B	160	40	200
Total	206	194	400

This table makes it clear that Surgeon A performs surgery mostly on high-risk patients, while Surgeon B performs surgery on mostly low-risk patients. The survival rate for high-risk patients is much lower than it is for low-risk patients. Since Surgeon A works mostly with high-risk patients, the fact that Surgeon A has an overall lower survival rate is to be expected, even though Surgeon A has the higher survival rate for both those with low and high risk.

A thorough discussion of Simpson's paradox involves *causal inference* a new and fascinating area of statistics.[8] In most cases, if a third variable (risk group in this example) is a plausible causal factor for survival, then the appropriate conclusions are those drawn from the disaggregated data. For the surgery data, possible causal models are illustrated in Figure 13.11. In the illustration on the left, only the surgeon affects survival. In the middle illustration, only the risk group affects survival. Finally, in the illustration on the right, both surgeon and risk group affect survival, although the surgeon may affect the survival in a different way than in the left-hand model, which excludes the risk group. Also in the rightmost figure, the risk group affects the surgeon. An analysis of the disaggregated data indicates that the rightmost figure is the most likely scenario.

Simpson's paradox often arises in cases of racial or gender bias in hiring or college admission decisions. The following example is based on a real case of college admissions at the University of California Berkeley (Bickel *et al.*, 1975). The aggregate admissions and gender table is real, but the program-level numbers are hypothetical. These hypothetical data demonstrate, in a small example, the issues that arose in the large data set that included many different programs.

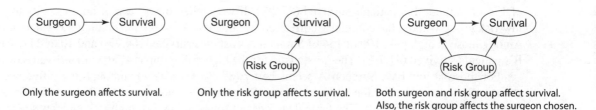

Only the surgeon affects survival.   Only the risk group affects survival.   Both surgeon and risk group affect survival.
                                                                                            Also, the risk group affects the surgeon chosen.

Figure 13.11 Possible causal diagrams for the surgery example.

---

[8]See the recent book by Pearl and Mackenzie (2018) entitled *The Book of Why: The New Science of Cause and Effect*.

■ **Example 13.11**   The aggregate counts for the admission decision (admit/deny) and gender (male/female) are:

		Admit	Deny	Total

		Decision		
		Admit	Deny	Total
Gender of	Male	3,738	4,704	8,442
applicant	Female	1,494	2,827	4,321
	Total	5,232	7,531	12,763

Investigate whether men and women had different acceptance rates.

Solution: Among the 8,442 men who applied, 3,738 were admitted, 44.3%, while among the 4,321 women who applied, 1,494 were admitted, 34.6%. Overall, men were admitted at a higher rate, by nearly 10 percentage points, than women. Thus, it looks like men received preferential treatment.

In reality, there were 101 programs to which students could apply. For simplicity, though, we will assume that there were only two: we'll call them Program A and Program B.[9] These two programs will illustrate the issue. The breakdown of applicants is

Program A

		Decision		
		Admit	Deny	Total
Gender of	M	3,000	2,000	5,000
applicant	F	600	300	900
	Total	3,600	2,300	5,900

Program B

		Decision		
		Admit	Deny	Total
Gender of	M	738	2,704	3,442
applicant	F	894	2,527	3,421
	Total	1,632	5,231	6,863

From these tables we can compute the admission rate for Programs A and B. For Program A,

$$\text{Admission rate for men in Program A} = \frac{3,000}{5,000} = 0.600 = 60.0\%$$

$$\text{Admission rate for women in Program A} = \frac{600}{900} = 0.667 = 66.7\%.$$

Thus, women have a slightly higher acceptance rate compared to men for Program A. For Program B,

$$\text{Admission rate for men in Program B} = \frac{738}{3,442} = 0.214 = 21.4\%$$

$$\text{Admission rate for women in Program B} = \frac{894}{3,421} = 0.261 = 26.1\%.$$

Again, women have a higher acceptance rate compared to men for Program B. Thus, neither program gives men preferential treatment; in fact, both programs admit women at a higher rate. But, at the university level, men are admitted at a higher rate. Simpson's paradox arises again.

If we look deeper, we see that men tended to apply to Program A, which has a much higher acceptance rate, while women tended to apply to Program B, which has a much lower acceptance rate. The 2 × 2 table of gender by program is:

		Applied to program		
		A	B	Total
Gender of	Male	5,000	3,442	8,442
applicant	Female	900	3,421	4,321
	Total	5,900	6,863	12,763

[9]Bickel *et al.* (1975) also presented a simplified two-program example. They called the programs machismatics and social warfare. We'll stick with Program A and Program B.

Thus, 59.2% of men applied to Program A, which had an acceptance rate of $3,600/5,900 = 61.0\%$, while only 20.8% of women applied to Program A. Similar numbers can be computed for men and women applying to Program B.

∎∎∎

13.9 Chapter Summary

Often we do not have direct measurements on each observation in our sample; rather, we observe which category the item belongs to. When there is a single variable, with k bins, we can test whether the probabilities of bin membership are as specified in some model. For example, we might hypothesize that observations are uniformly distributed across the various categories. When we have two variables, with r categories for one variable and c categories for the other, we can use all rc possible joint outcomes as the bins. From the $r \times c$ table, called a contingency table, we can test whether the two variables are independent. If the margins of either the row or column variables are fixed by the design of the study, then we can test whether the probabilities are the same across all categories. This is called a test of homogeneity. Tests of independence and homogeneity are carried out in exactly the same fashion, and are based on the χ^2 distribution.

When there are three variables, it is possible that we are led to different conclusions if one variable is excluded rather than separating (disaggregating) the variables by level. This is called Simpson's paradox, and it is important for a statistician or data scientist to understand the resolution of the paradox.

13.10 Problems

Problem 13.1 In the "nighttime" lottery run by the state of Texas, three numbers are selected from the digits 0 through 9. The frequencies of the first digit selected over a period of almost 30 years (from 1993 to 2023) are shown below for each of the 9,215 days:

0	1	2	3	4	5	6	7	8	9
918	905	908	916	900	911	963	948	937	909

(a) If each digit is equally likely, how many occurrences would we expect for each digit?

(b) Test whether each digit is equally likely. Use $\alpha = 0.05$.

(c) What is the P-value for the test in part (b)?

Problem 13.2 Consider the second digit selected in the Texas nighttime lottery described in Problem 13.1. Explain why the distribution for the second digit is discrete uniform over the digits 0 through 9. Do the same for the third digit.

Problem 13.3 The frequencies of the second digits of the Texas lottery (see Problem 13.1) are

0	1	2	3	4	5	6	7	8	9
886	911	915	929	942	917	897	955	920	943

Repeat Problem 13.1 for the second digit.

Problem 13.4 The frequencies of the third digits of the Texas lottery (see Problem 13.1) are

0	1	2	3	4	5	6	7	8	9
912	954	932	919	936	884	924	925	916	913

Repeat Problem 13.1 for the third digit.

Problem 13.5 Suppose a hospital monitors the daily number of admissions for COVID across three weeks. The numbers of admissions on these 21 days are:

5,	0,	1,	7,	5,	4,	8
2,	10,	5,	3,	3,	7,	2
7,	3,	2,	5,	9,	2,	6

(a) Assuming a POIS(λ) distribution for the number of daily admissions, estimate the parameter λ.

(b) Make a table of the number of times the admission counts were 0–2, 3–5, 6+.

(c) Find the MLE of λ.

(d) Use the χ^2 goodness-of-fit test to test whether the Poisson distribution is a reasonable model.

Problem 13.6 The following data give the failure times of ball bearings given by Lieblein and Zelen (1956). (See Caroni (2002) for suggested corrections to this data set.)

17.88,	28.92,	33.00,	41.52,	42.12,
45.60,	48.40,	51.84,	51.96,	54.12,
55.56,	67.80,	68.64,	68.64,	68.88,
84.12,	93.12,	98.64,	105.12,	105.84,
127.92,	128.04,	173.40		

(a) Make a table of the number of ball bearings that failed in the intervals 0–30, 30–60, 60–90, 90–120, and 120+.

(b) Assuming an EXP1(θ) distribution, estimate θ.

(c) Use the results of the table from part (a) and $\hat{\theta}$ from part (b) to test, using $\alpha = 0.05$, whether the exponential distribution is reasonable for this data set.

Problem 13.7 Examine the following R code:

```
n = 100
nsim = 1000
alpha = 0.05
rej = 0
chisq.cr = qchisq( 1 - alpha , 9 )
for( i in 1:nsim )
{
  u = runif( n )
  O = table( cut(u,seq(0,1,0.1)) )
  E = rep( 0.1*n , 10 )
  X2 = sum( (O-E)^2/E )
  if (X2 > chisq.cr ) rej = rej + 1
}
print( rej )
```

(a) Explain what line 5 (that calls the qchisq() function) is doing.

(b) At the R prompt, type ?cut to see what the cut function does. What is accomplished in the line that calls the cut() function?

(c) Explain what one loop over i accomplishes.

(d) Explain what the variable rej means.

Problem 13.8 Repeat Problem 13.7 using $\alpha = 0.01$. Explain what your final results mean.

Problem 13.9 Vermund *et al.* (1991) studied whether positivity rates for HIV and the human papillomavirus (HPV) were related. The HIV status of 96 women was classified as positive and symptomatic, positive and not symptomatic, and negative. HPV status was positive or negative. The contingency table for the results is shown in Table 13.16.

(a) Test whether positivity rates for HIV and HPV are independent. Do not use the continuity correction. Use $\alpha = 0.05$.

(b) Apply the test *with* the continuity correction.

(c) Compute the P-value for the test in part (a).

(d) Compute the P-value for the test in part (b).

Table 13.16 Data for Problem 13.9.

HIV	HPV Positive	Negative	
Pos & sympt	23	10	33
Pos & asympt	4	14	18
Negative	10	35	45
	37	59	96

Problem 13.10 In a study of drinking patterns among high school students in Ghana, Bondah *et al.* (2020) surveyed 762 students. A table of drinking habits (abstain or have ever drunk alcohol) against sex is given in Table 13.17.

Table 13.17 Data for Problem 13.10.

Drinking behavior	Sex Male	Female	
Never	23	269	455
Ever drank	124	183	307
	310	452	762

(a) Test whether sex and drinking behavior are independent. Use the continuity correction with $\alpha = 0.05$.
(b) What is the *P*-value for this test?
(c) Explain why this is a test of independence and not a test of homogeneity.

Problem 13.11 Suppose that in a situation like that described in Problem 13.10 the survey was designed to select 300 male students and 300 female students and the results were as shown in Table 13.18.

Table 13.18 Data for Problem 13.11.

Drinking behavior	Sex Male	Female	
Never	154	131	285
Ever drank	146	169	315
	300	300	600

(a) Use the continuity correction to test whether drinking behavior is the same across sexes. Use $\alpha = 0.05$.
(b) What is the *P*-value for this test?
(c) Explain why this is a test of homogeneity.

Problem 13.12 Wang *et al.* (2020) studied the intention of nurses to accept the COVID-19 vaccine. One of the predictors they studied was the respondent's age. The table of age against intention to vaccinate is given in Table 13.19.

Table 13.19 Data for Problem 13.12.

Age	Vaccine acceptance Accept	Reject	Und.	
18–29	73	26	75	174
30–39	116	40	95	251
40–49	78	44	96	218
50+	55	28	80	163
	322	138	346	

(a) Use the continuity correction to test whether age and vaccine intention (accept the vaccine, reject, or undecided) is independent of age.
(b) What is the *P*-value from the test in part (a)?
(c) Explain why this is a test of independence.

Problem 13.13 In a case–control study, individuals with some disease are identified; these are called *cases*. A number of people, similar in important demographics but without the disease are also identified. The numbers of people of each type are fixed in advance. In a study of acute pelvic inflammation (API), Scholes *et al.* (1992) studied the effect of income, smoking, and a number of other variables on the risk of API. The table of income against case/control is shown in Figure 13.20. We are interested in whether the risk of API is related to income.

Table 13.20 Data for Problem 13.13.

Income	Case	Control	
Low	55	57	112
Middle	47	122	169
High	28	114	142
	130	293	423

(a) Is this a test of independence or homogeneity? Explain.
(b) Using the continuity correction with $\alpha = 0.05$, carry out the appropriate test.
(c) What is the *P*-value for the test in part (b)?
(d) Explain the result in practical terms.

Problem 13.14 In a case–control study for gastric cancer, Nemati *et al.* (2012) looked at several diet and lifestyle variables, such as alcohol use. The contingency table of alcohol use versus case (gastric cancer) or control (no gastric cancer) is given in Table 13.21.

(a) Is this a test of independence or heterogeneity? Explain.
(b) Carry out the test using the continuity correction with $\alpha = 0.05$.

Table 13.21 Data for Problem 13.14.

Alcohol use	Case	Control	
Yes	5	10	15
No	37	76	113
	42	86	128

(c) Apply Fisher's exact test.

(d) Apply the method of simulation to approximate the P-value.

Problem 13.15 Schweizer *et al.* (1991) studied whether the drug carbamazepine could be used to assist patients in the process of discontinuing daily benzodiazepine therapy. In a clinical trial, patients were assigned either carbamazepine or a placebo and the benzodiazepine status was determined after five weeks. The data are summarized in Table 13.22.

Table 13.22 Data for Problem 13.15.

	Benzodiazepine use after 5 weeks		
Treatment	Yes	No	
Carbamazepine	1	18	19
Placebo	8	13	21
	9	31	40

(a) If we would like to test whether carbamazepine is effective in helping patients discontinue benzodiazepine therapy, would this be a test of independence or homogeneity? Explain.

(b) Apply the usual χ^2 test with $\alpha = 0.05$ without the continuity correction. Is there an issue in using this test?

(c) Apply Fisher's exact test.

(d) Apply the method of simulation to approximate the P-value.

Problem 13.16 Consider the vaccine intention survey described in Problem 13.12. Suppose that instead of taking a simple random sample of nurses, they took a stratified random sample of 200 from each age group. Explain why the resulting test would be a test of homogeneity rather than a test of independence.

Problem 13.17 Suppose that two imaging methods are used to detect the presence or absence of cancerous cells in the lung. Let's call these methods A and B. Investigators select 80 individuals suspected of having lung cancer. All 80 were giving both imaging tests. The results for methods A and B are summarized in Table 13.23.

Table 13.23 Data for Problem 13.17.

		Method B		
		Positive	Negative	
Method	Positive	26	6	32
A	Negative	14	34	48
		40	40	80

(a) Apply McNemar's test to compare the two methods.

(b) What are the null and alternative hypotheses for McNemar's test?

(c) What is the P-value for the test in part (a)?

(d) Explain your conclusions in plain language.

Problem 13.18 Eustace *et al.* (1997) studied two imaging methods for discovering locations where cancer has metastasized. They identified 175 locations on 25 patients. Magnetic resonance imaging (MRI) and a method called scintigraphy were used to classify each location. In all cases the ground truth about whether the location contained metastasized cells was known. Data for the MRI and scintigraphy are given in Table 13.24.

Table 13.24 Data for Problem 13.18. The top (bottom) table gives the results for the MRI (scintigraphy)

	Detected		
MRI	Yes	No	
Positive	57	0	57
Negative	2	116	118
	59	116	175

	Detected		
Scintigraphy	Yes	No	
Positive	43	16	59
Negative	2	114	116
	45	130	175

(a) For the MRI, estimate the sensitivity, specificity, and positive predictive value (PPV).

(b) Do the same for the method of scintigraphy.

(c) Suppose that the cross-tabulated counts for both methods (MRI and scintigraphy) are as given in Table 13.25. (The authors did not provide such a table; these values are therefore approximate.) Apply McNemar's test with $\alpha = 0.05$.

(d) What are the null and alternative hypothesis being tested here?

(e) What is the P-value for the test in part (a)?

Table 13.25 Cross tabulation of positive and negative results from MRI and scintigraphy.

	Scintigraphy		
MRI	Positive	Negative	Total
Positive	40	17	57
Negative	5	33	118
Total	45	130	175

(f) Explain your conclusions in plain language.

(g) Suppose that, after conducting the analysis, someone suggests that the observations may not be independent. What is it about this problem that suggests nonindependence?

Problem 13.19 Consider the study described in Problem 13.13. The three-way table with case/control, income, and smoking status is shown in Table 13.26.

Table 13.26 Data for Problem 13.19.

	Smoking	Case	Control	Total
	Low	29	41	70
Income	Middle	25	92	117
	High	20	85	105
	Total	74	218	292

	Nonsmoking	Case	Control	Total
	Low	26	16	42
Income	Middle	22	30	52
	High	8	29	37
	Total	56	75	131

(a) Separately for the smoking and nonsmoking groups, test whether the income group is related to the risk of API. Use $\alpha = 0.05$.

(b) What are the P-values for the tests in part (a)?

(c) Is this an example of Simpson's paradox?

Problem 13.20 Radelet (1981) presented data from Florida regarding the imposition of the death penalty for murder cases. For nonprimary murders (i.e., those

where the victim and the defendant did not know each other) the counts of death penalty (no, yes) against race of the defendant (white, black) are shown in Tables 13.27, 13.28, and 13.29.

Table 13.27 Data for Problem 13.20.

		Death penalty		
		No	Yes	Total
Race of	White	141	19	160
defendant	Black	149	17	166
	Total	290	36	326

Table 13.28 Data for Problem 13.20 when the victim is white.

White victim		Death penalty		
		No	Yes	Total
Race of	White	132	19	151
defendant	Black	52	11	63
	Total	184	30	214

Table 13.29 Data for Problem 13.20 when the victim is black.

Black victim		Death penalty		
		No	Yes	Total
Race of	White	9	0	9
defendant	Black	97	6	103
	Total	106	6	112

(a) Compare the overall percentages of white and black defendants who were sentenced to death.

(b) For the case where the victim was white, compare the percentages of white and black defendants who were sentenced to death.

(c) For the case where the victim was black, compare the percentages of white and black defendants who were sentenced to death.

(d) Construct a 2×2 table of race of victim and race of defendant.

(e) Is this an example of Simpson's paradox?

14 — Regression

14.1 Introduction

We are often interested in how one or more predictor variables are associated with some outcome or response. We might postulate that the outcome Y depends on the predictors x_1, x_2, ..., x_p through some function

$$Y = f(x_1, x_2, \ldots, x_n). \tag{14.1}$$

The objective is then to learn what this function f looks like.

There are two common objectives for learning about this function. First, we might want to use it to *predict* what the outcome might be for a particular set of predictor variables. Second, we may want to *explain* the relationship between the predictor variables and the outcome. There are situations where one of these is central to the study, and there are cases where both are important, but one of the objectives predominates.

Anyone who has run science experiments or analyzed data knows that data points never fit a relationship like that in (14.1) *perfectly*. There is always some noise that in inherent in the data. To allow for this noise, we reformulate the model as

$$Y = f(x_1, x_2, \ldots, x_n) + \varepsilon, \tag{14.2}$$

where we can think of ε as being the error between what we would expect to observe and what we actually observe. We will make some assumptions about ε in the next section. Normally, we think of ε as being a random variable. Since ε is random, the outcome Y is also a random variable. The predictor variables x_1, x_2, ...,x_p can be either random or fixed. Initially, we will treat them as fixed.

14.1.1 Prediction vs. Explanation

Let's look at a few situations where the objective might be prediction or explanation, and some where there is a combination.

■ **Example 14.1** A clinical trial is studying the effect of the dose of a new drug on some health outcome, such as blood pressure reduction. They have data on the dose and the outcome, as well as on a number of demographic variables.

Solution: In a case like this, the investigators would probably like to know the relationship between the dose and the outcome. If the relationship is nonlinear, there may be an optimal value of dose over the interval studied. In addition, they would like to know whether the drug works differently across genders, races, and age groups. They would like to learn about how all of these variables are related to the outcome. This is a problem of explanation.   ■ ■ ■

■ **Example 14.2**   A movie streaming company would like to suggest movies for you to watch based on your viewing history. For a particular movie, the outcome is $Y = 1$ if you watch it and 0 otherwise. The company has some demographic information about you and your family, such as age, number of children, etc., as well as geographic information about where you live.

Solution: This is a prediction problem. The company doesn't care much about whether or how age affects your movie preferences. They just want to know whether you'd like the movie. There is no desire to build and explain a model about what affects your movie choices.   ■ ■ ■

■ **Example 14.3**   Engineers are working to improve the reliability of an automobile transmission. They have identified several characteristics of the design of the transmission that might affect its reliability, including the shaft diameter and the annealing time, along with other variables. The outcome is the lifetime of the transmission.

Solution: The engineers would like to know which among the many variables studied have an effect on the transmission's lifetime. They would also like to know *how* these variables affect the outcome, so that they can choose the design characteristics that maximize the reliability. For example, the relationship between shaft diameter and lifetime may be nonlinear, where small and large diameters are unreliable, but diameters in the middle are more reliable. In this case, the optimal value would be somewhere in the middle. Finally, if some variables have no effect on the lifetime, then any setting would work satisfactorily. This example, which is similar to one described in Davis (1995) and is an exercise in Wu and Hamada (2000), involves both aspects of prediction and explanation.   ■ ■ ■

■ **Example 14.4**   Public health officials would like to know which groups of people are most hesitant to receive a COVID-19 vaccine. They develop a questionnaire with several demographic questions and several questions about feelings toward vaccines in general, and the COVID-19 vaccine in particular. From the latter set of questions they develop an index regarding vaccine hesitancy.

Solution: On the one hand, the investigators would like to understand the behavioral dynamics of people's vaccine hesitance. They would like to know whether some age groups (or genders or races) are more reluctant to receive a vaccine. This would contribute to a general understanding of vaccine hesitancy. On the other hand, the investigators would like to predict whether a particular person would be hesitant about receiving the vaccine so they can develop a strategy for encouraging vaccine uptake. This is a combination of explanation and prediction.   ■ ■ ■

In a prediction problem, the investigators do not necessarily need (or want) to know the rule used to predict the outcome. Attention is focused on making accurate predictions. The rule for prediction can be thought of as a "black box," as illustrated in Figure 14.1; the set of predictors (a vector) is put into the box, and the prediction comes out.

$x \longrightarrow$ ⬛ $\longrightarrow \hat{f}(x)$ **Figure 14.1** The predictor $\hat{f}(x)$.

In an explanation problem the investigators will often assume some sort of functional form for $f(\boldsymbol{x})$, such as linear or quadratic. More on this will come in the subsequent sections. For example, in Example 14.1 the model may be assumed to be of the form

$$\texttt{BP Reduction} = \beta_0 + \beta_1 \texttt{dose} + \beta_2 \texttt{dose}^2 + \beta_3 \texttt{age} + \beta_4 \texttt{Gender} + \varepsilon$$

where Gender is an indicator variable that identifies the gender of the subject (e.g., 0 = Female, 1 = Male). In cases like this, the investigators are often interested in whether particular βs are equal to zero, because if this is the case, then that variable has no effect on the outcome. Often, investigators are interested in the functional form so that they can identify an optimal value. In Example 14.1 the objective may be to identify the optimal dose by first learning what the functional relation is between the dose and the outcome, and how this may differ between demographic groups. A black box like the one shown in Figure 14.1 will not suffice.

14.1.2 Terminology

In many applications we are trying to predict or explain the variable Y based on one or more variables x_1, x_2, ..., x_p. The variable Y is called the *response*, *outcome*, or *dependent variable*. The xs are called *predictors*, *regressors*, *explanatory variables*, or *independent variables*. Some types of predictor variables are called *covariates*. In the machine learning literature, the term *feature* is often used. We will usually use the terms predictor variables for the xs and outcome for Y. In the model

$$Y = f(x_1, x_2, \ldots, x_n) + \varepsilon,$$

we will refer to f as the true regression equation and ε as the *error term*.

14.1.3 A Working Example

To make these ideas more concrete, let's consider a particular study that involves relating various variables measured in high school to the first-year GPA in college. We use an old data set[1] to illustrate the ideas contained in this chapter. The data set contains data from 219 students and includes the following variables:

GPA	First-year GPA in college
HSGPA	High school GPA
SATV	SAT verbal/critical reading score
SATM	SAT math score
Male	1 = male, 0 = female
HU	Number of credits earned in high school humanities courses
SS	Number of credits earned in high school social science courses
FirstGen	1 = student is first in family to attend college, 0 = otherwise
White	1 = white, 0 = other
CollegeBound	1 = attended a high school where at least half of all students intended to go on to college, 0 = otherwise

We would like to know how these variables are associated with GPA. For now, we'll investigate only the relationship between GPA and SATM. In subsequent sections we will study the effects of two or more variables on the outcome, GPA in this case.

The scatterplot of GPA and SATM, shown in Figure 14.2, shows that there is a lot of noise and that it will be difficult to predict with any certainty the GPA of a student with a particular SATM. We can see why the error term ε in (14.2) is needed! So, is this a prediction problem or an explanation problem?

---

[1] This data set is contained in the R package Rdatasets and can be obtained (along with hundreds of other interesting data sets) at https://vincentarelbundock.github.io/Rdatasets/articles/data.html.

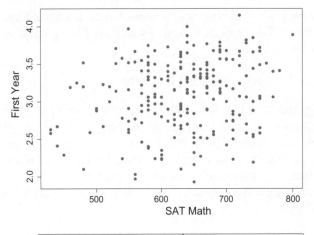

Figure 14.2 SATM score vs. first-year GPA.

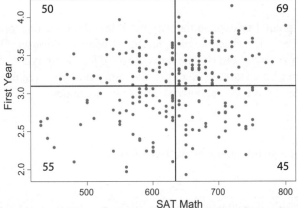

Figure 14.3 SATM score vs. first-year GPA
with crosshairs drawn at the mean for both
variables. The numbers give the count of points
in each quadrant.

Or is it a combination of both? This is primarily a prediction problem because the university would probably want to predict whether a student will be successful in college. There are, however, aspects of explanation that are present here. For example, we would like to know which predictor variable have an effect on GPA. Some variables may have virtually no effect on GPA. We might also be interested in whether high values of GPA are associated with small or large values of the predictors. (This is called the *direction* of the effect.)

From a second and more careful look at the scatterplot in Figure 14.2 you can probably see that large values of SATM tend to be associated with large values of GPA and small values of SATM tend to be associated with small values of GPA. If we were to draw horizontal and vertical lines through the means of both variables, as in Figure 14.3, we can see that there are slightly more points in the upper right and lower left quadrants than there are in the upper left and lower right quadrants. Thus, it looks like there is a relationship, but the relationship is very weak and it will be difficult to make an accurate prediction for a particular student.

We will use this example to illustrate many of the topics in this chapter. We begin in the next section with linear regression when there is a single predictor.

14.2 Simple Linear Regression

Suppose there is a single predictor variable and a single outcome or response. We assume a linear relationship of the form

$$y_i = \beta_0 + \beta_1 x_i + \varepsilon_i, \qquad \varepsilon_i \sim \text{i.i.d. } N(0, \sigma^2), \qquad i = 1, 2, \dots, n. \tag{14.3}$$

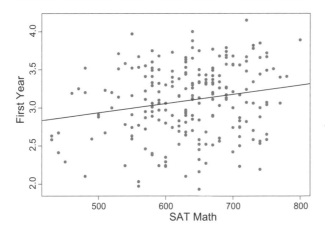

Figure 14.4 SAT math score vs. first-year GPA with least squares line.

Here, i is the index for the observation (ordered pair (x_i, y_i)). Notice that the error terms are independent and are assumed to be normally distributed with a mean of 0 and a common variance of σ^2. This is the *simple linear regression* model. We must always keep in mind that the true model (which we never know) is

$$y_i = f(x_i) + \varepsilon_i, \qquad \varepsilon_i \sim \text{i.i.d. } N(0, \sigma^2), \quad i = 1, 2, \ldots, n.$$

Here, we replaced $f(x_i)$ with $\beta_0 + \beta_1 x_i$, which may be a reasonable assumption is some cases, but unrealistic in others:

$$y_i = \underbrace{\beta_0 + \beta_1 x_i}_{\text{linear relationship}} + \underbrace{f(x_i) - (\beta_0 + \beta_1 x_i)}_{\text{model bias}} + \underbrace{\varepsilon_i,}_{\text{random error}} \qquad \varepsilon_i \sim \text{i.i.d. } N(0, \sigma^2). \qquad (14.4)$$

We must never forget that our linear assumption is rarely met in practice.

There are three parameters to estimate in the simple linear regression model in (14.3): β_0, β_1, and σ^2. Given a number of ordered pairs (x_i, y_i), we can estimate β_0 and β_1 using the method of *least squares*. We will address the issue of estimating σ^2 later in this section.

For fixed estimators $\hat{\beta}_0$ and $\hat{\beta}_1$, the *predicted value* for the outcome y_i is

$$\hat{y}_i = \hat{\beta}_0 + \hat{\beta}_1 x_i. \qquad (14.5)$$

The *residual* is then defined to be the difference between the observed value y_i and the predicted value $\hat{y}_i$:

$$r_i = y_i - \hat{y}_i. \qquad (14.6)$$

We would like to choose the estimators $\hat{\beta}_0$ and $\hat{\beta}_1$ so that the residuals are small in magnitude. The method of least squares selects the values of β_0 and β_1 that minimize the sum of squared residuals:

$$S(\beta_0, \beta_1) = \sum_{i=1}^{n} \left(y_i - (\beta_0 + \beta_1 x_i) \right)^2. \qquad (14.7)$$

For now, we will omit the "hat" on β_0 and β_1. Once we derive the least squares estimators, we will insert the hats to indicate these are *estimators*, not the true values.

We will use some calculus to find the values of β_0 and β_1 that minimize $S(\beta_0, \beta_1)$ in (14.7). We start by taking partial derivatives with respect to β_0 and β_1 and simplifying:

$$\frac{\partial S}{\partial \beta_0} = \sum_{i=1}^{n} 2\left(y_i - \beta_0 - \beta_1 x_i\right)^{2-1}(-1) \qquad \text{[the } -1 \text{ comes from the chain rule]}.$$

$$= 2\left(-\sum_{i=1}^{n} y_i + \sum_{i=1}^{n}\beta_0 + \beta_1\sum_{i=1}^{n} x_i\right)$$

$$= 2\left(-n\bar{y} + n\beta_0 + \beta_1 n\bar{x}\right)$$

and

$$\frac{\partial S}{\partial \beta_1} = \sum_{i=1}^{n} 2\left(y_i - \beta_0 - \beta_1 x_i\right)^{2-1}(-x_i) \qquad \text{[the } -x_i \text{ comes from the chain rule]}.$$

$$= 2\left(-\sum_{i=1}^{n} x_i y_i + \beta_0\sum_{i=1}^{n} x_i + \beta_1\sum_{i=1}^{n} x_i^2\right)$$

$$= 2\left(-\sum_{i=1}^{n} x_i y_i + \beta_0\, n\bar{x} + \beta_1\sum_{i=1}^{n} x_i^2\right).$$

We now equate these results to zero and solve the two equations in two unknowns. Note that these equations are linear in the *parameters*. After dividing both sides by 2 and rearranging, this yields

$$n\beta_0 + \beta_1 n\bar{x} = n\bar{y} \tag{14.8}$$

$$\beta_0\, n\bar{x} + \beta_1\sum_{i=1}^{n} x_i^2 = \sum_{i=1}^{n} x_i y_i. \tag{14.9}$$

We can solve the first equation for β_0 in terms of β_1 to obtain

$$\beta_0 = \bar{y} - \beta_1\bar{x}. \tag{14.10}$$

The second equation can then be written successively as

$$\beta_1\sum_{i=1}^{n} x_i^2 = \sum_{i=1}^{n} x_i y_i - n\beta_0\bar{x}$$

$$\beta_1\sum_{i=1}^{n} x_i^2 = \sum_{i=1}^{n} x_i y_i - n\left(\bar{y} - \beta_1\bar{x}\right)\bar{x}$$

$$\beta_1\sum_{i=1}^{n} x_i^2 = \sum_{i=1}^{n} x_i y_i - n\bar{x}\bar{y} + n\beta_1\bar{x}^2.$$

From this we can write

$$\beta_1\sum_{i=1}^{n} x_i^2 - n\beta_1\bar{x}^2 = \sum_{i=1}^{n} x_i y_i - n\bar{x}\bar{y}$$

$$\beta_1\left(\sum_{i=1}^{n} x_i^2 - n\bar{x}^2\right) = \sum_{i=1}^{n} x_i y_i - n\bar{x}\bar{y}$$

$$\beta_1 = \frac{\displaystyle\sum_{i=1}^{n} x_i y_i - n\bar{x}\bar{y}}{\displaystyle\sum_{i=1}^{n} x_i^2 - n\bar{x}^2}.$$

This can be rewritten in the equivalent form

$$\hat{\beta}_1 = \frac{\sum_{i=1}^{n}(x_i - \bar{x})(y_i - \bar{y})}{\sum_{i=1}^{n}(x_i - \bar{x})^2}. \tag{14.11}$$

This is now a closed-form expression for the estimator $\hat{\beta}_1$ because it does not depend on the other parameter estimator $\hat{\beta}_0$. Once the least squares estimator for β_1 is obtained, we can substitute this back into (14.10) to obtain the least squares estimator for the intercept parameter β_0:

$$\hat{\beta}_0 = \bar{y} - \hat{\beta}_1 \bar{x}. \tag{14.12}$$

An astute reader might recognize that we have found a stationary point, not necessarily the point that yields a minimum of $S(\beta_0, \beta_1)$. To demonstrate that indeed we've achieved a minimum, we can apply the second derivative test. We found that the first derivatives of $S(\beta_0, \beta_1)$ are

$$\frac{\partial S}{\partial \beta_0} = 2\left(-n\bar{y} + n\beta_0 + \beta_1 n\bar{x}\right)$$

$$\frac{\partial S}{\partial \beta_1} = 2\left(-\sum_{i=1}^{n} x_i y_i + \beta_0 n\bar{x} + \beta_1 \sum_{i=1}^{n} x_i^2\right).$$

The three second partials, including the mixed partial, are then

$$\frac{\partial^2 S}{\partial \beta_0^2} = 2n$$

$$\frac{\partial^2 S}{\partial \beta_0 \partial \beta_1} = 2n\bar{x}$$

$$\frac{\partial^2 S}{\partial \beta_1^2} = 2\sum_{i=1}^{n} x_i y_i.$$

The second-derivative test involves the determinant of the matrix of second derivatives:

$$\det\left(\begin{bmatrix} \frac{\partial^2 S}{\partial \beta_0^2} & \frac{\partial^2 S}{\partial \beta_0 \partial \beta_1} \\ \frac{\partial^2 S}{\partial \beta_0 \partial \beta_1} & \frac{\partial^2 S}{\partial \beta_1^2} \end{bmatrix}\right) = \det\left(\begin{bmatrix} 2n & 2n\bar{x} \\ 2n\bar{x} & 2\sum_{i=1}^{n} x_i^2 \end{bmatrix}\right)$$

$$= 4n\sum_{i=1}^{n} x_i^2 - 4n^2\bar{x}^2$$

$$= 4n\left(\sum_{i=1}^{n} x_i^2 - n\bar{x}^2\right)$$

$$= 4n\sum_{i=1}^{n}(x_i - \bar{x})^2 > 0.$$

Since the determinant of the second derivative matrix is positive and $\partial^2 S/\partial^2 \beta_0 = 2n > 0$, the function must be concave up at the stationary point; thus, the stationary point is one that yields a *minimum*.

Sums of squares (and cross products) are common in linear regression analysis, and we will find the following notation helpful. We define

$$S_{xx} = \sum_{i=1}^{n}(x_i - \bar{x})^2 \tag{14.13}$$

$$S_{xy} = \sum_{i=1}^{n}(x_i - \bar{x})(y_i - \bar{y}) \tag{14.14}$$

$$S_{yy} = \sum_{i=1}^{n}(y_i - \bar{y})^2. \tag{14.15}$$

With this notation we can write the least squares estimator for β_1 as

$$\hat{\beta}_1 = \frac{\sum_{i=1}^{n}(x_i - \bar{x})(y_i - \bar{y})}{\sum_{i=1}^{n}(x_i - \bar{x})^2} = \frac{S_{xy}}{S_{xx}}.$$

We will make further use of this notation when we develop the properties of these estimators in the sections that follow.

To illustrate the concepts presented here we've selected a subset of 10 of the 219 observations in the GPA vs. SATM scores data set. Figure 14.5 shows scatterplots of SATM vs. GPA. The plot also shows line segments from each point in the scatterplot to the line, projected vertically. The length of this line segment is $|y_i - \hat{y}_i|$. The method of least squares selects $\hat{\beta}_0$ and $\hat{\beta}_1$ so as to minimize the sum of these squared distances. The least squares estimators are

$$\hat{\beta}_1 = \frac{S_{xy}}{S_{xx}} = \frac{401.06}{100,760} \approx 0.003980$$

and

$$\hat{\beta}_0 = \bar{y} - \hat{\beta}_1 \bar{x} = 3.073 - 0.003980 \times 588 \approx 0.73276.$$

In Figure 14.5(a) and (b), arbitrary values for β_0 and β_1 were chosen (they are not the least squares estimators). Figure 14.5(a) shows the line with $\beta_0 = 2.8$ and $\beta_1 = 0.001$. Since the first eight points lie below the line and the next two points lie above the line, it seems like we could do better if we took a line with a steeper slope (i.e., one for which β_1 is larger). In Figure 14.5(b) it seems like the line has an intercept that is too high, since all of the points are below the line. Figure 14.5(c) shows the line with the intercept and slope parameters given by (14.12) and (14.11), respectively. In all three graphs, the sum of squared residuals is given in the bottom right corner. The one for which the sum of squared residuals is smallest is Figure 14.5(c). Try as we might, we cannot reduce the value of S, the sum of squared residuals in Figure 14.5(c). Any perturbation of the line in Figure 14.5(c) will increase the value of S. This is what the calculus showed us in the derivation of the least squares estimator.

Our simple linear regression model involved three parameters: β_0, β_1, and σ^2. We can use the method of least squares to find estimates for β_0 and β_1; we will address the estimation of σ^2 in Section 14.3.

Our criterion for selecting estimators $\hat{\beta}_0$ and $\hat{\beta}_1$ was to minimize the sum of *squared* residuals. This leads to the *least squares* estimators. The calculus for this method works out nicely, and we get the closed-form expressions for the estimators as given in (14.11) and (14.12).

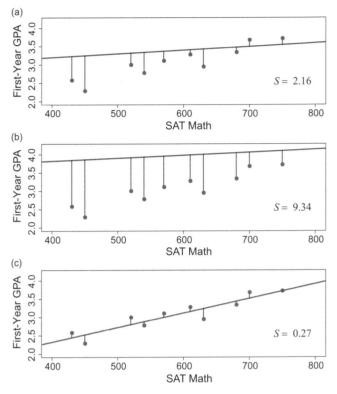

Figure 14.5 Scatterplot of a subset of points from the SAT math vs. GPA data set. (a) and (b) include lines with arbitrarily chosen slope and intercept; (c) shows the least squares line.

Another criterion is to minimize the sum of the absolute deviations:

$$A(\beta_0, \beta_1) = \sum_{i=1}^{n} |y_i - \hat{y}_i|. \tag{14.16}$$

The method that uses this criterion is called *least absolute deviations* and it is sometimes used in practice. See Problem 14.2 for a further analysis of least absolute deviations. Closed-form expressions for the estimators are not available, however, for least absolute deviations.

In Section 9.4 we described the method of maximum likelihood (ML). This is a general-purpose method that can be used in a variety of situations. Next we develop the MLEs for all three parameters in the simple linear regression model. We treat the response Y_i as a random variable, and the x_i as fixed numbers. The model from (14.3) can be rewritten in the equivalent form

$$Y_i \stackrel{\text{indep.}}{\sim} N(\beta_0 + \beta_1 x_i, \sigma^2). \tag{14.17}$$

Given that the Y_i are independent (but not identically distributed, unless $\beta_1 = 0$), the likelihood function is

$$L(\beta_0, \beta_1, \phi) = \prod_{i=1}^{n} \frac{1}{\sqrt{2\pi\phi}} \exp\left(\frac{1}{2\phi}(y_i - \beta_0 - \beta_1 x_i)^2\right)$$

$$= (2\pi)^{-n/2}\phi^{-n/2} \exp\left(\frac{1}{2\phi}\sum_{i=1}^{n}(y_i - \beta_0 - \beta_1 x_i)^2\right),$$

where $\phi = \sigma^2$. The log-likelihood function is therefore

$$\ell(\beta_0, \beta_1, \phi) = \underbrace{-\frac{n}{2}\log 2\pi}_{\text{constant}} - \underbrace{\frac{n}{2}\log \phi}_{\text{function of }\phi} - \underbrace{\frac{1}{2\phi}\sum_{i=1}^{n}(y_i - \beta_0 - \beta_1 x_i)^2}_{\text{function of }\beta_0, \beta_1}. \tag{14.18}$$

As in Section 9.4 we write $\phi = \sigma^2$ so there is no confusion in the meaning of the 2 in the exponent; that is, between *squared* superscript in σ^2 and the second derivative operator in $\partial^2/\partial\sigma^2$. The partial derivatives of $\ell(\beta_0, \beta_1, \phi)$ are then

$$\frac{\partial\ell}{\partial\beta_0} = -\frac{1}{2\phi}\sum_{i=1}^{n}2(y_i - \beta_0 - \beta_1 x_i)(-1)$$

$$\frac{\partial\ell}{\partial\beta_1} = -\frac{1}{2\phi}\sum_{i=1}^{n}2(y_i - \beta_0 - \beta_1 x_i)(-x_i)$$

$$\frac{\partial\ell}{\partial\phi} = -\frac{n}{2\phi} + \frac{1}{2\phi^2}\sum_{i=1}^{n}(y_i - \beta_0 - \beta_1 x_i)^2.$$

If we set the first two expressions equal to 0, we can simplify and then multiply both sides by $-\phi$, yielding two equations in two unknowns:

$$0 = \sum_{i=1}^{n}(y_i - \beta_0 - \beta_1 x_i)(-1) = -n\bar{y} + n\beta_0 + \beta_1 n\bar{x}$$

$$0 = \sum_{i=1}^{n}(y_i - \beta_0 - \beta_1 x_i)(-x_i) = -\sum_{i=1}^{n}x_i y_i + \beta_0\, n\bar{x} + \beta_1\sum_{i=1}^{n}x_i^2.$$

These are precisely the same two equations as in (14.8) and (14.9), which were derived in the search for the least squares estimators of β_0 and β_1. Thus, the least squares estimators from (14.12) and (14.11) are also the MLEs. This gives some additional justification to the use of least squares, as opposed to another method such as least absolute deviations. The result should be intuitive because in the log-likelihood function in (14.18) the β_0 and β_1 parameters appear only in the term containing the sum of squared residuals:

$$-\frac{1}{2\phi}\sum_{i=1}^{n}(y_i - \beta_0 - \beta_1 x_i)^2.$$

Because of the minus sign in front of the sum, maximizing this expression is equivalent to minimizing the sum of squared residuals, as in the method of least squares. (Note, too, that the constant $1/2\phi$ in front of the sum does not affect the values of β_0 and β_1 that yield the maximum.)

The MLE for ϕ can be obtained by solving for ϕ in

$$0 = \frac{\partial\ell}{\partial\phi} = -\frac{n}{2\phi} + \frac{1}{2\phi^2}\sum_{i=1}^{n}(y_i - \beta_0 - \beta_1 x_i)^2$$

$$0 \times 2\phi^2 = 2\phi^2 \times \left(-\frac{n}{2\phi} + \frac{1}{2\phi^2}\sum_{i=1}^{n}(y_i - \beta_0 - \beta_1 x_i)^2\right)$$

$$\phi = \frac{1}{n}\sum_{i=1}^{n}(y_i - \beta_0 - \beta_1 x_i)^2.$$

The sum of squared residuals is often denoted SSE (sum of squared errors[2]).

The MLEs for the three parameters β_0, β_1, and σ^2 are then

$$\hat{\beta}_1 = \frac{S_{xy}}{S_{xx}}$$

$$\hat{\beta}_0 = \bar{y} - \hat{\beta}_1 \bar{x}$$

$$\hat{\sigma}^2_{\text{MLE}} = \frac{1}{n}\sum_{i=1}^{n}(y_i - \hat{\beta}_0 - \hat{\beta}_1 x_i)^2 = \frac{1}{n}\text{SSE}.$$

Finally, note that the estimates must be obtained in this order, since each row in the above depends on the parameters in previous rows.

■ **Example 14.5** For the subset of 10 students from the GPA vs. SATM data set, estimate the parameter σ^2.

Solution: The MLE of σ^2 is

$$\hat{\sigma}^2_{\text{MLE}} = \frac{1}{10}\sum_{i=1}^{n}(y_i - 0.73276 - 0.00398x_i)^2 \approx 0.02687.$$ ■ ■ ■

In the next section we will show that the MLE of σ^2 is a biased estimator, and we will show how to use it to construct an unbiased estimator.

14.3 Properties of the Least Squares Estimators

Throughout, we have assumed that the response Y_i is a random variable; specifically,

$$Y_i \overset{\text{indep.}}{\sim} N(\beta_0 + \beta_1 x_i, \sigma^2).$$

The least squares estimators of β_0 and β_1 are functions of Y_i, $i = 1, 2, \ldots, n$ and so are random variables themselves. In this section we will derive some properties of these estimators. The next section will then use these properties to develop inference procedures (i.e., confidence intervals and hypothesis tests) for these parameters. In developing properties of these estimators, we will make use of several of the properties from Section 4.4, which we summarize here. For random variables Y_1, Y_2, $\ldots$, Y_n and constants c_1, c_2, $\ldots$, c_n:

$$\mathbb{E}\left(\sum_{i=1}^{n}c_i Y_i\right) = \sum_{i=1}^{n}c_i \mathbb{E}(Y_i)$$

$$\mathbb{V}(cY) = c^2 \mathbb{V}(Y)$$

$$\mathbb{V}\left(\sum_{i=1}^{n}c_i Y_i\right) = \sum_{i=1}^{n}c_i^2 \mathbb{V}(Y_i) \qquad [\text{provided } Y_i \text{ are independent}]$$

$$\text{cov}(c_1 Y_1, c_2 Y_2) = c_1 c_2 \text{cov}(Y_1, Y_2)$$

---

[2]"Sum of squared errors" is a bit of a misnomer, since the residuals $y_i - \hat{y}_i$ are distinct from the "error" term in the linear model $y_i = \beta_0 + \beta_1 x_i + \varepsilon_i$. Usually, ε_i is called the error term, which is different from the residual. We will use the term SSE for two reasons: (1) it is called SSE in most other books and software; and (2) the term SSR, which might stand for *sum of squared residuals*, is reserved for sum of squares due to regression. We will take this up in later sections of this chapter.

$$\mathbb{V}(Y_1 + Y_2) = \mathbb{V}(Y_1) + \mathbb{V}(Y_2) + 2\text{cov}(Y_1, Y_2)$$

$$\mathbb{V}(c_1 Y_1 + c_2 Y_2) = c_1^2 \mathbb{V}(Y_1) + c_2^2 \mathbb{V}(Y_2) + 2c_1 c_2 \text{cov}(Y_1, Y_2).$$

Theorem 14.3.1 — Distribution of $\hat{\beta}_1$. The distribution of the estimator $\hat{\beta}_1$ is

$$\hat{\beta}_1 \sim N\left(\beta_1, \frac{\sigma^2}{S_{xx}}\right).$$

Proof. We can write the estimator for β_1 as

$$\hat{\beta}_1 = \frac{1}{S_{xx}} \sum_{i=1}^{n} (x_i - \bar{x})(Y_i - \bar{Y})$$

$$= \frac{1}{S_{xx}} \left[\sum_{i=1}^{n} (x_i - \bar{x})Y_i + \sum_{i=1}^{n} (x_i - \bar{x})\bar{Y} \right]$$

$$= \frac{1}{S_{xx}} \left[\sum_{i=1}^{n} (x_i - \bar{x})Y_i + \bar{Y} \underbrace{\sum_{i=1}^{n} (x_i - \bar{x})}_{\text{equal to } 0} \right]$$

$$= \frac{1}{S_{xx}} \sum_{i=1}^{n} (x_i - \bar{x})Y_i.$$

This demonstrates that the random variable $\hat{\beta}_1$ is a linear combination of $Y_1, Y_2, \ldots, Y_n$, which are independent and normally distributed. Thus, $\hat{\beta}_1$ is normally distributed, but we don't yet know the mean and variance:

$$\hat{\beta}_1 \sim N\left(?, ?\right).$$

If we could find the mean and variance of $\hat{\beta}_1$, we would know precisely the distribution of $\hat{\beta}_1$. The mean, or expected value, can be found as follows:

$$\mathbb{E}(\hat{\beta}_1) = \mathbb{E}\left(\frac{1}{S_{xx}} \sum_{i=1}^{n} (x_i - \bar{x})Y_i \right)$$

$$= \frac{1}{S_{xx}} \sum_{i=1}^{n} (x_i - \bar{x})\mathbb{E}(Y_i)$$

$$= \frac{1}{S_{xx}} \sum_{i=1}^{n} (x_i - \bar{x})(\beta_0 + \beta_1 x_i)$$

$$= \frac{1}{S_{xx}} \left(\beta_0 \sum_{i=1}^{n} (x_i - \bar{x}) + \beta_1 \sum_{i=1}^{n} x_i(x_i - \bar{x}) \right)$$

$$= \frac{1}{S_{xx}} (\beta_0 \times 0 + \beta_1 S_{xx}) \qquad \text{[see Problem 14.1]}$$

$$= \beta_1.$$

Thus, $\hat{\beta}_1$ is an unbiased estimator of β_1. We can now summarize our knowledge of the distribution of $\hat{\beta}_1$ as

$$\hat{\beta}_1 \sim N\left(\beta_1, \, ?\right),$$

so all that is left is to find the variance of $\hat{\beta}_1$:

$$\begin{aligned}
\mathbb{V}(\hat{\beta}_1) &= \mathbb{V}\left(\frac{1}{S_{xx}}\sum_{i=1}^{n}(x_i-\bar{x})Y_i\right)\\
&= \left(\frac{1}{S_{xx}}\right)^2\sum_{i=1}^{n}\mathbb{V}((x_i-\bar{x})Y_i)\\
&= \frac{1}{S_{xx}^2}\sum_{i=1}^{n}(x_i-\bar{x})^2\mathbb{V}(Y_i)\\
&= \frac{1}{S_{xx}^2}\sum_{i=1}^{n}(x_i-\bar{x})^2\mathbb{V}(\beta_0+\beta_1 x_i+\varepsilon_i)\\
&= \frac{1}{S_{xx}^2}\sum_{i=1}^{n}(x_i-\bar{x})^2\mathbb{V}(\varepsilon_i) \qquad [\text{because } \beta_0+\beta_1 x_i \text{ is a constant}]\\
&= \frac{1}{S_{xx}^2}\sum_{i=1}^{n}(x_i-\bar{x})^2\sigma^2\\
&= \frac{1}{S_{xx}^2}\sigma^2 S_{xx}\\
&= \frac{\sigma^2}{S_{xx}}.
\end{aligned}$$

We can now write

$$\hat{\beta}_1 \sim N\left(\beta_1, \, \frac{\sigma^2}{S_{xx}}\right).$$

If we are willing to assume that ε_i are independent random variables with mean 0 and constant variance σ^2, but we're not willing to assume in addition that they are normally distributed, then it is still true that

$$\mathbb{E}(\hat{\beta}_1) = \hat{\beta}_1 \qquad \text{and} \qquad \mathbb{V}(\hat{\beta}_1) = \frac{\sigma^2}{S_{xx}},$$

but we are unable to conclude that the distribution of $\hat{\beta}_1$ is normal.

The result for $\hat{\beta}_0$ is similar, but is complicated by the fact that it depends on both $\overline{Y}$ and $\hat{\beta}_1$.

Theorem 14.3.2 — Distribution of $\hat{\beta}_0$. The distribution of the estimator $\hat{\beta}_0$ is

$$\hat{\beta}_0 \sim N\left(\beta_0, \, \sigma^2\frac{\sum_{i=1}^{n}x_i^2}{nS_{xx}}\right).$$

Proof. The expected value is determined as follows:

$$\mathbb{E}(\hat{\beta}_0) = \mathbb{E}(\overline{Y} - \hat{\beta}_1\overline{x})$$

$$= \mathbb{E}(\overline{Y}) - \overline{x}\mathbb{E}(\hat{\beta}_1)$$

$$= \mathbb{E}\left(\frac{1}{n}\sum_{i=1}^{n} Y_i\right) - \overline{x}\beta_1 \qquad [\text{because } \hat{\beta}_1 \text{ is unbiased}]$$

$$= \frac{1}{n}\mathbb{E}\left(\sum_{i=1}^{n} Y_i\right) - \overline{x}\beta_1$$

$$= \frac{1}{n}\sum_{i=1}^{n} \mathbb{E}(\beta_0 + \beta_1 x_i + \varepsilon_i) - \overline{x}\beta_1$$

$$= \frac{1}{n}\sum_{i=1}^{n} (\beta_0 + \beta_1 x_i + \mathbb{E}(\varepsilon_i)) - \overline{x}\beta_1$$

$$= \frac{1}{n}n\beta_0 + \frac{\beta_1}{n}n\overline{x} - \overline{x}\beta_1 \qquad [\text{because } \mathbb{E}(\varepsilon_i) = 0]$$

$$= \beta_0.$$

Thus, $\hat{\beta}_0$ is an unbiased estimator for β_0.

To find the variance of $\hat{\beta}_0$, we use the result for the variance of a difference of random variables:

$$V(\hat{\beta}_0) = \underbrace{\mathbb{V}(\overline{Y})}_{\text{①}} + \underbrace{\mathbb{V}(\overline{x}\hat{\beta}_1)}_{\text{②}} - \underbrace{2\,\text{cov}(\overline{Y}, \overline{x}\hat{\beta}_1)}_{\text{③}}.$$

Let's take these three terms one at a time. The first is

$$\text{①} = \mathbb{V}(\overline{Y})$$

$$= \mathbb{V}\left(\frac{1}{n}\sum_{i=1}^{n} Y_i\right)$$

$$= \frac{1}{n^2}\sum_{i=1}^{n} \mathbb{V}(Y_i) \qquad [\text{because the } Y_i \text{ are independent}]$$

$$= \frac{1}{n^2}\sum_{i=1}^{n} \mathbb{V}\left(\underbrace{\beta_0 + \beta_1 x_i}_{\text{not random}} + \underbrace{\varepsilon_i}_{\text{random}}\right)$$

$$= \frac{1}{n^2}\sum_{i=1}^{n} \mathbb{V}(\varepsilon_i)$$

$$= \frac{1}{n^2}\sum_{i=1}^{n} \sigma^2$$

$$= \frac{1}{n^2} n\,\sigma^2$$

$$= \frac{\sigma^2}{n}.$$

The second is

$$\text{②} = \mathbb{V}(\hat{\beta}_1 \bar{x}) = \bar{x}^2 \mathbb{V}(\hat{\beta}_1) = \frac{\bar{x}^2 \sigma^2}{S_{xx}}.$$

Finally, the third is

$$\text{③} = \text{cov}(\bar{Y}, \hat{\beta}_1 \bar{x})$$

$$= \text{cov}\left(\frac{1}{n} \sum_{i=1}^{n} Y_i, \bar{x} \frac{1}{S_{xx}} \sum_{j=1}^{n} (x_j - \bar{x}) Y_j \right)$$

$$= \frac{\bar{x}}{S_{xx}} \sum_{i=1}^{n} \sum_{j=1}^{n} \text{cov}(Y_i, (x_j - \bar{x}) Y_j)$$

$$= \frac{\bar{x}}{S_{xx}} \sum_{i=1}^{n} \sum_{j=1}^{n} (x_j - \bar{x}) \text{cov}(Y_i, Y_j).$$

The covariance between Y_i and Y_j is equal to zero unless $i = j$. Thus,

$$\text{③} = \frac{\bar{x}}{S_{xx}} \sum_{j=1}^{n} (x_j - \bar{x}) \text{cov}(Y_j, Y_j) = \frac{\bar{x}}{S_{xx}} \sum_{j=1}^{n} (x_j - \bar{x}) \sigma^2 = \frac{\bar{x}}{S_{xx}} \sigma^2 \underbrace{\sum_{j=1}^{n} (x_j - \bar{x})}_{\text{equal to } 0} = 0.$$

Now, let's put these all together:

$$\mathbb{V}(\hat{\beta}_0) = \text{①} + \text{②} - 2 \times \text{③}$$

$$= \frac{\sigma^2}{n} + \frac{\bar{x}^2 \sigma^2}{S_{xx}} - 2 \times 0$$

$$= \sigma^2 \left(\frac{1}{n} + \frac{\bar{x}^2}{S_{xx}} \right)$$

$$= \sigma^2 \left(\frac{S_{xx}}{n S_{xx}} + \frac{n \bar{x}^2}{n S_{xx}} \right)$$

$$= \frac{\sigma^2}{n S_{xx}} \left(S_{xx} + n \bar{x}^2 \right)$$

$$= \frac{\sigma^2}{n S_{xx}} \left(\sum_{i=1}^{n} (x_i - \bar{x})^2 + n \bar{x}^2 \right)$$

$$= \frac{\sigma^2}{n S_{xx}} \left(\sum_{i=1}^{n} x_i^2 - n \bar{x}^2 + n \bar{x}^2 \right)$$

$$= \frac{\sum_{i=1}^{n} x_i^2}{n S_{xx}} \sigma^2.$$

Finally, note that $\bar{Y}$ is normally distributed because it is the sum of independent (but not identically distributed) normal random variables. Also, from Theorem 14.3.1 we know that $\hat{\beta}_1$ is also normally distributed. Thus, $\hat{\beta}_0 = \overline{Y} - \overline{x}\hat{\beta}_1$ is normally distributed. This is the last piece we need to conclude that

$$\hat{\beta}_0 \sim N\left(\beta_0, \sigma^2 \frac{\sum_{i=1}^{n} x_i^2}{n S_{xx}}\right).$$

■

The last two theorems demonstrate that the marginal distributions for the estimators $\hat{\beta}_0$ and $\hat{\beta}_1$ are normal. The joint distribution of $\hat{\beta}_0$ and $\hat{\beta}_1$ is bivariate normal, which is discussed in Section 14.6.

The third parameter in the simple linear regression model is σ^2. Previously, we found that the MLE of σ^2 is

$$\sigma_{\text{MLE}}^2 = \frac{1}{n} \sum_{i=1}^{n} (Y_i - \overline{Y})^2 = \frac{\text{SSE}}{n}.$$

The MLE turns out to be a biased estimator of σ^2. In the next theorem, we find the expected value of SSE, which can be used to develop an unbiased estimator.

Theorem 14.3.3 For the simple linear regression model,

$$\mathbb{E}(\text{SSE}) = (n-2)\sigma^2.$$

Proof. We begin by manipulating the formula for SSE so that it is amenable to finding the expected value:

$$\text{SSE} = \sum_{i=1}^{n} (Y_i - \hat{Y})^2$$

$$= \sum_{i=1}^{n} (Y_i - \hat{\beta}_0 - \hat{\beta}_1 x_i)^2$$

$$= \sum_{i=1}^{n} \left(Y_i - (\overline{Y} - \hat{\beta}_1 \overline{x}) - \hat{\beta}_1 x_i\right)^2$$

$$= \sum_{i=1}^{n} \left((Y_i - \overline{Y}) - \hat{\beta}_1 (x_i - \overline{x})\right)^2$$

$$= \sum_{i=1}^{n} (Y_i - \overline{Y})^2 + \hat{\beta}_1^2 \sum_{i=1}^{n} (x_i - \overline{x})^2 - 2\hat{\beta}_1 \sum_{i=1}^{n} (x_i - \overline{x})(Y_i - \overline{Y})$$

$$= \sum_{i=1}^{n} Y_i^2 - n\overline{Y}^2 + \hat{\beta}_1^2 S_{xx} - 2\hat{\beta}_1 S_{xy}$$

$$= \sum_{i=1}^{n} Y_i^2 - n\overline{Y}^2 + \hat{\beta}_1^2 S_{xx} - 2(\hat{\beta}_1 S_{xx})$$

$$= \sum_{i=1}^{n} Y_i^2 - n\overline{Y}^2 - \hat{\beta}_1^2 S_{xx}.$$

Recall that for any random variable, say W, we can write $\mathbb{E}(W^2)$ as

$$\mathbb{E}(W^2) = \mathbb{V}(W) + \mathbb{E}(W)^2.$$

Using this for each of the three terms in the above formula for SSE, we obtain

$$
\begin{aligned}
\mathbb{E}(\mathrm{SSE}) &= \sum_{i=1}^{n} \mathbb{E}(Y_i)^2 - n\mathbb{E}(\overline{Y}^2) - S_{xx}\mathbb{E}(\hat{\beta}_1^2) \\
&= \sum_{i=1}^{n} [\mathbb{V}(Y_i) + \mathbb{E}(Y_i)^2] - n[\mathbb{V}(\overline{Y}) + \mathbb{E}(\overline{Y})^2] - S_{xx}\left[\mathbb{V}(\hat{\beta}_1) + \mathbb{E}(\hat{\beta}_1)^2\right] \\
&= \sum_{i=1}^{n} [\sigma^2 + (\beta_0 + \beta_1 x_i)^2] - n\left[\frac{\sigma^2}{n} + (\beta_0 + \beta_1\bar{x})^2\right] - S_{xx}\left[\frac{\sigma^2}{S_{xx}} + \beta_1^2\right] \\
&= n\sigma^2 + \sum_{i=1}^{n}(\beta_0 + \beta_1 x_i)^2 - \sigma^2 - n(\beta_0 + \beta_1\bar{x})^2 - \sigma^2 - S_{xx}\beta_1^2 \\
&= (n-2)\sigma^2 + \sum_{i=1}^{n}(\beta_0 + \beta_1 x_i)^2 - n(\beta_0 + \beta_1\bar{x})^2 - S_{xx}\beta_1^2 \\
&= (n-2)\sigma^2.
\end{aligned}
$$

We leave it as an exercise to show that the last three terms on the second-to-last line add to 0. Thus,

$$\mathbb{E}(\mathrm{SSE}) = (n-2)\sigma^2. \qquad \blacksquare$$

Exercise 14.1 Show that

$$\sum_{i=1}^{n}(\beta_0 + \beta_1 x_i)^2 - n(\beta_0 + \beta_1\bar{x})^2 - S_{xx}\beta_1^2 = 0,$$

thereby completing the proof of Theorem 14.3.3.

From the result of Theorem 14.3.3, we can see that the MLE $\hat{\sigma}_{\mathrm{MLE}}^2$ is a biased estimator since

$$\mathbb{E}\left(\hat{\sigma}_{\mathrm{MLE}}^2\right) = \mathbb{E}\left(\frac{1}{n}\mathrm{SSE}\right) = \frac{1}{n}(n-2)\sigma^2 = \frac{n-2}{n}\sigma^2.$$

We can construct an unbiased estimator with the right multiplicative constant. The estimator

$$\hat{\sigma}^2 = \frac{\mathrm{SSE}}{n-2}$$

is unbiased because

$$\mathbb{E}(\hat{\sigma}^2) = E\left(\frac{\mathrm{SSE}}{n-2}\right) = \frac{1}{n-2}(n-2)\sigma^2 = \sigma^2.$$

The expression $\mathrm{SSE}/(n-2)$ is called the mean squared error and is denoted

$$\mathrm{MSE} = \frac{\mathrm{SSE}}{n-2},$$

and it is an unbiased estimator of σ^2. Although the estimator is unbiased, it is model-dependent. If the model changes, say by adding more variables, then the estimator of σ^2 will change.

If we assume that the error terms in the simple linear regression model are normally distributed, then there is a distributional result that we will state, but not prove.

> **Theorem 14.3.4** If $Y_i = \beta_0 + \beta_1 x_i + \varepsilon_i$, where $\varepsilon_i \sim$ i.i.d. $N(0, \sigma^2)$, then $\dfrac{(n-2)\text{MSE}}{\sigma^2} \sim \chi^2(n-2)$.

Note that had we been given this result before Theorem 14.3.3, then the proof would have been considerably shorter. Since the expected value of the $\chi^2(v)$ distribution is equal to v, we can write

$$n - 2 = \mathbb{E}\left(\frac{(n-2)\text{MSE}}{\sigma^2}\right) = \mathbb{E}\left(\frac{\text{SSE}}{\sigma^2}\right) = \frac{1}{\sigma^2}\mathbb{E}(\text{SSE}).$$

From this we can conclude that

$$\mathbb{E}(\text{SSE}) = (n-2)\sigma^2.$$

In this section we derived some important distributional results for the estimators $\hat{\beta}_0, \hat{\beta}_1$, and $\hat{\sigma}^2 = \text{MSE}$. The next section uses these results to obtain confidence intervals and test statistics for these parameters.

14.4   Inference for Parameters of the Simple Linear Regression Model

In the last section we showed that the estimators $\hat{\beta}_0$ and $\hat{\beta}_1$ are unbiased for the parameters β_0 and β_1 of the simple linear regression model. We were also able to find the variances of these estimators. We can now standardize the results from Theorem 14.3.1 and 14.3.2 to obtain these distributional results:

$$\frac{\hat{\beta}_0 - \beta_0}{\sigma\sqrt{\sum_{i=1}^n x_i^2/(nS_{xx})}} \sim N(0,1)$$

and

$$\frac{\hat{\beta}_1 - \beta_1}{\sigma/\sqrt{S_{xx}}} \sim N(0,1).$$

From these results we could derive that

$$P\left(\hat{\beta}_0 - z_{\alpha/2}\sigma\sqrt{\frac{\sum_{i=1}^n x_i^2}{nS_{xx}}} < \beta_0 < \hat{\beta}_0 + z_{\alpha/2}\sigma\sqrt{\frac{\sum_{i=1}^n x_i^2}{nS_{xx}}}\right) = 1 - \alpha$$

and

$$P\left(\hat{\beta}_1 - z_{\alpha/2}\frac{\sigma}{\sqrt{S_{xx}}} < \beta_1 < \hat{\beta}_1 + z_{\alpha/2}\frac{\sigma}{\sqrt{S_{xx}}}\right) = 1 - \alpha.$$

Our goal in this section is to obtain confidence intervals and test statistics for the parameters β_0 and β_1. It might seem like the above expressions could be used to get confidence intervals, but the problem is that we do not know σ. If we were to use an estimator for σ or in place of σ as we did in Section 6.6, then we find that the statistics have t distributions rather than normal distributions.

> **Theorem 14.4.1 — t Distributions of Pivotal Quantities for $\hat{\beta}_0$ and $\hat{\beta}_1$.** If $Y_i \overset{\text{indep}}{\sim} N(\beta_0 + \beta_1 x_i, \sigma^2)$ and if $\hat{\beta}_0$ and $\hat{\beta}_1$ are the usual least squares estimators for β_0 and β_1, then
>
> $$T_0 = \frac{\hat{\beta}_0 - \beta_0}{S\sqrt{\sum_{i=1}^n x_i^2/(nS_{xx})}} \sim t(n-2)$$

and

$$T_1 = \frac{\hat{\beta}_1 - \beta_1}{S/\sqrt{S_{xx}}} \sim t(n-2),$$

where $S^2 = \dfrac{1}{n-2}\displaystyle\sum_{i=1}^{n}(Y_i - \hat{Y}_i)^2$.

Proof. For T_0 we have

$$
\begin{aligned}
T_0 &= \frac{\hat{\beta}_0 - \beta_0}{S\sqrt{\sum_{i=1}^{n} x_i^2/(nS_{xx})}} \\[2mm]
&= \frac{\hat{\beta}_0 - \beta_0}{\sigma\sqrt{\sum_{i=1}^{n} x_i^2/(nS_{xx})}}\frac{\sigma}{S} \\[2mm]
&= Z\sqrt{\frac{\sigma^2}{S^2}} \\[2mm]
&= \frac{Z}{\sqrt{\dfrac{S^2}{\sigma^2}\dfrac{(n-2)}{(n-2)}}} \\[2mm]
&= \frac{Z}{\sqrt{\dfrac{(n-2)S^2/\sigma^2}{n-2}}} \\[2mm]
&= \frac{Z}{\sqrt{\dfrac{W}{n-2}}},
\end{aligned}
$$

where $Z \sim N(0,1)$ and $W = (n-2)S^2/\sigma^2 \sim \chi^2(n-2)$. The random variables Z and W are independent, although we omit the proof here. Thus, by the definition of the t distribution in Section 6.6.2, we conclude that

$$T_0 \sim t(n-2).$$

We leave the second part of the proof (showing that $T_1 \sim t(n-2)$) as an exercise. ∎

Exercise 14.2 Complete the second half of the proof of the preceding theorem by showing that $\hat{\beta}_1$ can be written in the form

$$T_1 = \frac{Z}{\sqrt{\dfrac{W}{n-2}}},$$

where $Z \sim N(0,1)$ and $W \sim \chi^2(n-2)$. How are Z and W related to the data and the parameters?

With the results from Theorem 14.4.1, we are now able to obtain confidence intervals and test statistics for the parameters β_0 and β_1. We can write

$$\Pr\left(-t_{\alpha/2}(n-2) < \frac{\hat{\beta}_0 - \beta_0}{S\sqrt{\sum_{i=1}^{n} x_i^2/(nS_{xx})}} < t_{\alpha/2}(n-2) \right) = 1 - \alpha.$$

A little algebra on the inside of this probability statement yields

$$\Pr\left(\hat{\beta}_0 - t_{\alpha/2}(n-2)S\sqrt{\sum_{i=1}^{n} x_i^2/(nS_{xx})} < \beta_0 < \hat{\beta}_0 + t_{\alpha/2}(n-2)S\sqrt{\sum_{i=1}^{n} x_i^2/(nS_{xx})} \right) = 1 - \alpha.$$

A $100 \times (1-\alpha)\%$ confidence interval for β_0 is therefore

$$\left(\hat{\beta}_0 - t_{\alpha/2}(n-2)S\sqrt{\sum_{i=1}^{n} x_i^2/(nS_{xx})} \,,\; \hat{\beta}_0 + t_{\alpha/2}(n-2)S\sqrt{\sum_{i=1}^{n} x_i^2/(nS_{xx})} \;\right).$$

Similarly, we can obtain the $100 \times (1-\alpha)\%$ confidence interval for β_1, which is

$$\left(\hat{\beta}_1 - t_{\alpha/2}(n-2)\frac{S}{\sqrt{S_{xx}}} \,,\; \hat{\beta}_1 + t_{\alpha/2}(n-2)\frac{S}{\sqrt{S_{xx}}} \right).$$

We are more often interested in a confidence interval for β_1, since this tells us about the relationship between x and Y. Specifically, we are interested in whether the confidence interval includes (or excludes) 0. If the true value of β_1 is 0, then

$$Y_i = \beta_0 + 0 \times x_i + \varepsilon_i = \beta_0 + \varepsilon_i.$$

In other words, the distribution of Y_i does not depend on x_i in any way. Note that this assumes that the linear model is correct; this linearity is usually just an approximation to reality.

■ **Example 14.6** For the full GPA vs. SATM data, compute 95% confidence intervals for β_0 and β_1.

Solution: With $n = 219$, the degrees of freedom for the t distribution is $n - 2 = 217$. With this many degrees of freedom, the t distribution is very close to the normal. We can use the qt() function in R to get the 0.025 and 0.975 quantiles. This yields $t_{0.025}(217) = 1.971$, which is close to the normal quantile of $z_{0.025} = 1.96$. The following R code can be used to compute the necessary statistics in order to obtain the confidence intervals:

```
first = read.csv( "FirstYearGPA.csv" , header=TRUE )
x = first$SATM
y = first$GPA
ybar = mean(y)
xbar = mean(x)
Sxx = sum( (x-xbar)^2 )
Sxy = sum( (x-xbar)*(y-ybar) )
sx = sum( x^2 )
n = length( y )
b1.scr = Sxy / Sxx
b0.scr = ybar - b1.scr*xbar
yhat = b0.scr + b1.scr*x
SSE = sum( (y-yhat)^2 )
S = sqrt( SSE/(n-2) )
```

From this we can see that

$$S_{xx} = 1,233,965$$
$$S_{xy} = 1,483.705$$
$$\bar{y} = 3.0962$$
$$\bar{x} = 634.2922$$
$$\hat{\beta}_0 = 2.333499$$
$$\hat{\beta}_1 = 0.0012039$$
$$\sum_{i=1}^{n} x_i^2 = 89,343,500$$
$$\text{SSE} = 45.44959$$
$$S = 0.45765.$$

The confidence interval for β_0 is thus

$$\left(2.3335 - 1.971 \times 0.45765 \sqrt{\frac{89,343,500}{219 \times 1,233,965}} \;,\; 2.3335 + 1.971 \times 0.45765 \sqrt{\frac{89,343,500}{219 \times 1,233,965}} \right)$$

$$\left(2.3335 - 0.5187 \;,\; 2.3335 + 0.5187 \right)$$

$$\left(1.815 \;,\; 2.852 \right).$$

The confidence interval for β_1 is

$$\left(0.0012039 - 1.971 \frac{0.45765}{\sqrt{1,233,965}} \;,\; 0.0012039 + 1.971 \frac{0.45765}{\sqrt{1,233,965}} \right)$$

$$\left(0.00039 \;,\; 0.00202 \right).$$ ■■■

A confidence interval for σ^2 can be based on the pivotal quantity

$$\frac{(n-2)\text{SSE}}{\sigma^2} \sim \chi^2(n-2). \tag{14.19}$$

This result allows us to write

$$\Pr\left(\chi^2_{1-\alpha/2}(n-2) < \frac{(n-2)\text{SSE}}{\sigma^2} < \chi^2_{\alpha/2}(n-2) \right) = 1 - \alpha.$$

A little rearranging yields

$$\Pr\left(\frac{(n-2)\text{SSE}}{\chi^2_{\alpha/2}} < \sigma^2 < \frac{(n-2)\text{SSE}}{\chi^2_{1-\alpha/2}} \right) = 1 - \alpha,$$

which gives us the confidence interval

$$\left(\frac{(n-2)\text{SSE}}{\chi^2_{\alpha/2}} \;,\; \frac{(n-2)\text{SSE}}{\chi^2_{1-\alpha/2}} \right).$$

Exercise 14.3 Find a confidence interval for the error variance σ^2 in the GPA vs. SATM data set.

R has a built-in function `lm()` that can perform the calculations for linear regression. Once we construct vectors for the response y and the predictor x, we can fit the linear model $y_i = \beta_0 + \beta_1 x_i$ with the command `lm( y ~ x )`. This creates an object containing the important output. In our case, we have variables called GPA and SATM, so we could assign the linear model object to a variable called `lm1` and then apply the `summary()` function to view a summary of the output.

```
GPA = first$GPA
SATM = first$SATM
lm1 = lm( GPA ~ SATM )
summary( lm1 )
```

This yields the output

```
Call:
lm(formula = GPA ~ SATM)

Residuals:
Min      1Q  Median     3Q     Max
-1.1850 -0.3080  0.0409  0.3511  0.9752

Coefficients:
Estimate Std. Error t value Pr(>|t|)
(Intercept) 2.333499    0.263144    8.868 2.78e-16 ***
SATM        0.001202    0.000412    2.919  0.00389 **
---
Signif. codes:  0 '***' 0.001 '**' 0.01 '*' 0.05 '.' 0.1 ' ' 1

Residual standard error: 0.4577 on 217 degrees of freedom
Multiple R-squared:  0.03777,        Adjusted R-squared:  0.03334
F-statistic: 8.518 on 1 and 217 DF,  p-value: 0.003888
```

We can see the point estimates for β_0 and β_1 (2.333499 and 0.001202, respectively) as well as quite a bit more information which we will discuss shortly.

We can learn about the structure of the `lm1` object using the structure command `str()`:

```
str( lm1 )
```

Part of this output is contained below:

```
List of 12
$ coefficients : Named num [1:2] 2.3335 0.0012
..- attr(*, "names")= chr [1:2] "(Intercept)" "SATM"
$ residuals    : Named num [1:219] -0.1993 0.9508 0.3911 0.0348 0.413 ...
..- attr(*, "names")= chr [1:219] "1" "2" "3" "4" ...
$ effects      : Named num [1:219] -45.81905 -1.33566 0.46286 0.00113 0.45233 ...
..- attr(*, "names")= chr [1:219] "(Intercept)" "SATM" "" "" ...
$ rank         : int 2
$ fitted.values: Named num [1:219] 3.26 3.2 3.02 3.18 3.07 ...
..- attr(*, "names")= chr [1:219] "1" "2" "3" "4" ...
```

We can see that lm1 is a list of length 12, although only the first five are listed here. We can obtain the values contained in any of these 12 objects. For example, by running lm1$coefficients, we obtain the output

```
(Intercept)        SATM
2.333498796 0.001202388
```

You can usually obtain more accurate estimates of the parameter (in terms of additional decimal points) using this method, compared to reading the coefficients off of the output from summary(). You can also assign the estimates to variables if you need to use them subsequently. For example,

```
beta0hat = lm1$coefficients[1]
beta1hat = lm1$coefficients[2]
```

creates two new variables that contain the estimated coefficients.

The results of Theorem 14.4.1 can also be used to test a hypothesis about β_0 or β_1. Often, we are interested in testing $H_0 : \beta_1 = 0$ because if $\beta_1 = 0$ and if the linear model is correct, then the model becomes

$$Y_i = \beta_0 + 0 \times x_i = \beta_0,$$

in which case Y_i does not depend on the predictor variable x_i. We can, of course, test whether the parameter β_1 is equal to any particular value.

Let $\beta_0^{(0)}$ be any hypothesized value for β_0. To test $H_0 : \beta_0 = \beta_0^{(0)}$ against the alternative $H_1 : \beta_0 \neq \beta_0^{(0)}$ we use the test statistic

$$T = \frac{\hat{\beta}_0 - \beta_0^{(0)}}{S\sqrt{\sum_{i=1}^{n} x_i^2 / nS_{xx}}}, \tag{14.20}$$

which has a t distribution if H_0 is true. Similarly, we would base the test $H_0 : \beta_1 = \beta_1^{(0)}$ on the test statistic

$$T = \frac{\hat{\beta}_1 - \beta_1^{(0)}}{S/\sqrt{S_{xx}}}. \tag{14.21}$$

For both of these tests, with test statistics given in (14.20) and (14.21), the decision would be to reject H_0 if

$$T < -t_{\alpha/2}(n-2) \quad \text{or} \quad T > t_{\alpha/2}(n-2).$$

Exercise 14.4 Describe the rejection rule for one-sided tests involving β_0 or β_1.

■ **Example 14.7** For the data on first-year GPA and SATM score, test using $\alpha = 0.05$ whether the coefficient for SATM is zero against the alternative hypothesis that it is not zero.

Solution: For this data set $\hat{\beta}_0 = 2.3335$ and $\hat{\beta}_1 = 0.001202$. Also,

$$S_{xx} = \sum_{i=1}^{n}(x_i - \bar{x})^2 = 1,233,965$$

and

$$S^2 = \frac{\text{SSE}}{n-2} = \frac{45.44959}{219-2} = 0.20945.$$

The t statistic for testing $H_0 : \beta_1 = 0$ vs. $H_1 : \beta \neq 0$ is

$$t = \frac{\hat{\beta}_1 - 0}{S/\sqrt{S_{xx}}} = \frac{0.001202}{\sqrt{0.20945}/\sqrt{1,233,965}} \approx 2.919.$$

The critical value is $t_{0.025}(217) = 1.971$. We therefore reject H_0 if $t < -1.971$ or $t > 1.971$. Since $2.919 > 1.971$, we reject H_0 and conclude that there is evidence that the predictor SATM is associated with GPA. The P-value for this test is

$$P = 2P(T > 2.919) \approx 0.0039.$$

Thus, we would have rejected H_0 for smaller values of α; for example $\alpha = 0.01, \alpha = 0.005$, and even as low as $\alpha = 0.004$. This is strong evidence against $H_0 : \beta_1 = 0$. ■ ■ ■

We showed a scatterplot of GPA vs. SATM in Figure 14.2, and from this we concluded that there was large variability in the outcome GPA. For a student with a fixed value of SATM, the outcome GPA is difficult to predict because of this large variability. We will investigate this concept of prediction in Section 14.7. This seems to be at odds with the result of the previous example that concluded that there is strong evidence of a relationship between GPA and SATM. There is a key distinction to be made here between

1. the strength of the evidence of some relationship; and
2. the strength of the relationship.

In Example 14.7 we have strong evidence of a fairly weak relationship. This distinction is important in the era of big data. When there are a lot of data, even a weak relationship can become statistically significant. With the 219 data points in the previous example, we have plenty of information to say that first-year GPA is affected by the SATM score, but the relationship is not strong and it will be difficult to predict the GPA for a student having a particular SATM score.

14.5 Matrix Formulation of Simple Linear Regression

Matrices can be used to simplify some of the notation and results for linear regression, particularly when the number of predictor variables is large. We will first define the responses vector as the column vector:

$$y = \begin{bmatrix} y_1 \\ y_2 \\ \vdots \\ y_n \end{bmatrix}.$$

We will use capital letters if we think of the responses as random, and lower case if we think of them as fixed. The *model matrix* is defined as

$$X = \begin{bmatrix} 1 & x_1 \\ 1 & x_2 \\ \vdots & \vdots \\ 1 & x_n \end{bmatrix}.$$

The parameter vector for the linear model is

$$\beta = \begin{bmatrix} \beta_0 \\ \beta_1 \end{bmatrix}$$

and the vector of random errors is

$$\epsilon = \begin{bmatrix} \varepsilon_1 \\ \varepsilon_2 \\ \vdots \\ \varepsilon_n \end{bmatrix}.$$

With this notation, we can write the linear regression model

$$Y_i = \beta_0 + \beta_1 x_i + \varepsilon_i, \qquad \varepsilon_1, \varepsilon_2, \ldots, \varepsilon_n \sim \text{i.i.d. } N(0, \sigma^2)$$

as

$$Y = X\beta + \epsilon, \qquad \epsilon \sim N(0, \sigma^2 I),$$

where I is the $n \times n$ identity matrix, so the ϵ vector has mean $\mathbf{0}$ and a diagonal covariance matrix (with σ^2 along the diagonal). The least squares equations from (14.8) and (14.9) can then be written as

$$(X'X)\hat{\beta} = X'Y,$$

where X' indicates the transpose of X. Left multiplying both sides by $(X'X)^{-1}$ (assuming the inverse exists) yields

$$\hat{\beta} = (X'X)^{-1} X'Y.$$

When we study multiple regression later in this chapter, we will see that the estimator for the β parameter has this same form.

Many properties of the least squares and the ML estimators can be obtained directly using this vector and matrix notation. For example,

$$\mathbb{E}(Y) = \mathbb{E}(X\beta + \epsilon) = X\beta + E(\epsilon) = X\beta + \mathbf{0} = X\beta$$

and

$$\mathbb{E}(\hat{\beta}) = \mathbb{E}\left(X'X)^{-1} X'Y\right) = (X'X)^{-1} X' \mathbb{E}(Y) = (X'X)^{-1} X'X\beta = \beta,$$

since $(X'X)^{-1} X'X = I$, the identity matrix. It can be shown that

$$\text{cov}(\hat{\beta}) = \sigma^2 (X'X)^{-1}.$$

The predicted responses are contained in the vector

$$\hat{y} = X\hat{\beta},$$

and the residuals are contained in

$$r = y - \hat{y} = y - X\beta.$$

The sum of the squared residuals is then the scalar

$$\text{SSE} = r'r = \left(y - X\beta\right)'\left(y - X\beta\right),$$

so the unbiased estimator for σ^2 is

$$\hat{\sigma}^2 = \text{MSE} = \frac{\text{SSE}}{n-2} = \frac{1}{n-2}\left(y - X\beta\right)'\left(y - X\beta\right).$$

It can also be shown that the joint distribution of $\hat{\beta}$ is bivariate normal:

$$\hat{\beta} \sim N\left(\beta, \sigma^2 (X'X)^{-1}\right).$$

Finally, it can be shown that the random vector $\hat{\beta}$ and the random variable $\hat{\sigma}^2$ are independent.

14.6 Joint Confidence Regions

The results so far have given the marginal distributions for $\hat{\beta}_0$ and $\hat{\beta}_1$ and have led to methods for obtaining confidence intervals for β_0 and β_1. But each of these intervals must be evaluated in a one-at-a-time fashion. We are 95% confident that the interval for β_0 contains the true β_0, and we are 95% confident that the interval for β_1 contains the true β_1. We are *not* 95% confident that R contains both parameters. Next, we address the issue of a simultaneous confidence region, that is a region R for which we can say "we are 95% confident that both parameters are in R." The joint distribution of the vector $\hat{\beta} = [\hat{\beta}_0, \hat{\beta}_1]'$ is

$$\hat{\beta} = \begin{bmatrix} \hat{\beta}_0 \\ \hat{\beta}_1 \end{bmatrix} \sim N\left(\begin{bmatrix} \beta_0 \\ \beta_1 \end{bmatrix}, \sigma^2 (X'X)^{-1}\right),$$

or, using vector and matrix notation,

$$\hat{\beta} \sim N\left(\beta, \sigma^2 (X'X)^{-1}\right).$$

It can be shown that

$$(\hat{\beta} - \beta)'\left[\sigma^2 (X'X)^{-1}\right]^{-1}(\hat{\beta} - \beta) \sim \chi^2(2).$$

This can be written as

$$\frac{1}{\sigma^2}(\hat{\beta} - \beta)'(X'X)(\hat{\beta} - \beta) \sim \chi^2(2). \tag{14.22}$$

We also know that

$$\frac{(n-2)\text{MSE}}{\sigma^2} = \frac{\text{SSE}}{\sigma^2} \sim \chi^2(n-2). \tag{14.23}$$

Furthermore, since $\hat{\beta}$ and MSE are independent, we know that the random variables in (14.22) and (14.23) are independent. In Section 6.6.3 we defined an F random variable to be of the form

$$F = \frac{V_1/v_1}{V_2/v_2},$$

where $V_1 \sim \chi^2(v_1)$, $V_2 \sim \chi^2(v_2)$, and V_1 and V_2 are independent. Thus,

$$\frac{\frac{1}{\sigma^2}(\hat{\beta}-\beta)(X'X)(\hat{\beta}-\beta)/2}{\frac{\text{SSE}}{\sigma^2}/(n-2)} \sim F(2,n-2).$$

This allows us to write

$$\Pr\left(\frac{(\hat{\beta}-\beta)'(X'X)(\hat{\beta}-\beta)/2}{\text{SSE}/(n-2)} < F_\alpha(2,n-2)\right) = 1-\alpha. \tag{14.24}$$

A $100\times(1-\alpha)\%$ confidence region for β consists of all of those values of β for which

$$(\hat{\beta}-\beta)'(X'X)(\hat{\beta}-\beta) < \frac{2\,\text{SSE}}{n-2}F_\alpha(2,n-2). \tag{14.25}$$

■ **Example 14.8** For the GPA vs. SATM data, find a 95% joint confidence region for $\beta = [\beta_0,\beta_1]$.

Solution: For this data set,

$$X'X = \begin{bmatrix} 219 & 138,910 \\ 138,910 & 89,343,500 \end{bmatrix}$$

and

$$S^2 = \text{MSE} = \frac{\text{SSE}}{n-2} = 0.2095.$$

The confidence region therefore consists of those values of $\beta = [\beta_0,\beta_1]$ satisfying

$$[2.3335-\beta_0, 0.001202-\beta_1]'\begin{bmatrix} 219 & 138,910 \\ 138,910 & 89,343,500 \end{bmatrix}\begin{bmatrix} 2.3335-\beta_0 \\ 0.001202-\beta_1 \end{bmatrix}$$
$$\leq 2\times0.2095\times3.037 = 1.2725.$$

The region formed in this way is an ellipse (which includes the interior). Graphing the boundary of this by hand is a challenging task. Fortunately, the R package ellipse can be used to plot the confidence ellipse. The R code for this is given below, and the graph of the confidence region is shown in Figure 14.6:

```
first = read.csv( "FirstYearGPA.csv" , header=TRUE )
library( ellipse )

simpReg = lm( GPA ~ SATM , data=first )
confRegion = ellipse( simpReg, which = c(1, 2), level = 0.95,
t = sqrt(2 * qf(0.95, 2, simpReg$df.residual)) )

plot( confRegion , type="l" , xlim=c(1,3.5) , ylim=c(0,0.0025) )
polygon( confRegion , col="skyblue" )
```

The elliptical confidence region is quite narrow and shows a negative association. If β_0 is high, we would believe that the slope β_1 is small, and vice versa. The joint confidence region conveys much more information about the parameters than the two individual confidence intervals. Figure 14.6 also shows the confidence intervals for the two parameters individually superimposed on the joint confidence interval. ■ ■ ■

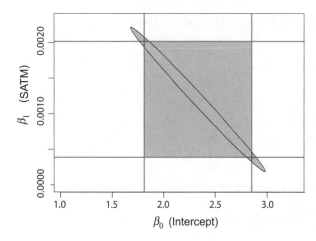

Figure 14.6 Joint confidence region for β_0 and β_1 in the first-year GPA and SATM data set.

14.7 Confidence and Prediction Intervals for Responses

In the linear model,

$$Y_i = \beta_0 + \beta_1 x_i + \varepsilon_i, \qquad \varepsilon_i \sim \text{i.i.d. } N(0, \sigma^2),$$

the expected value of the response (or outcome) is equal to

$$\mathbb{E}(Y_i) = \mathbb{E}(\beta_0 + \beta_1 x_i + \varepsilon_i) = \beta_0 + \beta_1 x_i + \mathbb{E}(\varepsilon_i) = \beta_0 + \beta_1 x_i.$$

A point estimate for this parameter[3] is

$$\widehat{E(Y_i)} = \hat{\beta}_0 + \hat{\beta}_1 x_i.$$

We would now like to obtain a confidence interval for $\widehat{E(Y_i)}$.

Theorem 14.7.1 Under the linear model with normally distributed error terms,

$$\frac{\widehat{E(Y_i)} - (\beta_0 + \beta_1 x)}{S\sqrt{\dfrac{1}{n} + \dfrac{(\bar{x} - x)^2}{S_{xx}}}} \sim t(n-2),$$

where x is any valid level of the predictor variable and $S^2 = \text{MSE}$ is the usual unbiased estimator for σ^2.

Proof. Using our rules for variances, we have

$$\mathbb{V}\left(\widehat{E(Y_i)}\right) = \mathbb{V}\left(\hat{\beta}_0 + \hat{\beta}_1 x\right)$$

$$= \mathbb{V}\left(\hat{\beta}_0\right) + x^2 \mathbb{V}\left(\hat{\beta}_1\right) + 2x \operatorname{cov}\left(\hat{\beta}_0, \hat{\beta}_1\right)$$

---

[3] Any function of parameters is a parameter, and this expression is a function of both β_0 and β_1 (as well as the predictor x_i). The important point here is that the expected response is a fixed quantity. By contrast, the response for a single subject or unit is a random variable. We will address this problem later in this section.

$$= \sigma^2 \frac{1}{n\,S_{xx}} \sum_{i=1}^{n} x_i^2 + x^2 \frac{\sigma^2}{S_{xx}} + 2x\sigma^2 \left(-\frac{\bar{x}}{S_{xx}}\right)$$

$$= \frac{\sigma^2}{S_{xx}} \left[\frac{1}{n} \sum_{i=1}^{n} x_i^2 + x^2 - 2x\bar{x} + \bar{x}^2 - \bar{x}^2\right]$$

$$= \frac{\sigma^2}{S_{xx}} \left[\frac{1}{n} \left(\sum_{i=1}^{n} x_i^2 - n\bar{x}^2\right) + (x-\bar{x})^2\right]$$

$$= \sigma^2 \left[\frac{1}{n} + \frac{(x-\bar{x})^2}{S_{xx}}\right]. \tag{14.26}$$

Thus,

$$\frac{\widehat{E(Y_i)} - (\beta_0 + \beta_1 x)}{\sigma \sqrt{\frac{1}{n} + \frac{(\bar{x}-x)^2}{S_{xx}}}} \sim N(0,1).$$

If we use S in place of σ, then

$$\frac{\widehat{E(Y_i)} - (\beta_0 + \beta_1 x)}{S \sqrt{\frac{1}{n} + \frac{(\bar{x}-x)^2}{S_{xx}}}} = \frac{\dfrac{\widehat{E(Y_i)} - (\beta_0 + \beta_1 x)}{\sigma \sqrt{\frac{1}{n} + \frac{(x-\bar{x})^2}{S_{xx}}}}}{\dfrac{S}{\sigma}}$$

$$= \frac{Z}{\sqrt{S^2/\sigma^2}}$$

$$= \frac{Z}{\sqrt{\frac{(n-2)S^2}{\sigma^2}/(n-2)}}$$

$$= \frac{Z}{\sqrt{V/v}},$$

where $v = n-2$, $Z \sim N(0,1)$, $V \sim \chi^2(v)$, and Z and V are independent. ∎

With the result from the preceding theorem, we can write

$$\Pr\left(-t_{\alpha/2} < \frac{\widehat{E(Y_i)} - (\beta_0 + \beta_1 x)}{S \sqrt{\frac{1}{n} + \frac{(\bar{x}-x)^2}{S_{xx}}}} < t_{\alpha/2}\right) = 1 - \alpha.$$

Substituting $\hat{\beta}_0 + \hat{\beta}_1 x$ for $\widehat{E(Y_i)}$, and doing a little algebra inside the probability, yields

$$\Pr\left(\hat{\beta}_0 + \hat{\beta}_1 x - t_{\alpha/2} S \sqrt{\frac{1}{n} + \frac{(x-\bar{x})^2}{S_{xx}}} < \beta_0 + \beta_1 x < \hat{\beta}_0 + \hat{\beta}_1 x + t_{\alpha/2} S \sqrt{\frac{1}{n} + \frac{(x-\bar{x})^2}{S_{xx}}}\right) = 1 - \alpha.$$

The value of x need not be one of the x values that occurred in the sample data set. To avoid confusion between x_i and the x for which the expected response is desired, we often use x^*. This leads to the $100 \times (1 - \alpha)\%$ confidence interval for $\beta_0 + \beta_1 x^*$:

$$\left(\hat{\beta}_0 + \hat{\beta}_1 x^* - t_{\alpha/2} S \sqrt{\frac{1}{n} + \frac{(x^* - \bar{x})^2}{S_{xx}}} \, , \, \hat{\beta}_0 + \hat{\beta}_1 x^* + t_{\alpha/2} S \sqrt{\frac{1}{n} + \frac{(x^* - \bar{x})^2}{S_{xx}}} \right). \tag{14.27}$$

■ **Example 14.9** For the first-year GPA vs. SATM data, find confidence intervals for the expected response (GPS) for values of SATM equal to 500, 600, and 700.

Solution: The following R code computes the confidence interval in (14.27):

```
first = read.csv( "FirstYearGPA.csv" , header=TRUE )
SATM = first$SATM
GPA = first$GPA
x = SATM
y = GPA
ybar = mean(y)
xbar = mean(x)
Sxx = sum( (x-xbar)^2 )
Sxy = sum( (x-xbar)*(y-ybar) )
lm1 = lm( GPA ~ SATM )
beta0.hat = lm1$coefficients[1]
names(beta0.hat) = ""
beta1.hat = lm1$coefficients[2]
names(beta1.hat) = ""
n = length( x )
yhat = beta0.hat + beta1.hat*x
SSE = sum( (y-yhat)^2 )
S = sqrt( SSE/(n-2) )

xstar = 500.0
t = qt(0.975,n-2)
LConfL = beta0.hat + beta1.hat*xstar - t*S*sqrt(1/n+(xstar-xbar)^2/Sxx)
UConfL = beta0.hat + beta1.hat*xstar + t*S*sqrt(1/n+(xstar-xbar)^2/Sxx)
print( c( LConfL , UConfL ) )
```

The first few lines of this code load in the data, compute the required quantities, and extract the estimated coefficients from the lm1 object. We then define xstar to be 500 and compute the confidence interval. The output is

```
2.809768 3.059618
```

If we change xstar to 600 and 700, we obtain the confidence intervals:

```
2.987920 3.121943
```

and

```
3.094165 3.256177
```

respectively. Given the 219 observations that we have, we can give a fairly narrow confidence interval for the expected or mean response.   ■ ■ ■

■ **Example 14.10** Compute and plot the confidence intervals for the expected outcome for values of SATM between 400 and 800.

Solution: We of course can't compute the confidence interval for *every* value of SATM between 400 and 800 (because there are infinitely many such values), but we can get a confidence interval for each value on a fine grid of values. Let's take x^* to be the vector of SATM values from 400 to 800 in increments of 1; that is, 400, 401, 402, ..., 799, 800. We can then compute and plot the confidence intervals for these values, and because there are so many, it will appear as a smooth curve.

We can use the predict() function in R to obtain the fitted value along with the corresponding confidence interval for a grid of x^* values. The predict() function takes as arguments the object that is the output from the lm() command, lm1 in this case. The second argument is a data frame consisting of one variable, the values of x^* for which we'd like the confidence interval. It is important that this be a data frame, not simply a vector of values for which we'd like the confidence interval. The third argument is either "confidence" or "prediction". For now we will use "confidence"; we will consider prediction intervals later in this section.

The following R code defines x^*, computes the confidence bands, and plots them:

```
xstar.vec = data.frame( SATM = 400:800 )
conf = predict( lm1 , xstar.vec , interval = "confidence" )
plot( SATM , GPA , pch=19 , col="red" , xlab="SAT Math" ,
ylab="First Year GPA" , main="Pointwise Confidence Intervals" )
lines( xstar.vec$SATM , conf[,"fit"] , lwd=2 , col="orange" )
lines( xstar.vec$SATM , conf[,"lwr"] , lwd=2 , col="orange")
lines( xstar.vec$SATM , conf[,"upr"] , lwd=2 , col="orange")
```

Figure 14.7 shows the scatterplot with the confidence intervals. Because we chose a grid of x^* to have so many values, the curves appear to be smooth, even though they are, strictly speaking, connected line segments. These are often called *confidence bands* for the mean response.

Note that these bands agree with the confidence intervals from the previous example. Also, note that the confidence bands are fairly narrow, indicating that we have a good estimate of the mean first-year GPA given a particular value of SATM.   ■ ■ ■

There are three important points to mention about the results of the last example.

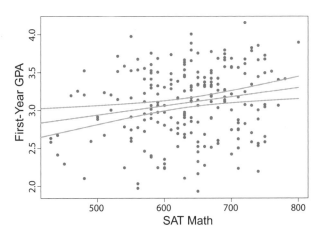

Figure 14.7 Ninety-five percent confidence bands for the expected response.

1. These are *pointwise* intervals. Thus, for a fixed value of SATM, the confidence bands provide a confidence interval for the mean outcome, first-year GPA, at that level of SATM. This is true for each individual value. For example, if we compute confidence intervals for the mean outcome for SATM= 500 and for SATM= 600, then we are 95% confident that the former interval (which, using the result from Example 14.9 was $[2.809768, 3.059618]$) contains the true mean response $\beta_0 + \beta_1 \times 500$. We are also 95% confident that the latter interval $[2.987920, 3.121943]$ contains the true mean response $\beta_0 + \beta_1 \times 600$. We are *not* 95% confident that simultaneously both intervals contain both mean responses. In other words, the intervals are *pointwise*, not *simultaneous*.

2. The interval is for the *mean response*, not the response of a particular student. We can see from Figure 14.7 that it would be very difficult to predict the outcome of a particular student whose SATM was 500. We have enough data to get an accurate estimate of the mean response, but there is too much variability in the outcome to give an accurate prediction of the GPA for a particular student.

3. Even though the fitted response is linear, the confidence bands are not. The confidence bands are narrowest near the center of the data, and widest at the low and high extremes of x^*. The formula for the variance of the estimated response,

$$\mathbb{V}\left(\widehat{E(Y_i)}\right) = \sigma^2 \left(\frac{1}{n} + \frac{(x^* - \bar{x})^2}{S_{xx}}\right),$$

involves the term $(x - \bar{x})^2$. This term is minimized when $x = \bar{x}$ and it increases as x moves away from $\bar{x}$. Thus, the confidence bands are thinnest at the mean $\bar{x}$ and are widest at the periphery of the data.

There are times when we would like to make inference for the mean response, $\mathbb{E}(Y|x)$, but there are also times when we want to make an inference for a future observation Y^* for a given value of x^*. In the GPA vs. SATM data set, it would probably be useful to predict the GPA of a student with a particular SATM score, say x^*. We would write the point prediction as

$$\hat{y}^* = \hat{\beta}_0 + \hat{\beta}_1 x^*.$$

At this point, we must introduce an important distinction in statistics. We *estimate* parameters and we *predict* random variables. An interval for a parameter is called a *confidence interval*, whereas an interval for a random variable is called a *prediction interval*. Thus, an interval for the mean response, $\beta_0 + \beta_1 x^*$ is a confidence interval. If we would like an interval for the outcome of a particular student with an SATM score of x^*, then we need a prediction interval.

In order to obtain a prediction interval for Y^* given an x^*, we must take into account the uncertainty in the estimates for β_0 and β_1, as well as the variability in the error term, ε^*. The error term ε^* for a future observation is independent of $\hat{\beta}_0$ and $\hat{\beta}_1$ because $\hat{\beta}_0$ and $\hat{\beta}_1$ are based on the original n observations and the error term for the future observation is independent of the data that made up the sample. The outcome for a future observation Y^* can be written as

$$Y^* = \hat{\beta}_0 + \hat{\beta}_1 x^* + \varepsilon^*,$$

where ε^* is the random error. The variance of the response $\hat{Y}^*$ is therefore

$$\mathbb{V}(\hat{Y}^*) = \mathbb{V}\left(\hat{\beta}_0 + \hat{\beta}_1 x^* + \varepsilon^*\right)$$

$$= \mathbb{V}\left(\hat{\beta}_0 + \hat{\beta}_1 x^*\right) + \mathbb{V}(\varepsilon^*) + 2 \operatorname{cov}\left(\hat{\beta}_0 + \hat{\beta}_1 x^*, \varepsilon^*\right)$$

$$= \sigma^2 \left(\frac{1}{n} + \frac{(x^* - \bar{x})^2}{S_{xx}}\right) + \sigma^2 + 2 \times 0$$

$$= \sigma^2 \left(1 + \frac{1}{n} + \frac{(x^* - \bar{x})^2}{S_{xx}}\right).$$

The variance of $\hat{\beta}_0 + \hat{\beta}_1 x^*$ (the first expression on the second line above) was derived earlier, in (14.26). Also, we've used the result that $\hat{\beta}_0 + \hat{\beta}_1 x^*$ and ε^* are independent, and therefore uncorrelated. The statistic

$$Z = \frac{Y^* - \hat{Y}^*}{\sigma\sqrt{1 + \dfrac{1}{n} + \dfrac{(x^* - \bar{x})^2}{S_{xx}}}}$$

is therefore a standard normal random variable. As we've seen several times before, replacing σ with $S = \sqrt{\text{SSE}/(n-2)}$ leads to the t distribution:

$$T = \frac{Y^* - \hat{Y}^*}{S\sqrt{1 + \dfrac{1}{n} + \dfrac{(x^* - \bar{x})^2}{S_{xx}}}} \sim t(n-2).$$

A $100 \times (1 - \alpha)\%$ prediction interval for Y^* is therefore

$$\left(\left(\hat{\beta}_0 + \hat{\beta}_1 x^*\right) - t_{\alpha/2}(n-2)S\sqrt{1 + \frac{1}{n} + \frac{(x^* - \bar{x})^2}{S_{xx}}} \, , \, \left(\hat{\beta}_0 + \hat{\beta}_1 x^*\right) + t_{\alpha/2}(n-2)S\sqrt{1 + \frac{1}{n} + \frac{(x^* - \bar{x})^2}{S_{xx}}} \right).$$

The formulas for a confidence interval for $E(Y^*|x^*) = \beta_0 + \beta_1 x^*$ and a prediction interval for the future observation Y^* are similar, and differ only in the presence of the "1" term inside the square root. This is, however, an important part of the formula for the prediction interval because it is what accounts for the variability in outcome for a fixed x^*.

If we were to have an infinite amount of data, giving us precise estimates of β_0 and β_1, we still would not be able to predict the response for a particular subject; this response has the inherent error ε^*, which has variance σ^2. Since we never have precise estimates of parameters, we must account for the uncertainty in the parameter estimates as well as the inherent variability in the response.

■ **Example 14.11** For the SAT vs. GPA data set, make 95% prediction intervals for students with SATM scores of 500, 600, and 700.

Solution: The following codes the required calculations:

```
xstar = 500.0
t = qt(0.975,n-2)
LPredL = beta0.hat + beta1.hat*xstar - t*S*sqrt(1+1/n+(xstar-xbar)^2/Sxx )
UPredL = beta0.hat + beta1.hat*xstar + t*S*sqrt(1+1/n+(xstar-xbar)^2/Sxx )
print( c( LPredL , UPredL ) )
```

The result from R is

```
2.024072 3.845314
```

giving $[2.024, 3.845]$ as the prediction interval. The prediction intervals for $x^* = 500$, 600, and 700 are shown in Table 14.1. Notice how for every value of x^* we are quite sure about the expected response, as evidenced by the narrow confidence interval, but we are quite *unsure* about the response for a particular student having that same value of x^*. From this data set we have strong evidence that first-year GPA is associated with SATM (recall the small P-value for testing $H_0 : \beta_1 = 0$), but because of the high variability in the response, we have a difficult time predicting the GPA for a particular student. ■ ■ ■

Table 14.1 Comparison of confidence interval for the mean first-year GPA and a prediction interval for an individual's first-year GPA.

x^*	Point estimate or prediction	95% confidence interval for mean response	95% prediction interval for future obs
500	2.935	(2.810, 3.060)	(2.024, 3.845)
600	3.055	(2.988, 3.122)	(2.150, 3.959)
700	3.175	(3.094, 3.256)	(2.270, 4.081)

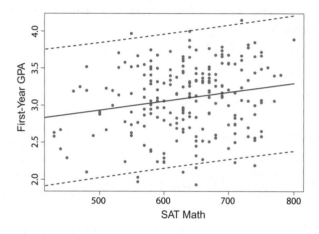

Figure 14.8 Ninety-five percent prediction bands for the expected response.

■ **Example 14.12** Plot 95% prediction bands for the GPA vs. SATM data set.

Solution: The R code necessary to compute and plot the prediction intervals follows the same pattern as for confidence intervals, as described in Example 14.10; the only difference is that we send the argument `interval="prediction"` when we call the `predict` function. The following R code computes and plots the prediction bands. The result is shown in Figure 14.8.

```
xstar.vec = data.frame( SATM = 400:800 )
pred = predict( lm1 , xstar.vec , interval = "prediction" )
plot( SATM , GPA , pch=19 , col="red" , xlab="SAT Math" ,
ylab="First Year GPA" , main="Pointwise Prediction Intervals" )
lines( xstar.vec$SATM , pred[,"fit"] , lwd=2 , col="darkgreen" )
lines( xstar.vec$SATM , pred[,"lwr"] , lwd=2 , col="darkgreen" , lty=2 )
lines( xstar.vec$SATM , pred[,"upr"] , lwd=2 , col="darkgreen" , lty=2 )
```

Notice that the prediction bands are substantially wider than the confidence intervals. This is because the prediction interval must account for the uncertainty in the parameter estimates *and* the inherent variability of the response. As was the case for confidence intervals for the mean response, these prediction bounds are narrowest at $\bar{x}$ and wider on the periphery of the x values.   ■ ■ ■

14.8 Optimal Selection of Levels of Predictor Variables

In some problems, the predictor variable is under the control of the investigators. For example, the outcome of an industrial experiment may depend on temperature, which can be set before the experiment is conducted. Choosing the settings for the experiment is a problem in the field of statistics called **design of experiments** (DOE). Is it possible to select the values of the predictor variables in an optimal way? We would first have to define a criterion to measure the "goodness" of a design. One possible criterion involves the variance of the estimator for β_1. Usually, β_1 is the more important parameter since it gives the slope of the linear relationship between x and Y.

■ **Example 14.13** Suppose that the outcome of an industrial experiment is a measure of the product quality and the predictor variable is temperature, which is constrained to be between 200°C and 300°C. If we want to choose the xs (temperature) to minimize the variance of the estimator $\hat{\beta}_1$, which of the following designs is best for a budget of $n = 4$ runs?

Design 1 : $x_1 = 200,$ $x_2 = 233,$ $x_3 = 267,$ $x_4 = 300$

Design 2 : $x_1 = 200,$ $x_2 = 250,$ $x_3 = 250,$ $x_4 = 300$

Design 3 : $x_1 = 200,$ $x_2 = 200,$ $x_3 = 300,$ $x_4 = 300$

Design 4 : $x_1 = 245,$ $x_2 = 245,$ $x_3 = 255,$ $x_4 = 255$

Solution: The variance of β_1 is

$$V(\hat{\beta}_1) = \frac{\sigma^2}{S_{xx}}.$$

The variance of the outcome σ^2 is not under the control of the experimenters, but S_{xx} is. Since S_{xx} is in the denominator of $V(\hat{\beta}_1)$, larger values of S_{xx} will lead to smaller values of $V(\hat{\beta}_1)$. Note that S_{xx} is equal to $(n-1)s_x^2$, where s_x^2 is the sample variance of the xs. Thus, the optimal design by the criterion of minimizing $V(\hat{\beta}_1)$ is the one that maximizes s_x^2. Some straightforward calculations in R give

$S_{xx} = 5,578$ for Design 1

$S_{xx} = 5,000$ for Design 2

$S_{xx} = 10,000$ for Design 3

$S_{xx} = 100$ for Design 4.

Note that the variances of $\hat{\beta}_1$ for the four designs are

$V(\hat{\beta}_1) = \dfrac{\sigma^2}{5,578}$ for Design 1

$V(\hat{\beta}_1) = \dfrac{\sigma^2}{5,000}$ for Design 2

$V(\hat{\beta}_1) = \dfrac{\sigma^2}{10,000}$ for Design 3

$V(\hat{\beta}_1) = \dfrac{\sigma^2}{100}$ for Design 4.

Each of these depends on the value of σ^2, which is unknown. Regardless of the value of σ^2, the variance of $\hat{\beta}_1$ will be least (best) for Design 3, and greatest (worst) for Design 4. Thus, we can find the best design among these four designs without knowing σ^2, since σ^2 is the same for all designs.

The best design among these four is Design 3, which puts two points at the left end of the design region and two points at the right end. (Remember, the temperature was constrained to be between 200°C and 300°C.) Design 1, which spreads the points roughly evenly between 200°C and 300°C, is the second best. The worst design is Design 4, which puts most points near the middle of the design region. ■ ■ ■

The optimal design using this criterion is the one that maximizes S_{xx}, or equivalently the one that maximizes the simple variance s_x^2. It can be shown that if n is even, then putting half of the design points at the left endpoint and the other half at the right endpoint will maximize s_x^2. If n is odd, then putting $(n-1)/2$ of them at the left endpoint and $(n+1)/2$ at the right endpoint (or vice versa) will maximize s_x^2.

We have barely scratched the surface of the field of optimal design of experiments. Our criterion was to minimize the variance of the estimate of the slope parameter $\hat{\beta}_1$. Here is a summary of other criteria:

1. D-optimal. Maximize the determinant of $X'X$. This minimizes the area of the joint confidence region for (β_0, β_1).
2. A-optimal. Minimize the average variance of the estimators $\hat{\beta}_0$ and $\hat{\beta}_1$. This is equivalent to minimizing the trace (sum of diagonal elements) of $(X'X)^{-1}$.
3. I-optimal. Minimize the average (or integrated) variance of a predicted value over some interval.
4. G-optimal. Minimize the maximum variance of predicted values over some interval.

There are others, and the interested reader is referred to Jones and Montgomery (2019).

14.9   The ANOVA Table for Simple Linear Regression

We can summarize the overall variability of the response y through the *total sum of squares* (SST), which is defined as

$$\text{SST} = \sum_{i=1}^{n} (y_i - \bar{y})^2.$$

This sum of squares can be decomposed into the variability explained by the regression and the variability of the residuals. These two sums of squares are the *sum of squares due to regression*,

$$\text{SSR} = \sum_{i=1}^{n} (\hat{y}_i - \bar{y})^2,$$

and the *sum of squared residuals*, which we called SSE, which is defined to be

$$\text{SSE} = \sum_{i=1}^{n} (y_i - \hat{y}_i)^2.$$

One of the fundamental results in the theory of linear regression is that SST is the sum of SSR and SSE.

Theorem 14.9.1 If $y_i = \beta_0 + \beta_1 x_i + \varepsilon_i$ and $\hat{\beta}_0$ and $\hat{\beta}_1$ are the least squares estimators for β_0 and β_1, then

$$\sum_{i=1}^{n} (y_i - \bar{y})^2 = \sum_{i=1}^{n} (y_i - \hat{y}_i)^2 + \sum_{i=1}^{n} (\hat{y}_i - \bar{y})^2.$$

In other words,

$$\text{SST} = \text{SSE} + \text{SSR}.$$

Proof. We begin by adding and subtracting $\hat{y}_i$ within the squared term, and then expanding the squared difference:

$$\text{SST} = \sum_{i=1}^{n}(y_i - \bar{y})^2$$

$$= \sum_{i=1}^{n}\left[(y_i - \hat{y}_i) + (\hat{y}_i - \bar{y})\right]^2$$

$$= \sum_{i=1}^{n}(y_i - \hat{y}_i)^2 + 2\sum_{i=1}^{n}(y_i - \hat{y}_i)(\hat{y}_i - \bar{y}) + \sum_{i=1}^{n}(\hat{y}_i - \bar{y})^2 \qquad (14.28)$$

$$= \text{SSR} + \underbrace{2\sum_{i=1}^{n}(y_i - \hat{y}_i)(\hat{y}_i - \bar{y})}_{\text{cross product term}} + \text{SSE}.$$

We leave it as an exercise to show that

$$\hat{y}_i - \bar{y} = \hat{\beta}_1(x_i - \bar{x})$$

and

$$y_i - \hat{y}_i = (y_i - \bar{y}) - \hat{\beta}_1(x_i - \bar{x}).$$

Given these results, we can write the cross product term from (14.28) as

$$\sum_{i=1}^{n}(y_i - \hat{y}_i)(\hat{y}_i - \bar{y}) = \sum_{i=1}^{n}\left((y_i - \bar{y}) - \hat{\beta}_1(x_i - \bar{x})\right)\left(\hat{\beta}_1(x_i - \bar{x})\right)$$

$$= \hat{\beta}_1\sum_{i=1}^{n}(x_i - \bar{x})(y_i - \bar{y}) - \hat{\beta}_1^2\sum_{i=1}^{n}(x_i - \bar{x})$$

$$= \frac{S_{xy}}{S_{xx}}S_{xy} - \frac{S_{xy}^2}{S_{xx}^2}S_{xx}$$

$$= 0.$$

Thus, the cross product in (14.28) is equal to zero, so we conclude that $\text{SST} = \text{SSR} + \text{SSE}$. ∎

Exercise 14.5 To complete the proof of Theorem 14.9.1, show that

$$\hat{y}_i - \bar{y} = \hat{\beta}_1(x_i - \bar{x}) \qquad \text{and} \qquad y_i - \hat{y}_i = (y_i - \bar{y}) - \hat{\beta}_1(x_i - \bar{x}).$$

The decomposition of the sums of squares in the previous theorem can be used to set up a table giving the various components of variance. A typical *analysis of variance* (ANOVA) table is shown below:

Source	Sum of squares	Degrees of freedom	Mean square
Regression	$\text{SSR} = \sum_{i=1}^{n}(\hat{y}_i - \bar{y})^2$	1	$\text{MSR} = \dfrac{\text{SSR}}{1}$
Error	$\text{SSE} = \sum_{i=1}^{n}(y_i - \hat{y}_i)^2$	$n-2$	$\text{MSE} = \dfrac{\text{SSE}}{n-2}$
Total	$\text{SST} = \sum_{i=1}^{n}(y_i - \bar{y})^2$	$n-1$	$\text{MST} = \dfrac{\text{SST}}{n-1}$

Notice that the two entries in the sums of squares column add to give the SST. *Degrees of freedom* is a slippery concept that is equal to the number of free, or unconstrained, terms in a sum. The entries in the degrees of freedom column also sum to the total degrees of freedom. The mean square entries are defined to be the corresponding sum of squares divided by the degrees of freedom.

Earlier, we noted that

$$\frac{(n-2)\text{MSE}}{\sigma^2} \sim \chi^2(n-2).$$

If the null hypothesis $H_0 : \beta_1 = 0$ is true (i.e., there is no relationship between the predictor x and the response y, given the assumption of a linear model), then $Y_i = \beta_0 + 0 \times x_i + \varepsilon_i = \beta_0 + \varepsilon_i$. In other words,

$$Y_1, Y_2, \ldots, Y_n \sim \text{ i.i.d. } N(\beta_0, \sigma^2).$$

The MST is just the sample variance of the Y_is:

$$\text{MST} = \frac{1}{n-1}\text{SST} = \frac{1}{n-1}\sum_{i=1}^{n}(Y_i - \bar{Y})^2.$$

We know that this is an unbiased estimator of σ^2 provided $H_0 : \beta_1 = 0$ is true. It can be shown that the MSR is also an unbiased estimator of σ^2 when H_0 is true. To summarize:

- If $H_0 : \beta_1 = 0$ is true, then $(n-1)\text{MST}/\sigma^2 \sim \chi^2(n-1)$ and so MST is an unbiased estimator of σ^2.
- If $H_0 : \beta_1 = 0$ is true, then $1 \times \text{MSR}/\sigma^2 \sim \chi^2(1)$ and MSR is an unbiased estimator of σ^2.
- Regardless of whether H_0 is true or false, $(n-2)\text{MSE}/\sigma^2 \sim \chi^2(n-2)$ and MSE is an unbiased estimator of σ^2.

It is also true that MSR and MSE are independent. Thus, we can construct the ratio

$$F = \frac{\text{MSR}}{\text{MSE}} = \frac{\dfrac{1 \times \text{MSR}/\sigma^2}{1}}{\dfrac{(n-2)\text{MSE}/\sigma^2}{n-2}}.$$

Note that this is of the form of a chi-square random variable over its degrees of freedom divided by an independent chi-square random variable over its degrees of freedom. Thus,

$$F = \frac{\text{MSR}}{\text{MSE}} \sim F(1, n-2).$$

If the null hypothesis is false, then SSR is a biased estimator for σ^2. Specifically, when $\beta_1 \neq 0$ we can show that the MSR estimates $\sigma^2 + g(\beta_1)$, where $g(\beta_1)$ is always positive. Consequently, we can use the F ratio as a test statistic in the ANOVA table shown previously. Thus, when H_0: $\beta_1 = 0$ is false, MSR *overestimates* σ^2, and this is true regardless of whether $\beta_1 > 0$ or $\beta_1 < 0$. Thus, the two-sided test of H_0: $\beta_1 = 0$ against H_1: $\beta_1 \neq 0$ leads to a rule that has a one-sided rejection region: We reject H_0 for sufficiently *large* values of F.

In Section 15.4 we developed a t-test for testing H_0: $\beta_1 = 0$. It can be shown that

$$\left(\frac{\hat{\beta}_1 - 0}{S/S_{xx}} \right)^2 = t^2 = F = \frac{\text{MSR}}{\text{MSE}}.$$

Thus, the F statistic described here is equivalent to the t-test described earlier. The advantage of the F-test is that it can be used in multiple regression to test whether *all* slope parameters are equal to 0. This is called testing for significance of regression.

14.10 Linear Models in More than One Predictor

So far, we have studied exclusively the problem where there is one predictor x and one response or predictor variable y. For example, a subject's response in a clinical trial may depend on age, gender, drug received, or a number of other variables. Our simple linear regression model can be extended to any number of predictor variables, as we demonstrate in this section. Suppose there are p predictor variables. Define

$$x_{ij} = \text{value of the } j\text{th predictor variable on the } i\text{th unit.}$$

For example, if there are three predictor variables, age, gender, and drug received, then $x_{100,1}$ would be the value of the first variable (age) for the 100th subject.[4] The multiple linear regression model is then

$$y_i = \beta_0 + \beta_1 x_{i1} + \beta_2 x_{i2} + \cdots + \beta_p x_{ip} + \varepsilon_i, \quad \varepsilon_i \sim \text{i.i.d. } N(0, \sigma^2). \tag{14.29}$$

We must always keep in mind that the linearity assumed in this model is just an approximation to reality. The true (unknown) functional relationship may be more complicated. Thus, we can decompose the model for the response y_i as

$$y_i = \underbrace{\beta_0 + \beta_1 x_{i1} + \cdots + \beta_p x_{ip}}_{\text{linear relationship}} + \underbrace{f(x_i) - (\beta_0 + \beta_1 x_{i1} + \cdots + \beta_p x_{ip})}_{\text{model bias}} + \underbrace{\varepsilon_i}_{\text{random error}}, \quad \varepsilon_i \sim \text{i.i.d. } N(0, \sigma^2),$$

$$\tag{14.30}$$

as we did for the simple linear regression model in (14.4).

The linear model in (14.29) can be written in matrix form as

$$\boldsymbol{y} = \mathbf{X}\boldsymbol{\beta} + \boldsymbol{\epsilon}, \tag{14.31}$$

---

[4]We will include commas in the subscript of x when necessary, and omit them when the meaning is clear. For example, we will write x_{ij} but when we use actual numbers we will write $x_{101,2}$ since the omission of commas would be vague. For example, does x_{1012} refer to variable 12 on subject 10 or variable 2 on subject 101?

where $\mathbf{X}$ is the *model matrix*,[5]

$$\mathbf{X} = \begin{bmatrix} 1 & x_{11} & x_{12} & \cdots & x_{1p} \\ 1 & x_{21} & x_{22} & \cdots & x_{2p} \\ 1 & x_{31} & x_{32} & \cdots & x_{3p} \\ \vdots & \vdots & \vdots & & \vdots \\ 1 & x_{n1} & x_{n2} & \cdots & x_{np} \end{bmatrix},$$

β is the vector of regression coefficients

$$\beta = \begin{bmatrix} \beta_0 \\ \beta_1 \\ \beta_2 \\ \vdots \\ \beta_p \end{bmatrix},$$

and ϵ is the vector of random errors, which are assumed to be i.i.d. $N(0,\sigma^2)$. We can write this as

$$\epsilon = \begin{bmatrix} \varepsilon_1 \\ \varepsilon_3 \\ \varepsilon_3 \\ \vdots \\ \varepsilon_n \end{bmatrix} \sim N_n \left(\begin{bmatrix} 0 \\ 0 \\ 0 \\ \vdots \\ 0 \end{bmatrix}, \begin{bmatrix} \sigma^2 & 0 & 0 & \cdots & 0 \\ 0 & \sigma^2 & 0 & \cdots & 0 \\ 0 & 0 & \sigma^2 & \cdots & 0 \\ \vdots & \vdots & \vdots & & \vdots \\ 0 & 0 & 0 & \cdots & \sigma^2 \end{bmatrix} \right).$$

The mean vector for ϵ can be written as $\mathbf{0}$ and the covariance matrix can be written as $\sigma^2 I$, where I is the $n \times n$ identity matrix.

There are now $p+2$ parameters to estimate: β_0, β_1, $\beta_2,\ldots,\beta_p$, and σ^2. We can estimate the β parameters using the method of least squares, using a similar method as we did for the simple linear regression model. We can define

$$S(\beta_0,\beta_1,\ldots,\beta_p) = \sum_{i=1}^n \left(y_i - (\beta_0 + \beta_1 x_{i1} + \cdots + \beta_p x_{ip}) \right)^2.$$

We can take the derivative with respect to β_j and obtain

$$\frac{\partial S}{\partial \beta_j} = \sum_{i=1}^n 2\left(y_i - (\beta_0 + \beta_1 x_{i1} + \cdots + \beta_p x_{ip}) \right)(-x_{ij}). \tag{14.32}$$

Following the usual convention, we define x_{i0} to be 1. With this definition, we see that (14.32) holds for β_0 as well as for β_1 through β_p. If we set the result in (14.32) equal to zero and simplify, we obtain

$$\beta_0 \sum_{i=1}^n 1 + \beta_1 \sum_{i=1}^n x_{i1}x_{ij} + \cdots + \beta_p \sum_{i=1}^n x_{ip}x_{ij} = \sum_{i=1}^n x_{ij}y_i.$$

[5]Some authors call this the *design* matrix. Here, we distinguish between the model matrix, which has the column of 1s for the constant term, and possibly squares or cross product terms if the model includes such terms (see Section 14.12 on polynomial regression). We use the term *design matrix* for the matrix that includes only those columns required to say what each of the levels of the predictors are. If you know the design matrix as well as the model, you can construct the model matrix.

Note that this is *linear in the parameters* in the sense that we have a constant that multiplies each β_j. (Note that the predictors x_{ij} are considered fixed, so functions of these are constants.) Thus the system of equations that must be solved (to get the least squares estimators) is linear. We can write this system as

$$n\beta_0 \; + \; \beta_1 \sum_{i=1}^n x_{i1} \; + \; \beta_2 \sum_{i=1}^n x_{i2} + \cdots + \beta_p \sum_{i=1}^n x_{ip} \; = \; \sum_{i=1}^n y_i$$

$$\beta_0 \sum_{i=1}^n x_{i1} + \beta_1 \sum_{i=1}^n x_{i1}^2 + \beta_2 \sum_{i=1}^n x_{i1}x_{i2} + \cdots + \beta_p \sum_{i=1}^n x_{i1}x_{ip} \; = \; \sum_{i=1}^n x_{i1}y_i$$

$$\beta_0 \sum_{i=1}^n x_{i2} + \beta_1 \sum_{i=1}^n x_{i1}x_{i2} + \beta_2 \sum_{i=1}^n x_{i2}^2 + \cdots + \beta_p \sum_{i=1}^n x_{i2}x_{ip} \; = \; \sum_{i=1}^n x_{i2}y_i \qquad (14.33)$$

$$\vdots \qquad\qquad\qquad\qquad\qquad \vdots$$

$$\beta_0 \sum_{i=1}^n x_{ip} + \beta_1 \sum_{i=1}^n x_{i1}x_{ip} + \beta_2 \sum_{i=1}^n x_{i2}x_{ip} + \cdots + \beta_p \sum_{i=1}^n x_{ip}^2 \; = \; \sum_{i=1}^n x_{ip}y_i.$$

Just as we did in Section 14.5, we can write this in matrix notation as

$$\begin{bmatrix} n\beta_0 & \beta_1 \sum_{i=1}^n x_{i1} & \beta_2 \sum_{i=1}^n x_{i2} & \cdots & \beta_p \sum_{i=1}^n x_{ip} \\ \beta_0 \sum_{i=1}^n x_{i1} & \beta_1 \sum_{i=1}^n x_{i1}^2 & \beta_2 \sum_{i=1}^n x_{i1}x_{i2} & \cdots & \beta_p \sum_{i=1}^n x_{i1}x_{ip} \\ \beta_0 \sum_{i=1}^n x_{i2} & \beta_1 \sum_{i=1}^n x_{i1}x_{i2} & \beta_2 \sum_{i=1}^n x_{i2}^2 & \cdots & \beta_p \sum_{i=1}^n x_{i2}x_{ip} \\ \vdots & \vdots & \vdots & & \vdots \\ \beta_0 \sum_{i=1}^n x_{ip} & \beta_1 \sum_{i=1}^n x_{i1}x_{ip} & \beta_2 \sum_{i=1}^n x_{i2}x_{ip} & \cdots & \beta_p \sum_{i=1}^n x_{ip}^2 \end{bmatrix} = \begin{bmatrix} \sum_{i=1}^n y_i \\ \sum_{i=1}^n x_{i1}y_i \\ \sum_{i=1}^n x_{i2}y_i \\ \vdots \\ \sum_{i=1}^n x_{ip}y_i \end{bmatrix}.$$

This can then be written as

$$\begin{bmatrix} n & \sum x_{i1} & \sum x_{i2} & \cdots & \sum x_{ip} \\ \sum x_{i1} & \sum x_{i1}^2 & \sum x_{i1}x_{i2} & \cdots & \sum x_{i1}x_{ip} \\ \sum x_{i2} & \sum x_{i1}x_{i2} & \sum x_{i2}^2 & \cdots & \sum x_{i2}x_{ip} \\ \vdots & \vdots & \vdots & & \vdots \\ \sum x_{ip} & \sum x_{i1}x_{ip} & \sum x_{i2}x_{ip} & \cdots & \sum x_{ip}^2 \end{bmatrix} \begin{bmatrix} \beta_0 \\ \beta_1 \\ \beta_2 \\ \vdots \\ \beta_p \end{bmatrix} = \begin{bmatrix} \sum y_i \\ \sum x_{i1}y_i \\ \sum x_{i2}y_i \\ \vdots \\ \sum x_{ip}y_i \end{bmatrix}.$$

This, in turn, can be written in matrix form as

$$\mathbf{X}'\mathbf{X}\boldsymbol{\beta} \; = \; \mathbf{X}\boldsymbol{y}. \qquad (14.34)$$

The estimator for the vector $\boldsymbol{\beta}$ is therefore[6]

$$\hat{\boldsymbol{\beta}} \; = \; (\mathbf{X}'\mathbf{X})^{-1}\mathbf{X}'\boldsymbol{y}. \qquad (14.35)$$

---

[6]Even though (14.35) provides a closed-form expression for the least squares estimator (which is also the MLE) for $\boldsymbol{\beta}$, it takes fewer computations to solve the system in (14.34) than it does to invert the matrix in (14.35).

Theorem 14.10.1 For the linear model in (14.31),

$$
\hat{\beta} \sim N\left(\begin{bmatrix} \beta_0 \\ \beta_1 \\ \beta_2 \\ \vdots \\ \beta_p \end{bmatrix},\ \sigma^2 \begin{bmatrix} c_{00} & c_{01} & c_{02} & \cdots & c_{0p} \\ c_{10} & c_{11} & c_{12} & \cdots & c_{1p} \\ c_{20} & c_{21} & c_{22} & \cdots & c_{2p} \\ \vdots & \vdots & \vdots & & \vdots \\ c_{p0} & c_{p1} & c_{p2} & \cdots & c_{pp} \end{bmatrix} \right),
$$

where c_{ij} is the (i, j) entry in the matrix $(\mathbf{X}'\mathbf{X})^{-1}$. Note that $(\mathbf{X}'\mathbf{X})^{-1}$ is a $(p+1) \times (p+1)$ matrix. We will label the top row as row 0 and the first column as column 0. By this convention, c_{00} is the top left element in $\mathbf{X}'\mathbf{X}$.

Proof. If we take the expected value of the estimator $\hat{\beta}$ we obtain

$$
\begin{aligned}
\mathbb{E}(\hat{\beta}) &= \mathbb{E}\left((\mathbf{X}'\mathbf{X})^{-1}\mathbf{X}'\boldsymbol{y}\right) \\
&= (\mathbf{X}'\mathbf{X})^{-1}\mathbf{X}'\,\mathbb{E}(\boldsymbol{y}) \\
&= (\mathbf{X}'\mathbf{X})^{-1}\mathbf{X}'\,\mathbb{E}(\mathbf{X}\boldsymbol{\beta}+\boldsymbol{\epsilon}) \\
&= (\mathbf{X}'\mathbf{X})^{-1}\mathbf{X}'\left(\mathbf{X}\boldsymbol{\beta}+\mathbb{E}(\boldsymbol{\epsilon})\right) \\
&= (\mathbf{X}'\mathbf{X})^{-1}(\mathbf{X}'\mathbf{X})\,\boldsymbol{\beta} \\
&= \boldsymbol{\beta},
\end{aligned}
$$

showing that $\mathbb{E}(\hat{\beta}) = \beta$. To determine the covariance matrix for $\hat{\beta}$, first note that

$$
\mathbb{V}(Y) = \mathbb{V}(\mathbf{X}\boldsymbol{\beta}+\boldsymbol{\epsilon}) = \mathbb{V}(\boldsymbol{\epsilon}) = \sigma^2 I.
$$

Obtaining the covariance matrix involves additional manipulation of matrices:

$$
\begin{aligned}
\mathbb{V}(\hat{\beta}) &= \mathbb{V}((\mathbf{X}'\mathbf{X})^{-1}\mathbf{X}'\,Y) \\
&= \left[(\mathbf{X}'\mathbf{X})^{-1}\mathbf{X}'\right]\mathbb{V}(Y)\left[(\mathbf{X}'\mathbf{X})^{-1}\mathbf{X}'\right]' \\
&= (\mathbf{X}'\mathbf{X})^{-1}\mathbf{X}'\,\sigma^2\,I(\mathbf{X}')'\left((\mathbf{X}'\mathbf{X})^{-1}\right)'.
\end{aligned}
$$

At this point, we notice that $(\mathbf{X}')' = \mathbf{X}$, and because $\mathbf{X}'\mathbf{X}$ is symmetric, so is $(\mathbf{X}'\mathbf{X})^{-1}$, and therefore $\left((\mathbf{X}'\mathbf{X})^{-1}\right)' = (\mathbf{X}'\mathbf{X})^{-1}$. Thus,

$$
\mathbb{V}(\hat{\beta}) = (\mathbf{X}'\mathbf{X})^{-1}\mathbf{X}'\,\sigma^2\,I\mathbf{X}(\mathbf{X}'\mathbf{X})^{-1} = \sigma^2(\mathbf{X}'\mathbf{X})^{-1}\mathbf{X}'\mathbf{X}(\mathbf{X}'\mathbf{X})^{-1} = \sigma^2(\mathbf{X}'\mathbf{X})^{-1}. \quad\blacksquare
$$

The variance of $\hat{\beta}_j$ is thus

$$
\mathbb{V}(\hat{\beta}_j) = \sigma^2 \times (j, j) \text{ entry in } (\mathbf{X}'\mathbf{X})^{-1}. \tag{14.36}
$$

As always, this is of no use to us for the purpose of deriving a confidence interval or a test statistic for making inference regarding β_j. We need one more theorem before we can get to that point. First, though, we turn our attention to the estimation of σ^2. The decomposition of the sums of squares follows the same pattern as for simple regression.

Theorem 14.10.2 If

$$y_i = \beta_0 + \beta_1 x_{i1} + \beta_2 x_{i2} + \cdots + \beta_p x_{pi} + \varepsilon_i$$

and $\hat{\beta}_0, \hat{\beta}_1, \hat{\beta}_2, \ldots, \hat{\beta}_p$ are the least squares estimators for $\beta_0, \beta_1, \beta_2, \ldots, \beta_p$, then

$$\underbrace{\sum_{i=1}^{n}(y_i - \bar{y})^2}_{\text{SST}} = \underbrace{\sum_{i=1}^{n}(y_i - \hat{y}_i)^2}_{\text{SSE}} + \underbrace{\sum_{i=1}^{n}(\hat{y}_i - \bar{y})^2}_{\text{SSR}}.$$

Proof. Before we begin the heart of the proof, we should point out a few aspects of matrix algebra. First,

$$(AB)' = B'A'.$$

Second, the transpose of a 1×1 matrix is itself. Thus, if $\boldsymbol{a}'AB\boldsymbol{b}$ turns out to be a 1×1 matrix, then

$$(\boldsymbol{a}'AB\boldsymbol{b})' = \boldsymbol{a}'AB\boldsymbol{b}.$$

Note that $(\mathbf{X}'\mathbf{X})$ is symmetric, so its inverse $(\mathbf{X}'\mathbf{X})^{-1}$ is also symmetric. Thus,

$$\left((\mathbf{X}'\mathbf{X})^{-1}\right)' = (\mathbf{X}'\mathbf{X})^{-1}.$$

Third,

$$\sum_{i=1}^{n}(y_i - \hat{y}_i) = 0.$$

The proof involves finding expressions for SST, SSR, and SSE, and then seeing that

$$\text{SST} = \text{SSR} + \text{SSE}.$$

We begin with SSE:

$$\begin{aligned}
\text{SSE} &= (\boldsymbol{y} - \hat{\boldsymbol{y}})'(\boldsymbol{y} - \hat{\boldsymbol{y}}) \\
&= \boldsymbol{y}'\boldsymbol{y} - \hat{\boldsymbol{y}}'\boldsymbol{y} - \boldsymbol{y}'\hat{\boldsymbol{y}} + \hat{\boldsymbol{y}}'\hat{\boldsymbol{y}} \\
&= \boldsymbol{y}'\boldsymbol{y} - 2\hat{\boldsymbol{y}}'\boldsymbol{y} + \hat{\boldsymbol{y}}'\hat{\boldsymbol{y}} \\
&= \boldsymbol{y}'\boldsymbol{y} - 2(\mathbf{X}\hat{\beta})'\boldsymbol{y} + (\mathbf{X}\hat{\beta})'(\mathbf{X}\hat{\beta}) \\
&= \boldsymbol{y}'\boldsymbol{y} - 2\hat{\beta}'\boldsymbol{X}'\boldsymbol{y} + \hat{\beta}'\boldsymbol{X}'\boldsymbol{X}\hat{\beta} \\
&= \boldsymbol{y}'\boldsymbol{y} - 2\hat{\beta}'\boldsymbol{X}'\boldsymbol{y} + \hat{\beta}'\boldsymbol{X}'\boldsymbol{X}\underbrace{(\boldsymbol{X}'\boldsymbol{X})^{-1}\boldsymbol{X}'\boldsymbol{y}}_{\hat{\beta}} \\
&= \boldsymbol{y}'\boldsymbol{y} - 2\hat{\beta}'\boldsymbol{X}'\boldsymbol{y} + \hat{\beta}'\underbrace{\boldsymbol{X}'\boldsymbol{X}(\boldsymbol{X}'\boldsymbol{X})^{-1}}_{\text{identity matrix}}\boldsymbol{X}'\boldsymbol{y} \\
&= \boldsymbol{y}'\boldsymbol{y} - 2\hat{\beta}'\boldsymbol{X}'\boldsymbol{y} + \hat{\beta}'\boldsymbol{X}'\boldsymbol{y} \\
&= \boldsymbol{y}'\boldsymbol{y} - \hat{\beta}'\boldsymbol{X}'\boldsymbol{y}.
\end{aligned}$$

This is as far as we will take the expression for SSE. Next, let's look at SSR:

$$\begin{aligned}
\mathrm{SSR} &= (\hat{y} - \bar{y}\mathbf{1})'(\hat{y} - \bar{y}\mathbf{1}) \\
&= \hat{y}'\hat{y} - \bar{y}\mathbf{1}'\hat{y} - \hat{y}'\bar{y}\mathbf{1} + (\bar{y}\mathbf{1})'(\bar{y}\mathbf{1}) \\
&= \hat{y}'\hat{y} - 2\bar{y}\mathbf{1}'\hat{y} + \bar{y}^2\mathbf{1}'\mathbf{1} \\
&= (\mathbf{X}\hat{\beta})'(\mathbf{X}\hat{\beta}) - 2\bar{y}\mathbf{1}'\hat{y} + n\bar{y}^2 \\
&= \hat{\beta}'(\mathbf{X}'\mathbf{X})\hat{\beta} - 2n\bar{y}^2 + n\bar{y}^2 \qquad [\text{since } \mathbf{1}'\hat{y} = n\bar{y}] \\
&= \hat{\beta}'(\mathbf{X}'\mathbf{X})(\mathbf{X}'\mathbf{X})^{-1}\mathbf{X}'y - n\bar{y}^2 \\
&= \hat{\beta}'\mathbf{X}'y - n\bar{y}^2.
\end{aligned}$$

The sum of SSE and SSR is therefore

$$\mathrm{SSE} + \mathrm{SSR} = \left(y'y - \hat{\beta}'\mathbf{X}'y\right) + \left(\hat{\beta}'\mathbf{X}'y - n\bar{y}^2\right) = y'y - n\bar{y}^2 = \mathrm{SST}. \qquad \blacksquare$$

We state, but do not prove, the following theorem regarding the statistics $(n - (p+1))\mathrm{SSE}/\sigma^2$ and its relationship to the statistic $\hat{\beta}_j$.

Theorem 14.10.3 For the linear model in (14.31),

$$\frac{(n - (p+1))\mathrm{MSE}}{\sigma^2} \sim \chi^2(n - (p+1))$$

and $\hat{\beta}_j$ and MSE are independent.

The results of the last three theorems can be used to develop a t-test for $H_0 : \beta_j = 0$, $j = 0, 1, \dots, p$. Theorem 14.10.1 tells us that the marginal distribution of $\hat{\beta}_j$ is

$$\hat{\beta}_j \sim N\left(\beta_j, \sigma^2 \times (j,j) \text{ entry in } (\mathbf{X}'\mathbf{X})^{-1}\right),$$

so we can conclude that

$$\frac{\hat{\beta}_j - 0}{\sigma\sqrt{(j,j) \text{ entry in } (\mathbf{X}'\mathbf{X})^{-1}}} \sim N(0,1). \qquad (14.37)$$

Remember that we label the rows and columns of $(\mathbf{X}'\mathbf{X})^{-1}$ beginning at 0. If we replace σ with the estimate $\sqrt{\mathrm{MSE}}$, then we have

$$\frac{\hat{\beta}_j - 0}{\sqrt{\mathrm{MSE}}\sqrt{(j,j) \text{ entry in } (\mathbf{X}'\mathbf{X})^{-1}}} = \frac{\dfrac{\hat{\beta}_j - 0}{\sqrt{(j,j) \text{ entry in } (\mathbf{X}'\mathbf{X})^{-1}}}}{\sqrt{\mathrm{MSE}}} \frac{\dfrac{1}{\sigma}}{\dfrac{1}{\sigma}}$$

$$= \frac{\dfrac{\hat{\beta}_j - 0}{\sigma\sqrt{(j,j) \text{ entry in } (\mathbf{X}'\mathbf{X})^{-1}}}}{\sqrt{\dfrac{\mathrm{MSE}}{\sigma^2}}}$$

$$= \frac{Z}{\sqrt{\dfrac{(n-(p+1))\mathrm{MSE}/\sigma^2}{n-(p+1)}}}$$

$$= \frac{Z}{\sqrt{V/(n-(p+1))}},$$

where $Z \sim N(0,1)$ and $V \sim \chi^2(n-(p+1))$. Thus,

$$\frac{\hat{\beta}_j - 0}{\sqrt{\mathrm{MSE}}\sqrt{(j,j) \text{ entry in } (\mathbf{X'X})^{-1}}} \sim t\big(n-(p+1)\big). \tag{14.38}$$

Here, we have used Theorem 14.10.3. The quantity

$$\sqrt{\mathrm{MSE}}\,\sqrt{(j,j) \text{ entry in } (\mathbf{X'X})^{-1}}$$

is often called the *standard error of the estimator*.[7]

■ **Example 14.14** In the first-year GPA, run a regression analysis in which GPA is the response variable and SATM and SATV are the predictor variables. Do we have evidence that either variable has an effect on first-year GPA?

Solution: It is a challenge to present multiple regression data graphically. When there are just two predictor variables, as we have here, it is possible to produce a three-dimensional scatterplot. Such a graph is shown in Figure 14.9. This plot shows the points along with a line from the *xy*-plane to the point. Without such a line, it is difficult to ascertain each point's location in a static 3D graph, unless the viewer has the capability of rotating the figure. The R code below can be used to create a dynamic graphic. After you install the `rgl` package you can run the code and a window will open with the 3D

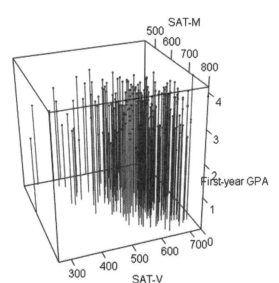

Figure 14.9 Three-dimensional scatterplot for first-year GPA data, with predictors SATM and SATV.

---

[7]Technically, the standard error is $\sigma\sqrt{(j,j) \text{ entry in } (\mathbf{X'X})^{-1}}$; when an estimator for σ is used in place of σ, this would more appropriately be called the *estimated standard error*. Many books and software do not make this distinction and call $\sqrt{\mathrm{MSE}}\sqrt{(j,j) \text{ entry in } (\mathbf{X'X})^{-1}}$ the standard error.

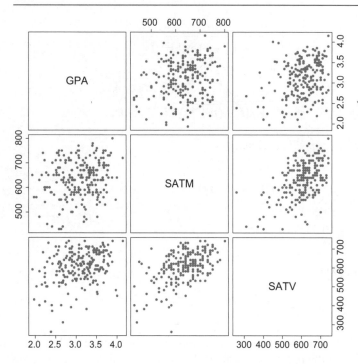

Figure 14.10 Matrix plot for first-year GPA data, with predictors SATM and SATV.

scatterplot. You should be able to use your mouse to rotate the figure so you can see the points from various angles. This should give you some insight into the data. If a user is unable to rotate the graph, then the lines from the *xy*-plane to the point are helpful.

```
library(rgl)
plot3d( x=SATM , y=SATV , z=GPA, type = 'h', col="blue" ,
   radius = 20, xlab="SAT-M", ylab="SAT-V", zlab="First-year GPA")
```

A *matrix plot* of the data consists of scatterplots of all possible pairs of variables (response and predictor variables). Such a scatterplot is shown in Figure 14.10. We can see that SATM and SATV are strongly associated; students who score high on one tend to score high on the other. Both variables are weakly associated with the response variable GPA.

We can run the multiple regression in R by separating the predictor variables with a plus sign. The R code below can perform the regression computations.

```
MultReg = lm( GPA ~ SATM + SATV , data=first )
summary( MultReg )
```

Part of the output is shown below:

```
Coefficients:
              Estimate Std. Error t value Pr(>|t|)
(Intercept) 1.9674974  0.2746617   7.163 1.22e-11 ***
SATM        0.0002911  0.0004714   0.618 0.537515
SATV        0.0015602  0.0004253   3.668 0.000307 ***
---
Signif. codes:  0 '***' 0.001 '**' 0.01 '*' 0.05 '.' 0.1 ' ' 1
```

From this output, we can read the coefficients (in the Estimate column underneath Coefficients) and find that our estimated regression equation is

$$\widehat{\text{GPA}} = 1.9674974 + 0.0002911\,\text{SATM} + 0.0015602\,\text{SATV}.$$

The output also contains the information needed to test $H_0 : \beta_j = 0$ for $j = 0, 1, 2$. The columns to the right of `Estimate` give the standard error of the estimate (the estimated standard deviation of the estimator); the t statistic as given in (14.38) is shown in the column labeled `t value`. Finally, the P-value for testing $H_0 : \beta_j = 0$ against the alternative $H_1 : \beta_j \neq 0$ is given in the last column, which is labeled `Pr(>|t|)`. Here we see that the hypothesis for β_1 (the coefficient of SATM) is not rejected for $\alpha = 0.05$ since the P-value exceeds α. The P-value for testing $H_0 : \beta_2 = 0$ is very small, P-value = 0.000307, giving us strong evidence that β_2, the coefficient for SATV is nonzero.

We see that this result seems to contradict the result from Example 14.7, where SATM had a significant effect on the response GPA. When SATV is taken into account, we see that SATV has a significant effect, but SATM does not. One possible explanation for this phenomenon is that SATV and SATM are associated, as can be seen in Figure 14.10. When a simple regression has a predictor that is significant,[8] the significance may disappear when other variables are introduced into the model in a multiple regression. Here, the true relationship may be between SATV and GPA, and SATM appears significant in a simple linear regression model only because it is associated with SATV.

The ANOVA table can be obtained using code `anova(MultReg)`:

```
Analysis of Variance Table

Response: GPA
           Df Sum Sq Mean Sq F value    Pr(>F)
SATM        1  1.784 1.78399  9.0067 0.0030063 **
SATV        1  2.666 2.66564 13.4578 0.0003073 ***
Residuals 216 42.784 0.19807
```

The SSR is equal to the sum of squares due to all of the predictors: SSR = $1.784 + 2.666 = 4.450$. In general, SSR has p degrees of freedom. We can test $H_0 : \beta_1 = \beta_2 = \cdots = \beta_p = 0$ using the F statistic

$$F = \frac{\text{SSR}/p}{\text{SSE}/(N-p-1)} = \frac{4.450/2}{42.784/216} = 11.23$$

yielding a P-value of 0.000023. This is evidence that the slope parameters are not both 0. Testing whether all slope parameters are equal to zero is called *testing for significance of regression*. ∎ ∎ ∎

14.11 Indicator Variables

Some of the variables in the first-year GPA data set contain only the numbers 0 or 1. For example, one variable is called `FirstGen`. This variable is equal to 1 for those students who are the first in their family to attend college (a first-generation college student), and 0 if they are not. Similarly, the variable `Male` is 1 for male students and 0 for female students. Variables such as these are called *indicator variables* because they indicate whether some condition is or is not met. Usually these variables are named for the characteristic that is assigned a 1. For example, `Male` is the variable name that takes on the value 1 for males. This practice is preferred over a more generic name, such as `Gender`, which can be vague; it raises the question "Which gender corresponds to a 1?"

Many variables are *dichotomous* like this, although you should be careful to assure that they are coded as 0/1. Other schemes, such as 1/2, can be used, but there are advantages to coding them as 0/1.

Let's look at how a linear model with one indicator and two other predictor variables would look. We could think of x_1 as the SATM score and x_2 as the SATV variable. The indicator variable `FirstGen` is x_3. The linear model is then

---

[8]We are abusing terminology a bit here. It is really the coefficient of the predictor variable that is significantly different from zero, but we will say that the *variable* is significant.

$$y_i = \beta_0 + \beta_1 x_{i1} + \beta_2 x_{i2} + \beta_3 x_{i3} + \varepsilon_i.$$

For those first-generation college students, $x_{i3} = 1$, so the model is

$$y_i = \beta_0 + \beta_1 x_{i1} + \beta_2 x_{i2} + \beta_3 \times 1 + \varepsilon_i = (\beta_0 + \beta_3) + \beta_1 x_{i1} + \beta_2 x_{i2} + \varepsilon_i, \qquad (14.39)$$

while for those who are not a first-generation college student the model is

$$y_i = \beta_0 + \beta_1 x_{i1} + \beta_2 x_{i2} + \beta_3 \times 0 + \varepsilon_i = \beta_0 + \beta_1 x_{i1} + \beta_2 x_{i2} + \varepsilon_i. \qquad (14.40)$$

Notice that the two models (for first-generation college students in (14.39) and non-first-generation college students in (14.40)) differ only in the intercept. Thus, as functions of x_1 and x_2, the response surfaces for the two groups are parallel planes.[9] The parameter β_3 then measures the effect of being a first-generation college student. If $\beta_3 = 0$, then there is no difference in the mean GPA for first-generation and non-first-generation college students.

The regression calculations are no different in the presence of one or more indicator variables than for the case where all predictors are continuous. All of the computations to obtain the least squares estimates, the standard errors, t statistics, and P-values are the same.

■ **Example 14.15** Run the multiple linear regression in R with GPA as the response and SATM, SATV, and FirstGen as the predictor variables. Is there evidence that FirstGen is associated with GPA?

Solution: Once we read in the data frame first, we can define the variables and run the multiple regression with the following commands:

```
SATM = first$SATM
SATV = first$SATV
FirstGen = first$FirstGen
GPA  = first$GPA
MultReg3 = lm( GPA ~ SATM + SATV + FirstGen )
summary( MultReg3 )
```

The output from the summary of MultReg3 is:

```
Call:
lm(formula = GPA ~ SATM + SATV + FirstGen)

Residuals:
     Min      1Q   Median      3Q      Max
-1.16201 -0.32598  0.04274  0.34587  1.11454

Coefficients:
              Estimate Std. Error t value Pr(>|t|)
(Intercept)  2.0626356  0.2851157   7.234 8.13e-12 ***
SATM         0.0002614  0.0004715   0.555 0.579794
SATV         0.0014567  0.0004331   3.363 0.000912 ***
FirstGen    -0.1199316  0.0978637  -1.225 0.221729
---
Signif. codes:  0 '***' 0.001 '**' 0.01 '*' 0.05 '.' 0.1 ' ' 1

Residual standard error: 0.4445 on 215 degrees of freedom
Multiple R-squared:  0.1005,      Adjusted R-squared:  0.08794
F-statistic: 8.006 on 3 and 215 DF,  p-value: 4.399e-05
```

---
[9]Had there been just a single nonindicator variable, the responses would be parallel lines.

The *P*-value for testing H_0: $\beta_3 = 0$ is 0.22, too large to provide evidence that being a first-generation college student is associated with first-year GPA. ∎ ∎ ∎

Building statistical models is part science and part art. How we handle variables that are not significant can be a point of contention. Almost everyone agrees that statistically significant variables should be retained in the model. Some argue that variables that are even marginally significant should be retained, while others argue that insignificant variables should be dropped.

Generally, leaving insignificant variables in a regression model inflates the MSE. This can result in wider confidence and prediction intervals, and statistical tests on model parameters that are less sensitive.

∎ **Example 14.16** Use (possibly) all of the variables in the first-year GPA example to build a predictive model for GPA.

Solution: The data set contains the following variables:

- GPA: First-year college GPA;
- HSGPA: High school;
- SATV: Verbal/critical reading SAT score;
- SATM: Math SAT score;
- Male: 1 = Male, 0 = Female;
- HU: Number of high school credits in humanities;
- SS: Number of high school credits in social sciences;
- FirstGen is 1 for students who are the first in their family to attend college, 0 otherwise;
- White is 1 for white students, 0 for others;
- CollegeBound is 1 for those who attended a high school where at least half of the students intended to go on to college, and 0 otherwise.

We start by running a multiple regression with all of these predictor variables, including the four indicator variables. This can be done with the following code:

```
MultRegFull = lm( GPA ~ SATV + SATM + HSGPA + Male + HU + SS +
                  FirstGen + White + CollegeBound , data = first )
summary( MultRegFull )
```

The result of summary(MultRegFull) is below:

```
Call:
lm(formula = GPA ~ SATV + SATM + HSGPA + Male + HU + SS + FirstGen +
White + CollegeBound, data = first)

Residuals:
    Min      1Q   Median      3Q     Max
-1.07412 -0.25827  0.05384  0.27675  0.85761

Coefficients:
              Estimate Std. Error t value Pr(>|t|)
(Intercept)  0.5268983  0.3487584   1.511  0.13235
SATV         0.0005919  0.0003945   1.501  0.13498
SATM         0.0000847  0.0004447   0.190  0.84912
HSGPA        0.4932945  0.0745553   6.616 3.03e-10 ***
```

```
Male          0.0482478  0.0570277   0.846  0.39850
HU            0.0161874  0.0039723   4.075  6.53e-05 ***
SS            0.0073370  0.0055635   1.319  0.18869
FirstGen     -0.0743417  0.0887490  -0.838  0.40318
White         0.1962316  0.0700182   2.803  0.00555 **
CollegeBound  0.0214530  0.1003350   0.214  0.83090
---
Signif. codes:  0 '***' 0.001 '**' 0.01 '*' 0.05 '.' 0.1 ' ' 1

Residual standard error: 0.3834 on 209 degrees of freedom
Multiple R-squared:  0.3496,        Adjusted R-squared:  0.3216
F-statistic: 12.48 on 9 and 209 DF,  p-value: 8.674e-16
```

Now, neither SATV nor SATM score is significant, with P-values of 0.135 and 0.849, respectively. High school GPA (HSGPA) is now strongly associated with the first-year college GPA. One explanation for this is that HSGPA affects both SATM and SATV, as well as GPA. By themselves, SATM and SATV might be associated with first-year college GPA, but this disappears once high school GPA is accounted for.

If we were to eliminate all variables that are nonsignificant, we would have the model with just HSGPA, White, and HU. This model can be run using the commands

```
MultRegSignificantVariables = lm( GPA ~ HSGPA + HU + White , data = first )
summary( MultRegSignificantVariables )
```

and yields the output

```
Call:
lm(formula = GPA ~ HSGPA + HU + White, data = first)

Residuals:
     Min      1Q   Median      3Q      Max
-1.09479 -0.27638  0.02287  0.25411  0.84538

Coefficients:
             Estimate Std. Error t value Pr(>|t|)
(Intercept) 0.933459   0.245673   3.800 0.000189 ***
HSGPA       0.507404   0.070197   7.228 8.42e-12 ***
HU          0.015328   0.003667   4.180 4.24e-05 ***
White       0.265644   0.064519   4.117 5.47e-05 ***
---
Signif. codes:  0 '***' 0.001 '**' 0.01 '*' 0.05 '.' 0.1 ' ' 1

Residual standard error: 0.3856 on 215 degrees of freedom
Multiple R-squared:  0.3231,        Adjusted R-squared:  0.3136
F-statistic: 34.21 on 3 and 215 DF,  p-value: < 2.2e-16
```

At this point we should address the purpose of the study. Is the goal to explain or to predict? Often such models are used to predict success in college, so that help may be provided to those who are admitted but are predicted to achieve a low GPA. The first model, which includes all of the predictors, may help us to identify the determinants of success, which may suggest racial or gender disparities. ■ ■ ■

14.12 Polynomial and Nonlinear Regression

The linear model $Y = X\beta + \varepsilon$ is a general model that can be used to fit any relationship that is linear in the unknown parameters β. This includes the important class of polynomial in regression models. For example, the second-degree polynomial in one variable,

Table 14.2 Average cost per unit (y) and production lot size (x).

y	1.81	1.70	1.65	1.55	1.48	1.40	1.30	1.26	1.24	1.21	1.20	1.18
x	20	25	30	35	40	50	60	65	70	75	80	80

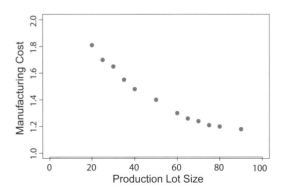

Figure 14.11 Scatterplot of cost versus lot size.

$$Y = \beta_0 + \beta_1 x + \beta_2 x^2 + \varepsilon,$$

is a linear regression model because it is linear in the unknown parameters $\beta = [\beta_0, \beta_1, \beta_2]'$. Similarly, the second-degree polynomial in two variables,

$$Y = \beta_0 + \beta_1 x_1 + \beta_2 x_2 + \beta_{11} x_1^2 + \beta_{22} x_2^2 + \beta_{12} x_1 x_2 + \varepsilon,$$

is also a linear regression model. Polynomial regression models are widely used when the response is curvilinear because the general techniques of fitting multiple regression models can be applied.

To illustrate fitting a polynomial regression model, consider the cost of manufacturing sidewall panels for the interior of an airplane, which are formed in a 1,500-ton press. The unit manufacturing cost varies with the production lot size. Table 14.2 shows the average cost per unit (in hundreds of dollars) for this product (y) and the production lot size (x). The scatter diagram, shown in Figure 14.11, indicates that a second-order polynomial may be appropriate. R code for loading the data and performing the linear regression is:

```
Cost = c( 1.81 , 1.70 , 1.65 , 1.55 , 1.48 , 1.40 , 1.30 ,
          1.26 , 1.24 , 1.21 , 1.20 , 1.18 )
LotSize = c( 20 , 25 , 30 , 35 , 40 , 50 , 60 , 65 , 70 , 75 , 80 , 90 )
regModel = lm( Cost ~ LotSize + I(LotSize^2) )
summary( regModel )
```

The second-order term `LotSize^2` must be placed inside the *insulate* function, called `I()`. This allows us to use arithmetic operations, such as exponentiation `^` in the formula for a linear model. The above code produces the output:

```
Call:
lm(formula = Cost ~ LotSize + I(LotSize^2))

Residuals:
       Min         1Q      Median         3Q        Max
-0.0174763 -0.0065087  0.0001297  0.0071482  0.0151887

Coefficients:
                Estimate Std. Error t value Pr(>|t|)
```

```
(Intercept)    2.198e+00   2.255e-02    97.48 6.38e-15 ***
LotSize       -2.252e-02   9.424e-04   -23.90 1.88e-09 ***
I(LotSize^2)   1.251e-04   8.658e-06    14.45 1.56e-07 ***
---
Signif. codes:  0 '***' 0.001 '**' 0.01 '*' 0.05 '.' 0.1 ' ' 1

Residual standard error: 0.01219 on 9 degrees of freedom
Multiple R-squared:  0.9975,        Adjusted R-squared:  0.9969
F-statistic:  1767 on 2 and 9 DF,  p-value: 2.096e-12
```

The output is interpreted much like that for multiple linear regression. The estimated equation is

$$\widehat{\text{Cost}} = 2.198 - 0.02252 \text{LotSize} + 0.0001251 \text{LotSize}^2.$$

We can test $H_0 : \beta_j = 0$ using a t-test. The t statistic is given in the t value column, and the P-value is shown in the last column. We can see that β_1 and β_2 are both significantly different from zero. Thus, LotSize and LotSize^2 should be included in the model.

There are several important considerations that arise when fitting a polynomial in one variable. Some of these are discussed below.

Order of the model It is important to keep the order of the model as low as possible. When the response function appears to be curvilinear, transformations should be tried to keep the model first order. If this fails, a second-order polynomial should be tried. As a general rule, the use of high-order polynomials ($k > 2$) should be avoided unless they can be justified for reasons outside the data. A low-order model in a transformed variable is almost always preferable to a high-order model in the original metric. Arbitrary fitting of high-order polynomials is a serious abuse of regression analysis. One should always maintain a sense of parsimony, that is, use the simplest possible model that is consistent with the data and knowledge of the problem environment. Remember that in an extreme case it is always possible to pass a polynomial of order $n - 1$ through n points so that a polynomial of sufficiently high degree can always be found that provides a "good" fit to the data. In most cases this would do nothing to enhance understanding of the unknown function, nor will it likely be a good predictor.

Model-building strategy Various strategies for choosing the order of an approximating polynomial have been suggested. One approach is to successively fit models of increasing order until the t-test for the highest-order term is nonsignificant. An alternative procedure is to appropriately fit the highest-order model and then delete terms one at a time, starting with the highest order, until the highest order remaining term has a significant t statistic. These two procedures are called forward selection and backward elimination, respectively. They do not necessarily lead to the same model. In light of the previous comment, these procedures should be used carefully. In most situations we should restrict our attention to first- and second-order polynomials.

Extrapolation Extrapolation with polynomial models can be extremely hazardous. For example, consider the second-order model in Figure 14.12. If we extrapolate beyond the range of the original data, the predicted response turns downward. This may be at odds with the true behavior of the system. In general, polynomial models may turn in unanticipated and inappropriate directions, both in interpolation and in extrapolation. Polynomial models, like all models, are just approximations of reality. A model may fit the data well in the region where we have data, but poorly outside that region. Figure 14.13 shows a situation where a second-order polynomial and an exponential model fit the data well over the region for which data exist. Outside of this region, the models behave quite differently. The second-order polynomial model reaches a maximum and then decreases, whereas the exponential model levels off at about 6.2. Predictions for $x > 5$ would be very different for the two models, even though both fit the data well from $x = 1$ to $x = 5$.

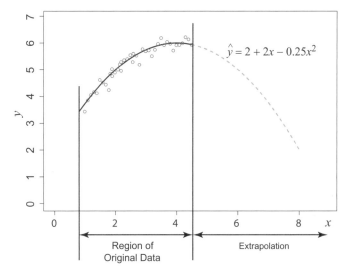

Figure 14.12 Extrapolation to make predictions outside the region where we have data.

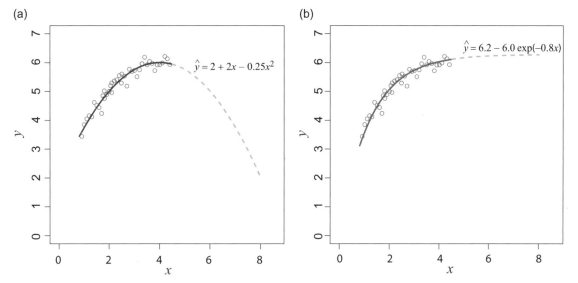

Figure 14.13 Danger of extrapolation.

Ill-conditioning I As the order of the polynomial increases, the $X'X$ matrix becomes ill-conditioned. This means that the matrix is "close to" a matrix that does not have an inverse. As a result, the matrix inversion calculations will be inaccurate, and considerable error may be introduced into the parameter estimates. Nonessential ill-conditioning caused by the arbitrary choice of origin can be removed by first centering the predictor variables (i.e., correcting x for its average), but even centering the data can still result in large sample correlations between some of the model coefficients.

Ill-conditioning II If the values of x are limited to a narrow range, there can be significant ill-conditioning or multicollinearity in the columns of the X matrix. For example, if x varies between 1 and 2, x^2 varies between 1 and 4, which could create strong multicollinearity between x and x^2.

Hierarchy The regression model

$$y = \beta_0 + \beta_1 x + \beta_2 x^2 + \beta_3 x^3 + \varepsilon$$

is said to be hierarchical because it contains all terms of order 3 and lower. By contrast, the model

$$y = \beta_0 + \beta_1 x + \beta_3 x^3 + \varepsilon$$

is not hierarchical. Only hierarchical models are invariant under linear transformation and many regression authorities suggest that all polynomial models should have this property. We have mixed feelings about this as a hard-and-fast rule. It is certainly attractive to have the model form preserved following a linear transformation (such as fitting the model in coded variables and then converting to a model in the natural variables), but it is purely a mathematical nicety. There are many mechanistic models that are not hierarchical; for example, Newton's law of gravity is an inverse square law, and the magnetic dipole law is an inverse cube law. Furthermore, there are many situations where using a polynomial regression model to represent the results of a designed experiment with a model such as

$$y = \beta_0 + \beta_1 x_1 + \beta_{12} x_1 x_2 + \varepsilon$$

would be supported by the data. Here the cross product term represents a two-factor interaction. A hierarchical model would require the inclusion of the other main effect x_2. However, this other term could really be entirely unnecessary from a statistical significance perspective. It may be perfectly logical from the viewpoint of the underlying science or engineering to have an interaction in the model without one (or even in some cases either) of the individual main effects. This occurs frequently when some of the variables involved in the interaction are categorical. The best advice is to fit a model that has all terms significant and to use discipline knowledge rather than an arbitrary rule as an additional guide in model formulation. Generally, a hierarchical model is easier to explain to a "customer" who is not familiar with statistical model-building, but a nonhierarchical model may produce better predictions of new data.

14.13 Inference for a Linear Combination of Model Parameters

Suppose we have a multiple linear regression model and we would like to estimate some linear combination of the β parameters, such as

$$c_0\beta_0 + c_1\beta_1 + c_2\beta_2 + \cdots + c_p\beta_p = [c_0, c_1, c_2, \ldots, c_p] \begin{bmatrix} \beta_0 \\ \beta_1 \\ \beta_2 \\ \vdots \\ \beta_p \end{bmatrix} = c'\beta.$$

We will see later in this section some applications of how and when we would use this result.

First, let's look at the distribution of $c'\hat{\beta}$. Since $\hat{\beta}$ is normally distributed (Theorem 14.10.1), any linear combination of the components of β is normally distributed; thus, $c'\hat{\beta} \sim N(?, ?)$. We now need to find the mean and variance. We will make use of these results for the random vector Y and constant matrix A:

$$\mathbb{E}(AY) = AE(Y)$$
$$\text{cov}(AY) = A\,\text{cov}(Y)A'. \tag{14.41}$$

Note that A could be a vector or a matrix, since a column vector is just a matrix with one column.

Applying the results in (14.41) to $c'\beta$ yields

$$\mathbb{E}(c'\hat{\beta}) = c'\mathbb{E}(\hat{\beta}) = c'\beta$$

and

$$\mathrm{cov}(c'\beta) = c'\mathrm{cov}(\hat{\beta})c = c'\sigma^2(X'X)^{-1}c = \sigma^2\,c'(X'X)^{-1}c.$$

Since $c'\beta$ is a scalar, its covariance is just $\mathbb{V}(c'\beta)$.

Thus,

$$c'\hat{\beta} \sim N\left(c'\beta,\,\sigma^2\,c'(X'X)^{-1}c\right). \tag{14.42}$$

The process of obtaining confidence intervals and test statistics should now be familiar: standardize the result, then substitute the estimate $\hat{\sigma} = \sqrt{\mathrm{MSE}}$ for σ and obtain a t statistic. Thus,

$$\frac{c'\hat{\beta} - c'\beta}{\sqrt{\mathrm{MSE}\left(c'(X'X)^{-1}c\right)}} \sim t(n-(p+1)). \tag{14.43}$$

See Problem 14.21.

Now that we've determined the distribution of $c'\hat{\beta}$, let's look at some applications. Suppose, we would like to make a point estimate and a confidence interval for the mean response for a particular set of predictor variables, say $x = [1,x_1,x_2,\ldots,x_p]'$. We are thus estimating $x'\beta$; the point estimate of the mean response at x is

$$\widehat{x'\beta} = x'\hat{\beta},$$

and its standard error is

$$\sqrt{\mathrm{MSE}\left(x'(X'X)^{-1}x\right)}.$$

A confidence interval for $x'\beta$ is therefore

$$\left(x'\hat{\beta} - z_{\alpha/2}\sqrt{\mathrm{MSE}\left(x'(X'X)^{-1}x\right)},\,x'\hat{\beta} + z_{\alpha/2}\sqrt{\mathrm{MSE}\left(x'(X'X)^{-1}x\right)}\right). \tag{14.44}$$

If we want a prediction interval for the response for a single subject or unit at the value x, we need an expression for the variance of the actual response Y^*. We will base our prediction interval on the statistic

$$T = \frac{Y^* - \hat{Y}^*}{\sqrt{\mathbb{V}(Y^* - \hat{Y}^*)}}.$$

Then the numerator has expectation

$$\mathbb{E}\left(Y^* - \hat{Y}^*\right) = \mathbb{E}\left(x'\hat{\beta} - (x'\beta + \varepsilon)\right) = \mathbb{E}\left(x'\hat{\beta}\right) - \mathbb{E}\left(x'\beta + \varepsilon\right) = x'\mathbb{E}\left(\hat{\beta}\right) - x'\beta - \mathbb{E}(\varepsilon)$$
$$= x'\beta - x'\beta - 0 = 0$$

and variance

$$\mathbb{V}\left(Y^* - \hat{Y}^*\right) = \mathbb{V}\left(x'\hat{\beta} - (x'\beta + \varepsilon)\right)$$
$$= \mathbb{V}\left(x'\hat{\beta}\right) + \mathbb{V}\left(x'\beta + \varepsilon\right)$$
$$= x'\mathrm{cov}(\hat{\beta})x + \mathbb{V}(\varepsilon)$$

$$= x'\left(\sigma^2(X'X)^{-1}\right)x + \sigma^2$$

$$= \sigma^2\left(1 + x'(X'X)^{-1}x\right).$$

Thus,

$$Z = \frac{Y^* - \hat{Y}^*}{\sigma\sqrt{1 + x'(X'X)^{-1}x}} \sim N(0,1).$$

When we substitute MSE in place of σ^2, we obtain a t statistic:

$$T = \frac{Y^* - \hat{Y}^*}{\sqrt{\text{MSE}}\sqrt{1 + x'(X'X)^{-1}x}} \sim t(n - (p+1)).$$

The prediction interval for y^* is therefore

$$\left[x'\hat{\beta} - t_{\alpha/2}(n - (p+1))\sqrt{\text{MSE}}\sqrt{1 + x'(X'X)^{-1}x} , \right.$$

$$\left. x'\hat{\beta} + t_{\alpha/2}(n - (p+1))\sqrt{\text{MSE}}\sqrt{1 + x'(X'X)^{-1}x} \right]. \tag{14.45}$$

The distributional result in (14.43) applies to *any* linear combination, so we can use it to compare two different slope parameters. For example, if we had three predictor variables we could choose

$$c = \begin{bmatrix} 0 \\ 1 \\ -1 \\ 0 \end{bmatrix}$$

so that

$$c'\beta = [0, 1, -1, 0] \begin{bmatrix} \beta_0 \\ \beta_1 \\ \beta_2 \\ \beta_3 \end{bmatrix} = \beta_1 - \beta_2.$$

We could obtain a confidence interval or a test statistic for testing H_0: $\beta_1 = \beta_2$, or equivalently H_0: $\beta_1 - \beta_2 = 0$. The applications for this are rather limited, and we would only want to raise this question when both variables are measured on the same scale.

■ **Example 14.17** Consider the linear regression from Example 14.15, where the predictor variable set included SATM, SATV, and FirstGen.

(a) Find a confidence interval for the mean response when SATM $= 600$, SATV $= 650$, and FirstGen $= 1$ (that is, for a first-generation college student).

(b) Find a prediction interval for a student with these same characteristics.

(c) Test whether the coefficients β_1 and β_2 are the same.

Solution: For parts (a) and (b) set

$$x = \begin{bmatrix} 1 \\ 600 \\ 650 \\ 1 \end{bmatrix}.$$

The computations can be done in R using these commands:

```
X = cbind( rep(1,219) , SATM , SATV , FirstGen )
print( t(X) %*% X )
XPXinv = solve( t(X) %*% X )
print( XPXinv )
x = matrix( c( 1 , 650 , 600 , 1) , ncol=1 )
betahat = XPXinv %*% t(X) %*% GPA
```

The first line in the above code defines the X matrix using cbind(); this command takes vectors and "binds" them to make columns of a matrix. The transpose function in R is t(). To perform matrix multiplication, we use the %*% command,[10] so to get $X'X$ we type t(X) %*% X. When we print this matrix, we obtain

```
             219    138910    132510       25
SATM      138910  89343500  84770800    14930
SATV      132510  84770800  81693700    13640
FirstGen      25     14930     13640       25
```

The R command to find the matrix inverse is solve(), so solve(t(X) %*% X), which is

```
                              SATM          SATV       FirstGen
          0.4113609360  -3.984195e-04  -2.473961e-04  -3.844551e-02
SATM     -0.0003984195   1.124885e-06  -5.230073e-07   1.199087e-05
SATV     -0.0002473961  -5.230073e-07   9.492496e-07   4.182553e-05
FirstGen -0.0384455105   1.199087e-05   4.182553e-05   4.846455e-02
```

The (scalar) value of $x'(X'X)^{-1}x$ is

```
0.04294068
```

This is the messiest part of the confidence interval from (14.44). We can find the MSE (the unbiased estimate for σ^2) through the following:

```
MultReg3 = lm( GPA ~ SATM + SATV + FirstGen )
MultReg3ANOVA= anova( MultReg3 )
print( MultReg3ANOVA )
MSE = MultReg3ANOVA$'Mean Sq'[4]
```

The output from the anova() function is

```
Analysis of Variance Table

Response: GPA
           Df Sum Sq Mean Sq F value    Pr(>F)
SATM        1  1.784 1.78399  9.0276 0.0029750 **
SATV        1  2.666 2.66564 13.4891 0.0003029 ***
FirstGen    1  0.297 0.29679  1.5018 0.2217294
Residuals 215 42.487 0.19761
---
Signif. codes:  0 '***' 0.001 '**' 0.01 '*' 0.05 '.' 0.1 ' ' 1
```

[10]If we have matrices A and B, the R command A*B computes the matrix whose entries equal the products of the corresponding entries of A and B. This assumes that A and B have the same number of rows and columns. If we want the usual matrix multiplication, we must use A%*%B; this assumes the number of columns of A equals the number of rows of B.

Here we can see that the MSE is 0.19761. Alternatively, we can extract the MSE from the `MultReg3ANOVA` object by pulling the fourth element of `MultReg3ANOVA$'Mean Sq'` with the command `MultReg3ANOVA$ 'Mean Sq'[4]`. The confidence interval can then be computed using

```
betahat = coef(MultReg3)
MSE = MultReg3ANOVA$'Mean Sq'[4]
LCL = t(x) %*% betahat - qt(0.975,219-(3+1))*sqrt(MSE*t(x) %*% XPXinv %*% x)
UCL = t(x) %*% betahat + qt(0.975,219-(3+1))*sqrt(MSE*t(x) %*% XPXinv %*% x)
print( c(LCL,UCL) )
```

The output from this is

```
[1] 2.805077 3.168217
```

This is a confidence interval for the mean response for the population of all first-generation college students (`FirstGen = 1`) whose SATM score is 650 and whose SATM is 600.

Part (b) asks for a prediction interval for one particular student with these characteristics. Computations are similar, but the expression in the square root now includes a $1+$ term. We can alter the lines above and rename them LPL and UPL to indicate that these are prediction limits:

```
LPL = t(x) %*% betahat - qt(0.975,219-(3+1)) *
                         sqrt( MSE * ( 1 + t(x) %*% XPXinv %*% x ) )
UPL = t(x) %*% betahat + qt(0.975,219-(3+1)) *
                         sqrt( MSE * ( 1 + t(x) %*% XPXinv %*% x ) )
print( c(LPL,UPL) )
```

The resulting prediction interval is

```
[1] 2.091820 3.881474
```

For part (c), we define $c = [0, 1, -1, 0]'$ so that

$$c'\beta = [0, 1, -1, 0] \begin{bmatrix} \beta_0 \\ \beta_1 \\ \beta_2 \\ \beta_3 \end{bmatrix} = \beta_1 - \beta_2.$$

From (14.43), we know that

$$\frac{c'\hat{\beta} - c'\beta}{\sqrt{\text{MSE}\left(c'(X'X)^{-1}c\right)}} \sim t(n - (p+1)).$$

Under H_0: $\beta_1 - \beta_2 = 0$, the numerator becomes simply $c'\hat{\beta}$. This test statistic can be computed in R by:

```
c.vec = c(0,1,-1,0)
t.stat = ( t(c.vec) %*% betahat ) / sqrt( MSE * t(c.vec) %*% XPXinv %*% c.vec )
```

This results in a t statistic of -1.315, which is insufficient to reject the null hypothesis. For $\alpha = 0.05$, the rejection region would be $t < -1.971$ or $t > 1.971$. In this example, the scales for SATM and SATV are the same. We are asking whether one additional point on the SATM score has the same effect on first-year GPA as one additional point on the SATV score. We have no evidence that these slope parameters differ. ∎ ■ ■ ■

In Example 14.9 we showed how to use the `predict()` function to obtain confidence or prediction intervals in the case of simple regression. This command also works for multiple regression if we create a data frame[11] whose columns are the variables `SATM`, `SATV`, and `FirstGen`, and whose rows consist of the sets of values of these variables at which we would like to obtain confidence or prediction intervals.

■ **Example 14.18** For Example 14.17, use R's built-in `predict()` command to obtain the confidence and prediction intervals for parts (a) and (b).

Solution: The required R commands are

```
x.pred = data.frame( SATM = 650 , SATV = 600 , FirstGen = 1 )
conf = predict( MultReg3 , x.pred , interval = "confidence" )
pred = predict( MultReg3 , x.pred , interval = "prediction" )
print( conf )
print( pred )
```

The results are

```
        fit      lwr      upr
1 2.986647 2.805077 3.168217
```

```
        fit      lwr      upr
1 2.986647 2.09182 3.881474
```

which agree with the results obtained by applying (14.44) and (14.45) directly, as in Example 14.17.

■ ■ ■

14.14 Correlation

So far in this chapter we have assumed that the predictor variables were constant and the only random variable was the response or outcome. There are cases, however, where we have two variables and both can be considered random. We assume that the ordered pair (X_i, Y_i) is a random vector having joint PDF $f(x, y)$.[12] The expected values of X_i and Y_i can then be computed as

$$\mathbb{E}(X_i) = \mu_X = \int_{-\infty}^{\infty} \int_{-\infty}^{\infty} x f(x,y) \, dy \, dx = \int_{-\infty}^{\infty} x f_X(x) \, dx$$

and

$$\mathbb{E}(Y_i) = \mu_Y = \int_{-\infty}^{\infty} \int_{-\infty}^{\infty} y f(x,y) \, dy \, dx = \int_{-\infty}^{\infty} y f_Y(y) \, dy,$$

where $f_X(x)$ and $f_Y(y)$ are the marginal distributions of X and Y, respectively. The variances are then defined similarly:

$$\mathbb{V}(X_i) = \sigma_X^2 = \int_{-\infty}^{\infty} \int_{-\infty}^{\infty} (x - \mu_X)^2 f(x,y) \, dy \, dx = \int_{-\infty}^{\infty} (x - \mu_X)^2 f_X(x) \, dx$$

---

[11]This must be a data frame. Even if we are making a confidence or prediction interval for one set of values for the predictors, this must be a data frame, not a vector. If we want to make a number of confidence or prediction intervals, the data frame would have multiple rows.

[12]Here we are assuming that both variables are continuous. If both X_i and Y_i are discrete, then $f(x,y)$ would be the joint PMF. Although there are cases where one random variable is continuous and one is discrete, we will not discuss that case here.

and

$$\mathbb{V}(Y_i) \;=\; \sigma_X^2 \;=\; \int_{-\infty}^{\infty}\!\!\int_{-\infty}^{\infty} (y-\mu_Y)^2\, f(x,y)\, dy\, dx \;=\; \int_{-\infty}^{\infty} (y-\mu_Y)^2\, f_Y(y)\, dy.$$

Note that both means and both variances can be obtained if we know the marginal distributions of X_i and Y_i. The covariance between X_i and Y_i is defined to be

$$\sigma_{XY} \;=\; \text{cov}(X,Y) \;=\; \mathbb{E}\big((X-\mu_X)(Y-\mu_Y)\big), \tag{14.46}$$

and it requires knowledge of the joint distribution $f(x,y)$.[13]

If we have a sample $(X_1,Y_1),\ (X_2,Y_2),\dots,(X_n,Y_n)$ from a bivariate normal distribution, we can construct the likelihood function and from it we can obtain the MLEs for the parameters

$$\mu = \begin{bmatrix} \mu_X \\ \mu_Y \end{bmatrix} \quad \text{and} \quad \Sigma = \begin{bmatrix} \sigma_X^2 & \sigma_{XY} \\ \sigma_{XY} & \sigma_Y^2 \end{bmatrix}.$$

The next theorem gives these results.

Theorem 14.14.1 If $(X_1,Y_1),\ (X_2,Y_2),\dots,(X_n,Y_n) \sim$ i.i.d. $N\left(\begin{bmatrix} \mu_X \\ \mu_Y \end{bmatrix}, \begin{bmatrix} \sigma_X^2 & \sigma_{XY} \\ \sigma_{XY} & \sigma_Y^2 \end{bmatrix}\right)$, then the MLEs of the parameters are

$$\hat\mu_X = \frac{1}{n}\sum_{i=1}^n X_i$$

$$\hat\mu_Y = \frac{1}{n}\sum_{i=1}^n Y_i$$

$$\hat\sigma_X^2 = \frac{1}{n}\sum_{i=1}^n (X_i-\overline{X})^2$$

$$\hat\sigma_Y^2 = \frac{1}{n}\sum_{i=1}^n (Y_i-\overline{Y})^2$$

$$\hat\sigma_{XY} = \frac{1}{n}\sum_{i=1}^n (X_i-\overline{X})(Y_i-\overline{Y}).$$

The MLEs for μ_X, μ_Y, σ_X^2, and σ_Y^2 can be found using the fact that the marginals for X and Y are $N(\mu_X,\sigma_X^2)$ and $N(\mu_Y,\sigma_Y^2)$, respectively. The MLE for the covariance σ_{XY} must be obtained from the joint PDF.

The MLE for the covariance matrix Σ can be written in matrix terms as

$$\hat\Sigma = \frac{1}{n}\sum_{i=1}^n \begin{bmatrix} x_i-\overline{x} \\ y_i-\overline{y} \end{bmatrix}\begin{bmatrix} x_i-\overline{x} & y_i-\overline{y} \end{bmatrix} = \sum_{i=1}^n \begin{bmatrix} (x_i-\overline{x})^2 & (x_i-\overline{x})(y_i-\overline{y}) \\ (x_i-\overline{x})(y_i-\overline{y}) & (y_i-\overline{y})^2 \end{bmatrix}. \tag{14.47}$$

The product of the 2×1 and the 1×2 vectors is the opposite of what we are used to seeing. Often we see a 1×2 vector times a 2×1, which results in a scalar; this operation is often called the *inner product*.

[13]If you know the joint distribution, you can find the marginals. If you know only the marginals, you cannot reconstruct the joint. Knowing the marginals is enough to get the mean and variance, but not enough to get the covariance!

Here, though, we have a 2×1 times a 1×2, which results in a 2×2 matrix, which is often called the *outer product*.

The result of Theorem 14.14.1 can be generalized to the multivariate normal distribution of any dimension.

Theorem 14.14.2 If $X_1, X_2, \ldots, X_p \sim$ i.i.d. $N(\mu, \Sigma)$, then the MLEs of μ and Σ are

$$\hat{\mu} = \frac{1}{n} \sum_{i=1}^{n} X_i$$

$$\hat{\Sigma} = \frac{1}{n} \sum_{i=1}^{n} \underbrace{(X_i - \hat{\mu})}_{p \times 1} \underbrace{(X_i - \hat{\mu})'}_{1 \times p}.$$

The *correlation* between two random variables X and Y is defined to be

$$\rho = \frac{\text{cov}(X, Y)}{\sigma_X \, \sigma_Y} = \frac{\sigma_{XY}}{\sigma_X \, \sigma_Y}. \tag{14.48}$$

The correlation is a dimensionless quantity, which means that its value has no units. For example, if X is measured in feet and Y is measured in seconds, then μ_X has units in feet, σ_X^2 has units in feet^2, and σ_{XY} has units in feet-seconds. The correlation, however, has no units.

By the invariance property of maximum likelihood estimators, the MLE of ρ, denoted R, is

$$
\begin{aligned}
R = \hat{\rho} &= \frac{\hat{\sigma}_{XY}}{\hat{\sigma}_X \, \hat{\sigma}_Y} \\
&= \frac{\frac{1}{n} \sum_{i=1}^{n} (X_i - \overline{X})(Y_i - \overline{Y})}{\sqrt{\frac{1}{n} \sum_{i=1}^{n} (X_i - \overline{X})^2} \sqrt{\frac{1}{n} \sum_{i=1}^{n} (Y_i - \overline{Y})^2}} \\
&= \frac{\sum_{i=1}^{n} (X_i - \overline{X})(Y_i - \overline{Y})}{\sqrt{\sum_{i=1}^{n} (X_i - \overline{X})^2 \times \sum_{i=1}^{n} (Y_i - \overline{Y})^2}} \\
&= \frac{(X - \overline{X})'(Y - \overline{Y})}{\sqrt{(X - \overline{X})'(X - \overline{X}) \times (Y - \overline{Y})'(Y - \overline{Y})}} \\
&= \frac{(X - \overline{X})'(Y - \overline{Y})}{\|X - \overline{X}\| \times \|Y - \overline{Y}\|}.
\end{aligned}
$$
$$\tag{14.49}$$
$$\tag{14.50}$$

Schwarz's inequality[14] implies that

$$|(X - \overline{X})'(Y - \overline{Y})| \leq \|X - \overline{X}\| \times \|Y - \overline{Y}\|,$$

---

[14]Schwarz's inequality (Rudin, 1976, section 1.35) states that for vectors u and v: $u \cdot v \leq \|u\| \, \|v\|$. Apply this to the vectors $(X - \overline{X})$ and $(Y - \overline{Y})$ to obtain the result.

which means that the estimate of the correlation must be between -1 and 1. It can also be shown that the population correlation ρ must be between -1 and 1.

Using the definitions in (14.13), (14.15), and (14.14) along with the result in (14.49), we can see that the estimate of ρ can be written as

$$R = \frac{S_{xy}}{\sqrt{S_{xx} S_{yy}}}.$$

This can also be written as

$$R = \frac{S_{xy}}{\sqrt{S_{xx} S_{yy}}} = \frac{S_{xy}}{S_{xx}} \sqrt{\frac{S_{xx}}{S_{yy}}} = \hat{\beta}_1 \sqrt{\frac{S_{xx}}{S_{yy}}},$$

where $\hat{\beta}_1 = S_{xy}/S_{xx}$.

■ **Example 14.19**   For the diabetes data in Example 10.8, give a scatterplot of (diastolic) blood pressure and BMI, and compute the correlation.

Solution: Figure 14.14(a) shows a scatterplot of diastolic blood pressure against BMI. The numerous points along the x-axis and the y-axis suggest that the data should first be cleaned, because a blood pressure or BMI of zero is not possible; these must be erroneous results. We can filter out those records with either a zero blood pressure or a zero BMI; we can also think of this as *keeping* those values for which blood pressure and BMI are both positive. This can be achieved with the `filter()` command and the pipe, both from the `dplyr` package:

```
diabetes = read.csv(Diabetes.csv)
library( dplyr )
diabetes = diabetes  %>% filter( BloodPressure > 0 & BMI > 0 )
```

Once this filtering is done, we obtain the scatterplot shown in Figure 14.14(b). The figure suggests that larger values of BMI tend to be associated with larger values of blood pressure. A useful trick for visualizing this is to draw "crosshairs" at the means for both variables, as shown in Figure 14.15. When you see more points in the upper right and lower left quadrants, this suggests a positive correlation. When you see more points in the upper left and lower right, this suggests a negative correlation.

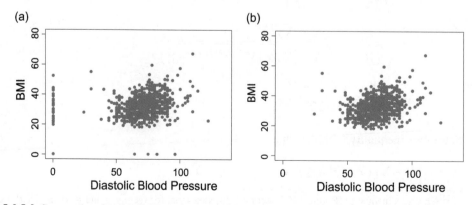

Figure 14.14 Scatterplots for the diabetes data set, before (a) and after (b) data cleaning.

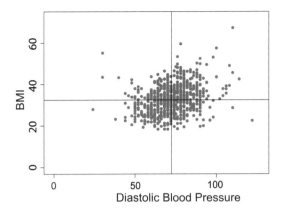

Figure 14.15 Scatterplots for diabetes data set with crosshairs drawn at the means.

The estimate of the correlation can be obtained using the cor() command:

```
cor( diabetes$BloodPressure , diabetes$BMI )
```

This yields the output

```
[1] 0.2892303
```

suggesting a small to moderate correlation between BMI and BloodPressure.   ■ ■ ■

Statisticians and data scientists should have a good idea of the strength of a relationship between variables given their correlation. If large values of one tend to be paired with large values of the other, and vice versa, then the correlation will be positive. If large values of one tend to be paired with small values of the other, then the correlation will be negative. Figure 14.16 shows scatterplots for nine different values of the estimated correlation r. As the correlation gets close to ± 1, the points get close to a straight line. This is what is meant when we say that the correlation measures the *linear relationship*.

It is certainly possible that there is a relationship between X and Y that is nonlinear. In this case, the correlation is not a useful measure of the association. Even when the relationship between X and Y is nonlinear, nothing prevents you from putting the numbers into the formula for the point estimate $\hat{\rho} = r$. While this would be considered inappropriate and bad practice, R is certainly willing to do the computations for you. Figure 14.17 shows several nonlinear relationships and the corresponding estimated correlations. The bottom right plot shows a strong circular relationship between X and Y, yet the correlation is zero.

We next present two theorems, which look similar. A closer look uncovers one of the common misconceptions in statistics.

Theorem 14.14.3 If X and Y are independent, then the correlation is equal to 0.

Proof. The covariance between X and Y is

$$
\begin{aligned}
\operatorname{cov}(X,Y) &= \mathbb{E}(XY) - \mathbb{E}(X)\mathbb{E}(Y) \\
&= \int_{-\infty}^{\infty} \int_{-\infty}^{\infty} x\,y\,f(x,y)\,dy\,dx - \mu_X \mu_Y \\
&= \int_{-\infty}^{\infty} \int_{-\infty}^{\infty} x\,y\,f_X(x)\,g_Y(y)\,dy\,dx - \mu_X \mu_Y \qquad \text{[because } X \text{ and } Y \text{ are independent]} \\
&= \int_{-\infty}^{\infty} x\,f_X(x) \int_{-\infty}^{\infty} y\,g_Y(y)\,dy\,dx - \mu_X \mu_Y
\end{aligned}
$$

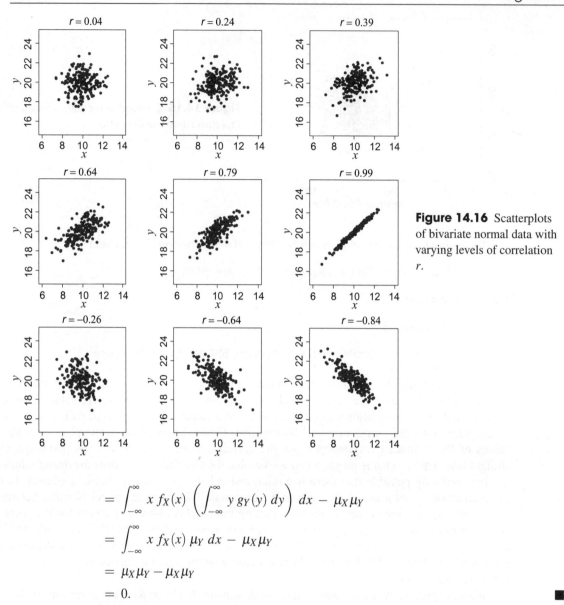

Figure 14.16 Scatterplots of bivariate normal data with varying levels of correlation r.

$$= \int_{-\infty}^{\infty} x\, f_X(x) \left(\int_{-\infty}^{\infty} y\, g_Y(y)\, dy \right) dx - \mu_X \mu_Y$$

$$= \int_{-\infty}^{\infty} x\, f_X(x)\, \mu_Y\, dx - \mu_X \mu_Y$$

$$= \mu_X \mu_Y - \mu_X \mu_Y$$

$$= 0. \qquad\qquad\qquad\qquad\qquad\qquad\qquad\qquad\qquad \blacksquare$$

Theorem 14.14.4 Suppose X and Y have the bivariate normal distribution

$$\begin{bmatrix} X \\ Y \end{bmatrix} \sim N\left(\begin{bmatrix} \mu_X \\ \mu_Y \end{bmatrix},\ \begin{bmatrix} \sigma_X^2 & \sigma_{XY} \\ \sigma_{XY} & \sigma_Y^2 \end{bmatrix} \right).$$

If $\rho_{XY} = 0$, then X and Y are independent.

This result was proved in Problem 6.27.

The two previous theorems may seem to indicate that random variables are independent if and only if their correlation (or covariance) is equal to zero. However, this is not true. The truth is a little more subtle. Independence implies a zero correlation regardless of the joint distribution of X and Y. The converse is, in general, not true. If X and Y are uncorrelated (i.e., their correlation ρ_{XY} is zero) then we

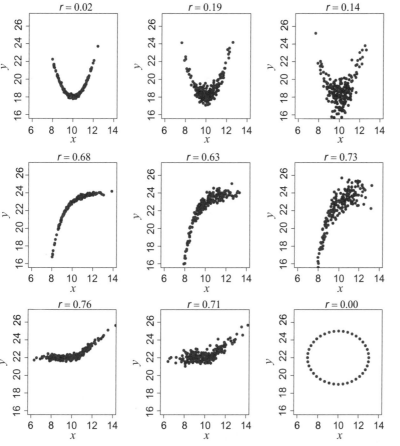

Figure 14.17 Scatterplots of nonlinear relationships with varying levels of correlation r.

cannot conclude that they are independent. If, however, the joint PDF is bivariate normal, then we can conclude that a zero correlation implies independence.

Often, we'd like to test whether the correlation ρ between random variables X and Y is zero against the alternative that it is not zero.[15] More generally, we can test H_0: $\rho = \rho_0$ vs. H_1: $\rho \neq \rho_0$ for any value of ρ_0 between -1 and 1. It is difficult to develop a test based directly on the statistic R as defined in (14.49) because the distribution of R is difficult to obtain. One approximation that is sometimes used is based on the Fisher transformation:

$$\log \frac{1+R}{1-R} \overset{\text{approx}}{\sim} N\left(\log \frac{1+\rho_0}{1-\rho_0} , \frac{4}{n-3} \right). \tag{14.51}$$

We could develop a z-test for H_0: $\rho = 0$ based on this result using the test statistic

$$Z = \frac{\log \dfrac{1+R}{1-R} - \log \dfrac{1-0}{1+0}}{2/\sqrt{n-3}} = \frac{\sqrt{n-3}}{2} \log \frac{1+R}{1-R}. \tag{14.52}$$

While this is sometimes used, there are better approaches. We describe one approach next, and defer the discussion of the other until Chapter 17.

---

[15]One-sided alternatives are also possible. The rejection regions for such tests should be obvious once a test statistic is derived.

If (X,Y) has the bivariate normal distribution, as described in Section 6.4.2,

$$\begin{bmatrix} X \\ Y \end{bmatrix} \sim N\left(\begin{bmatrix} \mu_X \\ \mu_Y \end{bmatrix}, \begin{bmatrix} \sigma_X^2 & \sigma_{XY} \\ \sigma_{XY} & \sigma_Y^2 \end{bmatrix} \right),$$

then

$$Y \mid (X = x) \sim N\left(\mu_Y + \frac{\sigma_Y}{\sigma_X}\rho(x - \mu_x), \, (1-\rho^2)\sigma_Y^2 \right).$$

Thus,

$$E(Y|X=x) = \mu_Y + \frac{\sigma_Y}{\sigma_X}\rho(x-\mu_X) = \underbrace{\mu_Y - \frac{\sigma_Y}{\sigma_X}\rho\,\mu_X}_{\beta_0} + \underbrace{\frac{\sigma_Y}{\sigma_X}\rho}_{\beta_1}\, x.$$

From this we can see that ρ and β_1 are related by

$$\beta_1 = \frac{\sigma_Y}{\sigma_X}\rho.$$

We conclude that $\rho = 0$ if and only if $\beta_1 = 0$. Thus, we could base a test of H_0 on the test statistic

$$t = \frac{\hat{\beta}_1 - 0}{S/\sqrt{S_{xx}}}, \tag{14.53}$$

where $S = \sqrt{\text{MSE}}$. This t statistic has a $t(n-2)$ distribution. The two-sided test with $H_1: \rho \neq 0$ would be rejected for t statistics in the left or right tail of the $t(n-2)$ distribution. One-sided tests can also be constructed. The result in (14.53) can be manipulated to be expressed in terms of the estimated correlation R. This result is proved in the theorem below, which relies on the result in the following exercise.

Exercise 14.6 Show that for the simple regression model

$$\text{SSE} = S_{yy} - \hat{\beta}_1 S_{xy}.$$

Theorem 14.14.5 The t statistic in (14.53) and the estimated correlation R are related by

$$t = \frac{R\sqrt{n-2}}{1-R^2}.$$

Proof.

$$t = \frac{\hat{\beta}_1 - 0}{S/\sqrt{S_{xx}}}$$

$$= \frac{\hat{\beta}_1 - 0}{\sqrt{\dfrac{\text{SSE}}{n-2}}\Big/\sqrt{S_{xx}}}$$

$$= \frac{\hat{\beta}_1 \sqrt{n-2}}{\sqrt{S_{yy} - \hat{\beta}_1 S_{xy}} / \sqrt{S_{xx}}}$$

$$= \frac{\frac{S_{xy}}{S_{xx}} \sqrt{n-2}}{\sqrt{S_{yy} - \frac{S_{xy}^2}{S_{xx}}} \sqrt{\frac{1}{S_{xx}}}}$$

$$= \frac{\frac{S_{xy}}{\sqrt{S_{xx}S_{yy}}} \sqrt{n-2}}{\sqrt{1 - \frac{S_{xy}^2}{S_{xx}S_{yy}}}}$$

$$= \frac{R \sqrt{n-2}}{\sqrt{1 - R^2}}.$$

A test of H_0: $\rho = 0$ vs. H_1: $\rho \neq 0$ can then be constructed using the statistic in Theorem 14.14.5: Reject H_0 if

$$t < -t_{\alpha/2}(n-2) \quad \text{or} \quad t > t_{\alpha/2}(n-2).$$

■ **Example 14.20** For the diabetes data described in Example 14.19, test whether the correlation is 0.

Solution: For this data set consisting of 729 ordered pairs (x_i, y_i) (after the 0 values for blood pressure and BMI are filtered out) we found the estimated correlation to be

$$r = 0.2892303.$$

We will use the statistic from Theorem 14.14.5 to test H_0: $\rho = 0$ versus H_1: $\rho \neq 0$:

$$t = \frac{r\sqrt{n-2}}{1 - r^2} = \frac{0.2892303\sqrt{727}}{1 - 0.2892303^2} = 8.51.$$

The rule is to reject H_0 if $t < -1.963$ or $t > 1.963$. Here, the observed t statistic is considerably larger than 1.963 so we reject H_0. The P-value is very small; R calculates this to be 9.87×10^{-17}, giving strong evidence against H_0. Even though the estimated correlation is only moderate (roughly 0.29), because there is so much data, the P-value is extremely small. We have strong evidence of a moderate correlation.

R is capable of testing H_0: $\rho = 0$ versus H_1: $\rho \neq 0$ through the `cor.test()` command. For example,

```
cor.test( diabetes$BloodPressure , diabetes$BMI )
```

This yields the output

```
        Pearson's product-moment correlation

data:  diabetes$BloodPressure and diabetes$BMI
t = 8.1467, df = 727, p-value = 1.63e-15
```

```
alternative hypothesis: true correlation is not equal to 0
95 percent confidence interval:
0.2212643 0.3544003
sample estimates:
cor
0.2892303
```

This test is based on the normal approximation from (14.52), which is often called Fisher's transformation. This accounts for the differing test statistic and *P*-value. The main conclusion that there is strong evidence of a nonzero correlation remains the same.   ∎ ∎ ∎

When we have several variables and we would like to study all possible correlations, we can estimate and test the correlations of all possible pairs of variables using R's rcorr() function in the Hmisc package. This will produce two matrices: a matrix of the estimates of all possible pairs of variables, and a matrix of *P*-values for testing whether each correlation is equal to zero.

∎ **Example 14.21**   For the diabetes data from Example 14.19, compute and test the correlations for all variables except the indicator variable Outcome.

Solution: First, let's load the Hmisc package. Second, we'll clean the data by removing the observations that have zero values for BMI or BloodPressure. Then, we will rename some of the variables. Shortening the variable names will make the output cleaner.

```
library( Hmisc )
diabetes = diabetes  %>% filter( BloodPressure > 0 & BMI > 0 )
diabetes = rename( diabetes , DiabetesPed = DiabetesPedigreeFunction ,
                   BloodPress = BloodPressure ,
                   SkinThick = SkinThickness )
```

We will then apply the rcorr function to compute and test the correlations of the first eight columns (variables) in the diabetes data set:

```
rcorr( as.matrix( diabetes[,1:8] ) , type = c("pearson") )
```

The resulting output is

	Pregnancies	Glucose	BloodPress	SkinThick	Insulin	BMI	DiabetesPed	Age
Pregnancies	1.00	0.14	0.21	-0.09	-0.08	0.02	-0.02	0.56
Glucose	0.14	1.00	0.22	0.06	0.34	0.21	0.14	0.26
BloodPress	0.21	0.22	1.00	0.01	-0.04	0.29	0.00	0.33
SkinThick	-0.09	0.06	0.01	1.00	0.42	0.40	0.18	-0.13
Insulin	-0.08	0.34	-0.04	0.42	1.00	0.19	0.18	-0.05
BMI	0.02	0.21	0.29	0.40	0.19	1.00	0.16	0.02
DiabetesPed	-0.02	0.14	0.00	0.18	0.18	0.16	1.00	0.03
Age	0.56	0.26	0.33	-0.13	-0.05	0.02	0.03	1.00

```
n= 729
```

```
P
```

	Pregnancies	Glucose	BloodPress	SkinThick	Insulin	BMI	DiabetesPed	Age
Pregnancies		0.0002	0.0000	0.0105	0.0342	0.6682	0.5295	0.0000
Glucose	0.0002		0.0000	0.1232	0.0000	0.0000	0.0002	0.0000
BloodPress	0.0000	0.0000		0.7566	0.2282	0.0000	0.9573	0.0000
SkinThick	0.0105	0.1232	0.7566		0.0000	0.0000	0.0000	0.0005
Insulin	0.0342	0.0000	0.2282	0.0000		0.0000	0.0000	0.2002

BMI	0.6682	0.0000	0.0000	0.0000	0.0000	0.0000	0.5057
DiabetesPed	0.5295	0.0002	0.9573	0.0000	0.0000	0.0000	0.4839
Age	0.0000	0.0000	0.0000	0.0005	0.2002	0.5057	0.4839

In the output above, the estimated correlations between variables are shown in the first matrix, and the P-value for testing whether the population correlation is zero is the corresponding entry in the second matrix. For example, the correlation between `Pregnancies` and `Glucose` is 0.14, and the P-value for testing whether the true correlation is zero is 0.0002. Both of these matrices are symmetric, because the correlation between X and Y is the same as the correlation between Y and X; the corresponding P-values will be the same too.

We can see a lot of highly significant correlations here. Care should be taken when interpreting these. It is always a good idea to present scatterplots of the variables to see more clearly what kind of relationship exists. Figure 14.18 shows a matrix plot of all possible scatterplots for the diabetes data set after the data cleaning. There are no strong linear relationships present among the eight variables. With 729 observations, though, many of the correlations will be statistically significant (i.e., different from 0) even if the relationship is small to moderate in magnitude. ∎∎∎

14.15 R^2 and Adjusted R^2

The partition (or decomposition) of sums of squares for multiple regression was found to be (Theorem 14.10.2)

$$\text{SST} = \text{SSR} + \text{SST};$$

in other words,

$$\sum_{i=1}^{n}(y_i - \bar{y})^2 = \sum_{i=1}^{n}(\hat{y}_i - \bar{y})^2 + \sum_{i=1}^{n}(y_i - \hat{y}_i)^2.$$

Each of these sums of squares measures a different aspect of the data and its fit to the regression model. The SST measures the variability of the data without taking into account any predictor variables. The SSE measures the variability after all of the predictor variables have been accounted for. We can therefore view the SSR as the amount that the SST is reduced once the predictor variables are included in the model. The *coefficient of determination*, defined to be the ratio

$$R^2 = \frac{\text{SSR}}{\text{SST}},$$

can be viewed as the variability in the data that can be explained by the regression, divided by the total amount of variability *before* any predictor variables are accounted for. This ratio cannot be less than zero, because both numerator and denominator are sums of squares and therefore must be nonnegative.[16] Also, since

$$R^2 = \frac{\text{SSR}}{\text{SST}} = \frac{\text{SSR}}{\text{SSR} + \text{SSE}} \le 1,$$

---

[16]It is possible that one of the sums of squares is zero. If all of the y_i are the same, then $y_i = \bar{y}$ for all i, in which case SST $= 0$. Also, if all of the estimated βs are zero, then the numerator, SSR, will be zero. These are extreme cases; for most data sets both numerator and denominator will be positive.

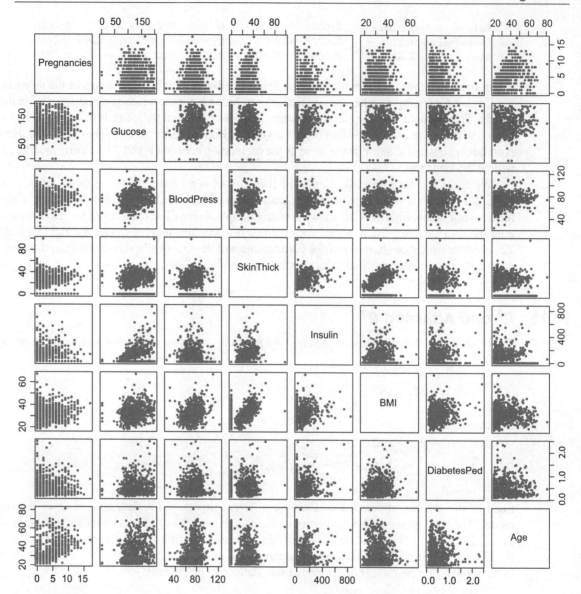

Figure 14.18 Matrix plot of all possible scatterplots for the diabetes data set.

we must have $R^2 \geq 0$. Thus,

$$0 \leq R^2 \leq 1.$$

The expression for R^2 can also be written as

$$R^2 = 1 - \frac{\text{SSE}}{\text{SST}}. \tag{14.54}$$

If the model fits all points perfectly, in the sense that $\hat{y}_i = y_i$ for all i, then SSE = 0. If SSE = 0, then SSR = SST, in which case $R^2 = 1$. If $\hat{\beta}_1 = \cdots = \hat{\beta}_p = 0$, then SST = SSR which implies that $R^2 = 1$. In general, values of R^2 nearer to 1 indicate a closer fit of the data to the model.

For the case of simple regression it is easy to show that the value of R^2 is equal to the square of the correlation R (justifying the use of the notation R^2). Before we prove this, we ask the reader to derive an expression for SSR.

Exercise 14.7 For simple linear regression, show that

$$SSR = \hat{\beta}_1^2 S_{xx}.$$

Hint: Write $\hat{y}_i - \bar{y} = \hat{y}_i - \hat{\beta}_1 x_i + \hat{\beta}_1 x_i - \bar{y}$.

Theorem 14.15.1 For simple linear regression, the value of R^2 is equal to the estimated correlation squared.

Proof.

$$SSR = \sum_{i=1}^n (\hat{y}_i - \bar{y})^2 = \hat{\beta}_1^2 S_{xx} = \left(\frac{S_{xy}}{S_{xx}}\right)^2 S_{xx} = \frac{S_{xy}^2}{S_{xx}}.$$

From this, we can write

$$R^2 = \frac{SSR}{SST} = \frac{S_{xy}^2/S_{xx}}{S_{yy}} = \left(\frac{S_{xy}^2}{S_{xx}S_{yy}}\right) = \left(\frac{S_{xy}}{\sqrt{S_{xx}S_{yy}}}\right)^2.$$

This last term is just the square of the estimated correlation between X and Y. ∎

For more than one predictor variable, the value of R^2 is a measure of the fit of the model. We can look at multiple regression as analyzing the variability in the response, given that we have accounted for one or more of the predictor variables. The variability in a model with no predictor variables can be described by SST. As we add predictor variables, we reduce the SSE, so from (14.54) we increase the R^2. Models with an R^2 near 1 explain more of the variability in the response than models with smaller R^2 values.

We can always add predictor variables to a model and the R^2 can only go up (or stay the same if the new variable has an estimated β that is exactly 0). It is tempting, therefore, to add predictor variables to the model to increase R^2 even if these variables have little or no effect on the response. To account for this, the *adjusted coefficient of determination* R_{adj} is often used to penalize models with more parameters. We will define this shortly, but to motivate the definition let's write R^2 as

$$R^2 = 1 - \frac{SSE}{SST} = 1 - \frac{SSE/(N-1)}{SST/(N-1)}. \tag{14.55}$$

This is almost the same as

$$1 - \frac{SSE/(N-p-1)}{SST/(N-1)} = 1 - \frac{MSE}{MST}, \tag{14.56}$$

except the denominator in the numerator is different. If we were to define the adjusted R^2 as in (14.56), then adding more predictor variables may not increase R^2. Using the denominator of $N - p$ rather than $N - 1$ means that adding a variable will certainly decrease the denominator in $SSE/(N-p)$, making the ratio larger. The SSE in the numerator can only decrease or remain the same. Thus, adding a predictor variable may increase or decrease $\text{SSE}/(N-p)$. Including a variable that has little effect on the response often leads to an increase in $\text{SSE}/(N-p-1)$. The denominator, SST, is unaffected by adding predictor variables.

We therefore define the adjusted R^2 to be

$$R_{\text{adj}}^2 = 1 - \frac{\text{MSE}}{\text{MST}}. \tag{14.57}$$

This can be written as

$$R_{\text{adj}}^2 = 1 - \frac{\text{SSE}/(N-p-1)}{\text{SST}/(N-1)} = 1 - \frac{N-1}{N-p-1}\left(1 - \frac{\text{SSR}}{\text{SST}}\right) = 1 - \frac{N-1}{N-p-1}(1-R^2). \tag{14.58}$$

Many books take (14.58) as the definition of R_{adj}^2. We take the definition in (14.57) because it is more intuitive.

■ **Example 14.22**   For the first-year GPA data, run all possible models with SATM, SATV, and HSGPA. For each model, compute R_{adj}^2. Which model gives the best R_{adj}^2?

Solution: The seven possible linear models (excluding the one with no predictors) can be run with the following commands:

```
lm100 = lm( GPA ~ SATV , data = first )
lm010 = lm( GPA ~ SATM , data = first )
lm001 = lm( GPA ~ HSGPA , data = first )
lm110 = lm( GPA ~ SATV + SATM , data = first )
lm101 = lm( GPA ~ SATV + HSGPA , data = first )
lm011 = lm( GPA ~ SATM + HSGPA , data = first )
lm111 = lm( GPA ~ SATV + SATM + HSGPA , data = first )
summary( lm100 )
summary( lm010 )
summary( lm001 )
summary( lm110 )
summary( lm101 )
summary( lm011 )
summary( lm111 )
```

The output from all seven of these models is long, so we include only the first one here.

```
Call:
lm(formula = GPA ~ SATV, data = first)

Residuals:
Min      1Q      Median   3Q      Max
-1.14057 -0.32756 0.03134  0.34259 1.16723

Coefficients:
            Estimate Std. Error t value Pr(>|t|)
(Intercept) 2.0684133 0.2204478  9.383  < 2e-16 ***
SATV        0.0016986 0.0003609  4.706  4.5e-06 ***
---
Signif. codes:  0 '***' 0.001 '**' 0.01 '*' 0.05 '.' 0.1 ' ' 1
```

Table 14.3 R^2 and R^2_{adj} from all seven possible models in the GPA example.

Predictor variables	R^2	R^2_{adj}
SATV	0.09261	0.08842
SATM	0.03777	0.03334
HSGPA	0.1997	0.196
SATV + SATM	0.0942	0.08582
SATV + HSGPA	0.246	0.239
SATM + HSGPA	0.216	0.2087
SATV + SATM + HSGPA	0.2464	0.2359

```
Residual standard error: 0.4444 on 217 degrees of freedom
Multiple R-squared:  0.09261,        Adjusted R-squared:  0.08842
F-statistic: 22.15 on 1 and 217 DF,  p-value: 4.499e-06
```

The values for R^2 and R^2_{adj} can be found on the second last line. Values of R^2 and R^2_{adj} for other regression models are shown in Table 14.3. The model with all three variables gives the greatest R^2_{adj}. ■ ■ ■

14.16 Model Checking

The linear regression model

$$y_i = \beta_0 + \beta_1 x_{i1} + \beta_2 x_{i2} + \cdots + \beta_p x_{ip} + \varepsilon_i, \qquad \varepsilon \sim \text{i.i.d. } N(0, \sigma^2)$$

involves a host of assumptions: linearity in each variable, independent normally distributed error terms, and constant variance of the error terms. As we've mentioned previously, particularly in the discussion regarding (14.4), the linearity fails to hold in many circumstances, and can be thought of as just an approximation to the true relationship. We can check many of these assumptions by looking at plots of the residuals. We defined the residuals to be the difference between the observed and the predicted responses:

$$r_i = y_i - \hat{y}_i.$$

We should emphasize that the residuals r_i are not the same as the error terms ε_i. The residuals satisfy

$$y_i = \hat{\beta}_0 + \hat{\beta}_1 x_{i1} + \cdots + \hat{\beta}_p x_{ip} + r_i,$$

whereas the responses themselves satisfy

$$y_i = \beta_0 + \beta_1 x_{i1} + \cdots + \beta_p x_{ip} + \varepsilon_i.$$

Thus, the r_i satisfy

$$r_i = y_i - \left(\hat{\beta}_0 + \hat{\beta}_1 x_{i1} + \cdots + \hat{\beta}_p x_{ip} \right),$$

whereas the ε satisfy

$$\varepsilon_i = y_i - (\beta_0 + \beta_1 x_{i1} + \cdots + \beta_p x_{ip}).$$

The ε_i are unobservable, whereas once we've collected data, the residuals r_i are observable. Finally, the ε are independent, whereas the residuals have a constraint; the residuals must sum to zero. To see this for the case of simple regression, note that

$$\sum_{i=1}^{n}(y_i - \hat{y}_i) = \sum_{i=1}^{n}\left(y_i - (\hat{\beta}_0 + \hat{\beta}_1 x_i)\right) = \sum_{i=1}^{n} y_i - n\hat{\beta}_0 - \hat{\beta}_1 \sum_{i=1}^{n} x_i = n\bar{y} - n\bar{y} + n\hat{\beta}_1\bar{x} - n\hat{\beta}_1\bar{x} = 0.$$

As mentioned previously, graphical analysis of residuals is an effective way to investigate the adequacy of the fit of a regression model and to check the underlying assumptions. In this section, we introduce and illustrate the basic residual plots. These plots are typically generated by regression computer software packages. They should be examined routinely in all regression modeling problems. We often plot externally studentized[17] residuals because they have constant variance.

14.16.1 Normal Probability Plots

Small departures from the normality assumption do not affect the model greatly, but gross nonnormality is potentially more serious because the t or F statistics and confidence and prediction intervals depend on the normality assumption. Furthermore, if the errors come from a distribution with thicker or heavier tails than the normal, the least-squares fit may be sensitive to a small subset of the data. Heavy-tailed error distributions often generate outliers that "pull" the least-squares fit too much in their direction and can lead to problems with nonconstant variance. In these cases, transformations or other estimation techniques such as robust regression methods should be considered.

A simple method of checking the normality assumption is to construct a normal probability plot of the residuals. We illustrate this using the regression model fit to the first-year GPA data that used only the three significant variables; HSGPA, HU, and White. The R code to perform the multiple regression is

```
first = read.csv( "FirstYearGPA.csv" , header=TRUE )
x1 = first$HSGPA
x2 = first$HU
x3 = first$White
y = first$GPA
mult.reg = lm( y ~ x1 + x2 + x3 )
summary( mult.reg )
```

The result of summary(mult.reg) is

```
Call:
lm(formula = y ~ x1 + x2 + x3)

Residuals:
Min        1Q    Median       3Q       Max
-1.09479  -0.27638   0.02287   0.25411   0.84538

Coefficients:
             Estimate Std. Error t value Pr(>|t|)
(Intercept) 0.933459   0.245673    3.800 0.000189 ***
x1          0.507404   0.070197    7.228 8.42e-12 ***
x2          0.015328   0.003667    4.180 4.24e-05 ***
x3          0.265644   0.064519    4.117 5.47e-05 ***
```

---

[17]*Studentizing* is similar to standardizing. While standardizing a random variable means subtracting the mean and dividing by the standard deviation, studentizing involves subtracting the mean and dividing by an estimate of the standard deviation. The term comes from the pseudonym "student" that Gossett used to publish his work on the t-test.

(a)

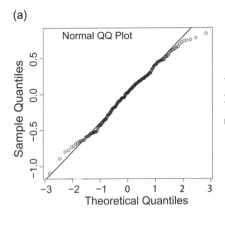

(b)

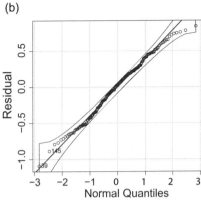

Figure 14.19 Normal probability plots for the residuals from a multiple linear regression using the GPA data.

```
---
Signif. codes:  0 '***' 0.001 '**' 0.01 '*' 0.05 '.' 0.1 ' ' 1

Residual standard error: 0.3856 on 215 degrees of freedom
Multiple R-squared:  0.3231,      Adjusted R-squared:  0.3136
F-statistic: 34.21 on 3 and 215 DF,  p-value: < 2.2e-16
```

The residuals are contained in the `mult.reg` object and are accessible as `mult.reg$residuals`. We can produce a normal probability plot using the base R function `qqnorm()`. A line can then be fitted to the data using `qqline()`. Bands can be placed around the points and line using the `qqPlot()` function in the `car` package:

```
resid = mult.reg$residuals
par( mfrow=c(1,2) )
qqnorm( resid ); qqline( resid , col="blue" , lwd=2 )
library( car )
qqPlot( resid )
```

Figure 14.19 shows the plot. Figure 14.19(a) shows the plot from the base R functions `qqnorm()` and `qqline()`; Figure 14.19(b) uses the package `car` with the function `qqPlot()`. Notice that the residuals lie approximately along a straight line, indicating that there is no serious problem with the normality assumption.

A common defect that shows up on the normal probability plot is the occurrence of one or two large residuals. Sometimes this is an indication that the corresponding observations are outliers. Outliers should be carefully investigated to see whether they are the result of an error in data collection or reporting, or if they are an unusual but real part of the process being studied. Finally, note that formal statistical tests for normality should not be applied to residuals. These tests generally assume that the observations are independent, and as we have noted previously, residuals are not independent.

14.16.2 Plot of Residuals against the Fitted Values

A plot of the residuals versus the corresponding fitted values is useful for detecting several common types of model inadequacies, such as unequal variance and independence of the observations. These violations of the assumptions are potentially more serious that nonnormality and can substantially impact the validity of the model. If this plot resembles Figure 14.20(a), which indicates that the residuals can be contained in a horizontal band, then there are no obvious model defects. Plots of the residuals versus the fitted values that resemble any of the patterns in panels (b)–(d) are symptomatic of model deficiencies.

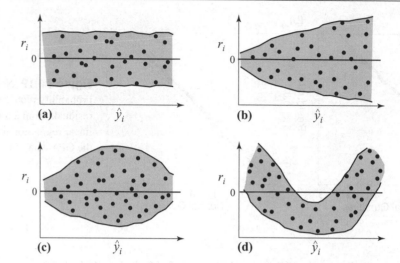

Figure 14.20 Patterns for residual plots: (a) satisfactory, (b) funnel, (c) double bow, and (d) nonlinear.

The patterns in panels (b) and (c) indicate that the variance of the errors is not constant. The outward-opening funnel pattern in panel (b) implies that the variance is an increasing function of y; an inward-opening funnel is also possible, indicating that $\mathbb{V}(\varepsilon)$ decreases as y increases. The double-bow pattern in panel (c) often occurs when y is a proportion between 0 and 1. The variance of a binomial proportion is greatest near 0.5 and decreases as you move away from 0.5. The usual approach for dealing with nonconstant variance is to apply a suitable transformation to either the regressor or the response variable or to use the method of weighted least squares. In practice, transformations on the response are generally employed to stabilize variance.

Plotting the residuals against the corresponding values of each regressor variable can also be helpful. These plots often exhibit patterns such as those in Figure 14.20, except that the horizontal scale is x_{ij} for the jth regressor rather than the fitted value. Once again, an impression of a horizontal band containing the residuals is desirable. The funnel and double-bow patterns in panels (b) and (c) indicate nonconstant variance. The curved band in panel (d) or a nonlinear pattern in general implies that the assumed relationship between y and the regressor x_j is not correct. Thus, either higher-order polynomial terms in x_j or a transformation should be considered.

In the simple linear regression case, it is not necessary to plot residuals versus both the fitted value and the regressor variable. The reason is that the fitted values are linear combinations of the regressor values, so the plots would only differ in the scale for the vertical axis.

Figure 14.21 shows plots of the residuals versus the fitted values and for each of the predictor variables for the first-year GPA data. These plots closely resemble the prototype in Figure 14.20(a), so we conclude that there is no significant problem with inequality of variance. The plot in Figure 14.21(c) suggests that the variability might be greater at the upper level (White=1) than at the lower level (White=0). This is an illusion because there are relatively few nonwhite students in the data set. The standard deviations of the residuals are nearly the same: 0.390 at the lower level and 0.382 at the upper level.

14.17 Chapter Summary

This chapter is an introduction to linear regression models. Linear regression models are often used as approximating functions in data analysis and modeling problems when the true functional relationships are unknown. The key concepts covered in the chapter include:

(a) using simple linear regression models to describe the relationship between a response or outcome variable y and a single predictor or regressor variable;

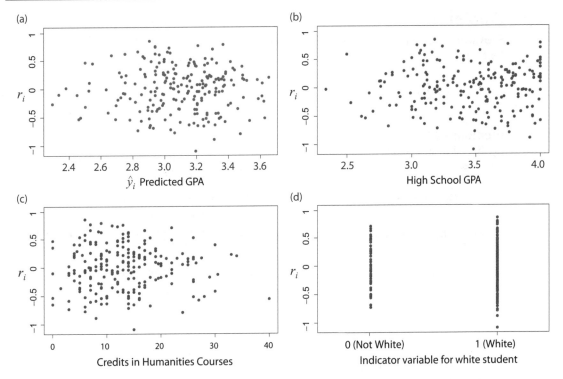

Figure 14.21 Residual plots for first-year GPA data.

(b) extending simple linear regression to multiple regression, a situation with several predictors;

(c) using the method of least squares to estimate the parameters in linear regression models;

(d) using regression models for prediction and explanation;

(e) maximum likelihood estimation and least squares estimation in linear regression are equivalent if model errors are normally and independently distributed;

(f) tools for assessing model adequacy;

(g) how to test hypotheses and construct confidence intervals on model parameters;

(h) how to construct confidence intervals on the mean response at specific values of the predictors and how to construct prediction intervals on new observations;

(i) how to build models that have categorical variables;

(j) how to build polynomial regression models;

(k) how to use and apply the normal correlation model.

14.18 Problems

Problem 14.1 Show that an alternative expression for S_{xx} is

$$S_{xx} = \sum_{i=1}^{n} (x_i - \bar{x}) x_i.$$

Problem 14.2 Suppose that the criterion for estimating β_0 and β_1 in a simple linear regression model is to minimize

$$A(\beta_0, \beta_1) = \sum_{i=1}^{n} |\hat{y}_i - y_i|. \tag{14.59}$$

Determine the equations that must be satisfied to find $\hat{\beta}_0$ and $\hat{\beta}_1$.

Problem 14.3 Suppose that the error terms in a linear regression model follow the double exponential distribution with PDF

$$f(\varepsilon_i) = \frac{1}{2\sigma} e^{-|\varepsilon_i|/\sigma}, \qquad -\infty < \varepsilon_i < \infty.$$

This distribution has heavier tails than the normal distribution. Show that if you find the maximum likelihood estimators of the parameters in a simple linear regression with double exponential errors that the MLEs must satisfy the least absolute deviations criterion described in Problem 14.2.

Problem 14.4 The purity of oxygen produced by a fractional distillation process is thought to be related to the percentage of hydrocarbons in the main condenser of the processing unit. Twenty samples are shown Table 14.4.

Table 14.4 Oxygen purity data.

Purity (%)	Hydro-carbon (%)	Purity (%)	Hydro-carbon (%)
86.91	1.02	96.73	1.46
89.85	1.11	99.42	1.55
90.28	1.43	98.66	1.55
86.34	1.11	96.07	1.55
92.58	1.01	93.65	1.40
87.33	0.95	87.31	1.15
86.29	1.11	95.00	1.01
91.86	0.87	96.85	0.99
95.61	1.43	85.20	0.95
89.86	1.02	90.56	0.98

(a) Fit a simple linear regression model to the data. Is there a significant linear relationship between purity and hydrocarbon percentage?

(b) Analyze the residuals from this model. Are there any potential problems with model adequacy?

Problem 14.5 Consider the oxygen plant data in Problem 14.4 and assume that purity and hydrocarbon percentage are jointly normally distributed random variables.

(a) What is the correlation between oxygen purity and hydrocarbon percentage?

(b) Test the hypothesis that $\rho = 0$.

(c) Construct a 95% confidence interval for ρ.

Problem 14.6 The number of pounds of steam used per month at a plant is thought to be related to the average monthly ambient temperature. The past year's usages and temperatures are shown in Table 14.5.

Table 14.5 Data for Problem 14.6.

Month	Temperature	Usage/1,000
January	21	185.79
February	24	214.47
March	32	288.03
April	47	424.84
May	50	454.68
June	59	539.03
July	68	621.55
August	74	675.06
September	62	562.03
October	50	452.93
November	41	369.95
December	30	273.98

(a) Fit a simple linear regression model to the data.

(b) Test for significance of regression.

(c) Plant management believes that an increase in average ambient temperature of 1 degree will increase average monthly steam consumption by 10,000 lb. Do the data support this statement?

(d) Construct a 99% prediction interval on steam usage in a month with average ambient temperature of 58°F.

Problem 14.7 Hsiue *et al.* (1995) studied the effect of the molar ratio of sebacic acid (the regressor) on the intrinsic viscosity of copolyesters (the response). Table 14.6 gives the data.

(a) Make a scatterplot of the data.

(b) Estimate the prediction equation.

(c) Perform a complete, appropriate analysis (statistical tests, calculation of R^2, and so forth).

(d) Calculate and plot the 95% confidence and prediction bands.

Table 14.6 Data for Problem 14.7.

Temperature (°C)	Viscosity (mPa·s)
24.9	1.1330
35.0	0.9772
44.9	0.8532
55.1	0.7550
65.2	0.6723
75.2	0.6021
85.2	0.5420
95.2	0.5074

Problem 14.8 Consider the maximum likelihood estimator of σ^2 in the simple linear regression model. We know that this estimator is a biased estimator for σ^2.

(a) Determine the amount of bias in this estimator.

(b) What happens to the bias as the sample size n becomes large?

Problem 14.9 A solid-fuel rocket propellant loses weight after it is produced. The data are shown in Table 14.7.

Table 14.7 Weight loss on solid rocket fuel.

Months since production, x	Weight loss, y (kg)
0.25	1.42
0.50	1.39
0.75	1.55
1.00	1.89
1.25	2.43
1.50	3.15
1.75	4.05
2.00	5.15
2.25	6.43
2.50	7.89

(a) Fit a second-order polynomial that expresses weight loss as a function of the number of months since production.

(b) Test for significance of regression; i.e., H_0: $\beta_1 = \beta_2 = 0$.

(c) Test the hypothesis $H_0 : \beta_2 = 0$. Comment on the need for the quadratic term in this model.

(d) Are there any potential hazards in extrapolating with this model?

Problem 14.10 Refer to Problem 14.7. Compute the residuals for the second-order model. Analyze the residuals and comment on the adequacy of the model.

Problem 14.11 Suppose that you are fitting the simple linear regression model

$$Y = \beta_0 + \beta_1 x + \varepsilon$$

but the true model is

$$Y = \beta_0 + \beta_1 x + \beta_{11} x^2 + \varepsilon.$$

Is the least squares estimator of the slope β_1 unbiased?

Problem 14.12 The carbonation level of a soft drink beverage is affected by the temperature of the product and the filler operating pressure. Twelve observations were obtained and the resulting data are shown in Table 14.8.

Table 14.8 Carbonation in soft drinks.

Carbonation, y	Temperature, x_1	Pressure, x_2
2.60	31.0	21.0
2.40	31.0	21.0
13.32	31.5	24.0
15.60	31.5	24.0
16.12	31.5	25.0
5.36	30.5	22.0
6.19	31.5	22.0
10.17	30.5	23.0
2.62	31.0	21.5
2.98	30.5	21.5
6.92	31.0	22.5
7.06	30.5	22.5

(a) Fit a full second-order polynomial.

(b) Test for significance of regression; i.e., H_0: $\beta_k = 0$ for all β_k except β_0.

(c) Does the interaction term contribute significantly to the model?

Problem 14.13 Shown in Table 14.9 are data on y = green liquor (g/L) and x = paper machine speed (ft/min) from a Kraft paper machine. (The data were read from a graph in an article in the *Tappi Journal*, March 1986.)

(a) Fit a second-order polynomial to the data.

(b) Test for significance of regression using $\alpha = 0.05$. What are your conclusions?

(c) Test the contribution of the quadratic term to the model and the contribution of the linear term. If $\alpha = 0.05$, what conclusion can you draw?

(d) Plot the residuals from the model. Does the model fit seem satisfactory?

Table 14.9 Green liquor vs. paper machine speed.

Green liquor (g/L), y	Paper machine speed (ft/min), x	Green liquor (g/L), y	Paper machine speed (ft/min), x
16.0	1,700	14.0	1,760
15.8	1,720	13.5	1,770
15.6	1,730	13.0	1,780
15.5	1,740	12.0	1,790
14.8	1,750	11.0	1,795

Problem 14.14 Reconsider the data from Table 14.9. Suppose that it is important to predict the response at the points $x = 1,750$ and $x = 1,775$.

(a) Find the predicted response at these points and the 95% prediction intervals for the future observed response at these points.

(b) Suppose that a first-order model is also being considered. Fit this model. Find the predicted response at these points. Calculate the 95% prediction intervals for the future observed response at these points. Does this give any insight about which model should be preferred?

Problem 14.15 Scientists studying the ozone layer over the Antarctic have developed a measure of the degree to which oceanic phytoplankton production is inhibited by exposure to ultraviolet radiation (UVB). The response is INHIBIT. The regressors are UVB and SURFACE, which is depth below the ocean's surface from which the sample was taken. The data are shown in Table 14.10. Perform an analysis of these data. Discuss your results.

Problem 14.16 The data in Table 14.11 come from a patient satisfaction study at a hospital. Fit a regression model to the satisfaction response using age, severity, and anxiety as the predictor variables. Are all of the model coefficients significant? Analyze the residuals and comment on model adequacy.

Problem 14.17 Consider the hospital patient satisfaction data in Problem 14.16. Fit a regression model to the satisfaction response using age, severity, and anxiety as the predictors and account for the medical versus surgical classification of each patient with an indicator variable. Has adding the indicator variable improved the model? Is that evidence to support the claim that medical and surgical patients differ in their satisfaction?

Problem 14.18 A chemical engineer is investigating how the amount of conversion of a product from a raw material (y) depends on reaction temperature

Table 14.10 Antarctic data.

Location	INHIBIT	UVB	SURFACE
1	0.0	0.00	Deep
2	1.0	0.00	Deep
3	6.0	0.01	Surface
4	7.0	0.01	Surface
5	7.0	0.02	Surface
6	7.0	0.03	Surface
7	9.0	0.04	Surface
8	9.5	0.01	Deep
9	10.0	0.00	Deep
10	11.0	0.03	Surface
11	12.5	0.03	Surface
12	14.0	0.01	Deep
13	20.0	0.03	Deep
14	21.0	0.04	Surface
15	25.0	0.02	Deep
16	39.0	0.03	Deep
17	59.0	0.03	Deep

Table 14.11 Hospital satisfaction data.

Satisfaction	Age	Severity	Surgical/ medical	Anxiety
68	55	50	0	2.1
77	46	24	1	2.8
96	30	46	1	3.3
80	35	48	1	4.5
43	59	58	0	2.0
44	61	60	0	5.1
26	74	65	1	5.5
88	38	42	1	3.2
75	27	42	0	3.1
57	51	50	1	2.4
56	53	38	1	2.2
88	41	30	0	2.1
88	37	31	0	1.9
102	24	34	0	3.1
88	42	30	0	3
70	50	48	1	4.2
82	58	61	1	4.6
43	60	71	1	5.3
46	62	62	0	7.2
56	68	38	0	1.8
59	70	41	1	7
26	79	66	1	6.2
52	63	31	1	4.1
83	39	42	0	3.5
75	49	40	1	2.1

(x_1) and reaction time (x_2). Two models have been developed:

$$\hat{y} = 100 + 0.2x_1 + 4x_2$$
$$\hat{y} = 95 + 0.15x_1 + 3x_1 + 1x_1x_2.$$

Both models have been built over the range $20 \leq x_1 \leq 50$ (°C) and $0.5 \leq x_2 \leq 10$ (hours).

(a) Using both models, what is the predicted value of conversion when $x_2 = 2$ in terms of x_1? Repeat this calculation for $x_2 = 8$. Draw a graph of the predicted values as a function of temperature (x_1) for both conversion models. Comment on the effect of the interaction term in model 2.

(b) Find the expected change in the mean conversion for a unit change in temperature x_1 for model 1 when $x_2 = 5$. Does this quantity depend on the specific value of reaction time selected? Why?

(c) Find the expected change in the mean conversion for a unit change in temperature x_1 for model 2 when $x_2 = 5$. Repeat this calculation for $x_2 = 2$ and $x_2 = 8$. Does the result depend on the value selected for x_2? Why?

Problem 14.19 Suppose we run a small clinical trial on a drug for cholesterol reduction. Those in group 1 receive an experimental drug, and those in group 2 receive a placebo. The data are shown in Table 14.12.

Table 14.12 Data for Problem 14.19.

Subject	Diff	Group	Age
1	−10	1	45
2	−22	1	57
3	−8	1	44
4	2	1	39
5	−30	1	63
6	−4	2	49
7	8	2	53
8	−10	2	70
9	0	2	63
10	−3	2	55

(a) Run a simple regression with Diff as the response and Group as the predictor.

(b) Run a multiple regression with Diff as the response and Group and Age as predictors.

(c) Explain the discrepancy between the results of parts (a) and (b).

Problem 14.20 Show that an equivalent way to perform the test for significance of regression in multiple linear regression is to base the test on R^2 using the following test statistic:

$$F_0 = \frac{R^2}{1-R^2} \frac{n-p-1}{p}.$$

Problem 14.21 Verify the t distribution in (14.43).

Problem 14.22 Suppose that a linear regression model with $k = 2$ predictor variables has been fit to $n = 25$ observations and $R^2 = 0.90$.

(a) Test for significance of regression at $\alpha = 0.05$. Use the results of the previous problem.

(b) What is the smallest value of R^2 that would lead to the conclusion of a significant regression if $\alpha = 0.05$? Are you surprised at how small this value of R^2 is?

Problem 14.23 Show that an alternate computing formula for the regression sum of squares in a linear regression model is

$$SSR = \sum_{i=1}^{n} \hat{y}_i^2 - n\bar{y}^2.$$

Problem 14.24 Use the Fisher transformation from (14.51) to show that the approximate $100 \times (1-\alpha)\%$ confidence interval for the correlation ρ for a sample from a bivariate normal distribution is

$$\left[\frac{\frac{1+R}{1-R} \exp\left(-2z_{\alpha/2}/\sqrt{n-3}\right) - 1}{\frac{1+R}{1-R} \exp\left(-2z_{\alpha/2}/\sqrt{n-3}\right) + 1}, \frac{\frac{1+R}{1-R} \exp\left(2z_{\alpha/2}/\sqrt{n-3}\right) - 1}{\frac{1+R}{1-R} \exp\left(2z_{\alpha/2}/\sqrt{n-3}\right) + 1} \right].$$

15 — Bayesian Methods

15.1 Introduction

Up to this point we have been talking about what are often called *frequentist* methods, because a statistical method is based on properties of its long-run relative frequency. With this approach, the probability of an event is defined as the proportion of times the event occurs in the long run. Parameters, that is values that characterize a distribution, such as the mean and variance of a normal distribution, are considered *fixed* but *unknown*.

The Bayesian approach departs from the frequentist approach at the very foundation: the definition of probability. To Bayesians, probability represents degree of belief. Parameters are treated as random variables having some distribution because they are *uncertain*, and uncertain things are expressed in probabilistic terms. In essence, uncertainty and randomness are the same in the Bayesian approach. This is not the case for the frequentist approach.

In the frequentist approach, we often collect data from some assumed distribution. We then use these data to find point estimates and confidence intervals for the unknown parameters. These confidence intervals have a rather subtle interpretation. For example, a 95% confidence interval for a parameter θ is an interval that has probability 0.95 of containing the true value of θ when viewed *before* the data were taken. In other words, before we take any data, but after we have chosen a rule for computing the confidence interval, the probability that the resulting interval will include the true value is 0.95. This seems odd to many users of statistics, who point out that we have already collected data, computed the relevant statistics, and computed the endpoints of the confidence interval, which we will denote C.

What should we make of this computed interval? With the frequentist or classical approach we cannot say that $\Pr(\theta \in C) = 0.95$, that is, the probability is 0.95 that the interval contains the true value. In the above probability there is nothing random. θ is not random in the frequentist sense and C is not random, since it has been computed from the observed data. (Once data are taken, C will be an interval with fixed endpoints, for example $C = (65, 67)$). We can only say that the probability that an interval *calculated in this way* will include the true value of θ is 0.95 when viewed before collecting the data. If we were repeatedly computing confidence intervals, 95% of the time we would get an interval that contains the true value of the parameter, and 5% of the time we would miss it. This follows from the Law of Large Numbers; it's as if we had a sequence of Bernoulli trials where the probability of success (obtaining a confidence interval that covers θ) is 0.95 on each trial. Why would we believe that our computed interval of (65,67) contains the true value of θ? Ninety-five percent of the time, when we compute such intervals, they contain θ, but whether this one does or doesn't, we do not know and will never know. In the frequentist paradigm, confidence intervals are probability statements about the *method* that produced the confidence interval, and its long-run properties.

The Bayesian approach, which treats unknown parameters as random variables, takes a more direct approach. Intervals are calculated and the interpretations are much more along the lines that many practitioners would like. They are direct probabilistic statements that a parameter is contained in the interval, conditioned on the observed data. This directness, however, comes at a price. Our interpretation of probability must change to the subjective, or degree of belief, interpretation, and we must specify what we believe about a parameter before we observe any data.

15.2 Bayes' Theorem

We saw Bayes' Theorem in Section 3.4.2, but we summarize the result here and explain how it forms the basis for a new system of statistical inference.

Suppose that the set of all possible outcomes for some random phenomenon, say taking a random sample from a population, is equal to the set S. Suppose also that the set S can be partitioned into sets $A_1, A_2, \ldots, A_k$, where these sets are *mutually exclusive* and *exhaustive*. Mutually exclusive means that if we take any two distinct sets, their intersection is the empty set. Exhaustive means that if we union up all of the A_is we end up with the entire sample space S; that is,

$$\bigcup_{i=1}^{k} A_i = S.$$

This means that every possible outcome is in one, and only one, of the sets $A_1, A_2, \ldots, A_k$.

Now let B be any set that is contained in the sample space S. Then every element in B will be in one and only one of the sets

$$A_1 \cap B, \quad A_2 \cap B, \quad \ldots, \quad A_k \cap B.$$

Since these events are mutually exclusive, we can write

$$\Pr(B) = \Pr(A_1 \cap B) + \Pr(A_2 \cap B) + \cdots + \Pr(A_k \cap B).$$

We can apply the laws of conditional probability to each term on the right side to obtain the Law of Total Probability:

$$\Pr(B) = \Pr(A_1)\Pr(B|A_1) + \Pr(A_2)\Pr(B|A_2) + \cdots + \Pr(A_k)\Pr(B|A_k) = \sum_{i=1}^{k} \Pr(A_i)\Pr(B|A_i). \quad (15.1)$$

This leaves us just short of Bayes' Theorem, which again is obtained from the definition of conditional probability. Bayes' Theorem says

$$\Pr(A_j|B) = \frac{\Pr(A_j)\Pr(B|A_j)}{\Pr(B)} = \frac{\Pr(A_j)\Pr(B|A_j)}{\sum_{i=1}^{k} \Pr(A_i)\Pr(B|A_i)}. \quad (15.2)$$

Notice how Bayes' Theorem *turns* conditional probabilities around. On the left side we have $\Pr(A_i|B)$ and on the right we have $\Pr(B|A_i)$. The Bayesian approach assumes $f(x|\theta)$ as a model for the distribution of the observation x (which is often a vector). What Bayes' Theorem tells us is how to determine the distribution of the parameter θ given the data, effectively swapping the events in the conditional probability.

■ **Example 15.1 — Disease testing.** Tests for a disease are never perfect. You could have the disease, but test negative; you could also be disease-free and test positive. The probability of testing positive given that you have the disease is called the *sensitivity*, and the probability that you test negative given that you don't have the disease is called the *specificity*. Suppose that for a particular test the sensitivity is 95% and the specificity is 90%. It is known that 1% of the population has the disease. Find the following:

(a) the probability that someone selected at random tests positive; and

(b) the probability that someone selected at random who tests positive actually has the disease.

Solution: First, notice that the sensitivity and specificity are defined as conditional probabilities. Let D denote the event that someone has the disease and ND denote the event that someone does not have the disease; also, let "+" denote the event that someone tests positive and "−" denote the event that someone tests negative. Then the sensitivity and specificity are

$$\text{sensitivity} = \Pr(+|D) \quad \text{and} \quad \text{specificity} = \Pr(-|ND).$$

The probability of having the disease is an *unconditional* probability, and is often called the *base rate*. The first part is answered by applying the law of total probability. Here, A_1 is the event that someone has the disease and A_2 is the event that the person does not have the disease. These events are clearly mutually exclusive and exhaustive. B is the event that someone tests positive. Thus,

$$
\begin{aligned}
\Pr(+) &= \Pr\big((+ \text{ and } D) \text{ or } (+ \text{ and } ND)\big) \\
&= \Pr(+ \text{ and } D) + \Pr(+ \text{ and } ND) \\
&= \Pr(D)\Pr(+|D) + \Pr(ND)\Pr(+|ND).
\end{aligned}
\tag{15.3}
$$

Remember that the logical "or" means the set union and the logical "and" means the set intersection. Each of the terms in this last expression is either a given in the problem or can be obtained using the complement rule. For example,

$$\Pr(ND) = 1 - \Pr(D) \quad \text{and} \quad \Pr(+|ND) = 1 - \Pr(-|ND) = 1 - \text{specificity}.$$

With this, everything in (15.3) is known. While the calculations could be done easily on a calculator, we show how this could be done in R:

```
sensitivity = 0.95
specificity = 0.90
baseRate = 0.01
probPositive = baseRate*sensitivity + (1-baseRate)*(1-specificity)
print(probPositive)
```

The result is 0.1085. Much of the work in answering the second part is already done once we have the probability of testing positive. Using Bayes' Theorem, we get

$$\Pr(D|+) = \frac{\Pr(D)\Pr(+|D)}{\Pr(+)} \approx \frac{0.01 \times 0.95}{0.1085} \approx 0.08756. \qquad \blacksquare\blacksquare\blacksquare$$

A few things about the solution are worth noting. First, about 11% of the population will test positive for the disease, even though only 1% have the disease. Second, if someone tests positive for the disease, the probability of actually having the disease is only 0.08756. Think about the importance of this. Fewer than 1 person in 10 who tests positive will have the disease. The value of $\Pr(D|+)$ is called the *positive predictive value* (PPV) and the value of $\Pr(ND|-)$ is called the *negative predictive value* (NPV). The

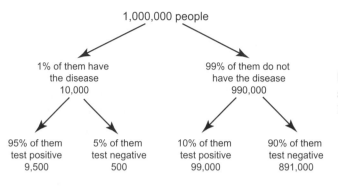

Figure 15.1 Tree diagram showing the status of a hypothetical population of 1,000,000 people.

PPV is very low even though the sensitivity and specificity are fairly high (0.95 and 0.90, respectively). This can be explained by the low base rate and the moderate probability (0.10) of those without the disease testing positive.

Figure 15.1 shows what would be expected in a population of 1,000,000 people. The tree first splits individuals on whether they do or don't have the disease. The second splits individuals on whether they test positive or negative. Thus, about $9,500 + 99,000 = 108,500$ people will test positive, and among these, only 9,500 will have the disease, a fraction of $9,500/108,500 = 0.08756$. The tree clearly shows why the PPV is so low: The number of false positives (99,000) is much greater than the number of true positives (9,500).

The second thing to note about this example is how it relates to the Bayesian approach to statistical inference. Think of the base rate as the prior probability of a person having the disease. If someone is selected at random from the population, we would, knowing nothing else, put the probability of having the disease at 0.01. Think of the result of the test for the disease as *data*. We use Bayes' Theorem to transform our assessment of how likely it is that a person has the disease as we move from *before* observing data to *after* observing data. We use the respective terms *prior* and *posterior* to denote our assessment before and after observing data. Note that if a person tests positive, our belief that a person has the disease has gone from 0.01 (the base rate) to 0.08756 (the PPV). Had the test been negative, the probability of having the disease would have gone from 0.01 to 0.000561; details are left to the reader.

Exercise 15.1 Verify in the previous example that $\Pr(D|-) = 0.000561$.

■ **Example 15.2** Suppose that someone has three coins. One coin has tails on both faces, so it will end up tails every time it is tossed. The second coin is fair, meaning that the probability of being heads is 0.5 (and the same for tails). The third coin is a two-headed coin, so it will always come up heads. One coin is selected at random, where each coin is equally likely, and tossed twice. We do not get to observe which coin was selected; we only see the results of the two tosses. Find the following:

(a) the probability that both tosses result in heads;
(b) the probability that the coin is the two-tailed coin given that both tosses result in heads;
(c) the probability that the coin is the fair one given that both tosses result in heads; and
(d) the probability that the coin is the two-headed coin given that both tosses result in heads.

Solution: To make the notation simpler, and to make this example a stepping stone to understanding the Bayesian approach, let θ denote the "head" probability for the selected coin. The possible values for θ are 0, $\frac{1}{2}$, and 1, and each of these occurs with probability $\frac{1}{3}$. We can summarize this prior distribution for θ:

θ	0	$\frac{1}{2}$	1
$\pi(\theta)$	$\frac{1}{3}$	$\frac{1}{3}$	$\frac{1}{3}$

Conditioned on the value of θ, the number of heads on two tosses then has a binomial distribution with $n = 2$ and probability of success θ. This is a conditional probability distribution, so we write $X|\theta \sim$ BIN$(2, \theta)$. To find the probability of getting exactly two heads, we apply the law of total probability:

$$
\begin{aligned}
\Pr\left(X = 2\right) &= \Pr\left(\theta = 0\right)\Pr\left(X = 2 \middle| \theta = 0\right) + \Pr\left(\theta = \frac{1}{2}\right)\Pr\left(X = 2 \middle| \theta = \frac{1}{2}\right)\Pr\left(\theta = 1\right)\Pr\left(X = 2 \middle| \theta = 1\right) \\
&= \frac{1}{3} \times 0 + \frac{1}{3} \times \left(\frac{1}{2}\right)^2 + \frac{1}{3} \times 1 \\
&= \frac{5}{12}.
\end{aligned}
$$

The next three parts are obtained by applying Bayes' Theorem:

$$
\begin{aligned}
\Pr\left(\theta = 0 \middle| X = 2\right) &= \frac{\pi(0)\,\Pr(X = 2|\theta = 0)}{\Pr(X = 2)} = \frac{\pi(0) \times 0}{5/12} = 0 \\
\Pr\left(\theta = \frac{1}{2} \middle| X = 2\right) &= \frac{\pi(\frac{1}{2})\,\Pr(X = 2|\theta = \frac{1}{2})}{\Pr(X = 2)} = \frac{\pi(\frac{1}{2}) \times \frac{1}{4}}{5/12} = \frac{1}{5} \\
\Pr\left(\theta = 1 \middle| X = 2\right) &= \frac{\pi(1)\,\Pr(X = 2|\theta = 1)}{\Pr(X = 2)} = \frac{\pi(1) \times 1}{5/12} = \frac{4}{5}.
\end{aligned}
$$

Given that $X = 2$ heads were observed, the posterior distribution for the parameter θ is

θ	0	$\frac{1}{2}$	1
$\pi(\theta|X = 2)$	0	$\frac{1}{5}$	$\frac{4}{5}$.

Again, a few important points should be made regarding this example. First, the posterior distribution is a valid (discrete) probability distribution, since all probabilities are nonnegative and sum to 1. Second, the probability that $\theta = 0$ is equal to 0, because if $\theta = 0$ then a head has probability 0 of occurring. Thus, on two tosses of the coin we would *always* observe $X = 0$ heads, so getting $X = 2$ is impossible. Third, if we observe both heads, we might believe it more likely that $\theta = 1$, that is, we have the two-headed coin. We might reason that if $\theta = \frac{1}{2}$, that is we selected the fair coin, then two heads is somewhat unlikely. Bayes' Theorem tells us exactly how we should change our minds about the value of θ from before data are taken to after, when we observe $X = 2$ heads.

■ **Example 15.3** Expand the previous experiment to one where the probability of getting a head is any number in

$$
\frac{0}{100}, \quad \frac{1}{100}, \quad \frac{2}{100}, \quad \ldots, \quad \frac{100}{100},
$$

where each of these 101 numbers is equally likely. As in the previous example, once a coin is selected it is tossed twice. This might arise in a situation where we are uncertain of the probability of some event,

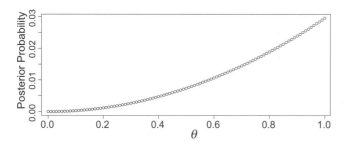

Figure 15.2 The (discrete) posterior distribution for the probability of success given $X = 2$ successes on $n = 2$ trials.

and we have the outcomes of two trials on which to estimate this probability. (This is a rather trivial problem, but soon we will give some more realistic numbers.) Before taking any data, any number (rounded to a multiple of $\frac{1}{100}$) between 0 and 1 is assumed to be equally likely. We then assign each of these 101 numbers the same prior probability, namely $\frac{1}{101}$. Apply the same reasoning as in the previous example to find the posterior probability for θ.

Solution: In principle, the approach is the same, just messier. We could run the following code in R to obtain and plot the posterior:

```
n = 2
x = 2
theta = (0:100)/100
prior = rep(1/101,101)
likelihood = dbinom(x,n,theta)
evidence = sum(prior*likelihood)
posterior = prior*likelihood/evidence
plot(theta , posterior , xlab="" , ylab="Posterior Probability" )
```

The plot produced is shown in Figure 15.2. ∎∎∎

Note that the most likely value of θ after observing two successes on two trials is $\theta = 1$, but even the most likely value is still rather unlikely. The posterior probability that $\theta = 1$ is approximately 0.03, which is small, but still the largest posterior probability.

To see what happens when the numbers are larger, consider the next example.

∎ **Example 15.4** Suppose we had $X = 36$ successes out of $n = 100$ trials. This could occur if we ran a clinical trial to test whether a medication was helpful in relieving pain. (This is a rather bad design, because there is no control group, but we will address this issue later.) Find the posterior probability distribution for the probability θ.

Solution: Apply similar code to what was used in the previous example:

```
x = 36
n = 100
theta = (0:100)/100
prior = rep(1/101,101)
likelihood = dbinom(x,n,theta)
evidence = sum(prior*likelihood)
posterior = prior*likelihood/evidence
plot(theta,posterior,xlab="",ylab="Posterior Probability")
```

The resulting plot of the posterior is shown in Figure 15.3. From this plot we can see that values between about 0.25 and 0.45 have a fairly high posterior probability, but values outside of this interval have posterior probability near zero. ∎∎∎

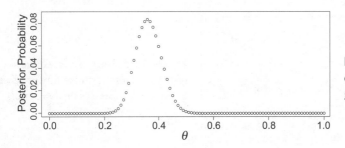

Figure 15.3 The (discrete) posterior distribution for the probability of success given $X = 2$ successes on $n = 2$ trials.

The major takeaway from this example is how we are interpreting probability. We can talk, for example, about the posterior probability of θ (which is equal to the probability of a success) being equal to 0.35 (which is about 0.08) or 0.70 (which is virtually 0.00). Such statements cannot be made in the frequentist or classical paradigm because parameters are considered fixed (not random) quantities. Therefore no probabilistic statements can be made about parameters, either before or after data are collected.

The second issue that one might raise about the previous example is the discreteness imposed on the parameter θ. Is it really reasonable to say that θ could be equal to 0.35, or 0.36, but θ could not be 0.355? It seems more reasonable to say that θ could be *any* value between 0 and 1. This means that we would put a *continuous* prior distribution on θ; we would then observe a *discrete* random variable with conditional distribution $X|\theta \sim \text{BIN}(100, \theta)$. The posterior would then be computed from the continuous analogue of Bayes' Theorem.

Consider the general case of observing X successes on n trials where the probability of success on any trial is θ. The conditional distribution is then $X|\theta \sim \text{BIN}(n, \theta)$. Let $\pi(\theta)$ denote the prior distribution, and let

$$f(x|\theta) = \binom{n}{x} \theta^x (1-\theta)^{n-x}, \quad x = 0, 1, \ldots, n$$

be the likelihood. For continuous parameters, Bayes' Theorem becomes

$$\pi(\theta|x) = \frac{\pi(\theta) f(x|\theta)}{\displaystyle\int_0^1 \pi(\theta) f(x|\theta) \, d\theta}. \tag{15.4}$$

Notice how the sum in the denominator of Bayes' Theorem becomes an integral (the limit of sums) when we move to a continuous prior.

■ **Example 15.5** Suppose we observed $x = 36$ successes on $n = 100$ trials as in the previous example. Suppose we take the prior on θ to be uniform on $[0, 1]$. Find and plot the posterior distribution.

Solution: The posterior would be

$$\pi(\theta|x) = \frac{1 \binom{100}{36} \theta^{36} (1-\theta)^{100-36}}{\displaystyle\int_0^1 \pi(\theta) f(x|\theta) \, d\theta} = c(x) \theta^{36} (1-\theta)^{100-36}, \quad 0 \leq \theta \leq 1.$$

Here, $c(x)$ is a function of x that includes $\binom{100}{36}$.[1] In the next section, we explain why the posterior distribution is the BETA(37,65) distribution. We can use the following R code to plot this posterior:

[1] Note that in the denominator the integral is over θ. Since the integrand involves both θ and x, the resulting integral will be a function of x only.

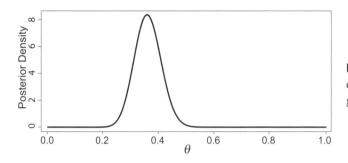

Figure 15.4 The (continuous) posterior distribution for the probability of success given $X = 36$ successes on $n = 100$ trials.

```
x = 36
n = 100
theta = (0:200)/200
prior = 1
posterior = dbeta( theta , 37 , 65 )
plot( theta , posterior , type="l" , xlab="" , lwd=2 ,
      ylab="Posterior Probability" )
```

The plot of the posterior is shown in Figure 15.4. Notice how similar this plot is to the plot in Figure 15.3. There is little practical difference between assuming a discrete uniform prior on $0, 0.01, 0.02, \ldots, 0.99, 1.00$ and a continuous prior on the interval $[0,1]$. One important difference, however is the vertical axis. In the former case, the vertical axis is the posterior probability of θ being equal to a particular value; here the sum of the posterior probabilities must be 1. In the latter, it is the probability density, and it must *integrate* to 1. ∎ ∎ ∎

15.3 The Bayesian Paradigm

In the Bayesian paradigm, unknown parameters are treated as *uncertain*, and uncertainty is stated in probabilistic terms. In other words, if we don't know the value of a parameter, we express our knowledge (or ignorance) in a probability distribution. Suppose θ is an unknown parameter (which could be a vector) and x is the data (usually a vector) whose distribution depends on the parameters. The Bayesian approach to statistics is, in principle, simple:

- assess what we believe about θ *before* observing any data, which is done through the probability distribution $\pi(\theta)$, called the *prior distribution*;
- apply Bayes' Theorem to obtain the conditional distribution of θ given the data x:

$$\pi(\theta|x) = \frac{\pi(\theta)f(x|\theta)}{\int \pi(\theta)f(x|\theta)\, d\theta},$$

which is called the *posterior distribution*.

The integral in the denominator is the integral over the entire θ space. For a scalar value, this will be an interval over the real line. For example, in the last example θ represented the unknown probability of a success, which must be in $[0,1]$. In other cases it might be $[0,\infty)$ or $(-\infty,\infty)$. If θ is a vector, then the integral will be a multiple integral whose dimension is equal to the dimension of θ. We will ordinarily write the integral with a single integral sign, whether it is a single or multiple integral, with the understanding that its dimension matches the dimension of θ. The two-step process described above is often called "turning the Bayesian crank." While in principle the process is straightforward, the computation of the denominator in Bayes' Theorem is usually a problem except in simple cases. Monte

Carlo simulation can be used to learn about the posterior in more complicated cases. We discuss this in Section 15.7.

The posterior distribution reflects what we believe about the parameter θ given the observed data. Notice how this is the reverse of the likelihood function $f(x|\theta)$. (Recall that Bayes' Theorem is all about swapping the events in a conditional probability.)

If we wanted a point estimate for θ, we could take some measure of the center of the posterior. Often the posterior mean

$$\hat{\theta} = E(\theta|\boldsymbol{x}) = \frac{\int \theta \, \pi(\theta) f(\boldsymbol{x}|\theta) \, d\theta}{\int \pi(\theta) f(x|\theta) \, d\theta} \tag{15.5}$$

is used as a point estimate. Sometimes the posterior median or mode is used as a point estimate for θ. If an interval estimate of a parameter is desired, then we could determine an interval I for the case of a scalar parameter, or a region R for the case of a vector parameter, with the property that

$$\Pr(\theta \in I|\boldsymbol{x}) = 1 - \alpha. \tag{15.6}$$

Such an interval is called a *credible interval*.

By contrast, the classical approach determines a random interval $I(\boldsymbol{X}) = (L, U)$ with the property that

$$\Pr(\theta \in I(\boldsymbol{X})) = \Pr(L < \theta < U) = 1 - \alpha. \tag{15.7}$$

These two formulations look similar, but they differ in a fundamental way. In (15.6) θ is the random variable and x is considered fixed, whereas in (15.7) the interval I is a random variable, depending on the random vector X, and θ is considered fixed.

■ **Example 15.6** For the data ($n = 100$, $x = 36$) from the previous example, find the posterior mean, median, and mode. Assume a (continuous) uniform prior over the interval $[0, 1]$. Find a 95% credible interval for θ.

Solution: Recall a few properties of the beta distribution from Section 6.7. The PDF is

$$f(\theta|a, b) = \frac{\Gamma(a+b)}{\Gamma(a)\Gamma(b)} \theta^{a-1}(1-\theta)^{b-1}, \qquad 0 < \theta < 1. \tag{15.8}$$

The mean of the BETA(a, b) distribution is

$$\mu = \frac{a}{a+b}$$

and the mode is

$$\text{mode} = \frac{a-1}{a+b-2}.$$

Given a uniform prior distribution for θ, which is a BETA$(1, 1)$ distribution, we find that the posterior distribution is

$$p(\theta|x) = \frac{\pi(\theta) \times \binom{100}{36} \theta^{36}(1-\theta)^{100-36}}{\int_0^1 \pi(\theta) \times \binom{100}{36} \theta^{36}(1-\theta)^{100-36} \, d\theta}.$$

Note that the denominator is a constant, say c. In general the constant will depend on the data but not the parameter. The posterior is therefore

$$p(\theta|x) = \frac{1}{c} \times \binom{100}{36} \theta^{36}(1-\theta)^{100-36} = c_1\theta^{37-1}(1-\theta)^{65-1}, \qquad 0 < \theta < 1.$$

We recognize this as a beta distribution with parameters $a - 37$ and $b - 65$.

There is no simple formula for the median, but it can easily be obtained by using the beta quantile function qbeta() in R. Since $\theta \mid (x = 36) \sim \text{BETA}(37,65)$, the posterior mean and mode are

$$\hat{\theta}_{\text{mean}} = E(\theta|x = 36) = \frac{37}{37+65} \approx 0.363$$

$$\hat{\theta}_{\text{mode}} = \frac{37-1}{37+65+2} = \frac{36}{100} = 0.36.$$

The median can be found using the qbeta() function in R. For example,

```
median = qbeta( 0.5 , 37 , 65 )
```

This yields $\hat{\theta}_{\text{median}} = 0.362$. You may be wondering why the mean and median are greater than the usual point estimate of $\frac{36}{100} = 0.36$. The prior puts uniform probability across the entire interval from 0 to 1. Thus, half of the prior probability is for $\theta > \frac{1}{2}$ and half for $\theta < \frac{1}{2}$. Thus, the posterior mean will be pulled slightly toward the prior mean, which is $\frac{1}{2}$. In this case, the point estimates are pulled very slightly from 0.36 toward 0.50.

To get a credible interval for θ, we find the 2.5 and 97.5 percentiles of the posterior distribution using R's qbeta() function:

```
lowerLimit = qbeta( 0.025 , 37 , 65 )
upperLimit = qbeta( 0.975 , 37 , 65 )
```

This yields the credible interval $(0.273, 0.458)$, which is interpreted as there is a 95% probability that θ lies in the interval $(0.273, 0.458)$ given the observed data. The two shaded tails are shown in Figure 15.5.

■ ■ ■

To summarize, here are the important quantities, terminology, and notation for Bayesian statistics.

- $\pi(\theta)$ is the *prior* distribution, which reflects what we believe about the parameter θ *prior* to observing any data.

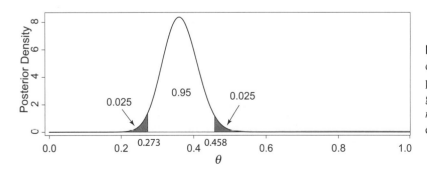

Figure 15.5 The posterior distribution for the probability of success given $X = 2$ successes on $n = 2$ trials, showing the equal-tail credible interval.

- $f(x|\theta)$ is the PDF (or PMF if the observed data are discrete) of the data x given the parameter θ. In classical statistics, the likelihood is often written as $L(\theta|x)$, or simply $L(\theta)$. Thus, the likelihood is functionally the same as the joint PDF; that is, $L(\theta|x) = f(x|\theta)$. The difference, to the extent there is one, is that in the joint PDF $f(x|\theta)$, we think of the parameter θ as *fixed* while the data x is the variable. Conversely, in the likelihood function $L(\theta|x)$ we think of the data x as fixed, and the parameter θ as the variable. In other words, the difference is a matter of perspective. Bayesians often ignore this difference and write $f(x|\theta)$.

- $\pi(\theta|x)$ is the *posterior distribution*, which says what we believe about θ given the observed data x. Bayes' Theorem tells us how to change our minds, going from the prior to the posterior once we have observed data. Bayes' Theorem says

$$\pi(\theta|x) = \frac{\pi(\theta)f(x|\theta)}{g(x)}.$$

- The denominator in Bayes' Theorem, $g(x)$, is called the *evidence*. It is the unconditional probability density of observing the data x. The evidence is obtained by integrating out the parameter θ in the joint PDF of x and θ:

$$g(x) = \int h(x,\theta)\, d\theta = \int \pi(\theta)f(x|\theta)\, d\theta$$

Note that Bayes' Theorem guarantees that the posterior integrates to 1 (as it must if it is to be a valid PDF), because

$$\int \pi(\theta|x)\, d\theta = \int \frac{\pi(\theta)f(x|\theta)}{\int \pi(\theta)f(x|\theta)\, d\theta} = \frac{\int \pi(\theta)f(x|\theta)\, d\theta}{\int \pi(\theta)f(x|\theta)\, d\theta} = 1.$$

Sometimes, mostly for simple problems, we can determine the family to which the posterior distribution belongs, without having to determine $g(x)$. For example, if $\theta \sim \text{UNIF}(0,1)$ and $X|\theta \sim \text{BIN}(n,\theta)$, then the posterior distribution is

$$\pi(\theta|x) = \frac{1\binom{n}{x}\theta^x(1-\theta)^{n-x}}{g(x)} = c(x)\theta^{x+1-1}(1-\theta)^{n+1-x-1}, \qquad 0 \le \theta \le 1,$$

where $c(x)$ is a function of x only. It does not depend on θ; this is important because it sometimes allows us to recognize the family of distributions from just the part that involves θ. The part of the posterior above that depends on just the parameter θ is called the *kernel* of the density. (Note that the kernel is not unique, so it might be better to call this "a" kernel, rather than "the" kernel.) In the above case the kernel is

$$\theta^{x+1-1}(1-\theta)^{n+1-x-1}.$$

In order to make this integrate to 1, the constant must be

$$\left(\int_0^1 \theta^{x+1-1}(1-\theta)^{n+1-x-1}\, d\theta\right)^{-1} = \left(\frac{\Gamma(x+1)\Gamma(n+1-x)}{\Gamma(n+2)}\right)^{-1} = \frac{\Gamma(n+2)}{\Gamma(x+1)\Gamma(n+1-x)}.$$

The posterior PDF is therefore

$$p(\theta|x) = \frac{\Gamma(n+2)}{\Gamma(x+1)\Gamma(n+1-x)}\, \theta^{(x+1)-1}(1-\theta)^{(n+1)-x-1}, \qquad 0 < \theta < 1.$$

This makes it clear that the posterior must be the $\text{BETA}(x+1, n+1-x)$ distribution. Often we can tell what the posterior is by looking at just the kernel. Here we would notice that $\theta^{x+1-1}(1-\theta)^{n+1-x-1}$ is the kernel of the $\text{BETA}(x+1, n+1-x)$ distribution, from which we would conclude that $\theta|x \sim \text{BETA}(x+1, n+1-x)$.

15.4 Two Paradoxes

In this section we describe some paradoxes involving Bayesian statistics, or the connection between Bayesian and frequentist methods. We give both Bayesian and frequentist solutions and indicate the inconsistencies with the frequentist approach.

The first paradox involves the nature of what is often called a noninformative prior, that is a prior that purportedly admits total ignorance about a parameter.

■ **Example 15.7** Suppose that $X|\theta \sim \text{BIN}(n, \theta)$ where θ is an unknown parameter. If, *a priori*, we know absolutely nothing about θ, we could place a $\text{UNIF}(0, 1)$ prior on θ. Does this mean that we know nothing about the log odds $\log(\theta/(1-\theta))$?

Solution: Let $\phi = \log(\theta/(1-\theta))$. If the prior is $\text{UNIF}(0,1)$, then the CDF for ϕ is

$$F(x) = \Pr(\phi < x) = \Pr\left(\log\frac{\theta}{1-\theta} < x\right) = \Pr\left(\theta < \frac{\exp(x)}{1+\exp(x)}\right) = \frac{\exp(x)}{1+\exp(x)}.$$

The last line follows because the CDF of the $\text{UNIF}(0,1)$ distribution is

$$F(x) = \begin{cases} 0 & \text{if } x \leq 0 \\ x & \text{if } 0 < x < 1 \\ 1 & \text{if } x \geq 1. \end{cases}$$

Note that the quantity $\exp(x)/(1+\exp(x))$ is *always* between 0 and 1, so the middle condition in the CDF above always holds. The PDF of ϕ is thus

$$f_\phi(x) = F'(x) = \frac{\exp(x)}{(1+\exp(x))^2}.$$

The graph of the PDF for the log odds $\phi = \log(\theta/(1-\theta))$ is shown in Figure 15.6. We might think that if we were completely ignorant about the value of θ, we should be equally ignorant about the log odds of θ, but this is not the case. The PDF shows that values of $\phi = \log(\theta/(1-\theta))$ below -5 or above 5 are unlikely, and values near 0 are most likely. A uniform prior in one parameterization does not imply a uniform prior in another parameterization. Care must be given to the choice of a parameterization in which we would like to supply a noninformative prior. We will discuss this more in Section 15.6. ■ ■ ■

The second paradox involves how, or even whether, the design for data collection affects the inference made. Let's begin by taking a frequentist approach.

■ **Example 15.8** Suppose that we want to test whether a pain reliever is more effective than a placebo. We select seven people at random and give them either the pain reliever or a placebo, chosen at random. After a few days, the person is given the other; for example, if they received the placebo first, they would get the real painkiller the second time, and vice versa. Each person is asked which drug worked better.

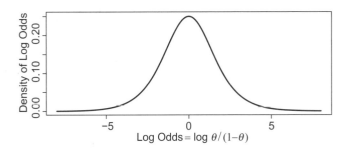

Figure 15.6 The prior for the log odds $\log(\theta/(1-\theta))$ given a uniform distribution for θ.

Participants had to select one drug or the other; ties were not allowed. We would like to know whether there is evidence that the painkiller works better than a placebo. In other words, if θ is the probability that someone would select the real painkiller (over the placebo) we would like to know whether $\theta = \frac{1}{2}$ or whether it is larger than $\frac{1}{2}$, so we set up the null and alternative hypotheses as $H_0 : \theta = \frac{1}{2}$ and $H_a : \theta > \frac{1}{2}$. Notice that this problem is exactly equivalent to tossing a coin and testing whether it is fair, or whether the probability of a head is greater than $\frac{1}{2}$. The outcomes of the seven cases are

D, D, D, D, D, D, P

where D indicates the drug was preferred and P indicates the placebo was preferred.

Solution: The test statistic here would be $X = $ number of times the drug was preferred out of the seven trials. Under the null hypothesis, $X \sim \text{BIN}(7, \frac{1}{2})$. The P-value would then be

$$
\begin{aligned}
P\text{-value} &= \Pr\left(X = 6 | \theta = \frac{1}{2}\right) + \Pr\left(X = 7 | \theta = \frac{1}{2}\right) \\
&= \binom{7}{6}\left(\frac{1}{2}\right)^6\left(1 - \frac{1}{2}\right)^{7-6} + \binom{7}{7}\left(\frac{1}{2}\right)^7\left(1 - \frac{1}{2}\right)^{7-7} \\
&= 0.0625.
\end{aligned}
$$

With a P-value exceeding the standard $\alpha = 0.05$ we would fail to reject $H_0 : \theta = \frac{1}{2}$. There is not enough evidence to say that the new painkiller is more effective than a placebo. ∎∎∎

■ **Example 15.9** Suppose now that the experimental design from the previous example was a bit different than that described. Suppose that, instead of selecting seven people to participate, we continued experimenting, one person after another, until we got one person who preferred the placebo. The observed data were

D, D, D, D, D, D, P

exactly as before, but the experimental protocol was different. Test whether the drug is effective.

Solution: The test statistic is $Y = $ number of trials needed to get one person to respond "placebo." Thus, $Y|\theta$ has a geometric distribution with parameter θ. In this case "as extreme or more extreme" means requiring seven or more trials to get one "placebo." The P-value is now

$$
P\text{-value} = \Pr\left(Y \geq 7 \mid \theta = \frac{1}{2}\right) = 1 - \Pr\left(Y \leq 6 \mid \theta = \frac{1}{2}\right) = 0.0078125.
$$

This P-value is well below 0.05, so we would reject $H_0 : \theta = \frac{1}{2}$. There *is* evidence that the drug is effective. ∎∎∎

So, if the plan is to test seven people, and we have six Ds followed by one P, then we have no evidence that the drug is effective. But if the plan is to test until we get one P, and it takes seven trials for this to occur, then we do have evidence that the drug is effective. But the data are exactly the same in both cases! Shouldn't the data itself determine whether there is or isn't evidence of the drug working? This is the *paradox* (a seeming contradiction). How would Bayesians approach this?

■ **Example 15.10** Apply a Bayesian analysis to Examples 15.8 and 15.9.

Solution: Let's take for θ a uniform prior across the interval $[0, 1]$. For the case of a fixed number of trials, the likelihood is

$$
f(x|\theta) = \binom{7}{x}\theta^x(1 - \theta)^{7-x}, \quad x = 0, 1, \dots, 7; \ 0 \leq \theta \leq 1.
$$

The posterior is therefore

$$
\begin{aligned}
\pi(\theta|x) &= c(x) \times 1 \times \binom{7}{x} \theta^x (1-\theta)^{7-x} \\
&= c_1(x)\, \theta^{x+1-1}(1-\theta)^{7+1-x-1}, \qquad 0 < \theta < 1.
\end{aligned}
$$

Here we combined $c(x)$ and $\binom{7}{x}$ into the new constant $c_1(x)$. We can see that this is the kernel of the $\mathrm{BETA}(x+1, 8-x)$ distribution, which for $x = 6$ is the $\mathrm{BETA}(7,2)$ distribution.

Now consider the second scenario where the experiment terminated after the first "placebo." Here we observe $Y =$ number of trials needed to get one "placebo." Thus, $Y \sim \mathrm{GEOM}(\theta)$, so the likelihood function is

$$
f(y|\theta) = (1-\theta)^{y-1}\theta, \qquad y = 1, 2, \ldots.
$$

With a uniform prior on θ as before, we find that the posterior is

$$
\pi(\theta|y = 7) = c \times 1 \times (1-\theta)^{7-1}\theta.
$$

We recognize this as the kernel of the $\mathrm{BETA}(7,2)$ distribution. This is the same conclusion that we reached when we assumed a fixed number of trials. ∎∎∎

Thus, for the Bayesian approach, it doesn't matter which experimental plan (fixed number of trials or fixed number of "placebos") is selected. The posterior distribution is exactly the same. Note that in both cases the posterior distribution is shown in Figure 15.7. A one-sided credible interval is $(0.539, 1)$ which lies entirely to the right of 0.5, giving evidence that the drug is effective. The frequentist approach had to account for the plan for data collection, whereas the Bayesian approach involved the data only.

This type of example has broader implications about stopping an experiment. For example, if we are running a clinical trial hoping to find evidence of a positive effect, can we put the trial on hold while we do a preliminary analysis of the data? Does this affect the inference? From the frequentist paradigm, yes, this matters. If we want to test the null hypothesis of no effect, and if we allow ourselves two chances to reject H_0, then these must be taken into account when determining α. If we don't account for this and select a decision rule for a fixed α, applying it in the interim analysis, and in the final analysis, then our true α will be a bit higher than we would like, because we are allowing two, not one, opportunities to reject H_0. It is possible to come up with a scenario (see Efron and Hastie, 2016: 31–32) where there was insufficient evidence to reject H_0 when we peeked at the data, but (assuming we don't account for the two-stage test) there was enough evidence to reject at the planned end of the experiment. But, had

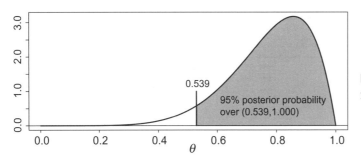

Figure 15.7 The posterior distribution for the probability of success.

we accounted for the two-stage nature of the test, the true α is inflated to the point where we could no longer reject H_0 at the planned end. Had we not peeked, the end result would be significant. Because we peeked at the data, the result is not significant. But why should the act of peeking at the data, which doesn't change the data at all, change the conclusion? In Bayesian statistics, only the data matters, so peeking at the data has no effect.

15.5 Conjugate Priors

In Section 15.2 we considered the problem of binomial data for which the probability of success had a uniform distribution over $(0,1)$. This suggests ignorance about the value of θ, but as the first example in Section 15.4 indicates, it might be an informative prior in a different parameterization. Suppose we do have prior information about θ, for example, θ might be the probability of having a particular disease. A prior that puts most probability between 0 and 0.1 might be better than spreading it over the whole interval $(0,1)$. In a case like this, we might search for a family of distributions whose support is the interval $(0,1)$ and is flexible to handle our belief about the parameter. The beta distribution, described in Section 6.7, may be able to fit these requirements. The BETA(a,b) distribution has mean

$$\mu = \frac{a}{a+b}$$

and standard deviation

$$\sigma = \sqrt{\frac{ab}{(a+b)^2(a+b+1)}} = \sqrt{\frac{a}{a+b}\left(1 - \frac{a}{a+b}\right)\frac{1}{a+b+1}} = \sqrt{\frac{\mu(1-\mu)}{a+b+1}}.$$

We can select the prior mean by appropriately choosing a and b, but there are infinitely many ways to do this. For example, if we wanted the prior mean to be 0.05 (for example, if we are sampling for a rather rare disease) we could select $a = 1$ and $b = 19$. Or we could select $a = 10$ and $b = 190$. The last part of the formula for the standard deviation suggests that for fixed values of $a/(a+b)$, the variance is proportional to $1/(a+b+1)$, so that larger values of a and b correspond to smaller variability. Figure 15.8 illustrates this. It shows three priors, each having a mean of 0.05, but the standard deviations are 0.04756 (for $a = 1, b = 19$), 0.02422 (for $a = 4, b = 76$), and 0.015373 (for $a = 10, b = 190$). Thus, the ratio $a/(a+b)$ controls the mean, and the magnitude of $a+b+1$ controls the scaling.

A family of distributions is said to be *rich* if it is able to accommodate a wide variety of prior beliefs. In this sense the beta distribution is rich because most forms of prior belief can be expressed (at least approximately) by a beta distribution with some a and b.

If we were to assume a beta prior distribution for the probability of success in a sequence of Bernoulli trials, then the posterior distribution will have a familiar form. To be specific, suppose we observe $X \sim$ BIN(n, θ) where θ is unknown and has prior distribution $\theta \sim$ BETA(a,b). Note that in a given problem, a and b would be *numbers* that express our belief about θ before observing any data; here we use arbitrary values a and b with the understanding that these will be known constants in our problem. The posterior is

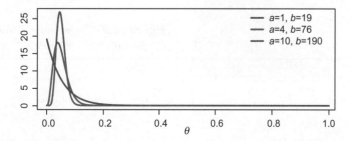

Figure 15.8 Three beta prior distributions. All have the same mean, $\mu = a/(a+b) = 0.05$, but the probability is more concentrated around the mean for larger values of $a+b$.

$$
\begin{aligned}
\pi(\theta|x) &= c(x)\pi(\theta)f(x|\theta) \\
&= c(x)\frac{\Gamma(a+b)}{\Gamma(a)\Gamma(b)}\theta^{a-1}(1-\theta)^{b-1}\binom{n}{x}\theta^x(1-\theta)^{n-x} \\
&= c_1(x)\theta^{a+x-1}(1-\theta)^{b+n-x-1}, \\
&= \underbrace{c_1(x)}_{\text{depends on } x \text{ only}}\ \underbrace{\theta^{a+x-1}(1-\theta)^{b+n-x-1}}_{\text{depends on }\theta\text{ and }x}, \qquad 0<\theta<1.
\end{aligned}
$$

Notice that this posterior distribution factors into a product of one expression that depends on the data x only, and a piece that depends on both the data x and the parameter θ. We recognize from the latter that this must be a beta distribution with parameters $a+x$ and $b+n-x$. Without doing any integration, we find that the normalizing constant must be

$$
c(x) = \frac{\Gamma(a+b+n)}{\Gamma(a+x)\Gamma(b+n-x)}.
$$

We call $c(x)$ a *constant* even though it depends on the data and the prior parameters a and b. Remember that in a particular problem a and b will both be fixed numbers. Also, the data x are considered to be fixed when we compute the posterior distribution. Thus, the posterior distribution of θ given the data x is the BETA$(a+x,b+n-x)$ distribution. This idea is illustrated in Figure 15.9.

Definition 15.5.1 — Conjugate Prior. If, for a given likelihood function, the posterior distribution is in the same family as the prior, then the prior is said to be a *conjugate prior* for the given likelihood.

We saw previously that the beta is the conjugate prior for the binomial likelihood. There are a number of others of these conjugate priors.

■ **Example 15.11** Suppose we observe a sample $X_1,X_2,\ldots,X_n$ from the EXP(λ) distribution, with PDF

$$
f(x|\lambda) = \lambda\exp(-\lambda x), \qquad x>0.
$$

Note that this is one of the two commonly used parameterizations of the exponential; the other uses the parameter $\theta = 1/\lambda$. What is a conjugate prior for λ?

Solution: The likelihood function is

$$
f(x_1,x_2,\ldots,x_n|\lambda) = \prod_{i=1}^{n}\lambda\exp(-\lambda x_i) = \lambda^n\exp\left(-\lambda\sum_{i=1}^{n}x_i\right).
$$

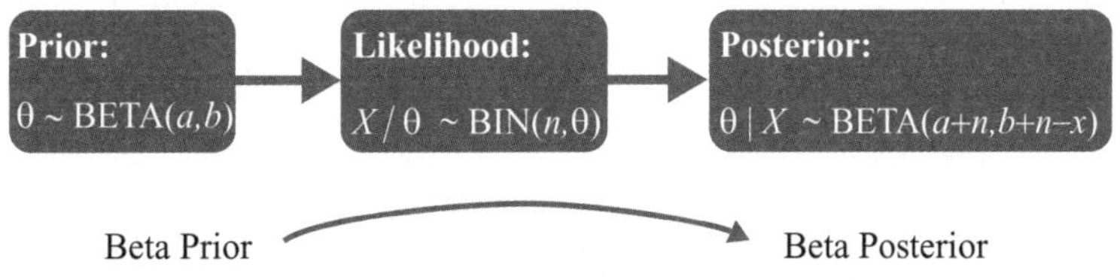

Beta Prior   Beta Posterior

Figure 15.9 If we observe binomial data, a beta prior distribution leads to a beta posterior distribution (but with different parameters). The beta distribution is a conjugate prior for the binomial.

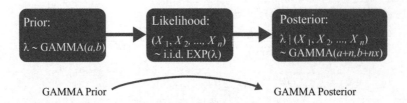

Figure 15.10 If we observe data from an exponential distribution, maybe lifetimes, then a gamma prior distribution will lead to a gamma posterior distribution (but with different parameters). This means that the gamma distribution is a conjugate prior for a sample from the exponential distribution.

In order to determine a conjugate prior, we have to look at the likelihood as a function of λ; that is, we have to view this as

$$f(x_1, x_2, \ldots, x_n | \lambda) = \lambda^A \exp(-B\lambda), \quad \lambda > 0,$$

where $A = n$ and $B = \sum_{i=1}^{n} x_i$. We then have to look for a prior that, when multiplied by this likelihood function, will yield a posterior in the same family. After a little searching, we would find that the $\text{GAM}(a, b)$ distribution fits the bill. If we take the prior to be

$$\pi(\lambda) = \frac{b^a}{\Gamma(a)} \lambda^{a-1} \exp(-b\lambda), \quad \lambda > 0,$$

then the posterior will be of the form

$$
\begin{aligned}
\pi(\lambda | x_1, x_2, \ldots, x_n) &= c\left(\lambda^{a-1} \exp(-b\lambda)\right)\left(\lambda^n \exp(-n\bar{x}\lambda)\right) \\
&= c\lambda^{a+n-1} \exp(-(b+n\bar{x})\lambda), \quad \lambda > 0.
\end{aligned}
$$

This is obviously the $\text{GAM}(a+n, b+n\bar{x})$ distribution, so the constant c must be

$$c = \frac{(b+n\bar{x})^{a+n}}{\Gamma(a+n)}.$$

A $\text{GAM}(a, b)$ prior for λ leads to a $\text{GAM}(a+n, b+n\bar{x})$ posterior. Since the prior and posterior are in the same family, the gamma distribution is a conjugate prior for the parameter λ of the exponential distribution. ∎ ∎ ∎

The next example is a bit contrived, because we assume that the variance of a normal is known and the mean is not. Just as we used the z-test for the mean of a normal distribution with known variance as a stepping stone to the t-test for the case when the variance was unknown, we use the Bayesian analysis of the known-variance case as a stepping stone to the unknown-variance case.

Bayesians usually prefer to work with the reciprocal of the variance, rather than the variance itself. This may seem odd, but the reason should be clear after working through the next example. A skeptical reader is invited to work this example using the variance rather than the precision.

Definition 15.5.2 — Precision. The reciprocal of the variance of a normal distribution is called the *precision*. Usually, we use σ^2 to denote the variance, and τ to represent the corresponding precision; that is,

$$\tau = \text{precision} = \frac{1}{\sigma^2}.$$

If there is a subscript on σ^2, we will use the same subscript on τ; for example, the precision for the $N(\mu_0, \sigma_0^2)$ distribution will be denoted τ_0.

■ **Example 15.12** Suppose that $X_1, X_2, \ldots, X_n \sim N(\mu, \sigma^2)$ where σ^2 is known. What is a conjugate prior for the parameter μ?

Solution: Let $x = [x_1, x_2, \ldots, x_n]'$ denote the observed data. The likelihood function is then

$$L(\mu|x) = \prod_{i=1}^{n} \frac{1}{\sqrt{2\pi}\sigma} \exp\left(-\frac{1}{2\sigma^2}(x_i - \mu)^2\right) = (2\pi)^{-n/2}\sigma^{-n}\exp\left(-\frac{1}{2\sigma^2}\sum_{i=1}^{n}(x_i - \mu)^2\right). \quad (15.9)$$

The question regarding a conjugate prior is really about what prior we might combine with this likelihood in order to obtain a recognizable posterior. The right side of (15.9), when viewed as a function of μ (not x) looks like it could be factored to be a normal distribution. The likelihood in (15.9) can be written in the form

$$L(\mu|x) = c\exp\left(-\frac{1}{2\sigma^2}(n\mu^2 - 2\mu n\bar{x})\right) = c\exp\left(-\frac{\tau}{2}(n\mu^2 - 2\mu n\bar{x})\right). \quad (15.10)$$

Note that the constant c can depend on the data x and τ, because $\tau = 1/\sigma$ was assumed known. We have absorbed the term $-(\tau/2)\sum_{i=1}^{n} x_i^2$ into the constant c. If we assume the $N(\mu_0, \sigma_0^2)$ prior distribution for the mean μ, then our prior is of the form

$$\pi(\mu) = c\exp\left(-\frac{1}{2\sigma_0^2}(\mu - \mu_0)^2\right) = c\exp\left(-\frac{\tau_0}{2}(\mu - \mu_0)^2\right).$$

It is important to keep the notation straight: μ is the unknown mean of the normal distribution from which our data $x_1, x_2, \ldots, x_n$ comes. With a Bayesian approach, we must specify a prior distribution for all unknown parameters; in this case μ is the only unknown parameter. We assume the normal prior with mean μ_0 and standard deviation σ_0. In a real application, μ_0 and σ_0 will be fixed numbers.

In terms of precision, the prior and likelihood function are

$$\pi(\mu) = \exp\left(-\frac{\tau_0}{2}(\mu - \mu_0)^2\right)$$

and

$$L(\mu|x) = c\exp\left(-\frac{\tau}{2}(n\mu^2 - 2\mu n\bar{x})\right).$$

The posterior distribution is therefore

$$\begin{aligned}
\pi(\mu|x) &= c\,\pi(\mu)\,L(\mu|x) \\
&= c\exp\left(-\frac{\tau_0}{2}(\mu - \mu_0)^2\right)\exp\left(-\frac{\tau}{2}(n\mu^2 - 2\mu n\bar{x})\right) \\
&= c\exp\left(-\frac{\tau_0}{2}(\mu^2 - 2\mu_0\mu + \mu_0^2) - \frac{\tau}{2}(n\mu^2 - 2n\bar{x}\mu)\right) \\
&= c_1\exp\left(-\frac{\tau_0 + n\tau}{2}\mu^2 + (\tau_0\mu_0 + \tau n\bar{x})\mu\right) \quad \left[\text{absorbing }\exp\left(-\frac{\tau_0}{2}\mu_0^2\right)\text{ into the constant}\right] \\
&= c_1\exp\left(-\frac{\tau_0 + n\tau}{2}\left(\mu^2 - 2\frac{\tau_0\mu_0 + \tau n\bar{x}}{\tau_0 + n\tau}\mu\right)\right).
\end{aligned}$$

Now comes the trick of completing the square. In order to make the exponent in this last expression be of the form (μ minus something) squared, we take half of the linear term, square it, and add it in the exponent. Because of the minus sign outside the parentheses, we are really subtracting this constant. Of course, if we subtract it out, we must also add it back in. This yields

$$\pi(\mu|x) = c_1 \exp\left(-\frac{\tau_0+n\tau}{2}\left(\mu^2 - 2\frac{\tau_0\mu_0+\tau n\bar{x}}{\tau_0+n\tau}\mu + \left(\frac{\tau_0\mu_0+\tau n\bar{x}}{\tau_0+n\tau}\right)^2\right) + \underbrace{\frac{\tau_0+n\tau}{2}\left(\frac{\tau_0\mu_0+\tau n\bar{x}}{\tau_0+n\tau}\right)^2}_{\text{adding the "square" back in}}\right).$$

Notice, however, that everything in the "complete the square" part (shown with the underbrace in the formula above) is *known*! (Remember, we've assumed that the variance, hence the precision τ, is known.) Thus, the entire "square" part can be absorbed in the constant, which we now call $c(x)$. The posterior distribution is therefore

$$\pi(\mu|x) = c_1(x) \exp\left(-\frac{\tau_0+n\tau}{2}\left(\mu - \left(\frac{\tau_0\mu_0+\tau n\bar{x}}{\tau_0+n\tau}\right)\right)^2\right).$$

This is the normal distribution with mean

$$\mu_1 = \frac{\tau_0\mu_0+\tau n\bar{x}}{\tau_0+n\tau}. \tag{15.11}$$

The precision is

$$\tau_1 = \tau_0+n\tau \tag{15.12}$$

and the variance is

$$\sigma_1^2 = \frac{1}{\frac{1}{\sigma_0^2}+\frac{n}{\sigma^2}}.$$

Thus, the normal distribution is the conjugate prior for the mean μ of the normal distribution when the precision τ (or the variance σ^2) is known. ∎

There are some interesting relationships to glean from (15.11) and (15.12). First, note that the posterior mean is a weighted average of the prior mean μ_0 and the sample mean $\bar{x}$; the weights are τ_0 (for the prior mean) and $n\tau$ for the sample mean. So the more data we have, the greater will be the weight on the sample mean. On the other hand, if the sample size is small, but the prior precision is large (large precision means small prior variance) then the prior precision will dominate. The other feature to notice about (15.12) is that the posterior precision is just the sum of the prior precision and n times the known precision of the distribution. Recall that the variance of the sample mean $\bar{X}$ is σ^2/n, so the precision of $\bar{X}$ is $n\tau$. The posterior precision is thus the sum of the prior precision and the precision of the sample mean. This is why Bayesians like to work with the precision, rather than the variance.

The conjugate prior for the case where both the mean μ and variance $\phi = \sigma^2$ is derived in a similar manner, but it is considerably messier. We give the result and then show how it meets the conjugacy requirement. For starters, we note that we are now dealing with a joint density, so the family for the joint posterior must match the family for the joint prior. Also, to avoid confusion about whether σ^2 is itself a parameter, or whether it is the square of a parameter, σ, we use ϕ to denote the variance.

The conjugate prior is developed by first assuming that the variance has the scaled inverse χ^2 distribution, denoted $\text{SInv}\chi^2(v_0, \phi_0)$, with PDF

$$p(\phi) = \frac{(v_0/2)^{v_0/2}\phi_0^{v_0/2}}{\Gamma(v_0/2)}\phi^{-v_0/2-1}\exp\left(-\frac{v_0\phi_0}{2\phi}\right), \qquad \phi > 0. \tag{15.13}$$

The mean of the $\mathrm{SInv}\chi^2(v_0, \phi_0)$ is

$$\mathbb{E}(\phi) = \frac{v_0}{v_0 - 2}\,\phi_0, \tag{15.14}$$

and the variance is

$$\mathbb{V}(\phi) = \frac{2v_0^2}{(v_0 - 2)^2(v_0 - 4)}\,\phi_0^2. \tag{15.15}$$

For the prior mean to exist we must have $v_0 > 2$, and for the variance to exist we must have $v_0 > 4$. We can incorporate prior information about ϕ by taking ϕ_0 to be near our prior estimate since the prior mean is $\frac{v_0}{v_0-2}\phi_0$, which is nearly ϕ_0 if v_0 is large. The more confident we are in our prior estimate, the larger we should choose v_0 because the prior variance goes to zero as $v_0 \to \infty$.

Given ϕ, we assign the normal prior $N(\mu_0, \phi/\kappa_0)$ to μ. Obviously, we would assign μ_0 to be our prior estimate of μ. If we have strong prior beliefs about μ we would choose κ_0 to be large, and if our prior beliefs are weak, we would choose a small κ_0. The parameter κ_0 can be thought of as the number of data points we believe our prior belief is worth. For example, suppose we believe that the mean μ of a population is about 100. When asked how strongly we feel about this, we might say that our belief is as strong as if we observed 5 data values from which we obtained a mean of 100; here we would select $\mu_0 = 100$ and $\kappa_0 = 5$.

The joint prior obtained this way is called the normal-inverse χ^2 distribution and it is denoted as $\mathrm{NInv}\chi^2(\mu_0, \kappa_0, \phi_0, v_0)$. If we observe $X_1, X_2, \ldots, X_n$ from the $N(\mu, \phi)$ distribution, it can be shown that the posterior distribution is the $\mathrm{NInv}\chi^2(\mu_1, \kappa_1, \phi_1, v_1)$ where

$$\mu_1 = \frac{\kappa_0}{\kappa_0 + n}\mu_0 + \frac{n}{\kappa_0 + n}\overline{X} \tag{15.16}$$

$$\kappa_1 = \kappa_0 + n \tag{15.17}$$

$$\phi_1 = \frac{v_0\phi_0}{v_0 + n} + \frac{(n-1)}{v_0 + n}S^2 + \frac{\kappa_0 n}{(v_0 + n)(\kappa_0 + n)}(\overline{X} - \mu_0)^2 \tag{15.18}$$

$$v_1 = v_0 + n, \tag{15.19}$$

where $\overline{X}$ and S^2 are the sample mean and variance, respectively. Several rather intuitive features can be seen in (15.16) through (15.19). First, (15.16) shows that the posterior mean is a weighted average of the prior mean μ_0 and the sample mean $\overline{X}$. Larger values of the sample size n make the sample mean $\overline{X}$ have a higher weight. Also, larger values of κ_0 give more weight to the prior mean μ_0. Since the prior parameters κ_0 and v_0 measure the strength of our prior beliefs, it shouldn't be surprising that our strength increases for larger sample sizes. This happens in a simple way, by adding n to each of κ_0 and v_0. Equation (15.18) is the messiest to interpret. The posterior estimate of the variance ϕ is dependent on the prior estimate ϕ_0, the sample variance S^2, and the squared difference between the prior mean μ_0 and the sample mean $\overline{X}$.

The marginal posterior distribution for μ is then

$$\mu \mid (\overline{X}, S^2) \sim t(v_1, \mu_1, \phi_1/\kappa_1), \tag{15.20}$$

where $\sim t_{v_1}(\mu_1, \phi_1/\kappa_1)$ is the t distribution with v_1 degrees of freedom, centered at μ_1 with scale parameter ϕ_1/κ_1; in other words,

$$\frac{\mu - \mu_1}{\sqrt{\phi_1/\kappa_1}} \mid (\overline{X}, S^2) \sim t(v_1). \tag{15.21}$$

Note a few important points here. First, the random variable on the left side of (15.21) is μ. Recall that in the Bayesian paradigm parameters are treated as random variables and the uncertainty is expressed in probabilistic terms. The values of μ_1 and ϕ_1 are functions of the data and the prior parameters, so once data are collected and recorded, μ_1 and ϕ_1 are fixed numbers. Second, this t distribution is centered at μ_1, but this is not what we call the noncentral t distribution. When we refer to the noncentral t distribution, we mean the distribution introduced in Chapter 11.

The marginal posterior for the population variance ϕ is

$$\phi \mid (\overline{X}, S^2) \sim \text{Inv}\chi^2(v_1, \phi_1). \tag{15.22}$$

Equivalently, we could say that

$$\frac{1}{\phi} \mid (\overline{X}, S^2) \sim \chi^2(v_1, \phi_1). \tag{15.23}$$

The derivations of the results described here are not difficult, but are messy. An interested reader should consult Gelman et al. (2014, chapter 3) or Colosimo and del Castillo (2006, chapter 1).

■ **Example 15.13 — Estimating the mean and variance of a normally distributed population.** Suppose that we wish to estimate the mean and variance of the body mass index (BMI) of a particular population. Before the study begins, we don't know whether to expect that the BMI is higher or lower than the country as a whole. For this case, develop reasonable priors for μ and $\phi = \sigma^2$. Then use these priors, along with the data

$$26, \quad 32, \quad 19, \quad 23, \quad 22, \quad 25, \quad 28, \quad 19, \quad 23, \quad 27, \quad 26, \quad 22, \quad 18, \quad 26, \quad 24$$

to find the joint posterior distribution for (μ, ϕ).

Solution: In the United States the mean BMI is about 29, with a standard deviation of approximately 5. For our prior for ϕ we choose $\phi_0 = 5^2$ and with little confidence in this initial estimate, we choose $v_0 = 1$. Since we don't know whether our population will have a higher or lower mean BMI, we take as our prior parameters

$$\mu_0 = 29, \qquad \phi_0 = 5^2, \qquad \text{and} \qquad \kappa_0 = 2.$$

Choosing $\kappa_0 = 2$ suggests little confidence in our initial estimate. Our joint prior distribution is therefore

$$(\mu, \phi) \sim \text{NInv}\chi^2(\mu_0 = 29, \kappa_0 = 2, \phi_0 = 5, v_0 = 1).$$

Figure 15.11 shows both the joint prior and the posterior for our hypothetical BMI data. Darker colors indicate higher values of the PDF. While the prior is mostly spread out over reasonable values of μ and $\phi = \sigma^2$, the posterior is concentrated near $\mu = 25$ and $\phi = 12$. The marginal posterior distributions are

$$\mu|\boldsymbol{x} \sim t(v_1, \mu_1, \phi_1/\kappa_1)$$

and

$$\phi|\boldsymbol{x} \sim \text{Inv}\chi^2(v_1, \phi_1).$$

If we take the posterior means as point estimates for the parameters, we have

$$\hat{\mu} = \mu_1 = \frac{\kappa_0}{\kappa_0 + n}\mu_0 + \frac{n}{\kappa_0 + n}\overline{X}$$

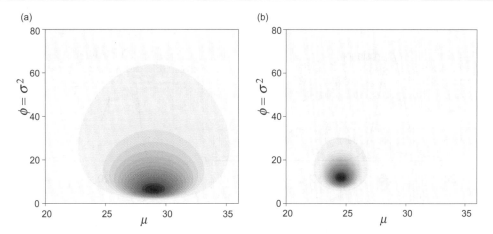

Figure 15.11 Joint $\mathrm{NInv}\chi^2$ prior (a) and posterior (b) for hypothetical data on BMI.

and

$$\hat{\sigma}^2 = \hat{\phi} = \frac{v_1}{v_1 - 2}\phi_1.$$

Often studies such as this focus on the population mean. The marginal prior distribution for μ is a $t(v_0, \mu_0, \phi_0/\kappa_0)$ distribution and the posterior is a $t(v_1, \mu_1, \phi_1/\kappa_1)$ distribution. Substituting the numbers for the parameters, we find that

Prior: $\quad t(1, 29, 5^2/2)$

Posterior: $\quad t(16, 24.59, 15.07/17)$.

Figure 15.12 shows the prior and posterior PDFs for the population mean μ and variance ϕ. ∎∎∎

In the next example we see that a larger sample size yields posteriors with much less variability.

∎ **Example 15.14** Consider the health data we studied previously from 768 pregnant women with diabetes. Using the same prior distribution for mean μ and variance ϕ of the BMI as given in Example 15.13, find the joint and marginal posterior distributions for μ and variance ϕ.

Solution: The joint prior is

$$(\mu, \phi) \sim \mathrm{NInv}\chi^2(\mu_0 = 29, \kappa_0 = 2, \phi_0 = 5, v_0 = 1). \tag{15.24}$$

The sample mean and sample variance are the sufficient statistics for this data set:

$$\bar{x} = 31.99$$
$$s^2 = 62.16.$$

The parameters of the normal-scaled inverse χ^2 posterior distribution are

$$\mu_1 = 31.98$$
$$\kappa_1 = 770$$
$$\phi_1 = 62.03$$
$$v_1 = 769,$$

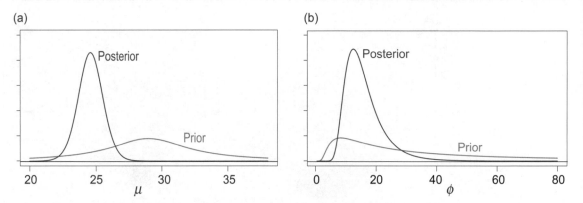

Figure 15.12 Marginal prior (red) assuming a joint NInvχ^2 prior and the posterior (blue) for the hypothetical BMI data. The prior and posterior for μ are shown in (a), and the prior and posterior for ϕ are shown in (b).

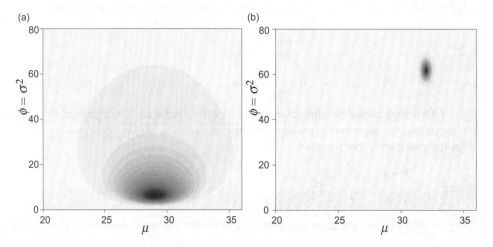

Figure 15.13 Joint NInvχ^2 prior (a) and posterior (b) for the data set on BMI for pregnant women.

yielding posterior point estimates (the means of the posterior distribution)

$$\hat{\mu} = \mu_1 = 31.98$$

$$\hat{\sigma}^2 = \hat{\phi} = \frac{v_1}{v_1 - 2}\phi_1 = 62.20.$$

Figure 15.13 shows the joint prior and posterior for the population mean and variance. Note that Figure 15.13(a) is the same as Figure 15.11(a); this occurs because the prior was taken to be the same for both cases. The joint posterior distribution, shown in Figure 15.13(b), indicates a very small region of high posterior density, much smaller than in Figure 15.11. This is due to the much larger sample size ($n = 768$ here compared to $n = 15$ in the previous example). The marginal priors and posteriors are shown in Figure 15.14. ∎ ∎ ∎

15.6 Noninformative Priors

If we have little or no information about a parameter, it seems that a prior that is uniform would reflect this. In Example 15.7 we showed that a uniform prior on the probability θ of the binomial distribution leads to a nonuniform prior on the log odds, $\log \theta/(1 - \theta)$. While we didn't show it, the converse also holds: A uniform prior on the log odds leads to a nonuniform prior on θ itself. This suggests that we should be careful in the choice of the parameterization and which parameterization gets the uniform

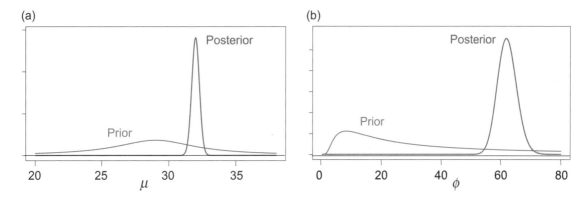

Figure 15.14 Marginal prior (red) assuming a joint NInvχ^2 prior and the posterior (green) for the BMI data for pregnant women. The prior and posterior for μ are shown in (a), and the prior and posterior for ϕ are shown in (b).

prior. The good news is that if we have a lot of data, the posterior is dominated by the data and the prior has little influence. When the sample size is small or moderate, the prior does exert a reasonable amount of influence on the posterior.

In 1946 Harold Jeffreys determined the form of a prior that is invariant under transformations. In other words, the probability of being in some region with one parameterization is the same as the probability in the transformed region with the transformed parameterization. Jeffreys showed that the prior

$$p(\theta) = c\sqrt{I(\theta)} \tag{15.25}$$

is invariant under transformations, where

$$I(\theta) = E\left(-\frac{\partial^2}{\partial\theta^2}\log f(x|\theta)\right) \tag{15.26}$$

is called the Fisher information. The definition given here assumes a single parameter, and as we have seen, most problems involve many parameters. Although there is a multivariate version of the definition of the Jeffreys prior, the prior is often determined by taking the Jeffreys prior for each parameter, one at a time, by assuming the other parameters are fixed.

■ **Example 15.15 — Jeffreys prior for the binomial.** Suppose we observe the number of successes on n Bernoulli trials. Let X denote the number of successes; then $X \sim \text{BIN}(n, \theta)$ where θ is the unknown probability of a success on any one trial. Find the Jeffreys prior.

Solution: The likelihood function is

$$f(x|\theta) = \binom{n}{x}\theta^x(1-\theta)^{n-x}, \qquad x = 0, 1, \ldots, n; \ 0 \le \theta \le 1.$$

Taking the logarithm and then the first two derivatives yields

$$\log f(x|\theta) = \log\binom{n}{x} + x\log\theta + (n-x)\log(1-\theta)$$

$$\frac{\partial\log f}{\partial\theta} = \frac{x}{\theta} + \frac{n-x}{1-\theta}(-1) = \frac{x}{\theta} - \frac{n-x}{1-\theta}$$

$$\frac{\partial^2\log f}{\partial\theta^2} = -\frac{x}{\theta^2} - \frac{n-x}{(1-\theta)^2}.$$

The Fisher information is therefore

$$I(\theta) = E\left(-\frac{\partial^2}{\partial\theta^2}\log f(X|\theta)\right) = \frac{1}{\theta^2}E(X|\theta) + \frac{1}{(1-\theta^2)}E(n-X|\theta) = \cdots = \frac{n}{\theta(1-\theta)}.$$

We've skipped a number of steps in the algebraic solution. Problem 15.25 asks you to fill in the required steps. The Jeffreys prior is therefore

$$p(\theta) = c\sqrt{I(\theta)} = c\sqrt{\frac{n}{\theta(1-\theta)}} = c_1\theta^{-1/2}(1-\theta)^{-1/2} \qquad 0 \le \theta \le 1. \tag{15.27}$$

This is the BETA$(\frac{1}{2},\frac{1}{2})$ distribution. ▪▪▪

In the last example, the Jeffreys prior is a proper PDF, in the sense that it integrates to 1, as all proper PDFs must. In the next example we see that sometimes the Jeffreys prior is improper, in the sense that it does not integrate to anything finite. Often, though not always, an improper prior still leads to a proper posterior.

■ **Example 15.16 — Jeffreys prior for the parameters of the normal distribution.**
Suppose we have a random sample $X_1, X_2, \ldots, X_n$ from the $N(\mu, \sigma^2)$ distribution. Find the Jeffreys priors for μ and for $\phi = \sigma^2$ assuming the other parameter is known.

Solution: We will obtain the Jeffreys prior for the two parameters one at a time by assuming that the other is known. We begin with the mean parameter μ. The likelihood function for a single observation is

$$f(x|\mu,\phi) = \frac{1}{\sqrt{2\pi}\sqrt{\phi}}\exp\left(-\frac{1}{2\phi}(x-\mu)^2\right).$$

Taking the logarithm and the first two derivatives yields

$$\log f(x|\mu,\phi) = -\frac{1}{2}\log 2\pi - \frac{1}{2}\log\phi - \frac{1}{2\phi}(x-\mu)^2$$

$$\frac{\partial\log f}{\partial\mu} = 0+0-\frac{1}{2\phi}2(X-\mu)(-1) = \frac{x-\mu}{\phi}$$

$$\frac{\partial^2\log f}{\partial\mu^2} = -\frac{1}{\phi}$$

$$I(\mu) = E\left(-\frac{\partial^2}{\partial\mu^2}\log f(X|\mu,\phi)\right) = -E\left(-\frac{1}{\phi}\right) = \frac{1}{\phi}.$$

The Jeffreys prior for μ is therefore

$$p(\mu) = c\sqrt{I(\mu)} = \frac{c}{\sqrt{\phi}}.$$

Note that the prior for μ is a constant. No matter how close to zero the constant is, the prior will always integrate to ∞. This is an example of an improper prior.

The partial derivatives of the PDF with respect to ϕ and the Fisher information are:

$$\frac{\partial \log f}{\partial \phi} = 0 - \frac{1}{2}\phi^{-1} + \frac{1}{2}\phi^{-2}(x-\mu)^2$$

$$\frac{\partial^2 \log f}{\partial \phi^2} = \frac{1}{2}\phi^{-2} - \phi^{-3}(x-\mu)^2$$

$$I(\phi) = E\left(-\frac{\partial^2}{\partial \phi^2}\log f(X|\mu,\phi)\right) = -\frac{1}{2}\phi^{-2} + \phi^{-3}E\left((X-\mu)^2\right) = \frac{1}{2}\phi^{-2}.$$

The Jeffreys prior for ϕ is therefore

$$p(\phi) = c\sqrt{\frac{1}{2}\phi^{-2}} = \frac{c_1}{\phi}.$$

This is also an improper prior.   ∎∎∎

> **Exercise 15.2** Rework Example 15.14 assuming the Jeffreys prior.
> *Answer*: Plots of the joint prior and posterior are shown in Figures 15.15 and 15.16.

15.7 Simulation Methods

As we have seen in the previous section, Bayesian calculations become difficult even for simple two-parameter problems like the problem of estimating the mean and variance of a normal distribution. For higher-dimensional problems, an exact solution for the posterior distribution and the marginal posterior distributions is often intractable. The problem usually lies in the denominator of Bayes' Theorem:

$$p(\boldsymbol{\theta}|\boldsymbol{x}) = \frac{p(\boldsymbol{\theta})L(\boldsymbol{\theta}|\boldsymbol{x})}{\int p(\boldsymbol{\theta})L(\boldsymbol{\theta}|\boldsymbol{x})\,d\boldsymbol{\theta}}.$$

This is a p-dimensional integral, where p is the number of unknown parameters in the vector $\boldsymbol{\theta}$.

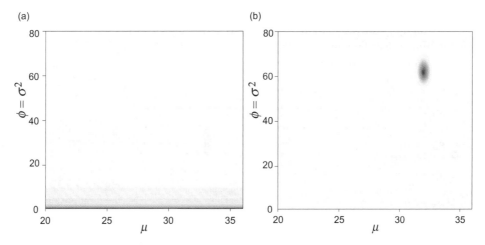

Figure 15.15 Marginal prior and posterior distributions for the mean μ (a) and variance ϕ (b) for the diabetes data. The priors are the Jeffreys priors.

(a)

(b)

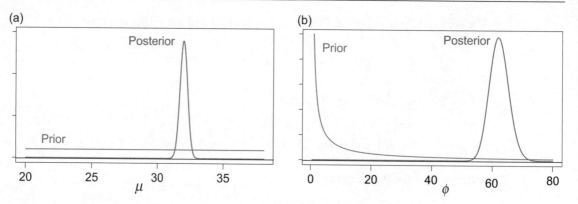

Figure 15.16 Marginal prior (a) and posterior (b) distributions for the mean μ and variance ϕ for the diabetes data. The priors are the Jeffreys priors.

For the case of an unknown mean and variance from a normal distribution with a $\text{NInv}\chi^2$ prior distribution, the posterior has PDF

$$p(\mu, \phi | x) = \frac{\phi^{-v_0/2 - n/2 - 3/2} \exp\left(-\frac{1}{2\phi}\left[v_0\phi_0^2 + \kappa_0(\mu - \mu_0)^2 + (n-1)s^2 + n(\bar{x} - \mu)^2\right]\right)}{\int_0^\infty \int_{-\infty}^\infty \phi^{-v_0/2 - n/2 - 3/2} \exp\left(-\frac{1}{2\phi}\left[v_0\phi_0^2 + \kappa_0(\mu\mu_0)^2 + (n-1)s^2 + n(\bar{x} - \mu)^2\right]\right) d\mu\, d\phi}.$$

(15.28)

This is a formidable equation indeed! Fortunately, in this case it works out to be the $\text{NInv}\chi^2$ distribution, whose marginals are known. In more complicated cases we're left with a complicated expression, and in dimensions higher than two it is difficult (impossible in some cases) to visualize what the posterior is telling us about the unknown parameters.

What if we could simulate random values from the joint posterior distribution? We could then plot them, either one parameter at a time if we're interested in a marginal posterior, or two at a time in a scatterplot if we're interested in a joint posterior. Figure 15.17 shows the scatterplot of 10,000 simulations from the posterior distribution of the mean and variance of the distribution of BMI in Example 15.14. Compare this with the contour plot of the posterior distribution given in Figure 15.13(b); these plots convey basically the same information.

Histograms for the marginals for μ and ϕ are shown in Figure 15.18. Compare these plots with the marginal distributions derived analytically and shown in Figure 15.14. Analogously, the marginal posteriors for μ and ϕ are shown in Figure 15.18. These convey much the same information as shown in Figure 15.14.

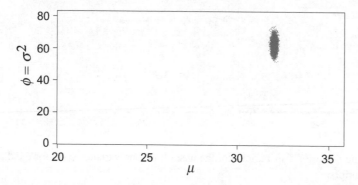

Figure 15.17 10,000 simulations from the joint posterior distribution.

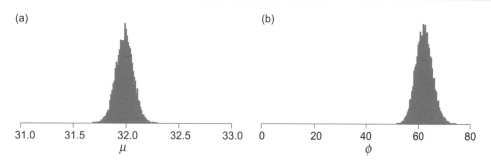

Figure 15.18 Histograms of the marginal posterior distributions for μ and ϕ based on 10,000 simulations.

In the case we have discussed at length in this chapter, namely the problem of estimating the posterior distribution of the mean and variance of a normal distribution, the exact solution is messy but tractable. The point of this section is that the tractability ends as problems get progressively harder, and in these cases some sort of simulation methods are required. The question you should ask yourself is this: Which gives me more useful information about the posterior, a messy formula such as that given in (15.28) or the scatterplot of 10,000 simulated values from the posterior? Usually, the answer is the latter.

15.7.1 Metropolis–Hastings Algorithm

The algorithm we describe here, called the Metropolis–Hastings algorithm, dates back to 1953 when Metropolis, Rosenbluth, Rosenbluth, Teller, and Teller (Metropolis *et al.*, 1953) developed a simulation method for a problem in physics. Nearly two decades later Hastings (1970) generalized the method. It remained an obscure method to statisticians until 1984 when brothers Donald Geman and Stuart Geman (Geman and Geman, 1984) applied the algorithm to the problem of image restoration using Bayesian methods. This led to the technique generally known as Markov chain Monte Carlo (MCMC). This method opened up a new world of applications of Bayesian statistics. Problems that were previously intractable were able to be solved using methods based on simulation.

Algorithm: Metropolis–Hastings Algorithm for Simulating from a Posterior Distribution

To simulate from the posterior distribution $p(\boldsymbol{\theta}|\boldsymbol{x})$:

Step 1: Starting Value: Begin with an initial guess for the value of $\boldsymbol{\theta}$ that is near the middle of the posterior distribution. Call this value $\boldsymbol{\theta}^{(0)}$. This must be a value for which $p(\boldsymbol{\theta}^{(0)}|\boldsymbol{x}) > 0$. Set $i = 1$.

Step 2a: Proposed Move: Let $g(\boldsymbol{\theta}|\boldsymbol{\theta}^{(i-1)})$ be some conditional PDF that has the same support as $p(\boldsymbol{\theta}|\boldsymbol{x})$. (This PDF may depend on the previous value $\boldsymbol{\theta}^{(i-1)}$ but it could be independent of $\boldsymbol{\theta}^{i-1}$.) Here, g is called the *proposal density*. Simulate one value from $g(\boldsymbol{\theta}|\boldsymbol{\theta}^{(i-1)})$ and call it $\boldsymbol{\theta}^{(\text{prop})}$.

Step 2b: Compute Acceptance Probability: Compute

$$\alpha = \frac{p(\boldsymbol{\theta}^{(\text{prop})}|\boldsymbol{x})\, g(\boldsymbol{\theta}^{(i-1)}|\boldsymbol{\theta}^{(\text{prop})})}{p(\boldsymbol{\theta}^{(i-1)}|\boldsymbol{x})\, g(\boldsymbol{\theta}^{(\text{prop})}|\boldsymbol{\theta}^{(i-1)})}. \tag{15.29}$$

Step 2c: Accept or Reject the Proposed Move to $\boldsymbol{\theta}^{(\text{prop})}$: If $\alpha \geq 1$, then accept the move to $\boldsymbol{\theta}^{(\text{prop})}$. If $\alpha < 1$ then accept the move to $\boldsymbol{\theta}^{(\text{prop})}$ with probability α. This can be done with the `sample()` command in R:

```
sample( c("Accept","Reject") , size=1 , prob=c(alpha,1-alpha) )
```

If we accept the move we set $\theta^{(i)} = \theta^{(\mathrm{prop})}$. Otherwise we stay at the same place, namely $\theta^{(i-1)}$, in which case we set $\theta^{(i)} = \theta^{(i-1)}$. In other words, we set

$$
\theta^{(i)} = \begin{cases} \theta^{(\mathrm{prop})} & \text{if the move is accepted} \\[2mm] \theta^{(i-1)} & \text{if the move is rejected} \end{cases} \tag{15.30}
$$

Step 3: Branch: If the required number of simulations has been reached, then stop. Otherwise, set $i \leftarrow i + 1$, go to Step 2a, and repeat.

The successive values generated by this process are not independent, partly because consecutive values might (or might not) be the same. The Metropolis–Hastings algorithm is a realization of a Markov chain, that is a sequence of random variables with the property that the probability distribution of the random variable at time i depends only on the value at time $i - 1$ and not on the value of the random variable at any time points before time $i - 1$. Under the right circumstances, Markov chains converge to a steady-state distribution where the unconditional distribution of the random variable at time i is the same regardless of i.[2]

If the proposal density is symmetric about the current value $\theta^{(i-1)}$, then

$$
g(\theta^{(\mathrm{prop})}|\theta^{(i-1)}) = g(\theta^{(i-1)}|\theta^{(\mathrm{prop})})
$$

so these cancel in the calculation of α, which then becomes

$$
\alpha = \frac{p(\theta^{(\mathrm{prop})}|x)}{p(\theta^{(i-1)}|x)} = \frac{p(\theta^{(\mathrm{prop})})\, f(x|\theta^{(\mathrm{prop})})}{p(\theta^{(i-1)})\, f(x|\theta^{(i-1)})}, \tag{15.31}
$$

where $p(\theta)$ is the prior for θ. When the proposal density is symmetric like this, the algorithm is usually called the Metropolis algorithm; when the proposal density is not symmetric, the algorithm is usually called the Metropolis–Hastings algorithm.

The theory says that the steady-state distribution of the Markov chain defined by the Metropolis–Hastings algorithm is the posterior distribution. This fact suggests the following method for learning about a posterior distribution.

Algorithm: Using the Metropolis–Hastings Algorithm to Learn about the Posterior

Burn-in: Apply the Metropolis–Hastings algorithm for enough iterations that we are convinced, for all practical purposes, that the steady state has been reached.

Simulate from the Posterior: Simulate a number of additional iterations from the Markov chain and use these as simulations from the desired posterior. Note, however, that these are not independent observations, so the information in, say, 1,000 iterations could be less that it would be if we had 1,000 independent simulations.

There are software engines that will perform the Metropolis–Hastings algorithm or its relative, Gibbs sampling (discussed below). We believe it is instructive to code the Metropolis–Hastings algorithm from scratch to see how it works. In the code below, we define three functions at the top: the beta PDF (excluding the $\Gamma(a+b)/\Gamma(a)\Gamma(b)$ constant), the likelihood (excluding the $\binom{n}{x}$ constant), and the function that proposes the next value in the Markov chain. The proposal distribution returns a normally distributed random variable with mean equal to the current value of θ and standard deviation of 0.1. Note that if this value is outside of the interval $[0, 1]$ the proposal is automatically rejected because the

---

[2]Note that this paragraph very briefly summarizes some deep theoretical notions. Interested readers are referred to books on Bayesian statistics for details that we have omitted here.

posterior distribution is equal to 0 outside this interval. We then set the random seed equal to 3210 so that repeated runs will give exactly the same output. If you'd rather get different results on each run, you can select a different seed or omit the `set.seed()` function. We define the data, which for this example is binomial with $n = 6$ and $x = 1$; that is, six trials with one success. The prior distribution is a BETA$(2,2)$ distribution. We run `nsim = 11000` simulations, expecting to discard several at the beginning to assure that we've reached the steady state. Then, inside the loop we propose a new θ called `thetap`, where the "p" stands for "proposed." The current value of θ is called `thetac`, where the "c" stands for "current." The heart of the Metropolis–Hastings algorithm is the line that computes `alpha`, the acceptance probability. Note that the proposal distribution, the normal in this case, is symmetric so the proposal density $g(|)$ drops out. The next line assigns the variable `decision` to be `"Acc"` or `"Rej"` with probabilities $\min(\alpha,1)$ and $1-\min(\alpha,1)$. Then, if the decision is `"Acc"` we set $\theta^{(i)} = \theta^{\text{prop}}$; otherwise we set $\theta^{(i)} = \theta^{(i-1)}$. Here is the R code:

```
pBeta = function(theta,a,b) {
  z = 0
  if (theta >= 0 && theta <= 1) { z = theta^(a-1) * (1-theta)^(b-1) }
  return(z)
}
thetaLik = function(x,n,theta) { theta^x * (1-theta)^(n-x) }
thetaprop = function(theta) { rnorm(n=1,mean=theta,sd=0.1) }
set.seed(3210)
n = 6
x = 1
a = 2;      b = 2
acc = 0
nsim = 11000
thetavec = rep(0,nsim)
thetavec[1] = 0.50
for (i in 2:nsim) {
  thetap = thetaprop( thetavec[i-1] )
  thetac = thetavec[i-1]
  alpha =    pBeta(thetap,a,b) * thetaLik(x,n,thetap) /
          ( pBeta(thetac,a,b) * thetaLik(x,n,thetac) )
  probAcc = min(1,alpha)
  decision = sample( c("Acc","Rej") , size=1 , prob=c(probAcc,1-probAcc) )
  if ( decision == "Acc" ) {  thetavec[i] = thetap; acc = acc + 1  }
  else thetavec[i] = thetac
}
```

The datum is the outcome $X \sim \text{BIN}(6,\theta)$ and the prior is BETA$(2,2)$. By the discussion in Section 15.5, since the beta is a conjugate prior for the binomial distribution, we could conclude immediately that the posterior distribution is BETA$(2+1, 2+6-1)$. We ran the MCMC simulations using the Metropolis algorithm (since the proposal distribution is symmetric, the Metropolis–Hastings reduces to the slightly simpler Metropolis algorithm) so that we could see how the algorithm works in a simple case.

The Metropolis–Hastings algorithm produces a Markov chain, in the sense that the distribution for the outcome at time t is dependent only on the outcome at time $t-1$, not on how we got to the outcome at time $t-1$. The theory says that the steady-state distribution of this Markov chain is the posterior distribution that we desire. We must, therefore, assure ourselves that we have reached the steady state. One way to assess whether the Markov chain has reached the steady state is to look at the sequence of values of the Markov chain. If these look like they have settled down to a single probability distribution,

(a)

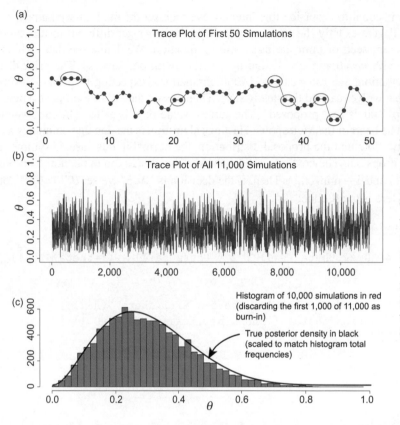

Figure 15.19 Results of the Metropolis algorithm applied to BIN$(6, \theta)$ data where we observed $x = 1$ success. The prior distribution was BETA$(2, 2)$. The first 50 simulations are shown in (a), where we can see several places where the proposal was rejected and the algorithm stayed at the same level. Near the beginning, we rejected the move twice in a row, leading to three consecutive $\theta^{(i)}$ that are identical. Places where the move was rejected are circled.

then we can infer that we have reached the steady state. There are more precise methods to make this inference, such as the Gelman–Rubin statistic, but we won't go into that here (see Gelman *et al.*, 1995).

The plot of successive values of the Markov chain is called a *trace plot* and for the simple case described here is shown in Figure 15.19. Figure 15.19(a) shows the first 50 simulations, and we can see that there were several cases in the first 50 where the algorithm rejected the proposal (as indicated by consecutive identical points). Points 3, 4, and 5 are identical, indicating that twice in a row we rejected the proposal. Figure 15.19(a) indicates rejections with a circle around consecutive identical points. Overall, in the 11,000 simulations we rejected about 21% of the time. From Figure 15.19(b), which shows all 11,000 simulations, it appears that the steady state is reached almost immediately. To be safe, we discard the first 1,000 simulations, called the burn-in, and keep simulations 1,001 through 11,000, for a total of 10,000 simulations. Figure 15.19(c) shows a histogram of these 10,000 simulations in red. This is what we believe the posterior distribution looks like. In this problem we know that the posterior is a BETA$(3,7)$ distribution so we can plot the (scaled) posterior distribution with the histogram. As Figure 15.19(c) shows, the histogram is a very close fit to the true posterior. If we wanted to estimate the probability θ by taking the posterior mean, we would simply average the values of the 10,000 simulations that we kept. This turns out to be 0.299, very close to the mean of the posterior we derived analytically since the mean of the BETA$(3,7)$ distribution is $3/(3+7) = 0.3$. This was from only 10,000 simulations. Keep in mind that in most realistic cases, the analytic form of the posterior will be unknown and we will have to rely solely on the simulated values.

15.7.2 The Gibbs Sampling Algorithm

Anther form of MCMC simulation is *Gibbs sampling*, which we explain in the context of a three-parameter problem; the extension from three to p dimensions should be apparent. To implement Gibbs sampling, we must know the conditional distribution of each parameter given all of the other parameters and the data.

Algorithm: Gibbs Sampling Algorithm for Simulating from a Posterior Distribution

Suppose we want to simulate from a posterior distribution $p(\theta_1, \theta_2, \theta_3 | x)$. Let $\left(\theta_1^{(0)}, \theta_2^{(0)}, \theta_3^{(0)}\right)$ denote a starting value for $(\theta_1, \theta_2, \theta_3)$. Set $i = 1$.

Step 1: Simulate one value from the conditional distribution,

$$p(\theta_1 | \theta_2^{(i-1)}, \theta_3^{(i-1)}, x),$$

and call this simulated value $\theta_1^{(i)}$.

Step 2: Simulate one value from the conditional distribution,

$$p(\theta_2 | \theta_1^{(i)}, \theta_3^{(i-1)}, x),$$

and call this simulated value $\theta_2^{(i)}$.

Step 3: Simulate one value from the conditional distribution,

$$p(\theta_3 | \theta_1^{(i)}, \theta_2^{(i)}, x),$$

and call this simulated value $\theta_3^{(i)}$.

Step 4: Stop when the required number of simulations has been reached. Otherwise, set $i \leftarrow i + 1$ and go to Step 1.

In the Gibbs sampling algorithm we cycle through all p of the parameters in the parameter space $\theta_1, \theta_2, \ldots, \theta_p$, each time conditioning on all of the other components. However, we always condition on the most recently computed value. For example, when we simulate the new $\theta_2^{(i)}$ we condition on $\theta_1^{(i)}$ because it has already been sampled, but we condition on $\theta_3^{(i-1)}$ since it was last updated on the previous step. The same theory applies to Gibbs sampling: If the Markov chain

$$(\theta_1^{(1)}, \theta_2^{(1)}, \theta_3^{(1)}), \ (\theta_1^{(2)}, \theta_2^{(2)}, \theta_3^{(2)}), \ (\theta_1^{(3)}, \theta_2^{(3)}, \theta_3^{(3)}), \ldots$$

converges to a steady state, then the steady state distribution is the desired posterior distribution.

Both Gibbs sampling and the Metropolis–Hastings algorithm, in theory, converge so that the steady-state distribution is the posterior distribution we are seeking. Thus, the choice of which to use is sometimes arbitrary and may depend on the complexity of the full conditional distributions. As we've mentioned, there are some software packages that will run the MCMC simulations automatically. Most of the older packages such as OpenBUGS, WinBUGS, and JAGS apply Gibbs sampling.[3]

More recent MCMC engines use enhancements of Gibbs sampling and the Metropolis–Hastings algorithm. Stan, named for the father of computer simulation, Stanislaw Ulam, uses what is called Hamiltonian dynamics, which borrows ideas from mechanics to jump more quickly to the region of high posterior probability. No matter what engine is used to perform the MCMC simulations, the important thing to keep in mind is that we have to sample sufficiently many times so that we can be confident that the steady state has been reached. Once this happens, we simulate additional runs to learn what the posterior distribution looks like. In simple models, the posterior distribution can be derived analytically, but as models get more complex we must rely on simulation in order to gain information about the posterior.

---

[3] The BUGS in OpenBUGS and WinBUGS stands for "Bayesian Using Gibbs Sampling." JAGS stands for "Just Another Gibbs Sampler."

15.8 Hierarchical Bayes Models

Most of what we've done so far involves data from some distribution and parameters of that distribution that must be estimated. This includes the i.i.d. case, such as

$$X_1, X_2, \ldots, X_n \sim \text{i.i.d. } N(\mu, \sigma^2),$$

and even the non i.i.d. case such as

$$Y_i \overset{\text{ind.}}{\sim} N(\beta_0 + \beta_1 x_i, \sigma^2).$$

The classical (frequentist) approach and the Bayesian approach diverge in the way we envision parameters. In the classical approach the parameters are fixed but unknown constants. Bayesians treat parameters as uncertain and assess the uncertainty in probabilistic terms.

There are situations where the observed variability is more complex than this simple data–parameters model would suggest. In many cases it is helpful to think of some process that is itself random and then view our data conditionally given this process. Such thinking leads us to hierarchical models, which we will treat from the Bayesian perspective. These ideas are presented in Figure 15.20.

To make these ideas concrete, let's consider a problem from baseball. All you need to know about baseball is that players attempt to get a hit every time they go to bat. Attempts are called "at bats" and successes are called "hits." A player will often get several hundred at bats in a given season. A player's "batting average" is the number of hits divided by the number of at bats. You should recognize that this sounds like the idea of Bernoulli trials we discussed in Chapter 6. The assumption of constant probability of success is questionable, however, because the chance of getting a hit depends on the strength of the opposing pitcher; good pitchers allow fewer hits. We will assume the usual conditions behind Bernoulli trials for each player.

Let θ_i denote the probability that player i will get a hit on any at bat. The problem is this: Given a very limited amount of data at the beginning of the season, estimate the (unknown) probability θ_i. This of course varies from player to player, with the number being higher for better players. An average player will have a batting average (H/AB) of about 0.250 (i.e., they get a hit about one time in four at bats), but good players will have a batting average of 0.300 or higher.

Let n_i denote the number of at bats (trials) and let X_i denote the number of hits (successes). Early in the season, players will not have had many at bats, so n_i will be small and, consequently, the simple estimate $\hat{\theta}_i = X_i/n_i$ has a lot of variability. We consider here the data from the 2018 Major League Baseball season. One player from the Pittsburgh Pirates named Austin Meadows had 18 hits in 44 at bats on May 31 for a 0.409 batting average. This was the highest batting average in Major League Baseball at that time. The at bat total of 44 is low even for this early in the season. Many players will have had 150–200 at bats at this point. Should we believe that he is a 0.400 hitter? No player in Major

Hyperparameters $g(\theta|\phi)$

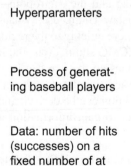

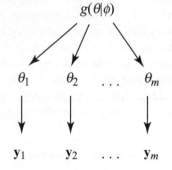

Process of generating baseball players $\theta_1 \quad \theta_2 \quad \ldots \quad \theta_m$ **Figure 15.20** The data–process–parameters paradigm.

Data: number of hits (successes) on a fixed number of at bats (trials) $\mathbf{y}_1 \quad \mathbf{y}_2 \quad \ldots \quad \mathbf{y}_m$

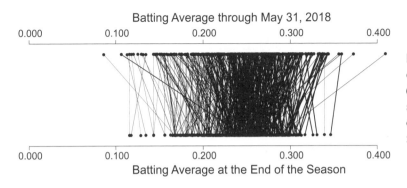

Figure 15.21 Batting averages of all players on May 31, 2018 (top) compared to the end-of-season averages (bottom). Lines connect the same player's two numbers.

League Baseball has hit 0.400 or higher since Ted Williams in 1941, so it is likely that his true θ_j is quite a bit less than 0.400.

As it turned out, his batting average at the end of the season was 0.287, higher than normal, but much closer to the average of all players. Conversely, a player who is at the low end on May 31 is likely to rebound and end up with a higher batting average. Thus, the end-of-season averages are shrunk toward the overall mean. Figure 15.21 illustrates the situation. The top shows the batting averages of all 419 players who had at least 25 at bats by May 31, and the bottom shows the end-of-season averages. Lines connect the players' May 31 averages with their end-of-season averages. We can clearly see this shrinkage toward the mean.

So, how would we model this situation? Let's envision a process that produces baseball players. Specifically, the process produces the probabilities of players getting hits; this is what we've called $\theta_1, \theta_2, \ldots, \theta_n$. These values are unobservable. Statisticians call these effects *latent*. We assume that these come from some distribution $p(\theta|\phi)$, where ϕ is itself unknown and possibly a vector of parameters. Even though we can't observe θ_j, we can observe the outcomes of random variables that depend on θ_j.

Since $\theta_1, \theta_2, \ldots, \theta_n$ are probabilities, we assume that they are selected from a BETA prior distribution. Since we don't know the parameters of this distribution, we assume a BETA(a,b) distribution and treat a and b as unknown parameters. These parameters will themselves have prior distributions if we take the Bayesian approach. The θ_is represent the *process*; the parameters, called hyperparameters in a hierarchical model like this, are then a and b. The data consist of the number of hits for player i out of his n_i at bats. If we denote this by X_i, then we can assume $X_i \sim \text{BIN}(n_i, \theta_i)$. Our model can then be described as follows:

$$
\begin{aligned}
(a,b) &\sim p(a,b) \\
(\theta_1, \theta_2, \ldots, \theta_n)|(a,b) &\sim i.i.d.\ \text{BETA}(a,b) \\
X_i|\theta_i &\overset{\text{ind.}}{\sim} \text{BIN}(n_i, \theta_i), \qquad i = 1,2,\ldots,419.
\end{aligned}
\tag{15.32}
$$

Here, $p(a,b)$ is the hyperprior for a and b which we took to be independent diffuse exponential distributions. We sent the code for the model into an MCMC engine (we used Nimble, described in the appendix to this chapter[4]) and turned it loose. The Nimble code is

```
library(nimble)
##   Likelihood
for (i in 1:n) {
```

---

[4]One R package that performs the MCMC simulations is called `nimble`. In R, variable names are case-sensitive, so the package name is in all lower case. The documentation on the `nimble` package describes the package as NIMBLE. There are other MCMC engines that have links to R, including WinBUGS (Spiegelhalter *et al.*, 2003) and JAGS (Plummer, 2012). The `nimble` package is described in NIMBLE Development Team (2022).

```
  H[i] ~ dbin( theta[i] , AB[i] )
}
##   Prior
for (i in 1:n){
  theta[i] ~ dbeta( a , b )
}
##   Hyper Priors
a ~ dexp(1/100)
b ~ dexp(1/100)
```

The R package `nimble` is described in the appendix to this chapter. We used Nimble to run the above code.

In hierarchical models like this, the hyperparameters are often the hardest to get to converge. Figure 15.22(a) and (b) shows the trace plots for a and b. As you can see, the successive values of the simulations are not independent. When a value is on the high side, the next few tend to also be on the high side. It is the same for lower values. This phenomenon is called *autocorrelation* or *serial correlation* and is discussed more thoroughly in Chapter 16. When we have high autocorrelation like this, it takes many more simulations to provide satisfactory information about the posterior. What is often done in cases like this is that users take every kth simulation and discard all those in between. This is called *thinning*, and in fact we thinned by taking every tenth simulation; had we not done this, we would have had even stronger autocorrelation.

Figure 15.22(c) shows the pairs of (a, b) and it shows our approximation to the joint posterior distribution. There is clearly a very high correlation between a and b. Most of the joint posterior probability for a and b lies roughly along the line $b = 3a$, suggesting that we are fairly confident that b is about three times a, but we are rather uncertain about whether this is $a = 60$, $b = 180$ or $a = 120$, $b = 360$, or something in the middle. Recall that the mean of the beta distribution is $a/(a+b)$, and if $b \approx 3a$, then

$$\mu \approx \frac{a}{a+b} = \frac{a}{a+3a} = \frac{1}{4} = 0.250.$$

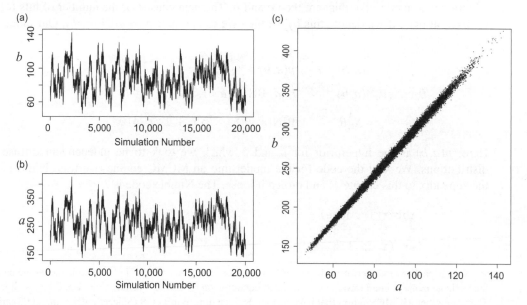

Figure 15.22 Hyperparameters in baseball data.

Thus, we're pretty confident that the average across all players is about 0.250, but there is some uncertainty about the spread of θs about this number.

We applied Gibbs sampling using Nimble in this 421-dimensional problem (419 θs and the two hyperparameters, a and b). This means that the algorithm cycled through all 421 parameters, simulating from the conditional distribution of each parameter given the most recently simulated value from the other 420 parameters. It did this 200,000 times, thinning by a factor of 10, to yield the simulated values of the Markov chain. This was done after a burn-in of 10,000 simulations. The estimates of the parameters were taken to be the posterior means, which were estimated by the average of the simulations after the burn-in. This yielded

$$\hat{a} = 86.7364 \quad \text{and} \quad \hat{b} = 261.0668.$$

The posterior means of the 20,000 saved values of the Markov chain were then used as point estimates for the θs. The May 31 batting averages and the hierarchical Bayes estimates are shown in Figure 15.23. Austin Meadows, mentioned earlier, had a batting average of 0.409 on May 31, and his Bayes estimate of the probability of getting a hit turned out to be 0.270.

If, on the morning of June 1, 2018, we were to make a prediction about what his average would be at the end of the year, we could go with one of two predictions: his current average of 0.409, or his hierarchical Bayes estimate of 0.270. If we use data from Austin Meadows' performance, and his performance only, we would have no other data except his current average of 0.409; this would be our prediction. If we take into account the variability of Major League Baseball players and the process that generates these, we could apply the hierarchical Bayes method and shrink this, and all other estimates of θ_i, toward the overall mean. This would yield a prediction of 0.270, much closer to the overall mean of 0.250. His actual batting average at the end of the year was 0.287, so our hierarchical Bayes estimate of 0.270 was much closer to his end-of-season average than his average of 0.409 on May 31.

Across all 419 players with 25 or more at bats on or before May 31, the mean squared prediction error using the hierarchical Bayes method is

$$\text{MSPE} = \sum_{i=1}^{419} (\hat{\theta}_i^{(\text{HB})} - \theta^{(\text{end})})^2 = 0.00097$$

and the mean squared prediction error using the May 31 average as the predictor is

$$\text{MSPE} = \sum_{i=1}^{419} (\hat{\theta}_i^{(\text{May 31})} - \theta^{(\text{end})})^2 = 0.00113.$$

Thus, using the hierarchical Bayes estimate to predict a player's average at the end of the season, given data through May 31, does a slightly better job of predicting the end-of-season average than the player's batting average on May 31.

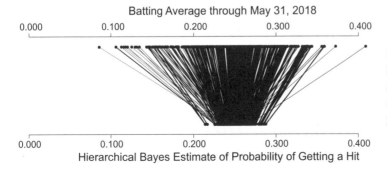

Figure 15.23 Results of the hierarchical Bayes model for batting averages. Point estimates for θ_i, $i = 1, 2, \ldots, 419$ are taken to be the posterior means.

■ **Example 15.17** A school district is studying the effect of the school and the teacher on students' performance. Suppose that a school district has m schools, and within school i, $i = 1, 2, \ldots, m$ there are t_i teachers. Within teacher j's class ($j = 1, 2, \ldots, t_i$) there are n_{ij} students. Describe a hierarchical model that might be used to model the effects due to schools and teachers on student performance.

Solution: Let $Y_{i,j,k}$ be the score of student k in the class of teacher j within school i. We might assume that, given the teacher and the school, the scores are normally distributed with some mean (which depends on the school and teacher) and a constant variance across all schools and teachers. That is,

$$Y_{i,j,k} \overset{\text{ind.}}{\sim} N(\delta_{jk}, \sigma^2).$$

Within school j we might assume that the δs make up a random sample:

$$\delta_{j1}, \delta_{j2}, \ldots, \delta_{j,t_j} \sim N(\theta_j, \sigma^2_{(t)}).$$

The schools then make up a random sample from a normal distribution:

$$\theta_1, \theta_2, \ldots, \theta_m \sim \text{i.i.d. } N(\mu, \sigma^2_{(s)}).$$

Finally, if we take a Bayesian approach, we would assign priors to the remaining parameters, σ^2 and $\sigma^2_{(t)}$, as well as the hyperparameters μ and $\sigma^2_{(s)}$. The model is illustrated in Figure 15.24.

This model could then be run using an MCMC engine such as Nimble. All of the parameters, including all of the δ_{jk}s and θ_is, which are called the *random effects* as well as the parameters μ, σ^2, $\sigma^2_{(t)}$, and $\sigma^2_{(s)}$, would be estimated in one overall model. The study might focus on the variability of the schools, measured by $\sigma^2_{(s)}$ and by the variability of the teachers within each school. ■ ■ ■

Appendix: Software for Bayesian Computations

MCMC algorithms that use the Metropolis–Hastings algorithm or Gibbs sampling are available in several R packages, or in standalone software packages to which R can link. One of the first was BUGS (Bayesian Using Gibbs Sampling; Gilks *et al.*, 1994) and its descendants, OpenBUGS (Gilks *et al.*, 1994; Spiegelhalter *et al.*, 2014) and WinBUGS (Spiegelhalter *et al.*, 2003). Another popular package is JAGS

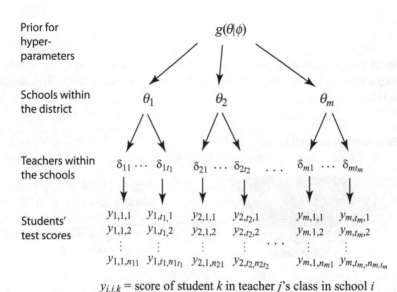

Figure 15.24 Hierarchical structure for students' achievement. There are m schools within a district and t_i teachers in school i. There are n_{ij} students in teacher j's class within school i.

$y_{i,j,k}$ = score of student k in teacher j's class in school i

(Just Another Gibbs Sampler; Plummer, 2012). When a user calls the WinBUGS or JAGS function, R essentially throws the important information "over the fence" to WinBUGS or JAGS, which does the required computations and throws the resulting simulations back "over the fence" to R. Typically, R is then used to summarize the results. Nimble works inside of R.

We must send to Nimble the following information:
1. the model;
2. the data;
3. all constants, including the sample sizes and predictor variables (in a regression);
4. a vector of strings that names the parameters to be monitored;
5. initial values for each of the parameters being monitored;
6. the number of simulations to use for burn in; and
7. the total number of simulations to run, counting the burn in.

We will illustrate these items in the following example. The code is available on the book's website.

■ **Example 15.18** Consider the 20 cholesterol measurements from 1952 given by Dixon and Massey (1983). Use `nimble` to approximate the posterior distributions of the population mean μ and precision τ (the reciprocal of the variance) assuming a $N(200, 20^2)$ prior distribution for μ and a $\text{GAM}(1, 0.01)$ prior for τ. Note that this is not the conjugate prior described in Section 15.5.

Solution: The code is essentially a string, with lines written in the Nimble language. Nimble, WinBUGS, and JAGS have very similar syntax. We must apply the `nimbleModel()` function with the code enclosed in braces. Within the braces, we must give the distribution for every random variable in our problem; in this example, the cholesterol measurements are 20 realizations of random variables. The tilde ($\sim$) is used to indicate "has distribution." Most distributions that we have studied in this book are *named* distributions in Nimble, and all begin with the letter "d." The first `for` loop, which has the same structure as the `for` loop in R, says that all 20 observations came from a normal distribution with mean `mu` and precision `tau`. Nimble, WinBUGS, and JAGS always use the precision, rather than the variance, for the normal distribution. The next two lines give the prior distributions for the parameters `mu` and `tau`. The very last line in the code snippet below closes the brace and the function's parenthesis.

```
code = nimbleCode({
        for (i in 1:N) {
          x[i] ~ dnorm( mu , tau )
        }
        mu ~ dnorm( 200 , 1/400 )
        tau ~ dgamma( 1 , 0.001 )
})
```

Next, we must define the data and the constants. Nimble makes a clear distinction between data and constants: Data are realizations of random variables, whereas constants are fixed numbers. Note that covariates or predictor variables are considered fixed numbers, therefore they are constants. The last line shown below combines all constants and all data into separate lists. Here there is only one constant and one vector of data, but in a more complicated problem there could be many.

```
N = 20
x = c( 240,243,250,254,264,279,284,285,290,298,302,
       310,312,315,322,337,348,384,386,520 )
constants = list( N = N )
data = list( x = x )
```

We must also tell Nimble which parameters to keep track of as it performs the simulations. We must also tell it how many iterations we want in total, as well as the number of burn-in samples. In the code below, we are asking for a burn-in of 500 and a total of 5,000, leaving just 4,500 simulations for inference.

```
parameters = c( "mu" , "tau" )
inits = list( mu = 200 , tau = 1 )
niter = 5000
nburn = 500
```

The next several steps involve a sequence of function calls to various Nimble functions in order to compile and run the model. We have defined a function called `run.nimble()` which does all of these in a single function.

```
run.nimble = function(code,data,constants,inits,parameters,niter,nburn)
{
  model = nimbleModel( code, data=data , constants=constants , inits=inits)
  compile1 = compileNimble( model )
  config = configureMCMC( compile1 , print = FALSE )
  config$addMonitors( parameters )
  mcmc = buildMCMC( config )
  compile2 = compileNimble( mcmc , project=model )
  samples = runMCMC( compile2 , niter = niter, nburnin = nburn,
                   inits = inits, nchains = 1, samplesAsCodaMCMC = TRUE )
  return( samples )
}
```

Once all of this is set up, it takes just one line to have Nimble run the Markov chain simulations:

```
samples = run.nimble(code,data,constants,inits,parameters,niter,nburn)
```

This stores the simulations in a matrix called `samples` whose columns have the names of the parameters. In this example, the columns are named `"mu"` and `"tau"`. We can check convergence by looking at time series plots of `samples[,"mu"]` and `samples[,"tau"]`. These plots, which are shown in Figure 15.25, suggest that the steady state has been reached, and we can treat these 4,500 samples as coming from the posterior distribution, although they are not *independent* samples. Histograms are shown in Figure 15.26. These plots tell us what we believe about the parameters μ and τ given the joint prior that we selected and the data. Point estimates of the parameters can be obtained by taking the mean of the 4,500 simulations:

$$\hat{\mu} = 291.01 \quad \text{and} \quad \hat{\tau} = 0.000247.$$ ■ ■ ■

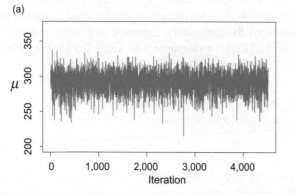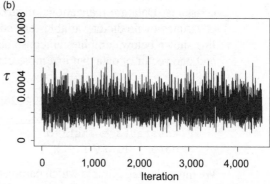

Figure 15.25 Trace plots for the model in Example 15.18.

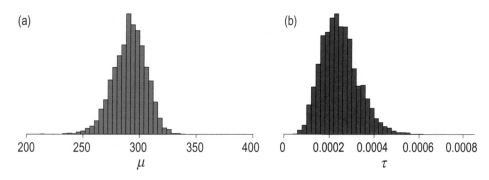

Figure 15.26 Posterior histograms for the model in Example 15.18.

15.9 **Chapter Summary**

Before this chapter, we've taken the position that parameters are fixed numbers. Even when we don't know these values, we still don't consider them to be random variables. The Bayesian approach is that we express uncertainty in terms of randomness. Thus, if we don't know the value of a parameter, we express our knowledge (or ignorance) of it in terms of a probability distribution. The Bayesian approach takes unknown parameters to be random variables.

The prior distribution is defined to be the distribution of our belief about a parameter *before* observing data. The posterior distribution is defined to be the distribution of our belief about a parameter *after* observing data. The two are related through Bayes' Theorem, which says that the posterior is proportional to the prior times the likelihood function.

In simple situations we can use what we know about distributions to express the posterior distribution exactly. In complicated situations this analytic approach is not possible and we must use recently developed approaches, such as MCMC, to *simulate* from the posterior distribution. The reasoning is that if you can't get a usable formula for the posterior distribution, the next best thing is to have many (thousands) of simulations from the posterior. We can then base our inference on these simulations.

15.10 Problems

Problem 15.1 Suppose that medical devices are made in four factories, which we'll call A, B, C, and D. These four factories make 10%, 40%, 20%, and 30% of the total output, respectively. The defect rates vary across the four factories and are given in the table below:

Factory	A	B	C	D
% of production	10%	40%	20%	30%
Defect rate	2%	2%	4%	3%

The outputs from these four factories are sent to a distribution center where they get mixed together. One item is selected at random from this distribution center.

(a) What is the probability that the item came from factory A?

(b) What is the probability that the item is defective?

(c) Given that the item is defective, what is the probability that it came from factory A?

(d) Repeat part (c) for factories B, C, and D.

(e) Must the "percent of production" values for the four factories add up to 100%? Explain.

(f) Must the "defect rate" values for the four factories add up to 100%? Explain.

Problem 15.2 Suppose that medical devices are made in four factories, which we'll call A, B, C, and D. These four factories make 50%, 30%, 5%, and 15% of the total output, respectively. The defect rates vary across the four factories and are given in the table below:

Factory	A	B	C	D
% of production	50%	30%	5%	15%
Defect rate	1%	4%	8%	3%

The outputs from these four factories are sent to a distribution center where they get mixed together. One item is selected at random from this distribution center.

(a) What is the probability that the item came from factory A?

(b) What is the probability that the item is defective?

(c) Given that the item is defective, what is the probability that it came from factory A?

(d) Repeat part (c) for factories B, C, and D.

(e) Must the "percent of production" values for the four factories add up to 100%? Explain.

(f) Must the "defect rate" values for the four factories add up to 100%? Explain.

Problem 15.3 Suppose that a test for a disease has a sensitivity of 0.98 and a specificity of 0.99. The base rate for the disease in the population is 0.004. A person is selected at random from the population.

(a) What is the probability that the person has the disease?

(b) What is the probability that the person tests positive for the disease?

(c) What is the probability that the person has the disease given that he tested positive?

(d) What is the probability that the person does not have the disease given that he tested negative?

Problem 15.4 Repeat Problem 15.3 assuming that the base rate is 0.15.

Problem 15.5 Suppose that the results of a test for a disease can be positive, inconclusive, or negative, which we label as $+$, 0, or $-$, respectively. If a person has the disease, then the probabilities of testing $+$, 0, or $-$ are

$$\Pr(+|D) = 0.80$$
$$\Pr(0|D) = 0.15$$
$$\Pr(-|D) = 0.05.$$

If a person does not have the disease, then the probabilities are

$$\Pr(+|ND) = 0.01$$
$$\Pr(0|ND) = 0.02$$
$$\Pr(-|ND) = 0.97.$$

Assume that the base rate for the disease in the population is 0.008.

(a) If a person tests positive, what is the probability that she has the disease?

(b) If a person tests inconclusive, what is the probability that she has the disease?

(c) If a person tests negative, what is the probability that she has the disease?

Problem 15.6 Repeat Problem 15.5 assuming that the base rate is 0.08.

Problem 15.7 Suppose I have three six-sided dice: one has the usual configuration of the numbers 1 through 6, the second has all 1s, and the third has three 1s and three 6s. Cutouts of the dice are shown below:

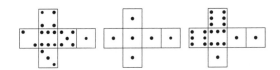

One die is selected at random and tossed. Find the following probabilities:

(a) The die was the one with the usual configuration (i.e., the first one) given that the roll was a 1.

(b) The die was the one with all 1s given that the roll was a 1.

(c) The die was the one with three 1s and three 6s given that the roll was a 1.

(d) The die was the one with the usual configuration given that the roll was a 2.

(e) The die was the one with all 1s given that the roll was a 2.

(f) The die was the one with three 1s and three 6s given that the roll was a 2.

(g) The die was the one with the usual configuration given that the roll was a 6.

(h) The die was the one with all 1s given that the roll was a 6.

(i) The die was the one with three 1s and three 6s given that the roll was a 6.

Problem 15.8 Suppose the die selected at random as described in Problem 15.7 is tossed twice and the result was $\{1, 6\}$. Find the probability that

(a) the die was the one with the usual configuration;

(b) the die was the one with all 1s;

(c) the die was the one with three 1s and three 6s.

Problem 15.9 Suppose I have two six-sided dice: one has the usual configuration of the numbers 1 through 6, the second has three 1s and three 6s Cutouts of the dice are shown below:

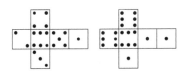

One die is selected at random and tossed four times. Find the probability that the die with the usual configuration (i.e., the first one) was selected given that the four tosses resulted in three 1s and one 6.

Problem 15.10 Consider the information in Problem 15.9 but suppose the outcomes of the four rolls were $\{3, 6, 1, 1\}$.

(a) Apply Bayes' Theorem to determine the probability that the die was the one with the usual configuration.

(b) Explain, without using any mathematics, why the correct answer is what you obtained in part (a).

Problem 15.11 Users arrive at a web site at a rate of λ hour. It is reasonable to assume that the times between events have an exponential distribution with PDF

$$f(t|\lambda) = \lambda e^{-\lambda t}, \quad t \geq 0,$$

where λ is the unknown rate. Before observing any data, we believe that λ will be 6 or 10 with equal probability. Suppose that the first five interarrival times were 0.3, 0.2, 0.1, 0.3, 0.1.

(a) Find the posterior distribution for λ.

(b) Find the posterior mean.

(c) Find the posterior standard deviation.

Problem 15.12 Repeat Problem 15.11 assuming the prior has equal weight on $\lambda = 2, 4, 6, 8, 10, 12$.

Problem 15.13 Repeat Problem 15.11 assuming the prior has a continuous distribution with PDF

$$\pi(\lambda) = \frac{1}{3}\lambda \exp(-\lambda/3), \quad \lambda > 0.$$

Problem 15.14 Repeat Problem 15.11 assuming the prior has a continuous distribution with PDF

$$\pi(\lambda) = \frac{1}{10^{10}}, \quad 0 < \lambda < 10^{10}.$$

Problem 15.15 Suppose that we have a sample of size $n = 10$ where we measured systolic blood pressure in patients with prostate cancer.

(a) If we assume a joint prior of the form

$$(\mu, \phi) \sim \text{NInv}\chi^2(\mu_0 = 130, \kappa_0 = 2, \phi_0 = 10, \nu_0 = 1),$$

what is the marginal prior for ϕ?

(b) Given ϕ, what is the conditional prior for μ?

(c) If the sample yields the data

156,	108,	129,	122,	128,
140,	135,	120,	139,	132

find the joint posterior for (μ, ϕ).

(d) Find and plot the marginal posterior distributions for μ and ϕ.

(e) Find the posterior means for μ and ϕ and use these as point estimates.

Problem 15.16 Suppose that in Problem 15.15 the sample size was $n = 200$ and it produced data with summary statistics $\bar{x} = 137.0$ and $s^2 = 115.0$. If we assume the same prior as in Problem 15.15, explain why this summary information is enough to find the joint and marginal posterior distributions. Then repeat parts (c), (d), and (e) of Problem 15.15.

Problem 15.17 Repeat Problem 15.15 with the same $n = 10$ data points, but this time assume the Jeffreys prior

$$p(\mu, \phi) = \frac{c}{\phi}, \quad \phi > 0.$$

Problem 15.18 Repeat Problem 15.15 with $n = 200$ and the summary statistics $\bar{x} = 137.0$ and $s^2 = 115.0$. Assume the Jeffreys prior as given in Problem 15.17

Problem 15.19 The number of visitors to a website is measured from 7:00 p.m. to 8:00 p.m. on 10 consecutive days. The data are

234, 197, 222, 260, 287,
178, 225, 214, 253, 209

(a) These numbers are constrained to be integers. Is the normal distribution a reasonable assumption here? Construct a normal probability plot to assess how reasonable this assumption is.

(b) For the rest of this problem, assume that the normal distribution is reasonable. With the $\text{NInv}\chi^2$ prior

$$(\mu, \phi) \sim \text{NInv}\chi^2(\mu_0 = 250, \kappa_0 = 2, \phi_0 = 20, \nu_0 = 1),$$

what is the prior for ϕ?

(c) Given ϕ, what is the conditional prior for μ?

(d) Find the joint posterior for (μ, ϕ).

(e) Find and plot the marginal posterior distributions for μ and ϕ.

(f) Find the posterior means for μ and ϕ and use these as point estimates.

Problem 15.20 Suppose that in Problem 15.19 the sample size was $n = 400$ and it produced data with summary statistics $\bar{x} = 230.0$ and $s^2 = 900$. If we assume the same prior as in Problem 15.19, explain why this summary information is enough to find the joint and marginal posterior distributions. Then repeat parts (c), (d), and (e) of Problem 15.19.

Problem 15.21 Repeat Problem 15.19 with the same $n = 10$ data points, but this time assume the Jeffreys prior

$$p(\mu, \phi) = \frac{c}{\phi}, \quad \phi > 0.$$

Problem 15.22 Repeat Problem 15.19 with $n = 400$ and the summary statistics $\bar{x} = 230.0$ and $s^2 = 900$. Assume the Jeffreys prior

$$p(\mu, \phi) = \frac{c}{\phi}, \quad \phi > 0.$$

Problem 15.23 Consider again the sample of size $n = 10$ where we measured systolic blood pressure in patients with prostate cancer. This data set was considered in Problem 15.15. This time apply MCMC using Nimble.

(a) Assume independent priors $\mu \sim N(140, 20^2)$ and $\tau \sim \text{GAM}(1, 0.01)$. Use Nimble to approximate the joint posterior distribution of μ and τ.

(b) Approximate the posterior means for μ and ϕ and use these as point estimates.

Problem 15.24 Repeat Problem 15.11 with MCMC using Nimble.

Problem 15.25 Refer to Example 15.15. Fill in the steps necessary to obtain the result that

$$E\left(-\frac{\partial^2}{\partial \theta^2} \log f(X|\theta)\right) = \frac{n}{\theta(1-\theta)}.$$

Problem 15.26 Find the Jeffreys prior for a sample from the Poisson distribution.

Problem 15.27 Sudden infant death syndrome (SIDS) occurs when a child under one year old suddenly dies, usually while sleeping. In a famous study (Symons *et al.*, 1983) researchers determined the number of births and the number of SIDS cases in each of the 100 counties in North Carolina in 1979. Part of the data set is shown below.

County	No. births	SIDS cases
Alamance	5,767	11
Alexander	1,683	2
⋮	⋮	⋮
Yancey	869	1

The SIDS rate varied across counties, and some counties, such as Ashe, even had a rate of zero because there were no SIDS cases in 1979. Suggest a hierarchical model for this data set as in Figure 15.20 and in (15.32).

Problem 15.28 Suppose that investigators are interested in testing a new drug to treat seizures. From among a cohort of $n = 100$ volunteers, 50 are given a new drug and the other 50 are given a placebo. The number of seizures in each week that the subject participated in the study was recorded. Note that the number of weeks the subjects participated in the study varied from person to person. A sample of the data is shown below. Suggest a hierarchical model for this data set as in Figure 15.20 and in (15.32). If you were interested in whether the new drug was effective in reducing the rate of seizures, what parameters or function of parameters would you be interested in?

Subject	Drug	No. seizures
1	Placebo	4,0,0,6,5,7,1,12,5
2	New drug	3,1,0,0,1,0
3	New drug	2,3,0,1,0
4	Placebo	7,8,7,2,6,11,6,2,4,9
⋮	⋮	⋮

16 — Time Series Methods

16.1 Introduction

Forecasting is an important problem that spans many fields, including business and industry, government, economics, environmental sciences, medicine, social science, politics, and finance. Forecasting problems are often classified as short term, medium term, and long term. Short-term forecasting problems involve predicting events only a few time periods (days, weeks, and months) into the future. Medium-term forecasts extend from 1 to 2 years into the future, and long-term forecasting problems can extend beyond that by many years. Short- and medium-term forecasts are required for activities that range from operations management to budgeting and selecting new research and development projects. Long-term forecasts impact issues such as strategic planning. Short- and medium-term forecasting is typically based on identifying, modeling, and extrapolating the patterns found in historical data. Because these historical data usually exhibit inertia and do not change very quickly, statistical methods are useful for short- and medium-term forecasting. This chapter is about the use of these statistical methods.

Most forecasting problems involve the use of time series data. A time series is a chronological sequence of observations on a variable of interest. For example, Figure 16.1 shows a graph of 100 hourly readings of the viscosity of a product from a chemical process. This graph is called a time series plot. The viscosity variable is collected at equally spaced time periods, as is typical in most time series modeling and forecasting applications. Time series plots are useful for displaying many common features of data, such as trends or seasonality. Figure 16.2 shows that the viscosity data varies around a long-term average of approximately 85 cP,[1] with a structured and not entirely random pattern. This pattern is typical of data that are positively autocorrelated; that is, a value above the long-run average tends to be followed by other values above the average, while a value below the average tends to be followed by other values below the average. We will discuss ways to measure autocorrelation later in this chapter. The viscosity data are shown in Table 16.1.

Many business applications of time series analysis utilize hourly, daily, weekly, monthly, quarterly, or annual data, but any reporting interval may be used. Furthermore, the data may be instantaneous, such as the viscosity of the chemical product in Figure 16.1, at the point in time where it is measured; it may be cumulative, such as the total sales of a product during the month; or it may be a statistic that in

---

[1]The unit of measurement for viscosity is usually taken to be the *centipoise*, abbreviated cP. This unit has dimensions of force × time/area.

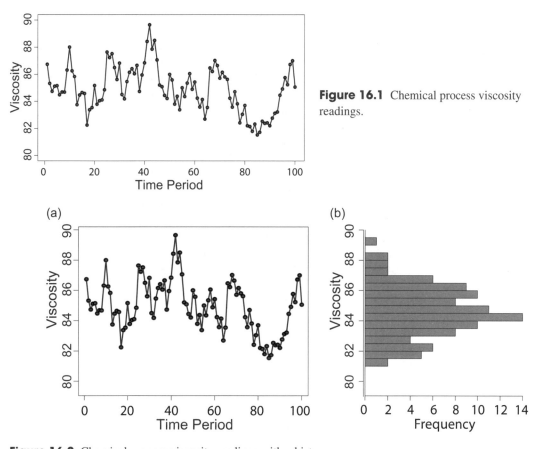

Figure 16.1 Chemical process viscosity readings.

Figure 16.2 Chemical process viscosity readings with a histogram

some way reflects the activity of the variable during the time period, such as the daily closing price of a specific stock on the New York Stock Exchange.

Some of the classical tools of descriptive statistics, such as the histogram and the stem-and-leaf plot, are not particularly useful for time series data because they do not take time order into account. To illustrate this, Figures 16.2 and 16.3 show time series plots for viscosity readings (shown in Table 16.1) and data on monthly beverage production shipments. At the right-hand side of each time series plot is a histogram of the data. Note that while these time series display very different characteristics, the histograms are remarkably similar. Essentially, the histogram summarizes the data across the time dimension, and in so doing, the key time-dependent features of the data are lost. Stem-and-leaf plots and box plots would have the same issues, losing time-dependent features.

Sometimes, it is useful to overlay a smoothed version of the original data on the original time series plot to help reveal patterns in the original data. There are several types of data smoothers that can be employed. One of the simplest and most widely used is the ordinary or simple moving average. A *simple moving average* of span N assigns weights $1/N$ to the most recent N observations $y_t, y_{t-1}, \ldots, y_{t-N+1}$, and weight zero to all other observations. If we let M_t be the moving average, then the *N-span moving average* at time period t is

$$ M_t = \frac{y_t + y_{t-1} + \cdots + y_{t-N+1}}{N} = \frac{1}{N} \sum_{s=t-N+1}^{t} y_s. $$

If the variance of an individual observation is σ^2 and the observations are uncorrelated, then the variance of the moving average is simply σ^2/N. Figure 16.4 shows a moving average of span $N = 5$ for the

Table 16.1 Chemical process viscosity data across 100 time periods.

Time	Viscosity	Time	Viscosity	Time	Viscosity	Time	Viscosity
1	86.7418	26	87.2397	51	85.5722	76	84.7052
2	85.3195	27	87.5219	52	83.7935	77	83.8168
3	84.7355	28	86.4992	53	84.3706	78	82.4171
4	85.1113	29	85.6050	54	83.3762	79	83.0420
5	85.1487	30	86.8293	55	84.9975	80	83.6993
6	84.4775	31	84.5004	56	84.3495	81	82.2033
7	84.6827	32	84.1844	57	85.3395	82	82.1413
8	84.6757	33	85.4563	58	86.0503	83	81.7961
9	86.3169	34	86.1511	59	84.8839	84	82.3241
10	88.0006	35	86.4142	60	85.4176	85	81.5316
11	86.2597	36	86.0498	61	84.2309	86	81.7280
12	85.8286	37	86.6642	62	83.5761	87	82.5375
13	83.7500	38	84.7289	63	84.1343	88	82.3877
14	84.4628	39	85.9523	64	82.6974	89	82.4159
15	84.6476	40	86.8473	65	83.5454	90	82.2102
16	84.5751	41	88.4250	66	86.4714	91	82.7673
17	82.2473	42	89.6481	67	86.2143	92	83.1234
18	83.3774	43	87.8566	68	87.0215	93	83.2203
19	83.5385	44	88.4997	69	86.6504	94	84.4510
20	85.1620	45	87.0622	70	85.7082	95	84.9145
21	83.7881	46	85.1973	71	86.1504	96	85.7609
22	84.0421	47	85.0767	72	85.8032	97	85.2302
23	84.1023	48	84.4362	73	85.6197	98	86.7312
24	84.8495	49	84.2112	74	84.2339	99	87.0048
25	87.6416	50	85.9952	75	83.5737	100	85.0572

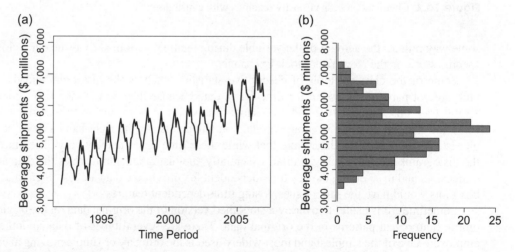

Figure 16.3 Time series plot and histogram of monthly beverage shipments.

viscosity data. The smoothed version of the data from the five-span moving average is shown as the red line without points in Figure 16.4. The plot's smooth curve does a reasonably good job of showing the pattern of points above and below the long-term average of about 85 cP.

The five-span moving average described here uses the current data value and four past data points. This means that we cannot calculate M_1, M_2, M_3, and M_4 because they involve values that we don't know:

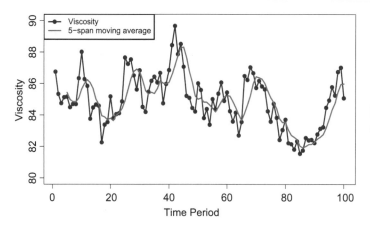

Figure 16.4 A moving average of span $N = 5$ for the viscosity data

specifically, the values of the time series at $t = -3, -2, -1$, and 0. On the other hand, we are able to calculate the moving average at the current time.

Another type of moving average is the *centered moving average*. This takes the average of the current data value and fixed number of values before and after the current value. For example, the five-span centered moving average is

$$CM_T = \frac{y_{T-2} + y_{T-1} + y_T + y_{T+1} + y_{T+2}}{5}.$$

For the centered moving average, it is clear that N must be odd.

The simple moving average is a *linear data smoother*, or a *linear filter*, because it replaces each observation y_t with a linear combination of the other data points that are near to it in time. The weights in the linear combination are equal, so the linear combination here is simply an average. Of course, unequal weights could be used. For example, the *Hanning filter* is a weighted, centered moving average, defined by

$$M_t^H = 0.25t_{t-1} + 0.5y_t + 0.25y_{t+1}.$$

Julius von Hann, a nineteenth-century Austrian meteorologist, used this filter to smooth weather data. An obvious disadvantage of a linear filter such as a moving average is that an unusual or erroneous data point or an outlier will dominate the moving averages that contain that observation, contaminating the moving averages for a length of time equal to the span of the filter. For example, consider the sequence of observations

15, 18, 13, 12, 16, 14 16, 17, 18, 15, 18, 200, 19, 14, 21, 24, 19, 25

which increases reasonably steadily from 15 to 25, except for the unusual value 200. Any reasonable smoothed version of the data should also increase steadily from 15 to 25 and not emphasize the value 200. Even if the value 200 is a legitimate observation, and not the result of a data recording or reporting error (perhaps it should be 20), it is so unusual that it deserves special attention and should likely not be analyzed along with the rest of the data.

Odd-span moving medians (or running medians) are effective data smoothers when a time series is contaminated with outliers or unusual vales. The *moving median of span N* is

$$m_t^N = \text{med}(y_{t-u}, y_{t-u+1}, \dots, y_{t+u}).$$

The median is the middle observation in rank order (or order of value). The moving median of span 3 is a popular and effective data smoother, where

$$m_t^3 = \text{med}(y_{t-1}, y_t, y_{t+1}).$$

This smoother would process the data three values at a time and replace the three original observations by their median. If we apply this smoother to the data above, we obtain

 __ , 15, 13, 13, 14, 16, 17, 17, 18, 18, 19, 19, 19, 21, 21, 24, __

These smoothed data are a reasonable representation of the original data, but they conveniently ignore the value 200. The first and last values are lost when using the moving median, and they are represented by "__".

In general, a moving median will pass monotone sequences of data unchanged. It will follow a step function in the data, but it will eliminate a spike or more persistent upset in the data that has duration of at most u consecutive observations. Moving medians can be applied more than once if desired to obtain an even smoother series of observations. If there are a lot of observations, the information loss from the missing end values is not serious. However, if it is necessary or desirable to keep the lengths of the original and smoothed data sets the same, a simple way to do this is to "copy on" or "add back" the end values from the original data. There are also other methods for smoothing the end values.

16.2 Using R for Time Series

R has a special object to describe time series. This object contains the values of the time series, together with the start time, end time, and frequency of the time series. A time series object can be created using the `ts()` function in R. For example, we can create a time series object for the viscosity data this way:

```
Viscosity = read.csv( "Viscosity.csv" )
n = nrow( Viscosity )
viscosity = ts( Viscosity , start=1 , end=n , frequency=1 )
```

When you deal with new classes of objects, we recommend that you always explore the properties of the object using R's `str()` (for structure) and `class()` functions. In this case, if we run the lines

```
class( viscosity )
str( viscosity )
```

we obtain the output

```
class( viscosity )
[1] "ts"
str( viscosity )
Time-Series [1:100, 1] from 1 to 100: 86.7 85.3 84.7 85.1 85.1 ...
- attr(*, "dimnames")=List of 2
..$ : NULL
..$ : chr "Viscosity"
```

From this we see that viscosity is a `ts` object and it is a list of two objects, the first of which is NULL. These `ts` objects can be sent to the generic `plot()` function in R and by default it creates a line plot:

```
plot( viscosity , type="l" , col="blue" , lwd=2 , xlab="Time Period" ,
      ylim=c(80 , 90) , ylab="Viscosity" )
```

Many of the optimal arguments in `plot()`, such as `col` and `ylim`, can be used when plotting a `ts` object. If you want points at each data value, you can add the statement

```
points( 1:n , viscosity , col="blue" , pch=19 )
```

The beverage shipment data set is a little more complex because it involves specific months in time. We must supply the `ts()` function with the start time, end time, and frequency:

```
beverage = read.csv( "BeverageShipments-1992-2006.csv" )
beverage.ts = ts(beverage,start=c(1992,1),end=c(2006,12),frequency=12)
```

The argument `start=c(1992,1)` indicates that the first time period is the first month of 1992. Similarly, `end=c(2006,12)` tells R that the end time is the 12th month in 2006. Finally, `frequency=12` indicates that there are 12 time periods per year.

16.3 Numerical Description of Time Series

An important type of time series is a stationary time series. A time series is said to be *strictly stationary* if its properties are not affected by a change in the time origin. That is, if for every $k > 0$ the joint probability distribution of the observations $y_t, y_{t+1}, \ldots, y_{t+n}$ is exactly the same as the joint probability distribution of the observations $y_{t+k}, y_{t+k+1}, \ldots, y_{t+k+n}$ then the time series is strictly stationary. When $n = 0$ the stationarity assumption means that the probability distribution of y_t is the same for all time periods and can be written as $f(y)$. The chemical viscosity time series data originally shown in Table 16.1 and Figure 16.1 is an example of a stationary time series. The time series plot shows that viscosity seems to vary around a fixed level. Based on the earlier definition, this is a characteristic of stationary time series. Now consider the Netflix stock price data in Table 16.2 and Figure 16.5. Stock data can be downloaded (assuming you have an internet connection) using `tidyquant` in R:

```
library(tidyquant)
getSymbols( "NFLX", from = '2020-01-01', to = "2022-06-30",warnings = FALSE,
            auto.assign = TRUE)
head(NFLX)
plot( NFLX$NFLX.Close )
```

The `getSymbols()` function uses the internet connection to download the data for the stock you choose. Here we chose Netflix, which has the abbreviation (also called the symbol) NFLX. We also give it the

Table 16.2 Netflix stock price from January 2, 2020 to June 29, 2022.

Date	Open	High	Low	Close	Volume	Adjusted
1/2/2020	326.10	329.98	324.78	329.81	4,485,800	329.81
1/3/2020	326.78	329.86	325.53	325.90	3,806,900	325.90
1/6/2020	323.12	336.36	321.20	335.83	5,663,100	335.83
1/7/2020	336.47	336.70	330.30	330.75	4,703,200	330.75
1/8/2020	331.49	342.70	331.05	339.26	7,104,500	339.26
1/9/2020	342.00	343.42	334.61	335.66	4,709,300	335.66
1/10/2020	337.13	338.50	327.27	329.05	4,718,300	329.05
1/13/2020	331.80	340.85	331.51	338.92	6,290,000	338.92
1/14/2020	344.40	345.38	335.52	338.69	7,199,400	338.69
1/15/2020	338.68	343.17	336.60	339.07	5,158,000	339.07
1/16/2020	343.50	343.56	335.85	338.62	5,016,000	338.62
1/17/2020	341.00	341.57	337.38	339.67	6,066,500	339.67
1/21/2020	340.00	341.00	332.59	338.11	14,350,300	338.11
1/22/2020	332.55	336.30	323.60	326.00	21,730,000	326.00
⋮	⋮	⋮	⋮	⋮	⋮	⋮
6/23/2022	180.50	182.76	175.87	181.71	9,538,200	181.71
6/24/2022	183.50	191.19	181.84	190.85	12,424,000	190.85
6/27/2022	191.77	191.78	182.39	189.14	8,363,900	189.14
6/28/2022	189.20	192.20	179.37	179.60	7,185,400	179.60
6/29/2022	179.55	180.67	175.10	178.36	5,766,800	178.36

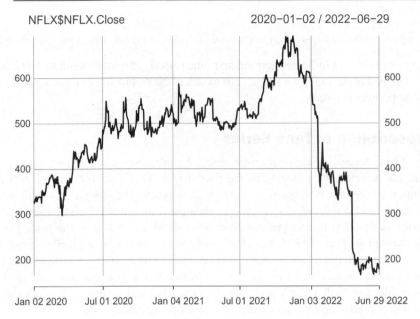

Figure 16.5 Time series plot of closing Netflix stock price from January 2, 2020 to June 29, 2022.

from and to dates for which we'd like to have data. The time series of the Netflix stock price tends to wander around or drift, with no obvious fixed level. This is behavior typical of a nonstationary time series.

Stationary implies a type of statistical equilibrium or stability in the data. Consequently, the time series has a constant mean defined in the usual way and estimated by the usual sample mean and variance. If a time series is stationary, this means that the joint probability distribution of any two observations, say, y_t and y_{t+k}, is the same for any two time periods t and $t+k$ that are separated by the same interval k. The interval k is called the *lag*. Looking closely at Figure 16.5 we observe that the pairs of adjacent observations exhibit positive correlation. That is, a small value of y tends to be followed in the next time period by another small value of y, and a large value of y tends to be followed immediately by another large value of y. The covariance between y_t and its value at another time period, say, y_{t+k}, is called the *autocovariance* at lag k, defined by

$$\gamma_k \,=\, \mathrm{cov}\,(y_t, y_{t+k}) \,=\, \mathbb{E}\big((y_t - \mu)(y_{t+k} - \mu)\big). \tag{16.1}$$

The collection of the values of γ_k, $k = 0, 1, \ldots$ is called the *autocovariance function*. Note that the autocovariance at lag $k = 0$ is just the variance of the time series; that is, $\gamma_0 = \sigma_y^2$, which is constant for a stationary time series. The *autocorrelation coefficient* at lag k for a stationary time series is

$$\rho_k \,=\, \frac{\mathrm{cov}\,(y_t, y_{t+k})}{\mathbb{V}(y_t)} \,=\, \frac{\gamma_k}{\gamma_0}. \tag{16.2}$$

The collection of the values of ρ_k, $k = 0, 1, 2, \ldots$ is called the *autocorrelation function* (ACF). Note that by definition $\rho_0 = 1$. Also, the ACF is independent of the scale of measurement of the time series, so it is a dimensionless quantity. Furthermore, $\rho_k = \rho_{-k}$ that is, the ACF is symmetric around zero, so it is only necessary to compute the positive (or negative) half.

If a time series has a finite mean and autocovariance function it is said to be second-order stationary (or weakly stationary of order 2). If, in addition, the joint probability distribution of the observations is multivariate normal, then that is sufficient to result in a time series that is strictly stationary. It is necessary to estimate the autocovariance and ACFs from a time series of finite length T. The usual estimate of the autocovariance function is

$$c_k = \hat{\gamma}_k = \frac{1}{T} \sum_{t=1}^{T-k} (y_t - \bar{y})(y_{t+k} - \bar{y}), \qquad k = 0, 1, 2, \ldots, K,$$

and the ACF is estimated by the *sample autocorrelation function* or *sample ACF*

$$r_k = \hat{\rho}_k = \frac{c_k}{c_0}, \qquad k = 0, 1, 2, \ldots, K.$$

A good general rule of thumb is that at least 50 observations are required to give a reliable estimate of the ACF, and the individual sample autocorrelations should be calculated up to lag K, where K is about $T/4$.

Often, we will need to determine whether the autocorrelation coefficient at a lag is zero. This is useful when examining the autocorrelations of the residuals from a fitted time series model. If a time series is uncorrelated (that is, it is *white noise*), the distribution of the sample autocorrelation coefficient at lag k in large samples is approximately normal with mean zero and variance $1/T$. Therefore, we could test the hypothesis H_0: $\rho_k = 0$ using the test statistic

$$Z_0 = r_k \sqrt{T}.$$

This procedure is a one-at-a-time test; that is, the significance level applies to the autocorrelations considered individually. We discuss this issue of large-scale hypothesis testing in Chapter 20. We are often interested in evaluating a set of autocorrelations jointly to determine whether they indicate that the time series is white noise. Box and Pierce (1970) have suggested such a procedure. A modification of this test that works better for small samples was devised by Ljung and Box (1978). The Ljung–Box goodness-of-fit statistic is

$$Q_{\mathrm{LB}} = T(T+2) \sum_{k=1}^{K} \left(\frac{1}{T-k} \right) r_k^2,$$

which is distributed approximately as chi-square with K degrees of freedom under the null hypothesis that the time series is white noise. Another useful way to describe a time series numerically is with the *partial autocorrelation* function. This function is related to the concept of partial correlation. Suppose that there are three random variables, X, Y, and Z. Then the correlation between X and Z given that Y is known is called the partial correlation of X and Z.

The partial autocorrelation function can be defined similarly. For details of calculating the partial autocorrelation function, see Montgomery *et al.* (2015). The partial autocorrelation function is useful in the identification phase of using a class of time series models called autoregressive integrated moving average (ARIMA) models, which we will discuss later in this chapter.

Two techniques that are useful for determining whether a time series is nonstationary are: (1) plotting a reasonably long series of the data to see if they drift or wander away from the mean for long periods of time; and (2) computing the sample ACF. In practice, however, there is often no clear demarcation between a stationary and a nonstationary process for many real-world time series. An additional diagnostic tool that is very useful is the *variogram*.

The variogram G_k measures variances of the differences between observations that are k lags apart, relative to the variance of the differences that are one time unit apart (or at lag 1). The variogram is defined mathematically as

$$G_k = \frac{\mathbb{V}(y_{t+k} - y_t)}{\mathbb{V}(y_{t+1} - y_t)},$$

and the values of G_k are plotted as a function of the lag k. If the time series is stationary, it turns out that

$$G_k = \frac{1 - \rho_k}{1 - \rho_1}. \tag{16.3}$$

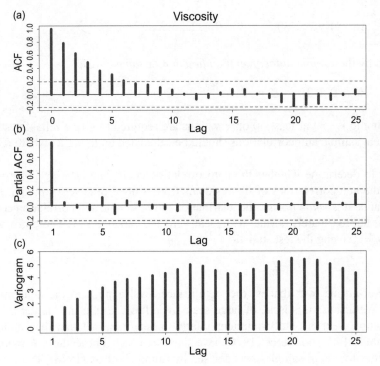

Figure 16.6 ACF, PACF, and variogram for the chemical process viscosity data.

However, for a stationary time series $\rho_k \to 0$ as the lag k increases, so when the variogram is plotted against lag k, G_k will reach the asymptote $1/(1 - \rho_1)$. If the time series is nonstationary, G_k will increase monotonically. The variogram is computed using the traditional sample variances for the quantities defined in the numerator and denominator. Many computer packages will compute and display the variogram. The variogram can be calculated in R using the result from (16.3):

```
acf.viscosity = acf(viscosity, lag.max=25)$acf
variogram.viscosity = (1-acf.viscosity[2:26])/(1-acf.viscosity[2])
```

Note that the first element of the vector `acf.viscosity` is the autocorrelation for lag 0, and in general the kth element is the autocorrelation for lag $k - 1$. Since the variogram begins at lag 1, we take `acf.viscosity[2:(kmax+1)]`, which gives the ACF for lags 1 through k. Similarly, the lag 1 ACF is `acf.viscosity[2]`.

Figure 16.6 presents the R output showing the sample autocorrelation function, the partial autocorrelation function, and the variogram for the chemical process viscosity data. Note the rate of decrease or decay in ACF values in Figure 16.6 from 0.78 to 0, followed by a sinusoidal pattern about 0. This ACF pattern is typical of stationary time series. The importance of ACF estimates exceeding the 5% significance limits shown in the display will be discussed later. By contrast, the plot of sample ACFs for a time series of random values with constant mean has a much different appearance. The sample ACFs for a white noise time series would appear with values distributed randomly positive or negative, with values near zero.

The Ljung–Box statistic can be computed in R using the `Box.test()` function:

```
LB = Box.test( viscosity, lag=1, type="Ljung-Box" )$statistic
```

The `type` argument can be either `"Ljung-Box"` or `"Box-Pierce"`. We can then loop over all desired lags k:

Table 16.3 Ljung–Box statistics for lags 1–25 for the viscosity data.

Lag	Ljung–Box statistic	Lag	Ljung–Box statistic
1	63.3639089	14	166.5691577
2	104.4186511	15	167.2048457
3	129.8297908	16	167.8127257
4	143.8208346	17	167.8129442
5	153.7794492	18	168.2233009
6	158.5182929	19	170.1323162
7	161.4797159	20	174.2884746
8	163.8039647	21	177.6964726
9	165.0075827	22	180.4831673
10	165.5096520	23	181.4954878
11	165.5114364	24	181.5134771
12	166.2027314	25	182.0562189
13	166.5192029		

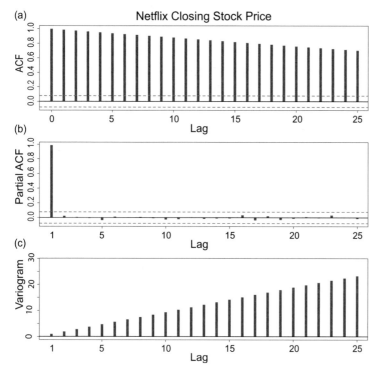

Figure 16.7 Sample autocorrelation function, sample partial autocorrelation function, and variogram for the Netflix stock price data.

```
LB = rep( -999 , 25 )
for(k in 1:25) LB[k] = Box.test(viscosity,lag=k,type="Ljung-Box")$statistic
```

Table 16.3 displays the Ljung–Box goodness-of-fit statistic. The *P*-values (not shown) are all very small, indicating that the autocorrelations considered as a group are not zero. Notice that the sample variogram generally converges to a stable level and then fluctuates around it. This is consistent with a stationary time series, and it provides additional evidence that the chemical process viscosity data are stationary. For the viscosity data, the variogram levels off to about 5 or 6 for lags 12 and beyond.

To illustrate what these numerical measures look like for a nonstationary time series, consider the Netflix closing stock price data in Figure 16.5. The R output for this time series is shown in Figure 16.7.

Notice that the sample autocorrelation function decays very slowly and the autocorrelations are very large (> 0.5) even at long lags. Also, the variogram is continuously increasing. This is typical behavior for a nonstationary time series.

16.4 Exponential Smoothing Methods

We can often think of a data set as consisting of two distinct components: *signal* and *noise*. Signal represents any pattern caused by the intrinsic dynamics of the process from which the data are collected. These patterns can take various forms, from a simple constant process to a more complicated structure that cannot be extracted visually or with any basic statistical tools. The constant process, for example, is represented as

$$y_t = \mu + \varepsilon_t, \tag{16.4}$$

where μ represents the underlying constant level of system response and ε_t is the noise at time t. The errors are often assumed to be uncorrelated with mean 0 and constant variance σ^2.

We have already discussed some basic data smoothers in Section 16.1. Smoothing is a technique to separate the signal and the noise as much as possible. A smoother is a filter that is used to obtain an "estimate" for the signal. The smoothers that we will discuss in this section achieve this by simply relating the current observation to the previous ones. For a given data set, one can devise forward- and/or backward-looking smoothers, but in this section we will only consider backward-looking smoothers. That is, at any given time, the observation will be replaced by a combination of observations at and before that time period. It does make intuitive sense to use some sort of an "average" of the current and the previous observations to smooth the data.

An important and widely used class of smoothers is *exponential smoothing*. This approach assigns geometrically decreasing weights to the past observations. The *first-order exponential smoothing equation* is

$$\tilde{y}_T = \lambda y_T + (1-\lambda)\tilde{y}_{T-1}.$$

This recursive equation shows that first-order exponential smoothing is the linear combination of the current observation and the smoothed observation at the previous time unit. Because we can write

$$
\begin{aligned}
\tilde{y}_T &= \lambda y_T + (1-\lambda)\tilde{y}_{T-1} \\
&= \lambda y_T + (1-\lambda)\Big(\lambda y_{T-1} + (1-\lambda)y_{T-2}\Big) \\
&= \cdots \\
&= \lambda y_T + \lambda(1-\lambda)y_{T-1} + \lambda(1-\lambda)^2 y_{T-2} + \cdots + \lambda(1-\lambda)^{T-1}y_1 + (1-\lambda)^{T+1}y_0,
\end{aligned}
$$

we see that the smoothed observation at time T is just the linear combination of the current observation and the discounted sum of all previous observations. This is also called *simple exponential* smoothing. We can see that the expected value of $\tilde{y}_T$ is

$$
\begin{aligned}
\mathbb{E}(\tilde{y}_T) &= \mathbb{E}\Big(\lambda y_T + \lambda(1-\lambda)y_{T-1} + \lambda(1-\lambda)^2 y_{T-2} + \cdots + \lambda(1-\lambda)^{T-1}y_1 + (1-\lambda)^{T+1}y_0\Big) \\
&= \lambda\mathbb{E}(y_T) + \lambda(1-\lambda)\mathbb{E}(y_{T-1}) + \lambda(1-\lambda)^2\mathbb{E}(y_{T-2}) + \cdots + \\
&\quad \lambda(1-\lambda)^{T-1}\mathbb{E}(y_1) + (1-\lambda)^T y_0
\end{aligned}
$$

$$= \mu\left(\lambda + \lambda(1-\lambda) + \lambda(1-\lambda)^2 + \cdots + \lambda(1-\lambda)^{T-1}\right) + (1-\lambda)^T y_0$$

$$= \frac{\mu\lambda\left(1 - (1-\lambda)^T\right)}{1 - (1-\lambda)} + (1-\lambda)^T y_0$$

$$= \mu\left(1 - (1-\lambda)^T\right) + (1-\lambda)^T y_0.$$

The second-last line follows because the summation is a geometric series. If T is large, in other words it's been a long time since the process of data collection started, then since $1 - \lambda$ is between 0 and 1, $(1-\lambda)^T \approx 0$. Thus,

$$\mathbb{E}(\tilde{y}_T) \approx \mu.$$

A similar argument shows that if in (16.4), $\mathbb{V}(\varepsilon_t) = \sigma^2$, then

$$\mathbb{V}(y_T) \approx \frac{\lambda}{2 - \lambda}\sigma^2. \tag{16.5}$$

> **Exercise 16.1** Show that if $y_t = \mu + \varepsilon_t$ where $\mathbb{E}(\varepsilon_t) = 0$ and $\mathbb{V}(\varepsilon_t) = \sigma^2$, then
>
> $$\mathbb{V}(y_T) \approx \frac{\lambda}{2 - \lambda}\sigma^2.$$

Useful references on exponential smoothing are the books by Brown (1963) and Montgomery *et al.* (2015).

The discount factor or smoothing constant, λ, represents the weight put on the current observation and $(1 - \lambda)$ represents the weight put on the smoothed value of the previous observations. Obviously, to compute the first smoothed value y_1 we would have to know the starting value y_0. It turns out that as T gets large the effect of the starting value dies out very quickly, so the choice of the starting value isn't terribly important. Typical choices are to set the starting value equal to the first observation in the time series or to use the average of a subset of the data as the starting value.

We must also choose the value of the discount factor λ. This determines how much smoothing is needed. Typically, λ values between 0.1 and 0.4 are recommended and do indeed perform well in practice.

Simple exponential smoothing is an optimal modeling and forecasting technique for a constant process; that is, a time series that varies around a fixed level. Figure 16.8 shows the result of using simple exponential smoothing with $\lambda = 0.2$ on the first 70 values of the chemical process viscosity data first shown in Figure 16.1. The smoothed value at any time period is an estimate of the average level of the time series based on the current and all previous observations. At the end of the graph, the last smoothed value is shown as a prediction of 25 future values through time 100. Obviously, since the time series is a constant process, the forecast of all future values is just the most recent smoothed value. This is certainly reasonable for perhaps the next few periods, but for longer forecast lead times there is more uncertainty in the prediction. Usually, a prediction interval is used to express this uncertainty. Figure 16.9 displays the 25 future values from time origin 100 along with the 95% prediction intervals. Notice that the prediction intervals widen as the lead time increases, reflecting the uncertainty in the prediction. For details on how these prediction intervals are constructed, see Montgomery *et al.* (2015).

Simple or first-order exponential smoothing can be applied to a time series that exhibits a linear trend, such as

$$y_t = a + bt + \varepsilon_t.$$

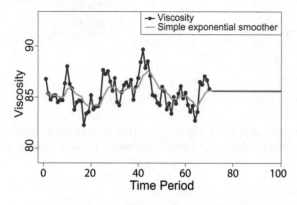

Figure 16.8 Viscosity data with simple exponential smoother using $\lambda = 0.2$.

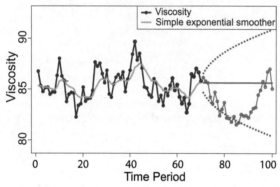

Figure 16.9 Forecasts and 95% prediction intervals for observations 71 through 100.

However, first-order exponential smoothing will consistently lag the trend. The amount of the lag, or the bias, can be shown to be $\left[-(1-\lambda)b/\lambda\right]$. At first glance a solution to this bias problem is to use a large value of the smoothing constant λ. A more general, and we think better, approach to model a linear trend is with second-order or double exponential smoothing, which applies simple exponential smoothing on $\tilde{y}_t$ resulting in

$$\tilde{y}_t^{(2)} = \lambda y_t^{(1)} + (1-\lambda)y_{t-1}^{(1)},$$

where $\tilde{y}_t^{(1)}$ and $\tilde{y}_t^{(2)}$ are the first-order and second-order smoothed statistics, respectively. Then, at the end of time period T, an unbiased estimate of y_T is

$$\hat{y}_T = 2\tilde{y}_t^{(1)} - \tilde{y}_t^{(2)}.$$

With simple exponential smoothing, we needed one value, y_0, to get started. For the second-order smoother we will need to specify two starting values. A simple and effective way to do this is to use $y_0^{(1)} = y_1$ and $y_0^{(2)} = \tilde{y}_1^{(1)}$. For other methods of obtaining starting values, see Montgomery *et al.* (2015). Double smoothing can be used to predict or forecast future values of the time series at lead time τ by using the equation

$$\hat{y}_{T+\tau}(T) = \tau\text{-step-ahead forecast made at time } T = \left(2 + \frac{\lambda}{1-\lambda}\tau\right)\tilde{y}_T^{(1)} - \left(1 + \frac{\lambda}{1-\lambda}\tau\right)\tilde{y}_T^{(2)}.$$

The idea of double smoothing can be extended to higher orders, and these higher-order smoothed statistics can be used to model and forecast time series that are represented as higher-order polynomials. For example, triple exponential smoothing would result in three smoothed statistics that could be used to estimate the parameters in a cubic polynomial. However, if a polynomial of higher order than two seems appropriate, a simpler approach may be to consider using one of the autoregressive integrated moving average (ARIMA) models presented later in this chapter.

Some time series data exhibit cyclical or seasonal patterns that cannot be effectively modeled using polynomials. Several approaches are available for the analysis of such data. We will now discuss exponential smoothing techniques that can be used in modeling seasonal time series. The methodology we will focus on was originally introduced by Holt (1957) and Winters (1960) and is generally known as Winters' method, where a seasonal adjustment is made to the linear trend model. Two types of adjustments are suggested: additive and multiplicative. We will focus on the additive version.

We assume that the seasonal time series can be represented by the following model:

$$y_t = L_t + S_t + \varepsilon_t,$$

where L_t represents the level or linear trend component and can in turn be represented by

$$y_t = a + bt.$$

Sometimes the level component is called the *permanent component*. The seasonal adjustment is S_t, with $S_t = S_{t+s} = S_{t+2s} = \cdots$ for $t = 1, 2, \ldots, s-1$, where s is the length of the season. One usual restriction on this model is that the seasonal adjustments add to zero during one period:

$$\sum_{t=1}^{T} S_t = 0.$$

In the model given above, for forecasting the future observations, we will employ first-order exponential smoothing with different discount factors. The procedure for updating the parameter estimates once the current observation y_T is obtained and for predicting the future observation at time period $T + \tau$ is as follows:

Step 1: Update the estimate of L_T using

$$\hat{L}_T = \alpha(y_T - \hat{S}_{T-s}) + (1 - \alpha)(\hat{L}_{T-1} + \hat{B}_{1,T-1}),$$

where $0 < \alpha < 1$. It should be noted that in this equation the first part is the "current" value for L_T and the second part is the forecast of L_T based on the estimates at $T - 1$.

Step 2: Update the estimate of the slope using

$$\hat{B}_{1,T} = \beta(\hat{L}_T - \hat{L}_{T-1}) + (1 - \beta)\hat{B}_{1,T-1}, \tag{16.6}$$

where $0 < \beta < 1$. As in Step 1, the estimate of B in (16.6) is the linear combination of the "current" value of B and its "forecast" at time $T - 1$.

Step 3: Update the estimate of the seasonal adjustment factor S_T using

$$\hat{S}_T = \gamma(y_T - \hat{L}_T) + (1 - \gamma)\hat{S}_{t-s},$$

where $0 < \gamma < 1$.

Step 4: The τ-step-ahead forecast made at the end of period T is

$$\hat{y}_{T+\tau} = \hat{L}_T + \hat{B}_{1,T}\tau + \hat{S}_T(\tau - s).$$

Estimating the initial values of the smoothed statistics can be done using standard regression modeling techniques, since the additive seasonal model is a linear regression model.

Table 16.4 Beginning and ending values of the beverage shipment data set.

Month	Shipments ($ million)
Jan-92	3,519
Feb-92	3,803
Mar-92	4,332
Apr-92	4,251
⋮	⋮
Oct-06	6,446
Nov-06	6,717
Dec-06	6,320

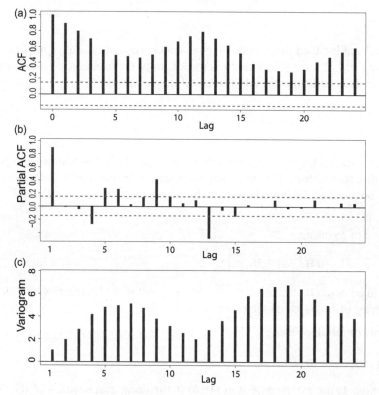

Figure 16.10 Autocorrelation, partial autocorrelation, and variogram plots for the beverage data.

We will illustrate using the additive Winters seasonal model using the annual beverage production data first shown in Figure 16.3. These data exhibit a strong seasonal pattern. The full data set is available in the file BeverageShipments-1992-2006.csv, and a subset is shown in Table 16.4.

The ACF, PACF, and the variogram for the beverage shipment data are shown in Figure 16.10. The ACF resembles a sinusoidal curve with period 12, suggesting that there is seasonality across the 12 months. The variogram seems to increase linearly, suggesting a nonstationary time series.

Once the data are read into R, we can apply Winters' method using the following code:

```
HW.out = HoltWinters( beverage.ts , seasonal="additive" )
HW.out.pred = predict( HW.out, 24, prediction.interval=TRUE, level=0.95 )
```

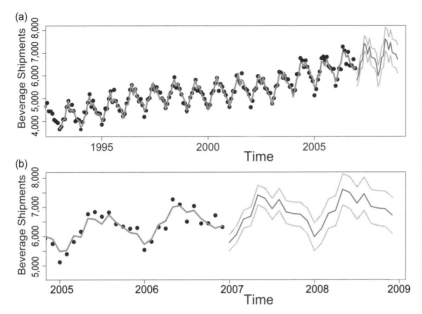

Figure 16.11 Beverage shipments from 1992 through 2006 with predictions made for 2007 through the end of 2008. Predictions are made using Winters' method. (a) shows all the data (dots), the fitted values (solid red curve), and the predictions (red with confidence bands in gray); (b) shows the more recent data along with the predictions.

The first line of code will select the optimal values for the smoothing constants α, β, and γ, and it will fit the model to the observations. The second line of code makes predictions for the next 24 months, including 95% prediction intervals. Figure 16.11 shows the observed data and these predictions for the next two years.

The estimated smoothing parameters are contained in the HW.out object, and can be extracted using

```
HW.out$alpha
HW.out$beta
HW.out$gamma
```

which yields

$$\hat{\alpha} = 0.342, \qquad \hat{\beta} = 0.016, \quad \text{and} \quad \hat{\gamma} = 0.489.$$

Winters' method is able to decompose the time series into (1) trend, (2) seasonal, and (3) random effects. Figure 16.12 shows this decomposition. The original time series is shown at the top. The second graph shows the trend, which seems to be increasing in a roughly linear fashion. The third graph shows the seasonal pattern. The fitted values are then obtained by adding the trend and the seasonal effect. Finally, the bottom graph shows the residuals, that is, the differences between the observed and fitted values. The residual plot (the bottom plot in Figure 16.12) does not exhibit any structure and the sample autocorrelations indicate that the residuals are white noise. We conclude that the additive seasonal model is a good fit to the data.

The bottom graph in Figure 16.12 shows the residuals from the additive seasonal model and Figure 16.13 shows the sample ACF and PACF function for the residuals. The residual plot does not exhibit any structure and the sample autocorrelations indicate that the residuals are white noise. We conclude that the additive seasonal model is a good fit to the data.

If the amplitude of the seasonal pattern is proportional to the average level of the seasonal time series, then a multiplicative seasonal model will be more appropriate. The usual form of the multiplicative season model is

$$y_t = L_t S_t + \varepsilon_t,$$

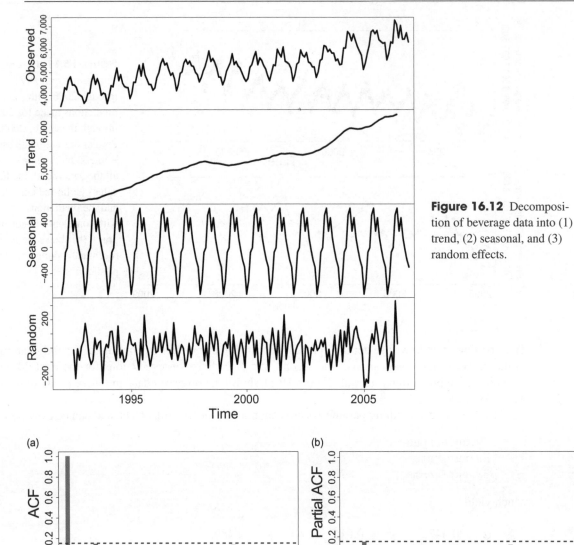

Figure 16.12 Decomposition of beverage data into (1) trend, (2) seasonal, and (3) random effects.

Figure 16.13 ACF and PACF functions for the residuals from the application of Winters' model to the beverage shipment data.

where L_t once again represents the permanent component (i.e., $a + bt$); S_t represents the seasonal adjustment with $S_t = S_{t+s} = S_{t+2s} = \cdots$ for $t = 1, 2, \ldots, s-1$, where s is the length of the period of the seasonal cycles and the errors are assumed to be uncorrelated with mean 0 and constant variance. As in the additive model, we will employ three exponential smoothers to estimate the parameters in the multiplicative seasonal model.

Step 1: Update the estimate of the permanent component L_T using

$$\hat{L}_T = \alpha \frac{y_T}{\hat{S}_{T-s}} + (1-\alpha)\left(\hat{L}_{T-1} + \hat{B}_{1,T-1}\right),$$

where $0 < \alpha < 1$. Similar interpretation as in the additive model can be made for this exponential smoothing operation.

Step 2: Update the estimate of the trend component using

$$\hat{B}_{1,T} = \beta\left(\hat{L}_T - \hat{L}_{T-1}\right) + (1-\beta)\hat{B}_{1,T-1},$$

where $0 < \beta < 1$.

Step 3: Update the estimate of the seasonal factor using

$$\hat{S}_T = \gamma\frac{y_T}{\hat{L}_T} + (1-\gamma)\hat{S}_{T-s},$$

where $0 < \gamma < 1$.

Step 4: The τ-step-ahead forecast made at the end of period T is

$$\hat{y}_{T+\tau}(T) = \left(\hat{L}_T + B_{1,T}\,\tau\right) \times \hat{S}_T(\tau - s).$$

To run a multiplicative model in R, simply use the argument `seasonal="multiplicative"` in the call to the function `HoltWinters()`.

> **Exercise 16.2** Apply a multiplicative model to the beverage data. Compare the fit with the additive model.

16.5 Autoregressive Integrated Moving Average (ARIMA) Models

In the previous chapter, we discussed forecasting techniques that, in general, were based on some variant of exponential smoothing. The general assumption for these models was that any time series data can be represented as the sum of two distinct components: deterministic and stochastic (random). The former is modeled as a function of time, whereas for the latter we assumed that some random noise that is added to the deterministic signal generates the stochastic behavior of the time series. One very important assumption is that the random noise is generated through independent shocks to the process. In practice, however, this assumption is often violated. That is, usually successive observations show serial dependence. Under these circumstances, forecasting methods based on exponential smoothing may be inefficient and sometimes inappropriate because they do not take advantage of the serial dependence in the observations in the most effective way. To formally incorporate this dependent structure, in this section we present a general class of models called *autoregressive integrated moving average* models or *ARIMA* models (also known as Box–Jenkins models, after the two authors who assembled most of the research literature on these models and popularized their use). The book by Box *et al.* (2008) is a useful and thorough reference on ARIMA models.

In statistical modeling, we are often engaged in an endless pursuit of finding the true relationship between certain inputs and the output. An assumption that often proves useful in time series modeling efforts is the linearity assumption. A general linear filter can be defined as

$$y_t = \mu + \sum_{i=0}^{\infty} \psi_i \varepsilon_{t-i},$$

where we assume that $\mathbb{E}(\varepsilon_t) = 0$, $\mathbb{V}(\varepsilon_t) = \sigma^2$, and the errors or "shocks" to the observations are uncorrelated. This general linear filter can be written as

$$y_t = \mu + \psi_0\varepsilon_t + \psi_1\varepsilon_{t-1} + \psi_2\varepsilon_{t-2} + \cdots = \mu + \sum_{i=0}^{\infty} \psi_i B^i \varepsilon_{t-i},$$

where B is the backshift operator defined as

$$B\varepsilon_t = \varepsilon_{t-1}$$

$$B^2\varepsilon_t = B(B\varepsilon_t) = B\varepsilon_{t-1} = \varepsilon_{t-2}$$

$$\vdots$$

$$B^i\varepsilon_t = B(B^{i-1}\varepsilon_t) = B(\varepsilon_{t-(i-1)}) = \varepsilon_{t-i}.$$

If we define the operator Ψ to be

$$\Psi(B) = \sum_{i=0}^{\infty} B^i,$$

then we can write the general linear filter compactly as

$$y_t = \Psi(B)\varepsilon_t.$$

The general linear filter is an infinite-order moving average process. It is stationary for any values of the ψ weights. Effectively, any stationary time series can be seen as the weighted sum of the present and all past random shocks that impact the observations.

The infinite-order moving average process is not very useful, because it would be tedious at best to estimate all the weights. In finite order moving average (MA) models, conventionally ψ_0 is set to 1 and the weights that are not set to 0 are represented by the Greek letter θ with a minus sign in front. Hence, a moving average process of order q (MA(q)) is given as

$$y_t = \mu + \varepsilon_t - \theta_1\varepsilon_{t-1} - \cdots - \theta_q\varepsilon_{t-q},$$

or more compactly as

$$y_t = \mu + \Theta(B)\varepsilon_t,$$

where

$$\Theta(B) = 1 - \sum_{i=1}^{q} \theta_i B^i.$$

The MA(q) process is stationary for any values of the θ weights.

The simplest finite-order MA model is obtained when $q = 1$:

$$y_t = \mu + \varepsilon_t - \theta_1\varepsilon_{t-1}.$$

The autocovariance function of the MA(1) model is very simple; there is a single large spike at lag 1 (which could either be positive or negative) and all other autocorrelations are zero. We say that the autocorrelation function "cuts off" after lag 1.

Another useful finite-order moving average process is MA(2):

$$y_t = \mu + \varepsilon_t - \theta_1\varepsilon_{t-1} - \theta_2\varepsilon_{t-2}.$$

The autocovariance function of the MA(2) model is also very simple; there are large spikes at lags 1 and 2, and all other autocorrelations are zero. The autocorrelation function for MA(2) "cuts off" after lag 2. Visual inspection of the time series plots for these models typically shows that the mean and variance remain stable while there are some short runs where successive observations tend to follow each other for very brief durations, suggesting that there is indeed some positive autocorrelation.

A special case of the general linear filter assumes that it can be adequately modeled by only estimating a finite number of weights and setting the rest equal to 0. Another interpretation of the finite order MA processes is that at any given time, of the infinitely many past disturbances, only a finite number of those disturbances "contribute" to the current value of the time series and that the time window of the contributors "moves" in time, making the "oldest" disturbance obsolete for the next observation. It is not unreasonable to think that many time series might have these intrinsic dynamics. However, for some others, we may be required to consider the "lingering" contributions of the disturbances that happened further back in the past. This will of course bring us back to square one in terms of our efforts in estimating infinitely many weights.

Another solution to this problem is through the autoregressive models in which the infinitely many weights are assumed to follow a distinct pattern and can be successfully represented with only a handful of parameters. We shall now consider some special cases of autoregressive processes. The first-order autoregressive process substitutes

$$y_{t-1} = \mu + \varepsilon_{t-1} + \phi\varepsilon_{t-2} + \phi^2\varepsilon_{t-3} + \cdots$$

into the general linear filter, resulting in

$$y_t = \delta + \phi y_{t-1} + \varepsilon_t,$$

where $\delta = (1 - \phi)\mu$. This process is called a first-order autoregressive process, AR(1), because it can be seen as a regression of y_t on y_{t-1} and hence the term autoregressive process. We will assume that $|\phi| < 1$, which results in weights that decay exponentially in time and guarantees that the sum of the weights in the underlying general linear filter is finite. This means that an AR(1) process is stationary if $|\phi| < 1$. For $|\phi| > 1$, past disturbances will get exponentially increasing weights as time goes on and the resulting time series will be explosive. Box et $al.$ (2008) argue that this type of process is of little practical interest and therefore only consider cases where $|\phi| \leq 1$. The mean of a stationary AR(1) process is

$$\mathbb{E}(y_t) = \mu = \frac{\delta}{1 - \phi}.$$

The autocovariance function of an AR(1) process has an exponential decay.

An obvious extension of the AR(1) model is the AR(2) model:

$$y_t = \delta + \phi_1 y_{t-1} + \phi_2 y_{t-2} + \varepsilon_t.$$

For the AR(2) model to be stationary the parameters must satisfy

$$\phi_1 + \phi_2 < 1$$
$$\phi_2 - \phi_1 < 1$$
$$|\phi_2| < 1.$$

The autocovariance function of AR(2) decays either as a mixture of exponentials or as a damped sinusoid. So using the sample autocorrelation function alone isn't usually adequate to identify a specific AR model. However, the sample partial autocorrelation function is helpful in this regard. It turns out that the autocorrelation function of an AR(p) process cuts off after lag p.

Sometimes we encounter a time series where it seems necessary to increase the order of the AR process beyond 2 or 3. Instead of increasing the order of the AR model to accommodate this, we can

Table 16.5 Behavior of theoretical ACF and PACF for stationary time series processes.

Model	ACF	PACF
MA(q)	Cuts off after lag q	Exponential decay or damped sinusoid
AR(p)	Exponential decay or damped sinusoid	Cuts off after lag p
ARMA(p,q)	Exponential decay or damped sinusoid	Exponential decay or damped sinusoid

instead add an MA(1) term. This results in a mixed autoregressive moving average, or ARMA(1,1), model. In general, an ARMA(p,q) model is given as

$$y_t = \delta + \sum_{i=1}^{p} \phi_i y_{t-i} - \sum_{i=1}^{q} \theta_i \varepsilon_{t-i},$$

or, using the backward operator,

$$\Psi(B) y_t = \delta + \Theta(B)\, \varepsilon_t.$$

It can be shown that the ACF and PACF of an ARMA(p,q) both exhibit exponential decay and/or damped sinusoid patterns, which makes the identification of the order of the ARMA(p,q) model relatively more difficult. For that, additional sample functions such as the extended sample ACF (ESACF), the generalized sample PACF (GPACF), the inverse ACF (IACF), and canonical correlations can be used. For further information, see Box *et al.* (2008). However, the availability of sophisticated statistical software packages makes it possible for the practitioner to consider several different models with various orders and compare them to identify the most appropriate model. A summary of the sample ACFs and sample PACFs is provided in Table 16.5.

Although the theoretical ACF and PACF functions involve unknown parameters, we can often look at the sample ACF and PACF and make an inference about which functions cut off after some lag, and which ones exhibit exponential decay or a damped sinusoidal pattern. To illustration the fitting of an ARMA model, consider the chemical process viscosity data originally shown in Figure 16.1 and Table 16.1. The sample ACF, PACF, and variogram are shown in Figure 16.6. The sample ACF exhibits exponential decay and the sample PACF cuts off after lag 1. These are characteristic of an AR(1) model. We fit this model to the viscosity data using R, resulting in the parameter estimates 0.7859 for the AR(1) parameter and 84.9827 for the mean. The needed R code is:

```
viscosity.arima = arima( viscosity , order=c(1,0,0) )
str( viscosity.arima )
viscosity.arima$coef
forecast1 = predict( viscosity.arima , 30 )
str( forecast1 )
```

The `arima()` function is part of the `stats` package, which is automatically loaded when you begin an R session. The required arguments are the time series object to be modeled and the order of the ARIMA model to be fit. The order should be a vector of length three, containing integers for p, d, and q. In our case we fit an AR(1) model, which is equivalent to an ARIMA(1,0,0) model. We can then use the `predict()` function to make predictions. The only required argument to `predict()` is the fitted object (`viscosity.arima` in this case), but if this is the only argument given, R will compute a prediction for the next time period only. In this case we want predictions over the next 30 time periods, so we must include this as the second argument. As always, we recommend that the user apply the structure function `str()` to learn what is included in a particular object. The output from the statement `str(viscosity.arima)` tells us that `viscosity.arima` is a list of 14 objects, the first of which is

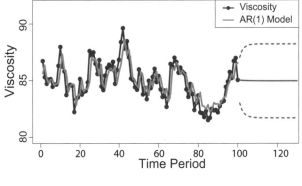

Figure 16.14 AR(1) or equivalently an ARIMA(1,0,0) model fit to the chemical process viscosity data. Predictions are made for the next 30 observations.

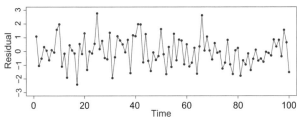

Figure 16.15 Residuals from AR(1), or equivalently an ARIMA(1,0,0), model fit to the chemical process viscosity data.

(a)

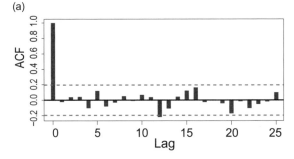

(b)

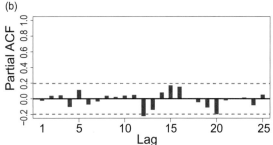

Figure 16.16 ACF and PACF for residuals from ARIMA(1,0,0) model applied to viscosity data.

`coef`, which gives the estimated coefficients. The `forecast1` object contains only two objects: the point predictions and the associated standard errors. We can make a $100 \times (1 - \alpha)\%$ prediction interval for an observation by taking the point prediction plus and minus $z_{\alpha/2}$ times the standard error.

Figure 16.14 shows the model fit and 30 predictions for times 101 through 130 along with 95% prediction limits. Figure 16.15 presents a plot of the residuals in time from this model. The impression is that the residuals are white noise. The sample ACF of the residuals is shown in Figure 16.16. None of the sample ACF coefficients are large, another indication that the residuals are structureless. The AR(1) model is a good fit for the viscosity data.

Exercise 16.3 Fit the AR(2) model to the viscosity data.

Often, we find that time series processes may not have a constant level. For example, consider the beverage shipment time series shown in Figure 16.3. These data exhibit strong seasonality and a linear trend. The Netflix stock price data in Table 16.2 and Figure 16.5 do not seem to have a stable mean level of the process. Similarly, processes may show nonstationary behavior in the slope as well. We will call a time series y_t *homogeneous nonstationary* if it is not stationary but its first difference, that is, $w_t = y_t - y_{t-1} = (1 - B)y_t$, or higher-order differences, $w_t = (1 - B)^d y_t$, produce a stationary time series. We will further call y_t an autoregressive integrated moving average (ARIMA) process of orders p, d,

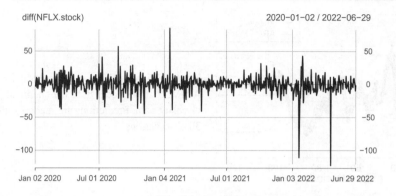

Figure 16.17 First differences of the Netflix stock price data.

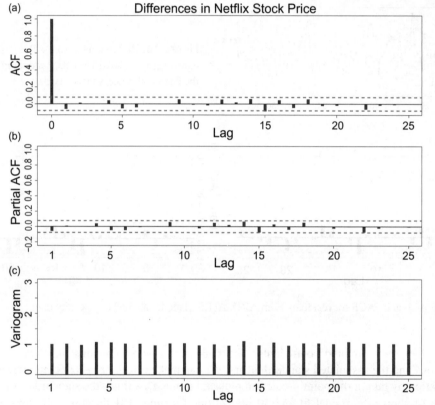

Figure 16.18 ACF, PACF, and variogram for the first differences of the Netflix stock price data.

and q – that is, ARIMA(p,d,q) – if its dth difference, denoted by $w_t = (1-B)^d y_t$, produces a stationary ARMA(p,q) process.

Figure 16.17 shows a time series plot of the first differences of the Netflix stock price data. The general impression is that the first differences fluctuate around a constant mean. The sample ACF, PACF, and variogram are shown in Figure 16.18. These displays indicate that there is no autocorrelation in the time series of first differences. Therefore, the model is the "trivial" ARIMA$(0,1,0)$ model $y_t = y_{t-1} + \varepsilon_t$. That is, the best model for the stock price on day t is the price on the previous day plus a "random shock." This is called the *random walk* model and many stock prices have similar behavior.

16.6 Chapter Summary

In much of this book we have assumed independence among random variables. Time series represent a situation where the independence assumption is usually not valid. Today's value of a stock is highly correlated with yesterday's; if the price was high yesterday, it will likely be high today. Time series usually exhibit a phenomenon called autocorrelation. We defined the lag k autocorrelation to be the correlation between two observations separated by k time units. The partial autocorrelation takes into account the outcomes of the time series between observations that are separated by k time units.

The autocorrelation and partial autocorrelation functions (ACF and PACF, respectively) can help us determine the type of time series model that might be reasonable. If the sample ACF exhibits exponential decay and the sample PACF cuts off after lag 1, then the AR(1) model is reasonable. If the PACF exhibits exponential decay and the sample ACF cuts off after lag 1, then the MA(1) model is reasonable. Other profiles of ACF and PACF can lead to both an AR and a MA component to the model. These models are called ARMA. For time series that are not stationary, it is often the case that first-differences of the time series *are* stationary. In a case like this, the procedure is usually to take the first differences and model these with an ARMA model. Such models are called ARIMA (autoregressive integrated moving average). Occasionally, a second difference (or even a third difference) is needed to reach a stationary process. Once the parameters of the model are estimated, we can use the resulting model to make predictions for future observations from the time series.

16.7 Problems

Problem 16.1 Consider the time series in Table 16.6.

Table 16.6 Data for Problem 16.1.

t	y_t	t	y_t	t	y_t
1	48.7	18	44.7	35	53.4
2	45.8	19	51.1	36	53.9
3	46.4	20	47.3	37	52.3
4	46.2	21	45.3	38	53
5	44	22	43.3	39	48.6
6	53.8	23	44.6	40	52.4
7	47.6	24	47.1	41	47.9
8	47	25	53.4	42	49.5
9	47.6	26	44.9	43	44
10	51.1	27	50.5	44	53.8
11	49.1	28	48.1	45	52.5
12	46.7	29	45.4	46	52
13	47.8	30	51.6	47	50.6
14	45.8	31	50.8	48	48.7
15	45.5	32	46.4	49	51.4
16	49.2	33	52.3	50	47.7
17	54.8	34	50.5		

(a) Construct a time series plot of the data.

(b) Use simple exponential smoothing with $\lambda = 0.2$ to smooth the first 40 periods of the data. How well does this procedure work?

(c) Make one-step-ahead predictions starting at period 40 for the last 10 observations. Determine the forecast errors.

Problem 16.2 Rework Problem 16.1 using simple exponential smoothing with $\lambda = 0.4$ to smooth the first 40 periods of the data. How well does this work in comparison to the results from Problem 16.1? Make one-step-ahead predictions starting at period 40 for the last 10 observations. Determine the forecast errors. Compare these results with those from Problem 16.1.

Problem 16.3 Find the sample ACF and sample PACF for the data in Table 16.6. Does this give you any insight about how the exponential smoothing procedures in Problems 16.1 and 16.2 work?

Problem 16.4 The data in Table 16.7 exhibit a linear trend.

(a) Verify that there is a trend by plotting the data.

(b) Using the first 12 observations, develop an appropriate procedure for forecasting.

(c) Forecast the last 12 observations and calculate the forecast errors. Does the forecasting procedure seem to be working satisfactorily?

Problem 16.5 Reconsider the linear trend data in Table 16.7. Take the first difference of this data and plot the time series of first differences. Has differencing removed the trend? Use exponential smoothing on the first 11 differences. Instead of forecasting the original data, forecast the first differences for the remaining data using exponential smoothing and use these forecasts of the first differences to obtain forecasts for the original data.

Table 16.7 Data for Problem 16.4.

y	y_t	y	y_t	y	y_t
1	315	9	420	17	458
2	195	10	410	18	570
3	310	11	485	19	520
4	316	12	420	20	400
5	325	13	460	21	420
6	335	14	395	22	580
7	318	15	390	23	475
8	355	16	450	24	560

Problem 16.6 Consider the Netflix stock price data in Table 16.2 and Figure 16.5. Use simple exponential smoothing with $\lambda = 0.4$ to smooth the first 40 periods of the data. How well does this procedure work?

Problem 16.7 Consider the Netflix stock price data in Table 16.2 and Figure 16.5. Use double exponential smoothing with $\lambda = 0.4$ to smooth the first 40 periods of the data. How well does this procedure work? Compare the results with those obtained in Problem 16.6.

Problem 16.8 Table 16.8 contains four years of monthly demand for a beverage product.

(a) Make a time series plot of the data and verify that it is seasonal. Why do you think seasonality is present in these data?

(b) Use Winters' additive method for the first three years to develop a forecasting method for this data. How well does this smoothing procedure work?

(c) Make one-step-ahead forecasts of the last 12 months. Determine the forecast errors. How well did your procedure work in forecasting the new data?

Problem 16.9 Suppose that simple exponential smoothing is being used to forecast a process. At the start of period t^*, the mean of the process shifts to a new level $\mu + \delta$. The mean remains at this new level for subsequent time periods. Show that the expected value of the exponentially smoothed statistic is

$$\mathbb{E}(\hat{y}_T) = \begin{cases} \mu, & T < t^* \\ \mu + \delta - \delta(1-\lambda)^{T-t^*+1}, & T \geq t^* \end{cases}.$$

Problem 16.10 Using the results of the previous problem, determine the number of periods following

Table 16.8 Data for Problem 16.8.

t	y_t	t	y_t	t	y_t
1	143	17	279	33	1,104
2	191	18	552	34	874
3	195	19	674	35	683
4	225	20	827	36	352
5	175	21	1,000	37	332
6	389	22	502	38	244
7	454	23	512	39	320
8	618	24	300	40	437
9	770	25	359	41	544
10	564	26	264	42	830
11	327	27	315	43	1,011
12	235	28	362	44	1,081
13	189	29	414	45	1,400
14	326	30	647	46	1,123
15	289	31	836	47	713
16	293	32	901	48	487

the step change for the expected value of the exponential smoothing statistic to be within 0.10δ of the new time series level $\mu + \delta$. Plot the number of periods as a function of the smoothing constant. What conclusions can you draw?

Problem 16.11 Suppose that simple exponential smoothing is being used to forecast the constant process. At the start of period t^*, the mean of the process experiences a transient; that is, it shifts to a new level $\mu + \delta$, but reverts to its original level μ at the start of the next period $t^* + 1$. The mean remains at this level for subsequent time periods. Show that the expected value of the exponentially smoothed statistic is

$$\mathbb{E}(\hat{y}_T) = \begin{cases} \mu, & T < t^* \\ \mu + \delta\lambda(1-\lambda)^{T-t^*}, & T \geq t^* \end{cases}.$$

Problem 16.12 Using the results of the previous problem, determine the number of periods that it will take following the impulse for the expected value of the exponential smoothing statistic to return to within 0.10δ of the original time series level μ. Plot the number of periods as a function of the smoothing constant. What conclusions can you draw?

Problem 16.13 Consider the data in Table 16.6.

(a) Fit an appropriate ARIMA model to the first 40 observations of this time series.

(b) Make one-step-ahead forecasts of the last 10 observations. Determine the forecast errors.

(c) In Problem 16.1 you used simple exponential smoothing with $\lambda = 0.2$ to smooth the first 40 time periods of these data and make forecasts of the last

10 observations. Compare the ARIMA forecasts with the exponential smoothing forecasts. How well do both techniques work?

Problem 16.14 Consider the time series model

$$y_t = 200 + 0.7y_{t-1} + \varepsilon_t.$$

(a) Is this a stationary time series process?
(b) What is the mean of the time series?
(c) If the current observation is $y_{100} = 750$, would you expect the next observation to be above or below the mean?

Problem 16.15 Consider the time series model

$$y_t = 50 - 0.8y_{t-1} + \varepsilon_t.$$

(a) Is this a stationary time series process?
(b) What is the mean of the time series?
(c) If the current observation is $y_{100} = 85$, would you expect the next observation to be above or below the mean?

Problem 16.16 Consider the time series model

$$y_t = 20 + \varepsilon_t + 0.25\varepsilon_{t-1}.$$

(a) Is this a stationary time series process?
(b) What is the mean of the time series?
(c) If the current observation is $y_{100} = 23$, would you expect the next observation to be above or below the mean? Explain your answer.

Problem 16.17 Consider the time series model

$$y_t = 50 + 0.8y_{t-1} + \varepsilon_t - 0.2\varepsilon_{t-1}.$$

(a) Is this a stationary time series process?
(b) What is the mean of the time series?
(c) If the current observation is $y_{100} = 270$, would you expect the next observation to be above or below the mean?

Problem 16.18 An ARIMA model has been fit to a time series, resulting in

$$y_t = 25 + 0.35y_{t-1} + \varepsilon_t.$$

Suppose that we are at time period $T = 100$ and $y_{100} = 31$. Determine forecasts for periods $101, 102, 103, \ldots$ from this model at origin 100.

(a) What is the shape of the forecast function from this model?

(b) Suppose that the observation for time period 101 turns out to be $y_{101} = 33$. Revise your forecasts for periods $102, 103, \ldots$ using period 101 as the new origin of time.

Problem 16.19 For each of the ARIMA models shown below, give the forecasting equation that evolves for lead times $\tau = 1, 2, \ldots, L$.

(a) AR(1)
(b) AR(2)
(c) MA(1)
(d) MA(2)
(e) ARMA(1, 1)
(f) IMA(1, 1)
(g) ARIMA(1, 1, 0)

Problem 16.20 Consider the AR(1) model in Problem 16.15. Use a random number generator to generate 100 observations from this time series. Assume that the errors are normally distributed with mean 0 and variance 1.

(a) Verify that your time series is AR(1).
(b) Generate 100 observations for an $N(0, 1)$ process and add these random numbers to the 100 AR(1) observations in part (a) to create a new time series that is the sum of AR(1) and "white noise."
(c) Find the sample autocorrelation and partial autocorrelation functions for the new time series created in part (b). Can you identify the new time series?
(d) Does this give you any insight about how the new time series might arise in practical settings?

Problem 16.21 Consider the AR(1) model in Problem 16.15.

(a) Sketch the theoretical ACF and PACF for this model.
(b) Generate 50 realizations of this AR(1) process and compute the sample ACF and PACF. Compare the sample ACF and the sample PACF to the theoretical ACF and PACF. How similar to the theoretical values are the sample values?
(c) Repeat part (b) using 200 realizations. How has increasing the sample size impacted the agreement between the sample and theoretical ACF and PACF? Does this give you any insight about the sample sizes required for model building, or the reliability of models built to short time series?

Problem 16.22 Consider the MA(1) model in Problem 16.15. Assume that the random shocks are normally and independently distributed with mean 0 and variance 1.

(a) Sketch the theoretical ACF and PACF for this model.
(b) Simulate 50 realizations of this AR(1) process and compute the sample ACF and PACF. Compare the sample ACF and the sample PACF to the theoretical ACF and PACF. How similar to the theoretical values are the sample values?
(c) Repeat part (b) using 200 realizations. How has increasing the sample size impacted the agreement between the sample and theoretical ACF and PACF? Does this give you any insight about the sample sizes required for model building, or the reliability of models built to short time series?

17 — Estimating the Standard Error: Analytic Approximations, the Jackknife, and the Bootstrap

17.1 Introduction

In Chapter 9 we discussed point estimation for a parameter or a vector of parameters. In Chapters 10 and 11, on confidence intervals and hypothesis testing, we needed the idea of the standard error of an estimator. For example, many of the confidence intervals you run across will be of the form

$$\hat{\theta} \pm z_{\alpha/2} \times \text{standard error},$$

where θ is the parameter being estimated and the standard error was given in Definition 10.3.1 as the standard deviation of the estimator $\hat{\theta}$.[1] We can also use the standard error to obtain a test statistic for a test of the parameter θ. Under some reasonable conditions,

$$\frac{\hat{\theta} - \theta_0}{\text{s.e.}(\hat{\theta})} \overset{\text{approx}}{\sim} N(0, 1), \tag{17.1}$$

provided H_0: $\theta = \theta_0$ is true and the sample size is large.

We saw a number of specific examples where the standard error could be computed directly. For example, if we have a sample $X_1, X_2, \ldots, X_n \sim$ i.i.d. $N(\mu, \sigma^2)$, where μ and σ^2 are both unknown, then

$$\mathbb{V}\left(\overline{X}\right) = \frac{\sigma^2}{n},$$

so

$$\text{s.e.}\left(\overline{X}\right) = \frac{\sigma}{\sqrt{n}}.$$

We cannot, however, use this to obtain a confidence interval for μ because σ^2 is unknown. Of course, we can use an estimator $\hat{\sigma}^2$ in place of σ^2. Using the sample variance S^2 in place of σ^2 really gives us an *estimated* standard error, denoted $\widehat{\text{s.e.}}(\hat{\mu})$. Thus,

---

[1]Here "standard deviation of $\hat{\theta}$" and "standard error of $\hat{\theta}$" are synonymous. If X is a random variable, but not an estimator of a parameter, we will still talk about "the standard deviation of X" but we will never refer to its "standard error." Estimators are the only random variables for which we will apply the term standard error.

$$\text{s.e.}(\hat{\mu}) = \frac{\sigma}{\sqrt{n}} \qquad \text{and} \qquad \widehat{\text{s.e.}}(\hat{\mu}) = \frac{S}{\sqrt{n}}.$$

The first is called the *standard error* of $\hat{\mu}$ and the second is called the *estimated standard error* of $\hat{\mu}$.

When we use a consistent estimator (recall Definition 9.2.2) for the parameter in the standard error instead of the true standard error (which usually involves some unknown parameters), the approximate normal distribution in (17.1) still holds.[2]

The standard error for estimators can be found directly in many problems involving the normal distribution. For example, for the one-sample normal problem described above, multiple linear regression, and one-factor ANOVA, we can find the standard error, and using an estimator for σ^2, we can obtain an estimated standard error. However, for most problems beyond these simple normal-distribution situations, there is no closed-form expression for the standard error or the estimated standard error.

17.2 Analytic Approximations to the Standard Error of an Estimator

An approximate standard error can often be obtained from the following theorem.

Theorem 17.2.1 — Asymptotic Normality of the MLE. Suppose that $X_1, X_2, \ldots, X_n \sim$ i.i.d. from some distribution with PDF $f(x \mid \theta)$. Let

$$\ell(\theta \mid x_1, \ldots, x_n) = \log L(\theta \mid x_1, \ldots, x_n) = \sum_{i=1}^{n} \log f(x_i \mid \theta)$$

denote the log-likelihood function. Under some regularity conditions, the MLE $\hat{\theta}$ is approximately normally distributed for large n, having mean θ and variance

$$\mathbb{V}(\hat{\theta}) \approx \left[\mathbb{E} \left(-\frac{\partial^2 \ell(\theta \mid x_1, \ldots, x_n)}{\partial \theta^2} \right) \right]^{-1}. \tag{17.2}$$

Also,

$$\frac{\hat{\theta} - \theta}{\sqrt{\left[\mathbb{E} \left(-\frac{\partial^2 \ell(\theta \mid x_1, \ldots, x_n)}{\partial \theta^2} \right) \right]^{-1}}} \overset{\text{approx}}{\sim} N(0,1). \tag{17.3}$$

We will not prove this theorem here, but we will point out one important facet of it. The variance involves the negative of the second derivative of the log-likelihood function $\ell(\theta \mid x_1, \ldots, x_n)$. Very often, the log-likelihood function is concave down throughout its domain. This means the second derivative is negative, so the expression inside the expectation in (17.2) is positive. (Remember, the MLE usually occurs when the first derivative is zero; a negative second derivative assures we have achieved a *maximum*.) This expectation is raised to the -1 power, so the larger the expectation, the smaller the variance. The curvature of the graph of a function $f(x)$ at $(a, f(a))$ is defined to be $\kappa = |f''(a)| / (1 + (f'(a))^2)$. When the first derivative of a function is equal to 0, the curvature is equal to the absolute value of the second derivative at that point. The osculating circle of a curve at $(a, f(a))$ is that circle passing through $(a, f(a))$ and matching the curvature of the curve at that point. The curvature of a circle is just the reciprocal of its radius. Thus, log-likelihood functions that are "sharper" at the

---

[2]An astute reader may recognize that in the case of estimating the mean of a normal distribution, we obtain a t distribution instead of a normal distribution. Remember, though, that (17.1) is an asymptotic result as $n \to \infty$, and as the degrees of freedom goes to infinity, the t distribution approaches the normal.

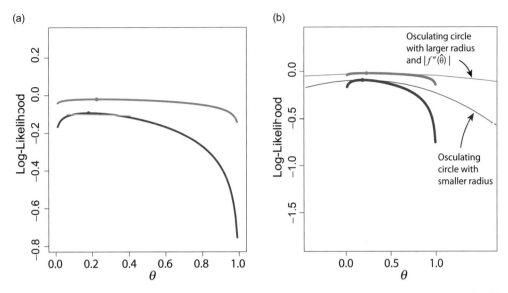

Figure 17.1 Two log-likelihood functions (a) with the osculating circles imposed (b). The blue log-likelihood is "sharper" at the MLE in the sense that the curvature is greater; in other words, the blue osculating circle has smaller radius. The red circle has a much greater radius.

MLE (i.e., the osculating circle has a smaller radius) have a greater $|f''(a)|$ and therefore lead to a smaller variance.

Figure 17.1 illustrates this concept. The blue log-likelihood function is "sharper" at the MLE in the sense that the curvature is greater; in other words, the blue osculating circle has smaller radius. The red circle has a much greater radius. The standard error for the parameter based on the blue log-likelihood function will be smaller than the standard error based on the red log-likelihood function.

Often the expectation in the denominator of (17.3) involves an unknown parameter. It can be shown that if a consistent estimator is used then the normal approximation still holds. Using the terminology introduced earlier in this section, the denominator in (17.3) is the standard error. If we use an estimator in place of one of the unknown parameters, it becomes the estimated standard error. A confidence interval based on the normal approximation in (17.3) is said to be a Wald interval.[3] Analogously, a test based on (17.3) is called a Wald test.

■ **Example 17.1** Suppose $X_1, X_2, \ldots, X_n \sim$ i.i.d. EXP1(θ). Find a confidence interval in two ways:

(a) Find a 95% confidence for θ using the pivotal quantity

$$\frac{2n\overline{X}}{\theta} \sim \chi^2(2n).$$

(b) Use the approximate normal distribution from (17.3) of Theorem 17.2.1 to approximate $\mathbb{V}(\hat{\theta})$ and thereby get an approximate confidence interval for θ.

(c) Apply both methods with $\alpha = 0.05$ to the data set
2.5   7.2   12.8   7.4   8.2   19.1   10.0   1.1   8.9   14.5

(d) Which confidence interval would you recommend?

Solution: Using the pivotal quantity, we can write

$$\Pr\left(\chi^2_{1-\alpha/2}(2n) < \frac{2n\overline{X}}{\theta} < \chi^2_{\alpha/2}(2n)\right) = 1 - \alpha.$$

---

[3]This is named after the Hungarian mathematician Abraham Wald.

A little algebra yields

$$\Pr\left(\frac{2n\overline{X}}{\chi^2_{\alpha/2}(2n)} < \theta < \frac{2n\overline{X}}{\chi^2_{1-\alpha/2}(2n)} \right) = 1 - \alpha.$$

Thus, a $100 \times (1 - \alpha)\%$ confidence interval for θ is

$$\left(\frac{2n\overline{X}}{\chi^2_{\alpha/2}(2n)}, \frac{2n\overline{X}}{\chi^2_{1-\alpha/2}(2n)} \right). \tag{17.4}$$

Now let's apply the normal approximation of the MLE. The likelihood function is

$$L(\theta \,|\, x_1, \ldots, x_n) = \prod_{i=1}^{n} \frac{1}{\theta}\, e^{-x_i/\theta} = \frac{1}{\theta^n}\, \exp\left(-\frac{1}{\theta} \sum_{i=1}^{n} x_i \right).$$

Thus, the log-likelihood is

$$\ell(\theta \,|\, x_1, \ldots, x_n) = -n\log\theta - \frac{1}{\theta} \sum_{i=1}^{n} x_i = -n\log\theta - \frac{n\overline{x}}{\theta}.$$

The denominator in (17.3) involves $\partial^2\ell/\partial\theta^2$, so let's begin finding derivatives:

$$\frac{\partial\ell}{\partial\theta} = -\frac{n}{\theta} + \frac{n\overline{x}}{\theta^2}.$$

Setting the first derivative equal to zero and solving yields the MLE for θ:

$$\hat{\theta} = \overline{x}.$$

The maximum likelihood *estimator* for θ is then formed by using the statistic $\overline{X}$ from the random sample. Viewed this way, the MLE is the random variable

$$\hat{\theta} = \overline{X}.$$

The second derivative is then

$$\frac{\partial^2\ell}{\partial\theta^2} = \frac{n}{\theta^2} - \frac{2n\overline{x}}{\theta^3}.$$

If (as above) we view the x_i as random variables (which we should really write as X_i), then the expected value in the denominator of (17.3) is:

$$\mathbb{E}\left(-\frac{\partial^2\ell(\theta \,|\, X_1, \ldots, X_n)}{\partial\theta^2} \right) = -\mathbb{E}\left(\frac{n}{\theta^2} - \frac{2n\overline{X}}{\theta^3} \right) = -\frac{n}{\theta^2} + \frac{2n}{\theta^3}\mathbb{E}\left(\overline{X} \right) = -\frac{n}{\theta^2} + \frac{2n}{\theta^3}\theta = \frac{n}{\theta^2}.$$

Thus, the Z statistic in (17.3) becomes

$$Z = \frac{\hat{\theta} - \theta}{\sqrt{n/\theta^2}} = \frac{\overline{X} - \theta}{\sqrt{n}/\theta} \stackrel{\text{approx}}{\sim} N(0, 1).$$

The denominator depends on the unknown θ. Since $\overline{X}$ is a consistent estimator of θ, we can substitute this estimator and obtain a further approximation

$$Z = \frac{\overline{X} - \theta}{\sqrt{n}/\overline{X}} \overset{\text{approx}}{\sim} N(0,1).$$

Using this as an approximate pivotal quantity yields

$$\Pr\left(-z_{\alpha/2} \leq \frac{\overline{X} - \theta}{\sqrt{n}/\overline{X}} \leq z_{\alpha/2}\right) = 1 - \alpha.$$

An approximate $100 \times (1 - \alpha)\%$ confidence interval for θ based on (17.3), in other words the Wald interval, is then

$$\left(\overline{X} - z_{\alpha/2}\frac{\overline{X}}{\sqrt{n}}, \overline{X} + z_{\alpha/2}\frac{\overline{X}}{\sqrt{n}}\right).$$

The two methods (exact and Wald) are coded in the R commands below:

```
x = c( 2.5,7.2,12.8,7.4,8.2,19.1,10.0,1.1,8.9,14.5)
n = length(x)
xbar = mean(x)
LCL.exact = 2*n*xbar / qchisq(0.025,2*n,lower.tail=FALSE)
UCL.exact = 2*n*xbar / qchisq(0.975,2*n,lower.tail=FALSE)
print( c(LCL.exact,UCL.exact) )
LCL.Wald = xbar - 1.96*xbar/sqrt(n)
UCL.Wald = xbar + 1.96*xbar/sqrt(n)
print( c(LCL.Wald,UCL.Wald) )
```

The confidence intervals are therefore:

$$(5.37, 19.12) \qquad \text{[exact]}$$

and

$$(3.49, 14.85) \qquad \text{[Wald]}.$$

The point estimate for θ is $\hat{\theta} = \overline{x} = 9.17$. Being based on a normal approximation, the Wald interval is symmetric about 9.17. The exact confidence interval is not symmetric; the upper confidence limit is quite a bit farther from 9.17 than the lower confidence limit. Figure 17.2 shows both of the confidence intervals on a number line, where we can clearly see the asymmetry of the exact interval and the symmetry of the Wald interval. An exact confidence interval is always preferred to one based on an approximation. ∎ ∎ ∎

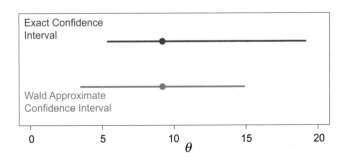

Figure 17.2 Comparison of the exact confidence interval based on the pivotal quantity $2n\overline{X}/\theta \sim \chi^2(2n)$ and the Wald interval based on the approximation in Theorem 17.2.1.

If we would like to estimate the standard error of some function of the parameter θ, say $g(\theta)$, we can use the following theorem, which is a generalization of Theorem 17.2.1.

Theorem 17.2.2 — Asymptotic Normality of a Function of the MLE. Suppose that $X_1, X_2, \ldots, X_n \sim$ i.i.d. from some distribution with PDF $f(x \mid \theta)$. Let

$$\ell(\theta \mid x_1, \ldots, x_n) = \log L(\theta \mid x_1, \ldots, x_n) = \sum_{i=1}^{n} \log f(x_i \mid \theta)$$

denote the log-likelihood function. Under some regularity conditions, the MLE of $g(\theta)$ is equal to $g(\hat{\theta})$, which has an approximate normal distribution for large n with mean $g(\theta)$ and variance

$$\mathbb{V}\big(g(\hat{\theta})\big) \approx \frac{\left(\dfrac{dg}{d\theta}\right)^2}{\mathbb{E}\left(-\dfrac{\partial^2 \ell(\theta \mid x_1, \ldots, x_n)}{\partial \theta^2}\right)}.$$

Also,

$$\frac{g(\hat{\theta}) - g(\theta)}{\sqrt{\big(g'(\theta)\big)^2 \left[\mathbb{E}\left(-\dfrac{\partial^2 \ell(\theta \mid x_1, \ldots, x_n)}{\partial \theta^2}\right)\right]^{-1}}} \overset{\text{approx}}{\sim} N(0, 1). \tag{17.5}$$

■ **Example 17.2**   Suppose that we observe X successes from a sequence of n independent trials, where the probability of success on any trial is p, which is unknown. Derive an approximate confidence interval for the log odds

$$\log \text{odds} = \log \frac{p}{1-p}.$$

Solution: Suppose we observe x successes on n trials. The likelihood function is

$$L(p \mid x) = \binom{n}{x} p^x (1-p)^{n-x}, \qquad x = 0, 1, \ldots, n; \ \ 0 \leq p \leq 1.$$

The log-likelihood function and its first two derivatives are

$$\ell(p \mid x) = \log \binom{n}{x} + x \log p + (n-x) \log(1-p)$$

$$\frac{\partial \ell}{\partial p} = \frac{x}{p} - \frac{n-x}{1-p} \tag{17.6}$$

$$\frac{\partial^2 \ell}{\partial p^2} = -\frac{x}{p^2} - \frac{n-x}{(1-p)^2}.$$

Some of the calculus and algebra has been omitted here; we leave it to the reader to work out these derivatives (see Problem 17.1). The expected value in the denominator of (17.5) is

$$\mathbb{E}\left(-\frac{\partial^2 \ell(p\,|\,X_1,\ldots,X_n)}{\partial p^2}\right) = \mathbb{E}\left(\frac{X}{p^2} + \frac{n-X}{(1-p)^2}\right)$$

$$= \frac{1}{p^2}\,np + \frac{1}{(1-p)^2}(n-np)$$

$$= \frac{n}{p} + \frac{n}{1-p}.$$

Let $g(p) = \log(p/(1-p))$. Then,

$$g'(p) = \frac{d}{dp}\log\frac{p}{1-p} = \frac{1}{\frac{p}{1-p}}\frac{d}{dp}\left(\frac{p}{1-p}\right) = \frac{1}{\frac{p}{1-p}}\frac{(1-p)(1)-p(-1)}{(1-p)^2} = \frac{1}{p(1-p)}.$$

The approximate variance of $\hat{p}$ is thus

$$\mathbb{V}(g(\hat{p})) \approx \frac{(g'(p))^2}{\mathbb{E}\left(-\frac{\partial^2 \ell(p\,|\,x_1,\ldots,x_n)}{\partial p^2}\right)} = \frac{\left(\frac{1}{p(1-p)}\right)^2}{\frac{n}{p}+\frac{n}{1-p}} = \frac{1}{np(1-p)}.$$

Using the consistent estimator $\hat{p} = X/n$ in place of p yields the estimated variance

$$\hat{\mathbb{V}}(g(\hat{p})) \approx \frac{1}{n\hat{p}(1-\hat{p})}.$$

The confidence interval for the log odds $\log(p/(1-p))$ is therefore

$$\left(\log\frac{\hat{p}}{1-\hat{p}} - z_{\alpha/2}\sqrt{\frac{1}{n\hat{p}(1-\hat{p})}}\,,\; \log\frac{\hat{p}}{1-\hat{p}} + z_{\alpha/2}\sqrt{\frac{1}{n\hat{p}(1-\hat{p})}}\right).$$

∎

For a vector of parameters, the following theorem gives the asymptotic normal distribution for the MLE.

Theorem 17.2.3 — Asymptotic Multivariate Normality of the MLE. Suppose that $X_1, X_2, \ldots, X_n \sim$ i.i.d. from some distribution with PDF $f(x\,|\,\theta)$. Let

$$\ell(\theta\,|\,x_1,\ldots,x_n) = \log L(\theta\,|\,x_1,\ldots,x_n) = \sum_{i=1}^{n}\log f(x_i\,|\,\theta)$$

denote the log-likelihood function, where $X_1,\ldots,X_n$ are random vectors and θ is a vector of unknown parameters. Under some regularity conditions, the MLE $\hat{\theta}$ is approximately multivariate normally distributed for large n, having mean θ. Define

$$\mathcal{I}_{ij} = \mathbb{E}\left(-\frac{\partial^2 \ell(\theta\,|\,x_1,\ldots,x_n)}{\partial \theta_i \partial \theta_j}\right). \tag{17.7}$$

The matrix of these expressions is

$$
\mathcal{I} = \begin{bmatrix} \mathcal{I}_{11} & \mathcal{I}_{12} & \cdots & \mathcal{I}_{1p} \\ \mathcal{I}_{21} & \mathcal{I}_{22} & \cdots & \mathcal{I}_{2p} \\ \vdots & \vdots & & \vdots \\ \mathcal{I}_{p1} & \mathcal{I}_{p2} & \cdots & \mathcal{I}_{pp} \end{bmatrix} = \begin{bmatrix} \mathbb{E}\left(-\frac{\partial^2 \ell}{\partial \theta_1^2}\right) & \mathbb{E}\left(-\frac{\partial^2 \ell}{\partial \theta_1 \partial \theta_2}\right) & \cdots & \mathbb{E}\left(-\frac{\partial^2 \ell}{\partial \theta_1 \partial \theta_p}\right) \\ \mathbb{E}\left(-\frac{\partial^2 \ell}{\partial \theta_2 \partial \theta_1}\right) & \mathbb{E}\left(-\frac{\partial^2 \ell}{\partial \theta_2^2}\right) & \cdots & \mathbb{E}\left(-\frac{\partial^2 \ell}{\partial \theta_2 \partial \theta_p}\right) \\ \vdots & \vdots & & \vdots \\ \mathbb{E}\left(-\frac{\partial^2 \ell}{\partial \theta_p \partial \theta_1}\right) & \mathbb{E}\left(-\frac{\partial^2 \ell}{\partial \theta_p \partial \theta_2}\right) & \cdots & \mathbb{E}\left(-\frac{\partial^2 \ell}{\partial \theta_p^2}\right) \end{bmatrix}.
$$

(17.8)

Then, for large n, the covariance matrix for the vector MLE $\hat{\theta}$ of θ is approximately

$$
\mathrm{cov}(\hat{\theta}) \approx \mathcal{I}^{-1}.
$$

Here p is the dimension of the vector θ. For large n,

$$
\hat{\theta} \overset{\mathrm{approx}}{\sim} N\left(\theta, \mathcal{I}^{-1}\right).
$$

(17.9)

The matrix $\mathcal{I}$ in Theorem 17.8 is called the *Fisher information matrix*. This theorem allows us to approximate the standard error of an estimator so long as the expected value of the negative of the second derivatives can be determined. There are situations where even this is difficult to achieve. This leads to the idea of the *observed Fisher information matrix*, which is defined to be

$$
\mathcal{J} = \begin{bmatrix} -\frac{\partial^2 \ell}{\partial \theta_1^2} & -\frac{\partial^2 \ell}{\partial \theta_1 \partial \theta_2} & \cdots & -\frac{\partial^2 \ell}{\partial \theta_1 \partial \theta_p} \\ -\frac{\partial^2 \ell}{\partial \theta_2 \partial \theta_1} & -\frac{\partial^2 \ell}{\partial \theta_1^2} & \cdots & -\frac{\partial^2 \ell}{\partial \theta_1 \partial \theta_p} \\ \vdots & \vdots & & \vdots \\ -\frac{\partial^2 \ell}{\partial \theta_p \partial \theta_1} & -\frac{\partial^2 \ell}{\partial \theta_p \partial \theta_2} & \cdots & -\frac{\partial^2 \ell}{\partial \theta_p^2} \end{bmatrix}\Bigg|_{\hat{\theta}}.
$$

(17.10)

Here, the vertical bar with the subscript $\hat{\theta}$ indicates that after taking the derivatives we evaluate each at the MLE $\hat{\theta}$. The approximate distribution for the estimator $\hat{\theta}$ is given in the next theorem.

Theorem 17.2.4 — Asymptotic Normality Using the Observed Fisher Information Matrix. Suppose that $X_1, X_2, \ldots, X_n \sim$ i.i.d. from some distribution with PDF $f(x \mid \theta)$. Let $\ell(\theta \mid x_1, \ldots, x_n)$ denote the log-likelihood function. Under some regularity conditions, for large n, the MLE of θ satisfies

$$
\hat{\theta} \overset{\mathrm{approx}}{\sim} N\left(\theta, \mathcal{J}^{-1}\right).
$$

(17.11)

Compare (17.9) and (17.11). Both give asymptotic normal distributions for the estimator $\hat{\theta}$. The former uses the Fisher information matrix and involves taking the expectations of the second derivatives of the negative of the log-likelihood function $\ell(\theta)$; since the result usually involves the unknown parameters, we substitute consistent estimators. The latter involves the observed Fisher information matrix; it skips the expectation step and simply plugs the MLE into the negative second derivative of

$\ell(\boldsymbol{\theta})$. In many cases the expectations are not tractable, leaving the observed Fisher information matrix as the only viable alternative.

One would think that the Fisher information would yield better results in terms of approximating the standard error of the estimator $\boldsymbol{\theta}$. There are cases where this is true, but for many one-dimensional problems the observed Fisher information matrix works as well or better than the Fisher information matrix. See Efron and Hinkley (1978) for a discussion of the issue.

■ **Example 17.3** Find the Fisher information matrix and the observed Fisher information matrix for a sample of size n from the $N(\mu, \phi)$ distribution, where both μ and ϕ (the variance, $\phi = \sigma^2$) are unknown. Then, find the asymptotic normal distribution for the MLEs of μ and ϕ.

Solution: The likelihood function for the $N(\mu, \phi)$ distribution is

$$L(\mu, \phi \,|\, x_1, \ldots, x_n) = (2\pi)^{-n/2} \phi^{-n/2} \exp\left(-\frac{1}{2\phi} \sum_{i=1}^{n} (x_i - \mu)^2\right),$$

and the log-likelihood function is

$$\ell(\mu, \phi \,|\, x_1, \ldots, x_n) = -\frac{n}{2}\log 2\pi - \frac{n}{2}\log\phi - \frac{1}{2\phi}\sum_{i=1}^{n}(x_i - \mu)^2. \tag{17.12}$$

The reader can verify (see Problem 17.11) that the first derivatives are:

$$\frac{\partial \ell}{\partial \mu} = n\bar{x}\phi^{-1} - n\mu\phi^{-1}$$

$$\frac{\partial \ell}{\partial \phi} = -\frac{n}{2}\phi^{-1} + \frac{1}{2}\phi^{-2}\sum_{i=1}^{n}(x_i - \mu)^2. \tag{17.13}$$

Now for the second derivatives (Problem 17.12):

$$\frac{\partial^2 \ell}{\partial \mu^2} = -n\phi^{-1}$$

$$\frac{\partial^2 \ell}{\partial \mu \,\partial \phi} = -(n\bar{x} - n\mu)\phi^{-2} \tag{17.14}$$

$$\frac{\partial^2 \ell}{\partial \phi^2} = \frac{n}{2}\phi^{-2} - \phi^{-3}\sum_{i=1}^{n}(x_i - \mu)^2.$$

To get the Fisher information, let's take the expected values of the negatives of these partial derivatives:

$$\mathbb{E}\left(-\frac{\partial^2 \ell}{\partial \mu^2}\right) = \mathbb{E}\left(n\phi^{-1}\right) = n\phi^{-1}$$

$$\mathbb{E}\left(-\frac{\partial^2 \ell}{\partial \mu \,\partial \phi}\right) = \mathbb{E}\left((n\bar{X} - n\mu)\phi^{-2}\right) = \phi^{-2}\left(n\mathbb{E}(\bar{X}) - n\mu\right) = 0 \tag{17.15}$$

$$\mathbb{E}\left(-\frac{\partial^2 \ell}{\partial \phi^2}\right) = -\frac{n}{2}\phi^{-2} + \phi^{-3}\sum_{i=1}^{n}\mathbb{E}\left((X_i - \mu)^2\right) = -\frac{n}{2}\phi^{-2} + \phi^{-3}n\phi = \frac{n}{2\phi^2}.$$

The Fisher information matrix is therefore

$$\mathcal{I} = \begin{bmatrix} \dfrac{n}{\phi} & 0 \\ 0 & \dfrac{n}{2\phi^2} \end{bmatrix}. \tag{17.16}$$

The MLEs are found to be

$$\hat{\mu} = \overline{X} \tag{17.17}$$

$$\hat{\phi} = \frac{1}{n} \sum_{i=1}^{n} (X_i - \mu)^2. \tag{17.18}$$

These are both consistent estimators. Substituting these into the expression in (17.16) yields

$$\mathcal{I} = \begin{bmatrix} \dfrac{n}{\hat{\phi}} & 0 \\ 0 & \dfrac{n}{2\hat{\phi}^2} \end{bmatrix}.$$

The asymptotic covariance matrix for the estimator $[\hat{\mu}, \hat{\phi}]$ is thus[4]

$$\hat{\mathcal{I}}^{-1} = \begin{bmatrix} \dfrac{\hat{\phi}}{n} & 0 \\ 0 & \dfrac{2\hat{\phi}^2}{n} \end{bmatrix}.$$

The observed Fisher information matrix is obtained by substituting the MLEs directly into (17.14):

$$\begin{aligned} -\frac{\partial^2 \ell}{\partial \mu^2}\bigg|_{[\hat{\mu},\hat{\phi}]} &= \frac{n}{\hat{\phi}} \\ -\frac{\partial^2 \ell}{\partial \mu \, \partial \phi}\bigg|_{[\hat{\mu},\hat{\phi}]} &= (n\bar{x} - n\hat{\mu})\hat{\phi}^{-2} = 0 \\ -\frac{\partial^2 \ell}{\partial \phi^2}\bigg|_{[\hat{\mu},\hat{\phi}]} &= -\frac{n}{2}\hat{\phi}^{-2} + \hat{\phi}^{-3}\sum_{i=1}^{n}(x_i - \hat{\mu})^2 = -\frac{n}{2\hat{\phi}^2} + \frac{n\hat{\phi}}{\hat{\phi}^3} = \frac{n}{2\hat{\phi}^2}. \end{aligned} \tag{17.19}$$

This yields

$$\mathcal{J} = \begin{bmatrix} \dfrac{n}{\hat{\phi}} & 0 \\ 0 & \dfrac{n}{2\hat{\phi}^2} \end{bmatrix}.$$

Thus, using the observed Fisher information matrix, we estimate the asymptotic covariance matrix as

$$\mathcal{J}^{-1} = \begin{bmatrix} \dfrac{\hat{\phi}}{n} & 0 \\ 0 & \dfrac{2\hat{\phi}^2}{n} \end{bmatrix}.$$

In this case the estimated Fisher information matrix $\hat{\mathcal{I}}$ and the observed Fisher information matrix are the same.   ∎ ∎ ∎

---

[4]For a diagonal matrix D, the inverse is the diagonal matrix whose entries on the main diagonal are the reciprocals of the diagonal entries in D. In general, though, the inverse of a matrix M is *not* the matrix of reciprocals.

■ **Example 17.4**   Suppose we have a sample $T_1, T_2, \ldots, T_n$ from the Weibull distribution with parameters θ and β. Investigate the practicality of using the Fisher information matrix and the observed Fisher information matrix to find 95% interval estimates for β and θ.

Solution: In Example 9.22 we derived the likelihood function and the first two derivatives of the log-likelihood function. Repeating the results in (9.11),

$$\frac{\partial \ell}{\partial \theta} = -\frac{n\beta}{\theta} + \sum_{i=1}^{n} \beta \theta^{-\beta-1} t_i^{\beta} = 0$$

$$\frac{\partial \ell}{\partial \beta} = \frac{n}{\beta} - n \log \theta + \sum_{i=1}^{n} \log t_i - \sum_{i=1}^{n} (t_i/\theta)^{\beta} \log \frac{t_i}{\theta} = 0. \tag{17.20}$$

These equations can be simplified to

$$\theta = \left(\frac{\sum_{i=1}^{n} t_i^{\beta}}{n} \right)^{1/\beta}$$

$$\beta = \frac{n \sum_{i=1}^{n} t_i^{\beta}}{n \sum_{i=1}^{n} t_i^{\beta} \log t_i - \left(\sum_{i=1}^{n} \log t_i \right) \left(\sum_{i=1}^{n} t_i^{\beta} \right)}. \tag{17.21}$$

The second equation must be solved numerically for β; once this is done, the result can be substituted in the first equation to find the MLE for θ.

Both the Fisher information and observed Fisher information matrices require the second derivatives. We leave it to the reader to verify the following results (Problem 17.8):

$$\frac{\partial^2 \ell}{\partial \theta^2} = \frac{n\beta}{\theta^2} - \frac{\beta(\beta+1)}{\theta^2} \sum_{i=1}^{n} \left(\frac{t_i}{\theta} \right)^{\beta}$$

$$\frac{\partial^2 \ell}{\partial \theta \partial \beta} = -\frac{n}{\theta} + \frac{1}{\theta} \sum_{i=1}^{n} \left(\frac{t_i}{\theta} \right)^{\beta} \left(1 + \beta \log \frac{t_i}{\theta} \right) \tag{17.22}$$

$$\frac{\partial^2 \ell}{\partial \beta^2} = -\frac{n}{\beta^2} - \sum_{i=1}^{n} \left(\frac{t_i}{\theta} \right)^{\beta} \left(\log \frac{t_i}{\theta} \right)^2.$$

Getting the expected values of these expressions presents serious challenges. It is not feasible to use the Fisher information matrix in this situation. Instead, we must use the observed Fisher information matrix:

$$\mathcal{J} = \left. \begin{bmatrix} -\left(\dfrac{n\beta}{\theta^2} - \dfrac{\beta(\beta+1)}{\theta^2} \displaystyle\sum_{i=1}^{n} \left(\dfrac{t_i}{\theta} \right)^{\beta} \right) & -\left(-\dfrac{n}{\theta} + \dfrac{1}{\theta} \displaystyle\sum_{i=1}^{n} \left(\dfrac{t_i}{\theta} \right)^{\beta} \left(1 + \beta \log \dfrac{t_i}{\theta} \right) \right) \\[20pt] -\left(-\dfrac{n}{\theta} + \dfrac{1}{\theta} \displaystyle\sum_{i=1}^{n} \left(\dfrac{t_i}{\theta} \right)^{\beta} \left(1 + \beta \log \dfrac{t_i}{\theta} \right) \right) & -\left(-\dfrac{n}{\beta^2} - \displaystyle\sum_{i=1}^{n} \left(\dfrac{t_i}{\theta} \right)^{\beta} \left(\log \dfrac{t_i}{\theta} \right)^2 \right) \end{bmatrix} \right|_{[\hat{\theta}, \hat{\beta}]}.$$

We would then approximate the MLE's $\hat{\theta}$ and $\hat{\beta}$, substitute them into this matrix, and finally invert the matrix to get the approximate covariance matrix for the estimators.   ■ ■ ■

■ Example 17.5   Assuming a Weibull distribution for the data in Example 9.22, use the observed Fisher information matrix to find approximate 95% confidence intervals for the parameters θ and β.

Solution: The observed lifetimes were 12, 17, 40, 71. In Example 9.22 we found that the MLEs were

$$\hat{\beta} = 1.577 \quad \text{and} \quad \hat{\theta} = 39.25.$$

The R code to evaluate the second derivative matrix for the log-likelihood function is:

```
loglike.2nd.deriv = function( theta , beta , x )
{
        n = length( x )
        s11 = sum( (x/theta)^beta )
        s12 = sum( (x/theta)^beta * ( 1 + beta*log(x/theta) ) )
        s22 = sum( (x/theta)^beta * (log((x/theta)))^2 )
        l.11 = n*beta/theta^2 - beta*(beta+1)*s11/theta^2
        l.12 = - n/theta + s12/theta
        l.22 = - n/beta^2 - s22
        H = matrix( c( l.11 , l.12 ,
                       l.12 , l.22 ) , nrow=2 , ncol=2 )
        return( H )
}
thetahat = 39.25
betahat = 1.577
t = c( 12 , 17 , 40 , 71 )
n = length( t )
ObsFisherInfMatrix = - loglike.2nd.deriv( thetahat , betahat , t )
```

The estimated variance of the estimators $\hat{\theta}$ and $\hat{\beta}$ are then the $(1,1)$ and $(2,2)$ entries in the inverse of `ObsFisherInfMatrix`. The following R code accomplishes this task and computes the 95% confidence interval for each of the parameters:

```
Sigma = solve( ObsFisherInfMatrix )
print( c(thetahat-1.96*sqrt(Sigma[1,1]),thetahat+1.96*sqrt(Sigma[1,1])) )
print( c(betahat-1.96*sqrt(Sigma[2,2]),betahat+1.96*sqrt(Sigma[2,2])) )
```

The output from these last two lines gives the confidence intervals for θ and β:

```
[1]  13.41285 65.08715
[1]  0.3597417 2.7942583
```

These confidence intervals are extremely wide, which should be expected from a sample of size $n = 4$.

                                                                              ■ ■ ■

17.3 The Jackknife

In Section 17.1 we looked at how we could use the log-likelihood to give us an approximation for the standard error of the MLE. This approach makes a particular distributional assumption for the data, and we must always keep in mind that these are *large sample approximations* based on the normal distribution. For small or moderate sample sizes, the approximation may be poor. We will now discuss two techniques that make no particular distributional assumptions. These are called the *jackknife* and the *bootstrap*.[5]

---

[5]These colorful terms have a history. The jackknife, a knife that folds up into a case to protect you from the blade, is a good all-purpose tool. Maybe it's not ideal for every situation, but it is widely useful. In statistics, the jackknife is

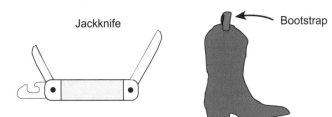

Figure 17.3 Common items that have meanings in statistics.

The jackknife and the bootstrap are computationally intensive methods and were not possible before computers. These methods can be used instead of the messy analytic approaches described in Section 17.1. Efron (1982), in his classic monograph *The Jackknife, the Bootstrap and Other Resampling Plans*, described these methods this way:

> An important theme of what follows is the substitution of computational power for theoretical analysis.

The jackknife can be used for two purposes: to reduce the bias in an estimator and to approximate the standard error of an estimator. The latter can then be used to approximate a confidence interval.

Suppose we have a random sample $X_1, X_2, \ldots, X_n$ from a distribution with CDF $F(x)$. We write this as $X_1, X_2, \ldots, X_n \sim F$. The jackknife approach involves successively leaving one of the observations out. Suppose that θ is some characteristic of the CDF F, such as the mean, median, variance, etc., and let $\hat{\theta}$ be an estimator of θ computed from the values in the sample $X_1, X_2, \ldots, X_n$. Define

$$\hat{\theta}_{(i)} = \text{the estimator } \hat{\theta} \text{ computed from the sample } X_1, \ldots, X_{i-1}, \; X_{i+1}, \ldots, X_n. \tag{17.23}$$

We define the jackknife estimator of θ to be the average of these "leave-one-out" estimators:

$$\hat{\theta}_{(\cdot)} = \frac{1}{n} \sum_{i=1}^{n} \hat{\theta}_{(i)}. \tag{17.24}$$

The bias in the estimator $\hat{\theta}$ can be estimated to be (see Efron, 1982: equation (2.7))

$$\widehat{\text{Bias}} = (n-1)(\hat{\theta}_{(\cdot)} - \hat{\theta}). \tag{17.25}$$

The bias-corrected jackknife estimator of θ is then

$$\hat{\theta}_{BC} = \hat{\theta} - \widehat{\text{Bias}} = \hat{\theta} - (n-1)\hat{\theta}_{(\cdot)} + (n-1)\hat{\theta} = n\hat{\theta} - (n-1)\hat{\theta}_{(\cdot)}. \tag{17.26}$$

The bias-corrected jackknife is not necessarily an unbiased estimator, because the bias from (17.25) is just an estimate, but the bias in the resulting estimator is usually less.

■ **Example 17.6** Suppose $X_1, X_2, \ldots, X_n \sim F$ where the distribution with CDF F has mean μ. Find the jackknife estimator of μ and the bias-corrected jackknife estimator.

---

believed to be a useful technique (although not always ideal) across a wide variety of circumstances. A bootstrap is a small loop found at the top of a boot used to pull the boot onto your foot (Figure 17.3). The phrase "pull yourself up by the bootstrap" implies that you could lift your body if you pulled hard enough. If this were possible (which it isn't!) you could lift yourself off the ground with no external help. In statistics, the bootstrap uses no external assumptions about the distribution of the data. While physical bootstrapping is not possible, statistical bootstrapping *is*, and we will describe it in Section 17.4

Solution: Let $\hat{\mu} = \overline{X}$. Then,

$$\hat{\mu}_{(i)} = \overline{X}_{(i)} = \frac{1}{n-1} \sum_{\substack{j=1 \\ j \neq i}}^{n} X_j.$$

Thus, the jackknife estimator is

$$
\begin{aligned}
\hat{\mu}_{(\cdot)} &= \frac{1}{n} \sum_{i=1}^{n} \hat{\mu}_{(i)} \\
&= \frac{1}{n} \sum_{i=1}^{n} \frac{1}{n-1} \sum_{\substack{j=1 \\ j \neq i}}^{n} X_j \\
&= \frac{1}{n} \frac{1}{n-1} \sum_{i=1}^{n} \sum_{\substack{j=1 \\ j \neq i}}^{n} X_j
\end{aligned}
$$

$$
= \frac{1}{n(n-1)} \big[\begin{array}{l} X_2 + X_3 + \cdots + X_{n-1} + X_n + \\ X_1 \quad\quad + X_3 + \cdots + X_{n-1} + X_n + \\ X_1 + X_2 \quad\quad + \cdots + X_{n-1} + X_n + \\ \cdots + \\ X_1 + X_2 + X_3 + \cdots \quad\quad + X_n + \\ X_1 + X_2 + X_3 + \cdots + X_{n-1} \big]
\end{array}
$$

$$= \frac{1}{n(n-1)} \big[(n-1)X_1 + (n-1)X_2 + (n-1)X_3 + \cdots + (n-1)X_{n-1} + (n-1)X_n\big]$$

$$= \overline{X}.$$

The estimated bias is therefore

$$\widehat{\text{Bias}} = (n-1)\big(\hat{\mu}_{(\cdot)} - \hat{\mu}\big) = (n-1)\big(\overline{X} - \overline{X}\big) = 0.$$

The bias-corrected jackknife estimator is thus

$$\hat{\mu}_{BC} = n\overline{X} - (n-1)\overline{X} = \overline{X}.$$

In this case, the bias-corrected jackknife estimate is the jackknife estimator, which is the same as the estimator $\hat{\mu} = \overline{X}$. The bias is zero, which we would expect since the sample mean $\overline{X}$ is an unbiased estimator for the population mean μ. ∎

■ **Example 17.7** Suppose $X_1, X_2, X_3 \sim$ from some distribution with mean μ and variance σ^2. Point estimates of the parameters are

$$\hat{\mu} = \overline{X} \quad \text{and} \quad \hat{\sigma}^2 = \frac{1}{3} \sum_{i=1}^{3} (x_i - \overline{x})^2 = \frac{1}{3}\left(\sum_{i=1}^{3} x_i^2 - 3\overline{x}^2\right).$$

Find the jackknife estimate of σ^2 and then find the jackknife estimate of the bias.

Solution: With only three observations, there are only three values of $\hat{\sigma}^2_{(i)}$ that must be computed; they are

$$\hat{\sigma}^2_{(1)} = \frac{1}{2}\left[\sum_{\substack{i=1\\j\neq 1}}^{3} x_i^2 - 2\bar{x}^2_{(1)}\right] = \frac{1}{2}\left[x_2^2 + x_3^2 - 2\left(\frac{x_2+x_3}{2}\right)^2\right] = \frac{1}{4}x_2^2 + \frac{1}{4}x_3^2 - \frac{1}{2}x_2 x_3$$

$$\hat{\sigma}^2_{(2)} = \frac{1}{2}\left[\sum_{\substack{i=1\\j\neq 2}}^{3} x_i^2 - 2\bar{x}^2_{(2)}\right] = \frac{1}{2}\left[x_1^2 + x_3^2 - 2\left(\frac{x_1+x_3}{2}\right)^2\right] = \frac{1}{4}x_1^2 + \frac{1}{4}x_3^2 - \frac{1}{2}x_1 x_3$$

$$\hat{\sigma}^2_{(3)} = \frac{1}{2}\left[\sum_{\substack{i=1\\j\neq 3}}^{3} x_i^2 - 2\bar{x}^2_{(3)}\right] = \frac{1}{2}\left[x_1^2 + x_2^2 - 2\left(\frac{x_1+x_2}{2}\right)^2\right] = \frac{1}{4}x_1^2 + \frac{1}{4}x_2^2 - \frac{1}{2}x_1 x_2.$$

Thus, the jackknife estimator of σ^2 is

$$\hat{\sigma}^2_{(\cdot)} = \frac{1}{3}\sum_{i=1}^{3}\hat{\sigma}^2_{(i)}$$

$$= \frac{1}{3}\left[\left(\frac{1}{4}x_2^2 + \frac{1}{4}x_3^2 - \frac{1}{2}x_2 x_3\right) + \left(\frac{1}{4}x_1^2 + \frac{1}{4}x_3^2 - \frac{1}{2}x_1 x_3\right) + \left(\frac{1}{4}x_1^2 + \frac{1}{4}x_2^2 - \frac{1}{2}x_1 x_2\right)\right]$$

$$= \frac{1}{3}\left[\frac{1}{2}x_1^2 + \frac{1}{2}x_2^2 + \frac{1}{2}x_3^2 - \frac{1}{2}x_1 x_2 - \frac{1}{2}x_1 x_3 - \frac{1}{2}x_2 x_3\right]$$

$$= \frac{1}{6}\left[x_1^2 + x_2^2 + x_3^2 - x_1 x_2 - x_1 x_3 - x_2 x_3\right]. \tag{17.27}$$

The MLE for σ^2 can be written as

$$\hat{\sigma}^2 = \frac{1}{3}\left[\sum_{i=1}^{3} x_i^2 - 3\left(\frac{x_1+x_2+x_3}{3}\right)^2\right]$$

$$= \frac{1}{3}\left[x_1^2 + x_2^2 + x_3^2 - \frac{1}{3}\left(x_1^2 + x_2^2 + x_3^2 + 2x_1 x_2 + 2x_1 x_3 + x_2 x_3\right)\right]$$

$$= \frac{2}{9}\left[x_1^2 + x_2^2 + x_3^2 - x_1 x_2 - x_1 x_3 - x_2 x_3\right].$$

The estimated bias is therefore

$$\widehat{\text{Bias}} = (3-1)\left(\hat{\sigma}^2_{(\cdot)} - \hat{\sigma}^2\right)$$

$$= 2\left(\frac{1}{6}\left[x_1^2 + x_2^2 + x_3^2 - x_1 x_2 - x_1 x_3 - x_2 x_3\right] - \frac{2}{9}\left[x_1^2 + x_2^2 + x_3^2 - x_1 x_2 - x_1 x_3 - x_2 x_3\right]\right)$$

$$= -\frac{1}{9}\left[x_1^2 + x_2^2 + x_3^2 - x_1 x_2 - x_1 x_3 - x_2 x_3\right]$$

$$= -\frac{1}{9} \times \frac{9}{2}\hat{\sigma}^2$$

$$= -\frac{1}{2}\hat{\sigma}^2.$$

The bias-corrected jackknife estimator is therefore

$$\hat{\sigma}^2_{BC} = \hat{\sigma}^2 - \widehat{\text{Bias}} = \hat{\sigma}^2 - \left(-\frac{1}{2}\,\hat{\sigma}^2\right) = \hat{\sigma}^2 + \frac{1}{2}\,\hat{\sigma}^2 = \frac{3}{2}\hat{\sigma}^2 = \frac{1}{3-1}\sum_{i=1}^{n}(x_i-\bar{x})^2.$$

Thus, the bias-corrected estimate of the MLE $\hat{\sigma}^2$ is the usual unbiased estimator

$$\hat{\sigma}^2_{BC} = \frac{1}{3-1}\sum_{i=1}^{n}(x_i-\bar{x})^2. \tag{17.28}$$

In other words, the bias-corrected estimator is the usual unbiased estimator of σ^2, which we have called the sample variance and denoted S^2. ∎ ∎ ∎

In the previous example, the sample size was $n=3$. The results for an arbitrary sample size of n hold as well. For example, the jackknife estimate of the variance, generalizing (17.27), is

$$\hat{\sigma}^2_{(\cdot)} = \frac{1}{n(n-1)}\left(\sum_{i=1}^{n}x_i^2 - \sum_{i<j}x_i x_j\right).$$

The MLE can be written as (see Problem 17.13)

$$\hat{\sigma}^2 = \frac{1}{n}\left[\sum_{i=1}^{n}x_i^2 - \frac{1}{n}\sum_{i=1}^{n}\sum_{j=1}^{n}x_i x_j\right] = \frac{1}{n^2}\left[(n-1)\sum_{i=1}^{n}x_i^2 - 2\sum_{i<j}x_i x_j\right]. \tag{17.29}$$

From (17.25) the estimated bias is therefore

$$\widehat{\text{Bias}} = -\frac{1}{n(n-1)}\sum_{i=1}^{n}(x_i-\bar{x})^2 = -\frac{1}{n-1}\hat{\sigma}^2,$$

and the bias-corrected jackknife estimator is

$$\hat{\sigma}^2_{BC} = \hat{\sigma}^2 - \widehat{\text{Bias}} = \hat{\sigma}^2 - \left(-\frac{1}{n-1}\hat{\sigma}^2\right) = \frac{n}{n-1}\hat{\sigma}^2 = \frac{1}{n-1}\sum_{i=1}^{n}(x_i-\bar{x})^2.$$

This is the usual unbiased estimator for the population variance.

The jackknife can also provide an estimate of the standard error of a statistic. This is particularly useful when the standard error cannot be easily calculated using the methods of Section 17.1. Let

$$\hat{\theta}_{(\cdot)} = \frac{1}{n}\sum_{i=1}^{n}\hat{\theta}_{(i)}$$

be the mean of the n jackknife samples. The jackknife estimate of the standard error if the statistic $\hat{\theta}$ is then

$$\widehat{\text{s.e.}}_{JK} = \sqrt{\frac{n-1}{n}\sum_{i=1}^{n}\left(\hat{\theta}_{(i)}-\hat{\theta}_{(\cdot)}\right)^2}.$$

■ **Example 17.8** Generalizing Example 17.7, suppose $X_1, X_2, \ldots, X_n \sim$ i.i.d. from some distribution with mean μ and variance σ^2. Point estimates of the parameters are

$$\hat{\mu} = \overline{X} \quad \text{and} \quad \hat{\sigma}^2 = \frac{1}{n}\sum_{i=1}^{n}(x_i - \bar{x})^2 = \frac{1}{n}\left(\sum_{i=1}^{n}x_i^2 - n\bar{x}^2\right).$$

Find the jackknife estimator of the standard error for $\hat{\mu} = \bar{x}$.

Solution: From the result of Example 17.7, we know that the jackknife estimator for μ is $\hat{\mu}_{(\cdot)} = \overline{X} = \hat{\mu}$. Since

$$\hat{\mu}_{(i)} = \frac{n\overline{X} - X_i}{n - 1},$$

we can write

$$\hat{\mu}_{(i)} - \hat{\mu}_{(\cdot)} = \hat{\mu}_{(i)} - \hat{\mu} = \frac{n\overline{X} - X_i}{n - 1} - \hat{\mu} = \frac{n\overline{X} - X_i}{n - 1} - \frac{(n-1)\overline{X}}{n - 1} = \frac{\overline{X} - X_i}{n - 1}.$$

The estimated standard error of $\hat{\mu} = \overline{X}$ is therefore

$$\widehat{\text{s.e.}}_{\text{JK}} = \sqrt{\frac{n-1}{n}\sum_{i=1}^{n}\left(\hat{\mu}_{(i)} - \hat{\mu}_{(\cdot)}\right)^2}$$

$$= \sqrt{\frac{n-1}{n}\sum_{i=1}^{n}\left(\frac{\overline{X} - X_i}{n-1}\right)^2}$$

$$= \sqrt{\underbrace{\frac{1}{n}\frac{1}{n-1}\sum_{i=1}^{n}(X_i - \overline{X})^2}_{S^2}}$$

$$= \frac{S}{\sqrt{n}},$$

which is the usual estimated standard error of $\overline{X}$. ■■■

In this example, the jackknife estimate of the standard error gives a familiar result. In many examples, like the next one, the jackknife can yield an estimate of the standard error when simple formulas don't exist.

■ **Example 17.9** Consider the GPA versus SATM score that played a prominent role in Chapter 14. Find the jackknife estimate of the correlation between GPA and SATM, and approximate its standard error. Use it to find an approximate 95% confidence interval.

Solution: The estimated correlation from the full sample is $\hat{\rho} = 0.1943439$. The following code reads the data and computes the jackknife estimate of the standard error of $\hat{\rho}$:

```
first = read.csv( "FirstYearGPA.csv" , header=TRUE )
xy = first[ , c("GPA","SATM") ]
n = nrow( xy )
rhohat. = rep( NA , n )
```

```
for( i in 1:n )
{
        xy. = xy[ -i , ]
        rhohat.[i] = cor( xy. )[1,2]
}
print( mean( rhohat. ) )
```

The jackknife estimate of ρ is thus

$$\hat{\rho}_{\text{JK}} = 0.1943393,$$

which is very close to the usual point estimate.

The jackknife estimate of the standard error is found from:

```
rhohat.jk = mean( rhohat. )
sehat = sqrt( ((n-1)/n) * sum( ( rhohat. - rhohat.jk )^2 ) )
print( sehat )
```

This yields the estimate

$$\widehat{\text{s.e.}}_{\text{JK}} = 0.06617.$$

An approximate confidence interval for ρ is therefore

$$\big(0.1943 - 1.96 \times 0.06617, 0.1943 + 1.96 \times 0.06617\big) = \big(0.0646, 0.324\big). \qquad \blacksquare\blacksquare\blacksquare$$

17.4 The Bootstrap

The standard error of the estimator $\hat{\theta}$ is defined to be its standard deviation. If we could repeatedly take samples from the population, we could compute $\hat{\theta}$ from each sample and thereby easily obtain simulated values of the sampling distribution. This could give us an estimate of the standard error of the estimator $\hat{\theta}$. But, of course, this is infeasible because we typically have only one sample.

As an alternative, we can sample from the sample. This is called *resampling*, and it is the essence of the bootstrap. While this may seem odd and fruitless, there is a sound reason to do this. Let's begin with the definition of the empirical cumulative distribution function.

Definition 17.4.1 — Empirical CDF. Suppose we have a sample $x_1, x_2, \ldots, x_n$ from a distribution with CDF F. The *empirical cumulative distribution function* (ECDF) is defined to be

$$\hat{F}(x) = \begin{cases} 0, & \text{if } x < x_{(1)} \\ \dfrac{1}{n}, & \text{if } x_{(1)} \leq x < x_{(2)} \\ \dfrac{2}{n}, & \text{if } x_{(2)} \leq x < x_{(3)} \\ \vdots & \\ \dfrac{n-1}{n} & \text{if } x_{(n-1)} \leq x < x_{(n)} \\ 1 & \text{if } x \geq x_{(n)}. \end{cases}$$

In other words,

$$\hat{F}(x) = \text{proportion of } x_i\text{s less than or equal to } x.$$

A bootstrap sample is a sample of size n selected from the CDF $\hat{F}$. The basic idea of the bootstrap is that we treat $\hat{F}$ as if it were CDF F of the population. This means that we are sampling from $\{x_1, x_2, \ldots, x_n\}$ *with replacement*. If we were to sample without replacement, we would always get the original sample, though probably in a different order. From each bootstrap sample, we compute the desired statistic; for example, if we were interested in the mean of the distribution, we would calculate the sample mean in each bootstrap sample. We will denote the original sample by the vector x and the ith bootstrap sample by $x^*_{(i)}$. The statistic that we are bootstrapping is $s(x)$, which, computed for the ith bootstrap sample, is $s(x^*_{(i)})$. In the present case, we're bootstrapping the sample mean, so we write

$$s(x^*_{(i)}) = \bar{x}^*_{(i)}.$$

The asterisk refers to the bootstrap sample, and the absence of an asterisk indicates the original sample.

The bootstrap concept is illustrated in Figure 17.4. Suppose we select one sample from some unknown distribution (Figure 17.4(a)). The empirical CDF is computed for the sample that we selected from the population. Figure 17.4(d) is a histogram of the sample means from all the resamples taken from the empirical CDF. Figure 17.4(c) shows, for comparison, what it would look like if we could resample

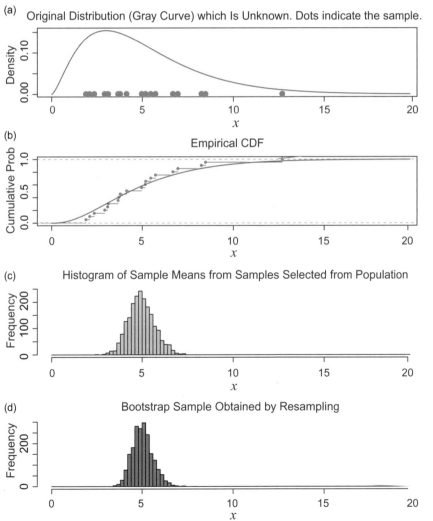

(a) Original Distribution (Gray Curve) which Is Unknown. Dots indicate the sample.

(b) Empirical CDF

(c) Histogram of Sample Means from Samples Selected from Population

(d) Bootstrap Sample Obtained by Resampling

Figure 17.4 Illustration of the bootstrap. A sample is selected from some unknown distribution (a). Since we cannot resample from the population, we resample from the empirical CDF (b). The statistic, $\bar{x}$, is computed for each sample selected from the population (c). The statistic, $\bar{x}$, is computed for each of the bootstrap samples (d).

from the population (which we can't in the real world). This would be the sampling distribution of the statistic, the sample mean in this case. Remember, the gray items in Figure 17.4 are unknown and so we can never observe them. The bootstrap concept is that the histogram of the statistic computed from each of the bootstrap resamples is an estimate of the actual sampling distribution of the statistic. Thus, the histogram in red in Figure 17.4(d) estimates the sampling distribution shown in Figure 17.4(c). There is some theory that backs up the assertions we've made here. Interested readers are referred to Efron (1982).

The histogram in Figure 17.4(d) is just an *estimate* of the sampling distribution. Depending on the original sample obtained (Figure 17.4(a)), the bootstrap distribution will sometimes be centered to the left of the true sampling distribution, and sometimes to the right. Also, the variability of the bootstrap distribution will sometimes be less than the variability of the true sampling distribution, and sometimes it will be greater. The code used to create these graphs is shown below. We invite the reader to run this code repeatedly (after deleting the set.seed() command on the first line). This will illustrate the idea that the bootstrap histogram *estimates* the sampling distribution. Note that the code performs 1,000 bootstrap samples, which is typically sufficient. A greater number of resamples will lead to a smoother looking histogram, but usually does little to affect the center or dispersion.

```
set.seed( 123 )
df = 5
n = 20
x = seq( 0 , 20 , 0.1 )
y = pchisq( x , df )
xRand = rchisq( n , df )

par( mfrow=c(4,1) )

plot( x , dchisq( x , df) , type="l" , col="darkgray" , lwd=2 ,
      xlim=c(0,15) , ylim=c(-0.02,0.20) , ylab="" ,
      main="Original Distribution (Gray Curve) which Is Unknown")
  points( xRand , rep(0,n) , cex=1.5 , pch=19 , col="red" )
  points( mean( xRand ) , -0.015 , pch=2 )
  abline( h=0 )
plot( ecdf(xRand) , xlim=c(0,15) , col="red" , lwd=2 ,
      main="Empirical CDF" , ylab="" )
  lines( x , y ,col="darkgray" , lwd=2 )

B = 1000
xMat = matrix( rchisq( B*n , df ) , nrow=B , ncol=n )
samp = apply( xMat , 1 , mean )
hist( samp , breaks=seq(0,15,0.2) , xlim=c(0,20) , ylab="" ,
      main="Histogram of Sample from Population" , col="lightgray" )

xBSMat = c()
for ( i in 1:B )
{
      xBSMat = rbind( xBSMat , sample(xRand,n,replace=TRUE) )
}
BSsamp = apply( xBSMat , 1 , mean )
hist( BSsamp , breaks=seq(0,15,0.2) , xlim=c(0,20) , ylab="" ,
      main="Bootstrap Sample Obtained by Resampling" , col="red" )
```

■ **Example 17.10**   Suppose our data consist of five values $x_1 = 2.0$, $x_2 = 3.1$, $x_3 = 3.7$, $x_4 = 4.5$, $x_5 = 6.2$ selected from a distribution with CDF F. Select three bootstrap samples and calculate the sample mean for each.

Solution: The R function `sample()` can be used to get a sample selected from the set of numbers $1, 2, 3, 4, 5$. We must tell R the set from which we want to sample, how many samples to select, and whether we want to sample with or without replacement. For example, one run of

```
sample( 1:5 , 5 , replace=TRUE )
```

yields

```
[1] 4 2 1 2 2
```

This means that we would select x_4, x_2, x_1, x_2, x_2 for our bootstrap sample. Here, x_2 got selected three times, while x_3 and x_5 were not selected at all. Second and third applications of the sample function yield

```
[1] 3 4 5 1 5
```

and

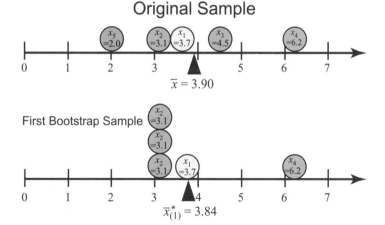

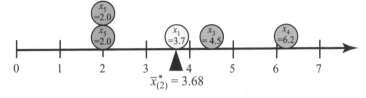

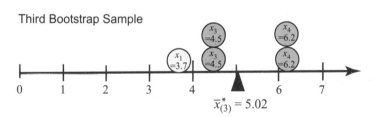

Figure 17.5 Illustration of the bootstrap sampling method for the data in Example 17.10.

```
[1] 3 4 4 3 1
```

These three bootstrap samples yield

$$\bar{x}_{(1)}^* = 3.84 \qquad \bar{x}_{(2)}^* = 3.68 \qquad \bar{x}_{(3)}^* = 5.02,$$

respectively. The bootstrap procedure is illustrated in Figure 17.5. We invite the reader to rerun the sample function several times to see the variability in the sample selected.

■ ■ ■

In practice, the number of bootstrap samples is $B = 1,000$ or higher.

■ **Example 17.11**   Using the small data set from Example 17.10, $x_1 = 2.0$, $x_2 = 3.1$, $x_3 = 3.7$, $x_4 = 4.5$, $x_5 = 6.2$, compute $B = 2,000$ bootstrap samples, compute the sample mean from each, and give a histogram of these sample means.

Solution: The following R code accomplishes these tasks:

```
x = c( 2.0 , 3.1 , 3.7 , 4.5, 6.2)
n = length( x )
B = 2000
BS.muhat = rep( NA , B )
for( i in 1:B )
{
        BSsamp = sample( x , n , replace=TRUE )
        BS.muhat[i] = mean( BSsamp )
}
hist( BS.muhat , breaks=seq(0,8,0.25) , xlab="x" , col="red" , main="" )
xL = quantile( BS.muhat , 0.025)
xU = quantile( BS.muhat , 0.975)
lines( c(xL,xL) , c(0,220) , lwd=2 , col="gray" )
lines( c(xU,xU) , c(0,220) , lwd=2 , col="gray" )
text( xL , 240 , paste0(round(xL,3)) )
text( xU , 240 , paste0(round(xU,3)) )
```

The histogram without the added lines and text is shown in Figure 17.6.   ■ ■ ■

Had we assumed in Example 17.11 that the sampling distribution of $\overline{X}$ was normal, then given values for the sample mean $\bar{x}$ and sample variance σ^2, our estimate of the sampling distribution would be $N(\bar{x}, s^2/n)$. For this data set we have

$$\bar{x} = 3.9 \qquad \text{and} \qquad \frac{s^2}{n} = 0.705.$$

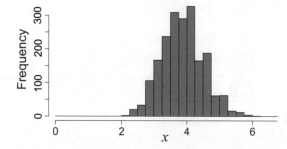

Figure 17.6 Histogram of $B = 2,000$ bootstrap samples from the small data set in Example 17.11.

Under the assumption of normality of the data, the 95% confidence interval for μ would be

$$\left(\bar{x} - z_{\alpha/2} \frac{s}{\sqrt{n}} \,,\, \bar{x} + z_{\alpha/2} \frac{s}{\sqrt{n}} \right) = (\, 2.581 \,,\, 5.282 \,).$$

Note that 2.581 and 5.282 are the 0.025 and 0.975 quantiles of the $N(3.9, 0.705)$ distribution, which is our estimate of the sampling distribution of $\overline{X}$.

The histogram in the previous example is our estimate of the sampling distribution for the sample statistic $\overline{X}$. Without assuming that the sampling distribution of $\overline{X}$ is normal, we can estimate the quantiles of the distribution of the statistic ($\overline{X}$ in this case) from the bootstrap samples. The interval found in this way is the confidence interval for the parameter that $\overline{X}$ estimates; in this example $\overline{X}$ estimates the population mean μ.

17.4.1 Bootstrap Confidence Intervals Based on Percentiles

This concept gives us a general method for approximating a confidence interval for a parameter: Take the $\alpha/2$ and $1 - \alpha/2$ quantiles of the distribution formed by determining the estimator based on each of the B bootstrap samples.

■ Example 17.12 Use the data from Example 17.11 to approximate a 95% confidence interval for the mean. Contrast this with the confidence interval based on the assumption that the data come from a normal distribution.

Solution: The 0.025 and 0.975 quantiles of the bootstrap distribution of $\bar{x}$s are 2.720 and 5.180, giving the bootstrap interval $(2.720, 5.180)$. If we assume that the data come from a normal distribution, then the confidence interval for μ is $(2.518, 5.282)$. Figure 17.7 shows the bootstrap distribution for $\overline{X}$ along with the $N(3.9, 0.705)$ distribution. Confidence limits based on the bootstrap are indicated by the gray lines extending upward, and confidence limits based on the normal distribution are the blue lines extending downward. **■ ■ ■**

■ Example 17.13 Find a 95% bootstrap confidence interval for the correlation between the SATM score and the first-year GPA for the data set described in Example 17.9.

Solution: The following R code reads the data, selects the bootstrap sample, and computes the correlation for the sample. Note that in this case, an "observation" is an ordered pair, so the bootstrap method selects a sample of size $n = 219$ with replacement from the set of 219 ordered pairs (GPA,SATM).

```
first = read.csv( "FirstYearGPA.csv" , header=TRUE )
x = first$SATM
y = first$GPA
xy = cbind( x , y )
```

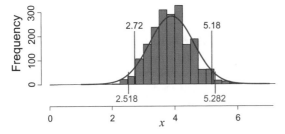

Figure 17.7 Histogram of $B = 2{,}000$ bootstrap samples from the small data set in Example 17.11. The bootstrap confidence interval is shown by the gray lines extended upward, and the confidence interval based on the assumption that the data come from a normal distribution is shown by the blue lines extending downward.

668 17 Analytic Approximations, the Jackknife, and the Bootstrap

```
n = length( x )
B = 2000
BS.rhohat = rep( NA , B )
for( i in 1:B )
{
  BSsamp = sample( 1:n , n , replace=TRUE )
  BS.rhohat[i] = cor( xy[ BSsamp , ])[1,2]
}
hist( BS.rhohat , breaks=seq(-0.2,0.6,0.02) , xlab="cor" ,
      col="red" , main="" )
xL = quantile( BS.rhohat , 0.025)
xU = quantile( BS.rhohat , 0.975)
lines( c(xL,xL) , c(0,200) , lwd=2 , col="gray" )
lines( c(xU,xU) , c(0,200) , lwd=2 , col="gray" )
text( xL , 220 , paste0(round(xL,3)) )
text( xU , 220 , paste0(round(xU,3)) )
```

The resulting histogram is shown in Figure 17.8. The confidence interval for ρ is $(0.061, 0.317)$. ∎ ∎ ∎

■ **Example 17.14** Consider a small clinical trial where two treatments, labeled A and B, are given to 357 patients, with 87 receiving Treatment A and 270 receiving Treatment B. The data, shown in Table 17.1, show us that Treatment A has an unsuccessful rate of 3.45% and Treatment B has an unsuccessful rate of 13.33%. Perform a χ^2 test of homogeneity for the proportions of unsuccessful cases. Estimate the relative risk, that is, the probability of being unsuccessful given Treatment A divided by the probability of being unsuccessful given Treatment B. Finally, use the bootstrap to obtain a confidence interval for the relative risk.

Solution: The relative risk is defined as

$$\text{Relative risk} = \frac{\Pr(\text{unsuccessful given Treatment A})}{\Pr(\text{unsuccessful given Treatment B})}.$$

Table 17.1 Results of a small clinical trial comparing two treatments, A and B.

	Treatment A	Treatment B	Total
Successful	84	234	318
Unsuccessful	3	36	39
Total	87	270	357

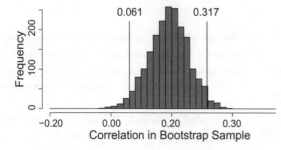

Figure 17.8 Histogram of $B = 2,000$ bootstrap samples from the small data set in Example 17.9. The bootstrap confidence interval is shown by the gray lines extended upward.

Table 17.2 Data structure for a small clinical trial comparing Treatments A and B. The variable `Unsuccessful` is coded as 1 for unsuccessful and 0 for successful.

Subject	Treatment	Unsuccessful	Subject	Treatment	Unsuccessful
1	A	0	180	B	0
2	B	0	181	A	0
3	A	0	182	B	1
⋮	⋮	⋮	⋮	⋮	⋮
178	B	0	357	B	0
179	A	0			

We can estimate this by using the estimated unsuccessful rates:

$$\widehat{\text{Relative risk}} = \frac{0.0345}{0.1333} = \frac{3/84}{36/234} = 0.23710.$$

A chi-square test of heterogeneity[6] yields

```
          Pearson's Chi-squared test with Yates' continuity correction

data:  SmallClinicalTrial
X-squared = 5.6304, df = 1, p-value = 0.01765
```

The small *P*-value indicates that there is evidence that the population unsuccessful rates differ for the two treatments. Interest then focuses on the relative risk of the treatment being unsuccessful.

Obtaining confidence intervals for ratios of parameters is a difficult problem in statistics, because there is no simple rule for approximating the standard error of a ratio given the standard errors for the numerator and denominator. This is a situation where the bootstrap can come to the rescue.

To apply the bootstrap to this data set, we must view the data as having the structure in Table 17.2. We view the data as having 357 rows, one for each person, and two columns, one for the treatment the patient received and another indicating whether the procedure was successful or unsuccessful. We coded the `Unsuccessful` variable as 1 for unsuccessful and 0 for successful.

Given this data structure, the bootstrap procedure proceeds by selecting a sample (with replacement) from the sample of 87 people who received Treatment A, and another sample (again, with replacement) from the sample of 270 people who received Treatment B. Then, the proportions ($\hat{p}_A^*$ and $\hat{p}_B^*$) of unsuccessful cases, or equivalently, the sum of the `Unsuccessful` variable divided by the number in each treatment group, are computed for the Treatment A group and the Treatment B group. Finally, the estimated relative risk

$$\widehat{\text{Relative risk}}^* = \frac{\hat{p}_A^*}{\hat{p}_B^*}$$

is computed for each of the pairs of bootstrap samples. The following R code defines the data, performs the χ^2 test of heterogeneity, and selects the 2,000 bootstrap samples:

```
SmallClinicalTrial = matrix(  c(  84 , 234 ,
                                   3 ,  36 ) ,
                              nrow=2 , ncol=2 , byrow=TRUE )
colnames( SmallClinicalTrial ) = c("Treatment A","Treatment B")
```

[6]This assumes that the margins for the two treatments are fixed ahead of time.

```
rownames( SmallClinicalTrial ) = c("Successful","Unsuccessful")
chisq.test( SmallClinicalTrial )
qchisq( 0.05 , 1 , lower.tail=FALSE )
x1 = c( rep(1,3) , rep(0,84) )
x2 = c( rep(1,36) , rep(0,234) )
n1 = 87
n2 = 270
B = 2000
rr.bss = rep( NA , B )
for( i in 1:B )
{
        samp1 = sample( 1:n1 , n1 , replace=TRUE )
        samp2 = sample( 1:n2 , n2 , replace=TRUE )
        p1hat.bss = mean( x1[samp1] )
        p2hat.bss = mean( x2[samp2] )
        rr.bss[i] = p1hat.bss / p2hat.bss
}
hist(rr.bss,col="red",breaks=seq(0.0,2.8,0.01),xlim=c(0.0,1.4),main="")
LCLp = quantile( rr.bss , 0.025 )
UCLp = quantile( rr.bss , 0.975 )
print( c( LCLp , UCLp ) )
```

The histogram of the 2,000 bootstrap samples is shown in Figure 17.9. The result of the `print()` statement is

```
      2.5%       97.5%
0.0000000 0.6206897
```

The confidence interval for the relative risk, based on the bootstrap method, is then $(0.000, 0.621)$. The most frequent estimate of the relative risk in the bootstrap sample is zero. This is because there is a low proportion of unsuccessful procedures in Treatment A (3 of 87), so it is often the case that all three unsuccessful cases are absent in the bootstrap sample, yielding a proportion of exactly 0 for the numerator in the relative risk.   ■ ■ ■

17.5 Parametric Bootstrap

The bootstrap described above is sometimes called the *nonparametric* bootstrap, since it makes no assumptions about the distribution of the data. In situations where it is reasonable to make particular assumptions about the distribution of the data (e.g., the data come from a normal distribution), then we can make use of this when we do bootstrapping.

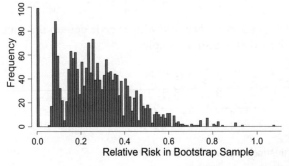

Figure 17.9 Bootstrap sample for the relative risk in Example 17.14.

If formulas exist for the standard error of the estimator, then, of course, we should use those. For example, if we sample from a normal distribution, we know that

$$\hat{\mu} \sim N(\mu, \sigma^2/n), \tag{17.30}$$

so there is no advantage in bootstrapping. When the approximate methods of Section 17.1 can be applied, then they should be used. There are cases, however, when none of the approximation methods yield a usable result for the standard error of an estimator. In these cases the parametric bootstrap may be used.

The parametric bootstrap differs from the nonparametric bootstrap in the following way. When we draw random samples and compute the statistic, we draw samples from the *estimated distribution* using parameter estimates (along with the particular distributional assumption being made) instead of drawing samples with replacement from the sample.

For example, if we were sampling from a normal distribution, we would estimate the mean and standard deviation of the normal distribution and then draw B random samples[7] from this continuous distribution. As in the nonparametric bootstrap, we compute the parameter estimate for each bootstrap sample. An astute reader may wonder why (if the original sample comes from the normal distribution) we bother sampling from the estimated distribution and computing the sample mean. After all, we know from (17.30) exactly what the sampling distribution is. This would be a valid concern, and as we mentioned previously we wouldn't bother bootstrapping in a problem where we knew from theory the form of the sampling distribution. The parametric bootstrap can be useful when a particular parametric distribution can be assumed, but the sampling distribution of the estimator is unknown.

The concept behind the parametric distribution is illustrated in Figure 17.10. Figure 17.10(a) shows the original data (red dots); (b) gives the estimated CDF. Note that this is the estimated CDF based on the assumption of a normal distribution (which from panel (a) is clearly inaccurate). Figure 17.10(c) shows the estimates of the population mean μ of samples taken from the true (unknown) distribution. Finally, Figure 17.10(d) shows the estimates of μ obtained from samples from the estimated distribution in the second panel. As with Figure 17.4, items in red (the dots, the empirical or estimated CDF, the histogram of bootstrap samples) are known, whereas those items in gray (the actual PDF and CDF, along with the histogram of samples from the population) are all unknown to the user.

■ **Example 17.15** Apply the nonparametric bootstrap to estimate the correlation between GPA and SATM.

Solution: We must first make an assumption about the parametric distribution we are assuming for the pairs (GPA, SATM). We will assume a bivariate normal distribution with mean vector μ and covariance matrix Σ. We can estimate these by their MLE:

$$\hat{\mu} = \begin{bmatrix} \hat{\mu}_{GPA} \\ \hat{\mu}_{SATM} \end{bmatrix} = \begin{bmatrix} 3.0962 \\ 634.29 \end{bmatrix}$$

and

$$\hat{\Sigma} = \begin{bmatrix} \hat{\sigma}^2_{GPA} & \hat{\sigma}_{GPA,SATM} \\ \hat{\sigma}_{GPA,SATM} & \hat{\sigma}^2_{SATM} \end{bmatrix} = \begin{bmatrix} 0.21568 & 6.7749 \\ 6.7749 & 5,634.5 \end{bmatrix}.$$

Each parametric bootstrap sample consists of $n = 219$ random pairs selected from the

$$N_2 \left(\begin{bmatrix} 3.0962 \\ 634.29 \end{bmatrix}, \begin{bmatrix} 0.21568 & 6.7749 \\ 6.7749 & 5,634.5 \end{bmatrix} \right)$$

---

[7]Because the distribution from which we are sampling is continuous, replacement of a sampled unit does not affect the probability distribution of the next one.

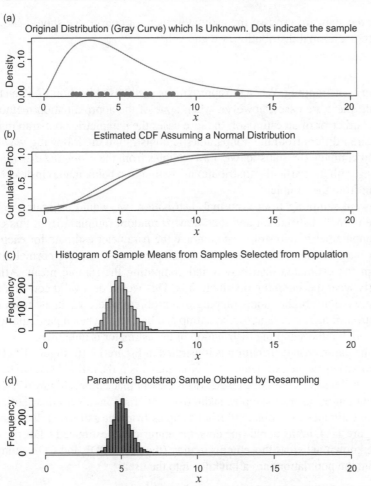

Figure 17.10 Illustration of the bootstrap. A sample is selected from some unknown distribution (a). Since we cannot resample from the population, we resample from the parametrically estimated CDF (b). The statistic, $\bar{x}$, is computed for each of the bootstrap samples (d).

distribution. This can be done with the `mvrnorm()` function in the `MASS` package. The following code generates $B = 2,000$ bootstrap samples, and from each it calculates the sample correlation coefficient:

```
library( MASS )
set.seed( 123 )
x = first$SATM
y = first$GPA
xy = cbind( x , y )
n = length( x )
muhat = colMeans( xy )
Sigmahat = cov( xy )
Sigmahat.MLE = ((n-1)/n) * cov( xy )
B = 2000
PBS.rhohat = rep( NA , B )
for( i in 1:B )
{
        PBSsamp = mvrnorm( n , muhat, Sigmahat.MLE )
        PBS.rhohat[i] = cor( PBSsamp )[1,2]
}
```

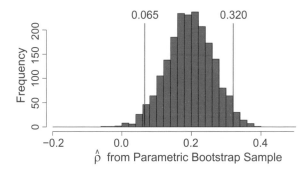

Figure 17.11 Estimated correlations from $B = 2,000$ parametric bootstrap simulations. 0.025 and 0.975 quantiles are shown by the gray lines.

A histogram of the $B = 2,000$ estimated correlations is shown in Figure 17.11. A 95% confidence interval for ρ is $[0.065, 0.320]$, which is nearly the same as the nonparametric bootstrap confidence interval. ∎∎∎

∎ **Example 17.16**   Compare the nonparametric and parametric bootstrap confidence intervals, and the confidence intervals based on the Fisher transformation given in (14.52).

Solution: The bootstrap confidence intervals are

Nonparametric bootstrap: $[0.061, 0.317]$

Parametric bootstrap: $[0.065, 0.320]$

Finally, using the confidence interval from Problem 14.24 that is derived from Fisher's transformation, we find that the confidence interval is

$$\left[\frac{\frac{1+R}{1-R} \exp\left(-2z_{\alpha/2}/\sqrt{n-3}\right) - 1}{\frac{1+R}{1-R} \exp\left(-2z_{\alpha/2}/\sqrt{n-3}\right) + 1}, \quad \frac{\frac{1+R}{1-R} \exp\left(2z_{\alpha/2}/\sqrt{n-3}\right) - 1}{\frac{1+R}{1-R} \exp\left(2z_{\alpha/2}/\sqrt{n-3}\right) + 1} \right].$$

Some straightforward calculations in R,

```
alpha = 0.05
z = qnorm(1-alpha/2)
R = cor( first$GPA , first$SATM )
LCL = ( ((1+R)/(1-R))*exp(-2*z/sqrt(n-3)) - 1 ) /
      ( ((1+R)/(1-R))*exp(-2*z/sqrt(n-3)) + 1 )
UCL = ( ((1+R)/(1-R))*exp(2*z/sqrt(n-3)) - 1 ) /
      ( ((1+R)/(1-R))*exp(2*z/sqrt(n-3)) + 1 )
print( c(LCL,UCL) )
```

yield

Fisher's transformation: $[0.06340, 0.31871]$.

Figure 17.12 shows that the confidence intervals from the three methods are nearly the same. ∎∎∎

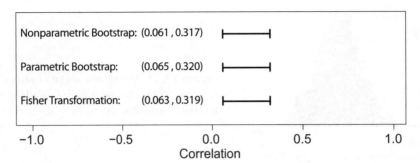

Figure 17.12 Confidence intervals for the correlation using the nonparametric bootstrap (top), the parametric bootstrap (middle), and Fisher's transformation (bottom).

17.6 Bootstrapping in R

In the previous sections we have shown how to code the bootstrap methods. Mostly we have coded the methods using `for` loops, which are inefficient in R. Fortunately, there are built-in functions in the `boot` package that automate the computations.

To begin, you will have to install the `boot` package if you haven't already done so. Then each R session that you use it, you will have to load the library by running `library(boot)`.

Before calling the `boot()` function in the `boot` package, you must first set up a function that computes the statistic of interest. One argument to this function should be the data frame or matrix containing your data. Another argument must be the arbitrary set of indices to use when evaluating your statistic. The `boot()` function needs this function in order to compute the statistic for those values in the bootstrap sample. Other arguments that are needed to compute the statistic can also be used.

In the code below, we define the function `cor1()`.[8] The arguments are `data` and `index`. The first line of code defines `d` to be the data frame or matrix evaluated at all those values in the array `index`.

```
cor1 = function( data , index ) {
        d = data[ index , ]
        rhohat = cor( d[,1] , d[,2] )
        return( rhohat )
}
```

After running this function, we should be able to recover the estimated correlation of the entire data set by defining `index = 1:219`, that is, the array of numbers from 1 through 219:

```
data = cbind( first$GPA , first$SATM )
index = 1:219
print( cor1( data , index ) )
```

This yields

```
[1] 0.1943439
```

If we wanted to take a sample of 219 ordered pairs, selected with replacement, we would run

```
index = sample( 1:219 , 219 , replace=TRUE )
```

Running

```
index = sample( 1:219 , 219 , replace=TRUE )
print( cor1( data , index ) )
```

---

[8]It is bad practice to define a new function whose name matches an existing R function. In this case, R already has a `cor()` function, so we make our function name something different: `cor1`.

yields

    0.1571041

With the function cor1() now ready, we can run the boot() function. In the code below, we send boot the data matrix by using the cbind command to make a 219 × 2 matrix containing the data. We also need to send it the name of the function that computes the desired statistic, and the number of bootstrap simulations to run. This can be done with the following call to the boot() function:

```
library(boot)
bs.results = boot(data=cbind(first$GPA,first$SATM),statistic=cor1,R=20000)
```

We can then display the results:

```
plot( bs.results )
boot.ci( bs.results , type="perc" )
```

The plot produced by the first line of code is shown in Figure 17.13. Figure 17.13(a) shows a density histogram, while Figure 17.13(b) shows a normal probability plot for the 20,000 bootstrap correlations.

A confidence interval can be calculated for the correlation using the boot.ci() function. This function takes as arguments the bootstrap object obtained from the call to the boot() function (bs.results in this case); the second argument is the method to be used to obtain the confidence interval. Here we have used the percentile method, type="perc". The full set of options for type is shown in Table 17.3. The bs.results variable is a complicated object that contains a lot of information. To see the structure, we can run the command

```
str( bs.results )
```

which results in

Table 17.3 Allowable types of bootstrap confidence intervals.

type=	Description
"perc"	Confidence interval based on percentiles
"norm"	Confidence interval based on a normal approximation
"bca"	Bias-corrected and accelerated confidence intervals
"stud"	Studentized confidence intervals
"basic"	Basic confidence interval
"all"	All of the above

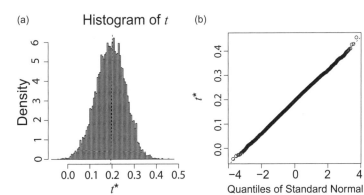

(a) Histogram of t (b)

Figure 17.13 Output from 20,000 bootstrap samples from GPA versus SATM data using the boot() function in R's boot package. (a) shows a density histogram, and (b) shows a normal probability plot for the 20,000 bootstrap correlations.

```
List of 11
$ t0        : num 0.194
$ t         : num [1:10000, 1] 0.152 0.182 0.258 0.315 0.208 ...
$ R         : num 10000
$ data      : num [1:219, 1:2] 3.06 4.15 3.41 3.21 3.48 2.95 3.6 2.87 3.67 3.49 ...
$ seed      : int [1:626] 10403 450 -1785717041 2043447706 1647338422 -48269213
-503457700 69659803 -1240200165 -1979698570 ...
$ statistic:function (data, index)
..- attr(*, "srcref")= 'srcref' int [1:8] 1 8 6 1 8 1 1 6
.. ..- attr(*, "srcfile")=Classes 'srcfilecopy',
'srcfile' <environment: 0x000002ac587e36c8>
$ sim       : chr "ordinary"
$ call      : language boot(data = cbind(first$GPA, first$SATM), statistic = cor1,
                    R = 10000)
$ stype     : chr "i"
$ strata    : num [1:219] 1 1 1 1 1 1 1 1 1 1 ...
$ weights   : num [1:219] 0.00457 0.00457 0.00457 0.00457 0.00457 ...
- attr(*, "class")= chr "boot"
- attr(*, "boot_type")= chr "boot"
```

We can access all bootstrap estimates of the correlation through bs.results$t. This feature can sometimes come in handy.

■ **Example 17.17** In Examples 12.14 and 12.15 we analyzed a small hypothetical data set similar to that described in an article in the *Journal of the American Medical Association*, where Thomas *et al.* (2021) describe a clinical trial to test the efficacy of high doses of zinc and/or ascorbic acid (vitamin C) on the recovery time for COVID-19. They studied a number of outcome variables, including the time to 50% reduction in symptoms; this is the response we will focus on. Volunteers were randomly assigned to one of four categories:

1. standard of care (no ascorbic acid, no zinc);

2. ascorbic acid only;

3. zinc only; and

4. ascorbic acid and zinc.

We will consider a data set that matches the means and standard deviations for each of the four treatments:

$$
\begin{aligned}
n_{\text{Standard}} &= 50 & \bar{y}_{\text{Standard}} &= 6.7 & s_{\text{Standard}} &= 4.4 \\
n_{\text{AscAcid}} &= 48 & \bar{y}_{\text{AscAcid}} &= 5.5 & s_{\text{AscAcid}} &= 3.7 \\
n_{\text{Zinc}} &= 58 & \bar{y}_{\text{Zinc}} &= 5.9 & s_{\text{Zinc}} &= 4.9 \\
n_{\text{AscAcid+Zinc}} &= 58 & \bar{y}_{\text{AscAcid+Zinc}} &= 5.5 & s_{\text{AscAcid+Zinc}} &= 3.4
\end{aligned}
$$

The data set is shown in Table 17.4. The data suggest that zinc had little effect, so we will combine the standard of care with the zinc only treatment and the ascorbic acid only with the ascorbic acid plus zinc; we call these treatments "Standard" and "AscAcid," respectively. Specifically, we would like to make an inference for the ratio of means $\mu_{\text{Standard}}/\mu_{\text{AscAcid}}$ while ignoring the effect of zinc. The statistics for the two groups are:

$$
\begin{aligned}
n_{\text{Standard}} &= 108 & \bar{y}_{\text{Standard}} &= 6.27 & s_{\text{Standard}} &= 4.67 \\
n_{\text{AscAcid}} &= 106 & \bar{y}_{\text{AscAcid}} &= 5.50 & s_{\text{AscAcid}} &= 3.52
\end{aligned}
$$

Table 17.4 Hypothetical data from clinical trial testing the effect of ascorbic acid (vitamin C) and zinc on the duration (time until 50% reduction) in symptoms. These data were simulated but the means and standard deviations match those given in Table 2 of Thomas *et al.* (2021).

Standard of care			Ascorbic acid			Zinc			Asc acid + zinc		
6.2	11.3	4.7	0.2	4.8	5.6	5.1	9.4	3.8	2.4	3.8	7.5
3.4	5.9	1.1	1.6	6.2	0.4	2.7	4.8	0.8	3.0	2.3	4.3
3.1	10.3	10.3	7.0	14.6	4.9	2.5	8.5	8.5	4.9	2.2	1.5
13.7	2.8	1.4	2.5	1.8	10.4	11.4	2.2	1.1	11.9	14.6	1.7
8.4	5.5	2.5	3.1	6.3	8.9	6.9	4.4	2.0	3.0	9.5	9.0
1.5	15.5	4.4	0.5	6.4	11.2	1.1	12.9	3.6	5.2	10.7	0.8
4.9	8.9	2.4	18.3	5.7	2.6	4.0	7.7	1.9	2.4	8.3	2.6
14.6	3.9	7.6	1.0	2.5	5	12.2	3.1	6.2	5.2	13.4	2.2
7.8	3.6	4.7	1.4	6.8	4.1	6.4	2.9		9.5	3.3	3.8
5.4	1.8	7.2	7.3	0.8	1.2	4.4	1.4		2.5	2.5	4.2
6.9	6.1		11.2	8.8	17.5	5.7	4.9		4.5	0.9	4.1
5.1	11.6		7.9	1.4	5.7	4.2	9.6		1.9	3.3	9.8
5.6	17.8		2.5	24.9	4.2	4.6	14.5		7.3	3.4	6.8
2.9	3.2		14.1	3.4	8.5	2.3	2.5		8.1	3.3	6.0
5.1	8.5		7.9	1.9	2.8	4.2	7.0		8.3	7.3	6.5
6.6	16.5		2.5	7.9	0.4	5.4	13.4		10.7	4.4	5.7
19.0	6.1		6.6	8.3	5.5	15.1	5.0		7.3	5.3	1.3
4.3	2.6		2.3	5.9	11.5	3.4	2.1		7.7	8.6	5.3
9.3	3.2		3.9	1.1		7.7	2.6		3.8	3.0	
4.6	5.2		4.0	6.5		3.7	4.3		13.3	3.3	

Apply the `boot()` function to compute a confidence interval for

$$\eta = \frac{\mu_{\text{Standard}}}{\mu_{\text{AscAcid}}}.$$

Solution: Define `Standard` to be the vector of outcomes (days until 50% reduction in symptoms) for the standard of care group, and similarly for `AscAcid`. For a two-sample problem like this, we would first, in separate steps, resample 108 of the observations in the standard group and 106 of the observations in the ascorbic acid group. Second, for bootstrap sample b ($b = 1, 2, \ldots, B$) we would form the statistics $\hat{\mu}_{\text{Standard}}^{(b)} = \bar{y}_{\text{Standard}}$ and $\hat{\mu}_{\text{AscAcid}}^{(b)} = \bar{y}_{\text{AscAcid}}$. Third, we would for each b in $1, 2, \ldots, B$ take the ratio of the bootstrap samples

$$\hat{\eta}^{(b)} = \frac{\hat{\mu}_{\text{Standard}}^{(b)}}{\hat{\mu}_{\text{AscAcid}}^{(b)}}, \quad b = 1, 2, \ldots, B.$$

The first two of these tasks can be performed using the `boot()` function in R. We must then extract the simulated bootstrap estimates of μ_{Standard} and of μ_{AscAcid} and compute the ratio.

The key to running the `boot()` function is to write the function that defines the statistic computed from the bootstrap sample. Since the data are a vector (as opposed to a matrix) the first line of the `sample.mean()` function is simply `d = data[index]`:

```
sample.mean = function( data , index ) {
        d = data[ index ]
        ybar = mean( d )
        return( ybar )
}
```

Once this function is defined, we can accomplish the first two tasks by sending this function in the `statistic=sample.mean` argument to the `boot()` function for both groups:

```
Thomas = read.csv("COVID-Zinc-AA-Full.csv")
Standard = c(Thomas[1:50,2],Thomas[1:58,3])
AscAcid = c(Thomas[1:48,4],Thomas[1:58,5])
bs.results.Standard = boot( data=Standard, statistic=sample.mean, R=10000 )
plot( bs.results.Standard )
boot.ci( bs.results.Standard , type="perc" )
bs.results.AscAcid = boot( data=AscAcid , statistic=sample.mean , R=10000 )
plot( bs.results.AscAcid )
boot.ci( bs.results.AscAcid , type="perc" )
```

Once we have run the bootstrap for each of the groups, we extract the simulated statistics $\hat{\mu}_{\text{Standard}}^{(b)}$ and $\hat{\mu}_{\text{AscAcid}}^{(b)}$, $b = 1, 2, \ldots, B$, and create the bootstrap plots:

```
bs.Standard = bs.results.Standard$t
bs.AscAcid  = bs.results.AscAcid$t
```

Figure 17.14 shows the results of these plots for the `Standard` group in (a) and (b), and the `AscAcid` group in (c) and (d). The recovery times seem to be a little longer on average for the `Standard` group. Finally, we compute the ratio of the estimated means:

```
bs.eta = bs.Standard / bs.AscAcid
hist( bs.eta )
```

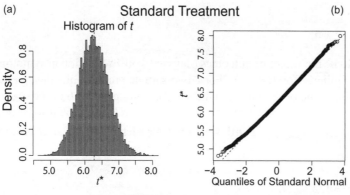

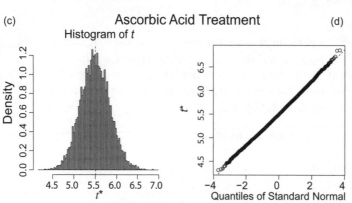

Figure 17.14 Separate plots of the bootstrap samples from the `Standard` group (a, b) and `AscAcid` group (c, d).

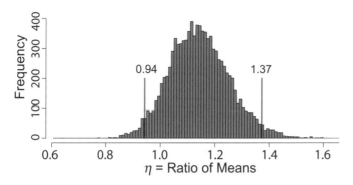

Figure 17.15 Histogram of 10,000 bootstrap estimates $\eta^{(b)}$ for the ratio $\hat{\eta}^{(b)} = \hat{\mu}_{\text{Standard}}^{(b)}/\hat{\mu}_{\text{AscAcid}}^{(b)}$, $b = 1, 2, \ldots, B$.

A plot of the $B = 10,000$ bootstrap estimates $\eta^{(b)}$ is shown in Figure 17.15. The bootstrap confidence interval for η is $(0.94, 1.37)$. The inclusion of 1 in this confidence interval indicates that 1 is a plausible value for the ratio of means. We therefore have no evidence that ascorbic acid affects time to recovery. ∎∎∎

When we bootstrap in a regression problem, we select rows of data from the data frame at random and with replacement. We then compute the set of regression coefficients, the intercept, and all of the slope parameters.

■ Example 17.18 Consider the cholesterol data given by Dixon and Massey (1983) and discussed in Example 10.13. We will investigate these data further by examining the relationship between the predictor variables:

(a) ageIn52 – the subject's age in 1952; and

(b) CoronaryEventBy62 – an indicator variable of whether the subject had a coronary event by 1962;

and the response variable diff, the difference between the 1962 and 1952 cholesterol levels. Use the bootstrap method to find 95% confidence intervals for the coefficients in the linear regression:

$$\texttt{diff} = \beta_0 + \beta_1\,\texttt{ageIn52} + \beta_2\,\texttt{CoronaryEventBy62} + \varepsilon.$$

Solution: We must first define a function that gives the estimated regression coefficients for an arbitrary set of rows of the data frame:

```
library( boot )
chol = read.csv("DixonMasseyCholesterol1952-1962.csv")
chol$diff = chol$chol62 - chol$chol52
bootReg = function( formula, data, i ) {
        d = data[i,]
        lm.fit = lm( formula, data=d )
        return(coef(lm.fit))
}
BootStrapResults = boot( statistic=bootReg ,
                         formula = diff ~ ageIn52 + CoronaryEventBy62 ,
                         data=chol , R=2000 )
summary( BootStrapResults )
BootStrapResults$t0
BootStrapResults$t
```

The fourth line of the code above, that begins bootReg, is the function that we will be bootstrapping. The first argument, formula, is the formula for the regression. The second argument is the data frame on which we will be operating. The third argument, i , is the set of indices that will be used to estimate the regression coefficients. The first line of the bootReg() function creates the data frame d by selecting the appropriate rows of the data frame. Note, in a bootstrap sample there will likely be several indices that are the same, and some indices that are missing. The call to the boot() function involves four arguments: (1) the statistic (really the function name that computes the statistic), bootReg in this case; (2) the regression formula; (3) the data frame on which we are operating, chol in this case; and (4) the number of bootstrap samples to select. The results of the bootstrap are contained in the object BootStrapResults.

We can obtain some information about the contents of BootStrapResults by using R's summary() function. The most informative objects within BootStrapResults are BootStrap Results$t0 and BootStrapResults$t. The former gives the result of applying the bootReg() function to the full set of data. The latter is a matrix with B rows and p columns, where p is the number of parameters to be estimated. In this example, $p = 3$. The code

```
summary( BootStrapResults )
BootStrapResults$t0
BootStrapResults$t
```

yields

```
        (Intercept)          ageIn52   CoronaryEventBy62
          27.154217        -1.839798          -22.921866
```

and

```
             [,1]             [,2]             [,3]
[1,]   -79.78687488  -0.093022336   12.0974699
[2,]   107.37973778  -2.890703218  -66.5994041
[3,]    94.41132424  -3.367691241  -42.0910562
[4,]   129.86940639  -3.807762557  -67.3155251
[5,]    55.17857143  -2.228571429  -45.3571429
. . . . . . . . . . . . . . . . . . . . . .
```

We can use the boot.ci() function in the boot package to compute confidence intervals for the three parameters. The code

```
boot.ci( BootStrapResults , conf = 0.95 , type = "perc" , index=1 )
boot.ci( BootStrapResults , conf = 0.95 , type = "perc" , index=2 )
boot.ci( BootStrapResults , conf = 0.95 , type = "perc" , index=3 )
```

produces confidence intervals for the three parameters β_0, β_1, and β_2. The output for the second command is

```
BOOTSTRAP CONFIDENCE INTERVAL CALCULATIONS
Based on 2000 bootstrap replicates

CALL :
boot.ci(boot.out = BootStrapResults, conf = 0.95, type = "perc", index = 2)

Intervals :
Level     Percentile
95%    (-5.548,  1.126 )
Calculations and Intervals on Original Scale
```

The output from the other two commands is similar. Confidence intervals for the three parameters are

$$(-137.55, 241.10) \quad \text{for } \beta_0$$
$$(-5.548, 1.126) \quad \text{for } \beta_1$$
$$(-120.87, 45.94) \quad \text{for } \beta_2$$

There is no evidence that either of the predictor variables affects the difference in cholesterol levels from 1952 to 1962, although we should keep in mind that this is a small study with only 20 subjects.   ∎∎∎

17.7 Chapter Summary

Once we obtain a point estimate for a parameter we are often interested in its standard error, because a confidence interval for the parameter depends on this standard error. We began this chapter with a review of the ideas of the standard error and the estimated standard error of a parameter. In some cases it is easy to estimate the standard error; in others it is very difficult. One approach, when no simple formula exists for the estimated standard error, is to use Theorem 17.2.1, which gives an approximation to the standard error of the MLE of a parameter. This depends on the expected value of the negative of the second derivative of the log-likelihood function. As you recall, the second derivative is related to the curvature of the graph of a function. Since the log-likelihood function is maximized at the MLE, the second derivative will be negative. When the magnitude of the second derivative is great, then the curvature is large and intuitively we have much confidence in the estimate of the parameter. This will translate into a narrow confidence interval.

Other approaches to the problem of estimating the standard error of an estimator are the jackknife and the bootstrap. The idea behind the bootstrap is this. If we could somehow repeatedly sample from the population and then get the parameter estimate for every randomly generated sample, we could obtain information about the sampling distribution from the simulated values; we could use this, sort of as a pivotal quantity, and obtain a confidence interval. But it is not possible to sample from the population itself because we don't know the parameters of the distribution. The idea behind the bootstrap is to sample not from the original population, but from the empirical CDF. In other words, we are sampling from the sample with replacement. The distribution of the bootstrap estimates of the parameter can then be used to obtain a confidence interval.

Instead of sampling from the empirical CDF, we could sample from the fitted distribution to the sample. For example, if we assume a normally distributed population, we would estimate μ and σ^2 and then sample from the fitted normal distribution. This is called the nonparametric bootstrap.

17.8 Problems

Problem 17.1 Verify the derivatives in (17.6).

Problem 17.2 Suppose $X \sim \text{BIN}(n, p)$. Find an approximate 99% confidence interval for $1/p$.

Problem 17.3 Suppose $X_1, X_2, \ldots, X_n \sim$ i.i.d. $\text{POIS}(\lambda)$.

(a) Find the likelihood function, the log-likelihood function, and the MLE $\hat{\lambda}$ for λ.

(b) Find $\mathbb{E}\left(-\frac{\partial^2}{\partial \lambda^2} \ell\right)$.

(c) Find the approximate standard error $\hat{\lambda}$.

(d) Given the following data from a sample of size $n = 5$ from the $\text{POIS}(\lambda)$ distribution

$$6, 14, 8, 5, 12$$

find the MLE $\hat{\lambda}$ and an approximate 90% confidence interval for λ.

Problem 17.4 Under the assumptions of Problem 17.3, use Theorem 17.2.2 to find an approximate confidence interval for $e^{-\lambda}$. Note, $\Pr(X = 0) = e^{-\lambda}$ so by estimating $e^{-\lambda}$ we are estimating the probability of observing a count of 0.

Problem 17.5 For the data in Example 17.1, find a 95% confidence interval for the variance θ^2 for the exponential distribution.

Problem 17.6 Suppose $X_1, X_2, \ldots, X_n \sim$ i.i.d. GAM $(2, \theta)$. This distribution, called the Rayleigh distribution, is sometimes used in fitting lifetimes.

(a) Find the MLE for θ.

(b) Is the MLE $\hat{\theta}$ an unbiased estimator? Explain.

(c) Use Theorem 17.2.1 to find an approximate confidence interval for θ.

Problem 17.7 Suppose $X \sim \text{BIN}(n, p)$. Use Theorem 17.2.2 to obtain a formula for an approximate confidence interval for $1/p$.

Problem 17.8 Verify the second derivatives for the Weibull likelihood function given in (17.22).

Problem 17.9 Suppose we have the following sample from the $\text{WEIB}(\beta, \theta)$.

| 115 | 99 | 75 | 31 | 127 | 33 | 24 | 64 | 68 | 167 |
| 126 | 132 | 61 | 98 | 51 | 84 | 58 | 9 | 98 | 50 |

(a) Find the MLEs for β and θ.

(b) Find the observed Fisher information matrix for this data set.

(c) Find approximate 95% confidence intervals for β and θ.

Problem 17.10 Consider the data in Example 9.21.

(a) Show that the second derivative of the likelihood function can be written as

$$\ell''(\theta) = -\frac{x_1}{(2+\theta)^2} - \frac{x_2 + x_3}{(1-\theta)^2} - \frac{x_4}{\theta^2}.$$

(b) Use the fact that the marginals of the multinomial are binomial (e.g., $X_1 \sim \text{BIN}(n, p_1)$ where $p_1 = (2+\theta)/4$) to find $\mathbb{E}\left(-\ell''(\theta)\right)$.

(c) Find an approximate confidence interval for θ.

Problem 17.11 Verify the first derivative calculation is in (17.13).

Problem 17.12 Verify the second derivative calculations in (17.14).

Problem 17.13 Verify that the MLE for σ^2 can be written as given in (17.29).

Problem 17.14 Suppose $X_1, X_2, \ldots, X_n \sim$ i.i.d. from a distribution with CDF F. Let $\hat{F}$ denote the ECDF. Fix the value x in the support of F.

(a) Show that $\mathbb{E}\left(\hat{F}(x)\right) = F(x)$.

(b) Show that $\mathbb{V}\left(\hat{F}(x)\right) = F(x)(1 - F(x))/n$.

(c) Explain why, for large n, $\hat{F}(x)$ has an approximate $N(F(y), F(x)(1 - F(x))/n)$ distribution.

(d) Explain why $\hat{F}(x)$ is a consistent estimator of $F(y)$.

Problem 17.15 Bootstrap samples are taken with replacement from the sample taken from the population. In this exercise we will investigate the probability that a particular observation in the sample, say the first one, will be contained in a bootstrap sample.

(a) Explain why, if the sample size is $n = 5$, the probability that the first observation is contained in a bootstrap sample is $(4/5)^5$.

(b) Explain why the probability is $(1 - 1/n)^n$ for a sample size of n.

(c) Compute the probability that the first observation is contained in the bootstrap sample if the sample size is $n = 100$.

(d) Repeat part (c) when $n = 1,000$.

(e) What is $\lim_{n \to \infty} (1 - 1/n)^n$? What does this have to do with parts (c) and (d)?

Problem 17.16 Table 17.5 gives the pre-intervention and post-intervention cholesterol levels for 10 subjects. Let μ_1 and μ_2 denote the population means for the pre- and post- populations.

(a) Use the bootstrap method to find a confidence interval for $\mu_1 - \mu_2$.

(b) Use the bootstrap method to find a confidence interval for μ_1 / μ_2.

Table 17.5 Data for Problem 17.16.

Subject	Pre	Post	Subject	Pre	Post
1	180	169	6	240	225
2	149	147	7	229	211
3	205	220	8	165	159
4	197	165	9	202	200
5	181	180	10	209	212

(c) Explain why you could apply the paired data confidence interval from Section 10.5.3. What assumptions are needed? Compute a confidence interval for $\mu_1 - \mu_2$ using this method.

(d) What problems would you encounter trying to find an exact confidence interval for μ_1/μ_2?

Problem 17.17 Consider the data in Problem 17.16 (Table 17.5), except now you'd like to estimate the correlation ρ between the pre- and post- measurements.

Table 17.6 Data for Problem 17.18.

	Exp.	Placebo	Total
Cured	625	633	1,258
Not cured	122	80	202
Total	747	713	1,460

(a) Use the jackknife method to obtain an approximate 95% confidence interval for ρ.

(b) Use the bootstrap method to obtain an approximate 95% confidence interval for ρ.

(c) Use the Fisher transformation method from Problem 14.24 to obtain an approximate 95% confidence interval for ρ.

Problem 17.18 Suppose a clinical trial is run to test an experimental drug against a placebo. The data on whether a patient was cured or not cured are shown in Table 17.6. Use the bootstrap method to get a confidence interval for the ratio $p_{\text{Exp}}/p_{\text{Placebo}}$.

Problem 17.19 Consider the clinical trial data in Problem 14.19 (Table 14.12). Apply the bootstrap to find confidence intervals for the intercept and the coefficients of Group and Age.

Problem 17.20 Consider the hospital satisfaction data described in Problem 14.16 (Table 14.11). Apply the bootstrap to find confidence intervals for the intercept and the coefficients of Age, Severity, SurgicalMedical, and Anxiety.

Problem 17.21 Consider the data in Example 12.11 and Table 12.3. Use the bootstrap method to find a confidence interval for the ratio of means $\mu_{\text{top}}/\mu_{\text{bottom}}$.

Problem 17.22 If our sample contains n distinct observations, the number of possible bootstrap samples is $\binom{2n-1}{n}$.

(a) Compute the number of possible bootstrap samples for $n = 5, 10$, and 15.

(b) Show that $\binom{2n-1}{n}$ is the number of possible bootstrap samples.

Problem 17.23 Consider the lifetimes given in Problem 9.19. Assume an exponential distribution. Apply the parametric bootstrap to estimate the mean θ of the exponential distribution.

Problem 17.24 Consider the data in Table 17.7 on the tensile strength of two types of aluminum. Assume that the tensile strengths are normally distributed with possibly different means and variances for the two groups.

Table 17.7 Data for Problem 17.24

Type of alum.	Tensile strength	Type of alum.	Tensile strength
1	46,200	2	40,100
1	48,100	2	41,900
1	40,300	2	43,300
1	43,100	2	43,000
1	44,000	2	39,900
1	47,400	2	40,200

(a) Use the parametric bootstrap to obtain an approximate confidence interval for $\mu_1 - \mu_2$.

(b) Use the parametric bootstrap to obtain an approximate confidence interval for μ_1/μ_2.

(c) Use the parametric bootstrap to obtain an approximate confidence interval for σ_1^2/σ_2^2.

Problem 17.25 Consider the first-year GPA data described in Example 12.1. Assume that the GPAs for first-generation college students and non-first-generation college students are normally distributed with means μ_1 and μ_2 and variances σ_1^2 and σ_2^2. Use the parametric bootstrap to find 95% confidence intervals for

(a) $\mu_1 - \mu_2$

(b) μ_1/μ_2

(c) σ_1/σ_2.

18 — Generalized Linear Models and Regression Trees

18.1 Logistic Regression

In Chapter 14 we studied multiple regression and polynomial regression and how these techniques can be used to determine the relationship between an outcome y and several predictor variables $x_1, x_2, \ldots, x_p$. One of the key assumptions was that conditioned on the values of the predictor variables, the outcome variable has a normal distribution whose mean is equal to the linear function of the parameters; that is,

$$y_i \sim N\left(\mu = \beta_0 + \beta_1 x_1 + \beta_2 x_2 + \cdots + \beta_p x_p, \sigma^2\right).$$

There are situations, however, where this normal distribution is unreasonable. Consider the case where the outcome variable is dichotomous – that is, it takes on one of two values, which we arbitrarily choose to be 0 and 1. For example, the predictor variable might be the amount of exposure to some chemical and the outcome is whether the subject developed cancer; here we might code y as

$$y_i = \begin{cases} 0, & \text{if the } i\text{th subject did not develop cancer} \\ 1, & \text{if the } i\text{th subject did develop cancer.} \end{cases}$$

Thus, y is an indicator variable that says whether the subject did develop cancer. Alternatively, we might study whether an email message is spam (unsolicited commercial email) or ham.[1] The predictor variables might include the presence or absence of certain words, punctuation, or phrases. For example, "sale," "free," or the $\$$ character might be indicators of spam, whereas my first name or my department name might be indicators of ham. Our outcome could then be

$$y_i = \begin{cases} 0, & \text{if the } i\text{th email message is spam} \\ 1, & \text{if the } i\text{th email message is ham.} \end{cases}$$

For cases like this, the linear regression model is clearly inappropriate. The outcome being either 0 or 1 precludes assuming a normally distributed response. Even if we attempt to model the probability

---

[1] "Ham" is often used to describe an email that is not spam; that is, there is some real "meat" to the message.

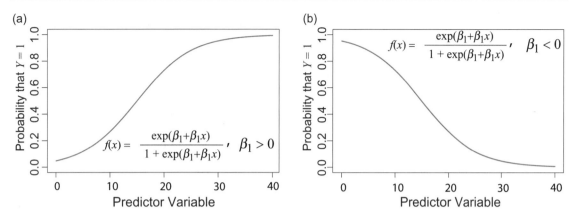

Figure 18.1 Logistic curves showing the effect of β_1. If $\beta_1 > 0$, the logistic curve approaches 1 as $x \to \infty$ and it approaches 0 as $x \to -\infty$. The reverse is true for $\beta_1 < 0$.

of $y = 1$ with linear regression, we run into the problem that, unless all slope coefficients are zero, the estimated probability will eventually exceed 1 or drop below 0, which is impossible for a probability. We can think of the outcome y_i as being the number of successes on a single Bernoulli trial; thus, $Y_i|(x_1, x_2, \ldots, x_p) \sim \text{BIN}(1, p)$ where p is some function of the predictor variables.

The model we develop in this section constrains the probability that $y = 1$ to be a number between 0 and 1. The most commonly used model that meets this constraint is the logistic model. For a single predictor variable, the logistic function is defined as

$$p = \frac{e^{\beta_0 + \beta_1 x_1}}{1 + e^{\beta_0 + \beta_1 x_1}}. \tag{18.1}$$

If $\beta_1 > 0$, then as $x_1 \to \infty$, we can see that p approaches 1 and as $x_1 \to -\infty$, p approaches 0. Thus, for a positive slope[2] the probability increases monotonically from 0 (as $x \to -\infty$) to 1 (as $x \to \infty$). The reverse is true if $\beta_1 < 0$. See Figure 18.1, which illustrates the effect of β_1 on the shape of the logistic curve.

When the outcome is an indicator variable, and the predictors are either continuous variables or indicators (or a mix), then the data can be summarized as follows for the case where $p = 3$ – that is, there are three predictor variables:

```
i     y      x1      x2      x3
-------------------------------
1     y1     x11     x12     x13
2     y2     x21     x22     x23
.  .  .  .  .  .  .  .  .  .  .
i     yi     xi1     xi2     xi3
.  .  .  .  .  .  .  .  .  .  .
n     yn     xn1     xn2     xn3
```

The y column will consist of all 0s and 1s, and the x columns will contain the values of the predictor variables. Under the assumptions we've laid out here, we can say that the outcome Y_i has a binomial distribution with $n = 1$ (which is also called the Bernoulli distribution) and p is equal to

$$p_i = \frac{e^{\beta_0 + \beta_1 x_{i1} + \cdots + \beta_p x_{ip}}}{1 + e^{\beta_0 + \beta_1 x_{i1} + \cdots + \beta_p x_{ip}}}. \tag{18.2}$$

---

[2] We call β_0 the "intercept" and $\beta_1, \beta_2, \ldots, \beta_p$ the "slope" parameters, even though they are certainly not the intercept and slope parameters for the logistic curve. They do, however, have these interpretations for the logit, or log odds, defined in (18.3).

It can be shown (Problem 18.2) that the logarithm of the odds ratio $p/(1-p)$, called the *logit* or *log odds*, can be written as

$$\text{logit}(p_i) = \log \frac{p_i}{1-p_i} = \beta_0 + \beta_1 x_{i1} + \beta_2 x_{i2} + \cdots + \beta_p x_{ip}. \tag{18.3}$$

This logit transformation maps the whole real number line to the open interval $(0,1)$, so it never gives a predicted probability in the prohibited region below 0 or above 1.

For logistic regression, we usually find the maximum likelihood estimators of the parameters. Since we assume that outcomes are independent, we can see that Y_1, Y_2, ..., Y_n are independent, but not identically distributed, random variables. The likelihood function can then be written as the product of binomial probabilities:

$$L(\beta_0, \beta_1, \beta_2, \ldots, \beta_p) = \prod_{i=1}^{n} \binom{1}{y_i} \left(\frac{e^{\beta_0 + \beta_1 x_{i1} + \cdots + \beta_p x_{ip}}}{1 + e^{\beta_0 + \beta_1 x_{i1} + \cdots + \beta_p x_{ip}}} \right)^{y_i} \left(1 - \frac{e^{\beta_0 + \beta_1 x_{i1} + \cdots + \beta_p x_{ip}}}{1 + e^{\beta_0 + \beta_1 x_{i1} + \cdots + \beta_p x_{ip}}} \right)^{1-y_i}. \tag{18.4}$$

Since y_i can only be 0 or 1, the binomial coefficient $\binom{1}{y_i}$ is always 1. We can use this fact and also get a common denominator for the second factor to simplify the likelihood somewhat:

$$L(\beta_0, \beta_1, \beta_2, \ldots, \beta_p) = \prod_{i=1}^{n} \left(\frac{e^{\beta_0 + \beta_1 x_{i1} + \cdots + \beta_p x_{ip}}}{1 + e^{\beta_0 + \beta_1 x_{i1} + \cdots + \beta_p x_{ip}}} \right)^{y_i} \left(\frac{1}{1 + e^{\beta_0 + \beta_1 x_{i1} + \cdots + \beta_p x_{ip}}} \right)^{1-y_i}. \tag{18.5}$$

Finally, note that the product in the numerator can be simplified by taking the product of the exponentials as the exponential of the sum, and the denominator can be simplified because $y_i + (1 - y_i) = 1$. Thus, the likelihood can be written as

$$
\begin{aligned}
L(\beta_0, \beta_1, \beta_2, \ldots, \beta_p) &= \frac{\exp\left(\sum_{i=1}^{n} \left(\beta_0 + \beta_1 x_{i1} + \cdots + \beta_p x_{ip} \right) \right)}{\prod_{i=1}^{n} \left(1 + \exp\left(\beta_0 + \beta_1 x_{i1} + \cdots + \beta_p x_{ip} \right) \right)} \\[2ex]
&= \frac{\exp\left(n\beta_0 + \beta_1 \sum_{i=1}^{n} x_{i1} + \cdots + \beta_p \sum_{i=1}^{n} x_{ip} \right)}{\prod_{i=1}^{n} \left(1 + \exp\left(\beta_0 + \beta_1 x_{i1} + \cdots + \beta_p x_{ip} \right) \right)}.
\end{aligned} \tag{18.6}
$$

The likelihood function cannot be simplified any further. We can take the partial derivatives with respect to each of the β parameters and set these results equal to zero, but there are no closed-form expressions for the solutions. The MLEs are thus found by approximating the solutions obtained by setting the derivatives equal to zero. Approximate standard errors are obtained using the observed Fisher information, and tests of hypotheses are usually based on Wald's approach. All of this is handled automatically in the `glm()` function in R using the `family="binomial"` argument.

■ **Example 18.1** Let's begin with a small hypothetical dose–response example. Suppose that 20 subjects received some dose or exposure from a potentially harmful chemical. The response is whether the person survived ($y = 0$) or died ($y = 1$). The data are shown in Table 18.1 and illustrated in Figure 18.2(a). Figure 18.2 suggests that higher doses lead to a greater chance of death. Fit a logistic regression model to these data.

Table 18.1 Hypothetical data from a dose–response study.

Dose	Response	Dose	Response
0	0	3	0
0	0	5	1
0	0	5	0
0	0	6	1
1	0	6	1
1	0	6	0
1	0	7	1
2	0	8	1
3	1	9	1
3	0	9	1

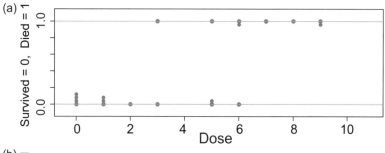

(a)

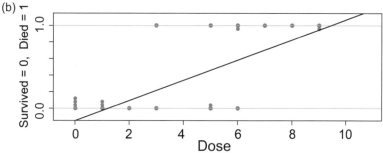

(b)

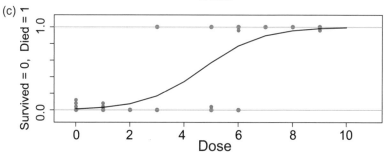

(c)

Figure 18.2 (a) Scatterplot of dose vs. response (1 = died, 0 = survived). (b) Linear fit to points in scatterplot. (c) Logistic fit to points in scatterplot.

Solution: Figure 18.2(b) shows a linear fit to the data, which is inappropriate for these kinds of data and model. We would like to model the probability of death as a function of the predictor variable, dose in this case. A linear fit clearly leads to estimated probabilities that are in the prohibited region, that is, outside of the interval from 0 to 1. We can use R to fit a logistic regression model to this simple example:

```
dose = c(0,0,0,0,1,1,1,2,3,3,3,5,5,6,6,6,7,8,9,9)
resp = c(0,0,0,0,0,0,0,0,1,0,0,1,0,1,1,0,1,1,1,1)
```

```
g0 = glm( resp ~ dose , family="binomial" )
summary( g0 )
```

The output from the `summary()` command is as follows:

```
Call:
glm(formula = resp ~ dose, family = "binomial")

Deviance Residuals:
     Min        1Q    Median        3Q       Max
 -1.7212   -0.2840   -0.1559    0.3395    1.8845

Coefficients:
             Estimate Std. Error z value Pr(>|z|)
(Intercept)   -4.4037     1.9656  -2.240   0.0251 *
dose           0.9379     0.3930   2.387   0.0170 *
---
Signif. codes:  0 '***' 0.001 '**' 0.01 '*' 0.05 '.' 0.1 ' ' 1

(Dispersion parameter for binomial family taken to be 1)

    Null deviance: 26.920  on 19  degrees of freedom
Residual deviance: 11.913  on 18  degrees of freedom
AIC: 15.913

Number of Fisher Scoring iterations: 6
```

Much of this output needn't concern us now, but let's take a look at some of the important facets of this analysis. Estimates of the intercept parameter and the one slope parameter are given in the table just below `Coefficients:` in the output. We see that

$$\hat{\beta}_0 = -4.4037 \quad \text{and} \quad \hat{\beta}_1 = 0.9379.$$

Thus, the estimated probability of death as a function of dose d is

$$\hat{p} = \frac{\exp(-4.4037 + 0.9379d)}{1 + \exp(-4.4037 + 0.9379d)}.$$

This function is plotted in Figure 18.2(c). The estimated standard errors of $\hat{\beta}_0$ and $\hat{\beta}_1$ are given under the `Std. Error` column in this table. The estimated parameter $\hat{\beta}_1$ and its corresponding standard error can be used to construct a Wald z statistic for testing whether the true parameter β_1 is equal to 0 using the test statistic

$$z = \frac{\hat{\beta}_k - 0}{\text{s.e.}(\hat{\beta}_k)}.$$

If the true β_1 parameter were equal to 0, then the dose would have no effect on the probability of death. The Wald z statistic for testing $H_0 : \beta_1 = 0$ is then

$$z = \frac{0.9379 - 0}{0.3930} = 2.387.$$

Since the absolute value of the z statistic exceeds 1.96 we would reject the null hypothesis. The P-value for this test is 0.0170, which is given in the last column of the table. Thus, there is evidence that the dose affects the probability of death. ∎∎∎

In a more realistic study there would ordinarily be many more subjects, and it is likely we would have demographic information on each of them. For example, we would probably have age, gender, race, education level, income, and many other variables. We could run a logistic regression model like the one in the previous example, but we would include these other variables called covariates. A covariate is really just a kind of predictor variable. In a study like this, the focus is on whether the *dose* affects the probability of death. The effects of these covariates is not the main focus of the study, but it may be of secondary interest. These covariates are used for adjustment so that we can make the strongest inference possible about whether the dose affects the outcome. Our model in R could look like this:

```
gFull = glm( resp ~ dose + age + gender + race + education + income ,
             family="binomial" )
summary( gFull )
```

■ **Example 18.2** In January 1986 the Space Shuttle Challenger exploded about a minute after takeoff, killing all seven members of the crew. After a lengthy investigation the cause was found to be a failure of an O-ring seal on the solid rocket booster. These O-rings were used at the connections between pieces of the cylindrical booster. There are three O-rings on each solid rocket booster and there are two solid rocket boosters. Failure was caused when one of the O-rings experienced a breach in the joint between pieces of the solid rocket booster, allowing gas to escape. This escaping gas then caused the catastrophic explosion. See Figure 18.3 for a time sequence of photos showing how the initially small leak led to the catastrophic explosion that destroyed the Challenger and killed the crew. It is believed that the low temperature (31°F) at the time of the launch caused the rubber O-rings to become stiff and inflexible, allowing the gas to escape.

Previous space shuttle launches had experienced at least some partial O-ring failure. The Challenger was the 25th space shuttle mission (although the Challenger itself had had only nine previous missions). Data regarding the temperature at launch and whether there was at least one partial O-ring failure are shown in Table 18.2. Use logistic regression to determine whether there is evidence that the temperature affects the probability of an O-ring failure.

Solution: The raw data are shown in the top scatterplot in Figure 18.4. Launches made at lower temperatures seem to have a greater chance of experiencing at least one O-ring failure. The temperature on January 28, 1986 was projected to be 31°F, well below the lowest temperature in the data set. We fit the logistic curve using R:

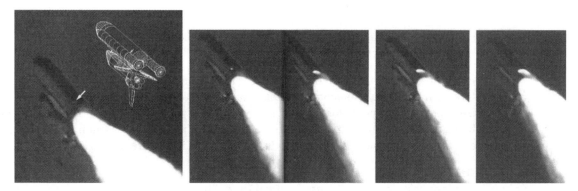

Figure 18.3 Time sequence of photos showing the moments before the explosion. The first frame shows the Challenger, along with a drawing to clarify the orientation, with a barely visible plume from the solid rocket booster on the right side. Successive frames show how this grew and eventually engulfed the spacecraft. Image is from the Report of the Presidential Commission on the Space Shuttle Challenger Accident: https://history.nasa.gov/rogersrep/v1p26.htm.

Table 18.2 Temperature in Fahrenheit and whether there was (1) at least one partial O-ring failure or (0) no failure.

Temperature	Failure	Temperature	Failure
66	0	78	0
70	1	67	0
69	0	53	1
80	0	67	0
68	0	75	0
67	0	70	0
72	0	81	0
73	0	76	0
70	0	79	0
57	1	75	1
63	1	58	1
70	1	76	0

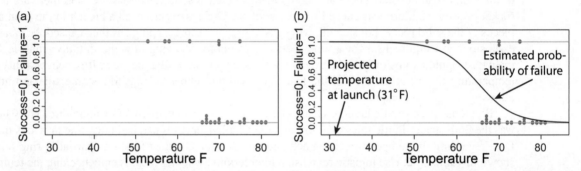

Figure 18.4 (a) Scatterplot of launch temperature vs. failure indicator (1 = failure, 0 = no failure). (b) Logistic fit to points in scatterplot.

```
tempF = c(66,70,69,80,68,67,72,73,70,57,63,70,
78,67,53,67,75,70,81,76,79,75,58,76)
fail = c(0,1,0,0,0,0,0,0,0,0,1,1,1,
0,0,1,0,0,0,0,0,0,1,1,0)
g1 = glm( fail ~ tempF , family="binomial" )
summary( g1 )
```

The key line here is the one containing the glm() command. This stands for generalized linear model. We need to tell R the outcome variable, fail in this case, and the predictor variable, tempF. If there were more than one predictor variable, they would be separated by + signs. The structure of the glm is much like the lm() command. For logistic regression we must add the option family="binomial" to indicate that the outcome has a binomial distribution. The results from R's glm() are:

```
Call:
glm(formula = fail ~ tempF, family = "binomial")

Deviance Residuals:

    Min      1Q   Median      3Q     Max
-1.0608  -0.7372  -0.3712  0.3948  2.2321
```

```
Coefficients:
            Estimate Std. Error z value Pr(>|z|)
(Intercept)  15.2968     7.3286   2.087   0.0369 *
tempF        -0.2360     0.1074  -2.198   0.0279 *
---
Signif. codes:  0 '***' 0.001 '**' 0.01 '*' 0.05 '.' 0.1 ' ' 1

(Dispersion parameter for binomial family taken to be 1)

    Null deviance: 28.975  on 23  degrees of freedom

Residual deviance: 20.371  on 22  degrees of freedom
AIC: 24.371

Number of Fisher Scoring iterations: 5
```

Whether the temperature affects the probability of at least one O-ring failure comes down to testing whether $\beta_1 = 0$. This can be done using a Wald test, which is a z-test based on the standard error of the estimator $\hat{\beta}_1$. From R's output, we see that

$$\hat{\beta} = -0.2360$$

$$\text{s.e.} = 0.1074$$

$$z = \frac{-0.2360 - 0}{0.1074} = -2.198.$$

The P-value for this test is 0.0279, providing evidence that temperature does affect the probability of a partial O-ring failure. The estimated probability for an O-ring failure at 31°F is

$$\hat{p} = \frac{e^{15.2968 - 0.2360 \times 31}}{1 + e^{15.2968 - 0.2360 \times 31}} = 0.99966.$$

This is the estimated probability of having at least a partial failure of an O-ring. Of course, previous shuttle missions had been successful even with partial failures, so this doesn't indicate that a catastrophic failure was imminent. This analysis does indicate that there was evidence of a relationship between temperature and partial O-ring failures. The launch temperature being 22 degrees colder than the coldest previous launch (31°F vs. 53°F) would (in hindsight) have made postponement of the Challenger's launch a prudent decision. ■■■

Any time the outcome variable is an indicator variable, the method of logistic regression can be used to assess the relationship between predictors (continuous, categorical, or a mix). Historically, logistic regression has been used when the predictor is a measure of dose or exposure and the outcome is a serious health condition like cancer or death. Today, logistic regression is used in innumerable situations that involve an indicator variable that is the response of some random mechanism.

■ **Example 18.3 — Spam filter.** This data set was created by Mark Hopkins, Erik Reeber, George Forman, and Jaap Suermondt from Hewlett-Packard Labs, and it involved 4,601 emails sent to George Forman. Each email was categorized by George Forman to be spam or ham (not spam); specifically, 1,813 were spam and 2,788 were ham. The purpose of this example is to build a *spam filter* – that is, a rule to predict whether a future incoming message is spam or ham. Data were collected in the summer of 1999. The data set consists of the frequency (measured as a percentage) that a particular word or character appeared in the email. The first 10 rows for some of the variables are shown in the list below:

y	word_freq_free	word_freq_000	word_freq_george	char_freq_$	capital_run_longest
1	0.32	0	0	0	61
1	0.14	0.43	0	0.18	101
1	0.06	1.16	0	0.184	485
1	0.31	0	0	0	40
1	0.31	0	0	0	40
1	0	0	0	0	15
1	0.96	0	0	0.054	4
1	0	0	0	0	11
1	0	0	0	0.203	445
1	0	0.19	0	0.081	43

The full list of variables can be obtained with the `names()` command in R.

"word_freq_make"	"word_freq_address"	"word_freq_all"
"word_freq_3d"	"word_freq_our"	"word_freq_over"
"word_freq_remove"	"word_freq_internet"	"word_freq_order"
"word_freq_mail"	"word_freq_receive"	"word_freq_will"
"word_freq_people"	"word_freq_report"	"word_freq_addresses"
"word_freq_free"	"word_freq_business"	"word_freq_email"
"word_freq_you"	"word_freq_credit"	"word_freq_your"
"word_freq_font"	"word_freq_000"	"word_freq_money"
"word_freq_hp"	"word_freq_hpl"	"word_freq_george"
"word_freq_650"	"word_freq_lab"	"word_freq_labs"
"word_freq_telnet"	"word_freq_857"	"word_freq_data"
"word_freq_415"	"word_freq_85"	"word_freq_technology"
"word_freq_1999"	"word_freq_parts"	"word_freq_pm"
"word_freq_direct"	"word_freq_cs"	"word_freq_meeting"
"word_freq_original"	"word_freq_project"	"word_freq_re"
"word_freq_edu"	"word_freq_table"	"word_freq_conference"
"char_freq_;"	"char_freq_("	"char_freq_["
"char_freq_!"	"char_freq_$"	"char_freq_#"
"capital_run_length_average"	"capital_run_length_longest"	"capital_run_length_total"

We will concentrate on just six of these variables (a more complete analysis could be based on all 58 variables):

```
"word_freq_free"
"word_freq_000"
"word_freq_george"
"word_freq_edu"
"char_freq_$"
"capital_run_length_longest"
```

Fit individual logistic regressions using the spam variable as the outcome with each of these six predictor variables. Also, run a multiple logistic regression model using all six variables.

Solution: Figure 18.5 shows plots of each of these six predictor variables against the response variable spam, which is coded as 1 = spam and 0 = not spam. The smooth curve in each panel shows the estimated logistic curve for these six one-variable-at-a-time models. All six of these variables appear to be highly associated with the email being spam. For the variables free, 000, char_freq_$, and capital_run_length_longest the association is positive, meaning that larger values tended to be associated with a higher likelihood of being spam. The email recipient's first name is George, so the word george (ignoring case) is associated with a smaller chance of being spam. Finally, edu is associated with

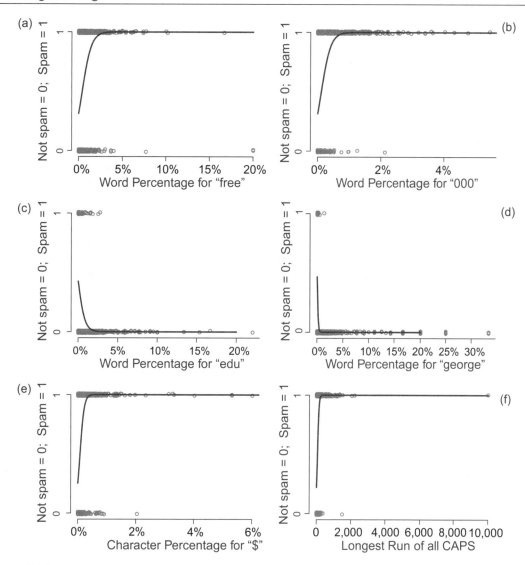

Figure 18.5 Plot of spam (1) or no spam (0) versus each of the six predictor variables. The fitted logistic curves are shown in blue.

a smaller likelihood of being spam. Folks in academia tend to receive emails from students or colleagues having an email address from an edu domain.

Figure 18.5 suggests that each of these predictors is associated with the outcome spam. In each of the one-variable-at-a-time models, the predictor is highly significant. For example, here is the output from R for the logistic regression using just the `capital_run_length_longest` variable. The P-value for the predictor is 2×10^{-16}.

```
Call:
glm(formula = y ~ capital_run_length_longest, family = "binomial",
data = spam)

Deviance Residuals:
    Min       1Q    Median       3Q      Max
-8.4904  -0.8085  -0.7329   1.0690   1.7232
```

```
Coefficients:
                          Estimate Std. Error z value Pr(>|z|)
(Intercept)               -1.25306    0.04571  -27.42   <2e-16 ***
capital_run_length_longest 0.02522    0.00119   21.19   <2e-16 ***
---
Signif. codes:  0 '***' 0.001 '**' 0.01 '*' 0.05 '.' 0.1 ' ' 1

(Dispersion parameter for binomial family taken to be 1)
    Null deviance: 6170.2  on 4600  degrees of freedom
Residual deviance: 5179.9  on 4599  degrees of freedom
AIC: 5183.9

Number of Fisher Scoring iterations: 8
```

The outputs from the other five one-predictor logistic regressions are similar. By combining several variables we might be able to build a strong spam filter that discriminates between spam and ham. If we run logistic regression with all six predictor variables, we obtain these results:

```
Call:
glm(formula = y ~ word_freq_free + word_freq_000 + word_freq_edu +
    word_freq_george + 'char_freq_$' + spam$capital_run_length_total,
    family = "binomial", data = spam )

Deviance Residuals:
    Min      1Q   Median      3Q      Max
-5.5780  -0.7379  -0.0019  0.3534   5.3804

Coefficients:
                              Estimate Std. Error z value Pr(>|z|)
(Intercept)                  -1.179e+00  5.418e-02 -21.755  < 2e-16 ***
word_freq_free                1.929e+00  1.344e-01  14.358  < 2e-16 ***
word_freq_000                 3.491e+00  4.217e-01   8.278  < 2e-16 ***
word_freq_edu                -2.365e+00  2.955e-01  -8.004 1.20e-15 ***
word_freq_george             -1.043e+01  1.758e+00  -5.929 3.05e-09 ***
'char_freq_$'                 9.564e+00  6.688e-01  14.300  < 2e-16 ***
spam$capital_run_length_total 6.158e-04  9.953e-05   6.187 6.12e-10 ***
---
Signif. codes:  0 '***' 0.001 '**' 0.01 '*' 0.05 '.' 0.1 ' ' 1

(Dispersion parameter for binomial family taken to be 1)

    Null deviance: 6170.2  on 4600  degrees of freedom
Residual deviance: 3440.0  on 4594  degrees of freedom
AIC: 3454

Number of Fisher Scoring iterations: 11
```

Of course, all of the variables are highly significant. Thus, there is strong evidence that each of these has an effect on the likelihood of an email message being spam, but this was expected. The real question for this spam filter is whether the algorithm can successfully discriminate between spam and ham, not to determine whether certain variables have an effect on spam. In Section 14.4 we pointed out the distinction between the strength of the evidence of some relationship and the strength of the relationship itself. Here we have strong evidence of a relationship, but the relation is only moderately strong in its ability to predict. We will take up this issue in Section 18.6. ∎ ∎ ∎

Often, the response that leads to $y = 1$ is an adverse outcome such as death or some serious condition. If one of the predictor variables besides the dose is an indicator variable, such as $1 =$ male and $0 =$ female, then many researchers will want to compare the risk of the adverse outcome for each level at a single dose. Suppose that variable x_2 is the dose and x_1 is the indicator variable. There may be additional covariates $x_3, x_4, \ldots, x_p$. The probability of an adverse outcome at level $x_1 = 0$ (and arbitrary values for the dose x_2 and all other covariates) is

$$p_0 = \frac{\exp(\beta_0 + \beta_1 \times 0 + \beta_2 x_2 + \beta_3 x_3 + \cdots + \beta_p x_p)}{1 + \exp(\beta_0 + \beta_1 \times 0 + \beta_2 x_2 + \beta_3 x_3 + \cdots + \beta_p x_p)},$$

and the probability of an adverse outcome at level $x_1 = 1$ is

$$p_1 = \frac{\exp(\beta_0 + \beta_1 \times 1 + \beta_2 x_2 + \beta_3 x_3 + \cdots + \beta_p x_p)}{1 + \exp(\beta_0 + \beta_1 \times 1 + \beta_2 x_2 + \beta_3 x_3 + \cdots + \beta_p x_p)}.$$

To compare the chance of an adverse outcome for the levels $x_1 = 0$ and $x_1 = 1$, researchers often look at the **odds ratio**, abbreviated **OR**, which is defined to be the ratio of these two odds:

$$\text{OR} = \frac{\dfrac{p_1}{1 - p_1}}{\dfrac{p_0}{1 - p_0}}. \tag{18.7}$$

Using (18.3) we see that the logits or log odds for these two conditions are equal to

$$\text{logit}(p_0) = \beta_0 + \beta_1 \times 0 + \beta_2 x_2 + \beta_3 x_3 + \cdots + \beta_p x_p$$

and

$$\text{logit}(p_1) = \beta_0 + \beta_1 \times 1 + \beta_2 x_2 + \beta_3 x_3 + \cdots + \beta_p x_p,$$

respectively. The log of the odds ratio in (18.7) is then

$$\log \text{OR} = \log \frac{p_1}{1 - p_1} - \log \frac{p_0}{1 - p_0} = \text{logit}(p_1) - \text{logit}(p_0),$$

which we can write as

$$\log \text{OR} = (\beta_0 + \beta_1 \times 1 + \beta_2 x_2 + \beta_3 x_3 + \cdots + \beta_p x_p) - (\beta_0 + \beta_1 \times 0 + \beta_2 x_2 + \beta_3 x_3 + \cdots + \beta_p x_p) = \beta_1. \tag{18.8}$$

Thus the log of the OR for variable x_1, an indicator variable, is equal to the parameter β_1. The OR is thus the exponential of the expression in (18.8):

$$\text{OR} = \exp(\log \text{OR}) = \exp(\beta_1). \tag{18.9}$$

The concept of the OR is often used in research to illustrate the effect of an indicator variable on the chance of an adverse outcome.

Table 18.3 Hypothetical data from a dose–response study.

Dose	Sex	Age	Response	Dose	Sex	Age	Response
0	1	4	0	3	1	4	0
0	1	6	0	5	1	4	1
0	1	5	0	5	0	3	0
0	1	6	0	6	0	5	1
1	1	4	0	6	0	7	1
1	0	3	0	6	1	6	0
1	1	4	0	7	0	5	1
2	1	4	0	8	0	3	1
3	0	7	1	9	0	3	1
3	1	6	0	9	0	3	1

■ **Example 18.4** Suppose that in Example 18.1 we also observed the sex variable x_2, coded as 1 = male and 0 = female, and the age of the animal x_3 measured in months. The augmented data set is shown in Table 18.3. Run the logistic regression model, compute the odds ratio for the sex variable, and interpret the results.

Solution: The R code to run the logistic regression is

```
dose = c(0,0,0,0,1,1,1,2,3,3, 3,5,5,6,6,6,7,8,9,9)
sex  = c(1,1,1,1,1,0,1,1,0,1, 1,1,0,0,0,1,0,0,0,0)
age  = c(4,6,5,6,4,3,4,4,7,6, 4,4,3,5,7,6,5,3,3,3)
resp = c(0,0,0,0,0,0,0,0,1,0, 0,1,0,1,1,0,1,1,1,1)
g1 = glm( resp ~ 1 + sex + dose + age , family="binomial" )
summary( g1 )
```

This yields the output:

```
Call:
glm(formula = resp ~ 1 + sex + dose + age, family = "binomial")

Coefficients:
            Estimate Std. Error z value Pr(>|z|)
(Intercept)  -6.2065     4.4629  -1.391   0.1643
sex          -2.3500     1.8249  -1.288   0.1978
dose          0.8820     0.4965   1.777   0.0756 .
age           0.6376     0.6201   1.028   0.3039
---
Signif. codes:  0 '***' 0.001 '**' 0.01 '*' 0.05 '.' 0.1 ' ' 1

(Dispersion parameter for binomial family taken to be 1)

    Null deviance: 26.9205  on 19  degrees of freedom
Residual deviance:  8.8041  on 16  degrees of freedom
AIC: 16.804

Number of Fisher Scoring iterations: 6
```

We see that the dose still has a significant effect, but neither sex nor age is significant. Larger values of dose are associated with an increased probability of death. The estimated coefficient for sex is

$\hat{\beta}_1 = -2.2325$, which is negative, suggesting that males (since sex is coded as 1 for male and 0 for female) have a lower risk of death. The odds ratio is

$$\text{OR} = e^{-2.2325} \approx 0.10726.$$

The coefficient for age is $\hat{\beta}_3 = 0.6350$, which is positive, indicating that the risk of death is higher for older animals, but the result is not statistically significant. ∎∎∎

When you perform logistic regression, remember that the OR is a ratio of *odds*, not *probabilities*, so care must be taken in explaining the results. The *relative risk* is a distinct, but related, term:

$$\text{Odds ratio} = \frac{\dfrac{p_1}{1 - p_1}}{\dfrac{p_0}{1 - p_0}} = e^{\beta_1}$$

$$\text{Relative risk} = \frac{p_1}{p_0}.$$

Some basic (but messy) algebra yields

$$
\begin{aligned}
\text{Relative risk} &= \frac{\exp\left(\beta_0 + \beta_1 + \sum_{j=2}^{p} \beta_j x_j\right) \Big/ \left[1 + \exp\left(\beta_0 + \beta_1 + \sum_{j=2}^{p} \beta_j x_j\right)\right]}{\exp\left(\beta_0 + \sum_{j=2}^{p} \beta_j x_j\right) \Big/ \left[1 + \exp\left(\beta_0 + \sum_{j=2}^{p} \beta_j x_j\right)\right]} \\[2em]
&= e^{\beta_1} \frac{1 + \exp\left(\beta_0 + \sum_{j=2}^{p} \beta_j x_j\right)}{1 + \exp\left(\beta_0 + \beta_1 + \sum_{j=2}^{p} \beta_j x_j\right)} \\[2em]
&= \text{Odds ratio} \times \frac{1 + \exp\left(\beta_0 + \sum_{j=2}^{p} \beta_j x_j\right)}{1 + \exp\left(\beta_0 + \beta_1 + \sum_{j=2}^{p} \beta_j x_j\right)}
\end{aligned}
$$

The most important thing to notice about the expressions for the OR and the relative risk is that the odds ratio, $\text{OR} = \exp(\beta_1)$, is independent of the other predictors, whereas the relative risk does depend on the other predictors. Note that for rare events, both $1 - p_0$ and $1 - p_1$ will both be nearly 1, causing the fraction in the last expression to be near 1; in this case, the relative risk will be nearly equal to the OR.

∎ **Example 18.5** Using the data from Example 18.4, plot the relative risk as a function of age and dose.

Solution: Figure 18.6 shows the relative risk for males compared to females. Based on the discussion above, the relative risk is not constant across other values of the predictor variables. (By contrast, the odds ratio *is* a constant, and is equal to 0.10726.) The relative risk is estimated to be

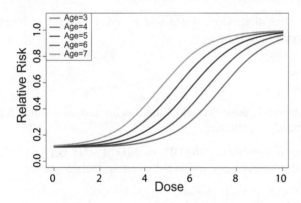

Figure 18.6 Relative risk as a function of dose and age.

$$\text{Estimated relative risk} = \text{Estimated OR} \times \frac{1 + \exp(\hat{\beta}_0 + \hat{\beta}_2 \text{dose} + \hat{\beta}_3 \text{age})}{1 + \exp(\hat{\beta}_0 + \hat{\beta}_1 + \hat{\beta}_2 \text{dose} + \hat{\beta}_3 \text{age})}$$

$$= 0.10726 \times \frac{1 + \exp(-6.5820 + 0.9436 \text{dose} + 0.6350 \text{age})}{1 + \exp(-6.5820 - 2.2325 + 0.9436 \text{dose} + 0.6350 \text{age})}.$$

■ ■ ■

18.2 Multinomial Logistic Regression

There are situations where each observation is classified into one of $m+1$ mutually exclusive outcomes (labeled 0, 1, …, m) and we would like to estimate the probability p_j of an observation falling into each category given a set of predictor variables. Two important things to note are: (1) if $m = 2$ then we are back in the familiar situation of logistic regression; and (2) we must assure that the sum of the probabilities is 1, that is $\sum_{j=0}^{m} p_j = 1$.

One approach that satisfies the condition where the probabilities sum to 1 is to assume the following:

$$P(Y_i = 0 | x_i) = \frac{1}{1 + \sum_{j=1}^{m} \exp(x_i' \beta^{(j)})}$$

$$P(Y_i = 1 | x_i) = \frac{\exp(x_i' \beta^{(1)})}{1 + \sum_{j=1}^{m} \exp(x_i' \beta^{(j)})}$$

$$P(Y_i = 2 | x_i) = \frac{\exp(x_i' \beta^{(2)})}{1 + \sum_{j=1}^{m} \exp(x_i' \beta^{(j)})} \tag{18.10}$$

$$\vdots$$

$$P(Y_i = m | x_i) = \frac{\exp(x_i' \beta^{(m)})}{1 + \sum_{j=1}^{m} \exp(x_i' \beta^{(j)})}$$

Notice that with this parameterization, the classification probabilities are guaranteed to sum to 1. Notice also that we can only model the classification probabilities for m of the $m+1$ categories because knowing the probabilities for m of the categories is enough to determine the probability for the last category. This "unmodeled" category is called the baseline, and we are really modeling the probability of being in category j relative to being in the baseline category.

It can be shown (see Problem 18.9) that

$$\log \frac{P(Y_i = 1|\boldsymbol{x}_i)}{P(Y_i = 0|\boldsymbol{x}_i)} = \boldsymbol{x}_i'\boldsymbol{\beta}^{(1)}$$

$$\log \frac{P(Y_i = 2|\boldsymbol{x}_i)}{P(Y_i = 0|\boldsymbol{x}_i)} = \boldsymbol{x}_i'\boldsymbol{\beta}^{(2)} \tag{18.11}$$

$$\vdots$$

$$\log \frac{P(Y_i = m|\boldsymbol{x}_i)}{P(Y_i = 0|\boldsymbol{x}_i)} = \boldsymbol{x}_i'\boldsymbol{\beta}^{(m)}.$$

■ **Example 18.6** The data set for this example comes from the study by Ayres-de Campos *et al.* (2000). The file contains 2,126 records from a cardiotocogram, a type of electronic fetal monitor which can take a number of different measurements related to fetal health. The results were classified by three obstetricians into one of three classes: normal (1), suspect (2), or pathological (3). The outcome is the class, and there are 22 predictor variables. Use the following predictor variables to classify the health of the fetus:

```
baseline.value              fetal_movement
uterine_contractions        severe_decelerations
prolongued_decelerations    abnormal_short_term_variability
```

Solution: Figure 18.7 shows scatterplots of these six predictors against the outcome class. We have jittered the class variable to aid in visualizing multiple points.

We take the normal category ($y = 1$) as the baseline and model the probabilities of being in class 2 or 3 relative to class 1:

```
library(nnet)
FetalHealth = read.csv("FetalHealth.csv")
FetalHealth$fetal_health_f = as.factor( FetalHealth$fetal_health )
fetal.multinom = multinom( fetal_health_f ~ baseline.value +
                                        fetal_movement +
                                        uterine_contractions +
                                        severe_decelerations +
                                        prolongued_decelerations +
                                        abnormal_short_term_variability ,
                           data = FetalHealth )
summary( fetal.multinom )
```

Because the class variable was coded as 1, 2, or 3, we had to convert it from numeric to a factor variable. The multinom() is a function in the nnet package that performs the multinomial logistic regression. The output from the multinom() function is summarized using the summary() command, and it yields the following:

```
Call:
multinom(formula = fetal_health_f ~ baseline.value + fetal_movement +
    uterine_contractions + severe_decelerations + prolongued_decelerations +
    abnormal_short_term_variability, data = FetalHealth)

Coefficients:
  (Intercept) baseline.value fetal_movement uterine_contractions
2   -18.66324    0.095263066       6.661813            -180.2777
3   -11.74142   -0.006165753       7.243815            -147.3874
```

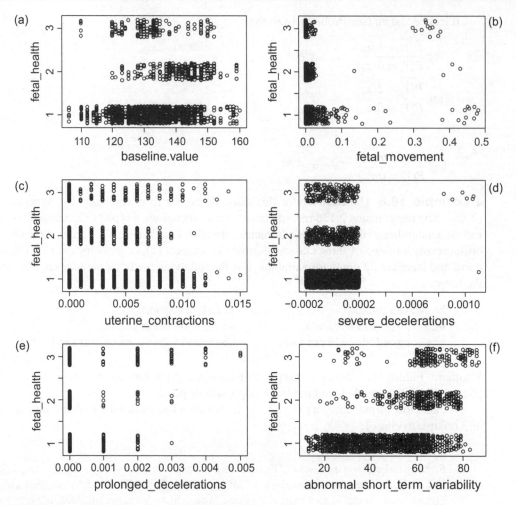

Figure 18.7 Scatterplots of fetal health against six of the possible predictors.

```
     severe_decelerations prolongued_decelerations
2              -6.964327                 51.96051
3              86.624135               2891.55480
     abnormal_short_term_variability
2                        0.0836883
3                        0.1690714

Std. Errors:
  (Intercept) baseline.value fetal_movement uterine_contractions
2    1.351122    0.009107235       1.728610           0.01115322
3    1.819566    0.013500228       2.401315           0.03705856
     severe_decelerations prolongued_decelerations
2            0.0001103281              0.002466545
3            0.0003668633              0.005887899
abnormal_short_term_variability
2                     0.006585948
3                     0.011228694
```

```
Residual Deviance: 1595.833
AIC: 1623.833
```

Notice how R gives the point estimates (the Coefficients) separately from the corresponding standard errors. We can extract the point estimates and standard errors in order to compute the Wald z statistic from the summary(fetal.multinom) object

```
summary.multinom.fetal = summary( fetal.multinom )
str( summary.multinom.fetal )
betahat = summary.multinom.fetal$coefficients
se.beta = summary.multinom.fetal$standard.errors
z = (betahat - 0)/se.beta
print( z )
```

The print() statement yields

```
   (Intercept) baseline.value fetal_movement uterine_contractions
2  -13.813149     10.4601520       3.853856           -16163.741
3   -6.452867     -0.4567147       3.016604            -3977.148
   severe_decelerations prolongued_decelerations
2             -63123.75                 21066.11
3             236120.99                491101.29
   abnormal_short_term_variability
2                        12.70710
3                        15.05708
```

In the printed output for the Wald z statistics, R retains the column names, making it easy to see which variables and which outcomes a particular z statistic applies to. For example, the z statistic for testing whether the baseline.value affects the probability of being assigned to category 3 is -0.4567147, which turns out to be the only z statistic that is not significant. All of the others exceed 3 in absolute value, indicating that the other variables have a significant effect in the probability of category membership.

We can use the estimated parameters to find the predicted probability of category membership. For example, let's consider the first fetus in the data set with

```
baseline.value = 120
fetal_movement = 0
uterine_contractions = 0
severe_decelerations = 0
prolongued_decelerations = 0
abnormal_short_term_variability = 73
```

For brevity, let's call these variables x_1, x_2, x_3, x_4, x_5, and x_6. The predicted probabilities for category membership can then be calculated using (18.10); this is accomplished in the following R code:

```
x1 = 120
x2 = 0
x3 = 0
x4 = 0
x5 = 0
x6 = 73
betahat2 = betahat[1,]
betahat3 = betahat[2,]
```

```
expnt2 = betahat2[1] + betahat2[2]*x1 + betahat2[3]*x2 + betahat2[4]*x3
                     + betahat2[5]*x4 + betahat2[6]*x5 + betahat2[7]*x6
p2 = exp(expnt2) / ( 1 + exp(expnt2) )
expnt3 = betahat3[1] + betahat3[2]*x1 + betahat3[3]*x2 + betahat3[4]*x3
                     + betahat3[5]*x4 + betahat3[6]*x5 + betahat3[7]*x6
p3 = exp(expnt3) / ( 1 + exp(expnt3) )
p1 = 1 - p2 - p3
print( c(p1,p2,p3) )
```

The result of the print() statement is

```
(Intercept)  (Intercept)  (Intercept)
9.992734e-01 7.227855e-04 3.796950e-06
```

Thus, the predicted probabilities of membership for this person are $p = [0.9993, 0.0007, 0.0000]$. In this case we observed a rare event, because the fetal health was category 2.

Figure 18.8 shows the same scatterplot as Figure 18.7, except the first point is highlighted with a larger symbol and plotted as a solid red circle. Figure 18.8(a) indicates that baseline.value = 120

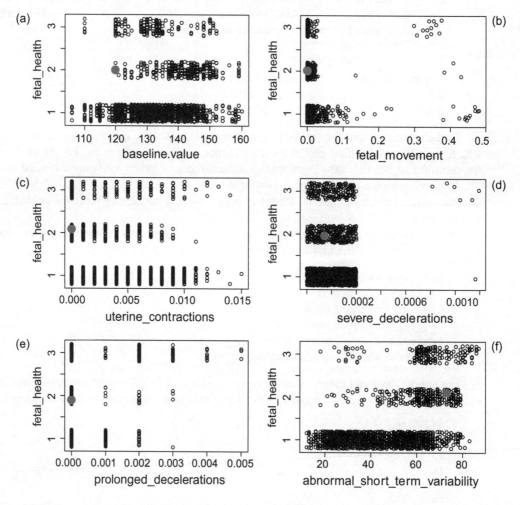

Figure 18.8 Scatterplots of fetal health against six of the possible predictors. The first data point is the large red circle.

is the smallest value that occurs with `fetal_health` = 2. Most of the other figures put the point in a cluster of points from the `fetal_health` category and don't distinguish it as an unusual point. Only Figure 18.8(d) gives an indication that the point could be in the category `fetal_health` = 2; the `abnormal_short_term_variability` would be on the high end if the point were in category `fetal_health` = 1; it is nearer the "pack" in category `fetal_health` = 2.   ∎∎∎

> **Exercise 18.1** Compare the scatterplot in Figure 18.7 with the coefficients, specifically the sign (positive or negative) of the coefficients, in the multinomial logistic regression.

18.3 Poisson Regression

In Section 18.1 we studied the situation where the outcome variable was an indicator variable that was equal to 1 if some event occurred and 0 if it did not occur. Various predictor variables can be used to estimate the probability of such an event occurring. Often the outcome is not *whether* an event occurred, but *how many times* an event occurred. For example, in a clinical trial for a seizure medication, the outcome may be how many seizures the patient suffered rather than whether a seizure occurred. In manufacturing of materials, the outcome may be the number of defects in a fixed region. For example, in large rolls of copper, the outcome may be the number of defects on each inspected square foot.

For cases like this, the Poisson distribution is often a reasonable assumption for the distribution of the outcome. If there are predictor variables $x_{i1}, x_{i2}, \ldots, x_{ip}$ on subject or unit i then we may assume that the mean of the Poisson distribution is

$$\mu_i = \exp(\beta_0 + \beta_1 x_{i1} + \beta_2 x_{i2} + \cdots + \beta_p x_{ip}). \tag{18.12}$$

Remember, the mean of the Poisson distribution must be a positive number, so we must guarantee that $\mu_i > 0$ for any possible values of the predictor. For this reason, some monotonic positive function of the predictors is needed; in this case, the exponential function serves this role. The model in (18.12) can be stated equivalently as

$$\log \mu_i = \beta_0 + \beta_1 x_{i1} + \beta_2 x_{i2} + \cdots + \beta_p x_{ip}. \tag{18.13}$$

The function of the μ_i that is linear in the predictors is called the **link** function. Thus we would say that for this model, the link function is the logarithm.

∎ **Example 18.7** Myers *et al.* (2010) describe a study conducted by the United States Navy during the Vietnam conflict. The outcome variable was the number of locations on the aircraft that suffered damage during low-altitude missions over Vietnam. These values ranged from 0 to 7, and the Poisson distribution seemed to be a reasonable model. The three predictor variables were:

- `Aircraft` This is the aircraft type, coded as 0 = McDonnell Douglas A-4 Skyhawk, and 1 = Grumman A-6 Intruder.
- `BombLoad` This is the bomb load in tons.
- `CrewExperience` This is the combined months of experience of the aircrew.

The data are shown in Table 18.4. Perform an exploratory analysis by plotting the response against each of the predictors.

Solution: The number of damage locations may be assumed to have a Poisson distribution whose mean depends on the levels of the three predictor variables: aircraft type, bomb load, and crew experience. A plot of the outcome variable `AircraftDamage` versus each of these predictor variables is shown in

Table 18.4 Aircraft damage.

Damage	Aircraft	BombLoad	CrewExp	Damage	Aircraft	BombLoad	CrewExp
0	0	4	91.5	3	1	7	116.1
1	0	4	84.0	1	1	7	100.6
0	0	4	76.5	1	1	7	85.0
0	0	5	69.0	1	1	10	69.4
0	0	5	61.5	2	1	10	53.9
0	0	5	80.0	0	1	10	112.3
1	0	6	72.5	1	1	12	96.7
0	0	6	65.0	1	1	12	81.1
0	0	6	57.5	2	1	12	65.6
2	0	7	50.0	5	1	8	50.0
1	0	7	103.0	1	1	8	120.0
1	0	7	95.5	1	1	8	104.4
1	0	8	88.0	5	1	14	88.9
1	0	8	80.5	5	1	14	73.7
2	0	8	73.0	7	1	14	57.8

Figure 18.9. Figure 18.9(a) indicates that damage is greater on the A-6 than on the A-4. Figure 18.9(b) shows a positive relationship, with higher bomb loads being associated with more damage. Finally, Figure 18.9(c) suggests that there could be a negative relationship between crew experience and damage. More experienced crews seem to suffer less damage.

These one-factor-at-a-time effects can be misleading, though. The plot character used in Figure 18.9 depends on the type of aircraft. The A-4 points are red circles, whereas the A-6 points are blue dots. The bomb loads are higher for the A-6 because it is a larger aircraft. In addition, the crew experience tends to be higher for the A-6. Thus, the predictor variables seem to be correlated with one another.

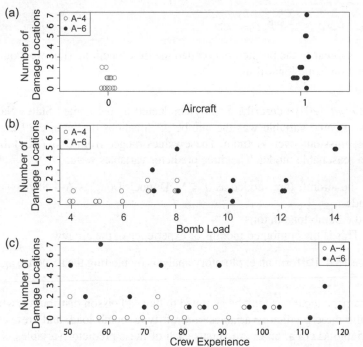

Figure 18.9 Aircraft damage plotted with each of the three predictor variables.

Since Poisson regression is a kind of generalized linear model (see the next section) we can use the glm() command along with the family='"poisson"' argument. We can run the full Poisson regression model, with all three predictor variables, using the code

```
damage = read.csv("AircraftDamage.csv")
DamageLocations = damage$DamageLocations
Aircraft = damage$Aircraft
BombLoad = damage$BombLoad
CrewExperience = damage$CrewExperience
glm1 = glm( DamageLocations ~ Aircraft + BombLoad + CrewExperience,
            family="poisson" )
summary( glm1 )
```

This yields the following output:

```
Call:
glm(formula = DamageLocations ~ Aircraft + BombLoad + CrewExperience,
    family = "poisson")
Deviance Residuals:
    Min       1Q   Median       3Q      Max
 -1.6418  -1.0064  -0.0180   0.5581   1.9094

Coefficients:
                Estimate Std. Error z value Pr(>|z|)
(Intercept)    -0.406023   0.877489  -0.463   0.6436
Aircraft        0.568772   0.504372   1.128   0.2595
BombLoad        0.165425   0.067541   2.449   0.0143 *
CrewExperience -0.013522   0.008281  -1.633   0.1025
---
Signif. codes:  0 '***' 0.001 '**' 0.01 '*' 0.05 '.' 0.1 ' ' 1

(Dispersion parameter for poisson family taken to be 1)

    Null deviance: 53.883  on 29  degrees of freedom
Residual deviance: 25.953  on 26  degrees of freedom
AIC: 87.649

Number of Fisher Scoring iterations: 5
```

We see that the only significant predictor variable is BombLoad, with a P-value of 0.0143. Despite the appearance in Figure 18.9, the aircraft type is not significant. The discrepancy between the A-4 and A-6 seen in Figure 18.9 can be explained by the difference in BombLoad values for the two types of aircraft. ∎∎∎

The Poisson distribution is a one-parameter distribution with the property that its mean and variance are the same. Both are equal to the parameter μ. This can sometimes impose an unrealistic assumption on the output since there are situations where the variance of the outcome, conditioned on the levels of the predictors, is larger than the mean, conditioned on the same levels of the predictors. This is called *overdispersion*. (While it is possible that the variance is less than the mean, situations like this are rare.) In cases like this, it is possible to model the outcome using a two-parameter distribution such as the negative binomial. See Hilbe (2011) for more details on negative binomial regression.

Another extension of Poisson regression involves the excess frequency of zeros. There are circumstances where the outcome is a mixture of "structural zeros" where the outcome *cannot* happen and "chance" zeros, where it is possible that the outcome occurs, but by chance it does not occur and a zero is observed. It is impossible to distinguish whether an observed count of zero is a structural zero or a chance zero. The *zero-inflated* Poisson model can be used to account for the overabundance of zeros in the data set. Lambert (1992) discusses the application of zero-inflated Poisson regression. It is even possible to construct a model for a zero-inflated negative binomial outcome (see Hilbe, 2011).

18.4 Generalized Linear Models

The usual linear regression model assumed that

$$Y_i \stackrel{\text{indep}}{\sim} N(\beta_0 + \beta_1 x_{i1} + \beta_2 x_{i2} + \cdots + \beta_p x_{ip}, \sigma^2).$$

The mean or expected outcome for predictors $x_{i1}, x_{i2}, \ldots, x_{ip}$ is therefore

$$\mu_i = \beta_0 + \beta_1 x_{i1} + \beta_2 x_{i2} + \cdots + \beta_p x_{ip}.$$

In the first section of this chapter we studied logistic regression, where the outcome was dichotomous with distribution

$$Y_i \stackrel{\text{indep}}{\sim} \text{BIN}(1, p_i),$$

where

$$\text{logit}\, p_i = \log \frac{p_i}{1 - p_i} = \beta_0 + \beta_1 x_{i1} + \beta_2 x_{i2} + \cdots + \beta_p x_{ip}.$$

Thus,

$$\mu_i = E(Y_i | x_{i1}, x_{i2}, \ldots, x_{ip}) = 1 \times p_i = \frac{\exp(\beta_0 + \beta_1 x_{i1} + \beta_2 x_{i2} + \cdots + \beta_p x_{ip})}{1 + \exp(\beta_0 + \beta_1 x_{i1} + \beta_2 x_{i2} + \cdots + \beta_p x_{ip})}.$$

This leads to

$$\text{logit}\,(\mu_i) = \beta_0 + \beta_1 x_{i1} + \beta_2 x_{i2} + \cdots + \beta_p x_{ip},$$

so a function of μ_i is linear in the parameters $\beta_0, \beta_1, \ldots, \beta_p$. In this case, the function is the logit function.

In Poisson regression, we assumed

$$Y_i \stackrel{\text{indep}}{\sim} \text{POIS}\big(\exp(\beta_0 + \beta_1 x_{i1} + \beta_2 x_{i2} + \cdots + \beta_p x_{ip})\big)$$

and since the mean of the Poisson is equal to its one (and only) parameter,

$$\mu_i = \exp(\beta_0 + \beta_1 x_{i1} + \beta_2 x_{i2} + \cdots + \beta_p x_{ip}).$$

Taking the log of both sides yields

$$\log \mu_i = \beta_0 + \beta_1 x_{i1} + \beta_2 x_{i2} + \cdots + \beta_p x_{ip}.$$

Again, some function of μ_i is linear. In this case the function is the natural log.

The *generalized linear model* assumes that some function of the mean outcome μ_i is linear; that is,

$$g(\mu_i) = g(E(Y_i)) = \beta_0 + \beta_1 x_{i1} + \beta_2 x_{i2} + \cdots + \beta_p x_{ip}. \tag{18.14}$$

The function g is called the *link* function. We have seen three link functions so far:

1. The identity link function, $g(\mu_i) = \mu_i$; this leads to the ordinary linear regression from Chapter 14.
2. The logit link function, $g(\mu_i) = \text{logit}\,\mu_i = \log\left[\mu_i/(1-\mu_i)\right]$; this leads to logistic regression, which we studied in Section 18.1.
3. The (natural) log link, $g(\mu_i) = \log\mu_i$; this leads to Poisson regression.

These three cases cover most of the generalized linear models that are used in practice. See Myers *et al.* (2010) for a rigorous coverage of generalized linear models.

18.5 Regression Trees

In Chapter 14 we studied regression models, such as the simple regression model shown below:

$$y_i = \beta_0 + \beta_1 x_i + \varepsilon_i, \qquad \varepsilon_1, \varepsilon_2, \ldots, \varepsilon_n \sim \text{i.i.d. } N(0, \sigma^2).$$

The assumptions behind this model are rather strong, and it is sometimes desired to make predictions for groups of predictor variables. In each of these groups, the point prediction is the same. For example, we might want to make the same prediction for any person who falls in a particular category. We can split the categories determined by the predictor variables to obtain homogeneous groups.

Consider, for example, the data set on diabetic women that we discussed in Section 10.8. Suppose the outcome is Glucose and the predictors include all other variables: Pregnancies, BloodPressure, SkinThickness, Insulin, BMI, DiabetesPedigreeFunction, and Age. We might want to make the same prediction for all women with exactly two prior pregnancies, or for all 30-year-old women with at least two pregnancies and a BMI of 30 or less.

We will illustrate the concept of a regression tree with a much-simplified version of the problem. Suppose that there are only 10 women in the study and there are just two predictor variables. The hypothetical data set is shown in Table 18.5.

We'll begin with the simplest kind of regression tree, one with just one predictor and one split. (We'll illustrate the concept of a split soon.) Suppose we would like to make one prediction of Glucose for women with two or fewer pregnancies and another for women with three or more pregnancies. There are six women in the former group and four in the latter group. If we wanted to make a prediction of Glucose for the six women in the two or fewer category, we could choose the mean Glucose level for those with two or fewer pregnancies. This is

Table 18.5 Small data set on glucose levels, prior pregnancies, and age.

Obs	Pregnancies	Age	Glucose
1	1	26	95
2	4	31	185
3	0	22	110
4	3	24	153
5	3	34	182
6	6	33	204
7	0	20	98
8	1	30	140
9	2	29	139
10	1	27	144

$$\bar{y}_- = \frac{95+110+98+140+139+144}{6} = \frac{726}{6} = 121.$$

Similarly, we would predict the Glucose level for women with three or more pregnancies by

$$\bar{y}_+ = \frac{185+153+182+204}{4} = \frac{724}{4} = 181.$$

Thus, our prediction model is simple: If the number of pregnancies is two or less, predict Glucose=121; otherwise, predict Glucose=181. This can be summarized in the following tree:

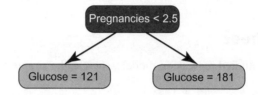

Note that we have used the midpoint of two consecutive possible values for Pregnancies. In other words, we set $c = 2.5$ as the number on which to split the prediction. The convention in reading trees like the one above is that if the condition in a node is true, we move to the node on the left; otherwise, we move to the node on the right. In this case, if the condition Pregnancies < 2.5 is true, we move down the tree and to the left and end up at a terminal node (called a *leaf*) that says Glucose=121. If the condition Pregnancies < 2.5 is false, we move down the tree and to the right, and end up at the leaf that says Glucose=181.

One might question our choice of using 2.5 as the split. Might a split of 1.5 or 5.5 have yielded better predictions? We can address this question by considering all possible splits. For this small data set, the numbers of pregnancies is an integer from 0 through 6. We could apply all of these splits, namely 0.5, 1.5, 2.5, 3.5, 4.5, and 5.5 (although since there were no records in the data set with Pregnancies=5, the last two splits would be the same). The latter two splits are equivalent to having one split at 5.0. In order to compare the fit of different splits, we must define a measure of fit. While there could be many definitions of fit, the most common is the sum of squared errors, much like the SSE in Chapter 14. We define

$$SS_k = \sum_{i:\, y_i < c_k} (y_i - \bar{y}_{-,k})^2 + \sum_{i:\, y_i > c_k} (y_i - \bar{y}_{+,k})^2,$$

where c_k is the split point and $y_{-,k}$ and $y_{+,k}$ are the mean Glucose levels for those values below and above c_k.

Figure 18.10 shows each of the possible splits for Pregnancies along with the resulting sums of squares. Notice that the SS is least when the split is $c = 2.5$. A plot of the SS as a function of the split point is shown in Figure 18.11.

Regression trees involve the consideration of several variables and possibly several splits on each variable. Before we pursue this, let's look at the connection between regression trees and the regression model that we studied in Chapter 14. Let x_i denote the ith observation on the one (and only, for now) predictor variable, and let y_k denote the kth possible split value. Define the indicator function

$$I(x_i < c_k) = \begin{cases} 1, & \text{if } x_i < c_k \\ 0, & \text{otherwise.} \end{cases}$$

We can then consider the regression problem

$$y_i = \beta_0 + \beta_1 I(x_i < c_k) + \varepsilon_i, \qquad \varepsilon_1, \varepsilon_2, \ldots, \varepsilon_n \sim \text{i.i.d. } N(0, \sigma^2),$$

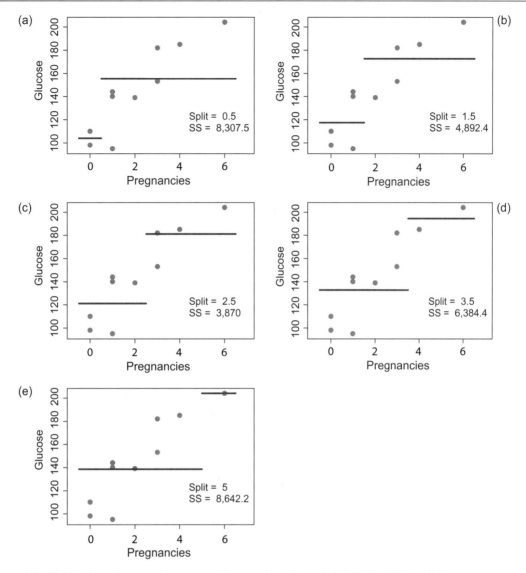

Figure 18.10 Number of pregnancies versus glucose with splits at 0.5, 1.5, 2.5, 3.5, and 5.0.

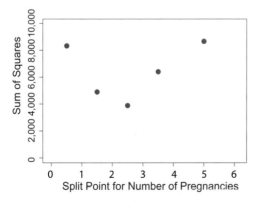

Figure 18.11 Sum of squares for the possible splits on the number of pregnancies. The minimum occurs when $c = 2.5$.

and estimate the parameters β_0 and β_1 by least squares. Define

$$S(c_k) = \sum_{i=1}^{n} \left(y_i - \beta_0 - \beta_1 I(x_i < c_k)\right)^2.$$

If we take the partial derivatives with respect to β_0 and β_1 and set the results equal to 0, we obtain the following two equations in two unknowns:

$$\frac{\partial S}{\partial \beta_0} = -2\sum_{i=1}^{n} \left(y_i - \beta_0 - \beta_1 I(x_i < c_k)\right) = 0 \tag{18.15}$$

$$\frac{\partial S}{\partial \beta_1} = -2\sum_{i=1}^{n} \left(y_i - \beta_0 - \beta_1 I(x_i < c_k)\right)I(x_i < c_k) = 0. \tag{18.16}$$

Some simple, but messy, calculations give the solutions

$$\hat{\beta}_0 = \bar{y}_{+,k}$$

$$\hat{\beta}_0 = \bar{y}_{-,k} - \bar{y}_{+,k}.$$

The predicted values $\hat{y}_i$ for the model with these estimates are, respectively, $\bar{y}_{-,k}$ and $\bar{y}_{+,k}$ for $x_i < c_k$ and $x_i > c_k$. See Problems 18.18 and 18.19. Thus, the regression tree can be seen in the context of linear regression using indicator variables for the predictors.

We've seen that splitting pregnancies between 2 and 3, that is with a split point of $c = 2.5$, minimizes the sum of squared deviations between the observed and the predicted. If there were only one predictor variable, then we would have determined the first split in the regression tree. But in our small (hypothetical) data set, there are two predictor variables for glucose: the number of previous pregnancies and age. Who's to say that we might not get a better fit (measured in terms of sums of squares) if we had chosen age?

So, let's consider age as a possible variable on which to split. The ordered values for Age are

    20,   22,   24,   26,   27,   29,   30,   31,   33,   34.

If we define the split points to be the midpoints of consecutive ordered ages, then

$$c_1 = 21, c_2 = 23, c_3 = 25, c_4 = 26.5, c_5 = 28, c_6 = 29.5, c_7 = 30.5, c_8 = 32, c_9 = 33.5.$$

Figure 18.12 shows scatterplots of Age vs. Glucose with fitted lines for all of these splits. The minimum sums of squares occurs at the split point 30.5, and the minimum SS is 3,702.38. This is smaller than the minimum for the number of pregnancies, which was 3,870. This indicates that splitting Age at 30.5 is a better fit than splitting Pregnancies at 2.5. The ending nodes in the tree, called *leaves*, give the mean glucose level for that branch of the tree. Here, the mean glucose level is 125.6 for those who are under 30.5 and 190.3 for those over 30.5. The first level of our tree is thus

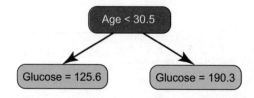

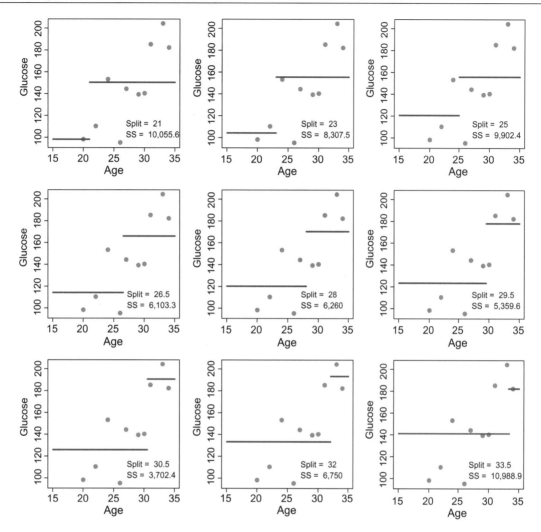

Figure 18.12 Age vs. Glucose with splits at 21.0, 23.0, 25.0, 26.5, 28.0, 29.5, 30.5, 32.0, and 33.5.

To recap, we have considered all possible splits of all possible variables (i.e., both of them), and found that the SS is minimized when we make the split on age at the value 30.5. We proceed with the rest of the regression tree in a similar manner, but this time using the already-defined split.

Consider the split when `Age < 30.5` is true; this is the left (true) branch of our tree. We should investigate all possible splits on pregnancies as well as all possible splits on `Age` again. (There is nothing wrong with making an additional split on the same variable.) The set of data values that satisfy `Age < 30` is

Age	Pregnancies	Glucose
26	1	95
22	0	110
24	3	153
20	0	98
30	1	140
29	2	139
27	1	144

We must therefore consider all of the following splits:

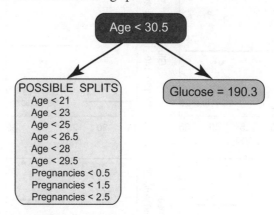

Table 18.6 shows the sums of squares for all of these possible splits. The splits `Age < 30.5 & Age < 23` and `Age < 30.5 & Pregnancies < 0.5` yield exactly the same split of the seven observations that satisfy the root node `Age < 30.5`. Either of these could be the split at this node.

Next, we would repeat this process on the right branch (the false branch) from the `Age < 30.5` node. This would yield a split that fills out the second level of our tree, but in our case there are only three observations that satisfy `Age > 30.5`, which is too few to make a meaningful split.

This process could be continued until there is only one observation in each node, or the observations are all identical at any given node. This process is not recommended, however, because there is too much variability in the predictions. We discuss the issue of overfitting in Section 19.1. The most common stopping rule is to stop when the size of all nodes, measured in terms of the number of observations that fall in it, drops below some threshold. This threshold is often set between 10 and 20. We tend to use larger thresholds for larger data sets. For example, if we set the threshold to be 5 in the small diabetes data set, then we would be done splitting our tree. This is because the largest node, `Age < 30.5` and `Pregnancies < 0.5` contains only five observations. Our final tree would look like this

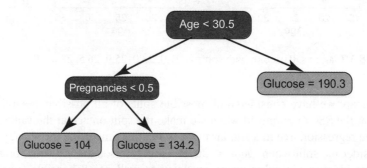

The procedure for making predictions is then to take all the predictor variables and follow the splits in the tree until you reach a leaf. The value in the leaf, which is the mean of all observations that satisfy the path through the tree, is then the prediction. The prediction function is therefore a step function that is constant across all values that satisfy a given path through the split. Figure 18.13 shows how the (`Age`,`Pregnancies`) plane is divided to make predictions. There is a vertical line at `Age = 30.5` that divides the observations into the sets below and above 30.5. Then there is a horizontal line across just the `Age < 30.5` line at `Pregnancies = 0.5` that divides this part of the plane. This creates three regions in the (`Age`,`Pregnancies`) plane; the predictions are constant across all points in a given region. Figure 18.13 shows the three regions in the `age-pregnancies` plane that are determined by the regression tree.

Table 18.6 Possible splits on the left branch of the Age < 30.5 node.

Split	Outcomes in split	Sum of squares
Age <30.5 & Age <21 Age <30.5 & Age <21	98 95, 110, 153, 140, 139, 144	SS = 0 SS = 2,530.8 Total SS = 2,530.8
Age <30.5 & Age <23 Age <30.5 & Age <23	110, 98 95, 153, 140, 139, 144	SS = 72 SS = 2,042.8 Total SS = 2,114.8
Age <30.5 & Age <25 Age <30.5 & Age <25	110, 153, 98 95, 140, 139, 144	SS = 1,672.7 SS = 1,601.0 Total SS = 3,273.7
Age <30.5 & Age <26.5 Age <30.5 & Age <26.5	95, 110, 153, 98 140, 139, 144	SS = 2,154 SS = 14 Total SS = 2,168
Age <30.5 & Age <28 Age <30.5 & Age <28	95, 110, 153, 98, 144 140, 139	SS = 2,874.0 SS = 0.5 Total SS = 2,874.5
Age <30.5 & Age <29.5 Age <30.5 & Age <29.5	95, 110, 153, 98, 139, 144 140	SS = 3,174.8 SS = 0 Total SS = 3,174.8
Age <30.5 & Pregnancies <0.5 Age <30.5 & Pregnancies >0.5	110, 98 95, 153, 140, 139, 144	SS = 72 SS = 2,042.8 Total SS = 2,114.8
Age <30.5 & Pregnancies <1.5 Age <30.5 & Pregnancies >1.5	95, 110, 98, 140, 144 153, 139	SS = 2,151.2 SS = 98.0 Total SS = 2,249.2
Age <30.5 & Pregnancies <2.5 Age <30.5 & Pregnancies >2.5	95, 110, 98, 140, 139, 144 153	SS = 2,540 SS = 0 Total SS = 2,540

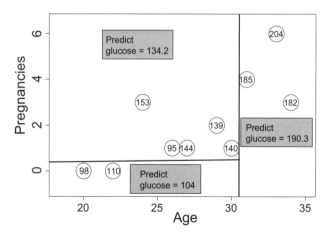

Figure 18.13 Results of regression tree analysis of the glucose data set.

For larger problems we will use an R package to construct the regression. While there are several R packages that can accomplish this task, we will use the `tree` package.

■ **Example 18.8** Construct a regression tree for the complete diabetes data set using the variables `Age` and `Pregnancies`.

Solution: You will have to install the `tree` package in R before proceeding. Once this is done, we can accomplish the tree fitting with the following code:

```
library( tree )
library( dplyr )
diabetes = read.csv( "PimaIndiansDiabetes.csv" )
diabetes0 = diabetes %>% filter( Age > 0 & BMI > 0 & Glucose > 0 )
RT = tree( Glucose ~ Pregnancies + Age , data = diabetes0 ,
           split="deviance" ,
           control=tree.control( nobs = 752 , mincut = 5) )
print( RT )
plot( RT )
text( RT )
```

The arguments to the `tree()` function are (1) the formula to be fit, (2) the data frame, (3) the rule for splitting, and (4) a method for terminating (optional). In the above code, the formula fits a regression tree with `Glucose` as the response variable, and `Pregnancies` and `Age` as the predictor variables. The second argument says to fit the model using the `diabetes0` data frame. The `control` statement says to use all 750 data points and use a minimum cut of 5 to split nodes. The textual output from R is

```
node), split, n, deviance, yval
      * denotes terminal node

1) root 752 703300 121.9
  2) Age < 28.5 356 260600 114.0
    4) Pregnancies < 0.5 81  68290 122.5 *
    5) Pregnancies > 0.5 275 184700 111.5 *
  3) Age > 28.5 396 399900 129.1
    6) Age < 48.5 304 299200 125.7 *
    7) Age > 48.5 92  85200 140.5 *
```

This output gives a description of the split, for example `Age` < 28.5, followed by the sum of squares for that split (called the deviance), and finally the mean response within this split, for example 114.0 for the `Age` < 28.5 split. The terminal nodes are marked with an asterisk.

The RT object can be plotted using the `plot(RT)` command. In order to get the text describing the tree to be placed on the plot, we must also run the command `text(RT)`. The resulting output is shown in Figure 18.14. The splits determined by the `tree()` command divide the `Age`–`Pregnancies` plane into four regions, as shown in Figure 18.15. The area of each circle is proportional to the `Glucose` level. ■ ■ ■

18.6 Discrimination and Classification

In Section 1.1 we emphasized that some models in data science are used to *explain* an observed phenomenon and some are used to *predict* future observations. In this section we look at prediction when the outcome is a categorical variable. Given a set of observations of several different groups, we would like to use some set of predictor variables to make a prediction about which class an observation

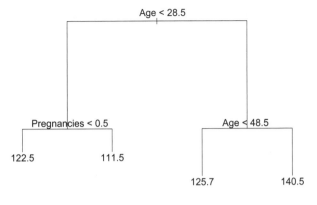

Figure 18.14 Output from the `tree()` function in R.

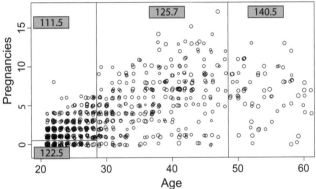

Figure 18.15 Division of the Age–Pregnancies plane into four regions, determined by the regression tree for the Glucose data. The area of each circle is proportional to the Glucose level. The number in the gray box is the mean Glucose level in that region of the plane.

falls in. This observation could be one from our available data set, or it could be some future observation for which we have available the predictor variables but not the group.

We will illustrate this idea of *discrimination* or *classification* by using a famous and old data set, the iris data set collected and analyzed by Edgar Anderson (Anderson, 1928, 1936), who was a botanist at the Missouri Botanical Garden in St. Louis.[3] The iris data set is available in base R and can be loaded using the `data(iris)` command.

■ Example 18.9   Load Anderson's iris data set, plot the data, and examine whether it is feasible to predict the species of an iris, given measurements of sepal length, sepal width, petal length, and petal length.

Solution: Let's begin by loading the data and exploring its structure. The code

```
data( iris )
head( iris )
```

yields

```
  Sepal.Length Sepal.Width Petal.Length Petal.Width Species
1          5.1         3.5          1.4         0.2  setosa
2          4.9         3.0          1.4         0.2  setosa
3          4.7         3.2          1.3         0.2  setosa
```

---

[3]The Missouri Botanical Garden, also known as Shaw's Garden to many native St. Louisans, was less than one block from the childhood home of one of the authors of this book. The author can recall many summer mornings spent as a child at the Missouri Botanical Garden learning about flowers, trees, insects, birds, etc. The data set was used by Fisher (1936) to illustrate the method of linear discrimination and is often referred to as Fisher's iris data.

4	4.6	3.1	1.5	0.2	setosa
5	5.0	3.6	1.4	0.2	setosa
6	5.4	3.9	1.7	0.4	setosa

There are four continuous variables, sepal length, sepal width, petal length, and petal width, and one categorical variable, the species, which is one of setosa, versicolor, or virginica. We would like to be able to take the four continuous measurements on one particular flower of unknown species and make an inference about what species it is. We begin by plotting each of the four measurements for the flower for the three species. This graph, which is shown in Figure 18.16, gives the four measurements for setosa (in red), versicolor (in blue), and virginica (in green). We can see that the two petal dimensions (petal length and width) for setosa are quite a bit smaller than for versicolor. Thus, for example, if a flower had a petal length less than 2 and a petal width less than 0.7, we could classify it as setosa, and we would be fairly confident in this inference. Versicolor and virginica look to be more difficult to distinguish because their distributions overlap considerably.

Maybe if we look at scatterplots of two variables at a time, we might see some separability between versicolor and virginica that is not visible in separate univariate plots. The R command `pairs()` produces an array of scatterplots for all pairs of variables. The following code

```
iris.jitter = data.frame( jitter( cbind(iris$Sepal.Length,iris$Sepal.Width,
                                    iris$Petal.Length,iris$Petal.Width) ) )
iris.jitter$spec = iris$spec
col1 = "red"
col2 = "blue"
col3 = "green3"
names( iris.jitter ) = c("Sepal.Length","Sepal.Width","Petal.Length",
                    "Petal.Width","spec")
```

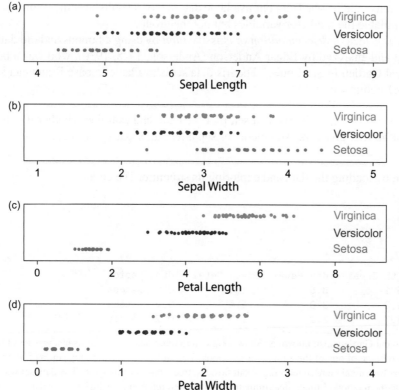

Figure 18.16 Fisher's iris data showing the sepal length, sepal width, and petal width for three groups of iris flowers. The observations are slightly jittered to show the extent of repeated observations.

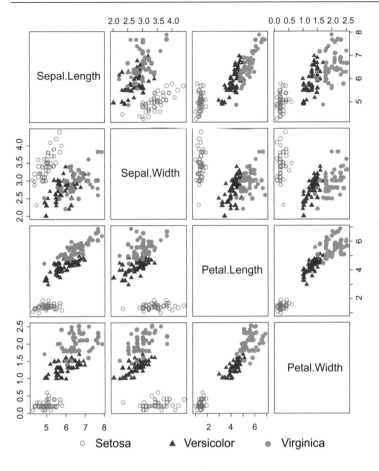

Figure 18.17 All possible pairs of scatterplots for Anderson's iris data.

```
pairs( iris.jitter[1:4] ,
       col=c(col1,col2,col3)[iris$spec] , pch=c(1,9,19)[iris$spec] ,
       cex=c(1.75,1.5,1.5) )
```

creates some jitter around each of the four continuous measurements and plots all pairs of scatterplots. The resulting set of scatterplots is shown in Figure 18.17. We must ask ourselves whether we could draw a boundary, such as a straight line, that would divide the different species. Again, setosa is clearly differentiated from versicolor and virginica. In the sepal length vs. sepal width scatterplot there is considerable overlap between versicolor and virginica, so it would be difficult to distinguish these species based solely on this plot. In some of the other plots it is possible to draw a line that separates most, but certainly not all, versicolor from virginica.

Three-dimensional scatterplots show individual points plotted in the (x, y, z) space. One of the disadvantages of ordinary 3D plots is that we are viewing the points from one particular angle, and we have no sense of depth. It is possible in R to create a 3D plot that you can rotate by simply moving the mouse. This added feature gives you the opportunity to see the data from many angles and discern patterns that you might otherwise miss in a static 3D plot. The R package plotly can be used to create these dynamic 3D scatterplots. The following R code can be used to create such a graph:

```
library( plotly )
plot_ly( data = iris , x = ~Sepal.Length , y = ~Petal.Length,
                        z = ~Sepal.Width , color = ~Species , size=5 )
```

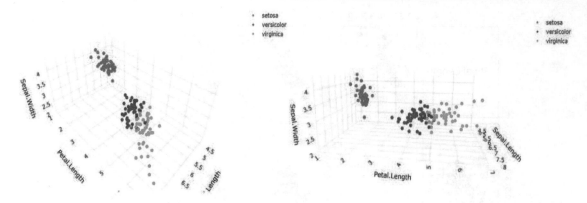

Figure 18.18 3D scatterplot from the `plotly` package in R. This plot can be rotated by clicking and dragging the mouse. The default view is shown in (a). The graph in (b) has been rotated to see as clearly as possible a plane that might separate the versicolor and virginica flowers.

The resulting graph should appear in the "Viewer" tab in the bottom right panel of R studio.[4] It is here where you can click and drag to rotate the plot. Figure 18.18(a) shows the view that is initially produced. The plot in Figure 18.18(b) has been rotated to see how a plane could be placed that would separate most versicolor flowers from most virginica flowers.

In summary, it looks like separating setosa from the other two species is easily done. Since there is some overlap between versicolor and virginica, even in the 3D plot, it may be difficult to discern these two species. Looking at the points in 3D space makes it feasible to discern the versicolor and virginica species, though not perfectly since there is some overlap.   ■ ■ ■

> **Exercise 18.2**   Rerun the 3D scatterplots in `plotly` for the other three combinations of measurements. Assess the feasibility of discriminating the three species, given the four continuous variables.

Four-dimensional scatterplots are not possible, so we are left wondering whether using all four dimensions at once will give a clear classification of the species given the four continuous variables. We will now pursue the problem of classification or discrimination for several special cases. We will return to Anderson's iris data once we have developed enough machinery to perform the classification.

18.6.1 *K* = 2 Groups and *p* = 1 Variable

Let's begin with the case of two groups or classes and a single measurement. We will let K denote the number of groups and p the number of variables, so we're starting with $K = 2$ and $p = 1$. We will assume that the group is a fixed variable (not random) and we assume that within each group we have a random sample of units from that group. Since we have $K = 2$ groups and $p = 1$ variable, we observe a random sample from each of the two groups. We will assume that both samples come from a normal distribution:

$$X_{11}, X_{12}, \ldots, X_{1n_1} \sim \text{ i.i.d. } N(\mu_1, \sigma_1^2)$$

$$X_{21}, X_{22}, \ldots, X_{2n_2} \sim \text{ i.i.d. } N(\mu_2, \sigma_2^2).$$

---

[4]If you run the above code you will get different colors for the points. Adjusting the colors is a bit tricky in `plotly`, so we have omitted the code we used to specify the colors.

Assume that $\mu_1 > \mu_2$, which implies that $\mu_1 - \mu_2 > 0$. Throughout this section we have assumed that all parameters are known, so assuming $\mu_1 - \mu_2 > 0$ is equivalent to labeling as group 1 the group with the larger mean.

We would like to assess how likely it is that an item with a measured value x came from group 1 and from group 2. To do this, we need to know how likely it is that an item came from each group *before* we observe x. This is the *prior distribution* for the group, which we will denote π_1 and π_2 for groups 1 and 2, respectively.[5] We will now use Bayes' Theorem to get the *posterior distribution* for the group given the single (because $p = 1$ for now) measurement on the variable x:

$$
\begin{aligned}
p_k(x) &= \Pr(\text{group } k | x) \\[6pt]
&= \frac{\pi_k f(x|\text{group } k)}{\pi_1 f(x|\text{group } 1) + \pi_2 f(x|\text{group } 2)} \\[6pt]
&= \frac{\pi_k \dfrac{1}{\sqrt{2\pi}\,\sigma_k} \exp\left(-(x-\mu_k)^2/2\sigma_k^2\right)}{\pi_1 \dfrac{1}{\sqrt{2\pi}\,\sigma_1} \exp\left(-(x-\mu_1)^2/2\sigma_1^2\right) + \pi_2 \dfrac{1}{\sqrt{2\pi}\,\sigma_2} \exp\left(-(x-\mu_2)^2/2\sigma_2^2\right)}.
\end{aligned} \tag{18.17}
$$

We will classify a unit with variable x to that group that has the highest posterior probability.[6] Let C denote the denominator in (18.17). Note that C depends on x but not on k, because we are summing over all possible k in the denominator, which for now is just $k = 1, 2$. The posterior probability for group 1 exceeds the posterior probability for group 2 provided

$$ p_1(x) > p_2(x) $$

$$ \frac{\pi_1 \dfrac{1}{\sqrt{2\pi}\,\sigma_1} \exp\left(-(x-\mu_1)^2/2\sigma_1^2\right)}{C} > \frac{\pi_2 \dfrac{1}{\sqrt{2\pi}\,\sigma_2} \exp\left(-(x-\mu_2)^2/2\sigma_2^2\right)}{C} $$

$$ \frac{\pi_1 \sigma_2}{\pi_2 \sigma_1} > \exp\left(-(x-\mu_2)^2/2\sigma_2^2 + (x-\mu_1)^2/2\sigma_1^2\right) $$

$$ \log \frac{\pi_1 \sigma_2}{\pi_2 \sigma_1} > -(x-\mu_2)^2/2\sigma_2^2 + (x-\mu_1)^2/2\sigma_1^2. $$

This reduces to inferring group 1 provided

$$ \left(\frac{1}{2\sigma_2^2} - \frac{1}{2\sigma_1^2}\right) x^2 + \left(\frac{\mu_1}{\sigma_1^2} - \frac{\mu_2}{\sigma_2^2}\right) x + \left(\frac{\mu_2^2}{2\sigma_2^2} - \frac{\mu_1^2}{2\sigma_1^2} + \log \frac{\pi_1 \sigma_2}{\pi_2 \sigma_1}\right) > 0. \tag{18.18} $$

Otherwise we infer that the item is in group 2. Note that the expression on the left side is quadratic in x. This method is called *quadratic discrimination analysis* (QDA).

We will make some further assumptions to simplify this result, starting with the assumption of equal variances in the two groups, that is, $\sigma_1^2 = \sigma_2^2 = \sigma^2$. With this assumption, the quadratic term in (18.18) becomes zero, and the result simplifies to

$$ (\mu_1 - \mu_2)x + \frac{1}{2}(\mu_1^2 - \mu_2^2) + \sigma^2 \log \frac{\pi_1}{\pi_2} > 0, $$

---

[5]Don't confuse π_k with the irrational mathematical constant $\pi \approx 3.1416$ that also appears in (18.17).

[6]The group with the highest posterior probability is the optimal decision here.

which is linear in the variable x. This method is called *linear discriminant analysis* (LDA). If the classification is made assuming equal variances, then the decision is to infer group 1 if[7]

$$x > \frac{1}{2}\frac{\mu_1^2 - \mu_2^2}{\mu_1 - \mu_2} - \frac{\log(\pi_1/\pi_2)}{\mu_1 - \mu_2}\sigma^2,$$

which can be simplified to

$$x > \frac{\mu_1 + \mu_2}{2} - \frac{\log(\pi_1/\pi_2)}{\mu_1 - \mu_2}\sigma^2. \tag{18.19}$$

If we make the additional assumption that $\pi_1 = \pi_2$, that is, an item to be classified came from either group with the same probability, then (18.19) becomes

$$x > \frac{\mu_1 + \mu_2}{2}. \tag{18.20}$$

This is a very intuitive result; it says that we would infer group 1 (the one with the larger mean) whenever the value of x exceeds the midpoint $(\mu_1 + \mu_2)/2$ of the two means.

In the development so far, we have assumed that the parameters of the distributions for each group are known. In practice, these are never known and they must be estimated from the available data. We will assume the usual unbiased estimators for the means and variances (and later covariances). Using estimates of the parameters, rather than the parameters themselves, means that the decision boundaries are estimated as well. We will address this issue of estimating parameters in Section 19.5.

■ **Example 18.10** For Anderson's iris data, consider just the versicolor and virginica species. What is the rule for choosing the species if we observed only the petal width?

Solution: We begin with an LDA, which assumes that the variances of the two groups are equal. The parameters μ_1, μ_2, σ_1, and σ_2 are unknown and must be estimated from the data. The sample means and standard deviations are

$$\hat{\mu}_1 = 2.026$$
$$\hat{\sigma}_1 = 0.2747$$
$$\hat{\mu}_2 = 1.326$$
$$\hat{\sigma}_2 = 0.1978.$$

Assuming that the two groups have the same standard deviation, we can use the pooled estimate of the common σ to get

$$\hat{\sigma} = \sqrt{\frac{(50-1) \times 0.2747^2 + (50-1) \times 0.1978^2}{50+50-2}} \approx 0.2393.$$

We will treat these estimates as if they were the actual parameters when we apply Bayes' Theorem to get the posterior distribution of the group given a continuous variable $x =$ petal width. If the two species are equally prevalent, we can assume that the prior probabilities are $\pi_1 = \pi_2 = 1/2$. Under these assumptions of equal prior probabilities and equal standard deviations, the rule becomes to infer group 1, the virginica species, provided

$$x > \frac{\hat{\mu}_1 + \hat{\mu}_2}{2} = 1.676.$$

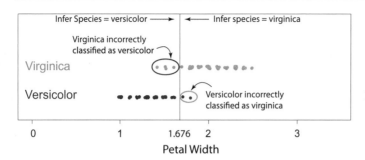

Figure 18.19 Iris data with only the petal width used to infer the species.

Figure 18.19 shows the petal widths for the virginica and versicolor flowers. Four of the virginica flowers were incorrectly classified as versicolor because their petal widths were less than the cutoff 1.676, and two of the versicolor flowers were classified as virginica because their petal widths were greater than 1.676. Table 18.7 shows all 100 observations, 50 from each species, along with the inferred group. The "Decision" column of this table indicates whether the inferred group was correct or incorrect. With LDA we made the correct inference 94% of the time.

Next, let's apply QDA, which does not make the assumption of equal variances. Using the estimates for the parameters we find using (18.18) that the rule for inferring the virginica species is

$$6.157293x^2 - 7.049339x - 5.055207 > 0,$$

where x is the petal width. The graph of this parabola is shown in Figure 18.20. We see that we would infer versicolor if the petal width x is between -0.499 and 1.644. In this case, the petal width cannot be negative, so we will never observe $x < -0.499$. The right cutoff point, 1.644, is nearly the same as in the LDA.   ∎∎∎

When we apply QDA we will infer one of the two groups for either large or small values of the (one and only) continuous variable x. If group 1 has the larger variability, then

$$\frac{1}{2\sigma_2^2} - \frac{1}{2\sigma_1^2} > 0,$$

so the parabola is concave up. This leads to inferring group 2 whenever x is between the lower and upper points where the parabola crosses the horizontal axis, and inferring group 1 whenever x is in one of the tails. If group 2 has larger variance, then the opposite is true.

■ **Example 18.11** Consider the data from two groups with a single predictor using the data as shown in Figure 18.21(a). Since group 2 clearly has a larger variance than group 1, apply QDA to discriminate the two groups.

Solution: The estimated means and standard deviations are

$$\hat{\mu}_1 = 41.03$$
$$\hat{\sigma}_1 = 3.84$$
$$\hat{\mu}_2 = 58.47$$
$$\hat{\sigma}_2 = 11.15.$$

---

[7]Note that we divided both sides by $\mu_1 - \mu_2$, which is assumed to be positive. Had it been negative, we would have had to reverse the inequality.

Table 18.7 Iris data with only the petal width to classify versicolor versus virginica.

Petal width	Species	Inferred species	Decision	Petal width	Species	Inferred species	Decision
1.4	versicolor	versicolor	Correct	2.5	virginica	virginica	Correct
1.5	versicolor	versicolor	Correct	1.9	virginica	virginica	Correct
1.5	versicolor	versicolor	Correct	2.1	virginica	virginica	Correct
1.3	versicolor	versicolor	Correct	1.8	virginica	virginica	Correct
1.5	versicolor	versicolor	Correct	2.2	virginica	virginica	Correct
1.3	versicolor	versicolor	Correct	2.1	virginica	virginica	Correct
1.6	versicolor	versicolor	Correct	1.7	virginica	virginica	Correct
1.0	versicolor	versicolor	Correct	1.8	virginica	virginica	Correct
1.3	versicolor	versicolor	Correct	1.8	virginica	virginica	Correct
1.4	versicolor	versicolor	Correct	2.5	virginica	virginica	Correct
1.0	versicolor	versicolor	Correct	2.0	virginica	virginica	Correct
1.5	versicolor	versicolor	Correct	1.9	virginica	virginica	Correct
1.0	versicolor	versicolor	Correct	2.1	virginica	virginica	Correct
1.4	versicolor	versicolor	Correct	2.0	virginica	virginica	Correct
1.3	versicolor	versicolor	Correct	2.4	virginica	virginica	Correct
1.4	versicolor	versicolor	Correct	2.3	virginica	virginica	Correct
1.5	versicolor	versicolor	Correct	1.8	virginica	virginica	Correct
1.0	versicolor	versicolor	Correct	2.2	virginica	virginica	Correct
1.5	versicolor	versicolor	Correct	2.3	virginica	virginica	Correct
1.1	versicolor	versicolor	Correct	1.5	virginica	versicolor	Incorrect
1.8	versicolor	virginica	Incorrect	2.3	virginica	virginica	Correct
1.3	versicolor	versicolor	Correct	2.0	virginica	virginica	Correct
1.5	versicolor	versicolor	Correct	2.0	virginica	virginica	Correct
1.2	versicolor	versicolor	Correct	1.8	virginica	virginica	Correct
1.3	versicolor	versicolor	Correct	2.1	virginica	virginica	Correct
1.4	versicolor	versicolor	Correct	1.8	virginica	virginica	Correct
1.4	versicolor	versicolor	Correct	1.8	virginica	virginica	Correct
1.7	versicolor	virginica	Incorrect	1.8	virginica	virginica	Correct
1.5	versicolor	versicolor	Correct	2.1	virginica	virginica	Correct
1.0	versicolor	versicolor	Correct	1.6	virginica	versicolor	Incorrect
1.1	versicolor	versicolor	Correct	1.9	virginica	virginica	Correct
1.0	versicolor	versicolor	Correct	2.0	virginica	virginica	Correct
1.2	versicolor	versicolor	Correct	2.2	virginica	virginica	Correct
1.6	versicolor	versicolor	Correct	1.5	virginica	versicolor	Incorrect
1.5	versicolor	versicolor	Correct	1.4	virginica	versicolor	Incorrect
1.6	versicolor	versicolor	Correct	2.3	virginica	virginica	Correct
1.5	versicolor	versicolor	Correct	2.4	virginica	virginica	Correct
1.3	versicolor	versicolor	Correct	1.8	virginica	virginica	Correct
1.3	versicolor	versicolor	Correct	1.8	virginica	virginica	Correct
1.3	versicolor	versicolor	Correct	2.1	virginica	virginica	Correct
1.2	versicolor	versicolor	Correct	2.4	virginica	virginica	Correct
1.4	versicolor	versicolor	Correct	2.3	virginica	virginica	Correct
1.2	versicolor	versicolor	Correct	1.9	virginica	virginica	Correct
1.0	versicolor	versicolor	Correct	2.3	virginica	virginica	Correct
1.3	versicolor	versicolor	Correct	2.5	virginica	virginica	Correct
1.2	versicolor	versicolor	Correct	2.3	virginica	virginica	Correct
1.3	versicolor	versicolor	Correct	1.9	virginica	virginica	Correct
1.3	versicolor	versicolor	Correct	2.0	virginica	virginica	Correct
1.1	versicolor	versicolor	Correct	2.3	virginica	virginica	Correct
1.3	versicolor	versicolor	Correct	1.8	virginica	virginica	Correct

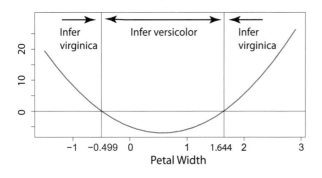

Figure 18.20 Parabola for discriminating versicolor and virginica using QDA.

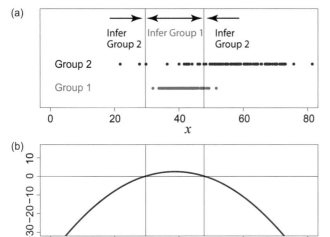

Figure 18.21 Data for Example 18.11. Notice that the variability from group 2 is clearly larger than the variability in group 1, so we infer group 2 when x is in either the left or right tail.

The estimated quadratic discrimination function is shown in Figure 18.21(b). For x between 29.6 and 47.7, we would infer group 1 and for x below 29.6 or greater than 47.6 we infer group 2. This two-tailed region for which we infer group 2 makes intuitive sense: since the variability is much greater in group 2 than in group 1, values that are far from the center of the data probably came from Group 2. Here there are observations in both the left and right tails that get categorized as Group 2.   ■ ■ ■

In the examples so far, we have assumed that the prior probabilities π_1 and π_2 are equal. This is not always the case. Often, units from one of the groups are much more prevalent than the other. If we assume that the units we have observed are selected at random from the combined population of groups 1 and 2, we could estimate π_1 and π_2 by the observed proportion. For example, if a random sample of 100 flowers from some region yielded 70 versicolor and 30 virginica, then we would estimate $\hat{\pi}_1 = 0.70$ and $\hat{\pi}_2 = 0.30$ and then use these estimates as the prior probabilities.

There are also situations where the combined data *cannot* be assumed to be a random sample from some population. For example, one common application of discriminant analysis is to infer whether a person with certain medical variables has or doesn't have a particular type of cancer. A common experimental plan is to collect data on those with cancer and those without, and run a discriminant analysis to see what variables are related to cancer. Often the plan is to obtain equal numbers of people with and without cancer. When this occurs, it is unreasonable to assume that half of the population has cancer and the other doesn't. Instead, we would have to use the rate of cancer in the general population for the prior probabilities.

18.6.2 *K* Groups and *p* = 1 Variable

Suppose now that we have an arbitrary number K of groups and one continuous variable x. The optimal inference procedure is still to assign the group based on the highest posterior probability. In the case of K groups, we have data of the form

$$X_{11}, X_{12}, \ldots, X_{1n_1} \sim \text{i.i.d. } N(\mu_1, \sigma_1^2)$$

$$X_{21}, X_{22}, \ldots, X_{2n_2} \sim \text{i.i.d. } N(\mu_2, \sigma_2^2)$$

$$\ldots$$

$$X_{K1}, X_{K2}, \ldots, X_{Kn_K} \sim \text{i.i.d. } N(\mu_K, \sigma_K^2).$$

The posterior probability is then

$$p_k(x) = \frac{\pi_k f(x|\text{group } k)}{\displaystyle\sum_{\ell=1}^{K} \pi_\ell f(x|\text{group } \ell)} = \frac{\pi_k \dfrac{1}{\sqrt{2\pi}\,\sigma_k} \exp\left(-(x-\mu_k)^2/2\sigma_k^2\right)}{\displaystyle\sum_{\ell=1}^{K} \pi_\ell \dfrac{1}{\sqrt{2\pi}\,\sigma_\ell} \exp\left(-(x-\mu_\ell)^2/2\sigma_\ell^2\right)}. \tag{18.21}$$

The procedure is similar to the case of $K = 2$ groups: Compute the posterior probability of each group given the value of x and assign the unit to the group with the highest posterior probability.

■ **Example 18.12** Apply LDA to Anderson's iris data using only the petal width for all three species, assuming the three species are equally likely to be presented for classification.

Solution: When the variances are assumed equal, the graphs of the group PDFs $f_k(x)$ are simply translations of one another, as shown in the second panel of Figure 18.22. The posterior probabilities $p_1(x), p_2(x),$ and $p_3(x)$ (each as a function of the petal width x) are shown in the third panel of Figure 18.22. These posterior probabilities are strictly between 0 and 1, and except for regions near 0.8 and 1.7, the curves are nearly 0 or 1. This means that we can do a good job classifying these observations except when the petal width is near 0.8 or 1.7. The cutoffs used to make the classification are the midpoints of the means of adjacent groups.[8] Thus, the classification is done as follows:

classify as setosa if $x < 0.786$;

classify as versicolor if $0.786 < x < 1.676$;

classify as virginica if $x > 1.676$.

Using this classification scheme, we can see that no setosa plants are misclassified, and no plants are misclassified as setosa. There are two versicolor plants that are classified as virginica, and there are four virginica plants that are classified as versicolor. We summarize the classification results with a *confusion matrix*, which shows a contingency table of the actual species and the predicted species; see Table 18.8.   ■ ■ ■

The LDA done in this example would seem to have a hard time distinguishing virginica from versicolor since the $f_k(x)$ functions, that is, the within-group PDFs, overlap considerably (Figure 18.22(b)). The raw data, shown in Figure 18.22(a), indicate that there is virtually no overlap between setosa and versicolor. By performing LDA, we are assuming equal variances among the groups, but this assumption seems to be violated; the variability in petal width x in setosa is less than the other groups. QDA might be a better method to apply in this case.

---

[8]By adjacent groups, we mean groups whose means are adjacent in a ordered list. For example, in the iris example, $\mu_1 < \mu_2 < \mu_3$, so μ_1 and μ_2 are adjacent, as are μ_2 and μ_3. Thus, the cutoffs are between the first group and the second group, and between the second group and the third group.

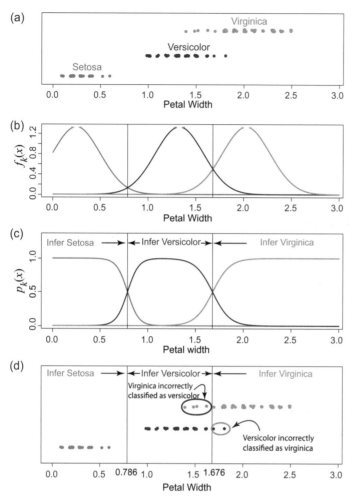

Figure 18.22 Linear discriminant analysis for iris data using only the variable petal width.

■ **Example 18.13**   Perform QDA on the petal width variable of the iris data set.

Solution: Figure 18.23 shows the estimated PDFs for the three groups. Since QDA does not assume equal variances, the PDFs are not translations of one another. We can see that the setosa species has smaller variability in petal width than the other two species. The variability for virginica is slightly larger than for versicolor. This makes it easier to distinguish between setosa and versicolor. There is little change in the ability to distinguish versicolor and virginica. Just as with LDA, QDA misclassifies four virginica plants as versicolor and two versicolor plants as virginica. The confusion matrix for QDA is the same as in Table 18.8.

■ ■ ■

18.6.3 *K* Groups and *p* Variables

Let's now address the general case of K groups and p variables. In Anderson's iris data, we have $K = 3$ groups (setosa, versicolor, and virginica) and $p = 4$ variables (petal length, petal width, sepal length, and sepal width). We will let X_{ijk} denote the measured value of the ith unit for variable j in group k. Here,

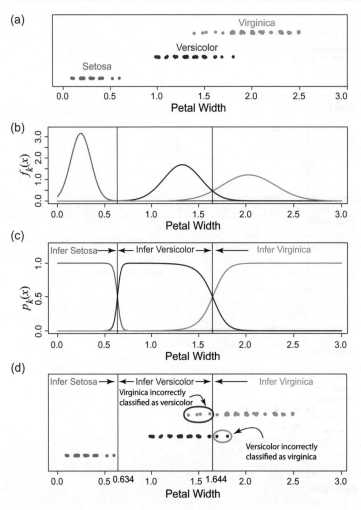

Figure 18.23 Quadratic discriminant analysis for iris data using only the variable petal width.

Table 18.8 Confusion matrix for LDA applied to the petal widths of the three iris species.

		Predicted species			
		Setosa	Versicolor	Virginica	Total
Actual	Setosa	50	0	0	50
species	Versicolor	0	48	2	50
	Virginica	0	4	46	50
	Total	50	52	48	150

$i = 1, 2, \ldots, n_k$, $j = 1, 2, \ldots, p$, and $k = 1, 2, \ldots, K$. Let $\boldsymbol{X}_{ik}$ denote the vector of all p variables for the ith item in the kth group. For the iris data, this would be the vector of petal length, petal width, sepal length, and sepal width for the ith flower ($i = 1, 2, \ldots, 50$) within species k ($k = 1, 2, 3$). That is,

$$\boldsymbol{X}_{ik} = \begin{bmatrix} X_{i1k} \\ X_{i2k} \\ \vdots \\ X_{ipk} \end{bmatrix}.$$

We then assume that

$$X_{11}, X_{21}, \ldots, X_{n_1 1} \sim N_p(\mu_1, \Sigma)$$
$$X_{12}, X_{22}, \ldots, X_{n_2 2} \sim N_p(\mu_2, \Sigma)$$
$$\vdots$$
$$X_{1K}, X_{2K}, \ldots, X_{n_K K} \sim N_p(\mu_K, \Sigma).$$

For the iris data, we assume

$$X_{11}, X_{21}, \ldots, X_{50,1} \sim N_4(\mu_1, \Sigma) \qquad \text{[species = setosa]}$$
$$X_{12}, X_{22}, \ldots, X_{50,2} \sim N_4(\mu_2, \Sigma) \qquad \text{[species = versicolor]}$$
$$X_{13}, X_{23}, \ldots, X_{50,3} \sim N_4(\mu_3, \Sigma) \qquad \text{[species = virginica]}.$$

The PDF of the p-variate normal PDF is

$$f(x|\mu, \Sigma) = \frac{1}{(2\pi)^{p/2} |\Sigma|^{1/2}} \exp\left(-\frac{1}{2}(x - \mu)' \Sigma^{-1}(x - \mu)\right),$$

so for a fixed x, and fixed prior probabilities $\pi_1, \pi_2, \ldots, \pi_K$ the posterior PDF for group r exceeds that of group s, provided

$$\frac{\pi_r}{(2\pi)^{p/2}|\Sigma|^{1/2}} \exp\left(-\frac{1}{2}(x - \mu_r)' \Sigma^{-1}(x - \mu_r)\right) > \frac{\pi_s}{(2\pi)^{p/2}|\Sigma|^{1/2}} \exp\left(-\frac{1}{2}(x - \mu_s)' \Sigma^{-1}(x - \mu_s)\right),$$

and since the covariance matrix Σ is assumed to be the same for all K groups, this simplifies as follows:

$$\pi_r \exp\left(-\frac{1}{2}(x - \mu_r)' \Sigma^{-1}(x - \mu_r)\right) > \pi_s \exp\left(-\frac{1}{2}(x - \mu_s)' \Sigma^{-1}(x - \mu_s)\right)$$

$$\log \pi_r - \frac{1}{2}(x - \mu_r)' \Sigma^{-1}(x - \mu_r) > \log \pi_s - \frac{1}{2}(x - \mu_s)' \Sigma^{-1}(x - \mu_s)$$

$$x' \Sigma^{-1} x - 2\mu_r' \Sigma^{-1} x + \mu_r' \Sigma^{-1} \mu_r < x' \Sigma^{-1} x - 2\mu_s' \Sigma^{-1} x + \mu_s' \Sigma^{-1} \mu_s - 2\log\frac{\pi_s}{\pi_r}$$

$$-2\mu_r' \Sigma^{-1} x + \mu_r' \Sigma^{-1} \mu_r < -2\mu_s' \Sigma^{-1} x + \mu_s' \Sigma^{-1} \mu_s - 2\log\frac{\pi_s}{\pi_r}$$

$$2\mu_s' \Sigma^{-1} x - 2\mu_r' \Sigma^{-1} x < \mu_s' \Sigma^{-1} \mu_s - \mu_r' \Sigma^{-1} \mu_r - 2\log\frac{\pi_s}{\pi_r}$$

$$\underbrace{(\mu_s - \mu_r)' \Sigma^{-1} x}_{\text{linear discriminant function}} < \frac{1}{2}\left(\mu_s' \Sigma^{-1} \mu_s - \mu_r' \Sigma^{-1} \mu_r\right) - \log\frac{\pi_s}{\pi_r}. \tag{18.22}$$

The left side of (18.22) is called the *linear discriminant function*. We will define

$$\ell_{rs} = (\mu_s - \mu_r)' \Sigma^{-1} \tag{18.23}$$

to be the $1 \times p$ matrix that transforms the predictor x to the decision variable $\ell_{rs}x$. In deciding between group r and s, we infer group r rather than group s if the inequality in (18.22) holds.

■ **Example 18.14**   Apply LDA to the iris data using only the two variables petal length and petal width. Assume $\pi_1 = \pi_2$.

Solution: The estimated mean vectors are

$$\hat{\mu}_1 = [1.462, 0.246]'$$
$$\hat{\mu}_2 = [4.260, 1.326]'$$
$$\hat{\mu}_3 = [5.552, 2.026]'$$

and the estimated covariance matrices are

$$\hat{\Sigma}_1 = \begin{bmatrix} 0.03016 & 0.00607 \\ 0.00607 & 0.01111 \end{bmatrix}$$

$$\hat{\Sigma}_2 = \begin{bmatrix} 0.22082 & 0.07310 \\ 0.07310 & 0.03911 \end{bmatrix}$$

$$\hat{\Sigma}_3 = \begin{bmatrix} 0.30459 & 0.04882 \\ 0.04882 & 0.07543 \end{bmatrix}.$$

Since LDA assumes equal covariance matrices, we estimate the common covariance matrix by pooling:

$$\hat{\Sigma} = \frac{(50-1)\hat{\Sigma}_1 + (50-1)\hat{\Sigma}_2 + (50-1)\hat{\Sigma}_3}{50+50+50-3} = \begin{bmatrix} 0.18519 & 0.04267 \\ 0.04267 & 0.04188 \end{bmatrix}.$$

Let's label the groups as 1 for setosa, 2 for versicolor, and 3 for virginica. From (18.23) we obtain

$$\ell_{12} = ([4.260 \quad 1.326] - [1.462 \quad 0.246]) \begin{bmatrix} 0.18519 & 0.04267 \\ 0.04267 & 0.04188 \end{bmatrix}^{-1}$$

$$= [11.98 \quad 13.58] \qquad \text{[setosa vs. versicolor]}$$

$$\ell_{13} = ([5.552 \quad 2.026] - [1.462 \quad 0.246]) \begin{bmatrix} 0.18519 & 0.04267 \\ 0.04267 & 0.04188 \end{bmatrix}^{-1}$$

$$= [16.06 \quad 26.14] \qquad \text{[setosa vs. virginica]}$$

$$\ell_{23} = ([5.552 \quad 2.026] - [4.260 \quad 1.326]) \begin{bmatrix} 0.18519 & 0.04267 \\ 0.04267 & 0.04188 \end{bmatrix}^{-1}$$

$$= [16.06 \quad 26.14] \qquad \text{[versicolor vs. virginica]}.$$

The decision boundaries are

$$b_{12} = \frac{1}{2} \left([4.260 \quad 1.326] \begin{bmatrix} 0.18519 & 0.04267 \\ 0.04267 & 0.04188 \end{bmatrix}^{-1} \begin{bmatrix} 4.260 \\ 1.326 \end{bmatrix} \right.$$

$$\left. -[1.462 \quad 0.246] \begin{bmatrix} 0.18519 & 0.04267 \\ 0.04267 & 0.04188 \end{bmatrix}^{-1} \begin{bmatrix} 1.462 \\ 0.246 \end{bmatrix} \right)$$

$$= 44.95$$

$$b_{13} = 86.03$$

$$b_{23} = 41.08.$$

The three lines in the x_3 (petal length) and x_4 (petal width) space that are used for discriminating between groups r and s are then determined by

$$\ell_{rs}\begin{bmatrix} x_3 \\ x_4 \end{bmatrix} = b_{rs}.$$

The three lines superimposed on a scatterplot of x_3 and x_4 are shown in Figure 18.24. As expected, setosa is easily distinguished from versicolor and virginica. The lines separating these groups have no misclassifications. The hardest discrimination is between versicolor and virginica. Here there are four virginica flowers classified as versicolor and one versicolor classified as virginica.

Figure 18.25 provides some intuition about how the LDA procedure works. If we project the dots for versicolor and virginica onto a line perpendicular to the discrimination line, we see that the inference of versicolor vs. virginica is made based on the location of the projected point. ■ ■ ■

So far we have done the LDA calculations "from scratch." When dealing with many groups and many predictor variables, the computations can become cumbersome and it is better to use an R package to do the computations for us. The MASS package contains the function `lda()` which performs linear discriminant analysis. The next example illustrates this package.

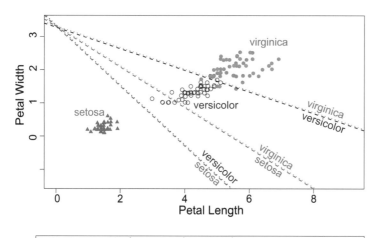

Figure 18.24 Scatterplot of petal length x_3 (horizontal axis) and petal width x_4 (vertical axis) along with the three lines to discriminate setosa vs. versicolor, setosa vs. virginica, and versicolor vs. virginica.

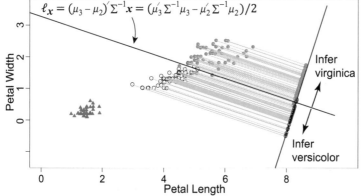

Figure 18.25 The LDA statistic $\ell_{23}x$ as the projection on the line perpendicular to the classification line.

■ **Example 18.15** Perform LDA on the iris data set using all three species and all four variables.

Solution: We begin by pulling in the MASS package along with the iris data set, which is part of base R:

```
library( MASS )
data( iris )
```

We send to the function lda() a model of the form

$$\text{outcome} \sim \text{predictor } 1 + \text{predictor } 2 + \cdots + \text{predictor } K$$

along with the data frame containing these variables. The following code accomplishes this:

```
LDA.model = lda( Species ~ Sepal.Length + Sepal.Width + Petal.Length +
                          Petal.Width , data = iris )
print( LDA.model )
```

The output from printing the result of LDA.model is

```
Call:
lda(Species ~ Sepal.Length + Sepal.Width + Petal.Length + Petal.Width,
data = iris)

Prior probabilities of groups:
    setosa versicolor  virginica
0.3333333  0.3333333  0.3333333

Group means:
           Sepal.Length Sepal.Width Petal.Length Petal.Width
setosa            5.006       3.428        1.462       0.246
versicolor        5.936       2.770        4.260       1.326
virginica         6.588       2.974        5.552       2.026

Coefficients of linear discriminants:
                    LD1         LD2
Sepal.Length  0.8293776  0.02410215
Sepal.Width   1.5344731  2.16452123
Petal.Length -2.2012117 -0.93192121
Petal.Width  -2.8104603  2.83918785

Proportion of trace:
LD1    LD2
0.9912 0.0088
```

The output gives quite a bit of information, including a statement of the model, the prior probabilities (assumed equal by default), the group means, the coefficients of linear determinants, and the proportion of trace. We can make predictions of the species using the predict() function. This can be done for the data set to which the model was fit or to a new data set if we wanted to infer the species of new plants. The syntax for predict() is

```
pred = predict( LDA.model , data = iris[,1:4] )
```

We must provide the fitted model for which we'd like prediction along with a set of values for all of the predictor variables. (Note that the first four columns of iris include the predictor variables; the fifth column is the species.) We can obtain the structure of the object pred using

```
str( pred )
```

which yields

```
List of 3
$ class    : Factor w/ 3 levels "setosa","versicolor",..: 1 1 1 1 1 1 ...
$ posterior: num [1:150, 1:3] 1 1 1 1 1 ...
..- attr(*, "dimnames")=List of 2
.. ..$ : chr [1:150] "1" "2" "3" "4" ...
.. ..$ : chr [1:3] "setosa" "versicolor" "virginica"
$ x        : num [1:150, 1:2] 8.06 7.13 7.49 6.81 8.13 ...
..- attr(*, "dimnames")=List of 2
.. ..$ : chr [1:150] "1" "2" "3" "4" ...
.. ..$ : chr [1:2] "LD1" "LD2"
```

This is a list of three objects, the first of which, class, gives the predicted class or group. The second object in pred is the posterior probability of a plant being in each category. The predicted classes can be seen by running pred$class, which yields

```
  [1] setosa     setosa     setosa     setosa     setosa     setosa     setosa
  [8] setosa     setosa     setosa     setosa     setosa     setosa     setosa
 [15] setosa     setosa     setosa     setosa     setosa     setosa     setosa
 [22] setosa     setosa     setosa     setosa     setosa     setosa     setosa
 [29] setosa     setosa     setosa     setosa     setosa     setosa     setosa
 [36] setosa     setosa     setosa     setosa     setosa     setosa     setosa
 [43] setosa     setosa     setosa     setosa     setosa     setosa     setosa
 [50] setosa     versicolor versicolor versicolor versicolor versicolor versicolor
 [57] versicolor versicolor versicolor versicolor versicolor versicolor versicolor
 [64] versicolor versicolor versicolor versicolor versicolor versicolor versicolor
 [71] virginica  versicolor versicolor versicolor versicolor versicolor versicolor
 [78] versicolor versicolor versicolor versicolor versicolor versicolor virginica
 [85] versicolor versicolor versicolor versicolor versicolor versicolor versicolor
 [92] versicolor versicolor versicolor versicolor versicolor versicolor versicolor
 [99] versicolor versicolor virginica  virginica  virginica  virginica  virginica
[106] virginica  virginica  virginica  virginica  virginica  virginica  virginica
[113] virginica  virginica  virginica  virginica  virginica  virginica  virginica
[120] virginica  virginica  virginica  virginica  virginica  virginica  virginica
[127] virginica  virginica  virginica  virginica  virginica  virginica  virginica
[134] versicolor virginica  virginica  virginica  virginica  virginica  virginica
[141] virginica  virginica  virginica  virginica  virginica  virginica  virginica
[148] virginica  virginica  virginica
Levels: setosa versicolor virginica
```

We can check which species predictions agreed with the actual species by running

```
pred$class == iris$Species
```

which yields a vector of 150 logical values (i.e., TRUE or FALSE), all of which are TRUE except observations 71, 84, and 134. We can count the number of correct classifications with the sum() command,[9] which returns the number of TRUEs:

---

[9]You might wonder how we sum up TRUEs and FALSEs. This is an example of what R calls *coercion*. Summing logicals doesn't make sense, so R must *coerce* the TRUEs and FALSEs to numbers, which it does by assigning 1 to TRUE and 0 to FALSE. Once these conversions are made, we are essentially counting how many TRUEs there are in the argument to sum.

```
sum( pred$class == iris$Species )
```

which returns 147. Thus, 147 of the 150 predicted species agree with the actual species.

The object returned from the function lda() is a list that contains the posterior probability of a plant's species. For the iris data set, most plants are easily classified; this can be seen in Figure 18.24. There are some plants near the boundary between versicolor and virginica that may be difficult to classify. It is here where some plants were classified. Thus, most of the iris plants have a posterior probability near 1 of being in a particular class. The posterior probabilities are in the posterior object within the object returned from lda(). For the iris data we can run

```
pred$posterior
```

Part of the output from rounding these probabilities is below:

```
      setosa versicolor virginica
1      1.000    0.0000    0.0000
2      1.000    0.0000    0.0000
3      1.000    0.0000    0.0000

49     1.000    0.0000    0.0000
50     1.000    0.0000    0.0000
51     0.000    0.9999    0.0001
52     0.000    0.9993    0.0007
53     0.000    0.9958    0.0042

70     0.000    1.0000    0.0000
71     0.000    0.2532    0.7468   ****
72     0.000    1.0000    0.0000
73     0.000    0.8155    0.1845

77     0.000    0.9983    0.0017
78     0.000    0.6892    0.3108
79     0.000    0.9925    0.0075
80     0.000    1.0000    0.0000
81     0.000    1.0000    0.0000
82     0.000    1.0000    0.0000
83     0.000    1.0000    0.0000
84     0.000    0.1434    0.8566   ****
85     0.000    0.9636    0.0364

100    0.000    0.9999    0.0001
101    0.000    0.0000    1.0000

126    0.000    0.0027    0.9973
127    0.000    0.1884    0.8116
128    0.000    0.1342    0.8658

133    0.000    0.0000    1.0000
134    0.000    0.7294    0.2706   ****
135    0.000    0.0660    0.9340
136    0.000    0.0000    1.0000
137    0.000    0.0000    1.0000
```

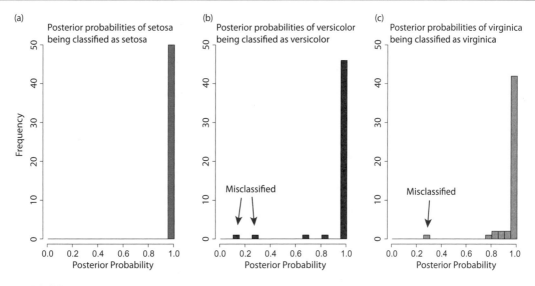

Figure 18.26 Histograms of the posterior probabilities of correctly classifying the three species.

138	0.000	0.0062	0.9938
139	0.000	0.1925	0.8075
140	0.000	0.0008	0.9992
149	0.000	0.0000	1.0000
150	0.000	0.0175	0.9825

We have removed many of the posterior probabilities that are near 0 or 1. We have also annotated the above output to flag the three plants that were misclassified. Figure 18.26 shows histograms of the 50 posterior probabilities of correctly classifying each of the three species. Most of the posterior probabilities of correct classification are near 1.

If we wanted, we could send `predict()` a set of measurements for a new set of plants – that is, plants that were not part of the fitted data set. This would return an inference for each plant in the new data set. ∎∎∎

The iris data set presents a fairly easy classification problem. When using all four variables we found that only three misclassifications occurred out of 150 plants. Other problems are often messier. We present one of these in the next example.

∎ **Example 18.16** The data set in `FetalHealth.csv` was presented by Ayres-de Campos *et al.* (2000) and gives a number of predictors along with a measure of fetal health for 2,126 fetuses. Fetal health is measured as an ordered categorical variable called `fetal_health` and coded as

    1 = normal
    2 = suspect
    3 = pathological

as determined by three obstetricians. Ayres-de Campos *et al.* (2000) used cardiotocograms to summarize image information and create a number of possible predictor variables. All of these predictor variables used data summarized in some way from the cardiotocograms. Use the variables

```
baseline.value , fetal_movement, uterine_contractions,
prolongued_decelerations, abnormal_short_term_variability,
histogram_width, histogram_min, histogram_max,
```

```
histogram_number_of_peaks, histogram_number_of_zeroes,
histogram_mode
```

to classify the level of fetal health.

Solution: To begin, let's load the data and subset the variables of interest:

```
FetalHealth = read.csv( "FetalHealth.csv" )
FH = FetalHealth[ , c("fetal_health","baseline.value","fetal_movement",
            "uterine_contractions","prolongued_decelerations",
            "abnormal_short_term_variability","histogram_width",
            "histogram_min","histogram_max","histogram_number_of_peaks",
            "histogram_number_of_zeroes","histogram_mode")]
n = length( FH$fetal_health )
```

Next, we can explore the data by creating all pairs of scatterplots using the pairs() function in R:

```
clr = c("green3","darkgoldenrod1","red")[FetalHealth$fetal_health]
pairs( FH , pch="." , cex=3 , col=clr )
```

The resulting plot, shown in Figure 18.27, shows the scatterplot with green indicating normal fetal health (fetal_health = 1), orange (really, a dark goldenrod) for suspect fetal health (fetal_health = 2), and red indicating pathological fetal health (fetal_health = 3). We can see that the red and orange points are usually in separate parts of the graph, but often in a nonlinear way. For example, in the histogram_min vs. histogram_max scatterplot (enlarged in Figure 18.28) the red points are clustered on the left side and the bottom right of the scatterplot. Linear discriminant analysis assumes equal covariance matrices, which is clearly violated in this example. We will apply LDA in hopes of obtaining a classification rule that predicts fetal health accurately. A more thorough analysis of this data set is warranted.[10]

The output from lda() is below:

```
Call:
lda(fetal_health ~ baseline.value + fetal_movement + uterine_contractions +
    prolongued_decelerations + abnormal_short_term_variability +
    histogram_width + histogram_min + histogram_max + histogram_number_of_peaks +
    histogram_number_of_zeroes + histogram_mode, data = FH)


Prior probabilities of groups:
         1          2          3
0.77845720 0.13875823 0.08278457


Group means:
  baseline.value fetal_movement uterine_contractions prolongued_decelerations
1       131.9819    0.007963142          0.004780665             5.135952e-05
2       141.6847    0.008332203          0.002389831             9.491525e-05
3       131.6875    0.025676136          0.003784091             1.272727e-03
  abnormal_short_term_variability histogram_width histogram_min histogram_max
1                        42.46586        73.40000      91.08580      164.4858
2                        61.90169        49.15932     113.29492      162.4542
3                        64.53977        78.34659      83.98295      162.3295
```

---

[10]This data set was the subject of a kaggle data science competition (see kaggle.com). Participating in these competitions is an excellent way to refine your data science skills.

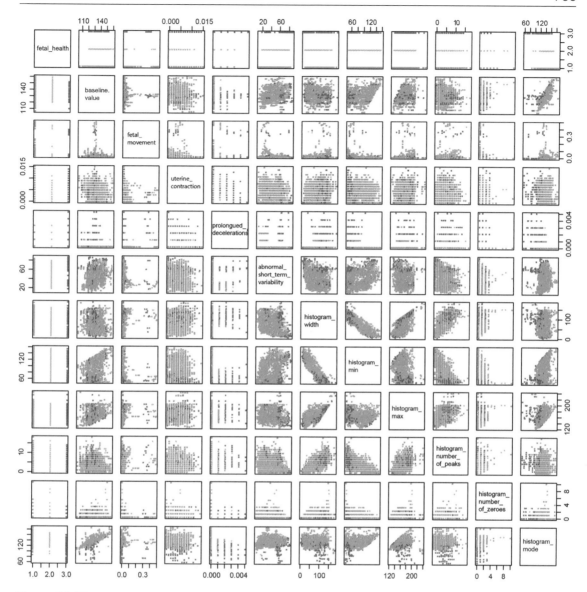

Figure 18.27 All possible pairs of scatterplots of variables in the fetal health data set.

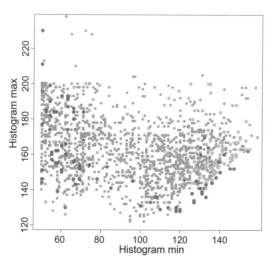

Figure 18.28 Scatterplot of `histogram_min` vs. `histogram_max` for fetal health data. Red points indicate pathological fetal health while orange points indicate suspect fetal health.

```
        histogram_number_of_peaks histogram_number_of_zeroes histogram_mode
1                        4.163142                   0.3353474      138.2586
2                        3.311864                   0.2440678      146.5559
3                        4.443182                   0.3465909      114.6080
```

```
Coefficients of linear discriminants:
                                       LD1            LD2
baseline.value                5.342136e-02   -0.034011121
fetal_movement                5.071733e-01   -0.976410218
uterine_contractions         -7.858397e+01  120.000849288
prolongued_decelerations      1.368291e+03  438.533117302
abnormal_short_term_variability 3.690142e-02  -0.018230944
histogram_width              -1.694109e-03    0.005472470
histogram_min                 3.222571e-03   -0.008492929
histogram_max                 4.911256e-04    0.003231217
histogram_number_of_peaks     3.802247e-03   -0.094488931
histogram_number_of_zeroes    7.021580e-02   -0.096276957
histogram_mode               -4.584064e-02   -0.020103266
```

```
Proportion of trace:
   LD1    LD2
0.8043 0.1957
```

Note that the `lda()` function estimates the prior probabilities using the observed frequencies in the data. In this case, the prior probabilities are estimated by $\hat{\pi}_1 = 0.778, \hat{\pi}_2 = 0.139$, and $\hat{\pi}_3 = 0.083$. We can make predictions for each element in the data set using the command

```
pred.fetal = predict( LDA.fetal , FH[,-1] )
```

We can determine the predicted fetal health group by using the `which.max` function applied to the set of posterior probabilities:

```
pred_FH = apply( pred.fetal$posterior , 1 , which.max )
cbind( round(pred.fetal$posterior,4) , pred_FH , FetalHealth$fetal_health )
```

Part of the output (the first 15 of the 2,126 observations) is shown below. We have annotated this output by flagging the misclassifications.

```
           1      2       3 pred_FH fetal_health
 1    0.7963 0.1805 0.0233       1            2   ****
 2    0.9980 0.0020 0.0000       1            1
 3    0.9989 0.0011 0.0000       1            1
 4    0.9976 0.0024 0.0000       1            1
 5    0.9985 0.0015 0.0000       1            1
 6    0.0001 0.0000 0.9999       3            3
 7    0.0000 0.0000 1.0000       3            3
 8    0.6683 0.2596 0.0721       1            3   ****
 9    0.7534 0.1873 0.0593       1            3   ****
10    0.7519 0.1895 0.0586       1            3   ****
11    0.6691 0.3253 0.0056       1            2   ****
12    0.6043 0.3919 0.0038       1            2   ****
13    0.9936 0.0064 0.0000       1            1
14    0.9950 0.0050 0.0000       1            1
15    0.9743 0.0252 0.0005       1            1
```

Table 18.9 Confusion matrix for fetal health data.

		\multicolumn{3}{c}{**Predicted fetal health**}		
		1 normal	2 suspect	3 pathological	Total
Actual	1 normal	1,584	53	18	1,655
fetal	2 suspect	146	139	10	295
health	3 pathological	24	53	99	176
	Total	1,754	245	127	2,126

Aside from one observation (number 29, not shown here), all of the predicted classes had a maximum posterior probability below 0.90, indicating that misclassifications usually occur on cases that are hard to predict – that is, where the highest posterior probability is not close to 1. The confusion matrix for the fetal health data set is shown in Table 18.9.   ■ ■ ■

18.6.4 Quadratic Discriminant Analysis

Linear discriminant analysis assumed equal covariance matrices for the K groups. If this assumption is not reasonable, we can relax this assumption and proceed assuming different covariance matrices for the groups. Let μ_r and Σ_r denote the mean and covariance matrix of all those units in group r. Assuming prior probabilities of $\pi_1, \pi_2, \ldots, \pi_K$ for the K groups, we can find that the posterior probability for group k is

$$p_k(\boldsymbol{x}) = C \frac{\pi_k}{(2\pi)^{p/2}|\Sigma_k|^{1/2}} \exp\left(-\frac{1}{2}(\boldsymbol{x}-\mu_k)'\Sigma_k^{-1}(\boldsymbol{x}-\mu_k)'\right), \tag{18.24}$$

where C is the reciprocal of the denominator in (18.21); in other words, C is the constant that assures $\sum_{k=1}^{K} p_k(x) = 1$. Thus, for a given x we compute (18.24) for each group $k = 1, 2, \ldots, K$. A bit of messy algebra shows that this is equivalent to selecting the group with the largest value of

$$q_k(\boldsymbol{x}) = \underbrace{-\frac{1}{2}\,\boldsymbol{x}'\,\Sigma_k^{-1}\,\boldsymbol{x}}_{\text{quadratic in } x} + \underbrace{\mu_k'\,\Sigma_k^{-1}\,\boldsymbol{x}}_{\text{linear in } x} - \underbrace{\frac{1}{2}\mu'\Sigma_k^{-1}\mu - \frac{1}{2}\log|\Sigma_k| - \log\pi_k}_{\text{constant in } x}. \tag{18.25}$$

This creates nonlinear boundaries in the p-dimensional data space. These boundaries are expressed algebraically as the solutions of a quadratic function in the components of $\boldsymbol{x}$. The next example shows what these regions look like when applied to two predictors.

■ **Example 18.17** Use the petal length and width to perform QDA for all three species in the iris data.

Solution: In Example 18.14 we performed LDA on the iris data using only petal length and width. We computed the sample means and covariances there, so we will not repeat these calculations. To obtain the boundaries for classification, we would compute $p_k(x_1, x_2)$ for $k = 1, 2, 3$, where x_1 and x_2 are the petal length and width, respectively. We would then find the solutions to

$$p_1(x_1, x_2) = p_2(x_1, x_2)$$
$$p_1(x_1, x_2) = p_3(x_1, x_2)$$
$$p_2(x_1, x_2) = p_3(x_1, x_2).$$

Each of these leads to an ellipse in the (x_1, x_2) plane. Figure 18.29(a) shows the three boundaries, between all pairs of the three species: setosa vs. versicolor, setosa vs. virginica, and versicolor vs. virginica. These boundaries are all given by ellipses.

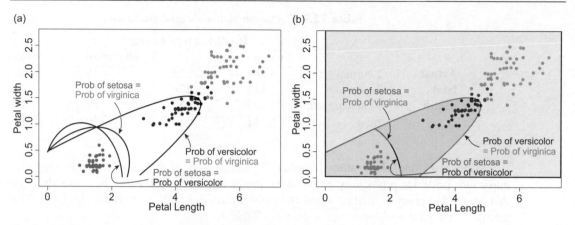

Figure 18.29 (a) Boundaries between the three comparisons: setosa vs. versicolor, setosa vs. virginica, and versicolor vs. virginica. (b) Shaded regions corresponding to predicted species.

Figure 18.29(b) shows the three regions for which we infer setosa, versicolor, or virginica. The boundary between setosa and virginica never comes into play because the versicolor species is in between.
■ ■ ■

■ **Example 18.18** Using R, apply QDA to the iris data using all four predictor variables.

Solution: The qda() function in R can be used to perform QDA:

```
iris.qda = qda( Species ~ Petal.Length + Petal.Width +
                          Sepal.Length + Sepal.Width ,
             data = iris )
```

Predictions for the entire iris data set can be obtained using the predict() function

```
pred.qda.iris = predict( iris.qda , iris[,c("Petal.Length",
            "Petal.Width","Sepal.Length","Sepal.Width")] )
```

We can find the entries in the confusion matrix by comparing the species and the predicted species. The following code accomplishes this for the $(1, 1)$ entry in the confusion matrix:

```
species = as.numeric( iris$Species )
pred_species = apply( pred.qda.iris$posterior , 1 , which.max )
cbind( round( pred.qda.iris$posterior , 4 ) , pred_species , species )
sum( (species == 1) & (pred_species == 1) )
```

Other entries in the confusion matrix can be found similarly. The confusion matrix is given in Table 18.10.
■ ■ ■

Table 18.10 Confusion matrix for QDA done in Example 18.18.

	setosa	versicolor	virginica
setosa	50	0	0
versicolor	0	48	2
virginica	0	1	49

18.6.5 **Dealing with Estimated Parameters**

So far we have assumed that the mean vectors and covariance matrices were all known. This is rarely the case in practice, and instead we must use samples selected from the population to estimate these parameters. We can then substitute these estimates into the decision rules we have derived above. Thus, our inferences about class membership are based on estimates of parameters, not the parameters themselves. An additional complication is that we often compare the predicted group with the actual group for members of the random sample. The catch here is that the observation x_i is used to estimate the parameters; we then turn around and compare the prediction to the actual group. It would be better to have a new data set that was not used in the estimation stage and to make inference to this new data set. What is often done in practice is to split the data into two groups, called a training data set and a validation data set. We use the training data to estimate the parameters and then make predictions for those observations in the validation data set. This way, we are not "double using" any observations. This avoids the concept of *overfitting*, where a complicated model is fit to too small of a data set. We will deal with the problem more generally in Chapter 19.

18.6.6 **Choosing Between Linear and Quadratic Discriminant Analysis**

The assumption of a common covariance matrix across all groups is a strong one that rarely holds in practice. Despite this, LDA often works better than QDA. This paradox arises because of the substantial variability in estimating the parameters. If there are p predictor variables, then for each group we need to estimate p means, p variances, and $p(p-1)/2$ covariances in order to perform QDA. This yields a total of

$$p + p + \frac{1}{2}p(p-1) = \frac{p^2 + 3p}{2} \text{ parameters for each group.}$$

If there are K groups, then the number of parameters to estimate when using QDA is

$$\frac{K(p^2 + 3p)}{2}. \tag{18.26}$$

When p and K are large, this is too many parameters to estimate with reasonable accuracy because the variability (and hence the uncertainty) in the estimates is too great. These are of course *unbiased* estimates, provided we use the usual sample mean vectors and sample covariance matrices.

If, on the other hand, we apply LDA, then we must estimate Kp means, but only p variances and $p(p-1)/2$ covariances, because a common covariance is assumed. For LDA, we must therefore estimate

$$Kp + p + \frac{1}{2}p(p-1) = \frac{(2K+1)p + p^2}{2} \text{ parameters.} \tag{18.27}$$

The biggest difference between the number of parameters that need to be estimated in (18.26) and (18.27) is that in (18.26) the number of groups K multiplies p^2, whereas it doesn't in (18.27). When K is moderately large this makes a big difference. Table 18.11 illustrates this effect. Notice how many more parameters must be estimate in QDA compared to LDA.

Assuming the multivariate normal distribution with differing covariance matrices is the correct model, the usual estimators of the group mean vectors and covariance matrices are unbiased, but because there are so many parameters to estimate, there is a lot of variance in these estimates. On the other hand, if a multivariate normal distribution with a common covariance matrix is assumed, then the estimates of covariance matrices (which are the same across all groups because we've assumed a common covariance matrix) are biased if we've assumed the wrong model, but there is much less variability in the estimates because fewer parameters are estimated. This is the classic trade-off between bias and variability that occurs in statistics. In LDA we are assuming a common covariance matrix, which is unreasonable in most cases, so we settle for a biased estimate of the covariance matrix. In exchange, we can estimate with more precision. We're trading unbiasedness for more precision.

Table 18.11 Number of parameters that need to be estimated for LDA and QDA for various values of K (the number of groups) and p (the number of predictor variables).

K	p	LDA	QDA	K	p	LDA	QDA
2	2	7	10	6	2	15	30
2	4	18	28	6	4	34	84
2	6	33	54	6	6	57	162
2	8	52	88	6	8	84	264
2	10	75	130	6	10	115	390
2	12	102	180	6	12	150	540
4	2	11	20	8	2	19	40
4	4	26	56	8	4	42	112
4	6	45	108	8	6	69	216
4	8	68	176	8	8	100	352
4	10	95	260	8	10	135	520
4	12	126	360	8	12	174	720

18.7 Logistic Regression for Classification

When there are $K = 2$ groups, we could perform logistic regression using the group as the random outcome and the values of the p predictors as the fixed predictor variables. When we perform LDA or QDA, the opposite is true; we treat the group as a fixed (not random) variable and the predictor variables as random within each group. If there are more than two groups, we can perform multinomial logistic regression.

It is possible to use logistic regression (binary or multinomial) to make group predictions, like we did earlier in this chapter with the Space Shuttle Challenger's O-ring data. We could fit a logistic regression and infer a group on the basis of which predicted probability is higher. In the case of the Challenger data, we treated the launch temperature as a fixed predictor and the presence or absence of O-ring damage as the random outcome. That made sense, because the performance of the O-ring depends on the temperature. It would seem a little awkward to look at the launches with no O-ring failures as a sample from a temperature distribution and those launches with O-ring failures as a sample from a possibly different temperature distribution.

In the iris data set, we assumed that for a fixed species, the sepal lengths and widths and the petal lengths and widths varied according to some distribution. The species was fixed and the measurements were random. Notice how the opposite is true for the Challenger O-ring data. Thus, the choice of whether to use logistic regression or discriminant analysis depends on which is more reasonable. Are the groups fixed or random? Are the measurements within each group fixed or random? Fixed groups with random measurements leads to discriminant analysis; random groups with fixed measurements leads to logistic regression. In some practical problems, the distinction becomes blurry and the selection of logistic regression or discriminant analysis requires some judgment.

If we have two predictor variables, x_1 and x_2, then the second-order logistic regression model might be assumed:[11]

$$\text{logit}\frac{p}{1-p} = \beta_0 + \beta_1 x_1 + \beta_2 x_2 + \beta_{11} x_1^2 + \beta_{22} x_2^2 + \beta_{12} x_1 x_2.$$

We could fit the logistic regression model and estimate the probability of a success (i.e., the outcome being 1) to be

$$\hat{p} = \text{logit}^{-1}\left(\hat{\beta}_0 + \hat{\beta}_1 x_1 + \hat{\beta}_2 x_2 + \hat{\beta}_{11} x_1^2 + \hat{\beta}_{22} x_2^2 + \hat{\beta}_{12} x_1 x_2\right) \tag{18.28}$$

---

[11]Extensions to higher (or lower) order models should be clear.

where

$$\text{logit}^{-1}(z) = \frac{e^z}{1+e^z}.$$

In cases where there are two predictor variables, we could plot contours in the (x_1, x_2) plane for various values of $\hat{p}$ in (18.28). We would decide what threshold to use to classify an observation (either a new observation or one of the observations in the original data set). Often this value is $1/2$, but it could differ from $1/2$ if the costs of misclassifying are unequal. For example, if group A consists of those with a disease and group B consists of those without the disease, it would be much more costly (not just monetarily) to classify someone with the disease in the no-disease group because this would withhold treatment from a person who truly has the disease. The other case, misclassifying someone without the disease as having the disease, has its own costs, which are certainly not zero, but usually the cost is lower. In this case, we might want to classify someone in the no-disease group if the estimated posterior probability of being in the no-disease group is 0.9 or higher.

■ **Example 18.19** Classification methods have been used to distinguish between genuine and forged banknotes. A number of banknotes in each of these categories (genuine and forged) was scanned at a resolution of 400 by 400 pixels. This data set is available on the University of California Irvine Machine Learning Repository (Dua and Graff, 2017; see also the paper by Gillich and Lohweg, 2010). The resulting images were run through a wavelet transformation. Details of how the wavelet transformation is used are not really important to the classification problem. Four characteristics of the output from the wavelet computations were extracted:

$x_1 = $ variance of the transformed image
$x_2 = $ skewness of the transformed image
$x_3 = $ kurtosis of the transformed image
$x_4 = $ entropy of the transformed image.

For the 1,372 banknotes available for the study, the ground truth regarding authenticity was known. The outcome variable is Y, which is 1 if the banknote is genuine and 0 if it is a forgery. The problem is to find a classification rule that could be used for future banknotes. Use logistic regression with only x_1 and x_2 to develop a model for whether a banknote is genuine or not. Plot contours for the probability and use $p = 1/2$ as the discriminator for whether a note is genuine or not.

Solution: We can read in the data and make a plot of all possible pairs of scatterplots using the code:

```
banknotes = read.csv("data_banknote_authentication.csv")
names( banknotes ) = c("X1","X2","X3","X4","Y")
clr = c("red","black")[banknotes$Y+1]
pairs( banknotes[ , 1:4 ] , col=clr )
X1 = banknotes$X1
X2 = banknotes$X2
Y = banknotes$Y
```

The set of all scatterplots is shown in Figure 18.30. We can see that the genuine notes are clustered together, as are the forged notes, but there is a lot of overlap. This example focuses on the first two variables, x_1 and x_2.

Figure 18.31(a) shows a scatterplot of x_1 and x_2 with black circles for genuine banknotes and red circles for notes that are not genuine. We can see that, generally, the genuine notes are in the bottom left part of the plot, while forged ones are in the upper right. We can run a logistic regression using Y, the indicator variable for genuine notes, as the response, and x_1 and x_2, the variance and skewness of the wavelet transformed image. The following code does a full second-order model in these two variables:

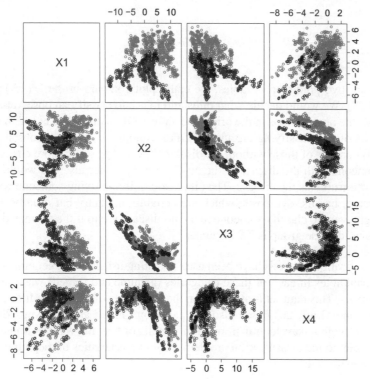

Figure 18.30 All possible scatterplots of the four variables obtained from the wavelet analysis of the images of the banknotes. Black circles indicate genuine notes and red circles indicate forged notes.

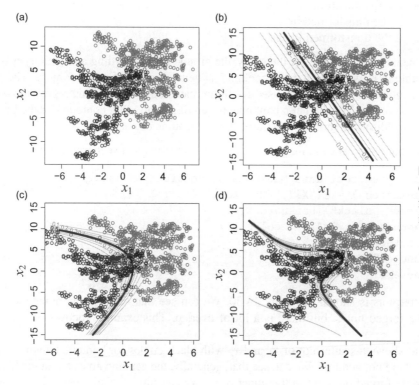

Figure 18.31 Banknote data. Red indicates $Y = 0$ (i.e., forged) banknote and black indicates $Y = 1$ (genuine).

```
glm2 = glm( Y ~ X1 + X2 + I(X1^2) + I(X2^2) + I(X1*X2), family="binomial" )
summary( glm2 )
```

The output from this model is

```
Call:
glm(formula = Y ~ X1 + X2 + X11 + X22 + X12, family = "binomial")

Deviance Residuals:
    Min       1Q    Median       3Q      Max
-3.4888  -0.2130  -0.0523   0.1557   2.2246

Coefficients:
              Estimate Std. Error z value Pr(>|z|)
(Intercept)   1.528499   0.168335   9.080  < 2e-16 ***
X1           -1.826396   0.139380 -13.104  < 2e-16 ***
X2           -0.047354   0.032249  -1.468    0.142
X11           0.051027   0.023930   2.132    0.033 *
X22          -0.054497   0.006148  -8.865  < 2e-16 ***
X12           0.138438   0.020945   6.610 3.85e-11 ***
---
Signif. codes:  0 '***' 0.001 '**' 0.01 '*' 0.05 '.' 0.1 ' ' 1

(Dispersion parameter for binomial family taken to be 1)

    Null deviance: 1883.94  on 1370  degrees of freedom
Residual deviance:  524.54  on 1365  degrees of freedom
AIC: 536.54

Number of Fisher Scoring iterations: 7
```

Our estimate for the probability that a banknote is genuine as a function of x_1 and x_2 is then

$$\hat{p} = \frac{\exp\left(1.528499 - 1.826396x_1 - 0.047354x_2 + 0.051027x_1^2 - 0.054497x_2^2 + 0.138438x_1x_2\right)}{1 + \exp\left(1.528499 - 1.826396x_1 - 0.047354x_2 + 0.051027x_1^2 - 0.054497x_2^2 + 0.138438x_1x_2\right)}.$$

The following code plots a contour plot for various values of p. The contour plot is plotted first; then the points (x_{1i}, x_{2i}) are added, and finally a thicker line is added for the $p = 1/2$ contour. We could use any of these contours as the discriminant function, but often the $p = 1/2$ contour is used.

```
logitinv = function(x) return( exp(x)/(1+exp(x)) )
betahat = glm2$coefficients
xx = seq( -6 , 6 , 0.01 )
yy = seq( -15 , 15 , 0.02 )
zz = matrix( 0 , nrow=length(xx) , ncol=length(yy) )
for (i in 1:length(xx))
{
        for (j in 1:length(yy))
        {
                zz[i,j] = logitinv(
                betahat[1] + betahat[2]*xx[i] + betahat[3]*yy[j] +
                betahat[4]*xx[i]*xx[i] + betahat[5]*yy[j]*yy[j] +
```

Table 18.12 Measures of model performance for first-, second-, and third-order models for the banknote data. Accuracy is the overall proportion of correct classifications, sensitivity is the proportion of correctly classified forged documents, and specificity is the proportion of correctly classified genuine documents.

Order	Accuracy	Sensitivity	Specificity
1	0.885	0.909	0.856
2	0.923	0.916	0.956
3	0.934	0.916	0.956

```
                        betahat[6]*xx[i]*yy[j] )
        }
     }
     contour( xx , yy , zz , col="gray" )
     points( banknotes$X1 , banknotes$X2 , col=clr )
     contour( xx , yy , zz , levels=0.5 , lwd=4 , add=TRUE )
```

Contours for first-order, second-order, and third-order logistic regression models are shown in Figure 18.31. We can see that the discriminating contours for higher order polynomials seem to do a better job at separating the points. Remember, though, that as the order of the model increases, the number of parameters that must be estimated increases, and the variability of the estimates increases. This is the bias–variance trade-off we discussed in Section 18.6.6.

We can evaluate the performance of the classification rules for models of order 1 through 3. The overall *accuracy* is the proportion of correct classifications. For the first-order model, there were 1,214 correct classifications out of the 1,371 banknotes. This ratio is $1,214/1,371 \approx 0.885$. The *sensitivity* is the probability that a forged document is classified correctly.[12] With the first-order model, 692 of the 761 forgeries were correctly classified, a proportion of approximately 0.909. Finally, with the first-order model, 522 of the 610 genuine notes were correctly identified, giving a specificity of $522/610 \approx 0.856$. Table 18.12 gives the accuracy of first-, second-, and third-order models. We can see that the proportions increase as the order of the model increases.

We must always keep in mind that we are estimating the parameters of the logistic regression. Thus, the fitted model will conform to the observed data. It would be better to make predictions on a new set of data. The idea of splitting the data into training and validation groups, called cross-validation, would provide a better assessment of the predictive performance of the model. This is discussed in Section 19.5.   ■ ■ ■

Exercise 18.3 Apply logistic regression to classify genuine and forged banknotes using variables x_3 and x_4.

We have discussed LDA, QDA, and logistic regression for classifying high-dimensional data. There are a number of other algorithms for accomplishing this task, but these are beyond the scope of this book. For a more complete treatment of the topic, see James *et al.* (2021).

---

[12]This is analogous to the case of disease testing, where the sensitivity is the probability of detecting the disease for someone with the disease. In this circumstance, we want the probability of detecting a forged document given that it is forged.

18.8 **Chapter Summary**

This chapter covered generalized linear models – that is, models for which some function of the mean or expected outcome is linear in the predictors. We focused mostly on two kinds of generalized linear models: logistic regression and Poisson regression. When the outcome is dichotomous (i.e., there are just two possible outcomes, which we usually call 0 and 1) we are led to logistic regression. We model the probability that the outcome is 1 as a function of the predictors. Specifically, we assume that the log odds $\log p/(1-p)$ is linear in the predictors. Logistic regression can be used to develop an explanatory model that helps us understand how the predictors affect the outcome. It can also be used to predict the category that an observation will fall in, given the values of the predictor. Poisson regression assumes that the outcome has a Poisson distribution with mean λ. We assume that $\log \lambda$ is linear in the predictor variables.

Regression trees provide an alternative to the usual regression model (see Chapter 14). They involve selecting a variable and a split point so that values of the variable below the split point lead to one prediction and values above the split point lead to another. We can then split each of these decision nodes and repeat the above procedure until some stopping criterion is met. Predictions for the outcome are then made by traversing down the tree until the final node is reached. This final node gives the predicted outcome.

Many problems involve predicting which class an object belongs to, given several measurements that can be used as predictors. Linear discriminant analysis leads to a line (or plane, or hyperplane) that divides the regions in the predictor space for the various categories. This division of the predictor space provides an inference for the category. Quadratic discriminant analysis leads to a more complicated division of the predictor space and it sometimes leads to better prediction rules. Logistic regression can also be used to classify objects as long as there are just two categories.

18.9 Problems

Problem 18.1 Consider the dose–response data shown in Table 18.13. Fit a logistic regression model with dose as the predictor variable. Using $\alpha = 0.05$, test whether the dose affects the response.

Table 18.13 Data for Problem 18.1.

Dose	Response	Dose	Response
16.08	0	34.93	0
50.45	1	20.40	0
49.17	1	13.96	0
31.99	1	16.14	0
67.33	1	20.49	0
38.04	1	24.29	1

Problem 18.2 Show that for the logistic model given in (18.1), the log odds is equal to

$$\log \frac{p}{1-p} = \beta_0 + \beta_1 x_{i1} + \beta_2 x_{i2} + \cdots + \beta_p x_{ip}.$$

Problem 18.3 Consider the data in Table 18.14.

Table 18.14 Data for Problem 18.3.

Dose	Age	Resp	Dose	Age	Resp
14.0	42	0	18.8	48	0
13.5	36	0	4.7	15	0
12.1	46	0	36.0	71	1
23.9	51	1	26.1	47	1
15.6	43	0	16.9	52	0
20.0	34	1	50.1	98	1
15.2	29	0	11.6	42	1
27.1	37	1	32.1	56	1
34.1	45	1	18.6	30	0
18.5	43	1	4.0	6	0

(a) Fit a logistic model using dose as the only predictor variable.
(b) Using $\alpha = 0.05$ test whether the coefficient of dose in this model is equal to 0 against the alternative that it is nonzero.
(c) Fit a logistic model using both dose and age as the predictor variables.
(d) Using $\alpha = 0.05$ test whether the coefficient of dose in this model is equal to 0 against the alternative that it is nonzero.
(e) Explain why the tests above reach different conclusions.

Problem 18.4 For the model you fit in Problem 18.3:

(a) What is the estimated probability that a 50 year-old person with a dose of 20 will have a response of 1?
(b) For a 30 year-old, make a plot of the probability of having a response of 1 as a function of the dose ($0 \leq$ dose ≤ 50).
(c) Repeat part (b) for a 40-year-old person.
(d) Create a graph showing (on the same axes) the probabilities of having a response of 1 for those who are 30, 40, 50, 60, and 70 years old.
(e) Based on the graph in part (d), describe the effect of age on the response curves.

Problem 18.5 Myers *et al.* (2010) describe a situation where families are surveyed about their intention to purchase a new car. Twenty families were surveyed, and for each the family income, age of their current car, and whether they actually purchased a new car within six months of the first survey. The data are contained in the file `PurchaseNewCar.csv`.

(a) Fit a logistic regression with predictor variables `IncomeThous` and `AgeOfCar`.
(b) Interpret the coefficients β_1 and β_2.
(c) Find 95% confidence intervals for β_1 and β_2.
(d) Estimate the probability that a family with a $50K income and a four-year-old car will purchase a new car.
(e) For families with a four-year-old car, estimate the probability that they will purchase a new car for each of a set of values for `IncomeThous` between 30 and 80. Plot the resulting curve.
(f) Repeat part (e) for families with a seven-year-old car.

Problem 18.6 Stokes *et al.* (2012) studied the factors that affect whether an offender is given a prison term or not. They considered the gender of the offender, whether the crime involved a home or a business, and whether the offender had a prior arrest record. Table 18.15 summarizes the data. Apply regression using indicator variables for `Gender`, `Type`, and `Prior`. You can use R's `glm()` function with the response `Prop` by giving R the weight for each observation, which is number of offenders `N`:

```
glm( Prop ~ Gender + Type + Prior ,
     family="binomial" , weights=N )
```

Problem 18.7 Create three contingency tables for each of the variables `Gender`, `Type`, and `Prior` against the `Prison` variable. Note that number *not* sentenced to a prison term is `N – Prison`. Apply a χ^2 test of independence for each of these tables.

Table 18.15 Data for Problem 18.6.

Gender	Type	Prior	Prison	N	Prop
Male	Bus.	Some	28	87	0.322
Male	Bus.	None	7	42	0.167
Male	Home	Some	19	94	0.202
Male	Home	None	27	227	0.119
Female	Bus.	Some	14	64	0.219
Female	Bus.	None	10	50	0.200
Female	Home	Some	14	114	0.123
Female	Home	None	27	186	0.145

Problem 18.8 Show that the logs of the ratios of the category probabilities for multinomial logistic regression satisfy (18.11).

Problem 18.9 Show that the assumptions in (18.10) imply that for $k > 0$

$$\log \frac{P(Y_i = k|\boldsymbol{x}_i)}{P(Y_i = 0|\boldsymbol{x}_i)} = \boldsymbol{x}_i'\boldsymbol{\beta}^{(k)}.$$

Problem 18.10 Myers *et al.* (2010) describe a situation where 202 customers were surveyed about their preferences for invoicing. Data were available for each company regarding whether they were a repeat customer, whether their account was small or large, and their preferred method of invoicing: (1) mail, (2) electronic, or (3) through a web portal. Data from these 202 customers are available in the file `CustInvoice.csv`. Apply multinomial logistic regression with the outcome being the preferred method of invoicing.

Problem 18.11 Refit the model in Problem 18.10 using only the factor that had a significant effect on the outcome.

Problem 18.12 The file `aneurysm.csv` contains data from a clinical trial comparing a new endovascular approach and a standard treatment (control) for the treatment of abdominal aortic aneurysms. The outcome is whether moderate to severe bleeding complications occur; this is an indicator variable called `comp30`. Other possible predictor variables include:

  `male`, an indicator for whether the patient was male;

  `control`, an indicator variable for the control (control = 1 , experimental group = 0);

  `early`, an indicator for whether the investigator had limited experience (limited experience = 1 , more experience = 0);

  `diameter`, a baseline measurement of the diameter of the aneurysm.

Build a model using logistic regression with `comp30` as the outcome. Is there evidence that the experimental treatment is effective at reducing the risk of complications?

Problem 18.13 The R package `palmerpenguins` contains data on three species of penguins: Adelie, Chinstrap, Gentoo. The data frame can be accessed by running the following R code:

```
library( palmerpenguins )
data( penguins )
```

For each of the 344 penguins in the data frame the following data are recorded:
```
species
island
bill_length_in_mm
bill_depth_in_mm
flipper_length_in_mm
body_mass_in_g
sex
```

(a) Filter the data using dplyr's `filter()` command to include only male penguins. Use multinomial logistic regression to obtain the posterior probability distribution for each of the three species using the variables `bill_length_in_mm`, `bill_depth_in_mm`, `flipper_length_in_mm`, and `body_mass_in_g`.

(b) For each male penguin in the data set, use the species with the highest posterior probability as the prediction for the species of that penguin.

(c) How many misclassifications are there?

Problem 18.14 Repeat Problem 18.13 for female penguins.

Problem 18.15 Suppose that the number of school absences due to illness is thought to be related to distance in miles from a chemical plant. Data for 14 students over one year are shown in Table 18.16. Assume that the number of absences has a Poisson distribution whose log mean is linearly related to distance from the plant. Fit a Poisson regression.

Table 18.16 Data for Problem 18.15.

Distance	Absent	Distance	Absent
4	4	11	1
9	0	2	5
9	2	6	2
8	1	9	1
9	1	7	3
3	0	4	3
5	1	9	1

Problem 18.16 Thall and Vail (1990) presented data from a clinical trial comparing the drug progabide

against a placebo for treatment of seizures. At baseline, the 59 subjects reported the number of seizures they had during the previous eight weeks. After being given the new drug (Treatment = 1) or the placebo (Treatment = 0), the number of seizures over the next eight weeks was observed. A portion of the data set is given in Table 18.17 (the full data set is in the file `epilepsyThallVail.csv`). Apply Poisson regression with the number of seizures as the outcome and treatment, age, and baseline as the predictor variables.

Table 18.17 Data for Problem 18.16.

Subject	Seizures	Treat	Age	Baseline
1	14	0	31	11
2	14	0	30	11
3	11	0	25	6
⋮	⋮	⋮	⋮	⋮
29	42	1	18	76
30	28	1	32	38
⋮	⋮	⋮	⋮	⋮
59	10	1	37	12

Problem 18.17 Consider Poisson regression with a single predictor variable, x.

(a) Determine the likelihood function.

(b) Determine the log-likelihood function.

(c) Differentiate the log-likelihood function with respect to β_0 and with respect to β_1, and set these results equal to zero.

(d) Show that it is possible to solve one of these equations for one of the parameters in terms of the other.

Problem 18.18 Show that for a regression tree the least squares estimators for β_0 and β_1 are those shown in (18.15) and (18.16).

Problem 18.19 Show that the predicted values $\hat{y}_i$ for the model in Problem 18.18 are, respectively, $\bar{y}_{-,k}$ and $\bar{y}_{+,k}$ for $x_i < c_k$ and $x_i > c_k$.

Problem 18.20 Give a detailed algorithm for constructing a regression tree with a suitable stopping criterion.

Problem 18.21 Consider the data on miles per gallon for recent makes of automobiles (Quinlan, 1993). This data set was downloaded from the University of California Irvine Machine Learning Repository (Dua and Graff, 2017). The file `auto-mpg.csv` contains the following variables

   mpg
   cylinders

   displacement
   horsepower
   weight
   acceleration
   model_year
   origin (1 = North America, 2 = Europe, 3 = Asia)
   car_name.

Use the variables `cylinders`, `displacement`, `horsepower`, `weight`, `acceleration`, `model_year` (as a continuous variable), and `origin` (as a three-level factor variable) to run a linear regression with `mpg` as the response variable. Which variables are significant predictors of `mpg`?

Problem 18.22 Consider using a regression tree to predict `mpg` for the data set in Problem 18.21.

(a) If `cylinders` is the only predictor, what is the optimal first split in the regression tree? What are the predicted values for `mpg` for values below and above this split?

(b) Repeat part (a) for the variable `displacement`.

(c) Repeat part (a) for the variable `horsepower`.

(d) Repeat part (a) for the variable `weight`.

(e) Repeat part (a) for the variable `acceleration`.

(f) Repeat part (a) for the variable `model_year` (used as a continuous variable).

Problem 18.23 Consider a regression tree for the data in Problem 18.22. Using the six variables, what is the optimal first split?

Problem 18.24 Consider again a regression tree for the data in Problem 18.22. Using the six variables, and the optimal first split from Problem 18.23, what is the optimal second split?

Problem 18.25 Use R's `tree` package and the `tree` function to create a regression tree for the `mpg` data from Problem 18.21. Use a minimum cut of 5 as the stopping criterion.

Problem 18.26 Consider the first-year college GPA for 219 students, discussed in Example 12.1. The file `FirstYearGPA.csv` includes the following variables:

  GPA, grade point average in college;

  HSGPA, high school GPA;

  SATV, SAT verbal score;

  SATM, SAT math score;

  Male, indicator variable for male, coded as 1 for male and 0 for female;

  HU, units of humanities courses in high school;

  SS, units of social science courses in high school;

  FirstGen, indicator for first-generation college student, coded as 1 for students who were the first in their family to attend college, and 0 otherwise;

White, indicator variable for white race, coded as 1 for white and 0 for otherwise;

CollegeBound, indicator variable coded as 1 if, at the student's high school, half or more of the students intended to go to college, and 0 otherwise.

(a) If HSGPA is the only predictor used in a regression tree, what is the optimal first split in the regression tree?

(b) If SATV is the only predictor used in a regression tree, what is the optimal first split in the regression tree?

(c) If SATM is the only predictor used in a regression tree, what is the optimal first split in the regression tree?

(d) If HU is the only predictor used in a regression tree, what is the optimal first split in the regression tree?

(e) If SS is the only predictor used in a regression tree, what is the optimal first split in the regression tree?

Problem 18.27 Consider a regression tree for the data in Problem 18.26. Using only the first five variables, what is the optimal first split?

Problem 18.28 Consider again a regression tree for the data in Problem 18.26. Using only the first five variables, and the optimal first split from Problem 18.27, what is the optimal second split?

Problem 18.29 Use R's tree package and the tree() function to create a regression tree for the GPA data from Problem 18.26. Use only the first five variables and a minimum cut of 5 as the stopping criterion.

Problem 18.30 The mtcars data set is available in base R and can be accessed by the command data (mtcars). You can learn about the variables in this data set by running ?mtcars. Suppose we would like to predict the time to drive one quarter-mile; this is in the qsec variable. Use R's tree package and the tree() function to create a regression tree for predicting qsec using the predictor variables mpg, cyl, disp, hp, drat, and wt.

Problem 18.31 Consider the mtcars data set from Problem 18.30. The variable am is an indicator variable for whether the car was built by an American manufacturer. Suppose we'd like to predict whether a car was built by an American manufacturer using the variables mpg, disp, hp, qsec, and wt. For parts (a) through (o), compute the confusion matrix after performing the LDA.

(a) Apply LDA with mpg only.

(b) Apply LDA with disp only.
(c) Apply LDA with hp only.
(d) Apply LDA with qsec only.
(e) Apply LDA with wt only.
(f) Apply LDA with mpg and disp.
(g) Apply LDA with mpg and hp.
(h) Apply LDA with mpg and qsec.
(i) Apply LDA with mpg and wt.
(j) Apply LDA with disp and hp.
(k) Apply LDA with disp and qsec.
(l) Apply LDA with disp and wt.
(m) Apply LDA with hp and qsec.
(n) Apply LDA with hp and wt.
(o) Apply LDA with qsec and wt.

Problem 18.32 Consider the penguin data from Problem 18.13. Consider the male penguins and the measurements bill length, bill depth, flipper length, and body mass. For parts (a) through (d), fit the given model and determine the confusion matrix.

(a) Apply LDA to infer the species using only the variable bill length.

(b) Apply LDA to infer the species using only the variable bill depth.

(c) Apply LDA to infer the species using only the variable flipper length.

(d) Apply LDA to infer the species using only the variable body mass.

(e) If you had to use just one of the four variables to predict the species, which variable would you choose? Explain.

Problem 18.33 Continuing with the penguin data from the previous problem, consider using a pair of the variables to predict the species. For parts (b) through (g), fit the given model and determine the confusion matrix.

(a) Make all six scatterplots of these variables taken two at a time. Use a different symbol or color for each of the three species.

(b) Apply LDA to infer the species using the bill length and bill depth.

(c) Apply LDA to infer the species using the bill length and flipper length.

(d) Apply LDA to infer the species using the bill length and body mass.

(e) Apply LDA to infer the species using the bill depth and flipper length.

(f) Apply LDA to infer the species using the bill depth and body mass.

(g) Apply LDA to infer the species using the flipper length and body mass.

(h) If you had to use just two variables to predict the species, which two variables would you choose? Explain.

Problem 18.34 For the penguin data, apply LDA using all four variables (bill length, bill depth, flipper length, and body mass) to infer the species. Give the confusion matrix. How many misclassifications are there?

Problem 18.35 Apply QDA to all parts of Problem 18.31.

Problem 18.36 Apply QDA to all parts of Problem 18.32.

Problem 18.37 The R package Stat2Data contains a data set called Hawks. You can access the data with the following code:

```
library( Stat2Data )
data( Hawks )
```

There is a lot of missing data in Hawks. The following code selects the species and five variables that might be used to predict a species. The code also selects all of those observations that do not contain missing values (i.e., those observations that are complete cases). Keeping only the complete cases reduces the number of observations from 908 to 891.

```
Hawks1=select(Hawks,Species,
        Wing,Weight,Culmen,Hallux,Tail)
Hawks1=Hawks1[complete.cases(Hawks1),]
```

The resulting data frame Hawks1 contains the following variables:

Species CH = Cooper's, RT = Red-tailed, SS = Sharp-Shinned;

Wing Length (in mm) of primary wing feather;

Weight Body mass (in g);

Culmen Length (in mm) of the upper bill;

Hallux Length (in mm) of the talon;

Tail Length (in mm) of the tail.

For parts (b) through (f), fit the LDA model and give the confusion matrix.

(a) Make all 10 scatterplots of the five predictor variables taken two at a time. Use a different symbol or color for each of the three species.

(b) Apply LDA for each variable taken one at a time.

(c) Apply LDA for all 10 pairs of variables.

(d) Apply LDA for all 10 groups of three variables.

(e) Apply LDA for all 5 groups of four variables.

(f) Apply LDA using all 5 variables.

Problem 18.38 Show that when the covariance matrices are not assumed equal, the group with the highest posterior probability is the one with the largest $q_k(x)$, where q_k is as given in (18.25).

Problem 18.39 Consider the data set from Problem 18.5 on families purchasing a new car. Fit a first-order logistic regression model with first-order terms for IncomeThous and AgeOfCar. Use this model to predict whether each family in the data frame will indeed purchase a new car. Give the confusion matrix.

Problem 18.40 For the mtcars data set described in Problem 18.31, apply logistic regression with first-order terms for the five predictor variables where the outcome is the indicator variable am. Give the confusion matrix.

Problem 18.41 For the mtcars data set described in Problem 18.31, apply logistic regression with second-order and all lower-order terms for the variables mpg, wt, and qsec where the outcome is the indicator variable am. Give the confusion matrix.

19 — Cross-Validation and Estimates of Prediction Error

19.1 Overfitting and Underfitting

Overfitting refers to the use of a model with more parameters than can be justified by the data. Models that are overfit are often poor at predicting the outcome of new observations, that is, observations that were not used in the construction of the model. The next example illustrates this concept.

■ **Example 19.1**   Weisberg (1986) considered a problem of determining the length of blugill (a small game fish found in the central and southern United States) from its age. We treat age as the predictor variable and length as a response. Given the data

Length	67	109	137	146	150
Age	1	2	3	4	5

fit polynomial regression models of order 1, 2, 3, and 4. (This is a subset of the data frame `lakemary` that is available in the `alr4` package. We will analyze this data set more fully later in this chapter.)

Solution: We can load the `alr4` package and the `lakemary` data using the following code. The five fish that we will analyze here are indexed as fish 1, 44, 12, 47, and 66.

```
library(alr4)
data(lakemary)
indx = c(1,44,12,47,66)
Age1 = lakemary$Age[ indx ]
Length1 = lakemary$Length[ indx ]
```

The code below also defines variables for the powers of `Length1` and runs the four polynomial regression models of order 1 through 4:

```
Age12 = Age1^2
Age13 = Age1^3
Age14 = Age1^4
mod1 = lm( Length1 ~ Age1 )
anova( mod1 )
```

```
mod2 = lm( Length1 ~ Age1 + Age12 )
anova( mod2 )
mod3 = lm( Length1 ~ Age1 + Age12 + Age13 )
anova( mod3 )
mod4 = lm( Length1 ~ Age1 + Age12 + Age13 + Age14 )
anova( mod4 )
```

The ANOVA tables from these four linear models are shown below:

```
Analysis of Variance Table

Response: Length1
          Df Sum Sq Mean Sq F value  Pr(>F)
Age1       1 5152.9  5152.9  13.124 0.03618 *
Residuals  3 1177.9   392.6

Analysis of Variance Table

Response: Length1
          Df Sum Sq Mean Sq F value  Pr(>F)
Age1       1 5152.9  5152.9 13.2787 0.06775 .
Age12      1  401.8   401.8  1.0354 0.41596
Residuals  2  776.1   388.1

Analysis of Variance Table

Response: Length1
          Df Sum Sq Mean Sq F value Pr(>F)
Age1       1 5152.9  5152.9  8.2577 0.2132
Age12      1  401.8   401.8  0.6439 0.5695
Age13      1  152.1   152.1  0.2437 0.7080
Residuals  1  624.0   624.0

Analysis of Variance Table

Response: Length1
          Df Sum Sq Mean Sq F value Pr(>F)
Age1       1 5152.9  5152.9     NaN    NaN
Age12      1  401.8   401.8     NaN    NaN
Age13      1  152.1   152.1     NaN    NaN
Age14      1  624.0   624.0     NaN    NaN
Residuals  0    0.0     NaN
Warning message: In anova.lm(mod4) :
ANOVA F-tests on an essentially perfect fit are unreliable
```

Some strange things seem to happen when we reach the fourth-order model. The SSE decreases as we add variables to the model, and it reaches 0 for the fourth-order model. The degrees of freedom for the SSE is also reduced to 0, making it impossible to compute the MSE. Because of this, the columns for F value and Pr(>F) are reported as NaN (not a number).

The estimated coefficients can be obtained from the linear model objects using the coefficients variable in the object; for example, mod1$coefficients:

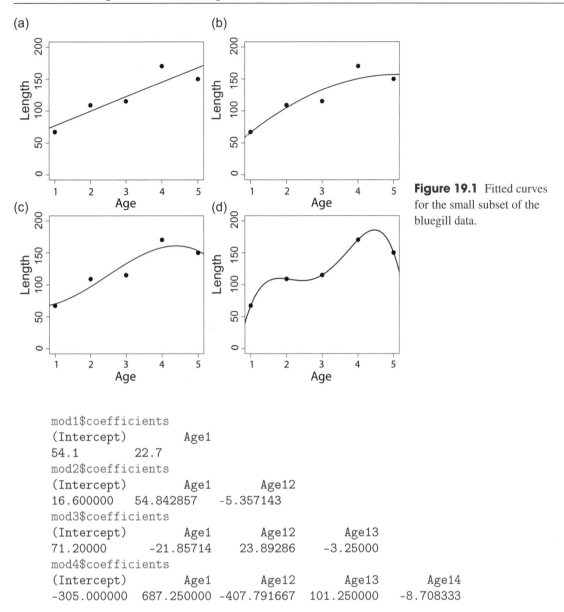

Figure 19.1 Fitted curves for the small subset of the bluegill data.

```
mod1$coefficients
(Intercept)        Age1
54.1         22.7
mod2$coefficients
(Intercept)        Age1         Age12
16.600000    54.842857    -5.357143
mod3$coefficients
(Intercept)        Age1         Age12         Age13
71.20000     -21.85714    23.89286      -3.25000
mod4$coefficients
(Intercept)        Age1         Age12         Age13         Age14
-305.000000  687.250000   -407.791667   101.250000   -8.708333
```

When we plot the estimated response curves, we obtain the plots shown in Figure 19.1. We see that the first-order model is probably inadequate because it fails to account for the curvature in the outcome. The second-order model seems to fit better because it accounts for the concave-down nature of the data. The response curve for the third-order model comes closer to the observed points, but the response seems to decrease when age exceeds 4, a feature that we would not expect. The fourth-order model passes through all five points, but gives a sharply decreasing response past age 5. In general, a set of n points in the plane can be fit by a polynomial of degree $n-1$ or less, so it is not surprising that the fourth-order model fits perfectly. We would say that the fourth-order model overfits the data. It is also likely that the third-order model overfits the data.   ■ ■ ■

■ **Example 19.2**   Analyze the full bluegill data using models of order 1 through 5.

Solution: Code very similar to that of the previous example can be used to load in the bluegill data from the alr4 package. This time we do not subset the fish given in the indx array; instead, we use the full data set. Otherwise, the code remains virtually unchanged.

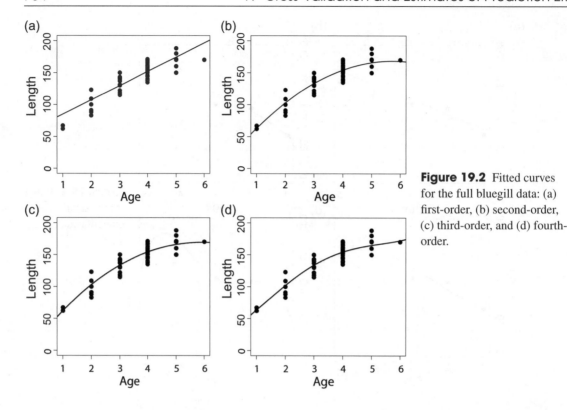

Figure 19.2 Fitted curves for the full bluegill data: (a) first-order, (b) second-order, (c) third-order, and (d) fourth-order.

Figure 19.2 shows the scatterplot of the raw data and the fitted curves. The first-order model in Figure 19.2(a) seems to underfit the data. Since the other three curves for polynomials of order 2, 3, and 4 are quite similar, we believe that the one of lowest order (the second-order polynomial in the upper right corner) is probably the best fit. One issue to note with the second-order model is that the estimated response curve is a parabola, which seems to reach its maximum near age 6. After age 6, the prediction curve will decrease. We should be careful about using this model to extrapolate beyond age 6.   ■ ■ ■

We might think that the SSE is a good measure of the fit (or lack of fit) of a model. There is a flaw in this thinking, though. The SSE

$$\text{SSE} = \sum_{i=1}^{n}(y_i - \hat{y}_i)^2$$

tends to underestimate the error because the ith observation y_i went into the calculation of the prediction $\hat{y}_i$. What would be ideal is if we had an additional data set that was not used in the calculation of the estimated parameters $\hat{\beta}_0, \hat{\beta}_1, \dots, \hat{\beta}_p$, and therefore not used in the calculation of each $\hat{y}_i$. Usually, there are no additional data available. Instead, we partition our available data into parts, fit the data with one part, and evaluate the fit with the other. This is the idea of cross-validation.

Exercise 19.1   Explain why the SSE cannot increase when we raise the order of the fitted polynomial by 1.

19.2 Cross-Validation

It is often possible to see how well a model fits by making predictions for new observations and then comparing how well the new data fit our predictions. Since we usually fit models with all of the available data, there are often no "new" data to fit. Instead, we will hold out a subset of data to be used to fit the model and then test it against the unused data. This idea of splitting data into a part used to fit the model and another part to test the model can help us guard against overfitting. This idea is called *cross-validation*.

> **Definition 19.2.1 — Cross-Validation.** *Cross-validation* involves splitting the data into two parts. One part, called the *training* data set, is used to fit the model. The other part, called the *validation* or *test* set, is used to test how well the model fits when it is applied to data that were not used in the fitting of the model.

As you might expect, there are many ways to perform cross-validation. In the next few subsections, we will look at (1) randomly splitting data, (2) leaving one observation out at time, and (3) *k*-fold cross-validation.

19.2.1 Splitting the Data at Random

Suppose we split the full data set 50/50, with half of the units being randomly assigned to the training data set and the remaining assigned to the validation data set. We can split the data with any fixed proportion, but an even 50/50 split is common. We then fit a model using the training data set and make predictions on the validation set. Let m denote the number of units in the validation data set and $n - m$ the number in the training data set. Let $\hat{y}_{(i)}$ denote the prediction of unit i in the validation set given the fitted model from the training data set. For an ordinary regression model we can measure the fit by the SSE

$$\text{SSE}_{\text{CV}} = \sum_{i=1}^{m} \left(y_i - \hat{y}_{(i)}\right)^2$$

or the MSE

$$\text{MSE}_{\text{CV}} = \frac{1}{m} \sum_{i=1}^{m} \left(y_i - \hat{y}_{(i)}\right)^2.$$

Note that we divide the SSE by m (not m minus the number of estimated β parameters) because the predictions are being made on observations that were *not* used in the fitting of the model.

We can then fit several different models on the training data set and get a measure of prediction accuracy using SSE_{CV} or MSE_{CV} by applying the estimated model to the validation set. The various models may involve varying degrees of the fitted polynomial or using various sets of predictors. We will illustrate the 50/50 cross-validation procedure on a data set on the compressive strength of concrete, given by Yeh (1998).

■ **Example 19.3** The concrete data set (`Concrete_Data.csv`) given by Yeh (1998) uses the response variable compression strength (`ComprStr`) and has several possible predictors. We will consider various polynomial models involving the age. The response is compressive strength (`ComprStr`) measured in MPa (mega pascals[1]). The age variable `Age` is measured in days. Apply the 50/50 cross-validation procedure and evaluate the fit based on the training data and applied to the validation data.

Solution: The scatterplot in Figure 19.3 shows that initially as concrete ages it becomes stronger, but after a while the strength reaches a maximum and then decreases, so some sort of nonlinearity in the response is needed. We will apply the 50/50 cross-validation procedure on models with $p = 1, 2, \ldots, 8$ and apply the fitted model to the validation data set. The R code to do this is:

---

[1]One pascal is one newton of force per square meter and one mega pascal, denoted 1 MPa, is one million pascals.

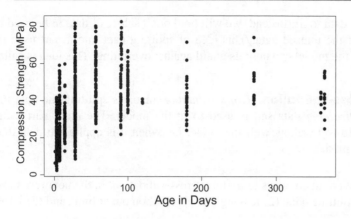

Figure 19.3 Compressive strength of concrete plotted against age (which is slightly jittered to show more detail).

```
concrete = read.csv( "Concrete_Data.csv" )
y = concrete$CompressiveStrength
age = concrete$Age_days
CV5050 = function( datafr , p , seed )
{
  set.seed( seed )
  n = nrow( datafr )
  predictors = names( datafr )[2:(p+1)]
  formla = paste( names(datafr)[1] , " ~ " ,
          paste( predictors , collapse=" + ") )
  smpl = sample(1:n,round((n+0.5)/2))
  training = datafr[ smpl , ]
  validation = datafr[ -smpl , ]
  df2 = datafr[ smpl , ]
  mod = lm( formla , data=df2 )
  pred = predict( mod , validation )
  MSE = sum( (validation[,1] - pred)^2 ) / round((n+0.5)/2)
  return( MSE )
}
ComprStr = concrete$CompressiveStrength
age = concrete$Age_days
datafr = data.frame( cbind(ComprStr,age,age^2,age^3,age^4,age^5,
                    age^6,age^7,age^8) )
CV5050( datafr , 2 , 12345 )
CV5050( datafr , 3 , 12345 )
```

After reading in the concrete data and extracting the important variables, we define the function CV5050. The function takes three arguments:

- the data frame, which is assumed to have columns for the response variables followed by each of the predictor variables (age raised to various powers in this example);
- the number of predictor variables to include, beginning with the first predictor variable in column 2; and
- the seed to use for the random selection.

Sending CV5050 a seed makes it possible to obtain exactly the same training/validation split for different models. For example, the last two lines call the CV5050 function with $p = 2$ and $p = 3$,

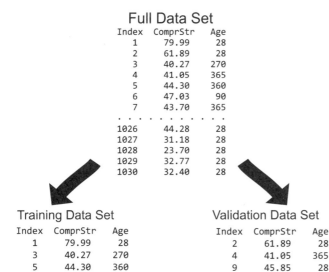

Full Data Set

Index	ComprStr	Age
1	79.99	28
2	61.89	28
3	40.27	270
4	41.05	365
5	44.30	360
6	47.03	90
7	43.70	365
. . .	. . . .	. . .
1026	44.28	28
1027	31.18	28
1028	23.70	28
1029	32.77	28
1030	32.40	28

Figure 19.4 The full concrete data set can be randomly split into two groups: the training data set (on which the model is fit) and the validation data set (on which the fitted model is assessed). Here the division was 50/50, with 515 units assigned to each group).

Training Data Set

Index	ComprStr	Age
1	79.99	28
3	40.27	270
5	44.30	360
6	47.03	90
7	43.70	365
. . .	. . . .	. . .
1019	37.27	28
1022	31.88	28
1024	39.46	28
1029	32.77	28

Validation Data Set

Index	ComprStr	Age
2	61.89	28
4	41.05	365
9	45.85	28
11	38.07	90
19	40.56	90
. . .	. . . .	. . .
1026	44.28	28
1027	31.18	28
1028	23.70	28
1030	32.40	28

but with the same random seed, and therefore the same split. The concrete data set contains 1,030 specimens, which can be divided into two groups of 515 each. One such division is shown in Figure 19.4.

Figure 19.5 shows the results of applying polynomial models of order $p = 1$, 2, 3, and 4 to the training data set. Figure 19.5(a) shows the training data set with the four fitted models to it. Figure 19.5(b) shows the corresponding validation set with the curves fitted from the training data set. The estimated response curves in blue are the same for the two curves in (a) and (b). Figures 19.5(c) and (d) show the results from a second 50/50 random assignment of units to the training and validation groups. Again, the curves in both parts are the same, but they are somewhat different than the curves in parts (a) and (b) because the model was fit on a different training data set.

We see that the linear model $p = 1$ is a poor fit, and this is exhibited by the large prediction MSE (MSE_{CV}) in the upper right corner of the validation sample #1. As the degree p of the polynomial increases, the MSE decreases, indicating a better predictive fit. Not shown in the figure are the curves for $p = 5, 6, 7$, and 8 (that would be too many curves on one graph!). It turns out that the MSE_{CV}s for $p = 5$, 6, 7, and 8 are *larger* than the MSE_{CV} for $p = 4$. This occurs because these models exhibit overfitting as described in Section 19.1. A similar phenomenon is seen in the second training/validation split, and again the MSE_{CV} is lowest for $p = 4$. Different random assignments of the split will give different parameter estimates and, of course, different predictions and MSE_{CV}. Thus, the 50/50 cross-validation procedure is usually run many times. While results will differ, the overall pattern is often the same in multiple runs.

We ran the 50/50 cross-validation with five random assignments using models with polynomials of order $p = 1, 2, \ldots, 8$ and for each we computed the MSE_{CV}. The results are shown in Figure 19.6. Note that the same random seed was used for orders of the polynomial for $p = 1$ through $p = 8$; this means that the same 50/50 split was made for points on a single curve, but different seeds were used for the five curves. While the levels differ across five different 50/50 splits of the data, the pattern is the same for all. MSE_{CV} is large for $p = 1$ and successively smaller for $p = 2, 3$, and 4. After $p = 4$ the MSE_{CV} is essentially flat. This suggests that a polynomial of order $p = 4$ is sufficient for making predictions in the validation set.

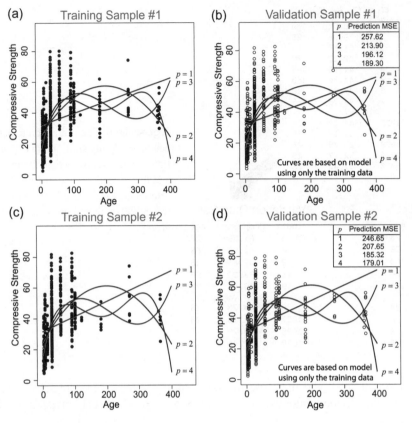

Figure 19.5 Two possible splits of training/validation sets. Polynomials of order 1, 2, 3, and 4 are fit to the training sample in (a) and (c). These models are applied to the validation set in (b) and (d).

Figure 19.5 shows that the predictions for compression strength using the various polynomial orders are similar for age between 0 and about 380. After that, the predictions vary drastically. For example, the $p = 3$ polynomial is increasing and concave-up for age > 400, while for $p = 2$ and $p = 4$ the estimated polynomial is decreasing and concave-down. While these models may be reasonable for predicting compression strength when age is between 0 and about 350, they should not be used to estimate strength when age exceeds 400. Extrapolation in this case is risky.   ■ ■ ■

Figure 19.3 shows a relationship between compression strength and age in which compression strength increases rapidly as age initially increases, then reaches a maximum strength, and finally decreases. This suggests that a transform may be applied to the predictor variable that "pulls back" the larger values more than it pulls back smaller values. The square root and the logarithm function can be used to accomplish this.

■ **Example 19.4**   Apply the 50/50 cross-validation method to the concrete compressive strength using the square root of the age.

Solution: The code from Example 19.3 can be adapted to consider the square root of age, rather than age itself. Figure 19.7 shows this relationship. After this transformation, the relationship looks much more symmetric about the point where strength is maximized, which is about 12 on the square root scale.

Let's follow the same patterns as in Example 19.3 and fit polynomials of various orders, but this time the predictor variables are powers of the square root of age. Figure 19.8 shows the results of two splits of testing and validation data. The estimated polynomial curves are shown in red, except for $p = 2$ which is shown in orange. Once this transformation is made, the second-order polynomial seems to be a reasonable fit.

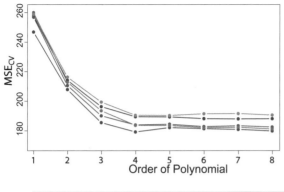

Figure 19.6 MSE_{CV} for the concrete data using five random 50/50 splits between the training and validation data sets applied to polynomial models of order $p = 1, 2, \ldots, 8$.

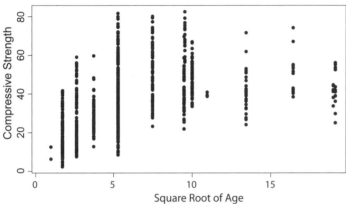

Figure 19.7 Compressive strength of concrete plotted against the square root of age (slightly jittered to show more detail).

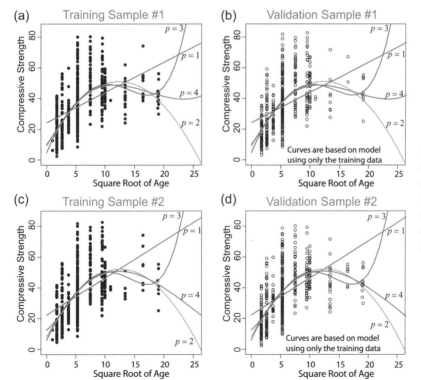

Figure 19.8 Polynomials of order 1, 2, 3, and 4 in the square root of age are fit to the training sample in (a) and (c). These models are applied to the validation set in (b) and (d).

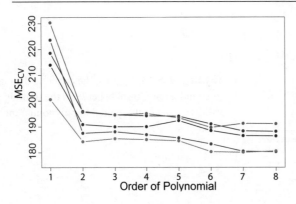

Figure 19.9 MSE_{CV} for the concrete data, with the square root of age, using five random 50/50 splits between the training and validation data sets applied to polynomial models of order $p = 1, 2, \ldots, 8$.

As in Example 19.3, we compute the MSE_{CV} for five different testing/validation data sets and for $p = 1, 2, \ldots, 8$. The results are shown in Figure 19.9. This time, we see that a first-order polynomial (a line) yields high MSE_{CV} values. The MSE_{CV} decreases substantially for $p = 2$ and remains fairly constant for larger values of p, although a small decrease is observed at $p = 6$.

In this case, the square root transformation results in a simpler model than in Example 19.3. Again, extrapolation is risky because the $p = 2$ curve drops precipitously after $\sqrt{age} > 20$, and even predicts negative strengths for sufficiently large $\sqrt{age}$.   ■ ■ ■

The idea of splitting the data into a training set and a validation step can be applied to almost any kind of modeling situation. This includes linear and polynomial regression, logistic regression, Poisson regression, and classification (which we will study in Section 19.5).

One disadvantage of splitting the data at random is that different random assignments will lead to different sets of MSE_{CV}. This difficulty can be overcome if we run multiple random splits. With multiple testing/validation data sets we can usually see a pattern. Another disadvantage, especially for smaller data sets, is that the training is done on only half the data, and consequently the degrees of freedom for error is not very large. In the next section we look at another cross-validation method that addresses these two issues.

19.3   Leave-One-Out Cross-Validation

Suppose we leave out just one observation, leaving the remaining data set with $n - 1$ observations. There are n ways to do this, and unless the data set is large we can perform this leave-one-out procedure for all possibilities. This is called *leave-one-out cross-validation* (LOOCV).

Suppose we have the observations

$$(\boldsymbol{x}_1, y_1), \ (\boldsymbol{x}_2, y_2), \ (\boldsymbol{x}_3, y_3), \ \ldots, \ (\boldsymbol{x}_{n-1}, y_{n-1}), \ (\boldsymbol{x}_n, y_n)$$

where the set of predictors for observation i, $\boldsymbol{x}_i$, could be a vector. In Example 19.3 there was only one predictor, age, and the various models used a set of powers of this variable, but in general there could be a number of predictor variables. The first split of the data is

Training	**Validation**
$(\boldsymbol{x}_2, y_2), \ (\boldsymbol{x}_3, y_3), \ \ldots, \ (\boldsymbol{x}_n, y_n)$	$(\boldsymbol{x}_1, y_1)$

We then use the $n - 1$ observations in the training set to fit the model and make the prediction $\hat{y}_1$ for the one excluded observation. The MSE (really, just the squared error) is

$$MSE_1 = (y_1 - \hat{y}_1)^2. \tag{19.1}$$

Next, we re-split the data leaving out the second observation $(\boldsymbol{x}_2, y_2)$ as follows:

Training	**Validation**
$(\boldsymbol{x}_1, y_1), \ (\boldsymbol{x}_3, y_3) \ldots, \ (\boldsymbol{x}_n, y_n)$	$(\boldsymbol{x}_2, y_2)$

Again, we fit the model on the training set of $n-1$ observations, make the prediction $\hat{y}_2$, and compute

$$\text{MSE}_2 = (y_2 - \hat{y}_2)^2.$$

We continue this process of excluding one observation, fitting the model on the remaining $n-1$ observations, making the prediction $\hat{y}_i$, and computing the MSE:

$$\text{MSE}_i = (y_i - \hat{y}_i)^2. \tag{19.2}$$

We therefore obtain n MSEs which we average to get the LOOCV measure:

$$\text{LOOCV}_{(n)} = \frac{1}{n} \sum_{i=1}^{n} \text{MSE}_i. \tag{19.3}$$

This method is completely deterministic, avoiding the randomness in selecting a sample of roughly half of the observations.

■ **Example 19.5** Apply the LOOCV method to the concrete data with compressive strength as the outcome and powers of age as the predictor variables.

Solution: The following code defines and calls the function LOOCV() for polynomials of degree 1 through 8:

```
LOOCV = function( datafr , p ) {
  n <<- nrow( datafr )
  MSE.sum = 0
  for (i in 1:n) {
    predictors = names( datafr )[2:(p+1)]
    formla = paste( names(datafr)[1] , " ~ " ,
            paste( predictors , collapse=" + ") )
    training = datafr[ -i , ]
    validation = datafr[ i , ]
    mod = lm( formla , data=training )
    pred = predict( mod , validation )
    MSE = ( validation[,1] - pred )^2
    MSE.sum = MSE.sum + MSE
  }
  return( MSE.sum/n )
}
LOOCV.vec = rep( NA , 8 )
for( i in 1:8 )  LOOCV.vec[i] = LOOCV( datafr , i )
```

The results of this computation are given in Table 19.1. We can see that MSE for the LOOCV method decreases to a level of about 183 for a fourth-order polynomial. Beyond that, the MSE is roughly constant. Thus, a fourth-order model seems reasonable for this data set. ■ ■ ■

Table 19.1 $\text{LOOCV}_{(n)}$ for concrete data set.

p	1	2	3	4	5	6	7	8
LOOCV	249.81	208.32	191.45	183.15	183.32	182.50	182.45	181.93

Table 19.2 $\text{LOOCV}_{(n)}$ for concrete data set with predictors being the powers of $\sqrt{\text{age}}$.

p	1	2	3	4	5	6	7	8
LOOCV	219.89	187.36	186.80	186.04	185.15	182.47	180.75	180.72

If a polynomial regression model with one predictor is assumed, it can be shown that

$$\text{LOOCV}_{(n)} = \frac{1}{n} \sum_{i=1}^{n} \left(\frac{y_i - \hat{y}_i}{1 - h_i} \right)^2, \tag{19.4}$$

where

$$h_i = \frac{1}{n} + \frac{(x_i - \bar{x})^2}{\sum_{j=1}^{n} (x_j - \bar{x})^2}.$$

The use of (19.4) means that we do not have to fit n separate models. Instead, we can fit the model on the full data set, make each prediction $\hat{y}_i$, compute h_i, and apply (19.4). In other cases, for example logistic regression, multiple regression with multiple predictor variables, etc., it is necessary to fit the model all n times and compute the $\text{LOOCV}_{(n)}$ from (19.3).

■ **Example 19.6**   Rerun the concrete strength analysis using LOOCV on the square root of age.

Solution: We can rerun the code from Example 19.5 if we simply add the line of code

```
age = sqrt( concrete$Age_days )
datafr = data.frame( cbind(y,age,age^2,age^3,age^4,age^5,age^6,age^7,age^8) )
names(datafr) = c("y","age","age2","age3","age4","age5","age6","age7","age8")
```

before running the LOOCV() function. The results are shown in Table 19.2.

We see that a polynomial model of order 1 has a high MSE. The MSE drops for a second-order model, and beyond that the MSE drops only slightly. We conclude that a second-order model based on the square root of age and its square[2] is a reasonable fit to the data.   ■ ■ ■

19.4 *k*-Fold Cross-Validation

Another method of splitting the data into training and validation sets is to divide randomly the full data set into k groups of roughly equal size. For example, if we have $n = 1,028$ observations and split them into $k = 5$ groups, we would take the group sizes to be 206, 206, 206, 205, and 205. In general, we let n_i denote the number of observations in group i. Typically, we divide the data into $k = 5$ to $k = 10$ groups. We then apply k separate training/validation sets as follows.

Initially, we use the first of the k groups as the validation data set and groups 2 through k as the training data set. As usual, we fit the model on the training data set and make predictions for the validation set. From these predictions in the validation data set, we compute

$$\text{MSE}_{[1]} = \frac{1}{n_1} \sum_{\text{all obs } i \text{ in group 1}} (y_i - \hat{y}_i)^2, \tag{19.5}$$

where y_i is the observed response on the ith observation in the validation set and $\hat{y}_i$ is the prediction for the ith observation made from the model fit on the training data set.

Next, we use the second group as the validation set and all the others as the training data set. As before, we fit the model on the training set and make predictions for the validation set. We then calculate $\text{MSE}_{[2]}$, defined analogously to (19.5).

---

[2]Note that the predictor variables are $\sqrt{\text{age}}$ and $\sqrt{\text{age}}^2 = \text{age}$.

We then continue this process to compute the MSE for the jth validation set, which is the jth group in the random splitting of the data. After computing $\text{MSE}_{[j]}$ for $j = 1, 2, \ldots, k$, we compute the average MSE as

$$\text{MSE}_{k\text{-fold}} = \frac{1}{k} \sum_{j=1}^{k} \text{MSE}_{[j]}.$$

> **Definition 19.4.1 — *k*-fold Cross-Validation.** The process of splitting the data into k groups and composing the training validation is called *k-fold cross-validation* and is illustrated (for the first two groups) in Figure 19.10.

There is, of course, some variation in the MSEs due to the random nature of the data splitting, but this is typically much less than the variable in the 50/50 split suggested in Section 19.2.

■ **Example 19.7** Apply five-fold cross-validation to the concrete data set.

Solution: The R code for performing five-fold cross-validation is:

```
k5Fold = function( datafr , p , seed )
{
  set.seed(seed)
  n = nrow( datafr )
  k = 5
  remainder = n - k*floor(n/k)
  m = rep( floor(n/k) , k )
  if (remainder > 0 ) for ( i in 1:remainder)  m[i] = m[i] + 1
  folds = matrix( NA , nrow=m[1] , ncol=k )
  cum.group = c()
  pool = 1:n
  pool0 = 1:n
  group = sort( sample( pool , m[1] ) )
  folds[1:m[1],1] = group
  cum.group = c( cum.group , group )
  for ( j in 2:k )
  {
      pool = pool0[ - cum.group ]
      group = sort( sample( pool , m[j] ) )
      folds[1:m[j],j] = group
      cum.group = sort( c( cum.group , group ) )
  }
  predictors = names( datafr )[2:(p+1)]
  formla = paste( names(datafr)[1] , " ~ " ,
  paste( predictors , collapse=" + ") )
  MSE.vec = rep( NA , k )
  for ( j in 1:k )
  {
      training = datafr[ - folds[,j] , 1:(p+1) ]
      validation = datafr[ folds[,j] , 1:(p+1) ]
      mod = lm( formla , data=training )
      pred = predict( mod , validation )
      MSE.vec[j] = sum( (validation[,1] - pred)^2 ) / m[j]
```

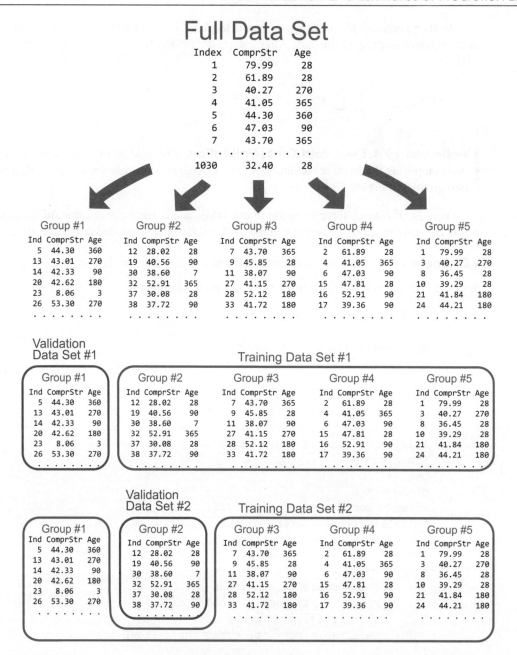

Figure 19.10 k-fold cross-validation. The full data set is divided into k data sets of roughly equal size. Here $k = 5$. The training data sets are obtained by successively leaving out group #1, group #2, etc., while the "left-out" group becomes the validation data set.

```
    }
    return( sum(MSE.vec)/k )
}
```

The function k5fold() takes arguments (1) the data frame on which to apply k-fold cross-validation; (2) the order of the polynomial; and (3) the random seed to set for the simulation. Being able to specify the seed is important because we must use the *seed* to get the same split of the data into groups.

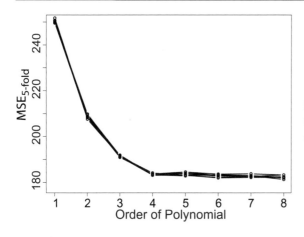

Figure 19.11 Five-fold cross-validation MSEs for the concrete data set from 10 random splits of the data.

The following code performs `nsim=10` random splits of the data, calls the `k5fold()` function, and plots the result. Note that the seed is defined once for each of the 10 simulations, assuring that the data splits are the same.

```
nsim = 10
MSE.mat = matrix( NA , nrow=nsim , ncol=8 )

for( i in 1:nsim ) {
seed = sample( 1:100000 , 1 )
for( j in 1:8 ) {
MSE.mat[i,j] = k5Fold( datafr , j , seed )
} }

plot( 1:8 , MSE.mat[1,] , type="b" , col=rgb(1,0,0.5) )
for( j in 1:8 ) points( 1:8, MSE.mat[j,], col=rgb(1,j/20,j/8), type="b" )
```

Figure 19.11 shows the MSE for 10 five-fold random splits of the data. We see that there is very little variability from the different data set splits. The MSE settles down to about 183 for polynomials of order $p = 4$ or more. ■ ■ ■

It is worth noting that the various cross-validation schemes (50/50, LOOCV, and k-fold cross-validation) lead to basically the same conclusion: that a polynomial of order $p = 4$ is adequate to predict the concrete compression strength.

19.5 Cross-Validation with Classification Data

In Section 18.6 we discussed the issue of classifying a continuous multivariate observation $(x_1, x_2, \ldots, x_p)$ into one of K groups. At the time, we used all of the data to make an inference about which group each observation was in. Comparing the inferred group to the actual group is problematic when the data point we are predicting was used to fit the model. In the language of this chapter, we should use the training data set to infer the group that contains each observation in the validation data set. To perform cross-validation with classification data, we need a measure of the fit of data to a model. The most common measure is the accuracy, defined to be the number of correct classifications divided by the number of predictions made. When there are only two categories, which we will call "positive" and "negative," we can alternatively use the sensitivity (the proportion of the positives that are correctly classified) or the sensitivity (the proportion of negatives that are correctly satisfied). We will assume for now that accuracy is the appropriate measure and define

$$\text{LOOCV}_{(n)} = \frac{1}{n} \sum_{i=1}^{n} \text{Accuracy}_i, \tag{19.6}$$

where Accuracy$_i$ is 1 if the "left out" observation is predicted correctly and 0 otherwise. When using LOOCV with regression data, we have available the shortcut described in (19.4), which obviated the need for rerunning the regression for each point removed. There is no shortcut procedure for classification problems, so the classification model must be run once for each point excluded.

Analogously, we can perform k-fold cross-validation by computing

$$\text{Accuracy}_{[j]} = \text{Proportion of correct classifications in validation group } j \tag{19.7}$$

for each of the training/validation splits of the data. We then average these accuracy values across all five splits or folds of the data:

$$\text{Accuracy}_{k\text{-fold}} = \frac{1}{k} \sum_{j=1}^{k} \text{Accuracy}_{[j]}. \tag{19.8}$$

Again, we could substitute sensitivity or specificity for accuracy if desired.

■ **Example 19.8**   Apply five-fold cross-validation to the banknote data from Example 18.19 using

1. linear discriminant analysis;
2. quadratic discriminant analysis;
3. first-order logistic regression; and
4. second-order logistic regression

Apply these techniques to the accuracy, sensitivity, and specificity.

Solution: We applied five-fold cross-validation by splitting the data into five groups. We fit each model on each of the five training data sets. The training data set for the first fold consisted of 1,096 banknotes, of which 606 were forged and 490 were genuine. The validation data set contained 275 banknotes, of which 155 were forged and 120 were genuine. After fitting the LDA model to the training data set, we made predictions to each of the 275 banknotes in the validation data set. Among the 155 forged documents, 146 were correctly classified as forged and 9 were incorrectly classified as genuine. Since the goal of the analysis is to detect fraudulent banknotes, we will define sensitivity to be the proportion of forged documents classified as forged. Similarly, we define specificity to be the proportion of genuine banknotes classified as genuine. For the first fold, the sensitivity is thus $146/155 \approx 0.942$. Among the 120 genuine banknotes, all 120 were correctly classified as genuine, yielding a specificity of 1.000. Overall, $146 + 120 = 266$ of the 275 documents were correctly classified, which gives an accuracy of 0.967. Figure 19.12 summarizes these results from the first fold. The results from all five folds are computed in a similar manner, and the results are summarized in Table 19.3.

All algorithms have a specificity that is close to 1, indicating that they can do an excellent job detecting genuine documents. The sensitivity values mostly exceed 0.95 so the method does a good job detecting forged documents. The best algorithm among the four seems to be the second-order logistic regression for which all three measures – accuracy, sensitivity, and specificity – are all 1.000.

Here we have used all four variables to make the classification of forged vs. genuine. In Example 18.19 we used only the first two predictors to make classifications and we found considerably lower accuracy, sensitivity, and specificity. This was true even though Example 18.19 used all of the data to make predictions. The results of the present example suggest that in four dimensions the genuine and forged documents are more easily classified. This demonstrates the power of multivariate analysis.

■ ■ ■

Table 19.3 LDA, QDA, first-order logistic regression, and second-order logistic regression applied to each of the five folds in the banknote data set.

Linear discriminant analysis

Measure	Fold 1	Fold 2	Fold 3	Fold 4	Fold 5
Accuracy	0.967	0.982	0.989	0.967	0.967
Sensitivity	0.942	0.965	0.979	0.942	0.945
Specificity	1.000	1.000	1.000	1.000	1.000

Quadratic discriminant analysis

Measure	Fold 1	Fold 2	Fold 3	Fold 4	Fold 5
Accuracy	0.978	0.993	1.000	0.978	0.978
Sensitivity	0.961	0.986	1.000	0.961	0.964
Specificity	1.000	1.000	1.000	1.000	1.000

First-order logistic regression

Measure	Fold 1	Fold 2	Fold 3	Fold 4	Fold 5
Accuracy	0.985	0.989	0.996	0.985	0.989
Sensitivity	1.000	0.986	1.000	0.981	0.982
Specificity	0.967	0.992	0.992	0.992	1.000

Second-order logistic regression

Measure	Fold 1	Fold 2	Fold 3	Fold 4	Fold 5
Accuracy	1.000	1.000	1.000	1.000	1.000
Sensitivity	1.000	1.000	1.000	1.000	1.000
Specificity	1.000	1.000	1.000	1.000	1.000

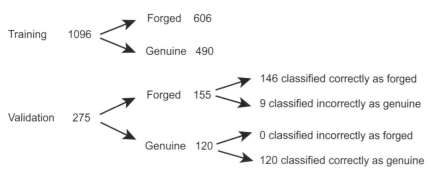

Figure 19.12 LDA applied to fold #1 in banknote data. Sensitivity is the proportion of forged documents classified as forged, and specificity is the proportion of genuine documents classified as genuine.

Exercise 19.2 Apply LOOCV to the banknote data using LDA, QDA, first-order logistic regression, and second-order logistic regression.

19.6 Chapter Summary

Models can be *overfit*, meaning that the number of predictors (that could include powers or products of other predictor variables) can be increased and the data will fit the model very well. In situations like this, predictions for future observations are often poor even though the fit for observations used to fit the model is very good. One remedy is to split the data into two parts:

1. a *training* set, which is used to fit the model; and

2. a *validation* set, which is used to test the predictive ability of the model.

We use the fitted model from the training set and use it to make predictions for the validation set. What makes this *cross-validation* work is that the data in the validation data set are not used to fit the model. Overfitting is thus penalized when predicting the observations in the validation data set. We can measure the fit in the validation data set by looking at the sum of squared prediction errors; smaller values indicate a better fit.

There are a number of strategies for performing this training/validation split. One way is to split the data roughly 50/50 into the training and validation data sets. Since this introduces variability we will often want to do this 50/50 split many times and average the sum of squared prediction errors from each. Another approach is to leave a single observation out, fit the model, and then make a prediction for the omitted observation. This is repeated for each observation in the data set and the sum of squared prediction errors can be computed. This is called leave-one-out cross-validation. A third approach is to divide the data into k different groups (called *folds*). Each fold is held out as the validation set and the model is fit on the remaining data. Predictions are then made for observations in the fold that is held out. This is done for each of the k folds.

Similar ideas can be applied when we are classifying objects using the methods from Chapter 18. The criteria for assessing the model fit can be the prediction accuracy (percent of correct predictions), the sensitivity, or the specificity.

19.7 Problems

Problem 19.1 Consider the data in Table 19.4.

Table 19.4 Data for Problem 19.1.

x	y
0	5
1	6
2	8
3	18
4	21

(a) Plot x and y along with predicted values for a linear model over the grid $x = 0, 0.1, 0.2, \ldots, 4.0$.

(b) Repeat part (a) for a second-order model.

(c) Repeat part (a) for a third-order model.

(d) Repeat part (a) for a fourth-order model.

(e) What happens if you try to fit a fifth-order model?

Problem 19.2 For each of the models in parts (a) through (d) in Problem 19.1:

(a) Make a point prediction for the outcome y when $x = 2$.

(b) Make a point prediction for the outcome y when $x = 3$.

(c) Make a point prediction for the outcome y when $x = 3.5$.

(d) For the predictions made in part (c) for models of order one through four, which prediction would you trust the most?

Problem 19.3 Consider the data in Table 19.5.

Table 19.5 Data for Problem 19.3.

x	y
5	0.10
8	0.05
9	0.05
12	0.09
14	0.20
15	0.22

(a) Plot x and y along with predicted values for a linear model over the grid $x = 5.0, 5.1, 5.2, \ldots, 15.0$.

(b) Repeat part (a) for a second-order model.

(c) Repeat part (a) for a third-order model.

(d) Repeat part (a) for a fourth-order model.

(e) Repeat part (a) for a fifth-order model.

(f) What happens if you try to fit a sixth-order model?

Problem 19.4 For each of the models in parts (a) through (e) in Problem 19.3:

(a) Make a point prediction for the outcome y when $x = 6$.

(b) Make a point prediction for the outcome y when $x = 8$.

(c) Make a point prediction for the outcome y when $x = 10$.

(d) For the predictions made in part (c) for models of order one through four, which prediction would you trust the most?

Problem 19.5 The file `FirstYearGPA.csv` contains data on first-year college GPA along with several high school measures of performance. Use high school GPA, found in the variable `HSGPA`, to predict the first-year college GPA, found in the variable `GPA`.

(a) Plot `HSGPA` and `GPA` along with predicted values for a linear model over the grid `HSGPA` = $2.0, 2.1, 2.2, \ldots, 4.0$.

(b) Repeat part (a) for a second-order model.

(c) Repeat part (a) for a third-order model.

(d) Repeat part (a) for a fourth-order model.

Problem 19.6 For each of the models in parts (a) through (d) in Problem 19.5:

(a) Make a point prediction of `GPA` for `HSGPA` = 2.5.

(b) Make a point prediction of `GPA` for `HSGPA` = 3.0.

(c) Make a point prediction of `GPA` for `HSGPA` = 3.5.

Problem 19.7 Consider the cost of manufacture and the lot size data described in Section 14.12.

(a) Plot `LotSize` and `Cost` along with predicted values for a linear model over the grid `LotSize` = $1, 2, \ldots, 100$.

(b) Repeat part (a) for a second-order model.

(c) Repeat part (a) for a third-order model.

(d) Repeat part (a) for a third-order model.

Problem 19.8 For each of the models in parts (a) through (d) in Problem 19.7:

(a) Make a point prediction of `Cost` for `LotSize` = 40.

(b) Make a point prediction of `Cost` for `LotSize` = 55.

(c) Make a point prediction of `Cost` for `LotSize` = 88.

Problem 19.9 For the data in Problem 19.3, apply five 50/50 splits of the data into training and validation data sets.

(a) For each of the five random splits, fit first- and second-order models and compute the SSE_{CV} for each.

(b) Compute the average of the SSE_{CV} for the five simulations in part (a).

(c) Which model is better at making predictions for the validation set?

(d) What happens if you try to fit a third-order model to the training data set?

Problem 19.10 For the Cost vs. LotSize data from Problem 19.7, apply ten 50/50 splits of the data into training and validation data sets.

(a) For each of the 10 random splits, fit first-, second-, third-, and fourth-order models and compute the SSE_{CV} for each.

(b) Compute the average of the SSE_{CV} for the 10 simulations in part (a).

(c) Which model is best at making predictions for the validation set?

Problem 19.11 Montgomery (2020a) describes a manufacturing process for a single-wafer etching tool. Energy is supplied by an radiofrequency (RF) generator and interest is focused in the relationship between the RF power setting and the etch rate. Data from a designed experiment in which the RF power was varied from 160 to 220 are shown in Table 19.6. Apply ten 50/50 splits of the data into training and validation data sets.

Table 19.6 Data for Problem 19.11.

RFPower	EtchRate	RFPower	EtchRate
160	575	200	600
160	542	200	651
160	530	200	610
160	539	200	637
160	570	200	629
180	565	220	725
180	593	220	700
180	590	220	715
180	579	220	685
180	610	220	710

(a) For each of the 10 random 50/50 splits, fit first-, second-, third-, and fourth-order models and compute the SSE_{CV} for each.

(b) Compute the average of the SSE_{CV} for the 10 simulations in part (a).

(c) Which model is best at making predictions for the validation set?

Problem 19.12 Consider the data from Table 18.5. This is a simplified data set from a much larger study on the effect of pregnancies and age on the glucose level. Fit the following models, and for each compute SSE_{CV} for five different 50/50 splits of the data.

(a) A first-order model in number of pregnancies.

(b) A first-order model in age

(c) A first-order model in age and number of pregnancies.

(d) A first-order model in the number of pregnancies and a second-order model in age.

(e) A first-order model in age and a second-order model in number of pregnancies.

(f) A first-order model in age and number of pregnancies with an added interaction term.

(g) What happens when you try to run a full second-order model in age and the number of pregnancies?

(h) Which model is preferred?

Problem 19.13 Consider again the hospital patient satisfaction data in Problem 14.16. Apply ten 13/12 splits of the data into training and validation data sets. For each of the following models, make predictions for the validation set and compute the SSE_{CV} for each.

(a) A first-order model using age.

(b) A first-order model using severity.

(c) A first-order model using anxiety.

(d) A first-order model using age and severity.

(e) A first-order model using age and anxiety.

(f) A first-order model using severity and anxiety.

(g) A first-order model using age, severity, and anxiety.

(h) A first-order model using age, severity, anxiety, and surgical/medical.

(i) A second-order model using age.

(j) A second-order model using anxiety.

(k) A second-order model using age and severity.

(l) A second-order model using age and anxiety.

(m) A second-order model using severity and anxiety.

(n) A second-order model using age, severity, and anxiety.

(o) Which model would you recommending for predicting hospital satisfaction?

Problem 19.14 Consider the data set from Problem 19.5. Apply ten 110/109 splits of the data into training and validation data sets. For each of the following models, make predictions for the validation set and compute the SSE_{CV} for each.

(a) Use the variables HSGPA, SATV, SATM, HU, and SS one at a time in a first-order linear model with GPA as the outcome.

(b) For each pair of variables, fit a full second-order model; i.e., a model of the form

$$y_i = \beta_0 + \beta_1 x_{i1} + \beta_2 x_{i2} + \beta_{11} x_{i1}^2 + \beta_{22} x_{i2}^2 + \beta_{12} x_{i1} x_{i2}.$$

(c) Fit the first-order model using all of the variables HSGPA, SATV, SATM, HU, and SS. Compute SSE_{CV} for this model.

(d) Which model is preferred?

Problem 19.15 Consider the data on seizures of 59 patients given in Problem 18.16. Apply Poisson regression to 10 splits of the data, with 30 chosen at random to be the training set and the remaining 29 to be the validation set. The outcome variable is the number of seizures in the eight-week follow-up period. For each split compute the SSE_{CV} for the following models.

(a) A first-order model with just the drug.

(b) A first-order model with just the patient's age.

(c) A first-order model with just the baseline number of seizures.

(d) A first-order model with the drug and age.

(e) A first-order model with the drug and baseline number of seizures.

(f) A first-order model with age and baseline number of seizures.

(g) A first-order model with all three predictors.

(h) A first-order model in the drug and a second-order model in age.

(i) A first-order model in the drug and age, and a second-order model in the baseline number of seizures.

(j) A first-order model in the drug and baseline number of seizures, and a second-order model in age.

(k) A first-order model in the drug and a full second-order model in age and baseline number of seizures.

(l) Which model is preferred?

Problem 19.16 Suppose you have five continuous variables and would like to consider models of third order and below. Recall the *hierarchy principle* from Section 12.6 which says that if a variable is included in a model, then all lower-order terms are included in a model. For example, if a model includes $x_1^2 x_2^2$, then x_1, x_2, $x_1 x_2$, x_1^2, x_2^2, $x_1^2 x_2$, and $x_1 x_2^2$ should be in the model. How many models must you fit?

Problem 19.17 Rework Problem 19.9 using the method of LOOCV.

Problem 19.18 Rework Problem 19.10 using the method of LOOCV.

Problem 19.19 Rework Problem 19.11 using the method of LOOCV.

Problem 19.20 Rework Problem 19.12 using the method of LOOCV.

Problem 19.21 Rework Problem 19.13 using the method of LOOCV.

Problem 19.22 Rework Problem 19.14 using the method of LOOCV.

Problem 19.23 Rework Problem 19.15 using the method of LOOCV.

Problem 19.24 Rework Problem 19.10 using three-fold cross-validation.

Problem 19.25 Rework Problem 19.11 using four-fold cross-validation.

Problem 19.26 Rework Problem 19.12 using five-fold cross-validation.

Problem 19.27 Rework Problem 19.14 using five-fold cross-validation.

Problem 19.28 Rework Problem 19.15 using five-fold cross-validation.

Problem 19.29 Consider the `mtcars` data set from Problem 18.31. In that problem you performed LDA to classify whether a car was built by an American company. Perform five 50/50 splits of the data and perform LDA on the training set. Compute the average accuracy, sensitivity, and specificity for the validation sets.

Problem 19.30 Consider the data on male penguins and the models given in Problem 18.32. Perform five 50/50 splits of the data and perform LDA on the training set. Compute the average accuracy for the validation sets.

Problem 19.31 Consider the data on male penguins and the models given in Problem 18.33. Perform five 50/50 splits of the data and perform LDA on the training set. Compute the average accuracy for the validation sets.

Problem 19.32 Consider the data on male penguins and the models given in Problem 18.34. Perform five 50/50 splits of the data and perform LDA on the training set. Compute the average accuracy for the validation sets.

Problem 19.33 Consider the data on hawks and the models given in Problem 18.37. Perform five 50/50 splits of the data and perform LDA on the training set. Compute the average accuracy for the validation sets.

Problem 19.34 Consider the `mtcars` data from Problem 19.29. Perform five 50/50 splits of the data and perform QDA on the training set to predict whether the car was built by an American manufacturer. Compute the average accuracy, sensitivity, and specificity for the validation sets.

Problem 19.35 Consider the male penguin data from Problem 19.30. Perform five 50/50 splits of the data and perform QDA on the training set to predict the species. Compute the average accuracy for the validation sets.

Problem 19.36 Consider the car purchasing data from Problems 18.5 and 18.39. Perform five 50/50 splits of the data and perform a first-order logistic regression on the training set. For each family in the validation data set, infer whether the family will purchase a new car by taking the category with the higher probability. Compute the average accuracy, sensitivity, and specificity for the validation sets.

Problem 19.37 Consider the mtcars data from Problem 19.29. Perform five 50/50 splits of the data and perform a first-order logistic regression on the training set. For each car in the validation data set, infer whether the car was made by an American company by taking the category with the higher probability. Compute the average accuracy, sensitivity, and specificity for the validation sets.

Problem 19.38 Consider mtcars data from Problem 19.29. Perform five 50/50 splits of the data. Use the variables hp and qsec to perform a second-order (with interaction term) logistic regression. For each car in the validation data set, infer whether the car was made by an American company by taking the category with the higher probability. Compute the average accuracy, sensitivity, and specificity for the validation sets.

Problem 19.39 Repeat Problem 19.29, except apply the method of LOOCV.

Problem 19.40 Repeat Problem 19.30, except apply the method of LOOCV.

Problem 19.41 Repeat Problem 19.31, except apply the method of LOOCV.

Problem 19.42 Repeat Problem 19.32, except apply the method of LOOCV.

Problem 19.43 Repeat Problem 19.33, except apply the method of LOOCV.

Problem 19.44 Repeat Problem 19.34, except apply the method of LOOCV.

Problem 19.45 Repeat Problem 19.35, except apply the method of LOOCV.

Problem 19.46 Repeat Problem 19.36, except apply the method of LOOCV.

Problem 19.47 Repeat Problem 19.37, except apply the method of LOOCV.

Problem 19.48 Repeat Problem 19.38, except apply the method of LOOCV.

Problem 19.49 Repeat Problem 19.29, except apply the method of four-fold cross-validation.

Problem 19.50 Repeat Problem 19.30, except apply the method of five-fold cross-validation.

Problem 19.51 Repeat Problem 19.31, except apply the method of five-fold cross-validation.

Problem 19.52 Repeat Problem 19.32, except apply the method of five-fold cross-validation.

Problem 19.53 Repeat Problem 19.33, except apply the method of five-fold cross-validation.

Problem 19.54 Repeat Problem 19.34, except apply the method of four-fold cross-validation.

Problem 19.55 Repeat Problem 19.35, except apply the method of five-fold cross-validation.

Problem 19.56 Repeat Problem 19.36, except apply the method of four-fold cross-validation.

Problem 19.57 Repeat Problem 19.37, except apply the method of four-fold cross-validation.

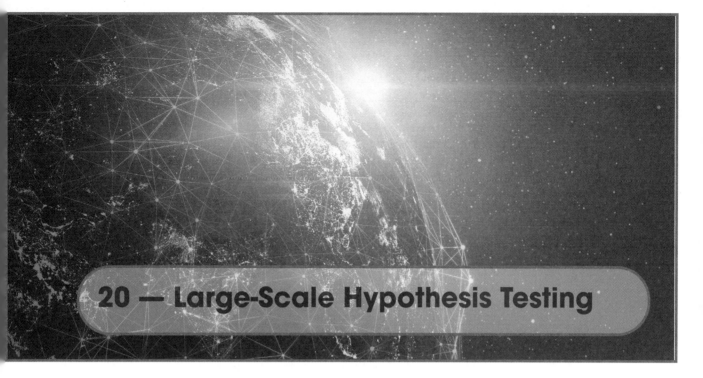

20 — Large-Scale Hypothesis Testing

20.1 Review of Hypothesis Testing

We begin this chapter with a review of hypothesis testing from Chapter 12. A hypothesis is a statement about one or more parameters of a model. The null hypothesis is usually a specific statement that encapsulates "no effect." For example, if we apply one of the two treatments, A or B, to volunteers[1] we may be interested in testing whether the population mean outcomes are equal. Label these mean outcomes as μ_A and μ_B. We might assume that the samples taken from the two groups are normally distributed:

$$X_1, X_2, \ldots, X_{n_1} \sim \text{i.i.d. } N(\mu_A, \sigma^2) \qquad \text{and} \qquad Y_1, Y_2, \ldots, Y_{n_2} \sim \text{i.i.d. } N(\mu_B, \sigma^2).$$

In some situations it may be more reasonable to allow the population variances to differ, say with values σ_A^2 and σ_B^2. It may also be the case that a transformation of the data is necessary to achieve something close to a normal distribution. These are important issues to address, but they divert attention from the topic of this section. For now, we'll assume that these assumptions are reasonable.

The null hypothesis is H_0: $\mu_A = \mu_B$ and we will take our alternative to be H_1: $\mu_A \neq \mu_B$. One-sided alternatives are possible, such as H_1: $\mu_A < \mu_B$ or H_1: $\mu_A > \mu_B$. Usually, the default choice is the two-sided alternative unless there is a compelling reason to use a one-sided alternative. The second step would involve choosing a test statistic. In this situation, the obvious test statistic is

$$T = \frac{\overline{X} - \overline{Y}}{S_p\sqrt{\dfrac{1}{n_1} + \dfrac{1}{n_2}}},$$

where $\overline{X}$ and $\overline{Y}$ are, respectively, the sample means from treatments A and B, and S_p^2 is the pooled variance:

---

[1]One of these treatments could be a placebo.

$$S_p^2 = \sqrt{\frac{\sum_{i=1}^{n_1}(X_i - \overline{X})^2 + \sum_{j=1}^{n_2}(Y_j - \overline{Y})^2}{n_1 + n_2 - 2}}.$$

We showed in Section 12.2 that $T \sim t(n_1 + n_2 - 2)$ when H_0 is true. We should use the computed value of the test statistic, along with the test statistic's distribution under H_0. If H_0 is true, we would expect $\overline{X}$ and $\overline{Y}$ to be close, and therefore we would expect T to be near 0. When the magnitude of the observed value of T is large, we are led to believe that H_0 is false. The question is, "How large is *large*?" We answer this question by fixing the probability of rejecting H_0 when it is true (a Type I error) to be some small predetermined number of α. We therefore choose the cutoff, or critical, values for "large" to be the lower and upper $\alpha/2$ quantiles of the t distribution. Our decision is thus:

$$\text{Reject } H_0 \text{ if } T < -t_{\alpha/2}(n_1 + n_2 - 2) \qquad \text{or} \qquad T > t_{\alpha/2}(n_1 + n_2 - 2).$$

The steps are summarized here.

Algorithm 1   Classical hypothesis testing for a fixed α.

1. Choose the null and alternative hypotheses to test.
2. Determine a test statistic T and its distribution when H_0 is true.
3. Determine whether small or large (or both) values of the test statistic give evidence against H_0. Choose the set $\mathcal{C}$ of test statistics so that $\Pr(T \in \mathcal{C}) = \alpha$.
4. Using the observed data, compute the test statistic.
5. If the computed test statistic falls in $\mathcal{C}$, reject H_0; otherwise, fail to reject H_0.

The P-value approach is different. Instead of selecting a value for α, we compute the P-value, defined to be the probability of getting a test statistic as extreme or more extreme than what was observed given H_0 is true. Alternatively, the P-value can be defined as the largest value of α for which we would fail to reject H_0. We then let the P-value stand as evidence against the null hypothesis, with smaller P-values giving stronger evidence against H_0. The P-value approach can be summarized as follows.

Algorithm 2 P-value approach to hypothesis testing.

1. Choose the null and alternative hypotheses to test.
2. Determine a test statistic T and its distribution when H_0 is true.
3. Using the data, compute the observed test statistic.
4. Compute the P-value.
5. Report the P-value.

■ **Example 20.1**   Two members of a research team have suggested different ways to report the results of their study. One member chose $\alpha = 0.05$, computed the test statistic, compared it to the critical region, and rejected the null hypothesis. The other member of the team computed the P-value to be 0.04 and reported this. What are the advantages and disadvantages of these two approaches?

Solution: These two approaches are consistent. A reader of the research report who is knowledgeable in statistics would know that a P-value of 0.04 would lead to rejection at the null hypothesis at the $\alpha = 0.05$ level. Both approaches would convey essentially the same information.   ■ ■ ■

■ **Example 20.2**   Two members of a research team have suggested different ways to report the results of their study. One member chose $\alpha = 0.05$, computed the test statistic, compared it to the critical region, and rejected the null hypothesis. The other member of the team computed the P-value to be 0.00008 and reported this. What are the advantages and disadvantages of these two approaches?

Solution: Again, both methods are consistent in the sense that a P-value of 0.00008 would lead to rejecting H_0 at the level $\alpha = 0.05$. In this case, however, the P-value approach has a distinct advantage. For this situation, the null hypothesis would also be to reject at the levels $\alpha = 0.01, 0.005$, and 0.001, and even $\alpha = 0.0001$. Rejecting H_0 at such small α levels indicates strong evidence against H_0. Reporting just that the null hypothesis was rejected at the $\alpha = 0.05$ is grossly understating the available evidence.

■ ■ ■

The last example illustrates an advantage of the P-value approach. Readers can make up their own mind about whether to reject H_0 if you report the P-value. If the P-value is smaller than the α they have in mind, they would reject H_0; otherwise they wouldn't reject H_0. Many researchers follow the P-value approach when reporting results of a study since it allows readers to apply their own α values when deciding on significance.

If we reject a null hypothesis that is false, then we have made a correct decision. Similarly, if we fail to reject a true null hypothesis we have made a correct decision. The other two possibilities lead to errors. Rejecting the null hypothesis when it is true is a Type I error, and failing to reject the null hypothesis when it is false is a Type II error. The four possibilities are shown below:

		Reality	
		H_0 is true	H_0 is false
Decision	Reject H_0	Type I error	No error
	Do not reject H_0	No error	Type II error

The usual procedure for hypothesis testing is to choose the decision rule so that

$$\Pr\left(\text{reject } H_0 \mid H_0 \text{ is true}\right) = \alpha.$$

In other words, the predetermined number α is the probability of making a Type I error. The probability of making a Type II error depends on how much the parameter(s) deviates from the null hypothesis. For example, if H_0: $\mu_A = \mu_B$, then the probability of failing to reject H_0 will be much greater if $\mu_A - \mu_B = 10$ than if $\mu_A - \mu_B = 1$. This leads to the topic of *power*, which was the topic of Section 11.3.

The important point here is that we are testing a *single* null hypothesis. Simultaneously testing multiple null hypotheses requires special attention.

20.2 Testing Multiple Hypotheses

When we test multiple null hypotheses we have the opportunity to make many Type I (and Type II) errors. We will illustrate the hazards of multiple testing with a famous data set on prostate cancer. The data set involves 50 men without cancer and 52 men with cancer. The study looked at gene activity for each man (with or without prostate cancer) for each of 6,033 genes. We define the outcome y_{ij} to be the activity of gene i on person j. This is a numeric variable that for most genes is between about -2 and $+5$. The goal is to discover which genes, among the many that were studied, act differently on men with and without prostate cancer. This might provide a genetic test to look for the likelihood of developing prostate cancer. If a man's gene activity (for a particular gene) was similar to the activity of men with cancer, we would infer that the man has a high chance of developing cancer. If the man's gene activity for the same gene was similar to those in the control group (no cancer), then we would infer a low chance of developing prostate cancer. If we could find several genes whose activity levels differ between those with cancer and those without, we might be able to do a better job of discriminating between men who are and are not likely to develop prostate cancer. In other words, these genes can give us *biomarkers* for prostate cancer.

Table 20.1 Table showing the number of Type I errors, Type II errors, and no errors in the terminology of rejecting and failing to reject $H_{0(i)}$, $i = 1, 2, \ldots, m$.

		Reality		
		H_0 is true	H_0 is false	Total
Decision	Reject H_0	V	S	R
	Fail to reject H_0	U	W	$m - R$
	Total	m_0	$m - m_0$	m

Suppose, for now, that we test with $\alpha = 0.05$ each $H_{0(i)}$: $\mu_{i1} = \mu_{i2}$, where μ_{i1} and μ_{i2} are the mean gene activity responses for the control and cancer groups, respectively, for gene i. For this data set we have 102 men (50 in the control group and 52 in the cancer group) and 6,033 genes. We are therefore testing 6,033 null hypotheses! If we were to test 6,033 null hypotheses, each with $\alpha = 0.05$, and if *all* of the null hypotheses are true, then we would expect to reject about 5% of them by chance alone. Thus, we'd expect to make about $6,033 \times 0.05 \approx 302$ Type I errors. To illustrate this conundrum, suppose the P-value for one particular gene was 0.04. We have to ask: Is this one of the many null hypotheses we reject by chance? Or is this gene really a biomarker for prostate cancer?

An obvious solution is that we have to raise the bar to make it harder to reject any particular null hypothesis.[2] One criterion for how much to lower the P-value threshold is to use the *family-wise error rate* (FWER).

Definition 20.2.1 — Family-Wise Error Rate. The FWER is

$$\text{FWER} = \Pr\left(\text{rejecting at least one of the true null hypotheses}\right). \tag{20.1}$$

We will find it helpful to consider the number of times each of the four possible outcomes occurs. These are shown in Table 20.1. Using the notation from this table, we can write the FWER as $\Pr(V \geq 1)$. Suppose that we are testing m null hypotheses, each with a fixed level of α. The FWER for this α is then

$$\text{FWER}(\alpha) = 1 - \Pr(V = 0) = 1 - \Pr(\text{we fail to reject each } H_0).$$

If the tests are independent (a strong assumption), and if we assume that all null hypotheses are true, then

$$\text{FWER}(\alpha) = 1 - \Pr\left(\text{we fail to reject } H_{0(1)} \text{ and } \ldots \text{ and we fail to reject } H_{0(m)}\right)$$

$$= 1 - \prod_{i=1}^{m}(1 - \alpha)$$

$$= 1 - (1 - \alpha)^m.$$

We would like the FWER to be small. For $\alpha = 0.05$ and $m = 10$, FWER $= 0.401$, and for $m = 100$, FWER $= 0.9941$. Thus, with $m = 100$ tests we expect to make about five Type I errors (since $m\alpha = 0.05$) and we're almost certain to make at least one Type I error. With $m = 6,033$, the probability of making at least one Type I error is FWER $= 1 - 0.95^{6,033} \approx 1 - 4.0 \times 10^{-135}$, which is virtually 1. Note that if $m = 1$ then the FWER and the Type I error rate α agree since $1 - (1 - \alpha)^1 = \alpha$.

---

[2]By "raising the bar" we mean increasing the critical T value to require stronger evidence to reject H_0. This is equivalent to *lowering* the P-value threshold for rejecting H_0.

■ **Example 20.3**   For the `prostate` microarray data, compute the t statistics for the first four genes and summarize the results. If you test each null hypothesis of no difference in gene activity between the control group and the cancer group using $\alpha = 0.05$, how many of the four null hypotheses will you reject?

Solution: The following R code loads in the data and prints parts of the first four rows:

```
prostate = read.csv( "prostmat.csv" )
head( prostate , 4 )
```

The data file has 6,033 rows, one for each gene whose activity is tested, and 102 columns, one for each man in the study. The `head()` command prints a lot of output, some of which is shown below:

```
       control  control.1   control.2   control.3 control.4   control.5
1 -0.93089516 -0.7518854 -0.545781228 -1.07851867 -0.9946767  0.01554727
2 -0.83999635 -0.8482712 -0.851687276 -0.15961030 -0.7519047 -0.51643585
3  0.06250802  0.1028951 -0.003043134  0.21534653 -1.1631092  1.02812986
4 -0.36159390  2.4210344 -0.122089373 -0.09627747 -1.1301438  0.45827185


     ...     ...      ...       ...       ...      ...     ...     ...     ...     ...

     cancer.46  cancer.47 cancer.48 cancer.49    cancer.50   cancer.51
1  0.4348645  2.0541222 2.7994978  1.294162  2.90558839  3.4345041
2  0.6758063 -0.4508528 1.3857200 -1.143997 -0.28211918 -1.1742330
3 -0.2353566 -1.0496002 1.1865992  0.962634 -0.02675318  1.5335323
4  1.4912386  0.5530220 0.1184763  1.220066 -1.13865102  0.1748307
```

Note that in this data frame, each row corresponds to a gene and each column to a person. This goes against the rule-of-thumb that, when we have data for human subjects, each row corresponds to a person. Notice also, the variable naming convention: The first column is `control`, the second is `control.1`, and the 52nd is `control.51`. Similarly for the cancer group. We can extract the control group's response to gene number one by taking the row 1 and columns 1 through 50; we can do the same for the cancer group by taking row 1 and columns 51–102. Here is the R code to accomplish this:

```
x1 = as.numeric( prostate[ 1 , 1:50 ] )
y1 = as.numeric( prostate[ 1 , 51:102 ] )
```

This can be done for each of the first four genes (or for all 6,033 of them). The two-sample t-test assuming equal variances is performed using the R command

```
t.test( x1 , y1 , var.equal=TRUE )
```

This yields the output for the first gene

```
        Two Sample t-test

data:  x1 and y1
t = -1.4812, df = 100, p-value = 0.1417
alternative hypothesis: true difference in means is not equal to 0
95 percent confidence interval:
-0.9222718  0.1338032
sample estimates:
mean of x  mean of y
-0.1860100  0.2082243
```

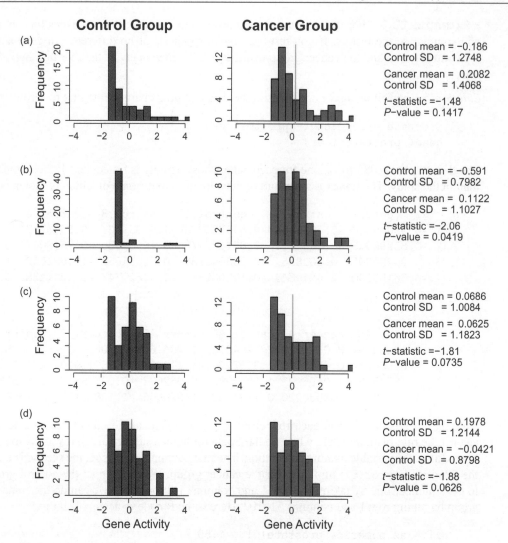

Figure 20.1 *t*-tests for each of the first four of 6,033 genes. The first column shows the histograms of the control group (for the four genes) and the second column shows the histograms for the cancer group. The column on the right gives the summary of the two-sample *t*-test.

With a *P*-value of 0.1417, we would fail to reject the null hypothesis of equal means for the first gene's activity across the two groups (cancer and control). Figure 20.1 shows histograms for the control and cancer groups on the activities of the first four genes. The information on the two-sample *t*-test is shown to the right of the four histograms. The only null hypothesis you would reject is the one on activity 2, which has a *P*-value of 0.0419. Some of the histograms in Figure 20.1 are distinctly nonnormal, and the assumption of equal variance seems to fail in some cases. The *t*-test is, however, fairly robust, especially for large sample sizes.   ∎ ∎ ∎

∎ **Example 20.4**   Compute the *t* statistics for all of the 6,033 genes and summarize.

Solution: The following R code assumes the `prostate` data frame has been loaded. It loops through all 6,033 genes, extracts the control and cancer group's responses to the gene, and computes the *t* statistic.

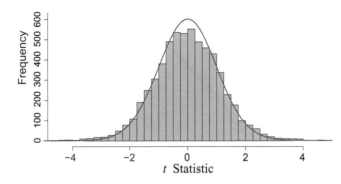

Figure 20.2 Histogram of 6,033 t statistics for comparing gene activity from the control group and the cancer group. The (scaled) $t(100)$ distribution is shown as the smooth curve.

```
m = nrow(prostate)
t.stat.vec = rep( NA , m )
for( i in 1:m )
{
        x = as.numeric( prostate[ i , 1:50 ] )
        y = as.numeric( prostate[ i , 51:102 ] )
        t.stat.vec[i] = t.test( x , y , var.equal=TRUE )$statistic
}
```

Once these 6,033 t statistics have been computed, we can display them in a histogram. This is accomplished by the following R code:

```
hist( t.stat.vec , breaks=seq(-6,6,0.25) , xlim=c(-4.5,4.5) ,
        ylim=c(0,700) , col="skyblue" )
xvec = seq(-5,5,0.1)
yvec = dt(xvec,100) * (0.25*m)
lines( xvec , yvec , lwd=2 , col="blue" )
```

Figure 20.2 superimposes the $t(100)$ distribution[3] on the histogram of the 6,033 t statistics. We see from Figure 20.2 that the t statistics roughly fit the $t(100)$ distribution, but there are some clear discrepancies. There are fewer t statistics near 0 than we would expect. There is an excess of t statistics in the left and right tails. ∎ ∎ ∎

The fact that there are more t statistics in the tails than we would expect is good news! It is these genes that might lead us to a genetic test for prostate cancer. How we determine which genes out of the 6,033 are "significant" is the topic of the rest of this chapter.

A useful result in what follows is that, under the right conditions, when we test a single null hypothesis the corresponding P-value has a uniform distribution.

Theorem 20.2.1 — P-Value is Uniformly Distributed. Suppose we are testing one null hypothesis H_0 where the distribution of the test statistic T is continuous and has CDF F_0 when H_0 is true. Suppose also that the test is one-sided, whereby we reject H_0 if T is sufficiently small. Let T_{obs} denote the observed value of the test statistic. If H_0 is true, then $P \sim \text{UNIF}(0,1)$.

Proof. Since the alternative is one-sided and we reject for small values of T_{obs}, the P-value is $P = F_0(T_{\text{obs}})$. Since T_{obs} is a random variable, so is $F_0(T_{\text{obs}}) = P$. Let p be some number between 0 and 1. Since the distribution of the test statistic T is continuous, its CDF F_0 under H_0 is monotonic increasing over its support, which implies the inverse function F_0^{-1} exists. Thus,

---

[3]The degrees of freedom here is $n_1 + n_2 - 2 = 50 + 52 - 2 = 100$. Also, note that the graph of the $t(100)$ distribution is scaled so that the area of the histogram bars equals the area under the plotted curve.

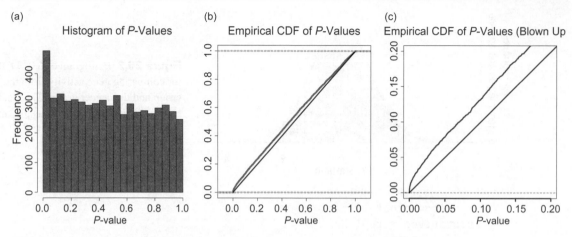

Figure 20.3 Histogram and empirical CDF of 6,033 P-values for prostate cancer microarray.

$$
\begin{aligned}
\Pr(P \le p) &= \Pr\left(F(T_{\text{obs}}) \le p | H_0 \text{ is true}\right) \\
&= \Pr\left(F_0^{-1}(F(T_{\text{obs}})) \le F_0^{-1}(p) | H_0 \text{ is true}\right) \\
&= \Pr\left(T_{\text{obs}} \le F_0^{-1}(p) | H_0 \text{ is true}\right) \\
&= F_0\left(F_0^{-1}(p)\right) \qquad \text{[because } F_0 \text{ is the CDF of } T_{\text{obs}} \text{ when } H_0 \text{ is true]} \\
&= p.
\end{aligned}
$$

Thus, $\Pr(P \le p) = p$ for any number $p \in [0,1]$. But this is the CDF of the UNIF$(0,1)$ distribution so $P \sim \text{UNIF}(0,1)$. ∎

This theorem is true for two-sided tests, but the proof is messier. If we look at all of the 6,033 P-values we see that there are noticeably more that are near zero than we would expect. Figure 20.3(a) shows a histogram of the 6,033 P-values and that there are more of them near zero than we would expect if the distribution were uniform. Figure 20.3(b) shows the empirical CDF (ECDF) of the 6,033 P-values. The ECDF is uniformly above the expected line ($y = p$). Figure 20.3(c) shows a blow-up of the graph in (b) for p between 0 and 0.2. This figure suggests that there are probably some genes that behave differently for the cancer group and the control group. The problem that we address in the rest of this chapter is how to discern which of these genes truly have an effect.

20.3 The FWER and the Bonferroni Correction

When we test multiple null hypotheses, it is often the case that the tests are not independent. Because of this, we must settle for a bound on the FWER. There are a number of ways of doing this, but we begin in this section with the most common method, which is based on Bonferroni's inequality. Before proving Bonferroni's inequality, we state and give a sketch of the proof of Boole's inequality.

> **Theorem 20.3.1 — Boole's Inequality.** If $A_1, A_2, \ldots, A_m$ are events, then
>
> $$
> \Pr\left(\bigcup_{i=1}^{m} A_i\right) \le \sum_{i=1}^{m} \Pr(A_i). \tag{20.2}
> $$

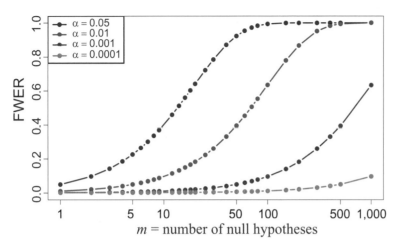

Figure 20.4 Family-wise error rate as a function of the number of (true) null hypotheses being tested and the level of α for each individual test.

Proof. This is actually a sketch of the proof. At the end, we will indicate how a rigorous proof could be given. Let's begin by looking at small values of m.

For $m = 1$, the left side of (20.2) is $\Pr(\bigcup_{i=1}^{1} A_i) = \Pr(A_1)$, which agrees with the right side, $\sum_{i=1}^{1} \Pr(A_i) = \Pr(A_1)$.

For $m = 2$, we begin with the left side of (20.2):

$$
\begin{aligned}
\Pr\left(\bigcup_{i=1}^{2} A_i\right) &= \Pr(A_1 \cup A_2) \\
&= \Pr(A_1) + \Pr(A_2) - \Pr(A_1 \cap A_2) \\
&\leq \Pr(A_1) + \Pr(A_2) \qquad [\text{because } \Pr(A_1 \cap A_2) \geq 0].
\end{aligned}
\tag{20.3}
$$

Thus, (20.2) is true for $m = 2$.

Next, consider $m = 3$. Again, beginning with the left side of (20.2) we have

$$
\begin{aligned}
\Pr\left(\bigcup_{i=1}^{3} A_i\right) &= \Pr(A_1 \cup A_2 \cup A_3) \\
&= \Pr\left((A_1 \cup A_2) \cup A_3\right) \\
&= \Pr(A_1 \cup A_2) + \Pr(A_3) - \Pr\left((A_1 \cap A_2) \cap A_3\right) \\
&\leq \Pr(A_1 \cup A_2) + \Pr(A_3) \qquad [\text{because } \Pr\left((A_1 \cap A_2) \cap A_3\right) \geq 0] \\
&\leq \Pr(A_1) + \Pr(A_2) + \Pr(A_3).
\end{aligned}
$$

The last line follows, because we showed in (20.3) that $\Pr(A_1 \cup A_2) \leq \Pr(A_1) + \Pr(A_2)$.

For $m = 4$, we would group the first three events and then proceed as above to show that

$$
\Pr(A_1 \cup A_2 \cup A_3 \cup A_4) = \Pr\left((A_1 \cup A_2 \cup A_3) \cup A_4\right) \leq \Pr(A_1) + \Pr(A_2) + \Pr(A_3) + \Pr(A_4),
$$

using along the way the result we showed above that $\Pr(A_1 \cup A_2 \cup A_3) \leq \Pr(A_1) + \Pr(A_2) + \Pr(A_3)$.

A rigorous proof would involve mathematical induction. The base case of $m = 1$ was shown above. We would then assume that the result was true for $m = k$ and prove that it is true for $m = k + 1$. As you might expect from the case above where $m = 4$, we would group the first m of the $m + 1$ sets and apply the inductive hypothesis. We leave this as an exercise for the reader (see Problem 20.17). ∎

> **Theorem 20.3.2 — Bonferroni's Inequality.** For any events $A_1, A_2, \ldots, A_m$, whether they are independent or not,
>
> $$\Pr\left(\bigcap_{i=1}^{m} A_i\right) \geq 1 - \sum_{i=1}^{m}\left(1 - \Pr(A_i)\right). \tag{20.4}$$

Proof. The proof is direct and it applies Boole's inequality (Theorem 20.3.1):

$$\Pr\left(\bigcap_{i=1}^{m} A_i\right) = 1 - \Pr\left(\left(\bigcap_{i=1}^{m} A_i\right)^c\right)$$

$$= 1 - \Pr\left(\bigcup_{i=1}^{m} A_i^c\right) \qquad \text{[by DeMorgan's Law for Sets]}$$

$$\geq 1 - \sum_{i=1}^{m} \Pr(A_i^c) \qquad \text{[by Boole's inequality].} \qquad \blacksquare$$

Suppose now that we are testing m true null hypotheses $H_{0(i)}$. We would like to make the FWER equal to a desired level α, for example, $\alpha = 0.05$. We will have to make it harder to reject each individual null hypothesis. If we let A_i be the event that we reject the true null hypothesis test $H_{0(i)}$ ($i = 1, 2, \ldots, m$) at the level α, then, by Boole's inequality (Theorem 20.3.1), we can write the FWER as

$$\text{FWER}(\alpha) = \Pr(R > 1) = \Pr\left(\bigcup_{i=1}^{m} A_i\right) \leq \sum_{i=1}^{m} \Pr(A_i), \tag{20.5}$$

where R is the top left entry in Table 20.1. If all null hypotheses are true, then the S in Table 20.1 is 0 and so $V = R$. Suppose we would like the FWER to be equal to some predetermined value α, say $\alpha = 0.05$. If we set the significance level of each individual test to α/m, then from (20.5) we have

$$\text{FWER} \leq m \times \frac{\alpha}{m} = \alpha. \tag{20.6}$$

This is the essence of the Bonferroni method.

■ **Example 20.5** Apply the two-sample t-test to the first four genes in the `prostate` microarray data, but use the Bonferroni method to control the FWER at $\alpha = 0.05$.

Solution: With four tests, we would take the significance level of each of the four tests to be

$$\frac{\alpha}{4} = \frac{0.05}{4} = 0.0125.$$

Thus, taking $\alpha = 0.0125$ for each of the four tests yields FWER = 0.05. None of the P-values fall below this level, so none of the four null hypotheses can be rejected. ■ ■ ■

■ **Example 20.6** Apply the two-sample t-test to all 6,033 genes in the `prostate` microarray data. Use FWER = 0.05, 0.10, and 0.20. How many of the genes are significant?

Solution: We would have to lower the significance level α for each of the 6,033 tests in order to achieve the desired FWER:

$$\text{For FWER} = 0.05, \alpha = \frac{0.05}{6,033} = 0.0000083 \text{ and } t_{0.0000083/2}(100) = 4.701$$

For FWER= 0.10, $\alpha = \dfrac{0.10}{6,033} = 0.0000166$ and $t_{0.0000166/2}(100) = 4.526$

For FWER= 0.20, $\alpha = \dfrac{0.20}{6,033} = 0.0000332$ and $t_{0.0000332/2}(100) = 4.348$.

These α levels for the 6,033 individual tests are very small, and the corresponding critical values are quite large. This is because we are testing so many null hypotheses.

Controlling FWER at 0.05 yields only two t statistics that exceed $t_{0.0000083}(100) = 4.701$; these are genes 610 and 1,720. At the level FWER= 0.10, seven genes are significant: 332, 364, 610, 914, 1,720, 3,940, and 4,546. Finally, at FWER= 0.20, nine genes are significant: 332, 364, 579, 610, 914, 1,068, 1,720, 3,940, and 4,546. Had we not controlled for multiple tests being performed with $\alpha = 0.05$, we would have $t_{0.025} = 1.984$ and 477 (almost 8%) of the genes would be significant. ∎∎∎

As the previous example shows, there is an inherent trade-off between setting the bar too high with a small α, in which case we might miss some genes that really do have a predictive effect on prostate cancer (Type II errors), and setting the bar too low with a high α, in which case we reject too many null hypotheses (Type I errors). For the example at hand, the loss is probably greater if we miss a significant gene than if we include an insignificant gene. But what is the relative loss from missing one gene that is truly related to prostate cancer compared to including 50 genes that have no relationship to prostate cancer? Using $\alpha = 0.05$ and not accounting for multiple tests yields 477 significant genes; this is probably too many. Taking FWER= 0.05 yields just two significant genes, while FWER= 0.10 yields seven and FWER= 0.20 yields nine. The choice of FWER is often subtle and based on the relative risks of making Type I and Type II errors.

20.4 Holm's Method

Holm's method is a more powerful alternative to Bonferroni's method. As with the Bonferroni method, the tests need not be independent. Holm's method will reject as many or more null hypotheses than Bonferroni's method and can be described in the following algorithm:

Algorithm 3 Holm's method
---
1. Determine the desired FWER α.
2. For each of the m hypotheses $H_{0(1)}, H_{0(2)}, \ldots, H_{0(m)}$, compute the respective P-value. Label these m P-values as $p_1, p_2, \ldots, p_m$.
3. Order the m P-values from smallest to largest, and call them $p_{(1)} \leq p_{(2)} \leq \cdots \leq p_{(m)}$.
4. Let L be the smallest index j for which $p_{(j)} > \dfrac{\alpha}{m+1-j}$.
5. Reject those null hypotheses for which $p_j < p_{(L)}$.

To visualize Holm's algorithm, it is helpful to set up a table with the ordered P-values, and the significance levels against which we compare these P-values. This is done in Table 20.2. The first row gives the ordered P-values from smallest to largest. The last row shows the null hypotheses that correspond to the ordered P-values. The middle row shows the cutoff against which the ordered P-value is compared. The method rejects all of the null hypotheses corresponding to the P-values that are smaller than $p_{(L)}$, where L is the smallest index j satisfying $p_{(j)} > \alpha/(m-1+j)$.

> **Theorem 20.4.1 — Holm's Method Controls the FWER.** Holm's method, outlined in Algorithm 3, controls the FWER to be at most α.

Table 20.2 Ordered P-values and cutoffs for Holm's method. Here, L is the smallest j for which $p_{(j)} > \alpha/(m - j + 1)$, so all of the null hypotheses $H_{0(1)}, H_{0(2)}, \ldots, H_{0,(L)}$ are rejected.

	Smallest PV							Largest PV
Ordered P-values	$p_{(1)}$	$p_{(2)}$	$\cdots$	$p_{(L-1)}$	$p_{(L)}$	$p_{(L+1)}$	$\cdots$	$p_{(m)}$
Cutoff	$\dfrac{\alpha}{m}$	$\dfrac{\alpha}{m-1}$	$\cdots$	$\dfrac{\alpha}{m-L+2}$	$\dfrac{\alpha}{m-L+1}$	$\dfrac{\alpha}{m-L}$	$\cdots$	$\dfrac{\alpha}{1}$
"Ordered" nulls	$H_{0(1)}$	$H_{0(2)}$	$\cdots$	$H_{0(L-1)}$	$H_{0(L)}$	$H_{0(L+1)}$		$H_{0(m)}$

If L is the smallest j such that $p_{(j)} > \dfrac{\alpha}{m-j+1}$, then all of *these* hypotheses are rejected.

Proof. Label the ordered P-values and the corresponding null hypotheses as indicated in the first and third rows of Table 20.2. Suppose that n_0 of the m null hypotheses are true; the remaining $m - n_0$ of them are false. Define the set I to be the indices of the null hypotheses that are true:

$$I = \left\{ i \,\middle|\, H_{0(i)} \text{ is true} \right\}.$$

The number of elements in I is therefore $|I| = n_0$. Let k be the first rejected true null hypothesis in the ordering $H_{0(1)}, H_{0(2)}, \ldots, H_{0(m)}$. If there is no such k (i.e., all of the true null hypotheses are correctly accepted), set $k = m + 1$. Then all $k - 1$ of the null hypotheses that come before $H_{0(k)}$ would be rejected, because the respective P-values are below $p_{(j)}$. Also, because k is the smallest index among the rejected true null hypotheses, we know that $H_{0(k)}$ is incorrectly rejected. The situation is illustrated below:

$$H_{0(1)} \qquad H_{0(2)} \qquad \cdots \qquad H_{0(k-1)} \qquad H_{0(k)} \qquad H_{0(k+1)} \qquad \cdots \qquad H_{0(m)}.$$

These are false null hypotheses that are correctly rejected

This is a true null hypothesis that is incorrectly rejected

We therefore know that the number of false null hypotheses, $m - n_0$, is at least $k - 1$. Thus,

$$m - n_0 \geq k - 1 \quad \Longrightarrow \quad n_0 \leq m - k + 1 \quad \Longrightarrow \quad \frac{1}{n_0} \geq \frac{1}{m - k + 1}.$$

Since we rejected $H_{0(k)}$, we know that

$$p_{(k)} < \frac{\alpha}{m - k + 1} \leq \frac{\alpha}{n_0}.$$

Recalling that $I = \left\{ i \,\middle|\, H_{0(i)} \text{ is true} \right\}$, we can write

$$
\begin{aligned}
\text{FWER} &= \Pr(\text{at least one false rejection}) \\
&= \Pr\left(\min_{i \in I} p_i \leq \frac{\alpha}{m - k + 1} \right) \\
&\leq \Pr\left(\min_{i \in I} p_i \leq \frac{\alpha}{n_0} \right)
\end{aligned}
$$

$$= \Pr\left(\bigcup_{i \in I}\left(p_i \leq \frac{\alpha}{n_0}\right)\right)$$

$$\leq \sum_{i \in I} \Pr\left(p_i \leq \frac{\alpha}{n_0}\right) \qquad \text{[by Boole's inequality]}$$

$$= \sum_{i \in I} \frac{\alpha}{n_0} \qquad \text{[because the probability of rejecting a true null equals the level of significance]}$$

$$= \alpha. \qquad\qquad\qquad\qquad\qquad\qquad\qquad\qquad\qquad\qquad\qquad\qquad \blacksquare$$

■ **Example 20.7** Apply Holm's method with $\alpha = 0.10$ to the full prostate cancer microarray data from Example 20.6.

Solution: We have $m = 6{,}033$ tests and $\alpha = 0.10$. If we run each test individually, order the P-values, and plot them, we obtain the graph in Figure 20.5(a). Figure 20.5, for which the vertical axis is on a log scale, shows 6,033 separate dots, although for much of the graph the dots are close enough together to give the impression that there is a smooth curve. At the bottom left of the curve, where the P-values are very small, we can see the individual dots.

The concept behind Holm's method is to compare the P-values to $\alpha/(6{,}033 - j + 1)$. Specifically, we look for the first P-value that falls above $\alpha/(6{,}033 - j + 1)$; call the first such index L. Then all hypotheses that come before $H_{0(L)}$ are rejected. Figure 20.5(b) shows the P-values as blue dots along with the values for $\alpha/(6{,}033 - j + 1)$. From this graph, we can see that the overwhelming majority of null hypotheses are not rejected, although there are a few in the very bottom left corner of the graph where the P-value falls below $\alpha/(6{,}033 - j + 1)$. Figure 20.5(c) in an enlargement of the leftmost part of Figure 20.5(b). This graph indicates that $L = 8$ is the smallest index j for which $p_{(j)} > \alpha/(6{,}033 - j + 1)$. Thus, the seven hypotheses that come before this one in the ordered list $H_{0(1)}, H_{0(2)}, \ldots, H_{0,(6{,}033)}$ are rejected. The first eight P-values with the corresponding null hypotheses are shown in Table 20.3.

We therefore see that by Holm's method, seven genes are significant: 332, 364, 610, 914, 1720, 3940, and 4546. These are the same genes that were significant using Bonferroni's method, so in this case the Bonferroni and Holm methods produce the same result, but in general Holm's method will lead to at least as many rejections as Bonferroni's method with no loss in power.

■ ■ ■

20.5 The False Discovery Rate

When we are testing a large number m of null hypotheses, it is unreasonable to control the probability of just one false rejection. When m is in the hundreds or thousands, it is nearly inevitable that we will reject at least one H_0. As a result, controlling the FWER means we must make it difficult to reject any particular null hypothesis. The FWER is just too restrictive for large-scale hypothesis testing. Instead, we might try to control the proportion of false positives, that is cases where we reject H_0 when it is true.

> **Definition 20.5.1 — False Discovery Proportion.** Using the notation from Table 20.1, we define the *false discovery proportion* (FDP) to be
>
> $$\text{FDP} = \frac{V}{V + S} = \frac{V}{R}.$$

In practice, a researcher has no way of knowing which null hypotheses are true and which are false, so it is not possible to control the FDP. It is possible, however, to control the expected value of the FDP, which is called the *false discovery rate*.

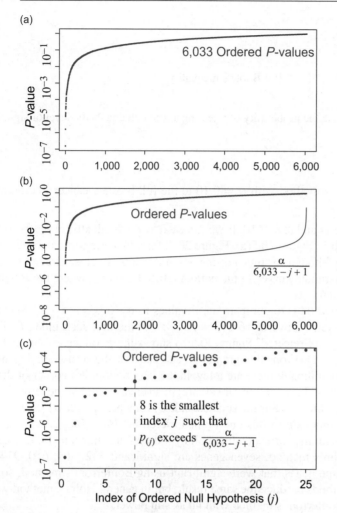

Figure 20.5 Holm's method applied to the prostate cancer data. Note that the P-value scale is logarithmic.

Table 20.3 First eight P-values and corresponding null hypotheses.

j	Hypothesis	P-value	$\dfrac{\alpha}{(6,033-j+1)}$	
1	610	1.544e-07	1.658e-05	
2	1,720	1.577e-06	1.658e-05	
3	364	9.398e-06	1.658e-05	
4	332	1.045e-05	1.658e-05	
5	914	1.210e-05	1.659e-05	
6	3,940	1.394e-05	1.659e-05	
7	4,546	1.573e-05	1.659e-05	
8	1,068	2.672e-05	1.659e-05	PV $> \alpha/(6,033-j+1)$

Definition 20.5.2 — False Discovery Rate. Suppose the breakdown of true/false null hypotheses and the reject/fail-to-reject decisions are as in Table 20.1. The *false discovery rate* (FDR) is.

$$\text{FDR} = \mathbb{E}(\text{FDP}) = c = \mathbb{E}\left(\frac{V}{V+S}\right) = \mathbb{E}\left(\frac{V}{R}\right).$$

It is possible, though unlikely (especially if m is very large), that none of the null hypotheses are rejected, that is $R = 0$. This creates a problem because R is in the denominator of the FDR. Some sources instead condition on $R > 0$ and define $\text{FDR} = \mathbb{E}(\text{FDP}|R > 0)\Pr(R > 0) = \mathbb{E}(V/R|R > 0)\Pr(R > 0)$. For large m, $\Pr(R > 0)$ will be near 1. Controlling the FDR at some rate, say $q = 0.10$, can be done using the method of Benjamini and Hochberg (1995). The algorithm for doing this is given below.

Algorithm 4   Benjamini–Hochberg method

---

1. Determine the desired FDR q.
2. For each of the m hypotheses $H_{0(1)}, H_{0(2)}, \ldots, H_{0(m)}$, compute the respective P-values and call them $p_1, p_2, \ldots, p_m$.
3. Order the m P-values from smallest to largest, and call them $p_{(1)} \leq p_{(2)} \leq \cdots \leq p_{(m)}$.
4. Let L be the smallest index j for which $p_{(j)} < \dfrac{qj}{m}$.
5. Reject those null hypotheses for which $p_j \leq p_{(L)}$.

■ **Example 20.8**   Apply the Benjamini–Hochberg algorithm to the prostate cancer microarray data from Example 20.6 using $q = 0.01$, 0.05, and 0.10.

Solution: The following R code applies the Benjamini–Hochberg algorithm:

```
prostate = read.csv( "prostmat.csv" )
m = nrow(prostate)
t.stat.vec = rep( NA , m )
pv = rep( NA , m )
for( i in 1:m ) {
        x = as.numeric( prostate[ i , 1:50 ] )
        y = as.numeric( prostate[ i , 51:102 ] )
        tTest = t.test( x , y , var.equal=TRUE )
        t.stat.vec[i] = tTest$statistic
        pv[i] = tTest$p.value
}
library( dplyr )
pv.sorted = arrange( data.frame( 1:6033 , pv ) , pv )
q = 0.05
BH.cutoff = q*(1:m)/m
plot( pv.sorted[,2] , pch="." , col="red" , log="y" ,cex=4 )
lines( 1:m , BH.cutoff )
pv.sorted[ pv.sorted[,2] < BH.cutoff , ][,1]
```

The first two lines read in the data. The next two statements and the `for` loop compute the t statistics from the 6,033 tests. The loop keeps track of the P-values in the vector pv. The line that defines `pv.sorted` uses the `dplyr` package to sort (using `dplyr`'s `arrange` function) the data frame constructed from the indices for the tests and the P-values. The cutoff for the Benjamini–Hochberg procedure is qj/m, $j = 1, 2, \ldots, 6,033$; this vector is defined to be `BH.cutoff`. The lines that begin `plot(...)` and

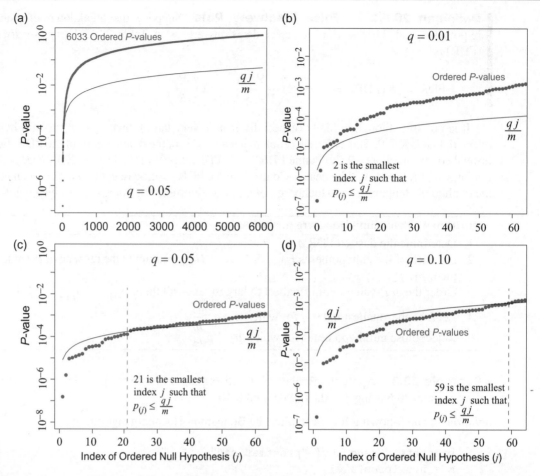

Figure 20.6 Results of Benjamini–Hochberg algorithm applied to the prostate cancer microarray data set. Panel (a) shows all 6,033 of the ordered P-values, while the other panels show the cutoffs for $q = 0.01, 0.05$, and 0.10.

lines(...) construct the plot of the ordered P-values against the index of the ordered null hypotheses. The last line tells us which null hypotheses would be rejected. The pv.sorted[,2] < BH.cutoff part of this line returns a vector of 6,033 TRUEs and FALSEs. The statement pv.sorted[pv.sorted[,2] < BH.cutoff , 1] selects the first column of the data frame formed with columns for the index j of the null hypothesis and the ordered P-value.

Figure 20.6(a) shows all 6,033 ordered P-values along with the cutoff qj/m. The figure suggests that all the action is taking place in the first 100 or so ordered P-values. The other three panels in Figure 20.6 show the ordered P-values along with the cutoffs for $q = 0.01, 0.05$, and 0.10 for the first 64 ordered P-values. Setting $q = 0.10$ leads to 59 rejections, $q = 0.05$ leads to 21 rejections, and $q = 0.01$ leads to 21 rejections. The output from the last line of the above code, which gives those null hypotheses that would be rejected with $q = 0.05$, is

```
pv.sorted[ pv.sorted[,2] < BH.cutoff , 1 ]
```

```
 [1]   610 1720  364  332  914 3940 4546 1068  579 4331 1089 3647 1113 1077
[15]  4518 1557 4088 3991 3375 4316 4073
```

∎ ∎ ∎

20.6 Simultaneous Confidence Intervals

The Bonferroni inequality can also be used to produce simultaneous confidence intervals for parameters or functions of parameters. Suppose we have parameters $\theta_1, \theta_2, \ldots, \theta_m$ (which could be, and often are, functions of other parameters) and we would like a set of intervals $I_1, I_2, \ldots, I_m$ that contain these parameters with high probability.[4] We would like to be able to say that all m of these intervals contain all m of the parameters. These are said to be *simultaneous* confidence intervals because they simultaneously contain all of the desired parameters.

Bonferroni's inequality can be written in the form[5]

$$\Pr\left(\bigcap_{i=1}^{m} B_i\right) \geq 1 - \sum_{i=1}^{m} \Pr\left(B_i^c\right). \tag{20.7}$$

If we let B_i denote the event that the ith interval contains the parameter θ_i, then the above becomes

$$\Pr\left((I_1 \text{ contains } \theta_1) \cap \cdots \cap (I_m \text{ contains } \theta_m)\right) \geq 1 - \sum_{i=1}^{m} \Pr\left((I_i \text{ contains } \theta_i)^c\right).$$

If the interval I_i is determined to be a $100 \times (1-\alpha)\%$ confidence interval for θ_i, then $\Pr\left((I_i \text{ contains } \theta_i)^c\right) = 1 - \alpha$. Then, (20.7) becomes

$$\Pr\left((I_1 \text{ contains } \theta_1) \cap \cdots \cap (I_m \text{ contains } \theta_m)\right) \geq 1 - \sum_{i=1}^{m} \left(1 - (1-\alpha)\right) = 1 - m\alpha. \tag{20.8}$$

Analogous to the FWER, we define the *family-wise confidence level* (FWCL) to be

$$\text{FWCL} = \Pr\left((I_1 \text{ contains } \theta_1) \cap \cdots \cap (I_m \text{ contains } \theta_m)\right).$$

Taking the α for each individual confidence interval to be α/m, we are guaranteeing that the FWCL satisfies

$$\text{FWCL} \geq 1 - m\frac{\alpha}{m} = 1 - \alpha.$$

Simultaneous confidence intervals for differences in means is often used in ANOVA when we reject the null hypothesis of equal means across all treatments. We are often interested in getting simultaneous confidence intervals for all possible differences of means $\mu_i - \mu_j$.

■ **Example 20.9** Consider the aluminum strength data in Table 12.4 and analyzed in Example 12.12. Find a set of simultaneous confidence intervals for the differences in means $\mu_1 - \mu_2$, $\mu_1 - \mu_3$, and $\mu_2 - \mu_3$ for which the FWCL is at least 0.90.

Solution: The ANOVA table is reproduced as Table 20.4. Since we are constructing three confidence intervals, $m = 3$ and we must set the confidence level for each confidence interval at $\alpha = 0.10/3 \approx 0.3333$. The degrees of freedom for the MSE is 13, so

$$t_{0.10/(3\times2)} = t_{0.3333/2}(13) = 2.380.$$

---

[4]This is seen from the point of view of *before* data are collected. These are *random* intervals.
[5]Here we use c to denote the complement of an event.

Table 20.4 ANOVA Table for aluminum data.

Source	Sum of squares	Degrees of freedom	Mean square	F	P
Factor A	123,187,500	2	61,593,750	14.032	0.00057
Error	57,062,500	13	4,389,423		
Total	180,250,000	15	12,016,667		

The confidence interval is therefore

$$\left((\bar{y}_{\cdot i_1} - \bar{y}_{\cdot i_2}) - t_{\alpha/(2\times3)}\sqrt{\mathrm{MSE}\left(\frac{1}{n_{i_1}} + \frac{1}{n_{i_2}}\right)}, \ (\bar{y}_{\cdot i_1} - \bar{y}_{\cdot i_2}) + t_{\alpha/(2\times3)}\sqrt{\mathrm{MSE}\left(\frac{1}{n_{i_1}} + \frac{1}{n_{i_2}}\right)} \right).$$

The difference between this and the formula for a one-at-a-time confidence interval from (12.40) is subtle, but important. The tail area for the t value in the set of simultaneous confidence intervals is $1/m$ times the one-at-a-time tail area.

The following R code performs the ANOVA and computes the confidence interval:

```
alum = read.csv("Aluminum.csv")
TypeOfAluminum = alum$TypeOfAluminum
TensileStrength = alum$TensileStrength
y1 = TensileStrength[ TypeOfAluminum == 1 ]
y2 = TensileStrength[ TypeOfAluminum == 2 ]
y3 = TensileStrength[ TypeOfAluminum == 3 ]
k = 3
n1 = length(y1)
n2 = length(y2)
n3 = length(y3)
N = n1 + n2 + n3
y1bar = mean(y1)
y2bar = mean(y2)
y3bar = mean(y3)
ybarbar = sum( c(y1,y2,y3) )/N
alum.lm = lm( TensileStrength ~ as.factor( TypeOfAluminum ) )
alum.anova = anova( alum.lm )
MSE = alum.anova$'Mean Sq'[2]   ## alum.anova$'Mean Sq' gives the MSEs
alpha = 0.10
m = choose(k,2)
df = N - k
t = qt( 1 - alpha/(2*m) , df )
LCL1 = ( y1bar - y2bar ) - t*sqrt( MSE*( 1/n1 + 1/n2 ) )
UCL1 = ( y1bar - y2bar ) + t*sqrt( MSE*( 1/n1 + 1/n2 ) )
print( c( LCL1 , UCL1 ) )
```

Code similar to the last three lines can be used to get confidence intervals for the other two differences of means: $\mu_1 - \mu_3$ and $\mu_2 - \mu_3$. The three confidence intervals are

```
(    571.64 ,  6328.36 )
(   3906.89 , 10343.11 )
(    456.89 ,  6893.11 )
```

Note that the confidence interval for $\mu_i - \mu_j$ uses information from *all* treatments, not just treatments i and j, in the sense that the common variance σ^2 is estimated (with 13 degrees of freedom) by the MSE, which uses the *whole* data set. All three of these confidence intervals exclude 0. ■■■

Exercise 20.1 If we chose $\alpha = 0.10$, what would be the t value if we wanted confidence intervals for all 6,033 differences in means for the microarray data set?

20.7 Tukey's Method

Suppose we observe

$$
\begin{aligned}
Y_{11}, Y_{12}, \ldots, Y_{1n_1} &\sim N(\mu_1, \sigma^2) \\
Y_{21}, Y_{22}, \ldots, Y_{2n_2} &\sim N(\mu_2, \sigma^2) \\
\vdots \qquad \vdots \quad \vdots & \\
Y_{k1}, Y_{k2}, \ldots, Y_{kn_k} &\sim N(\mu_k, \sigma^2).
\end{aligned}
\tag{20.9}
$$

We would like to construct simultaneous confidence intervals for all differences of means $\mu_i - \mu_j$, $i \neq j$. Thus, we would like to construct L_{ij} and U_{ij} such that

$$
\Pr\left(L_{ij} < \mu_i - \mu_j < U_{ij} \text{ for all } i, j, \ i \neq j\right) = 1 - \alpha,
$$

where α is some prespecified value for the simultaneous confidence intervals. Unlike Bonferroni's method, which makes no assumptions about the parameters involved in a confidence interval, Tukey's method takes advantage of the fact that we want to obtain confidence intervals for *differences of means*.

Let's begin by considering data from one of the treatments in (20.9), say the ith treatment. We therefore have random variables $Y_{i1}, Y_{i2}, \ldots, Y_{in_i}$. The range R_i is defined to be

$$
R_i = \max_{1 \leq j \leq n_i} Y_{ij} - \min_{1 \leq j \leq n_i} Y_{ij}.
$$

If SSE is the sum of squared errors from the ANOVA table, then

$$
\text{MSE} = \frac{\text{SSE}}{N - k}
$$

is an unbiased estimator of σ^2, where $N = \sum_{i=1}^{k} n_i$. The MSE also has the property that

$$
\frac{(N-k)\text{MSE}}{\sigma^2} \sim \chi^2(N-k).
$$

Definition 20.7.1 — Studentized Range. Suppose $Y_1, Y_2, \ldots, Y_n \sim$ i.i.d. $N(\mu, \sigma^2)$. Suppose also that S_ν^2 is an unbiased estimator of σ^2 (which could be based on information not contained in $Y_1, Y_2, \ldots, Y_n$) and is a function of a χ^2 random variable with ν degrees of freedom. Let

$$
R_n = \max_{1 \leq j \leq n} Y_i - \min_{1 \leq j \leq n} Y_i,
$$

then the *studentized range* is defined to be

$$
Q_{n,\nu} = \frac{R_n}{S_\nu}.
$$

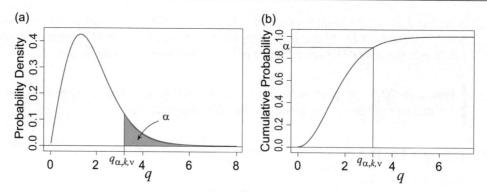

Figure 20.7 PDF (a) and CDF (b) of Tukey's distribution of the studentized range.

The distribution of the studentized range is messy and we will not give it here. We just mention that tables of percentiles of the distribution of the studentized range exist, and percentiles can be easily obtained in R. The functions `ptukey()` and `qtukey()` are available to obtain the CDF and the quantile function for the studentized range. Let $q_{\alpha,k,v}$ denote the $1-\alpha$ quantile of the distribution of the studentized range; that is,

$$\Pr\left(Q_{k,v} \leq q_{\alpha,k,v}\right) = 1-\alpha.$$

Figure 20.7 shows graphs of the PDF and CDF of the distribution of the studentized range.

The Tukey simultaneous confidence intervals for all $\binom{k}{2}$ differences $\mu_i - \mu_j$ are

$$\left((\overline{y}_{.i} - \overline{y}_{.j}) - q_{\alpha,k,N-k}\sqrt{\left(\frac{1}{n_1} + \frac{1}{n_2}\right)\frac{\text{MSE}}{2}}, (\overline{y}_{.i} - \overline{y}_{.j}) + q_{\alpha,k,N-k}\sqrt{\left(\frac{1}{n_1} + \frac{1}{n_2}\right)\frac{\text{MSE}}{2}} \right). \quad (20.10)$$

The next theorem gives the simultaneous coverage probability for the Tukey intervals.

Theorem 20.7.1 — Tukey's Method.   The Tukey confidence intervals from (20.10) cover all $\binom{k}{2}$ differences with probability $1-\alpha$.

Proof.   We will prove this for the special case where $n_1 = n_2 = \cdots = n_k = n$. Define

$$U_r = \sqrt{n}(\overline{Y}_{.r} - \mu_r) \sim \text{ i.i.d. } N(0,\sigma^2), \qquad r = 1, 2, \ldots, k.$$

Let MSE denote the mean squared error from the ANOVA table for all k groups. Then,

$$\frac{\displaystyle\max_{1\leq r\leq k} U_r - \min_{1\leq r\leq k} U_r}{\sqrt{\text{MSE}}} \sim Q_{k,N-k},$$

where $N = \sum_{i=1}^{n} n_i = nk$. This implies that

$$\Pr\left(\frac{\displaystyle\max_{1\leq r\leq k} U_r - \min_{1\leq r\leq k} U_r}{\sqrt{\text{MSE}}} < q_{\alpha,k,N-k} \right) = 1-\alpha$$

$$\Pr\left(\max_{1\leq r\leq k} U_r - \min_{1\leq r\leq k} U_r < q_{\alpha,k,N-k}\sqrt{\text{MSE}} \right) = 1-\alpha.$$

The next step is key. Notice that

$$\max_{1 \le r \le k} U_r - \min_{1 \le r \le k} U_r < c \quad \Longleftrightarrow \quad \left(|U_i - U_j| < c \text{ for all } i, j \right).$$

Thus,

$$
\begin{aligned}
1 - \alpha &= \Pr\left(\max_{1 \le i \le k} U_r - \min_{1 \le r \le k} U_r < q_{\alpha,k,N-k}\sqrt{\text{MSE}} \right) \\
&= \Pr\left(|U_i - U_j| < q_{\alpha,k,N-k}\sqrt{\text{MSE}} \text{ for all } i, j \right) \\
&= \Pr\left(-q_{\alpha,k,N-k}\sqrt{\text{MSE}} < U_i - U_j < q_{\alpha,k,N-k}\sqrt{\text{MSE}} \text{ for all } i, j \right) \\
&= \Pr\left(-q_{\alpha,k,N-k}\sqrt{\text{MSE}} < \sqrt{n}(\overline{Y}_{\cdot i} - \mu_i) - \sqrt{n}(\overline{Y}_{\cdot j} - \mu_j) < \right. \\
&\qquad \left. q_{\alpha,k,N-k}\sqrt{\text{MSE}} \text{ for all } i, j \right) \\
&= \Pr\left((\overline{Y}_{\cdot i} - \overline{Y}_{\cdot j}) - \frac{q_{\alpha,k,N-k}\sqrt{\text{MSE}}}{\sqrt{n}} < \mu_i - \mu_j < (\overline{Y}_{\cdot i} - \overline{Y}_{\cdot j}) + \frac{q_{\alpha,k,N-k}\sqrt{\text{MSE}}}{\sqrt{n}} \text{ for all } i, j \right).
\end{aligned}
$$

This agrees with (20.10) when $n_1 = n_2 = \cdots n_k = n$. ∎

■ **Example 20.10** For the aluminum data in Example 20.9, find 90% simultaneous Tukey confidence intervals for all possible differences in means.

Solution: From the ANOVA table in Table 20.4 we see that MSE $= 4,389,423$. We assume that the data are loaded and the code in Example 20.9 is run. We can then compute the Tukey simultaneous intervals using the following code:

```
alpha = 0.10
MSE = alum.anova$'Mean Sq'[2]
q = qtukey( 0.90 , k , N-3 )
LCLT12 = (y1bar - y2bar) - q * sqrt( (1/n1 + 1/n2) * MSE / 2 )
UCLT12 = (y1bar - y2bar) + q * sqrt( (1/n1 + 1/n2) * MSE / 2 )
print( c( LCLT12 , UCLT12 ) )
LCLT13 = (y1bar - y3bar) - q * sqrt( (1/n1 + 1/n3) * MSE / 2 )
UCLT13 = (y1bar - y3bar) + q * sqrt( (1/n1 + 1/n3) * MSE / 2 )
print( c( LCLT13 , UCLT13 ) )
LCLT23 = (y2bar - y3bar) - q * sqrt( (1/n2 + 1/n3) * MSE / 2 )
UCLT23 = (y2bar - y3bar) + q * sqrt( (1/n2 + 1/n3) * MSE / 2 )
print( c( LCLT23 , UCLT23 ) )
```

The resulting confidence intervals are

```
(  730.95 ,  6169.05 )
( 4085.01 , 10164.99 )
(  635.01 ,  6714.99 )
```

All of the Tukey intervals exclude 0, suggesting that all possible pairs of means are different. Figure 20.8 shows the extent of the Bonferroni and the Tukey confidence intervals. As is often the case, the Tukey intervals are narrower. The figure also shows the Scheffé confidence intervals, which will be discussed in Section 20.8. ■■■

When we reject the null hypothesis of equal means, the question of which means differ arises. Tukey's method is often used as a *post hoc* method when the null hypothesis is rejected.

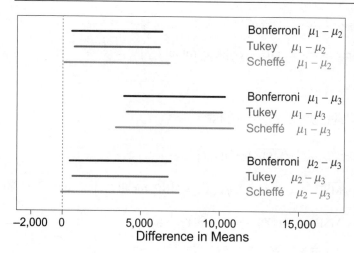

Figure 20.8 Comparison of Bonferroni, Tukey, and Scheffé simultaneous confidence intervals.

20.8   Scheffé's Method

Suppose we observe the data shown in (20.9). Scheffé's method deals with simultaneous confidence intervals for *contrasts* involving the means.

> **Definition 20.8.1 — Contrasts.** A *contrast* is some linear combination of the means,
>
> $$\eta = c_1\mu_1 + c_2\mu_2 + \cdots + c_k\mu_k, \tag{20.11}$$
>
> subject to the constraint that the constants c_j sum to 0, that is, $\sum_{j=1}^{k} c_j = 0$.

Differences in means can be represented by a contrast, with one of the c_j equal to 1 and another equal to -1, with the remaining c_j equal to 0. For example, if there are $k = 5$ groups, then

$$c_1 = 1; \quad c_2 = -1; \quad c_3 = 0; \quad c_4 = 0; \quad c_5 = 0$$

leads to the contrast

$$\eta = c_1\mu_1 + c_2\mu_2 + c_3\mu_3 + c_4\mu_4 + c_5\mu_5 = 1\mu_1 + (-1)\mu_2 + 0\mu_3 + 0\mu_4 + 0\mu_5 = \mu_1 - \mu_2.$$

For a fixed set of $c_1, c_2, \ldots, c_k$, we can estimate the contrast by

$$\hat{\eta} = c_1\overline{Y}_{\cdot 1} + c_2\overline{Y}_{\cdot 2} + \cdots + c_k\overline{Y}_{\cdot k}.$$

This estimator is unbiased because

$$\mathbb{E}(\hat{\eta}) = c_1\mathbb{E}\left(\overline{Y}_{\cdot 1}\right) + c_2\mathbb{E}\left(\overline{Y}_{\cdot 2}\right) + \cdots + c_k + \mathbb{E}\left(\overline{Y}_{\cdot k}\right) = c_1\mu_1 + c_2\mu_2 + \cdots + c_k\mu_k = \eta.$$

The variance of this estimator is

$$\mathbb{V}(\hat{\eta}) = c_1\mathbb{V}\left(\overline{Y}_{\cdot 1}\right) + c_2\mathbb{V}\left(\overline{Y}_{\cdot 2}\right) + \cdots + c_k\mathbb{V}\left(\overline{Y}_{\cdot k}\right)$$

$$= c_1^2\frac{\sigma^2}{n_1} + c_2^2\frac{\sigma^2}{n_2} + \cdots + c_k^2\frac{\sigma^2}{n_k}$$

$$= \sigma^2\sum_{j=1}^{k}\frac{c_j^2}{n_j}.$$

The MSE defined in Section 12.4 is an unbiased estimator of σ^2:

$$\hat{\sigma}^2 = \text{MSE} = \frac{\displaystyle\sum_{j=1}^{k}\sum_{i=1}^{n_k}(Y_{ij} - \overline{Y}_{\cdot j})^2}{\displaystyle\sum_{j=1}^{k}(n_j - 1)}.$$

The estimated variance of $\hat{\eta}$ is thus

$$\hat{\mathbb{V}}(\hat{\eta}) = \text{MSE}\sum_{j=1}^{k}\frac{c_j^2}{n_j}$$

and the MSE satisfies

$$\frac{(N-k)\text{MSE}}{\sigma^2} \sim \chi(N-k), \tag{20.12}$$

where $N = \sum_{j=1}^{k}n_j$ is the total sample size across all k groups. Since $\overline{X}_{\cdot j} \sim N(\mu_j, \sigma^2/n_j)$ we can standardize to obtain

$$\frac{\overline{X}_{\cdot j} - \mu_j}{\sigma/\sqrt{n_j}} \sim N(0,1).$$

Since the $\overline{X}_{\cdot j}$ are independent for $j = 1, 2, \ldots, k$,

$$\sum_{j=1}^{k}\left(\frac{\overline{X}_{\cdot j} - \mu_j}{\sigma/\sqrt{n_j}}\right)^2 = \sum_{j=1}^{k}\frac{(\overline{X}_{\cdot j} - \mu_j)^2}{\sigma^2/n_j} \sim \chi^2(k). \tag{20.13}$$

Furthermore, the means $\overline{X}_{\cdot j}$ are independent of the MSE, so

$$F = \frac{\displaystyle\sum_{j=1}^{k}\frac{(\overline{X}_{\cdot j} - \mu_j)^2}{\sigma^2/n_j}\Big/ k}{(N-k)\text{MSE}/\sigma^2 \Big/ (N-k)} = \frac{\dfrac{1}{k}\displaystyle\sum_{j=1}^{k}n_j(\overline{X}_{\cdot j} - \mu_j)^2}{\text{MSE}} \sim F(k, N-k). \tag{20.14}$$

We can therefore determine $F_\alpha(k, N-k)$ such that

$$\Pr\left(\frac{\dfrac{1}{k}\displaystyle\sum_{j=1}^{k}n_j(\overline{X}_{\cdot j} - \mu_j)^2}{\text{MSE}} \leq F_\alpha(k, N-k)\right) = 1 - \alpha,$$

or equivalently,

$$\Pr\left(\sum_{j=1}^{k}n_j(\overline{X}_{\cdot j} - \mu_j)^2 \leq \text{MSE}\, kF_\alpha(k, N-k)\right) = 1 - \alpha. \tag{20.15}$$

As we relate this to the Scheffé method for multiple confidence intervals, we will assume that there are just $k = 3$ groups, because the geometry for three groups can be easily visualized. The algebra for higher dimensions is only slightly more complicated and follows easily. We will also assume that the sample sizes n_j are all equal, and we will let n denote the common sample size. Although we will demonstrate Scheffé's method for $k = 3$ and $n_1 = n_2 = n_3 = n$, the method works for arbitrary dimension k and unequal sample sizes.

Consider the two points in $k = 3$D space: (μ_1, μ_2, μ_3) and $(\overline{X}_{\cdot 1}, \overline{X}_{\cdot 2}, \overline{X}_{\cdot 3})$, the latter being a random point. The distance between these two points is

$$D = \sqrt{(X_1 - \mu_1)^2 + (X_2 - \mu_2)^2 + (X_3 - \mu_3)^2}.$$

Note that the distance D is a random variable because $(\overline{X}_{\cdot 1}, \overline{X}_{\cdot 2}, \overline{X}_{\cdot 3})$ are random. Let (z_1, z_2, z_3) represent an arbitrary point in this 3D space. Consider the set of all planes in this space that pass through the point (μ_1, μ_2, μ_3); these planes can be written in the form

$$c_1(z_1 - \mu_1) + c_2(z_2 - \mu_2) + c_3(z_3 - \mu_3) = 0,$$

where at least one of the c_j is nonzero. (The relationship between one of these planes and the contrast $\sum_{j=1}^{3} c_j \mu_k$ will become apparent soon.) Figure 20.9(a) depicts the points and one of the infinitely many planes through (μ_1, μ_2, μ_3).

Using some methods from 3D geometry (Varberg *et al.*, 2007) we can say that the shortest distance between one of these planes and $(\overline{X}_{\cdot 1}, \overline{X}_{\cdot 2}, \overline{X}_{\cdot 3})$ is

$$\frac{\left(\sum_{j=1}^{3} c_j (\overline{X}_{\cdot j} - \mu_j) \right)^2}{\sum_{j=1}^{3} c_i^2}. \tag{20.16}$$

For most planes passing through (μ_1, μ_2, μ_3) the shortest distance between the point $(\overline{X}_{\cdot 1}, \overline{X}_{\cdot 2}, \overline{X}_{\cdot 3})$ and the plane is *less than* the distance between (μ_1, μ_2, μ_3) and $(\overline{X}_{\cdot 1}, \overline{X}_{\cdot 2}, \overline{X}_{\cdot 3})$. There is one plane, among the infinitely many that pass through (μ_1, μ_2, μ_3), for which the point closest to $(\overline{X}_{\cdot 1}, \overline{X}_{\cdot 2}, \overline{X}_{\cdot 3})$ is (μ_1, μ_2, μ_3). This occurs when the line segment from (μ_1, μ_2, μ_3) to $(\overline{X}_{\cdot 1}, \overline{X}_{\cdot 2}, \overline{X}_{\cdot 3})$ is perpendicular to the plane. Figure 20.9(b) shows the one plane having this property. The key to Scheffé's method is that for *every other* plane passing through (μ_1, μ_2, μ_3), the shortest distance between the plane and $(\overline{X}_{\cdot 1}, \overline{X}_{\cdot 2}, \overline{X}_{\cdot 3})$ is less than D. In other words,

$$\max_{c_1, c_2, c_3} \frac{\left(\sum_{j=1}^{3} c_j (\overline{X}_{\cdot j} - \mu_j) \right)^2}{\sum_{j=1}^{3} c_i^2} = \sum_{j=1}^{3} (X_j - \mu_j)^2 = D^2.$$

Thus,

$$\frac{\left(\sum_{j=1}^{3} c_j (\overline{X}_{\cdot j} - \mu_j) \right)^2}{\sum_{j=1}^{3} c_i^2} \leq \sum_{j=1}^{3} (X_j - \mu_j)^2 \qquad \text{for all } c_1, c_2, c_3.$$

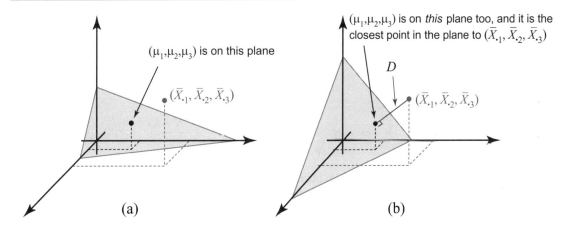

Figure 20.9 The 3D geometry behind Scheffé's method. Panel (a) shows the points (μ_1, μ_2, μ_3) and $(\overline{X}_{\cdot 1}, \overline{X}_{\cdot 2}, \overline{X}_{\cdot 3})$ and one of the infinitely many planes in gray that pass through (μ_1, μ_2, μ_3). Panel (b) shows the plane that is perpendicular to the line through (μ_1, μ_2, μ_3) and $(\overline{X}_{\cdot 1}, \overline{X}_{\cdot 2}, \overline{X}_{\cdot 3})$.

If A is any positive number, then

$$\left(D^2 \leq A \right) \quad \Longleftrightarrow \quad \left(\frac{\left(\sum_{j=1}^{3} c_j (\overline{X}_{\cdot j} - \mu_j) \right)^2}{\sum_{j=1}^{3} c_i^2} \leq A \quad \text{for all } c_1, c_2, c_3 \right).$$

See Problem 20.26. With this result we can claim that the following probabilities are equal:

$$\Pr\left(D^2 \leq A \right) = \Pr\left(\frac{\left(\sum_{j=1}^{3} c_j (\overline{X}_{\cdot j} - \mu_j) \right)^2}{\sum_{j=1}^{3} c_i^2} \leq A \quad \text{for all } c_1, c_2, c_3 \right). \tag{20.17}$$

Now let's take

$$A = \frac{k\, \text{MSE}\, F_\alpha(k, N-k)}{n},$$

which leads to

$$1 - \alpha = \Pr\left(\sum_{j=1}^{3} (\overline{X}_{\cdot j} - \mu_j)^2 \leq \frac{k\, \text{MSE}\, F_\alpha(k, N-k)}{n} \right)$$

$$= \Pr\left(\frac{\left(\sum_{j=1}^{3} c_j (\overline{X}_{\cdot j} - \mu_j) \right)^2}{\sum_{j=1}^{3} c_i^2} \leq \frac{k\, \text{MSE}\, F_\alpha(k, N-k)}{n} \quad \text{for all } c_1, c_2, c_3 \right).$$

This implies that

$$1 - \alpha = \Pr\left(\left(\sum_{j=1}^{3} c_j (\overline{X}_{\cdot j} - \mu_j) \right)^2 \leq \left(\sum_{j=1}^{3} c_j^2 \right) \frac{k\, \text{MSE}\, F_\alpha(k, N-k)}{n} \quad \text{for all } c_1, c_2, c_3 \right)$$

$$= \Pr\left(\left| \sum_{j=1}^{3} c_j \overline{X}_{\cdot j} - \sum_{j=1}^{3} c_j \mu_j \right| \leq \sqrt{\left(\sum_{j=1}^{3} c_j^2 \right) \frac{k\, \text{MSE}\, F_\alpha(k, N-k)}{n}} \quad \text{for all } c_1, c_2, c_3 \right).$$

Finally, this leads to

$$1 - \alpha = \Pr\left(\sum_{j=1}^{3} c_j \overline{X}_{\cdot j} - \sqrt{\left(\sum_{j=1}^{3} c_j^2 \right) \frac{k\,\mathrm{MSE}\,F_\alpha(k, N-k)}{n}} \leq \sum_{j=1}^{3} c_j \mu_j \leq \right.$$

$$\left. \sum_{j=1}^{3} c_j \overline{X}_{\cdot j} + \sqrt{\left(\sum_{j=1}^{3} c_j^2 \right) \frac{k\,\mathrm{MSE}\,F_\alpha(k, N-k)}{n}} \quad \text{for all } c_1, c_2, c_3 \right). \qquad (20.18)$$

This last result is quite remarkable, because the probability on the right side is for every possible c_1, c_2, c_3. It therefore covers every possible contrast $c_1\mu_1 + c_2\mu_2 + c_3\mu_3$ that we might ever want a confidence interval for.

Suppose we have a finite number, say r, of contrasts for which we would like confidence intervals. Let E_1 be the event that all of the confidence intervals calculated from (20.18) for the r contrasts contain their respective (population) contrasts. Similarly, let E_2 be the event that *all* possible contrasts, defined by c_1, c_2, c_3, where not all the c_j are zero, contain their respective (population) contrast. Since $E_2 \subset E_1$, $1 - \alpha = \Pr(E_2) \leq \Pr(E_1)$ we conclude that the probability that the Scheffé confidence intervals computed from (20.18) *simultaneously* cover their respective (population) contrasts is at least $1 - \alpha$.

The geometry and the algebra were simplified in the above derivation by assuming that $k = 3$ and that all sample sizes were the same: $n = n_1 = n_2 = n_3$. The case for an arbitrary number of groups and possibly unequal sample sizes in each group is

$$1 - \alpha = \Pr\left(\sum_{j=1}^{k} c_j \overline{X}_{\cdot j} - \sqrt{\left(\sum_{j=1}^{k} \frac{c_j^2}{n_j} \right) k\,\mathrm{MSE}\,F_\alpha(k, N-k)} \leq \sum_{j=1}^{k} c_j \mu_j \leq \right.$$

$$\left. \sum_{j=1}^{k} c_j \overline{X}_{\cdot j} + \sqrt{\left(\sum_{j=1}^{k} \frac{c_j^2}{n_j} \right) k\,\mathrm{MSE}\,F_\alpha(k, N-k)} \quad \text{for all } c_1, c_2, c_3 \right). \qquad (20.19)$$

■ **Example 20.11** Consider the aluminum strength data in Table 12.4, analyzed in Examples 12.12 and 20.9. Use the Scheffé method to find a set of 90% simultaneous confidence intervals for the differences in means $\mu_1 - \mu_2$, $\mu_1 - \mu_3$, and $\mu_2 - \mu_3$.

Solution: From the ANOVA table in Table 20.4, $\mathrm{MSE} = 4,389,423$. Once data are loaded and the code from Example 20.9 is run, we can compute the Scheffé confidence intervals using this code:

```
alum.lm = lm( TensileStrength ~ as.factor( TypeOfAluminum ) )
alum.anova = anova( alum.lm )
print( alum.anova )
alpha = 0.10
MSE = alum.anova$'Mean Sq'[2]
F = qf( 1-alpha , k , N-k )
k = 3
c1 = 1;  c2 = -1;  c3 = 0
sum1 = c1^2/n1 + c2^2/n2 + c3^2/n3
LCL1 = (y1bar - y2bar) - sqrt( sum1 * k * MSE * F )
UCL1 = (y1bar - y2bar) + sqrt( sum1 * k * MSE * F )
print( c( LCL1 , UCL1 ) )
c1 = 1;  c2 = 0;  c3 = -1
sum2 = c1^2/n1 + c2^2/n2 + c3^2/n3
```

```
LCL2 = (y1bar - y3bar) - sqrt( sum2 * k * MSE * F )
UCL2 = (y1bar - y3bar) + sqrt( sum2 * k * MSE * F )
print( c( LCL2 , UCL2 ) )
c1 = 0;  c2 = 1;  c3 = -1
sum3 = c1^2/n1 + c2^2/n2 + c3^2/n3
LCL3 = (y2bar - y3bar) - sqrt( sum3 * k * MSE * F )
UCL3 = (y2bar - y3bar) + sqrt( sum3 * k * MSE * F )
print( c( LCL3 , UCL3 ) )
```

The 90% Scheffé simultaneous confidence intervals calculated this way are

```
(   97.67 ,  6802.33 )
( 3376.98 , 10873.02 )
(  -73.02 ,  7423.02 )
```

Figure 20.8 compares the Bonferroni, Tukey, and Scheffé simultaneous confidence intervals. The Scheffé intervals are slightly wider, which is typical when the number of intervals is small. ■ ■ ■

■ **Example 20.12** For the aluminum data in Example 20.11, construct both Bonferroni and Scheffé 90% simultaneous confidence intervals for the following contrasts:

$$\mu_1 - \mu_2; \qquad \mu_1 - \mu_3; \qquad \mu_2 - \mu_3; \qquad \frac{\mu_1 + \mu_2}{2} - \mu_3; \qquad \frac{\mu_1 + \mu_3}{2} - \mu_2; \qquad \frac{\mu_2 + \mu_3}{2} - \mu_1.$$

Note that these are all pairwise comparisons and all possible combinations of two groups compared to the third group.

Solution: We have previously estimated the variance of the pairwise differences in sample means. The others are handled as follows:

$$\mathbb{V}\left(\frac{\overline{Y}_1 + \overline{Y}_2}{2} - \overline{Y}_3\right) = \mathbb{V}\left(\frac{\overline{Y}_1}{2} + \frac{\overline{Y}_2}{2} - \overline{Y}_3\right) = \frac{\sigma^2}{4n_1} + \frac{\sigma^2}{4n_2} + \frac{\sigma^2}{n_3} = \sigma^2\left(\frac{1}{4n_1} + \frac{1}{4n_2} + \frac{1}{n_3}\right).$$

The Bonferroni confidence interval for $(\mu_1 + \mu_2)/2 - \mu_3$ is therefore

$$\left(\left(\frac{\overline{y}_1 + \overline{y}_2}{2} - \overline{y}_3\right) - t_{\alpha/(2\times 6)}\sqrt{\text{MSE}\left(\frac{1}{4n_1} + \frac{1}{4n_2} + \frac{1}{n_3}\right)}, \right.$$

$$\left.\left(\frac{\overline{y}_1 + \overline{y}_2}{2} - \overline{y}_3\right) + t_{\alpha/(2\times 6)}\sqrt{\text{MSE}\left(\frac{1}{4n_1} + \frac{1}{4n_2} + \frac{1}{n_3}\right)}\right).$$

The confidence level for Bonferroni intervals must be adjusted because we're now obtaining simultaneous intervals for six contrasts, not three. The six Bonferroni intervals for the six contrasts $\mu_1 - \mu_2, \mu_1 - \mu_3, \mu_2 - \mu_3, \frac{\mu_1 + \mu_2}{2} - \mu_3, \frac{\mu_1 + \mu_3}{2} - \mu_2$, and $\frac{\mu_2 + \mu_3}{2} - \mu_1$ are, respectively,

```
(   128.50 ,  6771.50 )
(  3411.45 , 10838.55 )
(   -38.55 ,  7388.55 )
(  2078.50 ,  8721.50 )
( -3106.46 ,  2881.46 )
( -8281.46 , -2293.54 )
```

For the Scheffé method,

$$\sum_{j=1}^{3} \frac{c_j^2}{n_j} = \begin{cases} \frac{1}{n_1} + \frac{1}{n_2}, & \text{for the contrast } \mu_1 - \mu_2 \\ \frac{1}{n_1} + \frac{1}{n_3}, & \text{for the contrast } \mu_1 - \mu_3 \\ \frac{1}{n_2} + \frac{1}{n_3}, & \text{for the contrast } \mu_2 - \mu_3 \\ \frac{1}{4n_1} + \frac{1}{4n_2} + \frac{1}{n_3}, & \text{for the contrast } \frac{\mu_1 + \mu_2}{2} - \mu_3 \\ \frac{1}{4n_1} + \frac{1}{4n_3} + \frac{1}{n_2}, & \text{for the contrast } \frac{\mu_1 + \mu_3}{2} - \mu_2 \\ \frac{1}{4n_2} + \frac{1}{4n_3} + \frac{1}{n_1}, & \text{for the contrast } \frac{\mu_2 + \mu_3}{2} - \mu_3. \end{cases}$$

The first three intervals agree with those calculations from Example 20.11. These Scheffé intervals are unaffected by the fact that there are now six contrasts of interest. The six simultaneous confidence intervals are

```
(     97.67  ,  6802.33 )
(   3376.98  , 10873.02 )
(    -73.02  ,  7423.02 )
(   2047.67  ,  8752.33 )
(  -3134.25  ,  2909.25 )
(  -8309.25  , -2265.75 )
```

Figure 20.10 shows the Bonferroni and Scheffé confidence intervals for the six contrasts. In this case, the Bonferroni and Scheffé intervals are nearly the same. ■■■

Exercise 20.2 Suppose that for the data in Example 20.11, we would like to get 90% simultaneous confidence intervals for one million contrasts,[6] including the three pairwise comparison of means. What would be the two intervals for $\mu_1 - \mu_2$?

Answers: Bonferroni (-9270.78 , 16170.78); Scheffé is the same as in Example 20.11.

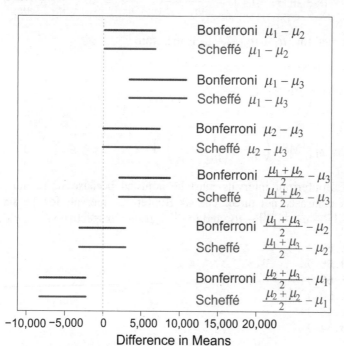

Figure 20.10 Comparison of Bonferroni and Scheffé simultaneous confidence intervals for six contrasts.

20.9 **Chapter Summary**

Before this chapter, the methods of inference were developed for one problem at a time. For example, a confidence interval for a parameter (Chapter 10) was developed as if this were the only confidence interval of interest. Similarly, when we developed a hypothesis test (Chapter 11), this single test was the only test being made. There are situations that call for testing multiple hypotheses (sometimes thousands) all at once. In this case the concept of α being the probability of rejecting a null hypothesis when it is true is blurred. For example, if we are testing 1,000 hypotheses based on independent data and each one uses $\alpha = 0.05$, then we would expect to reject about 5%, or 50, of the null hypotheses by chance. Controlling for this possibility of rejecting many null hypotheses by chance is challenging. We considered several approaches for this problem of multiple testing using a Bonferroni correction, Holm's method, and the Benjamini–Hochberg method.

Similar problems arise when we construct multiple confidence intervals. Often, we would like to obtain simultaneous confidence intervals. For example, we would like to say that we are 95% sure that *all* of the computed intervals contain *all* of the respective population parameters. This is much different than being 95% confident that each individual confidence interval contains the respective parameter. The Bonferroni method is the simplest method for obtaining simultaneous confidence intervals. Tukey's method can be used when we are comparing all possible differences of means in a one-factor ANOVA. We can apply Scheffé's method when we are interested in all possible contrasts of population means.

---

[6]There is no practical situation with $k = 3$ where this many confidence intervals would be needed, but this exercise makes the point about the width of Bonferroni and Scheffé intervals.

20.10 Problems

Problem 20.1 Suppose $X_1, X_2, \ldots, X_n \sim$ i.i.d. $N(\mu, \sigma^2)$. We want to test H_0: $\mu = 0$ against the alternative H_1: $\mu \neq 0$. Let P denote the P-value for this test (when viewed from before the data are collected). Given that H_0 is true, find:

(a) $\Pr(P < 0.05)$. (*Hint:* the probability that the P-value is less than 0.05 is equal to the probability that we reject H_0 when it is true for $\alpha = 0.05$.)

(b) $\Pr(P < 0.01)$

(c) $\Pr(P < 0.10)$

(d) $\Pr(P > 0.05)$

Problem 20.2 Continuation of Problem 20.1. Let w denote some number satisfying $0 < w < 1$. Given that H_0 is true, find $\Pr(P < w)$.

Problem 20.3 Suppose $X_1, X_2, \ldots, X_{n_1} \sim$ i.i.d. $N(\mu_1, \sigma^2)$ and $Y_1, Y_2, \ldots, Y_{n_1} \sim$ i.i.d. $N(\mu_2, \sigma^2)$. We want to test H_0: $\mu_1 = \mu_2$ against the alternative H_1: $\mu_1 \neq \mu_2$. Let P denote the P-value for this test (when viewed from before the data are collected). Given that H_0 is true, find

(a) $\Pr(P < 0.05)$

(b) $\Pr(P < 0.01)$

(c) $\Pr(P < 0.10)$

(d) $\Pr(P > 0.05)$

Problem 20.4 Continuation of Problem 20.3. Let w denote some number satisfying $0 < w < 1$. Given that H_0 is true, what is $\Pr(P < w)$?

Problem 20.5 Suppose $X_1, X_2, \ldots, X_n \sim$ i.i.d. $N(\mu, \sigma^2)$. We want to test H_0: $\mu = 0$ against the alternative H_1: $\mu > 0$. Let P denote the P-value for this test (when viewed from before the data are collected). Given that H_0 is true, find

(a) $\Pr(P < 0.05)$

(b) $\Pr(P < 0.01)$

(c) $\Pr(P < 0.10)$

(d) $\Pr(P > 0.05)$

Problem 20.6 Continuation of Problem 20.5. Let w denote some number satisfying $0 < w < 1$. Given that H_0 is true, what is $\Pr(P < w)$?

Problem 20.7 Suppose that every day the water utility in a city takes a sample of water from the local reservoir. They divide this sample into eight parts and send four to Lab 1 and four to Lab 2 for testing. Both labs determine the level of impurities in each of the four samples sent to it. This is done every day for 20 days. The first two day's worth of data are shown in Table 20.5. The full data set is given in the file LabSamples.csv.

Table 20.5 Data for Problem 20.7.

Day	Lab	Msmt	Day	Lab	Msmt
1	1	10.6	2	1	9.6
1	1	9.7	2	1	11.1
1	1	11.8	2	1	8.9
1	1	10.2	2	1	10.5
1	2	11.1	2	2	8.6
1	2	10.4	2	2	8.1
1	2	11.2	2	2	9.6
1	2	10.2	2	2	9.8

(a) Perform a two-sample t-test for the population mean measurements for Lab 1 and Lab 2 on Day 1. (You have four measurements from each.) Assume equal variance and $\alpha = 0.05$ when you do the t-test. What is the P-value?

(b) Repeat part (a) for the second day of data.

(c) Repeat the test in part (a) for days 3 through 20. Keep track of all the P-values.

(d) How many of the 20 P-values are below the level of $\alpha = 0.05$.

Problem 20.8 For the situation described in Problem 20.7, apply the Bonferroni method with $\alpha = 0.05$. With this new level, how many of the null hypotheses are rejected? Do you believe that for each sample the mean values of the two labs (in a hypothetically infinite sequence of measurements) are the same?

Problem 20.9 Refer to Problem 20.7. Suppose you have data for 365 days, and for each day you perform a t-test for the eight samples (four sent to Lab 1 and four to Lab 2). Suppose that all of the null hypotheses of equal means are true.

(a) If you fix $\alpha = 0.01$, how many of the 365 null hypotheses would you expect to reject?

(b) What is the probability that you will not reject any of the 365 null hypotheses?

(c) What is the probability that you will reject at least five of the null hypotheses?

Problem 20.10 Continuation of Problem 20.9. Apply the Bonferroni method to determine the α level for each of the 365 days if we want an FWER of 0.10.

Problem 20.11 Suppose that investigators test the genes of 860 cancer patients. The patients have one of four kinds of cancer, which we'll call type 1, type 2, type 3, and type 4. The numbers of patients with each type of cancer is below.

Type of cancer	1	2	3	4
No. of patients	200	210	150	300

A measure of gene activity is recorded for each person on each of the 500 genes. Data are contained in the file `Microarray_860_500.csv`. Use the group of 860 patients with type 2, 3, or 4 cancer as controls and compare the gene activation values between the type 1 group and the control 1 group for each of the 500 genes using a two-sample t-test. Record the P-value for each test.

(a) Apply the Bonferroni method with $\alpha = 0.10$ to control the FWER for the 500 tests. Which genes are associated with cancer type 1?

(b) Repeat part (a), except use Holm's method.

(c) Repeat part (b), except use the Benjamini–Hochberg method.

Problem 20.12 Repeat all parts of Problem 20.11, comparing type 2 cancer against the control group consisting of those with other kinds of cancers.

Problem 20.13 Repeat all parts of Problem 20.11, comparing type 3 cancer against the control group consisting of those with other kinds of cancers.

Problem 20.14 Repeat all parts of Problem 20.11, comparing type 4 cancer against the control group consisting of those with other kinds of cancers.

Problem 20.15 Ridgeway and MacDonald (2009) describe a study on racial bias in police stops in New York City in 2006. They modeled the racial makeup of those pedestrians stopped by 2,756 police officers who were regularly involved in these types of stops. Their model included a term for each officer's departure from the benchmark. These estimated effects, called $\hat{\theta}$, were standardized by dividing by the standard error of $\hat{\theta}$. This created 2,756 z statistics. If the omnibus hypothesis that all of the effects due to the 2,756 officers are zero, then $z_1, z_2, \ldots, z_n$ would be approximately i.i.d. $N(0, 1)$. Positive values of z_i indicate a tendency for officer i to stop more black and Hispanic pedestrians than expected, while negative values indicate a tendency to stop fewer black and Hispanic pedestrians than expected. The z statistics are contained in the file `Police.csv`. For each officer i, use the statistic z_i to compute the P-value for testing whether the $\hat{\theta}$ came from a population with mean zero against the alternative that the mean is greater than zero. The alternative hypothesis is therefore that the officer stops *more* black or Hispanic pedestrians than expected. This data set, also described and analyzed by Efron and Hastie (2016), is in the file `Police.csv`.

(a) Apply the Bonferroni method to achieve an FWER of $\alpha = 0.10$. How many officers would be flagged for stopping too many black or Hispanic pedestrians?

(b) Repeat part (a) using Holm's method.

(c) Repeat part (a) using the Benjamini–Hochberg Method.

Problem 20.16 Repeat all parts of Problem 20.15 if the alternative is that the officer is stopping *fewer* black or Hispanic pedestrians.

Problem 20.17 Use mathematical induction to provide a rigorous proof for Boole's inequality in Theorem 20.3.1.

Problem 20.18 Study the following function written in R code:

```
getPvalue = function(n)
{
  x = rnorm( n , 0 , 1 )
  xbar = mean(x)
  s = sd(x)
  t = (xbar - 0) / (s/sqrt(n))
  p = pt( t , n-1 , lower.tail=FALSE )
  return( p )
}
```

Explain what this code does, and how is it related to Problem 20.5.

Problem 20.19 Study the following code and then run it after you run the code from Problem 20.18.

```
Pvec = c()
for(i in 1:10000) Pvec[i]=getPvalue(20)
```

Make a histogram of the elements in the vector `Pvec`. Explain why the histogram looks flat.

Problem 20.20 Repeat Problem 20.19, except vary the argument to `getPvalue`. Explain why the resulting histogram is nearly flat regardless of the value sent to `getPvalue`.

Problem 20.21 The NIST's *e-Handbook of Statistical Methods* (NIST/SEMATECH, 2021) gives an example of a one-factor experiment with four levels and five observations per treatment. The data are shown in Table 20.6.

Table 20.6 Data for Problem 20.21.

Group	Outcomes					Mean
A	6.9	5.4	5.8	4.6	4.0	5.34
B	8.3	6.8	7.8	9.2	6.5	7.72
C	8.0	10.5	8.1	6.9	9.3	8.56
D	5.8	3.8	6.1	5.6	6.2	5.50

(a) Use ANOVA to test whether the four population means are equal.

(b) Use the Bonferroni method to find 95% simultaneous confidence intervals for all differences of means.

(c) Use Tukey's method to find 95% simultaneous confidence intervals for all differences of means.

(d) Use Scheffé's method to find 95% simultaneous confidence intervals for all differences of means.

Problem 20.22 In Problem 12.21 we looked at the tensile strength of a synthetic fiber and how it depends on the percent cotton. Five levels of percent cotton were selected and five units were constructed for each. Let $\mu_1, \mu_2, \mu_3, \mu_4$, and μ_5 be the population means for the five treatments.

(a) Find 95% confidence intervals for each of the 10 differences of means (i.e., $\mu_1 - \mu_2$, $\mu_1 - \mu_3$, etc.)

(b) Use the Bonferroni method to find 95% simultaneous confidence intervals for all differences of means.

(c) Use Tukey's method to find 95% simultaneous confidence intervals for all differences of means.

(d) Use Scheffé's method to find 95% simultaneous confidence intervals for all differences of means.

Problem 20.23 In Problem 19.11 we described a process for a single-wafer etching tool. The outcome etch rate depends on the RF power, which was varied from 160 to 220, and is shown in Table 19.6. Let μ_i denote the population mean for treatment i.

(a) Find 95% confidence intervals for each of the six differences of means (i.e., $\mu_1 - \mu_2$, $\mu_1 - \mu_3$, etc.).

(b) Use the Bonferroni method to find 95% simultaneous confidence intervals for all differences of means.

(c) Use Tukey's method to find 95% simultaneous confidence intervals for all differences of means.

(d) Use Scheffé's method to find 95% simultaneous confidence intervals for all differences of means.

Problem 20.24 For the situation described in Problem 20.22, use Bonferroni's and Scheffé's methods to find simultaneous confidence intervals for

$$\mu_i - \frac{1}{4}\sum_{j \neq i}\mu_j, \qquad i = 1, 2, 3, 4, 5,$$

in addition to all differences in means.

Problem 20.25 For the situation described in Problem 20.23, use Bonferroni's and Scheffé's methods to find simultaneous confidence intervals for

$$\mu_i - \frac{1}{3}\sum_{j \neq i}\mu_j, \qquad i = 1, 2, 3, 4,$$

in addition to all differences in means.

Problem 20.26 Suppose $f(c)$ is a function of one variable and $\max_c f(c) = d^2$; show that

$$\left(d^2 \leq A\right) \iff \left(f(c) \leq A \text{ for all } c\right).$$

Hint: Draw a picture. Then prove that if f is a function of $c_1, \ldots, c_k$ satisfying $\max_{c_1,\ldots,c_k} = d^2$, then

$$\left(d^2 \leq A\right) \iff \left(f(c_1,\ldots,c_k) \leq A \text{ for all } c_1,\ldots,c_k\right).$$

References

Anderson, E. 1928. The problem of species in the northern blue flags, Iris versicolor L. and Iris virginica L. *Annals of the Missouri Botanical Garden*, **15**(3), 241–332.

Anderson, E. 1936. The species problem in Iris. *Annals of the Missouri Botanical Garden*, **23**(3), 457–509.

Ayres-de Campos, D, Bernardes, J, Garrido, A, Marques-de Sa, J, & Pereira-Leite, L. 2000. SisPorto 2.0: a program for automated analysis of cardiotocograms. *Journal of Maternal–Fetal Medicine*, **9**(5), 311–318.

Bain, L.J., & Engelhardt, M. 1992. *Introduction to Probability and Mathematical Statistics*. 2nd edn. Duxbury Press.

Bartholomew, D.J. 1957. A problem in life testing. *Journal of the American Statistical Association*, **52**(279), 350–355.

Bartholomew, D.J. 1995. What is statistics? *Journal of the Royal Statistical Society, Series A (Statistics and Society)*, **158**, 3.

Benjamini, Y., & Hochberg, Y. 1995. Controlling the false discovery rate: a practical and powerful approach to multiple testing. *Journal of the Royal Statistical Society: Series B (Methodological)*, **57**(1), 289–300.

Bickel, P.J., Hammel, E.A., & O'Connell, J.W. 1975. Sex bias in graduate admissions: data from Berkeley. *Science*, **187**(4175), 398–404.

Bondah, E.K., Gren, L.H., & Talboys, S.L. 2020. Prevalence, drinking patterns, and risk factors of alcohol use and early onset among Ghanaian senior high school students. *American Journal of Preventive Medicine and Public Health*, **6**(1), 16–25.

Boslaugh, S. 2007. *Secondary Data Sources for Public Health*. Cambridge University Press.

Box, G.E.P., & Draper, N.R. 1987. *Empirical Model-Building and Response Surfaces*. Wiley.

Box, G.E.P., & Pierce, D.A. 1970. Distribution of residual autocorrelations in autoregressive-integrated moving average time series models. *Journal of the American Statistical Association*, **65**(332), 1509–1526.

Box, G.E.P., Jenkins, G.M., Reinsel, G.C., & Ljung, G.M. 2008. *Time Series Analysis: Forecasting and Control*. 4th edn. Wiley.

Brown, R. 1963. *Smoothing, Forecasting and Prediction of Discrete Time Series*. Prentice Hall.

Cannon, A.R., Cobb, G.W., Hartlaub, B.A., Legler, J.M., Lock, R.H., Moore, T.L., Rossman, A.J., & Witmer, J. 2018. *STAT2: Modeling with Regression and ANOVA*. W.H. Freeman.

Caroni, C. 2002. The correct "ball bearings" data. *Lifetime Data Analysis*, **8**, 395–399.

Cobb, G.W. 1998. *Introduction to Design and Analysis of Experiments*. Key College.

Colosimo, B.M., & Del Castillo, E. (eds.) 2006. *Bayesian Process Monitoring, Control and Optimization*. CRC Press.

Davis, T.P. 1995. Analysis of an experiment aimed at improving the reliability of transmission centre shafts. *Lifetime Data Analysis*, **1**(3), 275–306.

Dixon, W.J., & Massey, F.J. 1983. *Introduction to Statistical Analysis*. 4th edn. McGraw Hill.

Dua, D., & Graff, C. 2017. *UCI Machine Learning Repository*. UC Irvine.

Efron, B. 1982. *The Jackknife, the Bootstrap and Other Resampling Plans*. SIAM.

Efron, B., & Hastie, T. 2016. *Computer Age Statistical Inference: Algorithms, Evidence, and Data Science*. Cambridge University Press.

Efron, B., & Hinkley, D.V. 1978. Assessing the accuracy of the maximum likelihood estimator: observed versus expected Fisher information. *Biometrika*, **65**(3), 457–483.

Eustace, S., Tello, R., DeCarvalho, V., *et al.* 1997. A comparison of whole-body turboSTIR MR imaging and planar 99mTc-methylene diphosphonate scintigraphy in the examination of patients with suspected skeletal metastases. *American Journal of Roentgenology*, **169**(6), 1655–1661.

Fisher, R. A. 1935. *The Design of Experiments*. 9th edn. Macmillan.

Fisher, R. A. 1936. The use of multiple measurements in taxonomic problems. *Annals of Eugenics*, **7**(2), 179–188.

Fisher, R.A., & Balmukand, B. 1928. The estimation of linkage from the offspring of selfed heterozygotes. *Journal of Genetics*, **20**(1), 79–92.

Gelman, A., Carlin, J.B., Stern, H.S., Dunson, D.B, Vehtari, A., & Rubin, D.B. 2014. *Bayesian Data Analysis*. 3rd edn. Chapman and Hall/CRC.

Geman, S., & Geman, D. 1984. Stochastic relaxation, Gibbs distributions, and the Bayesian restoration of images. *IEEE Transactions on Pattern Analysis and Machine Intelligence*, **6**(6), 721–741.

Gilks, W.R., Thomas, A., & Spiegelhalter, D.J. 1994. A language and program for complex Bayesian modelling. *Journal of the Royal Statistical Society: Series D (The Statistician)*, **43**(1), 169–177.

Gillich, E., & Lohweg, V. 2010. Banknote authentication. *In: Bildverarbeitung in der Automation*. Springer.

Groves, R.M., & Lyberg, L. 2010. Total survey error: past, present, and future. *Public Opinion Quarterly*, **74**(5), 849–879.

Hare, L.B. 1988. In the soup: a case study to identify contributors to filling variability. *Journal of Quality Technology*, **20**(1), 36–43.

Hastings, W.K. 1970. Monte Carlo sampling methods using Markov chains and their applications. *Biometrika*, **57**, 97–109.

Hilbe, J.M. 2011. *Negative Binomial Regression*. Cambridge University Press.

Holt, C.C. 1957. *Forecasting Seasonals and Trends by Exponentially Weighted Moving Averages, Office of Naval Research Memorandum (Vol. 52)*. Carnegie Institute of Technology.

Hsiue, Lin-Tee, Ma, Chen-Chi M., & Tsai, Hong-Bing. 1995. Preparation and characterizations of thermotropic copolyesters of p-hydroxybenzoic acid, sebacic acid, and hydroquinone. *Journal of Applied Polymer Science*, **56**(4), 471–476.

Ioannidis, John P.A. 2005. Why most published research findings are false. *PLoS Medicine*, **2**(8), e124.

James, G., Witten, D., Hastie, T., & Tibshirani, R. 2021. *An Introduction to Statistical Learning: With Applications in R*. Springer.

Johnson, S.K, & Johnson, R.E. 1972. Tonsillectomy history in Hodgkin's disease. *New England Journal of Medicine*, **287**(22), 1122–1125.

Jones, B., & Montgomery, D.C. 2019. *Design of Experiments: A Modern Approach*. Wiley.

Lambert, D. 1992. Zero-inflated Poisson regression, with an application to defects in manufacturing. *Technometrics*, **34**(1), 1–14.

Lamberton, J.G., Arbogast, B.L., Deinzer, M.L., & Norris, L.A. 1976. A rapid method for the determination arsenic concentrations in urine at field locations. *American Industrial Hygiene Association Journal*, 3 418–422.

Larsen, R.J., & Marx, M.L. 2005. *An Introduction to Mathematical Statistics*. Prentice Hall.

Lieblein, J., & Zelen, M. 1956. Statistical investigation of the fatigue life of deep-groove ball bearings. *of Research of the National Bureau of Standards*, **57**(5), 273–316.

Lindsay, B.G., Kettenring, J., & Siegmund, D.O. 2004. A report on the future of statistics. *Statistical Science*, **19**(3), 387–413.

Ljung, G.M., & Box, G.E.P. 1978. On a measure of lack of fit in time series models. *Biometrika*, **65**, 197–303.

Lohr, S.L. 2010. *Sampling: Design and Analysis*. 2nd edn. Cengage.

Metropolis, N., Rosenbluth, A.W., Rosenbluth, M.N., Teller, A.H., & Teller, E. 1953. Equation of state calculations by fast computing machines. *The Journal of Chemical Physics*, **21**(6), 1087–1092.

Montgomery, D.C. 2020a. *Design and Analysis of Experiments*. 10th edn. Wiley.

Montgomery, D.C. 2020b. *Introduction to Statistical Quality Control*. 8th edn. Wiley.

Montgomery, D.C., Jennings, C.L., & Kulahci, M. 2015. *Introduction to Time Series Analysis and Forecasting*. 2nd edn. Wiley.

Myers, R.H., Montgomery, D.C., Vining, G.G., & Robinson, T.J. 2010. *Generalized Linear Models: With Applications in Engineering and the Sciences*. Wiley.

Nemati, A., Mahdavi, R., & Baghi, A.N. 2012. Case–control study of dietary pattern and other risk factors for gastric cancer. *Health Promotion Perspectives*, **2**(1), 20.

NIMBLE Development Team. 2022. *NIMBLE User Manual*. `https://r-nimble.org/manuals/NimbleUserManual.pdf`.

NIST/SEMATECH. 2021. *NIST/SEMATECH e-Handbook of Statistical Methods*. `www.itl.nist.gov/div898/handbook/`.

Patton, G.L., Bravman, J.C., & Plummer, J.D. 1986. Physics, technology, and modeling of polysilicon emitter contacts for VLSI bipolar transistors. *IEEE Transactions on Electron Devices*, **33**(11), 1754–1768.

Pearl, J., & Mackenzie, D. 2018. *The Book of Why: The New Science of Cause and Effect*. Basic Books.

Plummer, M. 2012. *JAGS Version 3.3. 0 User Manual*. International Agency for Research on Cancer.

Quinlan, J.R. 1993. Combining instance-based and model-based learning. In: *Proceedings of the Tenth International Conference on Machine Learning*. ACM.

Radelet, M.L. 1981. Racial characteristics and the imposition of the death penalty. *American Sociological Review*, **46**(6), 918–927.

Reintjes, R., de Boer, A., van Pelt, W., & Mintjes-de Groot, J. 2000. Simpson's paradox: an example from hospital epidemiology. *Epidemiology*, **11**(1), 81–83.

Ridgeway, G., & MacDonald, J.M. 2009. Doubly robust internal benchmarking and false discovery rates for detecting racial bias in police stops. *Journal of the American Statistical Association*, **104**(486), 661–668.

Rudin, W. 1976. *Principles of Mathematical Analysis*. McGraw Hill.

Salsburg, D. 2001. *The Lady Tasting Tea: How Statistics Revolutionized Science in the Twentieth Century*. Macmillan.

Scholes, D., Daling, J.R., & Stergachis, A.S. 1992. Current cigarette smoking and risk of acute pelvic inflammatory disease. *American Journal of Public Health*, **82**(10), 1352–1355.

Schweizer, E., Rickels, K., Case, W.G., & Greenblatt, D.J. 1991. Carbamazepine treatment in patients discontinuing long-term benzodiazepine therapy: effects on withdrawal severity and outcome. *Archives of General Psychiatry*, **48**(5), 448–452.

Serafini, M., Bugianesi, R., Maiani, G., Valtuena, S., De Santis, S., & Crozier, A. 2003. Plasma antioxidants from chocolate. *Nature*, **424**(6952), 1013–1013.

Shahar, D.R., Schultz, R., Shahar, A., & Wing, R.R. 2001. The effect of widowhood on weight change, dietary intake, and eating behavior in the elderly population. *Journal of Aging and Health*, **13**(2), 186–199.

Shmueli, G. 2010. To explain or to predict? *Statistical Science*, **25**(3), 289–310.

Soylak, M., Narin, I., de Almeida Bezerra, M., & Ferreira, S.L.C. 2005. Factorial design in the optimization of preconcentration procedure for lead determination by FAAS. *Talanta*, **65**(4), 895–899.

Spiegelhalter, D., Thomas, A., Best, N., & Lunn, D. 2003. *WinBUGS User Manual*.

Spiegelhalter, D., Thomas, A., Best, N., & Lunn, D. 2014. *OpenBUGS User Manual*. `www.openbugs.net/Manuals/Manual.html`.

Stokes, M.E., Davis, C.S., & Koch, G.G. 2012. *Categorical Data Analysis using SAS*. SAS Institute.

Symons, M.J., Grimson, R.C., & Yuan, Y.C. 1983. Clustering of rare events. *Biometrics*, **39**(1), 193–205.

Talalay, P., Fahey, J.W., Healy, Z.R., Wehage, S.L., Benedict, A.L., Min, C., & Dinkova-Kostova, A.T. 2007.

Sulforaphane mobilizes cellular defenses that protect skin against damage by UV radiation. *Proceedings of the National Academy of Sciences*, **104**(44), 17500–17505.

Thall, P.F., & Vail, S.C. 1990. Some covariance models for longitudinal count data with overdispersion. *Biometrics*, **46**(3) 657–671.

Theobald, C. 1981. The effect of nozzle design on the stability and performance of turbulent water jets. *Fire Safety Journal*, **4**(1), 1–13.

Thomas, S., Patel, D., Bittel, B., *et al.* 2021. Effect of high-dose zinc and ascorbic acid supplementation vs usual care on symptom length and reduction among ambulatory patients with SARS-CoV-2 infection: the COVID A to Z randomized clinical trial. *JAMA Network Open*, **4**(2), e210369–e210369.

Van Nood, E., Vrieze, A., Nieuwdorp, M., *et al.* 2013. Duodenal infusion of donor feces for recurrent *Clostridium difficile*. *New England Journal of Medicine*, **368**(5), 407–415.

Varberg, D., Purcell, E.J., & Rigdon, S.E. 2007. *Calculus Early Transcendentals*. Pearson Higher Education.

Vermund, S.H., Kelley, K.F., Klein, R.S., *et al.* 1991. High risk of human papillomavirus infection and cervical squamous intraepithelial lesions among women with symptomatic human immunodeficiency virus infection. *American Journal of Obstetrics and Gynecology*, **165**(2), 392–400.

Wackerly, D., Mendenhall, W., & Scheaffer, R.L. 2014. *Mathematical Statistics with Applications*. 7th edn. Brooks/Cole.

Wang, F.K. 2000. A new model with bathtub-shaped failure rate using an additive Burr XII distribution. *Reliability Engineering & System Safety*, **70**(3), 305–312.

Wang, K., Wong, E.L.Y., Ho, K.F., *et al.* 2020. Intention of nurses to accept coronavirus disease 2019 vaccination and change of intention to accept seasonal influenza vaccination during the coronavirus disease 2019 pandemic: a cross-sectional survey. *Vaccine*, **38**(45), 7049–7056.

Weisberg, S. 1986. A linear model approach to backcalculation of fish length. *Journal of the American Statistical Association*, **81**(396), 922–929.

White, C. 1990. Research on smoking and lung cancer: a landmark in the history of chronic disease epidemiology. *Yale Journal of Biology and Medicine*, **63**, 29–46.

Winters, P.R. 1960. Forecasting sales by exponentially weighted moving averages. *Management Science*, **6**(3), 324–342.

Wu, C.F.J., & Hamada, M. 2000. *Experiments: Planning, Analysis, and Parameter Design Optimization*. Wiley.

Yates, F. 1934. Contingency tables involving small numbers and the χ^2 test. *Supplement to the Journal of the Royal Statistical Society*, **1**(2), 217–235.

Yeh, I.-Cheng. 1998. Modeling of strength of high performance concrete using artificial neural networks. *Cement and Concrete Research*, **28**(12), 1797–1808.

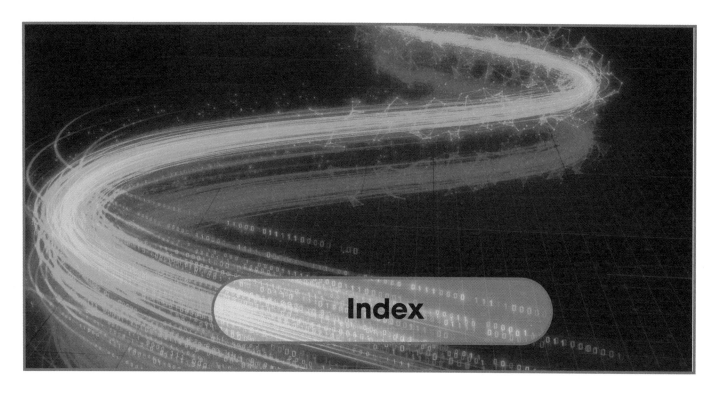

Index